Wavelet Methods for Time Series Analysis

CAMBRIDGE SERIES IN STATISTICAL AND PROBABILISTIC MATHEMATICS

This series of high quality upper-division textbooks and expository monographs covers all aspects of stochastic applicable mathematics. The topics range from pure and applied statistics to probability theory, operations research, optimization and mathematical programming. The books contain clear presentations of new developments in the field and also of the state of the art in classical methods. While emphasizing rigorous treatment of theoretical methods, the books also contain applications and discussions of new techniques made possible by advances in computational practice.

Already published

1. *Bootstrap Methods and Their Application*, by A. C. Davison and D. V. Hinkley
2. *Markov Chains*, by J. Norris
3. *Asymptotic Statistics*, by A. W. van der Vaart
4. *Wavelet Methods for Time Series Analysis*, by Donald B. Percival and Andrew T. Walden
5. *Bayesian Methods: An Analysis for Statisticians and Interdisciplinary Researchers*, by Thomas Leonard and John S. J. Hsu
6. *Empirical Processes in M-Estimation*, by Sara van de Geer

Wavelet Methods for Time Series Analysis

Donald B. Percival

UNIVERSITY OF WASHINGTON, SEATTLE

Andrew T. Walden

IMPERIAL COLLEGE OF SCIENCE,
TECHNOLOGY AND MEDICINE, LONDON

CAMBRIDGE UNIVERSITY PRESS
Cambridge, New York, Melbourne, Madrid, Cape Town, Singapore,
São Paulo, Delhi, Dubai, Tokyo, Mexico City

Cambridge University Press
32 Avenue of the Americas, New York, NY 10013-2473, USA

www.cambridge.org
Information on this title: www.cambridge.org/9780521685085

First published 2000
Reprinted 2000, 2002, 2003
First paperback edition 2006
Reprinted 2006, 2007, 2008

A catalog record for this publication is available from the British Library

Library of Congress Cataloging in Publication data

Percival, Donald B.
Wavelet methods for time series analysis / Donald B. Percival
and Andrew T. Walden.
p. cm.
Includes bibliographical references and indexes.
ISBN 0-521-64068-7
1. Time-series analysis. 2. Wavelets (Mathematics).
I. Walden, Andrew T. II. Title.
QA280.P47 2000
519.5′–dc21 00-029246

ISBN 978-0-521-64068-8 Hardback
ISBN 978-0-521-68508-5 Paperback

To Piroska and Hilary
for support, patience and endurance

Contents

Preface xiii
Conventions and Notation xvii

1. Introduction to Wavelets 1

1.0 Introduction 1
1.1 The Essence of a Wavelet 2
Comments and Extensions to Section 1.1 4
1.2 The Essence of Wavelet Analysis 5
Comments and Extensions to Section 1.2 12
1.3 Beyond the CWT: the Discrete Wavelet Transform 12
Comments and Extensions to Section 1.3 19

2. Review of Fourier Theory and Filters **20**

2.0 Introduction 20
2.1 Complex Variables and Complex Exponentials 20
2.2 Fourier Transform of Infinite Sequences 21
2.3 Convolution/Filtering of Infinite Sequences 24
2.4 Fourier Transform of Finite Sequences 28
2.5 Circular Convolution/Filtering of Finite Sequences 29
2.6 Periodized Filters 32
Comments and Extensions to Section 2.6 35
2.7 Summary of Fourier Theory 35
2.8 Exercises 39

3. Orthonormal Transforms of Time Series **41**

3.0 Introduction 41
3.1 Basic Theory for Orthonormal Transforms 41
3.2 The Projection Theorem 44
3.3 Complex-Valued Transforms 45
3.4 The Orthonormal Discrete Fourier Transform 46
Comments and Extensions to Section 3.4 52
3.5 Summary 53
3.6 Exercises 54

4. The Discrete Wavelet Transform **56**

4.0 Introduction 56
4.1 Qualitative Description of the DWT 57
Key Facts and Definitions in Section 4.1 67
Comments and Extensions to Section 4.1 68
4.2 The Wavelet Filter 68
Key Facts and Definitions in Section 4.2 74
Comments and Extensions to Section 4.2 75
4.3 The Scaling Filter 75
Key Facts and Definitions in Section 4.3 78
Comments and Extensions to Section 4.3 79
4.4 First Stage of the Pyramid Algorithm 80
Key Facts and Definitions in Section 4.4 86
Comments and Extensions to Section 4.4 87
4.5 Second Stage of the Pyramid Algorithm 88
Key Facts and Definitions in Section 4.5 93
4.6 General Stage of the Pyramid Algorithm 93
Key Facts and Definitions in Section 4.6 99
Comments and Extensions to Section 4.6 100
4.7 The Partial Discrete Wavelet Transform 104
4.8 Daubechies Wavelet and Scaling Filters: Form and Phase 105
Key Facts and Definitions in Section 4.8 116
Comments and Extensions to Section 4.8 117
4.9 Coiflet Wavelet and Scaling Filters: Form and Phase 123
4.10 Example: Electrocardiogram Data 125
Comments and Extensions to Section 4.10 134
4.11 Practical Considerations 135
Comments and Extensions to Section 4.11 145
4.12 Summary 150
4.13 Exercises 156

5. The Maximal Overlap Discrete Wavelet Transform **159**

5.0 Introduction 159
5.1 Effect of Circular Shifts on the DWT 160
5.2 MODWT Wavelet and Scaling Filters 163
5.3 Basic Concepts for MODWT 164
Key Facts and Definitions in Section 5.3 168
5.4 Definition of jth Level MODWT Coefficients 169
Key Facts and Definitions in Section 5.4 173
Comments and Extensions to Section 5.4 174
5.5 Pyramid Algorithm for the MODWT 174
Key Facts and Definitions in Section 5.5 177
Comments and Extensions to Section 5.5 177
5.6 MODWT Analysis of 'Bump' Time Series 179
5.7 Example: Electrocardiogram Data 182
5.8 Example: Subtidal Sea Level Fluctuations 185
5.9 Example: Nile River Minima 190
5.10 Example: Ocean Shear Measurements 193
5.11 Practical Considerations 195
5.12 Summary 200
5.13 Exercises 204

6. The Discrete Wavelet Packet Transform **206**

6.0 Introduction 206
6.1 Basic Concepts 207
Comments and Extensions to Section 6.1 217
6.2 Example: DWPT of Solar Physics Data 218
6.3 The Best Basis Algorithm 221
Comments and Extensions to Section 6.3 226
6.4 Example: Best Basis for Solar Physics Data 226
6.5 Time Shifts for Wavelet Packet Filters 229
Comments and Extensions to Section 6.5 231
6.6 Maximal Overlap Discrete Wavelet Packet Transform 231
6.7 Example: MODWPT of Solar Physics Data 234
6.8 Matching Pursuit 239
6.9 Example: Subtidal Sea Levels 243
Comments and Extensions to Section 6.9 247
6.10 Summary 247
6.11 Exercises 253

7. Random Variables and Stochastic Processes **255**

7.0 Introduction 255
7.1 Univariate Random Variables and PDFs 256
7.2 Random Vectors and PDFs 258
7.3 A Bayesian Perspective 264
7.4 Stationary Stochastic Processes 266
7.5 Spectral Density Estimation 269

Comments and Extensions to Section 7.5 278
7.6 Definition and Models for Long Memory Processes 279
Comments and Extensions to Section 7.6 285
7.7 Nonstationary $1/f$-Type Processes 287
Comments and Extensions to Section 7.7 289
7.8 Simulation of Stationary Processes 290
Comments and Extensions to Section 7.8 292
7.9 Simulation of Stationary Autoregressive Processes 292
7.10 Exercises 293

8. The Wavelet Variance 295

8.0 Introduction 295
8.1 Definition and Rationale for the Wavelet Variance 295
Comments and Extensions to Section 8.1 301
8.2 Basic Properties of the Wavelet Variance 304
Comments and Extensions to Section 8.2 306
8.3 Estimation of the Wavelet Variance 306
Comments and Extensions to Section 8.3 308
8.4 Confidence Intervals for the Wavelet Variance 311
Comments and Extensions to Section 8.4 315
8.5 Spectral Estimation via the Wavelet Variance 315
Comments and Extensions to Section 8.5 317
8.6 Example: Atomic Clock Deviates 317
8.7 Example: Subtidal Sea Level Fluctuations 324
8.8 Example: Nile River Minima 326
8.9 Example: Ocean Shear Measurements 327
8.10 Summary 335
8.11 Exercises 337

9. Analysis and Synthesis of Long Memory Processes 340

9.0 Introduction 340
9.1 Discrete Wavelet Transform of a Long Memory Process 341
Comments and Extensions to Section 9.1 350
9.2 Simulation of a Long Memory Process 355
Comments and Extensions to Section 9.2 361
9.3 MLEs for Stationary FD Processes 361
Comments and Extensions to Section 9.3 366
9.4 MLEs for Stationary or Nonstationary FD Processes 368
Comments and Extensions to Section 9.4 373
9.5 Least Squares Estimation for FD Processes 374
Comments and Extensions to Section 9.5 378
9.6 Testing for Homogeneity of Variance 379
Comments and Extensions to Section 9.6 382
9.7 Example: Atomic Clock Deviates 383
9.8 Example: Nile River Minima 386
9.9 Summary 388

9.10 Exercises 391

10. Wavelet-Based Signal Estimation 393

10.0 Introduction 393
10.1 Signal Representation via Wavelets 394
10.2 Signal Estimation via Thresholding 398
10.3 Stochastic Signal Estimation via Scaling 407
10.4 Stochastic Signal Estimation via Shrinkage 408
Comments and Extensions to Section 10.4 415
10.5 IID Gaussian Wavelet Coefficients 417
Comments and Extensions to Section 10.5 429
10.6 Uncorrelated Non-Gaussian Wavelet Coefficients 432
Comments and Extensions to Section 10.6 439
10.7 Correlated Gaussian Wavelet Coefficients 440
Comments and Extensions to Section 10.7 449
10.8 Clustering and Persistence of Wavelet Coefficients 450
10.9 Summary 452
10.10 Exercises 455

11. Wavelet Analysis of Finite Energy Signals 457

11.0 Introduction 457
11.1 Translation and Dilation 457
11.2 Scaling Functions and Approximation Spaces 459
Comments and Extensions to Section 11.2 462
11.3 Approximation of Finite Energy Signals 462
Comments and Extensions to Section 11.3 464
11.4 Two-Scale Relationships for Scaling Functions 464
11.5 Scaling Functions and Scaling Filters 469
Comments and Extensions to Section 11.5 472
11.6 Wavelet Functions and Detail Spaces 472
11.7 Wavelet Functions and Wavelet Filters 476
11.8 Multiresolution Analysis of Finite Energy Signals 478
11.9 Vanishing Moments 483
Comments and Extensions to Section 11.9 486
11.10 Spectral Factorization and Filter Coefficients 487
Comments and Extensions to Section 11.10 494
11.11 Summary 494
11.12 Exercises 500

Appendix. Answers to Embedded Exercises 501

References 552

Author Index 565

Subject Index 569

Preface

The last decade has seen an explosion of interest in wavelets, a subject area that has coalesced from roots in mathematics, physics, electrical engineering and other disciplines. As a result, wavelet methodology has had a significant impact in areas as diverse as differential equations, image processing and statistics. This book is an introduction to wavelets and their application in the analysis of discrete time series typical of those acquired in the physical sciences. While we present a thorough introduction to the basic theory behind the discrete wavelet transform (DWT), our goal is to bridge the gap between theory and practice by

- emphasizing what the DWT actually means in practical terms;
- showing how the DWT can be used to create informative descriptive statistics for time series analysts;
- discussing how stochastic models can be used to assess the statistical properties of quantities computed from the DWT; and
- presenting substantive examples of wavelet analysis of time series representative of those encountered in the physical sciences.

To date, most books on wavelets describe them in terms of continuous functions and often introduce the reader to a plethora of different types of wavelets. We concentrate on developing wavelet methods in discrete time via standard filtering and matrix transformation ideas. We purposely avoid overloading the reader by focusing almost exclusively on the class of wavelet filters described in Daubechies (1992), which are particularly convenient and useful for statistical applications; however, the understanding gained from a study of the Daubechies class of wavelets will put the reader in a excellent position to work with other classes of interest. For pedagogical purposes, this book in fact starts (Chapter 1) and ends (Chapter 11) with discussions of the continuous case. This organization allows us at the beginning to motivate ideas from a historical perspective and then at the end to link ideas arising in the discrete analysis to some of the widely known results for continuous time wavelet analysis.

Topics developed early on in the book (Chapters 4 and 5) include the DWT and the 'maximal overlap' discrete wavelet transform (MODWT), which can be regarded as a generalization of the DWT with certain quite appealing properties. As a whole, these two chapters provide a self-contained introduction to the basic properties of wavelets, with an emphasis both on algorithms for computing the DWT and MODWT and also on the use of these transforms to provide informative descriptive statistics for time series. In particular, both transforms lead to both a scale-based decomposition of the sample variance of a time series and also a scale-based additive decomposition known as a multiresolution analysis. A generalization of the DWT and MODWT that are known in the literature as 'wavelet packet' transforms, and the decomposition of time series via matching pursuit, are among the subjects of Chapter 6. In the second part of the book, we combine these transforms with stochastic models to develop wavelet-based statistical inference for time series analysis. Specific topics covered in this part of the book include

- the wavelet variance, which provides a scale-based analysis of variance complementary to traditional frequency-based spectral analysis (Chapter 8);
- the analysis and synthesis of 'long memory processes,' i.e., processes with slowly decaying correlations (Chapter 9); and
- signal estimation via 'thresholding' and 'denoising' (Chapter 10).

This book is written 'from the ground level and up.' We have attempted to make the book as self-contained as possible (to this end, Chapters 2, 3 and 7 contain reviews of, respectively, relevant Fourier and filtering theory; key ideas in the orthonormal transforms of time series; and important concepts involving random variables and stochastic processes). The text should thus be suitable for advanced undergraduates, but is primarily intended for graduate students and researchers in statistics, electrical engineering, physics, geophysics, astronomy, oceanography and other physical sciences. Readers with a strong mathematical background can skip Chapters 2 and 3 after a quick perusal. Those with prior knowledge of the DWT can make use of the Key Facts and Definitions toward the end of various sections in Chapters 4 and 5 to assess how much of these sections they need to study. This book – or drafts thereof – have been used as a textbook for a graduate course taught at the University of Washington for the past ten years, but we have also designed it to be a self-study work-book by including a large number of exercises embedded within the body of the chapters (particularly Chapters 2 to 5), with solutions provided in the Appendix. Working the embedded exercises will provide readers with a means of progressively understanding the material. For use as a course textbook, we have also provided additional exercises at the end of each chapter (instructors wishing to obtain a solution guide for the exercises should follow the guidance given on the Web site detailed below).

The wavelet analyses of time series that are described in Chapters 4 and 5 can readily be carried out once the basic algorithms for computing the DWT and MODWT (and their inverses) are implemented. While these can be immediately and readily coded up using the pseudo-code in the Comments and Extensions to Sections 4.6 and 5.5, links to existing software in `S-Plus`, `R`, `MATLAB` and Lisp can be found by consulting the Web site for this book, which currently is at

`http://faculty.washington.edu/dbp/wmtsa.html`

The reader should also consult this Web site to obtain a current errata sheet and to download the coefficients for various scaling filters (as discussed in Sections 4.8 and 4.9), the values for all the time series used as examples in this book, and certain computed values that can be used to check computer code. To facilitate preparation of overheads for courses and seminars, the Web site also allows access to pdf files with all the figures and tables in the book (please note that these figures and tables are the copyright of Cambridge University Press and must not be further distributed or used without written permission).

The book was written using Donald Knuth's superb typesetting system TEX as implemented by Blue Sky Research in their product TEXtures for Apple Macintosh™ computers. The figures in this book were created using either the plotting system GPL written by W. Hess (whom we thank for many years of support) or `S-Plus`, the commercial version of the `S` language developed by J. Chambers and co-workers and marketed by MathSoft, Inc. The computations necessary for the various examples and figures were carried out using either `S-Plus` or PITSSA (a Lisp-based object-oriented program for interactive time series and signal analysis that was developed in part by one of us (Percival)).

We thank R. Spindel and the late J. Harlett of the Applied Physics Laboratory, University of Washington, for providing discretionary funding that led to the start of this book. We thank the National Science Foundation, the National Institutes of Health, the Environmental Protection Agency (through the National Research Center for Statistics and the Environment at the University of Washington), the Office of Naval Research and the Air Force Office of Scientific Research for ongoing support during the writing of this book. Our stay at the Isaac Newton Institute for Mathematical Sciences (Cambridge University) during the program on Nonlinear and Nonstationary Signal Processing in 1998 contributed greatly to the completion of this book; we thank the Engineering and Physical Science Research Council (EPSRC) for the support of one of us (Percival) through a Senior Visiting Fellowship while at Cambridge.

We are indebted to those who have commented on drafts of the manuscript or supplied data to us, namely, G. Bardy, J. Bassingthwaighte, A. Bruce, M. Clyde, W. Constantine, A. Contreras Cristan, P. Craigmile, H.-Y. Gao, A. Gibbs, C. Greenhall, M. Gregg, M. Griffin, P. Guttorp, T. Horbury, M. Jensen, W. King, R. D. Martin, E. McCoy, F. McGraw, H. Mofjeld, F. Noraz, G. Raymond, P. Reinhall, S. Sardy, E. Tsakiroglou and B. Whitcher. We are also very grateful to the many graduate students who have given us valuable critiques of the manuscript and exercises and found numerous errors. We would like to thank E. Aldrich, C. Cornish, N. Derby, A. Jach, I. Kang, M. Keim, I. MacLeod, M. Meyer, K. Tanaka and Z. Xuelin for pointing out errors that have been corrected in reprintings of the book. For any remaining errors – which in a work of this size are inevitable – we apologize, and we would be pleased to hear from any reader who finds a mistake so that we can list them on the Web site and correct any future printings (our 'paper' and electronic mailing addresses are listed below). Finally we acknowledge two sources of great support for this project, Lauren Cowles and David Tranah, our editors at Cambridge University Press, and our respective families.

Don Percival
Applied Physics Laboratory
Box 355640
University of Washington
Seattle, WA 98195–5640
`dbp@apl.washington.edu`

Andrew Walden
Department of Mathematics
Imperial College of Science,
Technology and Medicine
London SW7 2BZ, UK
`a.walden@ic.ac.uk`

Conventions and Notation

- *Important conventions*

(83)	refers to the single displayed equation on page 83
(69a), (69b)	refers to different displayed equations on page 69
Figure 86	refers to the figure on page 86
Table 109	refers to the table on page 109
Exercise [72]	refers to the embedded exercise on page 72 (an answer is in the Appendix)
Exercise [4.9]	refers to the ninth exercise at the end of Chapter 4
$H(\cdot)$	refers to a function
$H(f)$	refers to the value of the function $H(\cdot)$ at f
$\{h_l\}$	refers to a sequence of values indexed by the integer l
h_l	refers to the lth value of a sequence

In the following lists, the numbers at the end of the brief descriptions are page numbers where more information about – or an example of the use of – an abbreviation or symbol can be found.

- *Abbreviations used frequently*

ACS	autocorrelation sequence	15, 266, 341
ACVS	autocovariance sequence	266
ANOVA	analysis of variance	19, 67
AR	autoregressive process	268
ARFIMA	autoregressive, fractionally integrated, moving average process	285
CWT	continuous wavelet transform	1, 10
dB	decibels, i.e., $10\log_{10}(\cdot)$	73

DFBM	discrete fractional Brownian motion	279
DFT	discrete Fourier transform	22
DHM	Davies–Harte method	290
DWPT	discrete wavelet packet transform	206, 209
DWT	discrete wavelet transform	1, 13, 56
ECG	electrocardiogram	125
EDOF	equivalent degrees of freedom	313
FBM	fractional Brownian motion	279
FD	fractionally differenced	281
FFT	fast Fourier transform	28
FGN	fractional Gaussian noise	279
GSSM	Gaussian spectral synthesis method	291
Hz	Hertz: 1 Hz = 1 cycle per second	48
IID	independent and identically distributed	262
LA	least asymmetric	107
LSE	least squares estimate or estimator	374, 378
MAD	median absolute deviation	420
MODWPT	maximal overlap discrete wavelet packet transform	207, 231
MODWT	maximal overlap discrete wavelet transform	159
ML	maximum likelihood	341, 361
MLE	maximum likelihood estimate or estimator	341, 361
MRA	multiresolution analysis	65, 461
MRC	mobile radio communications	436
NMR	nuclear magnetic resonance	420
NPES	normalized partial energy sequence	129, 394–5
ODFT	orthonormal discrete Fourier transform	41, 46
OLSE	ordinary least squares estimate or estimator	378
PDF	probability density function	256
PPL	pure power law	281
QMF	quadrature mirror filter	75, 474
RMSE	root mean square error	364, 436
RV	random variable	256
SDF	spectral density function	267
SURE	Stein's unbiased risk estimator	404
WLSE	weighted least squares estimate or estimator	374
WP	wavelet packet	209

- *Non-Greek notation used frequently*

A_j	integral of squared SDF $S_j^2(\cdot)$ for $\{\overline{W}_{j,t}\}$	307
$\mathcal{A}_j$	$N_j \times N_{j-1}$ matrix (rows have $\{g_l\}$ periodized to N_{j-1})	94
$\widetilde{\mathcal{A}}_j$	$N \times N$ matrix (rows have upsampled $\{\tilde{g}_l\}$ periodized to N)	176

Symbol	Description	Page
$\mathcal{A}_L(\cdot)$	squared gain function for low pass component of $\mathcal{H}^{(\mathrm{D})}(\cdot)$	106
$\{a_{n,t}\}$	nth order sine data taper	274
$\arg(z)$	argument of complex-valued number z	21
B	backward shift operator	283
B_t	discrete fractional Brownian motion (DFBM)	279
$B_H(\cdot)$	fractional Brownian motion (FBM)	279
$\mathcal{B}_j$	$N_j \times N_{j-1}$ matrix (rows have $\{h_l\}$ periodized to N_{j-1})	94
$\widetilde{\mathcal{B}}_j$	$N \times N$ matrix (rows have upsampled $\{\tilde{h}_l\}$ periodized to N)	176
C_j	average value of SDF $S_X(\cdot)$ over octave band $[\frac{1}{2^{j+1}}, \frac{1}{2^j}]$	343
$\widetilde{C}_j$	approximation to C_j for FD processes	344
$\{C_n\}$	normalized partial energy sequence (NPES)	129, 395
$\mathbf{C}$	N dimensional stochastic signal vector	393
$\operatorname{cov}\{\cdot,\cdot\}$	covariance operator	259
$\mathcal{D}(\cdot)$	squared gain function for difference filter	105
$\mathcal{D}_j$	jth level wavelet detail (DWT)	64
$\widetilde{\mathcal{D}}_j$	jth level wavelet detail (MODWT)	169, 171
$\mathbb{D}$	dictionary (collection of vectors) used in matching pursuit	239
$\mathbf{D}$	N dimensional deterministic signal vector	393
d	number of differencing operations	287
$d_{j,t}$	tth component on jth level of $\mathbf{d}$ for DWT	419
d_l	lth component of vector $\mathbf{d}$	398
$\mathbf{d}$	transform coefficients for deterministic signal $\mathbf{D}$	398
$\mathbf{d}_\gamma$	dictionary element (vector in matching pursuit dictionary $\mathbb{D}$)	239
$E\{\cdot\}$	expectation operator	256, 258
$E\{X_0 \mid X_1 = x_1\}$	conditional expectation of X_0 given $X_1 = x_1$	260
$\mathcal{E}_\mathbf{X}$	energy (squared norm) for vector $\mathbf{X}$	42, 72
e	$2.718281828459045\cdots$	3, 21
e^{ix}	complex exponential	21
$e_{j,t}$	tth component on jth level of $\mathbf{e}$ for DWT	419
e_l	lth component of vector $\mathbf{e}$	398
$\mathbf{e}$	transform coefficients for IID noise $\boldsymbol{\epsilon}$	398
$\{F_k\}$	orthonormal discrete Fourier transform (ODFT) coefficients	46
$\mathcal{F}$	$N \times N$ orthonormal discrete Fourier transform matrix	47
$\mathbf{F}$	vector containing ODFT coefficients $\{F_k\}$	47
f	frequency of a sinusoid	22
f_k	k/N or $k/(N\,\Delta t)$, the kth Fourier frequency	28, 87
$f_\mathcal{N}$	Nyquist frequency	87, 267
$f_X(\cdot)$	probability density function (PDF) for RV X	256
$f_{X_0,X_1}(\cdot,\cdot)$	joint PDF for RVs X_0 and X_1	258
$f_{X_0 \mid X_1 = x_1}(\cdot)$	conditional PDF for RV X_0 given $X_1 = x_1$	260
$G(\cdot)$	transfer function for $\{g_l\}$	76, 154

Symbol	Description	Pages
$\widetilde{G}(\cdot)$	transfer function for $\{\tilde{g}_l\}$	163, 202
$G_j(\cdot)$	transfer function for $\{g_{j,l}\}$, with $G_1(\cdot) \equiv G(\cdot)$	97, 154
$\widetilde{G}_j(\cdot)$	transfer function for $\{\tilde{g}_{j,l}\}$, with $\widetilde{G}_1(\cdot) \equiv \widetilde{G}(\cdot)$	169, 202
$\mathcal{G}(\cdot)$	squared gain function for $\{g_l\}$	76, 154
$\widetilde{\mathcal{G}}(\cdot)$	squared gain function for $\{\tilde{g}_l\}$	163, 202
$\mathcal{G}_j(\cdot)$	squared gain function for $\{g_{j,l}\}$, with $\mathcal{G}_1(\cdot) \equiv \mathcal{G}(\cdot)$	154
$\widetilde{\mathcal{G}}_j(\cdot)$	squared gain function for $\{\tilde{g}_{j,l}\}$, with $\widetilde{\mathcal{G}}_1(\cdot) \equiv \widetilde{\mathcal{G}}(\cdot)$	202
$\mathcal{G}^{(\mathrm{D})}(\cdot)$	squared gain function for Daubechies scaling filter $\{g_l\}$	105
$\{g_l\}$	DWT scaling filter	75, 154, 463
$\{\tilde{g}_l\}$	MODWT scaling filter	163, 202
$\{g_l^\circ\}$	$\{g_l\}$ periodized to length N	77
$\{\tilde{g}_l^\circ\}$	$\{\tilde{g}_l\}$ periodized to length N	168
$\{\bar{g}_l\}$	reversed scaling filter, i.e., $\bar{g}_l = g_{-l}$	463
$\{g_l^{(\mathrm{ep})}\}$	extremal phase (minimum delay) Daubechies scaling filter	106
$\{g_l^{(\mathrm{la})}\}$	least asymmetric (LA) Daubechies scaling filter	107
$\{g_{j,l}\}$	jth level DWT scaling filter, with $\{g_{1,l}\} \equiv \{g_j\}$	96, 154
$\{\tilde{g}_{j,l}\}$	jth level MODWT scaling filter, with $\{\tilde{g}_{1,l}\} \equiv \{\tilde{g}_j\}$	169, 202
$\{g_{j,l}^\circ\}$	$\{g_{j,l}\}$ periodized to length N	97
$\{\tilde{g}_{j,l}^\circ\}$	$\{\tilde{g}_{j,l}\}$ periodized to length N	170
H	Hurst coefficient	279, 286
$H(\cdot)$	transfer function for $\{h_l\}$	69, 154
$\widetilde{H}(\cdot)$	transfer function for $\{\tilde{h}_l\}$	163, 202
$H_j(\cdot)$	transfer function for $\{h_{j,l}\}$, with $H_1(\cdot) \equiv H(\cdot)$	96, 154
$\widetilde{H}_j(\cdot)$	transfer function for $\{\tilde{h}_{j,l}\}$, with $\widetilde{H}_1(\cdot) \equiv \widetilde{H}(\cdot)$	169, 202
$\mathcal{H}(\cdot)$	squared gain function for $\{h_l\}$	69, 154
$\widetilde{\mathcal{H}}(\cdot)$	squared gain function for $\{\tilde{h}_l\}$	163, 202
$\mathcal{H}_j(\cdot)$	squared gain function for $\{h_{j,l}\}$, with $\mathcal{H}_1(\cdot) \equiv \mathcal{H}(\cdot)$	154
$\widetilde{\mathcal{H}}_j(\cdot)$	squared gain function for $\{\tilde{h}_{j,l}\}$, with $\widetilde{\mathcal{H}}_1(\cdot) \equiv \widetilde{\mathcal{H}}(\cdot)$	202
$\mathcal{H}^{(\mathrm{D})}(\cdot)$	squared gain function for Daubechies wavelet filter $\{h_l\}$	105
$\{h_l\}$	DWT wavelet filter	68–9, 154, 474
$\{\tilde{h}_l\}$	MODWT wavelet filter	163, 202
$\{h_l^\circ\}$	$\{h_l\}$ periodized to length N	70–1
$\{\tilde{h}_l^\circ\}$	$\{\tilde{h}_l\}$ periodized to length N	167–8
$\{\bar{h}_l\}$	reversed wavelet filter, i.e., $\bar{h}_l = h_{-l}$	472, 474
$\{h_{j,l}\}$	jth level DWT wavelet filter, with $\{h_{1,l}\} \equiv \{h_j\}$	95, 154
$\{\tilde{h}_{j,l}\}$	jth level MODWT wavelet filter, with $\{\tilde{h}_{1,l}\} \equiv \{\tilde{h}_j\}$	169, 202
$\{h_{j,l}^\circ\}$	$\{h_{j,l}\}$ periodized to length N	96
$\{\tilde{h}_{j,l}^\circ\}$	$\{\tilde{h}_{j,l}\}$ periodized to length N	170
I_N	$N \times N$ identity matrix	42
$\Im(z)$	imaginary part of complex-valued number z	21
i	$\sqrt{-1}$	20

J — largest DWT level for sample size $N = 2^J$ 57

J_0 — level of partial DWT or of MODWT 104, 145, 169, 199

j — level (index) for scale usually (also used as generic index) 59

k — index for frequency usually (also used as generic index) 46

L — width of wavelet or scaling filter (unit scale) 68

L_j — width of jth level equivalent wavelet or scaling filter 96

L'_j — number of jth level DWT boundary coefficients 146

$L^2(\mathbb{R})$ — set of square integrable real-valued functions 458

$\log_{10}(\cdot)$, $\log(\cdot)$ — log base 10, log base e 73, 400–1

M_j — number of nonboundary jth level MODWT coefficients 306

$M(\mathbf{W}_{j,n})$ — cost of DWPT vector $\mathbf{W}_{j,n}$ 223

$m(\cdot)$ — additive cost functional 223

$m \bmod N$ — m modulo N 30

$m + n \bmod N$ — $(m + n)$ modulo N 30

N — sample size 28, 41

N_j — $N/2^j$, number of jth level DWT coefficients 94

$\mathcal{N}(\mu, \sigma^2)$ — Gaussian (normal) RV with mean μ and variance σ^2 257

n_l — lth component of vector $\mathbf{n}$ 403

$\mathbf{n}$ — transform coefficients for non-IID noise $\boldsymbol{\eta}$ 403

O_l — lth element of $\mathbf{O}$ 43, 398

$O_l^{(\mathrm{ht})}, O_l^{(\mathrm{st})}, O_l^{(\mathrm{mt})}$ — result of applying hard/soft/mid thresholding to O_l 399–400

$\mathcal{O}$ — $N \times N$ orthonormal transform matrix 42

$\mathbf{O}$ — transform coefficients obtained using $\mathcal{O}$ 43

$P_{\mathcal{F}}(f_k)$ — discrete Fourier empirical power spectrum 48

$P_{\mathcal{W}}(\tau_j)$ — discrete wavelet empirical power spectrum (DWT) 62

$P_{\widetilde{\mathcal{W}}}(\tau_j)$ — discrete wavelet empirical power spectrum (MODWT) 180

$\mathcal{P}_j$ — transform matrix for jth stage of DWT pyramid algorithm 94

$\widetilde{\mathcal{P}}_j$ — like $\mathcal{P}_j$, but for MODWT pyramid algorithm 176

$\mathbf{P}[A]$ — probability that the event A will occur 256

$Q_\eta(p)$ — $p \times 100\%$ percentage point of χ^2_η distribution 263–4

$R_{j,t}$ — tth component on jth level of $\mathbf{R}$ for DWT 424

R_l — lth component of vector $\mathbf{R}$ 407

$\mathcal{R}_j$ — jth level wavelet rough (DWT) 66

$\mathbb{R}$ — the entire real axis 457

$\mathbb{R}^N$ — space of real-valued N dimensional vectors 45

$\Re(z)$ — real part of complex-valued number z 21

$\mathbf{R}$ — transform coefficients for stochastic signal $\mathbf{C}$ 407

$S_j(\cdot)$ — SDF for $\{\overline{W}_{j,t}\}$ or for nonboundary part of $\{W_{j,t}\}$ 304, 348

$S_X(\cdot)$ — (power) spectral density function (SDF) 267

$\hat{S}_X^{(\mathrm{mt})}(\cdot)$ — multitaper SDF estimator 274

$\hat{S}_{X,n}^{(\mathrm{mt})}(\cdot)$ — nth eigenspectrum used to form $\hat{S}_X^{(\mathrm{mt})}(\cdot)$ 274

$\hat{S}_X^{(\mathrm{p})}(\cdot)$	periodogram	269
$\mathcal{S}_J$	Jth level wavelet smooth (DWT)	64
$\widetilde{\mathcal{S}}_{J_0}$	J_0th level wavelet smooth (MODWT)	169, 171
$\{s_{X,\tau}\}$	autocovariance sequence (ACVS)	266
$\{\hat{s}_{X,\tau}^{(\mathrm{p})}\}$	'biased' estimator of ACVS	269
$\mathcal{T}$	$N \times N$ circular shift matrix or unit delay operator	52, 457
t	actual time (continuous) or a unitless index (discrete)	5, 24
$U_{j,n}(\cdot)$	transfer function for $\{u_{j,n,l}\}$	215
$\widetilde{U}_{j,n}(\cdot)$	transfer function for $\{\tilde{u}_{j,n,l}\}$	232
$\{u_{j,n,l}\}$	DWPT filter for node (j,n)	214
$\{\tilde{u}_{j,n,l}\}$	MODWPT filter for node (j,n)	231
V_j	approximation subspace for functions of scale λ_j	462
$V_{j,t}$	tth element of $\mathbf{V}_j$	94
$\widetilde{V}_{j,t}$	tth element of $\widetilde{\mathbf{V}}_j$	169
$\mathcal{V}_j$	$N_j \times N$ matrix mapping $\mathbf{X}$ to $\mathbf{V}_j$	94
$\widetilde{\mathcal{V}}_j$	$N \times N$ matrix mapping $\mathbf{X}$ to $\widetilde{\mathbf{V}}_j$	171
$\mathbf{V}_j$	vector of jth level DWT scaling coefficients	94
$\widetilde{\mathbf{V}}_j$	vector of jth level MODWT scaling coefficients	169
$\mathrm{var}\{\cdot\}$	variance operator	259
W_j	detail subspace for functions of scale τ_j	472
$W_{j,n,t}$	tth element of $\mathbf{W}_{j,n}$	214
$\widetilde{W}_{j,n,t}$	tth element of $\widetilde{\mathbf{W}}_{j,n}$	231
$W_{j,t}$	tth element of $\mathbf{W}_j$	94
$\widetilde{W}_{j,t}$	tth element of $\widetilde{\mathbf{W}}_j$	169
$\{\overline{W}_{j,t}\}$	jth level MODWT coefficients for stochastic process $\{X_t\}$	296
W_n	nth DWT coefficient (nth element in $\mathbf{W}$)	57
$\mathcal{W}$	$N \times N$ discrete wavelet transform matrix	57
$\mathcal{W}_j$	$N_j \times N$ matrix mapping $\mathbf{X}$ to $\mathbf{W}_j$ (submatrix of $\mathcal{W}$)	94
$\widetilde{\mathcal{W}}_j$	$N \times N$ matrix mapping $\mathbf{X}$ to $\widetilde{\mathbf{W}}_j$	171
$\mathbf{W}$	vector containing DWT coefficients $\{W_n\}$	57, 150
$\mathbf{W}_j$	vector of jth level DWT wavelet coefficients (part of $\mathbf{W}$)	94
$\widetilde{\mathbf{W}}_j$	vector of jth level MODWT wavelet coefficients	169
$\mathbf{W}_{j,n}$	vector of DWPT coefficients at node (j,n)	209
$\widetilde{\mathbf{W}}_{j,n}$	vector of MODWPT coefficients at node (j,n)	232
$\mathrm{width_a}\{\cdot\}$	autocorrelation width	12, 103
$X_0, \ldots, X_{N-1}$	time series or portion of a stochastic process	41, 269
$\{X_t\}$	time series or stochastic process	41, 266, 295–6
$\overline{X}$	sample mean (arithmetic average) of $X_0, \ldots, X_{N-1}$	48
$\overline{X}_t(\lambda)$	sample mean of $X_{t-\lambda+1}, X_{t-\lambda+2}, \ldots, X_t$	58
$\{\mathcal{X}_k\}$	discrete Fourier transform of $\{X_t\}$	72
$\mathbf{X}$	vector containing $X_0, \ldots, X_{N-1}$	41–2

$Y^{(\mathrm{mt})}(f)$	log multitaper SDF estimate plus a constant	276
$Y^{(\mathrm{p})}(f)$	log periodogram plus a constant	271
Z	Gaussian (normal) RV with unit mean and zero variance	257

- *Greek notation used frequently*

α	exponent of power law spectral density function	279, 281, 286
α	significance level of a test	373, 434
β	slope in linear regression model related to FD parameter δ	374
$\hat{\beta}^{(\mathrm{wls})}$	WLSE of β	376
$\Gamma(\cdot)$	gamma function	257
γ	index for vectors in matching pursuit dictionary	239
γ	Euler's constant $(0.577215664901532\cdots)$	270, 432
$\gamma_{J_0}^{(G)}$, $\gamma_j^{(H)}$	index of coefficient earliest in time in $\mathbf{V}_{J_0}$, $\mathbf{W}_j$	137, 147
$\tilde{\gamma}_{J_0}^{(G)}$, $\tilde{\gamma}_j^{(H)}$	number of 'early' boundary coefficients in $\mathbf{V}_{J_0}$, $\mathbf{W}_j$	137, 147
γ_l^2	ratio of component variances in Gaussian mixture model	410
$\gamma(\cdot)$	real-valued function	457
$\gamma_{j,k}(\cdot)$	translated and dilated version of $\gamma(\cdot)$	459
Δt	sampling interval	48, 59
δ	generic threshold	223, 399
δ	long memory parameter for FD process	283–4, 286, 288
$\delta^{(\mathrm{s})}$	long memory parameter for stationary FD process	288, 368
$\delta^{(\mathrm{S})}$	threshold based on Stein's unbiased risk estimator	405
$\delta^{(\mathrm{u})}$	universal threshold	400
$\delta_{j,k}$	Kronecker delta function	42–3
$\hat{\delta}$	exact MLE of δ for stationary FD process	368
$\hat{\delta}^{(\mathrm{loocv})}$	threshold for leave-one-out cross-validation	402, 423
$\hat{\delta}^{(\mathrm{tfcv})}$	threshold for two-fold cross-validation	402, 422
$\hat{\delta}^{(\mathrm{wls})}$	WLSE of δ for stationary or nonstationary FD process	377
$\tilde{\delta}^{(\mathrm{s})}$	approximate MLE of δ for stationary FD process	363
$\tilde{\delta}^{(\mathrm{s/ns})}$	like $\tilde{\delta}^{(\mathrm{s})}$, but for nonstationary FD processes also	371
$\boldsymbol{\epsilon}$	N dimensional vector containing IID RVs	393
ϵ	a small positive number	2, 486
$\epsilon(f)$, $\epsilon(f_k)$	error term in frequency domain model (uncorrelated)	270, 432
ε_t	tth term in sequence of uncorrelated RVs (white noise)	268
ζ	intercept in linear regression model	374
$\boldsymbol{\eta}$	N dimensional vector containing non-IID RVs	393
η	degrees of freedom in a chi-square distribution	263, 313
$\eta_1, \eta_2, \eta_3, \eta_j$	EDOFs for wavelet variance estimator	313–4, 376
$\eta(f)$, $\eta(f_k)$	error term in frequency domain model (correlated)	276, 440
θ	argument in polar representation $z = \lvert z \rvert e^{i\theta}$	21
θ	parameter with prior distribution in Bayesian model	264

$\theta(\cdot)$	phase function for a filter	25
$\theta^{(G)}(\cdot)$	phase function for DWT scaling filter	106
$\theta^{(H)}(\cdot)$	phase function for DWT wavelet filter	112
$\theta_{c_{j,n,m}}(\cdot)$	component of phase function for DWPT wavelet filter	229
ϑ	degrees of freedom in a t distribution	257, 426
κ	scale parameter in a generalized t distribution	258, 414
κ	RV uniformly distributed over integers $0, 1, \ldots, N-1$	356
Λ_N	$N \times N$ diagonal covariance matrix	355
λ	scale (length of an interval or of an average)	6, 58
λ_j	2^j, unitless scale of jth level scaling coefficients ($j \geq 1$)	85, 481
μ	expected value of a random variable	256–7
ν	lag in frequency domain autocovariance	276–7
ν	advance for time series or filter	111–2
$\nu_j^{(G)}, \nu_j^{(H)}$	advance for scaling filter, wavelet filter	114
$\nu_{j,n}$	advance for wavelet packet filter	229
$\nu_X^2(\tau_j)$	wavelet variance at scale τ_j (time independent)	296
$\hat{\nu}_X^2(\tau_j)$	unbiased MODWT estimator of wavelet variance at scale τ_j	306
$\hat{\hat{\nu}}_X^2(\tau_j)$	unbiased DWT estimator of wavelet variance at scale τ_j	308
$\tilde{\nu}_X^2(\tau_j)$	biased MODWT estimator of wavelet variance at scale τ_j	306
$\tilde{\tilde{\nu}}_X^2(\tau_j)$	biased DWT estimator of wavelet variance at scale τ_j	308
π	$3.141592653589793\cdots$	3, 21
ρ	correlation between two RVs	259
$\rho_{X,\tau}$	autocorrelation sequence for stationary process at lag τ	266
$\hat{\rho}_{X,\tau}$	estimator of autocorrelation sequence at lag τ	16, 341
$\Sigma_{\mathbf{X}}$	covariance matrix for vector $\mathbf{X}$ of RVs	259, 262
$\widetilde{\Sigma}_{\mathbf{X}}$	wavelet-based approximation to covariance matrix $\Sigma_{\mathbf{X}}$	362
σ^2, σ_X^2	variance of a random variable	3, 257, 279
σ_ϵ^2	variance of an IID process	393
σ_ε^2	variance of a white noise process	268
$\sigma_{G_l}^2$	variance of one part of a Gaussian mixture model	410
$\sigma_{n_l}^2$	variance of a non-IID process	403
$\hat{\sigma}_X^2, \hat{\sigma}_Y^2$	sample variance formed using sample mean	48, 299
$\tilde{\sigma}_Y^2$	sample variance formed using process mean	299
$\hat{\sigma}_{(\mathrm{mad})}^2$	estimator of variance formed using MAD	420
$\tilde{\sigma}_{(\mathrm{mad})}^2$	like $\hat{\sigma}_{(\mathrm{mad})}^2$, but based on MODWT	429
τ	lag index in autocorrelation or autocovariance sequence	16, 266
τ_j	2^{j-1}, unitless scale of jth level wavelet coefficients ($j \geq 1$)	59
$\Upsilon(\cdot)$	function whose minimization yields SURE threshold $\delta^{(\mathrm{S})}$	405
υ, υ_l	inverse variance of Laplace distribution	257, 265, 413
$\phi(\cdot)$	scaling function	459
$\phi_{j,k}(\cdot)$	translated and dilated scaling function	460

$\phi^{(\mathrm{H})}(\cdot)$	Haar scaling function	460
$\phi_{p,1},\ldots,\phi_{p,p}$	coefficients of an AR(p) process	268, 292
χ^2_η	chi-square random variable with η degrees of freedom	263
$\psi(\cdot)$	wavelet function	2, 474
$\psi(\cdot)$, $\psi'(\cdot)$	digamma function, trigamma function	275, 376, 440
$\psi_{j,k}(\cdot)$	translated and dilated wavelet function	474
$\psi^{(\mathrm{H})}(\cdot)$	Haar wavelet function	2, 475
$\psi^{(\mathrm{Mh})}(\cdot)$	Mexican hat wavelet function	3
ω_0	parameter for Morlet wavelet function	4

- *Other mathematical conventions and symbols used frequently*

$\approx$	approximately equal to	83–4, 264, 297
$\{a \star a_t\}$	autocorrelation of real-valued sequence $\{a_t\}$	69
$\{a^* \star a_t\}$	autocorrelation of complex-valued sequence $\{a_t\}$	25, 30, 36–7
$\mathcal{O}^H$	complex conjugate (Hermitian) transpose of matrix $\mathcal{O}$	45
z^*	complex conjugate of z	21
$\{a^* \star b_t\}$	complex cross-correlation of sequences $\{a_t\}$ and $\{b_t\}$	24–5, 30
$\in$, $\notin$	contained in, not contained in	2, 398
$\{a * b_t\}$	convolution of sequences $\{a_t\}$ and $\{b_t\}$	24, 30, 36–7
$\lvert\Sigma_{\mathbf{X}}\rvert$	determinant of matrix $\Sigma_{\mathbf{X}}$	361
$\downarrow$	downsampling (removing values from a sequence)	70, 80, 92, 96
$\doteq$	equal at the stated precision (e.g., $\pi \doteq 3.14$ or $\pi \doteq 3.1416$)	3, 73
$\equiv$	equal by definition	20
$\stackrel{\mathrm{d}}{=}$	equal in distribution	257
$\hat{}$	estimator or estimate; e.g., $\hat{\nu}^2_X(\tau_j)$ is an estimator of $\nu^2_X(\tau_j)$	306
$\{a_t\} \longleftrightarrow \{A_k\}$	Fourier transform pair ($\{a_t\}$ is a finite sequence)	29, 36
$\{a_t\} \longleftrightarrow A(\cdot)$	Fourier transform pair ($\{a_t\}$ is an infinite sequence)	23, 35
$\lfloor x \rfloor$, $\lceil x \rceil$	greatest integer $\leq x$, smallest integer $\geq x$	50, 146
$1_{\mathcal{J}}(\cdot)$	indicator function for set $\mathcal{J}$	404
$\langle\cdot,\cdot\rangle$	inner product	42, 45
$\lvert z\rvert$	modulus (absolute value or magnitude) of z	21
$\mathbf{1}$	N dimensional vector of ones	50
$\{a_t^\circ\}$	periodized version (to length N) of infinite sequence $\{a_t\}$	33
$[a,b]$	set of values x such that $a \leq x \leq b$	22
(a,b)	set of values x such that $a < x < b$	2
$(a,b]$	set of values x such that $a < x \leq b$	465
$\lvert\lvert\cdot\rvert\rvert^2$	squared norm	42, 46
$\mathbf{X}^T$, $\mathcal{O}^T$	transpose of vector $\mathbf{X}$, transpose of matrix $\mathcal{O}$	42
$\uparrow$	upsampling (inserting zeros into a sequence)	82, 95, 201
$\mathcal{O}_{j\bullet}$	vector containing elements of jth row of $N \times N$ matrix $\mathcal{O}$	42
$\mathcal{O}_{\bullet k}$	vector containing elements of kth column of $N \times N$ matrix $\mathcal{O}$	42
$\mathbf{0}$, $\mathbf{0}_j$	vector of zeros	101

1

Introduction to Wavelets

1.0 Introduction

Wavelets are mathematical tools for analyzing time series or images (although not exclusively so: for examples of usage in other applications, see Stollnitz *et al.*, 1996, and Sweldens, 1996). Our discussion of wavelets in this book focuses on their use with time series, which we take to be any sequence of observations associated with an ordered independent variable t (the variable t can assume either a discrete set of values such as the integers or a continuum of values such as the entire real axis – examples of both types include time, depth or distance along a line, so a time series need not actually involve time). Wavelets are a relatively new way of analyzing time series in that the formal subject dates back to the 1980s, but in many aspects wavelets are a synthesis of older ideas with new elegant mathematical results and efficient computational algorithms. Wavelet analysis is in some cases complementary to existing analysis techniques (e.g., correlation and spectral analysis) and in other cases capable of solving problems for which little progress had been made prior to the introduction of wavelets.

Broadly speaking (and with apologies for the play on words!), there have been two main waves of wavelets. The first wave resulted in what is known as the continuous wavelet transform (CWT), which is designed to work with time series defined over the entire real axis; the second, in the discrete wavelet transform (DWT), which deals with series defined essentially over a range of integers (usually $t = 0, 1, \ldots, N-1$, where N denotes the number of values in the time series). In this chapter we introduce and motivate wavelets via the CWT. The emphasis is on conveying the ideas behind wavelet analysis as opposed to presenting a comprehensive mathematical development, which by now is available in many other places. Our approach will concentrate on what exactly wavelet analysis can tell us about a time series. We do not presume extensive familiarity with other common analysis techniques (in particular, Fourier analysis). After this introduction in Sections 1.1 and 1.2, we compare and contrast the DWT with the CWT and discuss why we feel the DWT is a natural tool for discrete time series analysis. The remainder of the book will then be devoted to presenting the DWT (and certain closely related transforms) from the ground level up (Chapters 2 to 6), followed by a discussion in Chapters 7 to 10 of the statistical analysis of time series via the DWT. We return to the CWT only in Chapter 11, where we deepen our understanding of the DWT by noting its connection to the CWT in the elegant theory of multiresolution analysis for functions defined over the entire real axis.

1.1 The Essence of a Wavelet

What is a wavelet? As the name suggests, a wavelet is a 'small wave.' A small wave grows and decays essentially in a limited time period. The contrasting notion is obviously a 'big wave.' An example of a big wave is the sine function, which keeps on oscillating up and down on a plot of $\sin(u)$ versus $u \in (-\infty, \infty)$. To begin to quantify the notion of a wavelet, let us consider a real-valued function $\psi(\cdot)$ defined over the real axis $(-\infty, \infty)$ and satisfying two basic properties.

[1] The integral of $\psi(\cdot)$ is zero:

$$\int_{-\infty}^{\infty} \psi(u)\, du = 0. \tag{2a}$$

[2] The square of $\psi(\cdot)$ integrates to unity:

$$\int_{-\infty}^{\infty} \psi^2(u)\, du = 1 \tag{2b}$$

(for the sine function, the above integral would be infinite, so $\sin^2(\cdot)$ cannot be renormalized to integrate to unity).

If Equation (2b) holds, then for any ϵ satisfying $0 < \epsilon < 1$, there must be an interval $[-T, T]$ of finite length such that

$$\int_{-T}^{T} \psi^2(u)\, du > 1 - \epsilon.$$

If we think of ϵ as being very close to zero, then $\psi(\cdot)$ can only deviate insignificantly from zero outside of $[-T, T]$: its nonzero activity is essentially limited to the finite interval $[-T, T]$. Since the length of the interval $[-T, T]$ is vanishingly small compared to the infinite length of the entire real axis $(-\infty, \infty)$, the nonzero activity of $\psi(\cdot)$ can be considered as limited to a relatively small interval of time. While Equation (2b) says $\psi(\cdot)$ has to make some excursions away from zero, Equation (2a) tells us that any excursions it makes above zero must be canceled out by excursions below zero, so $\psi(\cdot)$ must resemble a wave. Hence Equations (2a) and (2b) lead to a 'small wave' or wavelet.

Three such wavelets are plotted in Figure 3. Based on their definitions below, the reader can verify that these functions indeed satisfy Equations (2a) and (2b) (that they integrate to zero is evident from the plots). The first is called the Haar wavelet function:

$$\psi^{(\mathrm{H})}(u) \equiv \begin{cases} -1/\sqrt{2}, & -1 < u \le 0; \\ 1/\sqrt{2}, & 0 < u \le 1; \\ 0, & \text{otherwise} \end{cases} \tag{2c}$$

(a slightly different formulation of this wavelet is discussed in detail in Section 11.6). The above is arguably the oldest wavelet, being named after A. Haar, who developed

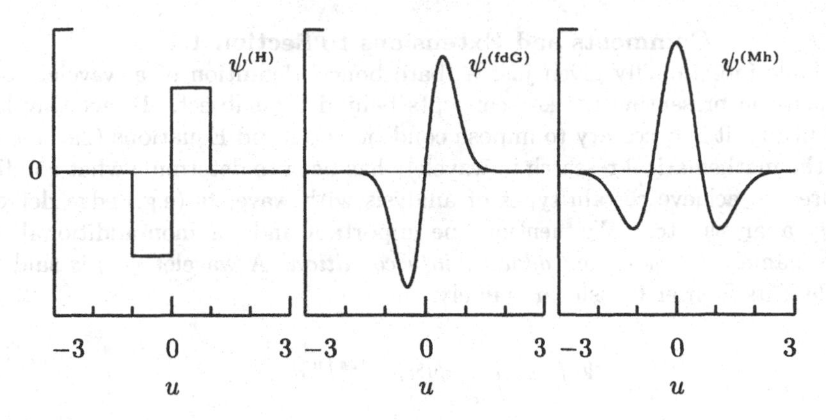

Figure 3. Three wavelets. From left to right, we have one version of the Haar wavelet; a wavelet that is related to the first derivative of the Gaussian probability density function (PDF); and the Mexican hat wavelet, which is related to the second derivative of the Gaussian PDF.

it as an analysis tool in an article in 1910. To form the other two wavelets, we start with the Gaussian (normal) probability density function (PDF) for a random variable with mean zero and variance σ^2:

$$\phi(u) \equiv \frac{e^{-u^2/2\sigma^2}}{\sqrt{2\pi\sigma^2}}, \quad -\infty < u < \infty.$$

The first derivative of $\phi(\cdot)$ is

$$\frac{d\phi(u)}{du} = -\frac{ue^{-u^2/2\sigma^2}}{\sigma^3\sqrt{2\pi}}.$$

If we renormalize the negative of the above to satisfy Equation (2b), we obtain the wavelet

$$\psi^{(\mathrm{fdG})}(u) \equiv \frac{\sqrt{2}\,ue^{-u^2/2\sigma^2}}{\sigma^{3/2}\pi^{1/4}}, \tag{3a}$$

which is shown in the middle of Figure 3 with $\sigma \doteq 0.44311$. With proper renormalization again, the negative of the second derivative of $\phi(\cdot)$ also yields a wavelet, usually referred to as the Mexican hat:

$$\psi^{(\mathrm{Mh})}(u) \equiv \frac{2\left(1 - \frac{u^2}{\sigma^2}\right)e^{-u^2/2\sigma^2}}{\pi^{1/4}\sqrt{3\sigma}}. \tag{3b}$$

The origin of its name should be apparent from a glance at the right-hand plot of Figure 3, in which $\sigma \doteq 0.63628$.

In summary, a wavelet by definition is any function that integrates to zero and is square integrable (see, however, item [1] in the Comments and Extensions below).

Comments and Extensions to Section 1.1

[1] We have intentionally given just a 'bare bones' definition of a wavelet so that we can focus on presenting the key concepts behind the subject. To get wavelets of practical utility, it is necessary to impose conditions beyond Equations (2a) and (2b). Much of the mathematical research in wavelets has been to determine what conditions are required to achieve certain types of analysis with wavelets (e.g., edge detection, singularity analysis, etc.). We mention one important and common additional condition here, namely, the so-called *admissibility condition.* A wavelet $\psi(\cdot)$ is said to be admissible if its Fourier transform, namely,

$$\Psi(f) \equiv \int_{-\infty}^{\infty} \psi(u) e^{-i2\pi f u}\, du,$$

is such that

$$C_\psi \equiv \int_0^{\infty} \frac{|\Psi(f)|^2}{f}\, df \text{ satisfies } 0 < C_\psi < \infty \tag{4a}$$

(Chapter 2 has a review of Fourier theory, including a summary of the key results for functions such as $\psi(\cdot)$ in the last part of Section 2.7). This condition allows the reconstruction of a function $x(\cdot)$ from its continuous wavelet transform (see Equation (11a)). For additional discussion on the admissibility condition, see, e.g., Daubechies (1992, pp. 24–6).

[2] To simplify our exposition in the main part of this chapter, we have assumed $\psi(\cdot)$ to be real-valued, but complex-valued wavelets are often used, particularly in geophysical applications (see, e.g., the articles in Foufoula–Georgiou and Kumar, 1994). One of the first articles on wavelet analysis (Goupillaud *et al.*, 1984) was motivated by Morlet's involvement in geophysical signal analysis for oil and gas exploration. He wanted to analyze signals containing short, high-frequency transients with a small number of cycles, as well as long, low-frequency transients. The examples given used the complex wavelet

$$\psi(u) = C e^{-i\omega_0 u} \left(e^{-u^2/2} - \sqrt{2} e^{-\omega_0^2/4} e^{-u^2} \right), \tag{4b}$$

where C and ω_0 are constants. It is well known that

$$\int_{-\infty}^{\infty} e^{-i\omega_0 u} e^{-u^2/2} du = \sqrt{2\pi} e^{-\omega_0^2/2}$$

(see e.g., Bracewell, 1978, p. 386, or Percival and Walden, 1993, p. 67), from which it follows that $\int_{-\infty}^{\infty} \psi(u) du = 0$, so that Equation (2a) is satisfied. For Morlet, the fact that Equation (2a) holds was not just a mathematical result, but also a physical necessity: the seismic reflection time series under analysis also 'integrate to zero' since compressions and rarefactions must cancel out. The constant C is chosen so that the complex-valued version of Equation (2b), namely, $\int |\psi(u)|^2\, du = 1$, holds for a particular choice of ω_0. For example, when $\omega_0 = 5$, we have $C \doteq 0.7528$. As ω_0 is increased further, the negative term in (4b) becomes negligible; when $\omega_0 = 10$ with $C = \pi^{-1/4}$, we have $\int_{-\infty}^{\infty} |\psi(u)|^2 du \doteq 1$ to nine decimal places accuracy. Hence for large ω_0,

$$\psi(u) \approx \psi_{\omega_0}^{(\mathrm{M})}(u) \equiv \pi^{-1/4} e^{-i\omega_0 u} e^{-u^2/2},$$

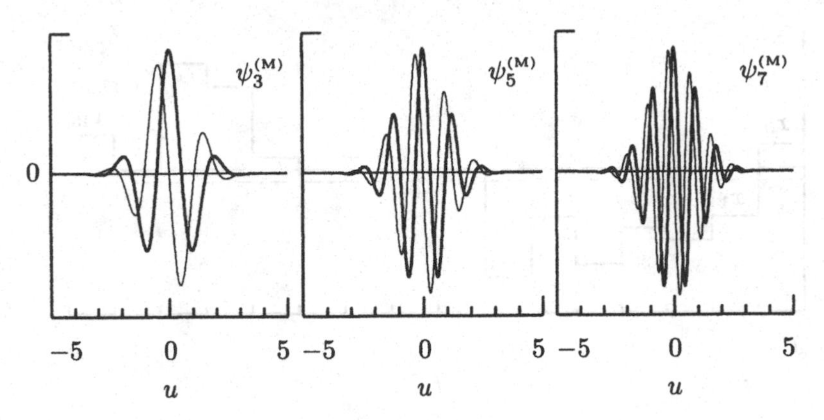

Figure 5. Three Morlet wavelets $\psi_{\omega_0}^{(M)}(\cdot)$. These wavelets are complex-valued, so their real and imaginary parts are plotted using, respectively, thick and thin curves. The parameter ω_0 controls the frequency of the complex exponential that is then modulated by a function whose shape is dictated by the standard Gaussian PDF. As ω_0 increases from 3 to 7, the number of oscillations within the effective width of the Gaussian PDF increases.

which is often called the Morlet wavelet (see, e.g., Kumar and Foufoula-Georgiou, 1994). Note, however, that because $\psi(\cdot)$ integrates exactly to zero, $\psi_{\omega_0}^{(M)}(\cdot)$ cannot also. Examples of $\psi_{\omega_0}^{(M)}(\cdot)$ are plotted in Figure 5 for $\omega_0 = 3$, 5 and 7.

The Morlet wavelet is essentially a complex exponential of frequency $f_0 = \omega_0/(2\pi)$ whose amplitude is modulated by a function proportional to the standard Gaussian PDF. As we argue in the next section, wavelets like those in Figure 3 provide a localized analysis of a function in terms of changes in averages over various scales. The Morlet wavelet yields an analysis that cannot be easily interpreted directly in terms of changes in averages over various scales (however, the similarity between the Morlet wavelet and the ones in Figure 3 is stronger in the frequency domain – all act as band-pass filters).

1.2 The Essence of Wavelet Analysis

We now have a definition for a wavelet and have some idea what one looks like, but for what is it useful? A quick answer is that wavelets such as those depicted in Figure 3 can tell us how weighted averages of certain other functions vary from one averaging period to the next. This interpretation of wavelet analysis is a key concept, so we will explore it in some depth in this section. To begin, let $x(\cdot)$ be a real-valued function of an independent variable t that we will refer to as 'time' (merely for convenience: t can actually have units of, for example, depth rather than time). We will refer to $x(\cdot)$ itself as a 'signal,' again merely for convenience. Let us consider the integral

$$\frac{1}{b-a}\int_a^b x(u)\,du \equiv \alpha(a,b), \tag{5}$$

where we assume that $a < b$ and that $x(\cdot)$ is such that the above integral is well defined. In elementary books on calculus, $\alpha(a,b)$ is called the average value of $x(\cdot)$ over the interval $[a,b]$. The above is related to the notion of a sample mean of a set of N observations. To see this connection, suppose for the moment that $x(\cdot)$ is a step function of the form

$$x(t) = x_j \text{ for } a + \frac{j}{N}(b-a) < t \leq a + \frac{j+1}{N}(b-a) \text{ and } j = 0, \ldots, N-1$$

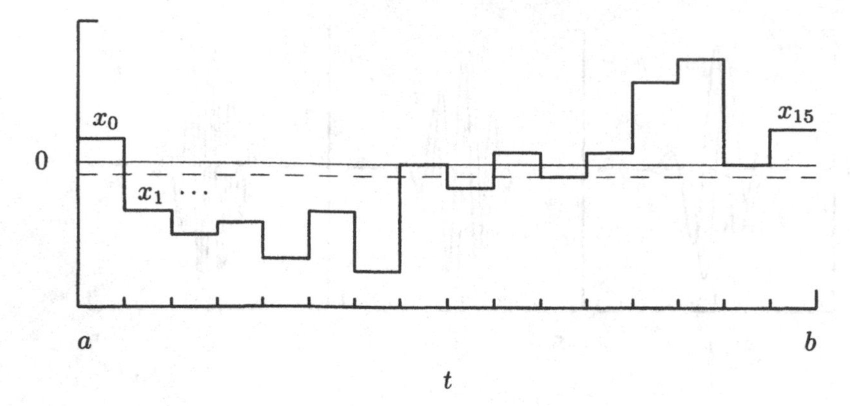

Figure 6. Step function $x(\cdot)$ successively taking on the values $x_0, x_1, \ldots, x_{15}$ over a partitioning of the interval $[a, b]$ into 16 equal subintervals. As defined by Equation (5), the average value of $x(\cdot)$ over $[a, b]$ is just the sample mean of all 16 x_j's (the dashed line indicates this average).

(see Figure 6). From the definition of the Riemann integral, it follows readily that

$$\frac{1}{b-a}\int_a^b x(u)\,du = \frac{1}{b-a}\sum_{j=0}^{N-1} x_j \frac{b-a}{N} = \frac{1}{N}\sum_{j=0}^{N-1} x_j.$$

Instead of regarding an average value $\alpha(a, b)$ as a function of the end points of the interval $[a, b]$ over which the integration is carried out, we can just as easily consider it to be a function of the length of the interval, i.e., $\lambda \equiv b - a$, and the center time of the interval, i.e., $t = (a + b)/2$. We refer to λ as the *scale* associated with the average. Using λ and t, we can define

$$A(\lambda, t) \equiv \alpha(t - \tfrac{\lambda}{2}, t + \tfrac{\lambda}{2}) = \frac{1}{\lambda}\int_{t-\frac{\lambda}{2}}^{t+\frac{\lambda}{2}} x(u)\,du.$$

We call $A(\lambda, t)$ the average value of the signal $x(\cdot)$ over a scale of λ centered about time t.

Average values of signals over various scales are of great interest in the physical sciences. Examples recently analyzed in the scientific literature include:

[1] one second averages of temperature and vertical velocity within and above a deciduous forest in Ontario, Canada (Gao and Li, 1993);
[2] ten minute and hourly rainfall rates of a severe squall line storm over Norman, Oklahoma (Kumar and Foufoula-Georgiou, 1993);
[3] monthly mean sea surface temperatures along the equatorial Pacific ocean (Mak, 1995);
[4] marine stratocumulus cloud inhomogeneities over a large range of scales (Gollmer *et al.*, 1995);
[5] thirty minute average vertical wind velocity profiles (Shimomai *et al.*, 1996);
[6] annual grades of wetness over regions of East China (Jiang *et al.*, 1997); and
[7] yearly average temperatures over central England (Baliunas *et al.*, 1997).

As a specific example to motivate the discussion in this section, consider the bottom part of Figure 8, which shows a series of measurements related to how well a particular cesium beam atomic clock (labeled as 'clock 571') keeps time on a daily basis as compared to a master clock maintained by the US Naval Observatory (this set of data is examined in more detail in Sections 8.6 and 9.7). The measurements are daily average fractional frequency deviates. A precise definition of these deviates is given in, for example, Rutman (1978), but suffice it to say here that they can be regarded as averages of a signal $x(\cdot)$ over intervals of the form $[t-\frac{1}{2}, t+\frac{1}{2}]$, where t is measured in days. The measurements shown at the bottom of Figure 8 are reported once per day, so we are essentially plotting $A(1,t)$ versus t over just integer values.

If $A(1,t) = 0$ at any time t, then clock 571 would have neither gained nor lost time with respect to the master clock over the one day period starting at time $t-\frac{1}{2}$ and ending at time $t+\frac{1}{2}$. When $A(1,t) < 0$ (as is always the case in Figure 8), then clock 571 has lost time, and the amount lost from $t-\frac{1}{2}$ to time $t+\frac{1}{2}$ is proportional to $A(1,t)$. If $A(1,t)$ were always equal to some constant negative value, it would be easy to correct for clock 571's tendency to lose time (similarly, if we know that a watch was set to the correct time on New Year's Day and if we also know that the watch loses exactly a minute each day, then we can get the correct time by looking at the watch and making a correction based upon the day of the year). A good indicator of the inherent ability of clock 571 to keep good time over a scale of a day is thus not $A(1,t)$ itself, but rather how much $A(1,t)$ *changes* from one time period to another. If we associate this change with the time point dividing the two averages, we are led to the quantity

$$D(1,t-\tfrac{1}{2}) \equiv A(1,t) - A(1,t-1) = \int_{t-\frac{1}{2}}^{t+\frac{1}{2}} x(u)\,du - \int_{t-\frac{3}{2}}^{t-\frac{1}{2}} x(u)\,du,$$

or, equivalently,

$$D(1,t) = A(1,t+\tfrac{1}{2}) - A(1,t-\tfrac{1}{2}) = \int_{t}^{t+1} x(u)\,du - \int_{t-1}^{t} x(u)\,du.$$

If $|D(1,t)|$ tends to be large as a function of t, then clock 571 performs poorly on a day to day basis; if $|D(1,t)|$ is close to zero, the daily performance of the clock is good. More generally, the timekeeping abilities of a clock over a scale λ can be judged by examining

$$D(\lambda,t) \equiv A(\lambda,t+\tfrac{\lambda}{2}) - A(\lambda,t-\tfrac{\lambda}{2}) = \frac{1}{\lambda}\int_{t}^{t+\lambda} x(u)\,du - \frac{1}{\lambda}\int_{t-\lambda}^{t} x(u)\,du. \tag{7}$$

For other processes, it is similarly the case that changes in averages over various scales are of more interest than the averages themselves (e.g., changes in daily average air temperature from one day to the next, changes in the temperature of the oceans when averaged over a year, etc.). A plot of $D(1,t)$ would tell us, for example, how quickly the daily average temperature is changing from one day to the next. Similarly, by increasing the scale λ up to a year, a plot of $D(\lambda,t)$ would tell us how much the yearly average temperature is changing from one year to the next.

We are now in a position to connect wavelets to the scheme of looking at changes in averages. Because the two integrals in Equation (7) involve adjacent nonoverlapping

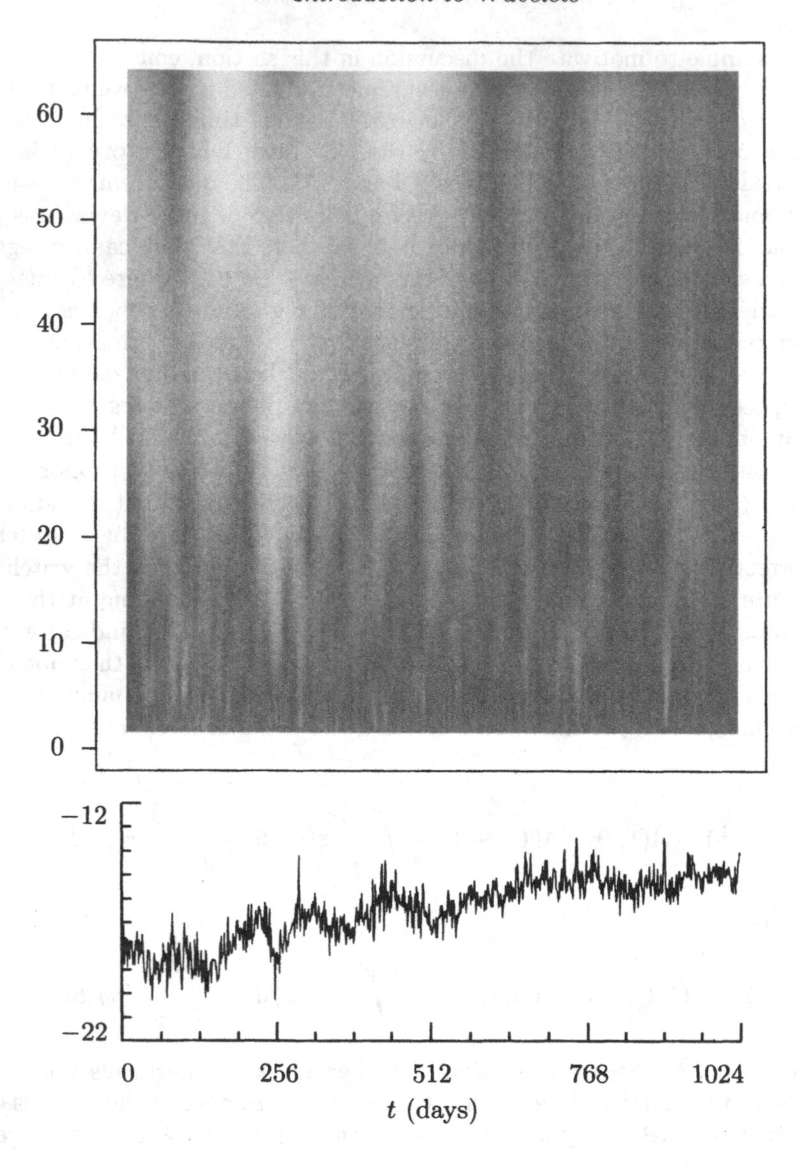

Figure 8. Average daily fractional frequency deviates for cesium beam atomic clock 571 (bottom plot) and its Mexican hat CWT. The fractional frequency deviates are recorded in parts in 10^{13} (a deviation of -15 parts in 10^{13} means that clock 571 lost about 129.6 billionths of a second in one day with respect to the US Naval Observatory master clock to which it was being compared). The vertical axis of the CWT plot is scale (ranging from 1 to 64 days), while the horizontal axis is the same as on the lower plot. The CWT plot is grey-scaled coded so that large magnitudes correspond to bright spots (regions where the plot is dark indicate scales and days at which the clock performed well).

intervals, it is easy to combine them into a single integral over the entire real axis to obtain

$$D(\lambda, t) = \int_{-\infty}^{\infty} x(u)\tilde{\psi}_{\lambda,t}(u)\,du,$$

where

$$\tilde{\psi}_{\lambda,t}(u) \equiv \begin{cases} -1/\lambda, & t-\lambda < u \le t; \\ 1/\lambda, & t < u \le t+\lambda; \\ 0, & \text{otherwise.} \end{cases}$$

Specializing to the case $\lambda = 1$ and $t = 0$ yields

$$\tilde{\psi}_{1,0}(u) \equiv \begin{cases} -1, & -1 < u \le 0; \\ 1, & 0 < u \le 1; \\ 0, & \text{otherwise.} \end{cases}$$

If we compare the above to the Haar wavelet $\psi^{(\mathrm{H})}(\cdot)$ of Equation (2c), we see that $\tilde{\psi}_{1,0}(u) = \sqrt{2}\psi^{(\mathrm{H})}(u)$. The scheme of looking at differences of averages on unit scale at time $t = 0$ is thus equivalent – to within a constant of proportionality – to integrating the product of the signal $x(\cdot)$ and the Haar wavelet. In effect, via the integral

$$\int_{-\infty}^{\infty} x(u)\psi^{(\mathrm{H})}(u)\,du \equiv W^{(\mathrm{H})}(1,0),$$

the Haar wavelet extracts information about how much difference there is between the two unit scale averages of $x(\cdot)$ bordering on time $t = 0$.

It is an easy matter to adjust the Haar wavelet to extract information about unit scale changes at other values of t: we merely need to shift the location of $\psi^{(\mathrm{H})}(\cdot)$. Accordingly, let us define

$$\psi_{1,t}^{(\mathrm{H})}(u) \equiv \psi^{(\mathrm{H})}(u-t) \text{ so that } \psi_{1,t}^{(\mathrm{H})}(u) = \begin{cases} -1/\sqrt{2}, & t-1 < u \le t; \\ 1/\sqrt{2}, & t < u \le t+1; \\ 0, & \text{otherwise.} \end{cases}$$

(See the top row of plots in Figure 10.) Because $\psi_{1,t}^{(\mathrm{H})}(\cdot)$ is just a shifted version of $\psi^{(\mathrm{H})}(\cdot)$, the function $\psi_{1,t}^{(\mathrm{H})}(\cdot)$ obviously satisfies the two basic properties of a wavelet (integration to zero and square-integrable to unity). Integration of the product of $x(\cdot)$ and $\psi_{1,t}^{(\mathrm{H})}(\cdot)$ yields

$$W^{(\mathrm{H})}(1,t) \equiv \int_{-\infty}^{\infty} x(u)\psi_{1,t}^{(\mathrm{H})}(u)\,du = \int_{t}^{t+1} x(u)\,du - \int_{t-1}^{t} x(u)\,du = D(1,t).$$

We can extract similar information about other scales λ and times t by considering

$$\psi_{\lambda,t}^{(\mathrm{H})}(u) \equiv \frac{1}{\sqrt{\lambda}}\psi^{(\mathrm{H})}\left(\frac{u-t}{\lambda}\right) = \begin{cases} -\frac{1}{\sqrt{2\lambda}}, & t-\lambda < u \le t; \\ \frac{1}{\sqrt{2\lambda}}, & t < u \le t+\lambda; \\ 0, & \text{otherwise.} \end{cases}$$

It is easy to see that $\psi_{\lambda,t}^{(\mathrm{H})}(\cdot)$ satisfies Equations (2a) and (2b). Using this wavelet, we obtain

$$W^{(\mathrm{H})}(\lambda,t) \equiv \int_{-\infty}^{\infty} x(u)\psi_{\lambda,t}^{(\mathrm{H})}(u)\,du \propto D(\lambda,t). \tag{9}$$

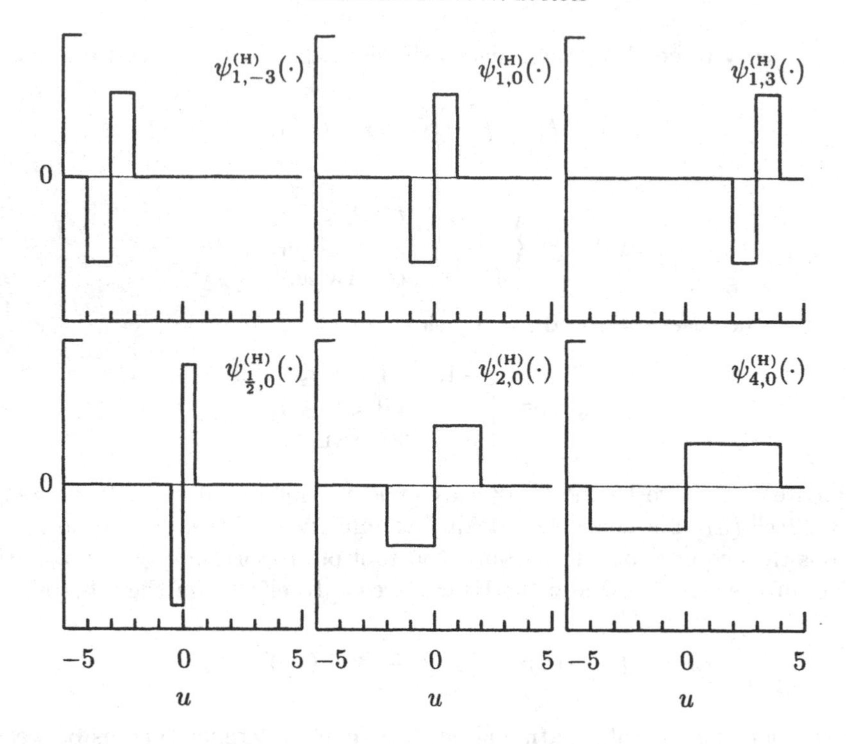

Figure 10. Shifted and rescaled versions of the Haar wavelet $\psi^{(H)}(\cdot)$. The plots above show $\psi^{(H)}_{\lambda,t}(\cdot)$, which can be used to measure how much adjacent averages of a signal $x(\cdot)$ over a scale of length λ change at time t. The top row of plots shows the effect of keeping λ fixed at unity while t is varied; in the bottom row t is fixed at zero while λ is varied.

The bottom row of Figure 10 shows three examples of $\psi^{(H)}_{\lambda,0}(\cdot)$.

By varying λ, we can build up a comprehensive picture of how averages of $x(\cdot)$ over many different scales are changing from one period of length λ to the next. The collection of variables $\{W^{(H)}(\lambda, t) : \lambda > 0, -\infty < t < \infty\}$ is known as the Haar continuous wavelet transform (CWT) of $x(\cdot)$. The interpretation of this transform is that $W^{(H)}(\lambda, t)$ is proportional to the difference between two adjacent averages of scale λ, the first average beginning at time t, and the second, ending at time t.

In a similar manner we can take any other wavelet $\psi(\cdot)$ (e.g., $\psi^{(fdG)}(\cdot)$ or $\psi^{(Mh)}(\cdot)$ of Figure 3) and construct a CWT based upon it by forming

$$W(\lambda, t) \equiv \int_{-\infty}^{\infty} x(u)\psi_{\lambda,t}(u)\, du, \text{ where } \psi_{\lambda,t}(u) \equiv \frac{1}{\sqrt{\lambda}}\psi\left(\frac{u-t}{\lambda}\right). \tag{10}$$

For the wavelets $\psi^{(fdG)}(\cdot)$ and $\psi^{(Mh)}(\cdot)$, we can make an interpretation of what $W(\lambda, t)$ is telling us about $x(\cdot)$ similar to what we built up for the Haar CWT. Considering first $\psi^{(fdG)}(\cdot)$ in the middle plot of Figure 3, we can argue that it yields the difference between adjacent *weighted* averages; i.e., whereas the Haar wavelet is essentially looking at differences of averages that are analogous to a sample mean like

$$\frac{1}{N}\sum_{j=0}^{N-1} x_j,$$

the $\psi^{(\mathrm{fdG})}(\cdot)$ wavelet involves differences of weighted averages analogous to

$$\frac{\sum_{j=0}^{N-1} w_j x_j}{\sum_{j=0}^{N-1} w_j}$$

(the weights for the two averages being differenced are reversed with respect to each other because of asymmetries in $\psi^{(\mathrm{fdG})}(\cdot)$). On the other hand, we can consider the Mexican hat wavelet $\psi^{(\mathrm{Mh})}(\cdot)$ as yielding a difference between a weighted average on unit scale and an average of two weighted averages surrounding it (see the Comments and Extensions below for a discussion as to why $\psi^{(\mathrm{fdG})}(\cdot)$ and $\psi^{(\mathrm{Mh})}(\cdot)$ can be considered to be associated with unit scale).

We now state a fundamental fact about the CWT, namely, that it preserves all the information in $x(\cdot)$. If $\psi(\cdot)$ satisfies the admissibility condition of Equation (4a) and if the signal $x(\cdot)$ satisfies

$$\int_{-\infty}^{\infty} x^2(t)\, dt < \infty$$

then we can recover $x(\cdot)$ from its CWT via

$$x(t) = \frac{1}{C_\psi} \int_0^\infty \left[\int_{-\infty}^{\infty} W(\lambda, u) \frac{1}{\sqrt{\lambda}} \psi\left(\frac{t-u}{\lambda}\right) du \right] \frac{d\lambda}{\lambda^2} \tag{11a}$$

(C_ψ is defined in Equation (4a)); moreover, we also have

$$\int_{-\infty}^{\infty} x^2(t)\, dt = \frac{1}{C_\psi} \int_0^\infty \left[\int_{-\infty}^{\infty} W^2(\lambda, t)\, dt \right] \frac{d\lambda}{\lambda^2} \tag{11b}$$

(Calderón, 1964; Grossmann and Morlet, 1984; Mallat, 1998, pp. 78–9). The left-hand side of the above equation is sometimes used to define the energy in the signal $x(\cdot)$; the above says that $W^2(\lambda, t)/\lambda^2$ essentially defines an energy density function that decomposes the energy across different scales and times.

The fundamental utility of the CWT is that it presents certain information about $x(\cdot)$ in a new manner, giving us the possibility of gaining insight about $x(\cdot)$ that is not readily available in, for example, a plot of $x(t)$ versus t. Because we can recover $x(\cdot)$ from its CWT, a function and its CWT are two representations for the same mathematical entity.

As an example of a CWT, the top plot of Figure 8 shows the CWT for the atomic clock data for scales λ from 1 to 64 days. In this figure the interval between tick marks on the time axis represents 64 days, the length of the longest scale in the CWT. Hence, at scale 64 (top of the top plot), the large magnitude bright spots indicate particularly big changes over the width of a tick mark interval at times around 170 and 480 days. At scale 32, the large magnitude bright spot indicates big changes over the width of half a tick mark interval at times around 256 days; this can also be appreciated from the time series plot. The grey-scaled plot of the CWT is generally 'darker' for the second half of the series, indicating more stable clock performance than in the first half; the dark bands stretching over nearly all scales around times 610 and 720 indicate particularly stable clock performance around these times.

Comments and Extensions to Section 1.2

[1] It is intuitively obvious that the Haar wavelet $\psi^{(\mathrm{H})}(\cdot)$ shown in Figure 3 measures changes on a unit scale, since the positive portion

$$\psi^{(\mathrm{H})+}(u) \equiv \begin{cases} \psi^{(\mathrm{H})}(u) = 1/\sqrt{2}, & \text{if } \psi^{(\mathrm{H})}(u) > 0, \text{ i.e., } 0 < u \leq 1; \\ 0, & \text{otherwise.} \end{cases}$$

and the negative portion, defined analogously, both have 'physical' or visual width of unity. However, another accepted measure of the width of a function $g(\cdot)$ is the autocorrelation width, which by definition is

$$\mathrm{width}_{\mathrm{a}}\{g(\cdot)\} = \frac{|\int g(u)\,du|^2}{\int g^2(u)\,du}.$$

The positive portion of the Haar wavelet has autocorrelation width

$$\mathrm{width}_{\mathrm{a}}\left\{\psi^{(\mathrm{H})+}(\cdot)\right\} = \frac{|\frac{1}{\sqrt{2}}\int_0^1 du|^2}{\frac{1}{2}\int_0^1 du} = 1,$$

and similarly for the negative portion. While there is no obvious 'physical' width for the positive portions of the wavelets $\psi^{(\mathrm{fdG})}(\cdot)$ and $\psi^{(\mathrm{Mh})}(\cdot)$ in Equations (3a) and (3b), we can look at the autocorrelation widths. Now $\psi^{(\mathrm{fdG})}(\cdot)$ and $\psi^{(\mathrm{Mh})}(\cdot)$ both contain an adjustable parameter σ. To make the dependence on σ explicit, we momentarily denote either of these wavelets by $\psi_\sigma(\cdot)$. Then we can write

$$\psi_\sigma(u) = \frac{1}{\sqrt{\sigma}}\psi_1\left(\frac{u}{\sigma}\right).$$

Let

$$\psi_\sigma^+(u) \equiv \begin{cases} \psi_\sigma(u), & \text{if } \psi_\sigma(u) > 0; \\ 0, & \text{otherwise.} \end{cases}$$

Then $\mathrm{width}_{\mathrm{a}}\{\psi_\sigma^+(u)\} = \sigma \times \mathrm{width}_{\mathrm{a}}\{\psi_1^+(u)\}$, where $\mathrm{width}_{\mathrm{a}}\{\psi_1^+(u)\}$ can be computed readily using numerical integration. By then setting $\sigma = 1/\mathrm{width}_{\mathrm{a}}\{\psi_1^+(u)\}$, we force $\mathrm{width}_{\mathrm{a}}\{\psi_\sigma^+(u)\}$, the width of the positive portion, to be unity, as for the Haar wavelet. This procedure yields $\sigma \doteq 0.44311$ for $\psi^{(\mathrm{fdG})}(\cdot)$ and $\sigma \doteq 0.63628$ for $\psi^{(\mathrm{Mh})}(\cdot)$, which are the values used to create the wavelets shown in Figure 3.

[2] For introductory expositions on the CWT with interesting applications in the physical sciences, see Meyers *et al.* (1993), Kumar and Foufoula–Georgiou (1994), Carmona *et al.* (1998) and Torrence and Compo (1998).

1.3 Beyond the CWT: the Discrete Wavelet Transform

An analysis of a signal based upon the CWT yields a potential wealth of information. When presented in the form of a plot such as the one in Figure 8, the CWT is essentially an exploratory data analysis tool that can help the human eye to pick out features of interest. To go beyond these plots, we are essentially faced with an image processing problem because of the two dimensional nature of the CWT. Since the two dimensional CWT depends on just a one dimensional signal, it is obvious that there is a lot of redundancy in the CWT. For example, as we move up into the larger scales in Figure 8, there is little difference in the CWT between adjacent scales (compare, for example, thin horizontal slices of the CWT at scales 40 and 42) and only slow variations across time at any fixed large scale (consider the variations in the CWT as we sweep from left to right along a slice at scale 40). We can thus advance beyond the CWT by considering subsamples that retain certain key features.

The main thrust of this book is on a specific version of the discrete wavelet transform (DWT) that can be applied directly to a time series observed over a discrete set of times, say, $t = 0, \ldots, N-1$. As discussed in Chapter 4, this DWT can be formulated entirely in its own right without explicitly connecting it to any CWT, but it can also be regarded as an attempt to preserve the key features of the CWT in a succinct manner (the precise connection between this DWT and a related CWT is discussed in Chapter 11). From this point of view, the DWT can be thought of as a judicious subsampling of $W(\lambda, t)$ in which we deal with just 'dyadic' scales (i.e., we pick λ to be of the form 2^{j-1}, $j = 1, 2, 3, \ldots$) and then, within a given dyadic scale 2^{j-1}, we pick times t that are separated by multiples of 2^j.

As an example, the left-hand portion of Figure 14 shows a DWT for the clock data based upon the Haar wavelet. The data consist of the $N = 1024$ values shown in the bottom left-hand plot (this is a replicate of the bottom of Figure 8). The DWT of these data also consists of $N = 1024$ values called 'DWT coefficients.' These coefficients can be organized into eight series, which are plotted in the left-hand column above the clock data. Seven of these series are called 'wavelet coefficients' – these are drawn as deviations from zero in the plots marked as 'scale 1' up to 'scale 64.' For scale 2^{j-1}, there are $N_j \equiv N/2^j$ wavelet coefficients, and the times associated with these coefficients are taken to be $(2n+1)2^{j-1} - \frac{1}{2}$, $n = 0, 1, \ldots, N_j - 1$ (for each scale, the product of the number of coefficients and their spacing in time is always N, i.e., the extent of the original data). With appropriate normalization, the nth wavelet coefficient at scale 2^{j-1} can be regarded as an approximation to $W^{(\mathrm{H})}(2^{j-1}, (2n+1)2^{j-1} - \frac{1}{2})$, i.e., the Haar CWT of Equation (9).

The wavelet coefficients for the seven scales account for 1016 of the DWT coefficients. The remaining eight coefficients are known as 'scaling coefficients' and are depicted via connected lines in the top left-hand plot of Figure 14 (these are plotted versus the same times associated with the wavelet coefficients for scale 64). The scaling coefficients are proportional to averages of the original data over a scale of 128 (in contrast to the wavelet coefficients, which are proportional to differences of averages). The scaling coefficients thus reflect long term variations, which is why they exhibit an upward trend similar to that of the data.

The interpretation of the wavelet coefficients in Figure 14 is quite similar to that of $W(\lambda, t)$. For example, the largest coefficient at scale 64 is the one indexed by $n = 1$ and is associated with time $(2n+1) \cdot 64 - \frac{1}{2} = 191.5$. This coefficient reflects the difference between averages of 64 values before and after time 191.5. The fact that this difference is large is an indication that, around $t = 191.5$, the clock performed relatively poorly over scale 64 as compared to other time periods. If we recall that Figure 8 shows the magnitudes of the Mexican hat CWT, we can see similar patterns between that CWT and the wavelet coefficients. For example, both figures indicate that the performance of the clock at scales 16 and 32 improved somewhat with the passage of time; on the other hand, its performance is fairly consistent across time at the smallest scales (see Section 9.6 for a discussion of a statistical test that allows us to assess objectively if wavelet coefficients for a particular scale can be regarded as

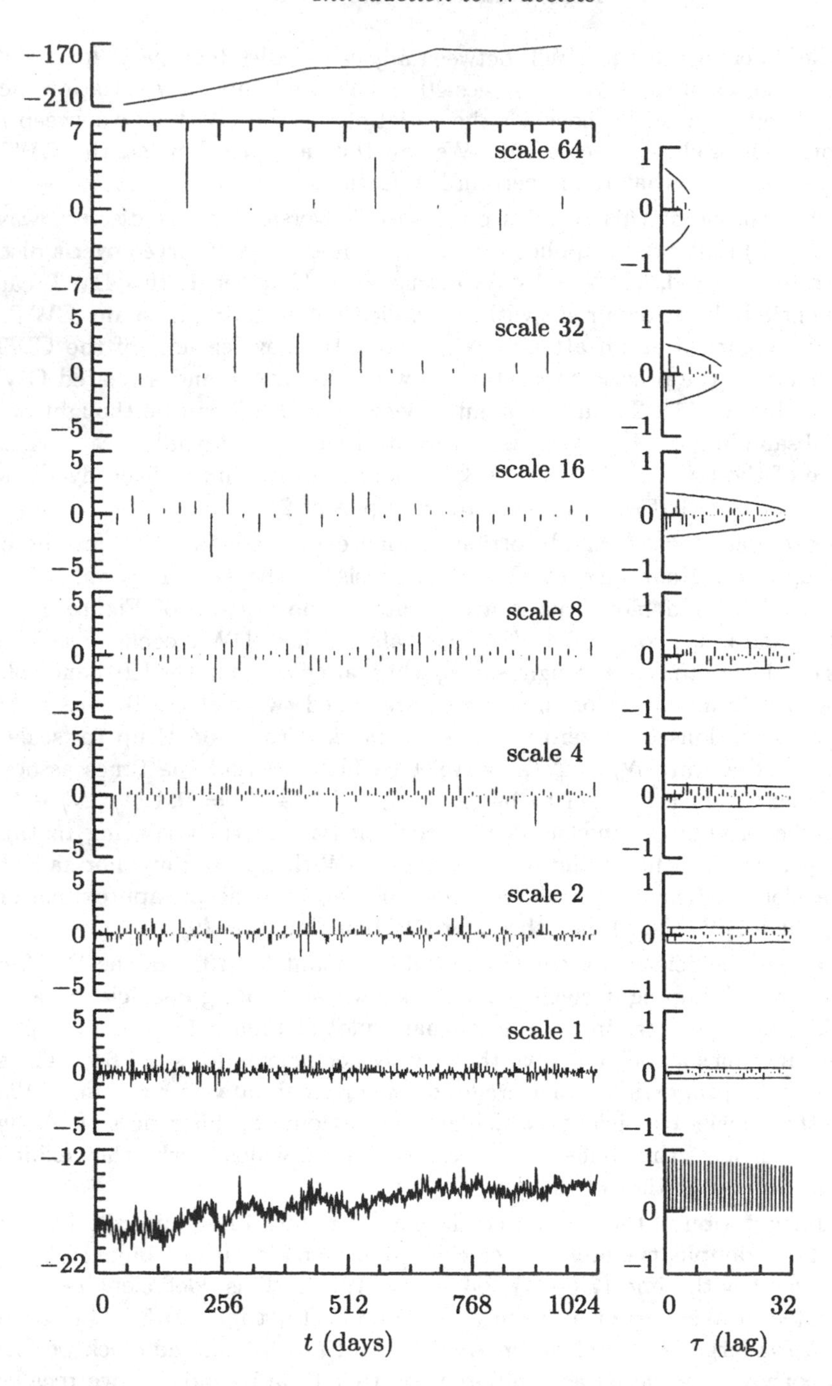

Figure 14. Haar DWT coefficients for clock 571 and sample ACSs.

having homogeneous variability across time).

Just as a signal $x(\cdot)$ can be recovered from its CWT (see Equation (11a)), it is possible to recover the clock data perfectly from its DWT coefficients (see Chapter 4 for details). Thus, while subsampling the CWT at just the dyadic scales might seem to be a drastic reduction, a time series and its DWT are actually two representations for the same mathematical entity. We can therefore claim that, when dealing with a sampled time series (the prevalent format for collecting time series these days), there is no loss of 'information' in dropping down from the CWT to the DWT. In fact the standardized set of dyadic scales is often sufficient to succinctly characterize physical processes, particularly in scientific areas where it has long been natural to describe the evolution of processes in terms of changes over various physical scales – examples abound in atmospheric science, oceanography, astronomy, etc. (If the standard dyadic scale summary of the CWT given by the DWT is not appropriate for certain problems, there are variations on these transforms – known as discrete wavelet packet transforms – that can provide a more appropriate analysis. These and related ideas are explored in Chapter 6.)

An additional aspect of the CWT that the DWT preserves is the ability to decompose the energy in a time series across scales. If we let X_t, $W_{j,n}$ and V_n represent, respectively, the tth data value, the nth wavelet coefficient for scale 2^{j-1} and the nth scaling coefficient, then we have

$$\sum_{t=0}^{1023} X_t^2 = \sum_{j=1}^{7} \sum_{n=0}^{N_j-1} W_{j,n}^2 + \sum_{n=0}^{7} V_n^2,$$

a result that is a discrete analog of the energy decomposition given in Equation (11b). The inner summation of the double summation above can be regarded as the contribution to the energy in the time series due to changes at scale 2^{j-1}. A plot of these contributions versus scale is shown by the connected lines in Figure 16. A clock analyst would deduce from this plot that clock 571 performs best over scales of eight to sixteen days (this would impact how to make best use of this clock in applications such as geodesy).

As discussed in Chapter 8, an energy decomposition is closely related to the concept of the wavelet variance, which partitions the variance of a process across dyadic scales. The wavelet variance is a succinct alternative to the power spectrum based on the Fourier transform, yielding a scale-based analysis that is often easier to interpret than the frequency-based spectrum. With the help of statistical models, we can assess the effect of sampling variability in various estimators of the wavelet variance in order to, for example, form confidence intervals for a hypothesized true wavelet variance (this allows us to assess the effect of sampling variability in plots quite similar to Figure 16). Additionally, as demonstrated by the examples in Chapter 8, we can make use of the localized nature of the wavelet coefficients to analyze the evolution of a process across time. This methodology allows us to study time series that, in contrast to the clock data, have scale-based characteristics that are not homogeneous over time.

Let us now consider another important aspect of the DWT that is illustrated by the right-hand column of plots in Figure 14. The bottom-most plot shows the first part of the sample autocorrelation sequence (ACS) for the time series, which by definition

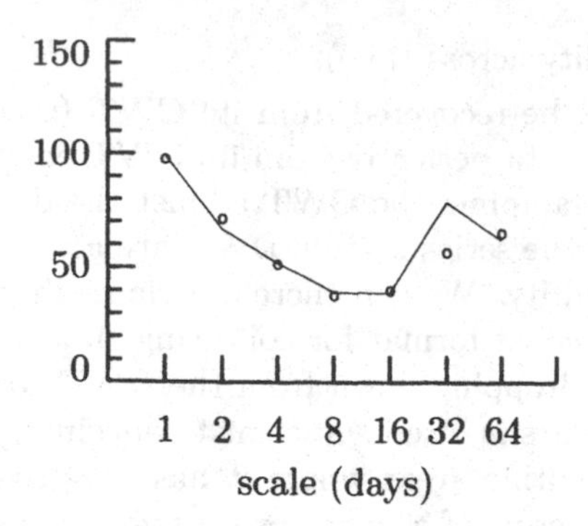

Figure 16. Energy analysis for clock 571 based on Haar DWT wavelet coefficients (curve) and Haar MODWT wavelet coefficients (o's).

is given by

$$\hat{\rho}_{X,\tau} \equiv \frac{\sum_{t=0}^{N-\tau-1}(X_t - \overline{X})(X_{t+\tau} - \overline{X})}{\sum_{t=0}^{N-1}(X_t - \overline{X})^2}, \quad \tau = 1, 2, \ldots, 32,$$

where $\overline{X}$ is the sample mean of the clock data. The statistic $\hat{\rho}_{X,\tau}$ is a measure of the linear association between values of the time series that are separated by τ units. The value of this statistic is constrained such that $-1 < \hat{\rho}_{X,\tau} < 1$. When $\hat{\rho}_{X,\tau}$ is close to unity, the values of X_t and $X_{t+\tau}$ tend to be close to one another (as compared to the total variability in the time series). The plot of $\hat{\rho}_{X,\tau}$ for the clock data indicates a high degree of correlation, even for deviates that are separated by 32 days. (In fact, the sample ACS beyond $\tau = 32$ decays very slowly toward zero. This slow rate of decay is typical of so-called 'long memory' processes – see Section 7.6 for a review of the basic concepts behind these processes.)

The wavelet coefficients for a given scale form a time series in their own right, so it is of interest to examine the sample ACSs for these coefficients. Portions of these ACSs are plotted as deviations from zero in the top seven plots of the right-hand column of Figure 14 (the entire sample ACS is shown for the three largest scales because of the limited number of wavelet coefficients at these scales). Here we see that the sample ACSs are much smaller in magnitude. Under certain mild assumptions that are reasonable here, statistical theory suggests that, if the wavelet coefficients actually had zero autocorrelation, then roughly 95% of the values in the sample ACS for scale 2^{j-1} should fall between $\pm\frac{2\sqrt{(N_j-\tau)}}{N_j}$ (for details, see Fuller, 1996, Corollary 6.3.6.2). These limits are plotted along with the sample ACSs. We see that, when sampling variability is taken into account, there is very little evidence of nonzero autocorrelation in the wavelet coefficients (a similar statement can be made with regard to the correlation between coefficients at different scales). Thus, while the original clock data exhibit a high degree of autocorrelation, their wavelet coefficients are approximately uncorrelated.

A central theme in the statistical analysis of time series has been the search for ways of re-expressing correlated series in terms of some combination of uncorrelated variables. The ACSs in Figure 14 suggest that the DWT effectively decorrelates even highly correlated series, which – combined with the fact that we can reconstruct a time series perfectly from its DWT coefficients – is a fundamental reason why the

DWT is a useful tool for time series analysis. The statistical methodology that we discuss in Chapters 8 (wavelet variance), 9 (analysis and synthesis of long memory processes) and 10 (signal estimation) all depend to a large extent on the perfect reconstruction and decorrelation properties of the DWT. Much of the recent excitement about wavelets can be traced to these two important properties and to the fact that each DWT coefficient depends on just a limited portion of a time series, leading to the possibility of effectively dealing with time series whose statistical characteristics evolve over time. (As discussed in Chapter 4, an additional 'selling point' for the DWT is purely computation. There exists an elegant way of computing the DWT via a 'pyramid' algorithm. By one measure of computational speed, this algorithm is actually faster than the fast Fourier transform algorithm that led to widespread use of the discrete Fourier transform starting in the 1960s.)

Finally, we note that, while the subsampling scheme of the DWT efficiently collapses the two dimensional CWT back into a one dimensional quantity, there are certain time series that can be more usefully analyzed using a less coarse sampling in time. For these series, there is merit in considering subsamples of the CWT that preserve some of its redundancy. In Chapter 5 we consider a variation on the DWT called the maximal overlap DWT (MODWT). Like the DWT, the MODWT can be thought of as a subsampling of the CWT at dyadic scales, but, in contrast to the DWT, we now deal with all times t and not just those that are multiples of 2^j. Retaining all possible times can lead to a more appropriate summary of the CWT because this can eliminate certain 'alignment' artifacts attributable to the way the DWT subsamples the CWT across time.

As an example, Figure 18 shows the Haar MODWT wavelet coefficients (middle seven plots) and scaling coefficients (top plot) for the clock data. In contrast to the DWT, the MODWT has 1024 wavelet coefficients at each scale, and there are also 1024 scaling coefficients. With appropriate normalization, the tth MODWT wavelet coefficient at scale 2^{j-1} can be regarded as an approximation to $W^{(\mathrm{H})}(2^{j-1}, t)$ of Equation (9). If we judiciously subsample and renormalize these MODWT coefficients, we can obtain all the DWT coefficients shown in Figure 14. We can readily reconstruct the clock data from their MODWT coefficients. If we let $\widetilde{W}_{j,t}$ and $\widetilde{V}_t$ represent, respectively, the tth MODWT wavelet coefficient for scale 2^{j-1} and the tth MODWT scaling coefficient, then we have the energy decomposition

$$\sum_{t=0}^{1023} X_t^2 = \sum_{j=1}^{7} \sum_{t=0}^{1023} \widetilde{W}_{j,t}^2 + \sum_{t=0}^{1023} \widetilde{V}_t^2.$$

As was true for the DWT, the inner summation of the double summation can be regarded as the contribution to the energy due to changes at scale 2^{j-1}. The values of these summations are plotted as the circles in Figure 16. This decomposition is put to good use in Chapter 8, where we use it to form a MODWT-based estimator of the wavelet variance. The resulting estimator is statistically more efficient than the corresponding DWT-based estimator, which is an important advantage to the MODWT (while this transform is slower to compute than the DWT, in fact there is an algorithm for computing it that has the same computational complexity as the fast Fourier transform algorithm).

In summary, while the DWT can be motivated as a subsampling of the CWT, it can be justified independently as an important practical tool for time series analysis. The basic reasons why the DWT is such an effective analysis tool are the following.

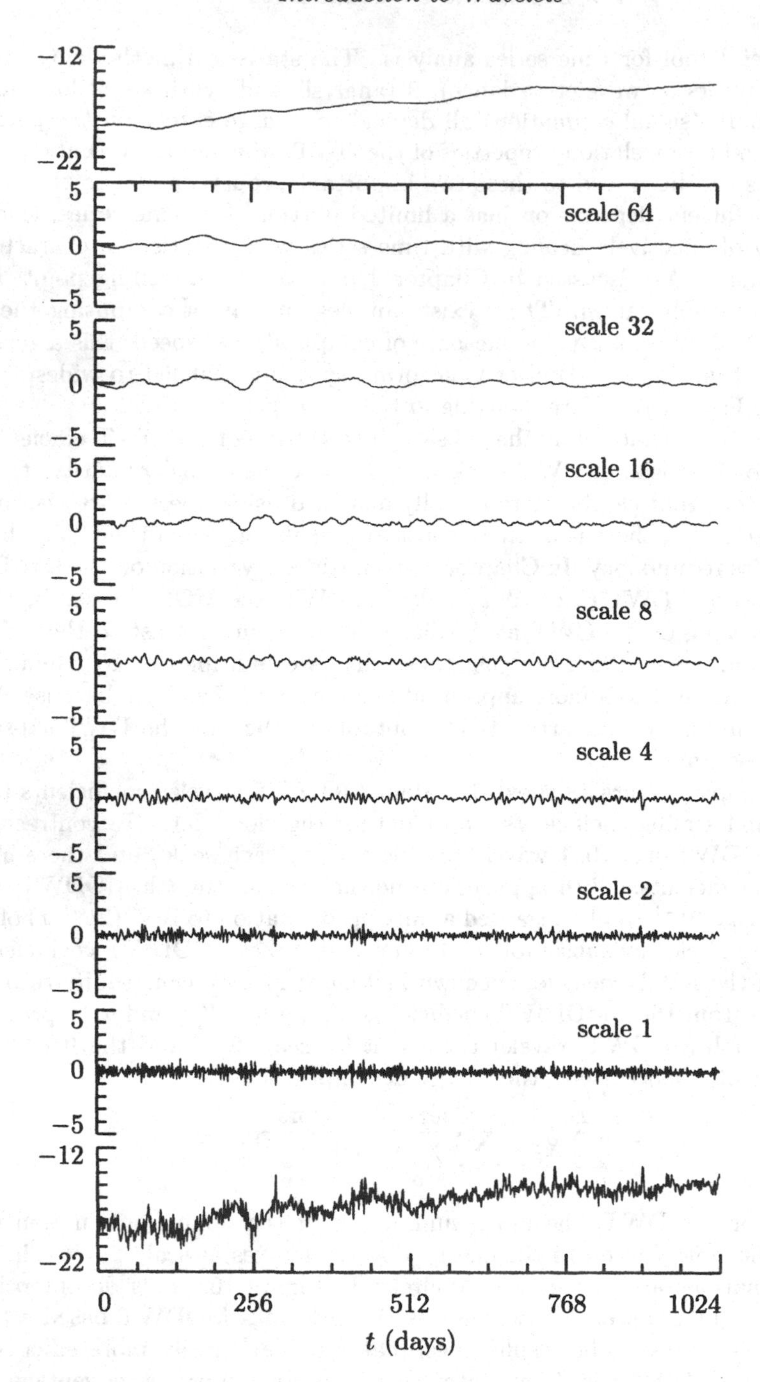

Figure 18. Haar MODWT coefficients for clock 571.

[1] The DWT re-expresses a time series in terms of coefficients that are associated with a particular time and a particular dyadic scale 2^{j-1}. These coefficients are fully equivalent to the original series in that we can perfectly reconstruct a time series from its DWT coefficients.
[2] The DWT allows us to partition the energy in a time series into pieces that are associated with different scales and times. An energy decomposition is very close to the statistical technique known as the analysis of variance (ANOVA), so the DWT leads to a scaled-based ANOVA that is quite analogous to the frequency-based ANOVA provided by the power spectrum.
[3] The DWT effectively decorrelates a wide variety of time series that occurs quite commonly in physical applications. This property is key to the use of the DWT in the statistical methodology discussed in Chapters 8, 9 and 10.
[4] The DWT can be computed using an algorithm that is faster than the celebrated fast Fourier transform algorithm.

A variation on the DWT is the MODWT, which introduces redundancy across time that can be put to good use in analyzing certain time series. The development of all the above - and related topics - from the ground level up is the main thrust of Chapters 4 to 10 of this book.

Comments and Extensions to Section 1.3

[1] For the sake of completeness, we need to give a few details about our analysis of the clock data (some of these depend on material in later chapters). The clock data (obtainable as per the instructions on page *xiv*) actually have 1025 values, but the analysis presented in this section is based on just the first 1024. As discussed in Section 4.7 and Chapter 5, the transforms that we made use of here are actually level $J_0 = 7$ partial DWT and MODWT transforms. In computing the MODWT, we have used reflection boundary conditions (see the discussion surrounding Equation (140)). For reasons discussed in Chapter 8, we can assume that the process mean for each series of wavelet coefficients is zero, which means that we estimated the τth element of the ACS for these coefficients using the formula

$$\frac{\sum_{t=0}^{N_j-\tau-1} W_{j,t}W_{j,t+\tau}}{\sum_{t=0}^{N_j-1} W_{j,t}^2}.$$

2

Review of Fourier Theory and Filters

2.0 Introduction

In subsequent chapters we will make substantial use of some basic results from the Fourier theory of sequences and – to a lesser extent – functions, and we will find that filters play a central role in the application of wavelets. This chapter is intended as a self-contained guide to some key results from Fourier and filtering theory. Our selection of material is intentionally limited to just what we will use later on. For a more thorough discussion employing the same notation and conventions adopted here, see Percival and Walden (1993). We also recommend Briggs and Henson (1995) and Hamming (1989) as complementary sources for further study.

Readers who have extensive experience with Fourier analysis and filters can just quickly scan this chapter to become familiar with our notation and conventions. We encourage others to study the material carefully and to work through as many of the embedded exercises as possible (answers are provided in the appendix). It is particularly important that readers understand the concept of periodized filters presented in Section 2.6 since we use this idea repeatedly in Chapters 4 and 5.

2.1 Complex Variables and Complex Exponentials

The most elegant version of Fourier theory for sequences and functions involves the use of complex variables, so here we review a few key concepts regarding them (see, for example, Brown and Churchill, 1995, for a thorough treatment). Let $i \equiv \sqrt{-1}$ so that $i^2 = -1$ (throughout the book, we take '$\equiv$' to mean 'equal by definition'). If x and y are real-valued variables, then $z \equiv x + iy$ defines a complex-valued variable whose real and imaginary components are, respectively, x and y. Essentially z is a two dimensional variable, and the use of the imaginary number i allows the delineation of its components. We note the following key properties and definitions (x, x_1, x_2, y, y_1 and y_2 are all assumed to be real-valued variables).

[1] Two complex-valued variables $z_1 \equiv x_1 + iy_1$ and $z_2 \equiv x_2 + iy_2$ are equal if and only if their real and imaginary components are equal; i.e., $z_1 = z_2$ if and only if $x_1 = x_2$ and $y_1 = y_2$.

[2] $z_1 + z_2 = x_1 + x_2 + i(y_1 + y_2)$ and $z_1 z_2 = x_1 x_2 - y_1 y_2 + i(x_1 y_2 + x_2 y_1)$.

[3] The complex conjugate of $z = x+iy$ is $z^* \equiv x-iy$. Note that $(z_1+z_2)^* = z_1^*+z_2^*$; $(z_1z_2)^* = z_1^*z_2^*$; and $x^* = x$ (i.e., the complex conjugate of a real-valued variable is the variable itself).

[4] The modulus (or absolute value, or magnitude) of $z = x + iy$ is the nonnegative number $|z| \equiv (x^2 + y^2)^{1/2}$; the squared modulus is $|z|^2 = x^2 + y^2$. Note that $zz^* = |z|^2$; that $|z^*| = |z|$; that $|z|^2 \neq z^2$ unless $y = 0$ (i.e., unless z can be regarded as real-valued); and that $|z_1z_2| = |z_1||z_2|$. If we plot the point (x, y) on a graph, then $|z|$ is the distance of the point from the origin $(0, 0)$.

[5] The complex exponential e^{ix} is defined to be the complex variable whose real and imaginary parts are $\cos(x)$ and $\sin(x)$:

$$e^{ix} \equiv \cos(x) + i\sin(x).$$

The above is known as the Euler relationship. Note that $|e^{ix}|^2 = \cos^2(x) + \sin^2(x) = 1$ for all x. Complex exponentials obey rules similar to those for ordinary exponentials, including:

(a) $e^{i(x+y)} = e^{ix}e^{iy}$;

(b) $(e^{ix})^n = e^{inx}$, where n is any integer (this result is known as de Moivre's theorem); and

(c) $\frac{de^{ix}}{dx} = ie^{ix}$ and $\int e^{ix}\,dx = e^{ix}/i$.

[6] Cosines and sines can be expressed in terms of complex exponentials as follows:

$$\cos(x) = \frac{e^{ix} + e^{-ix}}{2} \text{ and } \sin(x) = \frac{e^{ix} - e^{-ix}}{2i}.$$

[7] The Euler relationship implies that $i = e^{i\pi/2}$, $-i = e^{-i\pi/2}$ and $-1 = e^{\pm i\pi}$. These seemingly innocuous facts are handy for reducing various expressions.

[8] In addition to the Cartesian representation $z = x+iy$, any complex-valued variable has a polar representation, in which we write $z = |z|e^{i\theta}$, where, as before, $|z| \equiv (x^2 + y^2)^{1/2}$, while θ is called the argument of z (usually written as $\arg(z) = \theta$ and well defined only if $|z| > 0$). If we adopt the convention $-\pi < \theta \leq \pi$ and if we draw a line on a graph from the point (x, y) to the origin $(0, 0)$, then θ represents the smaller of the two angles that this line makes with the positive x axis (when $y > 0$, we have $\theta > 0$, while $\theta < 0$ when $y < 0$). We have $\tan(\theta) = y/x$, so θ can be computed from an arctangent function that pays attention to the signs of x and y (i.e., to which quadrant the point (x, y) lies in). The Cartesian representation for z in terms of the real-valued variables x and y is fully equivalent to the polar representation in terms of the nonnegative variable $|z|$ and the argument θ satisfying $-\pi < \theta \leq \pi$.

[9] For $z = x + iy$, denote $\Re(z) \equiv x$ and $\Im(z) \equiv y$ as the functions that extract the real and imaginary components of z.

2.2 Fourier Transform of Infinite Sequences

Let $\{a_t : t = \ldots, -1, 0, 1, \ldots\}$ denote an infinite sequence of real or complex-valued variables such that $\sum_{t=-\infty}^{\infty} |a_t|^2 < \infty$ (when the values that t can assume are clear from the context, we use the shorthand notation $\{a_t\}$). The assumption that the squared magnitudes sum to a finite number is a sufficient – but not necessary – condition ensuring that all the quantities we deal with in this section are well defined. The discrete Fourier transform (DFT) of $\{a_t\}$ is the complex-valued function defined by

$$A(f) \equiv \sum_{t=-\infty}^{\infty} a_t e^{-i2\pi ft},$$

where $-\infty < f < \infty$ is a real-valued variable known as frequency. Recall that, for $f > 0$, a function of u defined by $\cos(2\pi fu)$ or $\sin(2\pi fu)$ for $-\infty < u < \infty$ is called a sinusoid with frequency f, where the notion of frequency has a precise definition, namely, the number of cycles – or fractions thereof – the sinusoid goes through as the continuous variable u sweeps from 0 to 1. Since the Euler relationship states

$$e^{-i2\pi fu} = \cos(2\pi fu) - i\sin(2\pi fu),$$

it is natural to call the function of u defined by $e^{-i2\pi fu}$ a complex exponential with frequency f, where $|f|$ is how many cycles the real and imaginary components of the function each go through as u sweeps from 0 to 1. Note that, in the definition of $A(\cdot)$, we allow f to be negative: any negative frequency should be regarded as merely a mathematical construct that in fact will map to some positive frequency when a physical interpretation is required (consider, for example, Exercise [2.1] regarding the interpretation of negative frequencies in complex demodulation). While use of $e^{-i2\pi ft}$ in the definition of the DFT might worry the uninitiated, the reader can rest assured that the familiar cosine and sine functions are really all that are involved!

The function $A(\cdot)$ is sometimes called the Fourier analysis of $\{a_t\}$. Intuitively, $A(f)$ can be regarded as an attempt to see how well the sequences $\{a_t\}$ and $\{e^{-i2\pi ft}\}$ 'match' each other: if $|A(f)|$ is large (small), then the match is good (bad).

The next two exercises record two important properties of the DFT.

▷ **Exercise [22a]:** Show that $A(\cdot)$ is periodic with unit period; i.e., for any integer j, we have $A(f + j) = A(f)$. ◁

Because of its periodicity, we need only consider $A(\cdot)$ over any interval of unit length. A convenient choice is the closed interval $[-1/2, 1/2]$; i.e., for $|f| \leq 1/2$. The frequency 1/2 is often called the Nyquist frequency.

▷ **Exercise [22b]:** Show that, if $\{a_t\}$ is real-valued, then $A(-f) = A^*(f)$. ◁

Since $|A^*(f)| = |A(f)|$, it follows that the modulus of the DFT of a real-valued sequence is an even function; i.e., $|A(-f)| = |A(f)|$.

The inverse DFT is given by the following exercise.

▷ **Exercise [22c]:** Show that

$$\int_{-1/2}^{1/2} A(f)e^{i2\pi ft}\,df = a_t, \qquad t = \ldots, -1, 0, 1, \ldots.$$ ◁

The above says that we can synthesize (reconstruct) the sequence $\{a_t\}$ from its DFT $A(\cdot)$, so, in a very strong sense, $\{a_t\}$ and its DFT are two representations for the same mathematical entity. Roughly speaking, the inverse DFT states that $\{a_t\}$ can be re-expressed as a linear combination of sinusoids – the larger the magnitude $|A(f)|$

of $A(f)$, the more important sinusoids of frequency f are in reconstructing $\{a_t\}$. The Fourier relationship between $\{a_t\}$ and $A(\cdot)$ is noted as

$$\{a_t\} \longleftrightarrow A(\cdot),$$

by which it is meant that

$$A(f) = \sum_{t=-\infty}^{\infty} a_t e^{-i2\pi ft} \text{ and } a_t = \int_{-1/2}^{1/2} A(f) e^{i2\pi ft}\, df,$$

with the infinite summation and the integral both being well defined. The sequence $\{a_t\}$ and the function $A(\cdot)$ are said to constitute a Fourier transform pair. It is occasionally handy to use the alternative notation $\{a_t\} \longleftrightarrow A(f)$ even though this blurs our distinction between a function $A(\cdot)$ and its value $A(f)$ at f.

We note a particularly important result in the following exercise.

▷ **Exercise [23a]:** Suppose $\{a_t\} \longleftrightarrow A(\cdot)$ and $\{b_t\} \longleftrightarrow B(\cdot)$, where $\sum_{t=-\infty}^{\infty} |a_t|^2 < \infty$ and $\sum_{t=-\infty}^{\infty} |b_t|^2 < \infty$. Derive the 'two sequence' version of Parseval's theorem, namely,

$$\sum_{t=-\infty}^{\infty} a_t b_t^* = \int_{-1/2}^{1/2} A(f) B^*(f)\, df.$$ ◁

An immediate corollary of the above is the 'one sequence' version of Parseval's theorem:

$$\sum_{t=-\infty}^{\infty} |a_t|^2 = \int_{-1/2}^{1/2} |A(f)|^2\, df.$$

Parseval's theorem says that a key property of a sequence (the sum of its squared magnitudes) is preserved in a simple manner by its Fourier transform.

We will have occasion to refer to the following result, which relates the DFT of the even indexed variables in a sequence to the DFT of the entire sequence.

▷ **Exercise [23b]:** Suppose $\{a_t : t = \ldots, -1, 0, 1, \ldots\} \longleftrightarrow A(\cdot)$. Show that the DFT of the infinite sequence $\{a_{2n} : n = \ldots, -1, 0, 1, \ldots\}$ is the function defined by $\frac{1}{2}[A(\frac{f}{2}) + A(\frac{f}{2} + \frac{1}{2})]$. ◁

Finally, suppose that $\{a_t : t = 0, \ldots, N-1\}$ is a finite sequence of N variables, and suppose that we extend it to be an infinite sequence by defining $a_t = 0$ for $t \le -1$ and $t \ge N$. The DFT of the infinite sequence is then given by

$$A(f) \equiv \sum_{t=-\infty}^{\infty} a_t e^{-i2\pi ft} = \sum_{t=0}^{N-1} a_t e^{-i2\pi ft}.$$

We can thus take the notation

$$\{a_t : t = 0, \ldots, N-1\} \longleftrightarrow A(\cdot)$$

to mean that

$$A(f) \equiv \sum_{t=0}^{N-1} a_t e^{-i2\pi ft}, \qquad |f| \le 1/2;$$

i.e., we are implicitly assuming that a_t is zero for indices t outside of $0, \ldots, N-1$. If the range of indices over which a_t is explicitly defined is clear from the context, we use the shorthand notation $\{a_t\} \longleftrightarrow A(\cdot)$.

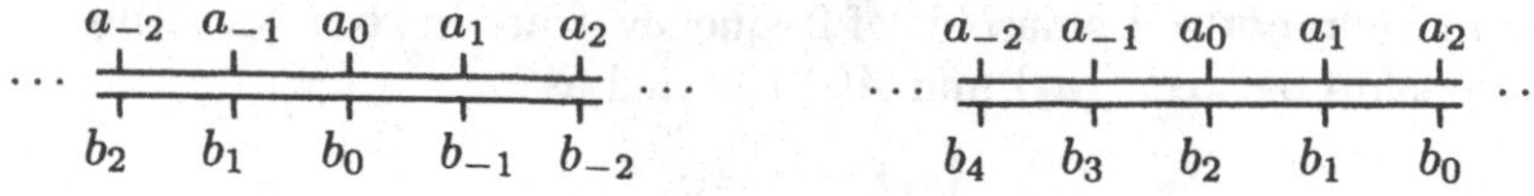

$$\cdots \begin{array}{ccccc} a_{-2} & a_{-1} & a_0 & a_1 & a_2 \\ b_2 & b_1 & b_0 & b_{-1} & b_{-2} \end{array} \cdots \qquad \cdots \begin{array}{ccccc} a_{-2} & a_{-1} & a_0 & a_1 & a_2 \\ b_4 & b_3 & b_2 & b_1 & b_0 \end{array} \cdots$$

$$a * b_0 = \sum_{u=-\infty}^{\infty} a_u b_{-u} \qquad\qquad a * b_2 = \sum_{u=-\infty}^{\infty} a_u b_{2-u}$$

Figure 24. Graphical illustration of convolution of the infinite sequences $\{a_t\}$ and $\{b_t\}$. The left-hand plot shows two lines. The upper line is labeled at equal intervals with elements of the infinite sequence $\{a_t\}$. The lower line is likewise labeled, but now with the *reverse* of $\{b_t\}$, i.e., $\{b_{-t}\}$ The zeroth element $a * b_0$ of the convolution of $\{a_t\}$ and $\{b_t\}$ is obtained by multiplying the a_t's and b_t's facing each other and then summing. In general, the tth element $a * b_t$ is obtained in a similar fashion *after* the lower line has been shifted to the right by t divisions – for example, the right-hand plot shows the alignment of the lines that yields the second element $a * b_2$ of the convolution.

2.3 Convolution/Filtering of Infinite Sequences

Suppose that $\{a_t\}$ and $\{b_t\}$ are two infinite sequences of real or complex-valued variables satisfying $\sum_{t=-\infty}^{\infty} |a_t|^2 < \infty$ and $\sum_{t=-\infty}^{\infty} |b_t|^2 < \infty$. Define the *convolution* of $\{a_t\}$ and $\{b_t\}$ as the infinite sequence whose tth element is

$$a * b_t \equiv \sum_{u=-\infty}^{\infty} a_u b_{t-u}, \qquad t = \ldots, -1, 0, 1, \ldots \tag{24}$$

(to simplify our discussion, we assume throughout this section that all convolutions $\{a * b_t\}$ satisfy $\sum_{t=-\infty}^{\infty} |a * b_t|^2 < \infty$, but it is important to note that this need not be true in general). Note that the second sequence $\{b_t\}$ is reversed and shifted with respect to the first prior to the element by element multiplication, a procedure that is illustrated in Figure 24.

The following exercise states probably the most important fact about convolutions.

▷ **Exercise [24]:** If $\{a_t\} \longleftrightarrow A(\cdot)$ and $\{b_t\} \longleftrightarrow B(\cdot)$, show that the DFT of the convolution $\{a * b_t\}$ is the function defined by $A(f)B(f)$; i.e.,

$$\sum_{t=-\infty}^{\infty} a * b_t e^{-i2\pi f t} = A(f)B(f), \qquad |f| \le 1/2. \qquad \triangleleft$$

A paraphrase for this result is 'convolution in the time domain is equivalent to multiplication in the frequency domain.' Using our notation for Fourier transform pairs, we can write

$$\{a * b_t\} \longleftrightarrow A(\cdot)B(\cdot),$$

where $A(\cdot)B(\cdot)$ refers to the function whose value at f is $A(f)B(f)$.

A concept that is closely related to convolution is complex cross-correlation, which is the infinite sequence defined by

$$a^* \star b_t \equiv \sum_{u=-\infty}^{\infty} a_u^* b_{u+t}, \qquad t = \ldots, -1, 0, 1, \ldots.$$

The complex cross-correlation is the convolution of $\{a_t^*\}$ with the time reverse of $\{b_t\}$. Exercise [2.2] says that $\{a^* \star b_t\} \longleftrightarrow A^*(\cdot)B(\cdot)$; i.e.,

$$\sum_{t=-\infty}^{\infty} a^* \star b_t e^{-i2\pi ft} = A^*(f)B(f). \tag{25a}$$

Letting $b_t = a_t$ in the above gives us the autocorrelation of $\{a_t\}$, which is the infinite sequence defined by

$$a^* \star a_t \equiv \sum_{u=-\infty}^{\infty} a_u^* a_{u+t}, \qquad t = \ldots, -1, 0, 1, \ldots. \tag{25b}$$

An immediate consequence of Exercise [2.2] is that

$$\{a^* \star a_t\} \longleftrightarrow A^*(\cdot)A(\cdot) = |A(\cdot)|^2, \tag{25c}$$

where $|A(\cdot)|^2$ denotes the function whose value at f is $|A(f)|^2$. Thus the DFT of an autocorrelation is not only real-valued but also necessarily nonnegative.

If we denote the components of a convolution properly, we are led to the engineering notion of filtering (technically, what we are calling a filter is actually a linear time-invariant filter; see, for example, Percival and Walden, 1993, for details). Thus, if we regard $\{a_t\}$ in Equation (24) as a filter and $\{b_t\}$ as a sequence to be filtered, then $\{a * b_t\}$ is the output from the filter, i.e., the filtered version of $\{b_t\}$. We can depict filtering in a flow diagram such as the following:

$$\{b_t\} \longrightarrow \boxed{\{a_t\}} \longrightarrow \{a * b_t\},$$

which says that inputing $\{b_t\}$ into the filter $\{a_t\}$ yields $\{a * b_t\}$ as its output. Since $\{a_t\} \longleftrightarrow A(\cdot)$, an equivalent representation for a filter is $A(\cdot)$. It is thus convenient to regard a filter as an abstract construct, two of whose representations are $\{a_t\}$ and $A(\cdot)$, so we can equally well refer to 'the filter $\{a_t\}$' or 'the filter $A(\cdot)$.' The 'time domain' representation $\{a_t\}$ is called the ***impulse response sequence*** for the filter, while its 'frequency domain' representation $A(\cdot)$ is called the ***transfer function*** (or ***frequency response function***). A flow diagram equivalent to the above is thus

$$\{b_t\} \longrightarrow \boxed{A(\cdot)} \longrightarrow \{a * b_t\}.$$

Since in general the transfer function $A(\cdot)$ for a filter is complex-valued, it is convenient to consider its polar representation, namely,

$$A(f) = |A(f)|e^{i\theta(f)}.$$

The functions defined by $|A(f)|$ and $\theta(f)$ are called, respectively, the ***gain function*** and the ***phase function*** for the filter (note that $\theta(f)$ is well defined only if $|A(f)| > 0$). Similarly, $|A(f)|^2$ defines the ***squared gain function.***

As our first concrete example of a filter, consider a filter whose impulse response sequence is given by

$$a_t = \begin{cases} 1/2, & t = 0; \\ 1/4, & t = \pm 1; \\ 0, & \text{otherwise.} \end{cases} \tag{25d}$$

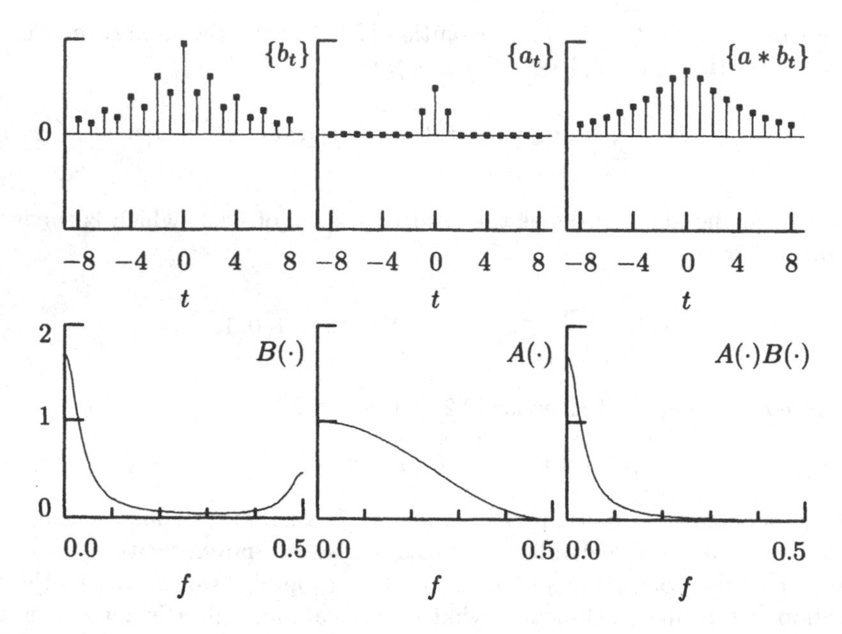

Figure 26. Example of filtering using a low-pass filter.

Exercise [2.3] is to show that its transfer function $A(\cdot)$ is given by $\cos^2(\pi f)$. We need an infinite sequence to filter, so consider

$$b_t \equiv \sum_{l=1}^{L} C_l \phi_l^{|t|}, \qquad t = \ldots, -1, 0, 1, \ldots, \tag{26a}$$

where the C_l's and ϕ_l's are all real-valued with $|\phi_l| < 1$. Exercise [2.4] is to show that the DFT of $\{b_t\}$ is given by

$$B(f) \equiv \sum_{l=1}^{L} C_l \frac{1 - \phi_l^2}{1 - 2\phi_l \cos(2\pi f) + \phi_l^2}. \tag{26b}$$

The DFT of the result of filtering $\{b_t\}$ with $\{a_t\}$ is thus given by $A(f)B(f) = \cos^2(\pi f)B(f)$.

An example is shown in Figure 26 for the case $L = 2$, $C_1 = 3/16$, $\phi_1 = 4/5$, $C_2 = 1/20$ and $\phi_2 = -4/5$ (the plots of the sequences are truncated at $t = \pm 8$). The DFT $B(\cdot)$ shows a low frequency peak at $f = 0$ and a high frequency peak at $f = 1/2$. The low frequency peak captures the overall decay of $\{b_t\}$ from its largest value of 19/80 at $t = 0$ to 0 as $t \to \pm\infty$. The high frequency peak is due to the 'choppy' variation between adjacent values. The transfer function for $\{a_t\}$ indicates that this filter is a 'low-pass' filter: it preserves low frequency components while attenuating high frequency components. The DFT $A(\cdot)B(\cdot)$ of the filtered series is thus very small at high frequencies, which is reflected in $\{a * b_t\}$ as a lack of 'choppiness' in comparison to $\{b_t\}$.

By way of contrast, let us consider the filter whose impulse response sequence is

$$a_t = \begin{cases} 1/2, & t = 0; \\ -1/4, & t = \pm 1; \\ 0, & \text{otherwise.} \end{cases} \tag{26c}$$

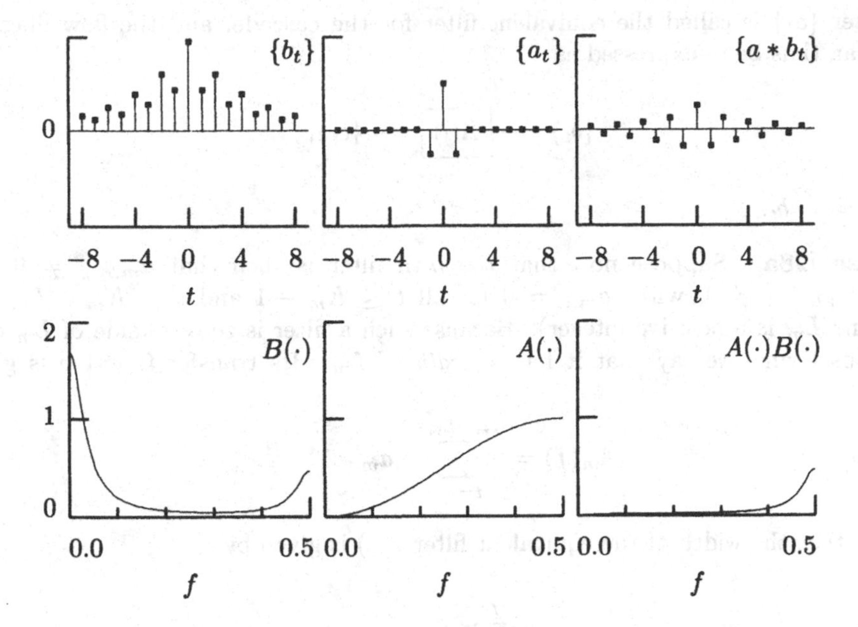

Figure 27. Example of filtering using a high-pass filter.

Exercise [2.5] says that its transfer function is given by $A(f) = \sin^2(\pi f)$. This transfer function is shown in the middle of the bottom row of plots in Figure 27 and indicates that the above is a 'high-pass' filter because it preserves high frequencies while attenuating low frequencies. The filtered series $\{a * b_t\}$ now consists of the choppy portion of $\{b_t\}$ that was eliminated by the low-pass filter in (25d).

Finally let us introduce the important notion of a *cascade of filters*, which we will encounter repeatedly in our study of wavelets. By definition a filter cascade is an arrangement of a set of M filters such that the output from the first filter is the input to the second filter and so forth.

▷ **Exercise [27]:** Suppose that $\{a_{m,t} : t = \ldots, -1, 0, 1, \ldots\}$, $m = 1, \ldots, M$, are M filters with corresponding transfer functions $A_m(\cdot)$. Suppose that these filters form a cascade whose mth filter is $\{a_{m,t}\}$. Let $\{b_t\}$ be the input to the cascade (i.e., the input to $\{a_{1,t}\}$), and let $\{c_t\}$ be the output (i.e., the output from $\{a_{M,t}\}$). In terms of a flow diagram we have

$$\{b_t\} \longrightarrow \boxed{A_1(\cdot)} \longrightarrow \boxed{A_2(\cdot)} \longrightarrow \cdots \longrightarrow \boxed{A_M(\cdot)} \longrightarrow \{c_t\}.$$

Show that we can write

$$c_t = \sum_{u=-\infty}^{\infty} a_u b_{t-u}, \quad t = \ldots, -1, 0, 1, \ldots,$$

where $\{a_t\}$ is a filter whose impulse response sequence is given by the successive convolution of the individual filters and whose transfer function is given by

$$A(f) \equiv \prod_{m=1}^{M} A_m(f).$$

(Recall our assumption that an impulse response sequence formed by convolution has squared magnitudes summing to a finite value.) ◁

The filter $\{a_t\}$ is called the equivalent filter for the cascade, and the flow diagram above can thus be re-expressed as

$$\{b_t\} \longrightarrow \boxed{A(\cdot)} \longrightarrow \{c_t\},$$

with $c_t = a * b_t$.

▷ **Exercise [28a]:** Suppose now that the mth filter is such that $a_{m,K_m} \neq 0$ and $a_{m,K_m+L_m-1} \neq 0$, while $a_{m,t} = 0$ for all $t \leq K_m - 1$ and $t \geq K_m + L_m$ (we assume L_m is a positive integer). Because such a filter is zero outside of L_m contiguous terms, we say that it has a *width* of L_m. Its transfer function is given by

$$A_m(f) = \sum_{t=K_m}^{K_m+L_m-1} a_{m,t} e^{-i2\pi f t}.$$

Show that the width of the equivalent filter $A(\cdot)$ is given by

$$L \equiv \sum_{m=1}^{M} L_m - M + 1. \tag{28a}$$ ◁

2.4 Fourier Transform of Finite Sequences

Suppose now that $\{a_t\} = \{a_t : t = 0, \ldots, N-1\}$ is a sequence of N real- or complex-valued variables. Its discrete Fourier transform (DFT) is the sequence $\{A_k\}$ of N variables given by

$$A_k \equiv \sum_{t=0}^{N-1} a_t e^{-i2\pi t k/N}, \qquad k = 0, \ldots, N-1. \tag{28b}$$

Note that A_k is associated with frequency $f_k \equiv k/N$. The right-hand side of the above can be used to define A_k for all integers k, which yields an infinite sequence with the following property.

▷ **Exercise [28b]:** Show that, if A_k is defined by the right-hand side of Equation (28b) for all k, then the resulting infinite sequence is periodic with a period of N terms; i.e., show that, for any nonzero integer n and any integer k such that $0 \leq k \leq N-1$, we have $A_{k+nN} = A_k$ (note the analogy between this and Exercise [22a]). ◁

Some authors call $\{A_k\}$ a 'finite Fourier transform' and use the name 'discrete Fourier transform' for infinite sequences only. We prefer to use 'discrete Fourier transform' for both finite and infinite sequences. Use of 'finite Fourier transform' leads to the unfortunate acronym 'FFT,' which is now commonly used for the 'fast Fourier transform' algorithm, a collection of procedures for computing $\{A_k\}$ quickly (see, for example, Oppenheim and Schafer, 1989).

We can reconstruct $\{a_t\}$ from its DFT $\{A_k\}$ using the following.

▷ **Exercise [29a]:** Show that

$$\frac{1}{N}\sum_{k=0}^{N-1} A_k e^{i2\pi tk/N} = a_t, \qquad t = 0, \ldots, N-1.$$

Hint: if z is any complex variable not equal to 1, show that

$$\sum_{u=0}^{N-1} z^u = \frac{1-z^N}{1-z},$$

and then note that we can write $\exp(i2\pi u/N) = z^u$ by letting $z = \exp(i2\pi/N)$. ◁

Since we can reconstruct $\{a_t\}$ from its DFT, these two sequences $\{a_t\}$ and $\{A_k\}$ can be considered equivalent representations of a common mathematical entity. The Fourier relationship between $\{a_t\}$ and $\{A_k\}$ is noted as

$$\{a_t\} \longleftrightarrow \{A_k\},$$

by which it is meant that

$$A_k = \sum_{t=0}^{N-1} a_t e^{-i2\pi tk/N} \text{ and } a_t = \frac{1}{N}\sum_{k=0}^{N-1} A_k e^{i2\pi tk/N},$$

where $\{a_t\}$ can now be called the inverse DFT of $\{A_k\}$. The right-hand summation above can be used to define a_t for $t \leq -1$ and $t \geq N$, yielding an infinite sequence that is periodic with period N (this result is a minor variation on Exercise [28b]).

The analog of Exercise [23a] for finite sequences is the following.

▷ **Exercise [29b]:** Suppose $\{a_t\} \longleftrightarrow \{A_k\}$ and $\{b_t\} \longleftrightarrow \{B_k\}$. Show that

$$\sum_{t=0}^{N-1} a_t b_t^* = \frac{1}{N}\sum_{k=0}^{N-1} A_k B_k^*.$$ ◁

By letting $b_t = a_t$, we obtain Parseval's theorem for finite sequences, namely,

$$\sum_{t=0}^{N-1} |a_t|^2 = \frac{1}{N}\sum_{k=0}^{N-1} |A_k|^2.$$

2.5 Circular Convolution/Filtering of Finite Sequences

Again let $\{a_t\}$ and $\{b_t\}$ be two sequences of length N such that $\{a_t\} \longleftrightarrow \{A_k\}$ and $\{b_t\} \longleftrightarrow \{B_k\}$. Let us define the convolution of $\{a_t\}$ and $\{b_t\}$ as the sequence of N numbers given by

$$a * b_t \equiv \sum_{u=0}^{N-1} a_u b_{t-u}, \qquad t = 0, \ldots, N-1,$$

where b_t is *defined* for $t \leq -1$ by periodic extension; i.e., $b_{-1} = b_{N-1}$; $b_{-2} = b_{N-2}$; and so forth. To remind ourselves that we are using a periodic extension, we can equivalently express the above as

$$a * b_t \equiv \sum_{u=0}^{N-1} a_u b_{t-u \bmod N}, \qquad t = 0, \ldots, N-1,$$

where '$t - u \bmod N$' stands for '$t - u$ modulo N,' which is defined as follows. If j is an integer such that $0 \leq j \leq N-1$, then $j \bmod N \equiv j$; if j is any other integer, then $j \bmod N \equiv j + nN$, where nN is the unique integer multiple of N such that $0 \leq j + nN \leq N-1$ (note carefully that, in expressions like $b_{10+s+t-[(u+5v)/2] \bmod N}$, the modulo operation is applied to what we get after computing $10+s+t-[(u+5v)/2]$). However we choose to express it, the convolution above is called *circular* or *cyclic.* Figure 31a illustrates the rationale for this nomenclature.

Analogous to the case of infinite sequences, the DFT of a circular convolution has a very simple relationship to the DFTs of $\{a_t\}$ and $\{b_t\}$.

▷ **Exercise [30]:** Show that the DFT of $\{a * b_t\}$ is given by

$$\sum_{t=0}^{N-1} a * b_t e^{-i2\pi tk/N} = A_k B_k, \qquad k = 0, \ldots, N-1;$$

i.e., $\{a * b_t\} \longleftrightarrow \{A_k B_k\}$. ◁

Two important related notions are complex cross-correlation, defined as

$$a^* \star b_t \equiv \sum_{u=0}^{N-1} a_u^* b_{u+t \bmod N}, \qquad t = 0, \ldots, N-1,$$

and autocorrelation, defined as

$$a^* \star a_t \equiv \sum_{u=0}^{N-1} a_u^* a_{u+t \bmod N}, \qquad t = 0, \ldots, N-1.$$

Figure 31b gives a graphical illustration of complex cross-correlation. Exercise [2.9] says that the DFT of $\{a^* \star b_t\}$ is given by

$$\sum_{t=0}^{N-1} a^* \star b_t e^{-i2\pi tk/N} = A_k^* B_k, \qquad k = 0, \ldots, N-1; \tag{30}$$

i.e., $\{a^* \star b_t\} \longleftrightarrow \{A_k^* B_k\}$. The DFT of an autocorrelation follows immediately:

$$\sum_{t=0}^{N-1} a^* \star a_t e^{-i2\pi tk/N} = A_k^* A_k = |A_k|^2, \qquad k = 0, \ldots, N-1;$$

i.e., $\{a^* \star a_t\} \longleftrightarrow \{|A_k|^2\}$.

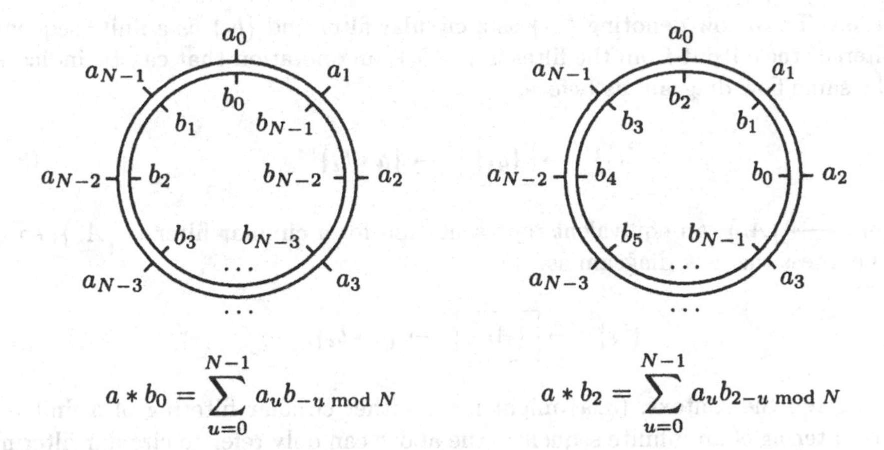

Figure 31a. Graphical illustration of circular convolution of the finite sequences $\{a_t\}$ and $\{b_t\}$. The left-hand plot shows two concentric circles. The outer circle is divided into N equal arcs, and the boundaries between the arcs are labeled clockwise from the top with $a_0, a_1, \ldots, a_{N-1}$. The inner circle is likewise divided, but now the boundaries are labeled counter-clockwise with $b_0, b_1, \ldots, b_{N-1}$. The zeroth element $a * b_0$ of the convolution of $\{a_t\}$ and $\{b_t\}$ is obtained by multiplying the a_t's and b_t's facing each other and then summing. In general, the tth element $a * b_t$ is obtained in a similar fashion *after* the inner circle has been rotated clockwise by t divisions – for example, the right-hand plot shows the alignment of the concentric circles that yields the second element $a * b_2$ of the convolution.

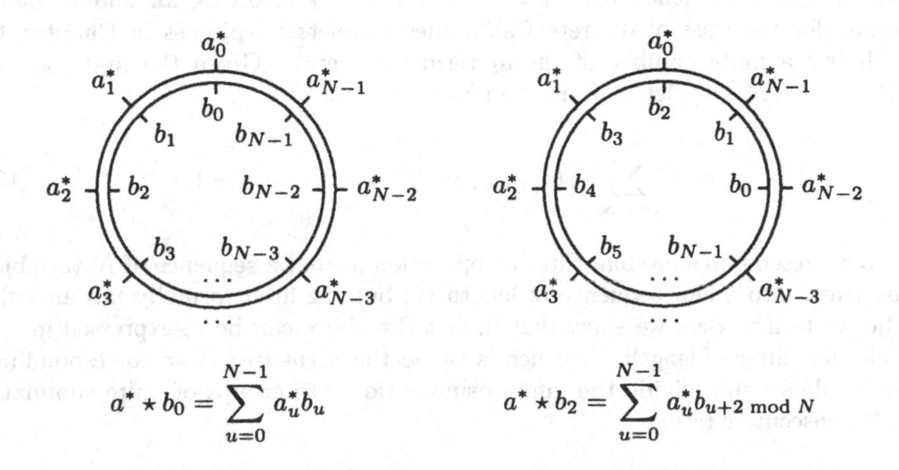

Figure 31b. Graphical illustration of circular complex cross-correlation of $\{a_t\}$ and $\{b_t\}$. The layout is similar to Figure 31a.

▷ **Exercise [31]:** Show that, with a proper definition for $\{\tilde{a}_t\}$, the sequence obtained by complex cross-correlating $\{a_t\}$ and $\{b_t\}$ is identical to the one obtained by convolving $\{\tilde{a}_t\}$ and $\{b_t\}$. Show also that, if $\{a_t\} \longleftrightarrow \{A_k\}$, then $\{\tilde{a}_t\} \longleftrightarrow \{A_k^*\}$; i.e., complex cross-correlating a sequence having DFT $\{A_k\}$ with $\{b_t\}$ is the same as convolving a sequence having DFT $\{A_k^*\}$ with $\{b_t\}$. Hint: compare Figures 31a and 31b. ◁

The notion of circular filtering in the engineering literature is the same as circular convolution. Thus, now denoting $\{a_t\}$ as a circular filter and $\{b_t\}$ as a finite sequence to be filtered, the output from the filter is $\{a * b_t\}$, an operation that can be indicated using the same flow diagram as before:

$$\{b_t\} \longrightarrow \boxed{\{a_t\}} \longrightarrow \{a * b_t\}. \tag{32a}$$

Since $\{a_t\} \longleftrightarrow \{A_k\}$, an equivalent representation for a circular filter is $\{A_k\}$, so we can also express the flow diagram as

$$\{b_t\} \longrightarrow \boxed{\{A_k\}} \longrightarrow \{a * b_t\}.$$

Whereas, given the context, (32a) might mean either circular filtering of a finite sequence or filtering of an infinite sequence, the above can only refer to circular filtering: the DFT of a circular filter must be a sequence like $\{A_k\}$, whereas the DFT of a filter defined over all integers must be a periodic function, say $A(\cdot)$, defined over the real axis.

Circular filtering can also lead to yet a third way of presenting the flow diagram as discussed in the next section.

2.6 Periodized Filters

In the study of wavelets, circular filters are usually constructed indirectly from an impulse response sequence $\{a_t : t = \ldots, -1, 0, 1, \ldots\}$ involving an infinite number of terms (for the class of discrete Daubechies wavelets we discuss in Chapter 4, in fact all but a finite number of the a_t terms are zero). Given the finite sequence $\{b_t : t = 0, \ldots, N-1\}$, let us form the finite sequence

$$c_t \equiv \sum_{v=-\infty}^{\infty} a_v b_{t-v \bmod N}, \qquad t = 0, \ldots, N-1. \tag{32b}$$

The above resembles a circular filtering operation in that a sequence of N variables is transformed into a new sequence of length N, but the filter formally has an infinite number of terms. Here we show that in fact the above can be re-expressed in terms of a circular filter of length N, which is called the *periodized filter* corresponding to $\{a_t\}$. To do so, first divide the infinite summation into groups of finite summations over N consecutive terms:

$$c_t = \sum_{n=-\infty}^{\infty} \sum_{u=nN}^{(n+1)N-1} a_u b_{t-u \bmod N} = \sum_{n=-\infty}^{\infty} \sum_{u=0}^{N-1} a_{u+nN} b_{t-(u+nN) \bmod N}.$$

Since $b_{t-(u+nN) \bmod N} = b_{t-u \bmod N}$ for all n, we can write

$$\begin{aligned} c_t = \sum_{n=-\infty}^{\infty} \sum_{u=0}^{N-1} a_{u+nN} b_{t-u \bmod N} &= \sum_{u=0}^{N-1} \left(\sum_{n=-\infty}^{\infty} a_{u+nN} \right) b_{t-u \bmod N} \\ &\equiv \sum_{u=0}^{N-1} a_u^{\circ} b_{t-u \bmod N}, \end{aligned}$$

where

$$a_t^\circ \equiv \sum_{n=-\infty}^{\infty} a_{t+nN}, \qquad t = 0, \ldots, N-1.$$

The above defines the periodized filter corresponding to $\{a_t\}$ (we also say that $\{a_t^\circ\}$ is $\{a_t\}$ periodized to length N). In essence, the filter $\{a_t\}$ is cut into blocks of N coefficients,

$$\ldots, \underbrace{a_{-N}, a_{-N+1}, \ldots, a_{-1}}_{\text{block } n=-1}, \underbrace{a_0, a_1, \ldots, a_{N-1}}_{\text{block } n=0}, \underbrace{a_N, a_{N+1}, \ldots, a_{2N-1}}_{\text{block } n=1}, \ldots$$

and these are overlain and then added to form the periodized filter $\{a_t^\circ\}$:

$$\begin{array}{lcccc}
\vdots & \vdots & \vdots & \vdots & \vdots \\
 & + & + & \cdots & + \\
\text{block } n=-1: & a_{-N} & a_{-N+1} & \cdots & a_{-1} \\
 & + & + & \cdots & + \\
\text{block } n=0: & a_0 & a_1 & \cdots & a_{N-1} \\
 & + & + & \cdots & + \\
\text{block } n=1: & a_N & a_{N+1} & \cdots & a_{2N-1} \\
 & + & + & \cdots & + \\
\vdots & \vdots & \vdots & \vdots & \vdots \\
 & \Downarrow & \Downarrow & \cdots & \Downarrow \\
\text{periodized filter:} & a_0^\circ & a_1^\circ & \cdots & a_{N-1}^\circ
\end{array}$$

Now let $\{A_k^\circ : k = 0, \ldots, N-1\}$ be the DFT of $\{a_t^\circ\}$, and let $A(\cdot)$ be the transfer function (i.e., DFT) corresponding to $\{a_t\}$. The following exercise shows that $\{A_k^\circ\}$ and $A(\cdot)$ are related in a very simple manner.

▷ **Exercise [33]:** Show that $A_k^\circ = A(\frac{k}{N})$. ◁

Thus the DFT $\{A_k^\circ\}$ for $\{a_t^\circ\}$ can be easily picked off from the transfer function $A(\cdot)$ for the original filter; i.e., $\{A_k^\circ\}$ is formed by sampling $A(\cdot)$ at a spacing of $1/N$.

In terms of a flow diagram, we can express the (implicit) circular filtering of Equation (32b) as

$$\{b_t\} \longrightarrow \boxed{\{A(\tfrac{k}{N})\}} \longrightarrow \{c_t\}.$$

For future reference, we note two minor variations on this flow diagram (see also item [1] in the Comments and Extensions below). First, if we place the elements of the sequences $\{b_t\}$ and $\{c_t\}$ into the N dimensional vectors $\mathbf{B}$ and $\mathbf{C}$, we can write

$$\mathbf{B} \longrightarrow \boxed{\{A(\tfrac{k}{N})\}} \longrightarrow \mathbf{C}.$$

Second, it is sometimes convenient to shorten $\{A(\frac{k}{N})\}$ (a sequence) to just $A(\frac{k}{N})$ (one element of that sequence):

$$\mathbf{B} \longrightarrow \boxed{A(\tfrac{k}{N})} \longrightarrow \mathbf{C}. \tag{33}$$

This form of the flow diagram is the one we will make most use of in subsequent chapters.

As an example of creating a periodized filter, consider the following infinite sequence:

$$a_t = \begin{cases} 1/2, & t = 0; \\ 1/4, & t = \pm 1; \\ 0, & \text{otherwise.} \end{cases}$$

This is just new notation for the impulse response sequence for the filter of Equation (25d), whose transfer function $A(\cdot)$ is given explicitly by Exercise [2.3]. For $N \geq 4$, the periodized filter is given by

$$a_0^\circ = \sum_{n=-\infty}^{\infty} a_{nN} = a_0 = 1/2$$

$$a_1^\circ = \sum_{n=-\infty}^{\infty} a_{1+nN} = a_1 = 1/4$$

$$a_t^\circ = \sum_{n=-\infty}^{\infty} a_{t+nN} = 0, \quad t = 2, \ldots, N-2$$

$$a_{N-1}^\circ = \sum_{n=-\infty}^{\infty} a_{N-1+nN} = a_{-1} = 1/4.$$

Now the DFT of $\{a_t^\circ\}$ is given by

$$\begin{aligned} A_k^\circ \equiv \sum_{t=0}^{N-1} a_t^\circ e^{-i2\pi tk/N} = \sum_{t=-1}^{1} a_{t \bmod N}^\circ e^{-i2\pi tk/N} &= \sum_{t=-1}^{1} a_t e^{-i2\pi tk/N} \\ &= \sum_{t=-\infty}^{\infty} a_t e^{-i2\pi tk/N} = A(\tfrac{k}{N}), \end{aligned}$$

as required (see Exercise [2.11] for a second example of periodization).

Finally we note an important result concerning a filter $\{a_t\}$ and its periodization $\{a_t^\circ\}$ (we will see an application of this result in Chapter 6).

▷ **Exercise [34]:** Suppose that the filter $\{a_t\}$ is normalized such that, say,

$$\sum_{t=-\infty}^{\infty} |a_t|^2 = 1.$$

If $\{a_t^\circ\}$ is $\{a_t\}$ periodized to length N, show that in general

$$\sum_{t=0}^{N-1} |a_t^\circ|^2 \neq 1;$$

i.e., the sums of squared magnitudes for a filter and its periodization need not be the same. (Hint: either construct a simple counter-example or use Parseval's theorem along with the relationship between the DFTs for $\{a_t\}$ and $\{a_t^\circ\}$.) ◁

Comments and Extensions to Section 2.6

[1] For the filtering operation of Equation (32b), a possible alternative to the flow diagram of Equation (33) is

$$\mathbf{B} \longrightarrow \boxed{A(\cdot)} \longrightarrow \mathbf{C}.$$

The reader should note carefully how the above is to be interpreted: the N elements of $\mathbf{B}$ are circularly filtered by the periodized filter whose frequency domain (Fourier) representation is $\{A(\frac{k}{N}) : k = 0, \ldots, N-1\}$, and the output from the filter is given by the N elements of $\mathbf{C}$. While flow diagrams like the above are quite prevalent in the wavelet literature, the sampling of $A(\cdot)$ is implicit, so we prefer (33) because it explicitly denotes the sampling operation.

[2] A well-known phenomenon in Fourier theory is aliasing, in which a discrete sampling in the time domain expresses itself in the frequency domain as a summation over equally spaced frequencies (see, for example, Percival and Walden, 1993, Section 3.8). Note that periodizing filters can be regarded as the 'dual' of aliasing in which the roles of time and frequency are interchanged: here summation over equally spaced time domain variables expresses itself as a discrete sampling in the frequency domain.

2.7 Summary of Fourier Theory

For later reference, we summarize here the most important results from the Fourier theory for an infinite sequence, a finite sequence and a function defined on the real axis.

- *Infinite real- or complex-valued sequence* $\{a_t\}$ *with* $\sum_{-\infty}^{\infty} |a_t|^2 < \infty$

(a) Fourier representation:

$$a_t = \int_{-1/2}^{1/2} A(f) e^{i2\pi f t}\, df, \qquad t = \ldots, -1, 0, 1, \ldots \tag{35a}$$

(synthesis equation), where

$$A(f) \equiv \sum_{t=-\infty}^{\infty} a_t e^{-i2\pi f t}, \qquad |f| \le 1/2 \tag{35b}$$

(analysis equation). The right-hand side of the above can be used to define $A(\cdot)$ for all f, yielding a periodic function with unit period. The Fourier relationship between $\{a_t\}$ and $A(\cdot)$ is noted as

$$\{a_t\} \longleftrightarrow A(\cdot).$$

(b) Parseval's theorems: The 'one sequence' version is

$$\sum_{t=-\infty}^{\infty} |a_t|^2 = \int_{-1/2}^{1/2} |A(f)|^2\, df; \tag{35c}$$

with $\{b_t\} \longleftrightarrow B(\cdot)$, the 'two sequence' version is

$$\sum_{t=-\infty}^{\infty} a_t^* b_t = \int_{-1/2}^{1/2} A^*(f) B(f)\, df. \tag{35d}$$

(c) Convolution and related theorems: The Fourier transform of the convolution

$$a * b_t \equiv \sum_{u=-\infty}^{\infty} a_u b_{t-u}, \qquad t = \ldots, -1, 0, 1, \ldots, \tag{36a}$$

is

$$\sum_{t=-\infty}^{\infty} a * b_t e^{-i2\pi ft} = A(f)B(f), \tag{36b}$$

so $\{a * b_t\} \longleftrightarrow A(\cdot)B(\cdot)$. The Fourier transform of the complex cross-correlation

$$a^* \star b_t \equiv \sum_{u=-\infty}^{\infty} a_u^* b_{u+t}, \qquad t = \ldots, -1, 0, 1, \ldots, \tag{36c}$$

is

$$\sum_{t=-\infty}^{\infty} a^* \star b_t e^{-i2\pi ft} = A^*(f)B(f), \tag{36d}$$

so $\{a^* \star b_t\} \longleftrightarrow A^*(\cdot)B(\cdot)$. Letting $b_t = a_t$ yields an autocorrelation, for which

$$\{a^* \star a_t\} \longleftrightarrow |A(\cdot)|^2. \tag{36e}$$

- *Finite real- or complex-valued sequence* $\{a_t : t = 0, \ldots, N-1\}$

(a) Fourier representation:

$$a_t = \frac{1}{N} \sum_{k=0}^{N-1} A_k e^{i2\pi tk/N}, \qquad t = 0, \ldots, N-1 \tag{36f}$$

(synthesis equation), where

$$A_k \equiv \sum_{t=0}^{N-1} a_t e^{-i2\pi tk/N}, \qquad k = 0, \ldots, N-1. \tag{36g}$$

(analysis equation). Note that A_k is associated with frequency $f_k \equiv k/N$. The right-hand sides of both equations above can be used to define a_t and A_k for all integers t and k, yielding periodic infinite sequences with period N. The Fourier relationship between $\{a_t\}$ and $\{A_k\}$ is noted as

$$\{a_t\} \longleftrightarrow \{A_k\}.$$

(b) Parseval's theorems: The 'one sequence' version is

$$\sum_{t=0}^{N-1} |a_t|^2 = \frac{1}{N} \sum_{k=0}^{N-1} |A_k|^2; \tag{36h}$$

with $\{b_t\} \longleftrightarrow \{B_k\}$, the 'two sequence' version is

$$\sum_{t=0}^{N-1} a_t^* b_t = \frac{1}{N} \sum_{k=0}^{N-1} A_k^* B_k. \tag{36i}$$

(c) Convolution and related theorems: The Fourier transform of the convolution

$$a * b_t \equiv \sum_{u=0}^{N-1} a_u b_{t-u \bmod N}, \qquad t = 0, \ldots, N-1, \tag{37a}$$

is

$$\sum_{t=0}^{N-1} a * b_t e^{-i2\pi tk/N} = A_k B_k, \qquad k = 0, \ldots, N-1, \tag{37b}$$

so $\{a * b_t\} \longleftrightarrow \{A_k B_k\}$. This type of convolution is called *circular* or *cyclic*. The Fourier transform of the complex cross-correlation

$$a^* \star b_t \equiv \sum_{u=0}^{N-1} a_u^* b_{u+t \bmod N}, \qquad t = 0, \ldots, N-1, \tag{37c}$$

is

$$\sum_{t=0}^{N-1} a^* \star b_t e^{-i2\pi tk/N} = A_k^* B_k, \qquad k = 0, \ldots, N-1, \tag{37d}$$

so $\{a^* \star b_t\} \longleftrightarrow \{A_k^* B_k\}$. Letting $b_t = a_t$ yields an autocorrelation, for which

$$\{a^* \star a_t\} \longleftrightarrow \{|A_k|^2\}. \tag{37e}$$

If $\{a_t : t = \ldots, -1, 0, 1, \ldots\} \longleftrightarrow A(\cdot)$, then

$$c_t \equiv \sum_{u=-\infty}^{\infty} a_u b_{t-u \bmod N}, \qquad t = 0, \ldots, N-1,$$

can be re-expressed as a circular convolution:

$$c_t = \sum_{u=0}^{N-1} a_u^\circ b_{t-u \bmod N}, \quad \text{where} \quad a_t^\circ \equiv \sum_{u=-\infty}^{\infty} a_{t+uN}.$$

The sequence $\{a_t^\circ\}$ of length N is said to be formed by periodizing $\{a_t\}$ to length N. We have

$$\{a_t^\circ : t = 0, \ldots, N-1\} \longleftrightarrow \{A(\tfrac{k}{N}) : k = 0, \ldots, N-1\}.$$

- *Real- or complex-valued function $a(\cdot)$ with $\int_{-\infty}^{\infty} |a(t)|^2 \, dt < \infty$*

(a) Fourier representation:

$$a(t) \overset{\text{ms}}{=} \int_{-\infty}^{\infty} A(f) e^{i2\pi ft} \, df, \qquad -\infty < t < \infty \tag{37f}$$

(synthesis equation), where

$$A(f) \equiv \int_{-\infty}^{\infty} a(t) e^{-i2\pi ft} \, dt, \qquad -\infty < t < \infty \tag{37g}$$

(analysis equation). The notation '$\stackrel{\text{ms}}{=}$' indicates 'equality in the mean square sense,' which says that, while the left- and right-hand sides of Equation (37f) need not be equal at any given t, we do have

$$\int_{-\infty}^{\infty} \left| a(t) - \int_{-\infty}^{\infty} A(f)e^{i2\pi ft}\,df \right|^2 dt = 0.$$

The Fourier relationship between $a(\cdot)$ and $A(\cdot)$ is noted as

$$a(\cdot) \longleftrightarrow A(\cdot).$$

(b) Parseval's theorems: The 'one function' version is

$$\int_{-\infty}^{\infty} |a(t)|^2\,dt = \int_{-\infty}^{\infty} |A(f)|^2\,df; \tag{38a}$$

with $b(\cdot) \longleftrightarrow B(\cdot)$, the 'two function' version is

$$\int_{-\infty}^{\infty} a^*(t)b(t)\,dt = \int_{-\infty}^{\infty} A^*(f)B(f)\,df. \tag{38b}$$

(c) Convolution and related theorems: The Fourier transform of the convolution

$$a * b(t) \equiv \int_{-\infty}^{\infty} a(u)b(t-u)\,du \tag{38c}$$

is

$$\int_{-\infty}^{\infty} a * b(t)e^{-i2\pi ft}\,dt = A(f)B(f), \tag{38d}$$

so $a * b(\cdot) \longleftrightarrow A(\cdot)B(\cdot)$. The Fourier transform of the complex cross-correlation

$$a^* \star b(t) \equiv \int_{-\infty}^{\infty} a^*(u)b(u+t)\,du \tag{38e}$$

is

$$\int_{-\infty}^{\infty} a^* \star b(t)e^{-i2\pi ft}\,dt = A^*(f)B(f), \tag{38f}$$

so $a^* \star b(\cdot) \longleftrightarrow A^*(\cdot)B(\cdot)$. Letting $b(f) = a(f)$ yields an autocorrelation, for which

$$a^* \star a(\cdot) \longleftrightarrow |A(\cdot)|^2. \tag{38g}$$

2.8 Exercises

[2.1] Suppose that $\{a_t : t = \ldots, -1, 0, 1, \ldots\}$ is a real-valued sequence whose DFT is $A(\cdot)$. For some f_0 such that $0 < f_0 < 1/2$, construct the complex-valued sequence $c_t \equiv a_t e^{-i2\pi f_0 t}$ (this procedure is known as *complex demodulation*). How is the DFT $C(\cdot)$ of $\{c_t\}$ related to the DFT $A(\cdot)$ of $\{a_t\}$? Suppose that ϵ denotes a variable such that $0 < f_0 - \epsilon < f_0 + \epsilon < 1/2$ and that the positive frequencies in $A(\cdot)$ can be related to some physical phenomenon. What physical interpretation can you make regarding the negative frequencies $-\epsilon \le f < 0$ in $C(f)$?

[2.2] Show that the DFT of the complex cross-correlation $\{a^* \star b_t\}$ of the infinite sequences $\{a_t\}$ and $\{b_t\}$ is given by Equation (25a).

[2.3] Show that the transfer function $A(\cdot)$ for the filter whose impulse response sequence is stated in Equation (25d) is given by $\cos^2(\pi f)$.

[2.4] Show that the DFT of the sequence $\{b_t\}$ defined by Equation (26a) is given by the function $B(\cdot)$ defined by Equation (26b). Hint: write $\phi_l^t e^{-i2\pi f t}$ as z^t with $z \equiv \phi_l e^{-i2\pi f}$.

[2.5] Show that the filter whose impulse response sequence is given by Equation (26c) has a transfer function $A(\cdot)$ given by $\sin^2(\pi f)$.

[2.6] Verify the contents of Figures 26 and 27.

[2.7] Show that the low-pass filter defined in Equation (25d) is the equivalent filter for a cascade of two filters $\{a_{1,t}\}$ and $\{a_{2,t}\}$, both with a width of two. Determine the transfer functions $A_1(\cdot)$ and $A_2(\cdot)$ for these filters, and verify that $A_1(f)A_2(f) = \cos^2(\pi f)$.

[2.8] If $\{a_t : t = 0, \ldots, N-1\} \longleftrightarrow A(\cdot)$, what are a_{-1} and a_{N+1} taken to be? If $\{a_t : t = 0, \ldots, N-1\} \longleftrightarrow \{A_k\}$, what are a_{-1} and a_{N+1} taken to be?

[2.9] Show that the DFT of the circular complex cross-correlation $\{a^* \star b_t\}$ of the finite sequences $\{a_t\}$ and $\{b_t\}$ is as stated in Equation (30).

[2.10] Let $\{a_t : t = \ldots, -1, 0, 1, \ldots\}$ be an infinite sequence whose DFT is $A(\cdot)$. Find the DFT of the infinite sequence $\{a_{2n+1} : n = \ldots, -1, 0, 1, \ldots\}$ (i.e., the odd indexed variables in $\{a_t\}$) in terms of $A(\cdot)$. (Exercise [23b] gives the corresponding DFT for the even index variables.)

[2.11] Let $\{b_t : t = 0, \ldots, 3\}$ be a finite sequence of length $N = 4$, and consider a filter defined by $a_t = \phi^{|t|}, t = \ldots, -1, 0, 1, \ldots$, where ϕ is a real-valued variable satisfying $|\phi| < 1$. Let

$$c_t \equiv \sum_{u=-\infty}^{\infty} a_u b_{t-u \bmod 4}, \qquad t = 0, \ldots, 3,$$

represent the result of filtering $\{b_t\}$ with $\{a_t\}$. Derive the periodized filter of length $N = 4$ from $\{a_t\}$, i.e., the filter $\{a_t^\circ : t = 0, \ldots, 3\}$ such that

$$c_t = \sum_{u=0}^{3} a_u^\circ b_{t-u \bmod 4}.$$

Determine the DFT $A(\cdot)$ of $\{a_t\}$ and the DFT $\{A_k^\circ\}$ of $\{a_t^\circ\}$, and verify explicitly that $A_k^\circ = A(\frac{k}{4})$ for $k = 0, \ldots, 3$.

[2.12] Consider the following filter defined over the integers:

$$a_t = \begin{cases} 1/2, & t = 0, 1, 2 \text{ or } 3; \\ 0, & t < 0 \text{ or } t \ge 4. \end{cases}$$

What are the corresponding periodized filters $\{a_t^\circ\}$ of lengths $N = 1, 2, 3, 4$ and 5? Answer the same question for the filter

$$a_t = \begin{cases} 1/2, & t = 0, 1; \\ -1/2, & t = 2, 3; \\ 0, & t < 0 \text{ or } t \geq 4. \end{cases}$$

(After Chapter 4 we can interpret the above filters as, respectively, Haar scaling and wavelet filters of second level.)

3

Orthonormal Transforms of Time Series

3.0 Introduction

The wavelet analysis of a time series can be defined in terms of an orthonormal transform, so here we briefly review the key ideas behind such transforms. We first review the basic theory for orthonormal transforms in Section 3.1. Section 3.2 discusses the important projection theorem, while 3.3 considers complex-valued transforms. Prior to introducing the discrete wavelet transform (DWT) in Chapter 4, we discuss the orthonormal discrete Fourier transform (ODFT) in Section 3.4 because it parallels and contrasts the DWT in a number of interesting ways. We summarize the key points of this chapter in Section 3.5 – readers who are already comfortable with orthonormal transforms can read this section simply to become familiar with our notation and conventions.

3.1 Basic Theory for Orthonormal Transforms

Orthonormal transforms are of interest because they can be used to re-express a time series in such a way that we can easily reconstruct the series from its transform. In a loose sense, the 'information' in the transform is thus equivalent to the 'information' in the original series; to put it another way, the series and its transform can be considered to be two representations of the same mathematical entity. Orthonormal transforms can be used to re-express a series in a standardized form (e.g., a Fourier series) for further manipulation, to reduce a series to a few values summarizing its salient features (compression), and to analyze a series to search for particular patterns of interest (e.g., analysis of variance). There are many examples of orthonormal transforms in the literature, including the orthonormal version of the discrete Fourier transform (the ODFT – see Section 3.4), various versions of the discrete cosine transform (see Exercise [3.3]), the Walsh transform (Beauchamp, 1984) and – as we shall see in Chapter 4 – the discrete wavelet transform.

Let $X_0, X_1, \ldots, X_{N-1}$ represent a time series of N real-valued variables (henceforth we will denote such a series as either $\{X_t : t = 0, \ldots, N-1\}$ or just $\{X_t\}$ if it is clear what values the dummy index t can assume). Two examples of such series are shown in Figure 42 for $N = 16$ values. Let $\mathbf{X}$ represent an N dimensional column

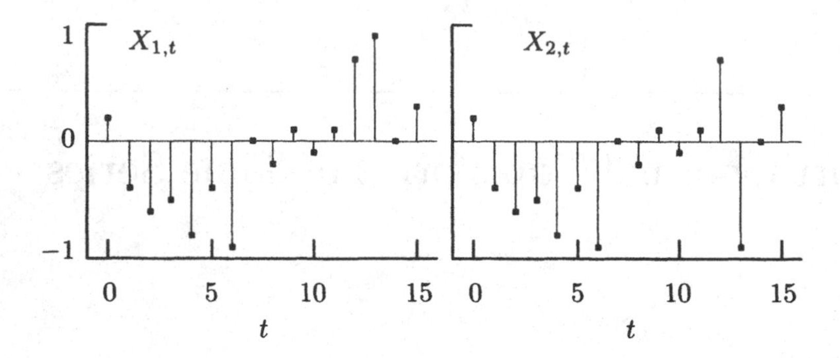

Figure 42. Two small time series $\{X_{1,t}\}$ and $\{X_{2,t}\}$, each with $N = 16$ values. The left-hand plot shows $X_{1,t}$ versus t for $t = 0, \ldots, 15$ as black dots, while the right-hand plot shows $X_{2,t}$. The two series differ only at the $t = 13$th value, for which $X_{2,13} = -X_{1,13}$. For the record, the 16 values for $\{X_{1,t}\}$ are 0.2, −0.4, −0.6, −0.5, −0.8, −0.4, −0.9, 0.0, −0.2, 0.1, −0.1, 0.1, 0.7, 0.9, 0.0 and 0.3.

vector whose tth element is X_t for $t = 0, \ldots, N-1$. If $\mathbf{Y}$ is another such vector containing $Y_0, \ldots, Y_{N-1}$, the inner product of $\mathbf{X}$ and $\mathbf{Y}$ is given by

$$\langle \mathbf{X}, \mathbf{Y} \rangle \equiv \mathbf{X}^T\mathbf{Y} = \sum_{t=0}^{N-1} X_t Y_t,$$

where $\mathbf{X}^T$ is the transpose of $\mathbf{X}$. The squared norm of $\mathbf{X}$ is given by

$$\|\mathbf{X}\|^2 \equiv \langle \mathbf{X}, \mathbf{X} \rangle = \mathbf{X}^T\mathbf{X} = \sum_{t=0}^{N-1} X_t^2.$$

We shall refer to the quantity $\mathcal{E}_{\mathbf{X}} \equiv \|\mathbf{X}\|^2$ as the energy in the time series $\{X_t\}$ (even though X_t must be measured in certain units for $\mathcal{E}$ to be called energy by a physicist).

Let $\mathcal{O}$ be an $N \times N$ matrix of real-valued variables satisfying the orthonormality property $\mathcal{O}^T\mathcal{O} = I_N$, where I_N is the $N \times N$ identity matrix. For $j, k = 0, \ldots, N-1$, let $\mathcal{O}_{j\bullet}^T$ and $\mathcal{O}_{\bullet k}$ refer to, respectively, the jth row vector and kth column vector of $\mathcal{O}$. For example,

$$\text{if } \mathcal{O} = \begin{bmatrix} 0 & \cos(\theta) & \sin(\theta) \\ 0 & -\sin(\theta) & \cos(\theta) \\ 1 & 0 & 0 \end{bmatrix}, \text{ then } \mathcal{O}_{2\bullet} = \begin{bmatrix} 1 \\ 0 \\ 0 \end{bmatrix}, \quad \mathcal{O}_{\bullet 2} = \begin{bmatrix} \sin(\theta) \\ \cos(\theta) \\ 0 \end{bmatrix}.$$

With this notation, we can write

$$\mathcal{O} = \begin{bmatrix} \mathcal{O}_{0\bullet}^T \\ \mathcal{O}_{1\bullet}^T \\ \vdots \\ \mathcal{O}_{N-1\bullet}^T \end{bmatrix} = [\mathcal{O}_{\bullet 0}, \mathcal{O}_{\bullet 1}, \ldots, \mathcal{O}_{\bullet N-1}].$$

The orthonormality property can be restated in terms of the inner product as

$$\langle \mathcal{O}_{\bullet k}, \mathcal{O}_{\bullet k'} \rangle = \delta_{k,k'} \equiv \begin{cases} 1, & \text{if } k = k'; \\ 0, & \text{otherwise,} \end{cases}$$

where $\delta_{j,k}$ is the Kronecker delta function. Orthonormality also implies that the inverse $\mathcal{O}^{-1}$ of the matrix $\mathcal{O}$ is just its transpose $\mathcal{O}^T$. Thus $\mathcal{O}\mathcal{O}^{-1} = \mathcal{O}\mathcal{O}^T = I_N$, from which it follows that $\langle \mathcal{O}_{j\bullet}, \mathcal{O}_{j'\bullet} \rangle = \delta_{j,j'}$ also. Hence the columns of $\mathcal{O}$ are an orthonormal set of vectors, as are its rows.

We can analyze (or decompose) our time series $\{X_t\}$ with respect to the orthonormal matrix $\mathcal{O}$ by premultiplying $\mathbf{X}$ by $\mathcal{O}$ to obtain

$$\mathbf{O} \equiv \mathcal{O}\mathbf{X} = \begin{bmatrix} \mathcal{O}_{0\bullet}^T \\ \mathcal{O}_{1\bullet}^T \\ \vdots \\ \mathcal{O}_{N-1\bullet}^T \end{bmatrix} \mathbf{X} = \begin{bmatrix} \mathcal{O}_{0\bullet}^T \mathbf{X} \\ \mathcal{O}_{1\bullet}^T \mathbf{X} \\ \vdots \\ \mathcal{O}_{N-1\bullet}^T \mathbf{X} \end{bmatrix} = \begin{bmatrix} \langle \mathbf{X}, \mathcal{O}_{0\bullet} \rangle \\ \langle \mathbf{X}, \mathcal{O}_{1\bullet} \rangle \\ \vdots \\ \langle \mathbf{X}, \mathcal{O}_{N-1\bullet} \rangle \end{bmatrix}, \tag{43a}$$

where we use the fact that $\mathcal{O}_{j\bullet}^T \mathbf{X} = \langle \mathcal{O}_{j\bullet}, \mathbf{X} \rangle = \langle \mathbf{X}, \mathcal{O}_{j\bullet} \rangle$. The N dimensional column vector $\mathbf{O}$ is referred to as the transform coefficients for $\mathbf{X}$ with respect to the orthonormal transform $\mathcal{O}$. The jth transform coefficient is O_j (i.e., the jth element of $\mathbf{O}$) and is given by the inner product $\langle \mathbf{X}, \mathcal{O}_{j\bullet} \rangle$. If we premultiply both sides of the above equation by $\mathcal{O}^T$ and recall that $\mathcal{O}^T\mathcal{O} = I_N$, we obtain the synthesis equation

$$\mathbf{X} = \mathcal{O}^T \mathbf{O} = [\mathcal{O}_{0\bullet}, \mathcal{O}_{1\bullet}, \ldots, \mathcal{O}_{N-1\bullet}] \begin{bmatrix} O_0 \\ O_1 \\ \vdots \\ O_{N-1} \end{bmatrix} = \sum_{j=0}^{N-1} O_j \mathcal{O}_{j\bullet}, \tag{43b}$$

which tells us how to reconstruct $\mathbf{X}$ from its transform coefficients $\mathbf{O}$ (the notation $O_j\mathcal{O}_{j\bullet}$ refers to the vector obtained by multiplying each element of the vector $\mathcal{O}_{j\bullet}$ by the scalar O_j). Since $O_j = \langle \mathbf{X}, \mathcal{O}_{j\bullet} \rangle$, we can re-express the synthesis of $\mathbf{X}$ as

$$\mathbf{X} = \sum_{j=0}^{N-1} \langle \mathbf{X}, \mathcal{O}_{j\bullet} \rangle \mathcal{O}_{j\bullet}. \tag{43c}$$

The above holds for arbitrary $\mathbf{X}$, so the $\mathcal{O}_{j\bullet}$ vectors form a basis for the finite dimensional space $\mathbb{R}^N$ of all N dimensional real-valued vectors; i.e., any real-valued N dimensional column vector can be expressed as a unique linear combination of $\mathcal{O}_{0\bullet}, \ldots, \mathcal{O}_{N-1\bullet}$.

▷ **Exercise [43]**: Show that a similar statement holds for the $\mathcal{O}_{\bullet k}$'s, i.e.,

$$\mathbf{X} = \sum_{k=0}^{N-1} \langle \mathbf{X}, \mathcal{O}_{\bullet k} \rangle \mathcal{O}_{\bullet k}. \tag{43d}$$

In words, if $\mathcal{O}$ is an $N \times N$ orthonormal matrix, then any N dimensional matrix $\mathbf{X}$ can be written as a linear combination of the N column vectors $\mathcal{O}_{\bullet k}$ of $\mathcal{O}$. ◁

An important fact about an orthonormal transform is that it preserves energy in the sense that the energy in the transform coefficients $\mathbf{O}$ is equal to the energy in the original series $\mathbf{X}$, as can be seen from the following argument:

$$\mathcal{E}_{\mathbf{O}} \equiv \|\mathbf{O}\|^2 = \mathbf{O}^T\mathbf{O} = (\mathcal{O}\mathbf{X})^T \mathcal{O}\mathbf{X} = \mathbf{X}^T\mathcal{O}^T\mathcal{O}\mathbf{X} = \mathbf{X}^T\mathbf{X} = \|\mathbf{X}\|^2 = \mathcal{E}_{\mathbf{X}}$$

(a transform that preserves energy is sometimes called *isometric*). Since $\mathcal{E}_{\mathbf{X}} = \sum X_t^2$ and $\mathcal{E}_{\mathbf{O}} = \sum O_j^2$, we can argue that the sequence $\{X_t^2\}$ describes how the energy in the time series is decomposed across time, whereas the sequence $\{O_j^2\}$ describes how the energy is decomposed across transform coefficients with different indices.

▷ **Exercise [44a]:** Use Equation (43b) to show that

$$\|\mathbf{X}\|^2 = \sum_{j=0}^{N-1} \|O_j \mathcal{O}_{j\bullet}\|^2;$$

i.e., the squared norm of $\mathbf{X}$ is equal to the sum of the squared norms for the vectors $O_j \mathcal{O}_{j\bullet}$ that synthesize $\mathbf{X}$. ◁

While there are an infinite number of possible orthonormal transforms, the ones of particular interest (such as the DWT and the ODFT) are those for which we can attach some physical significance to the transform coefficients O_j or the synthesis vectors $O_j \mathcal{O}_{j\bullet}$.

3.2 The Projection Theorem

One important property of orthonormal transforms is summarized in the *projection theorem*, which we can formulate as follows. Suppose that we wish to approximate $\mathbf{X}$ using a linear combination of the vectors $\mathcal{O}_{0\bullet}, \mathcal{O}_{1\bullet}, \ldots, \mathcal{O}_{N'-1\bullet}$, where $N' < N$. Our approximation $\widehat{\mathbf{X}}$ thus takes the form

$$\widehat{\mathbf{X}} = \sum_{j=0}^{N'-1} \alpha_j \mathcal{O}_{j\bullet}$$

for some set of coefficients $\{\alpha_j\}$. Suppose further that we choose the α_j's based upon a least squares criterion; i.e., the difference vector $\mathbf{e} \equiv \mathbf{X} - \widehat{\mathbf{X}}$ is to be made as small as possible in the sense that $\|\mathbf{e}\|^2$ should be minimized over all possible choices for the α_j's. The projection theorem states that setting $\alpha_j = O_j = \langle \mathbf{X}, \mathcal{O}_{j\bullet} \rangle$ yields the least squares choice of the α_j's.

▷ **Exercise [44b]:** Verify the key step in establishing the projection theorem, namely, that

$$\|\mathbf{e}\|^2 = \sum_{j=0}^{N'-1} (O_j - \alpha_j)^2 + \sum_{j=N'}^{N-1} O_j^2.$$ ◁

The projection theorem follows from the above by noting that the first sum of squares can obviously be minimized by letting $\alpha_j = O_j$ (we have no control over the second sum of squares because it involves the $\mathcal{O}_{j\bullet}$'s that are *not* used in forming the approximation $\widehat{\mathbf{X}}$).

With the least squares choice of the α_j's, i.e., $\alpha_j = O_j$ for $j = 0, \ldots, N' - 1$, we have

$$\|\mathbf{e}\|^2 = \sum_{j=N'}^{N-1} O_j^2,$$

which is a useful measure of how well $\widehat{\mathbf{X}}$ approximates $\mathbf{X}$. Note that increasing N' will typically result in a decrease in $\|\mathbf{e}\|^2$; i.e., using more $\mathcal{O}_{j\bullet}$'s to form the approximation cannot cause an increase in $\|\mathbf{e}\|^2$.

Finally let us review the following linear algebra concepts that we will have occasion to use in Chapter 4. While all N of the $\mathcal{O}_{j\bullet}$'s form a basis for the finite dimensional space $\mathbb{R}^N$, the set of vectors $\mathcal{O}_{j\bullet}$, $0 \le j \le N'-1$ with $N' < N$, forms a basis for a subspace of $\mathbb{R}^N$. We can refer to this subspace as the one *spanned* by the vectors $\mathcal{O}_{0\bullet}, \ldots, \mathcal{O}_{N'-1\bullet}$. For example,

$$\text{if } \mathcal{O} = \begin{bmatrix} 0 & \frac{1}{\sqrt{2}} & \frac{1}{\sqrt{2}} \\ 0 & -\frac{1}{\sqrt{2}} & \frac{1}{\sqrt{2}} \\ 1 & 0 & 0 \end{bmatrix} \text{ so } \mathcal{O}_{0\bullet} = \begin{bmatrix} 0 \\ \frac{1}{\sqrt{2}} \\ \frac{1}{\sqrt{2}} \end{bmatrix} \text{ and } \mathcal{O}_{1\bullet} = \begin{bmatrix} 0 \\ -\frac{1}{\sqrt{2}} \\ \frac{1}{\sqrt{2}} \end{bmatrix},$$

then the subspace of $\mathbb{R}^3$ that is spanned by $\mathcal{O}_{0\bullet}$ and $\mathcal{O}_{1\bullet}$ consists of the set of all vectors in $\mathbb{R}^3$ whose $j=0$ element is identically zero. There are obviously other bases for this subspace besides the one given by $\mathcal{O}_{0\bullet}$ and $\mathcal{O}_{1\bullet}$: for any θ, the two vectors

$$\begin{bmatrix} 0 \\ \cos(\theta) \\ \sin(\theta) \end{bmatrix} \text{ and } \begin{bmatrix} 0 \\ -\sin(\theta) \\ \cos(\theta) \end{bmatrix}$$

form a basis for the same subspace spanned by $\mathcal{O}_{0\bullet}$ and $\mathcal{O}_{1\bullet}$. With this terminology, we can refer to $\widehat{\mathbf{X}}$ above as the projection of $\mathbf{X}$ onto the subspace spanned by $\mathcal{O}_{0\bullet}, \ldots, \mathcal{O}_{N'-1\bullet}$.

3.3 Complex-Valued Transforms

It is of interest to generalize our development slightly to allow the elements of $\mathcal{O}$ to be complex-valued (this will be used in Section 3.4 in defining the ODFT). The orthonormality property then becomes $\mathcal{O}^H\mathcal{O} = I_N$, where the superscript H refers to Hermitian transpose – by definition, the Hermitian transpose $\mathcal{O}^H$ of a matrix is obtained by taking the ordinary transpose $\mathcal{O}^T$ and then replacing each of its elements by their complex conjugates. For example,

$$\text{if } \mathcal{O} = \begin{bmatrix} \mathcal{O}_{1,1} & \mathcal{O}_{1,2} \\ \mathcal{O}_{2,1} & \mathcal{O}_{2,2} \end{bmatrix}, \text{ then } \mathcal{O}^H = \begin{bmatrix} \mathcal{O}_{1,1}^* & \mathcal{O}_{2,1}^* \\ \mathcal{O}_{1,2}^* & \mathcal{O}_{2,2}^* \end{bmatrix},$$

where $\mathcal{O}_{j,k}^*$ denotes the complex conjugate of $\mathcal{O}_{j,k}$. Note that, if the elements of $\mathcal{O}$ are real-valued, then $\mathcal{O}^H$ is just equal to $\mathcal{O}^T$ because the complex conjugate of a real variable is just the variable itself. A matrix $\mathcal{O}$ satisfying $\mathcal{O}^H\mathcal{O} = I_N$ is sometimes called *unitary*. For $j, k = 0, \ldots, N-1$, let $\mathcal{O}_{j\bullet}^H$ and $\mathcal{O}_{\bullet k}$ refer to, respectively, the jth row vector and kth column vector of $\mathcal{O}$ (note that these definitions are consistent with the real-valued case because then $\mathcal{O}_{j\bullet}^H = \mathcal{O}_{j\bullet}^T$). We thus can write

$$\mathcal{O} = \begin{bmatrix} \mathcal{O}_{0\bullet}^H \\ \mathcal{O}_{1\bullet}^H \\ \vdots \\ \mathcal{O}_{N-1\bullet}^H \end{bmatrix} = [\mathcal{O}_{\bullet 0}, \mathcal{O}_{\bullet 1}, \ldots, \mathcal{O}_{\bullet N-1}].$$

The inner product of two complex-valued vectors $\mathbf{U}$ and $\mathbf{V}$ is defined as

$$\langle \mathbf{U}, \mathbf{V} \rangle = \sum_{t=0}^{N-1} U_t V_t^* = \mathbf{V}^H \mathbf{U}. \tag{45}$$

Note that $\langle \mathbf{V}, \mathbf{U} \rangle = (\langle \mathbf{U}, \mathbf{V} \rangle)^*$. The squared norm of $\mathbf{U}$ becomes

$$\|\mathbf{U}\|^2 = \langle \mathbf{U}, \mathbf{U} \rangle = \sum_{t=0}^{N-1} U_t U_t^* = \sum_{t=0}^{N-1} |U_t|^2.$$

With this definition for the inner product, the orthonormality property can be restated as before, namely, as either $\langle \mathcal{O}_{\bullet k}, \mathcal{O}_{\bullet k'} \rangle = \delta_{k,k'}$ or $\langle \mathcal{O}_{j\bullet}, \mathcal{O}_{j'\bullet} \rangle = \delta_{j,j'}$. Either the $\mathcal{O}_{j\bullet}$'s or the $\mathcal{O}_{\bullet k}$'s constitute a basis for the finite dimensional space $\mathbb{C}^N$ of all N dimensional complex-valued vectors.

Because the real-valued vector $\mathbf{X}$ is contained in $\mathbb{C}^N$, we can analyze it with respect to the unitary matrix $\mathcal{O}$. When we analyzed $\mathbf{X}$ with respect to a real-valued orthonormal matrix, we took $\langle \mathbf{X}, \mathcal{O}_{j\bullet} \rangle$ to be the definition of the jth transform coefficient O_j (see Equation (43a)). For a unitary matrix, let us again use $O_j \equiv \langle \mathbf{X}, \mathcal{O}_{j\bullet} \rangle$.

▷ **Exercise [46a]:** Show that

$$\mathbf{O} = \mathcal{O}\mathbf{X} \text{ and } \mathbf{X} = \mathcal{O}^H \mathbf{O}, \tag{46a}$$

which define, respectively, the analysis and synthesis equations for a transform based on a unitary matrix (these equations are still valid if we replace $\mathbf{X}$ with a complex-valued vector). ◁

Note that these equations are consistent with Equations (43a) and (43b) for the real-valued case.

Finally we note that energy preservation can be expressed as

$$\|\mathbf{X}\|^2 = \sum_{t=0}^{N-1} X_t^2 = \sum_{j=0}^{N-1} |O_j|^2 = \|\mathbf{O}\|^2$$

and that the projection theorem can be easily extended to complex-valued transforms.

3.4 The Orthonormal Discrete Fourier Transform

Let $\{F_k\}$ be the orthonormal discrete Fourier transform (ODFT) of the real-valued sequence $\{X_t\}$, which we define as

$$F_k \equiv \frac{1}{\sqrt{N}} \sum_{t=0}^{N-1} X_t e^{-i2\pi tk/N}, \qquad k = 0, \ldots, N-1 \tag{46b}$$

(the relationship between the ODFT and the DFT of Chapter 2 is discussed in item [1] of the Comments and Extensions at the end of this section). The quantity F_k is called the kth Fourier coefficient. Although the F_k's are N complex-valued variables, the $2N$ real values comprising their real and imaginary components are constrained such that only N real values are needed to construct all of the Fourier coefficients.

▷ **Exercise [46b]:** Let $\{X_t : t = 0, \ldots, N-1\}$ be a real-valued time series, and let F_k be its kth Fourier coefficient as defined in Equation (46b). Show that F_0 is real-valued, that $F_{N-k} = F_k^*$ for $1 \leq k < N/2$, and that $F_{N/2}$ is real-valued when N is even. ◁

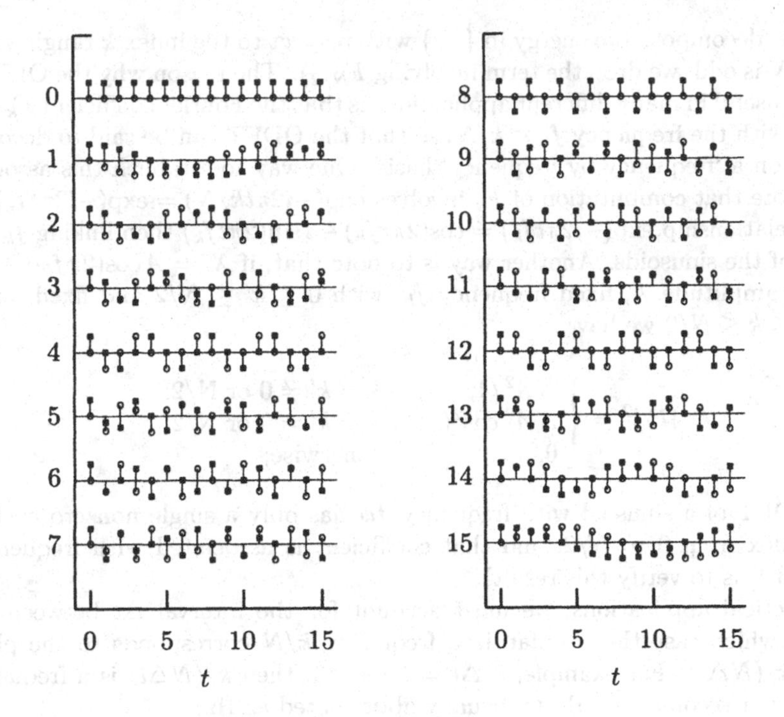

Figure 47. Row vectors $\mathcal{F}_{k\bullet}^H$ for the orthonormal discrete Fourier transform matrix $\mathcal{F}$ for $N = 16$ and $k = 0$ to 7 (top to bottom on left-hand plot) and $k = 8$ to 15 (right-hand plot). Most of the elements of $\mathcal{F}$ are complex-valued, so the real and imaginary components are represented, respectively, by solid squares and open circles. Note that the elements of $\mathcal{F}_{0\bullet}^H$ and $\mathcal{F}_{8\bullet}^H$ are real-valued and that, e.g., $\mathcal{F}_{15\bullet}^H = (\mathcal{F}_{1\bullet}^H)^*$.

From the above exercise, we can conclude that the F_k's are determined by N real values, namely, F_0 and $F_{N/2}$ (when N is even) and the real and imaginary components of the complex-valued variables F_k, $1 \le k < N/2$ (there are $N/2 - 1$ of these when N is even and $(N-1)/2$ when N is odd).

In vector notation we can write Equation (46b) as $\mathbf{F} = \mathcal{F}\mathbf{X}$, where $\mathbf{F}$ is a column vector of length N whose elements are $F_0, \ldots, F_{N-1}$, and $\mathcal{F}$ is an $N \times N$ matrix whose (k,t)th element is $\exp(-i2\pi tk/N)/\sqrt{N}$ for $0 \le k, t \le N-1$. Figure 47 depicts the matrix $\mathcal{F}$ for the case $N = 16$.

▷ **Exercise [47]:** Verify that $\mathcal{F}$ is a unitary matrix, i.e., that $\mathcal{F}^H\mathcal{F} = I_N$, where the superscript H refers to Hermitian transpose. Hint: use the hint for Exercise [29a]. ◁

It now follows that $\mathbf{X} = \mathcal{F}^H\mathbf{F}$ and that

$$\mathcal{E}_{\mathbf{F}} \equiv \|\mathbf{F}\|^2 = \|\mathbf{X}\|^2 \equiv \mathcal{E}_{\mathbf{X}}.$$

The above says that the ODFT is an energy preserving (isometric) transform. Because $F_{N-k} = F_k^*$, for even N we have

$$\|\mathbf{F}\|^2 = \sum_{k=0}^{N-1} |F_k|^2 = |F_0|^2 + |F_{N/2}|^2 + 2 \sum_{1 \le k < N/2} |F_k|^2.$$

We can thus decompose the energy in $\{X_t\}$ with respect to the index k ranging from 0 to $N/2$ (if N is odd, we drop the term involving $F_{N/2}$). The reason why the ODFT has been found useful in many different applications is that the Fourier coefficient F_k can be associated with the frequency $f_k \equiv k/N$, so that the ODFT can be said to decompose the energy on a 'frequency by frequency' basis. One way to establish this association is just to note that computation of F_k involves $\exp(-i2\pi tk/N) = \exp(-i2\pi t f_k)$, with the Euler relationship $\exp(-i2\pi t f_k) = \cos(2\pi t f_k) - i\sin(2\pi t f_k)$ then linking f_k to the frequency of the sinusoids. Another way is to note that, if $X_t = A\cos(2\pi f_{k'} t + \phi)$ for some fixed amplitude A, fixed frequency $f_{k'}$ with $0 \le k' \le N/2$ and fixed phase ϕ, then for $0 \le k \le N/2$ we have

$$|F_k|^2 = \begin{cases} NA^2/4, & k = k' \neq 0 \text{ or N/2;} \\ NA^2\cos^2(\phi), & k = k' = 0 \text{ or N/2;} \\ 0, & \text{otherwise;} \end{cases} \tag{48a}$$

i.e., the ODFT of a sinusoid with frequency $f_{k'}$ has only a single nonzero coefficient with an index from 0 to $N/2$, and that coefficient is associated with frequency $f_{k'}$ (Exercise [3.4] is to verify this result).

In practical applications, we must account for the interval Δt between observations, in which case the standardized frequency k/N corresponds to the physical frequency $k/(N\Delta t)$. For example, if $\Delta t = 1$ second, then $k/(N\Delta t)$ is a frequency of k/N cycles per second (i.e., Hertz, usually abbreviated as Hz).

Let

$$\overline{X} \equiv \frac{1}{N}\sum_{t=0}^{N-1} X_t$$

be the sample mean of the X_t's, and let

$$\hat{\sigma}_X^2 \equiv \frac{1}{N}\sum_{t=0}^{N-1}\left(X_t - \overline{X}\right)^2 = \frac{1}{N}\sum_{t=0}^{N-1} X_t^2 - \overline{X}^2 = \frac{1}{N}\|\mathbf{X}\|^2 - \overline{X}^2 \tag{48b}$$

be the sample variance. Since $F_0/\sqrt{N} = \overline{X}$, the sample variance can be decomposed for even N as

$$\hat{\sigma}_X^2 = \frac{1}{N}\|\mathbf{F}\|^2 - \overline{X}^2 = \frac{1}{N}|F_{N/2}|^2 + \frac{2}{N}\sum_{1\le k<N/2}|F_k|^2,$$

so that $2|F_k|^2/N$ represents the contribution to the sample variance of $\{X_t\}$ due to frequency f_k for $0 < k < N/2$ (for odd N, we must drop the term involving $F_{N/2}$). For even N, this decomposition of variance can be used to define a discrete Fourier empirical power (i.e., sample variance) spectrum $\{P_{\mathcal{F}}(f_k) : f_k \equiv k/N \text{ with } k = 1, \ldots, N/2\}$ for $\{X_t\}$ as follows:

$$P_{\mathcal{F}}(f_k) \equiv \begin{cases} 2|F_k|^2/N, & 1 \le k < N/2; \\ |F_k|^2/N, & k = N/2. \end{cases} \tag{48c}$$

With this definition, we have $\sum_{k=1}^{N/2} P_{\mathcal{F}}(f_k) = \hat{\sigma}_X^2$ (for odd N, we again drop the $F_{N/2}$ term, and the spectrum is defined over the frequencies $f_k = k/N$ with $k =$

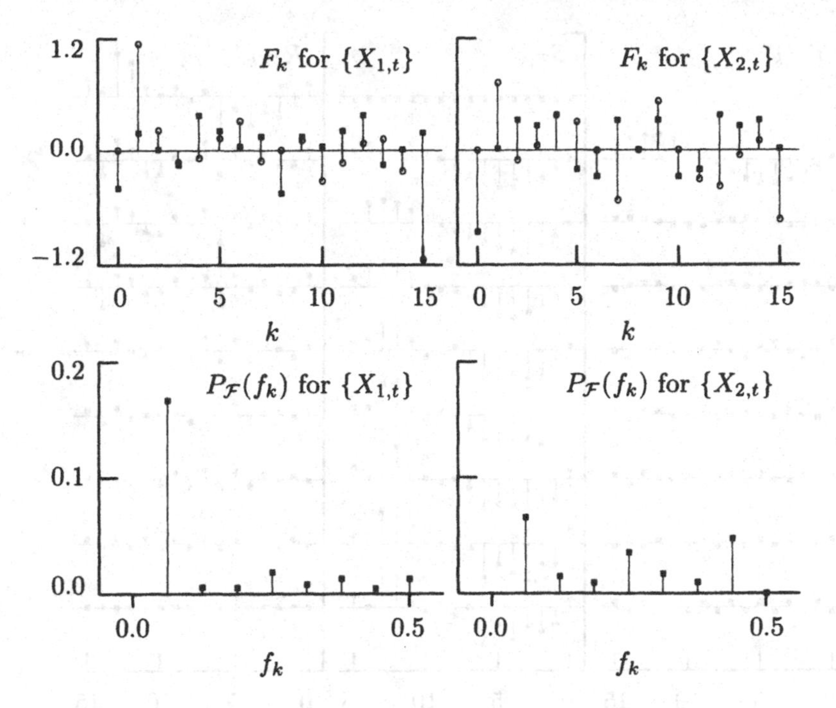

Figure 49. Orthonormal discrete Fourier transforms for the time series in Figure 42. The ODFTs $\mathbf{F}$ for $\{X_{1,t}\}$ and $\{X_{2,t}\}$ are plotted in the top row. The real and imaginary components of the ODFTs are represented by, respectively, solid squares and open circles. The discrete empirical power spectra corresponding to these ODFTs are shown in the bottom row.

$1, \ldots, (N-1)/2$). For even or odd N, the frequencies f_k over which $P_{\mathcal{F}}(f_k)$ is defined satisfy $1/N \leq f_k \leq 1/2$.

As an example, the top row of plots in Figure 49 shows the ODFTs for the two 16 point time series $\{X_{1,t}\}$ and $\{X_{2,t}\}$ in Figure 42. The corresponding discrete empirical power spectra are shown in the bottom row. Even though the two series differ only in their thirteenth value, their ODFTs and spectra differ at all points.

Let us now consider the Fourier synthesis of $\mathbf{X}$ indicated by Equation (46a):

$$\mathbf{X} = \mathcal{F}^H \mathbf{F} = \sum_{k=0}^{N-1} \langle \mathbf{X}, \mathcal{F}_{k\bullet} \rangle \mathcal{F}_{k\bullet} = \sum_{k=0}^{N-1} F_k \mathcal{F}_{k\bullet}.$$

By considering the tth element of this vector equation, we obtain the expression for the inverse ODFT corresponding to Equation (46b):

$$X_t = \frac{1}{\sqrt{N}} \sum_{k=0}^{N-1} F_k e^{i2\pi tk/N}, \qquad t = 0, \ldots, N-1.$$

Next we note that, for even N,

$$\sum_{k=0}^{N-1} F_k \mathcal{F}_{k\bullet} = F_0 \mathcal{F}_{0\bullet} + \sum_{1 \leq k < N/2} (F_k \mathcal{F}_{k\bullet} + F_{N-k} \mathcal{F}_{N-k\bullet}) + F_{N/2} \mathcal{F}_{N/2\bullet}$$

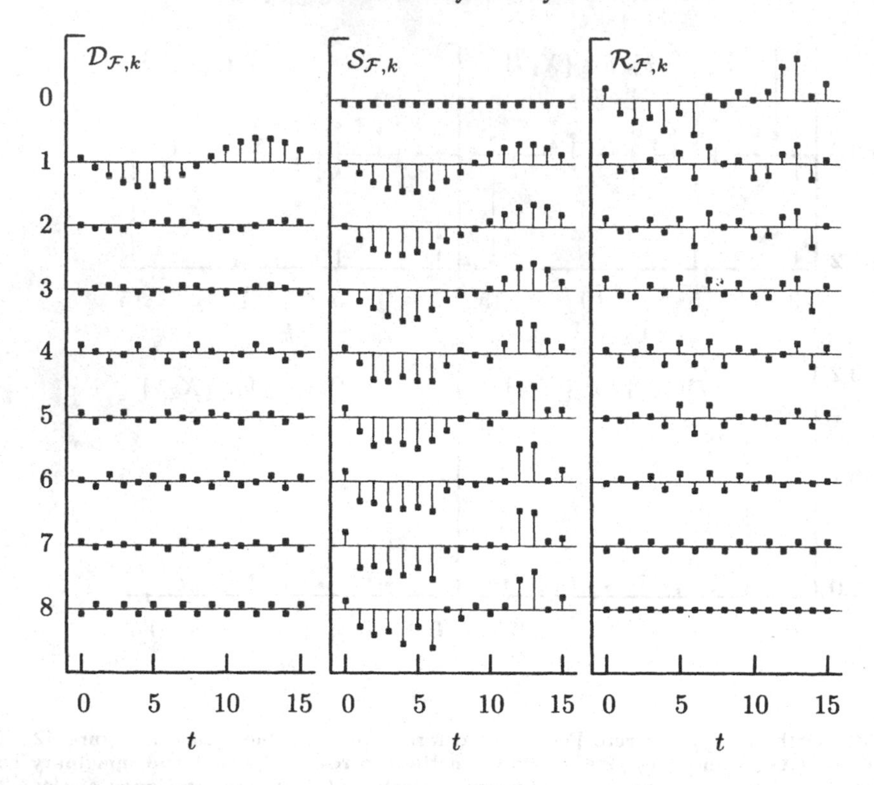

Figure 50. Fourier details $\mathcal{D}_{\mathcal{F},k}$, smooths $\mathcal{S}_{\mathcal{F},k}$ and roughs $\mathcal{R}_{\mathcal{F},k}$ for $\{X_{1,t}\}$ (left-hand plot of Figure 42) for $k = 0, \ldots, 8$ (top to bottom).

(again, we drop the term with $F_{N/2}$ if N is odd). Let us define

$$\mathcal{D}_{\mathcal{F},k} \equiv \begin{cases} F_k \mathcal{F}_{k\bullet} + F_{N-k}\mathcal{F}_{N-k\bullet}, & 1 \le k < N/2; \\ F_{N/2}\mathcal{F}_{N/2\bullet}, & k = N/2. \end{cases} \tag{50}$$

▷ **Exercise [50a]:** Show that $\mathcal{D}_{\mathcal{F},k} = 2\Re\left(F_k \mathcal{F}_{k\bullet}\right)$ for $0 < k < N/2$, where $\Re\left(F_k \mathcal{F}_{k\bullet}\right)$ is a vector whose kth component is the real part of the product of the complex variable F_k and the kth component of the vector $\mathcal{F}_{k\bullet}$. Hint: argue that $F_{N-k}\mathcal{F}_{N-k\bullet} = F_k^* \mathcal{F}_{k\bullet}^*$. ◁

Since $F_0 \mathcal{F}_{0\bullet} = \overline{X}\mathbf{1}$, where $\mathbf{1}$ is a vector of ones, we can now write

$$\mathbf{X} = \overline{X}\mathbf{1} + \sum_{k=1}^{\lfloor N/2 \rfloor} \mathcal{D}_{\mathcal{F},k},$$

where $\lfloor N/2 \rfloor$ is the largest integer less than or equal to $N/2$ (thus $\lfloor N/2 \rfloor = N/2$ if N is even, while $\lfloor N/2 \rfloor = (N-1)/2$ if N is odd). The above equation provides an additive decomposition of $\mathbf{X}$ in terms of the real-valued vectors $\mathcal{D}_{\mathcal{F},k}$, each of which is associated with a frequency f_k.

▷ **Exercise [50b]:** Let $F_k \equiv A_k - iB_k$, and let $\mathcal{D}_{\mathcal{F},k,t}$ be the tth element of $\mathcal{D}_{\mathcal{F},k}$ for $1 \le k < N/2$. Show that

$$\mathcal{D}_{\mathcal{F},k,t} = \frac{2}{\sqrt{N}}\left[A_k \cos(2\pi f_k t) + B_k \sin(2\pi f_k t)\right],$$

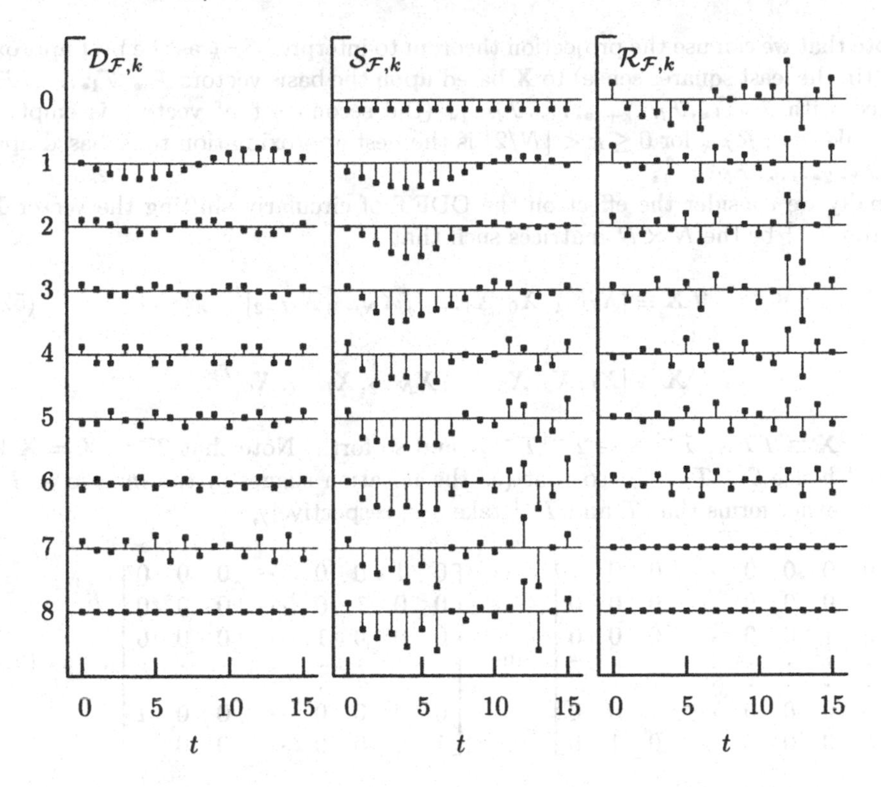

Figure 51. Fourier details $\mathcal{D}_{\mathcal{F},k}$, smooths $\mathcal{S}_{\mathcal{F},k}$ and roughs $\mathcal{R}_{\mathcal{F},k}$ for $\{X_{2,t}\}$ (right-hand plot of Figure 42) for $k = 0, \ldots, 8$ (top to bottom).

which verifies that $\mathcal{D}_{\mathcal{F},k}$ is real-valued and associated with f_k. ◁

We call $\mathcal{D}_{\mathcal{F},k}$ the kth order Fourier detail. The first columns of Figures 50 and 51 show the Fourier details for, respectively, $\{X_{1,t}\}$ and $\{X_{2,t}\}$ of Figure 42 (here $\lfloor N/2 \rfloor = 8$). Note that the $\mathcal{D}_{\mathcal{F},k}$'s are different for all k for the two series even though the series differ only at $t = 13$.

Let

$$\mathcal{S}_{\mathcal{F},k} \equiv \overline{X}\mathbf{1} + \sum_{j=1}^{k} \mathcal{D}_{\mathcal{F},j} \text{ and } \mathcal{R}_{\mathcal{F},k} \equiv \sum_{j=k+1}^{\lfloor N/2 \rfloor} \mathcal{D}_{\mathcal{F},j}$$

so that we have $\mathbf{X} = \mathcal{S}_{\mathcal{F},k} + \mathcal{R}_{\mathcal{F},k}$ for $0 \leq k \leq \lfloor N/2 \rfloor$ (here $\mathcal{S}_{\mathcal{F},0} \equiv \overline{X}\mathbf{1}$, and we define $\mathcal{R}_{\mathcal{F},\lfloor N/2 \rfloor}$ to be a vector of zeros). Since $\mathcal{S}_{\mathcal{F},k}$ is composed of the sample mean and the k lowest frequency components in the Fourier synthesis of $\mathbf{X}$ while $\mathcal{R}_{\mathcal{F},k}$ constitutes the $\lfloor N/2 \rfloor - k$ highest frequency components, the time series given by the elements of $\mathcal{S}_{\mathcal{F},k}$ should be 'smooth' in appearance compared to that of $\mathcal{R}_{\mathcal{F},k}$, which should be 'rough' in appearance (the terminologies 'smooth' and 'rough' were introduced by Tukey, 1977, in another context). We thus refer to $\mathcal{S}_{\mathcal{F},k}$ and $\mathcal{R}_{\mathcal{F},k}$ as, respectively, the kth order Fourier smooth and rough for $\mathbf{X}$ (engineers would say that $\mathcal{S}_{\mathcal{F},k}$ and $\mathcal{R}_{\mathcal{F},k}$ are, respectively, low-pass and high-pass filtered versions of $\mathbf{X}$). The second and third columns of Figures 50 and 51 show the Fourier smooths and roughs of all orders for, respectively, $\{X_{1,t}\}$ and $\{X_{2,t}\}$.

Note that we can use the projection theorem to interpret $\mathcal{S}_{\mathcal{F},k}$ as the best approximation (in the least squares sense) to $\mathbf{X}$ based upon the basis vectors $\mathcal{F}_{0\bullet}, \mathcal{F}_{1\bullet}, \ldots, \mathcal{F}_{k\bullet}$ combined with $\mathcal{F}_{N-k\bullet}, \mathcal{F}_{N-k+1\bullet}, \ldots, \mathcal{F}_{N-1\bullet}$ (the second set of vectors is empty if $k = 0$). Likewise, $\mathcal{R}_{\mathcal{F},k}$ for $0 \leq k < \lfloor N/2 \rfloor$ is the best approximation to $\mathbf{X}$ based upon $\mathcal{F}_{k+1\bullet}, \mathcal{F}_{k+2\bullet}, \ldots, \mathcal{F}_{N-k-1\bullet}$.

Finally we consider the effect on the ODFT of circularly shifting the vector $\mathbf{X}$. Let $\mathcal{T}$ and $\mathcal{T}^{-1}$ be the $N \times N$ matrices such that

$$\mathcal{T}\mathbf{X} \equiv [X_{N-1}, X_0, X_1, \ldots, X_{N-3}, X_{N-2}]^T \tag{52a}$$

and

$$\mathcal{T}^{-1}\mathbf{X} \equiv [X_1, X_2, X_3, \ldots, X_{N-2}, X_{N-1}, X_0]^T,$$

and let $\mathcal{T}^2\mathbf{X} \equiv \mathcal{T}\mathcal{T}\mathbf{X}$, $\mathcal{T}^{-2}\mathbf{X} \equiv \mathcal{T}^{-1}\mathcal{T}^{-1}\mathbf{X}$ and so forth. Note that $\mathcal{T}^{-1}\mathcal{T}\mathbf{X} = \mathbf{X}$ for all $\mathbf{X}$ and hence $\mathcal{T}^{-1}\mathcal{T} = I_N$ so that (as the notation suggests) the inverse of $\mathcal{T}$ is $\mathcal{T}^{-1}$. The exact forms that $\mathcal{T}$ and $\mathcal{T}^{-1}$ take are, respectively,

$$\begin{bmatrix} 0 & 0 & 0 & 0 & \cdots & 0 & 0 & 1 \\ 1 & 0 & 0 & 0 & \cdots & 0 & 0 & 0 \\ 0 & 1 & 0 & 0 & \cdots & 0 & 0 & 0 \\ \vdots & \vdots & \vdots & \vdots & \ddots & \vdots & \vdots & \vdots \\ 0 & 0 & 0 & 0 & \cdots & 1 & 0 & 0 \\ 0 & 0 & 0 & 0 & \cdots & 0 & 1 & 0 \end{bmatrix} \text{ and } \begin{bmatrix} 0 & 1 & 0 & 0 & \cdots & 0 & 0 & 0 \\ 0 & 0 & 1 & 0 & \cdots & 0 & 0 & 0 \\ 0 & 0 & 0 & 1 & \cdots & 0 & 0 & 0 \\ \vdots & \vdots & \vdots & \vdots & \ddots & \vdots & \vdots & \vdots \\ 0 & 0 & 0 & 0 & \cdots & 0 & 0 & 1 \\ 1 & 0 & 0 & 0 & \cdots & 0 & 0 & 0 \end{bmatrix}, \tag{52b}$$

from which it is evident that $\mathcal{T}^{-1} = \mathcal{T}^T$; i.e., $\mathcal{T}$ is in fact an orthonormal transform.

▷ **Exercise [52]:** For any integer m show that the kth Fourier coefficient for $\mathcal{T}^m\mathbf{X}$ is $F_k \exp(-i2\pi mk/N)$. ◁

Since $|F_k \exp(-i2\pi mk/N)| = |F_k|$, the discrete Fourier empirical power spectrum $\{P_{\mathcal{F}}(f_k)\}$ is invariant under circular shifts of $\mathbf{X}$.

Comments and Extensions to Section 3.4

[1] A comparison of the ODFT of Equation (46b) with the DFT of Equation (36g) shows that the two transforms differ only by a factor of $1/\sqrt{N}$. Thus, if we were to form an $N \times N$ dimensional matrix $\overline{\mathcal{F}}$ whose (k,t)th element is $\exp(-i2\pi tk/N)$, then $\overline{\mathcal{F}}\mathbf{X}$ would yield a vector containing the DFT of $\mathbf{X}$. Since $\overline{\mathcal{F}}^H\overline{\mathcal{F}}$ is not equal to the identity matrix I_N but rather is equal to N times it, we see that the DFT is not an orthonormal transform, but it can be called an orthogonal transform since all the off-diagonal elements of $\overline{\mathcal{F}}^H\overline{\mathcal{F}}$ are zero. The rationale for using the DFT rather than just sticking with the ODFT is that, as we have defined it, the DFT of a convolution is equal to the product of the individual DFTs (this is Equation (37b)), whereas the ODFT of a convolution is equal to $\sqrt{N}$ times the product of the individual ODFTs. Convolutions are used so extensively in wavelet theory and elsewhere that it is very convenient not to have to carry along the factor of $\sqrt{N}$ explicitly.

[2] The assumption of circularity that is inherent in the ODFT is not appropriate for many time series, particularly those for which $|X_0 - X_{N-1}|$ is much larger than the typical absolute difference between adjacent observations (i.e., $|X_t - X_{t-1}|$, $t = 2, \ldots, N$). Two procedures that have been found useful for lessening the impact of

circularity on the resulting analysis are tapering (sometimes called windowing) and extending $\mathbf{X}$ to twice its length by tacking on a reversed version of $\mathbf{X}$ at the end of $\mathbf{X}$ (this second procedure is discussed in Section 4.11). Tapering consists of multiplying the elements of $\mathbf{X}$ by a sequence $\{a_t\}$ that decreases to 0 as t approaches 0 or $N-1$ and that leaves the shape of the central part of $\mathbf{X}$ largely unaltered. By forcing the time series toward zero at its end points, we should find that $a_0 X_0 \approx a_{N-1} X_{N-1}$ so that the assumption of circularity is somewhat more reasonable.

As we shall see in Section 4.11, the discrete wavelet transform also assumes circularity, but the localized nature of the transform is such that the number of affected coefficients is much smaller than in the case of the ODFT.

3.5 Summary

Here we summarize the key points of this chapter. An $N \times N$ real-valued matrix $\mathcal{O}$ is said to be orthonormal if its transpose is its inverse, i.e., if $\mathcal{O}^T\mathcal{O} = I_N$, where I_N is the $N \times N$ identity matrix. Also of course $\mathcal{O}\mathcal{O}^T = I_N$. We can analyze (decompose) a time series $\mathbf{X}$ by premultiplying it by $\mathcal{O}$ to obtain $\mathbf{O} = \mathcal{O}\mathbf{X}$, an N dimensional column vector containing the transform coefficients; likewise, given $\mathbf{O}$, we can synthesize (reconstruct) $\mathbf{X}$ since $\mathcal{O}^T\mathbf{O} = \mathbf{X}$. If we let $\mathcal{O}_{j\bullet}$ be a column vector whose elements contain the jth row of $\mathcal{O}$, then the jth element O_j of $\mathbf{O}$ is given by the inner product $\langle \mathbf{X}, \mathcal{O}_{j\bullet}\rangle$, so we can express the synthesis of $\mathbf{X}$ as

$$\mathbf{X} = \mathcal{O}^T\mathbf{O} = \sum_{j=0}^{N-1} O_j \mathcal{O}_{j\bullet} = \sum_{j=0}^{N-1} \langle \mathbf{X}, \mathcal{O}_{j\bullet}\rangle \mathcal{O}_{j\bullet}.$$

An orthonormal transform is isometric, which means that it preserves energy in the sense that $\sum_t X_t^2 = \|\mathbf{X}\|^2 = \|\mathbf{O}\|^2$. The projection theorem states that the best approximation $\widehat{\mathbf{X}}$ to $\mathbf{X}$ that is formed using just the $N' < N$ vectors $\mathcal{O}_{0\bullet}, \ldots, \mathcal{O}_{N'-1\bullet}$ is given by

$$\widehat{\mathbf{X}} = \sum_{j=0}^{N'-1} O_j \mathcal{O}_{j\bullet},$$

where 'best' is to be interpreted in a least squares sense; i.e., the norm of the error vector $\mathbf{e} \equiv \mathbf{X} - \widehat{\mathbf{X}}$ is minimized.

Analogous results can be formulated for a matrix $\mathcal{O}$ whose elements are complex-valued. Such a matrix is said to be unitary if $\mathcal{O}^H\mathcal{O} = I_N$, where the superscript H denotes the Hermitian transpose. Also of course $\mathcal{O}\mathcal{O}^H = I_N$. As before, the transform coefficients for the real-valued vector $\mathbf{X}$ are given by $\mathbf{O} = \mathcal{O}\mathbf{X}$ with the jth element being $O_j = \langle \mathbf{X}, \mathcal{O}_{j\bullet}\rangle$, where $\mathcal{O}_{j\bullet}$ is the Hermitian transpose of the jth row of $\mathcal{O}$. Synthesis is achieved by

$$\mathbf{X} = \mathcal{O}^H\mathbf{O} = \sum_{j=0}^{N-1} O_j \mathcal{O}_{j\bullet} = \sum_{j=0}^{N-1} \langle \mathbf{X}, \mathcal{O}_{j\bullet}\rangle \mathcal{O}_{j\bullet}.$$

The transform $\mathcal{O}$ preserves energy: $\|\mathbf{X}\|^2 = \|\mathbf{O}\|^2 = \sum_t |O_j|^2$. The above results also hold if $\mathbf{X}$ is replaced with a complex-valued vector.

A prime example of a complex-valued orthonormal transform is the orthonormal discrete Fourier transform (ODFT), defined by the $N \times N$ matrix $\mathcal{F}$ whose (k, t)th element is $\exp(-i2\pi tk/N)/\sqrt{N}$ (the ODFT and the DFT differ by a factor of $1/\sqrt{N}$: cf. Equations (46b) and (36g)). Each coefficient for this transform can be associated with a particular frequency, namely, $f_k = k/N$, allowing a decomposition of energy across frequencies. By grouping the synthesis vectors by frequencies, it is possible to split $\mathbf{X}$ into a 'smooth' $\mathcal{S}_{\mathcal{F},k}$ (a low-pass version of $\mathbf{X}$) and a 'rough' $\mathcal{R}_{\mathcal{F},k}$ (a high-pass version of $\mathbf{X}$) so that $\mathbf{X} = \mathcal{S}_{\mathcal{F},k} + \mathcal{R}_{\mathcal{F},k}$. The detail $\mathcal{D}_{\mathcal{F},k}$ is associated with the frequency f_k and represents the difference between smooths (or roughs) of adjacent orders: $\mathcal{D}_{\mathcal{F},k} = \mathcal{S}_{\mathcal{F},k} - \mathcal{S}_{\mathcal{F},k-1}$ (or $\mathcal{D}_{\mathcal{F},k} = \mathcal{R}_{\mathcal{F},k-1} - \mathcal{R}_{\mathcal{F},k}$).

3.6 Exercises

[3.1] Show that an orthonormal transform $\mathcal{O}$ preserves inner products; i.e.,

$$\langle \mathcal{O}\mathbf{X}, \mathcal{O}\mathbf{Y}\rangle = \langle \mathbf{X}, \mathbf{Y}\rangle.$$

[3.2] Here we define a transform that is quite similar to the ODFT except that it uses cosines and sines directly rather than indirectly via complex exponentials. Let N be an even integer, and suppose that $\mathcal{O}$ is an $N \times N$ matrix whose (k,t)th element $\mathcal{O}_{k,t}$, $0 \le k,t \le N-1$, is given by

$$\mathcal{O}_{k,t} = \begin{cases} \sqrt{2/N}\cos\left([k+2]\pi t/N\right), & k = 0, 2, \ldots, N-4; \\ \sqrt{2/N}\sin\left([k+1]\pi t/N\right), & k = 1, 3, \ldots, N-3; \\ \cos(\pi t)/\sqrt{N}, & k = N-2; \\ 1/\sqrt{N}, & k = N-1. \end{cases}$$

Show that $\mathcal{O}$ is an orthonormal matrix. How would you define $\mathcal{O}$ for odd N?

[3.3] Here we consider the type II and type IV discrete cosine transforms (DCTs), which are orthonormal transforms that have been used extensively by engineers for, e.g., compression of images – see Rao and Yip (1990) for details.

(a) Let $\mathcal{O}$ be the $N \times N$ matrix whose (k,t)th element is given by

$$\frac{\sqrt{2}}{\sqrt{c_k N}} \cos\left(\pi k \frac{t+0.5}{N}\right), \qquad k,t = 0, \ldots, N-1,$$

where $c_0 = 2$ and $c_k = 1$ for $k = 1, \ldots, N-1$ (note that the elements of the $k = 0$th row are all $1/\sqrt{N}$). Show that $\mathcal{O}$ is an orthonormal matrix. This transform is known as the DCT-II transform.

(b) Let $\mathcal{O}$ be the $N \times N$ matrix whose (k,t)th element is given by

$$\frac{\sqrt{2}}{\sqrt{N}} \cos\left(\pi (k+0.5) \frac{t+0.5}{N}\right), \qquad k,t = 0, \ldots, N-1.$$

Show that $\mathcal{O}$ is an orthonormal matrix. This transform is known as the DCT-IV transform.

(Unlike the ODFT, DCT-II and DCT-IV are always real-valued; like the ODFT, each DCT coefficient can be associated with a frequency, but note that the spacing between adjacent frequencies is $1/2N$, which is twice as fine as the spacing between frequencies in the DFT, i.e., $1/N$. Like the ODFT, one of the basis vectors of DCT-II is constant, so one of the DCT-II coefficients is proportional to the sample mean $\overline{X}$; this does not hold for DCT-IV, which is thus most useful for series whose sample mean is zero.)

[3.4] As defined by Equation (46b), let F_k be the kth component of the orthonormal discrete Fourier transform of a sequence $X_0, \ldots, X_{N-1}$ of length N. Suppose that $X_t = A\cos(2\pi f_{k'} t + \phi)$, where A is a fixed amplitude, $f_{k'} \equiv k'/N$ is a fixed frequency with $0 \le k' \le N/2$, and ϕ is a fixed phase. Show that $|F_k|^2$ is given by Equation (48a).

[3.5] Show that $P_{\mathcal{F}}(f_k) = \|\mathcal{D}_{\mathcal{F},k}\|^2/N$ for $f_k \equiv k/N$ with $1 \le k \le N/2$, where $P_{\mathcal{F}}(f_k)$ is the kth component of the discrete Fourier empirical power spectrum (defined in Equation (48c)), while $\mathcal{D}_{\mathcal{F},k}$ is the kth order Fourier detail (defined in Equation (50)).

[3.6] Show that, if $\mathcal{D}_{\mathcal{F},k}$ is the kth order Fourier detail for $\mathbf{X}$, then $\mathcal{T}^m\mathcal{D}_{\mathcal{F},k}$ is the kth order Fourier detail for $\mathcal{T}^m\mathbf{X}$, where $\mathcal{T}$ is defined in Equation (52a), and m is any integer.

[3.7] Verify the contents of Figure 50 (the values for $\{X_{1,t}\}$ are given in the caption for Figure 42).

[3.8] Determine the circular filter $\{a_{k,t} : t = 0, \ldots, N-1\}$ such that filtering the elements of $\mathbf{X}$ with $\{a_{k,t}\}$ yields the elements of the Fourier smooth $\mathcal{S}_{\mathcal{F},k}$ as output. How can this filter be implemented using DFTs?

[3.9] Consider an orthonormal matrix $\mathcal{O}$ of the following form:

$$\mathcal{O} = \begin{bmatrix} -\frac{1}{\sqrt{2}} & \frac{1}{\sqrt{2}} & 0 & 0 & 0 & \cdots & 0 & 0 & 0 \\ 0 & 0 & -\frac{1}{\sqrt{2}} & \frac{1}{\sqrt{2}} & 0 & \cdots & 0 & 0 & 0 \\ \vdots & \vdots & \vdots & \vdots & \vdots & \ddots & \vdots & \vdots & \vdots \\ 0 & 0 & 0 & 0 & 0 & \cdots & 0 & -\frac{1}{\sqrt{2}} & \frac{1}{\sqrt{2}} \\ \frac{1}{\sqrt{2}} & \frac{1}{\sqrt{2}} & 0 & 0 & 0 & \cdots & 0 & 0 & 0 \\ 0 & 0 & \frac{1}{\sqrt{2}} & \frac{1}{\sqrt{2}} & 0 & \cdots & 0 & 0 & 0 \\ \vdots & \vdots & \vdots & \vdots & \vdots & \ddots & \vdots & \vdots & \vdots \\ 0 & 0 & 0 & 0 & 0 & \cdots & 0 & \frac{1}{\sqrt{2}} & \frac{1}{\sqrt{2}} \end{bmatrix}$$

(we will encounter such a matrix when we study the partial Haar discrete wavelet transform). Let $\mathbf{X}$ be a vector containing the 16 values of the time series $\{X_{1,t}\}$ of Figure 42 (the values for this series are given in the caption to that figure). Using a 16×16 version of $\mathcal{O}$, compute and plot the least squares approximation $\widehat{\mathbf{X}}_{0,\ldots,7}$ to $\mathbf{X}$ based upon $\mathcal{O}_{0\bullet}, \mathcal{O}_{1\bullet}, \ldots, \mathcal{O}_{7\bullet}$ (i.e., vectors formed from the elements of the first 8 rows of $\mathcal{O}$) – see the discussion concerning the projection theorem in Section 3.2. Repeat the above for the least squares approximation $\widehat{\mathbf{X}}_{8,\ldots,15}$ based upon $\mathcal{O}_{8\bullet}, \mathcal{O}_{9\bullet}, \ldots, \mathcal{O}_{15\bullet}$. How can $\mathbf{X}$ be reconstructed from $\widehat{\mathbf{X}}_{0,\ldots,7}$ and $\widehat{\mathbf{X}}_{8,\ldots,15}$?

4

The Discrete Wavelet Transform

4.0 Introduction

Here we introduce the discrete wavelet transform (DWT), which is the basic tool needed for studying time series via wavelets and plays a role analogous to that of the discrete Fourier transform in spectral analysis. We assume only that the reader is familiar with the basic ideas from linear filtering theory and linear algebra presented in Chapters 2 and 3. Our exposition builds slowly upon these ideas and hence is more detailed than necessary for readers with strong backgrounds in these areas. We encourage such readers just to use the Key Facts and Definitions in each section or to skip directly to Section 4.12 – this has a concise self-contained development of the DWT. For complementary introductions to the DWT, see Strang (1989, 1993), Rioul and Vetterli (1991), Press *et al.* (1992) and Mulcahy (1996).

The remainder of this chapter is organized as follows. Section 4.1 gives a qualitative description of the DWT using primarily the Haar and D(4) wavelets as examples. The formal mathematical development of the DWT begins in Section 4.2, which defines the wavelet filter and discusses some basic conditions that a filter must satisfy to qualify as a wavelet filter. Section 4.3 presents the scaling filter, which is constructed in a simple manner from the wavelet filter. The wavelet and scaling filters are used in parallel to define the pyramid algorithm for computing (and precisely defining) the DWT – various aspects of this algorithm are presented in Sections 4.4, 4.5 and 4.6. The notion of a 'partial' DWT is discussed in Section 4.7, after which we consider two specific classes of wavelet filters in Sections 4.8 and 4.9 (the Daubechies and coiflet filters, respectively). We then give an example of a DWT analysis using some electrocardiogram (ECG) measurements (Section 4.10). Section 4.11 discusses some practical considerations in using the DWT (choice of a particular wavelet, handling boundary conditions, handling time series whose lengths are not a power of two, and choice of the level of a partial DWT). The chapter concludes with the summary (Section 4.12).

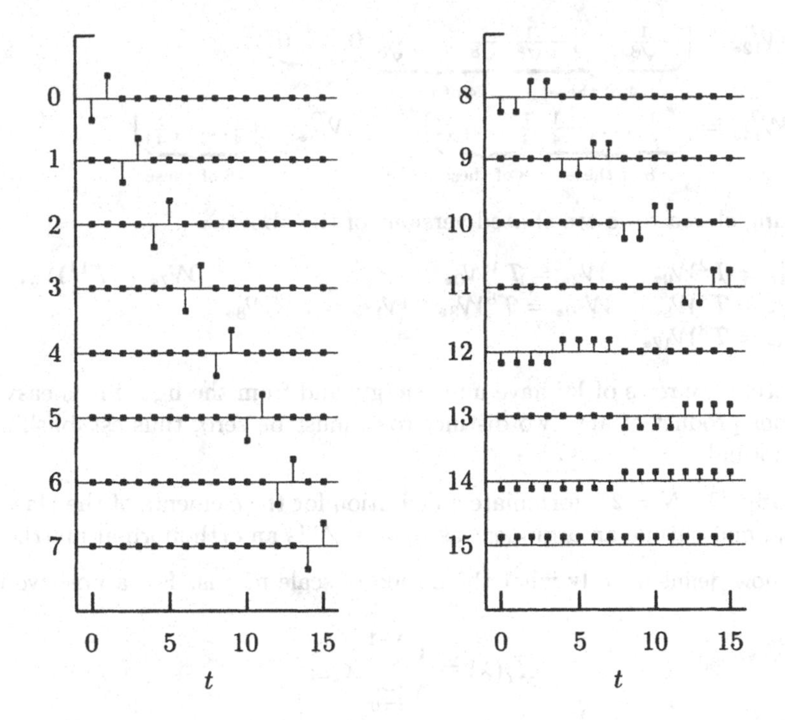

Figure 57. Row vectors $\mathcal{W}_{n\bullet}^T$ of the discrete wavelet transform matrix $\mathcal{W}$ based on the Haar wavelet for $N = 16$ and $n = 0$ to 7 (top to bottom on left plot) and $n = 8$ to 15 (right plot).

4.1 Qualitative Description of the DWT

Like the orthonormal discrete Fourier transform (ODFT) discussed in Section 3.4, the discrete wavelet transform (DWT) of $\{X_t\}$ is an orthonormal transform. If we let $\{W_n : n = 0, \ldots, N-1\}$ represent the DWT coefficients, then we can write $\mathbf{W} = \mathcal{W}\mathbf{X}$, where $\mathbf{W}$ is a column vector of length $N = 2^J$ whose nth element is the nth DWT coefficient W_n, and $\mathcal{W}$ is an $N \times N$ real-valued matrix defining the DWT and satisfying $\mathcal{W}^T\mathcal{W} = I_N$ (the condition that the length of $\mathbf{X}$ be a power of two is restrictive – see the discussions in Sections 4.7, 4.11 and 5.0). Similar to the ODFT, orthonormality implies that $\mathbf{X} = \mathcal{W}^T\mathbf{W}$ and $\|\mathbf{W}\|^2 = \|\mathbf{X}\|^2$. Hence W_n^2 represents the contribution to the energy attributable to the DWT coefficient with index n.

Whereas ODFT coefficients are associated with frequencies, the nth wavelet coefficient W_n is associated with a particular scale and with a particular set of times. To illustrate what is meant by this statement, Figure 57 depicts the elements of $\mathcal{W}$ in a row by row format for an $N = 16$ Haar DWT, whose simple structure makes it a convenient example to start with (historically the Haar DWT can be regarded as the 'first' DWT because it is analogous to a set of functions used by Haar, 1910, to construct an orthonormal basis for the set of all functions that are square integrable over the interval $(-\infty, \infty)$; see Chapters 1 and 11 for details). In this figure, solid squares falling on a thin horizontal line indicate elements of $\mathcal{W}$ that are zero, so we see that each row has a varying number of nonzero elements at different locations (related to the set of times). Explicitly, the rows of this matrix for $n = 0$, 8, 12, 14 and 15 are

$$\mathcal{W}_{0\bullet}^T = [-\tfrac{1}{\sqrt{2}}, \tfrac{1}{\sqrt{2}}, \underbrace{0, \ldots, 0}_{14 \text{ zeros}}], \qquad \mathcal{W}_{8\bullet}^T = [-\tfrac{1}{2}, -\tfrac{1}{2}, \tfrac{1}{2}, \tfrac{1}{2}, \underbrace{0, \ldots, 0}_{12 \text{ zeros}}]$$

$$\mathcal{W}_{12\bullet}^T = [\underbrace{-\tfrac{1}{\sqrt{8}}, \ldots, -\tfrac{1}{\sqrt{8}}}_{\text{4 of these}}, \underbrace{\tfrac{1}{\sqrt{8}}, \ldots, \tfrac{1}{\sqrt{8}}}_{\text{4 of these}}, \underbrace{0, \ldots, 0}_{\text{8 zeros}}]$$

$$\mathcal{W}_{14\bullet}^T = [\underbrace{-\tfrac{1}{4}, \ldots, -\tfrac{1}{4}}_{\text{8 of these}}, \underbrace{\tfrac{1}{4}, \ldots, \tfrac{1}{4}}_{\text{8 of these}}], \qquad \mathcal{W}_{15\bullet}^T = [\underbrace{\tfrac{1}{4}, \ldots, \tfrac{1}{4}}_{\text{16 of these}}].$$

The remaining eleven rows are shifted versions of the above:

$$\begin{array}{llll} \mathcal{W}_{1\bullet} = \mathcal{T}^2\mathcal{W}_{0\bullet} & \mathcal{W}_{2\bullet} = \mathcal{T}^4\mathcal{W}_{0\bullet} & \cdots & \mathcal{W}_{7\bullet} = \mathcal{T}^{14}\mathcal{W}_{0\bullet} \\ \mathcal{W}_{9\bullet} = \mathcal{T}^4\mathcal{W}_{8\bullet} & \mathcal{W}_{10\bullet} = \mathcal{T}^8\mathcal{W}_{8\bullet} & \mathcal{W}_{11\bullet} = \mathcal{T}^{12}\mathcal{W}_{8\bullet} & \\ \mathcal{W}_{13\bullet} = \mathcal{T}^8\mathcal{W}_{12\bullet} & & & \end{array}$$

By construction the rows of $\mathcal{W}$ have unit energy, and from the figure it is easy to see that the inner product of any two distinct rows must be zero, thus establishing that $\mathcal{W}$ is orthonormal.

▷ **Exercise [58]:** For $N = 2^J$, formulate a definition for the elements of the Haar DWT matrix $\mathcal{W}$, and outline an argument as to why $\mathcal{W}$ is an orthonormal matrix. ◁

Let us now define exactly what the notion of scale means. For a positive integer λ, let

$$\overline{X}_t(\lambda) \equiv \frac{1}{\lambda} \sum_{l=0}^{\lambda-1} X_{t-l}$$

represent the average of the λ contiguous data values with indices from $t - \lambda + 1$ to t (note that $\overline{X}_t(1) = X_t$, which we can regard as a 'single point average', and that $\overline{X}_{N-1}(N) = \overline{X}$, which is the sample average of all N values). We refer to $\overline{X}_t(\lambda)$ as the sample average for scale λ over the set of times $t - \lambda + 1$ to t. Since $\mathbf{W} = \mathcal{W}\mathbf{X}$, consideration of the rows of $\mathcal{W}$ shows that we can write

$$\mathbf{W} = \begin{bmatrix} W_0 \\ \vdots \\ W_7 \\ W_8 \\ \vdots \\ W_{11} \\ W_{12} \\ W_{13} \\ W_{14} \\ W_{15} \end{bmatrix} = \begin{bmatrix} \frac{1}{\sqrt{2}}(X_1 - X_0) \\ \vdots \\ \frac{1}{\sqrt{2}}(X_{15} - X_{14}) \\ \frac{1}{2}(X_3 + X_2 - X_1 - X_0) \\ \vdots \\ \frac{1}{2}(X_{15} + X_{14} - X_{13} - X_{12}) \\ \frac{1}{\sqrt{8}}(X_7 + \cdots + X_4 - X_3 - \cdots - X_0) \\ \frac{1}{\sqrt{8}}(X_{15} + \cdots + X_{12} - X_{11} - \cdots - X_8) \\ \frac{1}{4}(X_{15} + \cdots + X_8 - X_7 - \cdots - X_0) \\ \frac{1}{4}(X_{15} + \cdots + X_0) \end{bmatrix}$$

Using the definition for $\overline{X}_t(\lambda)$, we can rewrite the W_n's as

$$\begin{aligned} W_0 &= \tfrac{1}{\sqrt{2}}\left[\overline{X}_1(1) - \overline{X}_0(1)\right], \ldots, W_7 = \tfrac{1}{\sqrt{2}}\left[\overline{X}_{15}(1) - \overline{X}_{14}(1)\right] \\ W_8 &= \overline{X}_3(2) - \overline{X}_1(2), \ldots, W_{11} = \overline{X}_{15}(2) - \overline{X}_{13}(2) \\ W_{12} &= \sqrt{2}\left[\overline{X}_7(4) - \overline{X}_3(4)\right], W_{13} = \sqrt{2}\left[\overline{X}_{15}(4) - \overline{X}_{11}(4)\right] \\ W_{14} &= 2\left[\overline{X}_{15}(8) - \overline{X}_7(8)\right] \\ W_{15} &= 4\,\overline{X}_{15}(16). \end{aligned}$$

Note that the first eight DWT coefficients $W_0, \ldots, W_7$ are proportional to differences (changes) in adjacent averages of $\{X_t\}$ on a unit scale; the next four $W_8, \ldots, W_{11}$ are differences in adjacent averages on a scale of two; W_{12} and W_{13} are proportional to differences on a scale of four; W_{14} is proportional to a difference on a scale of eight; and the final coefficient W_{15} is proportional to the average of all the data.

For general $N = 2^J$ and for the Haar and other DWTs described in this chapter, the elements of $\mathcal{W}$ can be arranged such that the first $N/2$ DWT coefficients are associated with unit scale changes; the next $N/4$ coefficients, with changes on a scale of two; and so forth until we come to the coefficients W_{N-4} and W_{N-3}, which are associated with changes on a scale of $N/4$; the coefficient W_{N-2} is associated with a change on a scale of $N/2$; and finally W_{N-1} is again proportional to the average of all the data. There are thus exactly $N/(2\tau_j)$ coefficients in the DWT associated with changes on scale τ_j, where $\tau_j \equiv 2^{j-1}$ for $j = 1, \ldots, J$ (note that $\tau_1 = 1$ and $\tau_J = N/2$); additionally, there is one coefficient W_{N-1} associated with an average at scale N. The $N-1$ coefficients that are associated with changes on various scales are called *wavelet coefficients*, whereas W_{N-1} is called a *scaling coefficient*. The rows of $\mathcal{W}$ that produce the wavelet coefficients for a particular scale are circularly shifted versions of each other. The amount of the shift between adjacent rows for scale τ_j is $2\tau_j = 2^j$.

It is important to note that τ_j is a unitless standardized scale. In practical applications, we must account for the interval Δt between observations, in which case τ_j corresponds to a physical scale $\tau_j \, \Delta t$ with meaningful units. For example, in Section 4.10 we consider a time series of electrocardiogram measurements recorded at a rate of 180 samples per second, so the sampling interval is $\Delta t = 1/180$ of a second. The physical scales $\tau_j \, \Delta t$ are thus measured in seconds so that, for example, the standardized scale $\tau_3 = 4$ maps to $\tau_3 \, \Delta t = 1/45$ of a second.

Each wavelet coefficient at each scale is also localized in time. In the example above, W_0 involves just times $t = 0$ and 1, whereas W_8 involves times $t = 0$, 1, 2 and 3. In comparison, recall that the ODFT coefficients are not localized in time in any meaningful sense, which is an important distinction between a 'global' transform such as the ODFT and the 'localized' DWT. The notion that the wavelet coefficients are related to differences (of various orders) of (weighted) average values of portions of $\{X_t\}$ concentrated in time is not special just to the Haar wavelet transform, but rather is fundamental to all wavelet transforms (we argued this point also in Chapter 1).

As an example of a DWT other than the Haar DWT, Figure 60 shows the DWT matrix $\mathcal{W}$ for $N = 16$ corresponding to the D(4) wavelet, which is a 'four term' member of the class of discrete Daubechies wavelets (this class of wavelets is formally defined in Section 4.8; the Haar wavelet is the 'two term' member). Again, the first eight rows of this matrix correspond to unit scale (see item [2] of the Comments and Extensions for this section). Each of these rows is based upon four nonzero values, namely,

$$h_0 = \frac{1-\sqrt{3}}{4\sqrt{2}}, \; h_1 = \frac{-3+\sqrt{3}}{4\sqrt{2}}, \; h_2 = \frac{3+\sqrt{3}}{4\sqrt{2}} \text{ and } h_3 = \frac{-1-\sqrt{3}}{4\sqrt{2}}. \tag{59a}$$

For example, we have

$$\mathcal{W}_{0\bullet}^T = [h_1, h_0, \underbrace{0, \ldots, 0}_{\text{12 zeros}}, h_3, h_2] \text{ and } \mathcal{W}_{1\bullet}^T = [h_3, h_2, h_1, h_0, \underbrace{0, \ldots, 0}_{\text{12 zeros}}];$$

i.e., $\mathcal{W}_{1\bullet} = \mathcal{T}^2 \mathcal{W}_{0\bullet}$. Orthonormality of $\mathcal{W}$ requires that

$$\|\mathcal{W}_{0\bullet}\|^2 = h_0^2 + h_1^2 + h_2^2 + h_3^2 = 1 \text{ and } \langle \mathcal{W}_{0\bullet}, \mathcal{W}_{1\bullet} \rangle = h_0 h_2 + h_1 h_3 = 0, \tag{59b}$$

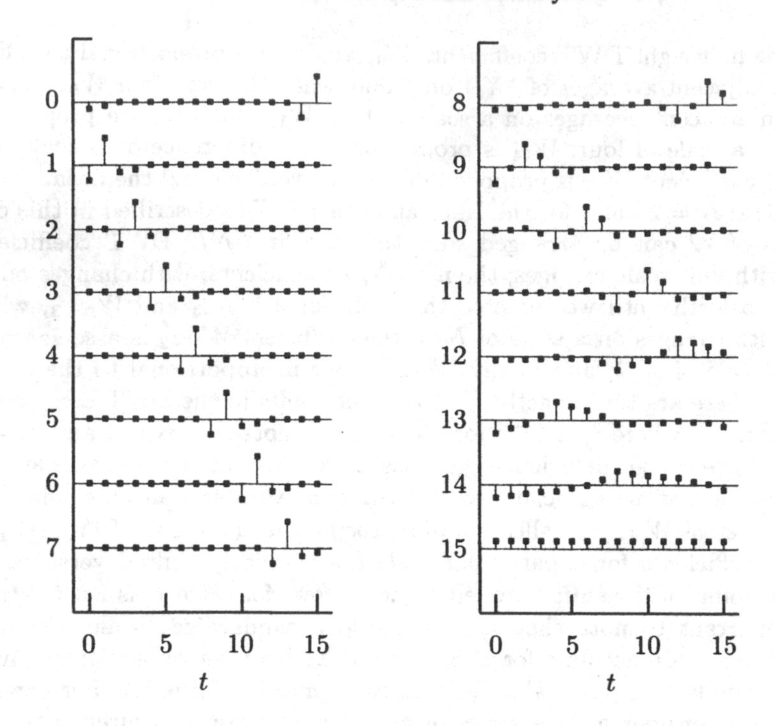

Figure 60. Row vectors $\mathcal{W}_{n\bullet}^T$ of the discrete wavelet transform matrix $\mathcal{W}$ based on the D(4) wavelet for $N = 16$ and $n = 0$ to 7 (top to bottom on left plot) and $n = 8$ to 15 (right plot).

two facts the reader can readily check.

Now the Haar DWT at unit scale is based upon two values, namely, $1/\sqrt{2}$ and $-1/\sqrt{2}$. As we noted above, each Haar wavelet coefficient at unit scale can be obtained by properly scaling the first backward difference between adjacent 'single point averages,' so that $W_n \propto X_{2n+1} - X_{2n}$ for $n = 0, \ldots, 7$. By comparison, each D(4) wavelet coefficient at unit scale can be obtained by properly scaling the second backward difference of contiguous two point weighted averages; i.e., we first form the weighted average $Y_t \equiv aX_t + bX_{t-1}$ and then form the second backward difference of the Y_t's, which by definition is the first backward difference of the first backward difference. Hence, if $Y_t^{(1)} \equiv Y_t - Y_{t-1}$ represents the first backward difference of $\{Y_t\}$, then

$$Y_t^{(2)} \equiv Y_t^{(1)} - Y_{t-1}^{(1)} = Y_t - 2Y_{t-1} + Y_{t-2}$$

is its second backward difference. With a particular choice of a and b, the D(4) wavelet coefficients at unit scale are given by

$$\begin{aligned} W_n &= Y_{2n+1} - 2Y_{2n} + Y_{2n-1} \\ &= aX_{2n+1} + bX_{2n} - 2(aX_{2n} + bX_{2n-1}) + aX_{2n-1} + bX_{2n-2} \\ &= aX_{2n+1} + (b - 2a)X_{2n} + (a - 2b)X_{2n-1} + bX_{2n-2} \\ &\equiv h_0 X_{2n+1} + h_1 X_{2n} + h_2 X_{2n-1} + h_3 X_{2n-2} \end{aligned} \tag{60}$$

for $n = 0, \ldots, 7$ (here we must make the definitions $X_{-1} \equiv X_{15}$ and $X_{-2} \equiv X_{14}$, i.e., a 'circularity' assumption). The two conditions of Equation (59b) can be used to find solutions for a and b, from which the values given for h_0, h_1, h_2 and h_3 in Equation (59a) emerge as one possibility (see Exercise [4.1]).

Let us now visually compare the rows of $\mathcal{W}$ corresponding to unit scale ($n = 0$ to 7 in Figure 57) with the rows for a scale of two ($n = 8$ to 11 in the same figure) for the Haar wavelet. The nonzero elements of the scale two rows appear to be 'stretched out and compressed down' versions of the nonzero elements of the scale one rows (this pattern persists as we go to higher scales). With this idea in mind (and a little imagination!), a look at Figure 60 shows the same pattern for the D(4) wavelet. As we shall see in Sections 4.5 and 4.6, the nonzero elements in the rows corresponding to scales greater than unity are indeed determined by 'stretching and compressing' the nonzero elements in the row for unit scale (the appropriate definition for 'stretching and compressing' is discussed in that section).

The pattern we have observed in the Haar and D(4) DWT continues to other Daubechies wavelets (again, see Section 4.8 for details). Thus, for even L and $L < N$, let $h_0, h_1, \ldots, h_{L-1}$ represent the nonzero elements of the rows of $\mathcal{W}$ associated with unit scale. Each wavelet coefficient at unit scale can be obtained by properly scaling the $(L/2)$th backward difference of contiguous $(L/2)$th point weighted averages. The $L/2$ weights – and hence the h_l's – can be determined by $L/2$ conditions analogous to Equation (59b), namely, that $\sum h_l^2 = 1$ and that

$$\begin{aligned} h_0 h_2 + h_1 h_3 + \cdots + h_{L-3} h_{L-1} &= 0 \\ h_0 h_4 + h_1 h_5 + \cdots + h_{L-5} h_{L-1} &= 0 \\ &\vdots \\ h_0 h_{L-2} + h_1 h_{L-1} &= 0 \end{aligned} \tag{61a}$$

(i.e., $\{h_l\}$ is orthogonal to its even shifts). The $L/2$ conditions do *not* yield a unique set of h_l's, so additional conditions (such as 'extremal phase' or 'least asymmetric', to be discussed in Section 4.8) must be imposed to yield uniqueness. The nonzero elements of the unit scale rows uniquely determine the nonzero elements for the higher scale rows by an appropriate 'stretching and compressing.'

Let us now decompose the elements of the vector $\mathbf{W}$ into $J + 1$ subvectors. The first J subvectors are denoted by $\mathbf{W}_j$, $j = 1, \ldots, J$, and the jth such subvector contains all of the DWT coefficients for scale τ_j. Note that $\mathbf{W}_j$ is a column vector with $N/2^j$ elements. The final subvector is denoted as $\mathbf{V}_J$ and contains just the scaling coefficient W_{N-1}. We can then write

$$\mathbf{W} = \begin{bmatrix} \mathbf{W}_1 \\ \mathbf{W}_2 \\ \vdots \\ \mathbf{W}_J \\ \mathbf{V}_J \end{bmatrix}. \tag{61b}$$

When $N = 2^J = 16$ so that $J = 4$, we have

$$\begin{aligned} \mathbf{W}_1^T &= [W_0, W_1, W_2, W_3, W_4, W_5, W_6, W_7] \\ \mathbf{W}_2^T &= [W_8, W_9, W_{10}, W_{11}] \\ \mathbf{W}_3^T &= [W_{12}, W_{13}] \\ \mathbf{W}_4^T &= [W_{14}] \\ \mathbf{V}_4^T &= [W_{15}]. \end{aligned} \tag{61c}$$

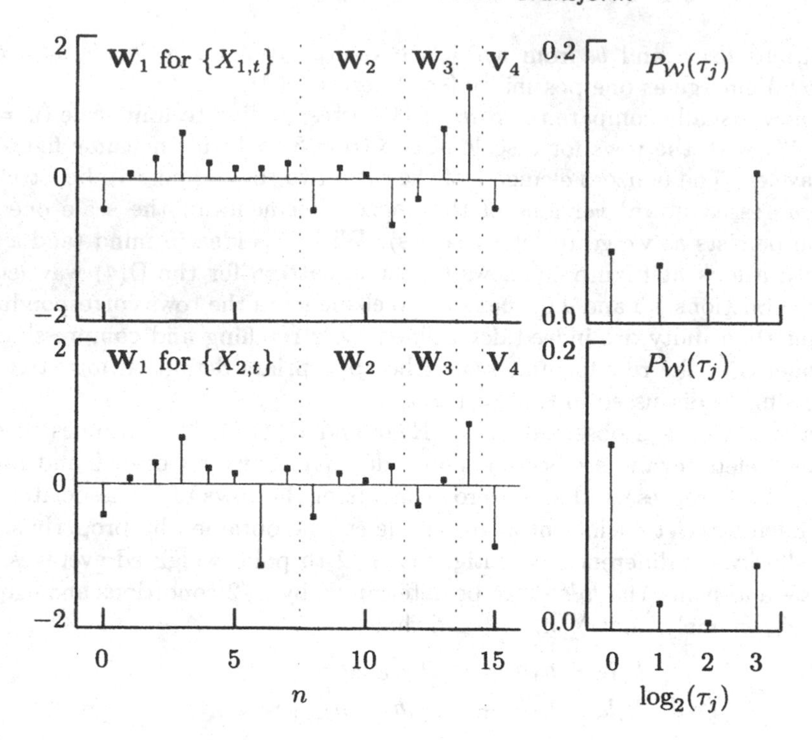

Figure 62. Haar DWTs for the two time series in Figure 42. The DWT coefficients $\mathbf{W}$ are shown in the left-hand plots (the corresponding discrete wavelet empirical power spectra are in the right-hand plots). The thin dotted lines delineate the subvectors $\mathbf{W}_1$, $\mathbf{W}_2$, $\mathbf{W}_3$, $\mathbf{W}_4$ and $\mathbf{V}_4$ (see Equation (61c); $\mathbf{W}_4$ is between $\mathbf{W}_3$ and $\mathbf{V}_4$ but is unlabeled due to lack of space).

We can now write the energy preserving condition as

$$\|\mathbf{X}\|^2 = \|\mathbf{W}\|^2 = \sum_{j=1}^{J} \|\mathbf{W}_j\|^2 + \|\mathbf{V}_J\|^2,$$

so that $\|\mathbf{W}_j\|^2$ represents the contribution to the energy of $\{X_t\}$ due to changes at scale τ_j. Exercise [97] establishes that $W_{N-1}/\sqrt{N} = \overline{X}$ for all DWTs formed using the Daubechies wavelets. Since this result implies that $\|\mathbf{V}_J\|^2 = N\overline{X}^2$, we can decompose the sample variance as

$$\hat{\sigma}_X^2 = \frac{1}{N}\|\mathbf{X}\|^2 - \overline{X}^2 = \frac{1}{N}\|\mathbf{W}\|^2 - \overline{X}^2 = \frac{1}{N}\sum_{j=1}^{J} \|\mathbf{W}_j\|^2, \tag{62}$$

so that $\|\mathbf{W}_j\|^2/N$ represents the contribution to the sample variance of $\{X_t\}$ due to changes at scale τ_j. This decomposition of sample variance can be used to define a discrete wavelet empirical power spectrum $\{P_\mathcal{W}(\tau_j) : \tau_j = 1, 2, 4, \ldots, N/2\}$ for $\{X_t\}$ as

$$P_\mathcal{W}(\tau_j) \equiv \frac{1}{N}\|\mathbf{W}_j\|^2, \text{ for which we have } \sum_{j=1}^{J} P_\mathcal{W}(\tau_j) = \hat{\sigma}_X^2,$$

which can be compared to the equivalent for the ODFT given in Equation (48c).

As an example, the left-hand plots of Figure 62 show the Haar DWTs for the two 16 point time series $\{X_{1,t}\}$ and $\{X_{2,t}\}$ of Figure 42. The corresponding discrete wavelet empirical power spectra are shown in the right-hand plots. Note that the difference between the two series in their thirteenth value affects only one coefficient in each of $\mathbf{W}_1$, $\mathbf{W}_2$ and $\mathbf{W}_3$ along with the single coefficients in $\mathbf{W}_4$ and $\mathbf{V}_4$; however, because the spectra are determined by all the points at a given scale, the spectra for the two series differ at all scales.

In contrast to the discrete Fourier empirical power spectrum, the discrete wavelet empirical power spectrum is in general *not* invariant under circular shifts of $\mathbf{X}$. As an example, consider the unit variance time series $\mathbf{X} = [0, 0, -2, 2, 0, 0, 0, 0]^T$, which has a discrete Haar wavelet empirical power spectrum given by

$$P_{\mathcal{W}}(\tau_j) = \begin{cases} 1, & \tau_j = 1; \\ 0, & \tau_j = 2 \text{ or } 4, \end{cases} \tag{63a}$$

whereas the circularly shifted series $[0, 0, 0, -2, 2, 0, 0, 0]^T$ has a spectrum given by

$$P_{\mathcal{W}}(\tau_j) = \begin{cases} 1/2, & \tau_j = 1; \\ 1/4, & \tau_j = 2 \text{ or } 4 \end{cases} \tag{63b}$$

(Exercise [4.3] is to verify the above).

Let us now consider the wavelet synthesis of $\mathbf{X}$ indicated by Equation (43b):

$$\mathbf{X} = \mathcal{W}^T\mathbf{W} = \sum_{n=0}^{N-1} W_n \mathcal{W}_{n\bullet} = \sum_{j=1}^{J} \mathcal{W}_j^T \mathbf{W}_j + \mathcal{V}_J^T \mathbf{V}_J, \tag{63c}$$

where we define the $\mathcal{W}_j$ and $\mathcal{V}_J$ matrices by partitioning the rows of $\mathcal{W}$ commensurate with the partitioning of $\mathbf{W}$ into $\mathbf{W}_1, \ldots, \mathbf{W}_J$ and $\mathbf{V}_J$. Thus the $\frac{N}{2} \times N$ matrix $\mathcal{W}_1$ is formed from the $n = 0$ up to $n = \frac{N}{2} - 1$ rows of $\mathcal{W}$; the $\frac{N}{4} \times N$ matrix $\mathcal{W}_2$ is formed from the $n = \frac{N}{2}$ up to $n = \frac{3N}{4} - 1$ rows; and so forth, until we come to the $1 \times N$ matrices $\mathcal{W}_J$ and $\mathcal{V}_J$, which are the last two rows of $\mathcal{W}$. We thus have

$$\mathcal{W} = \begin{bmatrix} \mathcal{W}_1 \\ \mathcal{W}_2 \\ \vdots \\ \mathcal{W}_J \\ \mathcal{V}_J \end{bmatrix}, \tag{63d}$$

where $\mathcal{W}_j$ is an $\frac{N}{2^j} \times N$ matrix for $j = 1, \ldots, J$, and $\mathcal{V}_J$ is a row vector of N elements (all of which are in fact equal to $1/\sqrt{N}$ – cf. row 15 of the matrices displayed in Figures 57 and 60, and see Exercise [97]). In our $N = 16$ DWT examples, $\mathcal{W}_1$ is the 8×16 matrix whose rows are the first eight rows of $\mathcal{W}$; i.e.,

$$\mathcal{W}_1 = [\mathcal{W}_{0\bullet}, \mathcal{W}_{1\bullet}, \mathcal{W}_{2\bullet}, \mathcal{W}_{3\bullet}, \mathcal{W}_{4\bullet}, \mathcal{W}_{5\bullet}, \mathcal{W}_{6\bullet}, \mathcal{W}_{7\bullet}]^T;$$

likewise, $\mathcal{W}_2$ is the 4×16 matrix given by

$$\mathcal{W}_2 = [\mathcal{W}_{8\bullet}, \mathcal{W}_{9\bullet}, \mathcal{W}_{10\bullet}, \mathcal{W}_{11\bullet}]^T,$$

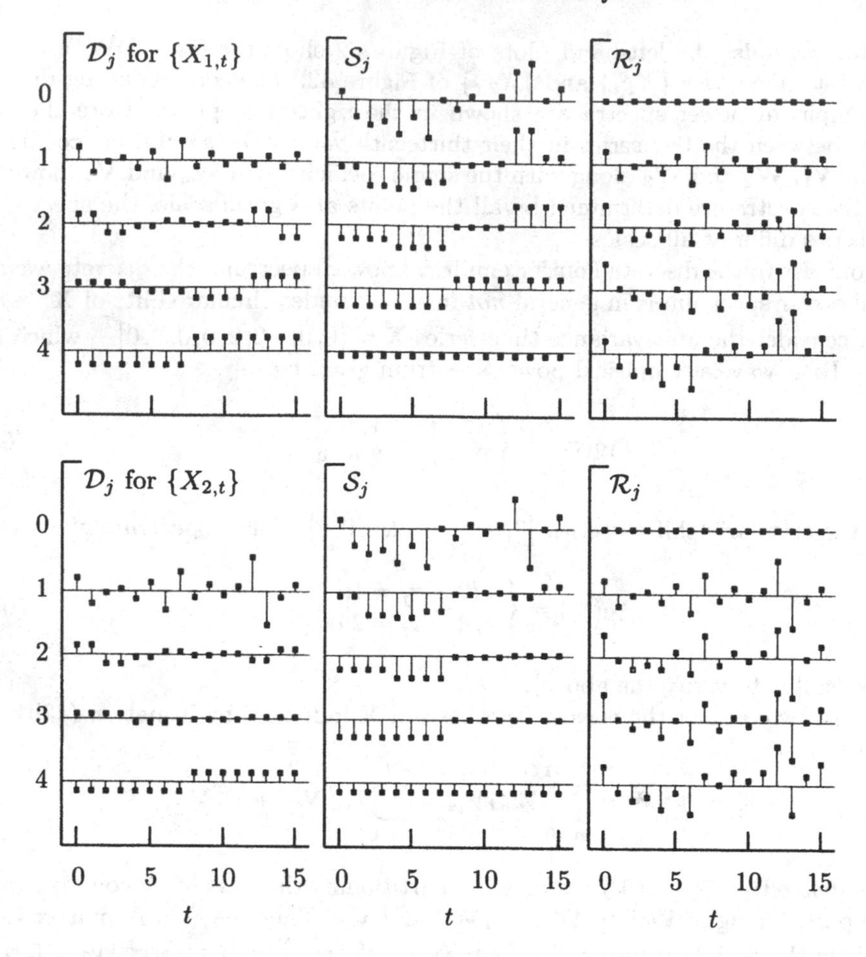

Figure 64. Haar wavelet details $\mathcal{D}_j$, smooths $\mathcal{S}_j$ and roughs $\mathcal{R}_j$ of levels $j = 0$ to 4 for $\{X_{1,t}\}$ (top plots) and $\{X_{2,t}\}$ (bottom). For any given j, we have $\mathcal{S}_j + \mathcal{R}_j = \mathbf{X}$. The jth detail can be interpreted as the difference between successive smooths or successive roughs: $\mathcal{D}_j = \mathcal{S}_{j-1} - \mathcal{S}_j$ and $\mathcal{D}_j = \mathcal{R}_j - \mathcal{R}_{j-1}$.

while $\mathcal{W}_3$, $\mathcal{W}_4$ and $\mathcal{V}_4$ are the 2×16, 1×16 and 1×16 matrices given by

$$\mathcal{W}_3 = \left[\mathcal{W}_{12\bullet}, \mathcal{W}_{13\bullet}\right]^T, \ \mathcal{W}_4 = \mathcal{W}_{14\bullet}^T \text{ and } \mathcal{V}_4 = \mathcal{W}_{15\bullet}^T.$$

Let us now define $\mathcal{D}_j \equiv \mathcal{W}_j^T \mathbf{W}_j$ for $j = 1, \ldots, J$, which is an N dimensional column vector whose elements are associated with changes in $\mathbf{X}$ at scale τ_j; i.e., $\mathbf{W}_j = \mathcal{W}_j \mathbf{X}$ represents the portion of the analysis $\mathbf{W} = \mathcal{W}\mathbf{X}$ attributable to scale τ_j, while $\mathcal{W}_j^T \mathbf{W}_j$ is the portion of the synthesis $\mathbf{X} = \mathcal{W}^T \mathbf{W}$ attributable to scale τ_j. Let $\mathcal{S}_J \equiv \mathcal{V}_J^T \mathbf{V}_J$, which – as indicated by Exercise [97] – has all of its elements equal to the sample mean $\overline{X}$. We can now write

$$\mathbf{X} = \sum_{j=1}^{J} \mathcal{D}_j + \mathcal{S}_J, \tag{64}$$

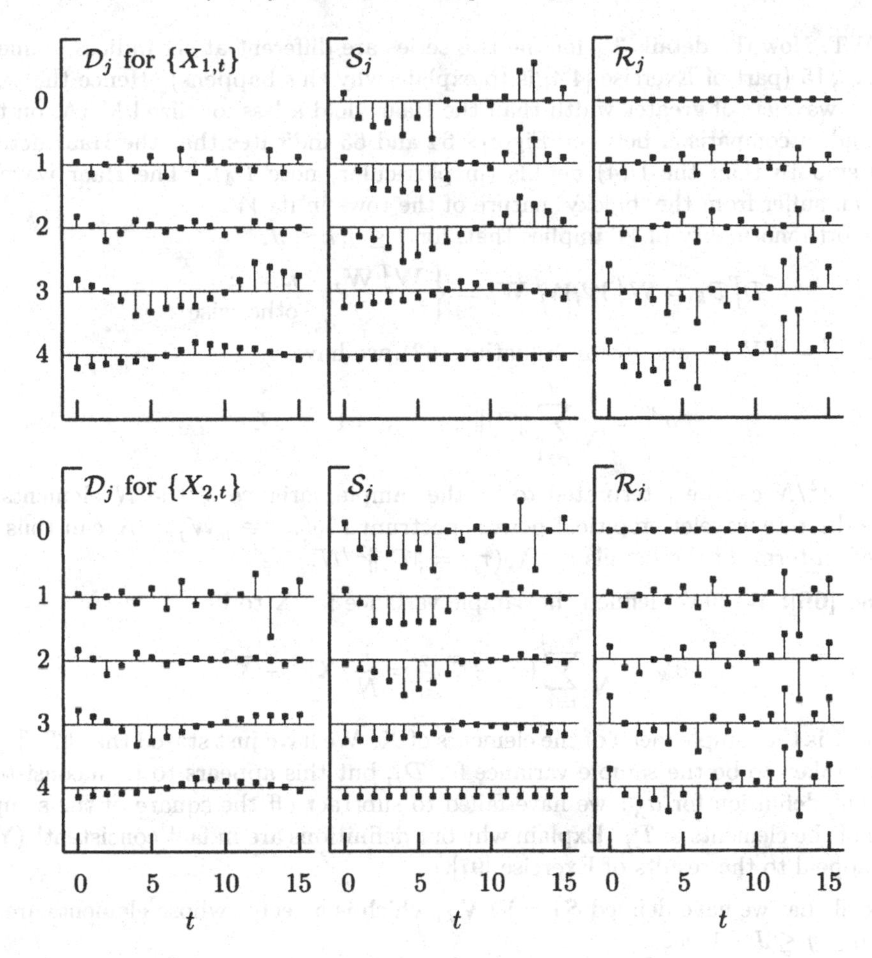

Figure 65. D(4) wavelet details $\mathcal{D}_j$, smooths $\mathcal{S}_j$ and roughs $\mathcal{R}_j$ for $\{X_{1,t}\}$ (top plots) and $\{X_{2,t}\}$ (bottom). Figure 64 has corresponding plots for the Haar wavelet. A comparison of these two figures shows that the Haar and D(4) smooths $\mathcal{S}_4$ agree perfectly for a given time series, which is a consequence of the fact that $\mathcal{V}_4$ is the same for both transforms (the roughs $\mathcal{R}_0$ also agree by definition).

which defines a *multiresolution analysis* (MRA) of $\mathbf{X}$; i.e., we express the series $\mathbf{X}$ as the sum of a constant vector $\mathcal{S}_J$ and J other vectors $\mathcal{D}_j$, $j = 1, \ldots, J$, each of which contains a time series related to variations in $\mathbf{X}$ at a certain scale. We refer to $\mathcal{D}_j$ as the jth level *wavelet detail*.

The top and bottom left-hand plots of Figure 64 show the Haar wavelet details for, respectively, $\{X_{1,t}\}$ and $\{X_{2,t}\}$ of Figure 42 (here $J = 4$ since $N = 16 = 2^4$). Let us compare their unit scale details $\mathcal{D}_1$. Note that the sole difference between the two series, namely, $X_{2,13} = -X_{1,13}$, is reflected as a difference in $\mathcal{D}_1$ at $t = 12$ and 13. To see why this happens, note that $\mathcal{D}_1 = \mathcal{W}_1^T \mathbf{W}_1 = \sum_{n=0}^{7} W_n \mathcal{W}_{n\bullet}$, and recall that the only difference between the unit scale wavelet coefficients for the two series is for W_6 (cf. the left-hand plots of Figure 62). The details $\mathcal{D}_1$ thus differ at $t = 12$ and 13 because $\mathcal{W}_{6\bullet}$ has nonzero elements only at those indices (cf. Figure 57). The top and bottom left-hand plots of Figure 65 show corresponding details using the

D(4) DWT. Now the details $\mathcal{D}_1$ for the two series are different at six indices, namely, $t = 10, \ldots, 15$ (part of Exercise [4.4] is to explain why this happens). Hence the D(4) and other wavelets of greater width than the Haar yield a less localized MRA; on the other hand, a comparison between Figures 64 and 65 indicates that the Haar details are less smooth than the D(4) details (in particular, note $\mathcal{D}_4$). The Haar wavelet details can suffer from the 'blocky' nature of the rows in its $\mathcal{W}$.

The orthonormality of $\mathcal{W}$ implies that, for $1 \le j, k \le J$,

$$\mathcal{D}_j^T \mathcal{D}_k = \mathbf{W}_j^T \mathcal{W}_j \mathcal{W}_k^T \mathbf{W}_k = \begin{cases} \mathbf{W}_j^T \mathbf{W}_j, & k = j; \\ 0, & \text{otherwise.} \end{cases}$$

Thus $\|\mathcal{D}_j\|^2 = \|\mathbf{W}_j\|^2$, and, using Equation (62), we have

$$\hat{\sigma}_X^2 = \frac{1}{N} \sum_{j=1}^{J} \|\mathcal{D}_j\|^2, \qquad j = 1, \ldots, J,$$

where $\|\mathcal{D}_j\|^2/N$ can be interpreted to be the sample variance of the N elements of $\mathcal{D}_j$. The discrete wavelet empirical power spectrum $P_{\mathcal{W}}(\tau_j) \equiv \|\mathbf{W}_j\|^2/N$ can thus be expressed in terms of the details as $P_{\mathcal{W}}(\tau_j) = \|\mathcal{D}_j\|^2/N$.

▷ **Exercise [66]:** We have defined the sample variance for $\mathbf{X}$ to be

$$\hat{\sigma}_X^2 = \frac{1}{N} \sum_{t=0}^{N-1} (X_t - \overline{X})^2 = \frac{1}{N} \|\mathbf{X}\|^2 - \overline{X}^2,$$

where $\overline{X}$ is the sample mean of the elements of $\mathbf{X}$. We have just stated that $\|\mathcal{D}_j\|^2/N$ can be taken to be the sample variance for $\mathcal{D}_j$, but this appears to be inconsistent with our definition for $\hat{\sigma}_X^2$: we have failed to subtract off the square of the sample mean of the elements of $\mathcal{D}_j$. Explain why our definitions are in fact consistent! (You may appeal to the results of Exercise [97].) ◁

Recall that we have defined $\mathcal{S}_J = \mathcal{V}_J^T \mathbf{V}_J$, which is a vector whose elements are all $\overline{X}$. For $0 \le j \le J - 1$, let

$$\mathcal{S}_j \equiv \sum_{k=j+1}^{J} \mathcal{D}_k + \mathcal{S}_J. \tag{66a}$$

Since for $j \ge 1$

$$\mathbf{X} - \mathcal{S}_j = \sum_{k=1}^{j} \mathcal{D}_k,$$

we can argue that $\mathcal{S}_j$ is a smoothed version of $\mathbf{X}$ since the difference between the two vectors involves only details at scale $\tau_j = 2^{j-1}$ and smaller – hence, as the index j increases, $\mathcal{S}_j$ should be smoother in appearance (as we have noted, when $j = J$, its elements are all the same). We refer to $\mathcal{S}_j$ as the jth level *wavelet smooth* for $\mathbf{X}$. Similarly, we define the jth level *wavelet rough* for $\mathbf{X}$ as

$$\mathcal{R}_j \equiv \begin{cases} \mathbf{0}, & j = 0; \\ \sum_{k=1}^{j} \mathcal{D}_k, & 1 \le j \le J, \end{cases} \tag{66b}$$

so that we have $\mathbf{X} = \mathcal{S}_j + \mathcal{R}_j$ for all j. Note that $\mathcal{S}_j - \mathcal{S}_{j+1} = \mathcal{D}_{j+1}$ and $\mathcal{R}_{j+1} - \mathcal{R}_j = \mathcal{D}_{j+1}$; i.e., the details are the differences both between adjacent smooths and between adjacent roughs. The middle and right-hand columns of Figure 64 show the Haar wavelet smooths and roughs of all levels for $\{X_{1,t}\}$ (top plots) and $\{X_{2,t}\}$ (bottom); the corresponding plots in Figure 65 show the smooths and roughs for the D(4) wavelet.

Key Facts and Definitions in Section 4.1

Under the restrictive assumption that our time series $\mathbf{X}$ has length $N = 2^J$, the rows of the $N \times N$ orthonormal DWT matrix $\mathcal{W}$ can be grouped into $J + 1$ submatrices, which in turn leads to a partitioning of the vector $\mathbf{W}$ of DWT coefficients:

$$\mathcal{W}\mathbf{X} = \begin{bmatrix} \mathcal{W}_1 \\ \mathcal{W}_2 \\ \vdots \\ \mathcal{W}_J \\ \mathcal{V}_J \end{bmatrix} \mathbf{X} = \begin{bmatrix} \mathcal{W}_1\mathbf{X} \\ \mathcal{W}_2\mathbf{X} \\ \vdots \\ \mathcal{W}_J\mathbf{X} \\ \mathcal{V}_J\mathbf{X} \end{bmatrix} = \begin{bmatrix} \mathbf{W}_1 \\ \mathbf{W}_2 \\ \vdots \\ \mathbf{W}_J \\ \mathbf{V}_J \end{bmatrix} = \mathbf{W},$$

where $\mathcal{W}_j$ has dimension $N/2^j \times N$; $\mathcal{V}_J$ is $1 \times N$; $\mathbf{W}_j$ is a column vector of length $N/2^j$; and $\mathbf{V}_J$ contains the last element of $\mathbf{W}$. Within each $\mathcal{W}_j$, the rows are circularly shifted versions of each other, but nevertheless are pairwise orthonormal (because of the orthonormality of $\mathcal{W}$). The wavelet coefficients in the vector $\mathbf{W}_j$ are associated with differences (of various orders) in adjacent (weighted) averages over a scale of $\tau_j = 2^{j-1}$, while the scaling coefficient in $\mathbf{V}_J$ is equal to $\sqrt{N}$ times the sample mean $\overline{X}$ of $\mathbf{X}$. Because of the orthonormality of $\mathcal{W}$ and the special form of $\mathbf{V}_J$, we can decompose (analyze) the sample variance (empirical power) of $\mathbf{X}$ into pieces that are associated with scales $\tau_1, \ldots, \tau_J$:

$$\hat{\sigma}^2_X \equiv \frac{1}{N}\|\mathbf{X}\|^2 - \overline{X}^2 = \frac{1}{N}\|\mathbf{W}\|^2 - \overline{X}^2 = \frac{1}{N}\sum_{j=1}^{J}\|\mathbf{W}_j\|^2 \equiv \sum_{j=1}^{J} P_{\mathcal{W}}(\tau_j),$$

where the sequence $\{P_{\mathcal{W}}(\tau_j)\}$ is the discrete wavelet empirical power spectrum. Using the same partitioning of $\mathcal{W}$ and $\mathbf{W}$, we can express the synthesis of $\mathbf{X}$ as the addition of $J+1$ vectors of length N, the first J of which are each associated with a particular scale τ_j, while the final vector $\mathcal{S}_J$ has all its elements equal to the sample mean:

$$\mathbf{X} = \mathcal{W}^T\mathbf{W} = [\mathcal{W}_1^T, \mathcal{W}_2^T, \ldots, \mathcal{W}_J^T, \mathcal{V}_J^T] \begin{bmatrix} \mathbf{W}_1 \\ \mathbf{W}_2 \\ \vdots \\ \mathbf{W}_J \\ \mathbf{V}_J \end{bmatrix} = \sum_{j=1}^{J} \mathcal{W}_j^T\mathbf{W}_j + \mathcal{V}_J^T\mathbf{V}_J \equiv \sum_{j=1}^{J} \mathcal{D}_j + \mathcal{S}_J.$$

Because $\|\mathcal{D}_j\|^2 = \|\mathbf{W}_j\|^2$ for $j = 1, \ldots, J$, the analysis of variance (ANOVA) on a scale-by-scale basis can be re-expressed as

$$\hat{\sigma}^2_X = \frac{1}{N}\sum_{j=1}^{J}\|\mathcal{D}_j\|^2.$$

The detail vectors can be added successively to form jth level smooths and roughs:

$$\mathcal{S}_j = \sum_{k=j+1}^{J} \mathcal{D}_k + \mathcal{S}_J \text{ and } \mathcal{R}_j = \sum_{k=1}^{j} \mathcal{D}_k,$$

in terms of which we have $\mathbf{X} = \mathcal{S}_j + \mathcal{R}_j$, $j = 1, \ldots, J$. The jth level smooth $\mathcal{S}_j$ is associated with scales τ_{j+1} and higher, while $\mathcal{R}_j$ is associated with scales τ_j and lower.

Comments and Extensions to Section 4.1

[1] A comparison of, say, Figures 50 and 64 shows that the indexing schemes for the ODFT and DWT roughs and smooths 'go in the opposite directions' in the sense that the ODFT smooths get smoother as j decreases, whereas the DWT smooths get rougher as j decreases.

[2] The notion of scale τ_j is naturally defined for the Haar wavelet because each Haar wavelet coefficient (with the exception of the very last one) is proportional to the difference of averages formed using $\tau_j = 2^{j-1}$ points. It is not so clear how to define scale for the D(4) and other wavelets, but it can be done in terms of a measure of the 'equivalent width' for a sequence – see item [3] of the Comments and Extensions to Section 4.6 for details. Suffice it to say here that this measure yields τ_j exactly for all the discrete Daubechies wavelets, so we are justified in using it.

[3] Note that the two stage construction of the D(4) wavelet coefficients is a simple example of an equivalent filter for a filter cascade, where here the cascade consists of two filters, one $\{a_{1,0} = a, a_{1,1} = b\}$ of width two forming the weighted average, and the other $\{a_{2,0} = 1, a_{2,1} = -2, a_{2,2} = 1\}$ of width three constituting the second backward difference (determination of a and b is the subject of Exercise [4.1]). If we let $\mathbf{X}$ and $\mathbf{Y}$ represent the input and output of the filter cascade, then the filtering operation depicted by the flow diagram

$$\mathbf{X} \longrightarrow \boxed{\{a, b\}} \longrightarrow \boxed{\{1, -2, 1\}} \longrightarrow \mathbf{Y}$$

is equivalent to one given by

$$\mathbf{X} \longrightarrow \boxed{\{h_0, h_1, h_2, h_3\}} \longrightarrow \mathbf{Y},$$

where $h_0, \ldots, h_3$ are given in Equation (59a).

4.2 The Wavelet Filter

With this section we begin our task of precisely defining the DWT (so far we have done so – via Exercise [58] – only for the Haar DWT). Our definition will be formulated as an algorithm that allows $\mathcal{W}$ to be factored in terms of very sparse matrices. This algorithm is known as the *pyramid algorithm* and was introduced in the context of wavelets by Mallat (1989b). It allows $\mathbf{W} = \mathcal{W}\mathbf{X}$ to be computed using only $O(N)$ multiplications, whereas brute force computation of the product of the $N \times N$ matrix $\mathcal{W}$ and the vector $\mathbf{X}$ involves N^2 multiplications (the notation $O(a_N)$ means that there exists a constant C such that the actual number of multiplications is less than or equal to Ca_N for all N). A related idea exists for the ODFT or DFT, where use of a fast Fourier transform algorithm reduces the number of multiplications to $O(N \log_2 N)$ – by this crude measure, the DWT pyramid algorithm is faster than the fast Fourier transform algorithm!

We shall describe the pyramid algorithm using both linear filtering operations and matrix manipulations. We begin with the filtering approach, which is built upon a real-valued *wavelet filter* $\{h_l : l = 0, \ldots, L-1\}$, where L is the width of the filter and must be an even integer (see Exercise [69]). For $\{h_l\}$ to have width L, we must have $h_0 \neq 0$ and $h_{L-1} \neq 0$. We define $h_l = 0$ for $l < 0$ and $l \geq L$ so that $\{h_l\}$ is actually an infinite sequence with at most L nonzero values. A wavelet filter must satisfy the following three basic properties:

$$\sum_{l=0}^{L-1} h_l = 0; \tag{69a}$$

$$\sum_{l=0}^{L-1} h_l^2 = 1; \tag{69b}$$

and

$$\sum_{l=0}^{L-1} h_l h_{l+2n} = \sum_{l=-\infty}^{\infty} h_l h_{l+2n} = 0 \tag{69c}$$

for all nonzero integers n. In words, a wavelet filter must sum to zero; must have unit energy; and must be orthogonal to its even shifts. The first condition is in keeping with the basic notion of a wavelet (cf. the discussion in Chapter 1), while the last two conditions arose previously in our qualitative description of the DWT (see the discussion surrounding Equation (61a)). We refer to Equations (69b) and (69c) collectively as the *orthonormality property* of wavelet filters. We have in fact already seen two wavelet filters in the previous section, namely, the Haar wavelet filter $\{h_0 = 1/\sqrt{2}, h_1 = -1/\sqrt{2}\}$ for which $L = 2$, and the D(4) wavelet filter of Equation (59a) for which $L = 4$.

▷ **Exercise [69]**: Suppose that $\{h'_l : l = 0, \ldots, L-1\}$ is a filter whose width L is *odd* (this assumes $h'_0 \neq 0$ and $h'_{L-1} \neq 0$). Define $h'_l = 0$ for $l < 0$ and $l \geq L$. Explain why this filter cannot satisfy Equation (69c). ◁

Let $H(\cdot)$ be the transfer function for $\{h_l\}$, i.e.,

$$H(f) \equiv \sum_{l=-\infty}^{\infty} h_l e^{-i2\pi fl} = \sum_{l=0}^{L-1} h_l e^{-i2\pi fl},$$

and let $\mathcal{H}(\cdot)$ denote the associated squared gain function:

$$\mathcal{H}(f) \equiv |H(f)|^2.$$

It is helpful to derive a rather simple condition that is equivalent to Equations (69b) and (69c) and is expressed in terms of the squared gain function $\mathcal{H}(\cdot)$, namely,

$$\mathcal{H}(f) + \mathcal{H}(f + \tfrac{1}{2}) = 2 \quad \text{for all } f. \tag{69d}$$

To establish this equivalence, first assume that $\{h_l\}$ is any real-valued filter whose squared gain function satisfies the above condition. Let

$$h \star h_j \equiv \sum_{l=-\infty}^{\infty} h_l h_{l+j}, \quad j = \ldots, -1, 0, 1, \ldots,$$

be the autocorrelation of this filter (see Equation (25b), recalling that $h_l^* = h_l$ because it is real-valued). Equation (25c) tells us that the Fourier transform of $\{h \star h_j\}$ is given by $|H(f)|^2 = \mathcal{H}(f)$. An appeal to Exercise [23b] says the DFT of $\{h \star h_{2n}\}$ (i.e., the infinite sequence consisting of just the even indexed variables in $\{h \star h_j\}$) is given by $\frac{1}{2}[\mathcal{H}(\frac{f}{2}) + \mathcal{H}(\frac{f}{2} + \frac{1}{2})]$. The inverse DFT (Equation (35a)) yields

$$h \star h_{2n} = \frac{1}{2} \int_{-1/2}^{1/2} \left[\mathcal{H}(\tfrac{f}{2}) + \mathcal{H}(\tfrac{f}{2} + \tfrac{1}{2})\right] e^{i2\pi fn}\, df. \tag{70a}$$

Because we assumed that Equation (69d) holds for all f, it must hold for all $f/2$, so the above yields

$$\sum_{l=-\infty}^{\infty} h_l h_{l+2n} = h \star h_{2n} = \int_{-1/2}^{1/2} e^{i2\pi fn}\, df = \begin{cases} 1, & n = 0; \\ 0, & n = \ldots, -2, -1, 1, 2, \ldots, \end{cases}$$

which shows that Equations (69b) and (69c) must hold.

▷ **Exercise [70]:** To complete the proof of equivalence, suppose now that $\{h_l\}$ satisfies Equations (69b) and (69c). Show that Equation (69d) must be true. ◁

To obtain the wavelet coefficients associated with unit scale, we circularly filter the time series $\{X_t : t = 0, \ldots, N-1\}$ with $\{h_l\}$ and retain every other value of the output, where $N \equiv 2^J$ for some positive integer J. Denote the result of circularly filtering $\{X_t\}$ with $\{h_l\}$ as

$$2^{1/2}\widetilde{W}_{1,t} \equiv \sum_{l=0}^{L-1} h_l X_{t-l \bmod N}, \quad t = 0, \ldots, N-1. \tag{70b}$$

We define the wavelet coefficients for unit scale as

$$W_{1,t} \equiv 2^{1/2}\widetilde{W}_{1,2t+1} = \sum_{l=0}^{L-1} h_l X_{2t+1-l \bmod N}, \quad t = 0, \ldots, \frac{N}{2} - 1. \tag{70c}$$

The first of the two subscripts on $W_{1,t}$ (and $\widetilde{W}_{1,t}$) keeps track of the scale $\tau_j = 2^{j-1}$ associated with these $N/2$ wavelet coefficients, so here $j = 1$ is the index for the unit scale. Note that the wavelet coefficients $\{W_{1,t}\}$ are given by the $N/2$ values with odd indices in the rescaled filter output $\{2^{1/2}\widetilde{W}_{1,t}\}$. The procedure of taking every other value of a filter output is called *subsampling* by two or *downsampling* by two (see item [1] of the Comments and Extensions for this section). The square root of two which appears in Equation (70c) is needed essentially to preserve energy following downsampling.

We can now connect our definition for $\{W_{1,t}\}$ to the matrix formulation $\mathbf{W} = \mathcal{W}\mathbf{X}$, where (as before) $\mathbf{W}$ is the N dimensional column vector of DWT coefficients, $\mathcal{W}$ is the $N \times N$ matrix defining the DWT, and $\mathbf{X}$ is the N dimensional column vector containing the time series $\{X_t\}$. The first $N/2$ elements of $\mathbf{W}$, i.e., the subvector $\mathbf{W}_1$, are defined to be $W_{1,t}$, $t = 0, \ldots, \frac{N}{2} - 1$. Since $\mathbf{W}_1 = \mathcal{W}_1\mathbf{X}$, where $\mathcal{W}_1$ is the $\frac{N}{2} \times N$ matrix consisting of the first $N/2$ rows of $\mathcal{W}$, our definition for $\mathbf{W}_1$ implies a definition for the rows of $\mathcal{W}_1$. To see precisely what these rows look like, let us rewrite Equation (70c) as

$$W_{1,t} = \sum_{l=0}^{N-1} h_l^\circ X_{2t+1-l \bmod N}, \quad t = 0, \ldots, \frac{N}{2} - 1,$$

where $\{h_l^\circ : l = 0, \ldots, N-1\}$ is $\{h_l\}$ periodized to length N (see Section 2.6). For $0 \le t \le \frac{N}{2} - 1$, the tth row $\mathcal{W}_{t\bullet}^T$ of $\mathcal{W}$ or $\mathcal{W}_1$ yields

$$W_{1,t} = \mathcal{W}_{t\bullet}^T \mathbf{X} = \sum_{l=0}^{N-1} h_l^\circ X_{2t+1-l \bmod N} = \sum_{l=0}^{N-1} h_{2t+1-l \bmod N}^\circ X_l. \tag{71a}$$

Letting $t = 0$, we obtain

$$W_{1,0} = \mathcal{W}_{0\bullet}^T \mathbf{X} = \sum_{l=0}^{N-1} h_{1-l \bmod N}^\circ X_l,$$

so we must have

$$\mathcal{W}_{0\bullet}^T = \left[h_1^\circ, h_0^\circ, h_{N-1}^\circ, h_{N-2}^\circ, \ldots, h_2^\circ\right]. \tag{71b}$$

It is evident from Equation (71a) that the remaining $\frac{N}{2} - 1$ rows of $\mathcal{W}_1$ can be expressed as circularly shifted versions of $\mathcal{W}_{0\bullet}^T$, namely,

$$\mathcal{W}_{t\bullet}^T = \left[\mathcal{T}^{2t} \mathcal{W}_{0\bullet}\right]^T, \quad t = 1, \ldots, \frac{N}{2} - 1,$$

where $\mathcal{T}$ is the $N \times N$ circular shift matrix described in Equation (52a). For example, when $t = 1$, we obtain

$$\mathcal{W}_{1\bullet}^T = \left[h_3^\circ, h_2^\circ, h_1^\circ, h_0^\circ, h_{N-1}^\circ, h_{N-2}^\circ, \ldots, h_4^\circ\right].$$

We are now in a position to prove that the rows of $\mathcal{W}_1$ constitute a set of $N/2$ orthonormal vectors. When $L \le N$, the periodized filter takes the simple form

$$h_l^\circ = \begin{cases} h_l, & 0 \le l \le L-1; \\ 0, & L \le l \le N-1, \end{cases}$$

so the first row of $\mathcal{W}_1$ looks like

$$\mathcal{W}_{0\bullet}^T = [h_1, h_0, \underbrace{0, \ldots, 0}_{N-L \text{ zeros}}, h_{L-1}, \ldots, h_2]. \tag{71c}$$

In this case, because $\mathcal{T}^{2t}$ is an orthonormal transform, we have

$$\langle \mathcal{W}_{t\bullet}, \mathcal{W}_{t\bullet} \rangle = \|\mathcal{W}_{t\bullet}\|^2 = \|\mathcal{T}^{2t} \mathcal{W}_{0\bullet}\|^2 = \|\mathcal{W}_{0\bullet}\|^2 = \sum_{l=0}^{L-1} h_l^2 = 1$$

because a wavelet filter has unit energy (Equation (69b)). For $t' \ne t$ we also have

$$\langle \mathcal{W}_{t'\bullet}, \mathcal{W}_{t\bullet} \rangle = \mathcal{W}_{t'\bullet}^T \mathcal{W}_{t\bullet} = \mathcal{W}_{0\bullet}^T \mathcal{T}^{-2t'} \mathcal{T}^{2t} \mathcal{W}_{0\bullet} = \sum_{l=0}^{L-1} h_l h_{l+2(t-t')} = 0$$

because a wavelet filter is orthogonal to its even shifts (Equation (69c)). Thus, when $L \le N$, proof of the orthonormality of the rows in $\mathcal{W}_1$ follows directly from two of the three basic properties of a wavelet filter (summation to zero – Equation (69a) – does not enter into play).

The above proof cannot be easily adapted to handle the $L > N$ case, but the following approach works both when $L \leq N$ and $L > N$. Because the last $\frac{N}{2} - 1$ rows of $\mathcal{W}_1$ are formed by circularly shifting $\mathcal{W}_{0\bullet}^T$ by multiples of 2 and because $\mathcal{W}_{0\bullet}^T$ is given by Equation (71b), we can establish orthonormality if we can show that

$$h^\circ \star h_l^\circ \equiv \sum_{n=0}^{N-1} h_n^\circ h_{n+l \bmod N}^\circ = \begin{cases} 1, & \text{if } l = 0; \\ 0, & \text{if } l = 2, 4, \ldots, N-2. \end{cases} \tag{72}$$

As discussed in Section 2.5, we have

$$\{h_l^\circ : l = 0, \ldots, N-1\} \longleftrightarrow \{H(\tfrac{k}{N}) : k = 0, \ldots, N-1\},$$

so $\{h^\circ \star h_l^\circ\} \longleftrightarrow \{|H(\frac{k}{N})|^2 = \mathcal{H}(\frac{k}{N})\}$ (see Equation (37e)). The inverse Fourier transform relationship yields

$$\begin{aligned} h^\circ \star h_l^\circ &= \frac{1}{N} \sum_{k=0}^{N-1} \mathcal{H}(\tfrac{k}{N}) e^{i2\pi lk/N} \\ &= \frac{1}{N} \left(\sum_{k=0}^{\frac{N}{2}-1} \mathcal{H}(\tfrac{k}{N}) e^{i2\pi lk/N} + \sum_{k=0}^{\frac{N}{2}-1} \mathcal{H}(\tfrac{k}{N} + \tfrac{1}{2}) e^{i2\pi l(\frac{k}{N} + \frac{1}{2})} \right). \end{aligned}$$

▷ **Exercise [72]:** Complete the proof of orthonormality by showing that this reduces to Equation (72) for $l = 2t$ with $t = 0, \ldots, \frac{N}{2} - 1$. ◁

Because we have defined the wavelet coefficients of unit scale in terms of the output from a filter, we can now attach an interpretation to these coefficients using some tools from filtering theory. Let

$$\mathcal{X}_k \equiv \sum_{t=0}^{N-1} X_t e^{-i2\pi tk/N}, \qquad k = 0, \ldots, N-1,$$

be the DFT of $\{X_t\}$. An application of Parseval's theorem (Equation (36h)) states that

$$\mathcal{E}_{\mathbf{X}} = \sum_{t=0}^{N-1} X_t^2 = \frac{1}{N} \sum_{k=0}^{N-1} |\mathcal{X}_k|^2,$$

so that $|\mathcal{X}_k|^2 / N$ defines an energy spectrum over frequencies k/N. Now consider the result of filtering $\{X_t\}$ with $\{h_t\}$ to produce

$$2^{1/2} \widetilde{W}_{1,t} = \sum_{l=0}^{L-1} h_l X_{t-l \bmod N} = \sum_{l=0}^{N-1} h_l^\circ X_{t-l \bmod N}, \quad t = 0, \ldots, N-1.$$

The sequence $\{2^{1/2} \widetilde{W}_{1,t}\}$ is thus obtained by circularly convolving $\{X_t\}$ with a sequence of length N, namely, $\{h_l^\circ\}$. The DFT of this sequence is $\{H(\frac{k}{N})\}$. The result stated in Equation (37b) regarding the DFT of a circular convolution tells us that

$$\{2^{1/2} \widetilde{W}_{1,t}\} \longleftrightarrow \{H(\tfrac{k}{N}) \mathcal{X}_k\},$$

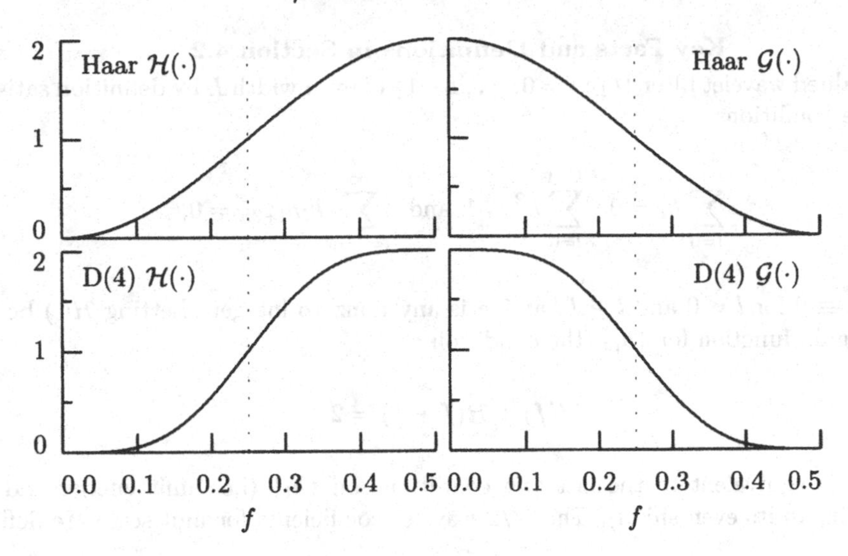

Figure 73. Squared gain functions for Haar wavelet filter (upper left-hand plot), Haar scaling filter (upper right), D(4) wavelet filter (lower left) and D(4) scaling filter (lower right). The dotted lines mark the frequency $f = 1/4$, which is the lower (upper) end of the nominal pass-band for the wavelet (scaling) filters.

and Parseval's theorem in turn yields

$$2\sum_{t=0}^{N-1} \widetilde{W}_{1,t}^2 = \frac{1}{N}\sum_{k=0}^{N-1} \left|H(\tfrac{k}{N})\mathcal{X}_k\right|^2 = \frac{1}{N}\sum_{k=0}^{N-1} \left|H(\tfrac{k}{N})\right|^2 |\mathcal{X}_k|^2 = \frac{1}{N}\sum_{k=0}^{N-1} \mathcal{H}(\tfrac{k}{N})\,|\mathcal{X}_k|^2 .$$

The above equation says that the energy spectrum at frequency k/N of the output from the filter is given by $\mathcal{H}(\frac{k}{N})\,|\mathcal{X}_k|^2/N$, which is just $\mathcal{H}(\frac{k}{N})$ times the energy spectrum at frequency k/N of the input to the filter.

The effect of filtering $\{X_t\}$ to obtain $\{2^{1/2}\widetilde{W}_{1,t}\}$ can thus be evaluated by studying $\mathcal{H}(\frac{k}{N})$ versus frequency k/N. The left-hand column of plots in Figure 73 shows $\mathcal{H}(\cdot)$ for, respectively, the Haar and D(4) wavelet filters (because the squared gain function satisfies $\mathcal{H}(-f) = \mathcal{H}(f)$ and is periodic with a period of unity, we need only plot it for $0 \le f \le \frac{1}{2}$). The plots indicate that these wavelet filters can be regarded as approximations to a high-pass filter with pass-band defined by $\frac{1}{4} \le |f| \le \frac{1}{2}$, with the D(4) wavelet filter being a somewhat better approximation (following standard engineering practice, we define the lower end of the pass-band as the frequency at which $\mathcal{H}(\cdot)$ is a factor of two smaller than its maximum value, i.e., '3 dB down' since $10 \cdot \log_{10}(2) \doteq 3$ dB; this is $f = \frac{1}{4}$ because for both filters $2\mathcal{H}(\frac{1}{4}) = 2 = \mathcal{H}(\frac{1}{2})$). This result holds for all the Daubechies wavelet filters (with the high-pass approximation improving as L increases – see Figure 107), so $\{2^{1/2}\widetilde{W}_{1,t}\}$ and hence the wavelet coefficients $\{W_{1,t}\}$ for unit scale are nominally associated with frequencies with absolute values in the interval $[\frac{1}{4}, \frac{1}{2}]$.

Key Facts and Definitions in Section 4.2

A real-valued wavelet filter $\{h_l : l = 0, \ldots, L-1\}$ of even width L by definition satisfies the three conditions

$$\sum_{l=0}^{L-1} h_l = 0, \quad \sum_{l=0}^{L-1} h_l^2 = 1 \text{ and } \sum_{l=-\infty}^{\infty} h_l h_{l+2n} = 0,$$

where $h_l \equiv 0$ for $l < 0$ and $l \geq L$, and n is any nonzero integer. Letting $\mathcal{H}(\cdot)$ be the squared gain function for $\{h_l\}$, the condition

$$\mathcal{H}(f) + \mathcal{H}(f + \tfrac{1}{2}) = 2$$

for all f is equivalent to the last two conditions on $\{h_l\}$ (i.e., unit energy and orthogonality to its even shifts). The $N/2$ wavelet coefficients for unit scale are defined as

$$W_{1,t} \equiv 2^{1/2}\widetilde{W}_{1,2t+1}, \quad t = 0, \ldots, \frac{N}{2} - 1,$$

with

$$2^{1/2}\widetilde{W}_{1,t} \equiv \sum_{l=0}^{L-1} h_l X_{t-l \bmod N}, \quad t = 0, \ldots, N-1;$$

i.e., we filter $\{X_t\}$ with $\{h_l\}$ to obtain $\{2^{1/2}\widetilde{W}_{1,t}\}$, which we then downsample to get the wavelet coefficients. We can also obtain $W_{1,t}$ directly without going through $2^{1/2}\widetilde{W}_{1,t}$:

$$W_{1,t} = \sum_{l=0}^{L-1} h_l X_{2t+1-l \bmod N} = \sum_{l=0}^{N-1} h_l^\circ X_{2t+1-l \bmod N}, \quad t = 0, \ldots, \frac{N}{2} - 1,$$

where $\{h_l^\circ\}$ is $\{h_l\}$ periodized to length N. These coefficients form the first $N/2$ elements of $\mathbf{W} = \mathcal{W}\mathbf{X}$, i.e., the elements of the subvector $\mathbf{W}_1 = \mathcal{W}_1\mathbf{X}$, where $\mathcal{W}_1$ is the $\frac{N}{2} \times N$ matrix consisting of the first $N/2$ rows of $\mathcal{W}$. The first row of $\mathcal{W}_1$ is given by

$$\mathcal{W}_{0\bullet}^T = \left[h_1^\circ, h_0^\circ, h_{N-1}^\circ, h_{N-2}^\circ, \ldots, h_2^\circ\right],$$

while the remaining $\frac{N}{2} - 1$ rows can be expressed as circularly shifted versions of $\mathcal{W}_{0\bullet}^T$, namely, $\mathcal{W}_{t\bullet}^T = \left[\mathcal{T}^{2t}\mathcal{W}_{0\bullet}\right]^T$, $t = 1, \ldots, \frac{N}{2} - 1$. Two of the defining properties of the wavelet filter (unit energy and orthogonality to even shifts) imply that the rows of $\mathcal{W}_1$ constitute a set of $N/2$ orthonormal vectors. Examination of two specific Daubechies wavelet filters (the Haar and the D(4)) leads to the interpretation that in practice these wavelet filters are high-pass filters with a nominal pass-band defined by $|f| \in [1/4, 1/2]$.

Comments and Extensions to Section 4.2

[1] In Equation (70c) we have subsampled the filter output $\{\widetilde{W}_{1,t}\}$ by retaining all values with odd indices. The more common convention in the engineering literature is to keep the even indexed values. If even indexed values are retained and if we specialize to the Haar wavelet filter $\{h_0, h_1\}$, then the first wavelet coefficient would be

$$2^{1/2}\widetilde{W}_{1,0} = h_0 X_0 + h_1 X_{N-1}$$

rather than our definition, namely,

$$W_{1,0} = 2^{1/2}\widetilde{W}_{1,1} = h_0 X_1 + h_1 X_0.$$

Since the Haar transform is often described as free of boundary effects, retaining the odd indexed values leads to a more appealing Haar DWT that does not explicitly treat $\{X_t\}$ as if it were a periodic sequence.

4.3 The Scaling Filter

In the previous section we used the wavelet filter $\{h_l\}$ to construct the first $N/2$ rows of the DWT matrix $\mathcal{W}$ – these rows constitute the matrix $\mathcal{W}_1$ in the decomposition of $\mathcal{W}$ shown in Equation (63d). In preparation for forming the last $N/2$ rows of $\mathcal{W}$ via the pyramid algorithm, we now define a second filter that will be used to construct an $\frac{N}{2} \times N$ matrix $\mathcal{V}_1$. We will show that this matrix spans the same subspace as the last $N/2$ rows of $\mathcal{W}$. Except for the case $N = 2$ so that $J = 1$ in Equation (63d), the rows of $\mathcal{V}_1$ are *not* equal to the last $N/2$ rows of $\mathcal{W}$, but we can obtain these rows by subsequent manipulation of $\mathcal{V}_1$.

The required second filter is the 'quadrature mirror' filter (QMF) $\{g_l\}$ that corresponds to $\{h_l\}$:

$$g_l \equiv (-1)^{l+1} h_{L-1-l}. \tag{75a}$$

For future reference, we note the inverse relationship

$$h_l = (-1)^l g_{L-1-l}. \tag{75b}$$

The filter $\{g_l\}$ is known as the *scaling filter.* Let us consider two examples. Since the Haar wavelet filter is given by $h_0 = 1/\sqrt{2}$ and $h_1 = -1/\sqrt{2}$, the nonzero portion of the corresponding scaling filter is

$$g_0 = -h_1 = 1/\sqrt{2} \text{ and } g_1 = h_0 = 1/\sqrt{2}. \tag{75c}$$

For the D(4) scaling filter, we have $g_0 = -h_3$, $g_1 = h_2$, $g_2 = -h_1$ and $g_3 = h_0$, so Equation (59a) yields

$$g_0 = \frac{1+\sqrt{3}}{4\sqrt{2}},\ g_1 = \frac{3+\sqrt{3}}{4\sqrt{2}},\ g_2 = \frac{3-\sqrt{3}}{4\sqrt{2}} \text{ and } g_3 = \frac{1-\sqrt{3}}{4\sqrt{2}}. \tag{75d}$$

Let us first note a basic fact about the transfer function for $\{g_l\}$.

▷ **Exercise [76a]:** Suppose that $\{h_l\}$ is a wavelet filter, and let $H(\cdot)$ be its transfer function. As defined in Equation (75a), let $\{g_l\}$ be the scaling filter corresponding to $\{h_l\}$. Show that the transfer function $G(\cdot)$ for $\{g_l\}$ is given by

$$G(f) \equiv \sum_{l=-\infty}^{\infty} g_l e^{-i2\pi fl} = \sum_{l=0}^{L-1} g_l e^{-i2\pi fl} = e^{-i2\pi f(L-1)} H(\tfrac{1}{2} - f) \tag{76a}$$

and hence

$$\mathcal{G}(f) = \mathcal{H}(\tfrac{1}{2} - f),$$

where $\mathcal{G}(f) \equiv |G(f)|^2$ is the squared gain function. ◁

Because $\mathcal{G}(f) = \mathcal{H}(\frac{1}{2} - f)$, it follows that alternative ways of stating the condition of Equation (69d) are

$$\mathcal{G}(f) + \mathcal{G}(f + \tfrac{1}{2}) = 2 \quad \text{or} \quad \mathcal{G}(f) + \mathcal{H}(f) = 2 \quad \text{for all } f \tag{76b}$$

(here we make use of the fact that $\mathcal{G}(\cdot)$ and $\mathcal{H}(\cdot)$ are even periodic functions with unit period). This second relationship is evident for the Haar and D(4) squared gain functions shown in Figure 73. It implies that, if the wavelet filter resembles a high-pass filter, the scaling filter should resemble a low-pass filter. Figure 73 indicates that indeed this is true for the Haar and D(4) scaling filters: these filters can be regarded as approximations to a low-pass filter with pass-band defined by $0 \leq |f| \leq 1/4$, with the D(4) scaling filter being a somewhat better approximation. The high-pass filter $\{h_l\}$ and the low-pass filter $\{g_l\}$ are sometimes called 'half-band' filters because they split the frequency band $[0, 1/2]$ in half.

We next establish some basic properties of the scaling filter.

▷ **Exercise [76b]:** Show that we must have either

$$\sum_{l=0}^{L-1} g_l = \sqrt{2} \text{ or } \sum_{l=0}^{L-1} g_l = -\sqrt{2}.$$

Show further that

$$\sum_{l=0}^{L-1} g_l^2 = 1 \text{ and } \sum_{l=0}^{L-1} g_l g_{l+2n} = \sum_{l=-\infty}^{\infty} g_l g_{l+2n} = 0$$

for all nonzero integers n. ◁

Henceforth we adopt the convention that $\sum_l g_l = \sqrt{2}$, which facilitates the interpretation of scaling coefficients as being localized weighted averages (and not the negative of such).

From the above exercise, we see that the wavelet and scaling filters are similar in that both satisfy the orthonormality property (i.e., unit energy and orthogonality to even shifts). Because the orthonormality property of $\{h_l\}$ is all we needed to establish the orthonormality of the rows of $\mathcal{W}_1$, if we use $\{g_l\}$ to construct a matrix $\mathcal{V}_1$ in exactly the same manner as we used $\{h_l\}$ to construct $\mathcal{W}_1$, then it follows immediately that the rows of $\mathcal{V}_1$ are pairwise orthonormal. Accordingly, we construct $\mathcal{V}_1$ as follows. Denote the result of circularly filtering $\{X_t\}$ with $\{g_l\}$ as

$$2^{1/2}\widetilde{V}_{1,t} \equiv \sum_{l=0}^{L-1} g_l X_{t-l \bmod N}, \quad t = 0, \ldots, N-1. \tag{77a}$$

Define the first level *scaling coefficients* as

$$V_{1,t} \equiv 2^{1/2}\widetilde{V}_{1,2t+1} = \sum_{l=0}^{L-1} g_l X_{2t+1-l \bmod N} = \sum_{l=0}^{N-1} g_l^\circ X_{2t+1-l \bmod N}, \tag{77b}$$

$t = 0, \ldots, \frac{N}{2} - 1$, where $\{g_l^\circ\}$ is $\{g_l\}$ periodized to length N. Let $\mathbf{V}_1$ be the vector of length $N/2$ whose tth element is $V_{1,t}$. Let $\mathcal{V}_1$ be the $\frac{N}{2} \times N$ matrix whose first row is given by

$$[g_1^\circ, g_0^\circ, g_{N-1}^\circ, g_{N-2}^\circ, \ldots, g_2^\circ] \equiv \mathcal{V}_{0\bullet}^T$$

(cf. Equation (71b)) and whose remaining $\frac{N}{2} - 1$ rows are $[\mathcal{T}^{2t}\mathcal{V}_{0\bullet}]^T$, $t = 1, \ldots, \frac{N}{2} - 1$. Then $\mathbf{V}_1 = \mathcal{V}_1\mathbf{X}$, and the rows of $\mathcal{V}_1$ constitute a set of $N/2$ orthonormal vectors.

Our next task is to show that the rows of $\mathcal{V}_1$ and $\mathcal{W}_1$ together constitute a set of N orthonormal vectors. Since the tth rows of $\mathcal{V}_1$ and $\mathcal{W}_1$ are, respectively, $[\mathcal{T}^{2t}\mathcal{V}_{0\bullet}]^T$ and $[\mathcal{T}^{2t}\mathcal{W}_{0\bullet}]^T$, we need to show that

$$\langle \mathcal{T}^{2t}\mathcal{V}_{0\bullet}, \mathcal{T}^{2t'}\mathcal{W}_{0\bullet} \rangle = 0 \text{ for } 0 \le t \le t' \le \frac{N}{2} - 1. \tag{77c}$$

Letting $n = t' - t$, we have, for $n = 0, \ldots, \frac{N}{2} - 1$,

$$\langle \mathcal{T}^{2t}\mathcal{V}_{0\bullet}, \mathcal{T}^{2t'}\mathcal{W}_{0\bullet} \rangle = \mathcal{V}_{0\bullet}^T \mathcal{T}^{-2t}\mathcal{T}^{2t'}\mathcal{W}_{0\bullet} = \mathcal{V}_{0\bullet}^T \mathcal{T}^{2n}\mathcal{W}_{0\bullet} = \sum_{l=0}^{N-1} g_l^\circ h_{l+2n \bmod N}^\circ.$$

When $L \le N$ so that $g_l^\circ = g_l$ and $h_l^\circ = h_l$, the above reduces to

$$\langle \mathcal{T}^{2t}\mathcal{V}_{0\bullet}, \mathcal{T}^{2t'}\mathcal{W}_{0\bullet} \rangle = \sum_{l=0}^{L-1} g_l h_{l+2n}.$$

▷ **Exercise [77]:** For the special case $L \le N$, establish Equation (77c) by showing (using preferably a time domain argument) that

$$\sum_{l=0}^{L-1} g_l h_{l+2n} = \sum_{l=-\infty}^{\infty} g_l h_{l+2n} = 0$$

for all integers n. ◁

In words, the above equation says that the scaling and wavelet filters and any of their even shifts are orthogonal to each other.

The above proof of orthonormality cannot be easily extended to include the $L > N$ case. An approach that works both for $L \le N$ and $L > N$ is the following. We need to show that

$$\sum_{l=0}^{N-1} g_l^\circ h_{l+2n \bmod N}^\circ = g^\circ \star h_{2n}^\circ = 0 \text{ for } n = 0, \ldots, \frac{N}{2} - 1, \tag{77d}$$

where $\{g^\circ \star h_l^\circ\}$ is the circular cross-correlation of $\{g_l^\circ\}$ and $\{h_l^\circ\}$ (see Equation (37c), recalling that $g_l^\circ = [g_l^\circ]^*$ because it is real-valued). Since $\{g_l^\circ\} \longleftrightarrow \{G(\frac{k}{N})\}$ and $\{h_l^\circ\} \longleftrightarrow \{H(\frac{k}{N})\}$, we have $\{g^\circ \star h_l^\circ\} \longleftrightarrow \{G^*(\frac{k}{N})H(\frac{k}{N})\}$ (see Equation (37d)).

▷ **Exercise [78]:** Use the inverse DFT relationship to argue that

$$g^\circ \star h^\circ_{2n} = \frac{1}{N} \sum_{k=0}^{\frac{N}{2}-1} \left[G^*(\tfrac{k}{N})H(\tfrac{k}{N}) + G^*(\tfrac{k}{N}+\tfrac{1}{2})H(\tfrac{k}{N}+\tfrac{1}{2})\right] e^{i4\pi nk/N},$$

and then show that, for all k,

$$G^*(\tfrac{k}{N})H(\tfrac{k}{N}) + G^*(\tfrac{k}{N}+\tfrac{1}{2})H(\tfrac{k}{N}+\tfrac{1}{2}) = 0,$$

which immediately establishes Equation (77d). ◁

In matrix notation, the fact that each row in $\mathcal{V}_1$ is orthogonal to each row in $\mathcal{W}_1$ can be expressed as

$$\mathcal{W}_1\mathcal{V}_1^T = \mathcal{V}_1\mathcal{W}_1^T = 0_{N/2},$$

where $0_{N/2}$ is an $\frac{N}{2} \times \frac{N}{2}$ matrix, all of whose entries are zero. Because we already know that $\mathcal{W}_1\mathcal{W}_1^T = I_{N/2}$ and $\mathcal{V}_1\mathcal{V}_1^T = I_{N/2}$, where $I_{N/2}$ is the $\frac{N}{2} \times \frac{N}{2}$ identity matrix, the $N \times N$ matrix

$$\mathcal{P}_1 \equiv \begin{bmatrix} \mathcal{W}_1 \\ \mathcal{V}_1 \end{bmatrix} \tag{78a}$$

is orthonormal because

$$\mathcal{P}_1\mathcal{P}_1^T = \begin{bmatrix} \mathcal{W}_1 \\ \mathcal{V}_1 \end{bmatrix} [\mathcal{W}_1^T \quad \mathcal{V}_1^T] = \begin{bmatrix} \mathcal{W}_1\mathcal{W}_1^T & \mathcal{W}_1\mathcal{V}_1^T \\ \mathcal{V}_1\mathcal{W}_1^T & \mathcal{V}_1\mathcal{V}_1^T \end{bmatrix} = I_N. \tag{78b}$$

Because the DWT matrix $\mathcal{W}$ is also orthonormal and because the first $N/2$ rows of $\mathcal{W}$ and $\mathcal{P}_1$ are identical, it follows that the last $N/2$ rows of $\mathcal{P}_1$, namely $\mathcal{V}_1$, must span the same subspace as the last $N/2$ rows of $\mathcal{W}$. Except in the case $N = 2$, the rows of $\mathcal{V}_1$ and the last $N/2$ rows of $\mathcal{W}$ are not identical, but we can manipulate $\mathcal{V}_1$ to obtain these rows – this is the subject of the next three sections.

Key Facts and Definitions in Section 4.3

Given the real-valued wavelet filter $\{h_l\}$, the scaling filter is defined as

$$g_l \equiv (-1)^{l+1} h_{L-1-l}.$$

This filter can be assumed to satisfy the conditions

$$\sum_{l=0}^{L-1} g_l = \sqrt{2}, \quad \sum_{l=0}^{L-1} g_l^2 = 1, \quad \sum_{l=-\infty}^{\infty} g_l g_{l+2n} = 0 \text{ and } \sum_{l=-\infty}^{\infty} g_l h_{l+2n'} = 0$$

for all nonzero integers n and all integers n'. Letting $G(\cdot)$ and $\mathcal{G}(\cdot)$ be, respectively, the transfer and squared gain functions for $\{g_l\}$, we have

$$G(f) = e^{-i2\pi f(L-1)} H(\tfrac{1}{2} - f), \quad \mathcal{G}(f) + \mathcal{G}(f + \tfrac{1}{2}) = 2 \text{ and } \mathcal{G}(f) + \mathcal{H}(f) = 2$$

for all f. In practice $\{g_l\}$ is a low-pass filter with nominal pass-band $[-1/4, 1/4]$.

The $N/2$ first level scaling coefficients are defined as

$$V_{1,t} \equiv 2^{1/2}\widetilde{V}_{1,2t+1}, \quad t = 0, \ldots, \frac{N}{2} - 1,$$

with

$$2^{1/2}\widetilde{V}_{1,t} \equiv \sum_{l=0}^{L-1} g_l X_{t-l \bmod N}, \quad t = 0, \ldots, N-1.$$

We can also obtain $V_{1,t}$ directly without going through $2^{1/2}\widetilde{V}_{1,t}$:

$$V_{1,t} = \sum_{l=0}^{L-1} g_l X_{2t+1-l \bmod N} = \sum_{l=0}^{N-1} g_l^\circ X_{2t+1-l \bmod N}, \quad t = 0, \ldots, \frac{N}{2} - 1,$$

where $\{g_l^\circ\}$ is $\{g_l\}$ periodized to length N. These coefficients form the $N/2$ elements of $\mathbf{V}_1 = \mathcal{V}_1\mathbf{X}$, where $\mathcal{V}_1$ is the $\frac{N}{2} \times N$ matrix whose first row is given by $\mathcal{V}_{0\bullet}^T = [g_1^\circ, g_0^\circ, g_{N-1}^\circ, g_{N-2}^\circ, \ldots, g_2^\circ]$, while the remaining $\frac{N}{2} - 1$ rows can be expressed as circularly shifted versions of $\mathcal{V}_{0\bullet}^T$, namely, $\left[\mathcal{T}^{2t}\mathcal{V}_{0\bullet}\right]^T$, $t = 1, \ldots, \frac{N}{2} - 1$. Two of the properties of the scaling filter (unit energy and orthogonality to even shifts) imply that the rows of $\mathcal{V}_1$ constitute a set of $N/2$ orthonormal vectors. The additional fact that the scaling filter is orthogonal to the wavelet filter and all its even shifts implies that $\mathcal{V}_1$ and $\mathcal{W}_1$ are orthogonal; i.e., when stacked together to form

$$\mathcal{P}_1 \equiv \begin{bmatrix} \mathcal{W}_1 \\ \mathcal{V}_1 \end{bmatrix},$$

the resulting $N \times N$ matrix $\mathcal{P}_1$ is orthonormal.

Comments and Extensions to Section 4.3

[1] There is a second common way of defining a QMF namely, $g_l = (-1)^{l-1}h_{1-l}$, which implies the inverse relationship $h_l = (-1)^l g_{1-l}$. This definition is used in, for example, Bruce and Gao (1996a). This second definition is particularly useful for filters with infinite width since the filter width L is absent from the formula. The definition we use (Equation (75a)) presumes that $\{h_l\}$ has finite width. A comparison with Equation (75b) shows that g_{L-1-l} in (75b) has become g_{1-l}, i.e., a shift of even length $-L+2$. As a result, given the scaling filter $\{g_l : l = 0, \ldots, L-1\}$, we obtain a wavelet filter whose nonzero coefficients are contained in $\{h_l : l = -L+2, \ldots, 0, 1\}$. By way of contrast, Equation (75a) gives $\{h_l : l = 0, \ldots, L-1\}$. Our preference for this latter definition is merely that it emphasizes the parallel role of the scaling and wavelet filters in the pyramid algorithm discussed below, and it also makes it somewhat easier to relate the matrix and filtering approaches to wavelets. Mathematically, the two definitions are equivalent ways of expressing the same important concept. (Note that, for the Haar wavelet filter, both definitions yield the same scaling filter.)

[2] Close examination of the proof that $\mathcal{P}_1$ of Equation (78b) is an orthonormal matrix shows that such a matrix is well defined and orthonormal as long as N is an even sample size; i.e., we need not assume that $N = 2^J$ to construct $\mathcal{P}_1$ as we do in order to construct the DWT matrix $\mathcal{W}$. This matter is discussed further in Section 4.7, where we will learn that $\mathcal{P}_1$ is an example of a 'partial DWT' matrix.

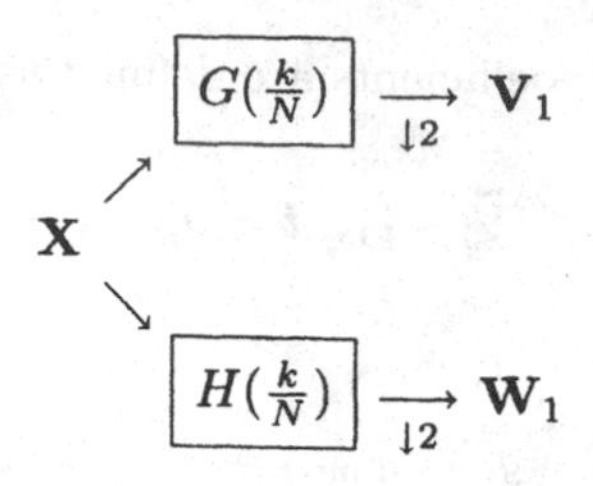

Figure 80. Flow diagram illustrating analysis of $\mathbf{X}$ into $\mathbf{W}_1$ and $\mathbf{V}_1$. The time series in the vector $\mathbf{X}$ of length N is circularly filtered using a wavelet filter $H(\cdot)$ periodized to length N (the frequency domain form of this filter is given by $\{H(\frac{k}{N}) : k = 0, \ldots, N-1\}$). All the odd indexed values of the filtered series are used to form the vector $\mathbf{W}_1$ of length $N/2$ containing the wavelet coefficients of level $j = 1$ ('$\downarrow 2$' indicates downsampling by two); in a similar manner, the vector $\mathbf{V}_1$ of length $N/2$ containing the scaling coefficients of level $j = 1$ is obtained by downsampling the output from filtering $\mathbf{X}$ with the scaling filter $G(\cdot)$ periodized to length N.

[3] The scaling filter is sometimes called the father wavelet filter ('le père' in French), while $\{h_l\}$ is sometimes called the mother wavelet filter ('la mère'). Strichartz (1994) comments that this terminology '... shows a scandalous misunderstanding of human reproduction; in fact, the generation of wavelets more closely resembles the reproductive life style of amoebas.' Needless to say, it is worth sticking with our less colorful terminology to avoid such a serious accusation!

4.4 First Stage of the Pyramid Algorithm

The first stage of the pyramid algorithm for computing the DWT simply consists of transforming the time series $\mathbf{X}$ of length $N = 2^J$ into the $N/2$ first level wavelet coefficients $\mathbf{W}_1$ and the $N/2$ first level scaling coefficients $\mathbf{V}_1$ (Figure 80 illustrates this transformation from a filtering point of view). There are $J-1$ subsequent stages to the pyramid algorithm. For $j = 2, \ldots, J$, the jth stage transforms the vector $\mathbf{V}_{j-1}$ of length $N/2^{j-1}$ into the vectors $\mathbf{W}_j$ and $\mathbf{V}_j$, each of length $N/2^j$. At the jth stage, we treat $\mathbf{V}_{j-1}$ in *exactly* the same manner as we treated $\mathbf{X}$ in the first stage: the elements of $\mathbf{V}_{j-1}$ are filtered separately with $\{h_l\}$ and $\{g_l\}$, and the filter outputs are subsampled to form, respectively, $\mathbf{W}_j$ and $\mathbf{V}_j$. The elements of $\mathbf{V}_j$ are called the scaling coefficients for level j, while $\mathbf{W}_j$ contains the desired wavelet coefficients for level j. At the end of the Jth stage, we can form the DWT coefficient vector $\mathbf{W}$ by concatenating the $J+1$ vectors $\mathbf{W}_1, \ldots, \mathbf{W}_J$ and $\mathbf{V}_J$ (cf. Equation (61b)).

While the above description of the pyramid algorithm tells us exactly how to compute the DWT, it does not give much insight as to what information the DWT is extracting about the original time series. To help understand the DWT, we will concentrate in this section and the next on examining and interpreting certain key points related to the first and second stages of the pyramid algorithm, after which we will be in a better position to understand this algorithm in its full generality in Section 4.6. The key aspects of the first stage we examine in this section are the synthesis of $\mathbf{X}$ from $\mathbf{W}_1$ and $\mathbf{V}_1$; the relationship between the DFT for $\mathbf{X}$ and the DFTs for $\mathbf{V}_1$ and $\mathbf{W}_1$; and the claim that the scale of the scaling coefficients $\mathbf{V}_1$ is given by $\lambda_1 = 2$.

Let us consider the synthesis of $\mathbf{X}$ first. Because we can use the matrix $\mathcal{P}_1$ in

Equation (78a) to express the first stage of the pyramid algorithm as

$$\mathcal{P}_1\mathbf{X} = \begin{bmatrix} \mathcal{W}_1 \\ \mathcal{V}_1 \end{bmatrix} \mathbf{X} = \begin{bmatrix} \mathcal{W}_1\mathbf{X} \\ \mathcal{V}_1\mathbf{X} \end{bmatrix} = \begin{bmatrix} \mathbf{W}_1 \\ \mathbf{V}_1 \end{bmatrix},$$

and because $\mathcal{P}_1$ is an orthonormal matrix, we can recover (synthesize) $\mathbf{X}$ using the equation

$$\mathbf{X} = \mathcal{P}_1^T \begin{bmatrix} \mathbf{W}_1 \\ \mathbf{V}_1 \end{bmatrix} = [\mathcal{W}_1^T \quad \mathcal{V}_1^T] \begin{bmatrix} \mathbf{W}_1 \\ \mathbf{V}_1 \end{bmatrix} = \mathcal{W}_1^T\mathbf{W}_1 + \mathcal{V}_1^T\mathbf{V}_1. \tag{81a}$$

Recall from Section 4.1 that the first level detail is by definition $\mathcal{D}_1 \equiv \mathcal{W}_1^T\mathbf{W}_1$ and that the first level smooth $\mathcal{S}_1$ is such that $\mathcal{S}_1 + \mathcal{D}_1 = \mathbf{X}$. Comparison with the above shows that we must have

$$\mathcal{S}_1 = \mathcal{V}_1^T\mathbf{V}_1 = \mathcal{V}_1^T\mathcal{V}_1\mathbf{X},$$

so the $N \times N$ matrix $\mathcal{V}_1^T\mathcal{V}_1$ can be regarded as an operator that extracts the first level wavelet smooth from $\mathbf{X}$. Likewise, we can write

$$\mathcal{D}_1 = \mathcal{W}_1^T\mathcal{W}_1\mathbf{X},$$

which shows us how to form $\mathcal{D}_1$ via a matrix operation on $\mathbf{X}$.

It is also of interest to see how $\mathcal{D}_1$ can be formed using filtering operations. For convenience, let us consider the form that $\mathcal{W}_1$ takes when $L = 4$ and $N > 4$:

$$\mathcal{W}_1 = \begin{bmatrix} h_1 & h_0 & 0 & 0 & 0 & \cdots & 0 & 0 & 0 & 0 & 0 & h_3 & h_2 \\ h_3 & h_2 & h_1 & h_0 & 0 & \cdots & 0 & 0 & 0 & 0 & 0 & 0 & 0 \\ \vdots & \vdots & \vdots & \vdots & \vdots & \ddots & \vdots & \vdots & \vdots & \vdots & \vdots & \vdots & \vdots \\ 0 & 0 & 0 & 0 & 0 & \cdots & 0 & h_3 & h_2 & h_1 & h_0 & 0 & 0 \\ 0 & 0 & 0 & 0 & 0 & \cdots & 0 & 0 & 0 & h_3 & h_2 & h_1 & h_0 \end{bmatrix}. \tag{81b}$$

Since $\mathcal{D}_1 = \mathcal{W}_1^T\mathbf{W}_1$, we have

$$\mathcal{D}_1 = \begin{bmatrix} h_1 & h_3 & 0 & \cdots & 0 & 0 \\ h_0 & h_2 & 0 & \cdots & 0 & 0 \\ 0 & h_1 & h_3 & \cdots & 0 & 0 \\ 0 & h_0 & h_2 & \cdots & 0 & 0 \\ \vdots & \vdots & \vdots & \ddots & \vdots & \vdots \\ 0 & 0 & 0 & \cdots & h_1 & h_3 \\ 0 & 0 & 0 & \cdots & h_0 & h_2 \\ h_3 & 0 & 0 & \cdots & 0 & h_1 \\ h_2 & 0 & 0 & \cdots & 0 & h_0 \end{bmatrix} \begin{bmatrix} W_{1,0} \\ W_{1,1} \\ W_{1,2} \\ \vdots \\ W_{1,\frac{N}{2}-2} \\ W_{1,\frac{N}{2}-1} \end{bmatrix}. \tag{81c}$$

If we let $\mathcal{D}_{1,t}$ be the tth element of $\mathcal{D}_1$, we can write

$$\mathcal{D}_{1,t} = \begin{cases} h_1 W_{1,\frac{t}{2}} + h_3 W_{1,\frac{t}{2}+1 \bmod \frac{N}{2}}, & t = 0, 2, \ldots, N-2; \\ h_0 W_{1,\frac{t-1}{2}} + h_2 W_{1,\frac{t-1}{2}+1 \bmod \frac{N}{2}}, & t = 1, 3, \ldots, N-1. \end{cases}$$

If we compare the above to Equation (37c), we see that half the elements of $\mathcal{D}_1$ are formed by circularly *cross-correlating* $\mathbf{W}_1$ with $\{h_1, h_3\}$, while the other half are formed by cross-correlating with $\{h_0, h_2\}$. Thus the elements of $\mathcal{D}_1$ are not formed from single cross-correlation operations but rather by interleaving the outputs from two cross-correlations of $\mathbf{W}_1$. Because $\mathcal{V}_1$ has a similar structure to $\mathcal{W}_1$ (with g_l's replacing h_l's), it follows that the first level smooth $\mathcal{S}_1$ can be interpreted as interleaving the outputs from two cross-correlations applied to $\mathbf{V}_1$. Since $\mathbf{X} = \mathcal{D}_1 + \mathcal{S}_1$, we can use these interpretations for $\mathcal{D}_1$ and $\mathcal{S}_1$ to write for general L

$$X_t = \sum_{l=0}^{\frac{L}{2}-1} h_{2l+1} W_{1,\frac{t}{2}+l \bmod \frac{N}{2}} + \sum_{l=0}^{\frac{L}{2}-1} g_{2l+1} V_{1,\frac{t}{2}+l \bmod \frac{N}{2}},$$

$t = 0, 2, \ldots, N-2$, and

$$X_t = \sum_{l=0}^{\frac{L}{2}-1} h_{2l} W_{1,\frac{t-1}{2}+l \bmod \frac{N}{2}} + \sum_{l=0}^{\frac{L}{2}-1} g_{2l} V_{1,\frac{t-1}{2}+l \bmod \frac{N}{2}},$$

$t = 1, 3, \ldots, N-1$.

There is, however, a second way of viewing the construction of $\mathcal{D}_1$ that involves just a single cross-correlation. Suppose we add $N/2$ columns to $\mathcal{W}_1^T$ (one to the left of each existing column) and insert $N/2$ corresponding zeros into $\mathbf{W}_1$ to allow us to write

$$\mathcal{D}_1 = \begin{bmatrix} h_0 & h_1 & h_2 & h_3 & 0 & 0 & \cdots & 0 & 0 & 0 \\ 0 & h_0 & h_1 & h_2 & h_3 & 0 & \cdots & 0 & 0 & 0 \\ 0 & 0 & h_0 & h_1 & h_2 & h_3 & \cdots & 0 & 0 & 0 \\ 0 & 0 & 0 & h_0 & h_1 & h_2 & \cdots & 0 & 0 & 0 \\ \vdots & \vdots & \vdots & \vdots & \vdots & \vdots & \cdots & \vdots & \vdots & \vdots \\ 0 & 0 & 0 & 0 & 0 & 0 & \cdots & h_1 & h_2 & h_3 \\ h_3 & 0 & 0 & 0 & 0 & 0 & \cdots & h_0 & h_1 & h_2 \\ h_2 & h_3 & 0 & 0 & 0 & 0 & \cdots & 0 & h_0 & h_1 \\ h_1 & h_2 & h_3 & 0 & 0 & 0 & \cdots & 0 & 0 & h_0 \end{bmatrix} \begin{bmatrix} 0 \\ W_{1,0} \\ 0 \\ W_{1,1} \\ 0 \\ W_{1,2} \\ \vdots \\ W_{1,\frac{N}{2}-2} \\ 0 \\ W_{1,\frac{N}{2}-1} \end{bmatrix}.$$

The procedure of inserting a zero before each of the elements of $\mathbf{W}_1$ is called *upsampling* (by two) in the engineering literature and, in a certain sense, is complementary to downsampling (see Exercise [4.9]). If we define

$$W_{1,t}^{\uparrow} \equiv \begin{cases} 0, & t = 0, 2, \ldots, N-2; \\ W_{1,\frac{t-1}{2}}, & t = 1, 3, \ldots, N-1, \end{cases}$$

we can write

$$\mathcal{D}_{1,t} = \sum_{l=0}^{L-1} h_l W_{1,t+l \bmod N}^{\uparrow} = \sum_{l=0}^{N-1} h_l^{\circ} W_{1,t+l \bmod N}^{\uparrow}, \quad t = 0, 1, \ldots, N-1.$$

If we compare the above to Equation (37c), we see that the elements of $\mathcal{D}_1$ are obtained by cross-correlating the filter $\{h_l^{\circ}\}$ with the upsampled version of $\mathbf{W}_1$. Since $\{h_l^{\circ}\} \longleftrightarrow \{H(\frac{k}{N})\}$, an alternative interpretation is that the elements of $\mathcal{D}_1$ are the

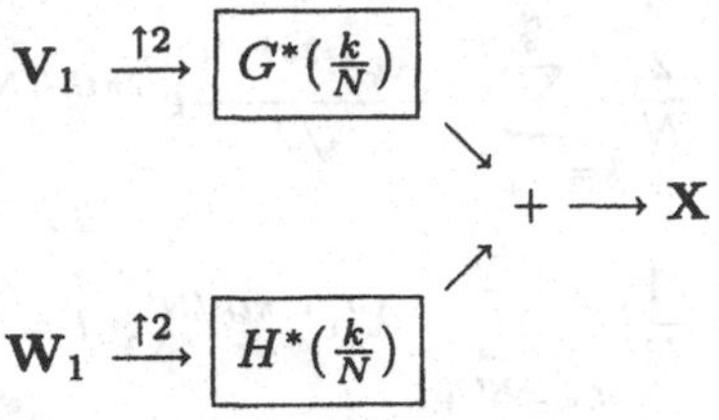

$$\mathbf{V}_1 \xrightarrow{\uparrow 2} \boxed{G^*(\tfrac{k}{N})} \searrow$$
$$+ \longrightarrow \mathbf{X}$$
$$\mathbf{W}_1 \xrightarrow{\uparrow 2} \boxed{H^*(\tfrac{k}{N})} \nearrow$$

Figure 83. Flow diagram illustrating synthesis of $\mathbf{X}$ from $\mathbf{W}_1$ and $\mathbf{V}_1$. The vector $\mathbf{W}_1$ of length $N/2$ is upsampled by two to form a vector of length N, whose contents are then circularly filtered using the filter $\{H^*(\frac{k}{N})\}$ (here upsampling $\mathbf{W}_1$ by two means adding a zero before each element in that vector and is indicated on the flow diagram by '$\uparrow 2$'). The vector $\mathbf{X}$ is formed by adding the output from this filter to a similar output obtained by filtering $\mathbf{V}_1$ (after upsampling) with the filter $\{G^*(\frac{k}{N})\}$.

output obtained by filtering the upsampled version of $\mathbf{W}_1$ with a circular filter whose frequency domain representation is $\{H^*(\frac{k}{N})\}$ (see Exercise [31]).

In a similar manner, the elements of $\mathcal{S}_1$ can be generated by filtering an upsampled version of $\mathbf{V}_1$ with the circular filter $\{G^*(\frac{k}{N})\}$. If we define $V^{\uparrow}_{1,t}$ in a similar way to $W^{\uparrow}_{1,t}$, we can now write

$$\begin{aligned} X_t &= \sum_{l=0}^{L-1} h_l W^{\uparrow}_{1,t+l \bmod N} + \sum_{l=0}^{L-1} g_l V^{\uparrow}_{1,t+l \bmod N} \\ &= \sum_{l=0}^{N-1} h^{\circ}_l W^{\uparrow}_{1,t+l \bmod N} + \sum_{l=0}^{N-1} g^{\circ}_l V^{\uparrow}_{1,t+l \bmod N}, \quad t = 0, 1, \ldots, N-1. \end{aligned} \tag{83}$$

Figure 83 shows the synthesis of $\mathbf{X}$ from $\mathbf{W}_1$ and $\mathbf{V}_1$ in a flow diagram.

Let us now consider the relationship between the DFT for $\mathbf{X}$ (the input to the first stage of the pyramid algorithm) and the DFTs for $\mathbf{V}_1$ and $\mathbf{W}_1$ (the outputs). Since $\{X_t\} \longleftrightarrow \{\mathcal{X}_k\}$, the inverse DFT (Equation (36f)) tells us that

$$X_t = \frac{1}{N}\sum_{k=0}^{N-1} \mathcal{X}_k e^{i2\pi tk/N} = \frac{1}{N}\sum_{k=-\frac{N}{2}+1}^{\frac{N}{2}} \mathcal{X}_k e^{i2\pi tk/N}, \quad t = 0, \ldots, N-1,$$

where we have made use of the fact that, because $\{\mathcal{X}_k\}$ and $\{e^{i2\pi tk/N}\}$ are periodic sequences with period N, so is their product $\{\mathcal{X}_k e^{i2\pi tk/N}\}$, and hence a sum over any N contiguous elements is invariant. Since $\{2^{1/2}\widetilde{V}_{1,t}\}$ is formed by filtering $\{X_t\}$ with the low-pass filter $\{g_l\}$ with nominal pass-band $[-1/4, 1/4]$ and since $\mathcal{X}_k$ is associated with frequency $f_k \equiv k/N$, it follows that

$$\widetilde{V}_{1,t} \approx \frac{1}{N}\sum_{k=-\frac{N}{4}+1}^{\frac{N}{4}} \mathcal{X}_k e^{i2\pi tk/N}, \qquad t = 0, \ldots, N-1.$$

Now consider the subsampled series: to the same degree of approximation,

$$V_{1,t} = 2^{1/2}\widetilde{V}_{1,2t+1} \approx \frac{\sqrt{2}}{N}\sum_{k=-\frac{N}{4}+1}^{\frac{N}{4}} \mathcal{X}_k e^{i2\pi(2t+1)k/N}, \quad t = 0, \ldots, \frac{N}{2} - 1,$$

$$= \frac{2}{N} \sum_{k=-\frac{N}{4}+1}^{\frac{N}{4}} \frac{\mathcal{X}_k e^{i2\pi k/N}}{\sqrt{2}} e^{i2\pi tk/(N/2)}$$

$$= \frac{1}{N'} \sum_{k=-\frac{N'}{2}+1}^{\frac{N'}{2}} \mathcal{X}'_k e^{i2\pi tk/N'}, \quad t = 0, \ldots, N'-1,$$

where $N' \equiv N/2$ and $\mathcal{X}'_k \equiv \mathcal{X}_k e^{i2\pi k/N}/\sqrt{2}$. The elements of the approximate Fourier transform $\{\mathcal{X}'_k\}$ of the subsampled series $\{V_{1,t}\}$ are associated with frequencies $f'_k \equiv k/N'$ ranging from

$$f'_{-\frac{N'}{2}+1} = -\frac{1}{2} + \frac{1}{N'} \quad \text{up to} \quad f'_{\frac{N'}{2}} = \frac{1}{2}.$$

Thus, whereas the series $\{\widetilde{V}_{1,t}\}$ can be called a 'half-band series' because it is the output of a low-pass filter with nominal pass-band $[-1/4, 1/4]$ and hence deficient in high frequency elements, the subsampled series $\{V_{1,t}\}$ is a 'full-band series' because it can have significant Fourier coefficients over all frequencies $f'_k \in [-1/2, 1/2]$. Note that $\mathcal{X}_k$ and $\mathcal{X}'_k$ correspond to each other; that $\mathcal{X}_k$ corresponds to frequency $f_k = k/N$; and that $\mathcal{X}'_k$ corresponds to frequency $f'_k = k/N' = 2k/N = 2f_k$. Hence the coefficients at frequencies $f_k \in [-1/4, 1/4]$ in the Fourier representation for $\{X_t\}$ map onto the coefficients at $f'_k \in [-1/2, 1/2]$ in the approximate Fourier representation for $\{V_{1,t}\}$.

Let us now follow a similar line of thought for the wavelet coefficients. Since $\{2^{1/2}\widetilde{W}_{1,t}\}$ is formed by filtering $\{X_t\}$ with the high-pass filter $\{h_l\}$ with nominal pass-band defined by $1/4 \le |f| \le 1/2$, it follows that

$$\widetilde{W}_{1,t} \approx \frac{1}{N} \left(\sum_{k=-\frac{N}{2}+1}^{-\frac{N}{4}} + \sum_{k=\frac{N}{4}+1}^{\frac{N}{2}} \right) \mathcal{X}_k e^{i2\pi tk/N},$$

for $t = 0, \ldots, N-1$, where we define

$$\left(\sum_{k=l}^{m} + \sum_{k=l'}^{m'} \right) A_k \equiv \sum_{k=l}^{m} A_k + \sum_{k=l'}^{m'} A_k.$$

For the subsampled series $\{W_{1,t} : t = 0, \ldots, \frac{N}{2} - 1\}$, we have

$$W_{1,t} = 2^{1/2}\widetilde{W}_{1,2t+1} \approx \frac{\sqrt{2}}{N} \left(\sum_{k=-\frac{N}{2}+1}^{-\frac{N}{4}} + \sum_{k=\frac{N}{4}+1}^{\frac{N}{2}} \right) \mathcal{X}_k e^{i2\pi(2t+1)k/N}$$

$$= \frac{2}{N} \left(\sum_{k=-\frac{N}{2}+1}^{-\frac{N}{4}} + \sum_{k=\frac{N}{4}+1}^{\frac{N}{2}} \right) \frac{\mathcal{X}_k e^{i2\pi k/N}}{\sqrt{2}} e^{i2\pi tk/(N/2)}$$

$$= \frac{1}{N'} \sum_{k=-\frac{N'}{2}+1}^{\frac{N'}{2}} \mathcal{X}'_k e^{i2\pi tk/N'},$$

where now

$$\mathcal{X}'_k \equiv \frac{\mathcal{X}_{k+\frac{N}{2}} e^{i2\pi(k+\frac{N}{2})/N}}{\sqrt{2}} = -\frac{\mathcal{X}_{k+\frac{N}{2}} e^{i2\pi k/N}}{\sqrt{2}}.$$

As before, the elements of the approximate Fourier transform $\{\mathcal{X}'_k\}$ of the subsampled series $\{W_{1,t}\}$ are associated with frequencies f'_k ranging from $-\frac{1}{2} + \frac{1}{N'}$ to $\frac{1}{2}$. Thus, whereas $\{\widetilde{W}_{1,t}\}$ is a half-band series that is deficient in low frequency elements (because it is the output of a high-pass filter), the subsampled series $\{W_{1,t}\}$ is full-band; i.e., it can have significant Fourier coefficients over all frequencies f'_k. Note that $\mathcal{X}_{k+\frac{N}{2}}$ and $\mathcal{X}'_k$ correspond to each other; $\mathcal{X}_{k+\frac{N}{2}}$ corresponds to frequency $f_{k+\frac{N}{2}} = \frac{k}{N} + \frac{1}{2}$; and $\mathcal{X}'_k$, to $f'_k = k/N' = 2k/N = 2f_k$. As k ranges from 0 to $N'/2$, f'_k ranges over the interval $[0, 1/2]$, while $f_{k+\frac{N}{2}}$ ranges from $\frac{1}{2}$ to $\frac{N'}{2N} + \frac{1}{2} = \frac{3}{4}$, i.e., over the interval $[1/2, 3/4]$. For a real-valued sequence, the Fourier coefficient at a frequency f_k in the interval $[1/2, 3/4]$ is the complex conjugate of the coefficient associated with frequency $1 - f_k$, which is in the interval $[1/4, 1/2]$. This mapping from $[1/2, 3/4]$ to $[1/4, 1/2]$ is in reverse order; i.e., as we sweep from left to right in $[1/2, 3/4]$, we sweep from right to left in $[1/4, 1/2]$. Hence the complex conjugates $\mathcal{X}^*_k$ of the coefficients at frequencies $f_k \in [1/4, 1/2]$ in the Fourier representation for $\{X_t\}$ map – *in reverse order* – onto $\mathcal{X}'_k$ at frequencies $f'_k \in [0, 1/2]$ in the approximate Fourier representation for $\{W_{1,t}\}$.

The first stage of the pyramid algorithm thus takes the full-band series $\{X_t\}$ of length N and transforms it into two new full-band series, namely, the first level scaling and wavelet coefficients $\{V_{1,t}\}$ and $\{W_{1,t}\}$, each of length $N/2$. The scaling coefficients capture approximately the low frequency content of $\{X_t\}$, whereas the wavelet coefficients capture the high frequency content (although with a reversal of the frequency ordering). Figure 86 illustrates these ideas for an artificial 32 point time series given, for $t = 0, \ldots, 31$, by

$$X_t = \sum_{k=0}^{31} \mathcal{X}_k e^{i2\pi tk/32} \text{ with } \mathcal{X}_k \equiv \begin{cases} 0.96 - 0.03k, & k = 0, \ldots, 13 \text{ or } 16; \\ 0.84, & k = 14; \\ 0.735, & k = 15; \\ \mathcal{X}_{32-k}, & k = 17, \ldots, 31. \end{cases} \tag{85}$$

Finally, let us consider the scale that should be attached to the scaling coefficients $\{V_{1,t}\}$. For the special case of the Haar DWT for which $g_0 = g_1 = 1/\sqrt{2}$, we have

$$V_{1,t} = 2^{1/2}\widetilde{V}_{1,2t+1} = g_0 X_{2t+1} + g_1 X_{2t} = \frac{X_{2t+1} + X_{2t}}{\sqrt{2}} \propto \overline{X}_{2t+1}(2);$$

i.e., $\{V_{1,t}\}$ is proportional to averages of $\{X_t\}$ on a scale of two. In a similar manner the series $\{V_{1,t}\}$ can be interpreted as being proportional to a weighted average at scale two for other members of the class of discrete Daubechies wavelets of finite width (see Section 4.8 for details). For example, with an appropriate choice of a and b, the D(4) scaling filter is equivalent to first smoothing $\{X_t\}$ using the two point filter $\{a, b\}$ to obtain the series $\{aX_t + bX_{t-1}\}$ and then smoothing this new series using a filter given by $\{1/4, 1/2, 1/4\}$. Note that $\{V_{1,t}\}$ is associated with averages on a scale of *two* while $\{W_{1,t}\}$ is associated with changes in averages on a scale of *one*. The scales associated with the outputs of the wavelet and scaling filters thus differ by a factor of two. In order to keep track of this important distinction in later stages of the pyramid algorithm, we use $\lambda_j \equiv 2^j$ to denote the scale of the output from the scaling filter, whereas we use $\tau_j \equiv 2^{j-1}$ to denote the scale associated with the output from the wavelet filter (thus at the first stage $\lambda_1 = 2$ while $\tau_1 = 1$).

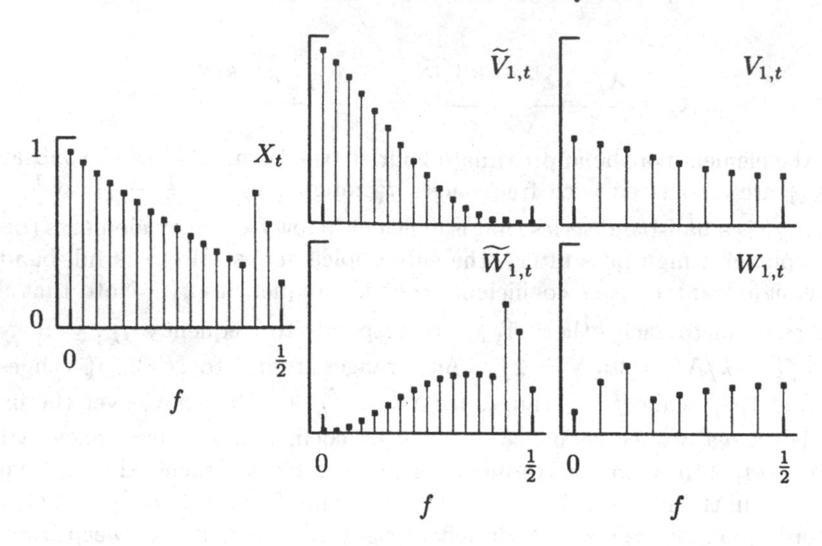

Figure 86. Magnitude squared DFT of the time series $\{X_t\}$ of Equation (85) (left-hand plot), along with the magnitude squared DFTs for the rescaled Haar scaling and wavelet filter outputs $\{\widetilde{V}_{1,t}\}$ and $\{\widetilde{W}_{1,t}\}$ (middle plots) and the Haar scaling and wavelet coefficients $\{V_{1,t}\}$ and $\{W_{1,t}\}$ (right-hand). Because $\{X_t\}$ and its filtered versions are real-valued, the squared magnitudes of all the DFTs are symmetric about zero, so just the values for the nonnegative Fourier frequencies are shown. The magnitude squared DFTs for $\{\widetilde{V}_{1,t}\}$ and $\{\widetilde{W}_{1,t}\}$ are obtained by multiplying the magnitude squared DFT for $\{X_t\}$ by values picked from the squared gain functions for $\{g_l/\sqrt{2}\}$ and $\{h_l/\sqrt{2}\}$ (these are defined by $\mathcal{G}(f)/2$ and $\mathcal{H}(f)/2$ – the shapes of these functions are shown in the top row of Figure 73). Note that $\{\widetilde{V}_{1,t}\}$ preserves the low frequency content of $\{X_t\}$ in its low frequencies, while $\{\widetilde{W}_{1,t}\}$ preserves the high frequency content of $\{X_t\}$ in its high frequencies. Whereas $\{\widetilde{V}_{1,t}\}$ and $\{\widetilde{W}_{1,t}\}$ are deficient in, respectively, high and low frequency content and hence are half-band series, the subsampled and rescaled series $\{V_{1,t}\}$ and $\{W_{1,t}\}$ are full-band series that preserve, respectively, the low and high frequency content of $\{X_t\}$. Whereas $\{V_{1,t}\}$ keeps the ordering of the frequencies in $\{X_t\}$, the ordering is reversed in $\{W_{1,t}\}$; e.g., the bulge at the high frequencies $f_{14} = \frac{14}{32}$ and $f_{15} = \frac{15}{32}$ in the DFT for $\{X_t\}$ appears in a reversed manner at the low frequencies $f'_1 = \frac{1}{16}$ and $f'_2 = \frac{2}{16}$ in the DFT for $\{W_{1,t}\}$.

Key Facts and Definitions in Section 4.4

The first stage of the pyramid algorithm orthogonally decomposes $\{X_t : t = 0, \ldots, N-1\}$ into two new series, namely, $\{W_{1,t} : t = 0, \ldots, \frac{N}{2} - 1\}$ and $\{V_{1,t} : t = 0, \ldots, \frac{N}{2} - 1\}$. The first level wavelet coefficients $\{W_{1,t}\}$ are associated with

[1] $\mathcal{W}_1$, an $\frac{N}{2} \times N$ matrix satisfying $\mathcal{W}_1 \mathcal{W}_1^T = I_{N/2}$ and consisting of the first $N/2$ rows of the DWT matrix $\mathcal{W}$ (each row of $\mathcal{W}_1$ contains the elements of $\{h_l^\circ\} \longleftrightarrow \{H(\frac{k}{N})\}$, i.e., the wavelet filter $\{h_l\}$ periodized to length N; the first row is $[h_1^\circ, h_0^\circ, h_{N-1}^\circ, h_{N-2}^\circ, \cdots, h_2^\circ]$, and the remaining $\frac{N}{2} - 1$ rows are formed by circularly shifting the first row to the right by, respectively, $2, 4, 6, \ldots, N-2$ units);

[2] the elements of the vector $\mathbf{W}_1 = \mathcal{W}_1 \mathbf{X}$ of length $N/2$;

[3] the first level detail $\mathcal{D}_1 = \mathcal{W}_1^T \mathbf{W}_1 = \mathcal{W}_1^T \mathcal{W}_1 \mathbf{X}$, which can be formed by upsampling $\mathbf{W}_1$ by two and then filtering with $\{H^*(\frac{k}{N})\}$;

[4] changes in averages of scale $\tau_1 = 1$;

[5] subsampling by two of a half-band process associated with the high frequencies $[1/4, 1/2]$ in the time series $\{X_t\}$; and
[6] the portion of the DFT of $\{X_t\}$ with frequencies in the interval $[1/4, 1/2]$, with a reversal in ordering with respect to frequency.

In contrast, the first level scaling coefficients $\{V_{1,t}\}$ go with

[1] $\mathcal{V}_1$, an $\frac{N}{2} \times N$ matrix that satisfies both $\mathcal{V}_1\mathcal{V}_1^T = I_{N/2}$ and $\mathcal{W}_1\mathcal{V}_1^T = \mathcal{V}_1\mathcal{W}_1^T = 0_{N/2}$ and whose rows span the same subspace of $\mathbb{R}^N$ as are spanned by the last $N/2$ rows of the DWT matrix $\mathcal{W}$ ($\mathcal{V}_1$ has the same structure as $\mathcal{W}_1$: we need only replace each h_l° with g_l°, where $\{g_l^\circ\}$ is the scaling filter $\{g_l\}$ periodized to length N, i.e., $\{g_l^\circ\} \longleftrightarrow \{G(\frac{k}{N})\}$);
[2] the elements of the vector $\mathbf{V}_1 = \mathcal{V}_1\mathbf{X}$ of length $N/2$;
[3] the first level smooth $\mathcal{S}_1 = \mathcal{V}_1^T\mathbf{V}_1 = \mathcal{V}_1^T\mathcal{V}_1\mathbf{X}$, which can be formed by upsampling $\mathbf{V}_1$ by two and then filtering with $\{G^*(\frac{k}{N})\}$;
[4] averages of scale $\lambda_1 = 2$;
[5] subsampling by two of a half-band process associated with the low frequencies $[0, 1/4]$ in the time series $\{X_t\}$; and
[6] the portion of the DFT of $\{X_t\}$ with frequencies in the interval $[0, 1/4]$.

The series $\{W_{1,t}\}$ constitutes the first half of the vector of wavelet coefficients $\mathbf{W}$; the remaining wavelet coefficients are obtained from $\{V_{1,t}\}$ at successive stages of the pyramid algorithm.

Comments and Extensions to Section 4.4

[1] We can gain more insight into the relationship between the frequencies in the DFTs for $\{X_t\}$ and $\{V_{1,t}\}$ if we use a variation on the DFT that explicitly incorporates the sampling interval Δt between observations. Accordingly, let us redefine the Fourier frequencies f_k to be $k/(N\,\Delta t)$ so that, if Δt has units of, say, seconds, then f_k would be measured in cycles per second. With Δt included, the DFT of $\{X_t\}$ and the corresponding inverse DFT are redefined to be

$$\mathcal{X}_k = \Delta t \sum_{t=0}^{N-1} X_t e^{-i2\pi f_k t\,\Delta t} \text{ and } X_t = \frac{1}{N\,\Delta t} \sum_{k=-\frac{N}{2}+1}^{\frac{N}{2}} \mathcal{X}_k e^{i2\pi f_k t\,\Delta t}.$$

The frequencies here satisfy $-f_\mathcal{N} < f_k \leq f_\mathcal{N}$, where $f_\mathcal{N} \equiv 1/(2\,\Delta t)$ is called the Nyquist frequency. The result of circularly filtering the time series with $\{g_l\}$ can be expressed approximately as

$$\widetilde{V}_{1,t} \approx \frac{1}{N\,\Delta t} \sum_{k=-\frac{N}{4}+1}^{\frac{N}{4}} \mathcal{X}_k e^{i2\pi f_k t\,\Delta t},$$

which involves frequencies satisfying $-f_\mathcal{N}/2 < f_k \leq f_\mathcal{N}/2$. Since the sampling interval for $\widetilde{V}_{1,t}$ is the same as for X_t, the subsampled series $\{V_{1,t}\}$ has a sampling interval of $2\,\Delta t \equiv \Delta t'$. Because $\{V_{1,t}\}$ is of length $N' \equiv N/2$, its DFT is related to Fourier frequencies $k/(N'\,\Delta t') = k/(N\,\Delta t) = f_k$, where now k ranges from $-\frac{N}{4} + 1$ to $\frac{N}{4}$. The Fourier frequencies for $\{X_t\}$ and $\{V_{1,t}\}$ are thus of the same form, but in the first case they satisfy $-f_{\mathcal{N}} < f_k \leq f_{\mathcal{N}}$, and in the second, $-f_{\mathcal{N}}/2 < f_k \leq f_{\mathcal{N}}/2$ (note that $f_{\mathcal{N}}/2 = 1/(2\,\Delta t')$, which is the Nyquist frequency for the subsampled series). We can now write

$$V_{1,t} \approx \frac{1}{N'\Delta t'} \sum_{k=-\frac{N'}{2}+1}^{\frac{N'}{2}} \mathcal{X}'_k e^{i2\pi f_k t\,\Delta t'}, \text{ where now } \mathcal{X}'_k = \sqrt{2}\mathcal{X}_k e^{i2\pi f_k\,\Delta t}.$$

Using DFTs involving frequencies with physically meaningful units, we see that frequencies $f_k \in [-f_{\mathcal{N}}/2, f_{\mathcal{N}}/2]$ in the Fourier representation for $\{X_t\}$ are the same as the frequencies in the approximate Fourier representation for $\{V_{1,t}\}$. Mapping the physical frequencies in the representations for $\{X_t\}$ and $\{V_{1,t}\}$ to the standard interval $[-1/2, 1/2]$ yields what we described in the main part of this section. (Likewise, we can argue that frequencies $f_k \in [f_{\mathcal{N}}/2, f_{\mathcal{N}}]$ in the Fourier representation for $\{X_t\}$ map – *in reverse order* – onto frequencies $f_k \in [0, f_{\mathcal{N}}/2]$ in the approximate Fourier representation for $\{W_{1,t}\}$.)

4.5 Second Stage of the Pyramid Algorithm

As described at the beginning of the previous section, the second stage of the pyramid algorithm consists of treating $\{V_{1,t}\}$ in the same way as $\{X_t\}$ was treated in the first stage. Intuitively this is a reasonable way to proceed because $\{X_t\}$ and $\{V_{1,t}\}$ are similar in the sense that $\{X_t\}$ can be regarded as an 'average' on unit scale, while the first level scaling coefficients can be regarded as an average over a scale of two. Thus, we circularly filter $\{V_{1,t}\}$ separately with $\{h_l\}$ and $\{g_l\}$ and subsample to produce two new series, namely,

$$W_{2,t} \equiv \sum_{l=0}^{L-1} h_l V_{1,2t+1-l \bmod \frac{N}{2}} \text{ and } V_{2,t} \equiv \sum_{l=0}^{L-1} g_l V_{1,2t+1-l \bmod \frac{N}{2}}, \tag{88}$$

$t = 0, \ldots, \frac{N}{4} - 1$. Note that the above is of the same filtering form as Equations (70c) and (77b) except 'mod N' is replaced by 'mod $\frac{N}{2}$.' The DWT wavelet coefficients for scale two and of level $j = 2$ are given by the $W_{2,t}$'s; i.e.,

$$\begin{aligned}\mathbf{W}_2 &\equiv \left[W_{\frac{N}{2}}, W_{\frac{N}{2}+1}, \ldots, W_{\frac{3N}{4}-1}\right]^T \\ &= \left[W_{2,0}, W_{2,1}, \ldots, W_{2,\frac{N}{4}-1}\right]^T.\end{aligned}$$

Define a vector containing the scaling coefficients of level $j = 2$ to be

$$\mathbf{V}_2 \equiv \left[V_{2,0}, V_{2,1}, \ldots, V_{2,\frac{N}{4}-1}\right]^T.$$

Let $\mathcal{B}_2$ and $\mathcal{A}_2$ be $\frac{N}{4} \times \frac{N}{2}$ matrices whose rows consist of circularly shifted versions of, respectively, $\{h_l\}$ and $\{g_l\}$ periodized to length $N/2$ such that we can express the transformation from $\mathbf{V}_1$ to $\mathbf{W}_2$ and $\mathbf{V}_2$ as

$$\begin{bmatrix}\mathbf{W}_2 \\ \mathbf{V}_2\end{bmatrix} = \mathcal{P}_2\mathbf{V}_1 = \begin{bmatrix}\mathcal{B}_2 \\ \mathcal{A}_2\end{bmatrix}\mathbf{V}_1 \text{ with } \mathcal{P}_2 \equiv \begin{bmatrix}\mathcal{B}_2 \\ \mathcal{A}_2\end{bmatrix}.$$

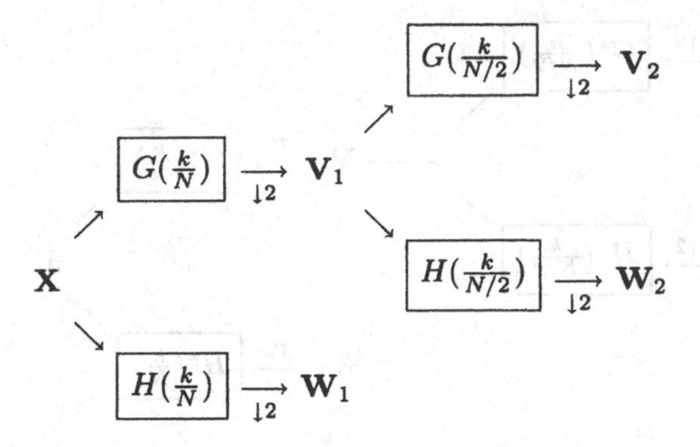

Figure 89. Flow diagram illustrating analysis of $\mathbf{X}$ into $\mathbf{W}_1$, $\mathbf{W}_2$ and $\mathbf{V}_2$.

As was the case with $\mathcal{W}_1$ and $\mathcal{V}_1$, the rows of $\mathcal{B}_2$ and $\mathcal{A}_2$ are orthonormal:

$$\mathcal{B}_2\mathcal{A}_2^T = \mathcal{A}_2\mathcal{B}_2^T = 0_{\frac{N}{4}} \text{ and } \mathcal{B}_2\mathcal{B}_2^T = \mathcal{A}_2\mathcal{A}_2^T = I_{\frac{N}{4}}.$$

The $\frac{N}{2} \times \frac{N}{2}$ matrix $\mathcal{P}_2$ is thus orthonormal, so we can recover $\mathbf{V}_1$ using the equation

$$\mathbf{V}_1 = \mathcal{P}_2^T \begin{bmatrix} \mathbf{W}_2 \\ \mathbf{V}_2 \end{bmatrix} = [\mathcal{B}_2^T \quad \mathcal{A}_2^T] \begin{bmatrix} \mathbf{W}_2 \\ \mathbf{V}_2 \end{bmatrix} = \mathcal{B}_2^T\mathbf{W}_2 + \mathcal{A}_2^T\mathbf{V}_2.$$

Substitution of the above into Equation (81a) yields

$$\mathbf{X} = \mathcal{W}_1^T\mathbf{W}_1 + \mathcal{V}_1^T\mathcal{B}_2^T\mathbf{W}_2 + \mathcal{V}_1^T\mathcal{A}_2^T\mathbf{V}_2 = \mathcal{B}_1^T\mathbf{W}_1 + \mathcal{A}_1^T\mathcal{B}_2^T\mathbf{W}_2 + \mathcal{A}_1^T\mathcal{A}_2^T\mathbf{V}_2,$$

where we let $\mathcal{A}_1 \equiv \mathcal{V}_1$ and $\mathcal{B}_1 \equiv \mathcal{W}_1$, a notation that will become convenient in describing the pyramid algorithm in its full generality. Comparison with Equation (63c) shows that we must have $\mathcal{W}_2 \equiv \mathcal{B}_2\mathcal{A}_1$ and hence that

$$\mathcal{D}_2 = \mathcal{W}_2^T\mathbf{W}_2 = \mathcal{A}_1^T\mathcal{B}_2^T\mathbf{W}_2.$$

Since $\mathcal{D}_1 = \mathcal{B}_1^T\mathbf{W}_1$, we have both

$$\mathbf{X} = \mathcal{D}_1 + \mathcal{D}_2 + \mathcal{A}_1^T\mathcal{A}_2^T\mathbf{V}_2 \text{ and } \mathbf{X} = \mathcal{D}_1 + \mathcal{D}_2 + \mathcal{S}_2,$$

so it follows that

$$\mathcal{S}_2 = \mathcal{A}_1^T\mathcal{A}_2^T\mathbf{V}_2 \equiv \mathcal{V}_2^T\mathbf{V}_2,$$

where $\mathcal{V}_2 \equiv \mathcal{A}_2\mathcal{A}_1$. Figures 89 and 90 depict, respectively, the analysis of $\mathbf{X}$ into $\mathbf{W}_1$, $\mathbf{W}_2$ and $\mathbf{V}_2$ via the pyramid algorithm and the synthesis of $\mathbf{X}$ from $\mathbf{W}_1$, $\mathbf{W}_2$ and $\mathbf{V}_2$ via the inverse of the pyramid algorithm.

Because the sequences $\{W_{2,t}\}$ and $\{V_{2,t}\}$ are obtained by filtering $\{V_{1,t}\}$ and subsampling and because $\{V_{1,t}\}$ in turn is obtained by filtering $\{X_t\}$ and subsampling, we can use the notion of a filter cascade to describe how $\{W_{2,t}\}$ and $\{V_{2,t}\}$ can be obtained by filtering $\{X_t\}$ directly if we can somehow take the subsampling operation into account. To do so, let us define $\{h_l^{\uparrow}\}$ to be the filter of width $2L-1$ with coefficients

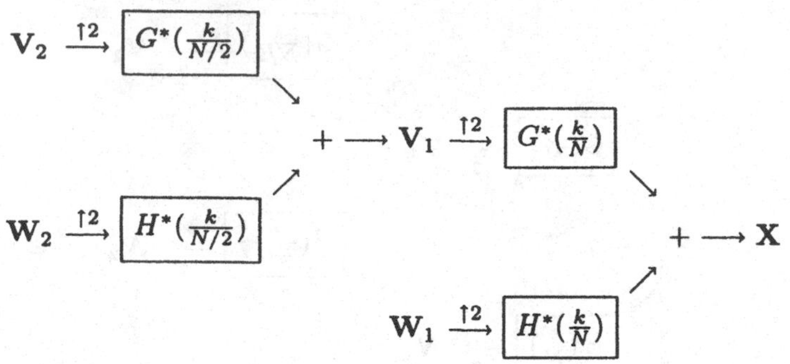

$$\mathbf{V}_2 \xrightarrow{\uparrow 2} \boxed{G^*(\tfrac{k}{N/2})} \searrow$$

$$+ \longrightarrow \mathbf{V}_1 \xrightarrow{\uparrow 2} \boxed{G^*(\tfrac{k}{N})} \searrow$$

$$\mathbf{W}_2 \xrightarrow{\uparrow 2} \boxed{H^*(\tfrac{k}{N/2})} \nearrow \qquad + \longrightarrow \mathbf{X}$$

$$\mathbf{W}_1 \xrightarrow{\uparrow 2} \boxed{H^*(\tfrac{k}{N})} \nearrow$$

Figure 90. Flow diagram illustrating synthesis of $\mathbf{X}$ from $\mathbf{W}_1$, $\mathbf{W}_2$ and $\mathbf{V}_2$.

$h_0, 0, h_1, 0, \ldots, h_{L-2}, 0, h_{L-1}$; i.e., we form $\{h_l^\uparrow\}$ by inserting one zero between each of the L elements of $\{h_l\}$. Let

$$2\widetilde{W}_{2,t} \equiv \sum_{l=0}^{2L-2} h_l^\uparrow 2^{1/2} \widetilde{V}_{1,t-l \bmod N}, \quad t = 0, \ldots, N-1. \tag{90a}$$

Equation (77a) tells us that $\{2^{1/2}\widetilde{V}_{1,t}\}$ is obtained by filtering $\{X_t\}$ with $\{g_l\}$, so $\{2\widetilde{W}_{2,t}\}$ is the output of a cascaded filter whose input $\{X_t\}$ is subject to filtering first by $\{g_l\}$ and then by $\{h_l^\uparrow\}$. We claim that

$$W_{2,t} = 2\widetilde{W}_{2,4t+3}, \quad t = 0, \ldots, \frac{N}{4} - 1, \tag{90b}$$

i.e., that $\{W_{2,t}\}$ can be formed by picking every fourth value of $\{2\widetilde{W}_{2,t}\}$ starting with $2\widetilde{W}_{2,3}$ (in engineering terminology, $\{W_{2,t}\}$ is the result of downsampling $\{2\widetilde{W}_{2,t}\}$ by four). To see that this is true, note that

$$\begin{aligned} 2\widetilde{W}_{2,4t+3} = \sum_{l=0}^{2L-2} h_l^\uparrow 2^{1/2} \widetilde{V}_{1,4t+3-l \bmod N} &= \sum_{l=0}^{L-1} h_l 2^{1/2} \widetilde{V}_{1,4t+3-2l \bmod N} \\ &= \sum_{l=0}^{L-1} h_l 2^{1/2} \widetilde{V}_{1,2(2t+1-l)+1 \bmod N} \\ &= \sum_{l=0}^{L-1} h_l V_{1,2t+1-l \bmod \frac{N}{2}} = W_{2,t}, \end{aligned}$$

where we have made use of Equations (77b) and (88). Let $\{h_{2,l}\} \equiv \{g * h_l^\uparrow\}$ be the convolution of $\{g_l\}$ and $\{h_l^\uparrow\}$. Since $\{g_l\}$ has width L while $\{h_l^\uparrow\}$ has width $2L-1$, an application of Exercise [28a] says that the width of $\{h_{2,l}\}$ is

$$L_2 \equiv L + (2L - 1) - 1 = 3L - 2.$$

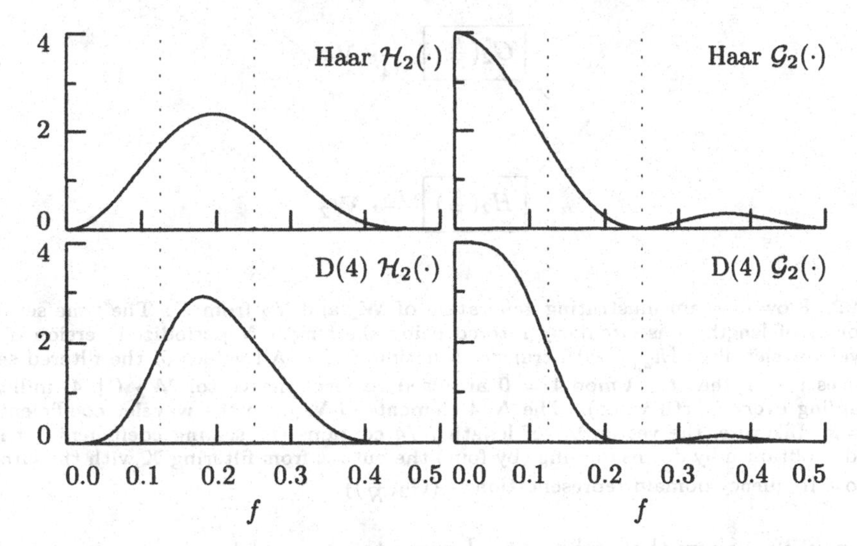

Figure 91. Squared gain functions for Haar $\{h_{2,l}\}$ and $\{g_{2,l}\}$ filters (upper left and right-hand plots, respectively) and D(4) $\{h_{2,l}\}$ and $\{g_{2,l}\}$ filters (lower left and right). The dotted lines mark the frequencies $f = 1/8$ and $f = 1/4$.

▷ **Exercise [91]:** Suppose that $\{h_l : l = 0, \ldots, L-1\}$ is a filter with transfer function $H(\cdot)$. Define a new filter by inserting m zeros between each of the elements of $\{h_l\}$:

$$h_0, \underbrace{0, \ldots, 0}_{m \text{ zeros}}, h_1, \ldots, h_{L-2}, \underbrace{0, \ldots, 0}_{m \text{ zeros}}, h_{L-1}.$$

Show that this filter has a transfer function given by $H([m+1]f)$. ◁

By letting $m = 1$ in the above, we can conclude that the transfer function for $\{h_{2,l}\}$ is given by

$$H_2(f) \equiv H(2f)G(f).$$

In a similar way, we can argue that the equivalent filter, say $\{g_{2,l}\}$, relating $\{V_{2,t}\}$ to $\{X_t\}$ is the filter of width L_2 obtained by convolving $\{g_l\}$ with

$$\{g_l^{\uparrow}\} \equiv \{g_0, 0, g_1, 0, \ldots, g_{L-2}, 0, g_{L-1}\}$$

and has a transfer function given by

$$G_2(f) \equiv G(2f)G(f).$$

Figure 91 shows plots of

$$\mathcal{H}_2(f) \equiv |H_2(f)|^2 \text{ and } \mathcal{G}_2(f) \equiv |G_2(f)|^2$$

versus f for the Haar and D(4) filters. Note that $\{h_{2,l}\}$ is an approximation to a band-pass filter with pass-band given by $1/8 \leq |f| \leq 1/4$, while $\{g_{2,l}\}$ is an approximation to a low-pass filter with pass-band given by $0 \leq |f| \leq 1/8$. A flow diagram depicting the direct generation of $\mathbf{W}_2$ and $\mathbf{V}_2$ from $\mathbf{X}$ is given in Figure 92.

For convenience, we call $\{h_{2,l}\}$ and $\{g_{2,l}\}$ the second level wavelet and scaling filters since they produce the second level wavelet and scaling coefficients. In the next section we generalize this notion to jth level wavelet and scaling filters. It is also convenient to define the first level filters $\{h_{1,l}\}$ and $\{g_{1,l}\}$ to be identical to $\{h_l\}$ and $\{g_l\}$.

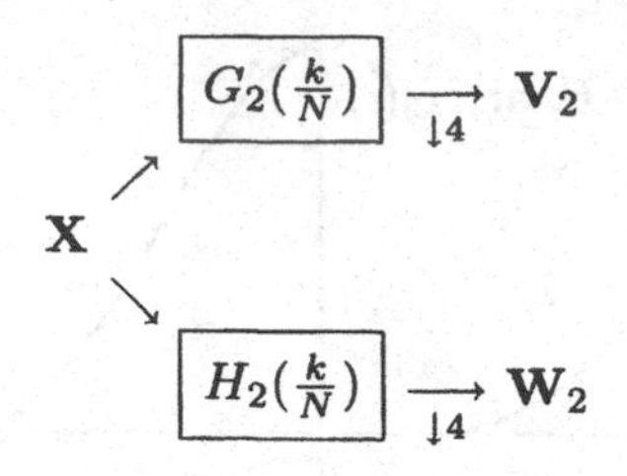

Figure 92. Flow diagram illustrating generation of $\mathbf{W}_2$ and $\mathbf{V}_2$ from $\mathbf{X}$. The time series in the vector $\mathbf{X}$ of length N is circularly filtered using the length N periodized version of the $j = 2$ level wavelet filter $\{h_{2,l}\}$ with transfer function $H_2(\cdot)$. All values of the filtered series with indices t such that $t + 1 \bmod 4 = 0$ are used to form the vector $\mathbf{W}_2$ ('↓ 4' indicates downsampling every fourth value). The $N/4$ elements of $\mathbf{W}_2$ are the wavelet coefficients of level $j = 2$. Likewise, the vector $\mathbf{V}_2$ of length $N/4$ contains the scaling coefficients of level $j = 2$ and is obtained by downsampling (by four) the output from filtering $\mathbf{X}$ with the circular filter whose frequency domain representation is $\{G_2(\frac{k}{N})\}$.

▷ **Exercise [92]:** Show that, while $\{h_{2,l}\}$ sums to zero and has unit energy, it does not satisfy the final defining property of a wavelet filter, namely, the orthogonality to even shifts of Equation (69c). Illustrate this result by considering the specific example of the second level Haar wavelet filter. Hint: consider the frequency domain formulation of the orthonormality property for wavelet filters. ◁

As the above exercise demonstrates, the second level wavelet filter in fact need not be a wavelet filter! A similar remark holds for second level scaling filters (and for the level $j \geq 3$ wavelet and scaling filters defined in the next section).

Because the elements of $\mathbf{W}_2$ are obtained by circularly filtering $\{X_t\}$ with $\{h_{2,l} : l = 0, \ldots, L_2 - 1\}$, they can also be obtained by circularly filtering with the periodized version of $\{h_{2,l}\}$, which we denote as $\{h^{\circ}_{2,l} : l = 0, \ldots, N-1\}$; i.e., we have

$$W_{2,t} = \sum_{l=0}^{N-1} h^{\circ}_{2,l} X_{4(t+1)-1-l \bmod N}, \qquad t = 0, \ldots, \frac{N}{4} - 1$$

(see Exercise [4.10]). Since we also have $\mathbf{W}_2 = \mathcal{W}_2\mathbf{X}$, the rows of $\mathcal{W}_2$ must contain circularly shifted versions of $\{h^{\circ}_{2,l}\}$. If we note that the zeroth element of $\mathbf{W}_2$ is

$$W_{2,0} = \sum_{l=0}^{N-1} h^{\circ}_{2,l} X_{3-l \bmod N},$$

it is evident that the first row of $\mathcal{W}_2$ (or, equivalently, row $N/2$ of $\mathcal{W}$) is given by

$$[h^{\circ}_{2,3}, h^{\circ}_{2,2}, h^{\circ}_{2,1}, h^{\circ}_{2,0}, h^{\circ}_{2,N-1}, h^{\circ}_{2,N-2}, \ldots, h^{\circ}_{2,5}, h^{\circ}_{2,4}] = \mathcal{W}^T_{\frac{N}{2}\bullet}; \tag{92}$$

the remaining $\frac{N}{4} - 1$ rows are given by $\mathcal{T}^{4k}\mathcal{W}^T_{\frac{N}{2}\bullet}, k = 1, \ldots, \frac{N}{4} - 1$, i.e., by circularly shifting the first row by multiples of 4. Also, we know from Exercise [33] that the DFT of $\{h^{\circ}_{2,l}\}$ is given by sampling the transfer function for $\{h_{2,l}\}$ at the frequencies $f_k = k/N, k = 0, \ldots, N-1$. Since this transfer function is defined by $H(2f)G(f)$, we have $\{h^{\circ}_{2,l}\} \longleftrightarrow \{H(\frac{k}{N/2})G(\frac{k}{N})\}$. In a similar manner, the periodized filter $\{g^{\circ}_{2,l}\} \longleftrightarrow \{G(\frac{k}{N/2})G(\frac{k}{N})\}$ provides the rows for $\mathcal{V}_2$.

Key Facts and Definitions in Section 4.5

The second stage of the pyramid algorithm amounts to rotating the $N/2$ basis vectors in $\mathcal{V}_1$ into two sets of $N/4$ basis vectors, namely, $\mathcal{W}_2 = \mathcal{B}_2\mathcal{V}_1 = \mathcal{B}_2\mathcal{A}_1$ and $\mathcal{V}_2 = \mathcal{A}_2\mathcal{V}_1 = \mathcal{A}_2\mathcal{A}_1$, where we define $\mathcal{A}_1 = \mathcal{V}_1$. Each row of $\mathcal{B}_2$ contains the elements of the wavelet filter $\{h_l\}$ periodized to length $N/2$ (i.e., the inverse DFT of $\{H(\frac{k}{N/2}) : k = 0, \ldots, \frac{N}{2} - 1\}$), with adjacent rows differing by a circular shift of 2 units; likewise, each row of $\mathcal{A}_2$ contains the scaling filter $\{g_l\}$ periodized to length $N/2$. This stage transforms $\{V_{1,t} : t = 0, \ldots, \frac{N}{2} - 1\}$ into two new series, namely, $\{W_{2,t} : t = 0, \ldots, \frac{N}{4} - 1\}$ and $\{V_{2,t} : t = 0, \ldots, \frac{N}{4} - 1\}$. Additionally, just as $\{X_t\}$ was orthogonally partitioned into a high-pass component $\{W_{1,t}\}$ and a low-pass component $\{V_{1,t}\}$ by the wavelet and scaling filters, $\{V_{1,t}\}$ is so partitioned into $\{W_{2,t}\}$ and $\{V_{2,t}\}$; however, high-pass and low-pass splitting of $\{V_{1,t}\}$ corresponding to splitting the low-pass part of $\{X_t\}$ into two parts, one corresponding to $1/8 \le |f| \le 1/4$ (associated with the wavelet coefficients for scale two) and the other, to $0 \le |f| \le 1/8$. The series of second level wavelet coefficients $\{W_{2,t}\}$ is thus associated with

[1] $\mathcal{W}_2 = \mathcal{B}_2\mathcal{A}_1$, an $\frac{N}{4} \times N$ matrix satisfying $\mathcal{W}_2\mathcal{W}_2^T = I_{N/4}$ and comprising rows $\frac{N}{2}$ to $\frac{3N}{4} - 1$ of the DWT matrix $\mathcal{W}$ (each row of $\mathcal{W}_2$ contains the elements of $\{h_{2,l}^\circ\} \longleftrightarrow \{H(\frac{k}{N/2})G(\frac{k}{N})\}$, i.e., the second level wavelet filter $\{h_{2,l}\}$ periodized to length N; the first row is displayed in Equation (92), and the remaining $\frac{N}{4} - 1$ rows are formed by circularly shifting the first row to the right by, respectively, $4, 8, 12, \ldots, N - 4$ units);
[2] the elements of the vector $\mathbf{W}_2 = \mathcal{W}_2\mathbf{X}$ of length $N/4$;
[3] the second level detail $\mathcal{D}_2 = \mathcal{W}_2^T\mathbf{W}_2 = \mathcal{W}_2^T\mathcal{W}_2\mathbf{X}$;
[4] changes in averages of scale $\tau_2 = 2$;
[5] subsampling by two of a half-band process associated with the high frequencies $[1/4, 1/2]$ in the first level scaling coefficients $\{V_{1,t}\}$; and
[6] the portion of the DFT of $\{X_t\}$ with frequencies in the interval $[1/8, 1/4]$, with a reversal in ordering with respect to frequency.

By comparison, the series of second level scaling coefficients $\{V_{2,t}\}$ is associated with

[1] $\mathcal{V}_2 = \mathcal{A}_2\mathcal{A}_1$, an $\frac{N}{4} \times N$ matrix that satisfies both $\mathcal{V}_2\mathcal{V}_2^T = I_{N/4}$ and $\mathcal{W}_2\mathcal{V}_2^T = \mathcal{V}_2\mathcal{W}_2^T = 0_{N/4}$ and whose rows span the same subspace of $\mathbb{R}^N$ as spanned by the last $N/4$ rows of the DWT matrix $\mathcal{W}$ ($\mathcal{V}_2$ has a construction similar to $\mathcal{W}_2$: we just need to replace each $h_{2,l}^\circ$ by $g_{2,l}^\circ$, where $\{g_{2,l}^\circ\}$ is $\{g_{2,l}\}$ periodized to length N, i.e., $\{g_{2,l}^\circ\} \longleftrightarrow \{G(\frac{k}{N/2})G(\frac{k}{N})\}$);
[2] the elements of the vector $\mathbf{V}_2 = \mathcal{V}_2\mathbf{X}$ of length $N/4$;
[3] the second level smooth $\mathcal{S}_2 = \mathcal{V}_2^T\mathbf{V}_2 = \mathcal{V}_2^T\mathcal{V}_2\mathbf{X}$;
[4] averages on scale $\lambda_2 = 4$;
[5] subsampling by two of a half-band process associated with the low frequencies $[0, 1/4]$ in the first level scaling coefficients $\{V_{1,t}\}$; and
[6] the portion of the DFT of $\{X_t\}$ with frequencies in the interval $[0, 1/8]$.

4.6 General Stage of the Pyramid Algorithm

Given the discussion in the previous two sections, we can readily state the general jth stage of the pyramid algorithm, where $j = 1, \ldots, J$ (recall that the sample size N is assumed to be equal to 2^J). With $V_{0,t}$ defined to be X_t, the jth stage input is

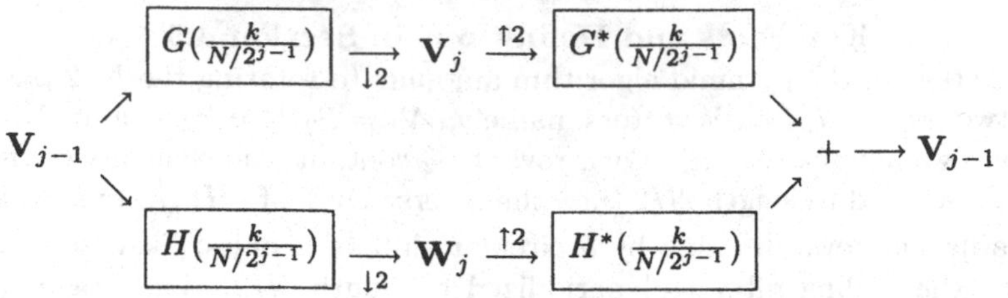

$$
\begin{array}{ccccccc}
 & \boxed{G(\frac{k}{N/2^{j-1}})} & \underset{\downarrow 2}{\longrightarrow} & \mathbf{V}_j & \xrightarrow{\uparrow 2} & \boxed{G^*(\frac{k}{N/2^{j-1}})} & \\
\nearrow & & & & & & \searrow \\
\mathbf{V}_{j-1} & & & & & & + \longrightarrow \mathbf{V}_{j-1} \\
\searrow & & & & & & \nearrow \\
 & \boxed{H(\frac{k}{N/2^{j-1}})} & \underset{\downarrow 2}{\longrightarrow} & \mathbf{W}_j & \xrightarrow{\uparrow 2} & \boxed{H^*(\frac{k}{N/2^{j-1}})} &
\end{array}
$$

Figure 94. Flow diagram illustrating analysis of $\mathbf{V}_{j-1}$ into $\mathbf{W}_j$ and $\mathbf{V}_j$, followed by synthesis of $\mathbf{V}_{j-1}$ from $\mathbf{W}_j$ and $\mathbf{V}_j$.

$\{V_{j-1,t} : t = 0, \ldots, N_{j-1} - 1\}$, where $N_j \equiv N/2^j$. This input is the scaling coefficients associated with averages over scale $\lambda_{j-1} = 2^{j-1}$. The jth stage outputs are the jth level wavelet and scaling coefficients:

$$W_{j,t} \equiv \sum_{l=0}^{L-1} h_l V_{j-1,2t+1-l \bmod N_{j-1}}, \quad V_{j,t} \equiv \sum_{l=0}^{L-1} g_l V_{j-1,2t+1-l \bmod N_{j-1}}, \tag{94a}$$

$t = 0, \ldots, N_j - 1$. The wavelet coefficients for scale $\tau_j = 2^{j-1}$ are given by the $W_{j,t}$'s; i.e.,

$$\mathbf{W}_j \equiv \left[W_{N-N_{j-1}}, W_{N-N_{j-1}+1}, \ldots, W_{N-N_j-1}\right]^T = \left[W_{j,0}, W_{j,1}, \ldots, W_{j,N_j-1}\right]^T.$$

If we define $\mathbf{V}_j^T$ similarly to contain the $V_{j,t}$'s, then the transformation from $\mathbf{V}_{j-1}$ to $\mathbf{W}_j$ and $\mathbf{V}_j$ can be expressed as

$$\begin{bmatrix} \mathbf{W}_j \\ \mathbf{V}_j \end{bmatrix} = \mathcal{P}_j \mathbf{V}_{j-1} = \begin{bmatrix} \mathcal{B}_j \\ \mathcal{A}_j \end{bmatrix} \mathbf{V}_{j-1} \text{ with } \mathcal{P}_j \equiv \begin{bmatrix} \mathcal{B}_j \\ \mathcal{A}_j \end{bmatrix}, \tag{94b}$$

where $\mathcal{B}_j$ and $\mathcal{A}_j$ are $N_j \times N_{j-1}$ matrices whose rows contain, respectively, circularly shifted versions of the wavelet and scaling filters $\{h_l\}$ and $\{g_l\}$ periodized to length N_{j-1}. The $N_{j-1} \times N_{j-1}$ matrix $\mathcal{P}_j$ is orthonormal, yielding

$$\mathcal{P}_j \mathcal{P}_j^T = \begin{bmatrix} \mathcal{B}_j \\ \mathcal{A}_j \end{bmatrix} [\, \mathcal{B}_j^T \quad \mathcal{A}_j^T \,] = \begin{bmatrix} \mathcal{B}_j \mathcal{B}_j^T & \mathcal{B}_j \mathcal{A}_j^T \\ \mathcal{A}_j \mathcal{B}_j^T & \mathcal{A}_j \mathcal{A}_j^T \end{bmatrix} = \begin{bmatrix} I_{N_j} & 0_{N_j} \\ 0_{N_j} & I_{N_j} \end{bmatrix} = I_{N_{j-1}}$$

and

$$\mathcal{P}_j^T \mathcal{P}_j = [\, \mathcal{B}_j^T \quad \mathcal{A}_j^T \,] \begin{bmatrix} \mathcal{B}_j \\ \mathcal{A}_j \end{bmatrix} = \mathcal{B}_j^T \mathcal{B}_j + \mathcal{A}_j^T \mathcal{A}_j = I_{N_{j-1}}.$$

If we recall that $\mathbf{V}_1 = \mathcal{A}_1 \mathbf{X}$, recursive application of $\mathbf{V}_j = \mathcal{A}_j \mathbf{V}_{j-1}$ yields

$$\mathbf{W}_j = \mathcal{B}_j \mathcal{A}_{j-1} \cdots \mathcal{A}_1 \mathbf{X} = \mathcal{W}_j \mathbf{X} \text{ and } \mathbf{V}_j = \mathcal{A}_j \mathcal{A}_{j-1} \cdots \mathcal{A}_1 \mathbf{X} = \mathcal{V}_j \mathbf{X},$$

where

$$\mathcal{W}_j \equiv \mathcal{B}_j \mathcal{A}_{j-1} \cdots \mathcal{A}_1 \text{ and } \mathcal{V}_j \equiv \mathcal{A}_j \mathcal{A}_{j-1} \cdots \mathcal{A}_1.$$

Since the $N_{j-1} \times N_{j-1}$ matrix $\mathcal{P}_j$ is orthonormal, we can recover $\mathbf{V}_{j-1}$ using the equation

$$\mathbf{V}_{j-1} = \mathcal{P}_j^T \begin{bmatrix} \mathbf{W}_j \\ \mathbf{V}_j \end{bmatrix} = [\mathcal{B}_j^T \quad \mathcal{A}_j^T] \begin{bmatrix} \mathbf{W}_j \\ \mathbf{V}_j \end{bmatrix} = \mathcal{B}_j^T \mathbf{W}_j + \mathcal{A}_j^T \mathbf{V}_j. \tag{95a}$$

In terms of filtering operations, the tth element of $\mathbf{V}_{j-1}$ can be reconstructed via an analog to Equation (83), namely,

$$V_{j-1,t} = \sum_{l=0}^{L-1} h_l W^{\uparrow}_{j,t+l \bmod N_{j-1}} + \sum_{l=0}^{L-1} g_l V^{\uparrow}_{j,t+l \bmod N_{j-1}}, \tag{95b}$$

$t = 0, 1, \ldots, N_{j-1} - 1$, where

$$W^{\uparrow}_{j,t} \equiv \begin{cases} 0, & t = 0, 2, \ldots, N_{j-1} - 2; \\ W_{j,\frac{t-1}{2}}, & t = 1, 3, \ldots, N_{j-1} - 1 \end{cases}$$

(we define $V^{\uparrow}_{j,t}$ in a similar manner). Equation (81a) along with recursive application of Equation (95a) yields

$$\mathbf{X} = \mathcal{B}_1^T \mathbf{W}_1 + \mathcal{A}_1^T \mathcal{B}_2^T \mathbf{W}_2 + \mathcal{A}_1^T \mathcal{A}_2^T \mathcal{B}_3^T \mathbf{W}_3 + \cdots$$
$$+ \mathcal{A}_1^T \cdots \mathcal{A}_{j-1}^T \mathcal{B}_j^T \mathbf{W}_j + \mathcal{A}_1^T \cdots \mathcal{A}_{j-1}^T \mathcal{A}_j^T \mathbf{V}_j.$$

The detail $\mathcal{D}_j$ and the smooth $\mathcal{S}_j$ are given by the last two terms above, namely,

$$\begin{aligned} \mathcal{D}_j &= \mathcal{A}_1^T \cdots \mathcal{A}_{j-1}^T \mathcal{B}_j^T \mathbf{W}_j = \mathcal{W}_j^T \mathbf{W}_j \\ \mathcal{S}_j &= \mathcal{A}_1^T \cdots \mathcal{A}_{j-1}^T \mathcal{A}_j^T \mathbf{V}_j = \mathcal{V}_j^T \mathbf{V}_j \end{aligned} \tag{95c}$$

The orthonormality of $\mathcal{P}_j$ tells us that

$$\|\mathbf{V}_{j-1}\|^2 = \|\mathbf{W}_j\|^2 + \|\mathbf{V}_j\|^2.$$

Since $\mathbf{V}_0 = \mathbf{X}$, recursive application of the above yields

$$\|\mathbf{X}\|^2 = \sum_{k=1}^{j} \|\mathbf{W}_k\|^2 + \|\mathbf{V}_j\|^2. \tag{95d}$$

▷ **Exercise [95]:** Using the above and Equation (95c), show that we also have

$$\|\mathbf{X}\|^2 = \sum_{k=1}^{j} \|\mathcal{D}_k\|^2 + \|\mathcal{S}_j\|^2. \quad \triangleleft$$

The equivalent filter $\{h_{j,l}\}$ relating $\{W_{j,t}\}$ to $\{X_t\}$ is formed by convolving together the following j filters:

$$\begin{aligned}
&\text{filter 1: } g_0, g_1, \ldots, g_{L-2}, g_{L-1}; \\
&\text{filter 2: } g_0, 0, g_1, 0, \ldots, g_{L-2}, 0, g_{L-1}; \\
&\text{filter 3: } g_0, 0, 0, 0, g_1, 0, 0, 0, \ldots, g_{L-2}, 0, 0, 0, g_{L-1}; \\
&\quad \vdots \\
&\text{filter } j-1\text{: } g_0, \underbrace{0, \ldots, 0}_{2^{j-2}-1 \text{ zeros}}, g_1, \underbrace{0, \ldots, 0}_{2^{j-2}-1 \text{ zeros}}, \ldots, g_{L-2}, \underbrace{0, \ldots, 0}_{2^{j-2}-1 \text{ zeros}}, g_{L-1}; \\
&\text{filter } j\text{: } h_0, \underbrace{0, \ldots, 0}_{2^{j-1}-1 \text{ zeros}}, h_1, \underbrace{0, \ldots, 0}_{2^{j-1}-1 \text{ zeros}}, \ldots, h_{L-2}, \underbrace{0, \ldots, 0}_{2^{j-1}-1 \text{ zeros}}, h_{L-1}.
\end{aligned} \tag{95e}$$

Note that, for $k = 2, \ldots, j-1$, the kth filter is obtained by inserting zeros between the elements of the previous filter.

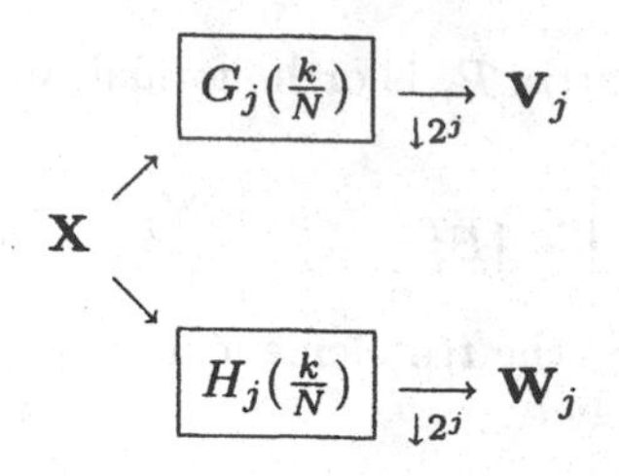

Figure 96. Flow diagram illustrating generation of $\mathbf{W}_j$ and $\mathbf{V}_j$ from $\mathbf{X}$. The time series in the vector $\mathbf{X}$ of length N is circularly filtered using a jth level wavelet filter $\{h_{j,l}\}$ with transfer function $H_j(\cdot)$, and all values of the filtered series with indices divisible by 2^j are used to form the vector $\mathbf{W}_j$ of length $N/2^j$ containing the wavelet coefficients of level j ('$\downarrow 2^j$' indicates downsampling every 2^jth value); in a similar manner, the vector $\mathbf{V}_j$ of length $N/2^j$ containing the scaling coefficients of level j is obtained by downsampling the output from filtering $\mathbf{X}$ using the jth level scaling filter with transfer function $G_j(\cdot)$.

▷ **Exercise [96]:** Show that the filter $\{h_{j,l}\}$ has width

$$L_j \equiv (2^j - 1)(L - 1) + 1 \tag{96a}$$

and a transfer function given by

$$H_j(f) \equiv H(2^{j-1} f) \prod_{l=0}^{j-2} G(2^l f) \tag{96b}$$ ◁

The filter $\{h_{j,l}\}$ has a nominal pass-band given by $1/2^{j+1} \le |f| \le 1/2^j$. With

$$2^{j/2}\widetilde{W}_{j,t} \equiv \sum_{l=0}^{L_j-1} h_{j,l} X_{t-l \bmod N}, \quad t = 0, \ldots, N-1, \tag{96c}$$

we have

$$W_{j,t} = 2^{j/2}\widetilde{W}_{j,2^j(t+1)-1} = \sum_{l=0}^{L_j-1} h_{j,l} X_{2^j(t+1)-1-l \bmod N}, \tag{96d}$$

$t = 0, 1, \ldots, N_j - 1$. Since we also have $\mathbf{W}_j = \mathcal{W}_j \mathbf{X}$, it is evident that the rows of $\mathcal{W}_j$ contain $\{h^\circ_{j,l}\}$, which by definition is $\{h_{j,l}\}$ periodized to length N.

In a similar manner, the equivalent filter $\{g_{j,l}\}$ relating $\{V_{j,t}\}$ to $\{X_t\}$ is formed by convolving together the following j filters:

$$\begin{aligned}
&\text{filter } 1\text{: } g_0, g_1, \ldots, g_{L-2}, g_{L-1};\\
&\text{filter } 2\text{: } g_0, 0, g_1, 0, \ldots, g_{L-2}, 0, g_{L-1};\\
&\text{filter } 3\text{: } g_0, 0, 0, 0, g_1, 0, 0, 0, \ldots, g_{L-2}, 0, 0, 0, g_{L-1};\\
&\quad\vdots\\
&\text{filter } j-1\text{: } g_0, \underbrace{0, \ldots, 0,}_{2^{j-2}-1 \text{ zeros}} g_1, \underbrace{0, \ldots, 0,}_{2^{j-2}-1 \text{ zeros}} \ldots, g_{L-2}, \underbrace{0, \ldots, 0,}_{2^{j-2}-1 \text{ zeros}} g_{L-1};\\
&\text{filter } j\text{: } g_0, \underbrace{0, \ldots, 0,}_{2^{j-1}-1 \text{ zeros}} g_1, \underbrace{0, \ldots, 0,}_{2^{j-1}-1 \text{ zeros}} \ldots, g_{L-2}, \underbrace{0, \ldots, 0,}_{2^{j-1}-1 \text{ zeros}} g_{L-1}.
\end{aligned} \tag{96e}$$

This equivalent filter has width L_j also, and its transfer function is given by

$$G_j(f) \equiv \prod_{l=0}^{j-1} G(2^l f). \tag{97}$$

The filter $\{g_{j,l}\}$ is an approximation to a low-pass filter with a pass-band given by $0 \le |f| \le 1/2^{j+1}$. With

$$2^{j/2}\widetilde{V}_{j,t} \equiv \sum_{l=0}^{L_j-1} g_{j,l} X_{t-l \bmod N}, \quad t = 0, \ldots, N-1,$$

we have

$$V_{j,t} = 2^{j/2}\widetilde{V}_{j,2^j(t+1)-1} = \sum_{l=0}^{L_j-1} g_{j,l} X_{2^j(t+1)-1-l \bmod N},$$

$t = 0, 1, \ldots, N_j - 1$. Since we also have $\mathbf{V}_j = \mathcal{V}_j\mathbf{X}$, it is evident that the rows of $\mathcal{V}_j$ contain $\{g^{\circ}_{j,l}\}$, which by definition is $\{g_{j,l}\}$ periodized to length N.

Let us see what the higher level scaling $\{g_{j,l}\}$ and wavelet $\{h_{j,l}\}$ filters look like as j increases beyond 1. Figure 98a shows plots of the D(4) $g_{j,l}$ (left-hand column) and $h_{j,l}$ (left-hand) versus $l = 0, \ldots, L_j - 1$ for scales indexed by $j = 1, 2, \ldots, 7$ (in these plots the individual values of the impulse response sequences are connected by lines). We see that $\{g_{j,l}\}$ and $\{h_{j,l}\}$ are converging to a characteristic shape as j increases (resembling a shark's fin in the case of $\{h_{j,l}\}$), a fact that we will explore in more detail in Sections 11.4 and 11.6. Similar plots for the 'least asymmetric' Daubechies filter of width $L = 8$ (see Section 4.8) are shown in Figure 98b – note that, as j increases, $\{g_{j,l}\}$ converges to something that is roughly Gaussian in shape, whereas $\{h_{j,l}\}$ is converging to a function that resembles somewhat the Mexican hat wavelet $\psi^{(\mathrm{Mh})}(\cdot)$ in the right-hand plot of Figure 3. The squared gain functions $\mathcal{H}_j(\cdot)$, $j = 1, \ldots, 4$, and $\mathcal{G}_4(\cdot)$ for these filters are shown in Figure 99, along with vertical lines delineating the associated nominal pass-bands.

Recalling our assumption that $N = 2^J$, we find that the Jth repetition of the pyramid algorithm yields output vectors $\mathbf{W}_J = [W_{J,0}]$ and $\mathbf{V}_J = [V_{J,0}]$, both of which are one dimensional. The algorithm terminates at this point, with $W_{J,0}$ and $V_{J,0}$ constituting the final two DWT coefficients W_{N-2} and W_{N-1} in the vector $\mathbf{W} = \mathcal{W}\mathbf{X}$. The coefficient W_{N-2} is associated with a change in $\{X_t\}$ at the scale $\tau_J = 2^{J-1} = N/2$, while the following exercise says that we always have $W_{N-1} = \overline{X}\sqrt{N}$, where $\overline{X}$ is the sample mean of $\{X_t\}$.

▷ **Exercise [97]:** Show that the elements of the final row $\mathcal{V}_J$ of the DWT matrix $\mathcal{W}$ are all equal to $1/\sqrt{N}$. Use this result to show explicitly what form $\mathcal{S}_J$ must take. Hint: consider the sum of the squares of the differences between the elements of $\mathcal{V}_J$ and $1/\sqrt{N}$, and use $\sum_l g_l = \sqrt{2}$ from Exercise [76b]. ◁

At the final stage, we also have an expression for all the rows of the DWT matrix,

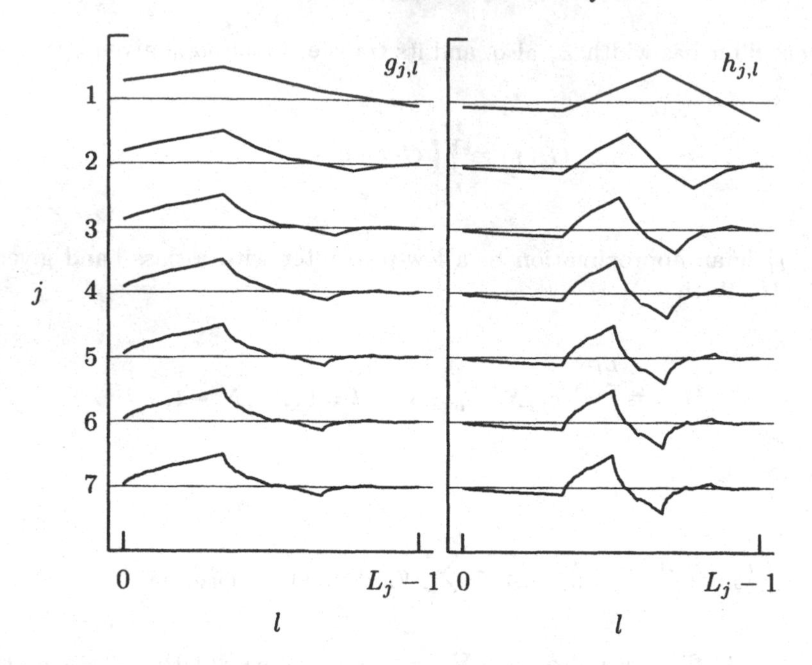

Figure 98a. D(4) scaling $\{g_{j,l}\}$ and wavelet $\{h_{j,l}\}$ filters for scales indexed by $j = 1, 2, \ldots, 7$ (here the individual values of the impulse response sequences are connected by lines).

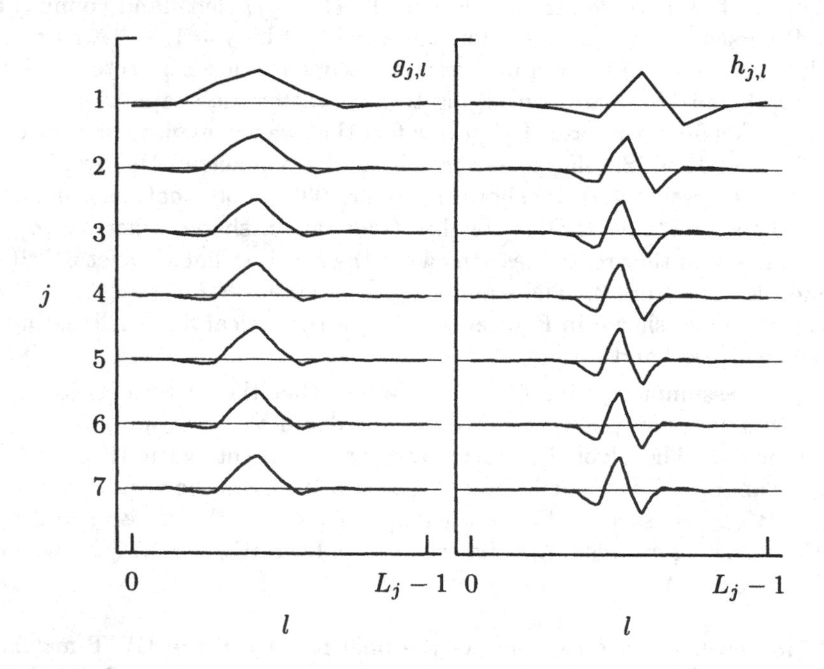

Figure 98b. LA(8) scaling $\{g_{j,l}\}$ and wavelet $\{h_{j,l}\}$ filters for scales indexed by $j = 1, 2, \ldots, 7$ (here the individual values of the impulse response sequences are connected by lines). These filters are defined in Section 4.8.

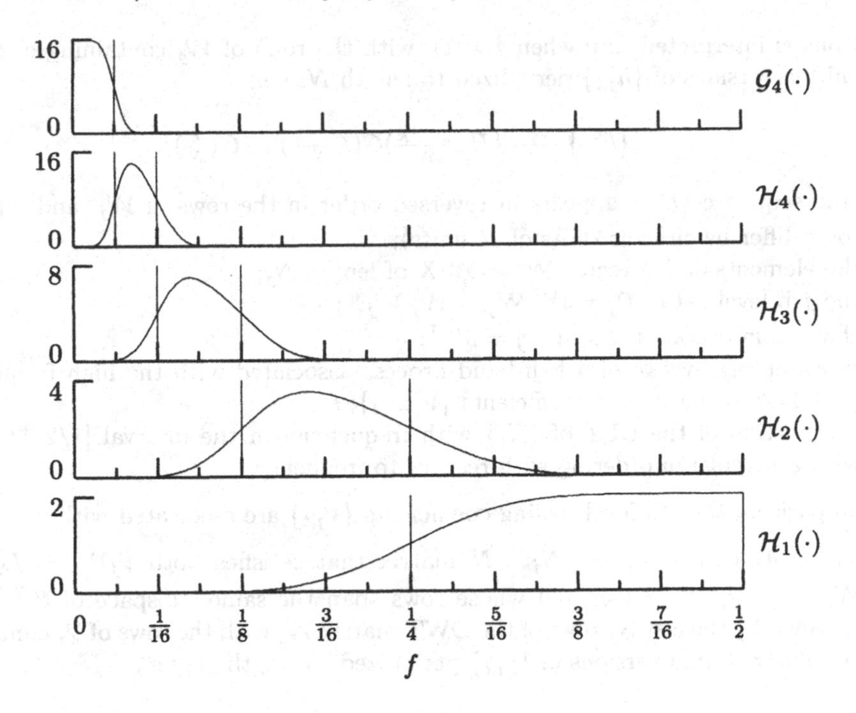

Figure 99. Squared gain functions for LA(8) filters $\{h_{j,l}\}, j = 1, \ldots, 4$, and $\{g_{4,l}\}$ (these filters are defined in Section 4.8).

namely,

$$\mathcal{W} = \begin{bmatrix} \mathcal{W}_1 \\ \mathcal{W}_2 \\ \vdots \\ \mathcal{W}_j \\ \vdots \\ \mathcal{W}_J \\ \mathcal{V}_J \end{bmatrix} = \begin{bmatrix} \mathcal{B}_1 \\ \mathcal{B}_2\mathcal{A}_1 \\ \vdots \\ \mathcal{B}_j\mathcal{A}_{j-1}\cdots\mathcal{A}_1 \\ \vdots \\ \mathcal{B}_J\mathcal{A}_{J-1}\cdots\mathcal{A}_1 \\ \mathcal{A}_J\mathcal{A}_{J-1}\cdots\mathcal{A}_1 \end{bmatrix}.$$

Key Facts and Definitions in Section 4.6

Recall the assumption that the sample size $N = 2^J$ for some integer J. For $j = 1, \ldots, J$ with $N_j \equiv N/2^j$, the jth stage of the pyramid algorithm rotates the N_{j-1} row vectors in the $N_{j-1} \times N$ matrix $\mathcal{V}_{j-1}$ into two sets of N_j row vectors, namely, $\mathcal{W}_j = \mathcal{B}_j\mathcal{V}_{j-1}$ and $\mathcal{V}_j = \mathcal{A}_j\mathcal{V}_{j-1}$, where $\mathcal{B}_j$ and $\mathcal{A}_j$ are $N_j \times N_{j-1}$ matrices containing, respectively, the wavelet and scaling filters $\{h_l\}$ and $\{g_l\}$ periodized to length N_{j-1} (here we define $\mathcal{V}_0 = I_N$). Jointly the row vectors of $\mathcal{A}_j$ and $\mathcal{B}_j$ form a set of N_{j-1} orthonormal vectors. With $V_{0,t} \equiv X_t$, the jth stage transforms $\{V_{j-1,t} : t = 0, \ldots, N_{j-1} - 1\}$ into two new series, namely, the jth level wavelet coefficients $\{W_{j,t} : t = 0, \ldots, N_j - 1\}$ and the jth level scaling coefficients $\{V_{j,t} : t = 0, \ldots, N_j - 1\}$. The jth level wavelet coefficients $\{W_{j,t}\}$ are associated with

[1] $\mathcal{W}_j = \mathcal{B}_j\mathcal{A}_{j-1}\cdots\mathcal{A}_1$, an $N_j \times N$ matrix satisfying $\mathcal{W}_j\mathcal{W}_j^T = I_{N_j}$ and comprising rows $\sum_{k=1}^{j-1} N_k$ to $\sum_{k=1}^{j} N_k - 1$ of the DWT matrix $\mathcal{W}$ (the first of these summa-

tions is interpreted as 0 when $j = 1$), with the rows of $\mathcal{W}_j$ containing circularly shifted versions of $\{h_{j,l}\}$ periodized to length N, i.e.,

$$\{h^{\circ}_{j,l}\} \longleftrightarrow \{H(\tfrac{2^{j-1}k}{N})G(\tfrac{2^{j-2}k}{N})\cdots G(\tfrac{k}{N})\}$$

(the sequence $\{h^{\circ}_{j,l}\}$ appears in reversed order in the rows of $\mathcal{W}_j$, and adjacent rows differ by circular shifts of 2^j units);

[2] the elements of the vector $\mathbf{W}_j = \mathcal{W}_j\mathbf{X}$ of length N_j;

[3] the jth level detail $\mathcal{D}_j = \mathcal{W}_j^T\mathbf{W}_j = \mathcal{W}_j^T\mathcal{W}_j\mathbf{X}$;

[4] changes in averages of scale $\tau_j = 2^{j-1}$;

[5] subsampling by two of a half-band process associated with the high frequencies $[1/4, 1/2]$ in the scaling coefficients $\{V_{j-1,t}\}$; and

[6] the portion of the DFT of $\{X_t\}$ with frequencies in the interval $[1/2^{j+1}, 1/2^j]$, with a reversal in ordering with respect to frequency.

By comparison, the jth level scaling coefficients $\{V_{j,t}\}$ are associated with

[1] $\mathcal{V}_j = \mathcal{A}_j\mathcal{A}_{j-1}\cdots\mathcal{A}_1$, an $N_j \times N$ matrix that satisfies both $\mathcal{V}_j\mathcal{V}_j^T = I_{N_j}$ and $\mathcal{W}_j\mathcal{V}_j^T = \mathcal{V}_j\mathcal{W}_j^T = 0_{N_j}$ and whose rows span the same subspace of $\mathbb{R}^N$ as are spanned by the last N_j rows of the DWT matrix $\mathcal{W}$, with the rows of $\mathcal{V}_j$ containing circularly shifted versions of $\{g_{j,l}\}$ periodized to length N, i.e.,

$$\{g^{\circ}_{j,l}\} \longleftrightarrow \{G(\tfrac{2^{j-1}k}{N})G(\tfrac{2^{j-2}k}{N})\cdots G(\tfrac{k}{N})\}$$

(the sequence $\{g^{\circ}_{j,l}\}$ appears in reversed order in the rows of $\mathcal{V}_j$);

[2] the elements of the vector $\mathbf{V}_j = \mathcal{V}_j\mathbf{X}$ of length N_j;

[3] the jth level smooth $\mathcal{S}_j = \mathcal{V}_j^T\mathbf{V}_j = \mathcal{V}_j^T\mathcal{V}_j\mathbf{X}$;

[4] averages on scale $\lambda_j = 2^j$;

[5] subsampling by two of a half-band process associated with the low frequencies $[0, 1/4]$ in the scaling coefficients $\{V_{j-1,t}\}$; and

[6] the portion of the DFT of $\{X_t\}$ with frequencies in the interval $[0, 1/2^{j+1}]$.

Comments and Extensions to Section 4.6

[1] Here we give pseudo-code that can be used to compute the DWT and its inverse using the elegant 'pyramid' algorithm introduced in the context of wavelets by Mallat (1989b). Let $V_{0,t} \equiv X_t$ for $t = 0, \ldots, N-1$, where $N = 2^J$. Let $\{h_n : n = 0, \ldots, L-1\}$ be a wavelet filter of even width L, and let $\{g_n\}$ be the corresponding scaling filter defined by Equation (75a). As in Section 4.1, partition the vector $\mathbf{W}$ of DWT coefficients into subvectors $\mathbf{W}_j$, $j = 1, \ldots, J$, and $\mathbf{V}_J$. Denote the elements of $\mathbf{W}_j$ and $\mathbf{V}_J$ as $W_{j,t}$, $t = 0, \ldots, N/2^j - 1$ and $V_{J,t}$, $t = 0, \ldots, N/2^J - 1$. Given the vector $\mathbf{V}_{j-1}$ of even length $M \equiv N/2^{j-1}$, the following pseudo-code computes the vectors $\mathbf{W}_j$ and $\mathbf{V}_j$, each of length $M/2$ (this is based on Equation (94a)):

For $t = 0, \ldots, M/2 - 1$, do the outer loop:
  Set u to $2t + 1$.
  Set $W_{j,t}$ to $h_0V_{j-1,u}$, and set $V_{j,t}$ to $g_0V_{j-1,u}$.
  For $n = 1, \ldots, L-1$, do the inner loop:
    Decrement u by 1.
    If $u < 0$, set u to $M - 1$.
    Increment $W_{j,t}$ by $h_nV_{j-1,u}$.

Increment $V_{j,t}$ by $g_n V_{j-1,u}$.
End of inner loop.
End of outer loop.

By using the pseudo-code above first for $j = 1$ (i.e., starting with $\mathbf{V}_0 = \mathbf{X}$) and then for $j = 2, \ldots, J$, we obtain the component vectors of the DWT, namely, $\mathbf{W}_1, \ldots, \mathbf{W}_J$ and $\mathbf{V}_J$, along with the intermediate vectors $\mathbf{V}_1, \ldots, \mathbf{V}_{J-1}$.

The task of the inverse DWT is to compute $\mathbf{X}$ given the subvectors $\mathbf{W}_1, \ldots, \mathbf{W}_J$ and $\mathbf{V}_J$ of $\mathbf{W}$. Given the vectors $\mathbf{W}_j$ and $\mathbf{V}_j$ of length $M' \equiv N/2^j$, the following pseudo-code computes the vector $\mathbf{V}_{j-1}$ of length $2M'$ (this is based on Equation (95b)):

Set l to -2, and set m to -1.
For $t = 0, \ldots, M' - 1$, do the outer loop:
Increment l by 2, and increment m by 2.
Set u to t; set i to 1; and set k to 0.
Set $V_{j-1,l}$ to $h_i W_{j,u} + g_i V_{j,u}$.
Set $V_{j-1,m}$ to $h_k W_{j,u} + g_k V_{j,u}$.
If $L > 2$, then, for $n = 1, \ldots, \frac{L}{2} - 1$, do the inner loop:
Increment u by 1.
If $u \geq M'$, set u to 0.
Increment i by 2, and increment k by 2.
Increment $V_{j-1,l}$ by $h_i W_{j,u} + g_i V_{j,u}$.
Increment $V_{j-1,m}$ by $h_k W_{j,u} + g_k V_{j,u}$.
End of the inner loop.
End of the outer loop.

By using the pseudo-code first with $\mathbf{W}_J$ and $\mathbf{V}_J$, we obtain $\mathbf{V}_{J-1}$. Let us depict this procedure as putting $\mathbf{W}_J$ and $\mathbf{V}_J$ into a box, out of which comes $\mathbf{V}_{J-1}$:

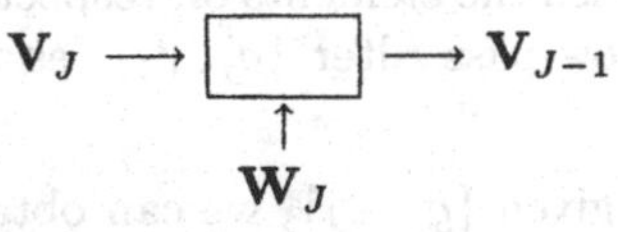

$$\mathbf{V}_J \longrightarrow \boxed{} \longrightarrow \mathbf{V}_{J-1} \quad (\uparrow \mathbf{W}_J)$$

By using the pseudo-code next with $\mathbf{W}_{J-1}$ and $\mathbf{V}_{J-1}$, we obtain $\mathbf{V}_{J-2}$. We can depict these two applications of the pseudo-code as

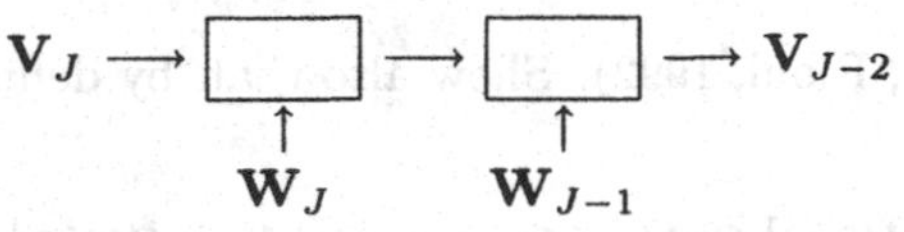

$$\mathbf{V}_J \longrightarrow \boxed{} \longrightarrow \boxed{} \longrightarrow \mathbf{V}_{J-2} \quad (\uparrow \mathbf{W}_J, \ \uparrow \mathbf{W}_{J-1})$$

If we continue in this manner, $J - 2$ more applications of the pseudo-code yields $\mathbf{V}_0 = \mathbf{X}$ (note that $\mathbf{V}_{J-3}, \ldots, \mathbf{V}_1$ are produced as intermediate calculations):

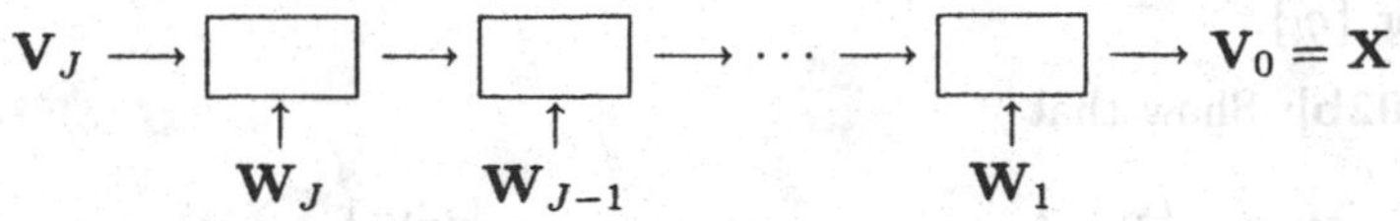

$$\mathbf{V}_J \longrightarrow \boxed{} \longrightarrow \boxed{} \longrightarrow \cdots \longrightarrow \boxed{} \longrightarrow \mathbf{V}_0 = \mathbf{X} \quad (\uparrow \mathbf{W}_J, \ \uparrow \mathbf{W}_{J-1}, \ \ldots, \ \uparrow \mathbf{W}_1)$$

The jth detail $\mathcal{D}_j$ is obtained by taking the inverse DWT of $\mathbf{0}_1, \ldots, \mathbf{0}_{j-1}, \mathbf{W}_j$ and $\mathbf{0}_j$, where $\mathbf{0}_k$, $k = 1, \ldots, j$, is a vector of $N/2^k$ zeros (when $j = 1$, we interpret $\mathbf{0}_1, \ldots, \mathbf{0}_{j-1}, \mathbf{W}_j$ as just $\mathbf{W}_1$). This detail can be computed by successively applying the inverse pyramid algorithm for $j, j-1, \ldots, 1$ starting with $\mathbf{W}_j$ and with $\mathbf{0}_j$ substituted for $\mathbf{V}_j$ in the above pseudo-code. At the end of j iterations, we obtain the desired $\mathcal{D}_j$:

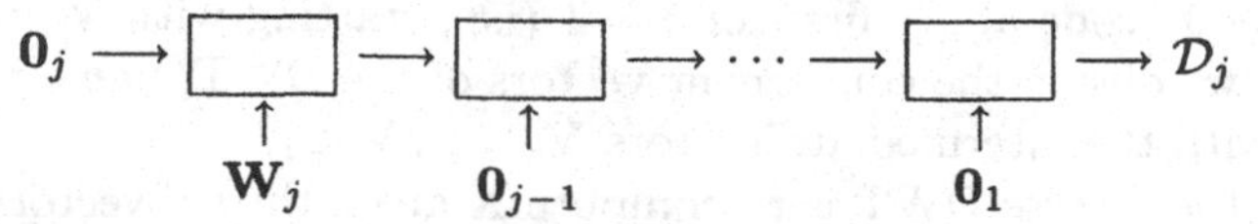

$$\mathbf{0}_j \longrightarrow \boxed{} \longrightarrow \boxed{} \longrightarrow \cdots \longrightarrow \boxed{} \longrightarrow \mathcal{D}_j$$
$$\uparrow \mathbf{W}_j \qquad \uparrow \mathbf{0}_{j-1} \qquad \uparrow \mathbf{0}_1$$

Likewise, the smooth $\mathcal{S}_J$ can be obtaining by applying the inverse DWT to $\mathbf{0}_1, \ldots, \mathbf{0}_J$ and $\mathbf{V}_J$:

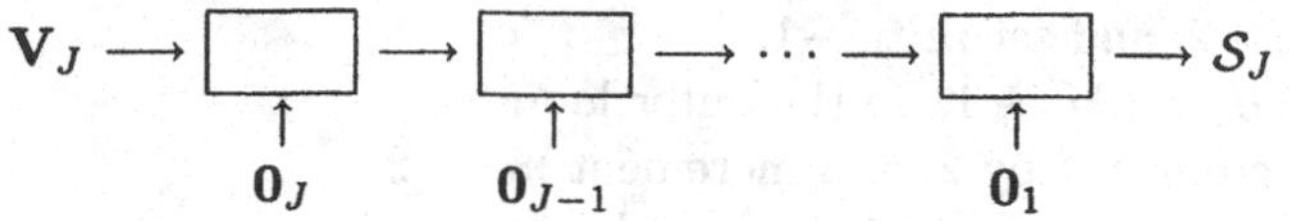

$$\mathbf{V}_J \longrightarrow \boxed{} \longrightarrow \boxed{} \longrightarrow \cdots \longrightarrow \boxed{} \longrightarrow \mathcal{S}_J$$
$$\uparrow \mathbf{0}_J \qquad \uparrow \mathbf{0}_{J-1} \qquad \uparrow \mathbf{0}_1$$

[2] Let us understand the nature of the higher level wavelet and scaling filters $\{h_{j,l}\}$ and $\{g_{j,l}\}$ a little better. We saw in Figure 73 that $\{g_{1,l}\} = \{g_l : l = 0, \ldots, L-1\}$ is a half-band low-pass filter (an approximation to an ideal low-pass filter with pass-band $0 \le |f| \le 1/4$), while we saw in Figure 91 that $\{g_{2,l}\}$ is a quarter-band low-pass filter (an approximation to an ideal low-pass filter with pass-band $0 \le |f| \le 1/8$). Continuing in this way, $\{g_{j,l}\}$ is a 2^{-j}th-band low-pass filter with ideal pass-band $0 \le |f| \le 1/2^{j+1}$. We also saw in Figure 73 that $\{h_{1,l}\} = \{h_l : l = 0, \ldots, L-1\}$ is a half-band high-pass filter, or band-pass filter, with ideal pass-band $1/4 \le |f| \le 1/2$, while we saw in Figure 91 that $\{h_{2,l}\}$ is a quarter-band band-pass filter with ideal pass-band $1/8 \le |f| \le 1/4$. Similarly, $\{h_{j,l}\}$ is a 2^{-j}th-band (octave-band) band-pass filter with ideal pass-band $1/2^{j+1} \le |f| \le 1/2^j$.

There are two equivalent routes by which we can consider $\{h_{j,l}\}$ and $\{g_{j,l}\}$ to be derived. In the first route, we generate the wavelet and scaling filters $\{h_{j,l}\}$ and $\{g_{j,l}\}$ by inserting $2^{j-1}-1$ zeros between the elements of, respectively, $\{h_l\}$ and $\{g_l\}$ and then applying the $2^{-(j-1)}$th-band low-pass filter $\{g_{j-1,l}\}$, as follows (see Equations (95e) and (96e)).

▷ **Exercise [102a]:** Show that, given $\{g_{j-1,l}\}$, we can obtain $\{h_{j,l}\}$ and $\{g_{j,l}\}$ via

$$h_{j,l} = \sum_{k=0}^{L-1} h_k g_{j-1,l-2^{j-1}k} \text{ and } g_{j,l} = \sum_{k=0}^{L-1} g_k g_{j-1,l-2^{j-1}k}, \tag{102}$$

$l = 0, \ldots, L_j - 1$ (e.g., Rioul, 1992). Show also that, by defining $G_1(f) \equiv G(f)$, we have for any $j \ge 2$

$$H_j(f) = H(2^{j-1}f)G_{j-1}(f) \text{ and } G_j(f) = G(2^{j-1}f)G_{j-1}(f).$$ ◁

In the second route, the wavelet and scaling filters at the next level are formed by inserting a single zero between $\{h_{j-1,l}\}$ or $\{g_{j-1,l}\}$ and then applying the half-band low-pass filter $\{g_l\}$.

▷ **Exercise [102b]:** Show that

$$h_{j,l} = \sum_{k=0}^{L_{j-1}-1} g_{l-2k} h_{j-1,k} \text{ and } g_{j,l} = \sum_{k=0}^{L_{j-1}-1} g_{l-2k} g_{j-1,k},$$

$l = 0, \ldots, L_j - 1$. ◁

Note that the creation of $\{h_{j,l}\}$ does not involve the convolution of $\{g_l\}$ and $\{h_{j-1,l}\}$ (it would if $2k$ were replaced by k in the formula above). The procedure that maps $\{h_{j-1,l}\}$ into $\{h_{j,l}\}$ is sometimes called an 'interpolation operator' in the engineering literature (e.g., Rioul and Duhamel, 1992). This terminology is appropriate here because we take the existing filter $\{h_{j-1,l}\}$ of scale τ_{j-1} and 'stretch it out' to scale $\tau_j = 2\tau_{j-1}$ by interpolating new filter coefficients using the low-pass filter $\{g_j\}$.

[3] Let us now justify our claim that the scaling filter $\{g_{j,l}\}$ is associated with averages over a scale of $\lambda_j = 2^j$. Although the width of $\{g_{j,l}\}$ is given by $L_j = (2^j-1)(L-1)+1$, a glance at Figures 98a and 98b indicates that, as j increases, many of the values of $g_{j,l}$ near $l = 0$ and $L_j - 1$ are quite close to zero, so L_j is a poor measure of the effective width of $\{g_{j,l}\}$. A better measure can be based upon the notion of the 'autocorrelation' width for a sequence, which we can take here to be defined as

$$\text{width}_a\{g_{j,l}\} \equiv \frac{\left(\sum_{l=-\infty}^{\infty} g_{j,l}\right)^2}{\sum_{l=-\infty}^{\infty} g_{j,l}^2} \tag{103}$$

(see, for example, Bracewell, 1978, or Percival and Walden, 1993, for details on this measure of width). Consider first the jth level Haar scaling filter:

$$g_{j,l} = \begin{cases} 1/2^{j/2}, & l = 0, \ldots, 2^j - 1; \\ 0, & \text{otherwise.} \end{cases}$$

This filter yields an output, each of whose values is proportional to a sample mean of 2^j input values, so the Haar $\{g_{j,l}\}$ can obviously be associated with scale $2^j = \lambda_j$.

▷ **Exercise [103]:** Show that the autocorrelation width for the Haar $\{g_{j,l}\}$ is given by 2^j, as is intuitively reasonable. ◁

For all scaling filters, we have

$$\sum_{l=-\infty}^{\infty} g_{j,l} = \sum_{l=0}^{L_j-1} g_{j,l} = G_j(0) = \prod_{k=0}^{j-1} G(0) = 2^{j/2}$$

since $G(0) = \sum_{l=0}^{L-1} g_l = \sqrt{2}$, while

$$\sum_{l=0}^{L_j-1} g_{j,l}^2 = 1$$

because, for any $N \geq L_j$, the filter coefficients $\{g_{j,l}\}$ are the nonzero elements in a row of $\mathcal{V}_j$, which satisfies the orthonormality property $\mathcal{V}_j \mathcal{V}_j^T = I_{N_j}$. We thus obtain $\text{width}_a\{g_{j,l}\} = 2^j = \lambda_j$ as claimed.

Moreover, because the wavelet filter $\{h_{j,l}\}$ has the same width L_j as $\{g_{j,l}\}$, is orthogonal to it and sums to zero, each wavelet coefficient can be interpreted as the difference between two generalized averages, each occupying half the effective width of λ_j; i.e., the scale associated with $\{h_{j,l}\}$ is $\lambda_j/2 = \tau_j$, as we have used throughout.

4.7 The Partial Discrete Wavelet Transform

For a time series $\{X_t : t = 0, \ldots, N-1\}$ with a sample size given by $N = 2^J$ (as we have assumed so far in this chapter), the pyramid algorithm yields the DWT coefficients $\mathbf{W} = \mathcal{W}\mathbf{X}$ after J repetitions. If we were to stop the algorithm after $J_0 < J$ repetitions, we would obtain a level J_0 partial DWT of $\mathbf{X}$, whose coefficients are given by

$$\begin{bmatrix} \mathbf{W}_1 \\ \mathbf{W}_2 \\ \vdots \\ \mathbf{W}_j \\ \vdots \\ \mathbf{W}_{J_0} \\ \mathbf{V}_{J_0} \end{bmatrix} = \begin{bmatrix} \mathcal{W}_1 \\ \mathcal{W}_2 \\ \vdots \\ \mathcal{W}_j \\ \vdots \\ \mathcal{W}_{J_0} \\ \mathcal{V}_{J_0} \end{bmatrix} \mathbf{X} = \begin{bmatrix} \mathcal{B}_1 \\ \mathcal{B}_2\mathcal{A}_1 \\ \vdots \\ \mathcal{B}_j\mathcal{A}_{j-1}\cdots\mathcal{A}_1 \\ \vdots \\ \mathcal{B}_{J_0}\mathcal{A}_{J_0-1}\cdots\mathcal{A}_1 \\ \mathcal{A}_{J_0}\mathcal{A}_{J_0-1}\cdots\mathcal{A}_1 \end{bmatrix} \mathbf{X}.$$

While $\mathbf{W}_j$, $j = 1, \ldots, J_0$, are all subvectors of the DWT coefficient vector $\mathbf{W}$, the final subvector $\mathbf{V}_{J_0}$ of $N/2^{J_0}$ scaling coefficients replaces the last $N/2^{J_0}$ coefficients of $\mathbf{W}$. These scaling coefficients represent averages over the scale of $\lambda_{J_0} = 2^{J_0}$ and hence capture the large scale (or low frequency) components in $\mathbf{X}$ (including its sample mean $\overline{X}$). By comparison, the 'full' DWT has but a single scaling coefficient, and this merely captures the sample mean (see Exercise [97]).

Partial DWTs are more commonly used in practice than the full DWT because of the flexibility they offer in specifying a scale beyond which a wavelet analysis into individual large scales is no longer of real interest. A partial DWT of level J_0 also allows us to relax the restriction that $N = 2^J$ for some J and to replace it with the condition that N be an integer multiple of 2^{J_0}; i.e., N need not be a power of two as required by the full DWT, but it does need to be a multiple of 2^{J_0} if we want to use a partial DWT of level J_0. When sample size is not an overriding consideration, the choice of J_0 can be set from the goals of the analysis for a particular application (for examples, see Sections 4.10 and 5.8).

For a level $J_0 < J$ partial DWT, we obtain the additive decomposition (i.e., MRA)

$$\mathbf{X} = \sum_{j=1}^{J_0} \mathcal{D}_j + \mathcal{S}_{J_0}, \tag{104a}$$

where, as before, the detail $\mathcal{D}_j$ represents changes on a scale of $\tau_j = 2^{j-1}$, while $\mathcal{S}_{J_0}$ represents averages of a scale of $\lambda_{J_0} = 2^{J_0}$. The energy decomposition is given by

$$\|\mathbf{X}\|^2 = \sum_{j=1}^{J_0} \|\mathbf{W}_j\|^2 + \|\mathbf{V}_{J_0}\|^2 = \sum_{j=1}^{J_0} \|\mathcal{D}_j\|^2 + \|\mathcal{S}_{J_0}\|^2, \tag{104b}$$

and, since the sample variance is given by $\hat{\sigma}_X^2 = \frac{1}{N}\|\mathbf{X}\|^2 - \overline{X}^2$, the decomposition of the sample variance is given by

$$\hat{\sigma}_X^2 = \frac{1}{N}\sum_{j=1}^{J_0} \|\mathbf{W}_j\|^2 + \frac{1}{N}\|\mathbf{V}_{J_0}\|^2 - \overline{X}^2 = \frac{1}{N}\sum_{j=1}^{J_0} \|\mathcal{D}_j\|^2 + \frac{1}{N}\|\mathcal{S}_{J_0}\|^2 - \overline{X}^2, \tag{104c}$$

where the last two terms can be regarded as the sample variance of the J_0 level smooth $\mathcal{S}_{J_0}$ (see Exercise [4.20]).

4.8 Daubechies Wavelet and Scaling Filters: Form and Phase

We have seen in the previous sections that we can construct an orthonormal DWT matrix $\mathcal{W}$ based on any filter satisfying the properties of a wavelet filter, namely, summation to zero (Equation (69a)) and orthonormality (Equations (69b) and (69c)). For the Haar and D(4) wavelet filters, we have noted through examination of Figure 73 that these are high-pass filters with nominal pass-bands given by $[1/4, 1/2]$. This single fact in turn implies that the associated scaling filters are low-pass filters and that the jth level wavelet coefficients can be associated with frequencies in the pass-band $[1/2^{j+1}, 1/2^j]$. It is important to understand, however, that this frequency domain interpretation of wavelet coefficients does *not* automatically follow from the definition of a wavelet filter, as the following exercise demonstrates.

▷ **Exercise [105a]:** Show that the filter $\{a_0 = -1/\sqrt{2}, a_1 = 0, a_2 = 0, a_3 = 1/\sqrt{2}\}$ is a wavelet filter. Compute its squared gain function, and argue that this wavelet filter is not a reasonable approximation to a high-pass filter. ◁

In addition to not being a reasonable high-pass filter, the wavelet filter of the above exercise does not act like the Haar and D(4) filters in giving us a wavelet transform that we can describe in terms of (generalized) differences of adjacent (weighted) averages. We thus must place additional constraints on a wavelet filter if the resulting wavelet transform is going to be interpretable in these terms.

By imposing an appealing set of regularity conditions (discussed in Section 11.9), Daubechies (1992, Section 6.1) came up with a useful class of wavelet filters, all of which yield a DWT in accordance with the notion of differences of adjacent averages. The definition for this class of filters can be expressed in terms of the squared gain function for the associated Daubechies scaling filters $\{g_l : l = 0, \ldots, L-1\}$:

$$\mathcal{G}^{(\mathrm{D})}(f) \equiv 2\cos^L(\pi f) \sum_{l=0}^{\frac{L}{2}-1} \binom{\frac{L}{2}-1+l}{l} \sin^{2l}(\pi f), \tag{105a}$$

where L is a positive even integer,

$$\binom{a}{b} \equiv \frac{a!}{b!(a-b)!} \text{ and } \sin^0(\pi f) = 1 \text{ for all } f \text{ (including } f = 0).$$

Using the relationship $\mathcal{H}^{(\mathrm{D})}(f) = \mathcal{G}^{(\mathrm{D})}(f+\frac{1}{2})$, we see that the corresponding Daubechies wavelet filters have squared gain functions satisfying

$$\mathcal{H}^{(\mathrm{D})}(f) \equiv 2\sin^L(\pi f) \sum_{l=0}^{\frac{L}{2}-1} \binom{\frac{L}{2}-1+l}{l} \cos^{2l}(\pi f). \tag{105b}$$

As the following suggests, we can interpret $\mathcal{H}^{(\mathrm{D})}(\cdot)$ as the squared gain function of the equivalent filter for a filter cascade.

▷ **Exercise [105b]:** Show that the squared gain function for the difference filter $\{a_0 = 1, a_1 = -1\}$ is given by $\mathcal{D}(f) \equiv 4\sin^2(\pi f)$. ◁

Using this result, we can now write $\mathcal{H}^{(D)}(f) = \mathcal{D}^{\frac{L}{2}}(f)\mathcal{A}_L(f)$, where

$$\mathcal{A}_L(f) \equiv \frac{1}{2^{L-1}} \sum_{l=0}^{\frac{L}{2}-1} \binom{\frac{L}{2}-1+l}{l} \cos^{2l}(\pi f).$$

Figure 107 plots the relative shapes of $\mathcal{D}^{\frac{L}{2}}(\cdot)$, $\mathcal{A}_L(\cdot)$ and $\mathcal{H}^{(D)}(\cdot)$ for $L = 2, 4, \ldots, 14$. For $L \geq 4$, we see that the filter $\mathcal{A}_L(\cdot)$ is approximately a low-pass filter and hence acts as a smoother or weighted average, in agreement with the interpretation of wavelet coefficients as differences of weighted averages (for $L = 2$, which corresponds to the Haar wavelet, the filter is an all-pass filter).

▷ **Exercise [106]:** Show that the squared gain function $\mathcal{H}^{(D)}(\cdot)$ of Equation (105b) satisfies $\mathcal{H}^{(D)}(f) + \mathcal{H}^{(D)}(f + \frac{1}{2}) = 2$ for all positive even L (recall that this condition ensures that any wavelet filter $\{h_l\}$ with squared gain $\mathcal{H}^{(D)}(\cdot)$ has unit energy and is orthogonal to its even shifts). Hint: this can be shown by induction. ◁

As L increases, $\mathcal{G}^{(D)}(\cdot)$ converges to the squared gain function for an ideal low-pass filter (Lai, 1995), and the number of distinct real-valued sequences of the form $\{g_l : l = 0, \ldots, L-1\}$ whose squared gain function is $\mathcal{G}^{(D)}(\cdot)$ increases (the exact number of such sequences is discussed in Section 11.10). The transfer functions $G(\cdot)$ for these $\{g_l\}$ are necessarily different, but the difference is only in their phase functions, i.e., $\theta^{(G)}(\cdot)$ in the polar representation

$$G(f) = [\mathcal{G}^{(D)}(f)]^{1/2} e^{i\theta^{(G)}(f)}.$$

Given $\mathcal{G}^{(D)}(\cdot)$, we can obtain all possible $\{g_l\}$ by a procedure known as *spectral factorization* (the subject of Section 11.10), with different factorizations leading to different phase functions $\theta(\cdot)$. The question then arises as to which particular factorization we should use. The factorization that Daubechies originally obtained corresponds to an *extremal phase* choice for the transfer function and produces what is known as a *minimum delay* filter in the engineering literature (Oppenheim and Schafer, 1989, Section 5.6). If we let $\{g_l^{(ep)}\}$ denote the extremal phase scaling filter and if $\{g_l\}$ is a filter corresponding to another factorization, then we have

$$\sum_{l=0}^{m} g_l^2 \leq \sum_{l=0}^{m} \left[g_l^{(ep)}\right]^2 \quad \text{for } m = 0, \ldots, L-1$$

(Oppenheim and Schafer, 1989, Equation 5.118). Note that equality occurs at $m = L-1$ because all scaling filters have unity energy. For $m < L-1$, a summation like the above defines a *partial energy sequence*, which builds up as rapidly as possible for a minimum delay filter.

We refer to the Daubechies extremal phase scaling filter of width L as the D(L) scaling filter, where $L = 2, 4, \ldots$ (these filters are called 'daublets' in Bruce and Gao, 1996a). The D(2) scaling filter is the same as the Haar scaling filter so that $g_0^{(ep)} = 1/\sqrt{2}$ and $g_1^{(ep)} = 1/\sqrt{2}$ for this case; the coefficients for the D(4) scaling filter are given in Equation (75d), while the coefficients for the D(6) and D(8) filters are given in Table 109 (the numbers in this table are taken from Table 6.1, p. 195, Daubechies, 1992). The Daubechies extremal phase filters for $L = 2, 4, \ldots, 20$ are plotted in Figure 108a, while the corresponding wavelet filters (obtained via Equation (75b)) are plotted in Figure 108b.

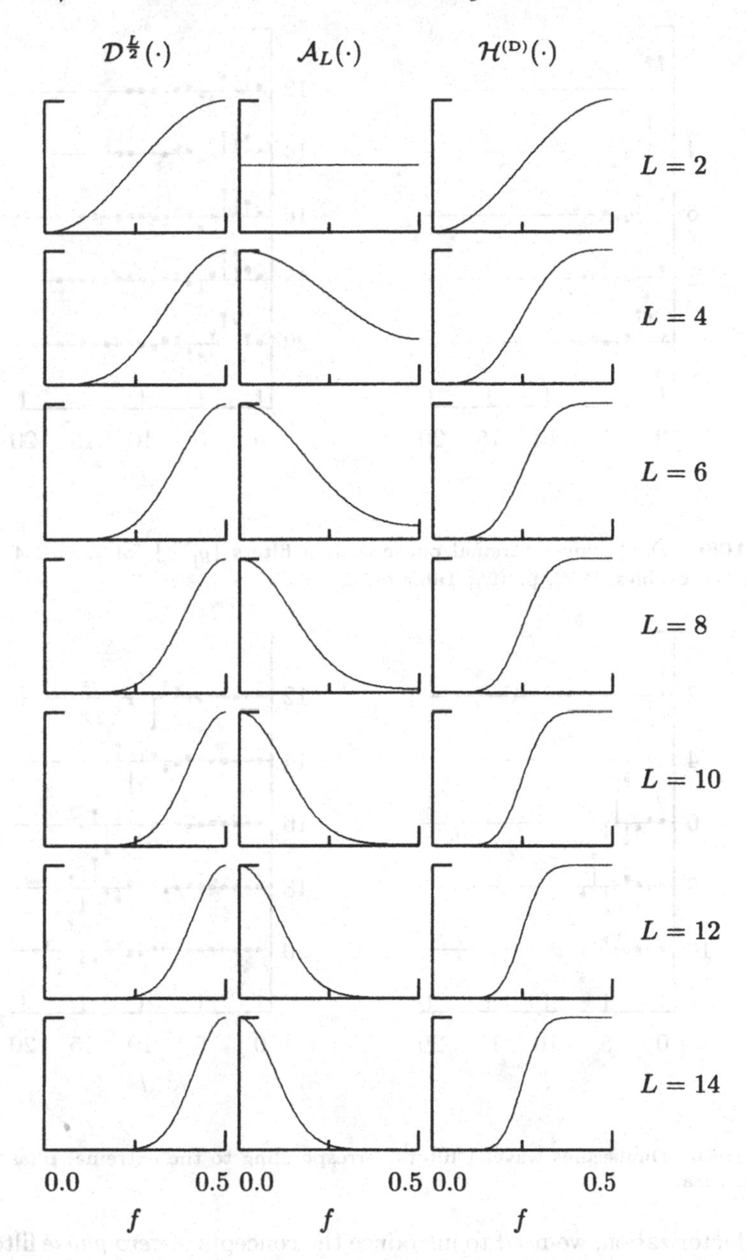

Figure 107. Squared gain functions $\mathcal{H}^{(D)}(\cdot)$ for Daubechies wavelet filters of widths $L = 2, 4, \ldots, 14$ (right-hand column). Each $\mathcal{H}^{(D)}(\cdot)$ is the product of two other squared gain functions, namely, $\mathcal{D}^{\frac{L}{2}}(\cdot)$ (left-hand column) and $\mathcal{A}_L(\cdot)$ (middle). The first corresponds to an $\frac{L}{2}$ order difference filter, while for $L \geq 4$ the second is associated with a weighted average (i.e., low-pass filter).

A second choice for factorizing $\mathcal{G}^{(D)}(\cdot)$ leads to the *least asymmetric* (LA) family of scaling filters, which we denote as $\{g_l^{(la)}\}$ (these are also called 'symmlets' – see, for example, Bruce and Gao, 1996a, or Härdle *et al.*, 1998). To understand the rationale

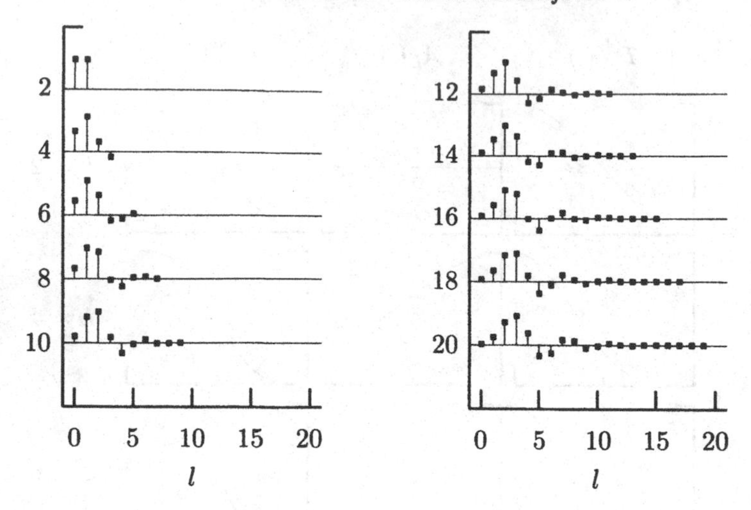

Figure 108a. Daubechies extremal phase scaling filters $\{g_l^{(\text{ep})}\}$ for $L = 2, 4, \ldots, 20$ (values based on Daubechies, 1992, p. 195, Table 6.1).

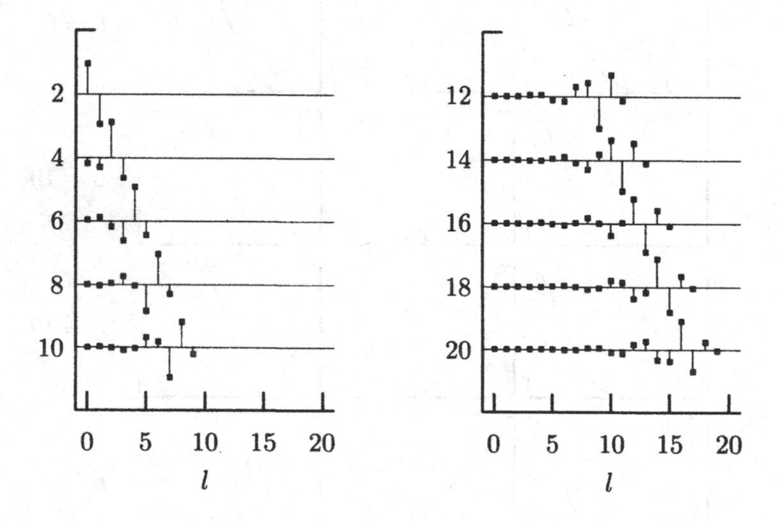

Figure 108b. Daubechies wavelet filters corresponding to the extremal phase scaling filters in Figure 108a.

for this factorization, we need to introduce the concepts of *zero phase* filters and *linear phase* filters. Consider a filter $\{u_l\}$ whose transfer function is given by

$$U(f) \equiv \sum_{l=\infty}^{\infty} u_l e^{-i2\pi fl}.$$

Let $\{u_l^\circ : l = 0, \ldots, N-1\}$ be $\{u_l\}$ periodized to length N, and let $\{U_k^\circ\}$ be its DFT:

$$U_k^\circ \equiv \sum_{l=0}^{N-1} u_l^\circ e^{-i2\pi kl/N} = U(f_k) \text{ with } f_k \equiv \frac{k}{N}.$$

l	g_l for D(6)	g_l for C(6)	g_l for D(8)
0	0.3326705529500827	−0.0156557285289848	0.2303778133074431
1	0.8068915093110928	−0.0727326213410511	0.7148465705484058
2	0.4598775021184915	0.3848648565381134	0.6308807679358788
3	−0.1350110200102546	0.8525720416423900	−0.0279837694166834
4	−0.0854412738820267	0.3378976709511590	−0.1870348117179132
5	0.0352262918857096	−0.0727322757411889	0.0308413818353661
6			0.0328830116666778
7			−0.0105974017850021

l	g_l for LA(8)	g_l for LA(12)	g_l for LA(16)
0	−0.0757657147893407	0.0154041093273377	−0.0033824159513594
1	−0.0296355276459541	0.0034907120843304	−0.0005421323316355
2	0.4976186676324578	−0.1179901111484105	0.0316950878103452
3	0.8037387518052163	−0.0483117425859981	0.0076074873252848
4	0.2978577956055422	0.4910559419276396	−0.1432942383510542
5	−0.0992195435769354	0.7876411410287941	−0.0612733590679088
6	−0.0126039672622612	0.3379294217282401	0.4813596512592012
7	0.0322231006040713	−0.0726375227866000	0.7771857516997478
8		−0.0210602925126954	0.3644418948359564
9		0.0447249017707482	−0.0519458381078751
10		0.0017677118643983	−0.0272190299168137
11		−0.0078007083247650	0.0491371796734768
12			0.0038087520140601
13			−0.0149522583367926
14			−0.0003029205145516
15			0.0018899503329007

Table 109. Coefficients for selected Daubechies scaling filters and for the coiflet scaling filter for $L = 6$ (the latter is discussed in Section 4.9). The coefficients in this table are derived from Daubechies (1992, 1993). These coefficients (and those for other $\{g_l\}$) are available on the Web site for this book (see page *xiv*).

Consider a time series $\{X_t : t = 0, \dots, N-1\}$ whose DFT is

$$\mathcal{X}_k \equiv \sum_{t=0}^{N-1} X_t e^{-i2\pi kt/N}, \quad \text{so that} \quad X_t = \frac{1}{N}\sum_{k=0}^{N-1} \mathcal{X}_k e^{i2\pi kt/N}.$$

Let

$$Y_t \equiv \sum_{l=0}^{N-1} u_l^\circ X_{t-l \bmod N}, \quad t = 0, \dots, N-1,$$

be the result of circularly filtering $\{X_t\}$ with $\{u_l^\circ\}$. Then the DFT for $\{Y_t\}$ is given by $\{U_k^\circ \mathcal{X}_k\}$, so the inverse DFT tells us that

$$Y_t = \frac{1}{N}\sum_{k=0}^{N-1} U_k^\circ \mathcal{X}_k e^{i2\pi kt/N}.$$

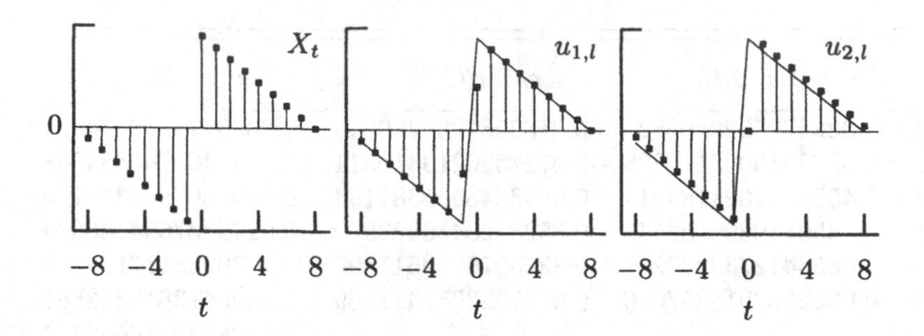

Figure 110. Example of filtering using a zero phase filter. The left-hand plot shows a time series $\{X_t\}$ with a discontinuity from $t = -1$ to $t = 0$. The solid points in the middle and right-hand plots show the results of filtering this series with, respectively, the filters $\{u_{1,l}\}$ (Equation (110a)) and $\{u_{2,l}\}$ (Equation (110b)), only the first of which has zero phase. The thin curves on these two plots show the original time series. Note that the discontinuity is spread out – but not shifted in location – using the zero phase filter $\{u_{1,l}\}$, whereas it is spread out and shifted forward in time using $\{u_{2,l}\}$.

Let us write the transfer function $U(\cdot)$ as $U(f) = |U(f)|e^{i\theta(f)}$, where $\theta(\cdot)$ is the phase function. If $\theta(f_k) = 0$ for all k, then $U(f_k) = |U(f_k)|$, and hence $U_k^\circ = |U_k^\circ|$, so we have

$$Y_t = \frac{1}{N}\sum_{k=0}^{N-1} |U_k^\circ|\mathcal{X}_k e^{i2\pi kt/N},$$

so that filtering $\{X_t\}$ with $\{u_l^\circ\}$ yields a series whose synthesis differs from that of $\{X_t\}$ only in the amplitudes of the sinusoids and not in their phases. The practical implication of $\theta(f_k) = 0$ is that events in $\{Y_t\}$ can be aligned to events in $\{X_t\}$, which is important for interpreting the meaning of $\{Y_t\}$ in physical applications. A filter with a phase function satisfying the property $\theta(f) = 0$ for all f is called a *zero phase* filter.

As an example of a filter with zero phase, consider

$$u_{1,l} = \begin{cases} 1/2, & l = 0; \\ 1/4, & l = \pm 1; \\ 0, & \text{otherwise.} \end{cases} \tag{110a}$$

From Exercise [2.3] we have $U_1(f) = \cos^2(\pi f)$, which is real-valued and nonnegative, and hence $\theta_1(f) = 0$ for all f. By way of contrast, consider

$$u_{2,l} = \begin{cases} 1/2, & l = 0, 1; \\ 0, & \text{otherwise,} \end{cases} \tag{110b}$$

whose transfer function is

$$U_2(f) = \frac{1}{2} + \frac{e^{-i2\pi f}}{2} = e^{-i\pi f}\cos(\pi f) = |U_2(f)|e^{-i\pi f}.$$

Since $\theta_2(f) = -\pi f$ here, this filter does not have zero phase. An example of filtering a time series with these two filters is shown in Figure 110.

To introduce linear phase filters, let us now consider the effect of circularly advancing the filter output $\{Y_t\}$ by ν units to form

$$Y_t^{(\nu)} \equiv Y_{t+\nu \bmod N}, \quad t = 0, \ldots, N-1,$$

where ν is an integer such that $1 \leq |\nu| \leq N-1$ (in vector notation, $\{Y_t^{(\nu)}\}$ forms the elements of $\mathcal{T}^{-\nu}\mathbf{Y}$ – see the discussion surrounding Equation (52a)). For example, if $\nu = 2$ and $N \geq 11$, then $Y_8^{(2)} = Y_{10}$ so that events in $\{Y_t^{(2)}\}$ occur two units in advance of events in $\{Y_t\}$. Now

$$\begin{aligned} Y_t^{(\nu)} = Y_{t+\nu \bmod N} = \sum_{l=0}^{N-1} u_l^\circ X_{t+\nu-l \bmod N} &= \sum_{l=-\nu}^{N-1-\nu} u_{l+\nu}^\circ X_{t-l \bmod N} \\ &= \sum_{l=0}^{N-1} u_{l+\nu \bmod N}^\circ X_{t-l \bmod N}. \end{aligned}$$

Hence advancing the filter output by ν units corresponds to using a filter whose coefficients have been advanced circularly by ν units. Now the circular filter $\{u_{l+\nu \bmod N}^\circ : l = 0, \ldots, N-1\}$ can be constructed by periodizing the filter $\{u_l^{(\nu)} \equiv u_{l+\nu} : l = \ldots, -1, 0, 1, \ldots\}$ to length N, so the phase properties of $\{u_{l+\nu \bmod N}^\circ\}$ can be picked off from the transfer function for $\{u_l^{(\nu)}\}$.

▷ **Exercise [111]:** Show that the transfer function for the advanced filter $\{u_l^{(\nu)}\}$ is given by

$$U^{(\nu)}(f) \equiv e^{i2\pi f\nu} U(f).$$ ◁

Thus, if $\{u_l\}$ has zero phase so that $U(f) = |U(f)|$, then $U^{(\nu)}(f) = |U(f)| \exp(i2\pi f\nu)$, and hence $\{u_l^{(\nu)}\}$ has a phase function given by

$$\theta(f) = 2\pi f\nu.$$

A filter with a phase function satisfying the above for some real-valued constant ν is called a *linear phase* filter. Note that, if the constant ν is an integer, a linear phase filter can be easily changed into a zero phase filter.

As an example, suppose that

$$u_{3,l} = \begin{cases} 1/2, & l = 1; \\ 1/4, & l = 0 \text{ or } 2; \\ 0, & \text{otherwise.} \end{cases}$$

The transfer function for this filter is

$$U_3(f) \equiv \frac{1}{4} + \frac{e^{-i2\pi f}}{2} + \frac{e^{-i4\pi f}}{4} = \cos^2(\pi f) e^{-i2\pi f},$$

which has a linear phase function with $\nu = -1$. Hence advancing (i.e., shifting to the left) $\{u_{3,l}\}$ by one unit turns it into the zero phase filter $\{u_{1,l}\}$ given by Equation (110a); i.e., $u_{3,l}^{(1)} = u_{1,l}$.

The definition for the Daubechies least asymmetric scaling filter of width L is as follows. For all possible sequences $\{g_l : l = 0, \ldots, L-1\}$ with squared gain function $\mathcal{G}^{(\mathrm{D})}(\cdot)$ and for a given shift $\tilde{\nu}$, we compute

$$\rho_{\tilde{\nu}}(\{g_l\}) \equiv \max_{-1/2 \leq f \leq 1/2} \left|\theta^{(G)}(f) - 2\pi f \tilde{\nu}\right|, \tag{112a}$$

where $\theta^{(G)}(\cdot)$ is the phase function for the $\{g_l\}$ under consideration (we must have $\rho_{\tilde{\nu}}(\{g_l\}) > 0$ because exact linear phase cannot be achieved with compactly supported scaling filters – see Daubechies, 1992, Theorem 8.1.4). For a given $\{g_l\}$, let ν be the shift that minimizes the above (in fact, because by assumption the nonzero values of $\{g_l\}$ can occur only at indices $l = 0$ to $L-1$, we must have $-(L-1) \leq \nu \leq 0$, so we need only compute $\rho_{\tilde{\nu}}(\{g_l\})$ for a finite number of shifts). The least asymmetric filter $\{g_l^{(\mathrm{la})}\}$ is the one such that $\rho_{\nu}(\{g_l\})$ is as small as possible. In words, the least asymmetric filter is the one whose phase function has the smallest maximum deviation in frequency from the best fitting linear phase function. We refer to the Daubechies least asymmetric scaling filter of width L as the LA(L) scaling filter, where $L = 8, 10, \ldots$. The coefficients for the LA(8), LA(12) and LA(16) filters are given in Table 109 (the values in this table were computed from the values in Table 6.3, p. 198, Daubechies, 1992, but, for reasons explained in item [1] of the Comments and Extensions below, the values in our table are not the same as those in Daubechies' table). The LA(8), LA(10), ..., LA(20) scaling filters are plotted in Figure 113a, while the corresponding wavelet filters are plotted in Figure 113b.

Since ν is the value of $\tilde{\nu}$ that minimizes (112a), we have

$$\theta^{(G)}(f) \approx 2\pi f \nu. \tag{112b}$$

▷ **Exercise [112]:** Use Equation (76a) to show that

$$H(f) = e^{-i2\pi f(L-1)+i\pi} G(\tfrac{1}{2} - f).$$ ◁

Hence the phase function for the corresponding wavelet filter is given by

$$\theta^{(H)}(f) = -2\pi f(L-1) + \pi + \theta^{(G)}(\tfrac{1}{2} - f).$$

Equation (112b) implies that

$$\theta^{(H)}(f) \approx -2\pi f(L-1+\nu) + \pi(\nu+1), \tag{112c}$$

so that the wavelet phase function is also approximately linear if ν is odd (recall that the phase function is defined modulo 2π). Computations indicate that ν is indeed odd for all the LA coefficients plotted in the figures, and thus all the LA scaling and wavelet phase functions can be taken to be approximately linear:

$$\theta^{(G)}(f) \approx 2\pi f \nu \quad \text{and} \quad \theta^{(H)}(f) \approx -2\pi f(L-1+\nu), \tag{112d}$$

where ν is always odd. In particular, computations indicate that

$$\nu = \begin{cases} -\frac{L}{2} + 1, & \text{if } L = 8, 12, 16 \text{ or } 20 \text{ (i.e., } L/2 \text{ is even);} \\ -\frac{L}{2}, & \text{if } L = 10 \text{ or } 18; \\ -\frac{L}{2} + 2, & \text{if } L = 14. \end{cases} \tag{112e}$$

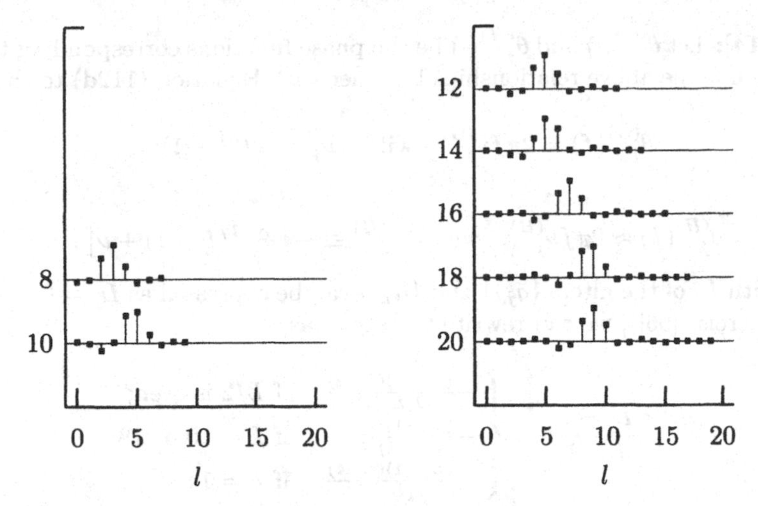

Figure 113a. Daubechies least asymmetric scaling filters $\{g_l^{(la)}\}$ for $L = 8, 10, \ldots, 20$ (values based on Daubechies, 1992, p. 198, Table 6.3, with modifications as noted in item [1] of the Comments and Extensions to this section).

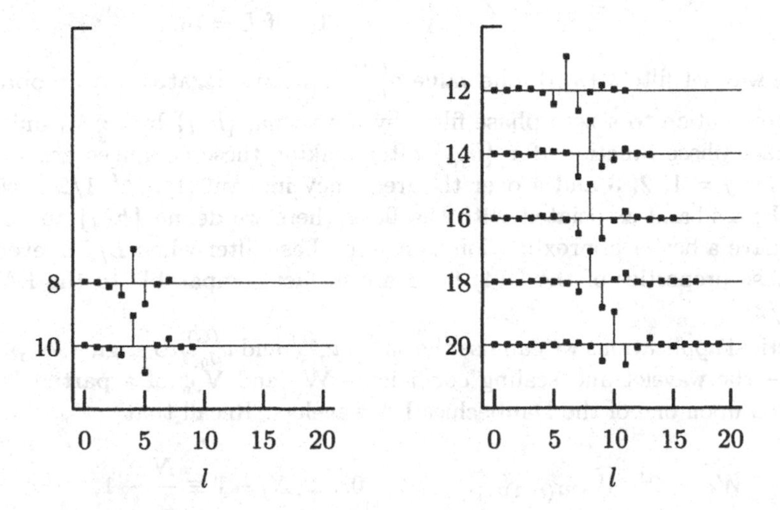

Figure 113b. Daubechies wavelet filters corresponding to the least asymmetric scaling filters shown in Figure 113a.

Note that ν is always negative, so that we can obtain an approximate zero phase scaling filter by advancing the filter output by $|\nu|$ units; likewise, because $-(L-1+\nu)$ is also always negative, we can obtain an approximate zero phase wavelet filter by advancing the filter output by $|L-1+\nu|$ units.

We now want to consider the phase functions of the scaling filters $\{g_{j,l}\}$ and wavelet filters $\{h_{j,l}\}$ at higher scales $\lambda_j = 2^j$ and $\tau_j = 2^{j-1}$ for $j = 2, 3, \ldots$. As noted in Section 4.6, the transfer functions $G_j(\cdot)$ and $H_j(\cdot)$ for these filters satisfy

$$G_j(f) = \prod_{l=0}^{j-1} G(2^l f) \quad \text{and} \quad H_j(f) = H(2^{j-1} f) G_{j-1}(f).$$

▷ **Exercise [114]:** Let $\theta_j^{(G)}(\cdot)$ and $\theta_j^{(H)}(\cdot)$ be the phase functions corresponding to $G_j(\cdot)$ and $H_j(\cdot)$. Use the above relationships together with Equation (112d) to show that

$$\theta_j^{(G)}(f) \approx 2\pi f \nu_j^{(G)} \quad \text{with} \quad \nu_j^{(G)} \equiv (2^j - 1)\nu \tag{114a}$$

and

$$\theta_j^{(H)}(f) \approx 2\pi f \nu_j^{(H)} \quad \text{with} \quad \nu_j^{(H)} \equiv -\left[2^{j-1}(L-1) + \nu\right]. \tag{114b}$$ ◁

Since the width L_j of the filters $\{g_{j,l}\}$ and $\{h_{j,l}\}$ can be expressed as $L_j = (2^j - 1)(L - 1) + 1$ (cf. Exercise [96]), we can rewrite the above as

$$\nu_j^{(G)} = \frac{L_j - 1}{L - 1}\nu = \begin{cases} -\frac{(L_j-1)(L-2)}{2(L-1)}, & \text{if } L/2 \text{ is even;} \\ -\frac{(L_j-1)L}{2(L-1)}, & \text{if } L = 10 \text{ or } 18; \\ -\frac{(L_j-1)(L-4)}{2(L-1)}, & \text{if } L = 14; \end{cases}$$

and

$$\nu_j^{(H)} = -\left(\frac{L_j}{2} + \frac{L}{2} + \nu - 1\right) = \begin{cases} -\frac{L_j}{2}, & \text{if } L/2 \text{ is even;} \\ -\frac{L_j}{2} + 1, & \text{if } L = 10 \text{ or } 18; \\ -\frac{L_j}{2} - 1, & \text{if } L = 14. \end{cases} \tag{114c}$$

For the wavelet filters $\{h_{j,l}\}$, the value $\nu_j^{(H)}$ is always negative, so we obtain the closest approximation to a zero phase filter by advancing $\{h_{j,l}\}$ by $|\nu_j^{(H)}|$ units. The resulting exact phase functions for $\{h_{j,l}\}$ after making these advances are shown in Figure 115 for $j = 1$, 2, 3 and 4 over the frequency interval $[1/2^{j+1}, 1/2^j]$, which is the nominal pass-band associated with the filter (here we define $\{h_{1,l}\}$ to be $\{h_l\}$). Clearly we have a better approximation to a zero phase filter when $L/2$ is even (note that the phase properties of the D(4) filter are in fact comparable to the LA filters with odd $L/2$).

In practical applications we can use the shifts $\nu_j^{(H)}$ and $\nu_{J_0}^{(G)}$ to align – for purposes of plotting – the wavelet and scaling coefficients $\mathbf{W}_j$ and $\mathbf{V}_{J_0}$ of a partial DWT of level J_0 based upon one of the Daubechies LA wavelets. Recall that

$$W_{j,t} = 2^{j/2}\widetilde{W}_{j,2^j(t+1)-1}, \qquad t = 0, \ldots, N_j - 1 \equiv \frac{N}{2^j} - 1,$$

where, for $t = 0, \ldots, N - 1$,

$$2^{j/2}\widetilde{W}_{j,t} \equiv \sum_{l=0}^{L_j-1} h_{j,l} X_{t-l \bmod N}.$$

Now we need to advance the jth level wavelet filter $\{h_{j,l}\}$ by $|\nu_j^{(H)}|$ units to achieve approximate zero phase; equivalently, we can advance the filter output $\{\widetilde{W}_{j,t}\}$ by $|\nu_j^{(H)}|$ units. If we let $\widetilde{\mathbf{W}}_j$ be the N dimensional vector containing the $\widetilde{W}_{j,t}$'s and if we recall that $\nu_j^{(H)} < 0$ so that we can write $\nu_j^{(H)} = -|\nu_j^{(H)}|$ to help to keep the direction in which things are moving straight, this advanced output can be described as the contents of the vector $\mathcal{T}^{-|\nu_j^{(H)}|}\widetilde{\mathbf{W}}_j$, the tth element of which is associated with X_t;

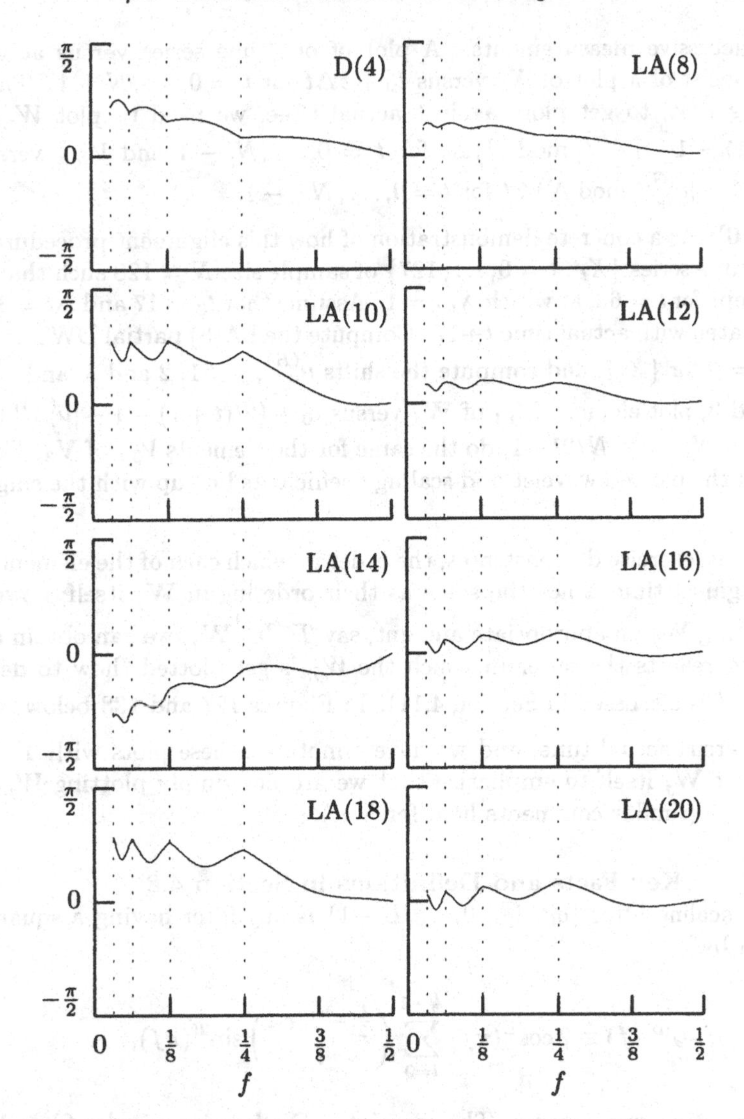

Figure 115. Exact phase functions of shifted LA wavelet filters $\{h_{j,l}\}$, $j = 1$, 2, 3 and 4 (the D(4) case is also shown, with shifts governed by setting $\nu = -1$). The phase functions for $\{h_{j,l}\}$ are plotted just over frequencies in the interval $[1/2^{j+1}, 1/2^j]$ – these intervals are indicated by the vertical dotted lines.

i.e., we associate the output $\widetilde{W}_{j,t+|\nu_j^{(H)}| \bmod N}$ with the input X_t, or, equivalently, we associate $\widetilde{W}_{j,t}$ with $X_{t-|\nu_j^{(H)}| \bmod N}$. Finally, since $W_{j,t} = 2^{j/2}\widetilde{W}_{j,2^j(t+1)-1}$, we see that $W_{j,t}$ should be associated with $X_{2^j(t+1)-1-|\nu_j^{(H)}| \bmod N}$. An analogous argument says that the scaling coefficient $V_{J_0,t}$ should be associated with $X_{2^{J_0}(t+1)-1-|\nu_{J_0}^{(G)}| \bmod N}$.

Suppose now that the tth value X_t of our time series is associated with actual time $t_0 + t\,\Delta t$, where t_0 is the time (in units of, say, seconds) at which the value X_0 was measured, and Δt is the time increment (in the same units as t_0) between subsequent successive measurements. A plot of our time series versus actual time would thus consist of a plot of X_t versus $t_0 + t\,\Delta t$ for $t = 0, \ldots, N-1$. The above argument says that, to get plots against actual time, we need to plot $W_{j,t}$ versus $t_0 + \left(2^j(t+1) - 1 - |\nu_j^{(H)}| \bmod N\right)\Delta t$ for $t = 0, \ldots, N_j - 1$ and $V_{J_0,t}$ versus $t_0 + \left(2^{J_0}(t+1) - 1 - |\nu_{J_0}^{(G)}| \bmod N\right)\Delta t$ for $t = 0, \ldots, N_{J_0} - 1$.

▷ **Exercise [116]:** As a concrete demonstration of how this alignment procedure works, consider a time series $\{X_t : t = 0, \ldots, 127\}$ of sample size $N = 128$ such that $X_t = 0$ for all t except for $t = 63$, at which $X_{63} = 1$. Assume that $t_0 = 17$ and $\Delta t = 1$ so that X_t is associated with actual time $t+17$. Compute the LA(8) partial DWT transform of level $J_0 = 3$ for $\{X_t\}$, and compute the shifts $\nu_j^{(H)}$, $j = 1$, 2 and 3, and $\nu_3^{(G)}$. For $j = 1$, 2 and 3, plot element $W_{j,t}$ of $\mathbf{W}_j$ versus $t_0 + \left(2^j(t+1) - 1 - |\nu_j^{(H)}| \bmod N\right)$ for $t = 0, \ldots, N_j - 1 \equiv N/2^j - 1$; do the same for the elements $V_{3,t}$ of $\mathbf{V}_3$. Comment on how well the plotted wavelet and scaling coefficients line up with the single spike in $\{X_t\}$. ◁

As the above exercise demonstrates, the order in which each of the elements of $\mathbf{W}_j$ gets plotted against time is not the same as their ordering in $\mathbf{W}_j$ itself; however, by circularly shifting $\mathbf{W}_j$ an appropriate amount, say $\mathcal{T}^{-\gamma_j^{(H)}}\mathbf{W}_j$, we can obtain a vector whose ordering reflects the order in which the $W_{j,t}$'s get plotted (how to determine the integer $\gamma_j^{(H)}$ is discussed in Section 4.11). In Figures 127 and 138 below, we show plots of $\mathbf{W}_j$ versus actual time, and we have annotated these plots with $\mathcal{T}^{-\gamma_j^{(H)}}\mathbf{W}_j$ rather than just $\mathbf{W}_j$ itself to emphasize that we are not simply plotting $W_{j,t}$ versus $t = 0, \ldots, N_j - 1$ (similar comments hold for $\mathbf{V}_{J_0}$).

Key Facts and Definitions in Section 4.8

A Daubechies scaling filter $\{g_l : l = 0, \ldots, L-1\}$ is any filter having a squared gain function given by

$$\mathcal{G}^{(\mathrm{D})}(f) \equiv 2\cos^L(\pi f) \sum_{l=0}^{\frac{L}{2}-1} \binom{\frac{L}{2} - 1 + l}{l} \sin^{2l}(\pi f),$$

where L is a positive even integer. The associated Daubechies wavelet filter $\{h_l : l = 0, \ldots, L-1\}$ then follows from the QMF relationship of Equation (75b) and has a squared gain function given by

$$\mathcal{H}^{(\mathrm{D})}(f) \equiv 2\sin^L(\pi f) \sum_{l=0}^{\frac{L}{2}-1} \binom{\frac{L}{2} - 1 + l}{l} \cos^{2l}(\pi f).$$

The above can be re-expressed as $\mathcal{H}^{(\mathrm{D})}(f) = \mathcal{D}^{\frac{L}{2}}(f)\mathcal{A}_L(f)$, where $\mathcal{D}(\cdot)$ is the squared gain function for the difference filter $\{1, -1\}$, while $\mathcal{A}_L(\cdot)$ is the squared gain for a low-pass filter (see Figure 107).

For a fixed L, the squared gain function $\mathcal{G}^{(\mathrm{D})}(\cdot)$ does not determine a unique scaling filter. The set of all $\{g_l\}$ having $\mathcal{G}^{(\mathrm{D})}(\cdot)$ as their squared gain function can be obtained by 'spectral factorization,' which amounts to 'taking the square root' of $\mathcal{G}^{(\mathrm{D})}(\cdot)$ in the sense of finding all possible transfer functions $G(\cdot)$ such that $|G(f)|^2 = \mathcal{G}^{(\mathrm{D})}(\cdot)$. For example, when $L = 2$, there are two possible factorizations, one yielding the Haar scaling filter $\{\frac{1}{\sqrt{2}}, \frac{1}{\sqrt{2}}\}$ and the other, its negative $\{-\frac{1}{\sqrt{2}}, -\frac{1}{\sqrt{2}}\}$; when $L = 4$, we obtain the D(4) scaling filter of Equation (75d) and three other filters that differ only in sign and direction. For $L \geq 6$, the multiple filters with squared gains $\mathcal{G}^{(D)}(\cdot)$ differ in nontrivial ways, so we can select amongst all possible factorizations by requiring some additional criterion to hold. One such criterion is the extremal phase choice that yields a minimum delay filter. This criterion picks the scaling filter $\{g_l^{(ep)}\}$ with squared gain $\mathcal{G}^{(D)}(\cdot)$ such that

$$\sum_{l=0}^{m} g_l^2 \leq \sum_{l=0}^{m} \left[g_l^{(ep)}\right]^2 \quad \text{for } m = 0, \ldots, L-1,$$

where $\{g_l\}$ is any other filter with squared gain $\mathcal{G}^{(D)}(\cdot)$. We denote the filters satisfying this criterion as the D(L) scaling filters, $L = 2, 4, \ldots$. These filters are plotted in Figure 108a (the corresponding wavelet filters are shown in Figure 108b).

A second criterion is to pick the scaling filter $\{g_l^{(la)}\}$ whose transfer function $G(f) = [\mathcal{G}^{(D)}(f)]^{1/2} e^{i\theta^{(G)}(f)}$ is such that the phase function $\theta^{(G)}(\cdot)$ is as close as possible (in a sense made precise by Equation (112a)) to that of a linear phase filter. We denote the filters satisfying this criterion as the LA(L) scaling filters, $L = 8, 10, \ldots$, where 'LA' stands for 'least asymmetric' (this name is appropriate because true linear phase filters are symmetric about some midpoint). These filters are plotted in Figure 113a (the LA wavelet filters are shown in Figure 113b). The phase functions for the LA scaling filters are such that $\theta^{(G)}(f) \approx 2\pi f \nu$, where ν depends on L as specified in Equation (112e).

Comments and Extensions to Section 4.8

[1] Table 6.3 of Daubechies (1992) lists the LA scaling filter coefficients for even filter widths L ranging from 8 to 20 with the normalization $\sum_l g_l = 2$ rather than the normalization that we use, namely, $\sum_l g_l = \sqrt{2}$. Computations indicate that ν in Equation (112c) is odd for $L = 8$, 12, 18 and 20, but is even for 10, 14, and 16. Thus, while the tabulated LA(8), LA(12), LA(18) and LA(20) scaling filter coefficients have approximately linear phase, the other three filters are said to have approximately *generalized linear phase*, which is defined to be 'linear phase plus a nonzero intercept term' (Oppenheim and Schafer, 1989); here the nonzero intercept term can be taken to be $\theta^{(H)}(0) = \pi$. The result of the following exercise can be used to get rid of this nonzero intercept.

▷ **Exercise [117]:** Show that, if the filter $\{g_0, g_1, \ldots, g_{L-2}, g_{L-1}\}$ has the transfer function $G(\cdot)$, the reversed filter $\{g_{L-1}, g_{L-2}, \ldots, g_1, g_0\}$ then has a transfer function defined by $\exp[-i2\pi f(L-1)]G^*(f)$. ◁

Hence, if the filter coefficients tabulated by Daubechies for these three cases are simply reversed, the transfer function for the reversed scaling filter will be given by $\exp[-i2\pi f(L-1)]\, G^*(f)$ and will thus have a phase function given by

$$-\theta^{(G)}(f) - 2\pi f(L-1) \approx 2\pi f(-\nu_b - [L-1]) \equiv 2\pi f \nu,$$

where ν_b is the even value before time reversal, and ν is the odd value after time reversal. Our plots for $L = 10$, 14, and 16 in Figure 113a (and the values given in

L	$\lvert\nu\rvert - e\{g_l\}$	L	$\lvert\nu\rvert - e\{g_l\}$
8	0.1536	14	−0.1615
10	0.4500	16	0.1546
12	0.1543	18	0.4471
		20	0.1547

Table 118. Comparison between advances $|p_j^{(H)}|$ and $|\nu_j^{(H)}|$ for Daubechies least asymmetric wavelet filters. Since $|p_j^{(H)}|-|\nu_j^{(H)}| = |\nu|-e\{g_l\}$ for all scale indices j, the above indicates that the two advances are the same when rounded to the nearest integer. To compare the advances for the corresponding scaling filters, the tabulated values must be multiplied by $2^j - 1$, so the advances diverge as j increases.

Table 109 for $L = 16$) thus show $\{g_l\}$ that are derived from Daubechies (1992) after a reverse in order.

[2] Wickerhauser (1994, pp. 171 and 341) and Hess–Nielsen and Wickerhauser (1996) gave advances $|p_j^{(G)}|$ and $|p_j^{(H)}|$ that can be used for both the extremal phase and least asymmetric filters (and other filter designs as well). These advances are given by

$$|p_j^{(G)}| = (2^j - 1)e\{g_l\} \text{ and } |p_j^{(H)}| = 2^{j-1}(e\{g_l\} + e\{h_l\}) - e\{g_l\},$$

where

$$e\{a_l\} \equiv \frac{\sum_{l=1}^{L-1} l a_l^2}{\sum_{l=0}^{L-1} a_l^2}$$

represents the 'center of energy' for a filter $\{a_l : l = 0, \ldots, L-1\}$.

▷ **Exercise [118]:** Show that, because the scaling $\{g_l\}$ and wavelet $\{h_l\}$ filters have unit energy and satisfy the QMF relationship of Equation (75a), we must have $e\{g_l\} + e\{h_l\} = L - 1$. ◁

For all wavelet and scaling filters, we thus have

$$|p_j^{(G)}| = (2^j - 1)e\{g_l\} \text{ and } |p_j^{(H)}| = 2^{j-1}(L-1) - e\{g_l\}.$$

We can compare the above to the advances we derived for the least asymmetric filters by using the right-hand parts of Equations (114a) and (114b):

$$|\nu_j^{(G)}| = (2^j - 1)|\nu| \text{ and } |\nu_j^{(H)}| = 2^{j-1}(L-1) - |\nu|,$$

where ν is given in Equation (112e) for various filter widths L. We thus see that

$$|\nu_j^{(G)}| - |p_j^{(G)}| = (2^j - 1)(|\nu| - e\{g_l\}), \text{ while } |p_j^{(H)}| - |\nu_j^{(H)}| = |\nu| - e\{g_l\}.$$

The differences $|\nu| - e\{g_l\}$ are given in Table 118 for $L = 8$ up to 20 and are seen to be always less than 0.5 in magnitude. This indicates that, when $|p_j^{(H)}|$ is rounded to the nearest integer, the advances for the wavelet coefficients are the same as determined by $|\nu_j^{(H)}|$ or $|p_j^{(H)}|$; on the other hand, the advances for the scaling coefficients diverge as j increases (we will discuss filter advances again in Chapter 6 when we consider the discrete wavelet packet transform).

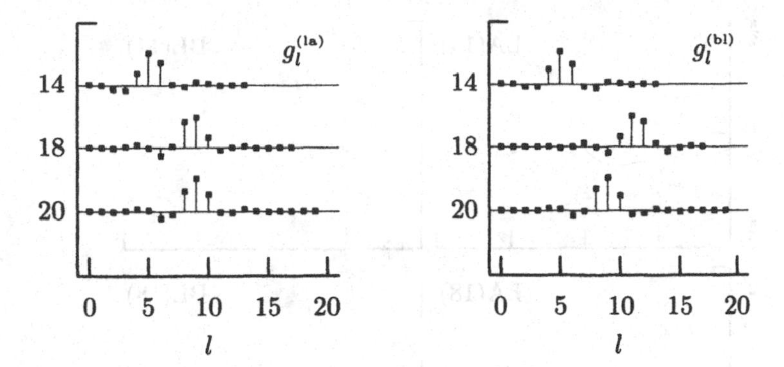

Figure 119. Least asymmetric scaling filters $\{g_l^{(la)}\}$ (left-hand column) and best localized scaling filters $\{g_l^{(bl)}\}$ (right) for $L = 14$, 18 and 20.

[3] Doroslovački (1998) proposed a 'best localized' (BL) factorization of the squared gain function for the Daubechies scaling filter – this scheme is an attempt to refine the least asymmetric idea. The LA factorization is the one that minimizes the maximum deviation of the phase function from linear phase over all frequencies; however, since the scaling filter is low-pass, the phases at high frequencies are not as important to control as the ones at low frequencies. In his work Doroslovački used a new measure of departure from linear phase that penalizes departures at the low frequencies more heavily than those at high frequencies. For $L = 8$, 10, 12 and 16, this new measure picks out the same factorization as the LA factorization; for $L = 14$, 18 and 20 the BL factorization is different. These new scaling filters are plotted in the right-hand column of Figure 119, while the corresponding LA scaling filters are given in the left-hand column. (It should be noted that the BL coefficients used here for $L = 14$ and 18 are in reverse order from those listed in Doroslovački, 1998; the rationale for this reversal is the same as that for reversing the LA(10), LA(14) and LA(16) coefficients.)

Computations indicate that the phase functions $\theta^{(G)}(\cdot)$ for these three filters are such that $\theta^{(G)}(f) \approx 2\pi f \nu$ if we take $\nu = -5$ for $L = 14$; $\nu = -11$ for $L = 18$; and $\nu = -9$ for $L = 20$. The right-hand column of Figure 120 shows the exact phase functions for the BL versions of $\{h_{j,l}\}$, $j = 1, 2, 3$ and 4, over the nominal pass-bands, while the left-column has corresponding plots for the LA filters (these are extracted from Figure 115). Since ideally the curves in these plots should be as close to zero as possible, we see that the BL(14) filter has markedly better phase properties than the LA(14) filter; however, the same cannot be said for the BL(18) and BL(20) filters, both of which have generally worse phase properties than the corresponding LA filters (since the phase functions have opposite signs for BL(18) and LA(18) filters, we have also plotted the negative of the exact phase functions for LA(18) for ease of comparison – this is the thicker curve in the middle of the left-hand column).

Taswell (2000) discusses other factorization criteria and a systematic approach for extracting the corresponding filters.

[4] Let us consider the phase properties of the LA wavelet filters in more detail. Koopmans (1974, pp. 95–6) defines a *time shift function* that transforms the angular displacement given by a phase function $\theta(\cdot)$ into a shift in terms of the index t:

$$\tau(f) \equiv \frac{\theta(f)}{2\pi f}, \quad f \neq 0.$$

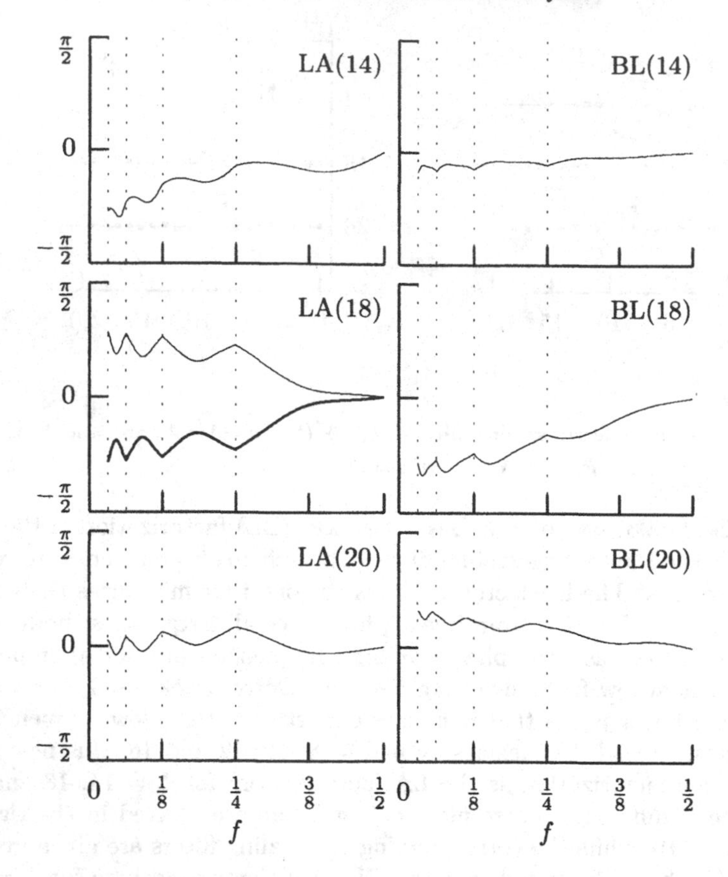

Figure 120. Exact phase functions of shifted least asymmetric and best localized wavelet filters (see the caption to Figure 115 and the text for details).

For example, suppose $\theta(f) = 2\pi f\nu$ for some integer ν; i.e., $\theta(\cdot)$ is the phase function for a linear phase filter. The time shift function is then $\tau(f) = \nu$, which tells us how much we need to shift (advance) the output from this filter in order to obtain a zero phase filter.

In Figure 115 we examined the phase functions for the LA wavelet filters after these have been advanced to achieve approximate zero phase. If these advanced filters were actually zero phase filters, all the plotted curves would be horizontal lines centered at zero. Since they are only approximately so, it is of interest to translate the angular displacements into shifts in terms of the index t. Figure 121 shows the exact time shift functions after rounding to the nearest integer. Note that, for unit scale (i.e., frequency band $[\frac{1}{4}, \frac{1}{2}]$), the rounded time shift functions are always zero – this is not unexpected since the advances are designed to achieve as good an approximation to a zero phase filter for unit scale as possible. As the scale increases, however, nonzero values of the rounded time shift functions start to appear, particularly for the filters in the left-hand column of Figure 121. These results indicate that we can actually obtain a somewhat better approximation to a zero phase filter over the nominal pass-bands $[\frac{1}{2^{j+1}}, \frac{1}{2^j}]$ by applying an additional advance dictated by the average value of the time shift function over the nominal pass-band, rounded to the nearest integer.

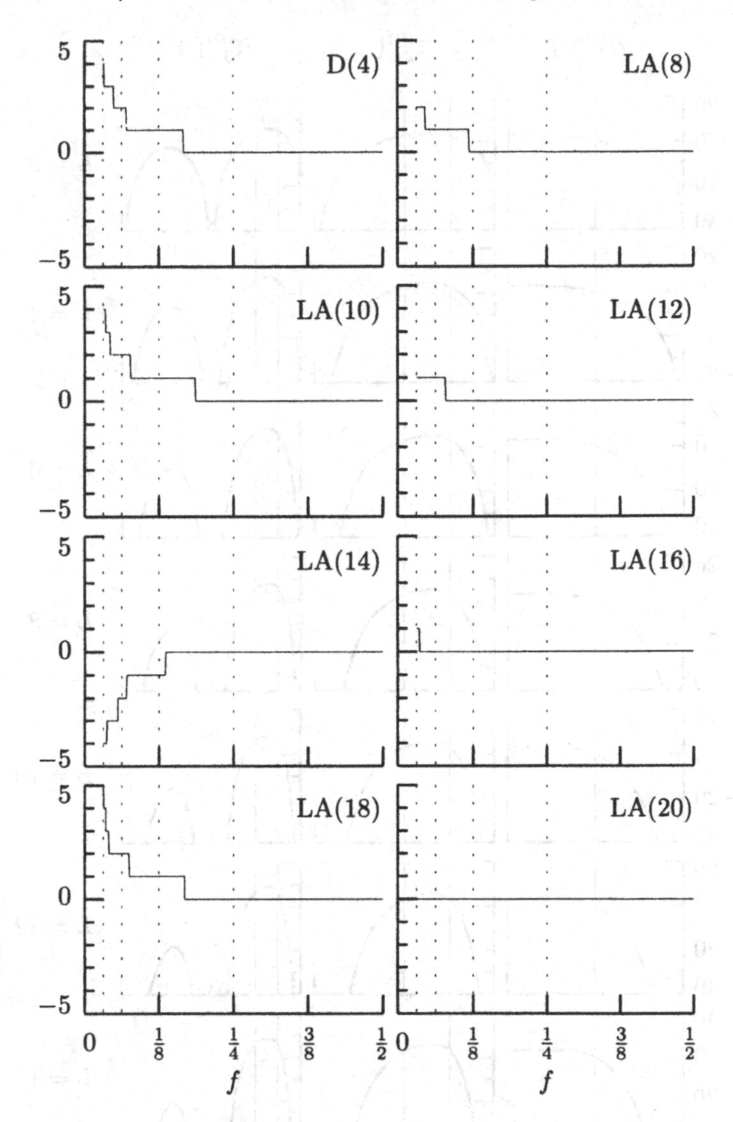

Figure 121. Time shift functions (rounded to the nearest integer) corresponding to the phase functions shown in Figure 115.

For example, we would need to advance the output for the LA(8) filter of level $j = 3$ by one additional unit.

[5] Let us consider the squared gain functions $\mathcal{H}_j(\cdot)$ for the Daubechies wavelet filters in a bit more detail. Figure 122 shows these functions for filter widths $L = 2, 4, \ldots, 14$ in combination with scale levels $j = 1, 2$ and 3. Thin vertical lines mark the nominal pass-bands $(1/2^{j+1}, 1/2^j]$, and the functions are plotted on a decibel (dB) scale; i.e., we are plotting $10 \cdot \log_{10}(\mathcal{H}_j(f))$ versus f. The distance between each of the tick marks on the vertical axis is 10 dB (i.e., one order of magnitude). Note that the band-pass approximation from these filters improves as L increases, but that, once we get past the Haar filter, the improvement in going from L to $L + 2$ is not dramatic. Since

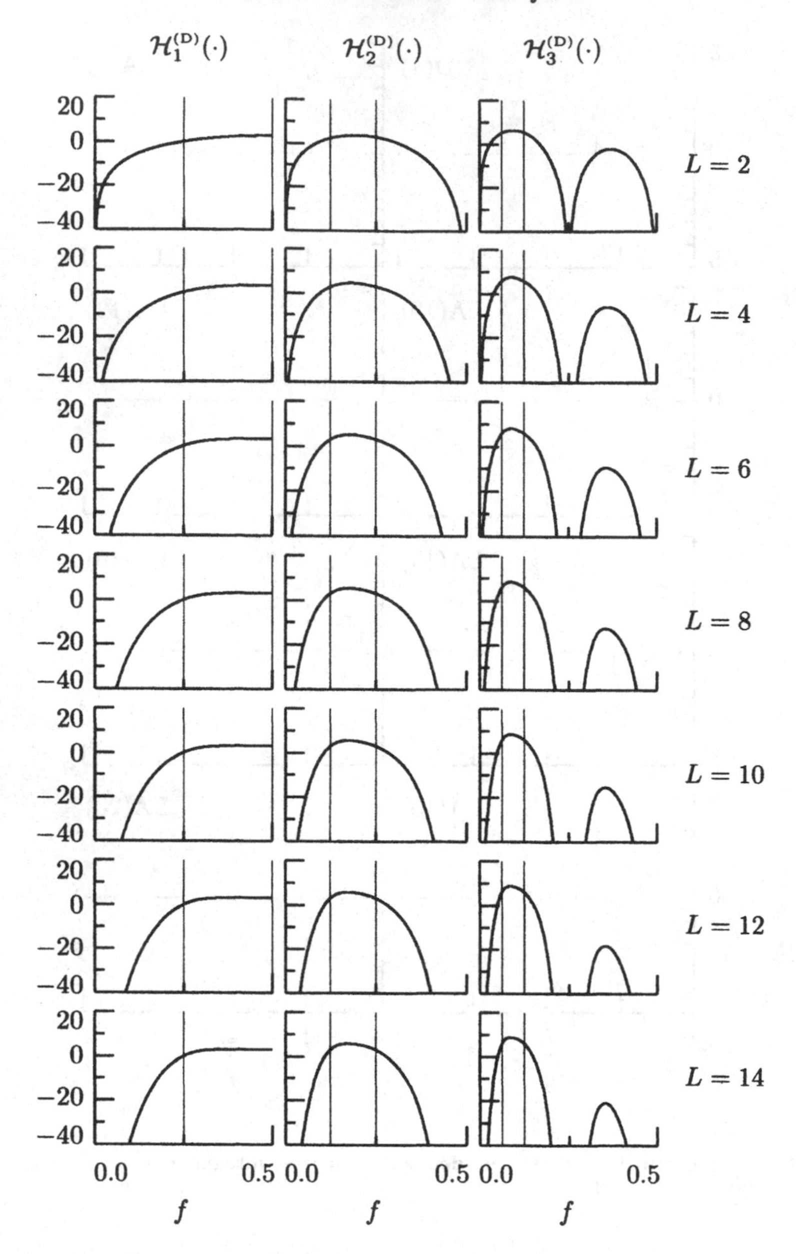

Figure 122. Squared gain functions $\mathcal{H}_j^{(D)}(\cdot)$, $j = 1, 2$ and 3 (left, middle and right columns, respectively), for Daubechies wavelet filters of widths $L = 2, 4, \ldots, 14$ (top to bottom rows, respectively). The two thin vertical lines in each plot delineate the nominal pass-band for the filter. The vertical axis is in decibels (i.e., we plot $10 \cdot \log_{10}(\mathcal{H}_j^{(D)}(f))$ versus f).

$\mathcal{H}_j(f) = \mathcal{H}(2^{j-1}f)\mathcal{G}_{j-1}(f)$ and since $\mathcal{G}_{j-1}(f) \approx 2^{j-1}$ for small f, the decay rate of $\mathcal{H}_j(\cdot)$ as we go from $f = 1/2^{j+1}$ (the left-hand side of its nominal pass-band) down to zero is dictated by the decay rate of $\mathcal{H}(\cdot)$. A study of Equation (105b) indicates that this decay rate is in turn controlled by the number of embedded differencing

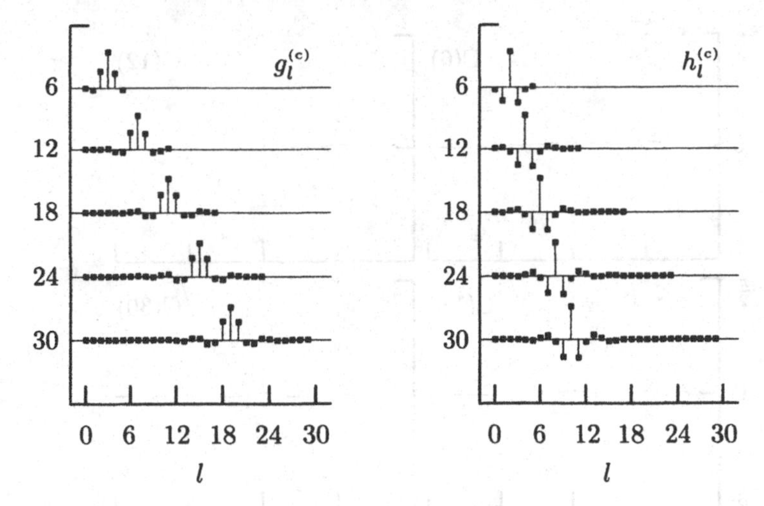

Figure 123. Coiflet scaling filters $\{g_l^{(c)}\}$ (left-hand column) and corresponding wavelet filters $\{h_l^{(c)}\}$ (right) of widths $L = 6$, 12, 18, 24 and 30 (Daubechies, 1992, p. 261, Table 8.1).

operations $L/2$, so that, for small f, we have $\mathcal{H}_j(f) \propto |f|^L$. Note also the prominent sidelobe in $\mathcal{H}_3(\cdot)$ to the right of its nominal pass-band. As we increase j, more of these sidelobes appear on the right: for $j \geq 3$, $\mathcal{H}_j(\cdot)$ has $2^{j-2} - 1$ such sidelobes (these are basically due to the fact that $\mathcal{G}(2^l f)$ in the expression above for $\mathcal{H}_j(f)$ defines a periodic function with period $1/2^l$). As we shall see in Section 8.9, the degree to which $\mathcal{H}_j(\cdot)$ approximates a band-pass filter impacts how we can interpret various estimates of the wavelet variance (the subject of Chapter 8).

4.9 Coiflet Wavelet and Scaling Filters: Form and Phase

It is important to realize the Daubechies wavelet filters are not the only ones yielding a DWT that can be described in terms of generalized differences of weighted averages. A second example of such filters is 'coiflets,' a term coined by Daubechies (1992, Section 8.2) to acknowledge the role of R. Coifman in suggesting the idea used to construct these filters. Coiflets provide an interesting contrast to the Daubechies scaling and wavelet filters. Although we have taken these latter filters to be defined via their squared gain functions (Equations (105a) and (105b)), in fact they were obtained by specifying certain 'vanishing moment' conditions on a wavelet function that is entirely determined by the associated scaling filter (see Section 11.9). The idea behind coiflets is also to specify vanishing moment conditions on the associated scaling function. This construction turns out to produce an appealing set of wavelet filters with remarkably good phase properties.

Figure 123 shows the coiflet scaling filters $\{g_l^{(c)}\}$ and the corresponding wavelet filters $\{h_l^{(c)}\}$ for widths $L = 6$, 12, 18, 24 and 30 as presented in Daubechies (1992, p. 261, Table 8.1) – henceforth we use the abbreviation C(L) to refer to a coiflet filter of width L. The coefficients for the C(6) scaling filter are also listed in Table 109. It is important to note that the ordering of the coefficients in Daubechies' table is here reversed for the same reason that we previously reversed the tabulated LA(10), LA(14) and LA(16) scaling coefficients (this is discussed in item [1] of the Comments and Extensions to Section 4.8). The squared gain functions for $\{h_l^{(c)}\}$ can be written

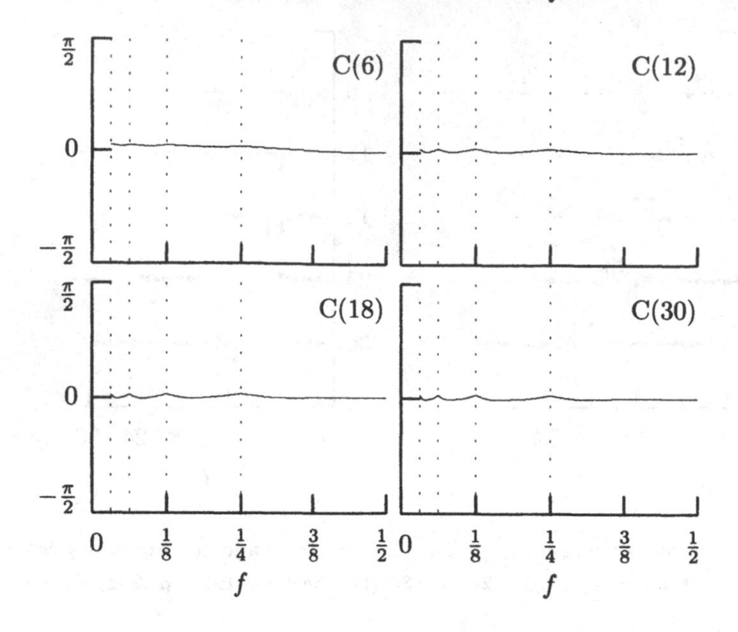

Figure 124. Exact phase functions of shifted coiflet wavelet filters (see the caption to Figure 115 and the text for details).

as

$$\mathcal{H}^{(c)}(f) = \mathcal{D}^{\frac{L}{3}}(f) \left(\sum_{l=0}^{\frac{L}{6}-1} \binom{\frac{L}{6}-1+l}{l} \cos^{2l}(\pi f) + \cos^{\frac{L}{3}}(\pi f) F(f) \right)^2,$$

where $F(\cdot)$ is a trigonometric polynomial chosen to force the condition of Equation (69d), namely, $\mathcal{H}^{(c)}(f) + \mathcal{H}^{(c)}(f + \frac{1}{2}) = 2$ for all f. Note that, whereas the Daubechies wavelet filters of width L have $L/2$ embedded differencing operations, the coiflet filters involve $L/3$ such differences, so there are more terms devoted to the averaging portion of the filter. Calculations indicate that the phase function for $\{g_l^{(c)}\}$ is approximately given by $2\pi f \nu$, with $\nu = -\frac{2L}{3} + 1$ (note that this is always odd when $L = 6$, 12, 18, 24 and 30). From Equations (114a) and (114b) we obtain the shift factors

$$\nu_j^{(G)} = -\frac{(L_j - 1)(2L - 3)}{3(L - 1)} \text{ and } \nu_j^{(H)} = -\frac{L_j}{2} + \frac{L}{6}. \tag{124}$$

In the same manner as in Figure 115 for the LA wavelet filters, Figure 124 shows exact phase functions for the shifted coiflet wavelet filters for $L = 6$, 12, 18 and 30 (the case $L = 24$ is not shown, but it is quite similar to the $L = 18$ and 30 cases). A comparison of Figures 115 and 124 indicate that coiflets provide a better approximation to zero phase filters than the LA filters do.

Figure 125 shows scaling $\{g_{j,l}\}$ and wavelet $\{h_{j,l}\}$ filters for levels $j = 1, \ldots, 7$ based upon the C(6) filter (Figures 98a and 98b show similar plots for the D(4) and LA(8) filters). Note that, as j increases, $\{g_{j,l}\}$ converges to an approximate triangular weighting scheme, but with a noticeable cusp at the top. Just as the shark's fin of the D(4) filters can introduce artifacts into an MRA, so can the shape of the C(6) filter. While this filter has better phase properties than the LA(8) filter, its triangular

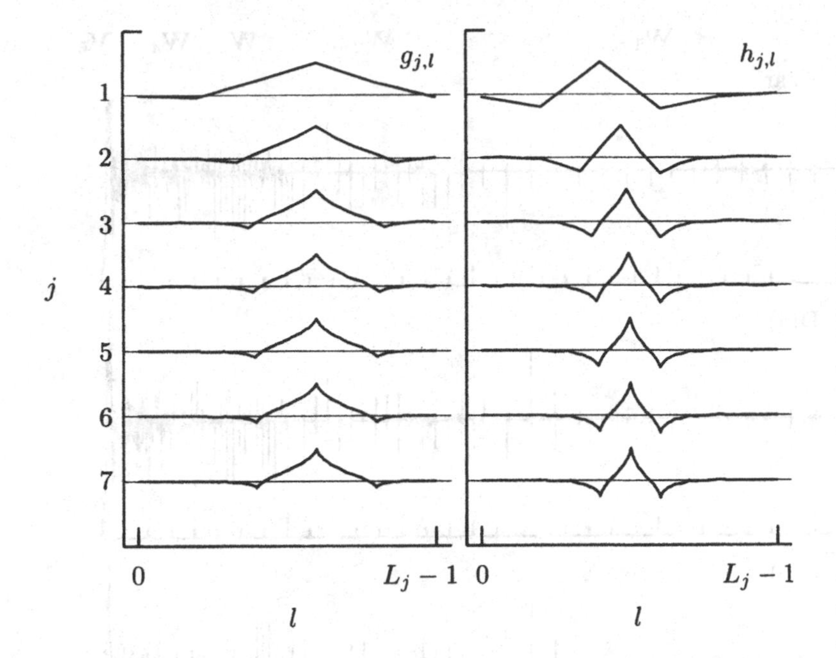

Figure 125. C(6) scaling $\{g_{j,l}\}$ and wavelet $\{h_{j,l}\}$ filters for scales indexed by $j = 1, 2, \ldots, 7$ (see Figures 98a and 98b for similar plots of the D(4) and LA(8) filters).

shape is not likely to be as good a match for the characteristic features in most time series as the smoother LA(8) shape. Thus the trade-off between the C(6) and LA(8) filters is one of good phase properties versus protection against artifacts (an additional consideration is that the C(6) filter uses two embedded differencing operations while the LA(8) filter uses four, which means that the gain function for the latter is a better approximation to an ideal band-pass filter in the sense of having reduced sidelobes outside the nominal pass-band).

4.10 Example: Electrocardiogram Data

As a first example of the use of the DWT on actual data, let us consider a time series $\mathbf{X}$ of electrocardiogram (ECG) measurements shown in the bottom panel of Figure 127. These data were measured during the normal sinus rhythm of a patient who occasionally experiences arhythmia. There are $N = 2048$ observations measured in units of millivolts and collected at a rate of 180 samples per second, so the sampling interval is $\Delta t = 1/180$ seconds, and the data cover an 11.37 second interval (merely for plotting purposes, we have taken the time of the first observation X_0 to be $t_0 =$ 0.31 seconds). This series is a good candidate for an MRA because its components are on different scales. For example, the large scale (low frequency) fluctuations – known as baseline drift – are due to patient respiration, while the prominent short scale (high frequency) intermittent fluctuations between 3 and 4 seconds are evidently due to patient movement. Neither of these fluctuations is directly related to the heart, but heart rhythm determines most of the remaining features in the series. The large spikes occurring about 0.7 seconds apart are the R waves of normal heart rhythm; the smaller – but sharp – peak coming just prior to an R wave is known as a P wave; and the broader peak that comes after the P and R waves is a T wave (one such PRT

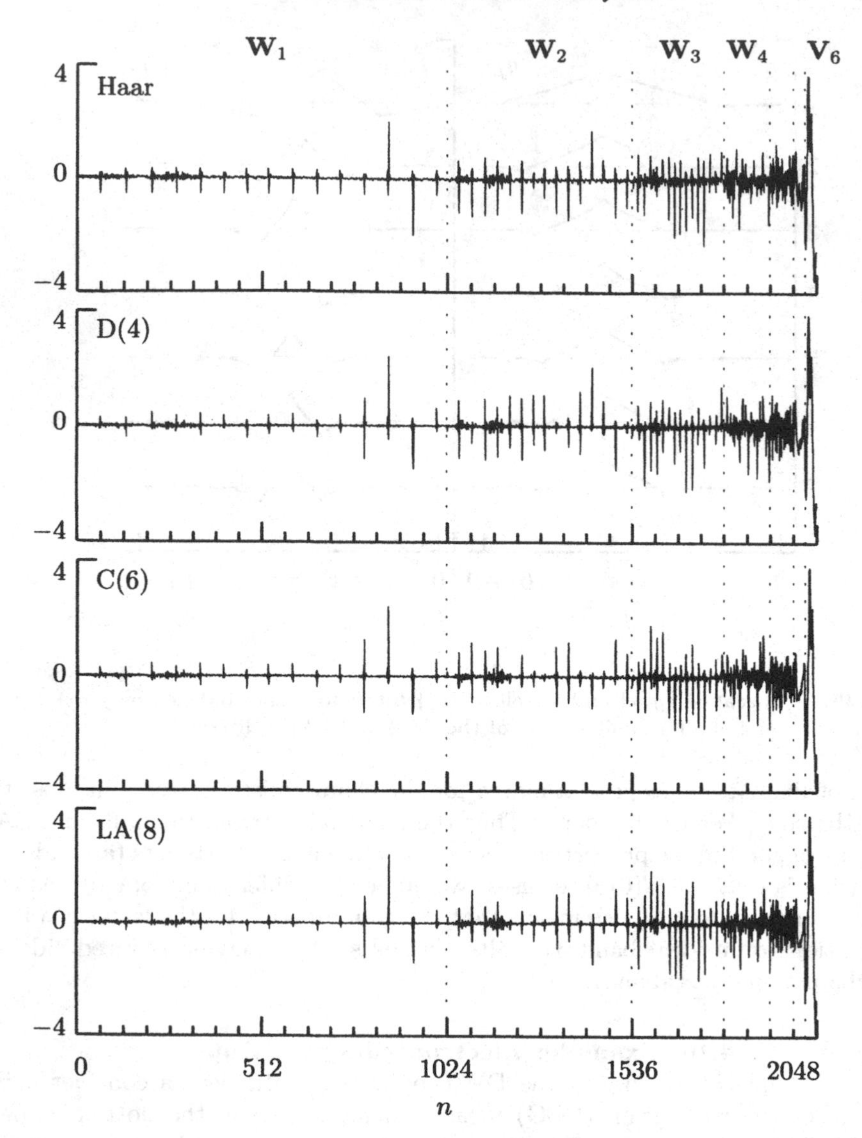

Figure 126. Partial DWT coefficients $\mathbf{W}$ of level $J_0 = 6$ for ECG time series using the Haar, D(4), C(6) and LA(8) wavelets. The elements W_n of $\mathbf{W}$ are plotted versus $n = 0, \ldots, N-1 = 2047$. The six vertical dotted lines delineate the seven subvectors of $\mathbf{W}$, namely, $\mathbf{W}_1, \ldots, \mathbf{W}_6$ and $\mathbf{V}_6$ ($\mathbf{W}_5$ and $\mathbf{W}_6$ are not labeled at the top of the figure due to lack of space).

complex is labeled on the bottom right-hand side of Figure 127).

Figure 126 shows the elements of the DWT coefficient vector $\mathbf{W}$ for $J_0 = 6$ partial DWTs based upon the Haar, D(4), C(6) and LA(8) wavelets. Vertical dotted lines delineate the subvectors $\mathbf{W}_1, \ldots, \mathbf{W}_6$ and $\mathbf{V}_6$. The number of coefficients in each subvector $\mathbf{W}_j$ of wavelet coefficients is $2048/2^j$, while there are $2048/2^6 = 32$ scaling coefficients in $\mathbf{V}_6$, for a total of 2048 coefficients in all. The sum of squares of all the elements in each $\mathbf{W}$ is equal to the sum of squares of the original ECG series. For each wavelet, the largest coefficients are at the end of $\mathbf{W}$, which corresponds to

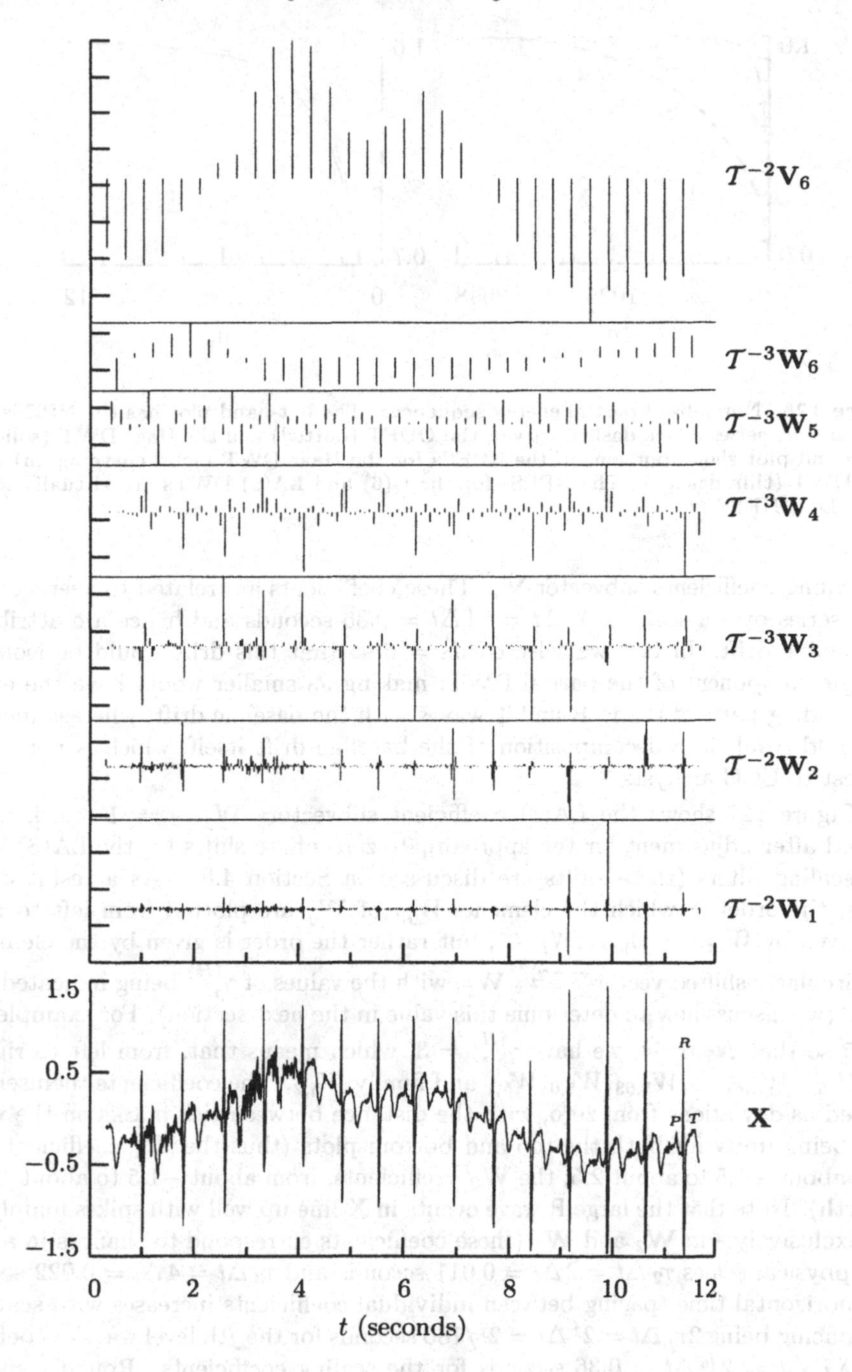

Figure 127. LA(8) DWT coefficients for ECG time series (data courtesy of Gust Bardy and Per Reinhall, University of Washington).

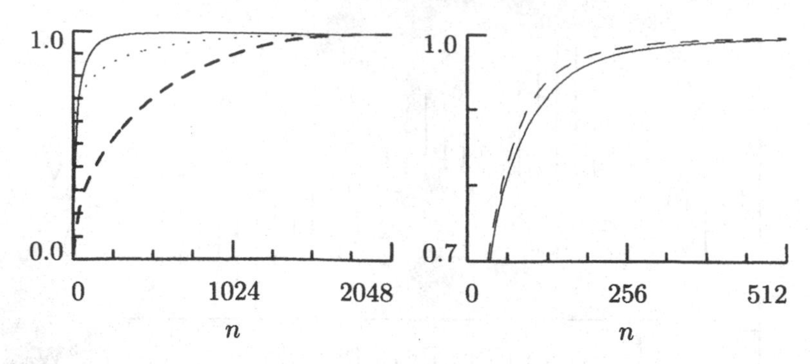

Figure 128. Normalized partial energy sequences. The left-hand plot has the NPESs for the original time series (thick dashed curve), the ODFT (dotted) and the Haar DWT (solid). The right-hand plot shows portions of the NPESs for the Haar DWT (solid curve again) and the D(4) DWT (thin dashed). The NPESs for the C(6) and LA(8) DWTs are virtually identical to the D(4) NPES.

the scaling coefficients subvector $\mathbf{V}_6$. These coefficients are related to averages of the ECG series over a scale of $\lambda_6 \, \Delta t = 64 \, \Delta t = 0.36$ seconds and hence are attributable to baseline drift. In fact we selected $J_0 = 6$ so that this drift would be isolated in a single component of the partial DWT: making J_0 smaller would have the effect of confounding parts of the P, R and T waves with the baseline drift, whereas increasing J_0 would result in a decomposition of the baseline drift itself, which is not of much interest to ECG analysts.

Figure 127 shows the LA(8) coefficient subvectors $\mathbf{W}_j$, $j = 1, \ldots, 6$, and $\mathbf{V}_6$ plotted after adjustment for the approximate zero phase shifts for the LA(8) wavelet and scaling filters (these shifts are discussed in Section 4.8). As a result of these shifts, the order in which the elements $W_{j,t}$ of $\mathbf{W}_j$ are plotted from left to right is not given by $W_{j,t}$, $t = 0, \ldots, N_j - 1$, but rather the order is given by the elements in the circularly shifted vector $\mathcal{T}^{-\gamma_j^{(H)}} \mathbf{W}_j$, with the values of $\gamma_j^{(H)}$ being indicated on the figure (we discuss how to determine this value in the next section). For example, when $j = 5$ so that $N_5 = 64$, we have $\gamma_5^{(H)} = 3$, which means that, from left to right, we plot $W_{5,3}, W_{5,4}, \ldots, W_{5,63}, W_{5,0}, W_{5,1}$ and finally $W_{5,2}$. The coefficients themselves are plotted as deviations from zero, with the distance between tick marks on the vertical axes being unity for both the top and bottom plots (thus the $\mathbf{W}_1$ coefficients range from about -1.5 to about 2.5; the $\mathbf{W}_2$ coefficients, from about -1.5 to about 1.5; and so forth). Note that the large R wave events in $\mathbf{X}$ line up well with spikes mainly – but not exclusively – in $\mathbf{W}_2$ and $\mathbf{W}_3$ (these coefficients correspond to changes in averages over physical scales $\tau_2 \, \Delta t = 2 \, \Delta t \doteq 0.011$ seconds and $\tau_3 \, \Delta t = 4 \, \Delta t \doteq 0.022$ seconds). The horizontal time spacing between individual coefficients increases with scale, with the spacing being $2\tau_j \, \Delta t = 2^j \, \Delta t = 2^j/180$ seconds for the jth level wavelet coefficients and $\lambda_{J_0} \, \Delta t = 2^{J_0} \, \Delta t = 0.36$ seconds for the scaling coefficients. Roughly speaking, the location of a particular coefficient plus or minus half the spacing between the coefficients indicates the portion of the ECG series that is summarized in some manner by the coefficients (either as differences of averages in the case of wavelet coefficients or as averages in the case of the scaling coefficients). Note that, because the scaling coefficients are proportional to averages over a scale of 0.36 seconds, their shape is roughly proportional to that of the baseline drift.

To study how well the various DWTs summarize the time series, Figure 128 shows plots of the normalized partial energy sequences (NPESs) for the original time series and for the wavelet transforms (along with the ODFT for the sake of comparison). For a sequence of real- or complex-valued variables $\{U_t : t = 0, \ldots, M-1\}$, an NPES is formed in the following manner. First, we form the squared magnitudes $|U_t|^2$, and then we order the squared magnitudes such that

$$|U_{(0)}|^2 \geq |U_{(1)}|^2 \geq \cdots \geq |U_{(M-2)}|^2 \geq |U_{(M-1)}|^2,$$

where $|U_{(0)}|^2$ represents the largest of the M squared magnitudes; $|U_{(1)}|^2$ is the next largest; and so forth until we come to $|U_{(M-1)}|^2$, which is the smallest observed squared magnitude. The NPES is defined as

$$C_n \equiv \frac{\sum_{u=0}^{n} |U_{(u)}|^2}{\sum_{u=0}^{M-1} |U_{(u)}|^2}, \qquad n = 0, 1, \ldots, M-1$$

(note that the denominator is equal to the sum of squared magnitudes of the original sequence $\{U_t\}$). By definition the NPES is a nondecreasing sequence such that $0 < C_n \leq 1$ for all n, with $C_{M-1} = 1$. If a particular orthonormal transform is capable of capturing the key features in a time series in a few coefficients, we would expect that C_n will become close to unity for relatively small n. The left-hand plot of Figure 128 indicates that, because of the combination of features in the ECG series, the Haar wavelet transform can summarize the energy in this series in a more succinct manner than either $\mathbf{X}$ itself or its ODFT. The right-hand plot indicates that the D(4) DWT is superior to the Haar DWT in summarizing the ECG series (NPESs for the C(6) and LA(8) DWTs are not shown, but they are virtually the same as the D(4) NPES). Evidently, the NPES for $\mathbf{X}$ itself increases slowly mainly due to the influence of the baseline drift; the NPES for the ODFT is better in that it can capture the global characteristics of the baseline drift in a few coefficients, but it cannot represent the localized P, R and T waves very well; and the NPESs for the partial DWTs demonstrate the ability of these transforms to capture succinctly both the baseline drift in the scaling coefficients and the localized P, R and T waves in the wavelet coefficients.

Figures 130, 131, 132 and 133 show multiresolution analyses of level $J_0 = 6$ for, respectively, the Haar, D(4), C(6) and LA(8) wavelets. The details $\mathcal{D}_j$ and the smooth $\mathcal{S}_6$ are plotted in a manner similar to that of Figure 127 in that the distance between adjacent tick marks on the vertical axes is unity in both plots on all four figures (however, in contrast to the wavelet coefficients, we do not have to apply any phase corrections, i.e., advances, to either $\mathcal{D}_j$ or $\mathcal{S}_6$). For all four wavelets, each detail $\mathcal{D}_j$ has a sample mean of zero, while the sample mean of each smooth $\mathcal{S}_6$ is equal to the sample mean of $\mathbf{X}$. The MRA for each wavelet satisfies the additive condition

$$\mathbf{X} = \sum_{j=1}^{6} \mathcal{D}_j + \mathcal{S}_6.$$

Qualitatively, the LA(8) MRA is visually more pleasing than the results from the other three wavelets: the Haar higher level details and smooth have a distinctive blocky look to them, whereas the D(4) and C(6) wavelets lead to some 'sharks' fins' and triangles

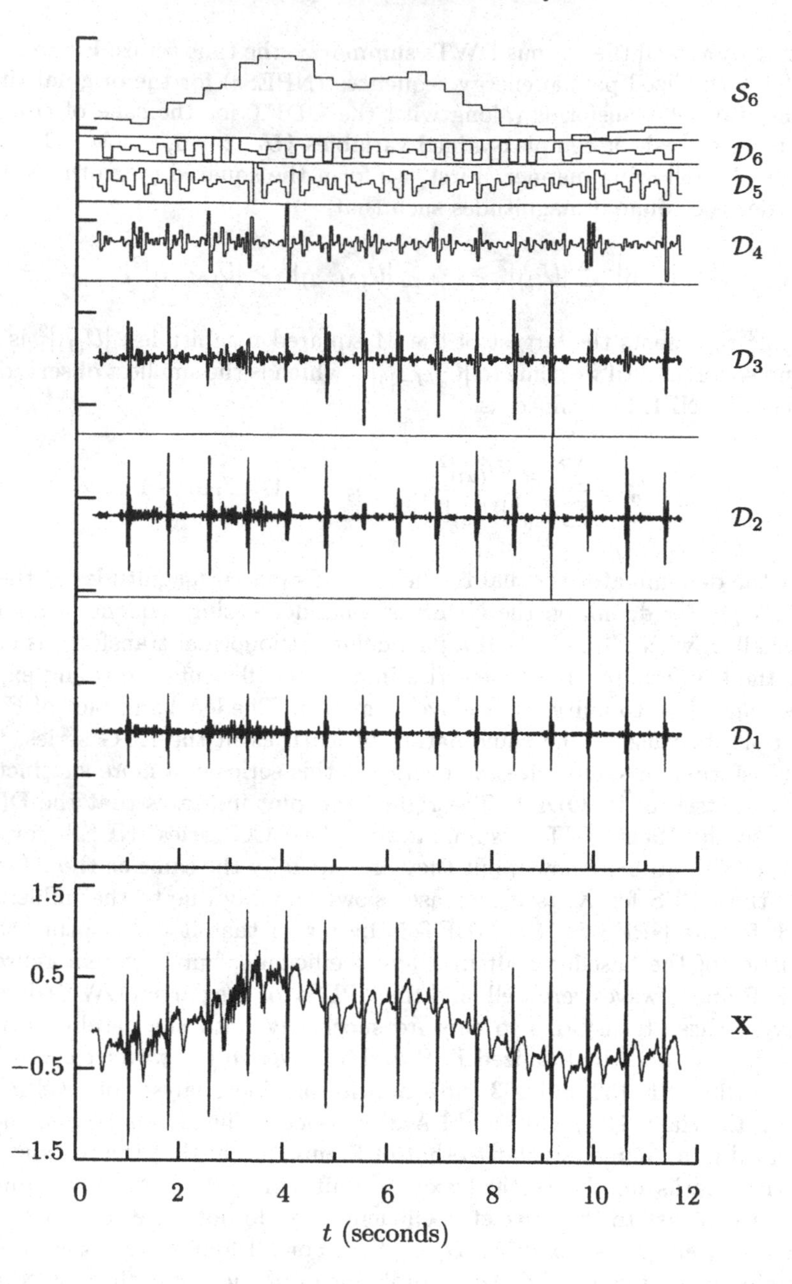

Figure 130. Haar DWT multiresolution analysis of ECG time series (see text for details).

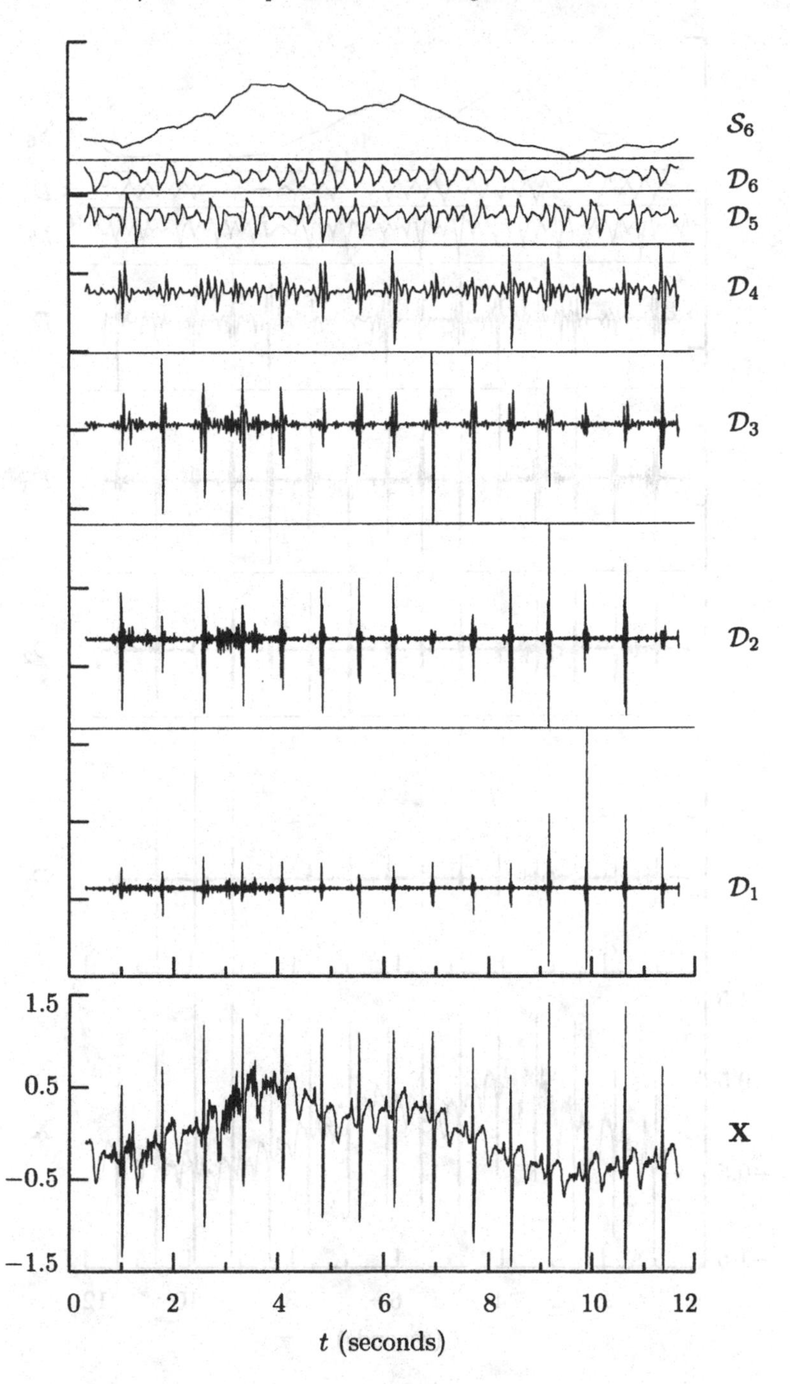

Figure 131. D(4) DWT multiresolution analysis of ECG time series (see text for details).

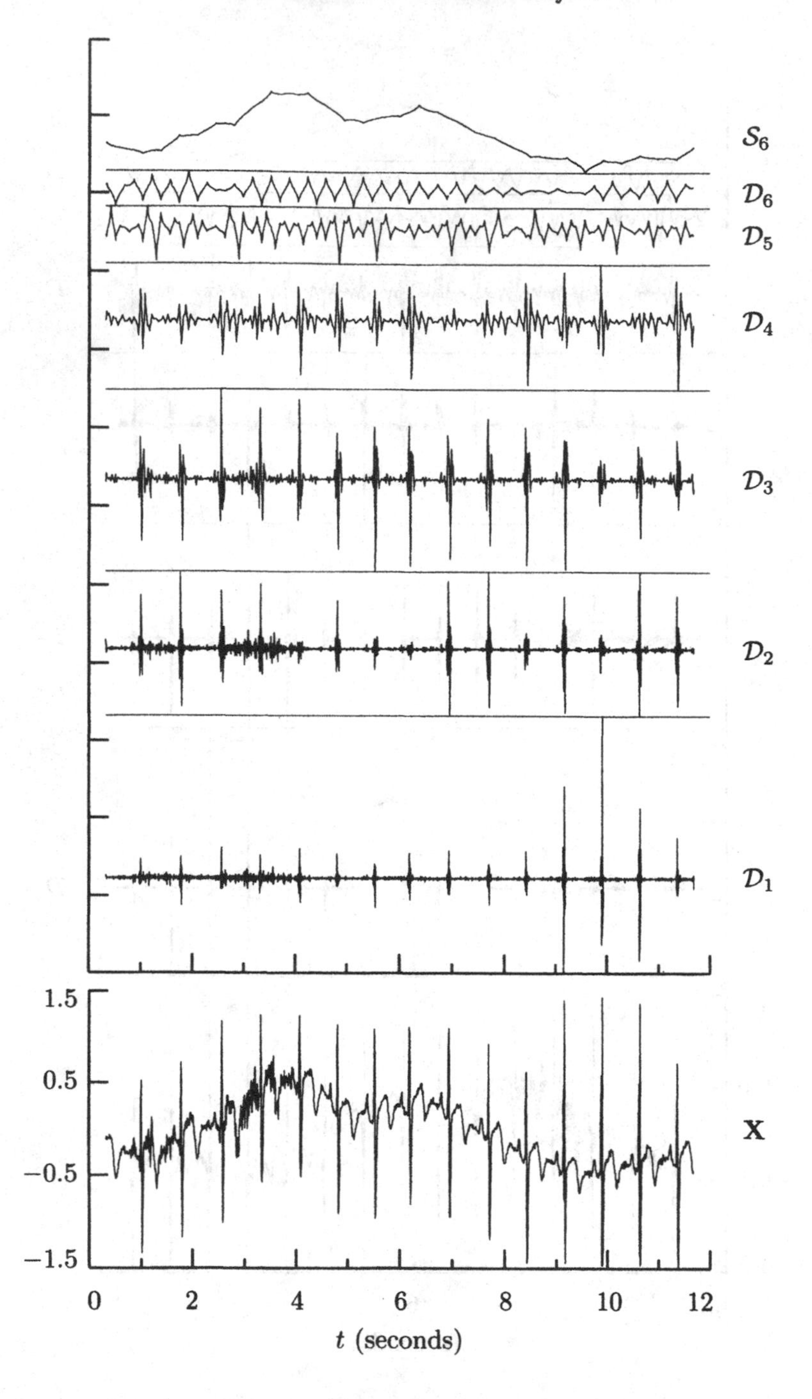

Figure 132. C(6) DWT multiresolution analysis of ECG time series (see text for details).

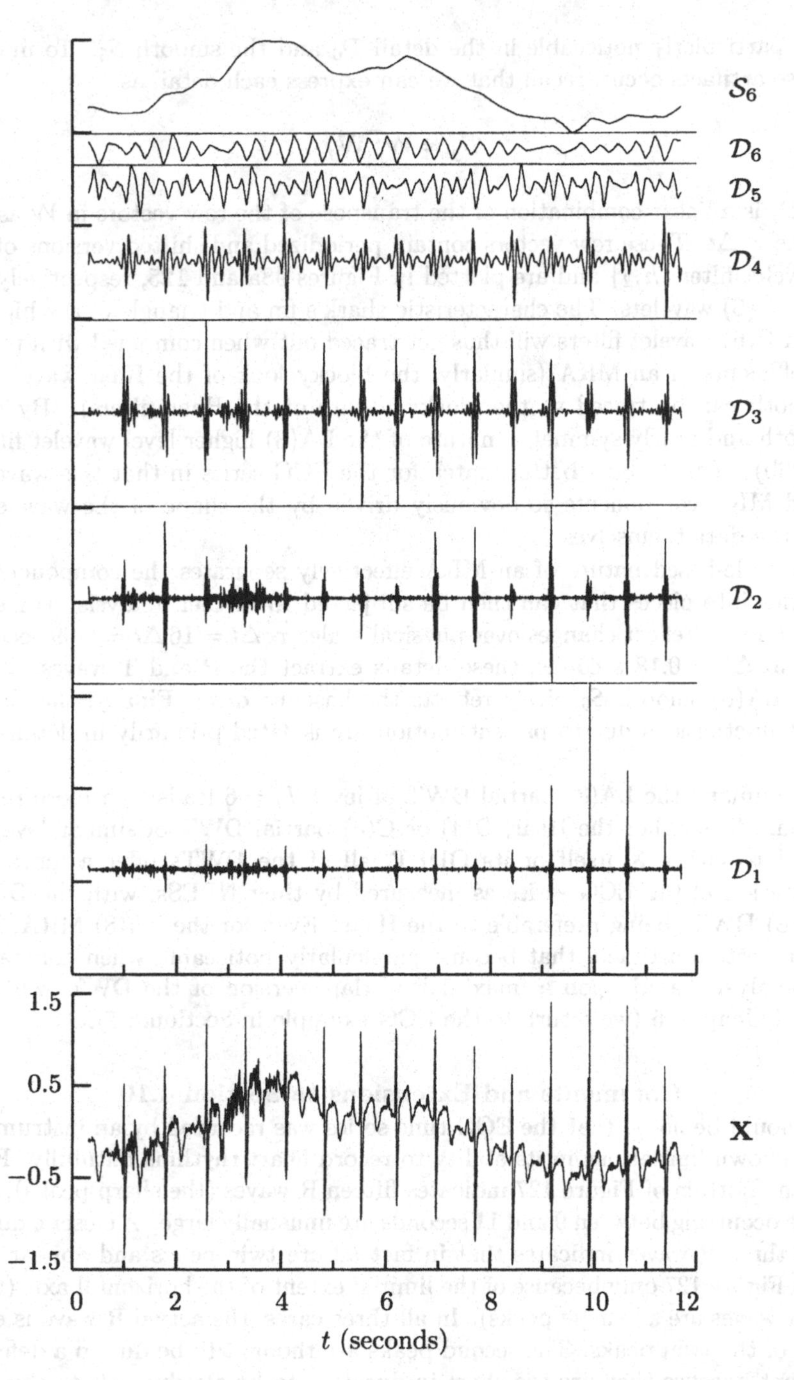

Figure 133. LA(8) DWT multiresolution analysis of ECG time series (see text for details).

that are particularly noticeable in the detail $\mathcal{D}_6$ and the smooth $\mathcal{S}_6$. To understand why these artifacts occur, recall that we can express each detail as

$$\mathcal{D}_j = \mathcal{W}_j^T \mathbf{W}_j,$$

so that $\mathcal{D}_j$ is a linear combination of the transpose of the row vectors in $\mathcal{W}$ associated with scale $\tau_j\,\Delta t$. These row vectors contain periodized and shifted versions of the jth level wavelet filter $\{h_{j,l}\}$ and are plotted in Figures 98a and 125, respectively, for the D(4) and C(6) wavelets. The characteristic shark's fin and triangle of the higher level D(4) and C(6) wavelet filters will thus get traced out when combined with prominent $W_{j,t}$ coefficients in an MRA (similarly, the blocky look of the Haar wavelet details and smooth can be traced to the blocky nature of the Haar filters). By contrast, the smooth and nearly symmetric nature of the LA(8) higher level wavelet filters (see Figure 98b) seems to be a better match for the ECG series in that this wavelet does not yield MRA components so obviously driven by the shape of the wavelet rather than by the data themselves.

The scale-based nature of an MRA effectively separates the components of the ECG series into pieces that can then be subjected to further analysis. For example, since $\mathcal{D}_5$ and $\mathcal{D}_6$ reflect changes over physical scales $\tau_5\,\Delta t = 16\,\Delta t \doteq 0.09$ seconds and $\tau_6\,\Delta t = 32\,\Delta t \doteq 0.18$ seconds, these details extract the P and T waves. Note also that the LA(8) smooth $\mathcal{S}_6$ nicely reflects the baseline drift. Finally, the short scale transient fluctuations due to patient motion are isolated primarily in details $\mathcal{D}_1$ and $\mathcal{D}_2$.

In summary, the LA(8) partial DWT of level $J_0 = 6$ leads to a more reasonable MRA than does either the Haar, D(4) or C(6) partial DWT of similar level. When compared to either $\mathbf{X}$ itself or its ODFT, all of the DWTs offer a more succinct representation of the ECG series as measured by their NPESs, with the D(4), C(6) and LA(8) DWTs being preferable to the Haar. Even for the LA(8) MRA, however, there are certain artifacts that become particularly noticeable when compared to a similar analysis based upon a 'maximal overlap' version of the DWT, which is the subject of Chapter 5 (we return to the ECG example in Section 5.7).

Comments and Extensions to Section 4.10

[1] It should be noted that the ECG time series was recorded by an instrument that has two known limitations in its ability to record heart rhythms faithfully. First, the plot at the bottom of Figure 127 indicates fifteen R waves (the sharp peaks), of which the three occurring between 9 and 11 seconds are unusually large. A closer examination of these three R waves indicates that in fact all are twin peaks and appear as single peaks in Figure 127 only because of the limited extent of the horizontal axis (the other twelve R waves are all single peaks). In all three cases, the actual R wave is evidently the first of the twin peaks. The second peaks are thought to be due to a defect in the instrument because they are too short in duration to be attributable to the patient's heart: the second peak consists of two similar outliers in the first case, and a single outlier in the other two cases. (It is interesting to note that, while all three twin peaks visibly influence the $\mathbf{W}_1$ component of the D(4), C(6) and LA(8) DWTs shown in Figure 126, only the latter two are evident in the Haar DWT. This is because the second peak in the first case consists of two similar outliers. The first (second) outlier has an even (odd) index, a pattern that makes this pair of outliers undetectable in the

Haar $\mathbf{W}_1$. These outliers would have been detected if the indexing had been odd/even instead of even/odd. In Chapter 5, we consider a variation on the DWT, part of the rationale for which is to offer us some protection against such alignment effects.)

Second, after the occurrence of each R wave, the ECG rapidly decreases into a sharp valley (the so-called 'S portion' of the 'QRS complex,' where Q and S are valleys that occur between the R wave and, respectively, the P and T waves – these waves are the peaks marked in the bottom right-hand of Figure 127). For technical reasons, data collected by the instrument are passed through a nonlinear filter that imposes a limit on how much the recorded ECG can change from one time point to the next. This filter makes it impossible to record a rapid decrease accurately. This defect in fact shows up in the $\mathbf{W}_1$ component of the Haar DWT in Figure 126, where there are fourteen valleys with almost identical magnitudes (the $\mathcal{D}_1$ component of the corresponding MRA in Figure 130 has a similar pathology). Thus, while we have argued that the Haar wavelet is not particularly well suited for analyzing ECG records, in fact it is useful for pointing out this subtle defect.

4.11 Practical Considerations

As the example in the previous section illustrated, there are a number of practical considerations that must be addressed in order to come up with a useful wavelet analysis of a time series. Here we comment upon some of the choices that must be made and some options that are available for handling boundary conditions and a time series whose length is not a power of two (see Bruce and Gao, 1996a, and Ogden, 1997, for additional discussions on practical considerations).

- *Choice of wavelet filter*

The first practical problem we face in constructing a wavelet analysis is how to choose a particular wavelet filter from amongst all those discussed in Sections 4.8 and 4.9. As is already apparent from the ECG example, a reasonable choice depends very much on the application at hand, so we will be reconsidering this problem as we present examples of wavelet analyses of actual time series in subsequent chapters. The choices we make there demonstrate the interplay between a specific analysis goal (such as isolation of transient events in a time series, signal estimation, estimation of parameters in a long memory process, testing for homogeneity of variance, estimation of the wavelet variance and so forth) and the properties we need in a wavelet filter to achieve that goal.

While studying the examples in the upcoming chapters is the best way to learn how to choose a wavelet filter (see, in particular, Sections 5.8, 5.9, 8.6, 8.9, 9.7 and 9.8), we can comment here on an overall strategy. Generally speaking, our choice is dictated by a desire to balance two considerations. On the one hand, wavelet filters of the very shortest widths (i.e., $L = 2, 4$ or 6) can sometimes introduce undesirable artifacts into the resulting analyses (while this was true in the ECG example for which Haar, D(4) and C(6) MRAs resulted in unrealistic blocks, sharks' fins and triangles (Figures 130, 131 and 132), it is also true that even the Haar wavelet is quite acceptable for certain applications – see Sections 8.6 and 9.6 for two examples). On the other hand, while wavelet filters with a large L can be a better match to the characteristic features in a time series, their use can result in (i) more coefficients being unduly influenced by boundary conditions, (ii) some decrease in the degree of localization of the DWT coefficients and (iii) an increase in computational burden. A reasonable overall strategy is thus to use the smallest L that gives reasonable results. In practice,

L	L'_1	L'_2	L'_3	L'_4	$L'_{j\geq5}$
2	0	0	0	0	0
4	1	2	2	2	2
6	2	3	4	4	4
8	3	5	6	6	6
10	4	6	7	8	8
12	5	8	9	10	10
14	6	9	11	12	12
16	7	11	13	14	14
18	8	12	14	15	16
20	9	14	16	17	18

Table 136. Number L'_j of boundary coefficients in $\mathbf{W}_j$ or $\mathbf{V}_j$ based on a wavelet filter of width L (here we assume $L'_j \leq N_j$, where $N_j = N/2^j$ is length of $\mathbf{W}_j$ or $\mathbf{V}_j$). The boundary coefficients are those that are influenced by boundary conditions at least to some degree.

this involves comparing a series of preliminary analyses in which we keep increasing L until we come to an analysis that is free of any artifacts attributable to the wavelet filter alone. When combined with a requirement that the resulting DWT coefficients be alignable in time (i.e., have phase shifts reasonably close to zero), this strategy often settles on the LA(8) wavelet filter as a good choice. (If alignment of wavelet coefficients is important, it is often best to use one of the LA wavelets or coiflets; however, if an LA wavelet is selected, it is better to pick one whose half width $L/2$ is an even integer – these have phase functions that are better approximations to linear phase, as indicated by Figure 115.)

- *Handling boundary conditions*

A partial or full DWT makes use of circular filtering, so a filtering operation near the beginning or end of the series (the 'boundaries') treats the time series $\mathbf{X}$ as if it were a portion of a periodic sequence with period N. With circular filtering we are essentially assuming that $X_{N-1}, X_{N-2}, \ldots$ are useful surrogates for the unobserved $X_{-1}, X_{-2}, \ldots$. This assumption is reasonable for some series if the sample size is chosen appropriately. For example, the subtidal sea level series discussed in Section 5.8 has a strong annual component, so it can be treated as a circular series with some validity if the number of samples covers close to an integer multiple of a year. For other series, circularity is a problematic assumption, particularly if there is a large discontinuity between X_{N-1} and X_0.

Because circularity can be a questionable assumption, we need to consider carefully exactly how it affects the wavelet analysis of a time series. To do so, let us first quantify the extent to which circularity influences the DWT coefficients and corresponding MRA. For the wavelet coefficients, we show in item [1] of the Comments and Extensions below that the number of 'boundary coefficients' in the N_j dimensional vector $\mathbf{W}_j$ (i.e., the coefficients that are affected to any degree by circularity) is given by $\min\{L'_j, N_j\}$, where L'_j is defined in Equation (146a) and listed in Table 136 for filter widths $L = 2, 4, \ldots, 20$ (a similar result holds for the scaling coefficients). As is apparent from this table, L'_j does not depend on N, increases as we increase L, is

L	$\bar{\gamma}_1^{(H)}, \gamma_1^{(H)}$	$\bar{\gamma}_2^{(H)}, \gamma_2^{(H)}$	$\bar{\gamma}_3^{(H)}, \gamma_3^{(H)}$	$\bar{\gamma}_4^{(H)}, \gamma_4^{(H)}$	$\bar{\gamma}_{j\geq 5}^{(H)}, \gamma_{j\geq 5}^{(H)}$
8	1,2	3,2	3,3	3,3	3,3
10	2,2	3,3	4,3	4,4	4,4
12	2,3	4,4	5,4	5,5	5,5
14	2,4	4,5	6,5	6,6	6,6
16	3,4	6,5	7,6	7,7	7,7
18	4,4	6,6	7,7	8,7	8,8
20	4,5	7,7	8,8	9,8	9,9

Table 137a. Number of LA boundary wavelet coefficients at the beginning and the end of $\mathcal{T}^{-\gamma_j^{(H)}}\mathbf{W}_j$ (assuming $L'_j \leq N_j$). The number at the beginning is given by $\bar{\gamma}_j^{(H)}$, and the number at the end, by $\gamma_j^{(H)}$. Note that $\bar{\gamma}_j^{(H)} + \gamma_j^{(H)} = L'_j$, which is the total number of boundary coefficients (see Table 136).

L	$\bar{\gamma}_1^{(G)}, \gamma_1^{(G)}$	$\bar{\gamma}_2^{(G)}, \gamma_2^{(G)}$	$\bar{\gamma}_3^{(G)}, \gamma_3^{(G)}$	$\bar{\gamma}_4^{(G)}, \gamma_4^{(G)}$	$\bar{\gamma}_{J_0\geq 5}^{(G)}, \gamma_{J_0\geq 5}^{(G)}$
8	2,1	3,2	4,2	4,2	4,2
10	2,2	3,3	3,4	4,4	4,4
12	3,2	5,3	5,4	6,4	6,4
14	4,2	6,3	7,4	8,4	8,4
16	4,3	6,5	7,6	8,6	8,6
18	4,4	6,6	7,7	7,8	8,8
20	5,4	8,6	9,7	9,8	10,8

Table 137b. As in Table 137a, but now for the LA scaling coefficients. Again we have $\bar{\gamma}_{J_0}^{(G)} + \gamma_{J_0}^{(G)} = L'_{J_0}$, where L'_{J_0} is given in Table 136.

nondecreasing as we increase j, and satisfies the inequality

$$\frac{L}{2} - 1 \leq L'_j \leq L - 2$$

(L'_j equals the lower bound when $j = 1$ and attains the upper bound for large enough j). When N_j grows as the result of making N larger, the proportion L'_j/N_j of boundary coefficients becomes arbitrarily small; on the other hand, for fixed N, this proportion increases as j increases (with the exception of the Haar wavelet ($L = 2$), which yields a DWT free of the circularity assumption). In addition, when $0 < L'_j \leq N_j$, the boundary coefficients in $\mathbf{W}_j$ or $\mathbf{V}_j$ are the very first L'_j elements in each vector.

Let us now concentrate on the LA and coiflet filters and (assuming $L'_j \leq N_j$) see what happens to the L'_j boundary wavelet coefficients when we follow the prescription described in Section 4.8 and apply circular advances to $\mathbf{W}_j$ for the purposes of plotting its elements $W_{j,t}$ versus actual time. We show in item [2] of the Comments and Extensions below that the order in which the coefficients are plotted corresponds to

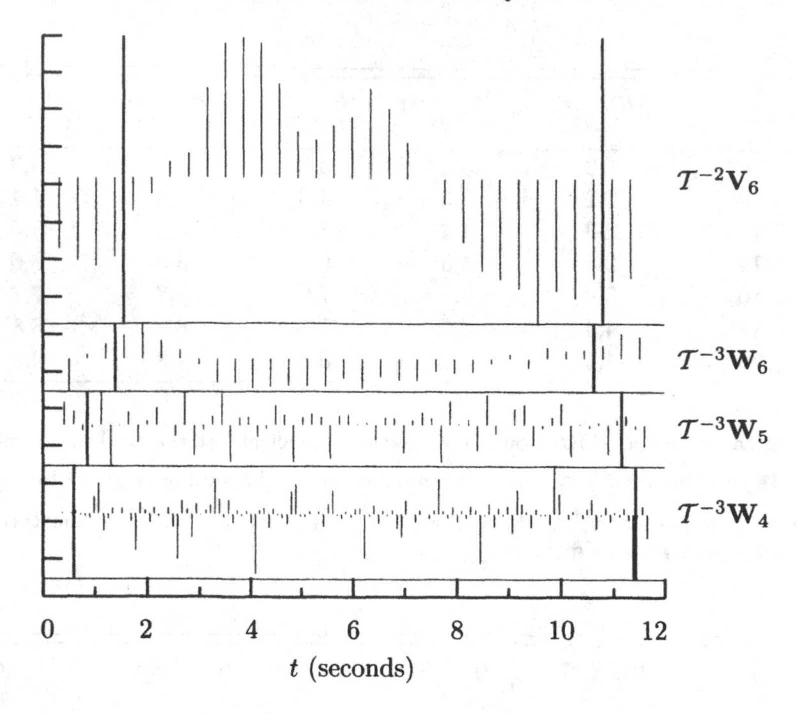

Figure 138. Four circularly advanced subvectors of the LA(8) DWT for the ECG time series (cf. Figure 127). The DWT coefficients plotted between the two thick vertical lines for a given subvector are unaffected by circularity, while those outside of the lines are the boundary coefficients. The number of plotted boundary coefficients agrees with the first rows of Tables 137a and 137b.

the ordering of the elements in $\mathcal{T}^{-\gamma_j^{(H)}}\mathbf{W}_j$, where $0 < \gamma_j^{(H)} < L'_j$. This means that $\gamma_j^{(H)}$ of the L'_j boundary coefficients are circularly shifted around so that they appear at the end of the plot, while the remaining $\tilde{\gamma}_j^{(H)} \equiv L'_j - \gamma_j^{(H)}$ boundary coefficients appear at the beginning. The values for $\tilde{\gamma}_j^{(H)}$ and $\gamma_j^{(H)}$ are listed in Table 137a for all LA filter widths L discussed in this book (similar quantities $\tilde{\gamma}_{J_0}^{(G)}$ and $\gamma_{J_0}^{(G)}$ can be defined for the scaling coefficients – these are given in Table 137b). Exercise [4.27] is to create similar tables for the coiflet filters.

A minimal way of dealing with the circularity assumption is to indicate on plots of the DWT coefficients exactly which values are the boundary coefficients. As an example, Figure 138 reproduces the LA(8) wavelet coefficients indexed by $j = 4, 5$ and 6 from Figure 127 for the ECG series (the scaling coefficients for $J_0 = 6$ are also shown). We have added thick vertical lines that delineate the portions of the plotted values outside of which circularity has some effect. For example, there are three boundary coefficients at the beginning and end of $\mathcal{T}^{-3}\mathbf{W}_6$, which agrees with the right-most entry in the first row of Table 137a; likewise, there are four boundary coefficients at the beginning and two at the end of $\mathcal{T}^{-2}\mathbf{V}_6$, in accordance with Table 137b.

Let us next consider how circularity influences the detail $\mathcal{D}_j$ (here we no longer restrict ourselves to just the LA and coiflet filters). We show in item [3] of the Comments and Extensions below that, except for $\mathcal{D}_j$ generated by the Haar wavelet, there are some elements at both the beginning and end of $\mathcal{D}_j$ that are affected in some manner by circularity. The indices of these elements are given by

$$t = 0, \ldots, 2^j L'_j - 1 \text{ along with } t = N - (L_j - 2^j), \ldots, N - 1; \qquad (139)$$

i.e, the first $2^j L'_j$ elements of $\mathcal{D}_j$ are affected by circularity, as are the final $L_j - 2^j$ (similar results hold for the smooth $\mathcal{S}_{J_0}$). Again we see that the number of boundary terms in $\mathcal{D}_j$ depends on L and j but not N, with more elements being affected as we increase L or j. When $j = 1$ so that $L'_1 = \frac{L}{2} - 1$ and $L_1 = L$, the lower and upper ranges become

$$t = 0, \ldots, (L-2) - 1 \text{ and } t = N - (L-2), \ldots, N-1,$$

so there are $L-2$ indices at both the beginning and end of $\mathcal{D}_j$ influenced by circularity. When j is large enough so that L'_j attains its maximum value of $L - 2$, we have $2^j L'_j = L_j - 2^j + L - 2$, so there are $L - 2$ more boundary elements at the beginning than at the end.

As an example, consider $\mathcal{D}_1$ of Equation (81c) for which $j = 1$ and $L = 4$. We saw above that $W_{1,0}$ is a boundary coefficient, so the first two elements of $\mathcal{D}_1$ will be affected through the matrix multiplication by $\mathcal{W}_1^T$. This is in agreement with Equation (139) since we can use Table 136 to obtain $2^j L'_j = 2L'_1 = 2$. In addition, the last two rows of $\mathcal{W}_1^T$ involve circularization also, so the last two elements of $\mathcal{D}_1$ will also be affected. Again, this agrees with Equation (139) because $L_j - 2^j = L - 2 = 2$.

As we did with the DWT coefficients, we can place vertical lines on plots of $\mathcal{D}_j$ and $\mathcal{S}_{J_0}$ to delineate the boundary regions. This is illustrated in Figure 140 (a reproduction of part of Figure 133). A comparison of the vertical markers on this figure with those in Figure 138 indicates that the proportion of affected terms is greater in the detail $\mathcal{D}_j$ than in the corresponding $\mathbf{W}_j$, as the following exercise quantifies.

▷ **Exercise [139]:** Show that, for large j and N, the ratio of the proportion of affected terms in $\mathcal{D}_j$ to the proportion in $\mathbf{W}_j$ is approximately two. ◁

A similar result holds for $\mathcal{S}_{J_0}$ and $\mathbf{V}_{J_0}$.

It should be emphasized that, even though we can now identify exactly which terms in $\mathbf{W}_j$, $\mathbf{V}_{J_0}$, $\mathcal{D}_j$ and $\mathcal{S}_{J_0}$ are affected to any degree by circularity, in fact the influence of circularity can be quite small, particularly when the discrepancy between the beginning and end of $\mathbf{X}$ is not too large. As an example, in the MRA depicted in Figure 140, the smooth $\mathcal{S}_6$ should reflect averages of the ECG series over a scale of $2^6 \Delta t = 0.36$ seconds. A comparison of this series and $\mathcal{S}_6$ before and after the vertical markers indicates that $\mathcal{S}_6$ follows the ECG series rather faithfully over 'eyeball' averages covering about a third of a second. The effect of circularity is evidently quite minimal over the bulk of the boundary regions. We need to realize that the placement of the vertical markers is largely dictated by the width L_j of $\{g_{j,l}\}$, but in fact this equivalent filter has many small terms (see, for example, Figure 98b). Unless there is a large difference between X_0 and X_{N-1}, the regions that are seriously influenced by circularity will be smaller than what the vertical markers indicate. The marked boundary regions should thus be regarded as a quite conservative measure of the influence of circularity.

Now that we have seen how to determine what portions of a wavelet analysis are influenced by boundary conditions, let us now consider ways of reducing the impact

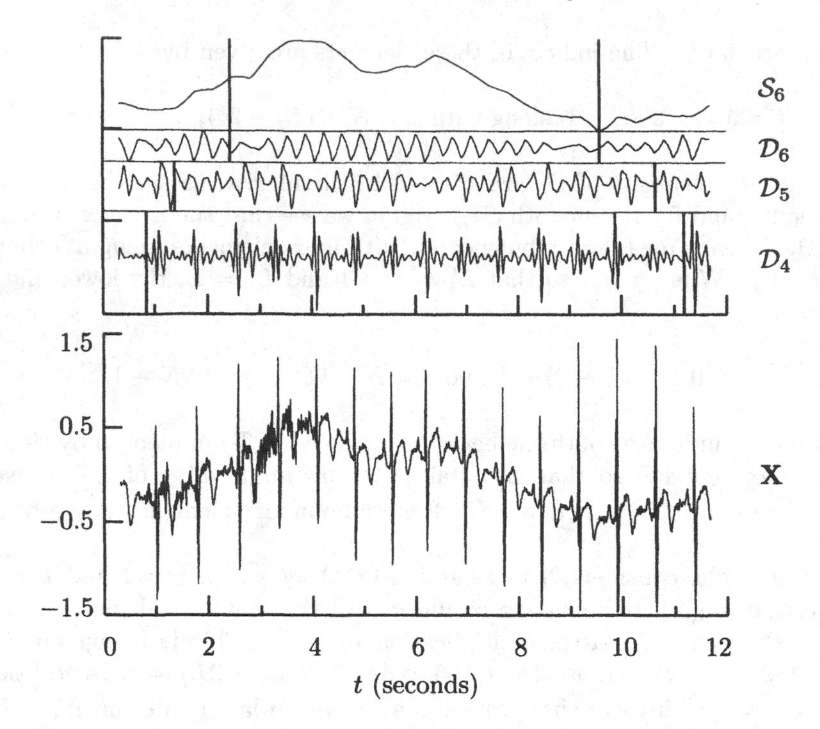

Figure 140. Portion of the LA(8) DWT multiresolution analysis for the ECG time series (the full analysis is shown in Figure 133). The thick vertical lines delineate the boundary regions in the details $\mathcal{D}_j$ and smooth $\mathcal{S}_6$ (i.e., those parts influenced to some degree by circularity).

of circularity. Because the discrete Fourier transform also treats a time series as if it were circular, we can make use of some of the techniques that have been successful in Fourier analysis. One simple – but very effective – technique is to replace **X** by a series of length $2N$ that we construct by appending a reversed version of **X** onto the end of **X**; i.e., instead of analyzing **X**, we analyze

$$X_0, X_1, \ldots, X_{N-2}, X_{N-1}, X_{N-1}, X_{N-2}, \ldots, X_1, X_0. \tag{140}$$

Note that this series of length $2N$ has the same sample mean and variance as **X**. Circular filtering of **X** itself in effect uses $X_{N-1}, X_{N-2}, \ldots$ as surrogates for X_{-1}, $X_{-2}, \ldots$; with the constructed series, $X_0, X_1, \ldots$ are now the surrogates for X_{-1}, $X_{-2}, \ldots$; i.e., we have eliminated effects due to a serious mismatch between X_0 and X_{N-1}. When **X** itself is subject to a DWT, we refer to an 'analysis of **X** using circular boundary conditions;' when we use the $2N$ series in place of **X**, we refer to an 'analysis of **X** using reflection boundary conditions.' The price we pay in using reflection boundary conditions is an increase in computer time and memory, but this is often quite acceptable.

As an example, Figure 142 shows an LA(8) MRA of the ECG time series using reflection boundary conditions. Here we computed the details and smooth based upon a series of length $N = 4096$ values constructed from the original ECG series of $N = 2048$ values as per Equation (140). We then truncated back the resulting details and smooth to $N = 2048$ values to create Figure 142, which should be compared with Figure 133 (computed using circular boundary conditions). There is virtually no difference in the two analyses except at the very beginnings and ends of the details and smooth, where use of reflection boundary conditions reduces some small blips due to a mismatch between the beginning and end of $\mathbf{X}$. (A comparison of the top part of Figure 142 with Figure 140 shows that – except at the very extremes – there is no visible change in the details and smooth over the boundary region. This reemphasizes the point we made earlier that the substantive influence of circularity can be restricted to a considerably smaller region than the actual boundary region.)

There are several other ways to cope with circularity, including a polynomial extrapolation of the time series at both ends, specially designed 'boundary wavelets,' and a variation on the reflection scheme in which a series is differenced, extended as in Equation (140) and then cumulatively summed (first differences are useful for a time series with a pronounced linear trend; second differences, with a quadratic trend; and so forth). For details on these procedures, see Chapter 14 of Bruce and Gao (1996a), Taswell and McGill (1994), Cohen *et al.* (1993) and Greenhall *et al.* (1999).

- *Handling sample sizes that are not a power of two*

The 'full' DWT which we described in the early sections of this chapter is designed to work with a time series whose sample size N is a power of two, i.e., $N = 2^J$ for some positive integer J. We noted in Section 4.7 that use of a partial DWT of level J_0 reduces this sample size restriction somewhat in that we now just need N to be an integer multiple of 2^{J_0}. When N is not a multiple of 2^{J_0}, it is sometimes possible to use one of the three simple methods presented below to obtain a DWT-like transform valid for general N. While these methods are acceptable for many applications, they are admittedly somewhat *ad hoc*. We note here that a reasonable alternative to the DWT that works for all N is the 'maximal overlap' discrete wavelet transform (MODWT) presented in Chapter 5. While the MODWT is not an orthonormal transform, some of its properties are similar to such a transform (e.g., it yields an exact ANOVA and MRA), and it is arguably superior to the DWT in certain aspects (e.g., its MRA is associated with zero phase filters, thus ensuring an alignment between $\mathbf{X}$ and its MRA that need not hold for a DWT-based MRA).

Suppose now that we want to compute a partial DWT of level J_0 for a time series whose sample size N is not a multiple of 2^{J_0}. Let N' be the smallest integer that is greater than N and is an integer multiple of 2^{J_0}. The first method is to 'pad' $\{X_t\}$ in some manner to create a new time series, say $\{X'_t\}$, of length $N' > N$ and then to take the DWT of the padded series (this trick is commonly used in Fourier analysis, where padding a time series up to a power of two is routinely used in conjunction with standard fast Fourier transform algorithms). One natural definition for X'_t is to set the padded values equal to the sample mean $\overline{X}$ of $\{X_t\}$; i.e., we take

$$X'_t \equiv \begin{cases} X_t, & t = 0, \ldots, N-1; \\ \overline{X}, & t = N, \ldots, N'-1. \end{cases}$$

Because the sample mean of $\{X'_t\}$ is also equal to $\overline{X}$, the sample variances $\hat{\sigma}^2_X$ and $\hat{\sigma}^2_{X'}$ of, respectively, $\{X_t\}$ and $\{X'_t\}$ are related in a simple manner:

$$\hat{\sigma}^2_{X'} \equiv \frac{1}{N'} \sum_{t=0}^{N'-1} (X'_t - \overline{X})^2 = \frac{1}{N'} \sum_{t=0}^{N-1} (X_t - \overline{X})^2 = \frac{N}{N'} \hat{\sigma}^2_X.$$

If we let $\mathbf{W}'_j$ and $\mathbf{V}'_{J_0}$ represent the subvectors of the DWT for $\{X'_t\}$, we can use the above in conjunction with Equation (104c) to obtain an ANOVA for $\{X_t\}$ based upon

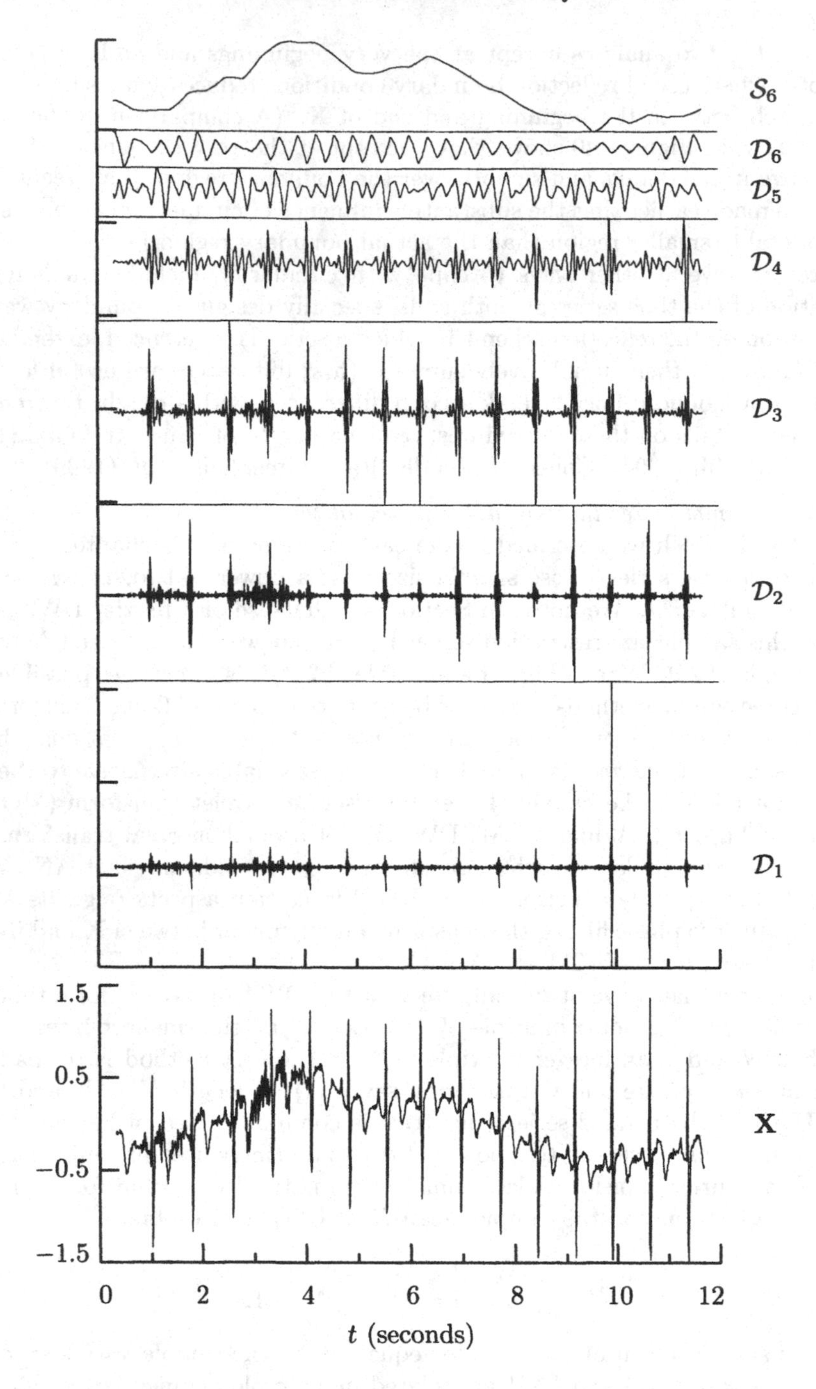

Figure 142. LA(8) DWT multiresolution analysis of ECG time series using reflection boundary conditions (cf. Figure 133).

the ANOVA for $\{X'_t\}$:

$$\hat{\sigma}_X^2 = \frac{1}{N}\sum_{j=1}^{J_0} \|\mathbf{W}'_j\|^2 + \frac{1}{N}\|\mathbf{V}'_{J_0}\|^2 - \frac{N'}{N}\overline{X}^2.$$

In the above we would interpret $\frac{1}{N}\|\mathbf{W}'_j\|^2$ as the contribution to the sample variance $\hat{\sigma}_X^2$ due to changes on scale τ_j. Letting $\mathbf{X}'$ be the vector with elements $\{X'_t\}$, we can use the $\mathbf{W}'_j$ and $\mathbf{V}'_{J_0}$ to form the MRA

$$\mathbf{X}' = \sum_{j=1}^{J_0} \mathcal{D}'_j + \mathcal{S}'_{J_0}$$

in the usual manner. Because this additive decomposition is valid on a pointwise basis, we can obtain an MRA of $\mathbf{X}$ by just using the first N elements of the $\mathcal{D}'_j$ and $\mathcal{S}'_{J_0}$ vectors:

$$\mathbf{X} = \sum_{j=1}^{J_0} I_{N,N'}\mathcal{D}'_j + I_{N,N'}\mathcal{S}'_{J_0}, \tag{143}$$

where $I_{N,N'}$ is an $N \times N'$ matrix whose (i,j)th element is either unity when $i = j$ or zero when $i \neq j$. (It is also possible to pad $\{X_t\}$ with values other than the sample mean. If, for example, we were to pad using a linear extrapolation of the time series, we could still obtain an MRA using Equation (143), but the procedure for obtaining an ANOVA outlined above would no longer work in general.)

A second method for dealing with general N is to truncate the time series by at most $2^{J_0} - 1$ observations to obtain shorter series whose lengths are a multiple of 2^{J_0} and hence amenable to the usual DWT. If we let N'' be the largest integer that is smaller than N and is an integer multiple of 2^{J_0}, we can define two shortened series

$$\mathbf{X}^{(1)} \equiv [X_0, \ldots, X_{N''-1}]^T \text{ and } \mathbf{X}^{(2)} \equiv [X_{N-N''}, \ldots, X_{N-1}]^T,$$

both of length N''. Letting $\mathbf{W}_j^{(l)}$ and $\mathbf{V}_{J_0}^{(l)}$ represent the subvectors of the DWT for $\mathbf{X}^{(l)}$, we can combine the two analyses to achieve an inexact ANOVA – but an exact MRA – as follows. Since we have both

$$\sum_{t=0}^{N''-1} X_t^2 = \sum_{j=1}^{J_0} \|\mathbf{W}_j^{(1)}\|^2 + \|\mathbf{V}_{J_0}^{(1)}\|^2$$

and

$$\sum_{t=N-N''}^{N-1} X_t^2 = \sum_{j=1}^{J_0} \|\mathbf{W}_j^{(2)}\|^2 + \|\mathbf{V}_{J_0}^{(2)}\|^2,$$

we can average the above to obtain

$$\frac{1}{2}\sum_{t=0}^{N-N''-1} X_t^2 + \sum_{t=N-N''}^{N''-1} X_t^2 + \frac{1}{2}\sum_{t=N''}^{N-1} X_t^2$$
$$= \sum_{j=1}^{J_0} \frac{1}{2}\left(\|\mathbf{W}_j^{(1)}\|^2 + \|\mathbf{W}_j^{(2)}\|^2\right) + \frac{1}{2}\left(\|\mathbf{V}_{J_0}^{(1)}\|^2 + \|\mathbf{V}_{J_0}^{(2)}\|^2\right).$$

This yields a scale-based decomposition of a weighted sum of squares, with half weights being attached to the $N - N''$ points at the beginning and end of the series. To obtain a corresponding MRA, we form

$$\mathbf{X}^{(l)} = \sum_{j=1}^{J_0} \mathcal{D}_j^{(l)} + \mathcal{S}_{J_0}^{(l)}, \quad l = 1 \text{ and } 2,$$

and then combine these by keeping the pointwise decompositions for $X_0, \ldots, X_{N-N''-1}$ (provided by the MRA of $\mathbf{X}^{(1)}$) and also for $X_{N''}, \ldots, X_{N-1}$ (provided by the MRA of $\mathbf{X}^{(2)}$) and by averaging the two pointwise decompositions for $X_{N-N''}, \ldots, X_{N''-1}$. Mathematically, we can write the resulting MRA as

$$\mathbf{X} = \sum_{l=1}^{2} \left(\sum_{j=1}^{J_0} I_{N,N''}^{(l)} \mathcal{D}_j^{(l)} + I_{N,N''}^{(l)} \mathcal{S}_{J_0}^{(l)} \right),$$

where the $I_{N,N''}^{(l)}$ matrices are $N \times N''$ with elements that are best deduced by studying an example, say, $N = 6$ and $N'' = 4$:

$$I_{6,4}^{(1)} = \begin{bmatrix} 1 & 0 & 0 & 0 \\ 0 & 1 & 0 & 0 \\ 0 & 0 & \frac{1}{2} & 0 \\ 0 & 0 & 0 & \frac{1}{2} \\ 0 & 0 & 0 & 0 \\ 0 & 0 & 0 & 0 \end{bmatrix} \quad \text{while} \quad I_{6,4}^{(2)} = \begin{bmatrix} 0 & 0 & 0 & 0 \\ 0 & 0 & 0 & 0 \\ \frac{1}{2} & 0 & 0 & 0 \\ 0 & \frac{1}{2} & 0 & 0 \\ 0 & 0 & 1 & 0 \\ 0 & 0 & 0 & 1 \end{bmatrix}.$$

(A variation on the above scheme would be to combine the $N - N'' + 1$ analyses for $[X_l, \ldots, X_{N''-1+l}]^T$, $l = 0, \ldots, N - N''$, which is quite close in spirit to the notion of 'cycle spinning' discussed in Coifman and Donoho, 1995.)

A third method is to reformulate the basic step of the pyramid algorithm to preserve at most one more scaling coefficient at each level j. To see how this works, suppose that we have $V_{j-1,t}$, $t = 0, \ldots, M - 1$, where M is odd. We would then apply the usual pyramid algorithm to just $V_{j-1,t}$, $t = 1, \ldots, M - 1$ to obtain the $M - 1$ coefficients $V_{j,t}$ and $W_{j,t}$, $t = 0, \ldots, \frac{M-1}{2} - 1$. The sum of squares decomposition then takes the form

$$\sum_{t=0}^{M-1} V_{j-1,t}^2 = \sum_{t=0}^{\frac{M-1}{2}-1} V_{j,t}^2 + \sum_{t=0}^{\frac{M-1}{2}-1} W_{j,t}^2 + V_{j-1,0}^2;$$

i.e., the first piece $V_{j-1,0}^2$ is left alone and does not enter into the next level. This scheme leads to a valid partitioning of $\sum X_t^2$ in terms of sums of squares of wavelet and scaling coefficients (however, the scaling coefficients will now no longer be on a single scale as they are with the usual DWT: for a J_0 level transform, there can be at most one scaling coefficient at each level from $j = 1$ to $J_0 - 1$). The inverse algorithm reconstructs $\{V_{j-1,t}\}$ using $\{W_{j,t}\}$, $\{V_{j,t}\}$ and the preserved $V_{j-1,0}$. We define the detail $\mathcal{D}_j$ in a manner analogous to the usual procedure: we apply the inverse transform to $\{W_{j,t}\}$, with $\{V_{j,t}\}$, $\{W_{j-1,t}\}$, $\{W_{j-2,t}\}$, $\{W_{1,t}\}$ all set to zero (any preserved scaling coefficients at lower levels are also set to zero). In a similar manner, we define the smooth $\mathcal{S}_{J_0}$ by what we get by inverse transforming with all the lower order wavelet coefficients set to zero (in this case all of the preserved scaling coefficients are retained).

As an example, let us consider a partial DWT of level $J_0 = 3$ for a time series $X_0, \ldots, X_{36}$ of length $N = 37$. The first level of the transform would preserve X_0 and decompose $X_1, \ldots, X_{36}$ into $W_{1,t}$ and $V_{1,t}$, $t = 0, \ldots, 17$. Since $\{V_{1,t}\}$ is of even length, there is no need to preserve any of these first level scaling coefficients, so we just decompose $\{V_{1,t}\}$ into $W_{2,t}$ and $V_{2,t}$, $t = 0, \ldots, 8$. Since $\{V_{2,t}\}$ has nine values, we preserve $V_{2,0}$ and decompose $V_{2,1}, \ldots, V_{2,8}$ into $W_{3,t}$ and $V_{3,t}$, $t = 0, \ldots, 3$. The sum of squares decomposition thus takes the form

$$\sum_{t=0}^{36} X_t^2 = \sum_{t=0}^{17} W_{1,t}^2 + \sum_{t=0}^{8} W_{2,t}^2 + \sum_{t=0}^{3} W_{3,t}^2 + \sum_{t=0}^{3} V_{3,t}^2 + V_{2,0}^2 + X_0^2.$$

To obtain, for example, $\mathcal{D}_1$, we inverse transform from level one using $W_{1,t}$, $t = 0, \ldots, 17$, with $V_{1,t}$, $t = 0, \ldots, 17$ and X_0 all being replaced by zeros; likewise, to obtain $\mathcal{S}_1$, we inverse transform from level one using $V_{1,t}$, $t = 0, \ldots, 17$ and X_0, with $W_{1,t}$, $t = 0, \ldots, 17$ all being replaced by zeros. This resulting MRA satisfies $\mathbf{X} = \mathcal{D}_1 + \mathcal{S}_1$. (Higher level MRAs can be obtained in an analogous manner.)

- *Choice of level J_0 of partial DWT*

As was true for the choice of the wavelet filter, a reasonable selection of the level J_0 for a partial DWT must take into account the application at hand. In the ECG example, we noted that setting $J_0 = 6$ yields scaling coefficients $\mathbf{V}_6$ associated with a physical scale of $\lambda_6 \, \Delta t = 0.36$ seconds. This choice effectively isolates large scale fluctuations ('baseline drift') of little interest to ECG analysts into the scaling coefficients so that the wavelet coefficients give a decomposition of the heart rhythm across time scales naturally associated with this phenomenon. In examples that we present later (in particular, see Sections 5.8, 5.9 and 5.10), we also use physical considerations to help to pick J_0.

It should also be noted that the width L of the wavelet filter can impact the choice of J_0 somewhat. Since a large L will lead to larger filter widths L_j for the higher level wavelet and scaling filters, a default upper bound for J_0 is to set it such that $L_{J_0} \le N < L_{J_0+1}$. This choice ensures that at least some of the coefficients in $\mathbf{W}_{J_0}$ and $\mathbf{V}_{J_0}$ are unaffected by boundary conditions.

Comments and Extensions to Section 4.11

[1] Here we quantify the extent to which the circularity assumption influences the DWT coefficients. Recall that the jth level wavelet coefficients can be expressed as

$$W_{j,t} = 2^{j/2} \widetilde{W}_{j,2^j(t+1)-1}, \qquad t = 0, \ldots, N_j - 1 \equiv \frac{N}{2^j} - 1,$$

where, for $t = 0, \ldots, N - 1$,

$$2^{j/2} \widetilde{W}_{j,t} \equiv \sum_{l=0}^{L_j - 1} h_{j,l} X_{t-l \bmod N}$$

(see Equations (96c) and (96d)). From Exercise [96] we know that $\{h_{j,l}\}$ has width $L_j \equiv (2^j - 1)(L - 1) + 1$. Thus, in the circular filtering of $\{X_t\}$ with $\{h_{j,l}\}$ that yields

$\{2^{j/2}\widetilde{W}_{j,t}\}$, the output values that are indexed by $t = 0, \ldots, L_j - 2$ make explicit use of the circularity assumption, whereas those indexed by $t = L_j - 1$ to $t = N - 1$ are unaffected by circularity. The index t of the first wavelet coefficient $W_{j,t}$ that is unaffected by circularity is the smallest t satisfying

$$2^j(t+1) - 1 \geq L_j - 1$$

or, equivalently (after some reduction),

$$t \geq (L-2)\left(1 - \frac{1}{2^j}\right), \text{ i.e., } t = \left\lceil (L-2)\left(1 - \frac{1}{2^j}\right)\right\rceil \equiv L'_j, \qquad \text{(146a)}$$

where $\lceil x \rceil$ refers to the smallest integer greater than or equal to x. Note that

$$L'_1 = \frac{L}{2} - 1 \text{ and } \frac{L}{2} - 1 \leq L'_j \leq L - 2. \qquad \text{(146b)}$$

Moreover, because $L'_j = L - 2$ happens if and only if

$$(L-2)\left(1 - \frac{1}{2^j}\right) > L - 3,$$

we see that $L'_j = L - 2$ for all j such that $2^j > L - 2$.

Under the assumption that $L'_j \leq N_j = N/2^j$, we can interpret L'_j as the number of jth level wavelet coefficients $W_{j,t}$ that are directly influenced by the circularity assumption – more generally, the number of affected coefficients is given by $\min\{L'_j, N_j\}$. We refer to these as 'boundary coefficients.' Note that the values of $\widetilde{W}_{j,t}$ indexed by $t = L_j - 1$ to $N - 1$ are unaffected by circularity; in other words, only the beginning of this sequence is affected. After we subsample and rescale to obtain the elements of $\mathbf{W}_j$, this is still true: only the first L'_j are boundary coefficients. Table 136 lists L'_j for filter widths $L = 2, \ldots, 20$.

As an example, for $j = 1$ and $L = 4$ only the first row of the matrix $\mathcal{W}_1$ of Equation (81b) is affected by circularity; hence only $W_{1,0}$ is a boundary coefficient. Table 136 correctly gives $L'_1 = 1$ for $L = 4$.

A similar line of argument says that the number of boundary scaling coefficients in a partial DWT of level J_0 is $\min\{L'_{J_0}, N_{J_0}\}$. Exercise [97] suggests, however, that there is an interesting deviation from this rule when $N = 2^J$ and $J_0 = J$: for a full DWT, the single scaling coefficient is proportional to the sample mean of the time series, and hence is free of the circularity assumption!

[2] Here we determine what happens to the L'_j boundary coefficients in $\mathbf{W}_j$ when we use an LA or coiflet wavelet filter and plot the coefficients versus actual time as described in Section 4.8. When we take into account the circular advances appropriate for a particular LA or coiflet filter, the order in which the $W_{j,t}$ are plotted from left to right on a graph is generally not the same as their ordering in $\mathbf{W}_j$. To see this, consider the case $j = 1$. From the discussion in Section 4.8 (and assuming $t_0 = 0$ and $\Delta t = 1$ for convenience), it follows that $W_{1,t}$ is associated with actual time $2t + 1 - |\nu_1^{(H)}| \bmod N$ for $t = 0, \ldots, \frac{N}{2} - 1$, where we can use Equation (114c) to write, for the LA filters,

$$|\nu_1^{(H)}| = \frac{L}{2} + \delta, \text{ where } \delta = \begin{cases} 0, & \text{if } L/2 \text{ is even;} \\ -1, & \text{if } L = 10 \text{ or } 18; \\ 1, & \text{if } L = 14; \end{cases}$$

for the coiflet filters, we have $|\nu_1^{(H)}| = \frac{L}{3}$. Since L is usually small compared to N, the first LA or coiflet coefficient $W_{1,0}$ of unit scale is associated with actual time, respectively,

$$1 - \frac{L}{2} - \delta \bmod N = N + 1 - \frac{L}{2} - \delta \text{ or } 1 - \frac{L}{3} \bmod N = N + 1 - \frac{L}{3},$$

so $W_{1,0}$ is plotted near the end of the actual times. The coefficient $W_{1,t}$ that is associated with the earliest actual time is the one with the smallest index t satisfying, for the LA filters,

$$2t + 1 - \frac{L}{2} - \delta \geq 0; \text{ i.e., } t = \left\lceil \frac{L + 2\delta - 2}{4} \right\rceil \equiv \gamma_1^{(H)}$$

or, for the coiflet filters, for which L is always a multiple of six,

$$2t + 1 - \frac{L}{3} \geq 0; \text{ i.e., } t = \left\lceil \frac{L-3}{6} \right\rceil = \frac{L}{6} \equiv \gamma_1^{(H)}.$$

The order in which the elements of $\mathbf{W}_1$ are plotted is thus given by the elements of

$$\mathcal{T}^{-\gamma_1^{(H)}} \mathbf{W}_1 = \left[W_{1,\gamma_1^{(H)}}, W_{1,\gamma_1^{(H)}+1}, \ldots, W_{1,\frac{N}{2}-1}, W_{1,0}, \ldots, W_{1,\gamma_1^{(H)}-1} \right]^T.$$

Now the first $L'_1 = \frac{L}{2} - 1$ coefficients in $\mathbf{W}_1$ are boundary coefficients, i.e., those with indices $t = 0, \ldots, \frac{L}{2} - 2$. Since the index of the first element in $\mathcal{T}^{-\gamma_1^{(H)}}$ is $\gamma_1^{(H)}$ and since $\gamma_1^{(H)} \leq \frac{L}{2} - 2 < L'_1$ for all the LA and coiflet filters, it follows that the last $\gamma_1^{(H)}$ coefficients in $\mathcal{T}^{-\gamma_1^{(H)}} \mathbf{W}_1$ are all affected by circularity, as are the first $\frac{L}{2} - 1 - \gamma_1^{(H)} \equiv \tilde{\gamma}_1^{(H)}$. A similar analysis for higher scales indicates that, under the assumption that $L'_j \leq N_j$, the order in which the elements of $\mathbf{W}_j$ are plotted is given by $\mathcal{T}^{-\gamma_j^{(H)}} \mathbf{W}_j$, where we have, for the LA and coiflet filters respectively,

$$\gamma_j^{(H)} \equiv \left\lceil \frac{L_j + 2\delta + 2}{2^{j+1}} - 1 \right\rceil \text{ and } \gamma_j^{(H)} \equiv \left\lceil \frac{3L_j - L + 6}{3 \cdot 2^{j+1}} - 1 \right\rceil.$$

For both filters, we always have $0 < \gamma_j^{(H)} < L'_j$. Thus the first $\gamma_j^{(H)}$ coefficients in $\mathbf{W}_j$ are associated with the $\gamma_j^{(H)}$ latest actual times, all of which are boundary coefficients. There are $\tilde{\gamma}_j^{(H)} \equiv L'_j - \gamma_j^{(H)}$ other boundary coefficients, and these are associated with the $\tilde{\gamma}_j^{(H)}$ earliest actual times. The values of $\tilde{\gamma}_j^{(H)}$ and $\gamma_j^{(H)}$ for the LA filters are given in Table 137a. A similar analysis for the scaling coefficients leads to the analogous quantities $\gamma_{J_0}^{(G)}$ and $\tilde{\gamma}_{J_0}^{(G)}$, which are given in Table 137b. Exercise [4.27] is to create similar tables for the coiflet filters.

[3] Here we consider how circularity influences the detail series. We can write

$$\mathcal{D}_j = \mathcal{W}_j^T \mathbf{W}_j = \sum_{n=0}^{N_j - 1} W_{j,n} \mathcal{W}_{j,n\bullet},$$

where $\mathcal{W}_{j,n\bullet}$ is an N dimensional column vector whose elements comprise the nth row of $\mathcal{W}_j$. When $L'_j \leq N_j$, we know that the boundary coefficients are $W_{j,n}, n = 0, \ldots, L'_j - 1$, which in turn influence $\mathcal{D}_j$ via the nonzero portions of the corresponding vectors $\mathcal{W}_{j,n\bullet}$; i.e., we can write

$$\mathcal{D}_j = \sum_{n=0}^{L'_j - 1} W_{j,n}\mathcal{W}_{j,n\bullet} + \sum_{n=L'_j}^{N_j - 1} W_{j,n}\mathcal{W}_{j,n\bullet},$$

where the first summation depends explicitly on circularity, and the second does not. For example, consider the case $j = 1$ and $L = 4 < N$, for which only $W_{1,0}$ is a boundary coefficient. We can deduce from Equation (81b) that $\mathcal{D}_1$ is affected by circularity via

$$\sum_{n=0}^{L'_1 - 1} W_{1,n}\mathcal{W}_{1,n\bullet} = W_{1,0}\mathcal{W}_{1,0\bullet} = W_{1,0}[h_1, h_0, 0, \ldots, 0, h_3, h_2]^T$$

so the boundary elements of $\mathcal{D}_1$ are those with indices $n = 0, 1, N-2$ and $N-1$. For $j = 1$ and $4 \leq L < N$ so that $L'_1 = \frac{L}{2} - 1 > 0$, the boundary coefficients $W_{1,0}, W_{1,1}, \ldots, W_{1,L'_1 - 1}$ influence $\mathcal{D}_1$ via

$$\sum_{n=0}^{L'_1 - 1} W_{1,n}\mathcal{W}_{1,n\bullet} = W_{1,0}\begin{bmatrix} h_1 \\ h_0 \\ 0 \\ \vdots \\ 0 \\ h_{L-1} \\ \vdots \\ h_2 \end{bmatrix} + \cdots + W_{1,L'_1 - 1}\begin{bmatrix} h_{L-3} \\ \vdots \\ h_0 \\ 0 \\ \vdots \\ 0 \\ h_{L-1} \\ h_{L-2} \end{bmatrix}.$$

Note that the collection of affected indices can be deduced from the nonzero portions of the first and last terms in the above sum: because $\mathcal{W}_{1,n\bullet}$ is formed by circularly shifting $\mathcal{W}_{1,0\bullet}$ by $2n$ units, for every nonzero element in $\mathcal{W}_{1,1\bullet}, \ldots, \mathcal{W}_{1,L'_1 - 2\bullet}$, there is a corresponding nonzero element in either $\mathcal{W}_{1,0\bullet}$ or $\mathcal{W}_{1,L'_1 - 1\bullet}$. The indices that are affected in $\mathcal{D}_1$ are thus $n = 0, \ldots, L-3$ (this is deduced from $\mathcal{W}_{1,L'_1 - 1\bullet}$) combined with $n = N - (L-2), \ldots, N-1$ (from $\mathcal{W}_{1,0\bullet}$).

In a similar manner, when $0 < L'_j \leq N_j$, we can deduce the affected indices in $\mathcal{D}_j$ by just determining the indices of the nonzero entries in $\mathcal{W}_{j,0\bullet}$ along with those in $\mathcal{W}_{j,L'_j - 1\bullet}$. Recalling that $\langle \mathbf{X}, \mathcal{W}_{j,n\bullet} \rangle = W_{j,n}$, we can use Equation (96d) to obtain

$$\langle \mathbf{X}, \mathcal{W}_{j,0\bullet} \rangle = \sum_{l=0}^{L_j - 1} h_{j,l} X_{2^j - 1 - l \bmod N}$$

and

$$\langle \mathbf{X}, \mathcal{W}_{j,L'_j - 1\bullet} \rangle = \sum_{l=0}^{L_j - 1} h_{j,l} X_{2^j L'_j - 1 - l \bmod N}.$$

The desired indices are the ones taken on by X_t in these two summations. From the first we get the indices

$$t = 0, \ldots, 2^j - 1 \text{ and } t = N + 2^j - L_j, \ldots, N - 1; \tag{149a}$$

from the second,

$$t = 0, \ldots, 2^j L_j' - 1 \text{ and } t = N + 2^j L_j' - L_j, \ldots, N - 1. \tag{149b}$$

Since $L_j' \geq 1$, the combination of these two sets of indices yields what is reported in Equation (139). (Similar results hold for $\mathcal{S}_{J_0}$, with again the case $J_0 = J$ being special.)

▷ **Exercise [149]:** If there are to be affected values at the end of $\mathcal{D}_j$, we must have either $L_j > 2^j$ in Equation (149a) or $L_j > 2^j L_j'$ in Equation (149b). Show that in fact both these inequalities hold under our assumption $L_j' > 0$. ◁

[4] It should be noted that there is no unique definition in the literature for *any* of the DWT transforms $\mathcal{W}$ based upon the discrete Daubechies wavelets, so it is important to check what conventions have been adopted if, say, a figure in this book is compared to computations from a software package for wavelets. For example, the Haar DWT as defined by Bruce and Gao (1996a) differs from our definition $\mathcal{W}$ in how the rows are ordered: their orthonormal matrix is

$$\begin{bmatrix} \mathcal{V}_J \\ \mathcal{W}_J \\ \vdots \\ \mathcal{W}_2 \\ \mathcal{W}_1 \end{bmatrix} \text{ rather than } \mathcal{W} = \begin{bmatrix} \mathcal{W}_1 \\ \mathcal{W}_2 \\ \vdots \\ \mathcal{W}_J \\ \mathcal{V}_J \end{bmatrix}.$$

The two definitions are essentially the same in that the elements of the Haar wavelet coefficients for the two transforms are merely reordered.

For wavelets other than the Haar, the differences are not so easy to reconcile. For example, Figure 150 shows the row vectors of the D(4) DWT matrix for $N = 16$ as defined in Bruce and Gao (1996a), with the vectors reordered to match our convention. A careful comparison with Figure 60 for our $\mathcal{W}$ indicates that, even if the arbitrary ordering of the rows is ignored, the row vectors for the two transforms are shifted with respect to each other (to go from Figure 60 to 150 requires shifting each row associated with τ_j in Figure 60 to the right by $2\tau_j - 1$ units). As a result, the DWT coefficients for the two D(4) transforms in general will not agree except for the single coefficient that is proportional to the sample mean!

The source of discrepancies such as these is the arbitrariness with which certain parts of the wavelet transform can be defined. For example, there is a downsampling (or subsampling) operation inherent in the wavelet transform, and an implementor can choose to subsample the even or odd indexed values of a sequence (indeed the time series itself is often indexed from 1 to N rather than 0 to $N - 1$ as we have chosen). Additionally, we implement the wavelet transform in terms of filtering operations (i.e., convolutions), but it is equally valid to implement it in terms of dot products (i.e., autocorrelations). The choice is largely 'cultural:' electrical engineers tend to implement wavelets as filters, whereas mathematicians tend to use dot products. There is no fundamental way of resolving these differences, so practitioners must beware!

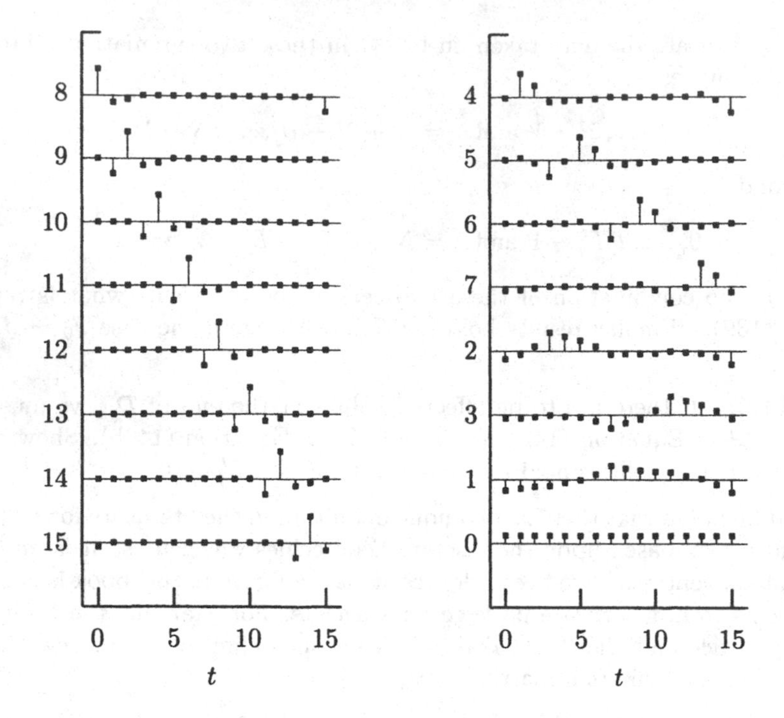

Figure 150. Row vectors of the discrete wavelet transform matrix based on the D(4) wavelet as defined in Bruce and Gao (1996a) – cf. Figure 60.

4.12 Summary

Let $\mathbf{X}$ be an N dimensional vector whose elements are the real-valued time series $\{X_t : t = 0, \ldots, N-1\}$, where the sample size N is taken to be an integer multiple of 2^{J_0}. The partial DWT of level J_0 of $\mathbf{X}$ is an orthonormal transform given by $\mathbf{W} = \mathcal{W}\mathbf{X}$, where $\mathbf{W}$ is an N dimensional vector of DWT coefficients, and $\mathcal{W}$ is an $N \times N$ real-valued matrix defining the DWT (if $N = 2^J$ and $J_0 = J$, we obtain a full DWT). The DWT coefficients $\mathbf{W}$ and matrix $\mathcal{W}$ can be partitioned such that

$$\mathbf{W} = \begin{bmatrix} \mathbf{W}_1 \\ \mathbf{W}_2 \\ \vdots \\ \mathbf{W}_{J_0} \\ \mathbf{V}_{J_0} \end{bmatrix} \text{ and } \mathcal{W} = \begin{bmatrix} \mathcal{W}_1 \\ \mathcal{W}_2 \\ \vdots \\ \mathcal{W}_{J_0} \\ \mathcal{V}_{J_0} \end{bmatrix} \tag{150}$$

so that $\mathbf{W}_j = \mathcal{W}_j\mathbf{X}$ and $\mathbf{V}_{J_0} = \mathcal{V}_{J_0}\mathbf{X}$. Here $\mathbf{W}_j$ is an $N_j \equiv N/2^j$ dimensional vector of wavelet coefficients associated with changes on scale $\tau_j \equiv 2^{j-1}$; $\mathcal{W}_j$ is an $N_j \times N$ dimensional matrix; $\mathbf{V}_{J_0}$ is an N_{J_0} dimensional vector of scaling coefficients associated with averages on scale $\lambda_{J_0} \equiv 2^{J_0}$; and $\mathcal{V}_{J_0}$ is an $N_{J_0} \times N$ dimensional matrix. The vector $\mathbf{X}$ can be synthesized from $\mathbf{W}$ via

$$\mathbf{X} = \mathcal{W}^T\mathbf{W} = \sum_{j=1}^{J_0} \mathcal{W}_j^T\mathbf{W}_j + \mathcal{V}_{J_0}^T\mathbf{V}_{J_0} \equiv \sum_{j=1}^{J_0} \mathcal{D}_j + \mathcal{S}_{J_0},$$

which defines a multiresolution analysis (MRA) of $\mathbf{X}$, i.e., an additive decomposition in terms of the N dimensional vectors $\mathcal{D}_j \equiv \mathcal{W}_j^T \mathbf{W}_j$ (the jth level detail) and $\mathcal{S}_{J_0} \equiv \mathcal{V}_{J_0}^T \mathbf{V}_{J_0}$ (the J_0th level smooth), each of which can be associated with a particular scale (τ_j in the case of $\mathcal{D}_j$ and λ_{J_0} in the case of $\mathcal{S}_{J_0}$). A scale-by-scale ANOVA can be based upon the energy decomposition

$$\|\mathbf{X}\|^2 = \|\mathbf{W}\|^2 = \sum_{j=1}^{J_0} \|\mathbf{W}_j\|^2 + \|\mathbf{V}_{J_0}\|^2 = \sum_{j=1}^{J_0} \|\mathcal{D}_j\|^2 + \|\mathcal{S}_{J_0}\|^2.$$

In practice the DWT matrix $\mathcal{W}$ is not formed explicitly, but rather $\mathbf{W}$ is computed using a 'pyramid' algorithm that makes use of a wavelet filter and a scaling filter (this algorithm in fact precisely defines the DWT). By definition, a filter $\{h_l : l = 0, \ldots, L-1\}$ of even width L (implying $h_0 \neq 0$ and $h_{L-1} \neq 0$) is called a wavelet filter if

$$\sum_{l=0}^{L-1} h_l = 0 \text{ and } \sum_{l=0}^{L-1} h_l h_{l+2n} = \begin{cases} 1, & \text{if } n = 0; \\ 0, & \text{if } n \text{ is a nonzero integer,} \end{cases}$$

where the second summation expresses the orthonormality property of a wavelet filter (in the above $h_l \equiv 0$ for $l < 0$ and $l \geq L$, so we actually consider $\{h_l\}$ to be an infinite sequence with at most L nonzero values). The scaling filter is defined in terms of the wavelet filter via thc 'quadrature mirror' relationship

$$g_l \equiv (-1)^{l+1} h_{L-1-l}.$$

This filter satisfies the conditions

$$\sum_{l=0}^{L-1} g_l g_{l+2n} = \begin{cases} 1, & \text{if } n = 0; \\ 0, & \text{otherwise;} \end{cases} \quad \text{and} \quad \sum_{l=0}^{L-1} g_l h_{l+2n} = 0 \text{ for all } n.$$

Without loss of generality, we can also assume that $\sum_l g_l = \sqrt{2}$. If we let $H(\cdot)$ be the transfer function for $\{h_l\}$, i.e.,

$$H(f) \equiv \sum_{l=-\infty}^{\infty} h_l e^{-i2\pi fl} = \sum_{l=0}^{L-1} h_l e^{-i2\pi fl},$$

and if we use $\mathcal{H}(f) \equiv |H(f)|^2$ to define the associated squared gain function, then the orthonormality property is equivalent to

$$\mathcal{H}(f) + \mathcal{H}(f + \tfrac{1}{2}) = 2 \quad \text{for all } f.$$

If we let $G(\cdot)$ and $\mathcal{G}(\cdot)$ be the transfer and squared gain functions for the scaling filter, we have $\mathcal{G}(f) = \mathcal{H}(\frac{1}{2} - f)$, from which it follows that

$$\mathcal{G}(f) + \mathcal{G}(f + \tfrac{1}{2}) = 2 \text{ and } \mathcal{G}(f) + \mathcal{H}(f) = 2 \quad \text{for all } f.$$

In practice, $\{h_l\}$ is nominally a high-pass filter with a pass-band given by $\frac{1}{4} \leq |f| \leq \frac{1}{2}$, while $\{g_l\}$ is nominally a low-pass filter with pass-band $0 \leq |f| \leq \frac{1}{4}$. Because each filter has a nominal pass-band covering half the full band of frequencies, both $\{h_l\}$ and $\{g_l\}$ can be called 'half-band' filters.

With $\{h_l\}$ and $\{g_l\}$ so defined, the pyramid algorithm for stage j consists of circularly filtering the N_{j-1} elements of

$$\mathbf{V}_{j-1} \equiv \left[V_{j-1,0}, V_{j-1,1}, \ldots, V_{j-1,N_{j-1}-1}\right]^T$$

and retaining the filtered values with odd indices – this yields the jth level wavelet and scaling coefficients, namely,

$$W_{j,t} \equiv \sum_{l=0}^{L-1} h_l V_{j-1,2t+1-l \bmod N_{j-1}}, \quad V_{j,t} \equiv \sum_{l=0}^{L-1} g_l V_{j-1,2t+1-l \bmod N_{j-1}},$$

$t = 0, \ldots, N_j - 1$ (these are the elements of $\mathbf{W}_j$ and $\mathbf{V}_j$). The procedure of retaining just every other filtered value is called 'downsampling by two.' With $\mathbf{V}_0 \equiv \mathbf{X}$, we start the pyramid algorithm with $j = 1$ and, after repeating the algorithm with $j = 2, 3, \ldots, J_0$, we obtain all the vectors of coefficients needed to form $\mathbf{W}$, namely, the J_0 vectors of wavelet coefficients $\mathbf{W}_1, \ldots, \mathbf{W}_{J_0}$ and a single vector of scaling coefficients $\mathbf{V}_{J_0}$ (the other vectors $\mathbf{V}_1, \ldots, \mathbf{V}_{J_0-1}$ of scaling coefficients can be regarded as intermediate computations).

Although in practice $\mathbf{W}_j$ and $\mathbf{V}_j$ are computed using the pyramid algorithm, it is of theoretical interest to note that we could obtain their elements directly from $\mathbf{X}$ via

$$W_{j,t} = \sum_{l=0}^{L_j-1} h_{j,l} X_{2^j(t+1)-1-l \bmod N}, \quad V_{j,t} = \sum_{l=0}^{L_j-1} g_{j,l} X_{2^j(t+1)-1-l \bmod N},$$

where $\{h_{j,l}\}$ and $\{g_{j,l}\}$ are the jth level equivalent wavelet and scaling filters, each having width $L_j \equiv (2^j - 1)(L-1) + 1$ (here $h_{1,l} \equiv h_l$ and $g_{1,l} \equiv g_l$). These filters have transfer functions

$$H_j(f) \equiv H(2^{j-1} f) \prod_{l=0}^{j-2} G(2^l f) \text{ and } G_j(f) \equiv \prod_{l=0}^{j-1} G(2^l f)$$

(again, $H_1(f) \equiv H(f)$ and $G_1(f) \equiv G(f)$). The filter $\{h_{j,l}\}$ is nominally a band-pass filter with pass-band given by $1/2^{j+1} \leq |f| \leq 1/2^j$, while $\{g_{j,l}\}$ is nominally a low-pass filter with pass-band $0 \leq |f| \leq 1/2^{j+1}$. Table 154 summarizes the most important properties of the wavelet and scaling filters and the outputs they generate.

Given $\mathbf{W}_j$ and $\mathbf{V}_j$, we can reconstruct (synthesize) the elements of $\mathbf{V}_{j-1}$ via the jth stage of the inverse pyramid algorithm, namely,

$$V_{j-1,t} = \sum_{l=0}^{L-1} h_l W^{\uparrow}_{j,t+l \bmod N_{j-1}} + \sum_{l=0}^{L-1} g_l V^{\uparrow}_{j,t+l \bmod N_{j-1}},$$

$t = 0, 1, \ldots, N_{j-1} - 1$, where

$$W^{\uparrow}_{j,t} \equiv \begin{cases} 0, & t = 0, 2, \ldots, N_{j-1} - 2; \\ W_{j,\frac{t-1}{2}}, & t = 1, 3, \ldots, N_{j-1} - 1 \end{cases}$$

($V^{\uparrow}_{j,t}$ is defined similarly). The procedure of inserting a single zero between the elements of $\{W_{j,t}\}$ to form $\{W^{\uparrow}_{j,t}\}$ is called 'upsampling by two.'

The filtering operations that make up the pyramid algorithm completely determine the $N \times N$ matrix $\mathcal{W}$ that defines the DWT. Recalling the partitioning of $\mathcal{W}$ given in Equation (150), we can write

$$\mathcal{W}_j = \mathcal{B}_j \mathcal{A}_{j-1} \cdots \mathcal{A}_1 \text{ and } \mathcal{V}_{J_0} = \mathcal{A}_{J_0} \mathcal{A}_{J_0-1} \cdots \mathcal{A}_1,$$

where $\mathcal{B}_j$ and $\mathcal{A}_j$ are $N_j \times N_{j-1}$ matrices satisfying $\mathcal{B}_j \mathcal{A}_j^T = \mathcal{A}_j \mathcal{B}_j^T = 0_{N_j}$ and $\mathcal{B}_j \mathcal{B}_j^T = \mathcal{A}_j \mathcal{A}_j^T = I_{N_j}$ (here 0_{N_j} is an $N_j \times N_j$ matrix, all of whose entries are zero). The first row of $\mathcal{B}_j$ is given by

$$\left[h_1^\circ, h_0^\circ, h_{N_{j-1}-1}^\circ, h_{N_{j-1}-2}^\circ, \ldots, h_3^\circ, h_2^\circ\right]^T,$$

where $\{h_l^\circ\}$ is the wavelet filter periodized to length N_{j-1} (and hence satisfies $\{h_l^\circ\} \longleftrightarrow \{H(\frac{k}{N_{j-1}})\}$). Subsequent rows are formed by circularly shifting the above to the right by $2k$ with $k = 1, \ldots, N_j - 1$; in particular, the final row is given by

$$\left[h_{N_{j-1}-1}^\circ, h_{N_{j-1}-2}^\circ, \ldots, h_1^\circ, h_0^\circ\right]^T.$$

A similar construction holds for $\mathcal{A}_j$, but now we use $\{g_l\}$ instead of $\{h_l\}$.

The rows of $\mathcal{W}_j$ can also be described directly in terms of the jth level wavelet filter $\{h_{j,l}\}$ periodized to length N. This periodized filter is denoted as $\{h_{j,l}^\circ\}$ and satisfies

$$\{h_{j,l}^\circ\} \longleftrightarrow \{H(\tfrac{2^{j-1}k}{N}) G(\tfrac{2^{j-2}k}{N}) \cdots G(\tfrac{k}{N})\}.$$

The first row of $\mathcal{W}_j$ is given by

$$\left[h_{j,2^j-1}^\circ, h_{j,2^j-2}^\circ, \ldots, h_{j,1}^\circ, h_{j,0}^\circ, h_{j,N-1}^\circ, h_{j,N-2}^\circ, \ldots, h_{j,2^j+1}^\circ, h_{j,2^j}^\circ\right]^T.$$

Subsequent rows are formed by circularly shifting the above to the right by $k2^j$ with $k = 1, \ldots, N_j - 1$; in particular, the final row is given by

$$\left[h_{j,N-1}^\circ, h_{j,N-2}^\circ, \ldots, h_{j,1}^\circ, h_{j,0}^\circ\right]^T.$$

A similar construction holds for $\mathcal{V}_{J_0}$ using $\{g_{J_0,l}^\circ\}$ instead of $\{h_{j,l}^\circ\}$.

The defining properties of a wavelet filter (summation to zero and orthonormality) are not sufficient to yield a DWT whose wavelet coefficients can reasonably be interpreted in terms of changes in adjacent weighted averages over particular scales. By considering certain regularity conditions (see Section 11.9), Daubechies (1992, Section 6.1) defined a useful class of wavelet filters, all of which yield wavelet coefficients in accordance with such an interpretation. By definition, a Daubechies wavelet filter of even width L has a squared gain function given by

$$\mathcal{H}^{(\mathrm{D})}(f) \equiv \mathcal{D}^{\frac{L}{2}}(f) \mathcal{A}_L(f),$$

Wavelet filter	Scaling filter
$\{h_l\} \longleftrightarrow H(\cdot)$	$\{g_l\} \longleftrightarrow G(\cdot)$
$h_l = (-1)^l g_{L-1-l}$	$g_l \equiv (-1)^{l+1} h_{L-1-l}$
$H(f) = -e^{-i2\pi f(L-1)} G(\frac{1}{2} - f)$	$G(f) = e^{-i2\pi f(L-1)} H(\frac{1}{2} - f)$
$\sum_l h_l = H(0) \equiv 0$	$\sum_l g_l = G(0) = \sqrt{2}$
$\sum_l h_l^2 \equiv 1$	$\sum_l g_l^2 = 1$
$\sum_l h_l h_{l+2n} \equiv 0, n \neq 0$	$\sum_l g_l g_{l+2n} = 0, n \neq 0$
$\sum_l g_l h_{l+2n} = 0$	
$\mathcal{H}(f) \equiv \|H(f)\|^2$	$\mathcal{G}(f) \equiv \|G(f)\|^2$
$\mathcal{H}(f) + \mathcal{H}(f + \frac{1}{2}) = 2$	$\mathcal{G}(f) + \mathcal{G}(f + \frac{1}{2}) = 2$
$\mathcal{G}(f) + \mathcal{H}(f) = 2$	
$W_{1,t} \equiv \sum_l h_l X_{2t+1-l \bmod N}$	$V_{1,t} \equiv \sum_l g_l X_{2t+1-l \bmod N}$
$W_{j,t} \equiv \sum_l h_l V_{j-1,2t+1-l \bmod N_{j-1}}$	$V_{j,t} \equiv \sum_l g_l V_{j-1,2t+1-l \bmod N_{j-1}}$
$h_{1,l} \equiv h_l, \ H_1(f) \equiv H(f)$	$g_{1,l} \equiv g_l, \ G_1(f) \equiv G(f)$
$H_j(f) \equiv H(2^{j-1} f) \prod_{l=0}^{j-2} G(2^l f)$	$G_j(f) \equiv \prod_{l=0}^{j-1} G(2^l f)$
$H_j(f) = H(2^{j-1} f) G_{j-1}(f)$	$G_j(f) = G(2^{j-1} f) G_{j-1}(f)$
$\{h_{j,l}\} \longleftrightarrow H_j(\cdot)$	$\{g_{j,l}\} \longleftrightarrow G_j(\cdot)$
$\sum_l h_{j,l} = H_j(0) = 0$	$\sum_l g_{j,l} = G_j(0) = 2^{j/2}$
$\sum_l h_{j,l}^2 = 1$	$\sum_l g_{j,l}^2 = 1$
$\sum_l h_{j,l} h_{j,l+2^j n} = 0, n \neq 0$	$\sum_l g_{j,l} g_{j,l+2^j n} = 0, n \neq 0$
$\sum_l g_{j,l} h_{j,l+2^j n} = 0$	
$\mathcal{H}_j(f) \equiv \|H_j(f)\|^2$	$\mathcal{G}_j(f) \equiv \|G_j(f)\|^2$
$W_{j,t} = \sum_l h_{j,l} X_{2^j(t+1)-1-l \bmod N}$	$V_{j,t} = \sum_l g_{j,l} X_{2^j(t+1)-1-l \bmod N}$

Table 154. Key relationships involving wavelet and scaling filters. Because $h_l = g_l = 0$ for all $l < 0$ and $l \geq L$, summations involving h_l or g_l can be taken to range from either $l = 0$ to $l = L - 1$ or over all integers; likewise, summations involving $h_{j,l}$ or $g_{j,l}$ can range either from $l = 0$ to $l = L_j - 1$ or over all integers (recall that $L_j \equiv (2^j - 1)(L - 1) + 1$ and that $N_j \equiv N/2^j$).

where $\mathcal{D}(f) \equiv 4\sin^2(\pi f)$ defines the squared gain function for the difference filter $\{1,-1\}$, and

$$\mathcal{A}_L(f) \equiv \frac{1}{2^{L-1}} \sum_{l=0}^{\frac{L}{2}-1} \binom{\frac{L}{2}-1+l}{l} \cos^{2l}(\pi f),$$

which constitutes the squared gain function of a low-pass filter. It can be shown (Exercise [106]) that $\mathcal{H}^{(D)}(f) + \mathcal{H}^{(D)}(f+\frac{1}{2}) = 2$ for all f, whereas $\mathcal{H}^{(D)}(0) = 0$ implies $\sum h_l = 0$, so any filter $\{h_l\}$ with squared gain function $\mathcal{H}^{(D)}(\cdot)$ is indeed a wavelet filter. The above decomposition of $\mathcal{H}^{(D)}(\cdot)$ indicates that the Daubechies wavelet filters can be interpreted as the equivalent filter for a filter cascade consisting of $\frac{L}{2}$ difference filters (yielding the overall differencing operation) along with a low-pass filter (yielding the weighted average). The scaling filter $\{g_l\}$ that corresponds to a Daubechies wavelet filter has a squared gain function given by

$$\mathcal{G}^{(D)}(f) \equiv \mathcal{H}^{(D)}(\tfrac{1}{2} - f) = 2\cos^L(\pi f) \sum_{l=0}^{\frac{L}{2}-1} \binom{\frac{L}{2}-1+l}{l} \sin^{2l}(\pi f).$$

Without loss of generality, the Daubechies wavelet filters of widths $L = 2$ and $L = 4$ can be taken to be, respectively, the Haar wavelet filter $\{h_0 = \frac{1}{\sqrt{2}}, h_1 = -\frac{1}{\sqrt{2}}\}$ and the D(4) wavelet filter $\{h_0, h_1, h_2, h_3\}$ displayed in Equation (59a). In general, there are several real-valued wavelet filters of the form $\{h_l : l = 0, \ldots, L-1\}$ with the same squared gain function $\mathcal{H}^{(D)}(\cdot)$. As L increases, we can impose some additional criterion to select a unique wavelet filter or, equivalently, a unique scaling filter. Daubechies (1992, Section 6.4) discusses two such criteria. The first is to pick the scaling filter $\{g_l^{(ep)}\}$ with squared gain $\mathcal{G}^{(D)}(\cdot)$ such that

$$\sum_{l=0}^{m} g_l^2 \le \sum_{l=0}^{m} \left[g_l^{(ep)}\right]^2 \quad \text{for } m = 0, \ldots, L-1,$$

where $\{g_l\}$ is any other filter with squared gain $\mathcal{G}^{(D)}(\cdot)$. This 'extremal phase' choice yields a scaling filter with minimum delay. We denote the filters satisfying this criterion as the D(L) filters, $L = 2, 4, \ldots$. The second criterion is to pick the scaling filter whose transfer function $G(f) = [\mathcal{G}^{(D)}(f)]^{1/2} e^{i\theta^{(G)}(f)}$ has a phase function $\theta^{(G)}(\cdot)$ that is as close as possible to that of a linear phase filter (here closeness is defined via Equation (112a)). We denote the filters satisfying this criterion as the LA(L) filters, $L = 8, 10, \ldots$, where 'LA' stands for 'least asymmetric.' The advantage of the LA filters is that we can use the value ν of $\tilde{\nu}$ that minimizes the above to align both the scaling and wavelet coefficients such that they can be regarded as approximately the output from zero phase filters. This approximate zero phase property is important because it allows us to meaningfully relate DWT coefficients to various events in the original time series. In particular, if we assume that X_t is associated with the actual time $t_0 + t\,\Delta t$, then the phase properties of the LA filters dictate associating the wavelet coefficient $W_{j,t}$ with actual time

$$t_0 + (2^j(t+1) - 1 - |\nu_j^{(H)}| \bmod N)\Delta t, \quad t = 0, \ldots, N_j - 1,$$

where

$$|\nu_j^{(H)}| = \frac{L_j}{2} + \frac{L}{2} + \nu - 1,$$

and

$$\nu = \begin{cases} -\frac{L}{2}+1, & \text{if } L = 8, 12, 16 \text{ or } 20 \text{ (i.e., } L/2 \text{ is even);} \\ -\frac{L}{2}, & \text{if } L = 10 \text{ or } 18; \\ -\frac{L}{2}+2, & \text{if } L = 14. \end{cases}$$

For the scaling coefficient $V_{j,t}$, we obtain a similar expression by replacing $|\nu_j^{(H)}|$ with

$$|\nu_j^{(G)}| = \frac{L_j - 1}{L - 1}|\nu|.$$

The coiflet wavelet filters – denoted as C(L) with $L = 6, 12, 18, 24$ or 30 – are alternatives to the Daubechies filters that provide better approximations to zero phase filters than the LA filters do; however, they have a less appealing filter shape and fewer embedded differencing operations for a given filter width ($L/3$ versus $L/2$ for the Daubechies filters). We can associate coiflet wavelet and scaling coefficients with actual times using the same formulae as in the LA case, but now we need to use

$$\nu = -\frac{2L}{3} + 1.$$

4.13 Exercises

[4.1] Using the conditions of Equation (59b), solve for a and b in Equation (60), and show that one set of solutions leads to the values given for h_0, h_1, h_2 and h_3 in Equation (59a). How many other sets of solutions are there, and what values of h_0, h_1, h_2 and h_3 do these yield?

[4.2] Verify the upper left-hand plot of Figure 62, which shows the Haar wavelet transform for $\{X_{1,t}\}$ (the values for this time series are given in the caption to Figure 42).

[4.3] Verify the discrete Haar wavelet empirical power spectra calculated in Equations (63a) and (63b).

[4.4] The $\mathcal{D}_1$ details in the top and bottom left-hand plots of Figure 65 for the two sixteen point time series $\{X_{1,t}\}$ and $\{X_{2,t}\}$ in Figure 42 were constructed using the D(4) wavelet filter. Given that the two time series differ only at $t = 13$, explain why the corresponding $\mathcal{D}_1$ details differ at $t = 10, \ldots, 15$. At what indices t would the $\mathcal{D}_1$ details differ if we used a wavelet filter of width $L = 6$ instead?

[4.5] Verify Equation (69d) for the Haar wavelet filter $\{h_0 = 1/\sqrt{2}, h_1 = -1/\sqrt{2}\}$.

[4.6] Suppose that $\{h_l : l = 0, \ldots, 11\}$ is a wavelet filter of width $L = 12$ and that we construct the DWT matrix $\mathcal{W}$ based upon this filter. If $N = 4$, what are the elements of the first two rows $\mathcal{W}_{0\bullet}^T$ and $\mathcal{W}_{1\bullet}^T$ of $\mathcal{W}$ in terms of the h_l's? If $N = 8$, what do these rows now become?

[4.7] As a concrete example that the rows of $\mathcal{W}_1$ are orthonormal when $L > N$, construct this matrix explicitly for the case $N = 4$ and $L = 6$. Use Equations (69b) and (69c) directly to verify that $\mathcal{W}_1\mathcal{W}_1^T = I_2$, where I_2 is the 2×2 identity matrix.

[4.8] Show that the squared gain function for any wavelet filter $\{h_l\}$ satisfies $\mathcal{H}(0) = 0$ and $\mathcal{H}(\frac{1}{2}) = 2$ (cf. the left-hand column of Figure 73).

[4.9] Let $\mathbf{X} = [X_0, X_1, X_2, X_3]^T$, and suppose that we downsample $\mathbf{X}$ by two to obtain, say, $\mathbf{X}_\downarrow$. What vector do we end up with if we now upsample $\mathbf{X}_\downarrow$ by two?

[4.10] Based upon the definition for $2\widetilde{W}_{2,t}$ in Equation (90a), explain why

$$2\widetilde{W}_{2,t} = \sum_{l=0}^{L_2-1} h_{2,l} X_{t-l \bmod N}, \quad t = 0, \ldots, N-1,$$

and why

$$W_{2,t} = \sum_{l=0}^{L_2-1} h_{2,l} X_{4(t+1)-1-l \bmod N}, \quad t = 0, \ldots, \tfrac{N}{4} - 1.$$

[4.11] For the D(4) wavelet filter, compute the elements of the filters $\{h_{2,l}\}$ and $\{h_{3,l}\}$ that can be used to filter $\{X_t\}$ to obtain, respectively, $\{W_{2,t}\}$ and $\{W_{3,t}\}$ (the wavelet coefficients for scales $\tau_2 = 2$ and $\tau_3 = 4$). Verify that the squared gain function for $\{h_{2,l}\}$ is correctly plotted in Figure 91, and create a similar plot for the squared gain function for $\{h_{3,l}\}$.

[4.12] Let $\mathcal{H}_j(\cdot)$ and $\mathcal{G}_j(\cdot)$ be the squared gain functions corresponding to the transfer functions of Equations (96b) and (97). What is $\int_{-1/2}^{1/2} \mathcal{H}_j(f)\,df$ equal to? What is $\int_{-1/2}^{1/2} \mathcal{G}_j(f)\,df$ equal to?

[4.13] Verify that use of the pyramid algorithm with the Haar wavelet and scaling filters yields a transform that agrees with the Haar DWT matrix $\mathcal{W}$ described in Section 4.1 for $N = 16$. In particular, describe the contents of $\mathcal{B}_j$ and $\mathcal{A}_j$ for $j = 1, 2, 3$ and 4, and show that $\mathcal{B}_2\mathcal{A}_1$, $\mathcal{B}_3\mathcal{A}_2\mathcal{A}_1$, $\mathcal{B}_4\mathcal{A}_3\mathcal{A}_2\mathcal{A}_1$ and $\mathcal{A}_4\mathcal{A}_3\mathcal{A}_2\mathcal{A}_1$ yield the bottom 8 rows of $\mathcal{W}$ as shown in the right-hand column of Figure 57.

[4.14] Plot the squared gain functions $\mathcal{G}_j^{(\mathrm{D})}(\cdot)$ and $\mathcal{H}_j^{(\mathrm{D})}(\cdot)$ corresponding to the jth level LA(8) scaling and wavelet filters in Figure 98b.

[4.15] Let $H_j(\cdot)$ and $G_j(\cdot)$ be the transfer functions corresponding to, respectively, the jth level wavelet and scaling filters $\{h_{j,l}\}$ and $\{g_{j,l}\}$. Under what circumstances is $\mathbf{Y} = \mathbf{X}$ in the following flow diagram?

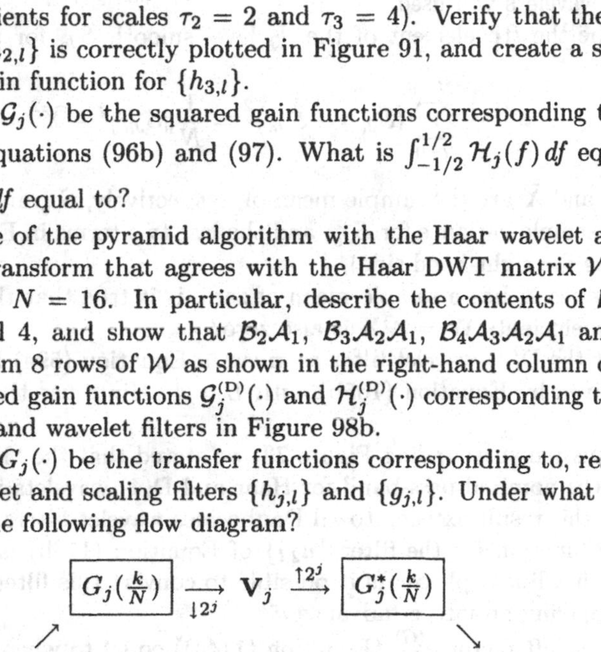

$$\mathbf{X} \nearrow \boxed{G_j(\tfrac{k}{N})} \xrightarrow[\downarrow 2^j]{} \mathbf{V}_j \xrightarrow{\uparrow 2^j} \boxed{G_j^*(\tfrac{k}{N})} \searrow + \longrightarrow \mathbf{Y}$$

$$\mathbf{X} \searrow \boxed{H_j(\tfrac{k}{N})} \xrightarrow[\downarrow 2^j]{} \mathbf{W}_j \xrightarrow{\uparrow 2^j} \boxed{H_j^*(\tfrac{k}{N})} \nearrow + \longrightarrow \mathbf{Y}$$

[4.16] Show that $\mathcal{W}_1^T\mathcal{W}_1 + \mathcal{V}_1^T\mathcal{V}_1 = I_N$, an equation that can be referred to as a $j = 1$ order resolution of the Nth order identity matrix. State and prove a corresponding resolution of identity for $j > 1$.

[4.17] Using the pseudo-code in item [1] of the Comments and Extensions to Section 4.6, implement the DWT pyramid algorithm and its inverse in your favorite computer language. Using the D(4) wavelet and scaling filters (see Equations (59a) and (75d)), verify your implementation by computing the DWT and its inverse for the 16 point time series $\{X_{1,t}\}$ listed in the caption to Figure 42.

[4.18] Using the pyramid algorithm and its inverse along with the D(4) wavelet, compute $\mathcal{D}_3$ for the 16 point time series $\{X_{1,t}\}$ listed in the caption to Figure 42.

[4.19] For $t = 0, 1, \ldots, 1023$, define a 'chirp' time series by

$$X_t = \cos\left(\tfrac{3\pi}{64}\left[1 + \tfrac{t}{1023}\right] t + 0.82\right) = \cos(2\pi f_t t + 0.82),$$

where $f_t \equiv \frac{3}{128} + \frac{3}{128}\left(\frac{t}{1023}\right)$ so that f_t ranges from $f_0 = \frac{3}{128}$ up to $f_{1023} = \frac{3}{64}$. If we regard the argument to the cosine above as a continuous function of t, we can differentiate it to obtain

$$\frac{d(2\pi f_t t + 0.82)}{dt} = 2\pi\left[\tfrac{3}{128} + \tfrac{3}{128}\left(\tfrac{2t}{1023}\right)\right] \equiv 2\pi \bar{f}_t,$$

where $\bar{f}_t$ is known as the instantaneous frequency. For this chirp series, compute and plot the DWT details $\mathcal{D}_k$ for $k = 1$ to 6 using any two wavelets of the following four wavelets: the Haar, D(4), C(6) and LA(8). Explain the appearance of the detail series with regard to (a) the instantaneous frequency for $\{X_t\}$ and (b) the particular wavelets you used.

[4.20] Let $\mathcal{S}_{J_0,t}$ be the tth element of the J_0 level smooth $\mathcal{S}_{J_0}$ for the time series $\mathbf{X}$. Show that

$$\frac{1}{N}\sum_{t=0}^{N-1}\left(\mathcal{S}_{J_0,t} - \overline{\mathcal{S}}_{J_0}\right)^2 = \frac{1}{N}\|\mathcal{S}_{J_0}\|^2 - \overline{X}^2,$$

where $\overline{\mathcal{S}}_{J_0}$ and $\overline{X}$ are the sample mean of, respectively, $\mathcal{S}_{J_0}$ and $\mathbf{X}$ (the left-hand side is the sample variance for $\mathcal{S}_{J_0}$, and the last two terms in Equation (104c) are the same as the right-hand side).

[4.21] If a time series $\mathbf{X}$ has a sample mean of zero, is it true that the sample mean of its DWT coefficients $\mathbf{W} = \mathcal{W}\mathbf{X}$ is also zero?

[4.22] Verify that the D(4) wavelet filter as given in Equation (59a) has a squared gain function given by Equation (105b) with $L = 4$. Hint: use the notion of a filter cascade.

[4.23] In the discussion surrounding Figure 73, we stated that the '3 dB down' point is such that the nominal pass-band for Haar and D(4) wavelets is $1/4 \le |f| \le 1/2$. Show that this result extends to all Daubechies wavelet filters.

[4.24] The phase function for the filter $\{u_{2,l}\}$ of Equation (110b) is given by $\theta_2(f) = -\pi f$, so it has linear phase. Is it possible to convert this filter into a zero phase filter by applying an integer advance ν?

[4.25] What is the shift factor $\nu_J^{(G)}$ (Equation (114a)) equal to when $N = 2^J$?

[4.26] Plot the phase functions $\theta^{(H)}(\cdot)$ for $\{h_l\}$ and $\theta_j^{(H)}(\cdot)$ for $\{h_{j,l}\}$, $j = 2$, 3 and 4 for the Haar and D(4) wavelet filters with the following advances incorporated:

	$\{h_l\}$	$\{h_{2,l}\}$	$\{h_{3,l}\}$	$\{h_{4,l}\}$
Haar	0	1	3	7
D(4)	2	5	11	23

(these are the advances that make these filters as close to zero phase as possible). Note: if $z = x + iy$, then θ in the polar representation $z = |z|e^{i\theta}$ can be computed in most computer languages using a 'two argument' version of the arctangent, which computes $\arctan(y/x)$ by paying attention to the signs of y and x.

[4.27] Create tables similar to Tables 137a and 137b that give the number of boundary coefficients for the coiflet wavelet and scaling filters with $L = 6$, 12, 18, 24 and 30.

5

The Maximal Overlap Discrete Wavelet Transform

5.0 Introduction

In this chapter we describe a modified version of the discrete wavelet transform called the maximal overlap DWT (MODWT). The name comes from the literature on the Allan variance (see Section 8.6), where the MODWT based upon the Haar wavelet filter has been used since the 1970s (a good entry point to this body of literature is Greenhall, 1991; see also Percival and Guttorp, 1994). Transforms that are essentially the same as – or have some similarity to – the MODWT have been discussed in the wavelet literature under the names 'undecimated DWT' (Shensa, 1992), 'shift invariant DWT' (Beylkin, 1992; Lang *et al.*, 1995), 'wavelet frames' (Unser, 1995), 'translation invariant DWT' (Coifman and Donoho, 1995; Liang and Parks, 1996; Del Marco and Weiss, 1997), 'stationary DWT' (Nason and Silverman, 1995), 'time invariant DWT' (Pesquet *et al.*, 1996) and 'non-decimated DWT' (Bruce and Gao, 1996a). While all these names are lacking somewhat as an accurate description of the transform, we prefer 'maximal overlap DWT' mainly because it leads to an acronym that is easy to say ('mod WT') and carries the connotation of a 'modification of the DWT.'

In contrast to the orthonormal partial DWT, the MODWT of level J_0 for a time series $\mathbf{X}$ is a highly redundant nonorthogonal transform yielding the column vectors $\widetilde{\mathbf{W}}_1, \widetilde{\mathbf{W}}_2, \ldots, \widetilde{\mathbf{W}}_{J_0}$ and $\widetilde{\mathbf{V}}_{J_0}$, each of dimension N. The vector $\widetilde{\mathbf{W}}_j$ contains the MODWT wavelet coefficients associated with changes in $\mathbf{X}$ on a scale of $\tau_j = 2^{j-1}$, while $\widetilde{\mathbf{V}}_{J_0}$ contains the MODWT scaling coefficients associated with variations at scales $\lambda_{J_0} = 2^{J_0}$ and higher. Like the DWT, the MODWT is defined in terms of a computationally efficient pyramid algorithm. There are five important properties that distinguish the MODWT from the DWT:

[1] While the partial DWT of level J_0 restricts the sample size to an integer multiple of 2^{J_0}, the MODWT of level J_0 is well defined for any sample size N. When N is an integer multiple of 2^{J_0}, the partial DWT can be computed using $O(N)$ multiplications, whereas the corresponding MODWT requires $O(N \log_2 N)$ multiplications. There is thus a computational price to pay for using the MODWT, but its computational burden is the same as the widely used fast Fourier transform algorithm and hence is usually quite acceptable.

[2] As is true for the DWT, the MODWT can be used to form a multiresolution analysis (MRA). In contrast to the usual DWT, the details $\widetilde{\mathcal{D}}_j$ and smooth $\widetilde{\mathcal{S}}_{J_0}$ of this MRA are such that circularly shifting the time series by any amount will circularly shift each detail and smooth by a corresponding amount.
[3] In contrast to the DWT, the MODWT details and smooths are associated with zero phase filters, thus making it easy to line up features in an MRA with the original time series meaningfully.
[4] As is true for the DWT, the MODWT can be used to form an analysis of variance (ANOVA) based upon the wavelet and scaling coefficients; in contrast to the DWT, the MODWT details and smooths cannot be used to form such an analysis.
[5] Whereas a time series and a circular shift of the series can have different DWT-based empirical power spectra (as demonstrated by Equations (63a) and (63b)), the corresponding MODWT-based spectra are the same. In fact, we can obtain the MODWT of a circularly shifted time series by just applying a similar shift to each of the components $\widetilde{\mathbf{W}}_j$ and $\widetilde{\mathbf{V}}_{J_0}$ of the MODWT of the original series; by contrast, the DWT of a circular series cannot in general be obtained by any circular shift of the components $\mathbf{W}_j$ and $\mathbf{V}_{J_0}$ of the DWT of the original series.

After we illustrate in Section 5.1 the problems that can arise with the ordinary DWT when we circularly shift $\mathbf{X}$, we define the basic MODWT wavelet and scaling filters in Section 5.2 and use these to motivate and prove the basic results for the MODWT in Sections 5.3 and 5.4. We present a pyramid algorithm for efficiently computing the MODWT in Section 5.5, followed by a simple example in Section 5.6 of how the MODWT treats a circularly shifted time series. We then illustrate the use of the MODWT through several examples in Sections 5.7 to 5.10. We close with a discussion of some practical considerations (Section 5.11) and a summary of the key results in this chapter (Section 5.12).

5.1 Effect of Circular Shifts on the DWT

Our interest in the MODWT stems from the fact that an analysis of a time series using the DWT can depend critically on where we 'break into' the series, i.e., what we take as a starting point or origin for analysis. We have noted the sensitivity of the DWT to such effects briefly in the context of the discrete Haar wavelet empirical power spectrum (see Equations (63a) and (63b)), but here we wish to show how the starting point can affect the DWT and its associated MRA (for comparison, we will revisit the example presented below in Section 5.6 using the MODWT).

The bottom row of Figure 161 has duplicate plots of both a time series $\mathbf{X}$ (first and third columns) and a version of the series circularly shifted by five units, i.e., $\mathcal{T}^5\mathbf{X}$ (second and fourth columns). The time series has a sample size of $N = 128$ and consists of a 'bump' extending over six values surrounded by zeros:

$$X_t = \begin{cases} \frac{1}{2}\cos(\frac{3\pi t}{16} + 0.08), & t = 40, \ldots, 45; \\ 0, & t = 0, \ldots, 39 \text{ and } 46, \ldots, 127. \end{cases}$$

The top five rows of Figure 161 show partial DWTs of order $J_0 = 4$ for $\mathbf{X}$ and $\mathcal{T}^5\mathbf{X}$ (first two columns) along with the related MRAs (last two columns). The DWTs are all based upon the LA(8) wavelet filter. The DWT coefficients $\mathbf{W}_j$ and $\mathbf{V}_4$ are plotted as described in Section 4.8 so that they are aligned with the original time series based upon the phase properties of the LA(8) filter. Note that, because these coefficients are

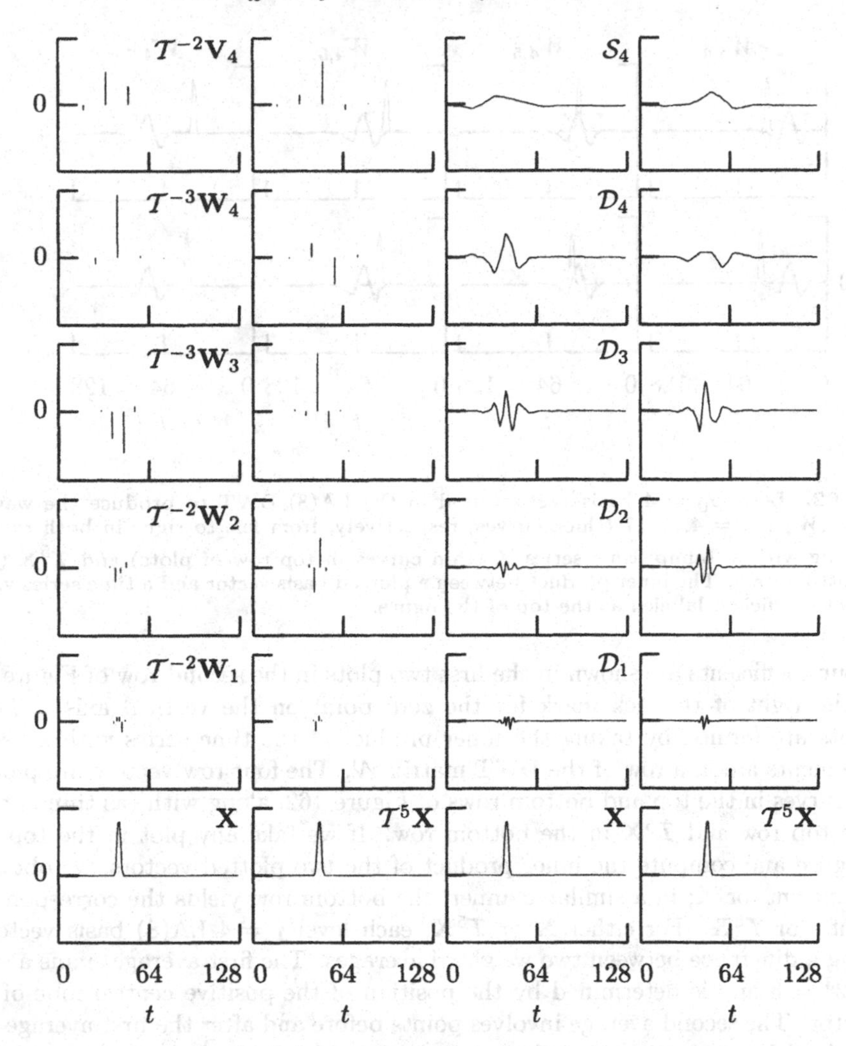

Figure 161. LA(8) DWTs of level $J_0 = 4$ (top five rows, first two columns) and corresponding multiresolution analyses (last two columns) for a time series $\mathbf{X}$ and its circularly shift $\mathcal{T}^5\mathbf{X}$ (bottom row of plots). A comparison of the first and second columns shows that circularly shifting a time series can yield substantial changes in its DWT; likewise, the third and fourth columns indicate the same is true for the corresponding MRAs. This figure should be compared with Figure 181, which uses the MODWT in place of the DWT.

plotted as deviations from the zero line, any coefficients that are identically zero show up as blank space (there are quite a few such coefficients due to the localized natures of the bump and the DWT). Although the sum of squares of all the DWT coefficients for either $\mathbf{X}$ and $\mathcal{T}^5\mathbf{X}$ is equal to $\|\mathbf{X}\|^2$, a comparison of the first two columns shows that shifting the time origin of the bump has altered the DWT considerably. An examination of the last two columns shows that the same is true for the MRAs. Although in each case the summation of the details and smooth yields the time series in the bottom row, corresponding details for $\mathbf{X}$ and $\mathcal{T}^5\mathbf{X}$ can be quite different.

To understand why shifting the time origin yields such different DWTs, let us concentrate on elements $W_{4,j}$, $j = 4, \ldots, 7$, of the subvectors $\mathbf{W}_4$ for $\mathbf{X}$ and $\mathcal{T}^5\mathbf{X}$

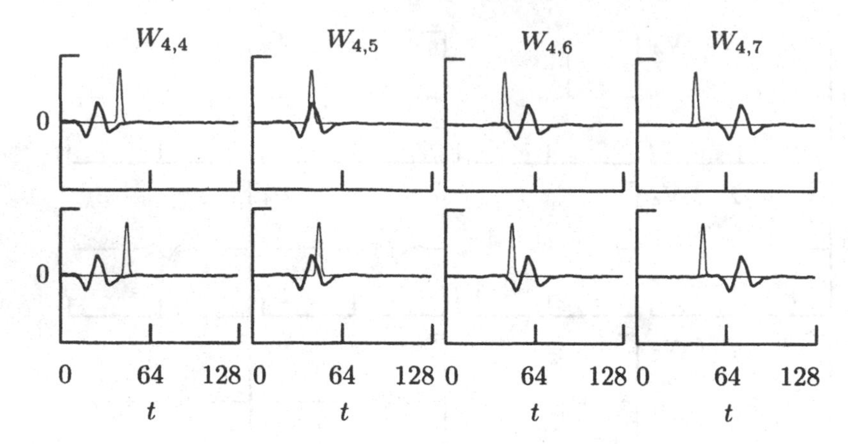

Figure 162. Level $J_0 = 4$ basis vectors used in the LA(8) DWT to produce the wavelet coefficients $W_{4,j}$, $j = 4, \ldots, 7$ (thick curves, respectively, from left to right in both rows of plots), along with a 'bump' time series $\mathbf{X}$ (thin curves in top row of plots) and $\mathcal{T}^5\mathbf{X}$ (thin curves, bottom row). The inner product between a plotted basis vector and a time series yields the wavelet coefficient labeled at the top of the figure.

(these four coefficients are shown in the first two plots in the second row of Figure 161 just to the right of the tick mark for the zero point on the vertical axis). These coefficients are formed by taking the inner product of the time series with a vector whose elements are in a row of the DWT matrix $\mathcal{W}$. The four row vectors are plotted as thick curves in the top and bottom rows of Figure 162, along with (as thin curves) $\mathbf{X}$ in the top row and $\mathcal{T}^5\mathbf{X}$ in the bottom row. If we take any plot in the top row of this figure and compute the inner product of the two plotted vectors, we obtain a $W_{4,j}$ coefficient for $\mathbf{X}$; in a similar manner, the bottom row yields the corresponding coefficients for $\mathcal{T}^5\mathbf{X}$. For either $\mathbf{X}$ or $\mathcal{T}^5\mathbf{X}$, each level $j = 4$ LA(8) basis vector is extracting a difference between two weighted averages. The first average spans a scale of $\tau_4 = 2^3 = 8$ and is determined by the position of the positive central lobe of the basis vector. The second average involves points before and after the first average and is determined by the positions of the two negative sidelobes. If we consider the first row of Figure 162, we see that $W_{4,5}$ is large because the bump in $\mathbf{X}$ is contained wholly within the central lobe of the LA(8) basis vector, whereas the two negative sidelobes cover portions of $\mathbf{X}$ that are zero; on the other hand, $W_{4,4}$, $W_{4,6}$ and $W_{4,7}$ are all close to zero because the bump is not well aligned with either the central lobe or the sidelobes. The second row of Figure 162 shows that shifting the bump to obtain $\mathcal{T}^5\mathbf{X}$ changes the nature of $W_{4,5}$ and $W_{4,6}$. For $W_{4,5}$, the shifted bump covers part of the central lobe and part of one sidelobe, so the resulting coefficient is much smaller than in the case of $\mathbf{X}$; on the other hand, for $W_{4,6}$, the shifted bump now matches up with one of the sidelobes, resulting in a negative coefficient.

In summary, while the wavelet coefficients for the DWT can be interpreted as a difference between two weighted averages, the intervals over which these averages are made are rigidly fixed *a priori* and hence might not line up well with interesting features in a time series. A change in the starting point for a time series can thus yield quite different results due to the juxtaposition of the time series with the averaging intervals predefined by the DWT. As we shall see, the MODWT is an attempt to get away from effects attributable to the choice of a starting time by essentially including all possible placements of averaging intervals.

5.2 MODWT Wavelet and Scaling Filters

In order to easily make connections between the DWT and the MODWT when we formulate the latter in the next two sections, it is convenient to define a MODWT wavelet filter $\{\tilde{h}_l\}$ via $\tilde{h}_l \equiv h_l/\sqrt{2}$ and a MODWT scaling filter $\{\tilde{g}_l\}$ via $\tilde{g}_l \equiv g_l/\sqrt{2}$. As $\{\tilde{h}_l\}$ is simply a rescaled version of the wavelet filter, it readily follows from Equations (69a), (69b) and (69c) that

$$\sum_{l=0}^{L-1} \tilde{h}_l = 0, \quad \sum_{l=0}^{L-1} \tilde{h}_l^2 = \tfrac{1}{2} \text{ and } \sum_{l=-\infty}^{\infty} \tilde{h}_l \tilde{h}_{l+2n} = 0 \tag{163a}$$

for all nonzero integers n. If we let $\widetilde{H}(\cdot)$ and $\widetilde{\mathcal{H}}(\cdot)$ denote, respectively, the transfer and squared gain functions for $\{\tilde{h}_l\}$, we have $\widetilde{H}(f) = H(f)/\sqrt{2}$ and $\widetilde{\mathcal{H}}(f) = \mathcal{H}(f)/2$, so Equation (69d) leads to the relationship

$$\widetilde{\mathcal{H}}(f) + \widetilde{\mathcal{H}}(f + \tfrac{1}{2}) = 1 \text{ for all } f. \tag{163b}$$

Similarly, for the MODWT scaling filter, it follows from Exercise [76b] that

$$\sum_{l=0}^{L-1} \tilde{g}_l = 1, \quad \sum_{l=0}^{L-1} \tilde{g}_l^2 = \tfrac{1}{2} \text{ and } \sum_{l=-\infty}^{\infty} \tilde{g}_l \tilde{g}_{l+2n} = 0$$

for all nonzero integers n, while it follows from Exercise [77] that

$$\sum_{l=-\infty}^{\infty} \tilde{g}_l \tilde{h}_{l+2n} = 0$$

for all integers n. If we let $\widetilde{G}(\cdot)$ and $\widetilde{\mathcal{G}}(\cdot)$ denote, respectively, the transfer and squared gain functions for $\{\tilde{g}_l\}$, we have $\widetilde{G}(f) = G(f)/\sqrt{2}$ and $\widetilde{\mathcal{G}}(f) = \mathcal{G}(f)/2$, so Equation (76b) tells us that

$$\widetilde{\mathcal{G}}(f) + \widetilde{\mathcal{G}}(f + \tfrac{1}{2}) = 1 \text{ and } \widetilde{\mathcal{G}}(f) + \widetilde{\mathcal{H}}(f) = 1 \quad \text{for all } f. \tag{163c}$$

Note also that the quadrature mirror relationships of Equations (75a) and (75b) also hold for the MODWT filters:

$$\tilde{g}_l = (-1)^{l+1} \tilde{h}_{L-1-l} \text{ and } \tilde{h}_l = (-1)^l \tilde{g}_{L-1-l}. \tag{163d}$$

Using the above, we can re-express Equations (70b) and (77a) as

$$\widetilde{W}_{1,t} \equiv \sum_{l=0}^{L-1} \tilde{h}_l X_{t-l \bmod N} \text{ and } \widetilde{V}_{1,t} \equiv \sum_{l=0}^{L-1} \tilde{g}_l X_{t-l \bmod N}, \tag{163e}$$

$t = 0, \ldots, N-1$. Thus the sequences $\{\widetilde{W}_{1,t}\}$ and $\{\widetilde{V}_{1,t}\}$ are obtained by circularly filtering $\{X_t\}$ with, respectively, the MODWT wavelet and scaling filters. (As we shall see in the following section, these two sequences in fact constitute the MODWT of level $J_0 = 1$.)

5.3 Basic Concepts for MODWT

The motivation for formulating the MODWT is essentially to define a transform that acts as much as possible like the DWT, but does not suffer from the DWT's sensitivity to the choice of a starting point for a time series. This sensitivity is entirely due to downsampling (subsampling) the outputs from the wavelet and scaling filters at each stage of the pyramid algorithm. To obtain a degree of insensitivity to the starting point, we need somehow to eliminate this downsampling while preserving the ability to carry out an ANOVA and an MRA.

As a first step in this direction, note that the filter outputs that are usually discarded at the first stage of the DWT pyramid algorithm can be obtained by applying the DWT pyramid algorithm to the circularly shifted vector $\mathcal{T}\mathbf{X}$ rather than $\mathbf{X}$. This suggests the following procedure for defining the first stage of the pyramid algorithm for the MODWT when N is an even sample size (an assumption that we will in fact drop later on). The idea is to apply the usual DWT pyramid algorithm twice, once to $\mathbf{X}$ and once to $\mathcal{T}\mathbf{X}$, after which we merge the two sets of DWT coefficients together. The first application yields the usual DWT, namely,

$$\begin{bmatrix} \mathbf{W}_1 \\ \mathbf{V}_1 \end{bmatrix} = \mathcal{P}_1\mathbf{X} = \begin{bmatrix} \mathcal{B}_1 \\ \mathcal{A}_1 \end{bmatrix}\mathbf{X},$$

where (appealing to Equation (81b) since $\mathcal{W}_1 = \mathcal{B}_1$)

$$\mathcal{B}_1 = \begin{bmatrix} h_1 & h_0 & 0 & 0 & 0 & \cdots & 0 & 0 & 0 & 0 & 0 & h_3 & h_2 \\ h_3 & h_2 & h_1 & h_0 & 0 & \cdots & 0 & 0 & 0 & 0 & 0 & 0 & 0 \\ \vdots & \vdots & \vdots & \vdots & \vdots & \ddots & \vdots & \vdots & \vdots & \vdots & \vdots & \vdots & \vdots \\ 0 & 0 & 0 & 0 & 0 & \cdots & 0 & h_3 & h_2 & h_1 & h_0 & 0 & 0 \\ 0 & 0 & 0 & 0 & 0 & \cdots & 0 & 0 & 0 & h_3 & h_2 & h_1 & h_0 \end{bmatrix},$$

and $\mathcal{A}_1$ has the same structure as the above with each h_l replaced by g_l (in displaying the elements of various matrices in this section, we specialize to the case $L = 4$ and $N > L$ for clarity, but our mathematical treatment holds in general). In view of Equations (70c) and (77b), we can denote the elements of $\mathbf{W}_1$ and $\mathbf{V}_1$ by

$$\mathbf{W}_1 = [2^{1/2}\widetilde{W}_{1,1}, 2^{1/2}\widetilde{W}_{1,3}, \ldots, 2^{1/2}\widetilde{W}_{1,N-1}]^T$$

and

$$\mathbf{V}_1 = [2^{1/2}\widetilde{V}_{1,1}, 2^{1/2}\widetilde{V}_{1,3}, \ldots, 2^{1/2}\widetilde{V}_{1,N-1}]^T;$$

i.e., $\mathbf{W}_1$ and $\mathbf{V}_1$ contain all the odd indexed elements of the length N sequences $\{2^{1/2}\widetilde{W}_{1,t}\}$ and $\{2^{1/2}\widetilde{V}_{1,t}\}$, which are formed by circularly convolving the time series $\mathbf{X}$ with, respectively, the wavelet filter $\{h_l\}$ and the scaling filter $\{g_l\}$. The second application consists of substituting the circularly shifted vector $\mathcal{T}\mathbf{X}$ for $\mathbf{X}$ to obtain

$$\begin{bmatrix} \mathbf{W}_{\mathcal{T},1} \\ \mathbf{V}_{\mathcal{T},1} \end{bmatrix} \equiv \mathcal{P}_1\mathcal{T}\mathbf{X}.$$

Equivalently, if we define

$$\mathcal{P}_{\mathcal{T},1} \equiv \mathcal{P}_1\mathcal{T} = \begin{bmatrix} \mathcal{B}_1 \\ \mathcal{A}_1 \end{bmatrix}\mathcal{T} = \begin{bmatrix} \mathcal{B}_1\mathcal{T} \\ \mathcal{A}_1\mathcal{T} \end{bmatrix} \equiv \begin{bmatrix} \mathcal{B}_{\mathcal{T},1} \\ \mathcal{A}_{\mathcal{T},1} \end{bmatrix},$$

we can write

$$\begin{bmatrix} \mathbf{W}_{\mathcal{T},1} \\ \mathbf{V}_{\mathcal{T},1} \end{bmatrix} = \mathcal{P}_{\mathcal{T},1}\mathbf{X},$$

where

$$\mathcal{B}_{\mathcal{T},1} = \begin{bmatrix} h_0 & 0 & 0 & 0 & 0 & \cdots & 0 & 0 & 0 & 0 & h_3 & h_2 & h_1 \\ h_2 & h_1 & h_0 & 0 & 0 & \cdots & 0 & 0 & 0 & 0 & 0 & 0 & h_3 \\ \vdots & \vdots & \vdots & \vdots & \vdots & \ddots & \vdots & \vdots & \vdots & \vdots & \vdots & \vdots & \vdots \\ 0 & 0 & 0 & 0 & 0 & \cdots & h_3 & h_2 & h_1 & h_0 & 0 & 0 & 0 \\ 0 & 0 & 0 & 0 & 0 & \cdots & 0 & 0 & h_3 & h_2 & h_1 & h_0 & 0 \end{bmatrix},$$

and $\mathcal{A}_{\mathcal{T},1}$ has a similar structure with each h_l being replaced by g_l. A comparison of the contents of $\mathcal{B}_1$ and $\mathcal{B}_{\mathcal{T},1}$ says that, because $\mathcal{B}_1\mathbf{X}$ forms the odd indexed values of the filter output $\{2^{1/2}\widetilde{W}_{1,t}\}$, then $\mathcal{B}_{\mathcal{T},1}\mathbf{X}$ must form the even indexed values. Hence the elements of $\mathbf{W}_{\mathcal{T},1}$ are given by

$$\mathbf{W}_{\mathcal{T},1} = [2^{1/2}\widetilde{W}_{1,0}, 2^{1/2}\widetilde{W}_{1,2}, \ldots, 2^{1/2}\widetilde{W}_{1,N-2}]^T,$$

and by a similar argument those of $\mathbf{V}_{\mathcal{T},1}$ are given by

$$\mathbf{V}_{\mathcal{T},1} = [2^{1/2}\widetilde{V}_{1,0}, 2^{1/2}\widetilde{V}_{1,2}, \ldots, 2^{1/2}\widetilde{V}_{1,N-2}]^T.$$

Next we define

$$\widetilde{\mathbf{W}}_1 \equiv [\widetilde{W}_{1,0}, \widetilde{W}_{1,1}, \widetilde{W}_{1,2}, \ldots, \widetilde{W}_{1,N-1}]^T$$

and

$$\widetilde{\mathbf{V}}_1 \equiv [\widetilde{V}_{1,0}, \widetilde{V}_{1,1}, \widetilde{V}_{1,2}, \ldots, \widetilde{V}_{1,N-1}]^T;$$

i.e., $\widetilde{\mathbf{W}}_1$ is formed by rescaling the interleaved elements of $\mathbf{W}_1$ and $\mathbf{W}_{\mathcal{T},1}$, and $\widetilde{\mathbf{V}}_1$ is constructed in a similar fashion from $\mathbf{V}_1$ and $\mathbf{V}_{\mathcal{T},1}$. Note that the elements of $\widetilde{\mathbf{W}}_1$ and $\widetilde{\mathbf{V}}_1$ are exactly the filter outputs $\{\widetilde{W}_{1,t}\}$ and $\{\widetilde{V}_{1,t}\}$ obtained by using the MODWT filters $\{\tilde{h}_l\}$ and $\{\tilde{g}_l\}$ – see Equation (163e). Let us define $\widetilde{\mathcal{B}}_1$ as the $N \times N$ matrix formed by interleaving the rows of $\mathcal{B}_{\mathcal{T},1}$ and $\mathcal{B}_1$ and replacing each h_l by $\tilde{h}_l$; i.e.,

$$\widetilde{\mathcal{B}}_1 \equiv \begin{bmatrix} \tilde{h}_0 & 0 & 0 & \cdots & 0 & 0 & 0 & 0 & \tilde{h}_3 & \tilde{h}_2 & \tilde{h}_1 \\ \tilde{h}_1 & \tilde{h}_0 & 0 & \cdots & 0 & 0 & 0 & 0 & 0 & \tilde{h}_3 & \tilde{h}_2 \\ \tilde{h}_2 & \tilde{h}_1 & \tilde{h}_0 & \cdots & 0 & 0 & 0 & 0 & 0 & 0 & \tilde{h}_3 \\ \vdots & \vdots & \vdots & \ddots & \vdots & \vdots & \vdots & \vdots & \vdots & \vdots & \vdots \\ 0 & 0 & 0 & \cdots & 0 & \tilde{h}_3 & \tilde{h}_2 & \tilde{h}_1 & \tilde{h}_0 & 0 & 0 \\ 0 & 0 & 0 & \cdots & 0 & 0 & \tilde{h}_3 & \tilde{h}_2 & \tilde{h}_1 & \tilde{h}_0 & 0 \\ 0 & 0 & 0 & \cdots & 0 & 0 & 0 & \tilde{h}_3 & \tilde{h}_2 & \tilde{h}_1 & \tilde{h}_0 \end{bmatrix}. \tag{165a}$$

We then have $\widetilde{\mathbf{W}}_1 = \widetilde{\mathcal{B}}_1\mathbf{X}$. With an analogous definition for $\widetilde{\mathcal{A}}_1$, we have $\widetilde{\mathbf{V}}_1 = \widetilde{\mathcal{A}}_1\mathbf{X}$. Finally, we can represent the first stage of the MODWT pyramid algorithm as

$$\begin{bmatrix} \widetilde{\mathbf{W}}_1 \\ \widetilde{\mathbf{V}}_1 \end{bmatrix} = \begin{bmatrix} \widetilde{\mathcal{B}}_1 \\ \widetilde{\mathcal{A}}_1 \end{bmatrix}\mathbf{X} = \widetilde{\mathcal{P}}_1\mathbf{X} \quad \text{with} \quad \widetilde{\mathcal{P}}_1 \equiv \begin{bmatrix} \widetilde{\mathcal{B}}_1 \\ \widetilde{\mathcal{A}}_1 \end{bmatrix}. \tag{165b}$$

Because $\mathcal{P}_1^T\mathcal{P}_1 = I_N$ and $\mathcal{T}^T\mathcal{T} = I_N$, it follows that

$$\mathcal{P}_{\mathcal{T},1}^T\mathcal{P}_{\mathcal{T},1} = \mathcal{T}^T\mathcal{P}_1^T\mathcal{P}_1\mathcal{T} = I_N$$

and hence that $\mathcal{P}_{\mathcal{T},1}$ is an orthonormal matrix. Hence we have both

$$\|\mathbf{X}\|^2 = \|\mathbf{W}_1\|^2 + \|\mathbf{V}_1\|^2 \text{ and } \|\mathbf{X}\|^2 = \|\mathbf{W}_{\mathcal{T},1}\|^2 + \|\mathbf{V}_{\mathcal{T},1}\|^2.$$

Since

$$\|\mathbf{W}_1\|^2 + \|\mathbf{W}_{\mathcal{T},1}\|^2 = 2\|\widetilde{\mathbf{W}}_1\|^2 \text{ and } \|\mathbf{V}_1\|^2 + \|\mathbf{V}_{\mathcal{T},1}\|^2 = 2\|\widetilde{\mathbf{V}}_1\|^2,$$

it follows that

$$\|\mathbf{X}\|^2 = \|\widetilde{\mathbf{W}}_1\|^2 + \|\widetilde{\mathbf{V}}_1\|^2,$$

so that $\|\widetilde{\mathbf{W}}_1\|^2$ and $\|\widetilde{\mathbf{V}}_1\|^2$ decompose $\|\mathbf{X}\|^2$, which is the basic result needed to define an ANOVA using the MODWT.

Since we have both

$$\mathbf{X} = [\mathcal{B}_1^T, \mathcal{A}_1^T]\begin{bmatrix}\mathbf{W}_1\\ \mathbf{V}_1\end{bmatrix} \text{ and } \mathbf{X} = [\mathcal{B}_{\mathcal{T},1}^T, \mathcal{A}_{\mathcal{T},1}^T]\begin{bmatrix}\mathbf{W}_{\mathcal{T},1}\\ \mathbf{V}_{\mathcal{T},1}\end{bmatrix},$$

we can also recover $\mathbf{X}$ by averaging the two right-hand sides:

$$\begin{aligned}\mathbf{X} &= \tfrac{1}{2}[\mathcal{B}_1^T, \mathcal{A}_1^T]\begin{bmatrix}\mathbf{W}_1\\ \mathbf{V}_1\end{bmatrix} + \tfrac{1}{2}[\mathcal{B}_{\mathcal{T},1}^T, \mathcal{A}_{\mathcal{T},1}^T]\begin{bmatrix}\mathbf{W}_{\mathcal{T},1}\\ \mathbf{V}_{\mathcal{T},1}\end{bmatrix}\\ &= \tfrac{1}{2}\left(\mathcal{B}_1^T\mathbf{W}_1 + \mathcal{A}_1^T\mathbf{V}_1 + \mathcal{B}_{\mathcal{T},1}^T\mathbf{W}_{\mathcal{T},1} + \mathcal{A}_{\mathcal{T},1}^T\mathbf{V}_{\mathcal{T},1}\right).\end{aligned}$$

▷ **Exercise [166]:** Argue that

$$\tfrac{1}{2}\left(\mathcal{B}_1^T\mathbf{W}_1 + \mathcal{B}_{\mathcal{T},1}^T\mathbf{W}_{\mathcal{T},1}\right) = \widetilde{\mathcal{B}}_1^T\widetilde{\mathbf{W}}_1$$

and

$$\tfrac{1}{2}\left(\mathcal{A}_1^T\mathbf{V}_1 + \mathcal{A}_{\mathcal{T},1}^T\mathbf{V}_{\mathcal{T},1}\right) = \widetilde{\mathcal{A}}_1^T\widetilde{\mathbf{V}}_1.$$ ◁

With the above, we can write

$$\mathbf{X} = \widetilde{\mathcal{B}}_1^T\widetilde{\mathbf{W}}_1 + \widetilde{\mathcal{A}}_1^T\widetilde{\mathbf{V}}_1 \equiv \widetilde{\mathcal{D}}_1 + \widetilde{\mathcal{S}}_1,$$

where $\widetilde{\mathcal{D}}_1 \equiv \widetilde{\mathcal{B}}_1^T\widetilde{\mathbf{W}}_1$ is the first level maximal overlap detail, while $\widetilde{\mathcal{S}}_1 \equiv \widetilde{\mathcal{A}}_1^T\widetilde{\mathbf{V}}_1$ is the corresponding smooth. The above equation is the basic result needed to define an MRA of level $J_0 = 1$ using the MODWT. (Exercise [5.1] explores an interesting interpretation of $\widetilde{\mathcal{D}}_1$ and $\widetilde{\mathcal{S}}_1$. We can in fact obtain $\widetilde{\mathcal{D}}_1$ by averaging $\mathcal{D}_1$ and $\mathcal{T}\mathcal{D}_{\mathcal{T},1}$, where $\mathcal{D}_1$ and $\mathcal{D}_{\mathcal{T},1}$ are, respectively, first level DWT details for $\mathbf{X}$ and $\mathcal{T}\mathbf{X}$. An analogous statement holds for $\widetilde{\mathcal{S}}_1$.)

It is important to note that, whereas we have the relationship $\|\mathbf{W}_1\|^2 = \|\mathcal{D}_1\|^2$ for the usual DWT, we do *not* in general have equality between $\|\widetilde{\mathbf{W}}_1\|^2$ and $\|\widetilde{\mathcal{D}}_1\|^2$, as the following exercise indicates.

▷ **Exercise [167a]:** Show that

$$\|\widetilde{\mathcal{D}}_1\|^2 = \tfrac{1}{2}\left(\|\widetilde{\mathbf{W}}_1\|^2 + \mathbf{W}_1^T \mathcal{B}_1 \mathcal{B}_{\mathcal{T},1}^T \mathbf{W}_{\mathcal{T},1}\right). \quad ◁$$

Thus, whereas we can define the discrete wavelet empirical power spectrum at scale τ_1 using the usual DWT as either $\|\mathbf{W}_1\|^2/N$ or $\|\mathcal{D}_1\|^2/N$, we can only use $\|\widetilde{\mathbf{W}}_1\|^2/N$ when dealing with the MODWT. (Exercise [5.7] indicates that in fact $\|\widetilde{\mathcal{D}}_1\|^2 \le \|\widetilde{\mathbf{W}}_1\|^2$, but this exercise is best tackled after studying Section 5.4.)

The above derivation of the MODWT pyramid algorithm assumes that N is even so that we can apply the usual DWT pyramid algorithm to $\mathbf{X}$ and to $\mathcal{T}\mathbf{X}$. The following exercise says that the first stage of this algorithm in fact works for all N.

▷ **Exercise [167b]:** For any sample size $N \ge L$, let $\widetilde{\mathcal{B}}_1$ be defined as in Equation (165a), and define $\widetilde{\mathcal{A}}_1$ analogously by substituting $\tilde{g}_l$'s for $\tilde{h}_l$'s. Suppose that $\widetilde{\mathbf{W}}_1$ and $\widetilde{\mathbf{V}}_1$ are defined by Equation (165b). Show that

$$\mathbf{X} = \widetilde{\mathcal{B}}_1^T \widetilde{\mathbf{W}}_1 + \widetilde{\mathcal{A}}_1^T \widetilde{\mathbf{V}}_1 \quad \text{and} \quad \|\mathbf{X}\|^2 = \|\widetilde{\mathbf{W}}_1\|^2 + \|\widetilde{\mathbf{V}}_1\|^2.$$

Hint: show that $\widetilde{\mathcal{B}}_1^T \widetilde{\mathcal{B}}_1 + \widetilde{\mathcal{A}}_1^T \widetilde{\mathcal{A}}_1 = I_N$. ◁

In fact, as indicated by the discussion in Section 5.4, the restriction $N \ge L$ can be relaxed.

With $\widetilde{\mathcal{P}}_1$ defined as in Equation (165b), we can re-express $\mathbf{X} = \widetilde{\mathcal{B}}_1^T \widetilde{\mathbf{W}}_1 + \widetilde{\mathcal{A}}_1^T \widetilde{\mathbf{V}}_1$ as

$$\mathbf{X} = \begin{bmatrix} \widetilde{\mathcal{B}}_1 \\ \widetilde{\mathcal{A}}_1 \end{bmatrix}^T \begin{bmatrix} \widetilde{\mathbf{W}}_1 \\ \widetilde{\mathbf{V}}_1 \end{bmatrix} = \widetilde{\mathcal{P}}_1^T \begin{bmatrix} \widetilde{\mathbf{W}}_1 \\ \widetilde{\mathbf{V}}_1 \end{bmatrix}; \text{ furthermore, since } \begin{bmatrix} \widetilde{\mathbf{W}}_1 \\ \widetilde{\mathbf{V}}_1 \end{bmatrix} = \widetilde{\mathcal{P}}_1 \mathbf{X},$$

it is evident that $\mathbf{X} = \widetilde{\mathcal{P}}_1^T \widetilde{\mathcal{P}}_1 \mathbf{X}$ for any time series and hence that $\widetilde{\mathcal{P}}_1^T$ can be considered as an inverse to the MODWT transform $\widetilde{\mathcal{P}}_1$. Because $\widetilde{\mathcal{P}}_1$ is a $2N \times N$ matrix, there are in fact an infinite number of ways of recovering $\mathbf{X}$ from its MODWT coefficients $\widetilde{\mathbf{W}}_1$ and $\widetilde{\mathbf{V}}_1$, so $\widetilde{\mathcal{P}}_1^T$ is by no means a unique inverse for $\widetilde{\mathcal{P}}_1$; however, Exercise [5.2] is to show that $\widetilde{\mathcal{P}}_1^T$ is in fact the (unique) Moore–Penrose generalized inverse for $\widetilde{\mathcal{P}}_1$.

For the DWT, we argued in Section 4.4 that the detail $\mathcal{D}_1$ can be constructed by cross-correlating the periodized filter $\{h_l^\circ\}$ with an upsampled version of $\mathbf{W}_1$. Here we show that the MODWT detail $\widetilde{\mathcal{D}}_1$ can be interpreted in terms of a filtering operation involving $\mathbf{X}$ and a zero phase filter. Let $\widetilde{\mathcal{D}}_{1,t}$ be the tth element of $\widetilde{\mathcal{D}}_1$. Expanding out $\widetilde{\mathcal{D}}_1 = \widetilde{\mathcal{B}}_1^T \widetilde{\mathbf{W}}_1$ yields

$$\widetilde{\mathcal{D}}_1 = \begin{bmatrix} \tilde{h}_0 & \tilde{h}_1 & \tilde{h}_2 & \tilde{h}_3 & 0 & \cdots & 0 & 0 & 0 & 0 \\ 0 & \tilde{h}_0 & \tilde{h}_1 & \tilde{h}_2 & \tilde{h}_3 & \cdots & 0 & 0 & 0 & 0 \\ \vdots & \vdots & \vdots & \vdots & \vdots & \ddots & \vdots & \vdots & \vdots & \vdots \\ 0 & 0 & 0 & 0 & 0 & \cdots & \tilde{h}_0 & \tilde{h}_1 & \tilde{h}_2 & \tilde{h}_3 \\ \tilde{h}_3 & 0 & 0 & 0 & 0 & \cdots & 0 & \tilde{h}_0 & \tilde{h}_1 & \tilde{h}_2 \\ \tilde{h}_2 & \tilde{h}_3 & 0 & 0 & 0 & \cdots & 0 & 0 & \tilde{h}_0 & \tilde{h}_1 \\ \tilde{h}_1 & \tilde{h}_2 & \tilde{h}_3 & 0 & 0 & \cdots & 0 & 0 & 0 & \tilde{h}_0 \end{bmatrix} \begin{bmatrix} \widetilde{W}_{1,0} \\ \widetilde{W}_{1,1} \\ \widetilde{W}_{1,2} \\ \vdots \\ \widetilde{W}_{1,N-2} \\ \widetilde{W}_{1,N-1} \end{bmatrix},$$

from which we have

$$\widetilde{\mathcal{D}}_{1,t} = \sum_{l=0}^{L-1} \tilde{h}_l \widetilde{W}_{1,t+l \bmod N} = \sum_{l=0}^{N-1} \tilde{h}_l^\circ \widetilde{W}_{1,t+l \bmod N}, \quad t = 0, 1, \ldots, N-1,$$

where $\{\tilde{h}_l^\circ : l = 0, \ldots, N-1\}$ is $\{\tilde{h}_l : l = 0, \ldots, L-1\}$ periodized to length N. Thus the detail series can be obtained by circularly cross-correlating $\{\widetilde{W}_{1,t}\}$ with $\{\tilde{h}_l^\circ\}$ (cf. Equation (37c)). Since $\widetilde{H}(\cdot)$ is the transfer function for $\{\tilde{h}_l\}$, an equivalent statement is that the detail series can be obtained by circularly filtering $\{\widetilde{W}_{1,t}\}$ with a filter whose DFT is $\{\widetilde{H}^*(\frac{k}{N}) : k = 0, \ldots, N-1\}$. Since $\{\widetilde{W}_{1,t}\}$ was formed by filtering $\{X_t\}$ using $\{\tilde{h}_l^\circ\}$, the detail series can equivalently be formed by filtering $\{X_t\}$ with a composite filter obtained by convolving $\{\tilde{h}_l^\circ\}$ with the filter whose DFT is $\{\widetilde{H}^*(\frac{k}{N})\}$. Now the DFT for the composite circular filter is given by the product of the DFTs for the individual filters, which yields

$$\widetilde{H}^*(\tfrac{k}{N})\widetilde{H}(\tfrac{k}{N}) = |\widetilde{H}(\tfrac{k}{N})|^2 = \widetilde{\mathcal{H}}(\tfrac{k}{N}).$$

Because $\widetilde{\mathcal{H}}(\cdot)$ is real-valued and nonnegative, its phase function is zero for all frequencies (if we define it as such for $f = 0$, where $\widetilde{\mathcal{H}}(0) = 0$), which establishes the important result that the composite filter associated with $\widetilde{\mathcal{D}}_1$ has zero phase. Using a similar argument, it follows that $\widetilde{\mathcal{S}}_1$ can be expressed as filtering $\{\widetilde{V}_{1,t}\}$ with a filter whose DFT is $\{\widetilde{G}^*(\frac{k}{N})\}$, where $\widetilde{G}(\cdot)$ is the transfer function for $\{\tilde{g}_l\}$. The composite filter we would apply to $\mathbf{X}$ to obtain $\widetilde{\mathcal{S}}_1$ has a DFT composed of samples from a zero phase transfer function defined by $\widetilde{\mathcal{G}}(f) \equiv |\widetilde{G}(f)|^2$.

Key Facts and Definitions in Section 5.3

To summarize, the N dimensional vector $\widetilde{\mathbf{W}}_1$ contains the MODWT wavelet coefficients of level $j = 1$. The tth element of $\widetilde{\mathbf{W}}_1$ is $\widetilde{W}_{1,t}$, $t = 0, 1, \ldots, N-1$, and is formed by circularly filtering the time series $\{X_t\}$ with the MODWT wavelet filter $\{\tilde{h}_l \equiv h_l/\sqrt{2}\}$ (see Equation (163e)); equivalently, we can regard $\{\widetilde{W}_{1,t}\}$ as the result of circularly filtering $\{X_t\}$ with the filter $\{\tilde{h}_l^\circ\}$, which is $\{\tilde{h}_l\}$ periodized to length N. Likewise, the MODWT scaling coefficients $\widetilde{\mathbf{V}}_1$ are obtained by filtering $\{X_t\}$ with the MODWT scaling filter $\{\tilde{g}_l \equiv g_l/\sqrt{2}\}$; equivalently, we can consider $\{\widetilde{V}_{1,t}\}$ to be the result of filtering $\{X_t\}$ with $\{\tilde{g}_l^\circ\}$, which is the periodized version of $\{\tilde{g}_l\}$). Each row of the $N \times N$ matrix $\widetilde{\mathcal{B}}_1$ contains the periodized filter $\{\tilde{h}_l^\circ\}$, with the difference between rows given by circular shifts; likewise, $\widetilde{\mathcal{A}}_1$ has rows whose elements are the periodized filter $\{\tilde{g}_l^\circ\}$. With these definitions, we have

$$\|\mathbf{X}\|^2 = \|\widetilde{\mathbf{W}}_1\|^2 + \|\widetilde{\mathbf{V}}_1\|^2 \text{ and } \mathbf{X} = \widetilde{\mathcal{B}}_1^T\widetilde{\mathbf{W}}_1 + \widetilde{\mathcal{A}}_1^T\widetilde{\mathbf{V}}_1 \equiv \widetilde{\mathcal{D}}_1 + \widetilde{\mathcal{S}}_1,$$

which yield, respectively, an energy decomposition and an additive decomposition in terms of the first level MODWT detail $\widetilde{\mathcal{D}}_1$ and the first level MODWT smooth $\widetilde{\mathcal{S}}_1$. The elements of $\widetilde{\mathcal{D}}_1$ can be obtained by filtering $\widetilde{\mathbf{W}}_1$ with a filter whose transfer function is $\widetilde{H}^*(\cdot)$, i.e., the complex conjugate of the transfer function $\widetilde{H}(\cdot)$ of $\{\tilde{h}_l\}$; likewise, the elements of $\widetilde{\mathcal{S}}_1$ are the result of filtering $\widetilde{\mathbf{V}}_1$ with a filter whose transfer function is $\widetilde{G}^*(\cdot)$, which is the complex conjugate of the transfer function for $\{\tilde{g}_l\}$. A flow diagram depicting the analysis and synthesis of $\mathbf{X}$ in terms of MODWT coefficients is shown in Figure 169.

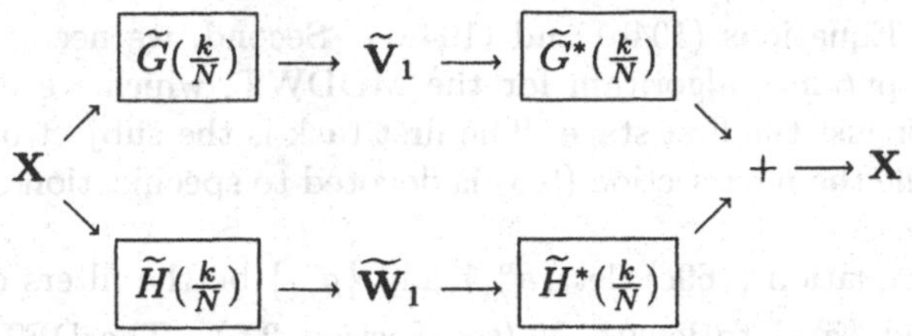

$$\mathbf{X} \nearrow \boxed{\widetilde{G}(\tfrac{k}{N})} \longrightarrow \widetilde{\mathbf{V}}_1 \longrightarrow \boxed{\widetilde{G}^*(\tfrac{k}{N})} \searrow + \longrightarrow \mathbf{X}$$
$$\mathbf{X} \searrow \boxed{\widetilde{H}(\tfrac{k}{N})} \longrightarrow \widetilde{\mathbf{W}}_1 \longrightarrow \boxed{\widetilde{H}^*(\tfrac{k}{N})} \nearrow + \longrightarrow \mathbf{X}$$

Figure 169. Flow diagram illustrating analysis of $\mathbf{X}$ into the MODWT wavelet and scaling coefficients $\widetilde{\mathbf{W}}_1$ and $\widetilde{\mathbf{V}}_1$ of first level, followed by the synthesis of $\mathbf{X}$ from $\widetilde{\mathbf{W}}_1$ and $\widetilde{\mathbf{V}}_1$.

5.4 Definition of jth Level MODWT Coefficients

For arbitrary sample size N, we now *define* the jth level MODWT wavelet and scaling coefficients to be the N dimensional vectors $\widetilde{\mathbf{W}}_j$ and $\widetilde{\mathbf{V}}_j$ whose elements are, respectively,

$$\widetilde{W}_{j,t} \equiv \sum_{l=0}^{L_j-1} \tilde{h}_{j,l} X_{t-l \bmod N} \text{ and } \widetilde{V}_{j,t} \equiv \sum_{l=0}^{L_j-1} \tilde{g}_{j,l} X_{t-l \bmod N}, \tag{169a}$$

$t = 0, \ldots, N-1$, where $\tilde{h}_{j,l} \equiv h_{j,l}/2^{j/2}$ and $\tilde{g}_{j,l} \equiv g_{j,l}/2^{j/2}$. The filters $\{\tilde{h}_{j,l}\}$ and $\{\tilde{g}_{j,l}\}$ are called the jth level MODWT wavelet and scaling filters. Note that, because $\{h_{j,l}\}$ and $\{g_{j,l}\}$ each have width

$$L_j \equiv (2^j - 1)(L-1) + 1$$

(cf. Exercise [96]), so do $\{\tilde{h}_{j,l}\}$ and $\{\tilde{g}_{j,l}\}$. Because the transfer functions for $\{\tilde{h}_l\}$ and $\{\tilde{g}_l\}$ are given by $\widetilde{H}(f) \equiv H(f)/\sqrt{2}$ and $\widetilde{G}(f) \equiv G(f)/\sqrt{2}$, it is evident from Equations (96b) and (97) that the transfer functions for $\{\tilde{h}_{j,l}\}$ and $\{\tilde{g}_{j,l}\}$ are given, respectively, by

$$\widetilde{H}_j(f) \equiv \widetilde{H}(2^{j-1}f)\prod_{l=0}^{j-2}\widetilde{G}(2^l f) \text{ and } \widetilde{G}_j(f) \equiv \prod_{l=0}^{j-1}\widetilde{G}(2^l f). \tag{169b}$$

For $j = 1$, the above definitions yield $L_1 \equiv L$ and $\widetilde{G}_1(f) \equiv \widetilde{G}(f)$; we also take $\widetilde{H}_1(f) \equiv \widetilde{H}(f)$, $\tilde{h}_{1,l} \equiv \tilde{h}_l$ and $\tilde{g}_{1,l} \equiv \tilde{g}_l$.

We now have two tasks in front of us. First, given any integer $J_0 \geq 1$ and any sample size N, we need to prove that our definition for the MODWT yields an energy decomposition

$$\|\mathbf{X}\|^2 = \sum_{j=1}^{J_0} \|\widetilde{\mathbf{W}}_j\|^2 + \|\widetilde{\mathbf{V}}_{J_0}\|^2 \tag{169c}$$

and an additive decomposition

$$\mathbf{X} = \sum_{j=1}^{J_0} \widetilde{\mathcal{D}}_j + \widetilde{\mathcal{S}}_{J_0}, \tag{169d}$$

where $\widetilde{\mathcal{D}}_j$ and $\widetilde{\mathcal{S}}_{J_0}$ depend on $\mathbf{X}$ only through, respectively, $\widetilde{\mathbf{W}}_j$ and $\widetilde{\mathbf{V}}_{J_0}$ in a yet to be defined way. Note that the above results are similar to the ones derived in Section 4.7

for the DWT (see Equations (104b) and (104a)). Second, we need to complete our description of the pyramid algorithm for the MODWT, which we described in the previous section for just the first stage. The first task is the subject of the remainder of this section, while the next section (5.5) is devoted to specification of the MODWT pyramid algorithm.

To establish Equation (169c), let $\{\tilde{h}_{j,l}^{\circ}\}$ and $\{\tilde{g}_{j,l}^{\circ}\}$ be the filters obtained by periodizing $\{\tilde{h}_{j,l}\}$ and $\{\tilde{g}_{j,l}\}$ to length N (see Section 2.6). The DFTs of $\{\tilde{h}_{j,l}^{\circ}\}$ and $\{\tilde{g}_{j,l}^{\circ}\}$ are thus given, respectively, by $\{\widetilde{H}_j(\frac{k}{N})\}$ and $\{\widetilde{G}_j(\frac{k}{N})\}$. Because circularly filtering $\{X_t\}$ with $\{\tilde{h}_{j,l}\}$ is equivalent to filtering it with $\{\tilde{h}_{j,l}^{\circ}\}$, we can re-express Equation (169a) as

$$\widetilde{W}_{j,t} = \sum_{l=0}^{N-1} \tilde{h}_{j,l}^{\circ} X_{t-l \bmod N} \text{ and } \widetilde{V}_{j,t} = \sum_{l=0}^{N-1} \tilde{g}_{j,l}^{\circ} X_{t-l \bmod N}, \tag{170a}$$

$t = 0, \ldots, N-1$. Letting $\{\mathcal{X}_k\}$ be the DFT of $\{X_t\}$, we have

$$\{\widetilde{W}_{j,t}\} \longleftrightarrow \{\widetilde{H}_j(\tfrac{k}{N})\mathcal{X}_k\} \text{ and } \{\widetilde{V}_{j,t}\} \longleftrightarrow \{\widetilde{G}_j(\tfrac{k}{N})\mathcal{X}_k\} \tag{170b}$$

(see Equations (37a) and (37b)). Parseval's theorem (Equation (36h)) now tells us that

$$\|\widetilde{\mathbf{W}}_j\|^2 = \frac{1}{N}\sum_{k=0}^{N-1} |\widetilde{H}_j(\tfrac{k}{N})|^2 |\mathcal{X}_k|^2 \text{ and } \|\widetilde{\mathbf{V}}_j\|^2 = \frac{1}{N}\sum_{k=0}^{N-1} |\widetilde{G}_j(\tfrac{k}{N})|^2 |\mathcal{X}_k|^2, \tag{170c}$$

which in turn yields

$$\|\widetilde{\mathbf{W}}_j\|^2 + \|\widetilde{\mathbf{V}}_j\|^2 = \frac{1}{N}\sum_{k=0}^{N-1} |\mathcal{X}_k|^2 \left(|\widetilde{H}_j(\tfrac{k}{N})|^2 + |\widetilde{G}_j(\tfrac{k}{N})|^2\right).$$

For any $j \geq 2$, we can use Equation (169b) to reduce the terms in the parentheses above as follows:

$$\begin{aligned}
|\widetilde{H}_j(\tfrac{k}{N})|^2 + |\widetilde{G}_j(\tfrac{k}{N})|^2 &= |\widetilde{H}(2^{j-1}\tfrac{k}{N})|^2 \prod_{l=0}^{j-2} |\widetilde{G}(2^l \tfrac{k}{N})|^2 + \prod_{l=0}^{j-1} |\widetilde{G}(2^l \tfrac{k}{N})|^2 \\
&= \left(|\widetilde{H}(2^{j-1}\tfrac{k}{N})|^2 + |\widetilde{G}(2^{j-1}\tfrac{k}{N})|^2\right) \prod_{l=0}^{j-2} |\widetilde{G}(2^l \tfrac{k}{N})|^2 \\
&= \left(\widetilde{\mathcal{H}}(2^{j-1}\tfrac{k}{N}) + \widetilde{\mathcal{G}}(2^{j-1}\tfrac{k}{N})\right) |\widetilde{G}_{j-1}(\tfrac{k}{N})|^2 \\
&= |\widetilde{G}_{j-1}(\tfrac{k}{N})|^2,
\end{aligned} \tag{170d}$$

where we have used the fact that $\widetilde{\mathcal{H}}(f) + \widetilde{\mathcal{G}}(f) = 1$ for all f (this is Equation (163c)). We now have

$$\|\widetilde{\mathbf{W}}_j\|^2 + \|\widetilde{\mathbf{V}}_j\|^2 = \frac{1}{N}\sum_{k=0}^{N-1} |\mathcal{X}_k|^2 |\widetilde{G}_{j-1}(\tfrac{k}{N})|^2 = \|\widetilde{\mathbf{V}}_{j-1}\|^2 \tag{170e}$$

from an application of Equation (170c). Since the above holds for any $j \geq 2$, a proof by induction says that

$$\|\widetilde{\mathbf{V}}_1\|^2 = \sum_{j=2}^{J_0} \|\widetilde{\mathbf{W}}_j\|^2 + \|\widetilde{\mathbf{V}}_{J_0}\|^2$$

for any $J_0 \geq 2$ also. The energy decomposition of Equation (169c) is now established if we can show that $\|\mathbf{X}\|^2 = \|\widetilde{\mathbf{W}}_1\|^2 + \|\widetilde{\mathbf{V}}_1\|^2$; in fact, Exercise [167b] states that this is true when $N \geq L$, but this restriction is removed in the following exercise.

▷ **Exercise [171a]:** Show that

$$\|\mathbf{X}\|^2 = \|\widetilde{\mathbf{W}}_1\|^2 + \|\widetilde{\mathbf{V}}_1\|^2$$

by using an argument that parallels the one leading to (170e). ◁

We can use the energy decomposition of Equation (169c) to provide a MODWT-based analysis of the sample variance of $\mathbf{X}$ as follows. Since $\hat{\sigma}_X^2 = \frac{1}{N}\|\mathbf{X}\|^2 - \overline{X}^2$ (as indicated in Equation (48b)), it follows that

$$\hat{\sigma}_X^2 = \frac{1}{N}\sum_{j=1}^{J_0} \|\widetilde{\mathbf{W}}_j\|^2 + \frac{1}{N}\|\widetilde{\mathbf{V}}_{J_0}\|^2 - \overline{X}^2 \tag{171a}$$

(the above closely parallels the DWT-based ANOVA given in Equation (104c)). The following exercise shows that the last two terms above can be regarded as the sample variance of $\widetilde{\mathbf{V}}_{J_0}$.

▷ **Exercise [171b]:** Show that the sample mean of $\widetilde{\mathbf{V}}_{J_0}$ is equal to $\overline{X}$ and hence that the sample variance of $\widetilde{\mathbf{V}}_{J_0}$ is given by $\frac{1}{N}\|\widetilde{\mathbf{V}}_{J_0}\|^2 - \overline{X}^2$. ◁

In preparation for establishing the MODWT-based MRA of Equation (169d), we first need to define $\widetilde{\mathcal{D}}_j$ and $\widetilde{\mathcal{S}}_j$. To do so, first note that, in matrix notation, we can express the transforms from $\mathbf{X}$ to $\widetilde{\mathbf{W}}_j$ and from $\mathbf{X}$ to $\widetilde{\mathbf{V}}_j$ as

$$\widetilde{\mathbf{W}}_j = \widetilde{\mathcal{W}}_j\mathbf{X} \text{ and } \widetilde{\mathbf{V}}_j = \widetilde{\mathcal{V}}_j\mathbf{X}, \tag{171b}$$

where each row of the $N \times N$ matrix $\widetilde{\mathcal{W}}_j$ has values dictated by $\{\tilde{h}_{j,l}^\circ\}$, while $\widetilde{\mathcal{V}}_j$ has values dictated by $\{\tilde{g}_{j,l}^\circ\}$; i.e., in view of Equation (170a), we must have

$$\widetilde{\mathcal{W}}_j \equiv \begin{bmatrix} \tilde{h}_{j,0}^\circ & \tilde{h}_{j,N-1}^\circ & \tilde{h}_{j,N-2}^\circ & \tilde{h}_{j,N-3}^\circ & \cdots & \tilde{h}_{j,3}^\circ & \tilde{h}_{j,2}^\circ & \tilde{h}_{j,1}^\circ \\ \tilde{h}_{j,1}^\circ & \tilde{h}_{j,0}^\circ & \tilde{h}_{j,N-1}^\circ & \tilde{h}_{j,N-2}^\circ & \cdots & \tilde{h}_{j,4}^\circ & \tilde{h}_{j,3}^\circ & \tilde{h}_{j,2}^\circ \\ \tilde{h}_{j,2}^\circ & \tilde{h}_{j,1}^\circ & \tilde{h}_{j,0}^\circ & \tilde{h}_{j,N-1}^\circ & \cdots & \tilde{h}_{j,5}^\circ & \tilde{h}_{j,4}^\circ & \tilde{h}_{j,3}^\circ \\ \vdots & \vdots & \vdots & \vdots & \cdots & \vdots & \vdots & \vdots \\ \tilde{h}_{j,N-2}^\circ & \tilde{h}_{j,N-3}^\circ & \tilde{h}_{j,N-4}^\circ & \tilde{h}_{j,N-5}^\circ & \cdots & \tilde{h}_{j,1}^\circ & \tilde{h}_{j,0}^\circ & \tilde{h}_{j,N-1}^\circ \\ \tilde{h}_{j,N-1}^\circ & \tilde{h}_{j,N-2}^\circ & \tilde{h}_{j,N-3}^\circ & \tilde{h}_{j,N-4}^\circ & \cdots & \tilde{h}_{j,2}^\circ & \tilde{h}_{j,1}^\circ & \tilde{h}_{j,0}^\circ \end{bmatrix}, \tag{171c}$$

while $\widetilde{\mathcal{V}}_j$ is expressed as above with each $\tilde{h}_{j,l}^\circ$ replaced by $\tilde{g}_{j,l}^\circ$. In analogy to Equation (95c), the desired definitions are

$$\widetilde{\mathcal{D}}_j \equiv \widetilde{\mathcal{W}}_j^T\widetilde{\mathbf{W}}_j \text{ and } \widetilde{\mathcal{S}}_j \equiv \widetilde{\mathcal{V}}_j^T\widetilde{\mathbf{V}}_j.$$

With these definitions, we can now prove the MRA of Equation (169d) as follows.

In view of what the rows of $\widetilde{\mathcal{W}}_j^T$ and $\widetilde{\mathcal{V}}_j^T$ look like (cf. Equation (171c)), we can express the elements of $\widetilde{\mathcal{D}}_j$ and $\widetilde{\mathcal{S}}_j$ as

$$\widetilde{\mathcal{D}}_{j,t} = \sum_{l=0}^{N-1} \tilde{h}_{j,l}^{\circ} \widetilde{W}_{j,t+l \bmod N} \text{ and } \widetilde{\mathcal{S}}_{j,t} = \sum_{l=0}^{N-1} \tilde{g}_{j,l}^{\circ} \widetilde{V}_{j,t+l \bmod N},$$

$t = 0, 1, \ldots, N-1$. Using an argument analogous to the one in Section 5.3 establishing that the elements of $\widetilde{\mathcal{D}}_1$ are the output from a zero phase filter, we know that $\{\widetilde{\mathcal{D}}_{j,t}\}$ is the result of circularly filtering $\{\widetilde{W}_{j,t}\}$ with a filter whose DFT is $\{\widetilde{H}_j^*(\frac{k}{N})\}$; likewise, $\{\widetilde{W}_{j,t}\}$ is obtained by circularly filtering $\{X_t\}$ with the filter $\{\tilde{h}_{j,l}^{\circ}\}$ whose DFT is $\{\widetilde{H}_j(\frac{k}{N})\}$. Letting $\{\mathcal{X}_k\}$ represent the DFT of $\{X_t\}$ again, it follows that

$$\{\widetilde{\mathcal{D}}_{j,t}\} \longleftrightarrow \{\widetilde{H}_j^*(\tfrac{k}{N})\widetilde{H}_j(\tfrac{k}{N})\mathcal{X}_k\} = \{|\widetilde{H}_j(\tfrac{k}{N})|^2 \mathcal{X}_k\}; \tag{172a}$$

likewise, we have

$$\{\widetilde{\mathcal{S}}_{j,t}\} \longleftrightarrow \{|\widetilde{G}_j(\tfrac{k}{N})|^2 \mathcal{X}_k\}. \tag{172b}$$

Since $\{a_t\} \longleftrightarrow \{A_k\}$ and $\{b_t\} \longleftrightarrow \{B_k\}$ implies $\{a_t + b_t\} \longleftrightarrow \{A_k + B_k\}$ (an easy exercise!), we have

$$\{\widetilde{\mathcal{D}}_{j,t} + \widetilde{\mathcal{S}}_{j,t}\} \longleftrightarrow \{(|\widetilde{H}_j(\tfrac{k}{N})|^2 + |\widetilde{G}_j(\tfrac{k}{N})|^2)\mathcal{X}_k\}.$$

Equation (170d) says that $|\widetilde{H}_j(\frac{k}{N})|^2 + |\widetilde{G}_j(\frac{k}{N})|^2 = |\widetilde{G}_{j-1}(\frac{k}{N})|^2$ when $j \geq 2$, so the above becomes

$$\{\widetilde{\mathcal{D}}_{j,t} + \widetilde{\mathcal{S}}_{j,t}\} \longleftrightarrow \{|\widetilde{G}_{j-1}(\tfrac{k}{N})|^2 \mathcal{X}_k\};$$

however, Equation (172b) says that $\{\widetilde{\mathcal{S}}_{j-1,t}\} \longleftrightarrow \{|\widetilde{G}_{j-1}(\frac{k}{N})|^2 \mathcal{X}_k\}$ also, so we must have $\widetilde{\mathcal{S}}_{j-1,t} = \widetilde{\mathcal{D}}_{j,t} + \widetilde{\mathcal{S}}_{j,t}$ for all t, establishing that

$$\widetilde{\mathcal{S}}_{j-1} = \widetilde{\mathcal{D}}_j + \widetilde{\mathcal{S}}_j \tag{172c}$$

for all $j \geq 2$. An easy induction argument says that

$$\widetilde{\mathcal{S}}_1 = \sum_{j=2}^{J_0} \widetilde{\mathcal{D}}_j + \widetilde{\mathcal{S}}_{J_0}$$

for any $J_0 \geq 2$. The MRA of Equation (169d) is now established if we can show that $\mathbf{X} = \widetilde{\mathcal{S}}_1 + \widetilde{\mathcal{D}}_1$; although Exercise [167b] says that this holds when $N \geq L$, we can remove this restriction in the following exercise.

▷ **Exercise [172]:** Show that

$$\mathbf{X} = \widetilde{\mathcal{S}}_1 + \widetilde{\mathcal{D}}_1$$

by using an argument that parallels the one leading to Equation (172c). ◁

Key Facts and Definitions in Section 5.4

For any sample size N, the MODWT wavelet and scaling coefficients are defined, respectively, by

$$\widetilde{W}_{j,t} \equiv \sum_{l=0}^{L_j-1} \tilde{h}_{j,l} X_{t-l \bmod N} \text{ and } \widetilde{V}_{j,t} \equiv \sum_{l=0}^{L_j-1} \tilde{g}_{j,l} X_{t-l \bmod N},$$

$t = 0, \ldots, N-1$, where $\{\tilde{h}_{j,l} : l = 0, \ldots, L_j - 1\}$ and $\{\tilde{g}_{j,l} : l = 0, \ldots, L_j - 1\}$ are, respectively, the jth level MODWT wavelet and scaling filters, which are defined in terms of the jth level wavelet and scaling filters $\{h_{j,l}\}$ and $\{g_{j,l}\}$ via $\tilde{h}_{j,l} \equiv h_{j,l}/2^{j/2}$ and $\tilde{g}_{j,l} \equiv g_{j,l}/2^{j/2}$ (here $L_j \equiv (2^j - 1)(L-1) + 1$ as before). In matrix notation, we can express the above as

$$\widetilde{\mathbf{W}}_j = \widetilde{\mathcal{W}}_j \mathbf{X} \text{ and } \widetilde{\mathbf{V}}_j = \widetilde{\mathcal{V}}_j \mathbf{X},$$

where the rows of $\widetilde{\mathbf{W}}_j$ contain circularly shifted versions of $\{\tilde{h}^{\circ}_{j,l}\}$ (i.e., $\{\tilde{h}_{j,l}\}$ periodized to length N), while the rows of $\widetilde{\mathbf{V}}_j$ contain circularly shifted versions of $\{\tilde{g}^{\circ}_{j,l}\}$. The transfer functions for the jth MODWT wavelet and scaling filters are given by, respectively,

$$\widetilde{H}_j(f) \equiv \widetilde{H}(2^{j-1} f) \prod_{l=0}^{j-2} \widetilde{G}(2^l f) \text{ and } \widetilde{G}_j(f) \equiv \prod_{l=0}^{j-1} \widetilde{G}(2^l f),$$

where, $\widetilde{H}(\cdot)$ and $\widetilde{G}(\cdot)$ are the transfer functions for $\{\tilde{h}_l\} \equiv \{h_l/\sqrt{2}\}$ and $\{\tilde{g}_l\} \equiv \{g_l/\sqrt{2}\}$. With the definition $\widetilde{\mathbf{V}}_0 = \mathbf{X}$, we have

$$\|\widetilde{\mathbf{W}}_j\|^2 + \|\widetilde{\mathbf{V}}_j\|^2 = \|\widetilde{\mathbf{V}}_{j-1}\|^2$$

for $j = 1, 2, \ldots$, leading to the MODWT energy decomposition

$$\|\mathbf{X}\|^2 = \sum_{j=1}^{J_0} \|\widetilde{\mathbf{W}}_j\|^2 + \|\widetilde{\mathbf{V}}_{J_0}\|^2$$

for any $J_0 \geq 1$. The jth level MODWT detail and smooth are defined by, respectively,

$$\widetilde{\mathcal{D}}_j \equiv \widetilde{\mathcal{W}}_j^T \widetilde{\mathbf{W}}_j \text{ and } \widetilde{\mathcal{S}}_j \equiv \widetilde{\mathcal{V}}_j^T \widetilde{\mathbf{V}}_j,$$

in terms of which we can express the MODWT additive decomposition

$$\mathbf{X} = \sum_{j=1}^{J_0} \widetilde{\mathcal{D}}_j + \widetilde{\mathcal{S}}_{J_0}$$

for any $J_0 \geq 1$.

Comments and Extensions to Section 5.4

[1] If the sample size N is a power of two, the DWT of the time series $\mathbf{X}$ can be extracted from the MODWT because of the relationship $W_{j,t} = 2^{j/2}\widetilde{W}_{j,2^j(t+1)-1}$. It is also true that, for any m, the DWT of the circularly shifted series $\mathcal{T}^m\mathbf{X}$ can be extracted from the MODWT. Liang and Parks (1996) used this fact in defining an orthonormal 'translation invariant' DWT (similar ideas are considered by Del Marco and Weiss, 1997). They considered the DWTs of $\mathcal{T}^m\mathbf{X}$, $m = 0, \ldots, N-1$, and picked the one that maximizes a certain energy concentration measure. This procedure yields the shifted DWT that is best matched to $\mathbf{X}$ in the sense of concentrating a large portion of the DWT energy in as few coefficients as possible. This transform is indeed insensitive to circular shifts because it takes into account all possible circular shifts of $\mathbf{X}$ and hence will yield the same result if we were to use any $\mathcal{T}^m\mathbf{X}$ as our original time series in place of $\mathbf{X}$.

Since the DWT requires $O(N)$ multiplications, a brute force approach to computing all N of the DWTs needed to find the translation invariant DWT would take $O(N^2)$ multiplications, but Liang and Parks noted that the MODWT can compute all the required coefficients in just $O(N \log_2 N)$ multiplications. The fundamental reason for this reduction is that the DWTs for particular collections of $\mathcal{T}^m\mathbf{X}$'s have the same coefficients at particular scales. For example, $\mathcal{T}^m\mathbf{X}$ for $m = 0, 2, \ldots, N-2$ all have the same $j = 1$ coefficients, as do $\mathcal{T}^m\mathbf{X}$ for $m = 1, 3, \ldots, N-1$; $\mathcal{T}^m\mathbf{X}$ for $m = 0, 4, \ldots, N-4$ all share the same $j = 2$ coefficients, as do $\mathcal{T}^m\mathbf{X}$ for either $m = 1, 5, \ldots, N-3$, $m = 2, 6, \ldots, N-2$ or $m = 3, 7, \ldots, N-1$; and so forth.

[2] In Section 4.6, we argued that it is reasonable to associate the DWT scaling coefficients $\{V_{j,t}\}$ with scale $\lambda_j \equiv 2^j$ because the autocorrelation width for the scaling filter $\{g_{j,l}\}$ is always equal to λ_j:

$$\text{width}_\text{a}\,\{g_{j,l}\} \equiv \frac{\left(\sum_{l=-\infty}^{\infty} g_{j,l}\right)^2}{\sum_{l=-\infty}^{\infty} g_{j,l}^2} = \lambda_j$$

(this is Equation (103)). Since $\tilde{g}_{j,l} \equiv g_{j,l}/2^{j/2}$, it is easy to see that $\text{width}_\text{a}\,\{\tilde{g}_{j,l}\} = \lambda_j$ also because the factor $1/2^{j/2}$ cancels out in the ratio above. We can thus claim that the MODWT scaling coefficients $\{\widetilde{V}_{j,t}\}$ are associated with averages over scale λ_j and, using the same line of reasoning as in the DWT case, that the MODWT wavelet coefficients $\{\widetilde{W}_{j,t}\}$ are associated with changes in averages over scale $\tau_j \equiv 2^{j-1}$.

5.5 Pyramid Algorithm for the MODWT

Here we describe an efficient algorithm for computing the jth level MODWT wavelet and scaling coefficients $\widetilde{\mathbf{W}}_j$ and $\widetilde{\mathbf{V}}_j$ based upon the scaling coefficients $\widetilde{\mathbf{V}}_{j-1}$ of level $j-1$ (pseudo-code for this algorithm is given in item [1] of the Comments and Extensions). The key to this algorithm is to note the relationship between the filters used to compute the coefficients of levels $j-1$ and j. Equation (170a) states that the elements of $\widetilde{\mathbf{W}}_j$, $\widetilde{\mathbf{V}}_j$ and $\widetilde{\mathbf{V}}_{j-1}$ are obtained by circularly filtering $\{X_t\}$ with, respectively, the periodized filters $\{\tilde{h}_{j,l}^\circ\}$, $\{\tilde{g}_{j,l}^\circ\}$ and $\{\tilde{g}_{j-1,l}^\circ\}$. The DFTs of these filters are, respectively, $\{\widetilde{H}_j(\frac{k}{N})\}$, $\{\widetilde{G}_j(\frac{k}{N})\}$ and $\{\widetilde{G}_{j-1}(\frac{k}{N})\}$. An application of Equation (169b) tells us that the first two of these DFTs can be expressed in terms of the third DFT and the transfer functions $\widetilde{H}(\cdot)$ and $\widetilde{G}(\cdot)$ for the basic wavelet and scaling filters $\{\tilde{h}_l\}$ and $\{\tilde{g}_l\}$ in the following way:

$$\widetilde{H}_j(\tfrac{k}{N}) = \widetilde{G}_{j-1}(\tfrac{k}{N})\widetilde{H}(2^{j-1}\tfrac{k}{N}) \text{ and } \widetilde{G}_j(\tfrac{k}{N}) = \widetilde{G}_{j-1}(\tfrac{k}{N})\widetilde{G}(2^{j-1}\tfrac{k}{N}).$$

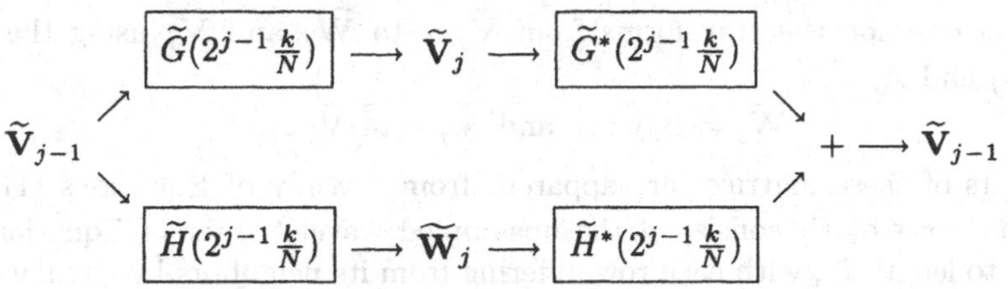

$$\widetilde{G}(2^{j-1}\tfrac{k}{N}) \longrightarrow \widetilde{\mathbf{V}}_j \longrightarrow \widetilde{G}^*(2^{j-1}\tfrac{k}{N})$$

$$\widetilde{\mathbf{V}}_{j-1} \qquad\qquad + \longrightarrow \widetilde{\mathbf{V}}_{j-1}$$

$$\widetilde{H}(2^{j-1}\tfrac{k}{N}) \longrightarrow \widetilde{\mathbf{W}}_j \longrightarrow \widetilde{H}^*(2^{j-1}\tfrac{k}{N})$$

Figure 175. Flow diagram illustrating analysis of $\widetilde{\mathbf{V}}_{j-1}$ into $\widetilde{\mathbf{W}}_j$ and $\widetilde{\mathbf{V}}_j$, followed by the synthesis of $\widetilde{\mathbf{V}}_{j-1}$ from $\widetilde{\mathbf{W}}_j$ and $\widetilde{\mathbf{V}}_j$.

Now the right-hand sides of the above equations describe filter cascades consisting of two circular filters. The first filter in both cases is the filter that – when applied to $\mathbf{X}$ – yields $\widetilde{\mathbf{V}}_{j-1}$ as its output. If we were to take this output and feed it into the second circular filters, namely, those with DFTs $\{\widetilde{H}(2^{j-1}\frac{k}{N})\}$ and $\{\widetilde{G}(2^{j-1}\frac{k}{N})\}$, we would obtain $\widetilde{\mathbf{W}}_j$ and $\widetilde{\mathbf{V}}_j$ as outputs. Thus, we can obtain $\widetilde{\mathbf{W}}_j$ and $\widetilde{\mathbf{V}}_j$ by filtering $\widetilde{\mathbf{V}}_{j-1}$ with the circular filters described by $\{\widetilde{H}(2^{j-1}\frac{k}{N})\}$ and $\{\widetilde{G}(2^{j-1}\frac{k}{N})\}$.

To see precisely what form these two filters take, recall the result of Exercise [91]: if the filter $\{\tilde{h}_l : l = 0, \ldots, L-1\}$ of width L has transfer function $\widetilde{H}(\cdot)$, then the filter of width $2^{j-1}(L-1)+1$ with impulse response sequence

$$\tilde{h}_0,\ \underbrace{0,\ldots,0,}_{2^{j-1}-1 \text{ zeros}}\ \tilde{h}_1,\ \underbrace{0,\ldots,0,}_{2^{j-1}-1 \text{ zeros}}\ \ldots, \tilde{h}_{L-2},\ \underbrace{0,\ldots,0,}_{2^{j-1}-1 \text{ zeros}}\ \tilde{h}_{L-1} \tag{175a}$$

has a transfer function defined by $\widetilde{H}(2^{j-1}f)$. Thus we can obtain the elements $\{\widetilde{W}_{j,t}\}$ from $\{\widetilde{V}_{j-1,t}\}$ using the formula

$$\widetilde{W}_{j,t} = \sum_{l=0}^{L-1} \tilde{h}_l \widetilde{V}_{j-1,t-2^{j-1}l \bmod N}, \qquad t = 0, 1, \ldots, N-1; \tag{175b}$$

using a similar argument, we also have

$$\widetilde{V}_{j,t} = \sum_{l=0}^{L-1} \tilde{g}_l \widetilde{V}_{j-1,t-2^{j-1}l \bmod N}, \qquad t = 0, 1, \ldots, N-1. \tag{175c}$$

These two equations constitute the MODWT pyramid algorithm. Note that, if we define $\widetilde{V}_{0,t} = X_t$, the above two equations yield the first level MODWT wavelet and scaling coefficients $\widetilde{\mathbf{W}}_1$ and $\widetilde{\mathbf{V}}_1$ (see Equation (163e)).

Since it is possible to obtain $\widetilde{\mathbf{W}}_j$ and $\widetilde{\mathbf{V}}_j$ by filtering $\widetilde{\mathbf{V}}_{j-1}$, it is not surprising that we can reconstruct $\widetilde{\mathbf{V}}_{j-1}$ from $\widetilde{\mathbf{W}}_j$ and $\widetilde{\mathbf{V}}_j$, as the following exercise indicates.

▷ **Exercise [175]:** Show that, by using filters whose DFTs are given by $\{\widetilde{H}^*(2^{j-1}\frac{k}{N})\}$ and $\{\widetilde{G}^*(2^{j-1}\frac{k}{N})\}$, the inverse MODWT can be computed via an inverse pyramid algorithm described by the following equation:

$$\widetilde{V}_{j-1,t} = \sum_{l=0}^{L-1} \tilde{h}_l \widetilde{W}_{j,t+2^{j-1}l \bmod N} + \sum_{l=0}^{L-1} \tilde{g}_l \widetilde{V}_{j,t+2^{j-1}l \bmod N} \tag{175d}$$

for $t = 0, 1, \ldots, N-1$. ◁

Figure 175 depicts a flow diagram for the MODWT pyramid algorithm and its inverse.

We can describe the transforms from $\widetilde{\mathbf{V}}_{j-1}$ to $\widetilde{\mathbf{W}}_j$ and $\widetilde{\mathbf{V}}_j$ using the $N \times N$ matrices $\widetilde{\mathcal{B}}_j$ and $\widetilde{\mathcal{A}}_j$:

$$\widetilde{\mathbf{W}}_j = \widetilde{\mathcal{B}}_j \widetilde{\mathbf{V}}_{j-1} \text{ and } \widetilde{\mathbf{V}}_j = \widetilde{\mathcal{A}}_j \widetilde{\mathbf{V}}_{j-1}.$$

The contents of these matrices are apparent from a study of Equations (175b) and (175c). The rows of $\widetilde{\mathcal{B}}_j$ consist of the upsampled wavelet filter of Equation (175a) periodized to length N, with each row differing from its neighbors by circular shifts of one unit either forward or backward; likewise, the rows of $\widetilde{\mathcal{A}}_j$ contain values dictated by the upsampled scaling filter. For example, for $j = 2$ and $N = 12$ with $L = 4$, we have

$$\widetilde{\mathcal{B}}_2 \equiv \begin{bmatrix} \tilde{h}_0 & 0 & 0 & 0 & 0 & 0 & \tilde{h}_3 & 0 & \tilde{h}_2 & 0 & \tilde{h}_1 & 0 \\ 0 & \tilde{h}_0 & 0 & 0 & 0 & 0 & 0 & \tilde{h}_3 & 0 & \tilde{h}_2 & 0 & \tilde{h}_1 \\ \tilde{h}_1 & 0 & \tilde{h}_0 & 0 & 0 & 0 & 0 & 0 & \tilde{h}_3 & 0 & \tilde{h}_2 & 0 \\ \vdots & \vdots & \vdots & \vdots & \vdots & \vdots & \vdots & \vdots & \vdots & \vdots & \vdots & \vdots \\ 0 & 0 & 0 & \tilde{h}_3 & 0 & \tilde{h}_2 & 0 & \tilde{h}_1 & 0 & \tilde{h}_0 & 0 & 0 \\ 0 & 0 & 0 & 0 & \tilde{h}_3 & 0 & \tilde{h}_2 & 0 & \tilde{h}_1 & 0 & \tilde{h}_0 & 0 \\ 0 & 0 & 0 & 0 & 0 & \tilde{h}_3 & 0 & \tilde{h}_2 & 0 & \tilde{h}_1 & 0 & \tilde{h}_0 \end{bmatrix}, \quad (176)$$

while $\widetilde{\mathcal{A}}_2$ is obtained by replacing the $\tilde{h}_l$'s in the above with $\tilde{g}_l$'s. Note that this description is consistent with our formulation of $\widetilde{\mathcal{B}}_1$ and $\widetilde{\mathcal{A}}_1$ in Section 5.3. With $\widetilde{\mathcal{B}}_j$ and $\widetilde{\mathcal{A}}_j$, we can express the MODWT transform from $\widetilde{\mathbf{V}}_{j-1}$ to $\widetilde{\mathbf{W}}_j$ and $\widetilde{\mathbf{V}}_j$ in a manner analogous to the DWT transform. Thus we have

$$\begin{bmatrix} \widetilde{\mathbf{W}}_j \\ \widetilde{\mathbf{V}}_j \end{bmatrix} = \widetilde{\mathcal{P}}_j \widetilde{\mathbf{V}}_{j-1} = \begin{bmatrix} \widetilde{\mathcal{B}}_j \\ \widetilde{\mathcal{A}}_j \end{bmatrix} \widetilde{\mathbf{V}}_{j-1} \text{ with } \widetilde{\mathcal{P}}_j \equiv \begin{bmatrix} \widetilde{\mathcal{B}}_j \\ \widetilde{\mathcal{A}}_j \end{bmatrix},$$

which is seen to be identical to Equation (94b) with the placement of tildes over all the components, indicating use of the MODWT filters. Furthermore, it is apparent from Equation (175d) that the synthesis of $\widetilde{\mathbf{V}}_{j-1}$ from $\widetilde{\mathbf{W}}_j$ and $\widetilde{\mathbf{V}}_j$ can also be expressed in terms of $\widetilde{\mathcal{B}}_j$ and $\widetilde{\mathcal{A}}_j$:

$$\widetilde{\mathbf{V}}_{j-1} = \widetilde{\mathcal{B}}_j^T \widetilde{\mathbf{W}}_j + \widetilde{\mathcal{A}}_j^T \widetilde{\mathbf{V}}_j.$$

If we recall that $\widetilde{\mathbf{V}}_0 \equiv \mathbf{X}$, then recursive application of the above out to stage J_0 yields

$$\begin{aligned} \mathbf{X} = \widetilde{\mathcal{B}}_1^T \widetilde{\mathbf{W}}_1 + \widetilde{\mathcal{A}}_1^T \widetilde{\mathcal{B}}_2^T \widetilde{\mathbf{W}}_2 + \widetilde{\mathcal{A}}_1^T \widetilde{\mathcal{A}}_2^T \widetilde{\mathcal{B}}_3^T \widetilde{\mathbf{W}}_3 + \cdots + \widetilde{\mathcal{A}}_1^T \cdots \widetilde{\mathcal{A}}_{J_0-1}^T \widetilde{\mathcal{B}}_{J_0}^T \widetilde{\mathbf{W}}_{J_0} \\ + \widetilde{\mathcal{A}}_1^T \cdots \widetilde{\mathcal{A}}_{J_0-1}^T \widetilde{\mathcal{A}}_{J_0}^T \widetilde{\mathbf{V}}_{J_0}. \end{aligned}$$

We can use the above to identify expressions for the jth MODWT detail $\widetilde{\mathcal{D}}_j$ and the J_0th MODWT smooth $\widetilde{\mathcal{S}}_{J_0}$, namely,

$$\widetilde{\mathcal{D}}_j = \widetilde{\mathcal{A}}_1^T \cdots \widetilde{\mathcal{A}}_{j-1}^T \widetilde{\mathcal{B}}_j^T \widetilde{\mathbf{W}}_j \text{ and } \widetilde{\mathcal{S}}_{J_0} = \widetilde{\mathcal{A}}_1^T \cdots \widetilde{\mathcal{A}}_{J_0-1}^T \widetilde{\mathcal{A}}_{J_0}^T \widetilde{\mathbf{V}}_{J_0}.$$

Comparison of the above with the definitions $\widetilde{\mathcal{D}}_j \equiv \widetilde{\mathcal{W}}_j^T \widetilde{\mathbf{W}}_j$ and $\widetilde{\mathcal{S}}_{J_0} \equiv \widetilde{\mathcal{V}}_{J_0}^T \widetilde{\mathbf{V}}_{J_0}$ yields expressions for $\widetilde{\mathcal{W}}_j$ and $\widetilde{\mathcal{V}}_j$ in terms of the $\mathcal{B}_j$ and $\mathcal{A}_j$ matrices:

$$\widetilde{\mathcal{W}}_j = \widetilde{\mathcal{B}}_j \widetilde{\mathcal{A}}_{j-1} \cdots \widetilde{\mathcal{A}}_1 \text{ and } \widetilde{\mathcal{V}}_j = \widetilde{\mathcal{A}}_j \widetilde{\mathcal{A}}_{j-1} \cdots \widetilde{\mathcal{A}}_1.$$

Note that Equation (171b) says that we can obtain $\widetilde{\mathbf{W}}_j$ and $\widetilde{\mathbf{V}}_j$ by premultiplying $\mathbf{X}$ individually by the above $N \times N$ matrices.

Key Facts and Definitions in Section 5.5

Although by definition

$$\widetilde{W}_{j,t} \equiv \sum_{l=0}^{L_j-1} \tilde{h}_{j,l} X_{t-l \bmod N} \text{ and } \widetilde{V}_{j,t} \equiv \sum_{l=0}^{L_j-1} \tilde{g}_{j,l} X_{t-l \bmod N},$$

$t = 0, \ldots, N-1$, we can also obtain these jth level MODWT coefficients from the MODWT scaling coefficients of level $j-1$ via the recursions

$$\widetilde{W}_{j,t} = \sum_{l=0}^{L-1} \tilde{h}_l \widetilde{V}_{j-1,t-2^{j-1}l \bmod N} \text{ and } \widetilde{V}_{j,t} = \sum_{l=0}^{L-1} \tilde{g}_l \widetilde{V}_{j-1,t-2^{j-1}l \bmod N}$$

(here we define $\widetilde{V}_{0,t} \equiv X_t$). We can describe the above in matrix notation as

$$\widetilde{\mathbf{W}}_j = \widetilde{\mathcal{B}}_j \widetilde{\mathbf{V}}_{j-1} \text{ and } \widetilde{\mathbf{V}}_j = \widetilde{\mathcal{A}}_j \widetilde{\mathbf{V}}_{j-1},$$

where the rows of $\widetilde{\mathcal{B}}_j$ contain circularly shifted versions of $\{\tilde{h}_l\}$ after it has been upsampled to width $2^{j-1}(L-1)+1$ and then periodized to length N, with a similar construction for $\widetilde{\mathcal{A}}_j$ based upon $\{\tilde{g}_l\}$ (here upsampling consists of inserting $2^{j-1}-1$ zeros between each of the L values of the original filter). We can reconstruct $\{\widetilde{V}_{j-1,t}\}$ from $\{\widetilde{W}_{j,t}\}$ and $\{\widetilde{V}_{j,t}\}$ via

$$\widetilde{V}_{j-1,t} = \sum_{l=0}^{L-1} \tilde{h}_l \widetilde{W}_{j,t+2^{j-1}l \bmod N} + \sum_{l=0}^{L-1} \tilde{g}_l \widetilde{V}_{j,t+2^{j-1}l \bmod N},$$

which can be expressed in matrix notation as $\widetilde{\mathbf{V}}_{j-1} = \widetilde{\mathcal{B}}_j^T \widetilde{\mathbf{W}}_j + \widetilde{\mathcal{A}}_j^T \widetilde{\mathbf{V}}_j$. The above relationships allow us to express the MODWT details and smooths as

$$\widetilde{\mathcal{D}}_j = \widetilde{\mathcal{A}}_1^T \cdots \widetilde{\mathcal{A}}_{j-1}^T \widetilde{\mathcal{B}}_j^T \widetilde{\mathbf{W}}_j \text{ and } \widetilde{\mathcal{S}}_{J_0} = \widetilde{\mathcal{A}}_1^T \cdots \widetilde{\mathcal{A}}_{J_0-1}^T \widetilde{\mathcal{A}}_{J_0}^T \widetilde{\mathbf{V}}_{J_0},$$

which in turn says that the matrices $\widetilde{\mathcal{W}}_j$ and $\widetilde{\mathcal{V}}_j$ can be expressed as

$$\widetilde{\mathcal{W}}_j = \widetilde{\mathcal{B}}_j \widetilde{\mathcal{A}}_{j-1} \cdots \widetilde{\mathcal{A}}_1 \text{ and } \widetilde{\mathcal{V}}_j = \widetilde{\mathcal{A}}_j \widetilde{\mathcal{A}}_{j-1} \cdots \widetilde{\mathcal{A}}_1.$$

(these matrices yield $\widetilde{\mathbf{W}}_j = \widetilde{\mathcal{W}}_j \mathbf{X}$ and $\widetilde{\mathbf{V}}_j = \widetilde{\mathcal{V}}_j \mathbf{X}$).

Comments and Extensions to Section 5.5

[1] Here we give pseudo-code that can be used to compute the MODWT and its inverse. Define $\widetilde{V}_{0,t} \equiv X_t$ for $t = 0, \ldots, N-1$, where N is any positive integer. Let $\{\tilde{h}_l : l = 0, \ldots, L-1\}$ be a MODWT wavelet filter of even width L. Let $\{\tilde{g}_l\}$ be the corresponding MODWT scaling filter constructed from $\{\tilde{h}_l\}$ via Equation (163d). Denote the elements of $\widetilde{\mathbf{W}}_j$ and $\widetilde{\mathbf{V}}_{J_0}$ as $\widetilde{W}_{j,t}$ and $\widetilde{V}_{J_0,t}$, $t = 0, \ldots, N-1$. Given $\widetilde{\mathbf{V}}_{j-1}$, the following pseudo-code computes $\widetilde{\mathbf{W}}_j$ and $\widetilde{\mathbf{V}}_j$ (this is based on Equations (175b) and (175c)):

For $t = 0, \ldots, N-1$, do the outer loop:
  Set k to t.
  Set $\widetilde{W}_{j,t}$ to $\tilde{h}_0 \widetilde{V}_{j-1,k}$.
  Set $\widetilde{V}_{j,t}$ to $\tilde{g}_0 \widetilde{V}_{j-1,k}$.
  For $n = 1, \ldots, L-1$, do the inner loop:
    Decrement k by 2^{j-1}.
    If $k < 0$, set k to $k \bmod N$.
    Increment $\widetilde{W}_{j,t}$ by $\tilde{h}_n \widetilde{V}_{j-1,k}$.
    Increment $\widetilde{V}_{j,t}$ by $\tilde{g}_n \widetilde{V}_{j-1,k}$.
  End of inner loop.
End of outer loop.

Note that, if j is such that $2^{j-1} \leq N$ (as will usually be the case in practical applications), then the single '$k \bmod N$' in the above code can be replaced by '$k + N$.' By using the pseudo-code above first for $j = 1$ (i.e., starting with $\widetilde{\mathbf{V}}_0 = \mathbf{X}$) and then for $j = 2, \ldots, J_0$, we obtain the component vectors of the MODWT, namely, $\widetilde{\mathbf{W}}_1, \ldots, \widetilde{\mathbf{W}}_{J_0}$ and $\widetilde{\mathbf{V}}_{J_0}$, along with the intermediate vectors $\widetilde{\mathbf{V}}_1, \ldots \widetilde{\mathbf{V}}_{J_0-1}$.

The task of the inverse MODWT is to compute $\mathbf{X}$ given the vectors $\widetilde{\mathbf{W}}_1, \ldots, \widetilde{\mathbf{W}}_{J_0}$ and $\widetilde{\mathbf{V}}_{J_0}$. Given $\widetilde{\mathbf{W}}_j$ and $\widetilde{\mathbf{V}}_j$, the following pseudo-code computes $\widetilde{\mathbf{V}}_{j-1}$ (this is based on Equation (175d)):

For $t = 0, \ldots, N-1$, do the outer loop:
  Set k to t.
  Set $\widetilde{V}_{j-1,t}$ to $\tilde{h}_0 \widetilde{W}_{j,k} + \tilde{g}_0 \widetilde{V}_{j,k}$.
  For $n = 1, \ldots, L-1$, do the inner loop:
    Increment k by 2^{j-1}.
    If $k \geq N$, set k to $k \bmod N$.
    Increment $\widetilde{V}_{j-1,t}$ by $\tilde{h}_n \widetilde{W}_{j,k} + \tilde{g}_n \widetilde{V}_{j,k}$.
  End of the inner loop.
End of the outer loop.

Again, if j is such that $2^{j-1} \leq N$, then the single '$k \bmod N$' in the above code can be replaced by '$k - N$.' By using the pseudo-code first with $\widetilde{\mathbf{W}}_{J_0}$ and $\widetilde{\mathbf{V}}_{J_0}$, we obtain $\widetilde{\mathbf{V}}_{J_0-1}$. We can depict this procedure as putting $\widetilde{\mathbf{W}}_{J_0}$ and $\widetilde{\mathbf{V}}_{J_0}$ into a box, out of which comes $\widetilde{\mathbf{V}}_{J_0-1}$:

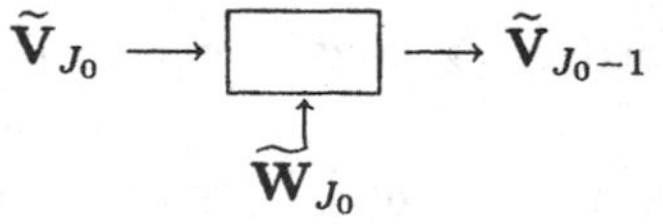

$$\widetilde{\mathbf{V}}_{J_0} \longrightarrow \boxed{} \longrightarrow \widetilde{\mathbf{V}}_{J_0-1}$$
$$\uparrow \; \widetilde{\mathbf{W}}_{J_0}$$

Using the pseudo-code next with $\widetilde{\mathbf{W}}_{J_0-1}$ and $\widetilde{\mathbf{V}}_{J_0-1}$ yields $\widetilde{\mathbf{V}}_{J_0-2}$:

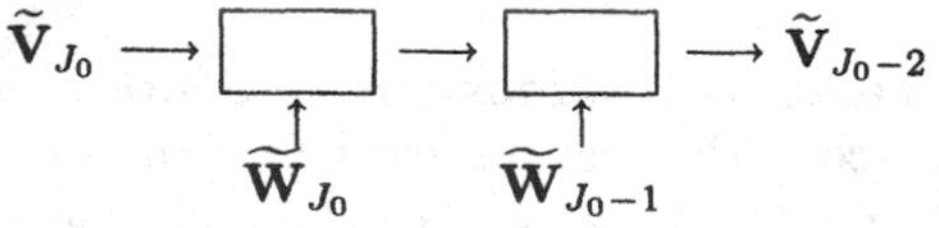

$$\widetilde{\mathbf{V}}_{J_0} \longrightarrow \boxed{} \longrightarrow \boxed{} \longrightarrow \widetilde{\mathbf{V}}_{J_0-2}$$
$$\uparrow \; \widetilde{\mathbf{W}}_{J_0} \qquad \uparrow \; \widetilde{\mathbf{W}}_{J_0-1}$$

After $J_0 - 2$ more applications of the pseudo-code, we obtain $\widetilde{\mathbf{V}}_0 = \mathbf{X}$:

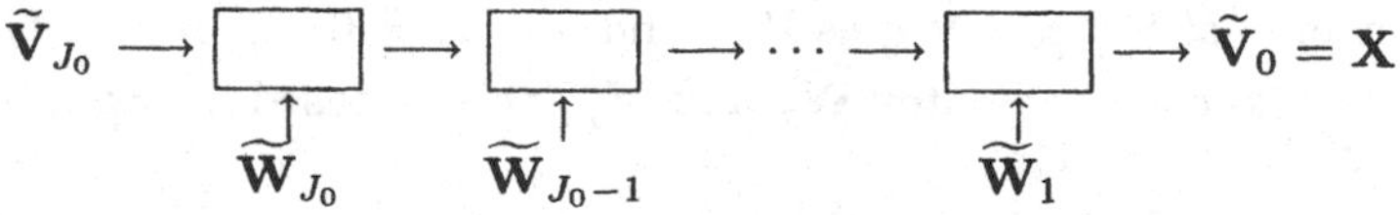

$$\widetilde{\mathbf{V}}_{J_0} \longrightarrow \boxed{} \longrightarrow \boxed{} \longrightarrow \cdots \longrightarrow \boxed{} \longrightarrow \widetilde{\mathbf{V}}_0 = \mathbf{X}$$
$$\uparrow \; \widetilde{\mathbf{W}}_{J_0} \qquad \uparrow \; \widetilde{\mathbf{W}}_{J_0-1} \qquad \qquad \uparrow \; \widetilde{\mathbf{W}}_1$$

($\widetilde{\mathbf{V}}_{J_0-3}, \ldots, \widetilde{\mathbf{V}}_1$ are produced as intermediate calculations).

The jth detail $\widetilde{\mathcal{D}}_j$ is obtained by taking the inverse MODWT of $\widetilde{\mathbf{0}}_1, \ldots, \widetilde{\mathbf{0}}_{j-1}$, $\widetilde{\mathbf{W}}_j$ and $\widetilde{\mathbf{0}}_j$, where each $\widetilde{\mathbf{0}}_k$, $k = 1, \ldots, j$, is a vector of N zeros. This detail can be computed by successively applying the inverse MODWT algorithm for $j, j-1, \ldots, 1$ starting with $\widetilde{\mathbf{W}}_j$ and with $\widetilde{\mathbf{0}}_j$ substituted for $\widetilde{\mathbf{V}}_j$ in the above pseudo-code. At the end of j iterations, we obtain the desired $\widetilde{\mathcal{D}}_j$:

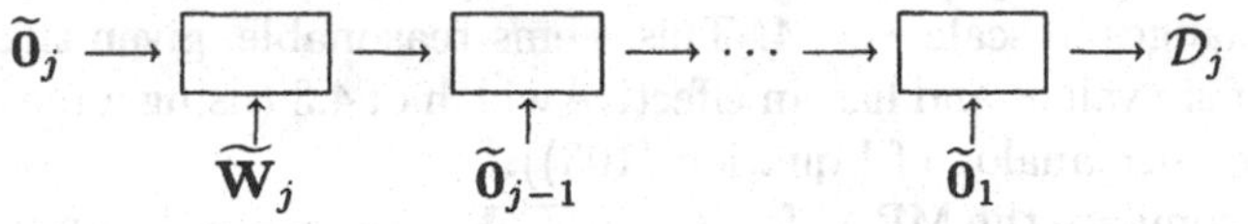

$$\widetilde{\mathbf{0}}_j \longrightarrow \boxed{}_{\uparrow \widetilde{\mathbf{W}}_j} \longrightarrow \boxed{}_{\uparrow \widetilde{\mathbf{0}}_{j-1}} \longrightarrow \cdots \longrightarrow \boxed{}_{\uparrow \widetilde{\mathbf{0}}_1} \longrightarrow \widetilde{\mathcal{D}}_j$$

Likewise, the smooth $\widetilde{\mathcal{S}}_{J_0}$ can be obtaining by applying the inverse MODWT to $\widetilde{\mathbf{0}}_1, \ldots, \widetilde{\mathbf{0}}_{J_0}$ and $\widetilde{\mathbf{V}}_{J_0}$:

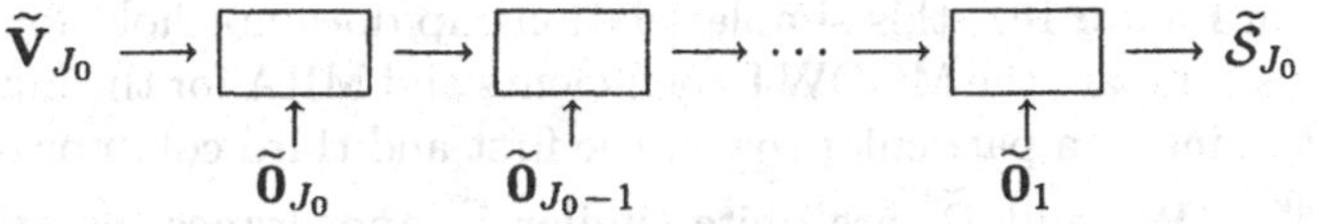

$$\widetilde{\mathbf{V}}_{J_0} \longrightarrow \boxed{}_{\uparrow \widetilde{\mathbf{0}}_{J_0}} \longrightarrow \boxed{}_{\uparrow \widetilde{\mathbf{0}}_{J_0-1}} \longrightarrow \cdots \longrightarrow \boxed{}_{\uparrow \widetilde{\mathbf{0}}_1} \longrightarrow \widetilde{\mathcal{S}}_{J_0}$$

[2] The fact that the MODWT details and smooths are associated with zero phase filters is an example of a well-known 'trick' for obtaining a zero phase output based upon any arbitrary circular filter $\{a_l^\circ\} \longleftrightarrow \{A(\frac{k}{N})\}$. The trick (described in, for example, Hamming, 1989, p. 252) is to filter the time series $\mathbf{X}$ with $\{a_l^\circ\}$; reverse the output from this filter; and then filter the reversed output with $\{a_l^\circ\}$ again to obtain an output, which – when reversed again to obtain, say, $\mathbf{Y}$ – is associated with zero phase. Because filtering the reversed output, followed by reversal, is the same thing as cross correlation of the original output, this procedure is equivalent to the following flow diagram:

$$\mathbf{X} \longrightarrow \boxed{A(\tfrac{k}{N})} \longrightarrow \boxed{A^*(\tfrac{k}{N})} \longrightarrow \mathbf{Y};$$

i.e., the equivalent circular filter for the two stage cascade has a frequency domain representation of $\{|A(\frac{k}{N})|^2\}$, which necessarily has zero phase.

5.6 MODWT Analysis of 'Bump' Time Series

Let us now revisit the example discussed in Section 5.1, in which we looked at a DWT analysis of a 'bump' time series $\mathbf{X}$ and its circular shift $\mathcal{T}^5\mathbf{X}$. The bottom six rows of Figure 181 show the same quantities as in Figure 161, but now we use the MODWT instead of the DWT. Let us first consider the MODWT coefficients for $\mathbf{X}$ (first column, middle five rows). In contrast to the DWT, each of the MODWT coefficient vectors $\widetilde{\mathbf{W}}_j$ and $\widetilde{\mathbf{V}}_4$ has the same number of elements as $\mathbf{X}$. As indicated by the labels in the first column, we have plotted each coefficient vector after advancing it (i.e., circularly shifting it to the left) by the amount indicated by the absolute value of the exponent for the shift operator $\mathcal{T}$ (see Equation (52b); the first and last values in, for example, $\mathcal{T}^{-53}\widetilde{\mathbf{W}}_4$ are, respectively, $\widetilde{W}_{4,53}$ and $\widetilde{W}_{4,52}$). Because we are using the LA(8) wavelet filter in this example, these circular shifts advance the MODWT coefficient vectors so that they are approximately the outputs from a zero phase filter. The particular values of the advances for the wavelet and scaling coefficients are determined by the absolute values of, respectively, $\nu_j^{(H)}$ in Equation (114b) and $\nu_4^{(G)}$ in Equation (114a) (in these equations, $\nu = -3$ for the LA(8) filter). The corresponding MODWT coefficient vectors for the shifted series $\mathcal{T}^5\mathbf{X}$ are shown in the second column. A careful comparison of the first and second columns indicates that shifting $\mathbf{X}$ has the effect of shifting each of the MODWT coefficient vectors by a similar amount; i.e., if $\widetilde{\mathbf{W}}_j$ and $\widetilde{\mathbf{V}}_{J_0}$ constitute the MODWT of $\mathbf{X}$, then $\mathcal{T}^m\widetilde{\mathbf{W}}_j$ and $\mathcal{T}^m\widetilde{\mathbf{V}}_{J_0}$ constitute the MODWT of $\mathcal{T}^m\mathbf{X}$. In contrast to the DWT, circularly shifting $\mathbf{X}$ modifies the MODWT in an intuitively reasonable manner. Note also that the largest MODWT coefficients occur in $\widetilde{\mathbf{W}}_3$, i.e., at changes in scale $\tau_3 = 4$. This seems reasonable, given that the bump is nonzero over just six values and has an effective width of 4.3 (using a measure provided by a finite sample size analog of Equation (103)).

Let us next compare the MRAs for $\mathbf{X}$ and $\mathcal{T}^5\mathbf{X}$ – these are shown in, respectively, the third and fourth columns of Figure 181 (middle five rows). The MODWT details and smooth for the two series are again related in a simple and reasonable manner: if $\widetilde{\mathcal{D}}_j$ and $\widetilde{\mathcal{S}}_{J_0}$ form the MRA for $\mathbf{X}$, then $\mathcal{T}^m\widetilde{\mathcal{D}}_j$ and $\mathcal{T}^m\widetilde{\mathcal{S}}_{J_0}$ form the MRA for $\mathcal{T}^m\mathbf{X}$. As is evident from Figure 161, this simple relationship does not hold for the DWT.

Finally, let us compare the MODWT coefficients and MRA for the bump series. As can be seen by looking at a particular row in the first and third columns of Figure 181, the vectors $\mathcal{T}^{-|\nu_j^{(H)}|}\widetilde{\mathbf{W}}_j$ and $\widetilde{\mathcal{D}}_j$ are quite similar in appearance, as are $\mathcal{T}^{-|\nu_4^{(G)}|}\widetilde{\mathbf{V}}_4$ and $\widetilde{\mathcal{S}}_4$. By contrast, the corresponding portion of Figure 161 shows that the DWT detail $\mathcal{D}_j$ can look quite different from $\mathbf{W}_j$ (partly because the former has 2^j more elements than the latter). To understand why $\widetilde{\mathbf{W}}_j$ and $\widetilde{\mathcal{D}}_j$ look so similar, recall that both vectors arise as outputs from circularly filtering $\mathbf{X}$. In the case of $\widetilde{\mathbf{W}}_j$, the circular filter has a frequency domain representation of $\{H_j(\frac{k}{N})\}$. Because $\mathcal{T}^{-|\nu_j^{(H)}|}\widetilde{\mathbf{W}}_j$ is approximately the output from a zero phase filter, its associated circular filter is approximately $\{|H_j(\frac{k}{N})|\}$. On the other hand, we know from Equation (172a) that the circular filter yielding $\mathcal{D}_j$ is given by $\{|H_j(\frac{k}{N})|^2\}$. The circular filters yielding $\mathcal{T}^{-|\nu_j^{(H)}|}\widetilde{\mathbf{W}}_j$ and $\widetilde{\mathcal{D}}_j$ thus have similar phase and gain properties (the fact that $H_j(\cdot)$ is approximately a band-pass filter means that $|H_j(\cdot)|^2$ must be also – it can be argued that the latter is a better approximation by certain measures). In the DWT case, we do not have such a tight correspondence, mainly because filtering is now accompanied by downsampling to form $\mathbf{W}_j$ and then upsampling to form $\mathcal{D}_j$.

Although the MODWT details and shifted wavelet coefficients are outputs from filters with similar phase and gain properties, the ways in which they can be put to good use are quite different. Let us demonstrate this using the bump series $\mathbf{X}$ and $\mathcal{T}^5\mathbf{X}$. First, whereas Equation (169c) says that we can decompose the energy in $\mathbf{X}$ or $\mathcal{T}^5\mathbf{X}$ using the MODWT coefficients, the same is not true for the details and smooth. To show this, we computed level $J_0 = 7$ MODWTs based again upon the LA(8) wavelet filter and used the wavelet coefficients to construct discrete wavelet empirical power spectra:

$$P_{\widetilde{\mathcal{W}}}(\tau_j) \equiv \frac{1}{N}\|\widetilde{\mathbf{W}}_j\|^2 = \frac{1}{N}\|\mathcal{T}^5\widetilde{\mathbf{W}}_j\|^2, \quad j = 1, \ldots, 7.$$

Since there are $N = 128 = 2^7$ points in the bump series, we can use Exercise [5.9a] to claim that these spectra provide an analysis of the sample variance for $\mathbf{X}$ or $\mathcal{T}^5\mathbf{X}$:

$$\hat{\sigma}_X^2 = \sum_{j=1}^{7} P_{\widetilde{\mathcal{W}}}(\tau_j)$$

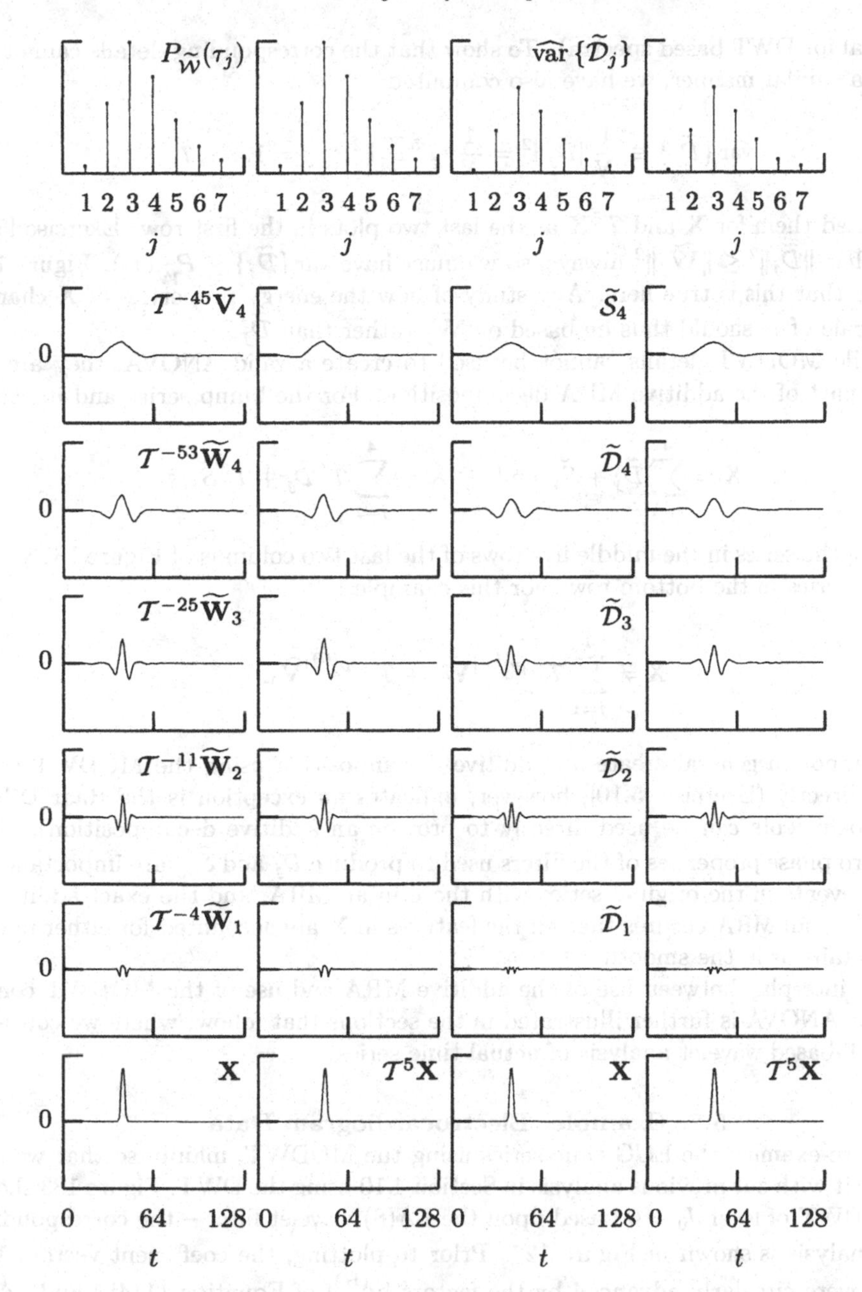

Figure 181. Bottom six rows as in Figure 161, with MODWT replacing DWT. The top row shows the discrete wavelet empirical power spectra (first two columns) and a corresponding quantity based upon level $j = 1, \ldots, 7$ MODWT details (last two columns) – see text for details.

(the analysis of $\hat{\sigma}_X^2$ in terms of MODWT coefficients is studied further in Chapter 8). The spectra for $\mathbf{X}$ and $\mathcal{T}^5\mathbf{X}$ are shown in the first two plots in the first row in Figure 181 and are seen to be identical – circularly shifting a series does not change its MODWT-based power spectrum (Equations (63a) and (63b) demonstrate that this is not true in general for DWT-based spectra). To show that the corresponding details cannot be used in a similar manner, we have also computed

$$\widehat{\operatorname{var}}\{\widetilde{\mathcal{D}}_j\} \equiv \frac{1}{N}\|\widetilde{\mathcal{D}}_j\|^2 = \frac{1}{N}\|\mathcal{T}^5\widetilde{\mathcal{D}}_j\|^2, \quad j = 1, \ldots, 7,$$

and plotted them for $\mathbf{X}$ and $\mathcal{T}^5\mathbf{X}$ as the last two plots in the first row. Exercise [5.7] claims that $\|\widetilde{\mathcal{D}}_j\|^2 \leq \|\widetilde{\mathbf{W}}_j\|^2$ always, so we must have $\widehat{\operatorname{var}}\{\widetilde{\mathcal{D}}_j\} \leq P_{\widetilde{\mathcal{W}}}(\tau_j)$. Figure 181 indicates that this is true here. Any study of how the energy properties of $\mathbf{X}$ change over a scale of τ_j should thus be based on $\widetilde{\mathbf{W}}_j$ rather than $\widetilde{\mathcal{D}}_j$.

While MODWT details cannot be used to create a valid ANOVA, they are an integral part of the additive MRA decomposition. For the bump series and its shift, we have

$$\mathbf{X} = \sum_{j=1}^{4} \widetilde{\mathcal{D}}_j + \widetilde{\mathcal{S}}_4 \text{ and } \mathcal{T}^5\mathbf{X} = \sum_{j=1}^{4} \mathcal{T}^5\widetilde{\mathcal{D}}_j + \mathcal{T}^5\widetilde{\mathcal{S}}_4,$$

so adding the series in the middle five rows of the last two columns of Figure 181 yields the time series in the bottom row. For this example,

$$\mathbf{X} \neq \sum_{j=1}^{4} \mathcal{T}^{-|\nu_j^{(H)}|}\widetilde{\mathbf{W}}_j + \mathcal{T}^{-|\nu_4^{(G)}|}\widetilde{\mathbf{V}}_4,$$

so we cannot in general create an additive decomposition using the MODWT coefficients directly (Exercise [5.10], however, indicates an exception is the Haar DWT, whose coefficients *can* be used directly to provide an additive decomposition). The exact zero phase properties of the filters used to produce $\widetilde{\mathcal{D}}_j$ and $\widetilde{\mathcal{S}}_{J_0}$ are important for aligning events in the original series with those in an MRA, and the exact additivity provided by an MRA ensures that all the features in $\mathbf{X}$ are accounted for either in one of the details or in the smooth.

The interplay between use of the additive MRA and use of the MODWT coefficients for ANOVA is further illustrated in the sections that follow, where we consider MODWT-based wavelet analysis of actual time series.

5.7 Example: Electrocardiogram Data

Here we re-examine the ECG time series using the MODWT, mainly so that we can compare it with our previous analysis in Section 4.10 using the DWT. Figure 183 shows the MODWT of level $J_0 = 6$ based upon the LA(8) wavelet filter – the corresponding DWT analysis is shown in Figure 127. Prior to plotting, the coefficient vectors $\widetilde{\mathbf{W}}_j$ and $\widetilde{\mathbf{V}}_6$ were circularly advanced by the factors $|\nu_j^{(H)}|$ of Equation (114b) and $|\nu_6^{(G)}|$ of Equation (114a), respectively, to align the coefficients properly as much as possible with the original time series (the amounts of the shifts are indicated by the exponent of each $\mathcal{T}$ recorded on the plot). The two thick vertical lines crossing a given $\mathcal{T}^{-|\nu_j^{(H)}|}\widetilde{\mathbf{W}}_j$ or $\mathcal{T}^{-|\nu_6^{(G)}|}\widetilde{\mathbf{V}}_6$ delineate regions at the beginning and end of the shifted vector that are affected at least to some degree by the assumption of circularity (we discuss how to compute the locations of these lines in Section 5.11). Judicious subsampling and rescaling of the MODWT coefficients yield the DWT coefficients. If we compare the coefficients for larger scales in Figure 127 with those in Figure 183, we see that the

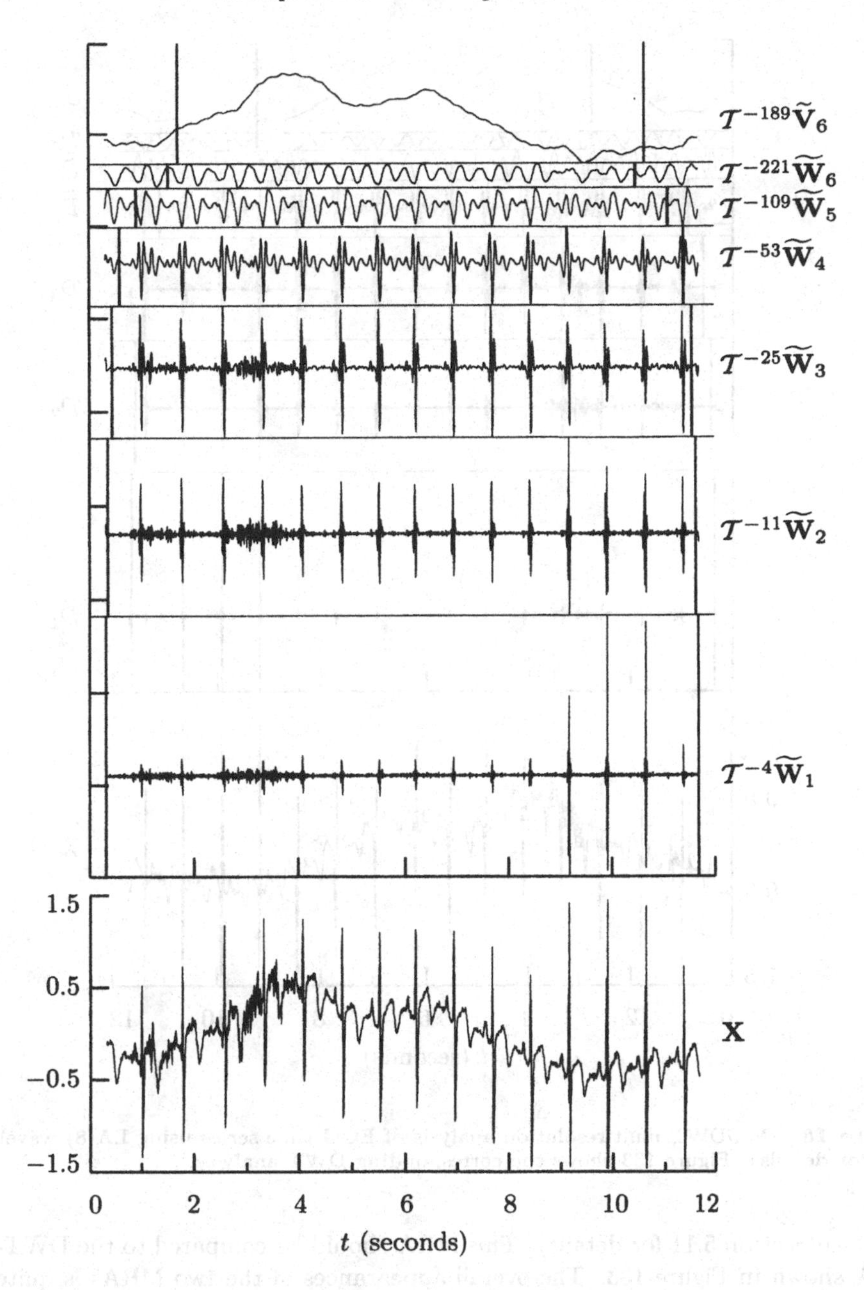

Figure 183. MODWT wavelet analysis of ECG time series using LA(8) wavelet. The above should be compared to Figure 127 for the corresponding DWT analysis (see text for details).

subsampling inherent in the DWT significantly hampers our ability to interpret its coefficients in terms of events in the original ECG series.

The MODWT multiresolution analysis is shown in Figure 184. As before, the thick vertical lines crossing a given $\widetilde{\mathcal{D}}_j$ or $\widetilde{\mathcal{S}}_6$ delineate regions at the beginning and end of the vector that are affected at least to some degree by the circularity assump-

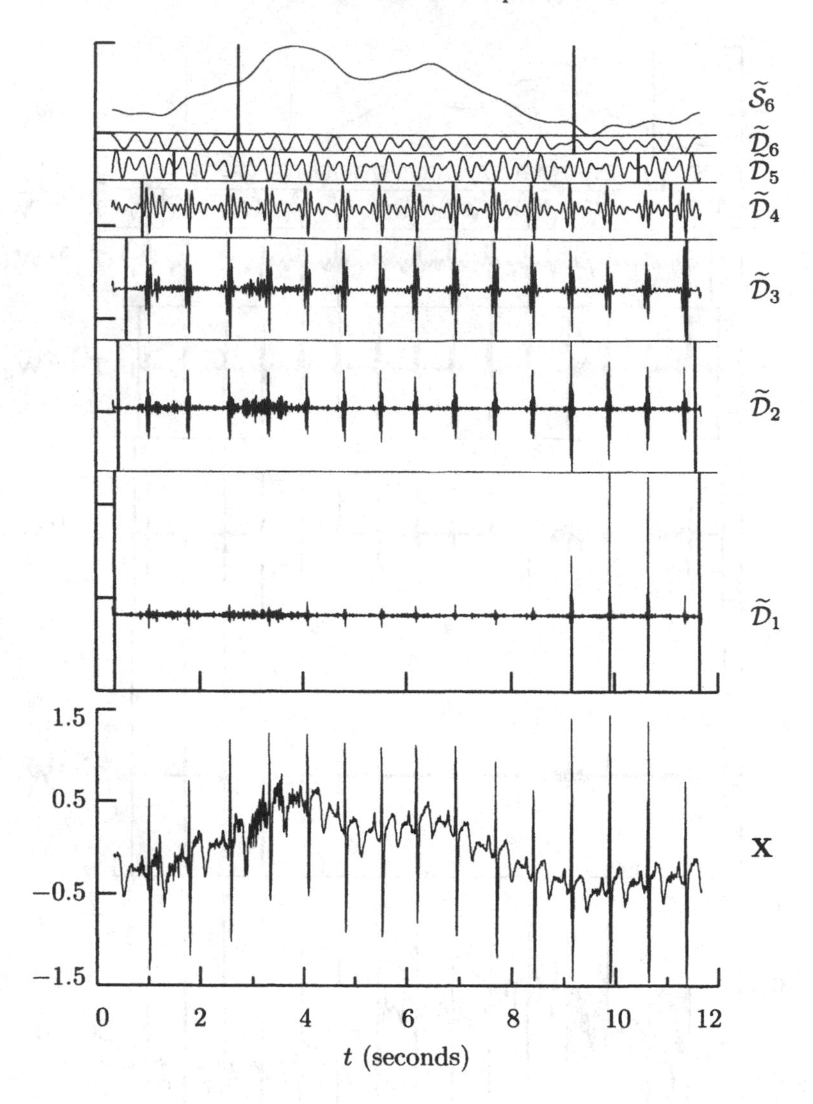

Figure 184. MODWT multiresolution analysis of ECG time series using LA(8) wavelet (see text for details). Figure 133 shows the corresponding DWT analysis.

tion (see Section 5.11 for details). This MRA should be compared to the DWT-based MRA shown in Figure 133. The overall appearances of the two MRAs is quite comparable, but note the difference between the DWT details $\mathcal{D}_3$, $\mathcal{D}_4$ and $\mathcal{D}_5$ and the corresponding MODWT details $\widetilde{\mathcal{D}}_3$, $\widetilde{\mathcal{D}}_4$ and $\widetilde{\mathcal{D}}_5$: in each case, the MODWT detail is more consistent across time, indicating that the time variations in the DWT details are in part attributable to the particular alignment assumed by the DWT. Note also that the MODWT detail $\widetilde{\mathcal{D}}_6$ does not tend to fade in and out nearly as much as the DWT detail $\mathcal{D}_6$ (particularly at the beginning of the series), again undoubtedly due to alignment effects in the DWT. The asymmetric appearance of the characteristic 'bumps' in $\mathcal{D}_6$ is also eliminated in the MODWT detail $\widetilde{\mathcal{D}}_6$ due to the zero phase property associated with this detail.

Finally let us compare the MODWT wavelet and scaling coefficients in Figure 183 with the MODWT details and smooth in Figure 184. While the former preserves energy and the latter preserves additivity, the overall appearances of the two scale-by-scale decompositions are quite similar. The details and smooth are obtained by filtering the wavelet and scaling coefficients, so each $\widetilde{\mathcal{D}}_j$ or $\widetilde{\mathcal{S}}_6$ is necessarily smoother in appearance than the corresponding $\widetilde{\mathbf{W}}_j$ or $\widetilde{\mathbf{V}}_6$. This is particularly evident in a comparison of $\widetilde{\mathcal{S}}_6$ with $\widetilde{\mathbf{V}}_6$: the latter contains small ripples that are almost entirely absent in the former. While $\widetilde{\mathcal{S}}_6$ is a quite useful estimate of the baseline drift here (and, likewise, $\widetilde{\mathcal{S}}_{J_0}$ is a useful estimate of trend in other time series), any meaningful partitioning of the energy in the ECG series across scales and/or time must be done using the MODWT coefficients. Both types of analysis (the $\widetilde{\mathbf{W}}_j$'s with $\widetilde{\mathbf{V}}_{J_0}$ and the $\widetilde{\mathcal{D}}_j$'s with $\widetilde{\mathcal{S}}_{J_0}$) thus have an important role to play in the analysis of this and other time series.

5.8 Example: Subtidal Sea Level Fluctuations

In this section we illustrate MODWT-based MRA and ANOVA using a time series of subtidal sea levels for Crescent City, which is located on the open coast of Northern California. Inside its harbor, a permanent tide gauge is maintained by the National Ocean Service (NOS). The gauge measures water levels every 6 minutes within a stilling well, which suppresses high-frequency fluctuations due to wind waves and swell. Periodic leveling surveys ensure that the reference level of the gauge remains constant relative to the surrounding land. The time series considered here is based on a segment of the water level series from the beginning of 1980 to the end of 1991. This segment contains one gap two weeks in length in addition to a few relatively minor short gaps. The two week gap occurred in the summer of 1990 and was filled in using predicted tides plus local means, whereas interpolated values supplied by NOS were used to bridge the shorter gaps. The edited series was then subsampled every tenth value to produce hourly values and low-pass filtered using a Kaiser filter to remove the diurnal (daily) and semidiurnal (twice daily) tides. The 3 dB down (half-power) point of the squared gain function for the filter occurs at a frequency corresponding to a 1.82 day period, while frequencies corresponding to a 2.68 day period and longer are in the pass-band of the low-pass filter and are attenuated in power by less than 1%. Finally the filtered series was resampled every $\Delta t = 1/2$ day, yielding a time series $\{X_t\}$ measured in centimeters with $N = 8746$ values. This series reflects subtidal variations in the coastal sea level at Crescent City (i.e., variations after removal of the predictable tidal component) and is plotted at the bottom of Figure 186. (For more details about this application, see Percival and Mofjeld, 1997.)

Eight other series depicting a level $J_0 = 7$ LA(8) MODWT-based MRA are plotted above the Crescent City sea level data in Figure 186. The uppermost series is the MODWT smooth $\widetilde{\mathcal{S}}_7$ (corresponding to averages over physical scales of $\lambda_7 \Delta t = 64$ days) while the seven series below it are the MODWT detail series $\widetilde{\mathcal{D}}_1, \widetilde{\mathcal{D}}_2, \ldots, \widetilde{\mathcal{D}}_7$ (corresponding to changes in physical scales of, from bottom to top, $1/2, 1, \ldots, 32$ days). As usual, we have $\mathbf{X} = \sum_j \widetilde{\mathcal{D}}_j + \widetilde{\mathcal{S}}_7$. As explained in Section 5.11, the two solid vertical lines intersecting a given $\widetilde{\mathcal{D}}_j$ or $\widetilde{\mathcal{S}}_7$ delineate the regions of the MRA component affected to some degree (however small) by the assumption that $\{X_t\}$ is circularly defined for $t < 0$ and $t \geq N$; however, a circular extension of the series is reasonable here, given that we are looking at a phenomenon with an annual

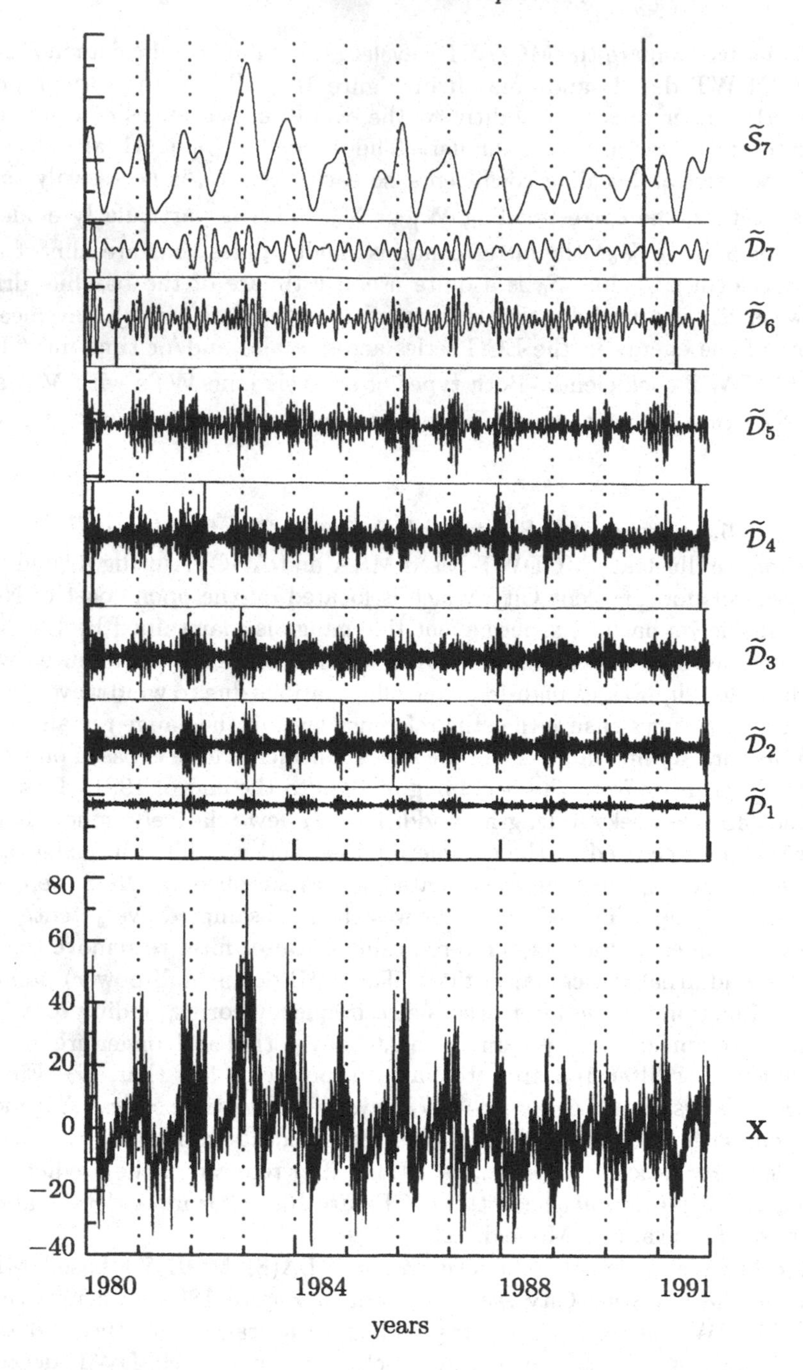

Figure 186. LA(8) MODWT multiresolution analysis for Crescent City subtidal variations (see text for details). This series is measured in centimeters.

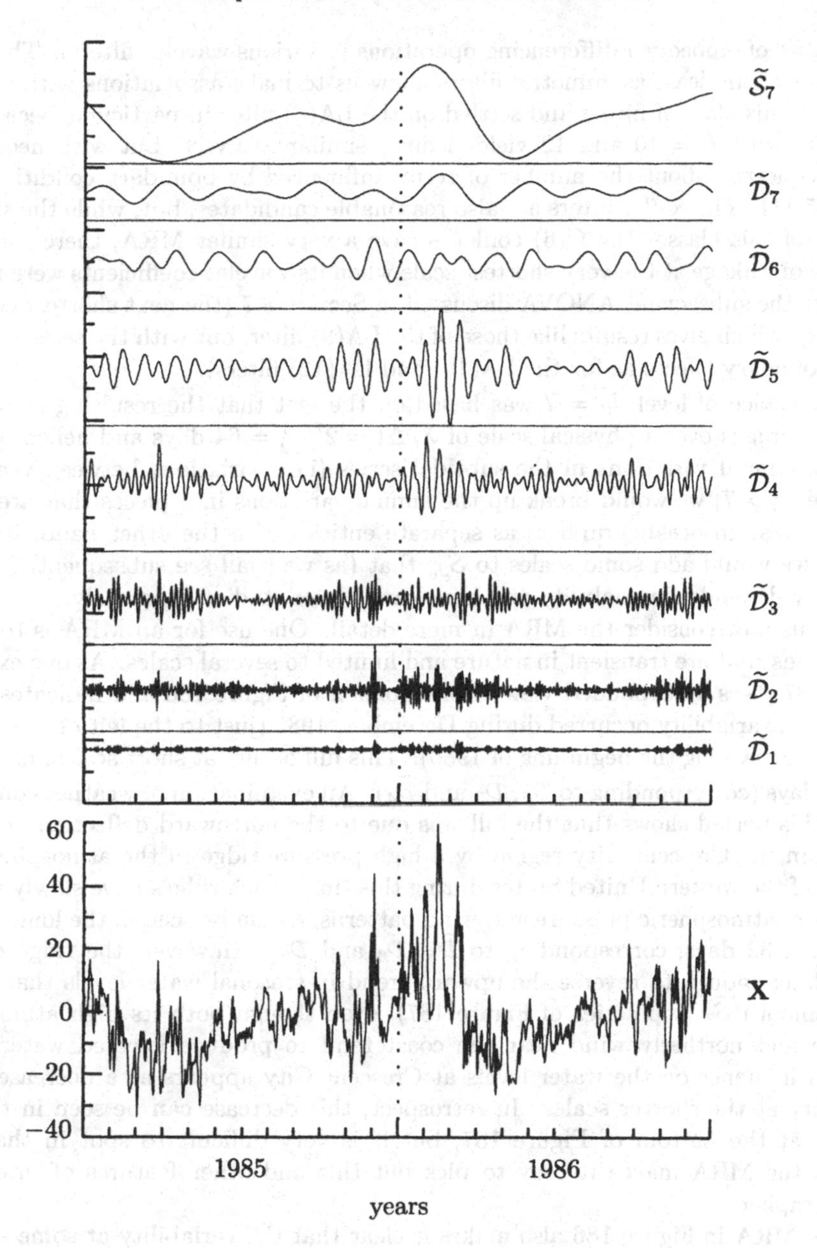

Figure 187. Expanded view of 1985 and 1986 portion of Figure 186.

component and that the amount of data we have is very close to 12 complete years.

Before we consider this MRA in more detail, however, let us comment upon why we decided to use the LA(8) wavelet filter and to set $J_0 = 7$. This particular filter was chosen because we eventually want to associate the MODWT coefficients themselves with actual times (had we just been interested in an MRA, a D(4) filter would have yielded nearly the same results, but detail series for small scales based on a Haar filter suffer from a form of 'leakage' in that they are noticeably tracking fluctuations from higher scales – see Section 8.9 for a discussion of leakage and how it is influenced by the number of embedded differencing operations in various wavelet filters). The phase properties of the least asymmetric filters allow us to make associations with time, so we picked this class of filters and settled on the LA(8) filter in particular because the LA filters with $L = 10$ and 12 yielded quite similar analyses, but with necessarily greater concerns about the number of terms influenced by boundary conditions (see Section 5.11). The coiflet filters are also reasonable candidates, but, while the shortest member of this class – the C(6) coiflet – gave a very similar MRA, there was some evidence of leakage at the very shortest scale when its wavelet coefficients were used to carry out the subsequent ANOVA discussed in Section 8.7 (the next shortest coiflet is the C(12), which gives results like those of the LA(8) filter, but with the same concerns about boundary effects as for the LA(10) and LA(12) filters).

Our choice of level $J_0 = 7$ was based on the fact that the resulting smooth $\widetilde{\mathcal{S}}_7$ reflects averages over a physical scale of $\lambda_7\,\Delta t = 2^7 \cdot \frac{1}{2} = 64$ days and hence contains the intra-annual variations in the subtidal series (i.e., variations between years). If we made $J_0 > 7$, we would break up the annual variations into pieces that are of not much interest to oceanographers as separate entities. On the other hand, if we set $J_0 < 7$, we would add some scales to $\widetilde{\mathcal{S}}_{J_0}$ that (as we shall see subsequently) have a seasonally dependent variability and hence are better studied separately.

Let us now consider the MRA in more detail. One use for an MRA is to detect fluctuations that are transient in nature and limited to several scales. As one example, Figure 187 gives an expanded view of one portion of Figure 186 and indicates that a brief lull in variability occurred during December, 1985 (just to the left of the vertical dotted line marking the beginning of 1986). This lull occurs at short scales, namely, 1, 2 and 4 days (corresponding to $\widetilde{\mathcal{D}}_2$, $\widetilde{\mathcal{D}}_3$ and $\widetilde{\mathcal{D}}_4$). An examination of weather conditions during this period shows that the lull was due to the northward deflection of storms away from the Crescent City region by a high pressure ridge in the atmosphere that formed off the western United States during this time. Such ridges have slowly varying patterns of atmospheric pressure and wind patterns, as can be seen in the longer scales (8, 16 and 32 days, corresponding to $\widetilde{\mathcal{D}}_5$, $\widetilde{\mathcal{D}}_6$ and $\widetilde{\mathcal{D}}_7$). However, the ridge did not persist long enough to reverse the upward trend in seasonal water levels that is seen in the smooth $\widetilde{\mathcal{S}}_7$ (top series of Figure 187), even though both its high atmospheric pressure and northerly wind near the coast tend to produce lowered water levels. Its main influence on the water levels at Crescent City appears as a decrease in the variability at the shorter scales. In retrospect, this decrease can be seen in the plot of $\{X_t\}$ at the bottom of Figure 187, but it is very difficult to spot in that plot, whereas the MRA makes it easy to pick out this and other features of interest to oceanographers.

The MRA in Figure 186 also makes it clear that the variability at some scales is seasonally dependent. It is, however, problematic to quantitatively study the scale-dependent variance properties of $\{X_t\}$ from this analysis, because here

$$\|\mathbf{X}\|^2 \neq \sum_{j=1}^{7} \|\widetilde{\mathcal{D}}_j\|^2 + \|\widetilde{\mathcal{S}}_7\|^2.$$

We can instead examine the MODWT coefficients, for which

$$\|\mathbf{X}\|^2 = \sum_{j=1}^{7} \|\widetilde{\mathbf{W}}_j\|^2 + \|\widetilde{\mathbf{V}}_7\|^2.$$

We can qualitatively study the time dependence of the variance on a scale-by-scale basis by creating rotated cumulative variance plots, defined as follows. First we circularly advance the elements of $\widetilde{\mathbf{W}}_j$ by $|\nu_j^{(H)}|$ units so that $\mathcal{T}^{-|\nu_j^{(H)}|}\widetilde{\mathbf{W}}_j$ is aligned with $\mathbf{X}$ (see Equation (114b) and the discussion surrounding it). We then cumulatively sum the squares of the elements of $\mathcal{T}^{-|\nu_j^{(H)}|}\widetilde{\mathbf{W}}_j$ and normalize by $1/N$ to form

$$C_{j,t} \equiv \frac{1}{N}\sum_{u=0}^{t} \widetilde{W}^2_{j,u+|\nu_j^{(H)}| \bmod N}, \quad t = 0, \ldots, N-1.$$

Since $C_{j,N-1} = \frac{1}{N}\|\mathcal{T}^{-|\nu_j^{(H)}|}\widetilde{\mathbf{W}}_j\|^2 = \frac{1}{N}\|\widetilde{\mathbf{W}}_j\|^2$ can be regarded as the contribution to the sample variance of $\mathbf{X}$ due to physical scale $\tau_j\,\Delta t$, the series $\{C_{j,t}\}$ tracks the build up of the sample variance for the jth scale across time. Because $\{C_{j,t}\}$ is a nondecreasing sequence, it is sometimes difficult to deduce variations in the manner in which the sample variance is building up by plotting $C_{j,t}$ versus t, but we can overcome this difficulty by instead plotting

$$C'_{j,t} \equiv C_{j,t} - t\frac{C_{j,N-1}}{N-1} \quad \text{versus } t, \tag{189}$$

which we refer to as a rotated cumulative variance plot (the rotation essentially allows us to see deviations from a uniform build up of $\frac{C_{j,N-1}}{N-1}$ per increment in t; this idea is exploited again in Section 9.6 to estimate the location of a change in variance in a time series).

As an example, Figure 190 shows rotated cumulative variance plots for physical scales $\tau_2\,\Delta t = 1$ day and $\tau_7\,\Delta t = 32$ days (since $\nu = -3$ for the LA(8) wavelet filter, Equation (114b) tells us that $\nu_2^{(H)} = -11$ and $\nu_7^{(H)} = -445$). The bottom plot in this figure clearly indicates that the variability in changes on a scale of a day is seasonally dependent, with higher variability occurring during the winter (i.e., close to the vertical dotted lines). The linear appearance of the rotated cumulative variance plot during the summer months shows that the variability at mid-summer is quite stable and predictable from year to year. The point at which the stable summer pattern shifts into the more erratic winter pattern can change from one year to the next by up to a month; moreover, the shift appears to be fairly abrupt some years (1985) and diffuse in others (1988). On the other hand, the top plot of Figure 190 shows a corresponding plot for variability in changes on a scale of 32 days. The seasonal dependence at this scale is much more erratic. For example, the rotated cumulative variance indicates a marked increase in variability around the beginning of 1980, 1982, 1983, 1986, 1987 and 1991, but quite constant variability of a roughly 20 month span from about March 1989 to January 1991. (We consider the time varying aspect of this rotated cumulative variance further in Section 8.7.)

Finally, as we noted in Section 5.3, the details and smooths constructed by the MODWT can be regarded as outputs from zero phase filters. This zero phase property is important if we want to relate events in the details to events in the original series and if we want to understand the temporal relationships between events at different scales. To illustrate what this property of zero phase means in practice, we show in the top plot of Figure 191 portions of the MODWT detail $\widetilde{\mathcal{D}}_5$ (thin curve) and $\mathbf{X}$ (thick) around the first part of 1986. This detail should reflect changes in $\mathbf{X}$ on a scale of

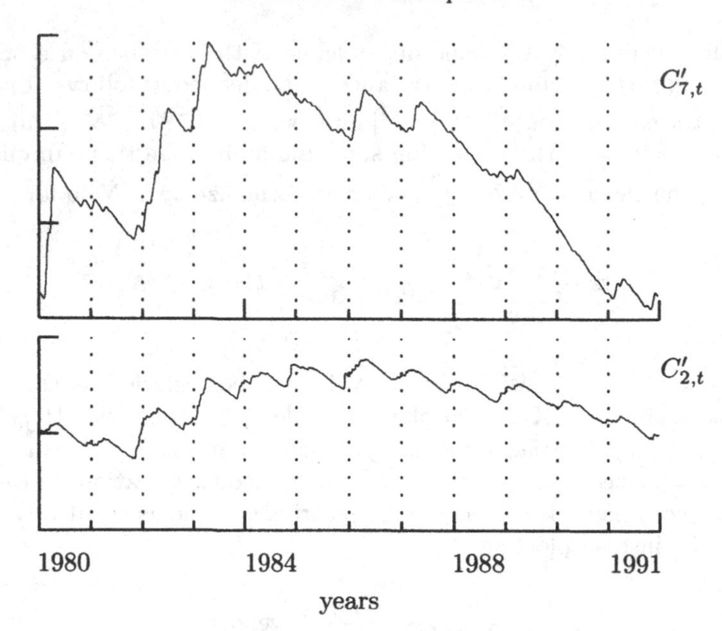

Figure 190. Rotated cumulative variance plots for subtidal sea level variations at physical scales $\tau_2\,\Delta t = 1$ day (bottom plot) and $\tau_7\,\Delta t = 32$ days (top).

8 days, so we would expect values in $\widetilde{\mathcal{D}}_5$ to be large when an average over roughly 8 days differs markedly from values surrounding it. This time period was picked because there are three peaks in $\mathbf{X}$ of such a duration – these are centered at about 32, 48 and 72 days from the beginning of 1986 (note that the distance between minor tick marks on the horizontal axis is 8 days). The MODWT detail picks out these events nicely, and the locations of the peaks in $\widetilde{\mathcal{D}}_5$ roughly match those in $\mathbf{X}$. Because $\widetilde{\mathcal{D}}_5$ circularly shifts when we apply circular shifts to $\mathbf{X}$, the top plot will remain the same no matter what origin we pick for $\mathbf{X}$.

By way of contrast, the thin curves in the middle and bottom plots of Figure 191 show DWT details $\mathcal{D}_5$ for two different shifts in the origin of $\mathbf{X}$, namely, 12 and 8 values (corresponding to 6 and 4 days). Note that changing the origin of $\mathbf{X}$ by 2 days markedly changes the $\mathcal{D}_5$'s. Because DWT details do not possess the zero phase property, the locations of peaks in $\mathcal{D}_5$ need not match up very well with those in $\mathbf{X}$. For example, consider the location of the peak in $\mathcal{D}_5$ corresponding to the one in $\mathbf{X}$ near 32 days. Relative to $\widetilde{\mathcal{D}}_5$ in the upper plot, the peak in $\mathcal{D}_5$ occurs markedly later in the middle plot and markedly earlier in the bottom plot. The ability of DWT details to locate transient events in $\mathbf{X}$ properly is thus limited, whereas MODWT details perform well because of their zero phase property.

5.9 Example: Nile River Minima

Here we present an MRA of a time series consisting of measurements of the minimum yearly water level of the Nile River over the years 622 to 1284 as recorded at the Roda gauge near Cairo (this analysis is adapted from an unpublished 1995 report by Simon Byers, Department of Statistics, University of Washington). The time series, with water levels in meters, is plotted in the bottom of Figure 192 and is actually a portion

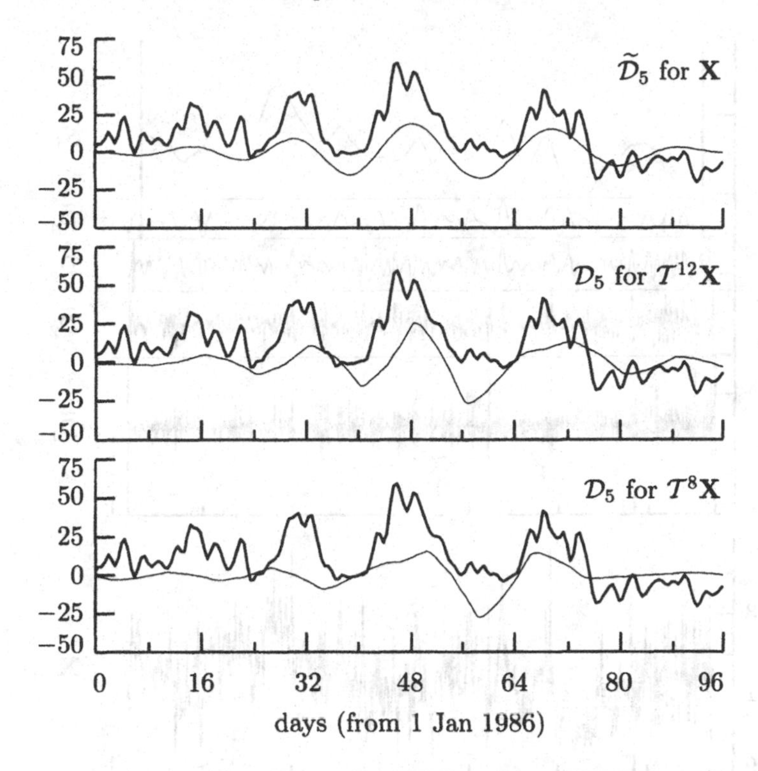

Figure 191. Demonstration of zero phase property of MODWT details. The thick curve in each plot is a 96 day portion of the Crescent City series starting at the beginning of 1986. The thin curves (from top to bottom) are the LA(8) MODWT detail $\widetilde{\mathcal{D}}_5$ for the original time series $\mathbf{X}$, the LA(8) DWT detail $\mathcal{D}_5$ for $\mathcal{T}^{12}\mathbf{X}$ (the time series delayed 12 data values) and the LA(8) DWT detail $\mathcal{D}_5$ for $\mathcal{T}^{8}\mathbf{X}$ (the time series delayed 8 data values). The physical scale associated with these details is $\tau_5 \, \Delta t = 8$ days; i.e., the details should be associated with changes of averages on a scale of 8 days (this is the distance between minor tick marks on the horizontal axis).

of a much longer historical record of Nile River minima compiled from various sources by Prince Omar Toussoun and published as part of a lengthy monogram entitled *Mémoire sur L'Histoire du Nil* in multiple volumes of *Mémoires de l'Institut d'Égypte* in 1925. Toussoun's record extends up to 1921, with the years 622 to 1284 representing the longest segment without interruptions (after 1284 there are a few scattered missing values up until 1470, after which coverage is much less complete: for example, there are only 17 measurements during the entire sixteenth century). The building of the Aswan Dam (1902) and the Aswan High Dam (1960–70) ended the primary role of nature in controlling the flow of the Nile River.

While the phenomena primarily responsible for the time series in Figure 192 are no longer operable, the historical record of Nile River minima has developed an interesting history in its own right. The hydrologist Hurst (1951) used this and other time series in a study that demonstrated the ineffectiveness of traditional statistical methodology for time series exhibiting long term persistence (i.e., a slowly decaying autocorrelation sequence – see Section 7.4). After Hurst's pioneering work, Mandelbrot and Wallis (1968) proposed to account for Hurst's empirical results by modeling the various series he studied as fractional Gaussian noise (this is defined in Section 7.6).

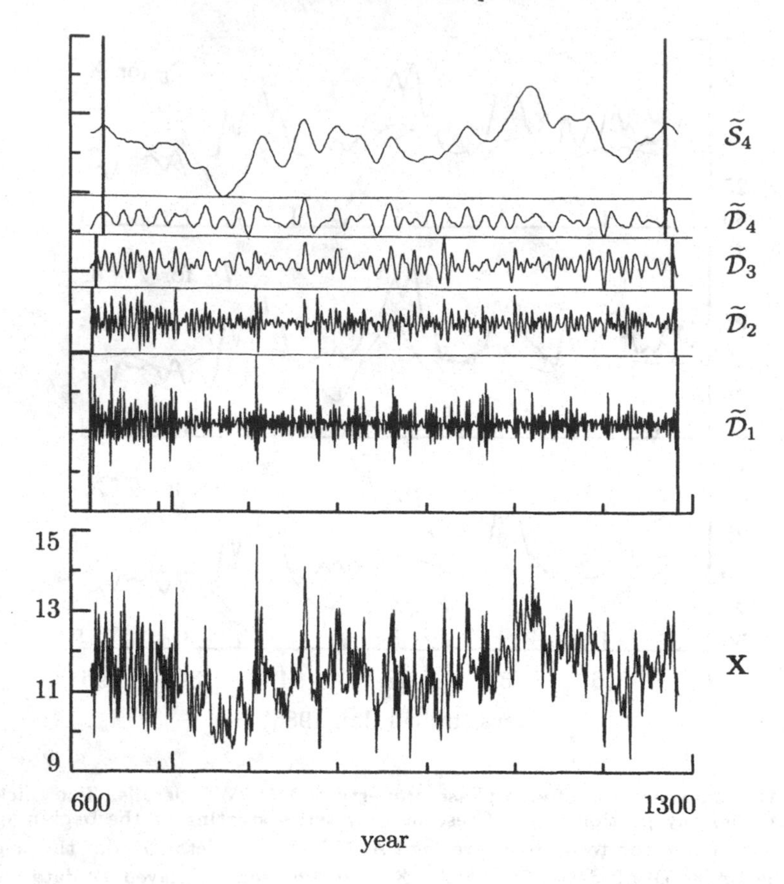

Figure 192. Level $J_0 = 4$ MODWT multiresolution analysis of Nile River minima (in meters) using the Haar wavelet filter.

This model was the first thorough attempt at characterizing what are now known as stationary long memory processes (Chapter 9 looks at various ways that wavelets can be used to study such processes). Beran (1994) used the Nile River series extensively as an example in his pioneering book on the statistics of long memory processes. More recently Eltahir and Wang (1999) studied this series as an indirect means of determining how often the El Niño phenomenon has occurred historically.

The top portion of Figure 192 shows a MODWT MRA of level $J_0 = 4$ for the Nile River minima. We used the Haar wavelet filter because MRAs from filters with $L = 4$ or larger were qualitatively similar to the Haar MRA – since there was no discernible benefit in using the wider filters, the Haar filter is preferable here because it minimizes concerns about the effects of boundary conditions (we have marked the boundary regions of each MRA component via a pair of thick vertical lines – see Section 5.11). We chose $J_0 = 4$ because we are primarily interested in pointing out a certain time dependency across scales τ_1 and τ_2 and contrasting it to a lack of such dependency at higher scales. Since the data were recorded annually (i.e., $\Delta t = 1$ year), the components of the MRA are associated with, from bottom to top, changes in averages on physical scales of 1, 2, 4 and 8 years ($\widetilde{\mathcal{D}}_1$ up to $\widetilde{\mathcal{D}}_4$, respectively) and averages over 16 years ($\widetilde{\mathcal{S}}_4$). The most interesting aspect of the MRA is the apparent inhomogeneity of variance in details $\widetilde{\mathcal{D}}_1$ and $\widetilde{\mathcal{D}}_2$: visually it appears that there is greater variability in these two details up until the early 700s. The details for higher scales ($\widetilde{\mathcal{D}}_3$ and $\widetilde{\mathcal{D}}_4$) show no evidence of increased variability during the early part of the record. While the increased variability is apparent – to a somewhat lesser degree – in the plot of $\{X_t\}$ itself and has been investigated using other methods (Beran, 1994), wavelet-based MRA picks it out quite clearly and nicely demonstrates that it is localized in scale. An examination of Toussoun's monogram and subsequent historical studies by Popper (1951) and Balek (1977) all indicate the construction in 715 (or soon thereafter) of a 'nilometer' in a mosque on Roda Island in the Nile River near Cairo (this year is marked by an elongated tick mark on the horizontal axes in both plots of Figure 192). The yearly minimum water levels from then to 1284 were measured using this device (or a reconstruction of it done in 861). The precise source of measurements for 622 to 714 is unknown, but they were most likely made at different locations around Cairo with possibly different types of measurement devices of less accuracy than the one in the Roda Island mosque. It is thus quite reasonable that this new nilometer led to a reduction in variability at the very smallest scales (it would be difficult to explain this change in variability on a geophysical basis).

We return to this example in Sections 8.8 and 9.8, where we make use of informal and formal statistical tests to verify that the variability is significantly different at scales of 1 and 2 days before and after about 715.

5.10 Example: Ocean Shear Measurements

Here we examine a 'time' series of vertical ocean shear measurements. These data were collected by an instrument that is dropped over the side of a ship and designed then to descend vertically into the ocean. As it descends, the probe collects measurements concerning the ocean as a function of depth. The ordering variable of our 'time' series is thus depth. One of the measurements is the x component of the velocity of water. This velocity is collected every 0.1 meters, first differenced over an interval of 10 meters, and then low-pass filtered to obtain a series related to vertical shear in the ocean. The bottom panel of Figure 194 shows the series of vertical shear measurements used here (this series has units of inverse seconds). The data extend from a depth of 350.0 meters down to 1037.4 meters in increments of $\Delta t = 0.1$ meters (there are $N = 6875$ data values in all). There are two thin vertical lines marked on the plot, between which there are 4096 values ranging from 489.5 meters to 899.0 meters (this subseries is discussed further in Section 8.9).

Figure 194 shows an MRA of level $J_0 = 6$ constructed using the LA(8) wavelet filter. We chose this filter primarily because, when we return to this example in Section 8.9, we will be interested both in looking at how the wavelet coefficients at different scales are related across time and also in using a wavelet filter that offers good protection against leakage (the concept of leakage is discussed in Section 8.9 also). The natural choices to achieve alignment in time are the least asymmetric or coiflet filters. Within these two classes, the LA(8) is the shortest filter with acceptable leakage properties (had we been interested solely in an MRA, the D(4), D(6) and C(6) filters would yield MRAs that are quite comparable to that of the LA(8) filter, but the small scale details $\widetilde{\mathcal{D}}_j$ for the Haar MRA look considerably different due to leakage). We set $J_0 = 6$ because this is the index of the smallest scale that is visibly free of bursts (to be discussed in what follows). This choice yields an MRA that isolates the bursts into the detail series and also yields – for the sake of comparison – one detail

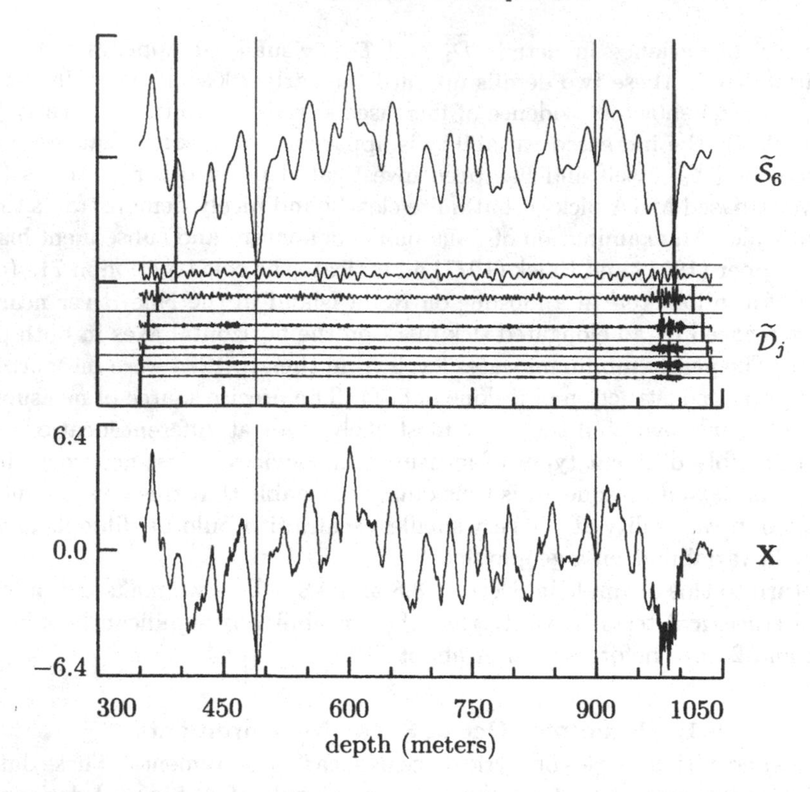

Figure 194. Level $J_0 = 6$ MODWT multiresolution analysis using LA(8) wavelet of vertical shear measurements (in inverse seconds) versus depth (in meters). This series was collected and supplied by Mike Gregg, Applied Physics Laboratory, University of Washington.

series that is burst free.

The MRA in Figure 194 is dominated by the smooth $\widetilde{\mathcal{S}}_6$. This smooth appears to be a reasonable extraction of the large scale variations in $\{X_t\}$ except at the very beginning and end of the series, where the periodic assumption inherent in the MODWT arguably causes distortions in $\widetilde{\mathcal{S}}_6$ (as before, the pair of thick vertical lines intersecting a given $\widetilde{\mathcal{D}}_j$ or $\widetilde{\mathcal{S}}_6$ delineates the portions of the MRA influenced at least in part by boundary conditions). If so desired, we could use reflection boundary conditions to alleviate these slight distortions (see the discussion surrounding Equation (140)). Because the smooth dominates the MRA to the point that the details are not distinguishable, we have plotted the $\widetilde{\mathcal{D}}_j$'s separately in Figure 195. The details nicely pick out two bursts, one near 450 meters, and the other just after 950 meters (the second burst is clearly visible in the plot of $\{X_t\}$ in Figure 194, but the first is only apparent in retrospect). Since the physical scale associated with $\widetilde{\mathcal{D}}_j$ is $\tau_j \, \Delta t = 0.1\tau_j$ meters, we see that the first burst is prominent at scales of 0.8 meters and smaller (i.e., $\widetilde{\mathcal{D}}_1$ up to $\widetilde{\mathcal{D}}_4$), while there is still prominent evidence of the second burst at scale 1.6 meters (i.e., $\widetilde{\mathcal{D}}_5$). An examination of MRAs for $J_0 \geq 7$ shows no indication of these bursts for physical scales longer than 3.2 meters. If we examine the first burst more closely, we see that it shuts down rapidly in $\widetilde{\mathcal{D}}_4$ (0.8 meters), but the burst continues beyond this shut down point in smaller scale details ($\widetilde{\mathcal{D}}_3$, $\widetilde{\mathcal{D}}_2$ and $\widetilde{\mathcal{D}}_1$). A physical interpretation might be that turbulence at a 0.8 meter scale is driving shorter scale turbulence at

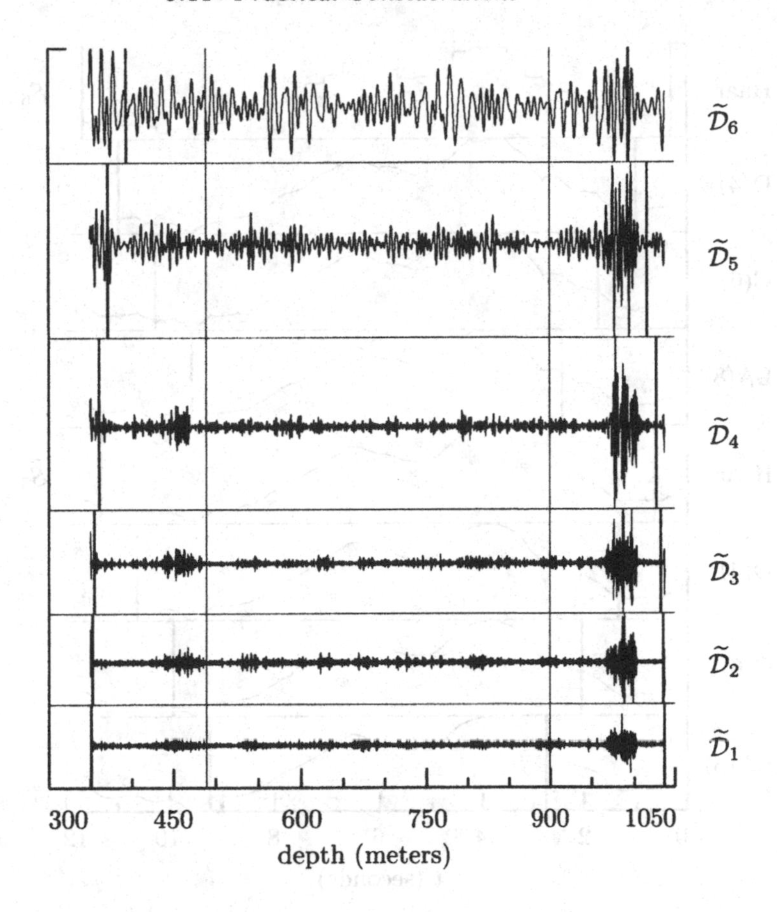

Figure 195. Expanded view of details series in MRA shown in Figure 194.

greater depths. By contrast, the larger second burst seems to be colocated in details $\widetilde{\mathcal{D}}_1$ up to $\widetilde{\mathcal{D}}_4$, but there seems to be hints of increased variability in $\widetilde{\mathcal{D}}_5$ and $\widetilde{\mathcal{D}}_6$ prior to the main portion of the burst. A MODWT MRA facilitates studies such as this one in which the depth (or time) relationships between events at different scales are important because its details and smooth are associated with zero phase filters. (While the additive decomposition given by the MRA gives a useful basis description of the shear series, we can say more about its properties in Section 8.9 after we have discussed ways of assessing a wavelet-based ANOVA with simple statistical models.)

5.11 Practical Considerations

In Section 4.11 we discussed several practical considerations that must be taken into account when carrying out a DWT analysis of a time series. Some of the remarks made there are also appropriate for the MODWT; however, because there are significant differences between the DWT and MODWT, each of the following topics merits more discussion.

• *Choice of wavelet filter*

For the most part, the remarks made in Section 4.11 about selecting a particular wavelet filter for the DWT hold also for the MODWT, but use of the MODWT for

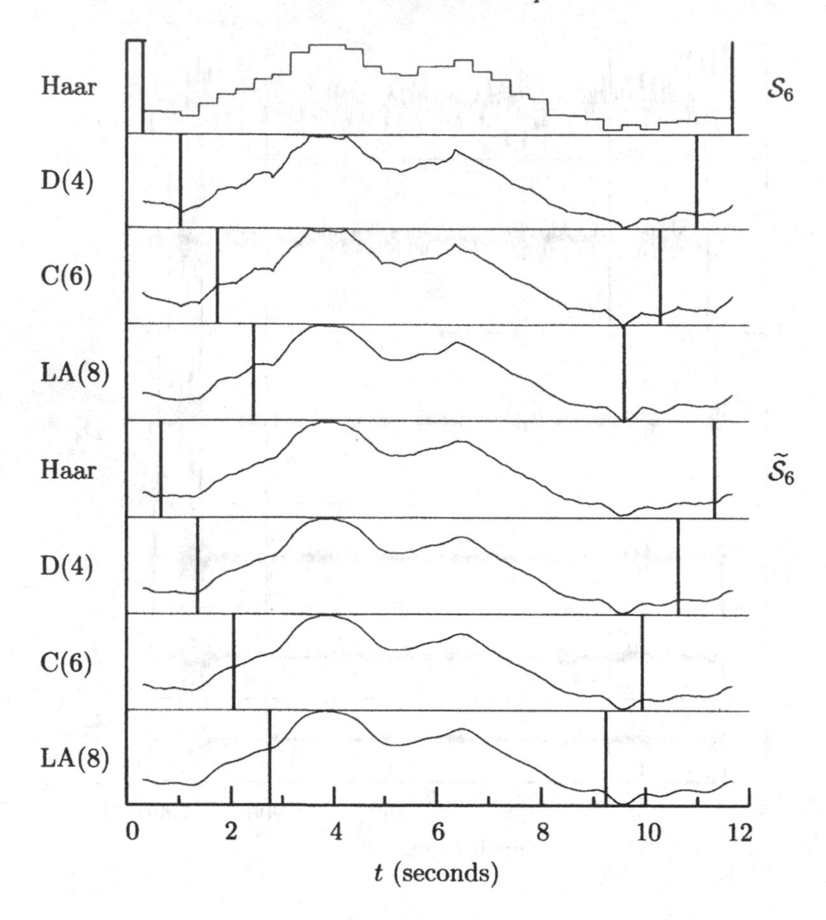

Figure 196. Comparison of DWT smooths $\mathcal{S}_6$ (top four series) and MODWT smooths $\widetilde{\mathcal{S}}_6$ (bottom four) for the ECG time series using, from top to bottom within each group, the Haar, D(4), C(6) and LA(8) wavelet filters. The thick vertical lines mark the boundary regions of the smooths, i.e., the parts of the smooths that are influenced (at least to some degree) by the assumption of circularity.

creating an MRA makes the choice of the wavelet filter less critical, as we illustrate in Figure 196. The top four series there show the DWT-based smooths $\mathcal{S}_6$ from the Haar, D(4), C(6) and LA(8) MRAs for the ECG time series (these are taken from Figures 130 to 133). The smooths based on the Haar, D(4) and C(6) wavelet filters contain unappealing blocks, sharks' fins and triangles, which are artifacts due to the filters rather than pertinent features in the ECG data. On the other hand, the LA(8) smooth shows no obvious artifacts and hence can be regarded as a more realistic depiction of the baseline drift in this series. These considerations would lead us to recommend against using Haar, D(4) or C(6) DWT-based MRAs for most time series since there are not many series whose characteristic features are well matched to blocks, sharks' fins and triangles.

Let us now see what happens when we switch from the DWT to the MODWT. The bottom four series in Figure 196 show MODWT smooths $\widetilde{\mathcal{S}}_6$ based upon the same four wavelet filters as before. Qualitatively, the three smooths look quite similar, although the Haar $\widetilde{\mathcal{S}}_6$ shows evidence of very small ripples that are not present in the other three MODWT smooths. A similar examination of the MODWT details $\widetilde{\mathcal{D}}_j$ for the ECG series again shows only quite small differences due to the different wavelet filters. We thus cannot as easily dismiss the Haar, D(4) and C(6) MRAs as we could in the DWT case. In general, we must resort to application-dependent considerations in order to make a reasonable choice of wavelet filters (for example, if $\widetilde{\mathcal{S}}_6$ is meant to capture the baseline drift and if we assume for the sake of argument that the notion of this drift precludes small ripples, an ECG analyst might prefer the D(4), C(6) or LA(8) wavelet filters over the Haar).

The key to understanding why MODWT-based MRAs are less dependent on the wavelet filter than DWT-based MRAs is given in Exercise [5.1], which is to show that MODWT details and smooths can be generated by averaging certain circularly shifted DWT details and smooths created from circularly shifted versions of a time series. This extra layer of averaging acts to blur out blocks, sharks' fins and triangles, thus eliminating the obvious defects in the DWT-based MRAs. From the point of view of filtering theory, the desirable properties of the MODWT detail $\widetilde{\mathcal{D}}_j$ arise because it is the output from a filter cascade with two stages (see Section 5.4). The first stage involves an 'analysis' filter with transfer function $\widetilde{H}_j(\cdot)$, and the second stage, a 'synthesis' filter $\widetilde{H}_j^*(\cdot)$. The equivalent filter $|\widetilde{H}_j(\cdot)|^2$ has a phase function that is always zero, no matter what wavelet filter we choose to use. This interpretation suggests that we can expect to find more marked differences between MODWT wavelet coefficients $\widetilde{\mathbf{W}}_j$ from different wavelet filters than between the corresponding details: the $\widetilde{\mathbf{W}}_j$ are formed using $\widetilde{H}_j(\cdot)$, whose phase properties depend upon the particular wavelet filter. Additionally, the squared gain functions associated with the MODWT wavelet coefficients and details are, respectively, $|\widetilde{H}_j(\cdot)|^2$ and $|\widetilde{H}_j(\cdot)|^4$, so we can expect the latter to provide a decent approximation to a band-pass filter in certain cases where the former might be inadequate. This point is illustrated by the example in Section 8.9, where we show that use of the Haar $\widetilde{\mathbf{W}}_j$ leads to an ANOVA that suffers from 'leakage' when compared to $\widetilde{\mathbf{W}}_j$ based on wavelet filters of length $L \geq 4$.

To summarize, as compared to DWT-based MRAs, a MODWT-based MRA is less dependent upon our choice of wavelet filter, but not so much so that we can recommend always using a particular filter. A careful study of the differences between MRAs based on different wavelet filters is still needed to see which filter is best matched to a particular application. We can expect the wavelet filter to have a greater impact on the MODWT wavelet coefficients because these depend more highly on the characteristics of individual wavelet filters.

• *Handling boundary conditions*

Like the DWT, the MODWT makes use of circular filtering, so the comments we made in Section 4.11 about circular boundary conditions and the simple alternative of reflection boundary conditions apply equally well to the MODWT. In what follows, we consider the extent to which circularity affects the MODWT coefficients and MRAs – our discussion closely parallels the one for the DWT, but the relevant formulae are quite different.

Let us first determine exactly which MODWT coefficients are affected by the assumption of circularity. To do so, we just have to look at the definitions for $\widetilde{W}_{j,t}$ and $\widetilde{V}_{j,t}$ in Equation (169a) and see for which values of t we can eliminate the 'mod' operation in $X_{t-l \bmod N}$ without changing the way in which these coefficients are defined. In forming either coefficient, the dummy index l ranges from 0 to $L_j - 1 = (2^j - 1)(L - 1)$.

Since $t - l \bmod N = t - l$ for all l as long as $t \geq L_j - 1$, the MODWT coefficients that are affected by circularity are

$$\widetilde{W}_{j,t} \text{ and } \widetilde{V}_{j,t} \text{ for } t = 0, \ldots, \min\{L_j - 2, N - 1\}. \tag{198a}$$

We refer to the above as the MODWT boundary coefficients. As was true for the DWT coefficients in $\mathbf{W}_j$ and $\mathbf{V}_j$, the boundary coefficients all occur at the beginning of $\widetilde{\mathbf{W}}_j$ and $\widetilde{\mathbf{V}}_j$. Since the dimensions of these vectors are always N and since L_j increases with j, the proportion of boundary coefficients increases with j and reaches unity when $L_j - 1 \geq N$, i.e., when $j \geq \log_2(\frac{N}{L-1} + 1)$.

Let us now specialize to the LA and coiflet wavelet filters and assume that $L_j \leq N$ so that $\min\{L_j - 2, N - 1\} = L_j - 2$ (i.e., we have at least one coefficient that is not a boundary coefficient). The discussion in Sections 4.8 and 4.9 says that, in order to obtain filter outputs that are aligned with $\mathbf{X}$, we must circularly shift the MODWT coefficient vectors by amounts that are dictated by the phase properties of the LA or coiflet filters. With $\nu_j^{(H)}$ and $\nu_j^{(G)}$ defined as in Equations (114b) and (114a), this procedure yields the circularly shifted vectors $\mathcal{T}^{-|\nu_j^{(H)}|}\widetilde{\mathbf{W}}_j$ and $\mathcal{T}^{-|\nu_j^{(G)}|}\widetilde{\mathbf{V}}_j$, whose elements can then be plotted versus the same actual times associated with the corresponding elements of $\mathbf{X}$. Thus some of the MODWT boundary coefficients will be transferred to the ends of the circularly shifted vectors. The indices of the boundary wavelet coefficients in $\mathcal{T}^{-|\nu_j^{(H)}|}\widetilde{\mathbf{W}}_j$ are given by

$$t = 0, \ldots, L_j - 2 - |\nu_j^{(H)}| \text{ along with } t = N - |\nu_j^{(H)}|, \ldots, N - 1; \tag{198b}$$

likewise, the indices of the boundary coefficients in $\mathcal{T}^{-|\nu_j^{(G)}|}\widetilde{\mathbf{V}}_j$ are given by

$$t = 0, \ldots, L_j - 2 - |\nu_j^{(G)}| \text{ along with } t = N - |\nu_j^{(G)}|, \ldots, N - 1 \tag{198c}$$

(Exercise [5.11] is to check that one aspect of these equations is sensible, namely, that we always have $L_j - 2 \geq |\nu_j^{(H)}|$ and $L_j - 2 \geq |\nu_j^{(G)}|$). These results can be used to add vertical lines to plots of the circularly shifted vectors, with the interpretation that these lines delineate the portions of the plotted coefficients outside of which circularity has some effect. As an example, these lines appear as part of Figure 183 showing the LA(8) MODWT of the ECG series.

Finally let us study how circularity affects the MODWT details $\widetilde{\mathcal{D}}_j$ and smooths $\widetilde{\mathcal{S}}_j$ (this discussion applies to all wavelet filters; i.e., we are no longer restricting ourselves to the LA or coiflet filters). As indicated by Equation (172a), the detail $\widetilde{\mathcal{D}}_j$ can be obtained directly from $\mathbf{X}$ by filtering the latter with a circular filter given by $\{|\widetilde{H}_j(\frac{k}{N})|^2\}$. This filter is a periodized filter formed by the cross-correlation of the filter $\{\tilde{h}_{j,l}\}$ with itself, i.e., the autocorrelation of $\{\tilde{h}_{j,l}\}$ (see Equation (36e)). If we denote this filter by $\{\tilde{h}_j \star \tilde{h}_{j,l}\}$, we can use Equation (36c) to write

$$\tilde{h}_j \star \tilde{h}_{j,l} = \sum_{u=-\infty}^{\infty} \tilde{h}_{j,u}\tilde{h}_{j,u+l} = \sum_{u=0}^{L_j-1} \tilde{h}_{j,u}\tilde{h}_{j,u+l}, \quad l = \ldots, -1, 0, 1, \ldots,$$

where we have used the fact that $\tilde{h}_{j,l} = 0$ for $l < 0$ and $l \geq L_j$ (in comparing the above to (36c), recall that $\tilde{h}_{j,l}$ is real-valued and hence $\tilde{h}_{j,l}^* = \tilde{h}_{j,l}$). From the above, we can argue that $\tilde{h}_j \star \tilde{h}_{j,l} = 0$ when $|l| \geq L_j$; i.e., the filter $\{\tilde{h}_j \star \tilde{h}_{j,l}\}$ is of odd width $2L_j - 1$ and is centered about $l = 0$. We can thus write the elements of $\widetilde{\mathcal{D}}_j$ as

$$\widetilde{\mathcal{D}}_{j,t} = \sum_{l=-(L_j-1)}^{L_j-1} \tilde{h}_j \star \tilde{h}_{j,l} X_{t-l \bmod N}, \quad t = 0, \ldots, N-1,$$

from which it is evident that, as long as we have both $t-(L_j-1) \geq 0$ and $t+(L_j-1) \leq N-1$, the element $\widetilde{\mathcal{D}}_{j,t}$ is unaffected by circularity; i.e., we can write

$$\widetilde{\mathcal{D}}_{j,t} = \sum_{l=-(L_j-1)}^{L_j-1} \tilde{h}_j \star \tilde{h}_{j,l} X_{t-l} \text{ for } t = L_j - 1, \ldots, N - L_j.$$

The elements of $\widetilde{\mathcal{D}}_j$ that are affected by circularity are thus those with indices t satisfying either $t = 0, \ldots, L_j - 2$ or $t = N - L_j + 1, \ldots, N - 1$. The same argument also holds for $\widetilde{\mathcal{S}}_j$.

We can use the above results to place vertical lines on plots of $\widetilde{\mathcal{D}}_j$ and $\widetilde{\mathcal{S}}_j$ to delineate the boundary regions, i.e., the regions influenced by circularity. This has been done in Figures 183, 184, 186, 192, 194 and 195 and also on the MODWT smooths $\widetilde{\mathcal{S}}_6$ plotted in Figure 196 (these are the bottom four series). From this latter figure, we see that, as the wavelet filter width L increases, the extent of the boundary region increases. For Figure 196 the boundary regions are also delineated on the corresponding DWT smooths $\mathcal{S}_6$ (the top four plotted series). A comparison between $\mathcal{S}_6$ and $\widetilde{\mathcal{S}}_6$ for a given wavelet filter shows that the MODWT boundary regions are somewhat larger than the DWT regions, a price that we must unavoidably pay for the appealing properties of MODWT MRAs. As we emphasized in Section 4.11, we must regard the boundary regions as very conservative measures of the influence of circularity – the regions over which circularity has a deleterious effect are usually much smaller.

- *Handling sample sizes that are not a power of two*

As we have noted earlier in this chapter, the MODWT is defined naturally for all sample sizes. Unlike the DWT, there is thus no need for special adaptations to handle certain sample sizes.

- *Choice of level J_0 of MODWT*

Our discussion on selecting J_0 for the DWT noted that a sensible choice depends primarily on the time series at hand – the same also holds for the MODWT. It is interesting to note, however, that, if $N = 2^J$, we cannot set J_0 to be greater than J for the DWT (because the DWT pyramid algorithm must terminate when there is but a single scaling coefficient left), whereas in theory there is nothing to stop us from picking $J_0 > J$ for the MODWT (because the MODWT pyramid algorithm always yields N scaling coefficients for use at the next level). Setting $J_0 > J$ would superficially seem to offer a way of obtaining information about variations in our times series over scales greater than the entire extent of the series, which is counter-intuitive. Exercise [5.9c] indicates that in fact we gain nothing by setting $J_0 > J$: when $N = 2^J$, all the elements of $\widetilde{\mathcal{S}}_{J_0}$ must be equal to the sample mean $\overline{X}$ for all $J_0 \geq J$. Since Equation (172c) says that $\widetilde{\mathcal{D}}_{J_0+1} = \widetilde{\mathcal{S}}_{J_0} - \widetilde{\mathcal{S}}_{J_0+1}$, all MODWT details for levels $J_0 > J$

are constrained to be zero. For sample sizes that are not a power of two, in practice $\widetilde{\mathcal{D}}_{J_0}$ rapidly becomes quite close to zero as $2^{J_0}/N$ increases beyond about 1.5.

If, as in the DWT discussion, we insist that there be at least one MODWT coefficient in $\widetilde{\mathbf{V}}_{J_0}$ that is not a boundary coefficient, then a rule of thumb is to require the condition $J_0 < \log_2(\frac{N}{L-1}+1)$; if this rule is deemed to be too conservative, alternative upper bounds are $J_0 \le \log_2(N)$ or, in rare cases, $J_0 \le \log_2(1.5N)$. All three upper bounds should be regarded as just guides for initially picking J_0 – the final choice of J_0 must come from application-dependent considerations.

5.12 Summary

Let $\mathbf{X}$ be an N dimensional vector whose elements are the real-valued time series $\{X_t : t = 0, \ldots, N-1\}$, where the sample size N is any positive integer. For any positive integer J_0, the level J_0 MODWT of $\mathbf{X}$ is a transform consisting of the J_0+1 vectors $\widetilde{\mathbf{W}}_1, \ldots, \widetilde{\mathbf{W}}_{J_0}$ and $\widetilde{\mathbf{V}}_{J_0}$, all of which have dimension N. The vector $\widetilde{\mathbf{W}}_j$ contains the MODWT wavelet coefficients associated with changes on scale $\tau_j \equiv 2^{j-1}$, while $\widetilde{\mathbf{V}}_{J_0}$ contains the MODWT scaling coefficients associated with averages on scale $\lambda_{J_0} \equiv 2^{J_0}$. Conceptually, based upon the definition for the MODWT coefficients (see below), we can write $\widetilde{\mathbf{W}}_j = \widetilde{\mathcal{W}}_j\mathbf{X}$ and $\widetilde{\mathbf{V}}_{J_0} = \widetilde{\mathcal{V}}_{J_0}\mathbf{X}$, where $\widetilde{\mathcal{W}}_j$ and $\widetilde{\mathcal{V}}_{J_0}$ are $N \times N$ matrices (in practice, however, the MODWT coefficient vectors are generated via an efficient 'pyramid' algorithm summarized below). The time series $\mathbf{X}$ can be recovered from its MODWT via

$$\mathbf{X} = \sum_{j=1}^{J_0} \widetilde{\mathcal{W}}_j^T \widetilde{\mathbf{W}}_j + \widetilde{\mathcal{V}}_{J_0}^T \widetilde{\mathbf{V}}_{J_0} \equiv \sum_{j=1}^{J_0} \widetilde{\mathcal{D}}_j + \widetilde{\mathcal{S}}_{J_0},$$

which defines a MODWT-based multiresolution analysis (MRA) of $\mathbf{X}$ in terms of jth level MODWT details $\widetilde{\mathcal{D}}_j \equiv \widetilde{\mathcal{W}}_j^T \widetilde{\mathbf{W}}_j$ and the J_0 level MODWT smooth $\widetilde{\mathcal{S}}_{J_0} \equiv \widetilde{\mathcal{V}}_{J_0}^T \widetilde{\mathbf{V}}_{J_0}$. A scale-by-scale ANOVA can be based upon the energy decomposition

$$\|\mathbf{X}\|^2 = \sum_{j=1}^{J_0} \|\widetilde{\mathbf{W}}_j\|^2 + \|\widetilde{\mathbf{V}}_{J_0}\|^2.$$

Unlike the DWT, the MODWT is *not* an orthonormal transform, but it is nevertheless capable of producing the above MRA and energy decomposition, which are similar to what the DWT yields. A primary advantage of the MODWT is that, unlike the DWT, it is not as critically dependent upon the assumed starting point for a time series. In particular, if $\widetilde{\mathbf{W}}_1, \ldots, \widetilde{\mathbf{W}}_{J_0}$ and $\widetilde{\mathbf{V}}_{J_0}$ are the MODWT for $\mathbf{X}$, then the MODWT for the circularly shifted series $\mathcal{T}^m\mathbf{X}$ is given by $\mathcal{T}^m\widetilde{\mathbf{W}}_1, \ldots, \mathcal{T}^m\widetilde{\mathbf{W}}_{J_0}$ and $\mathcal{T}^m\widetilde{\mathbf{V}}_{J_0}$. This fact implies that the MRA for $\mathcal{T}^m\mathbf{X}$ has details $\mathcal{T}^m\widetilde{\mathcal{D}}_j$ and smooth $\mathcal{T}^m\widetilde{\mathcal{S}}_{J_0}$; on the other hand, the energy decomposition remains the same because $\|\mathcal{T}^m\widetilde{\mathbf{W}}_j\|^2 = \|\widetilde{\mathbf{W}}_j\|^2$ and $\|\mathcal{T}^m\widetilde{\mathbf{V}}_{J_0}\|^2 = \|\widetilde{\mathbf{V}}_{J_0}\|^2$.

By definition, the elements of $\widetilde{\mathbf{W}}_j$ and $\widetilde{\mathbf{V}}_j$ are outputs obtained by filtering $\mathbf{X}$, namely,

$$\widetilde{W}_{j,t} \equiv \sum_{l=0}^{L_j-1} \tilde{h}_{j,l} X_{t-l \bmod N} \text{ and } \widetilde{V}_{j,t} \equiv \sum_{l=0}^{L_j-1} \tilde{g}_{j,l} X_{t-l \bmod N},$$

$t = 0, \ldots, N-1$, where $\{\tilde{h}_{j,l}\}$ and $\{\tilde{g}_{j,l}\}$ are the jth level MODWT wavelet and scaling filters. These filters are defined in terms of the jth level equivalent wavelet and scaling filters $\{h_{j,l}\}$ and $\{g_{j,l}\}$ for the DWT as $\tilde{h}_{j,l} = h_{j,l}/2^{j/2}$ and $\tilde{g}_{j,l} = g_{j,l}/2^{j/2}$. Each of the MODWT filters has width $L_j \equiv (2^j - 1)(L - 1) + 1$ and can be generated once a basic MODWT wavelet filter $\tilde{h}_{1,l} \equiv \tilde{h}_l \equiv h_l/\sqrt{2}$ and its related MODWT scaling filter $\tilde{g}_{1,l} \equiv \tilde{g}_l \equiv (-1)^{l+1}\tilde{h}_{L-1-l}$ have been specified. Specifically, if $\widetilde{H}(\cdot)$ and $\widetilde{G}(\cdot)$ are the transfer function for, respectively, $\{\tilde{h}_l\}$ and $\{\tilde{g}_l\}$, then the transfer functions for $\{\tilde{h}_{j,l}\}$ and $\{\tilde{g}_{j,l}\}$ are given by

$$\widetilde{H}_j(f) \equiv \widetilde{H}(2^{j-1}f)\prod_{l=0}^{j-2}\widetilde{G}(2^l f) \text{ and } \widetilde{G}_j(f) \equiv \prod_{l=0}^{j-1}\widetilde{G}(2^l f).$$

Because the MODWT filters are just rescaled versions of the corresponding equivalent filters for the DWT, all the relationships we have derived for the former hold for the latter after obvious adjustments – the most important ones are given in Table 202. If we let $\{\tilde{h}^{\circ}_{j,l}\}$ and $\{\tilde{g}^{\circ}_{j,l}\}$ be the periodization of $\{\tilde{h}_{j,l}\}$ and $\{\tilde{g}_{j,l}\}$ to circular filters of length N, we can use the definitions for the MODWT coefficients to obtain

$$\widetilde{W}_{j,t} = \sum_{l=0}^{N-1} \tilde{h}^{\circ}_{j,l}X_{t-l \bmod N} \text{ and } \widetilde{V}_{j,t} = \sum_{l=0}^{N-1} \tilde{g}^{\circ}_{j,l}X_{t-l \bmod N}.$$

These can be used to deduce the elements of the matrices $\widetilde{\mathcal{W}}_j$ and $\widetilde{\mathcal{V}}_{J_0}$. In particular, Equation (171c) shows how each row of $\widetilde{\mathcal{W}}_j$ is constructed using the N elements of $\{\tilde{h}^{\circ}_{j,l}\}$ (the matrix $\widetilde{\mathcal{V}}_{J_0}$ is formed similarly using $\{\tilde{g}^{\circ}_{j,l}\}$ instead).

In practice the matrices $\widetilde{\mathcal{W}}_j$ and $\widetilde{\mathcal{V}}_{J_0}$ are not used explicitly, but rather the vectors constituting the MODWT are formed using a 'pyramid' algorithm that uses just the basic MODWT wavelet and scaling filters. The MODWT pyramid algorithm at stage j consists of circularly filtering the N elements of $\widetilde{\mathbf{V}}_{j-1} \equiv [\widetilde{V}_{j-1,0}, \widetilde{V}_{j-1,1}, \ldots, \widetilde{V}_{j-1,N-1}]^T$ to obtain the jth level MODWT wavelet and scaling coefficients via

$$\widetilde{W}_{j,t} = \sum_{l=0}^{L-1} \tilde{h}_l \widetilde{V}_{j-1,t-2^{j-1}l \bmod N} \text{ and } \widetilde{V}_{j,t} = \sum_{l=0}^{L-1} \tilde{g}_l \widetilde{V}_{j-1,t-2^{j-1}l \bmod N},$$

$t = 0, \ldots, N-1$ (these are the elements of $\widetilde{\mathbf{W}}_j$ and $\widetilde{\mathbf{V}}_j$). With $\widetilde{\mathbf{V}}_0 \equiv \mathbf{X}$, we start the pyramid algorithm with $j = 1$ and, after repeating the algorithm with $j = 2, 3, \ldots, J_0$, we obtain all J_0+1 of the MODWT coefficient vectors (the other vectors $\widetilde{\mathbf{V}}_1, \ldots, \widetilde{\mathbf{V}}_{J_0-1}$ are MODWT scaling coefficients for scales smaller than λ_{J_0} and can be regarded as intermediate computations). The pyramid algorithm can be expressed in matrix notation as $\widetilde{\mathbf{W}}_j = \widetilde{\mathcal{B}}_j\widetilde{\mathbf{V}}_{j-1}$ and $\widetilde{\mathbf{V}}_j = \widetilde{\mathcal{A}}_j\widetilde{\mathbf{V}}_{j-1}$, where $\widetilde{\mathcal{B}}_j$ and $\widetilde{\mathcal{A}}_j$ are $N \times N$ matrices. The elements of these matrices are taken from circular filters of length N obtained by periodizing upsampled versions of $\{\tilde{h}_l\}$ and $\{\tilde{g}_l\}$. In particular, suppose we upsample $\{\tilde{h}_l\}$ by a factor of 2^{j-1} (i.e., we insert $2^{j-1} - 1$ zeros between each element of the filter). If we denote the result of periodizing this upsampled filter to length N as $\{\tilde{h}_l^{(\uparrow 2^{j-1})\circ}\}$, then the first row of $\widetilde{\mathcal{B}}_j$ is given by

$$[\tilde{h}_0^{(\uparrow 2^{j-1})\circ}, \tilde{h}_{N-1}^{(\uparrow 2^{j-1})\circ}, \tilde{h}_{N-2}^{(\uparrow 2^{j-1})\circ}, \ldots, \tilde{h}_2^{(\uparrow 2^{j-1})\circ}, \tilde{h}_1^{(\uparrow 2^{j-1})\circ}],$$

and the $N-1$ remaining rows are obtained by circularly shifting the above to the right by, respectively, $1, 2, \ldots, N-1$ units.

$$\tilde{h}_l \equiv h_l/\sqrt{2} \qquad \tilde{g}_l \equiv g_l/\sqrt{2}$$

$$\{\tilde{h}_l\} \longleftrightarrow \widetilde{H}(\cdot) = \tfrac{1}{\sqrt{2}} H(\cdot) \qquad \{\tilde{g}_l\} \longleftrightarrow \widetilde{G}(\cdot) = \tfrac{1}{\sqrt{2}} G(\cdot)$$

$$\tilde{h}_l = (-1)^l \tilde{g}_{L-1-l} \qquad \tilde{g}_l \equiv (-1)^{l+1} \tilde{h}_{L-1-l}$$

$$\widetilde{H}(f) = -e^{-i2\pi f(L-1)} \widetilde{G}(\tfrac{1}{2} - f) \qquad \widetilde{G}(f) = e^{-i2\pi f(L-1)} \widetilde{H}(\tfrac{1}{2} - f)$$

$$\textstyle\sum_l \tilde{h}_l = \widetilde{H}(0) \equiv 0 \qquad \sum_l \tilde{g}_l = \widetilde{G}(0) = 1$$

$$\textstyle\sum_l \tilde{h}_l^2 \equiv \tfrac{1}{2} \qquad \sum_l \tilde{g}_l^2 = \tfrac{1}{2}$$

$$\textstyle\sum_l \tilde{h}_l \tilde{h}_{l+2n} \equiv 0, n \neq 0 \qquad \sum_l \tilde{g}_l \tilde{g}_{l+2n} = 0, n \neq 0$$

$$\textstyle\sum_l \tilde{g}_l \tilde{h}_{l+2n} = 0$$

$$\widetilde{\mathcal{H}}(f) \equiv |\widetilde{H}(f)|^2 = \tfrac{1}{2}\mathcal{H}(f) \qquad \widetilde{\mathcal{G}}(f) \equiv |\widetilde{G}(f)|^2 = \tfrac{1}{2}\mathcal{G}(f)$$

$$\widetilde{\mathcal{H}}(f) + \widetilde{\mathcal{H}}(f + \tfrac{1}{2}) = 1 \qquad \widetilde{\mathcal{G}}(f) + \widetilde{\mathcal{G}}(f + \tfrac{1}{2}) = 1$$

$$\widetilde{\mathcal{G}}(f) + \widetilde{\mathcal{H}}(f) = 1$$

$$\widetilde{W}_{1,t} \equiv \textstyle\sum_l \tilde{h}_l X_{t-l \bmod N} \qquad \widetilde{V}_{1,t} \equiv \sum_l \tilde{g}_l X_{t-l \bmod N}$$

$$\widetilde{W}_{j,t} = \textstyle\sum_l \tilde{h}_l \widetilde{V}_{j-1,t-2^{j-1}l \bmod N} \qquad \widetilde{V}_{j,t} = \sum_l \tilde{g}_l \widetilde{V}_{j-1,t-2^{j-1}l \bmod N}$$

$$\tilde{h}_{1,l} \equiv \tilde{h}_l,\ \widetilde{H}_1(f) \equiv \widetilde{H}(f) \qquad \tilde{g}_{1,l} \equiv \tilde{g}_l,\ \widetilde{G}_1(f) \equiv \widetilde{G}(f)$$

$$\widetilde{H}_j(f) \equiv \widetilde{H}(2^{j-1}f) \prod_{l=0}^{j-2} \widetilde{G}(2^l f) \qquad \widetilde{G}_j(f) \equiv \prod_{l=0}^{j-1} \widetilde{G}(2^l f)$$

$$\widetilde{H}_j(f) = \tfrac{1}{2^{j/2}} H_j(f) \qquad \widetilde{G}_j(f) = \tfrac{1}{2^{j/2}} G_j(f)$$

$$\widetilde{H}_j(f) = \widetilde{H}(2^{j-1}f)\widetilde{G}_{j-1}(f) \qquad \widetilde{G}_j(f) = \widetilde{G}(2^{j-1}f)\widetilde{G}_{j-1}(f)$$

$$\{\tilde{h}_{j,l}\} \longleftrightarrow \widetilde{H}_j(\cdot) \qquad \{\tilde{g}_{j,l}\} \longleftrightarrow \widetilde{G}_j(\cdot)$$

$$\textstyle\sum_l \tilde{h}_{j,l} = \widetilde{H}_j(0) = 0 \qquad \sum_l \tilde{g}_{j,l} = \widetilde{G}_j(0) = 1$$

$$\textstyle\sum_l \tilde{h}_{j,l}^2 = \tfrac{1}{2^j} \qquad \sum_l \tilde{g}_{j,l}^2 = \tfrac{1}{2^j}$$

$$\textstyle\sum_l \tilde{h}_{j,l} \tilde{h}_{j,l+2^j n} = 0, n \neq 0 \qquad \sum_l \tilde{g}_{j,l} \tilde{g}_{j,l+2^j n} = 0, n \neq 0$$

$$\textstyle\sum_l \tilde{g}_{j,l} \tilde{h}_{j,l+2^j n} = 0$$

$$\widetilde{\mathcal{H}}_j(f) \equiv |\widetilde{H}_j(f)|^2 = \tfrac{1}{2^j}\mathcal{H}_j(f) \qquad \widetilde{\mathcal{G}}_j(f) \equiv |\widetilde{G}_j(f)|^2 = \tfrac{1}{2^j}\mathcal{G}_j(f)$$

$$\widetilde{W}_{j,t} \equiv \textstyle\sum_l \tilde{h}_{j,l} X_{t-l \bmod N} \qquad \widetilde{V}_{j,t} \equiv \sum_l \tilde{g}_{j,l} X_{t-l \bmod N}$$

Table 202. Key relationships involving MODWT wavelet and scaling filters (the conventions in Table 154 for limits on sums over l apply here too).

Given $\widetilde{\mathbf{W}}_j$ and $\widetilde{\mathbf{V}}_j$, we can recover the elements of $\widetilde{\mathbf{V}}_{j-1}$ via the jth stage of the inverse MODWT pyramid algorithm, which is given by

$$\widetilde{V}_{j-1,t} = \sum_{l=0}^{L-1} \tilde{h}_l \widetilde{W}_{j,t+2^{j-1}l \bmod N} + \sum_{l=0}^{L-1} \tilde{g}_l \widetilde{V}_{j,t+2^{j-1}l \bmod N},$$

$t = 0, 1, \ldots, N-1$. The above can be expressed in matrix notation as

$$\widetilde{\mathbf{V}}_{j-1} = \widetilde{\mathcal{B}}_j^T \widetilde{\mathbf{W}}_j + \widetilde{\mathcal{A}}_j^T \widetilde{\mathbf{V}}_j.$$

If we start with $\widetilde{\mathbf{V}}_0 = \mathbf{X}$ and combine the results of the above for $j = 1, \ldots, J_0$, we obtain

$$\mathbf{X} = \widetilde{\mathcal{B}}_1^T \widetilde{\mathbf{W}}_1 + \widetilde{\mathcal{A}}_1^T \widetilde{\mathcal{B}}_2^T \widetilde{\mathbf{W}}_2 + \widetilde{\mathcal{A}}_1^T \widetilde{\mathcal{A}}_2^T \widetilde{\mathcal{B}}_3^T \widetilde{\mathbf{W}}_3 + \cdots + \widetilde{\mathcal{A}}_1^T \cdots \widetilde{\mathcal{A}}_{J_0-1}^T \widetilde{\mathcal{B}}_{J_0}^T \widetilde{\mathbf{W}}_{J_0} + \widetilde{\mathcal{A}}_1^T \cdots \widetilde{\mathcal{A}}_{J_0-1}^T \widetilde{\mathcal{A}}_{J_0}^T \widetilde{\mathbf{V}}_{J_0}.$$

From the above, we can identify expressions for the MODWT details and smooth as $\widetilde{\mathcal{D}}_j = \widetilde{\mathcal{A}}_1^T \cdots \widetilde{\mathcal{A}}_{j-1}^T \widetilde{\mathcal{B}}_j^T \widetilde{\mathbf{W}}_j$ and $\widetilde{\mathcal{S}}_{J_0} = \widetilde{\mathcal{A}}_1^T \cdots \widetilde{\mathcal{A}}_{J_0-1}^T \widetilde{\mathcal{A}}_{J_0}^T \widetilde{\mathbf{V}}_{J_0}$. Since $\widetilde{\mathcal{D}}_j = \widetilde{\mathcal{W}}_j^T \widetilde{\mathbf{W}}_j$ and $\widetilde{\mathcal{S}}_{J_0} = \widetilde{\mathcal{V}}_{J_0}^T \widetilde{\mathbf{V}}_{J_0}$ also, we get $\widetilde{\mathcal{W}}_j = \widetilde{\mathcal{B}}_j \widetilde{\mathcal{A}}_{j-1} \cdots \widetilde{\mathcal{A}}_1$ and $\widetilde{\mathcal{V}}_j = \widetilde{\mathcal{A}}_j \widetilde{\mathcal{A}}_{j-1} \cdots \widetilde{\mathcal{A}}_1$.

Given the structure of $\widetilde{\mathcal{W}}_j$ and $\widetilde{\mathcal{V}}_{J_0}$, we can express the elements of $\widetilde{\mathcal{D}}_j$ and $\widetilde{\mathcal{S}}_{J_0}$ explicitly as

$$\widetilde{\mathcal{D}}_{j,t} = \sum_{l=0}^{N-1} \tilde{h}_{j,l}^{\circ} \widetilde{W}_{j,t+l \bmod N} \text{ and } \widetilde{\mathcal{S}}_{J_0,t} = \sum_{l=0}^{N-1} \tilde{g}_{J_0,l}^{\circ} \widetilde{V}_{J_0,t+l \bmod N},$$

$t = 0, 1, \ldots, N-1$. By combining the above with expressions for $\widetilde{W}_{j,t}$ and $\widetilde{V}_{J_0,t}$ involving $\tilde{h}_{j,l}^{\circ}$ and $\tilde{g}_{J_0,l}^{\circ}$, we can argue that $\widetilde{\mathcal{D}}_j$ and $\widetilde{\mathcal{S}}_{J_0}$ can be obtained directly by filtering the time series $\mathbf{X}$ with filters whose transfer functions are given by $|\widetilde{H}_j(\cdot)|^2$ and $|\widetilde{G}_{J_0}(\cdot)|^2$. The phase functions for both these filters are identically zero, so each component of a MODWT-based MRA is associated with a zero phase filter. The zero phase property is important because it allows us to line up events in the details and smooths meaningfully with events in $\mathbf{X}$.

Finally, if the sample size for $\mathbf{X}$ is such that we can compute a level J_0 DWT, we note the following relationships between the jth level DWT and MODWT wavelet coefficients for any $j \leq J_0$:

$$W_{j,t} = 2^{j/2} \widetilde{W}_{j,2^j(t+1)-1}, \quad t = 0, \ldots, N_j - 1,$$

where $N_j \equiv N/2^j$ (the above is Equation (96d)). Similarly, the scaling coefficients are related by $V_{J_0,t} = 2^{J_0/2} \widetilde{V}_{J_0,2^{J_0}(t+1)-1}$.

5.13 Exercises

[5.1] Suppose that $\mathbf{X}$ contains a time series of even sample size N. Let $\mathcal{D}_1$ and $\mathcal{S}_1$ be the detail and smooth of an MRA of $\mathbf{X}$ based upon a $J_0 = 1$ level partial DWT. We thus have $\mathbf{X} = \mathcal{D}_1 + \mathcal{S}_1$. Let $\mathcal{D}_{\mathcal{T},1}$ and $\mathcal{S}_{\mathcal{T},1}$ be corresponding quantities for an MRA of $\mathcal{T}\mathbf{X}$ (i.e., a circular shift of $\mathbf{X}$ to the right by one unit as per Equation (52a)). We thus have $\mathcal{T}\mathbf{X} = \mathcal{D}_{\mathcal{T},1} + \mathcal{S}_{\mathcal{T},1}$ and hence $\mathbf{X} = \mathcal{T}^{-1}\mathcal{D}_{\mathcal{T},1} + \mathcal{T}^{-1}\mathcal{S}_{\mathcal{T},1}$. Let $\widetilde{\mathcal{D}}_1$ and $\widetilde{\mathcal{S}}_1$ be the detail and smooth of a level $J_0 = 1$ MODWT-based MRA of $\mathbf{X}$ so that we have $\mathbf{X} = \widetilde{\mathcal{D}}_1 + \widetilde{\mathcal{S}}_1$. Show that

$$\widetilde{\mathcal{D}}_1 = \tfrac{1}{2}\left(\mathcal{D}_1 + \mathcal{T}^{-1}\mathcal{D}_{\mathcal{T},1}\right) \text{ and } \widetilde{\mathcal{S}}_1 = \tfrac{1}{2}\left(\mathcal{S}_1 + \mathcal{T}^{-1}\mathcal{S}_{\mathcal{T},1}\right).$$

Argue that we can also write

$$\widetilde{\mathcal{D}}_1 = \frac{1}{N}\sum_{n=0}^{N-1} \mathcal{T}^{-n}\mathcal{D}_{\mathcal{T}^n,1} \text{ and } \widetilde{\mathcal{S}}_1 = \frac{1}{N}\sum_{n=0}^{N-1} \mathcal{T}^{-n}\mathcal{S}_{\mathcal{T}^n,1},$$

where $\mathcal{D}_{\mathcal{T}^n,1}$ and $\mathcal{S}_{\mathcal{T}^n,1}$ constitute the DWT-based MRA for $\mathcal{T}^n\mathbf{X}$. (This interpretation of the MODWT details and smooth as averages over shifted DWT details and smooth for shifted versions of $\mathbf{X}$ holds in general in that, when $N = 2^J$ and $1 \le j \le J_0 \le J$, we have both

$$\widetilde{\mathcal{D}}_j = \frac{1}{2^j}\sum_{n=0}^{2^j-1} \mathcal{T}^{-n}\mathcal{D}_{\mathcal{T}^n,j} \text{ and } \widetilde{\mathcal{S}}_{J_0} = \frac{1}{2^{J_0}}\sum_{n=0}^{2^{J_0}-1} \mathcal{T}^{-n}\mathcal{S}_{\mathcal{T}^n,J_0} \quad (204)$$

and also

$$\widetilde{\mathcal{D}}_j = \frac{1}{N}\sum_{n=0}^{N-1} \mathcal{T}^{-n}\mathcal{D}_{\mathcal{T}^n,j} \text{ and } \widetilde{\mathcal{S}}_{J_0} = \frac{1}{N}\sum_{n=0}^{N-1} \mathcal{T}^{-n}\mathcal{S}_{\mathcal{T}^n,J_0}.$$

The idea of forming DWTs for the various shifts of a time series and then averaging together properly shifted syntheses based upon these DWTs is the key concept behind 'cycle spinning' as discussed in Coifman and Donoho, 1995.)

[5.2] Show that $\widetilde{\mathcal{P}}_1^T$ is the Moore–Penrose generalized inverse for $\widetilde{\mathcal{P}}_1$ (Peterson, 1996). Hint: you may use the result that, if $\widetilde{\mathcal{M}}$ is an $N \times 2N$ matrix such that $\widetilde{\mathcal{M}}\widetilde{\mathcal{P}}_1 = I_N$ and $\widetilde{\mathcal{P}}_1\widetilde{\mathcal{M}}$ is symmetric, then $\widetilde{\mathcal{M}}$ is the Moore–Penrose inverse for $\widetilde{\mathcal{P}}_1$ (Rao and Mitra, 1971).

[5.3] Equation (176) displays $\widetilde{\mathcal{B}}_2$ for the case $N = 12$ and $L = 4$. What would $\widetilde{\mathcal{B}}_3$ look like for this case?

[5.4] Using the pseudo-code in item [1] of the Comments and Extensions to Section 5.5, implement the MODWT pyramid algorithm and its inverse in your favorite computer language. Using the D(4) wavelet and scaling filters, compute and plot (or list the values for) $\widetilde{\mathbf{W}}_j$ and $\widetilde{\mathbf{V}}_j$, $j = 1$, 2 and 3, for the 16 point time series $\{X_{1,t}\}$ (see the caption to Figure 42). Verify your implementation of the inverse MODWT pyramid algorithm by showing that you can reconstruct $\widetilde{\mathbf{V}}_{j-1}$ from $\widetilde{\mathbf{W}}_j$ and $\widetilde{\mathbf{V}}_j$ for $j = 1$, 2 and 3 (here we define $\widetilde{\mathbf{V}}_0 = \mathbf{X}$). Verify that, by appropriately subsampling and rescaling $\widetilde{\mathbf{W}}_j$ and $\widetilde{\mathbf{V}}_j$, you can obtain the DWT $\mathbf{W}_j$ and $\mathbf{V}_j$ computed in Exercise [4.17].

[5.5] For the same time series $\{X_t\}$ considered in Exercise [4.19], compute and plot the MODWT details $\widetilde{\mathcal{D}}_k$ for $k = 1$ to 6 using any two wavelets of the following four wavelets: the Haar, D(4), C(6) and LA(8). Explain the appearance of the detail series with regard to (a) how $\{X_t\}$ is defined and (b) the particular wavelets used.

[5.6] For the 16 point time series $\mathbf{X} = [X_{1,0}, X_{1,1}, \ldots, X_{1,15}]^T$ listed in the caption to Figure 42, compute a level $J_0 = 2$ MRA based upon the Haar DWT, and compute corresponding MRAs for the circularly shifted series $\mathcal{T}^n\mathbf{X}$, $n = 1, 2$ and 3. Let $\mathcal{D}_{\mathcal{T}^n,1}$, $\mathcal{D}_{\mathcal{T}^n,2}$ and $\mathcal{S}_{\mathcal{T}^n,2}$ denote the details and smooth of the MRAs for $\mathcal{T}^n\mathbf{X}$, $n = 0, \ldots, 3$. Compute also the MODWT details $\widetilde{\mathcal{D}}_1$, $\widetilde{\mathcal{D}}_2$ and smooth $\widetilde{\mathcal{S}}_2$ for $\mathbf{X}$. Verify numerically that Equation (204) holds.

[5.7] Show that $\|\widetilde{\mathcal{D}}_j\|^2 \leq \|\widetilde{\mathbf{W}}_j\|^2$ and that $\|\widetilde{\mathcal{S}}_j\|^2 \leq \|\widetilde{\mathbf{V}}_j\|^2$ for all j.

[5.8] Show that $\widetilde{\mathcal{B}}_j^T\widetilde{\mathcal{B}}_j + \widetilde{\mathcal{A}}_j^T\widetilde{\mathcal{A}}_j = I_N$ holds for all j (this type of result is sometimes called a 'resolution of identity').

[5.9] Here we consider three complements to Exercise [171b].

(a) Show that, if the sample size is a power of two, i.e., $N = 2^J$, and if we select $J_0 \geq J$, then all the MODWT scaling coefficients in $\widetilde{\mathbf{V}}_{J_0}$ are in fact equal to $\overline{X}$, whereas, if N is not a power of two, then there can be some variation amongst the elements of $\widetilde{\mathbf{V}}_{J_0}$. We can use this to argue that Equation (171a) reduces to

$$\hat{\sigma}_X^2 = \frac{1}{N}\sum_{j=1}^{J_0} \|\widetilde{\mathbf{W}}_j\|^2 \text{ when } N = 2^J \text{ and } J_0 \geq J.$$

(b) Show that, for any sample size N and any level J_0, the sample mean of the elements of the MODWT smooth $\widetilde{\mathcal{S}}_{J_0}$ is equal to $\overline{X}$, whereas the sample mean of any MODWT detail $\widetilde{\mathcal{D}}_j$ is zero for any level j.

(c) Show that, if $N = 2^J$ and if we select $J_0 \geq J$, then all the elements of $\widetilde{\mathcal{S}}_{J_0}$ are in fact equal to $\overline{X}$, whereas, if N is not a power of two, then there can be some variation amongst the elements of $\widetilde{\mathcal{S}}_{J_0}$.

[5.10] Show that, for the Haar wavelet filter, we have $\widetilde{H}(f)+\widetilde{G}(f) = 1$. Demonstrate that this resolution of identity can be used to formulate an additive decomposition of a time series based directly on the MODWT Haar wavelet coefficients (C. A. Greenhall, private communication). What are the advantages and disadvantages of this approach as compared to the usual Haar MODWT MRA?

[5.11] Show that, for all the LA and coiflet wavelet filters discussed in Section 4.8 and Section 4.9, respectively, we always have $L_j - 2 \geq |\nu_j^{(H)}|$ and $L_j - 2 \geq |\nu_j^{(G)}|$, where $\nu_j^{(H)}$ and $\nu_j^{(G)}$ are defined in Equations (114b) and (114a) (this exercise verifies that the limits for t given in Equations (198b) and (198c) make sense).

6

The Discrete Wavelet Packet Transform

6.0 Introduction

In Chapter 4 we discussed the discrete wavelet transform (DWT), which essentially decomposes a time series $\mathbf{X}$ into coefficients that can be associated with different scales and times. We can thus regard the DWT of $\mathbf{X}$ as a 'time/scale' decomposition. The wavelet coefficients for a given scale $\tau_j \equiv 2^{j-1}$ tell us how localized weighted averages of $\mathbf{X}$ vary from one averaging period to the next. The scale τ_j gives us the effective width in time (i.e., degree of localization) of the weighted averages. Because the DWT can be formulated in terms of filters, we can relate the notion of scale to certain bands of frequencies. The equivalent filter that yields the wavelet coefficients for scale τ_j is approximately a band-pass filter with a pass-band given by $[1/2^{j+1}, 1/2^j]$. For a sample size $N = 2^J$, the $N-1$ wavelet coefficients constitute – when taken together – an octave band decomposition of the frequency interval $[1/2^{J+1}, 1/2]$, while the single scaling coefficient is associated with the interval $[0, 1/2^{J+1}]$. Taken as a whole, the DWT coefficients thus decompose the frequency interval $[0, 1/2]$ into adjacent individual intervals.

In this chapter we consider the *discrete wavelet packet transform* (DWPT), which can be regarded as any one of a collection of orthonormal transforms, each of which can be readily computed using a very simple modification of the pyramid algorithm for the DWT. Each DWPT is associated with a level j, and the jth level DWPT decomposes the frequency interval $[0, 1/2]$ into 2^j equal and individual intervals. Because the $J-1$th level decomposition splits $[0, 1/2]$ into $N/2 = 2^{J-1}$ equal intervals, there is a DWPT that mimics – but is not the same as – the decomposition of $[0, 1/2]$ given by the discrete Fourier transform (DFT). When $j = 1, \ldots, J-1$, the resulting DWPT yields what can be called a 'time/frequency' decomposition because each DWPT coefficient can be localized to a particular band of frequencies and a particular interval of time (this is similar in spirit to the so-called short-time Fourier transform, which essentially is computed by applying the DFT to subseries extracted from $\mathbf{X}$). We can also create an even larger collection of orthonormal transforms by grouping together carefully selected basis vectors from different DWPTs. In fact the DWT and all partial DWTs can be formed using basis vectors from different DWPTs, so this scheme leads to a flexible collection of transforms that serve as a bridge between time/scale and time/frequency decompositions.

The remainder of this chapter is organized as follows. First we discuss the basic concepts behind the DWPT in Section 6.1, noting in particular that, like the DWT, this transform depends on just a wavelet filter $\{h_l\}$ and its associated scaling filter $\{g_l\}$. In Section 6.2 we give an example of a DWPT analysis using a solar physics time series. Because DWPTs can be used to generate an entire collection of orthonormal transforms, we consider in Section 6.3 a method for defining an 'optimal' transform for a given time series – in Section 6.4 we then demonstrate the use of this method on the solar physics series. When we create a DWPT using any one of the Daubechies least asymmetric (LA) or coiflet filters, we obtain transform coefficients that can be regarded approximately as the output from a zero phase filter. In Section 6.5 we discuss the time shifts that are needed to achieve this approximate zero phase property. In Section 6.6 we define a maximal overlap DWPT (MODWPT) that is analogous to the maximal overlap DWT, and in Section 6.7 we again use the solar physics series to provide an example of a MODWPT analysis. We next discuss the idea of a 'matching pursuit' in Section 6.8, which allows us to approximate a time series via a linear combination of a small number of vectors chosen from a large set of vectors (including ones formed by merging distinct DWPTs or MODWPTs). We apply the matching pursuit idea to the Crescent City subtidal sea level series in Section 6.9. Finally, in Sections 6.10 and 6.11, we give a summary of the material in this chapter and some exercises.

6.1 Basic Concepts

As in Chapter 4, let $\mathbf{W} = \mathcal{W}\mathbf{X}$ represent the wavelet coefficients obtained by transforming $\mathbf{X}$ using the $N \times N$ orthonormal DWT matrix $\mathcal{W}$ (for convenience, we assume that $N = 2^J$ for some integer J). In practice the matrix $\mathcal{W}$ is not actually generated except implicitly through the pyramid algorithm. The first stage of the algorithm can be described in matrix notation as

$$\mathcal{P}_1\mathbf{X} = \begin{bmatrix} \mathcal{B}_1 \\ \mathcal{A}_1 \end{bmatrix} \mathbf{X} = \begin{bmatrix} \mathbf{W}_1 \\ \mathbf{V}_1 \end{bmatrix} \equiv \begin{bmatrix} \mathbf{W}_{1,1} \\ \mathbf{W}_{1,0} \end{bmatrix},$$

where $\mathbf{W}_{1,1} \equiv \mathbf{W}_1$ and $\mathbf{W}_{1,0} \equiv \mathbf{V}_1$. The matrix $\mathcal{P}_1$ is orthonormal because

$$\mathcal{P}_1\mathcal{P}_1^T = \begin{bmatrix} \mathcal{B}_1 \\ \mathcal{A}_1 \end{bmatrix} [\mathcal{B}_1^T, \mathcal{A}_1^T] = \begin{bmatrix} \mathcal{B}_1\mathcal{B}_1^T & \mathcal{B}_1\mathcal{A}_1^T \\ \mathcal{A}_1\mathcal{B}_1^T & \mathcal{A}_1\mathcal{A}_1^T \end{bmatrix} = \begin{bmatrix} I_{\frac{N}{2}} & 0_{\frac{N}{2}} \\ 0_{\frac{N}{2}} & I_{\frac{N}{2}} \end{bmatrix} = I_N$$

since we have constructed the $\frac{N}{2} \times N$ matrices $\mathcal{B}_1$ and $\mathcal{A}_1$ such that

$$\mathcal{B}_1\mathcal{B}_1^T = \mathcal{A}_1\mathcal{A}_1^T = I_{\frac{N}{2}} \text{ and } \mathcal{B}_1\mathcal{A}_1^T = \mathcal{A}_1\mathcal{B}_1^T = 0_{\frac{N}{2}}$$

(here $I_{\frac{N}{2}}$ is the $\frac{N}{2} \times \frac{N}{2}$ identity matrix, while $0_{\frac{N}{2}}$ represents an $\frac{N}{2} \times \frac{N}{2}$ matrix, all of whose elements are zero). Recall that, to a certain degree of approximation, the wavelet coefficients in $\mathbf{W}_{1,1}$ represent the frequency content in $\mathbf{X}$ for frequencies f such that $|f| \in [\frac{1}{4}, \frac{1}{2}]$, whereas $\mathbf{W}_{1,0}$ represents frequencies such that $|f| \in [0, \frac{1}{4}]$. The transform $\mathcal{P}_1$ corresponds to a level $J_0 = 1$ partial DWT (see Section 4.7).

Similarly, at the end of the second stage of the pyramid algorithm, we have a transform (a level $J_0 = 2$ partial DWT) that can be expressed as

$$\begin{bmatrix} \mathcal{B}_1 \\ \mathcal{B}_2\mathcal{A}_1 \\ \mathcal{A}_2\mathcal{A}_1 \end{bmatrix} \mathbf{X} = \begin{bmatrix} \mathbf{W}_1 \\ \mathbf{W}_2 \\ \mathbf{V}_2 \end{bmatrix} \equiv \begin{bmatrix} \mathbf{W}_{1,1} \\ \mathbf{W}_{2,1} \\ \mathbf{W}_{2,0} \end{bmatrix},$$

where $\mathbf{W}_{1,1} \equiv \mathbf{W}_1$ as before, while $\mathbf{W}_{2,1} \equiv \mathbf{W}_2$, and $\mathbf{W}_{2,0} \equiv \mathbf{V}_2$. By construction, we have

$$\mathcal{B}_2\mathcal{B}_2^T = \mathcal{A}_2\mathcal{A}_2^T = I_{\frac{N}{4}} \text{ and } \mathcal{B}_2\mathcal{A}_2^T = \mathcal{A}_2\mathcal{B}_2^T = 0_{\frac{N}{4}},$$

so it follows readily that the transform is orthonormal because

$$\begin{aligned}
\begin{bmatrix} \mathcal{B}_1 \\ \mathcal{B}_2\mathcal{A}_1 \\ \mathcal{A}_2\mathcal{A}_1 \end{bmatrix} \left[\mathcal{B}_1^T, \mathcal{A}_1^T\mathcal{B}_2^T, \mathcal{A}_1^T\mathcal{A}_2^T\right] &= \begin{bmatrix} \mathcal{B}_1\mathcal{B}_1^T & \mathcal{B}_1\mathcal{A}_1^T\mathcal{B}_2^T & \mathcal{B}_1\mathcal{A}_1^T\mathcal{A}_2^T \\ \mathcal{B}_2\mathcal{A}_1\mathcal{B}_1^T & \mathcal{B}_2\mathcal{A}_1\mathcal{A}_1^T\mathcal{B}_2^T & \mathcal{B}_2\mathcal{A}_1\mathcal{A}_1^T\mathcal{A}_2^T \\ \mathcal{A}_2\mathcal{A}_1\mathcal{B}_1^T & \mathcal{A}_2\mathcal{A}_1\mathcal{A}_1^T\mathcal{B}_2^T & \mathcal{A}_2\mathcal{A}_1\mathcal{A}_1^T\mathcal{A}_2^T \end{bmatrix} \\
&= \begin{bmatrix} I_{\frac{N}{2}} & 0_{\frac{N}{2}}\mathcal{B}_2^T & 0_{\frac{N}{2}}\mathcal{A}_2^T \\ \mathcal{B}_2 0_{\frac{N}{2}} & \mathcal{B}_2\mathcal{B}_2^T & \mathcal{B}_2\mathcal{A}_2^T \\ \mathcal{A}_2 0_{\frac{N}{2}} & \mathcal{A}_2\mathcal{B}_2^T & \mathcal{A}_2\mathcal{A}_2^T \end{bmatrix} \\
&= \begin{bmatrix} I_{\frac{N}{2}} & 0_{\frac{N}{2},\frac{N}{4}} & 0_{\frac{N}{2},\frac{N}{4}} \\ 0_{\frac{N}{4},\frac{N}{2}} & I_{\frac{N}{4}} & 0_{\frac{N}{4}} \\ 0_{\frac{N}{4},\frac{N}{2}} & 0_{\frac{N}{4}} & I_{\frac{N}{4}} \end{bmatrix} = I_N,
\end{aligned}$$

where $0_{\frac{N}{2},\frac{N}{4}}$ is an $\frac{N}{2} \times \frac{N}{4}$ matrix of zeros. In terms of the frequency domain, $\mathbf{W}_{1,1}$ is related to the frequency interval $[\frac{1}{4}, \frac{1}{2}]$; $\mathbf{W}_{2,1}$, to the interval $[\frac{1}{8}, \frac{1}{4}]$; and $\mathbf{W}_{2,0}$, to $[0, \frac{1}{8}]$.

Note that the second stage orthonormal transform is derived from the first by leaving the vectors in $\mathcal{B}_1$ alone and 'rotating' the vectors in $\mathcal{A}_1$ using $\mathcal{B}_2$ and $\mathcal{A}_2$. What would happen if instead we left the vectors in $\mathcal{A}_1$ alone and rotated the vectors in $\mathcal{B}_1$? We would then have a transform whose coefficients are given by

$$\begin{bmatrix} \mathbf{W}_{2,3} \\ \mathbf{W}_{2,2} \\ \mathbf{W}_{1,0} \end{bmatrix} \equiv \begin{bmatrix} \mathcal{A}_2\mathcal{B}_1 \\ \mathcal{B}_2\mathcal{B}_1 \\ \mathcal{A}_1 \end{bmatrix} \mathbf{X}.$$

▷ **Exercise [208]:** Show that the above transform is orthonormal. ◁

Furthermore, suppose that we combine together all the second level results:

$$\begin{bmatrix} \mathbf{W}_{2,3} \\ \mathbf{W}_{2,2} \\ \mathbf{W}_{2,1} \\ \mathbf{W}_{2,0} \end{bmatrix} = \begin{bmatrix} \mathcal{A}_2\mathcal{B}_1 \\ \mathcal{B}_2\mathcal{B}_1 \\ \mathcal{B}_2\mathcal{A}_1 \\ \mathcal{A}_2\mathcal{A}_1 \end{bmatrix} \mathbf{X}. \tag{208a}$$

Exercise [6.1] is to verify that this transform is orthonormal also. Orthonormality implies that

$$\begin{aligned}
\mathbf{X} &= \left[\mathcal{B}_1^T\mathcal{A}_2^T, \mathcal{B}_1^T\mathcal{B}_2^T, \mathcal{A}_1^T\mathcal{B}_2^T, \mathcal{A}_1^T\mathcal{A}_2^T\right] \begin{bmatrix} \mathbf{W}_{2,3} \\ \mathbf{W}_{2,2} \\ \mathbf{W}_{2,1} \\ \mathbf{W}_{2,0} \end{bmatrix} \\
&= \mathcal{B}_1^T\mathcal{A}_2^T\mathbf{W}_{2,3} + \mathcal{B}_1^T\mathcal{B}_2^T\mathbf{W}_{2,2} + \mathcal{A}_1^T\mathcal{B}_2^T\mathbf{W}_{2,1} + \mathcal{A}_1^T\mathcal{A}_2^T\mathbf{W}_{2,0},
\end{aligned} \tag{208b}$$

an additive decomposition of $\mathbf{X}$ that mimics the multiresolution analysis of the usual DWT; likewise, we have

$$\|\mathbf{X}\|^2 = \|\mathbf{W}_{2,0}\|^2 + \|\mathbf{W}_{2,1}\|^2 + \|\mathbf{W}_{2,2}\|^2 + \|\mathbf{W}_{2,3}\|^2,$$

from which we can produce an analysis of variance (ANOVA) similar to that based on the DWT. Figure 210a depicts the analysis of $\mathbf{X}$ into $\mathbf{W}_{2,0}$, $\mathbf{W}_{2,1}$, $\mathbf{W}_{2,2}$ and $\mathbf{W}_{2,3}$.

The ordering of the $\mathbf{W}_{2,n}$ in Equation (208a) and Figure 210a is particularly useful, as n corresponds to a frequency ordering: $\mathbf{W}_{2,0}$ is nominally related to the frequency interval $[0, \frac{1}{8}]$; $\mathbf{W}_{2,1}$, to $[\frac{1}{8}, \frac{1}{4}]$; $\mathbf{W}_{2,2}$, to $[\frac{1}{4}, \frac{3}{8}]$; and $\mathbf{W}_{2,3}$, to $[\frac{3}{8}, \frac{1}{2}]$. Such an ordering is called *sequency* ordering in Wickerhauser (1994). Note that the second filtering step shown in Figure 210a is low-pass, high-pass, high-pass and low-pass for $\mathbf{W}_{2,n}$, $n = 0, \ldots, 3$, respectively. The explanation for this ordering of the low-pass and high-pass filters is given in Section 4.4, where we noted the following two important facts:

- on the one hand, circularly filtering $\mathbf{X}$ with the scaling filter $\{g_l\}$ and then downsampling by 2 yields a series $\mathbf{V}_1 = \mathbf{W}_{1,0}$ whose frequency content over $f \in [0, 1/2]$ is related to the frequency content in $\mathbf{X}$ over $f \in [0, 1/4]$;
- on the other, circularly filtering $\mathbf{X}$ with the wavelet filter $\{h_l\}$ and then downsampling by 2 yields a series $\mathbf{W}_1 = \mathbf{W}_{1,1}$ whose frequency content over $f \in [0, 1/2]$ is related to the frequency content in $\mathbf{X}$ over $f \in [1/4, 1/2]$ *but in reversed order* (see Figure 86).

It is this reversal of the ordering of the frequencies in the representation for $\mathbf{W}_{1,1}$ as compared to the representation for $\mathbf{X}$ that accounts for the need to reverse the high-pass and low-pass filtering operations to get the proper correspondence of frequencies in $\mathbf{W}_{2,j}, j = 0, \ldots, 3$. Figure 211 depicts the magnitude squared DFTs that would arise at various stages of the flow diagram in Figure 210a if the wavelet and scaling filters were perfect high- and low-pass filters.

Figure 210b shows an alternative ordering of the filtering operations that might seem to be more intuitive because the upper branch of each split in the flow diagram always goes into $G(\cdot)$, while the lower branch always goes into $H(\cdot)$. Wickerhauser (1994) in fact refers to this as the *natural* ordering, but the frequency ordering is no longer directly reflected in the second index.

The results above can be generalized in an obvious way. In order to generate the physically useful sequency ordering, we must obey the following two part rule.

[1] If n in $\mathbf{W}_{j-1,n}$ is even, we use the low-pass filter $G(\cdot)$ to obtain $\mathbf{W}_{j,2n}$ and the high-pass filter $H(\cdot)$ to obtain $\mathbf{W}_{j,2n+1}$. In matrix notation, we have

$$\mathbf{W}_{j,2n} = \mathcal{A}_j \mathbf{W}_{j-1,n} \text{ and } \mathbf{W}_{j,2n+1} = \mathcal{B}_j \mathbf{W}_{j-1,n}. \tag{209a}$$

[2] If n is odd, we use $H(\cdot)$ to obtain $\mathbf{W}_{j,2n}$ and $G(\cdot)$ to obtain $\mathbf{W}_{j,2n+1}$:

$$\mathbf{W}_{j,2n} = \mathcal{B}_j \mathbf{W}_{j-1,n} \text{ and } \mathbf{W}_{j,2n+1} = \mathcal{A}_j \mathbf{W}_{j-1,n}. \tag{209b}$$

With this rule, we can construct 2^j vectors at level j, namely, $\mathbf{W}_{j,n}$, $n = 0, \ldots, 2^j - 1$. The vector $\mathbf{W}_{j,n}$ is nominally associated with frequencies in the interval $\mathcal{I}_{j,n} \equiv [\frac{n}{2^{j+1}}, \frac{n+1}{2^{j+1}}]$. The rule is illustrated in Figure 212a, which takes the analysis out to level 3. This figure is an example of a *wavelet packet table* (WP table) or *wavelet packet tree*. Since the starting point for our analysis is the time series $\mathbf{X}$, it is convenient to define $\mathbf{W}_{0,0} \equiv \mathbf{X}$ so that $\mathbf{X}$ is associated with a (j, n) doublet, namely, $(0, 0)$.

The transform that takes $\mathbf{X}$ to $\mathbf{W}_{j,n}$, $n = 0, \ldots, 2^j - 1$, for any j between 0 and J is called a discrete wavelet packet transform (DWPT). Any such transform is

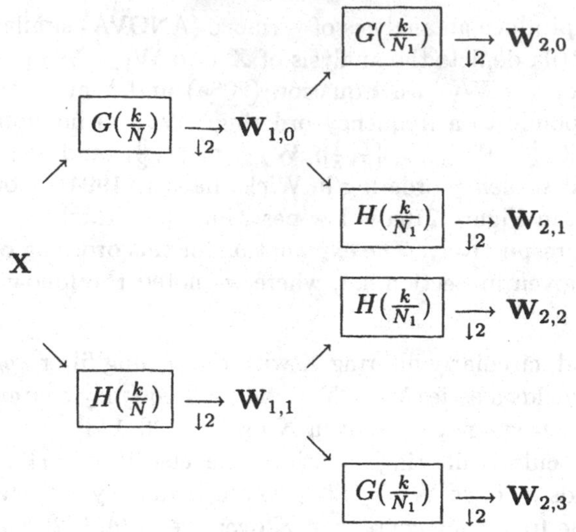

Figure 210a. Flow diagram illustrating the analysis of $\mathbf{X}$ into $\mathbf{W}_{2,0}$, $\mathbf{W}_{2,1}$, $\mathbf{W}_{2,2}$ and $\mathbf{W}_{2,3}$ (sequency ordering). In the above recall that $N_1 \equiv N/2$.

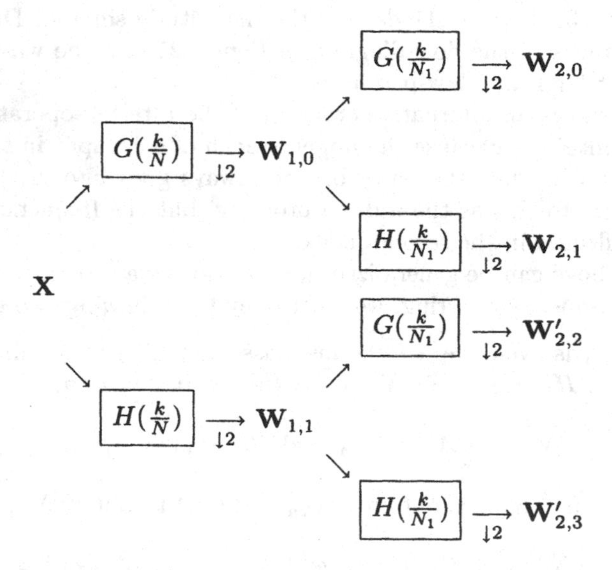

Figure 210b. Flow diagram illustrating the analysis of $\mathbf{X}$ into $\mathbf{W}_{2,0}$, $\mathbf{W}_{2,1}$, $\mathbf{W}'_{2,2}$ and $\mathbf{W}'_{2,3}$ (natural ordering).

orthonormal. To see why this is so, let us call $\mathbf{W}_{j-1,n}$ the 'parent' of its 'children' $\mathbf{W}_{j,2n}$ and $\mathbf{W}_{j,2n+1}$. In going from a level $j-1$ DWPT to one of level j, we essentially replace each parent by its two children. As described by Equations (209a) and (209b), the splitting at a parent node is an orthonormal transform. Since this is true at every parent node on level $j-1$, the orthonormality of the level $j-1$ DWPT implies the

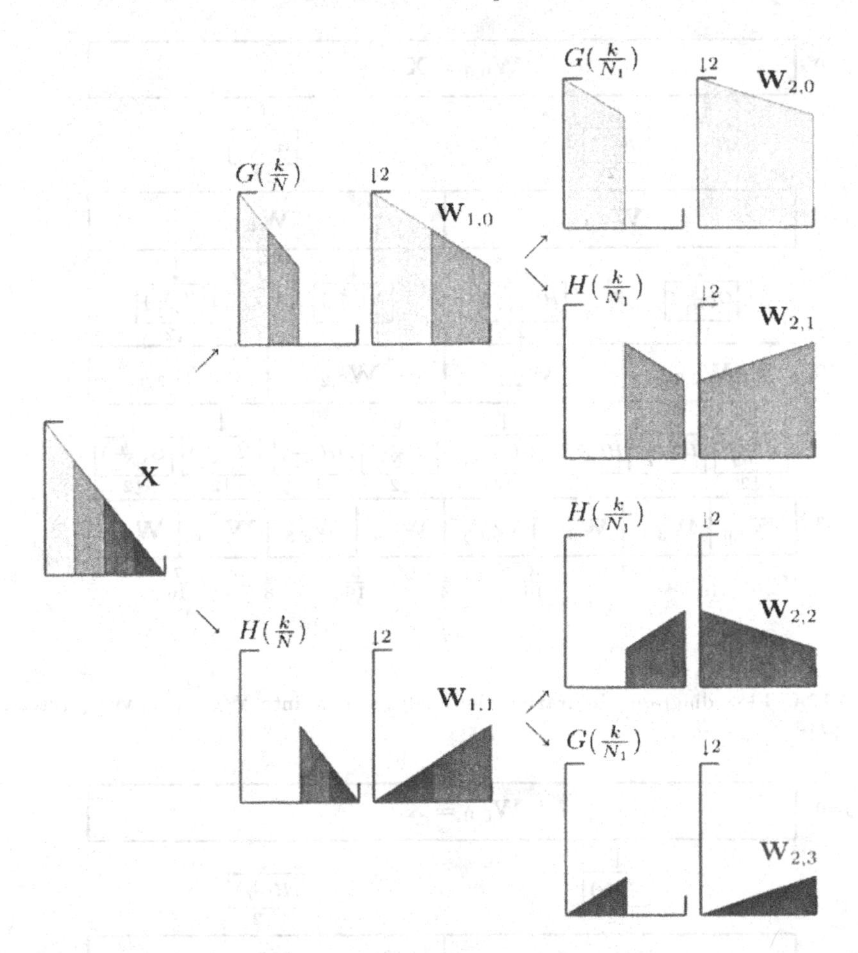

Figure 211. Illustration of effect in the frequency domain of filtering with ideal low- and high-pass filters $G(\cdot)$ and $H(\cdot)$, followed by downsampling (this figure parallels the flow diagram in Figure 210a). The magnitude squared DFT for the input $\mathbf{X}$ of length N is assumed to roll off linearly as f sweeps from 0 up to the Nyquist frequency. Filtering $\mathbf{X}$ with either $G(\cdot)$ or $H(\cdot)$ yields a half-band series that, after downsampling, becomes a full-band series (either $\mathbf{W}_{1,0}$ or $\mathbf{W}_{1,1}$, each with $N/2$ points). A second round of filtering and downsampling yields four new full-band series, namely, $\mathbf{W}_{2,n}$, $n = 0, 1, 2$ and 3. Each $\mathbf{W}_{2,n}$ is of length $N/4$ and is related to a single quarter-band in $\mathbf{X}$. Note that, each time that we create a half-band series by filtering with $H(\cdot)$, the subsequent downsampling creates a full-band series whose frequency content is reversed with respect to that of the half-band series; on the other hand, filtering with $G(\cdot)$ causes no such reversal (for an explanation of the reversal induced by $H(\cdot)$, see Section 4.4).

orthonormality of the level j transform.

The collection of doublets (j, n) forming the indices of the table nodes will be denoted by $\mathcal{N} \equiv \{(j, n) : j = 0, \ldots, J_0; n = 0, \ldots, 2^j - 1\}$, where we are free to pick any J_0 satisfying $J_0 \leq J$ (as is also true for partial DWTs, if $J_0 < J$, we can in fact relax the stipulation $N = 2^J$ and allow N to be any integer multiple of 2^{J_0}). The doublets (j, n) that form the indices of the WP coefficients corresponding to an orthonormal transform will be denoted $\mathcal{C}$; for example, we have $\mathcal{C} = \{(3, n) : n = 0, \ldots, 7\}$ for the DWPT transform of level $j = 3$ giving $\mathbf{W}_{3,0}, \ldots, \mathbf{W}_{3,7}$. Clearly $\mathcal{C} \subset \mathcal{N}$.

It is interesting to note that, in addition to the DWPTs, we can extract many

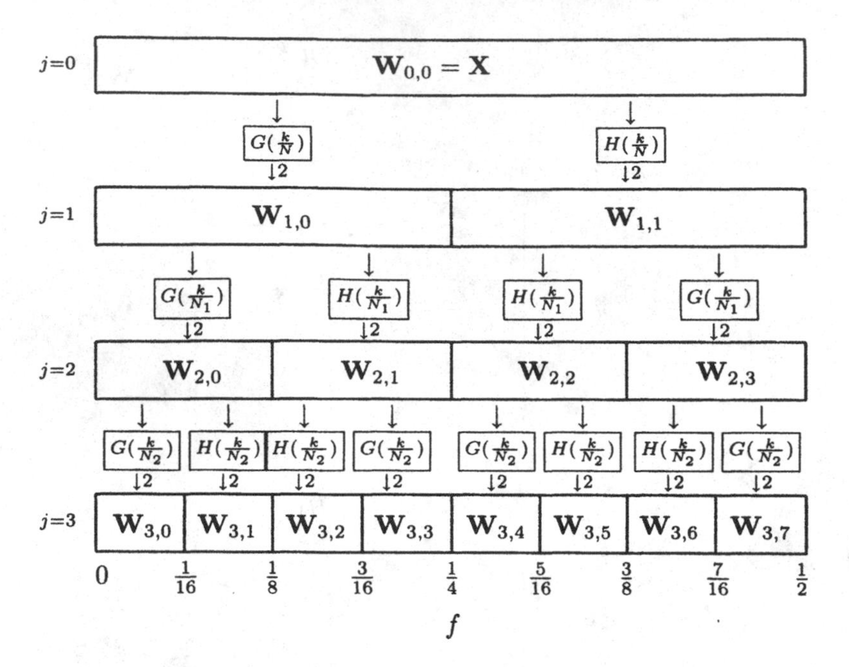

Figure 212a. Flow diagram illustrating the analysis of $\mathbf{X}$ into $\mathbf{W}_{3,0}, \ldots, \mathbf{W}_{3,7}$ (recall that $N_j \equiv N/2^j$).

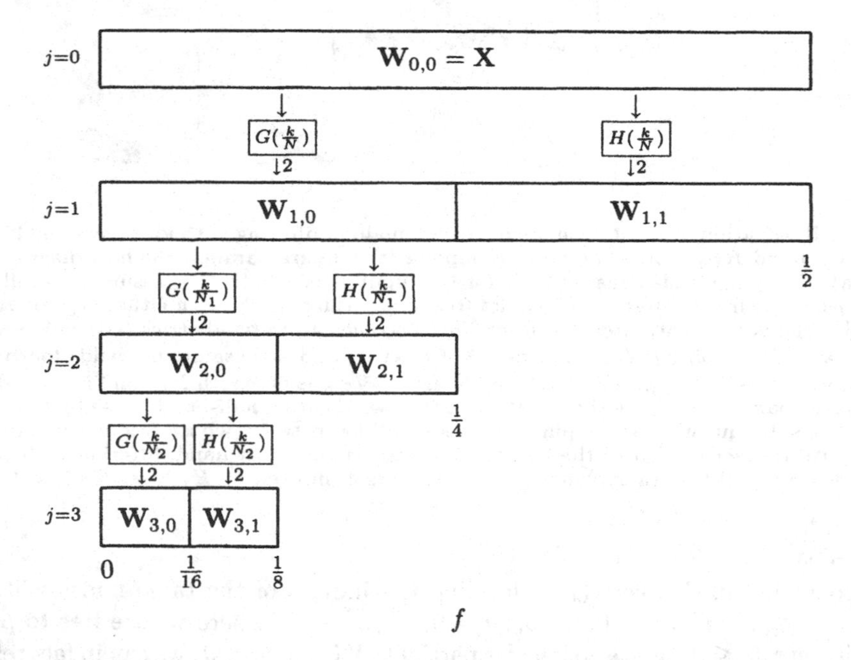

Figure 212b. Flow diagram illustrating the analysis of $\mathbf{X}$ into $\mathbf{W}_{3,0}$, $\mathbf{W}_{3,1}$, $\mathbf{W}_{2,1}$ and $\mathbf{W}_{1,1}$, which is identical to a partial DWT of level $J_0 = 3$.

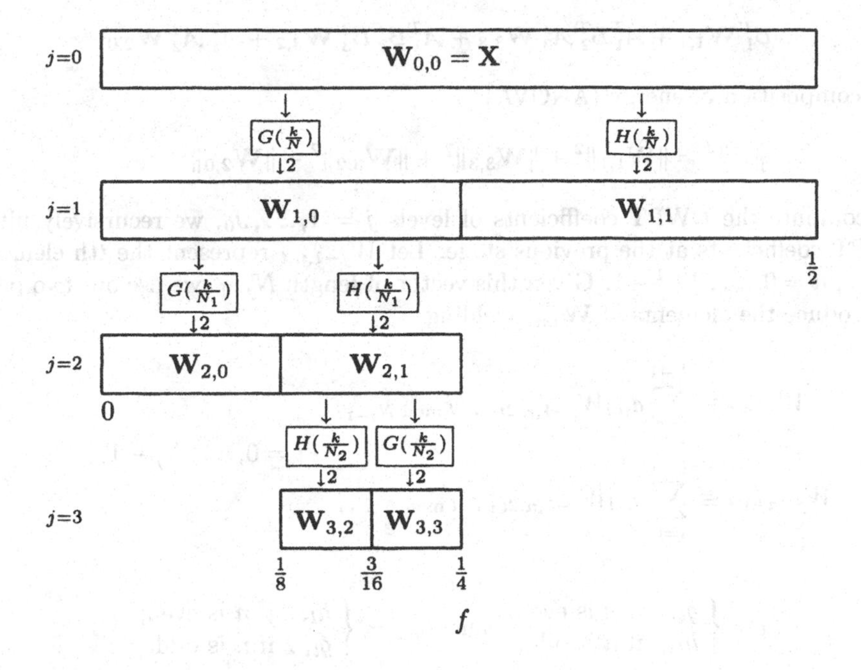

Figure 213. Flow diagram illustrating the analysis of $\mathbf{X}$ into $\mathbf{W}_{2,0}$, $\mathbf{W}_{3,2}$, $\mathbf{W}_{3,3}$ and $\mathbf{W}_{1,1}$, an arbitrary disjoint dyadic decomposition.

other orthonormal transforms from a WP table. For example, all partial DWTs can be pieced together from such a table. This is illustrated in Figure 212b, where the level $J_0 = 3$ partial DWT is shown to consist of four nodes from the WP table, namely, $\mathcal{C} = \{(3,0),(3,1),(2,1),(1,1)\}$. Note that the nominal frequency intervals for these four nodes form a complete partition of $[0, \frac{1}{2}]$. More generally, let us define a *disjoint dyadic decomposition* as one in which, at each splitting point (potential parent node) in the table structure of Figure 212a, we either carry out the splitting using both $G(\cdot)$ and $H(\cdot)$, as shown, or do not split at all. Such a decomposition is also orthonormal, and it is also associated with a complete partition of $[0, \frac{1}{2}]$. Figure 213 shows an arbitrary disjoint dyadic decomposition, where now $\mathcal{C}$ contains the doublets $(2,0)$, $(3,2)$, $(3,3)$ and $(1,1)$.

A WP table is fully specified once we have selected the wavelet filter $\{h_l\}$ (the corresponding scaling filter $\{g_l\}$ is determined by its quadrature mirror relationship to $\{h_l\}$ – see Equation (75a)). Any orthonormal transform that can be extracted from a WP table is specified by the doublets in $\mathcal{C}$. For the orthonormal transform of Figure 213 we have

$$\begin{bmatrix} \mathbf{W}_{1,1} \\ \mathbf{W}_{3,3} \\ \mathbf{W}_{3,2} \\ \mathbf{W}_{2,0} \end{bmatrix} = \begin{bmatrix} \mathcal{B}_1 \\ \mathcal{A}_3\mathcal{B}_2\mathcal{A}_1 \\ \mathcal{B}_3\mathcal{B}_2\mathcal{A}_1 \\ \mathcal{A}_2\mathcal{A}_1 \end{bmatrix} \mathbf{X}.$$

Because of the orthonormality of this transform, we have an additive decomposition

$$\mathbf{X} = [\mathcal{B}_1^T, \mathcal{A}_1^T\mathcal{B}_2^T\mathcal{A}_3^T, \mathcal{A}_1^T\mathcal{B}_2^T\mathcal{B}_3^T, \mathcal{A}_1^T\mathcal{A}_2^T] \begin{bmatrix} \mathbf{W}_{1,1} \\ \mathbf{W}_{3,3} \\ \mathbf{W}_{3,2} \\ \mathbf{W}_{2,0} \end{bmatrix}$$

$$= \mathcal{B}_1^T \mathbf{W}_{1,1} + \mathcal{A}_1^T \mathcal{B}_2^T \mathcal{A}_3^T \mathbf{W}_{3,3} + \mathcal{A}_1^T \mathcal{B}_2^T \mathcal{B}_3^T \mathbf{W}_{3,2} + \mathcal{A}_1^T \mathcal{A}_2^T \mathbf{W}_{2,0}$$

and a decomposition of energy (ANOVA):

$$\|\mathbf{X}\|^2 = \|\mathbf{W}_{1,1}\|^2 + \|\mathbf{W}_{3,3}\|^2 + \|\mathbf{W}_{3,2}\|^2 + \|\mathbf{W}_{2,0}\|^2.$$

To compute the DWPT coefficients of levels $j = 1, \ldots, J_0$, we recursively filter the DWPT coefficients at the previous stage. Let $W_{j-1,n,t}$ represent the tth element of $\mathbf{W}_{j-1,n}$, $n = 0, \ldots, 2^{j-1} - 1$. Given this vector of length N_{j-1}, we use our two part rule to produce the elements of $\mathbf{W}_{j,n}$, yielding

$$\begin{aligned} W_{j,2n,t} &\equiv \sum_{l=0}^{L-1} a_{n,l} W_{j-1,n,2t+1-l \bmod N_{j-1}}, \\ W_{j,2n+1,t} &\equiv \sum_{l=0}^{L-1} b_{n,l} W_{j-1,n,2t+1-l \bmod N_{j-1}}, \end{aligned} \qquad t = 0, \ldots, N_j - 1,$$

where

$$a_{n,l} = \begin{cases} g_l, & \text{if } n \text{ is even;} \\ h_l, & \text{if } n \text{ is odd,} \end{cases} \quad \text{and } b_{n,l} = \begin{cases} h_l, & \text{if } n \text{ is even;} \\ g_l, & \text{if } n \text{ is odd.} \end{cases}$$

▷ **Exercise [214]:** Show that an equivalent – but more compact – way of writing the above is

$$W_{j,n,t} = \sum_{l=0}^{L-1} u_{n,l} W_{j-1,\lfloor \frac{n}{2} \rfloor, 2t+1-l \bmod N_{j-1}}, \quad t = 0, \ldots, N_j - 1, \tag{214a}$$

where

$$u_{n,l} \equiv \begin{cases} g_l, & \text{if } n \bmod 4 = 0 \text{ or } 3; \\ h_l, & \text{if } n \bmod 4 = 1 \text{ or } 2, \end{cases} \tag{214b}$$

and $\lfloor \cdot \rfloor$ denotes the 'integer part' operator. ◁

We can also write $\mathbf{W}_{j,n}$ in terms of filtering $\mathbf{X}$ followed by an appropriate downsampling (the argument for constructing the filter that takes us from $\mathbf{X}$ to $\mathbf{W}_{j,n}$ closely parallels a similar one for the DWT in Section 4.6 – a study of that section will indicate how to justify the statements we make here). We must first formulate the correct form of the filter $\{u_{j,n,l} : l = 0, \ldots, L_j - 1\}$ corresponding to the node (j, n) in the WP table (as in Equation (96a), we have $L_j \equiv (2^j - 1)(L - 1) + 1$). Suppose that we let $u_{1,0,l} \equiv g_l$ and $u_{1,1,l} \equiv h_l$. For nodes (j, n) with $j > 1$ (and in parallel to Equation (102)), let

$$u_{j,n,l} \equiv \sum_{k=0}^{L-1} u_{n,k} u_{j-1,\lfloor \frac{n}{2} \rfloor, l-2^{j-1}k}, \quad l = 0, \ldots, L_j - 1 \tag{214c}$$

(as usual, we define $u_{j,n,l} = 0$ for $l < 0$ and $l \geq L_j$). For example, for $j = 2$,

$$u_{2,0,l} = \sum_{k=0}^{L-1} u_{0,k} u_{1,0,l-2k} = \sum_{k=0}^{L-1} g_k g_{l-2k}$$

$$u_{2,1,l} = \sum_{k=0}^{L-1} u_{1,k}u_{1,0,l-2k} = \sum_{k=0}^{L-1} h_k g_{l-2k}$$
$$u_{2,2,l} = \sum_{k=0}^{L-1} u_{2,k}u_{1,1,l-2k} = \sum_{k=0}^{L-1} h_k h_{l-2k}$$
$$u_{2,3,l} = \sum_{k=0}^{L-1} u_{3,k}u_{1,1,l-2k} = \sum_{k=0}^{L-1} g_k h_{l-2k}.$$

Then for $j = 1, \ldots, J_0$, we can write the elements of $\mathbf{W}_{j,n}$ in terms of a filtering of $\mathbf{X}$ with appropriate downsampling, namely,

$$W_{j,n,t} = \sum_{l=0}^{L_j-1} u_{j,n,l} X_{2^j[t+1]-1-l \bmod N}, \quad t = 0, 1, \ldots, N_j - 1. \tag{215a}$$

Let $U_{j,n}(\cdot)$ denote the transfer function for $\{u_{j,n,l}\}$. This transfer function depends only on the transfer functions $G(\cdot)$ and $H(\cdot)$ for the scaling and wavelet filters. To express this dependence, suppose that we let $M_0(\cdot) \equiv G(\cdot)$ and $M_1(\cdot) \equiv H(\cdot)$ as in Coifman *et al.* (1992). Suppose also that, to each node (j, n) in the WP table, we attach a vector $\mathbf{c}_{j,n}$ of length j, all of whose elements are either one or zero. For $j = 1$, we define $\mathbf{c}_{1,0} \equiv [0]$ and $\mathbf{c}_{1,1} \equiv [1]$. For $j > 1$, we define $\mathbf{c}_{j,n}$, $n = 0, \ldots, 2^j - 1$, recursively in terms of $\mathbf{c}_{j-1,n'}$, $n' = 0, \ldots, 2^{j-1} - 1$, via the following two part rule.

[1] If $n \bmod 4$ is either 0 or 3, we create $\mathbf{c}_{j,n}$ by appending a zero to the end of $\mathbf{c}_{j-1,\lfloor \frac{n}{2} \rfloor}$.

[2] If $n \bmod 4$ is either 1 or 2, we create $\mathbf{c}_{j,n}$ by appending a one to $\mathbf{c}_{j-1,\lfloor \frac{n}{2} \rfloor}$.

To see how to use this rule, suppose, for example, that we want to construct $\mathbf{c}_{j,n}$ for $j = 4$ and $n = 7$. Since $7 \bmod 4 = 3$, part [1] of the rule says to construct $\mathbf{c}_{4,7}$ by appending 0 to $\mathbf{c}_{3,3}$. To construct $\mathbf{c}_{3,3}$ in turn, note that $3 \bmod 4 = 3$, so another application of part [1] says to append 0 to $\mathbf{c}_{2,1}$. Finally, since $1 \bmod 4 = 1$, we construct $\mathbf{c}_{2,1}$ using part [2] of the rule, i.e., we must append 1 to $\mathbf{c}_{1,0} \equiv [0]$. Collecting these results together, we have $\mathbf{c}_{4,7} = [0, 1, 0, 0]^T$.

Let $c_{j,n,m}$ denote the mth element of $\mathbf{c}_{j,n}$. Then

$$U_{j,n}(f) = \prod_{m=0}^{j-1} M_{c_{j,n,m}}(2^m f). \tag{215b}$$

The vectors $\mathbf{c}_{j,n}$ are illustrated for the sequency ordered WP table in Figure 216. As an example, consider $U_{3,3}(f)$, which is associated with $\mathbf{c}_{3,3} = [c_{3,3,0}, c_{3,3,1}, c_{3,3,2}]^T = [0, 1, 0]^T$. We then have that

$$U_{3,3}(f) = M_0(f)M_1(2f)M_0(4f) = G(f)H(2f)G(4f).$$

The squared gain functions $|U_{j,n}(\cdot)|^2$ are shown in Figure 217 for $j = 3$ and $n = 0, \ldots, 7$ for LA(8) filters. The nominal 'ideal' pass-bands are marked by vertical lines. For example, the nominal pass-band for $U_{3,3}(\cdot)$ is $[\frac{3}{16}, \frac{1}{4}]$.

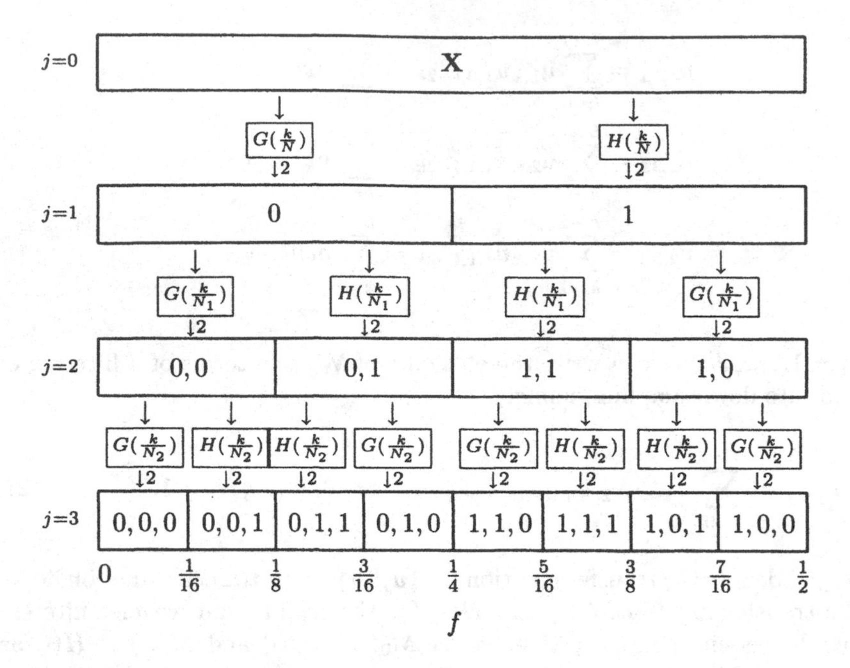

Figure 216. Illustration of the vectors $\mathbf{c}_{j,n}$ for the sequency ordered WP table for levels $j = 1, 2$ and 3 (0 or 1 indicate use of, respectively, a scaling (low-pass) or wavelet (high-pass) filter). Note that, if, in going from a parent node at level $j-1$ to a child at level j, we require use of $G(\cdot)$, then we append a zero to the parent's $\mathbf{c}_{j-1,\lfloor \frac{n}{2} \rfloor}$ to obtain the child's $\mathbf{c}_{j,n}$; on the other hand, if we use $H(\cdot)$, then we append a one. Note also that, as we sweep from left to right across either row $j = 2$ or 3 and pick out the last element of each $\mathbf{c}_{j,n}$, we obtain the pattern '0, 1, 1, 0' or this pattern followed by a replicate (for general $j \geq 2$, collecting the last elements of $\mathbf{c}_{j,n}$, $n = 0, \ldots, 2^j - 1$, will yield 2^{j-2} replications of '0, 1, 1, 0').

Finally, recall that each $\mathbf{W}_{j,n}$ in a jth level DWPT is nominally associated with bandwidth $1/2^{j+1}$, i.e., the width of its associated frequency interval $\mathcal{I}_{j,n}$. Note also $\mathbf{W}_{j,0}$ is exactly the same as the vector $\mathbf{V}_j$ of scaling coefficients in a level j partial DWT. These scaling coefficients are associated with scale $\lambda_j = 2^j$ (this measure can be justified using the notion of the 'autocorrelation' width of a sequence – see Equation (103) and the discussion surrounding it). We can thus associate a nominal 'time width' of λ_j with $\mathbf{W}_{j,0}$. Since the equivalent filters used to create all $\mathbf{W}_{j,n}$ on level j have exactly the same width L_j and same nominal bandwidth, we can arguably attach the same time width λ_j to each $\mathbf{W}_{j,n}$ (note that, since there are $N_j = N/2^j = N/\lambda_j$ coefficients in each $\mathbf{W}_{j,n}$, a nominal width of λ_j for each coefficient yields a total time coverage of $N_j \times \lambda_j = N$; i.e., collectively the coefficients in $\mathbf{W}_{j,n}$ cover the entire time over which $\mathbf{X}$ was recorded, as is intuitively reasonable). For $j = 0$, the time width is unity, whereas the bandwidth is 1/2, which is sensible given that a level $j = 0$ DWPT is the identity transform. If we consider the other extreme case, i.e., a level J DWPT of a series with length $N = 2^J$, the time width is N, whereas the bandwidth is $1/(2N)$. This time width is reasonable since none of the equivalent filters $\{u_{J,n,l}\}$ is localized in any sense. For all levels j, the product of the time width and the bandwidth is a constant, namely, 1/2, in keeping with a number of 'reciprocity relationships' between the time and frequency domains (see, for example, Bracewell, 1978, Chapter 8, or Percival and Walden, 1993, Chapter 3).

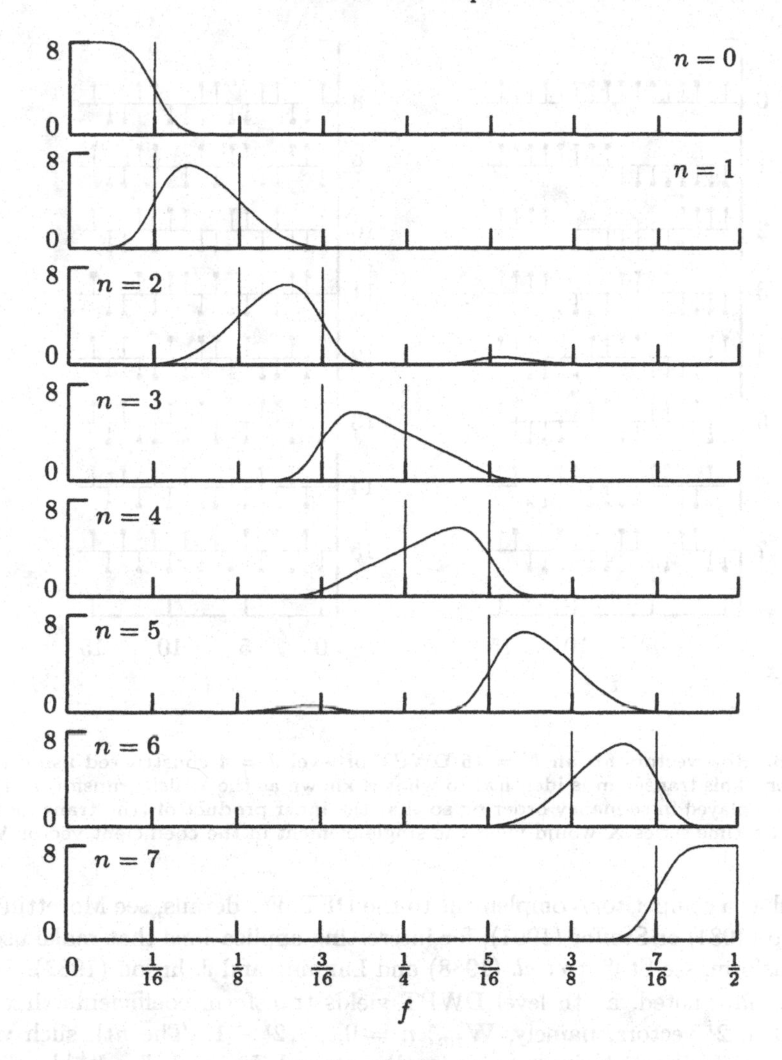

Figure 217. Squared gain functions $|U_{j,n}(\cdot)|^2$ for $j = 3$ and $n = 0, \ldots, 7$ based upon the LA(8) wavelet and scaling filters. The nominal 'ideal' pass-bands are marked by vertical lines.

Comments and Extensions to Section 6.1

[1] The two part rule that we gave above for constructing the binary-valued vector $\mathbf{c}_{j,n}$ allows us to determine which sequence of convolved filters with transfer functions of the form $G(2^m f)$ and/or $H(2^m f)$ lead to a filter with a squared gain function that is nominally band-pass over the frequency interval $\mathcal{I}_{j,n} \equiv [\frac{n}{2^{j+1}}, \frac{n+1}{2^{j+1}}]$. For a more formal proof that $\mathcal{I}_{j,n}$ is connected to $\mathbf{c}_{j,n}$ as the rule states, see Walden and Contreras Cristan (1998a).

[2] Suppose that we have a time series of length $N = 2^J$ and that we construct a DWPT of level J using the Haar wavelet. This transform is in fact identical to what is known in the literature as the (orthonormal) *Walsh transform.* The row vectors for the $N = 16$ case of this transform are shown in Figure 218. Note that, like the DFT (see Figure 47) but unlike DWPTs of levels $j < J$, the basis vectors for the Walsh transform are not localized in time. The Walsh transform has been investigated in

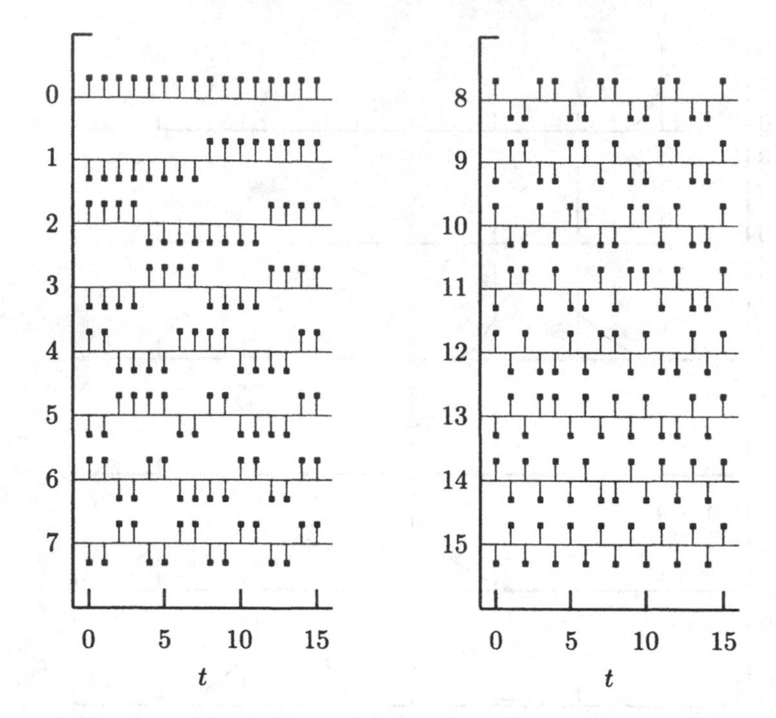

Figure 218. Row vectors for an $N = 16$ DWPT of level $J = 4$ constructed using the Haar wavelet filter. This transform is identical to what is known as the Walsh transform. The basis vectors are displayed in sequency ordering so that the inner product of (the transpose of) the nth row and a time series $\mathbf{X}$ would yield the single element in the coefficient vector $\mathbf{W}_{4,n}$.

some detail as a competitor/complement to the DFT. For details, see Morettin (1981), Beauchamp (1984) or Stoffer (1991); for interesting applications that make use of the Walsh transform, see Stoffer *et al.* (1988) and Lanning and Johnson (1983).

[3] As we have noted, a jth level DWPT yields transform coefficients that can be partitioned in 2^j vectors, namely, $\mathbf{W}_{j,n}$, $n = 0, \ldots, 2^j - 1$. The nth such vector is associated nominally with frequencies in the interval $\mathcal{I}_{j,n} = [\frac{n}{2^{j+1}}, \frac{n+1}{2^{j+1}}]$. Since the width of each $\mathcal{I}_{j,n}$ is the same, the DWPT effectively partitions the frequency range $[0, \frac{1}{2}]$ into 2^j regularly spaced intervals. Because of this partitioning and the fact that the DWPT is computed in a manner quite similar to the DWT, some investigators refer to the DWPT as a 'regular DWT' (see, for example, Medley *et al.*, 1994).

6.2 Example: DWPT of Solar Physics Data

The physics of magnetic fields in the heliosphere of the Sun consists of a complex superposition of processes and phenomena that occur on the Sun and in the solar wind over different length scales, as well as more locally near the point of observation. The Ulysses space mission provided, for the first time, observations over the polar regions of the Sun. These regions were measured during a minimum phase of the solar activity cycle and were thus dominated by fast solar wind streams from the polar coronal holes. The unprecedented high rate of data coverage (in excess of 95%) – along with the solar polar orbit of Ulysses – provided a unique high quality record of the heliospheric magnetic field.

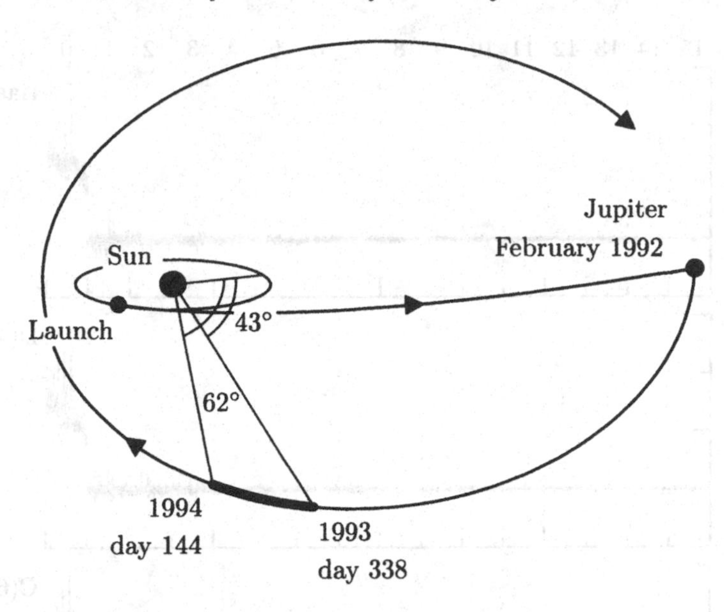

Figure 219. Path of the Ulysses spacecraft. After its launch, the spacecraft traveled in the plane of the Earth's orbit to near Jupiter, whose gravitational field then flung it into polar orbit about the Sun. The segment of data plotted at the bottom of Figure 222 was collected as the spacecraft traveled from heliographic latitude 43° S to 62° S south of the Sun – in changing latitudes, the spacecraft also decreased its distance from the Sun. After collecting the data of interest here, the spacecraft continued on its solar polar orbit, but now travelling north of the Sun. (Illustration courtesy of T. Horbury, Queen Mary and Westfield College, London.)

The data available for analysis consist of a gap-free segment of $N = 4096$ hourly averages of magnetic field magnitude recorded from 21 hours Universal Time (UT) on day 338 of 1993 (4 December) until 12 hours UT on day 144 of 1994 (24 May) and cover a period when Ulysses was recording emissions from the solar south pole (see Figure 219). The series is plotted in the bottom of Figure 222 (an expanded version of this same plot is shown at the bottom of Figure 235). For convenience we have taken the time variable to be measured in days so that the sampling interval is $\Delta t = 1/24$ day (frequencies are thus measured in cycles per day and range from 0 up to a Nyquist frequency of 12 cycles per day). The units of measurement are nanoteslas, i.e., 10^{-9} tesla, denoted by nT. As indicated in Figure 219, in the interval over which this time series was recorded, Ulysses changed its heliographic latitude from about 43° S to about 62° S, and its heliocentric range from approximately 4 to 3 astronomical units (by definition an astronomical unit is equal to the mean distance from the earth to the Sun). These southern hemisphere data show some large scale structures. The four shock waves we concentrate on are marked *a* (12 February, 1994), *b* (26 February 1994), *c* (10 March 1994) and *d* (3 April 1994) in Figure 222. We will examine the time/frequency nature of these high-latitude features.

Interplanetary shock waves are typically classified as 'corotating' shocks or 'transient' shocks (e.g., González–Esparza *et al.*, 1996). Corotating shocks are produced by the interaction of fast solar wind (originating from coronal holes on the Sun) overtaking slow solar wind in the interplanetary medium. Interaction regions that are recurrent are called corotating interaction regions (CIRs) and recur every solar ro-

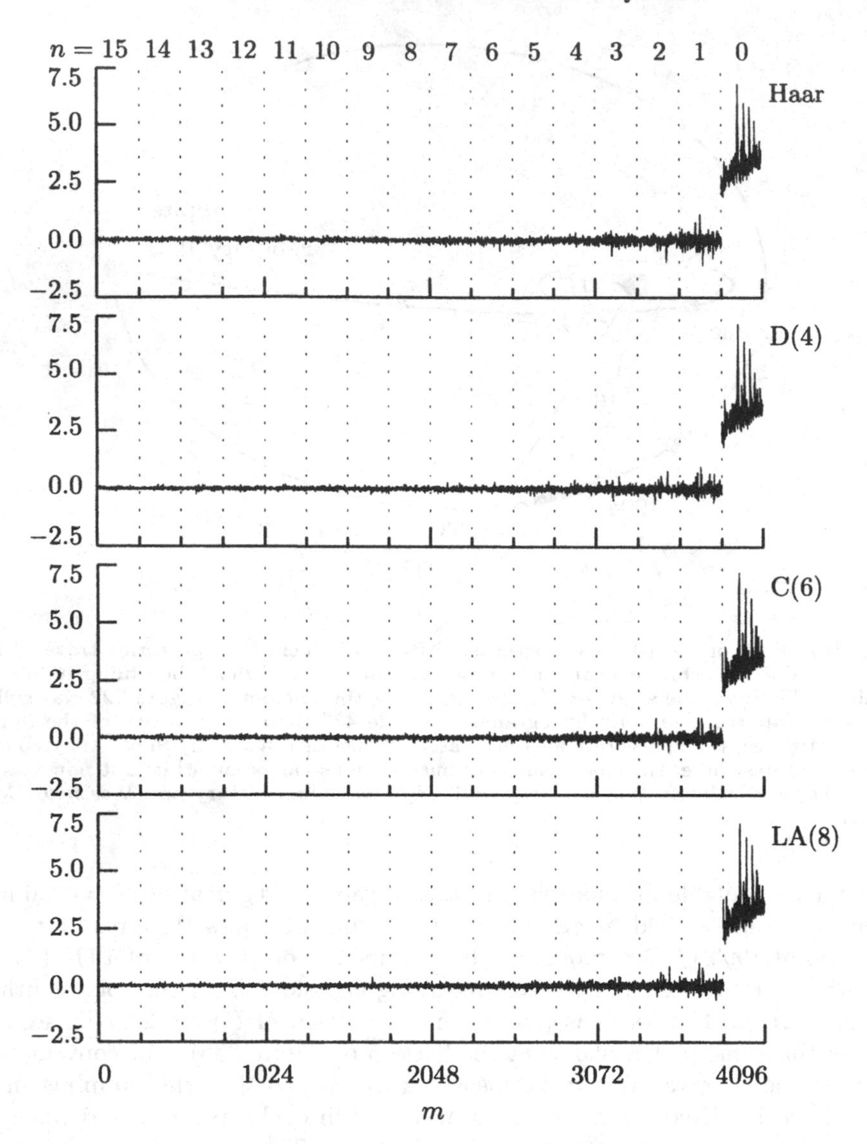

Figure 220. DWPT coefficients $\mathbf{W}$ of level $j = 4$ for solar physics time series using the Haar, D(4), C(6) and LA(8) wavelets. The elements W_m of $\mathbf{W}$ are plotted versus $m = 0, \ldots, N-1 = 4095$. The 15 vertical dotted lines delineate the 16 subvectors of $\mathbf{W}$, namely, from left to right, $\mathbf{W}_{4,n}$, $n = 15, 14, \ldots, 0$.

tation. Transient shocks are believed to be produced by fast coronal mass ejections (CMEs). Details of classification methods are given in González–Esparza *et al.* (1996). Balogh *et al.* (1995) classified the first two shock waves in Figure 222 as transient, associated with CMEs, and the latter two with recurrent CIRs. However, Zurbuchen *et al.* (1996) treated the first event as recurrent, in addition to the third and fourth.

Because this series has time-varying components, it is a good candidate to illustrate the capabilities of a DWPT analysis. Figure 220 shows the elements W_m of the DWPT coefficient vectors $\mathbf{W}$ of level $j = 4$ based upon the Haar, D(4), C(6) and

LA(8) wavelets. Vertical dotted lines indicate the partitioning of each $\mathbf{W}$ into its subvectors $\mathbf{W}_{4,n}$, $n = 0, \ldots, 2^j - 1 = 15$ (note that $\mathbf{W}_{4,15}$ is left-most on the plot, while $\mathbf{W}_{4,0}$ is right-most). The coefficient vector $\mathbf{W}$ has the same number of elements as the time series $\mathbf{X}$. We have $\|\mathbf{W}\|^2 = \|\mathbf{X}\|^2$ because DWPTs are orthonormal transforms. Each subvector $\mathbf{W}_{4,n}$ has the same number of elements, namely, $N_j = N/2^j = 256$. The subvector $\mathbf{W}_{4,n}$ tracks variations over time in the interval of physical frequencies $\frac{1}{\Delta t}\mathcal{I}_{4,n} = [\frac{n}{32\,\Delta t}, \frac{n+1}{32\,\Delta t}]$. The figure indicates that the magnitudes of the coefficients generally increase with decreasing frequency, with the largest coefficients being in $\mathbf{W}_{4,0}$ (associated with the lowest interval of frequencies). Note that, while the coefficients in $\mathbf{W}_{4,n}$ for $n > 0$ fluctuate about zero, those in $\mathbf{W}_{4,0}$ fluctuate about a positive value. The latter pattern can be explained by appealing to the main result of Exercise [6.6], which says that the sample mean of $\mathbf{W}_{4,0}$ is equal to $4\overline{X}$, where, as usual, $\overline{X}$ is the the sample mean for $\mathbf{X}$ (this is markedly positive, as Figure 222 indicates).

Within a given $\mathbf{W}_{4,n}$ for $n \geq 3$, the variations across time in Figure 220 appear to be fairly homogeneous and qualitatively the same for all four DWPTs, but there are some noticeable bursts in the subvectors associated with the three lowest frequency bands, particularly in $\mathbf{W}_{4,0}$. We explore the time relationships between the bursts in different bands in the top part of Figure 222, where each element of the LA(8) DWPT subvectors is plotted versus a time that takes into account the approximate zero phase properties of the LA(8) filters. These adjustments depend on shifts $\nu_{4,n}$ that are discussed in Section 6.5, which are then used to connect the tth element $W_{4,n,t}$ of $\mathbf{W}_{4,n}$ to time $t_0 + (2^4(t+1) - 1 - |\nu_{4,n}| \bmod N)\Delta t$, where t_0 is the actual time associated with X_0 (this formula exactly parallels one derived in Section 4.8 for the DWT). The circular shift of $\mathbf{W}_{4,n}$ yielding the ordering in which its elements actually get plotted is indicated to the right of Figure 222 – for example, $\mathcal{T}^{-2}\mathbf{W}_{4,0}$ indicates that elements $W_{4,0,2}$ and $W_{4,0,1}$ are associated with, respectively, the earliest and latest times. The thick vertical lines at the beginning and end of each subvector delineate the boundary coefficients: coefficients plotted outside of these two lines are influenced to some degree by the circularity assumed by the DWPT. We can determine which elements of $\mathbf{W}_{4,n}$ are boundary coefficients in exactly the same manner as described for the DWT in Section 4.11 (for each $\mathbf{W}_{4,n}$, the number of boundary coefficients is given by L_4', which Table 136 says is 6 for the LA(8) filter). Dotted vertical lines in the upper plot indicate the locations of the four shock wave structures as determined by Balogh *et al.* (1995) – these timings were determined from the raw (not hourly averaged) records by visually looking for the correct character in the data. With the phase alignment made possible by using the LA(8) wavelet, we can see the colocation of large coefficients in several frequency bands around the times of these shock waves; however, as we shall see in Section 6.7, the downsampling inherent in the DWPT compromises our ability to pick out these broad-band features.

6.3 The Best Basis Algorithm

As we noted in Section 6.1, we can extract many different orthonormal transforms from a WP table taken out to some level J_0. Every possible orthonormal transform is a disjoint dyadic decomposition, which by definition is a partitioning of the frequency interval $[0, \frac{1}{2}]$ (these transforms include – but are not limited to – partial DWTs and DWPTs of levels $j = 1, \ldots, J_0$). How can we define which one of these many transforms is in some sense optimal for a particular time series $\mathbf{X}$? Equivalently, given the choice of wavelet filter, which $\mathcal{C} \subset \mathcal{N}$ is in some sense optimal?

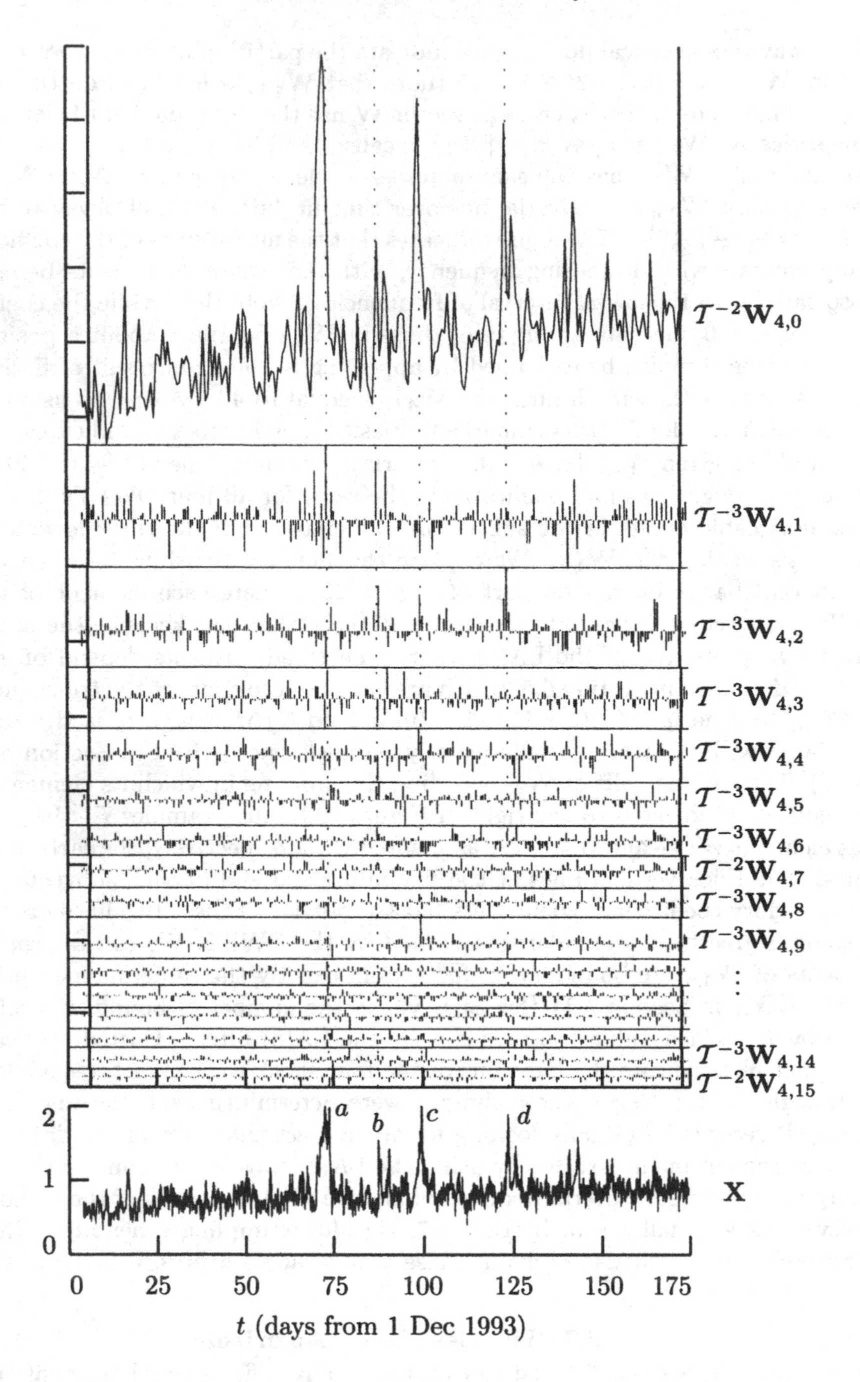

Figure 222. Level $j = 4$ LA(8) DWPT coefficients for solar physics time series (see text for details).

Coifman and Wickerhauser (1992) tackled this problem under the name of the 'best basis algorithm,' which consists of the following two basic steps.

[1] Given a WP table to level J_0, then for every $(j,n) \in \mathcal{N}$ we associate with $\mathbf{W}_{j,n}$ a cost $M(\mathbf{W}_{j,n})$, where $M(\cdot)$ is an additive cost functional of the form

$$M(\mathbf{W}_{j,n}) \equiv \sum_{t=0}^{N_j-1} m(|W_{j,n,t}|),$$

and $m(\cdot)$ is a real-valued function defined on $[0,\infty)$ with $m(0) = 0$.

[2] The 'optimal' orthonormal transform that can be extracted from the WP table is the solution of

$$\min_{\mathcal{C}} \sum_{(j,n)\in\mathcal{C}} M(\mathbf{W}_{j,n}),$$

i.e., we seek the orthonormal transform specified by $\mathcal{C} \subset \mathcal{N}$ such that the cost summed over all doublets $(j,n) \in \mathcal{C}$ is minimized.

A number of additive cost functionals have been proposed (see, for example, Wickerhauser, 1994, Chapter 8). Here are three examples.

[1] The $-\ell^2 \log(\ell^2)$ norm, also called the 'entropy' information cost functional, where

$$m(|\overline{W}_{j,n,t}|) = \begin{cases} -\overline{W}^2_{j,n,t} \log(\overline{W}^2_{j,n,t}), & \text{if } W_{j,n,t} \neq 0; \\ 0, & \text{if } W_{j,n,t} = 0, \end{cases}$$

and $\overline{W}_{j,n,t} \equiv W_{j,n,t}/\|\mathbf{X}\|$. This quantity has a monotonic relationship with the entropy of a sequence.

[2] The threshold functional, i.e., the number of terms exceeding a specified threshold δ, with

$$m(|W_{j,n,t}|) = \begin{cases} 1, & \text{if } |W_{j,n,t}| > \delta; \\ 0, & \text{otherwise.} \end{cases}$$

[3] The ℓ_p information cost functional, defined by

$$m(|W_{j,n,t}|) = |W_{j,n,t}|^p,$$

for $0 < p < 2$. If this cost functional is used, then $M^{\frac{1}{p}}(\mathbf{W}_{j,n})$ gives the ℓ_p norm of the sequence.

▷ **Exercise [223]:** Explain why the cost functional $m(|W_{j,n,t}|) = |W_{j,n,t}|^2$ is not of much use for selecting an 'optimal' orthonormal transform. ◁

To get some idea of what these cost functionals are trying to measure, let us consider the following simple example. Suppose we use the Haar wavelet to form DWPT coefficients of levels $j = 1, 2$ and 3 from a time series $\mathbf{X}$ of length $N = 8$:

$$\begin{bmatrix} \mathbf{W}_{1,1} \\ \mathbf{W}_{1,0} \end{bmatrix} = \begin{bmatrix} \mathcal{B}_1 \\ \mathcal{A}_1 \end{bmatrix} \mathbf{X} \qquad \begin{bmatrix} \mathbf{W}_{2,3} \\ \mathbf{W}_{2,2} \\ \mathbf{W}_{2,1} \\ \mathbf{W}_{2,0} \end{bmatrix} = \begin{bmatrix} \mathcal{A}_2\mathcal{B}_1 \\ \mathcal{B}_2\mathcal{B}_1 \\ \mathcal{B}_2\mathcal{A}_1 \\ \mathcal{A}_2\mathcal{A}_1 \end{bmatrix} \mathbf{X} \quad \text{and} \quad \begin{bmatrix} \mathbf{W}_{3,7} \\ \mathbf{W}_{3,6} \\ \mathbf{W}_{3,5} \\ \mathbf{W}_{3,4} \\ \mathbf{W}_{3,3} \\ \mathbf{W}_{3,2} \\ \mathbf{W}_{3,1} \\ \mathbf{W}_{3,0} \end{bmatrix} = \begin{bmatrix} \mathcal{A}_3\mathcal{A}_2\mathcal{B}_1 \\ \mathcal{B}_3\mathcal{A}_2\mathcal{B}_1 \\ \mathcal{B}_3\mathcal{B}_2\mathcal{B}_1 \\ \mathcal{A}_3\mathcal{B}_2\mathcal{B}_1 \\ \mathcal{A}_3\mathcal{B}_2\mathcal{A}_1 \\ \mathcal{B}_3\mathcal{B}_2\mathcal{A}_1 \\ \mathcal{B}_3\mathcal{A}_2\mathcal{A}_1 \\ \mathcal{A}_3\mathcal{A}_2\mathcal{A}_1 \end{bmatrix} \mathbf{X},$$

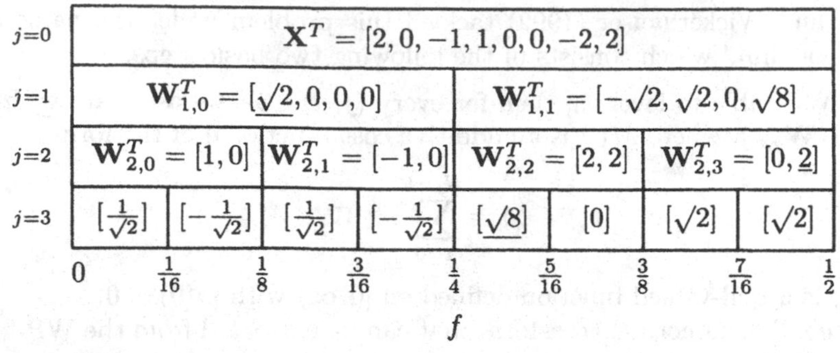

<table>
<tr><td>j=0</td><td colspan="8">$\mathbf{X}^T = [2, 0, -1, 1, 0, 0, -2, 2]$</td></tr>
<tr><td>j=1</td><td colspan="4">$\mathbf{W}_{1,0}^T = [\underline{\sqrt{2}}, 0, 0, 0]$</td><td colspan="4">$\mathbf{W}_{1,1}^T = [-\sqrt{2}, \sqrt{2}, 0, \sqrt{8}]$</td></tr>
<tr><td>j=2</td><td colspan="2">$\mathbf{W}_{2,0}^T = [1, 0]$</td><td colspan="2">$\mathbf{W}_{2,1}^T = [-1, 0]$</td><td colspan="2">$\mathbf{W}_{2,2}^T = [2, 2]$</td><td colspan="2">$\mathbf{W}_{2,3}^T = [0, \underline{2}]$</td></tr>
<tr><td>j=3</td><td>$[\frac{1}{\sqrt{2}}]$</td><td>$[-\frac{1}{\sqrt{2}}]$</td><td>$[\frac{1}{\sqrt{2}}]$</td><td>$[-\frac{1}{\sqrt{2}}]$</td><td>$[\underline{\sqrt{8}}]$</td><td>$[0]$</td><td>$[\sqrt{2}]$</td><td>$[\sqrt{2}]$</td></tr>
</table>

$0 \quad \frac{1}{16} \quad \frac{1}{8} \quad \frac{3}{16} \quad \frac{1}{4} \quad \frac{5}{16} \quad \frac{3}{8} \quad \frac{7}{16} \quad \frac{1}{2}$

f

Figure 224. Haar DWPT coefficients $\mathbf{W}_{j,n}$ of levels $j = 1, 2$ and 3 for a time series $\mathbf{X}$ of length $N = 8$. The series $\mathbf{X}$ was constructed from a linear combination of three basis vectors, one from each of the three levels (see Equation (224)). In the above, the DWPT coefficients corresponding to the vectors used in formation of $\mathbf{X}$ are underlined (one each in $\mathbf{W}_{1,0}$, $\mathbf{W}_{2,3}$ and $\mathbf{W}_{3,4}$).

where $\mathcal{A}_j$ and $\mathcal{B}_j$ are of dimension $\frac{8}{2^j} \times \frac{8}{2^{j-1}}$. We construct $\mathbf{X}$ from a linear combination of (the transposes of) the first row vector in $\mathcal{A}_1$, the second row vector in $\mathcal{A}_2\mathcal{B}_1$ and the row vector $\mathcal{A}_3\mathcal{B}_2\mathcal{B}_1$:

$$\mathbf{X} = \sqrt{2}\begin{bmatrix} \frac{1}{\sqrt{2}} \\ \frac{1}{\sqrt{2}} \\ 0 \\ 0 \\ 0 \\ 0 \\ 0 \\ 0 \end{bmatrix} + 2\begin{bmatrix} 0 \\ 0 \\ 0 \\ 0 \\ -\frac{1}{2} \\ \frac{1}{2} \\ -\frac{1}{2} \\ \frac{1}{2} \end{bmatrix} + \sqrt{8}\begin{bmatrix} \frac{1}{\sqrt{8}} \\ -\frac{1}{\sqrt{8}} \\ -\frac{1}{\sqrt{8}} \\ \frac{1}{\sqrt{8}} \\ \frac{1}{\sqrt{8}} \\ -\frac{1}{\sqrt{8}} \\ -\frac{1}{\sqrt{8}} \\ \frac{1}{\sqrt{8}} \end{bmatrix} = \begin{bmatrix} 2 \\ 0 \\ -1 \\ 1 \\ 0 \\ 0 \\ -2 \\ 2 \end{bmatrix}. \tag{224}$$

As can be seen by inspection, the three vectors used to form $\mathbf{X}$ are orthogonal to each other, but each comes from a DWPT of a different level. Figure 224 shows the contents of $\mathbf{W}_{j,n}$ for the Haar-based DWPTs out to level $J_0 = 3$ – here we have underlined the coefficients formed using the three vectors from which we constructed $\mathbf{X}$ (one each in $\mathbf{W}_{1,0}$, $\mathbf{W}_{2,3}$ and $\mathbf{W}_{3,4}$).

An intuitive notion of what would constitute a 'best basis' for $\mathbf{X}$ would surely contain the three orthonormal vectors from which we formed $\mathbf{X}$ – if these were included, we would be able to express $\mathbf{X}$ succinctly using three nonzero coefficients. Consider the coefficient vector $\mathbf{W}_{1,0}$, the first element of which is formed using the first vector displayed in Equation (224). Since $\mathbf{W}_{1,0}$ corresponds to the frequency interval $[0, \frac{1}{4}]$, its competitors in forming a disjoint dyadic decomposition are either $\mathbf{W}_{2,0}$ together with $\mathbf{W}_{2,1}$ or $\mathbf{W}_{3,0}$ together with $\mathbf{W}_{3,1}$, $\mathbf{W}_{3,2}$ and $\mathbf{W}_{3,3}$. Thus a reasonable additive cost functional should assign a lower cost to $[\sqrt{2}, 0, 0, 0]$ than to either $[1, 0]$ and $[-1, 0]$ (i.e., $\mathbf{W}_{2,0}$ and $\mathbf{W}_{2,1}$ grouped together) or $[\frac{1}{\sqrt{2}}], [-\frac{1}{\sqrt{2}}], [\frac{1}{\sqrt{2}}]$ and $[-\frac{1}{\sqrt{2}}]$ (i.e., the competing $j = 3$ coefficients grouped together). Qualitatively, the property that distinguishes $[\sqrt{2}, 0, 0, 0]$ from the other two sets of competing coefficients is that one of its coefficients stands out in magnitude above the rest; in other words, its coefficients are inhomogeneous. Let us thus see how the three cost functionals treat a simple case

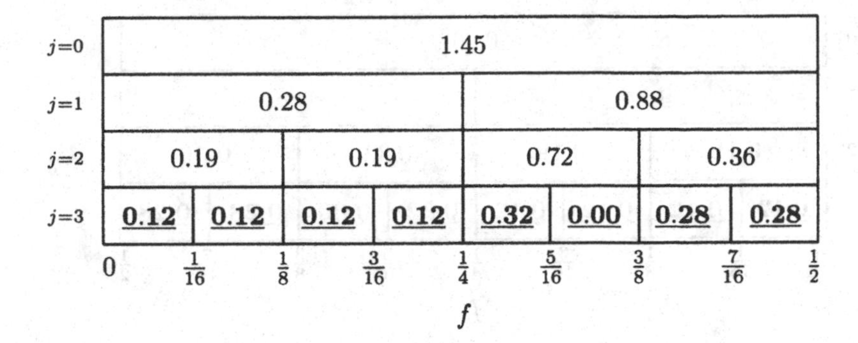

Figure 225. Cost table for the WP table shown in Figure 224. The entries in the above are the values of $M(\mathbf{W}_{j,n})$ computed using the $-\ell^2 \log(\ell^2)$ norm as a cost functional. The initial step of the best basis algorithm dictates that we mark the values in the bottom row in some manner – here we have done so by underlining and also by using a bold font.

of homogeneous coefficients, say, $\mathbf{W}_h \equiv [\pm\frac{\epsilon}{2}, \pm\frac{\epsilon}{2}, \pm\frac{\epsilon}{2}, \pm\frac{\epsilon}{2}]^T$ versus a vector of inhomogeneous coefficients, say, $\mathbf{W}_i \equiv [\pm\epsilon, 0, 0, 0]^T$ (note that the energy of both vectors is the same, as would be true for any sets of coefficients in competition for a given frequency interval $\mathcal{I}_{j,n}$).

[1] Using the $-\ell^2 \log(\ell^2)$ norm (and assuming for convenience that $\mathbf{W}_h$ and $\mathbf{W}_i$ are derived from a series such that $\|\mathbf{X}\| = 1$), we have

$$M(\mathbf{W}_h) = -4 \cdot \tfrac{\epsilon^2}{4} \log(\tfrac{\epsilon^2}{4}) = -\epsilon^2 (\log(\epsilon^2) - \log(4)) > -\epsilon^2 \log(\epsilon^2) = M(\mathbf{W}_i),$$

so, as desired, $\mathbf{W}_i$ is assigned a lower cost than $\mathbf{W}_h$.

[2] If we use, say, $\delta = \frac{\epsilon}{4}$ as the threshold for the threshold functional, then $M(\mathbf{W}_h) = 4$, whereas $M(\mathbf{W}_i) = 1$, so again $\mathbf{W}_i$ has a lower cost; on the other hand, if we set, say, $\delta = \frac{3\epsilon}{4}$, then $M(\mathbf{W}_h) = 0$ while $M(\mathbf{W}_i) = 1$, which is the reverse of what we want. The threshold functional can thus be tricky to use because of the problem of selecting an appropriate δ.

[3] If we use the ℓ_p information cost functional, then

$$M(\mathbf{W}_h) = 4|\tfrac{\epsilon}{2}|^p > |\epsilon|^p = M(\mathbf{W}_i)$$

for any $0 < p < 2$, again in agreement with what we want.

Thus, assuming that we can properly set δ in the threshold functional, all three of the cost functionals assign a lower cost to the vector with inhomogeneous coefficients.

Once a cost functional has been chosen, we can assign a cost to each vector $\mathbf{W}_{j,n}$ in the WP table. An example of such a 'cost table' (based upon the $-\ell^2 \log(\ell^2)$ norm) is given in Figure 225 for the WP table of Figure 224. The best basis algorithm (Coifman and Wickerhauser, 1992; Wickerhauser, 1994) proceeds in a simple way, as illustrated in Figures 225 and 226.

[1] We mark all costs in the nodes at the bottom of the table in some way (they are underlined and printed in bold in Figure 225). We start by examining this bottom row of nodes.

[2] We compare the costs of the sum of each pair of children nodes with their parent node and then do one of the following:

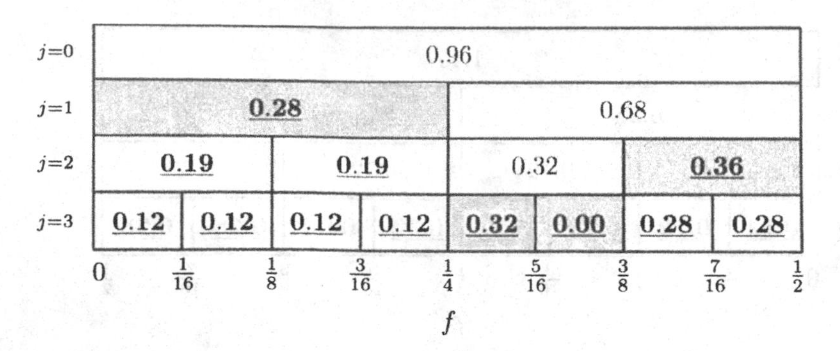

Figure 226. Final step of the best basis algorithm. The best basis transform is indicated by the shaded boxes. As is intuitively reasonable, the selected transform includes the three basis vectors used to form $\mathbf{X}$.

(a) if the parent node has a lower cost than the sum of the costs of the children nodes, we mark the parent node (i.e., cost is underlined and printed in bold in the figures), while

(b) if the sum of the costs of the children nodes is lower than the cost of the parent node, we replace the cost of the parent node by the sum of the costs of the children nodes (no new marking is done).

[3] We then repeat step [2] for each level as we move up the table. The end result is shown in Figure 226.

[4] Once we have reached the top of the table, we look back down the table at the marked nodes. The top-most marked nodes that correspond to a disjoint dyadic decomposition define the best basis transform.

For our simple example, the best basis transform is shown by the shaded boxes in Figure 226 and consists of the basis vectors corresponding to the DWPT coefficients $\mathbf{W}_{1,0}$, $\mathbf{W}_{3,4}$, $\mathbf{W}_{3,5}$ and $\mathbf{W}_{2,3}$. This choice is intuitively reasonable because $\mathbf{X}$ was formed using basis vectors from $\mathbf{W}_{1,0}$, $\mathbf{W}_{3,4}$ and $\mathbf{W}_{2,3}$ (these three sets of basis vectors do not form a disjoint dyadic decomposition, but such a decomposition is formed upon inclusion of the additional vector used in forming $\mathbf{W}_{3,5}$).

Comments and Extensions to Section 6.3

[1] The best basis algorithm is a 'bottom up' algorithm and hence requires computation of an entire WP table out to level J_0. Taswell (1996) discusses a 'top down' approach that can require considerable less computation because only a portion of the WP table needs actually to be computed. Although this approach selects a basis that is suboptimal in theory, computer experiments indicate that there is little or no practical difference between the suboptimal and best bases as measured by various criteria.

6.4 Example: Best Basis for Solar Physics Data

Here we apply the best basis algorithm to the solar magnetic field magnitude data (these are described in Section 6.2 and plotted in Figure 222). Since the sampling interval for this series is $\Delta t = 1/24$ day, we map the standardized frequency intervals $\mathcal{I}_{j,n} = [\frac{n}{2^{j+1}}, \frac{n+1}{2^{j+1}}]$ over to physically meaningful frequency intervals by multiplying both end points by $1/\Delta t = 24$, yielding $[\frac{n}{2^{j+1}\Delta t}, \frac{n+1}{2^{j+1}\Delta t}]$, where any frequency in this

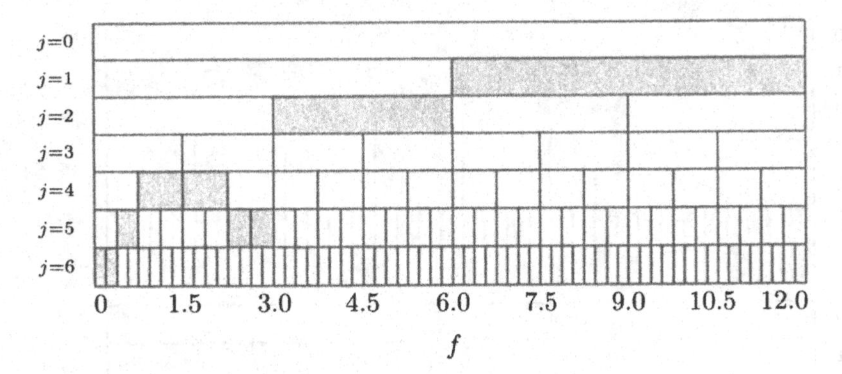

Figure 227. The best basis transform for the solar magnetic field magnitude data of Figure 222. This transform is based on the LA(8) wavelet filter and a cost functional based upon the $-\ell^2 \log(\ell^2)$ norm.

interval is measured in cycles per day. If we construct a WP table out to level $J_0 = 6$ for this series using the LA(8) wavelet filter, and if we then apply the best basis algorithm using a cost functional based upon the $-\ell^2 \log(\ell^2)$ norm, we obtain the best basis transform depicted in Figure 227. The transform consists of $N/2$ vectors from the level $j = 1$ DWPT; $N/4$ vectors from level $j = 2$; $N/8$ from level $j = 4$; $3N/32$ from level $j = 5$; and $N/32$ from level $j = 6$. Note that, in a manner somewhat similar to the DWT, this transform generally increases its frequency resolution as the frequency decreases – the frequency interval $[2.25, 3.0]$ cycles/day is an exception to this general pattern (we will refer back to this transform in Section 6.7 in conjunction with picking the level for a MODWPT analysis of this series that seeks to track frequency changes as a function of time).

Let us now consider two refinements to the best basis algorithm that might be appropriate for the solar magnetic field magnitude data. First, as can be seen from the plot at the bottom of Figure 222, there is a noticeable upward drift in this time series, which is problematic because, like the DWT, the DWPT makes use of circular filtering. The mismatch between the beginning and end of the time series can lead to few quite large boundary coefficients. Since the best basis algorithm makes use of measures that are intended to select sets of inhomogeneous coefficients, these anomalous coefficients are potentially troublesome. To alleviate such concerns, let us recompute a best basis transform for this time series using reflection boundary conditions; i.e., as described by Equation (140), we extend the series to twice its original length and analyze this augmented series. The selected basis is shown in the top of Figure 228. If we compare this to the basis in Figure 227 selected using periodic boundary conditions, we see that the basis vectors are the same up to level $j = 3$, beyond which there are some noticeable differences. The choice of boundary conditions can thus influence the best basis algorithm.

A second refinement attempts to alleviate the dependence of the DWPT on where we 'break into' the time series (this problem is analogous to one we noted for the DWT in Section 5.1). This dependence impacts the best basis algorithm in that, if we circularly shift a time series, the best basis transform for the shifted series need not be the same as the one for the original series. This consideration has led some researchers to propose a form of 'translation invariance' that effectively selects amongst different

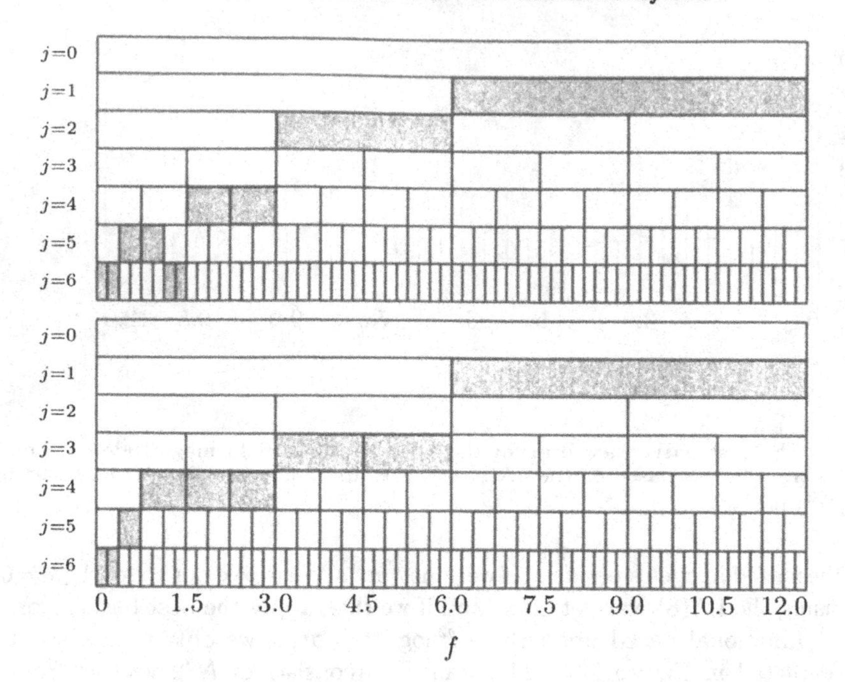

Figure 228. Best basis transform for the solar magnetic field magnitude data based upon reflection boundary conditions (top), and the best shift basis transform for the same data (bottom), again using reflection boundary conditions. Both transforms are based on the same wavelet filter and cost functional as used in Figure 227. The best shift basis algorithm selected the shift $m = 52$ (i.e., $\mathcal{T}^{52}$) as the best shift.

circular shifts by making use of the same cost functional as is used in the best basis algorithm (see, for example, Del Marco and Weiss, 1997, and references therein). This leads to the following modification of the best basis algorithm, which we can call the 'best shift basis algorithm.' We first form a WP table out to level J_0 for $\mathbf{X}$. We then apply the best basis algorithm and note the total cost associated with the selected basis (for the example shown in Figure 226, this amounts to summing the numbers in the four shaded boxes). Next we repeat the above for each of the circularly shifted series $\mathcal{T}^m\mathbf{X}$, $m = 1, \ldots, 2^{J_0} - 1$ (note that we need only consider these shifts: any $m < 0$ or $m \geq 2^{J_0}$ will yield a WP table whose components $\mathbf{W}_{j,n}$ will be merely circularly shifted versions of the components for a shift of $m \bmod 2^{J_0}$ – the resulting cost tables for $\mathcal{T}^m\mathbf{X}$ and $\mathcal{T}^{m \bmod 2^{J_0}}\mathbf{X}$ are thus the same and hence lead to the same best basis transform). The best shift basis transform is defined as the best basis transform associated with the shift yielding the smallest total cost.

When applied to the solar magnetic field magnitude data, the best shift basis transform is associated with shift $m = 52$ and is shown in the bottom of Figure 228 (as in the top part of this figure, we have used the LA(8) wavelet, a cost functional based on the $-\ell^2 \log(\ell^2)$ norm and symmetric boundary conditions). Note that the frequency resolution of the selected transform increases with decreasing frequency (a similar pattern almost holds for the transforms depicted in Figure 227 and the top of Figure 228). For $J_0 = 6$, there are a total of 64 shifts (including $m = 0$) that can potentially lead to different best basis transforms. For the time series under study, all

64 best basis transforms are quite similar, being at most minor variations on the three depicted in the figures.

6.5 Time Shifts for Wavelet Packet Filters

In Section 4.8 we looked at the phase properties of the Daubechies least asymmetric (LA) wavelet and scaling filters, and we did the same thing in Section 4.9 for the coiflet filters. In particular we derived the time shifts that enable us to plot the DWT coefficients as approximately the (downsampled) output from a zero phase filter (we made use of these shifts again in Chapter 5 when plotting the MODWT coefficients). Here we extend this approach to the filters used to create the DWPT (and also a 'maximal overlap' version of this transform – the MODWPT – to be discussed in Section 6.6).

The key to obtaining the proper time shifts is Equation (215b), which gives the transfer function $U_{j,n}(\cdot)$ of the filter $\{u_{j,n,l}\}$. Each factor in the product defining $U_{j,n}(\cdot)$ can be written as

$$M_{c_{j,n,m}}(2^m f) = |M_{c_{j,n,m}}(2^m f)|e^{i\theta_{c_{j,n,m}}(2^m f)},$$

where

$$\theta_{c_{j,n,m}}(f) = \begin{cases} \theta^{(G)}(f), & \text{if } c_{j,n,m} = 0; \\ \theta^{(H)}(f), & \text{if } c_{j,n,m} = 1. \end{cases}$$

From Equation (112d)

$$\theta^{(G)}(f) \approx 2\pi f\nu \text{ and } \theta^{(H)}(f) \approx -2\pi f(L-1+\nu),$$

where, from Equation (112e) and the discussion in Section 4.9,

$$\nu = \begin{cases} -\frac{L}{2}+1, & \text{for the LA filters with even } \frac{L}{2}; \\ -\frac{L}{2}, & \text{for the LA(10) and LA(18) filters;} \\ -\frac{L}{2}+2, & \text{for the LA(14) filter; and} \\ -\frac{2L}{3}+1, & \text{for the coiflet filters } (L = 6, 12, \ldots, 30). \end{cases} \tag{229a}$$

Note that ν is always negative (we can remind ourselves of this fact by writing $-|\nu|$ in place of ν). Hence the phase function corresponding to $U_{j,n}(\cdot)$ can be written as

$$\sum_{m=0}^{j-1} \theta_{c_{j,n,m}}(2^m f) \approx 2\pi f[\nu S_{j,n,0} - (L-1+\nu)S_{j,n,1}],$$

where

$$S_{j,n,0} \equiv \sum_{m=0}^{j-1}(1-c_{j,n,m})2^m \text{ and } S_{j,n,1} \equiv \sum_{m=0}^{j-1} c_{j,n,m}2^m.$$

If we note that we always have

$$S_{j,n,0} + S_{j,n,1} = \sum_{m=0}^{j-1} 2^m = 2^j - 1, \tag{229b}$$

then we can write

$$\sum_{m=0}^{j-1} \theta_{c_{j,n,m}}(2^m f) \approx 2\pi f[\nu(2^j - 1 - S_{j,n,1}) - (L-1+\nu)S_{j,n,1}] \equiv 2\pi f\nu_{j,n},$$

where $\nu_{j,n} \equiv \nu(2^j - 1) - S_{j,n,1}(2\nu + L - 1)$.

j	n	$\mathbf{c}_{j,n}$	$S_{j,n,0}$	$S_{j,n,1}$	$\lvert\nu_{j,n}\rvert$	LA(8)	$\lvert p_{j,n}\rvert$	LA(8)
1	0	0	1	0	$\frac{L_1}{2} - 1$	3	$e\{g_l\}$	2.8
	1	1	0	1	$\frac{L_1}{2} + 0$	4	$e\{h_l\}$	4.2
2	0	0,0	3	0	$\frac{L_2}{2} - 2$	9	$3e\{g_l\}$	8.5
	1	0,1	1	2	$\frac{L_2}{2} + 0$	11	$e\{g_l\} + 2e\{h_l\}$	11.2
	2	1,1	0	3	$\frac{L_2}{2} + 1$	12	$3e\{h_l\}$	12.5
	3	1,0	2	1	$\frac{L_2}{2} - 1$	10	$2e\{g_l\} + e\{h_l\}$	9.8
3	0	0,0,0	7	0	$\frac{L_3}{2} - 4$	21	$7e\{g_l\}$	19.9
	1	0,0,1	3	4	$\frac{L_3}{2} + 0$	25	$3e\{g_l\} + 4e\{h_l\}$	25.2
	2	0,1,1	1	6	$\frac{L_3}{2} + 2$	27	$e\{g_l\} + 6e\{h_l\}$	27.8
	3	0,1,0	5	2	$\frac{L_3}{2} - 2$	23	$5e\{g_l\} + 2e\{h_l\}$	22.5
	4	1,1,0	4	3	$\frac{L_3}{2} - 1$	24	$4e\{g_l\} + 3e\{h_l\}$	23.8
	5	1,1,1	0	7	$\frac{L_3}{2} + 3$	28	$7e\{h_l\}$	29.1
	6	1,0,1	2	5	$\frac{L_3}{2} + 1$	26	$2e\{g_l\} + 5e\{h_l\}$	26.5
	7	1,0,0	6	1	$\frac{L_3}{2} - 3$	22	$6e\{g_l\} + e\{h_l\}$	21.2

Table 230. Form of the advances $|\nu_{j,n}|$ required to achieve approximate zero phase output using a filter $\{u_{j,n,l}\}$ based upon an LA filter with $L/2$ even. Here we let $j = 1, 2$ and 3 and $n = 0, \ldots, 2^j - 1$. The seventh column gives specific values for the LA(8) filter (see Comments and Extensions to Section 6.5 for a discussion on the last two columns).

▷ **Exercise [230]:** Show that $\nu_{j,n} < 0$ for all the LA and coiflet filters we have considered. ◁

We hence obtain the closest approximation to a zero phase filter by advancing $\{u_{j,n,l}\}$ by $|\nu_{j,n}| = -\nu_{j,n}$ units.

Now the filter $\{u_{j,n,l}\}$ has width $L_j \equiv (2^j - 1)(L - 1) + 1$, so we can write

$$\nu_{j,n} = \left(\frac{L_j - 1}{L - 1}\right)\nu - S_{j,n,1}[2\nu + L - 1].$$

Using the above and the expressions for ν given in Equation (229a) for the various filters, we obtain (after some algebraic manipulations)

$$\nu_{j,n} = \begin{cases} -\frac{L_j}{2} + (2^{j-1} - S_{j,n,1}), & \text{for LA filters, even } \frac{L}{2}; \\ -\frac{L_j}{2} - (2^{j-1} - S_{j,n,1}) + 1, & \text{for LA(10) and LA(18);} \\ -\frac{L_j}{2} + 3(2^{j-1} - S_{j,n,1}) - 1, & \text{for LA(14); and} \\ -\frac{L_j}{2} - \frac{L-3}{3}(2^{j-1} - S_{j,n,1} - \frac{1}{2}) + \frac{1}{2}, & \text{for the coiflet filters.} \end{cases} \tag{230}$$

The above formulation is convenient for tabulation since it expresses $\nu_{j,n}$ in terms of deviations from half the filter width. The algebraic form of the advances $|\nu_{j,n}|$ required to make the filter $\{u_{j,n,l}\}$ closest to zero phase is given in the sixth column of Table 230 for the case of the LA filters with even $\frac{L}{2}$ and for $j = 1, \ldots, 3$ with $n = 0, \ldots, 2^j - 1$. The seventh column gives the actual numerical values of $|\nu_{j,n}|$ for the LA(8) filter (the final two columns are discussed in the Comments and Extensions below).

Comments and Extensions to Section 6.5

[1] As noted in item [2] of the Comments and Extensions to Section 4.8, Wickerhauser (1994, pp. 171 and 341) and Hess–Nielsen and Wickerhauser (1996) formulated advances applicable to a wide range of filter designs using the notion of the 'center of energy' for a filter (these designs include the Daubechies extremal phase and least asymmetric filters and the coiflet filters). These ideas can also be used with the DWPT filters $\{u_{j,n,l}\}$. The appropriate advances $|p_{j,n}|$ are given by

$$|p_{j,n}| = S_{j,n,0}e\{g_l\} + S_{j,n,1}e\{h_l\},$$

where, because they have unit energy, the centers of energy for $\{g_l\}$ and $\{h_l\}$ are given by, respectively,

$$e\{g_l\} = \sum_{l=0}^{L-1} lg_l \text{ and } e\{h_l\} = \sum_{l=0}^{L-1} lh_l.$$

The form of $|p_{j,n}|$ for LA filters with even $L/2$ is shown in the next to last column of Table 230, and values specific to the LA(8) filter are shown in the last column (for this filter, calculations show that $e\{g_l\} \doteq 2.8464$ and $e\{h_l\} \doteq 4.1536$). Note that, after $|p_{j,n}|$ is rounded to the nearest integer, it differs at most by unity from $|\nu_{j,n}|$, so the two methods of determining advances are quite comparable.

6.6 Maximal Overlap Discrete Wavelet Packet Transform

In Chapter 5 we explored the maximal overlap discrete wavelet transform (MODWT) as an alternative to the DWT. The MODWT has several desirable characteristics that the DWT lacks, including less sensitivity to the assumed starting point for a time series, applicability to any sample size N and a multiresolution analysis that is associated with zero phase filters (the chief prices we pay in switching to the MODWT are an extra computational burden and a loss of orthonormality). It is straightforward to extend the maximal overlap idea to define a maximal overlap discrete wavelet packet transform (MODWPT). Here we sketch the main ideas and results for this transform (formal proofs are just sketched or omitted since they are close copies of those already presented in Chapter 5 for the MODWT).

We start by defining the MODWPT filters $\{\tilde{u}_{n,l}\}$ and $\{\tilde{u}_{j,n,l}\}$ in terms of their DWPT counterparts, namely

$$\tilde{u}_{n,l} \equiv \frac{u_{n,l}}{\sqrt{2}} \text{ and } \tilde{u}_{j,n,l} \equiv \frac{u_{j,n,l}}{2^{j/2}},$$

where $u_{n,l}$ and $u_{j,n,l}$ are given in Equations (214b) and (214c). Using these filters, we can define level j coefficients $\widetilde{W}_{j,n,t}$ for the MODWPT in one of two equivalent ways. The first involves filtering $\mathbf{X}$ directly using $\{\tilde{u}_{j,n,l}\}$:

$$\widetilde{W}_{j,n,t} \equiv \sum_{l=0}^{L_j-1} \tilde{u}_{j,n,l}X_{t-l \bmod N}, \quad t = 0, 1, \ldots, N-1. \tag{231a}$$

The second is a recursive scheme for which we assume that we already have the level $j-1$ coefficients:

$$\widetilde{W}_{j,n,t} \equiv \sum_{l=0}^{L-1} \tilde{u}_{n,l}\widetilde{W}_{j-1,\lfloor \frac{n}{2} \rfloor, t-2^{j-1}l \bmod N}, \quad t = 0, \ldots, N-1 \tag{231b}$$

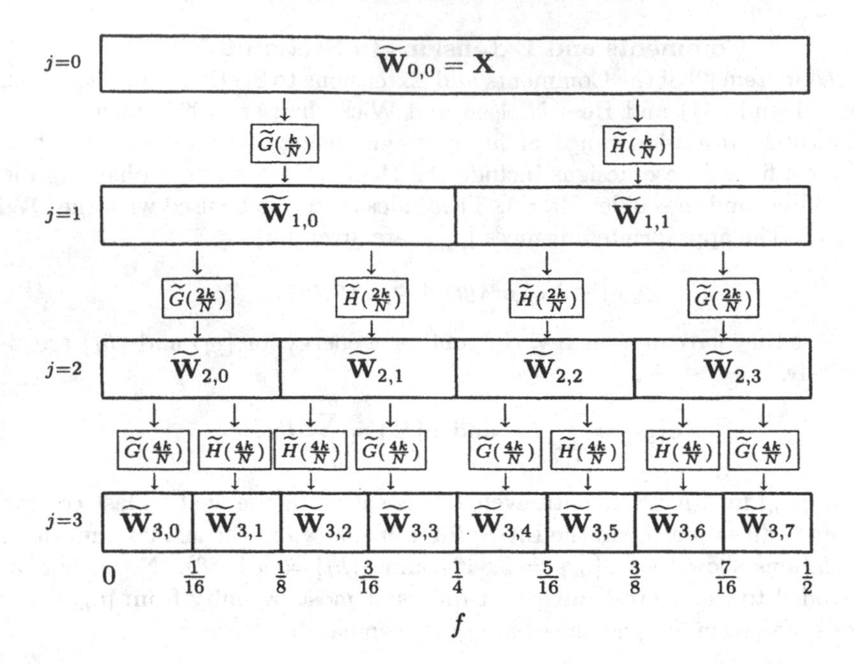

Figure 232. Flow diagram illustrating the analysis of $\mathbf{X}$ into MODWPT coefficients $\widetilde{\mathbf{W}}_{j,n}$ of levels $j = 1, 2$ and 3. Collectively, all the coefficients in the jth row constitute the level j MODWPT of $\mathbf{X}$. Note that, within the jth level, the frequency index n ranges from 0 to $2^j - 1$ and that each $\widetilde{\mathbf{W}}_{j,n}$ has length N (by contrast, in Figure 212a for the DWPT, each $\mathbf{W}_{j,n}$ has length $N/2^j$).

(this formulation leads to a MODWPT pyramid algorithm that closely parallels the MODWT algorithm – see Section 5.5). We initiate this scheme by defining the zeroth level coefficients to be the time series itself, i.e., $\widetilde{W}_{0,0,t} \equiv X_t$. There are N MODWPT coefficients corresponding to each node (j, n), and we place these in the N dimensional vector $\widetilde{\mathbf{W}}_{j,n}$ (hence $\widetilde{\mathbf{W}}_{0,0} \equiv \mathbf{X}$). As with the DWPT, we can collect together all the MODWPT coefficients out to, say, level J_0 and form a *maximal overlap WP table.* Figure 232 is an example of such a table to level $J_0 = 3$.

A comparison of Equations (231a) and (231b) with the corresponding equations for the DWPT, namely, Equations (215a) and (214a), shows that, other than the use of renormalized filters, the essential difference between the MODWPT and DWPT is the elimination of downsampling. Since essentially the same filters are used to create both $\mathbf{W}_{j,n}$ and $\widetilde{\mathbf{W}}_{j,n}$, the MODWPT coefficients are nominally related to the same frequency interval as the DWPT coefficients, namely, $\mathcal{I}_{j,n} = [\frac{n}{2^{j+1}}, \frac{n+1}{2^{j+1}}]$. The transfer function for the filter $\{\tilde{u}_{j,n,l}\}$ – call it $\widetilde{U}_{j,n}(\cdot)$ – is simply related to the transfer function $U_{j,n}(\cdot)$ for $\{u_{j,n,l}\}$ via $\widetilde{U}_{j,n}(f) = U_{j,n}(f)/2^{j/2}$. With $\widetilde{M}_0(f) \equiv M_0(f)/\sqrt{2} = G(f)/\sqrt{2}$ and $\widetilde{M}_1(f) \equiv M_1(f)/\sqrt{2} = H(f)/\sqrt{2}$, it follows from Equation (215b) that we can write

$$\widetilde{U}_{j,n}(f) = \prod_{m=0}^{j-1} \widetilde{M}_{c_{j,n,m}}(2^m f), \tag{232}$$

where, as before, $c_{j,n,m}$ is the mth element of the binary-valued vector $\mathbf{c}_{j,n}$ – this vector is the same as used for the DWPT, and examples of its structure are shown in

Figure 216. Because rescaling a filter does not alter its phase function, if we construct a MODWPT based upon (rescaled) LA or coiflet filters, we can then take advantage of the advances $|\nu_{j,n}|$ developed in Section 6.5 to circularly shift the elements of $\widetilde{\mathbf{W}}_{j,n}$ so that they are associated – to a good approximation – with the same times as are the elements of $\mathbf{X}$. For example, the tth element of the circularly shifted vector

$$\mathcal{T}^{-|\nu_{j,n}|}\widetilde{\mathbf{W}}_{j,n} = [\widetilde{W}_{j,n,|\nu_{j,n}|}, \widetilde{W}_{j,n,|\nu_{j,n}|+1}, \ldots, \widetilde{W}_{j,n,N-1}, \widetilde{W}_{j,n,0}, \ldots, \widetilde{W}_{j,n,|\nu_{j,n}|-1}]^T$$

can be associated with the same time as X_t.

Let us conclude our discussion by showing that, as is true for the DWPT, at any level $j \geq 1$ the MODWPT coefficients can be used to form an energy decomposition for $\mathbf{X}$ and an additive decomposition similar in spirit to an MRA based upon the MODWT (however, each of the components of the additive decomposition are now associated with a frequency interval $\mathcal{I}_{j,n}$ rather than a scale).

- *Energy decomposition*

Recall that the 'energy' in a time series is given by the sum of the squares of its elements, i.e., $\sum_{t=0}^{N-1} X_t^2 = \|\mathbf{X}\|^2$. For any level $j \geq 1$, we claim that

$$\|\mathbf{X}\|^2 = \sum_{n=0}^{2^j-1} \|\widetilde{\mathbf{W}}_{j,n}\|^2.$$

In other words, energy is preserved in a MODWPT of level j. To see this, let

$$\mathcal{X}_k = \sum_{t=0}^{N-1} X_t e^{-i2\pi tk/N}, \quad k = 0, \ldots, N-1,$$

be the DFT of $\mathbf{X}$. Since Equation (231a) tells us that $\widetilde{\mathbf{W}}_{j,n}$ is the result of circularly convolving $\mathbf{X}$ with the filter $\{\tilde{u}_{j,n,l}\}$ implicitly periodized to length N, Parseval's theorem states that

$$\|\widetilde{\mathbf{W}}_{j,n}\|^2 = \frac{1}{N}\sum_{k=0}^{N-1} |\widetilde{U}_{j,n}(\tfrac{k}{N})|^2 |\mathcal{X}_k|^2$$

(cf. Equation (170c)). Hence we have

$$\sum_{n=0}^{2^j-1} \|\widetilde{\mathbf{W}}_{j,n}\|^2 = \sum_{n=0}^{2^j-1} \frac{1}{N}\sum_{k=0}^{N-1} |\widetilde{U}_{j,n}(\tfrac{k}{N})|^2 |\mathcal{X}_k|^2 = \frac{1}{N}\sum_{k=0}^{N-1} |\mathcal{X}_k|^2 \sum_{n=0}^{2^j-1} |\widetilde{U}_{j,n}(\tfrac{k}{N})|^2.$$

▷ **Exercise [233]**: Show that

$$\sum_{n=0}^{2^j-1} |\widetilde{U}_{j,n}(f)|^2 = 1 \text{ for all } f. \quad (233)$$ ◁

The result thus follows from another application of Parseval's theorem:

$$\sum_{n=0}^{2^j-1} \|\widetilde{\mathbf{W}}_{j,n}\|^2 = \frac{1}{N}\sum_{k=0}^{N-1} |\mathcal{X}_k|^2 = \sum_{t=0}^{N-1} X_t^2 = \|\mathbf{X}\|^2.$$

- *Additive decomposition*

At any level $j \geq 1$, we claim that we can reconstruct $\mathbf{X}$ by circularly cross-correlating $\{\widetilde{W}_{j,n,t}\}$ with $\{\tilde{u}_{j,n,l}\}$ for each frequency band n and summing:

$$X_t = \sum_{n=0}^{2^j-1} \sum_{l=0}^{L_j-1} \tilde{u}_{j,n,l} \widetilde{W}_{j,n,t+l \bmod N} \quad t = 0, \ldots, N-1. \tag{234}$$

To show this, we will show that an equivalent statement is true, namely, one formed by applying the DFT to both sides of the above:

$$\mathcal{X}_k = \sum_{n=0}^{2^j-1} \sum_{t=0}^{N-1} \left(\sum_{l=0}^{L_j-1} \tilde{u}_{j,n,l} \widetilde{W}_{j,n,t+l \bmod N} \right) e^{-i2\pi tk/N}, \quad k = 0, \ldots, N-1.$$

Now the DFT of the cross-correlation is given by the complex conjugate of the DFT of $\{\tilde{u}_{j,n,l}\}$ periodized to length N, namely, $\{\widetilde{U}^*_{j,n}(\frac{k}{N})\}$, multiplied by the DFT of $\{\widetilde{W}_{j,n,t}\}$, namely, $\{\widetilde{U}_{j,n}(\frac{k}{N})\mathcal{X}_k\}$ (this follows from Equation (231a), which says that $\{W_{j,n,t}\}$ is the circular convolution of $\{X_t\}$ with $\{\tilde{u}_{j,n,l}\}$ periodized to length N). Hence the above is equivalent to

$$\mathcal{X}_k = \sum_{n=0}^{2^j-1} \widetilde{U}^*_{j,n}(\tfrac{k}{N}) \widetilde{U}_{j,n}(\tfrac{k}{N}) \mathcal{X}_k = \mathcal{X}_k \sum_{n=0}^{2^j-1} |\widetilde{U}_{j,n}(\tfrac{k}{N})|^2, \quad k = 0, \ldots, N-1;$$

however, Equation (233) says that the summation on the right-hand side is unity, which leads to the required result.

Let $\widetilde{\mathcal{D}}_{j,n}$ be the N dimensional vector containing the sequence formed by circular cross-correlation, and let $\widetilde{\mathcal{D}}_{j,n,t}$ denote its tth element. We then have

$$\widetilde{\mathcal{D}}_{j,n,t} = \sum_{l=0}^{L_j-1} \tilde{u}_{j,n,l} \widetilde{W}_{j,n,t+l \bmod N} \text{ and hence } X_t = \sum_{n=0}^{2^j-1} \widetilde{\mathcal{D}}_{j,n,t}.$$

We can say that at time t the value X_t is the sum of the tth terms $\widetilde{\mathcal{D}}_{j,n,t}$ in the 'detail' sequence in each frequency band n. Since the DFT of $\widetilde{\mathcal{D}}_{j,n}$ is given by $\{|\widetilde{U}_{j,n}(\frac{k}{N})|^2 \mathcal{X}_k\}$, we notice that, as was true for the MODWT, the transfer function of the composite filter acting on $\{X_t\}$ is given by $|\widetilde{U}_{j,n}(f)|^2$, which is real and positive, and hence has zero phase. The details $\widetilde{\mathcal{D}}_{j,n}$ in each frequency band at time t thus line up *perfectly* with features in the time series $\mathbf{X}$ at the same time. This appealing characteristic is not shared by the raw MODWPT coefficients $\widetilde{\mathbf{W}}_{j,n}$.

6.7 Example: MODWPT of Solar Physics Data

Here we give an example of MODWPT analysis by applying this technique to the solar physics time series (this is described in Section 6.2 and replotted at the bottom of Figure 235). To do so, we compute a level $j = 4$ MODWPT based on the LA(8) wavelet filter. In view of the transforms picked out by the best basis algorithm and variations thereof (see Figures 227 and 228), this choice of level provides a reasonable trade-off between resolution in the time and frequency domains: frequencies below

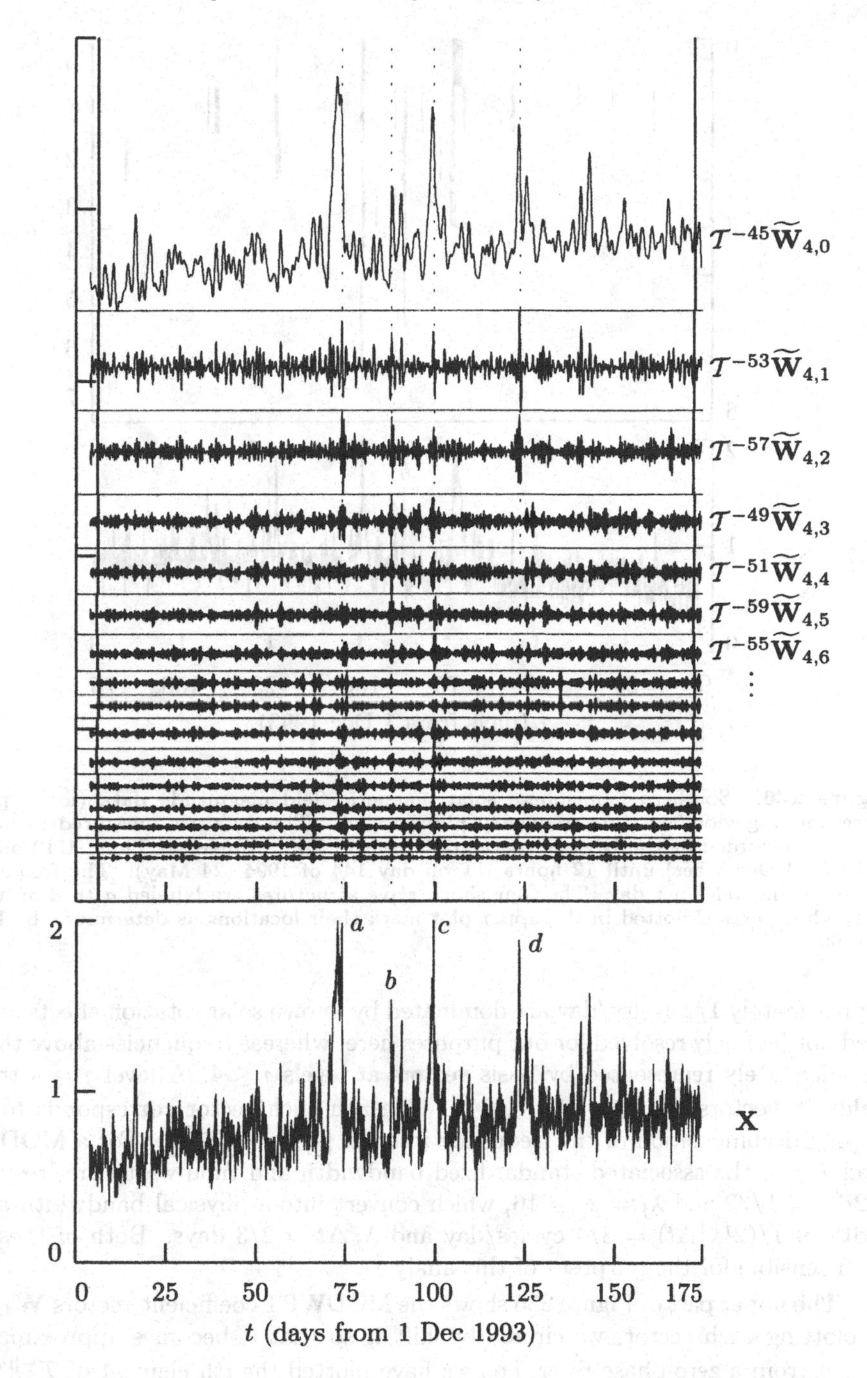

Figure 235. Level $j = 4$ LA(8) MODWPT coefficients for solar physics time series (see text for details).

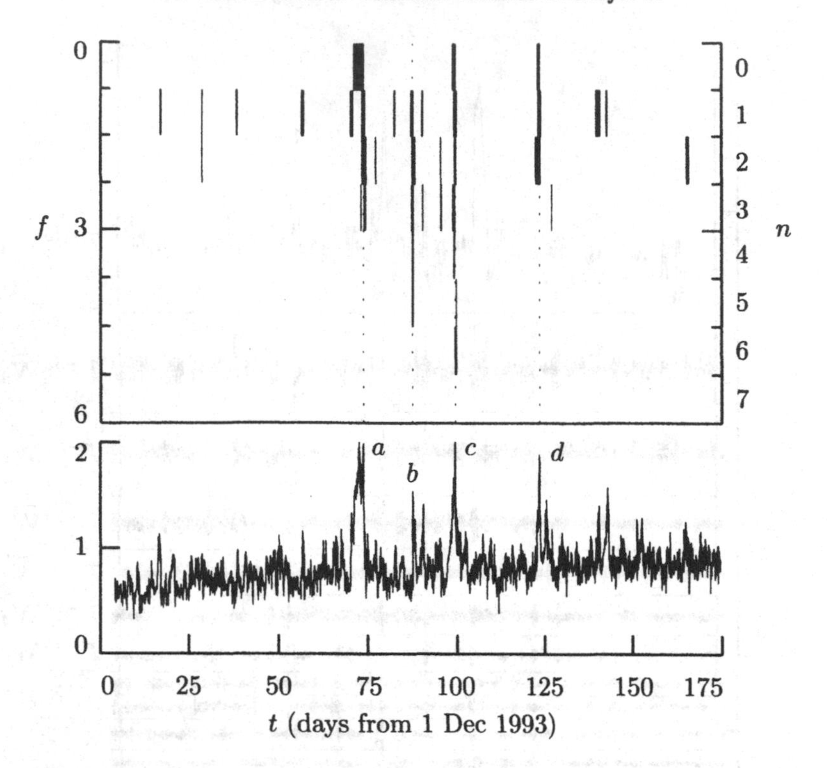

Figure 236. Southern hemisphere solar magnetic field magnitude data (lower plot) and corresponding modified time/frequency plot (upper). The data are measured in nanoteslas and were recorded by the Ulysses space craft from 21 hours Universal Time (UT) on day 338 of 1993 (4 December) until 12 hours UT on day 144 of 1994 (24 May). The frequency f is measured in cycles per day. The four shock wave structures are labeled a to d on the lower plot, while vertical dotted in the upper plot mark their locations as determined by Balogh *et al.* (1995).

approximately 1/2 cycles/day are dominated by known solar rotation effects and hence need not be finely resolved for our purposes here, whereas frequencies above this cutoff are adequately represented by basis vectors at levels $j \leq 4$. A level $j = 4$ transform yields 16 vectors $\widetilde{\mathbf{W}}_{4,n}$, $n = 0, \ldots, 15$. The nth such vector corresponds to a time-varying decomposition of the frequency interval $[\frac{n}{2^{j+1}\Delta t}, \frac{n+1}{2^{j+1}\Delta t}]$. For a MODWPT of level $j = 4$, the associated standardized bandwidth and time width are, respectively, $1/2^{j+1} = 1/32$ and $\lambda_j = 2^j = 16$, which convert into a physical bandwidth and time width of $1/(2^{j+1}\Delta t) = 3/4$ cycles/day and $\lambda_j \, \Delta t = 2/3$ days. Both of these values seem sensible for the purposes of this analysis.

The upper part of Figure 235 shows the MODWPT coefficient vectors $\widetilde{\mathbf{W}}_{4,n}$. Prior to plotting each vector, we circularly shift it so that it becomes approximately the output from a zero phase filter; i.e., we have plotted the tth element of $\mathcal{T}^{-|\nu_{4,n}|}\widetilde{\mathbf{W}}_{4,n}$ versus the time associated with X_t, where $\nu_{4,n}$ is the shift dictated by the phase properties of the LA(8) filters (see Equation (230)). The actual values of $|\nu_{4,n}|$ for $n = 0, \ldots, 6$ are displayed on the right-hand of Figure 235 (the values for $n > 7$ cannot all be listed due to space restrictions, so their determination is left to the reader as Exercise [6.8]). As in Figure 222, the thick vertical lines intersecting the beginning and end of each $\mathcal{T}^{-|\nu_{4,n}|}\widetilde{\mathbf{W}}_{4,n}$ delineate the MODWPT coefficients that are affected by circular boundary conditions. Placement of these lines is dictated by an argument exactly the same as the one we made for the MODWT. This led up to Equation (198b), which we can adapt here to tell us that the indices of the boundary coefficients in $\mathcal{T}^{-|\nu_{j,n}|}\widetilde{\mathbf{W}}_{j,n}$ are given by

$$t = 0, \ldots, L_j - 2 - |\nu_{j,n}| \text{ along with } t = N - |\nu_{j,n}|, \ldots, N - 1.$$

Dotted lines mark the location of the four shock waves as given in Balogh *et al.* (1995). It is interesting to compare the MODWPT coefficients in this figure with the DWPT coefficients plotted in Figure 222 – due to the downsampling inherent in the DWPT, it is much more difficult to see how the shock waves spread out over different frequency bands.

The upper part of Figure 236 is a modified time/frequency plot showing the position of the 100 largest squared time-advanced coefficients in $\mathcal{T}^{-|\nu_{4,0}|}\widetilde{\mathbf{W}}_{4,0}$, along with the 100 largest in $\mathcal{T}^{-|\nu_{4,1}|}\widetilde{\mathbf{W}}_{4,1}$ and the 100 largest in $\mathcal{T}^{-|\nu_{4,n}|}\widetilde{\mathbf{W}}_{4,n}$, $n = 2, \ldots, 15$, taken as one group. In this plot the left-hand y axis shows the frequency interval in physical units corresponding to each WP index n (these are indicated on the right-hand y axis). The 100 largest squared coefficients in $\mathcal{T}^{-|\nu_{4,n}|}\widetilde{\mathbf{W}}_{4,n}$, $n = 2, \ldots, 15$, all occur in the five lowest frequency bands (i.e., those indexed by $n = 2, \ldots, 6$), so we have limited the frequency range shown in Figure 236 to 0 to 6 cycles/day. This figure clearly shows the time localization of the shock wave features. The time shift adjustments are important here. A comparison of $\mathcal{T}^{-|\nu_{4,n}|}\widetilde{\mathbf{W}}_{4,n}$ with the original time series shows that time localized events in the MODWPT coefficients occur at the same physical time as in $\mathbf{X}$ and also can be readily traced across frequency bands. The results are much harder to interpret when the time advances are not used (see Exercise [6.10]). For each event the picked times coincide with the times at which the time/frequency decomposition is coherently broad-band (i.e., extends across many frequency bands). Since a sharply defined pulse has a broad-band Fourier decomposition, it is reasonable that the locations of events coincide with broad-band features.

Of the four marked events, the most interesting is event a. Figure 236 shows low frequency power (0 to 0.75 cycles per day) for about 2.5 days before the development of the broad-band characteristic corresponding to the picked time. One possible explanation is that the event at the picked time for a is one event (possibly a CIR), while the earlier low frequency energy derives from another event (possibly a CME), which arrived at Ulysses at nearly the same time. These events need not have occurred at the same longitude on the Sun, which would be most unlikely, since they would travel at different speeds towards Ulysses. Event b is obviously not recurrent, occurring roughly half-way through the solar cycle linking a and c.

The top plot of Figure 238 is a stacked plot of the details $\widetilde{\mathcal{D}}_{4,n}$ for frequency bands indexed by $n = 0, \ldots, 15$ (covering the entire frequency range from 0 to 12 cycles per day). While the details have been offset vertically to stack them, they all use the same relative scaling as is used in the bottom plot of the time series. Shock wave locations can easily be traced across several frequency bands. Because these details were constructed using zero phase filters, this plot helps to confirm the timings of the events in Figures 235 and 236 (the striking similarity between this plot and Figure 235 indicates that the LA(8) filters are quite good approximations to zero phase filters here). For example, the detail series $\widetilde{\mathcal{D}}_{4,0}$ shows the presence of low frequency energy arriving before the more characteristic broad-band features in the higher frequency

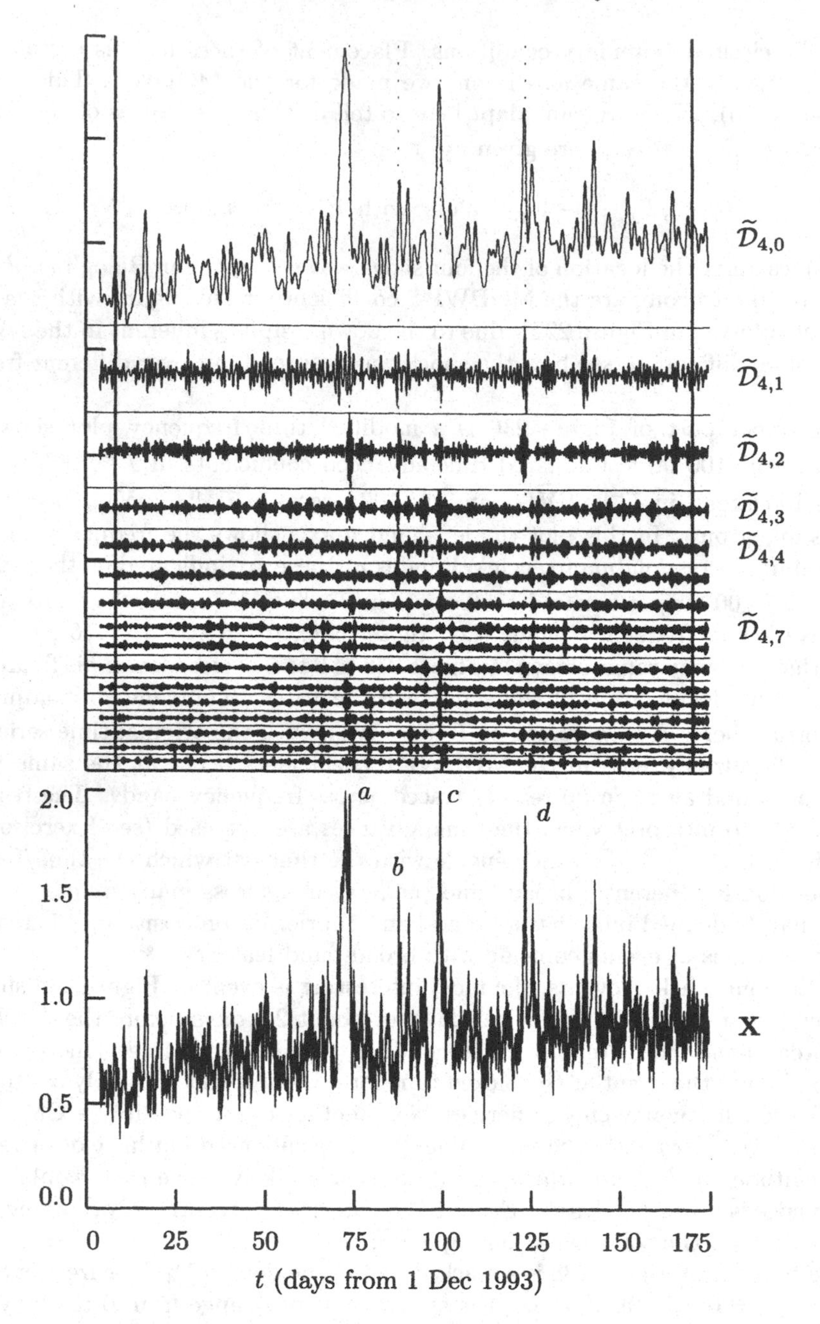

Figure 238. Solar magnetic field magnitude data (lower plot) and corresponding plot of the MODWPT details $\widetilde{\mathcal{D}}_{4,n}$, $n = 0, \ldots, 15$ (upper). The nth detail corresponds to the frequency band from $0.75n$ cycles per hour to $0.75(n+1)$ cycles per hour.

details $\widetilde{\mathcal{D}}_{4,n}$, $n > 0$. (For more details about this time series, see Walden and Contreras Cristan, 1998a.)

6.8 Matching Pursuit

The original idea behind matching pursuit was to approximate a function using linear combinations of a small number of more primitive functions (Mallat and Zhang, 1993; Davis, Mallat and Zhang, 1994). If we assume that these primitive functions (or waveforms) can each be associated with a particular time and frequency, we can obtain an adaptive time/frequency decomposition of a function by approximating it using a sum of waveforms whose localizations in time and frequency match those of pertinent structures in the function; likewise, we can obtain a time/scale decomposition by using waveforms that can individually be associated with a particular time and scale. The waveforms used for either purpose are drawn from a large and redundant collection called a *dictionary*. Here we consider this approach in the context of analyzing a time series observed only at discrete times.

As usual, let $\mathbf{X}$ be an N dimensional vector containing the elements of the time series $\{X_t : t = 0, \ldots, N-1\}$. We wish to expand $\mathbf{X}$ into linear combinations of vectors contained in a 'dictionary' $\mathbb{D}$:

$$\mathbb{D} \equiv \{\mathbf{d}_\gamma : \gamma \in \Gamma\},$$

where $\mathbf{d}_\gamma = \left[d_{\gamma,0}, d_{\gamma,1}, \ldots, d_{\gamma,N-1}\right]^T$ is called a 'dictionary element.' The subscript γ is in general a vector of parameters connecting $\mathbf{d}_\gamma$ to time and frequency (or time and scale), while Γ denotes the set of all possible values for these parameters and is assumed for simplicity to be a finite set (for example, if $\mathbf{d}_\gamma$ were one of the basis vectors for a DWPT, then γ would be a three dimensional vector $[j, n, t]^T$, the first two elements of which would map to frequency, while the last element would map to time). Each dictionary element is normalized such that

$$\|\mathbf{d}_\gamma\|^2 = \sum_{l=0}^{N-1} d_{\gamma,l}^2 = 1.$$

It is further assumed that some subset of the elements of $\mathbb{D}$ can be used to form a basis of the N dimensional vector space $\mathbb{R}^N$, of which $\mathbf{X}$ is assumed to be an element. The smallest possible dictionary $\mathbb{D}$ is thus a basis for $\mathbb{R}^N$, but dictionaries are intended to be highly redundant: the redundancy allows increased freedom in constructing expansions well matched to the time/frequency or time/scale structure of $\mathbf{X}$.

A matching pursuit is an algorithm that successively approximates $\mathbf{X}$ with orthogonal projections onto elements of $\mathbb{D}$. For any $\mathbf{d}_{\gamma_0} \in \mathbb{D}$, we project $\mathbf{X}$ onto this vector to form the approximation

$$\mathbf{X}^{(1)} \equiv \langle \mathbf{X}, \mathbf{d}_{\gamma_0} \rangle \mathbf{d}_{\gamma_0}$$

(as usual, $\langle \cdot, \cdot \rangle$ denotes inner product). We then construct the residual vector

$$\mathbf{R}^{(1)} \equiv \mathbf{X} - \mathbf{X}^{(1)} \text{ so that } \mathbf{X} = \mathbf{X}^{(1)} + \mathbf{R}^{(1)} = \langle \mathbf{X}, \mathbf{d}_{\gamma_0} \rangle \mathbf{d}_{\gamma_0} + \mathbf{R}^{(1)}.$$

The following exercise indicates that in fact this yields an orthogonal decomposition.

▷ **Exercise [240]:** Show that $\langle \mathbf{X}^{(1)}, \mathbf{R}^{(1)} \rangle = 0$ and hence that

$$\|\mathbf{X}\|^2 = \|\mathbf{X}^{(1)}\|^2 + \|\mathbf{R}^{(1)}\|^2 = \left|\langle \mathbf{X}, \mathbf{d}_{\gamma_0} \rangle\right|^2 + \|\mathbf{R}^{(1)}\|^2. \quad \triangleleft$$

In order to minimize $\|\mathbf{R}^{(1)}\|^2$, we choose $\gamma_0 \in \Gamma$ such that

$$\left|\langle \mathbf{X}, \mathbf{d}_{\gamma_0} \rangle\right| = \max_{\gamma \in \Gamma} \left|\langle \mathbf{X}, \mathbf{d}_{\gamma} \rangle\right|.$$

The next step is to decompose $\mathbf{R}^{(1)}$ by means of projecting it onto the vector of $\mathbb{D}$ that best matches $\mathbf{R}^{(1)}$:

$$\mathbf{R}^{(1)} = \langle \mathbf{R}^{(1)}, \mathbf{d}_{\gamma_1} \rangle \mathbf{d}_{\gamma_1} + \mathbf{R}^{(2)},$$

where $\left|\langle \mathbf{R}^{(1)}, \mathbf{d}_{\gamma_1} \rangle\right| = \max_{\gamma \in \Gamma} \left|\langle \mathbf{R}^{(1)}, \mathbf{d}_{\gamma} \rangle\right|$. This leads to the decomposition $\mathbf{X} = \mathbf{X}^{(2)} + \mathbf{R}^{(2)}$, where (letting $\mathbf{R}^{(0)} \equiv \mathbf{X}$)

$$\mathbf{X}^{(2)} \equiv \mathbf{X}^{(1)} + \langle \mathbf{R}^{(1)}, \mathbf{d}_{\gamma_1} \rangle \mathbf{d}_{\gamma_1} = \sum_{k=0}^{1} \langle \mathbf{R}^{(k)}, \mathbf{d}_{\gamma_k} \rangle \mathbf{d}_{\gamma_k}.$$

We repeat this iteratively to get after m steps

$$\mathbf{X} = \mathbf{X}^{(m)} + \mathbf{R}^{(m)}, \quad \text{where } \mathbf{X}^{(m)} \equiv \sum_{n=0}^{m-1} \langle \mathbf{R}^{(n)}, \mathbf{d}_{\gamma_n} \rangle \mathbf{d}_{\gamma_n}. \tag{240}$$

Energy is conserved because

$$\|\mathbf{X}\|^2 = \sum_{n=0}^{m-1} \|\langle \mathbf{R}^{(n)}, \mathbf{d}_{\gamma_n} \rangle \mathbf{d}_{\gamma_n}\|^2 + \|\mathbf{R}^{(m)}\|^2 = \sum_{n=0}^{m-1} |\langle \mathbf{R}^{(n)}, \mathbf{d}_{\gamma_n} \rangle|^2 + \|\mathbf{R}^{(m)}\|^2.$$

As we increase m, the residual energy $\|\mathbf{R}^{(m)}\|^2$ cannot increase. Conditions under which $\|\mathbf{R}^{(m)}\|^2$ decreases to zero as m increases are discussed in Mallat and Zhang (1993) and Mallat (1998).

• *DWT and other orthonormal dictionaries*

As a first example of a dictionary $\mathbb{D}$, let us consider a level J_0 partial DWT, for which we must assume that $N = q2^{J_0}$, where J_0 and q are positive integers. As before, let $\mathcal{W}$ represent the $N \times N$ orthonormal matrix defining this DWT, and let $\mathbf{W} \equiv \mathcal{W}\mathbf{X}$ be the DWT coefficients. If we let $\mathcal{W}_{j\bullet}$ be a column vector containing the elements in the jth row of $\mathcal{W}$, we can write

$$\mathbf{X} = \sum_{j=0}^{N-1} \langle \mathbf{X}, \mathcal{W}_{j\bullet} \rangle \mathcal{W}_{j\bullet} = \sum_{j=0}^{N-1} W_j \mathcal{W}_{j\bullet},$$

where W_j is the jth element of $\mathbf{W}$. Because $\mathcal{W}$ is an orthonormal transform, we must have $\|\mathcal{W}_{j\bullet}\|^2 = 1$ for all j. The DWT dictionary $\mathbb{D}^{(\mathrm{dwt})}$ can thus be taken to be the N vectors $\mathcal{W}_{j\bullet}$, $j = 0, \ldots, N-1$. The vectors in $\mathbb{D}^{(\mathrm{dwt})}$ form a basis for $\mathbb{R}^N$, and there is no redundancy in $\mathbb{D}^{(\mathrm{dwt})}$. At the first stage, the matching pursuit algorithm just picks out the vector from $\mathbb{D}^{(\text{dwt})}$ corresponding to the largest $|W_j|$; at stage m, it picks out the vector corresponding to the mth largest $|W_j|$. This same pattern holds for any orthonormal transform, so – for such transforms – the matching pursuit algorithm just consists of adding in vectors according to the magnitudes of the transform coefficients. It can be argued that, if we run the matching pursuit algorithm all the way out to stage N, we must have $\|\mathbf{R}^{(N)}\|^2 = 0$ (this is part of Exercise [6.11]).

- *Wavelet packet table dictionary*

As an example of a redundant dictionary, we can consider one that includes all the basis vectors used to create a WP table out to level J_0 based on a particular wavelet filter $\{h_l\}$. Note that this collection of wavelet-based vectors is exactly the same as those under consideration in the best basis algorithm. For example, when $J_0 = 2$, this dictionary would consist of the $2N$ column vectors contained in $\mathcal{B}_1^T$, $\mathcal{A}_1^T$, $\mathcal{B}_1^T\mathcal{A}_2^T$, $\mathcal{B}_1^T\mathcal{B}_2^T$, $\mathcal{A}_1^T\mathcal{B}_2^T$ and $\mathcal{A}_1^T\mathcal{A}_2^T$.

While the matching pursuit and best basis algorithms both involve the same basis vectors, the end results can be quite different. Using the $J_0 = 2$ example again, the best basis algorithm will end up describing $\mathbf{X}$ in terms of one of the following five orthonormal transforms: $\mathcal{O}^{(1)} \equiv I_N$,

$$\mathcal{O}^{(2)} \equiv \begin{bmatrix} \mathcal{B}_1 \\ \mathcal{A}_1 \end{bmatrix}, \mathcal{O}^{(3)} \equiv \begin{bmatrix} \mathcal{B}_1 \\ \mathcal{B}_2\mathcal{A}_1 \\ \mathcal{A}_2\mathcal{A}_1 \end{bmatrix}, \mathcal{O}^{(4)} \equiv \begin{bmatrix} \mathcal{A}_2\mathcal{B}_1 \\ \mathcal{B}_2\mathcal{B}_1 \\ \mathcal{A}_1 \end{bmatrix} \text{ or } \mathcal{O}^{(5)} \equiv \begin{bmatrix} \mathcal{A}_2\mathcal{B}_1 \\ \mathcal{B}_2\mathcal{B}_1 \\ \mathcal{B}_2\mathcal{A}_1 \\ \mathcal{A}_2\mathcal{A}_1 \end{bmatrix}.$$

If the selected transform is, say, $\mathcal{O}^{(5)}$, we could then approximate $\mathbf{X}$ by making use of the m coefficients in $\mathbf{O}^{(5)} \equiv \mathcal{O}^{(5)}\mathbf{X}$ with largest magnitudes (this would be equivalent to applying m steps of the matching pursuit algorithm with $\mathcal{O}^{(5)}$ as the dictionary). The approximation to $\mathbf{X}$ would thus be expressed entirely in terms of N vectors in the transpose of $\mathcal{O}^{(5)}$. By contrast, if we apply the matching pursuit algorithm using the same WP table, we could end up selecting some vectors from the transpose of $\mathcal{O}^{(5)}$ and some from the transpose of $\mathcal{O}^{(2)}$. The fact that we now have twice as many approximating vectors to choose from suggests that a level m matching pursuit approximation to $\mathbf{X}$ should have smaller residuals (as measured by $\|\mathbf{R}^{(m)}\|^2$) than a similar approximation extracted from the best basis transform. This point is discussed further in Section 10.4 of Bruce and Gao (1996a), which gives an example where matching pursuit does indeed outperform the best basis algorithm.

- *Combined WP table and ODFT dictionary*

Here we augment the dictionary in the previous example by including basis vectors from the orthonormal discrete Fourier transform (ODFT), which we discussed in Section 3.4. Inclusion of these vectors opens up the possibility of succinctly representing time series that have both broad- and narrow-band time/frequency characteristics. Because the ODFT is complex-valued, we need to modify the matching pursuit algorithm slightly if we want the approximations $\mathbf{X}^{(m)}$ and hence the residuals $\mathbf{R}^{(m)}$ to be real-valued. Accordingly, suppose that at step $n+1$ the vector $\mathbf{d}_\gamma \in \mathbb{D}$ maximizing $|\langle\mathbf{R}^{(n)}, \mathbf{d}_\gamma\rangle|$ is the ODFT vector $\mathcal{F}_{k\bullet}$ and that this vector is associated with a frequency $f_k \equiv k/N$ satisfying $0 < f_k < 1/2$. If we let $F_k \equiv \langle\mathbf{R}^{(n)}, \mathcal{F}_{k\bullet}\rangle$ and if we recall the constraint $F_{N-k}^* = F_k$, we see that, if the choice $\mathbf{d}_\gamma = \mathcal{F}_{k\bullet}$ maximizes $|\langle\mathbf{R}^{(n)}, \mathbf{d}_\gamma\rangle|$, then so does $\mathbf{d}_\gamma = \mathcal{F}_{N-k\bullet}$. This suggests that we should include both basis vectors so that the nth term in Equation (240) for the approximation $\mathbf{X}^{(m)}$ is given by

$$\langle \mathbf{R}^{(n)}, \mathcal{F}_{k\bullet}\rangle \mathcal{F}_{k\bullet} + \langle \mathbf{R}^{(n)}, \mathcal{F}_{N-k\bullet}\rangle \mathcal{F}_{N-k\bullet},$$

which is necessarily real-valued (see Equation (50) and Exercise [50a]). (Note that there is no need for this modification if the maximizing vector is either $\mathcal{F}_{0\bullet}$ or $\mathcal{F}_{\frac{N}{2}\bullet}$ since both of these are real-valued.)

- *MODWT dictionary*

While $\mathbf{X}$ can be represented exactly in terms of the (transposed) rows of $\mathcal{W}$ weighted by the wavelet coefficients $\mathbf{W} = \mathcal{W}\mathbf{X}$, we have seen in Section 5.1 that these coefficients can depend critically on the assumed starting point for the time series. In addition, we would like to be able to treat easily a series whose length is not necessarily a multiple of a power of two, so we here consider using a dictionary $\mathbb{D}^{(\text{modwt})}$ based upon the MODWT discussed in Chapter 5. Accordingly, after selection of a wavelet filter and a level J_0, we define $\mathbb{D}^{(\text{modwt})}$ to be the set of $(J_0 + 1)N$ vectors given by the transposes of the rows of $\widetilde{\mathcal{W}}_j$, $j = 1, \ldots, J_0$, and $\widetilde{\mathcal{V}}_{J_0}$ with an appropriate renormalization, determined as follows. Equation (171c) indicates that each row vector in $\widetilde{\mathcal{W}}_j$ consists of elements picked from the MODWT wavelet filter $\{\tilde{h}_{j,l}\}$ after it has been periodized to length N. Now the width of this filter is given by $L_j \equiv (2^j - 1)(L - 1) + 1$. If $L_j \leq N$, the periodized filter $\{\tilde{h}^{\circ}_{j,l}\}$ has the same nonzero elements as $\{\tilde{h}_{j,l}\}$, so we can use the fact that

$$2^j \sum_{j=0}^{L_j - 1} \tilde{h}^2_{j,l} = 1$$

to renormalize the rows of $\widetilde{\mathcal{W}}_j$ to have unit norm – in fact, since $\tilde{h}_{j,l} \equiv h_{j,l}/2^{j/2}$, this renormalization implicitly leads us back to the DWT filter. If we pick J_0 such that $L_{J_0} \leq N$, i.e.,

$$J_0 < \log_2\left(\frac{N}{L-1} + 1\right) \text{ or, equivalently, } J_0 \leq \left\lfloor \log_2\left(\frac{N-1}{L-1} + 1\right)\right\rfloor, \tag{242}$$

then it is a trivial task to renormalize the rows of $\widetilde{\mathcal{W}}_j$ and $\widetilde{\mathcal{V}}_{J_0}$. If we pick J_0 such that $L_j > N$ for some $j \leq J_0$, we can still multiply by $2^{j/2}$ to renormalize the rows of $\widetilde{\mathcal{W}}_j$ if N is a sample size amenable to a level j DWT (this follows because, after renormalization and a possible circular shift, each row in the MODWT matrix $\widetilde{\mathcal{W}}_j$ is the same as some row in the DWT matrix $\mathcal{W}_j$, for which we have $\mathcal{W}_j \mathcal{W}_j^T = I_{N_j}$). If N is not such, then our only recourse is to compute explicitly

$$C_j \equiv \sum_{j=0}^{N-1} [\tilde{h}^{\circ}_{j,l}]^2$$

and to multiple the rows of $\widetilde{\mathcal{W}}_j$ by $1/\sqrt{C_j}$ (similar remarks hold for $\widetilde{\mathcal{V}}_{J_0}$ when $L_{J_0} > N$).

6.9 Example: Subtidal Sea Levels

Here we illustrate the ideas behind matching pursuit by using the dictionary $\mathbb{D}^{(\text{modwt})}$ to analyze the subtidal sea level fluctuations discussed in Section 5.8. This time series (plotted at the bottom of Figure 246) has $N = 8746$ observations that are spaced $\Delta t = 1/2$ day apart and that cover very close to 12 complete years. As discussed in Chapters 4 and 5, the DWT and MODWT assume a circular time series so that, for example, the first observation X_0 is regarded as being preceded by the last observation X_{N-1}. Since subtidal sea level fluctuations have a strong annual component, they can be regarded – at least to some degree – as a circular series if the number of samples covers an integer number of years, which is approximately true here. The sample size is a multiple of two but not four, so only a partial DWT of level $J_0 = 1$ is possible with the standard formulation of the DWT. The MODWT requires no such restriction. In addition, since we could have just as easily looked at data beginning and ending in, say, June rather than January, we would like our analysis to be relatively insensitive to the assumed starting point. Given these considerations and the fact that subtidal sea level fluctuations are influenced by time-dependent phenomena operating over a range of different scales, the MODWT is a suitable framework for the structural decomposition of this series.

As in Section 5.8, we use the LA(8) wavelet filter to carry out our analysis and hence $L = 8$. We set J_0 to be the largest integer satisfying the inequality of Equation (242), i.e.,

$$J_0 = \left\lfloor \log_2 \left(\frac{N-1}{L-1} + 1 \right) \right\rfloor = 10.$$

This choice implies that the nonzero elements of the level j MODWT dictionary vectors are picked from the DWT filters $\{h_{j,l}\}$, $j = 1, \ldots, J_0$, or $\{g_{J_0,l}\}$ (these are what we obtain when we renormalize the MODWT filters to have unit energy). The dictionary thus contains $(J_0 + 1)N = 96,206$ renormalized rows from the $\widetilde{\mathcal{W}}_j$ and $\widetilde{\mathcal{V}}_{J_0}$. As usual, we can associate the rows of $\widetilde{\mathcal{W}}_j$ with changes in averages over a physical scale of $\tau_j \,\Delta t \equiv 2^{j-1}\,\Delta t$ and rows of $\widetilde{\mathcal{V}}_{J_0}$ with averages over a physical scale of $\lambda_{J_0}\,\Delta t \equiv 2^{J_0}\,\Delta t = 512$ days.

The first twenty basis vectors $\mathbf{d}_{\gamma_n}$ selected by the matching pursuit algorithm are shown in Figures 244 and 245. The heights of the vectors in the plots have been adjusted to fill up the available range – they have not been weighted by $\langle \mathbf{R}^{(n)}, \mathbf{d}_{\gamma_n} \rangle$ as they would need to be to form $\mathbf{X}^{(m)}$ in Equation (240); however, each vector has been multiplied by $+1$ or -1 according to the sign of this inner product to make it easier to see what feature in $\mathbf{X}$ is being extracted. The first ten vectors are all associated with large scales ranging from changes over $\tau_8\,\Delta t = 64$ days to averages over $\lambda_{10}\,\Delta t = 512$ days. The first chosen is a large scale average that draws our attention to a lengthy overall increase in the fluctuations spanning 1982–3. Seven of the first twenty selected vectors are associated with changes on physical scale $\tau_9\,\Delta t = 128$ days, which are needed to account for seasonal changes. Two of these ($n = 3$ and $n = 8$) are inverted with respect to the other five. It is problematic to interpret the $n = 8$ inversion because this vector explicitly makes use of the circularity assumption, but the $n = 3$ inversion points out a more gradual dip in the fluctuations in the Spring of 1984 as compared to rapid dips in the Springs of 1981, 1983, 1985, 1987 and 1988 ($n =$ 2, 1, 18, 6 and 11, respectively). This anomaly is also evident upon careful study of the $\widetilde{\mathcal{S}}_7$ component of the MODWT MRA in Figure 186. As discussed in

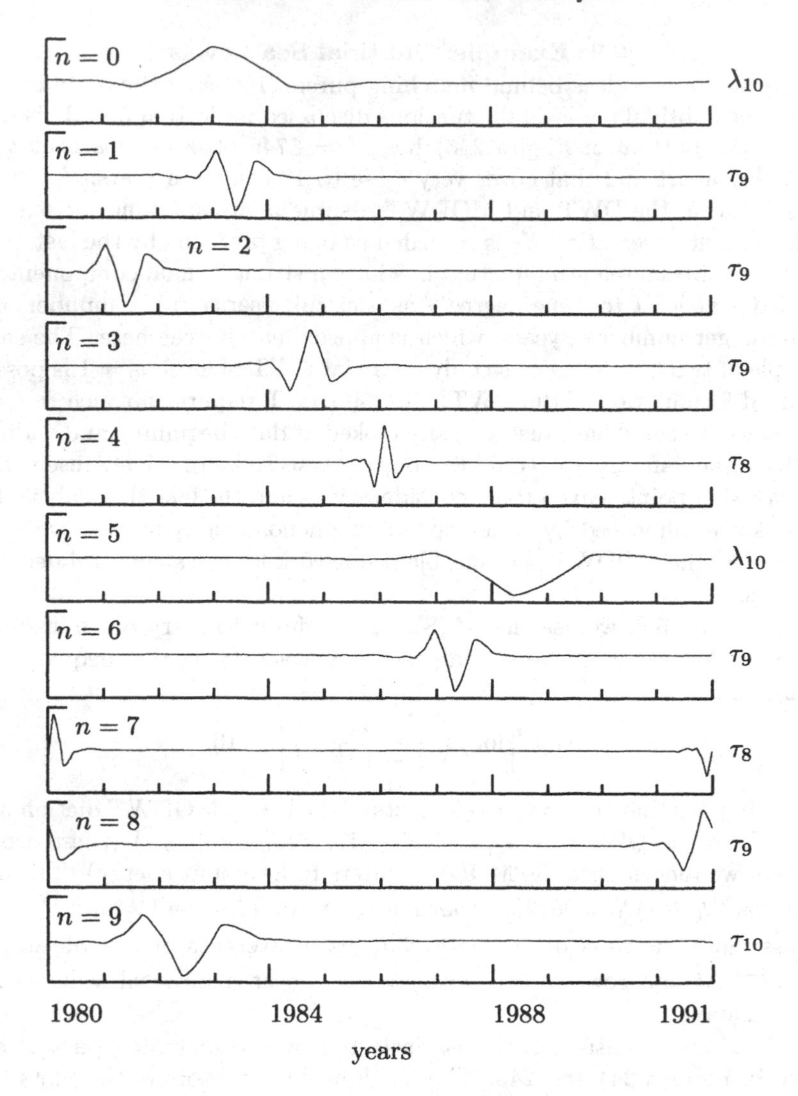

Figure 244. First ten MODWT vectors selected using the matching pursuit algorithm applied to the subtidal sea level fluctuations. The dictionary is $\mathbb{D}^{(\text{modwt})}$ constructed using the LA(8) wavelet filter. Each vector has been multiplied by +1 or −1 according to the sign of its inner product with $\mathbf{R}^{(n)}$, thus enabling visual correlation with the original series. The standardized scale for each vector is written to the right of its plot (τ_j for a vector from $\widetilde{\mathcal{W}}_j$ and λ_{10} for one from $\widetilde{\mathcal{V}}_{10}$).

Percival and Mofjeld (1997), these dips in sea level in March/April are due to the spring transition in the atmosphere over the eastern North Pacific Ocean, following high winter sea levels due to low atmospheric pressure and northward winds. Note also that, in the second ten selected vectors (Figure 245), there are six that are associated with prominent changes on scales of $\tau_5\,\Delta t = 8$ and $\tau_6\,\Delta t = 16$ days. These are all located in Winter or Spring and can be attributed to storm systems at that time of the year.

The top plot in Figure 246 shows the approximation $\mathbf{X}^{(20)}$ to $\mathbf{X}$ based upon the

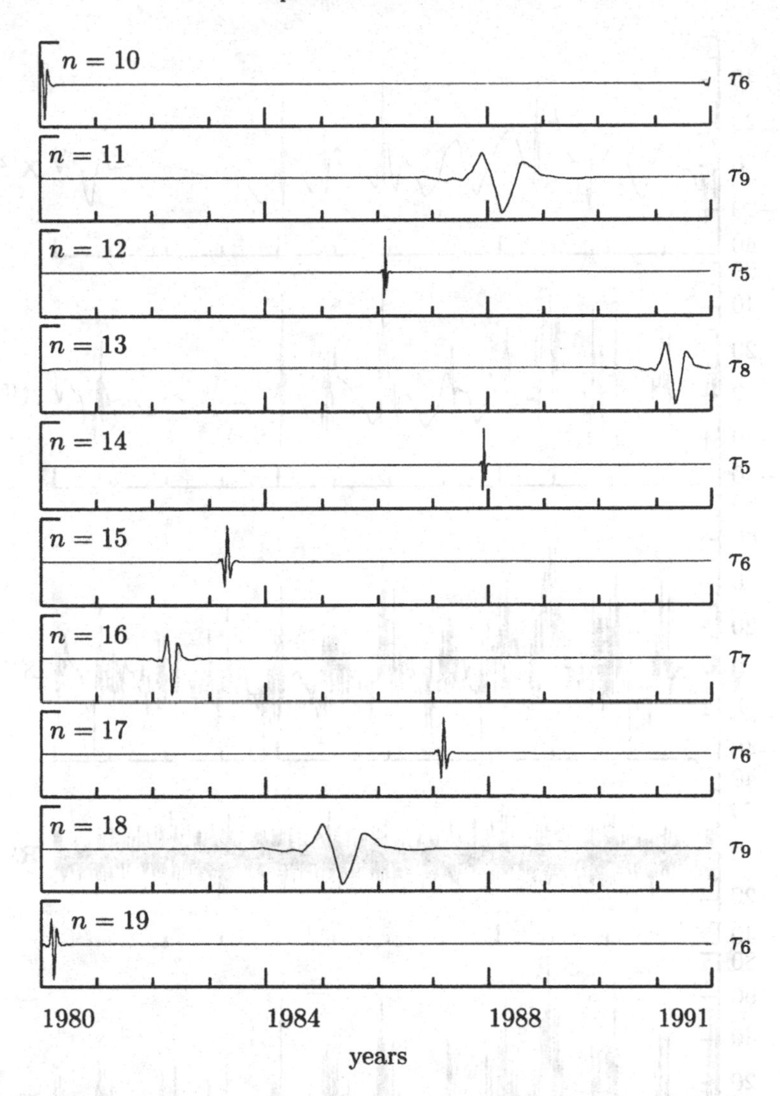

Figure 245. MODWT vectors selected eleventh to twentieth using the matching pursuit algorithm (see Figure 244 for details).

first twenty selected vectors ($\mathbf{X}$ itself is shown in the bottom plot). In addition to the characteristic March/April dips and storm-related features, this approximation points out that the subtidal fluctuations for 1989–90 were relatively quiescent compared to the other years in this twelve year stretch. If we move up to an approximation $\mathbf{X}^{(50)}$ based on fifty vectors, we see some structure appearing for that two year period. The final two plots in Figure 246 show the $m = 200$ order approximation $\mathbf{X}^{(200)}$ and the associated residuals $\mathbf{R}^{(200)}$. Note that these residuals are much more homogeneous than the original series (however, there is still a pronounced seasonally dependent variability).

This step-by-step matching pursuit approach clearly allows us to appreciate how the series is constructed and in particular emphasizes differences in construction for

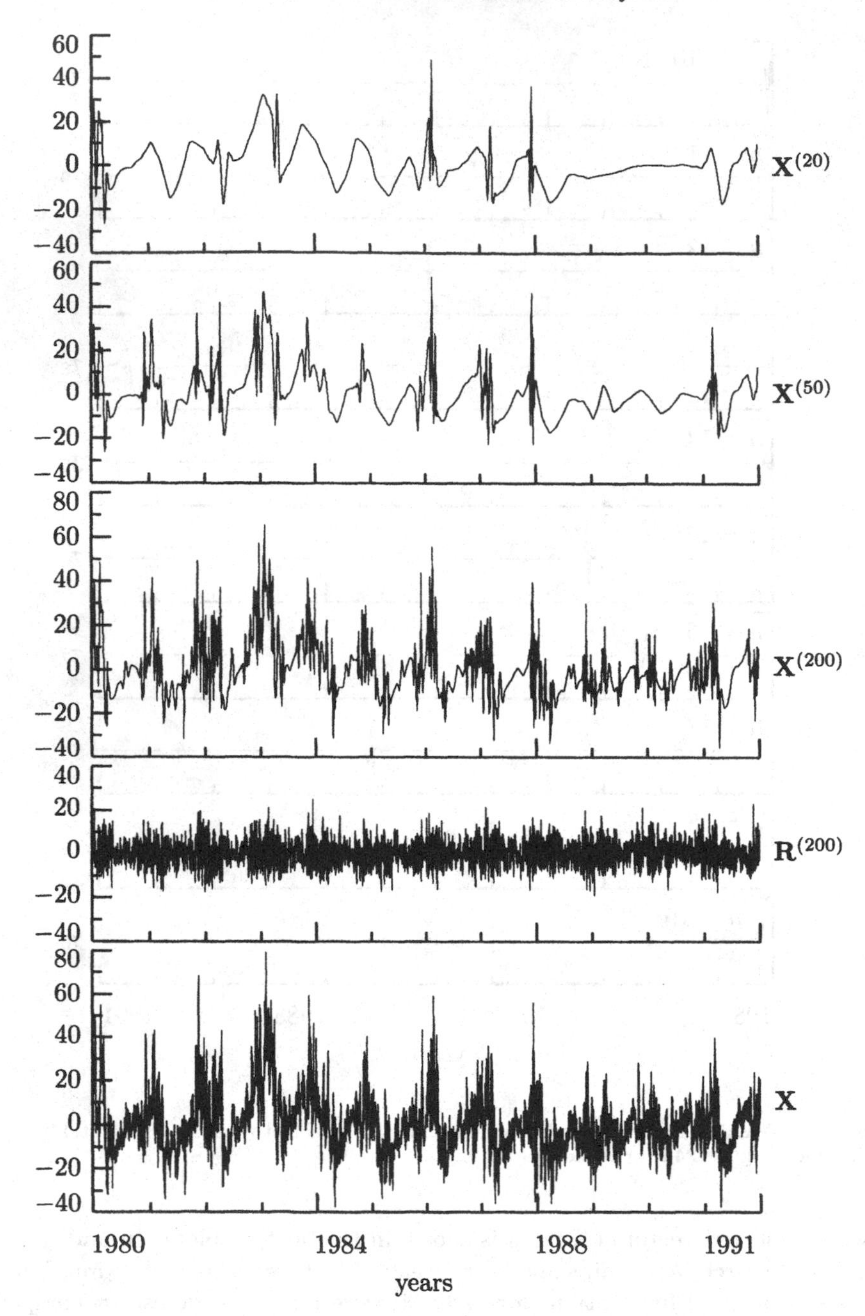

Figure 246. Matching pursuit approximations $\mathbf{X}^{(m)}$ of the subtidal sea level fluctuations $\mathbf{X}$ using $m = 20$, 50 and 200 vectors from a MODWT dictionary. The residuals $\mathbf{R}^{(200)}$ corresponding to $\mathbf{X}^{(200)}$ are shown above the plot of $\mathbf{X}$.

various parts of the series. It conveys information distinct from that gleaned using a MODWT multiresolution analysis (see Section 5.8).

Comments and Extensions to Section 6.9

[1] Walden and Contreras Cristan (1998b) present a complementary matching pursuit analysis of a shorter (5 year) segment of subtidal sea level fluctuations. The values for this shorter segment were measured with respect to a reference level that is significantly larger than the one for the series we studied here. This change in level results in a sample mean that is large compared to fluctuations in the series (the sample mean is very small for $\mathbf{X}$ in Figure 246). As a result, they augment the dictionary $\mathbb{D}^{(\text{modwt})}$ by including a single constant vector, namely, one whose elements are all equal to $1/\sqrt{N}$. If we let $\mathbf{1}$ be a vector whose elements are all unity, the constant vector is then $\frac{1}{\sqrt{N}}\mathbf{1}$, so we have $\langle \mathbf{X}, \frac{1}{\sqrt{N}}\mathbf{1}\rangle \frac{1}{\sqrt{N}}\mathbf{1} = \overline{X}\mathbf{1}$, where $\overline{X}$ is the sample mean of $\mathbf{X}$. Note that, because this vector is invariant under circular shifts, using it along with $\mathbb{D}^{(\text{modwt})}$ yields an overall dictionary with the key property that, if a vector is in this dictionary, then so are all of its possible circular shifts. Inclusion of the constant vector expands the overdetermined basis in a physically sensible way while maintaining this key property. With this augmented dictionary, Walden and Contreras Cristan (1998b) find the first vector chosen by the matching pursuit algorithm for their shorter series to be $\frac{1}{\sqrt{N}}\mathbf{1}$; i.e., the first approximation is just the sample mean, and the corresponding residuals are deviations from the sample mean.

[2] Because of the prominent annual variations in the subtidal sea level fluctuations, it might seem that we would obtain a better matching pursuit approximation to these data if we considered a dictionary consisting of $\mathbb{D}^{(\text{modwt})}$ combined with vectors from the ODFT. An analysis using this augmented dictionary shows that, while the first vector picked out is the same as the one shown in Figure 244, the second vector is from the ODFT and has an associated period quite close to one cycle per year; however, if we continue running the matching pursuit algorithm out to step $m = 200$, only three of the next 198 vectors are from the ODFT – all of the rest are from $\mathbb{D}^{(\text{modwt})}$ (the other three ODFT vectors are picked at steps $m = 65$, 84 and 192 and are associated with, respectively, zero frequency, 26.7 cycles/year and 36.2 cycles/year – these latter two might be attributable to weather patterns on a scale of one to two weeks). The approximations $\mathbf{X}^{(m)}$ for $m = 20$, 50 and 200 are shown in Figure 248a and are seen to be fairly similar to the ones in Figure 246 – the chief difference is that the annual variations are more fully developed in $\mathbf{X}^{(20)}$ in Figure 248a. In addition, Figure 248b shows the normalized residual sum of squares $\|\mathbf{R}^{(m)}\|^2/\|\mathbf{X}\|^2$ plotted versus m using both $\mathbb{D}^{(\text{modwt})}$ (thick curve) and this dictionary augmented with the ODFT vectors (thin curve). While we gain a substantial reduction when the ODFT vector is picked at the second step, the $\mathbb{D}^{(\text{modwt})}$ approximations close the gap around step 50. Thus the bulk of the features in this series is more suitably described in terms of localized time/scale variations than global frequency variations.

6.10 Summary

Let $\mathbf{X}$ be an N dimensional vector containing a real-valued time series whose sample size N is an integer multiple of 2^{J_0} for some integer J_0. For $0 \le j \le J_0$, the level j DWPT of $\mathbf{X}$ is an orthonormal transform yielding an N dimensional vector of

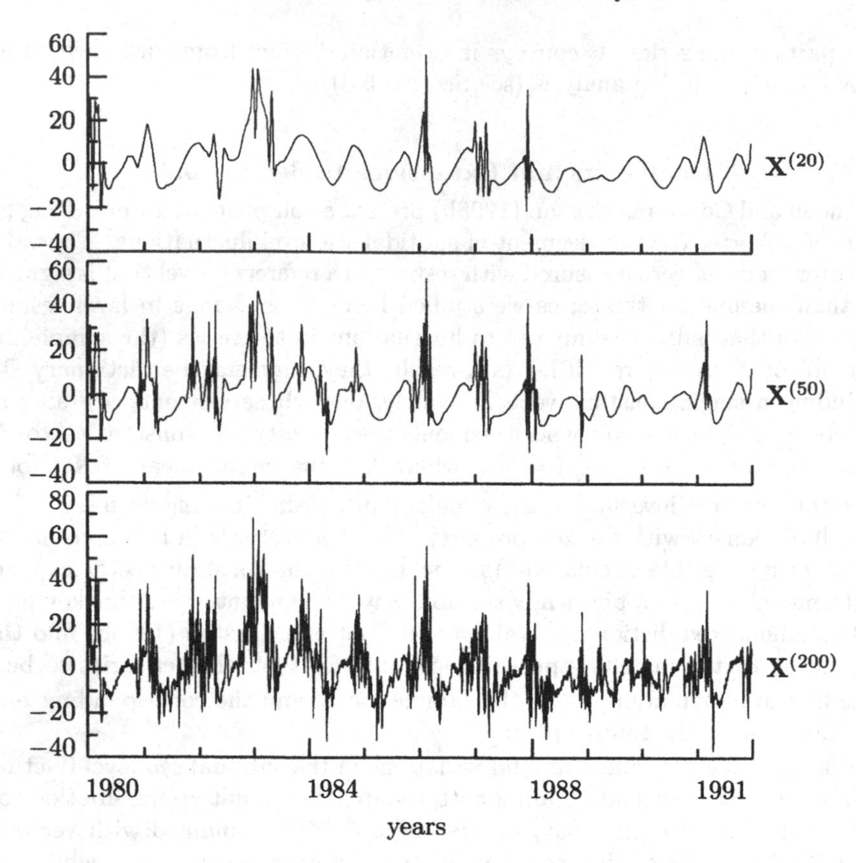

Figure 248a. Matching pursuit approximations $\mathbf{X}^{(m)}$ of the subtidal sea level fluctuations $\mathbf{X}$ using $m = 20$, 50 and 200 vectors, but now from a dictionary composed of both MODWT and ODFT vectors.

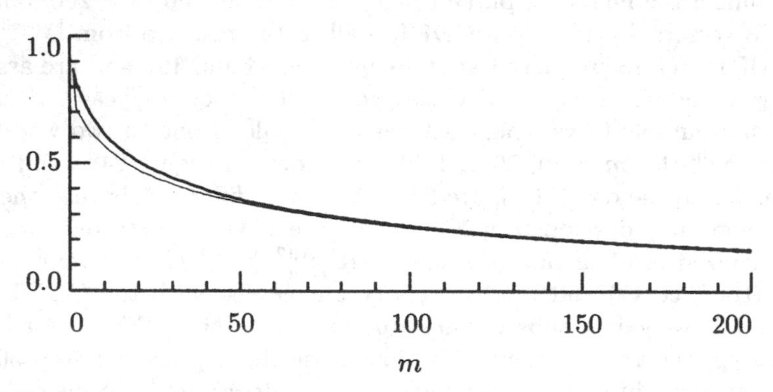

Figure 248b. Normalized residual sum of squares $\|\mathbf{R}^{(m)}\|^2/\|\mathbf{X}\|^2$ versus the number of terms m in the matching pursuit approximation using the MODWT dictionary $\mathbb{D}^{(\text{modwt})}$ (thick curve) and this dictionary combined with vectors from the ODFT (thin curve).

coefficients that can be partitioned as

$$\begin{bmatrix} \mathbf{W}_{j,2^j-1} \\ \mathbf{W}_{j,2^j-2} \\ \vdots \\ \mathbf{W}_{j,1} \\ \mathbf{W}_{j,0} \end{bmatrix},$$

where each $\mathbf{W}_{j,n}$ has dimension $N_j \equiv N/2^j$ and is nominally associated with the frequency interval $\mathcal{I}_{j,n} \equiv [\frac{n}{2^{j+1}}, \frac{n+1}{2^{j+1}}]$ (together these 2^j vectors divide up the standard frequency interval $[0, \frac{1}{2}]$ into 2^j intervals of equal width, so the bandwidth associated with each jth level DWPT coefficient is $1/2^{j+1}$). In practice a level j DWPT is formed by filtering a level $j-1$ DWPT using the wavelet and scaling filters $\{h_l\}$ and $\{g_l\}$ defined in Sections 4.2 and 4.3 (we initiate this recursive scheme by defining the DWPT for level $j = 0$ as the time series itself, i.e., $\mathbf{W}_{0,0} \equiv \mathbf{X}$). If we let $W_{j,n,t}$ be the tth element of $\mathbf{W}_{j,n}$, then we can write

$$W_{j,n,t} = \sum_{l=0}^{L-1} u_{n,l} W_{j-1,\lfloor \frac{n}{2} \rfloor, 2t+1-l \bmod N_{j-1}}, \quad t = 0, \ldots, N_j - 1.$$

where

$$u_{n,l} \equiv \begin{cases} g_l, & \text{if } n \bmod 4 = 0 \text{ or } 3; \\ h_l, & \text{if } n \bmod 4 = 1 \text{ or } 2, \end{cases}$$

and $\lfloor \frac{n}{2} \rfloor$ is $\frac{n}{2}$ if n is even – otherwise it is $\frac{n-1}{2}$ (the above are Equations (214a) and (214b)). Note that, in going from the level $j-1$ transform to the one of level j, each 'parent' vector $\mathbf{W}_{j-1,n'}$ gets filtered twice (once with the wavelet filter and once with the scaling filter), yielding two 'children' vectors $\mathbf{W}_{j,n}$ indexed by $n = 2n'$ and $2n' + 1$; however, the wavelet filter produces the first (even indexed) of these children only if $n \bmod 4$ is 2 – if it is 1 instead, it produces the second (odd indexed) child. Note also that each child is half the length of its parent. The coefficients in each child cover the entire time span of the original series $\mathbf{X}$ but are localized in time because of the localization properties of the wavelet and scaling filters (this fact supports the contention that the time width associated with each jth level DWPT coefficient is 2^j).

Each $\mathbf{W}_{j,n}$ on level j is the result of a unique ordered set of filtering operations applied to a subset of coefficients on levels $0, 1, \ldots, j-1$. If we use 0 as a symbol for the scaling filter and 1 for the wavelet filter, we can determine the order in which we need to apply these filters to construct a given $\mathbf{W}_{j,n}$ by looking at the elements of a binary-valued vector $\mathbf{c}_{j,n}$ of length j, constructed as follows. When $j = 1$, we define the one dimensional vectors $\mathbf{c}_{1,0} \equiv [0]$ and $\mathbf{c}_{1,1} \equiv [1]$. When $j > 1$, we construct $\mathbf{c}_{j,n}$ in terms of the level $j-1$ vector $\mathbf{c}_{j-1,\lfloor \frac{n}{2} \rfloor}$ by the simple rule of appending to the end of $\mathbf{c}_{j-1,\lfloor \frac{n}{2} \rfloor}$ either a zero (if $n \bmod 4$ is either 0 or 3) or a one (if $n \bmod 4$ is either 1 or 2). This presumes that we already know $\mathbf{c}_{j-1,\lfloor \frac{n}{2} \rfloor}$ – this is true for $j = 2$, but for $j > 2$ we can obtain it in terms of a level $j-2$ vector by applying the simple rule again (and so forth if levels $j-2$ and smaller are unknown). As an example, this construction procedure yields $\mathbf{c}_{4,9} = [1, 1, 0, 1]^T$, which tells us that we can create $\mathbf{W}_{4,9}$ by, first, filtering $\mathbf{X}$ with $\{h_l\}$ to obtain (after downsampling) $\mathbf{W}_{1,1}$; second, filtering $\mathbf{W}_{1,1}$ with $\{h_l\}$ to get $\mathbf{W}_{2,2}$; third, filtering $\mathbf{W}_{2,2}$ with $\{g_l\}$ to get $\mathbf{W}_{3,4}$; and finally filtering $\mathbf{W}_{3,4}$ with $\{h_l\}$ to get $\mathbf{W}_{4,9}$.

Although in practice a level j DWPT is usually computed recursively in terms of lower level DWPTs, it is of interest to note that the elements $W_{j,n,t}$ of $\mathbf{W}_{j,n}$ can be obtained directly from $\mathbf{X}$ via Equation (215a), namely,

$$W_{j,n,t} = \sum_{l=0}^{L_j-1} u_{j,n,l} X_{2^j[t+1]-1-l \bmod N}, \quad t = 0, 1, \ldots, N_j - 1,$$

where $\{u_{j,n,l}\}$ is the impulse response sequence for the filter whose transfer function is given by

$$U_{j,n}(f) = \prod_{m=0}^{j-1} M_{c_{j,n,m}}(2^m f)$$

(this is Equation (215b)); here $M_0(\cdot) \equiv G(\cdot)$ and $M_1(\cdot) \equiv H(\cdot)$ (i.e., the transfer functions for $\{g_l\}$ and $\{h_l\}$), while $c_{j,n,m}$ is the mth element of $\mathbf{c}_{j,n}$.

We can also formulate the DWPT in terms of matrix operations. Thus we have

$$\mathbf{W}_{j,n} = \begin{cases} \mathcal{A}_j \mathbf{W}_{j-1,\lfloor \frac{n}{2} \rfloor}, & \text{if } n \bmod 4 = 0 \text{ or } 3; \\ \mathcal{B}_j \mathbf{W}_{j-1,\lfloor \frac{n}{2} \rfloor}, & \text{if } n \bmod 4 = 1 \text{ or } 2; \end{cases}$$

where $\mathcal{A}_j$ and $\mathcal{B}_j$ are $N_j \times N_{j-1}$ matrices whose rows contain elements formed by periodizing, respectively, $\{g_l\}$ and $\{h_l\}$ to length N_{j-1} (Section 4.12 summarizes how to construct these matrices). We can apply the above recursively to obtain a matrix expression that directly relates $\mathbf{W}_{j,n}$ to $\mathbf{X}$. To do this, we just need to consider the elements of $\mathbf{c}_{j,n}$. If $c_{j,n,m} = 0$, then we use $\mathcal{A}_m$ to go from level $m - 1$ to m; on the other hand, if $c_{j,n,m} = 1$, we use $\mathcal{B}_m$. For example, since $\mathbf{c}_{4,9} = [1, 1, 0, 1]^T$, we can write $\mathbf{W}_{4,9} = \mathcal{B}_4 \mathcal{A}_3 \mathcal{B}_2 \mathcal{B}_1 \mathbf{X}$.

Since the DWPT is an orthonormal transform, we can use it to partition the energy in $\mathbf{X}$ via

$$\|\mathbf{X}\|^2 = \sum_{n=0}^{2^j-1} \|\mathbf{W}_{j,n}\|^2,$$

where $\|\mathbf{W}_{j,n}\|^2$ can be interpreted as the contribution to the energy due to frequencies in the band $\mathcal{I}_{j,n}$. Similarly, since $\mathbf{X}$ can be synthesized from its DWPT coefficients, we can express $\mathbf{X}$ as the addition of 2^j vectors, each of which is associated with a particular frequency band $\mathcal{I}_{j,n}$ (Equation (208b) shows such a decomposition for $j = 2$).

Because a level j DWPT is formed by taking each 'parent' vector in a level $j - 1$ DWPT and creating two 'children' vectors from it, we can organize all the DWPTs for levels $j = 0, 1, \ldots, J_0$ into a WP table, an example of which is shown in Figure 212a for $J_0 = 3$. The doublets (j, n) that form the indices for the DWPT coefficient vectors $\mathbf{W}_{j,n}$ in this table can be collected together to form a set $\mathcal{N} \equiv \{(j, n) : j = 0, \ldots, J_0,\ n = 0, \ldots, 2^j - 1\}$. Each $(j, n) \in \mathcal{N}$ is associated with a frequency interval, namely, $\mathcal{I}_{j,n}$. The WP table can be used to form a large collection of orthonormal transforms known as disjoint dyadic decompositions. By definition, each such decomposition is associated with a subset, say $\mathcal{C}$, of doublets from $\mathcal{N}$ that satisfies two properties. First, the union of all the frequency intervals $\mathcal{I}_{j,n}$ for each $(j, n) \in \mathcal{C}$ is exactly the interval $[0, \frac{1}{2}]$, and, second, if (j, n) and (j', n') are any two distinct elements of $\mathcal{C}$, then $\mathcal{I}_{j,n}$ and $\mathcal{I}_{j',n'}$ have no frequencies in common. The disjoint dyadic decompositions include the level j DWPTs (for which $\mathcal{C} = \{(j, 0), (j, 1), \ldots, (j, 2^j - 1)\}$), and the level j partial DWTs (for which $\mathcal{C} = \{(j, 0), (j, 1), (j - 1, 1), (j - 2, 1), \ldots, (1, 1)\}$).

Because so many disjoint dyadic decompositions can be extracted from a WP table, it is of interest to define an 'optimal' discrete WP table transform for a given time series $\mathbf{X}$. The best basis algorithm is one attempt to do so by assigning a 'cost' $M(\mathbf{W}_{j,n})$ to each vector $\mathbf{W}_{j,n}$ in the table in terms of an additive cost functional of the form

$$M(\mathbf{W}_{j,n}) \equiv \sum_{t=0}^{N_j-1} m(|W_{j,n,t}|),$$

where $m(\cdot)$ is a real-valued function defined on $[0,\infty)$ with $m(0)=0$. Three examples of $m(\cdot)$ are given in Section 6.3. As discussed in that section, this algorithm picks out the disjoint dyadic decomposition $\mathcal{C} \in \mathcal{N}$ such that

$$\sum_{(j,n)\in\mathcal{C}} M(\mathbf{W}_{j,n}) \le \sum_{(j,n)\in\mathcal{C}'} M(\mathbf{W}_{j,n})$$

for any other disjoint dyadic decomposition $\mathcal{C}' \in \mathcal{N}$.

As is true for the DWT, the DWPT can be sensitive to the assumed starting point for a time series. In the case of the DWT, this consideration led to the definition of the MODWT; for the DWPT, we are led to the analogous 'maximal overlap' DWPT (MODWPT). Using the MODWPT filters

$$\tilde{u}_{n,l} \equiv \frac{u_{n,l}}{\sqrt{2}} \text{ and } \tilde{u}_{j,n,l} \equiv \frac{u_{j,n,l}}{2^{j/2}},$$

we can define the jth level MODWPT coefficients either directly in terms of the time series $\mathbf{X}$ via

$$\widetilde{W}_{j,n,t} \equiv \sum_{l=0}^{L_j-1} \tilde{u}_{j,n,l} X_{t-l \bmod N}, \quad t=0,1,\ldots,N-1$$

(this is Equation (231a)) or recursively in terms of previously determined level $j-1$ coefficients via

$$\widetilde{W}_{j,n,t} \equiv \sum_{l=0}^{L-1} \tilde{u}_{n,l} \widetilde{W}_{j-1,\lfloor \frac{n}{2} \rfloor, t-2^{j-1}l \bmod N}, \quad t=0,\ldots,N-1,$$

where $W_{0,0,t} \equiv X_t$ (the above is Equation (231b)). If we place these coefficients into the N dimensional vector $\widetilde{\mathbf{W}}_{j,n}$, we can organize all the MODWPTs for levels $j = 0,1,\ldots,J_0$ into a MODWPT WP table, an example of which is shown in Figure 232 for $J_0 = 3$.

As is true for the MODWT, the MODWPT leads to both an energy decomposition and an additive decomposition for a time series $\mathbf{X}$. We can partition the energy in $\mathbf{X}$ via

$$\|\mathbf{X}\|^2 = \sum_{n=0}^{2^j-1} \|\widetilde{\mathbf{W}}_{j,n}\|^2,$$

where, analogous to the DWPT case, $\|\widetilde{\mathbf{W}}_{j,n}\|^2$ can be interpreted as the contribution to the energy due to frequencies in the band $\mathcal{I}_{j,n}$. We can form an additive decomposition for $\mathbf{X}$ using 'detail' vectors $\widetilde{\mathcal{D}}_{j,n}$, whose tth elements are

$$\widetilde{\mathcal{D}}_{j,n,t} = \sum_{l=0}^{L_j-1} \tilde{u}_{j,n,l} \widetilde{W}_{j,n,t+l \bmod N}, \text{ from which we get } \mathbf{X} = \sum_{n=0}^{2^j-1} \widetilde{\mathcal{D}}_{j,n}.$$

The vector $\widetilde{\mathcal{D}}_{j,n}$ is associated with the frequency interval $\mathcal{I}_{j,n}$ and has an alternative formulation in terms of filtering $\mathbf{X}$ with a zero phase filter – this allows us easily to relate broad-band features in $\widetilde{\mathcal{D}}_{j,n}$ to features in the original series. The MODWPT coefficients $\widetilde{\mathbf{W}}_{j,n}$ themselves do not have this property, but, as discussed in Section 6.5, we can adjust these to be approximately the outputs from zero phase filters if we use either the LA or coiflet filters and then circularly advance the coefficients via $\mathcal{T}^{-|\nu_{j,n}|}\widetilde{\mathbf{W}}_{j,n}$, where $\nu_{j,n}$ is given in Equation (230).

Finally, we come to the matching pursuit algorithm, which is designed to approximate a time series via a linear combination of a small number of vectors chosen from a large – but finite – set of vectors with unit norm. This set of vectors is called a dictionary and is denoted as $\mathbb{D}$. Although in the context of the current chapter an obvious dictionary would be a WP table (or the MODWPT version thereof), in fact $\mathbb{D}$ can be formed from a variety of sources, including – but not limited to – various combinations of (possibly renormalized) vectors from DWTs, MODWTs, DWPTs and MODWPTs and the ODFT. In order to form an approximation, say $\mathbf{X}^{(m)}$, to $\mathbf{X}$ using a linear combination of m vectors from $\mathbb{D}$, we form a series of approximations $\mathbf{X}^{(n)}$, $n = 0, \ldots, m-1$, along with associated residuals $\mathbf{R}^{(n)}$. The approximation $\mathbf{X}^{(n)}$ is based on n vectors from $\mathbb{D}$ (we define $\mathbf{X}^{(0)} \equiv \mathbf{X}$), and the residual vectors are such that $\mathbf{X}^{(n)} + \mathbf{R}^{(n)} = \mathbf{X}$. Given the nth order approximation, we obtain the approximation of order $n+1$ by computing the inner product between $\mathbf{R}^{(n)}$ and each vector $\mathbf{d}_{\boldsymbol{\gamma}} \in \mathbb{D}$ (the subscript $\boldsymbol{\gamma}$ is typically used to relate a vector to time/frequency or time/scale characteristics – for example, if $\mathbb{D}$ contains vectors related to a MODWT, then $\boldsymbol{\gamma}$ would specify the scale and location in time for $\mathbf{d}_{\boldsymbol{\gamma}}$). To form $\mathbf{X}^{(n+1)}$, we pick any vector – call it $\mathbf{d}_{\boldsymbol{\gamma}_n}$ – satisfying

$$\left|\langle \mathbf{R}^{(n)}, \mathbf{d}_{\boldsymbol{\gamma}_n} \rangle\right| \leq \left|\langle \mathbf{R}^{(n)}, \mathbf{d}_{\boldsymbol{\gamma}} \rangle\right| \quad \text{for all } \mathbf{d}_{\boldsymbol{\gamma}} \in \mathbb{D}$$

(usually there is only one such vector satisfying the above, but there is no reason for there to be just one). The new approximation is defined to be

$$\mathbf{X}^{(n+1)} \equiv \mathbf{X}^{(n)} + \langle \mathbf{R}^{(n)}, \mathbf{d}_{\boldsymbol{\gamma}_n} \rangle \mathbf{d}_{\boldsymbol{\gamma}_n},$$

and we set $\mathbf{R}^{(n+1)} = \mathbf{X} - \mathbf{X}^{(n+1)}$. After m such steps, we have the approximation

$$\mathbf{X}^{(m)} = \sum_{n=0}^{m-1} \langle \mathbf{R}^{(n)}, \mathbf{d}_{\boldsymbol{\gamma}_n} \rangle \mathbf{d}_{\boldsymbol{\gamma}_n}$$

along with the energy decomposition

$$\|\mathbf{X}\|^2 = \sum_{n=0}^{m-1} |\langle \mathbf{R}^{(n)}, \mathbf{d}_{\boldsymbol{\gamma}_n} \rangle|^2 + \|\mathbf{R}^{(m)}\|^2$$

(since $|\langle \mathbf{R}^{(n)}, \mathbf{d}_{\boldsymbol{\gamma}_n} \rangle|^2 = \|\langle \mathbf{R}^{(n)}, \mathbf{d}_{\boldsymbol{\gamma}_n} \rangle \mathbf{d}_{\boldsymbol{\gamma}_n}\|^2$, we can regard the nth term in the above summation as the energy associated with the nth approximating vector in $\mathbf{X}^{(m)}$).

6.11 Exercises

[6.1] Show that the transform given in Equation (208a) is orthonormal.

[6.2] Suppose that we accidently interchanged the wavelet and scaling filters $\{h_l\}$ and $\{g_l\}$ when computing the DWT out to level J_0. In terms of a level J_0 WP table, what transform would this lead to?

[6.3] Compute and plot the squared gain functions $|U_{3,n}(\cdot)|^2$ for $n = 0, \ldots, 7$ based upon the D(4) and LA(16) filters. Compare these to the LA(8) case shown in Figure 217.

[6.4] What sequence of filtering operations would we need to perform on a time series $\mathbf{X}$ in order to obtain DWPT coefficients $\mathbf{W}_{j,n}$ that nominally represent fluctuations in $\mathbf{X}$ within the frequency interval $[\frac{19}{128}, \frac{20}{128}]$? Based upon the LA(8) filter, compute and plot the impulse response sequence $\{u_{j,n,l}\}$ and the squared gain function $|U_{j,n}(\cdot)|^2$ for the associated equivalent filter.

[6.5] Figure 216 displays the contents of the binary-valued vectors $\mathbf{c}_{j,n}$ for the levels $j = 1, 2$ and 3. Determine these vectors for levels $j = 4$ and 5.

[6.6] Suppose that we compute a jth level DWPT $\mathbf{W}_{j,n}$, $n = 0, \ldots, 2^j - 1$, of a time series $\mathbf{X}$ with sample mean $\overline{X}$. Show that the sample mean of $\mathbf{W}_{j,0}$ is equal to $2^{j/2}\overline{X}$. Is it necessarily true that the sample mean of $\mathbf{W}_{j,n}$ for $n > 0$ is zero?

[6.7] As in Figures 225 and 226, apply the best basis algorithm to the DWPT coefficients in the WP table of Figure 224, but now use (a) the ℓ_p information cost functional with $p = 1$ and (b) the threshold cost functional with the threshold δ set to the median of the absolute values of all the coefficients in the WP table (including those at level $j = 0$, i.e., $\mathbf{W}_{0,0} = \mathbf{X}$; this method of setting δ is used in Bruce and Gao, 1996a). How do the best basis transforms picked out by these two cost functionals compare to what we found using the $-\ell^2 \log \ell^2$ norm?

[6.8] For $j = 1$ to 3, Table 230 lists the advances $|\nu_{j,n}|$ required to shift the filter $\{u_{j,n,l}\}$ (based on the LA(8) wavelet filter) so that it has approximately zero phase. Figure 235 gives $|\nu_{4,n}|$ for $n = 0, \ldots, 6$ in the labels $\mathcal{T}^{-|\nu_{4,n}|}\widetilde{\mathbf{W}}_{4,n}$ on that figure. Compute the remaining advances for level $j = 4$, i.e., $|\nu_{j,n}|, n = 7, \ldots, 15$.

[6.9] Create plots of the phase functions for the filters $\{u_{3,n,l}\}$, $n = 0, \ldots, 7$, similar to those for $\{h_{j,l}\}$ in Figures 115 and 124; i.e., for each $\{u_{3,n,l}\}$ based upon a Daubechies or coiflet wavelet filter in these figures, compute and plot the exact phase function over the nominal pass-band $\mathcal{I}_{3,n}$ after this function has been adjusted to take into account the advance $|\nu_{3,n}|$.

[6.10] Create a figure similar to Figure 235 for the MODWPT coefficients $\widetilde{\mathbf{W}}_{4,n}$ *without* making use of the advances $|\nu_{4,n}|$. How does this figure compare to Figures 235 and 238?

[6.11] Suppose that we have a time series $\mathbf{X}$ of length $N = 2^J$ that we wish to analyze using the matching pursuit algorithm. Show that, if we apply N steps of this algorithm using any DWT dictionary, the residuals $\mathbf{R}^{(N)}$ at the last step must satisfy $\|\mathbf{R}^{(N)}\|^2 = 0$. Similarly, show that, if we apply $(J_0 + 1)N$ steps of the algorithm using any MODWT dictionary based upon a level J_0 transform, the residuals at the last step must satisfy $\|\mathbf{R}^{([J_0+1]N)}\|^2 = 0$.

[6.12] For the four point time series $\mathbf{X} = [2, 4, 8, 10]^T$, apply four steps of the matching pursuit algorithm using the MODWT dictionary $\mathbb{D}^{(\text{modwt})}$ for level $J_0 = 2$ based upon the Haar wavelet. In particular, compute the approximations $\mathbf{X}^{(m)}$ and corresponding residuals $\mathbf{R}^{(m)}$ for $m = 1, 2, 3$ and 4.

[6.13] Using a dictionary $\mathbb{D}$ that consists of the vectors

$$\begin{bmatrix} \frac{1}{\sqrt{2}} \\ \frac{1}{\sqrt{2}} \end{bmatrix}, \begin{bmatrix} 1 \\ 0 \end{bmatrix} \text{ and } \begin{bmatrix} 0 \\ 1 \end{bmatrix},$$

apply the matching pursuit algorithm to the two point time series $\mathbf{X} = [5, 3]^T$. How many steps m are needed in order to obtain a residual vector $\mathbf{R}^{(m)}$ with zero norm?

[6.14] Using the eight point time series of Equation (224), apply three steps of the matching pursuit algorithm using a dictionary consisting of all the basis vectors for Haar DWPTs of levels $j = 1$, 2, and 3. What vectors does the algorithm pick out at each step, and what are the values for $\|\mathbf{R}^{(m)}\|^2$ at each step?

7

Random Variables and Stochastic Processes

7.0 Introduction

In the previous three chapters we developed the basic theory for the discrete wavelet transform (DWT) and transforms related to it, including the maximal overlap DWT (MODWT) and the discrete wavelet packet transform (DWPT). We described how these transforms can be applied to a time series $\{X_t\}$, where – up to now – we have taken each X_t in the series to be a real-valued variable. We have shown that these transforms lead to useful additive decompositions that re-express the series as the sum of a small number of other series, each of which is associated with an independent variable of physical interest (scale in the case of a DWT or MODWT and intervals of frequencies in the case of a DWPT). We can also use each transform to partition the sample variance of $\{X_t\}$ into pieces that can be associated with this same independent variable. While potentially quite useful, these additive decompositions and analyses of variance are inherently descriptive in nature and do not take into account the potentially important effects of sampling variability. For example, in the MODWT additive decomposition of the Nile River minima series shown in Figure 192, our eyes easily pick out an increased variability at the beginning of $\widetilde{\mathcal{D}}_1$ and $\widetilde{\mathcal{D}}_2$; however, if we were unaware of the history behind this series, we might well question whether or not we have merely latched onto an artifact that might occur occasionally in series generated by purely random mechanisms. To answer questions such as this, we must determine how stochastic variability in a time series affects wavelet-based transforms and quantities computed from such transforms. This requires a change in how we regard a time series: we must now think of it as a realization of a stochastic process $\{X_t\}$ (this concept is defined in Section 7.4). If we model our series using such processes, we can begin our task of assessing the effect of sampling variability on wavelet-based time series analysis.

The use of statistical models to address questions about time series has blossomed over the last half century (some classic texts on the subject are Anderson, 1971; Blackman and Tukey, 1958; Bloomfield, 1976; Box *et al.*, 1994; Brillinger, 1981; Fuller, 1996; Hannan, 1970; Koopmans, 1974; and Priestley, 1981). In the next three chapters we will be exploring the use of wavelet methods in conjunction with statistical models (for complementary treatments, see Ogden, 1997, Carmona *et al.*, 1998, and Vidakovic,

1999). In preparation for this discussion, here we briefly summarize various results in probability, statistics and stochastic process theory. Our treatment is a summary only and is intentionally limited to just what we will actually need in the chapters that follow. Readers with a solid background in statistics and time series analysis can skim this chapter to familiarize themselves with our notation. Readers who want a more detailed treatment on statistical analysis should consult, for example, Casella and Berger (1990), Mood *et al.* (1974), Papoulis (1991) or Priestley (1981).

7.1 Univariate Random Variables and PDFs

A real-valued random variable (RV) is a function, or mapping, from the sample space of possible outcomes of a random experiment to the real line (or a subset thereof). Let X denote an RV, and let x denote any of its possible outcomes or realizations i.e., a particular value in the sample space. If x can assume only a set of discrete values (e.g., the integers), we say that X is a discrete RV. For such an RV, we denote the probability of the event that X assumes the value x by $\mathbf{P}[X = x]$. By contrast, an RV for which the outcomes form a continuum (e.g., the entire real axis) is known as a continuous RV (we concentrate our attention on such RVs). In this case it makes no sense to talk about the probability of the event $X = x$, and interest focuses on the probability of obtaining an outcome in a specified interval, i.e., $\mathbf{P}[x < X < x+\Delta] = f_X(x)\Delta$, where $f_X(\cdot)$ is the probability density function (PDF) for X, and Δ is an infinitesimal. A PDF has two fundamental properties: it must be nonnegative, and it must integrate to unity.

- *Mean and variance*

If X is a continuous RV, its mean or expected value is the constant defined as

$$E\{X\} = \int_{-\infty}^{\infty} x f_X(x)\, dx$$

(here, and elsewhere in this chapter, we assume the finite existence of necessary quantities – in this case, we assume that the above integral exists and is finite). The mean of a function of X, say $g(X)$, is the constant

$$E\{g(X)\} = \int_{-\infty}^{\infty} g(x) f_X(x)\, dx.$$

The variance of X is the constant given by

$$\operatorname{var}\{X\} \equiv E\{(X - E\{X\})^2\} = \int_{-\infty}^{\infty} (x - E\{X\})^2 f_X(x)\, dx.$$

- *Gaussian PDF*

The best known PDF is the Gaussian (or normal), which takes the form

$$f_X(x;\mu,\sigma) = \frac{1}{\sqrt{(2\pi\sigma^2)}} e^{-(x-\mu)^2/(2\sigma^2)}, \quad -\infty < x, \mu < \infty, \ \sigma > 0, \tag{256}$$

where μ and σ are location and scale parameters (although similar in some ways, a scale parameter for a PDF should be carefully distinguished from the scale for a scaling or wavelet filter: the former gives us an idea about the range of values about μ an RV is likely to assume, whereas the latter tells us the effective range in time over which these filters are forming localized weighted averages). The function $f_X(\cdot;\mu,\sigma)$ defines a symmetric bell-shaped curve (see the thick curves in Figure 276). Note that $f_X(\cdot;\mu,\sigma)$ is often simply abbreviated to $f_X(\cdot)$; i.e., the parameters are suppressed. If X is a Gaussian RV, we in fact have $E\{X\} = \mu$ and $\text{var}\{X\} = \sigma^2$. As a shorthand for saying 'X has a Gaussian PDF with mean μ and variance σ^2,' we shall write

$$X \stackrel{\text{d}}{=} \mathcal{N}(\mu,\sigma^2), \tag{257a}$$

which means that X has the same distribution as a Gaussian RV with parameters μ and σ^2 (we interpret $\mathcal{N}(\mu,\sigma^2)$ to be such an RV); the above is read as 'X is equal in distribution to $\mathcal{N}(\mu,\sigma^2)$.' The RV defined by $Z = (X-\mu)/\sigma$ is called a standard Gaussian RV or standard normal RV; it has zero mean and unit variance. Note that $\mathbf{P}[X \le x] = \mathbf{P}[Z \le z]$, where $z \equiv (x-\mu)/\sigma$. Let

$$\Phi(z) \equiv \mathbf{P}[Z \le z] = \int_{-\infty}^{z} f_X(x;0,1)\,dx = \int_{-\infty}^{z} \frac{1}{\sqrt{(2\pi)}} e^{-x^2/2}\,dx.$$

The function $\Phi(\cdot)$ is called the Gaussian cumulative distribution function and is an increasing function of z satisfying $0 < \Phi(z) < 1$ for all finite z. This function has a well-defined inverse $\Phi^{-1}(\cdot)$ such that $\Phi(\Phi^{-1}(p)) = p$ for all $0 \le p \le 1$. We have

$$\mathbf{P}[Z \le \Phi^{-1}(p)] = p. \tag{257b}$$

The value $\Phi^{-1}(p)$ is called the $p \times 100\%$ percentage point for the standard Gaussian PDF and is tabulated for selected p in the last line of Table 263. The Gaussian PDF is used on many occasions in subsequent chapters, often with $\mu = 0$, for which it is symmetric and bell-shaped about zero.

- *Some other symmetric PDFs*

The Laplace or two-sided exponential PDF is symmetric, but cusp-like, about its location parameter μ and takes the form

$$f_X(x;\mu,v) = \frac{1}{\sqrt{(2/v)}} e^{-(2v)^{1/2}|x-\mu|}, \quad -\infty < x,\mu < \infty, \; v > 0. \tag{257c}$$

The mean of X is $E\{X\} = \mu$, and $\text{var}\{X\} = 1/v$.

The Gaussian and Laplace PDFs belong to the generalized Gaussian class, which have PDFs of the form

$$f_X(x;\mu,\alpha,\beta) = \frac{\alpha}{2\beta\Gamma(1/\alpha)} e^{-(|x-\mu|/\beta)^\alpha}, \quad -\infty < x,\mu < \infty, \; \alpha,\beta > 0 \tag{257d}$$

($\Gamma(\cdot)$ is the gamma function – see, for example, Abramowitz and Stegun, 1964, Chapter 6). Here α is a shape parameter such that the PDF is identical to the Gaussian PDF when $\alpha = 2$, and to the Laplace PDF when $\alpha = 1$. The generalized Gaussian PDF becomes increasingly peaked about its mean μ as α decreases toward zero.

The t distribution with ϑ degrees of freedom has a PDF given by

$$f_X(x;\vartheta) = \frac{\Gamma([\vartheta+1]/2)}{\Gamma(\vartheta/2)\sqrt{(\pi\vartheta)}} \left(1 + \frac{x^2}{\vartheta}\right)^{-(\vartheta+1)/2}, \quad -\infty < x < \infty, \; \vartheta > 0.$$

This PDF can be generalized to include a location parameter μ and a scale parameter κ, yielding a PDF given by

$$f_X(x;\mu,\kappa,\vartheta) = \frac{\Gamma([\vartheta+1]/2)}{\Gamma(\vartheta/2)\sqrt{(\pi\kappa^2\vartheta)}}\left(1+\frac{(x-\mu)^2}{\kappa^2\vartheta}\right)^{-(\vartheta+1)/2}, \quad (258a)$$

where $-\infty < x, \mu < \infty$, $\kappa > 0$ and $\vartheta > 0$. This PDF is symmetric and smoothly peaked about μ. If X is an RV with this PDF, we have $E\{X\} = \mu$ if $\vartheta > 1$, and $\text{var}\{X\} = \kappa^2\vartheta/(\vartheta-2)$ if $\vartheta > 2$. The above is a member of the so-called generalized Cauchy class of distributions (see, for example, Johnson *et al.*, 1994, p. 327). (The t and Laplace PDFs are used in Sections 10.4 and 10.5.)

- *Transformation of an RV*

Suppose that $u(\cdot)$ is a differentiable and single-valued function whose inverse is $v(\cdot)$; i.e., $v(u(x)) = x$ for all x. Given the RV X, suppose we construct $Y \equiv u(X)$. If the PDF of X is given by $f_X(\cdot)$, then the PDF for Y is given by

$$f_Y(y) = |J(y)| \cdot f_X(v(y)), \text{ where } J(y) \equiv \frac{dv(y)}{dy}; \quad (258b)$$

the function $J(\cdot)$ is known as the Jacobian of the (inverse) transformation. For example, suppose that $Y = X^3$, where X is a standard Gaussian RV so that $X \equiv Z$. Here we have $u(x) = x^3$ and $v(y) = y^{1/3}$, so the Jacobian is $\frac{1}{3}y^{-2/3}$. we have

$$f_Y(y) = \frac{1}{3y^{2/3}} f_X(y^{1/3}) = \frac{1}{3y^{2/3}\sqrt{(2\pi)}} e^{-y^{2/3}/2}.$$

(Exercise [7.1] gives a second example of the use of Equation (258b).)

7.2 Random Vectors and PDFs

Let X_0 and X_1 be two real-valued continuous RVs. Let $f_{X_0}(\cdot)$ be the PDF for X_0, and let $f_{X_1}(\cdot)$ be the PDF for X_1. We denote their joint PDF at the point (x_0, x_1) as $f_{X_0,X_1}(x_0, x_1)$. The function $f_{X_0,X_1}(\cdot,\cdot)$ must be nonnegative over the entire real plane (its domain of definition), and the (double) integral of $f_{X_0,X_1}(\cdot,\cdot)$ over this plane must be unity.

- *Marginal PDFs and independence*

The individual PDFs of X_0 and X_1 are known as the marginal PDFs. Each marginal PDF can be obtained from the joint PDF by integrating over values of the other variable:

$$f_{X_0}(x_0) = \int_{-\infty}^{\infty} f_{X_0,X_1}(x_0,x_1)\,dx_1; \quad f_{X_1}(x_1) = \int_{-\infty}^{\infty} f_{X_0,X_1}(x_0,x_1)\,dx_0.$$

The RVs X_0 and X_1 are by definition independent if their joint PDF is the product of the marginal PDFs; i.e., $f_{X_0,X_1}(x_0,x_1) = f_{X_0}(x_0)f_{X_1}(x_1)$ for all x_0, x_1.

- *Expected value of a function of two RVs*

Let $g(\cdot,\cdot)$ be a function of two variables that we use to create a new RV from X_0 and X_1, namely, $g(X_0, X_1)$ (a simple example would be $g(X_0, X_1) = X_0 + X_1$). The mean or expected value of $g(X_0, X_1)$ is then

$$E\{g(X_0,X_1)\} = \int_{-\infty}^{\infty}\int_{-\infty}^{\infty} g(x_0,x_1) f_{X_0,X_1}(x_0,x_1)\,dx_0\,dx_1.$$

- *Covariance and correlation*

Suppose X_0 and X_1 have expected values $E\{X_0\} = \mu_0$ and $E\{X_1\} = \mu_1$. By definition, the covariance between X_0 and X_1 is

$$\begin{aligned}\operatorname{cov}\{X_0, X_1\} &\equiv E\{(X_0 - \mu_0)(X_1 - \mu_1)\} \\ &= \int_{-\infty}^{\infty}\int_{-\infty}^{\infty} (x_0 - \mu_0)(x_1 - \mu_1) f_{X_0,X_1}(x_0, x_1)\, dx_0\, dx_1.\end{aligned}$$

Note that $\operatorname{cov}\{X_0, X_1\} \equiv \operatorname{cov}\{X_1, X_0\}$. If we use these two RVs to form a vector $\mathbf{X} = [X_0, X_1]^T$, then the expected value of $\mathbf{X}$ and its covariance matrix are defined to be

$$E\{\mathbf{X}\} \equiv \begin{bmatrix} E\{X_0\} \\ E\{X_1\} \end{bmatrix} \text{ and } \Sigma_{\mathbf{X}} \equiv \begin{bmatrix} \operatorname{var}\{X_0\} & \operatorname{cov}\{X_0, X_1\} \\ \operatorname{cov}\{X_1, X_0\} & \operatorname{var}\{X_1\} \end{bmatrix},$$

where $\operatorname{var}\{X_0\}$ is the variance of X_0:

$$\operatorname{var}\{X_0\} \equiv \operatorname{cov}\{X_0, X_0\} = E\{(X_0 - \mu_0)^2\} \equiv \sigma_0^2, \text{ say}$$

(similarly, let $\sigma_1^2 \equiv \operatorname{var}\{X_1\}$). Since $\operatorname{cov}\{X_0, X_1\} = \operatorname{cov}\{X_1, X_0\}$ the covariance matrix is symmetric. The correlation ρ between X_0 and X_1 is a standardized version of the covariance:

$$\rho \equiv \frac{\operatorname{cov}\{X_0, X_1\}}{\sigma_0 \sigma_1}.$$

Some additional properties of covariances are explored in Exercise [7.2].

- *Jointly Gaussian RVs*

The RVs X_0 and X_1 are said to be jointly Gaussian (or bivariate Gaussian) if their joint PDF is given by

$$\begin{aligned}&f_{X_0,X_1}(x_0, x_1; \mu_0, \mu_1, \sigma_0, \sigma_1, \rho) \\ &\quad = \frac{1}{2\pi\sigma_0\sigma_1\sqrt{(1-\rho^2)}} \times \\ &\quad\quad \exp\left(-\frac{1}{2(1-\rho^2)}\left[\frac{(x_0-\mu_0)^2}{\sigma_0^2} - 2\rho\frac{(x_0-\mu_0)(x_1-\mu_1)}{\sigma_0\sigma_1} + \frac{(x_1-\mu_1)^2}{\sigma_1^2}\right]\right)\end{aligned}$$

for $-\infty < x_0, x_1, \mu_0, \mu_1 < \infty$; $\sigma_0, \sigma_1 > 0$; and $-1 < \rho < 1$. We can write

$$E\{\mathbf{X}\} \equiv \begin{bmatrix} \mu_0 \\ \mu_1 \end{bmatrix} \text{ and } \Sigma_{\mathbf{X}} \equiv \begin{bmatrix} \sigma_0^2 & \sigma_0\sigma_1\rho \\ \sigma_0\sigma_1\rho & \sigma_1^2 \end{bmatrix}.$$

Note that, if jointly Gaussian RVs are uncorrelated, i.e., $\rho = 0$, then the joint PDF becomes

$$\frac{1}{2\pi\sigma_0\sigma_1}\exp\left(-\frac{(x_0-\mu_0)^2}{2\sigma_0^2} - \frac{(x_1-\mu_1)^2}{2\sigma_1^2}\right) = f_{X_0}(x_0; \mu_0, \sigma_0) f_{X_1}(x_1; \mu_1, \sigma_1),$$

where the PDFs $f_{X_0}(\cdot; \mu_0, \sigma_0)$ and $f_{X_1}(\cdot; \mu_1, \sigma_1)$ are as given in Equation (256). Since the joint PDF can be written as the product of the marginal PDFs, it follows that X_0 and X_1 are independent. (It is important to note that, if the RVs X_0 and X_1 are not jointly Gaussian, then uncorrelatedness does not necessarily mean that X_0 and X_1 are also independent.)

• *Conditional PDFs, means and medians*
Another way of defining independence is via conditional PDFs. The conditional PDF of X_0 given $X_1 = x_1$ is defined as

$$f_{X_0|X_1=x_1}(x_0) = \frac{f_{X_0,X_1}(x_0, x_1)}{f_{X_1}(x_1)}. \tag{260}$$

The symbol '|' should be read as 'given' (or 'conditional on'), so that $f_{X_0|X_1=x_1}(\cdot)$ defines the PDF of X_0 given that $X_1 = x_1$. Using Equation (260), it follows that independence of X_0 and X_1 means that $f_{X_0|X_1=x_1}(x_0) \equiv f_{X_0}(x_0)$ and $f_{X_1|X_0=x_0}(x_1) \equiv f_{X_1}(x_1)$; i.e., knowledge of the outcome of one RV does not influence the distribution of the other. Note that we can obtain a bivariate PDF from a marginal PDF and a conditional PDF:

$$f_{X_0,X_1}(x_0, x_1) = f_{X_0|X_1=x_1}(x_0) f_{X_1}(x_1)$$

or, by symmetry,

$$f_{X_0,X_1}(x_0, x_1) = f_{X_1|X_0=x_0}(x_1) f_{X_0}(x_0).$$

Associated with the idea of conditional distributions is that of conditional means or expectations (and variances). The conditional mean of X_0 given $X_1 = x_1$, is

$$E_{X_0|X_1}\{X_0|X_1 = x_1\} = \int_{-\infty}^{\infty} x_0 f_{X_0|X_1=x_1}(x_0)\, dx_0,$$

and likewise for X_1 given $X_0 = x_0$. It can be shown that

$$E_{X_1}\{E_{X_0|X_1}\{X_0|X_1\}\} = E\{X_0\}$$

and

$$E_{X_0}\{E_{X_1|X_0}\{X_1|X_0\}\} = E\{X_1\},$$

i.e., if we average the conditional mean over values of the conditioning RV, we obtain the (unconditional) mean. When a conditional mean is obvious, we shall sometimes write $E_{X_0|X_1}\{X_0|X_1 = x_1\}$ as $E\{X_0|X_1 = x_1\}$ to simplify notation.

An important application of conditional means is the following. Let X_0 and X_1 be two related RVs, of which we can only observe one, say X_0. Suppose that we want to approximate the unobserved X_1 via an RV that depends on the observed X_0. Let $U_2(X_0)$ denote this new RV, where $U_2(\cdot)$ is some nonlinear function. Suppose that we measure the quality of this approximation by looking at the mean square difference $E\{(X_1 - U_2(X_0))^2\}$ between the unobserved X_1 and $U_2(X_0)$. We claim that $E\{(X_1 - U_2(X_0))^2\}$ is minimized by setting $U_2(X_0) = E\{X_1|X_0\}$, the conditional mean of X_1 given X_0. To see this, note that

$$E\{(X_1 - U_2(X_0))^2\} = \int_{-\infty}^{\infty}\int_{-\infty}^{\infty} (x_1 - U_2(x_0))^2 f_{X_0,X_1}(x_0, x_1)\, dx_0\, dx_1;$$

however, $f_{X_0,X_1}(x_0, x_1) = f_{X_1|X_0=x_0}(x_1) f_{X_0}(x_0)$, so we have

$$E\{(X_1 - U_2(X_0))^2\} = \int_{-\infty}^{\infty} f_{X_0}(x_0) \left[\int_{-\infty}^{\infty} (x_1 - U_2(x_0))^2 f_{X_1|X_0=x_0}(x_1)\, dx_1\right] dx_0.$$

The integral in the square brackets is nonnegative, as is $f_{X_0}(\cdot)$, so the expectation is minimized if we make $\int (x_1 - U_2(x_0))^2 f_{X_1|X_0=x_0}(x_1)\, dx_1$ as small as possible for each outcome x_0. For a given value x_0 this integral is the second moment of the conditional PDF $f_{X_1|X_0=x_0}(\cdot)$ about the constant $U_2(x_0)$.

▷ **Exercise [261]:** Suppose that X is an RV whose PDF is $f_X(\cdot)$. If a is a constant, show that

$$E\{(X-a)^2\} = \int_{-\infty}^{\infty} (x-a)^2 f_X(x)\, dx$$

is minimized when $a = E\{X\}$. ◁

Hence the mean square error in approximating an RV by a constant is minimized when the constant is set equal to the mean of the RV (a physical interpretation is that the moment of inertia with respect to the center of gravity (the mean) is smaller than with respect to any other point). The result of Exercise [261] means that

$$\int (x_1 - E\{X_1|X_0 = x_0\})^2 f_{X_1|X_0=x_0}(x_1)\, dx_1 \le \int (x_1 - U_2(x_0))^2 f_{X_1|X_0=x_0}(x_1)\, dx_1,$$

so the choice $U_2(x_0) = E\{X_1|X_0 = x_0\}$, i.e., $U_2(X_0) = E\{X_1|X_0\}$, minimizes $E\{(X_1 - U_2(X_0))^2\}$, as claimed. For future use, it is handy to write the fact that $E\{X_1|X_0 = x_0\}$ is the value (argument) of $U_2(x_0)$ minimizing the above integral as follows:

$$E\{X_1|X_0 = x_0\} = \arg\min_{U_2(x_0)} \int (x_1 - U_2(x_0))^2 f_{X_1|X_0=x_0}(x_1)\, dx_1. \tag{261a}$$

The conditional mean estimator arises from minimizing a mean square error. Another possibility is to choose to minimize the mean absolute error $E\{|X_1 - U_1(X_0)|\}$. In this case the estimator, say $U_1(X_0)$, is the conditional median rather than the conditional mean, (see, for example, Van Trees, 1968, p. 56). When $X_0 = x_0$ we need to solve

$$\int_{-\infty}^{U_1(x_0)} f_{X_1|X_0=x_0}(x_1)\, dx_1 = 0.5 \tag{261b}$$

to determine $U_1(x_0)$.

- *Transformation of RVs*

Here we generalize Equation (258b) so that we can handle transformations depending on more than one RV. Suppose that $u_0(\cdot,\cdot)$ and $u_1(\cdot,\cdot)$ are differentiable and single-valued functions of two variables. This implies that, if $y_0 = u_0(x_0, x_1)$ and $y_1 = u_1(x_0, x_1)$, there exist inverse functions $v_0(\cdot,\cdot)$ and $v_1(\cdot,\cdot)$ such that $x_0 = v_0(y_0, y_1)$ and $x_1 = v_1(y_0, y_1)$. Suppose that we take the RVs X_0 and X_1 and create transformed RVs via

$$Y_0 = u_0(X_0, X_1) \text{ and } Y_1 = u_1(X_0, X_1).$$

We can readily determine the joint PDF $f_{Y_0,Y_1}(\cdot,\cdot)$ for Y_0 and Y_1 in terms of the joint PDF $f_{X_0,X_1}(\cdot,\cdot)$ for X_0 and X_1 if the Jacobian $J(\cdot,\cdot)$ of the (inverse) transformation is non-null. Thus, if

$$J(y_0, y_1) \equiv \begin{vmatrix} \dfrac{\partial v_0(y_0,y_1)}{\partial y_0} & \dfrac{\partial v_0(y_0,y_1)}{\partial y_1} \\ \dfrac{\partial v_1(y_0,y_1)}{\partial y_0} & \dfrac{\partial v_1(y_0,y_1)}{\partial y_1} \end{vmatrix}$$
$$= \frac{\partial v_0(y_0,y_1)}{\partial y_0} \cdot \frac{\partial v_1(y_0,y_1)}{\partial y_1} - \frac{\partial v_0(y_0,y_1)}{\partial y_1} \cdot \frac{\partial v_1(y_0,y_1)}{\partial y_0} \neq 0,$$

then we have

$$f_{Y_0,Y_1}(y_0, y_1) = |J(y_0, y_1)| \cdot f_{X_0,X_1}(v_0(y_0, y_1), v_1(y_0, y_1)).$$

- *PDF of a sum*

Let us now use the above results to derive an expression for the PDF of a sum of two RVs.

▷ **Exercise [262a]:** Show that, if $Y_0 = X_0$ and $Y_1 = X_0 + X_1$, then

$$f_{Y_0,Y_1}(y_0, y_1) = f_{X_0,X_1}(y_0, y_1 - y_0).$$ ◁

The desired PDF $f_{Y_1}(\cdot)$ is just the marginal PDF for the sum Y_1 in the above joint PDF. Hence

$$f_{Y_1}(y_1) = \int_{-\infty}^{\infty} f_{Y_0,Y_1}(y_0, y_1)\, dy_0 = \int_{-\infty}^{\infty} f_{X_0,X_1}(y_0, y_1 - y_0)\, dy_0.$$

If in addition X_0 and X_1 are independent, then $f_{X_0,X_1}(y_0, y_1 - y_0) = f_{X_0}(y_0) f_{X_1}(y_1 - y_0)$, so we have

$$f_{Y_1}(y_1) = \int_{-\infty}^{\infty} f_{X_0}(y_0) f_{X_1}(y_1 - y_0)\, dy_0, \tag{262a}$$

i.e., if two RVs are independent, the PDF of their sum is the convolution of their PDFs.

- *Multivariate random vectors*

The results on bivariate random vectors can generally be extended easily to higher dimensional vectors $\mathbf{X} = [X_0, X_1, \ldots, X_{N-1}]^T$. We mention a few additional results of interest to us.

Let $\mathcal{M}$ be an $M \times N$ matrix. Then the mean of $\mathbf{Y} = \mathcal{M}\mathbf{X}$ is given by

$$\mu_{\mathbf{Y}} = E\{\mathbf{Y}\} = \mathcal{M} E\{\mathbf{X}\} = \mathcal{M} \mu_{\mathbf{X}}, \tag{262b}$$

and its covariance matrix is given by

$$\Sigma_{\mathbf{Y}} = \mathcal{M} \Sigma_{\mathbf{X}} \mathcal{M}^T. \tag{262c}$$

▷ **Exercise [262b]:** Suppose that $\mathcal{M} = \mathcal{O}$, where $\mathcal{O}$ is an orthonormal transform (and hence is $N \times N$). Show that the total variance is preserved in the sense that

$$\sum_{t=0}^{N-1} \operatorname{var}\{Y_t\} = \sum_{t=0}^{N-1} \operatorname{var}\{X_t\}.$$ ◁

If in addition we assume that the N RVs in $\mathbf{X}$ are multivariate Gaussian, then the M RVs in $\mathbf{Y} = \mathcal{M}\mathbf{X}$ are also such. Since uncorrelatedness implies independence (and *vice versa*) for Gaussian RVs, the following exercise indicates that an orthonormal transform of independent and identically distributed – often abbreviated to 'IID' – Gaussian RVs with zero mean yields a new set of IID RVs with the same joint distribution.

	p					
η	0.005	0.025	0.05	0.95	0.975	0.995
1	0.00004	0.0010	0.0039	3.8415	5.0239	7.8794
1.5	0.0015	0.0131	0.0332	4.9802	6.2758	9.3310
2	0.0100	0.0506	0.1026	5.9915	7.3778	10.5966
2.5	0.0321	0.1186	0.2108	6.9281	8.3923	11.7538
3	0.0717	0.2158	0.3518	7.8147	9.3484	12.8382
3.5	0.1301	0.3389	0.5201	8.6651	10.2621	13.8696
4	0.2070	0.4844	0.7107	9.4877	11.1433	14.8603
4.5	0.3013	0.6494	0.9201	10.2882	11.9985	15.8183
5	0.4117	0.8312	1.1455	11.0705	12.8325	16.7496
5.5	0.5370	1.0278	1.3845	11.8376	13.6486	17.6583
6	0.6757	1.2373	1.6354	12.5916	14.4494	18.5476
6.5	0.8268	1.4584	1.8967	13.3343	15.2369	19.4201
7	0.9893	1.6899	2.1673	14.0671	16.0128	20.2777
7.5	1.1621	1.9306	2.4463	14.7912	16.7783	21.1222
$\Phi^{-1}(p)$	−2.5758	−1.9600	−1.6449	1.6449	1.9600	2.5758

Table 263. Percentage points $Q_\eta(p)$ for χ^2_η distribution for $\eta = 1$ to 7.5 in steps of 0.5. The bottom row gives percentage points $\Phi^{-1}(p)$ for the standard Gaussian distribution.

▷ **Exercise [263]:** Suppose that $\mathcal{O}$ is an orthonormal transform. Show that, if $\mathbf{X}$ is IID Gaussian with zero mean (i.e., $\boldsymbol{\mu}_{\mathbf{X}} = \mathbf{0}$, a vector of zeros) and covariance $\sigma^2 I_N$, then $\mathbf{Y} \equiv \mathcal{O}\mathbf{X}$ is also IID Gaussian with zero mean and covariance $\sigma^2 I_N$; i.e., the statistical properties of $\mathbf{Y}$ and $\mathbf{X}$ are *identical.* Does this statement still hold if $\boldsymbol{\mu}_{\mathbf{X}} \neq \mathbf{0}$? (You may assume the result stated above, namely, that, if the RVs in $\mathbf{X}$ are multivariate Gaussian, then so are the RVs in $\mathbf{Y}$.) ◁

• *Chi-square random variables*

Let us now review the properties of a chi-square RV, which can be introduced via the sum of squares of IID standard Gaussian RVs $Z_1, \ldots, Z_\eta$, i.e.,

$$\sum_{k=1}^{\eta} Z_k^2 \equiv \chi^2_\eta. \tag{263a}$$

The RV χ^2_η is said to obey a chi-square distribution with η degrees of freedom. The PDF for χ^2_η takes the form

$$f_{\chi^2_\eta}(x) = \frac{1}{2^{\eta/2}\Gamma(\eta/2)} x^{(\eta/2)-1} e^{-x/2}, \quad x \geq 0, \ \eta > 0 \tag{263b}$$

(proof of the above is the burden of Exercise [7.3]). The above function is nonnegative and integrates to unity for all $\eta > 0$, so we can use it to extend the definition of a chi-square RV to degrees of freedom other than just the nonnegative integers. The mean and variance of χ^2_η are given by $E\{\chi^2_\eta\} = \eta$ and $\operatorname{var}\{\chi^2_\eta\} = 2\eta$.

Let $Q_\eta(p)$ represent the $p \times 100\%$ percentage point for the χ^2_η PDF:

$$\mathbf{P}[\chi^2_\eta \le Q_\eta(p)] = p.$$

The percentage points required to compute 90%, 95% and 99% confidence intervals are listed in Table 263 for degrees of freedom η from 1 to 7.5 in steps of 0.5; these values are utilized in Section 8.4. For larger values of η, we can compute $Q_\eta(p)$ using the following approximation (Chambers *et al.*, 1983):

$$Q_\eta(p) \approx \tilde{Q}_\eta(p) \equiv \eta \left(1 - \frac{2}{9\eta} + \Phi^{-1}(p)\left(\frac{2}{9\eta}\right)^{1/2}\right)^3, \tag{264}$$

where $\Phi^{-1}(p)$ is the $p \times 100\%$ percentage point of the standard Gaussian distribution given in the bottom line of the table. For all $\eta \ge 8$, the relative absolute error $|\tilde{Q}_\eta(p) - Q_\eta(p)|/Q_\eta(p)$ is less than 0.05 for all six values of p used in the table. An algorithm for computing $Q_\eta(p)$ to high accuracy for arbitrary η and p is given by Best and Roberts (1975) and is the source of the values in the table.

The chi-square distribution is often used as a means of approximating the distribution of a nonnegative linear combination of squared Gaussian RVs. An example of how to formulate this approximation is given in Section 8.4 in the discussion surrounding Equation (313a).

7.3 A Bayesian Perspective

Perhaps the simplest way to summarize some Bayesian ideas and nomenclature needed later on (Chapter 10) is through the following situation. Suppose that we observe an RV X with a PDF that depends on θ; i.e., θ is a parameter of the PDF. As an example, suppose that θ is the mean of a Gaussian RV with unit variance. Viewed from a *frequentist* or classical statistics viewpoint, this could be expressed as (see Equation (257a))

$$X \stackrel{\mathrm{d}}{=} \mathcal{N}(\theta, 1),$$

where θ is regarded as a fixed but unknown value. From a Bayesian perspective, however, θ is regarded as random in the sense that we have certain beliefs about its value, so-called *prior information*, and the model is typically expressed as

$$X|\theta \stackrel{\mathrm{d}}{=} \mathcal{N}(\theta, 1).$$

Since θ is now regarded as an RV, it too will have a PDF, known as the *prior PDF*, e.g.,

$$\theta \stackrel{\mathrm{d}}{=} \mathcal{N}(0, \sigma^2_\theta).$$

Here σ^2_θ is a parameter of the prior distribution – a *hyperparameter* in the Bayesian context. (These ideas extend readily to multivariate random vectors $\mathbf{X}$ whose joint distribution involves a vector of hyperparameters.)

- *A variance mixture*

In a Bayesian context the Laplace PDF arises as a 'variance mixture' of the Gaussian distribution for which the variance has a (one-sided) exponential PDF (see Sections 10.4 and 10.5). Let

$$X|\sigma^2 \stackrel{\mathrm{d}}{=} \mathcal{N}(0, \sigma^2),$$

and suppose that the RV σ^2 has the exponential PDF

$$f_{\sigma^2}(\sigma_0^2; v) = v e^{-\sigma_0^2 v}, \quad v > 0.$$

Here v is a hyperparameter. Then the *variance mixture of the Gaussian* is found by integrating over all values of σ_0^2 and is given by

$$\int_0^\infty f_{X|\sigma^2=\sigma_0^2}(x) f_{\sigma^2}(\sigma_0^2; v)\, d\sigma_0^2 = \frac{1}{\sqrt{(2/v)}} e^{-(2v)^{1/2}|x|} \tag{265a}$$

(Teichroew, 1957). This integral gives precisely the Laplace PDF or two-sided exponential PDF in Equation (257c) with $\mu = 0$. If the Gaussian distribution is centered about μ instead of zero, Equation (257c) exactly is obtained. In view of Equation (260), we can interpret the integrand above as the joint PDF of the RVs X and σ^2, and hence the integral yields the marginal PDF of X, namely, the Laplace PDF.

- *Bayesian risk and Bayes rule*

Suppose that, based on an outcome $X = x$, a decision $d(x)$ is made resulting in a loss $\ell(d(x), \theta_0)$ when the outcome $\theta = \theta_0$ occurs. If we take the expectation over all possible outcomes of X (with θ regarded as fixed at outcome θ_0), we obtain a risk function

$$r_d(\theta_0) = E_{X|\theta=\theta_0}\{\ell(d(X), \theta_0)\} = \int \ell(d(x), \theta_0) f_{X|\theta=\theta_0}(x)\, dx.$$

The Bayes risk R_d associated with $d(\cdot)$ is the expectation of $r_d(\theta_0)$ over all possible values of θ_0; i.e., the risk function is weighted by the prior PDF $f_\theta(\cdot)$ of θ and integrated.

▷ **Exercise [265]:** Show that

$$R_d \equiv E_\theta\{r_d(\theta)\} = \int \left(\int \ell(d(x), \theta_0) f_{\theta|X=x}(\theta_0)\, d\theta_0 \right) f_X(x)\, dx. \qquad \triangleleft$$

The integral within the parentheses is called the posterior expected loss – this function of x is then weighted by the PDF $f_X(\cdot)$ to obtain the Bayes risk. By definition, the Bayes rule is the selection of $d(\cdot)$ that minimizes the Bayes risk. This risk is minimized if $d(\cdot)$ is chosen so that the posterior expected loss is a minimum for all x.

The Bayes rule can be used to find a point estimate of θ, i.e., a single number derived from the data. A Bayesian decision rule in such a case is called a Bayes estimator. Let us consider what the Bayes estimator is for squared error loss; i.e., $\ell(d(x), \theta) = (d(x) - \theta)^2$. The posterior expected loss is then the mean square error:

$$\int_{-\infty}^\infty \ell(d(x), \theta_0) f_{\theta|X=x}(\theta_0)\, d\theta_0 = \int_{-\infty}^\infty (d(x) - \theta_0)^2 f_{\theta|X=x}(\theta_0)\, d\theta_0.$$

Exercise [261] says that the right-hand integral is minimized by setting $d(x)$ equal to the expected value of an RV with PDF $f_{\theta|X=x}(\cdot)$, so the Bayes rule – call it $B_2(\cdot)$ – is given by

$$B_2(x) \equiv E\{\theta | X = x\}. \tag{265b}$$

Under squared error loss, the Bayes estimator of θ is thus the posterior mean of θ. Note that, if we think of X and θ as bivariate RVs like X_0 and X_1 in Equation (261a), then $B_2(x)$ is the conditional mean of θ given $X = x$.

As a second example, let us consider what the Bayes estimator is for absolute error loss; i.e., $\ell(d(x), \theta) = |d(x) - \theta|$. The posterior expected loss is then the mean absolute error:

$$\int_{-\infty}^{\infty} \ell(d(x), \theta_0) f_{\theta|X=x}(\theta_0)\, d\theta_0 = \int_{-\infty}^{\infty} |d(x) - \theta_0| f_{\theta|X=x}(\theta_0)\, d\theta_0.$$

The Bayes rule, say $B_1(\cdot)$, minimizing the above is the solution of

$$\int_{-\infty}^{B_1(x)} f_{\theta|X=x}(\theta_0)\, d\theta_0 = 0.5, \tag{266}$$

so that the Bayes estimator in this case is the posterior median of θ. Again, if we think of X and θ as bivariate RVs as in Equation (261b), then $B_1(x)$ is the conditional median of θ given $X = x$.

There are many excellent textbooks on Bayesian statistics; particularly accessible accounts are given in Box and Tiao (1973), Lee (1989) and Gelman *et al.* (1995).

7.4 Stationary Stochastic Processes

Now, let $\{X_t : t = \ldots, -1, 0, 1, \ldots\}$ be a discrete parameter real-valued stochastic process, which by definition is a sequence of RVs indexed over the integers (when the independent variable t is instead taken to vary over the entire real axis, the stochastic process is said to have a continuous parameter). A process such as $\{X_t\}$ can serve as a stochastic model for a sequence of observations of some physical phenomenon. We assume that these observations are recorded at a sampling interval of Δt, which is assumed to have physically meaningful units (e.g., seconds or years).

- *Stationarity and autocovariance sequence*

The process $\{X_t\}$ is said to be (second order) stationary if it satisfies the following two properties:

[1] $E\{X_t\} = \mu_X$ for all integers t; i.e., the expected value of the tth component X_t of the process $\{X_t\}$ is equal to a finite constant μ_X which does not depend on t; and

[2] $\operatorname{cov}\{X_t, X_{t+\tau}\} = s_{X,\tau}$ for all integers t and τ; i.e., the covariance between any two components X_t and $X_{t+\tau}$ of the stationary process $\{X_t\}$ is equal to a finite constant $s_{X,\tau}$ which depends only on the separation τ between the indices t and $t + \tau$ of the components.

An important consequence of property [2] is that the variance of the tth component X_t of the stationary process $\{X_t\}$ is a constant independent of t because

$$\operatorname{var}\{X_t\} = \operatorname{cov}\{X_t, X_t\} = E\{(X_t - \mu_X)^2\} = s_{X,0}.$$

The sequence $\{s_{X,\tau} : \tau = \ldots, -1, 0, 1, \ldots\}$ is called the *autocovariance sequence* (ACVS). This sequence is symmetric about $\tau = 0$ in the sense that $s_{X,-\tau} = s_{X,\tau}$ for all τ. If we let $\rho_{X,\tau} \equiv s_{X,\tau}/s_{X,0}$, we obtain a sequence $\{\rho_{X,\tau}\}$ of correlation coefficients known as the autocorrelation sequence (ACS).

- *Spectral density function*

For many real-valued stationary processes of interest as models in the physical sciences, the information about $\{X_t\}$ that is contained in the ACVS can be re-expressed in terms of a *spectral density function* $S_X(\cdot)$ (SDF), which is also commonly called the power spectrum (or sometimes just the spectrum). If the ACVS is square summable, i.e.,

$$\sum_{\tau=-\infty}^{\infty} s_{X,\tau}^2 < \infty, \tag{267a}$$

we have the relationship

$$S_X(f) = \Delta t \sum_{\tau=-\infty}^{\infty} s_{X,\tau} e^{-i2\pi f\tau\,\Delta t} \text{ for } |f| \le f_{\mathcal{N}} \equiv \frac{1}{2\,\Delta t}, \tag{267b}$$

where $f_{\mathcal{N}}$ is called the Nyquist frequency and has units of, say, cycles per second (square summability of the ACVS is a sufficient – but not necessary – condition for the SDF to exist). When Δt is taken to be unity, a comparison of the above with Equation (35b) shows that $S_X(\cdot)$ is the Fourier transform of $\{s_{X,\tau}\}$ and is often referred to as such even when $\Delta t \neq 1$. The SDF is an even function of frequency f (i.e., $S_X(-f) = S_X(f)$) and satisfies

$$\int_{-f_{\mathcal{N}}}^{f_{\mathcal{N}}} S_X(f) e^{i2\pi f\tau\,\Delta t}\,df = s_{X,\tau}, \quad \tau = \ldots, -1, 0, 1, \ldots \tag{267c}$$

(if $\{s_{X,\tau}\}$ is square summable and Δt is unity, this relationship is just an inverse Fourier transform; however, even if the SDF exists but cannot be expressed in terms of the ACVS via Equation (267b), the above relationship is still valid). In particular, for $\tau = 0$ we have the fundamental result

$$\int_{-f_{\mathcal{N}}}^{f_{\mathcal{N}}} S_X(f)\,df = s_{X,0} = \text{var}\{X_t\}; \tag{267d}$$

i.e., $S_X(\cdot)$ decomposes the process variance with respect to frequency. The quantity $S_X(f)\,\Delta f$ can be thought of as the contribution to the process variance due to frequencies in a small interval of width Δf containing f. The requirements on a real-valued function $S_X(\cdot)$ such that it is the SDF for some real-valued stationary process with sampling time Δt are quite simple. We merely require that $S_X(f) \ge 0$ for all f, that $S_X(f) = S_X(-f)$, and that $0 \le \int_{-f_{\mathcal{N}}}^{f_{\mathcal{N}}} S_X(f)df < \infty$, this integral being the variance of the process (the case in which the integral is zero is thus of little practical interest).

- *Linear filtering*

Let us now assume that $\Delta t = 1$ for convenience and consider the effect of filtering the stationary process $\{X_t\}$. Suppose that $\{a_t : t = 0, \ldots, M-1\}$ is a linear filter of width M with transfer function defined by

$$A(f) \equiv \sum_{t=0}^{M-1} a_t e^{-i2\pi ft}.$$

If we then construct the stochastic process

$$U_t \equiv \sum_{l=0}^{M-1} a_l X_{t-l}, \tag{268a}$$

then it follows that $\{U_t\}$ is a stationary process with an SDF given by

$$S_U(f) \equiv \mathcal{A}(f) S_X(f), \tag{268b}$$

where $\mathcal{A}(f) \equiv |A(f)|^2$ defines the squared gain function corresponding to $A(\cdot)$. Since the integral of the SDF is always equal to the process variance, we see that

$$\operatorname{var}\{U_t\} = \int_{-1/2}^{1/2} S_U(f)\, df = \int_{-1/2}^{1/2} \mathcal{A}(f) S_X(f)\, df. \tag{268c}$$

Equation (268b) is a stochastic version of the following result discussed in Section 4.2: if we circularly filter a finite sequence $\{X_t : t = 0, \ldots, N-1\}$ of (nonrandom) variables with DFT $\{\mathcal{X}_k : k = 0, \ldots, N-1\}$ and energy spectrum $\{|\mathcal{X}_k|^2/N\}$ using a filter whose squared gain function is $\mathcal{H}(\cdot)$, then the energy spectrum for the output from the filter is given by $\{\mathcal{H}(\frac{k}{N})|\mathcal{X}_k|^2/N\}$. Similarly, Equation (268c) is a stochastic version of the fact that the energy of the output sequence is given by $\sum_{k=0}^{N-1} \mathcal{H}(\frac{k}{N})|\mathcal{X}_k|^2/N$.

The result of filtering a stationary process with a filter of infinite width again yields a stationary process as long as the integrals in Equation (268c) yield a finite value for $\operatorname{var}\{U_t\}$.

- *White noise*

A stationary process of particular interest is the *white noise process* $\{\varepsilon_t\}$ (also called a *purely random process*). By definition, $\{\varepsilon_t\}$ is a sequence of pairwise uncorrelated RVs, each with the same mean (usually taken to be zero) and the same variance σ_ε^2. Since uncorrelatedness means that $\operatorname{cov}\{\varepsilon_t, \varepsilon_{t+\tau}\} = 0$ for all t and $\tau \neq 0$, we see that the ACVS for $\{\varepsilon_t\}$ is

$$s_{\varepsilon,\tau} = \operatorname{cov}\{\varepsilon_t, \varepsilon_{t+\tau}\} = \begin{cases} \sigma_\varepsilon^2, & \tau = 0; \\ 0, & \text{otherwise.} \end{cases} \tag{268d}$$

Since this ACVS is obviously square summable, it follows from Equation (267b) that $\{\varepsilon_t\}$ has an SDF given by $S_\varepsilon(f) = \sigma_\varepsilon^2\, \Delta t$ for $|f| \leq f_{\mathcal{N}}$.

- *Autoregressive processes*

The stochastic process $\{X_t\}$ is said to be autoregressive (AR) of order p if the RVs in the process are related by

$$X_t = \sum_{n=1}^{p} \phi_{p,n} X_{t-n} + \varepsilon_t, \tag{268e}$$

where $\{\varepsilon_t\}$ is a white noise process with mean zero and variance σ_ε^2. We refer to the above as an AR(p) process. If the AR coefficients $\phi_{p,n}$ satisfy certain restrictive conditions (see, for example, Priestley, 1981, Section 3.5.4), then $\{X_t\}$ is a stationary process with an SDF given by

$$S_X(f) = \frac{\sigma_\varepsilon^2\, \Delta t}{|1 - \sum_{n=1}^{p} \phi_{p,n} e^{-i2\pi f n\, \Delta t}|^2}, \quad |f| \leq f_{\mathcal{N}}. \tag{268f}$$

We discuss simulation of stationary AR processes in Section 7.9.

7.5 Spectral Density Estimation

The estimation of the SDF has been the subject of a number of textbooks (see, for example, Blackman and Tukey, 1958; Jenkins and Watts, 1968; Koopmans, 1974; Bloomfield, 1976; Priestley, 1981; Marple, 1987; Kay, 1988; and Percival and Walden, 1993). Here we give a brief introduction to the subject as background for material in Chapters 8 to 10. We first review the properties of the periodogram, which is a 'raw' estimator in the sense that it is an inconsistent estimator of the SDF (i.e., its variance does not decrease with sample size). We next sketch a well-known scheme for producing a consistent estimator by averaging (i.e., smoothing or filtering) the periodogram across frequencies. We note that the resulting estimator is characterized by a smoothing bandwidth measuring the effective range of frequencies over which values of the periodogram are averaged together. A critique of this estimator is that it has a fixed smoothing bandwidth over the entire SDF, even though different parts of the SDF can reflect a rapidly or slowly changing spectral shape. We then discuss a second 'raw' SDF estimator known as the multitaper estimator. This estimator plays an important role in Section 10.7, where we discuss an attractive wavelet-based scheme for constructing a variable bandwidth smoother.

• *The periodogram*

Given a time series that can be regarded as a realization of a portion $X_0, X_1, \ldots, X_{N-1}$ of a real-valued stationary process $\{X_t\}$ with zero mean, ACVS $\{s_{X,\tau}\}$ and SDF $S_X(\cdot)$ satisfying Equation (267b), it is possible to estimate $s_{X,\tau}$ for $|\tau| = 0, 1, \ldots, (N-1)$ – but *not* for $|\tau| \geq N$ – using the so-called biased ACVS estimator, namely,

$$\hat{s}^{(\mathrm{p})}_{X,\tau} = \frac{1}{N} \sum_{t=0}^{N-|\tau|-1} X_t X_{t+|\tau|}$$

(this makes explicit use of the assumption that the process mean is known to be zero). It thus seems natural to replace $s_{X,\tau}$ in Equation (267b) by $\hat{s}^{(\mathrm{p})}_{X,\tau}$ for $|\tau| \leq N-1$ and to truncate the summation over τ at the points $\pm(N-1)$ – this amounts to *defining* $\hat{s}^{(\mathrm{p})}_{X,\tau} = 0$ for $|\tau| \geq N$. Now

$$\begin{aligned}
\Delta t \sum_{\tau=-(N-1)}^{N-1} \hat{s}^{(\mathrm{p})}_{X,\tau} e^{-i2\pi f\tau\,\Delta t} &= \frac{\Delta t}{N} \sum_{\tau=-(N-1)}^{N-1} \sum_{t=0}^{N-|\tau|-1} X_t X_{t+|\tau|} e^{-i2\pi f\tau\,\Delta t} \\
&= \frac{\Delta t}{N} \sum_{j=0}^{N-1} \sum_{k=0}^{N-1} X_j X_k e^{-i2\pi f(k-j)\,\Delta t} \\
&= \frac{\Delta t}{N} \left| \sum_{t=0}^{N-1} X_t e^{-i2\pi ft\,\Delta t} \right|^2 \equiv \hat{S}^{(\mathrm{p})}_X(f), \qquad \text{(269a)}
\end{aligned}$$

after a change of variables in the double summation. The function $\hat{S}^{(\mathrm{p})}_X(\cdot)$ defined above is known as the *periodogram* (even though it is a function of frequency and not period). Like $S_X(\cdot)$, it is defined over the interval $[-f_{\mathcal{N}}, f_{\mathcal{N}}]$. It can be shown (e.g., Brillinger, 1981, Section 5.2) that – subject to the finiteness of certain high order moments –

$$\hat{S}^{(\mathrm{p})}_X(f) \stackrel{\mathrm{d}}{=} \begin{cases} S_X(f)\chi^2_2/2, & \text{for } 0 < f < f_{\mathcal{N}}; \\ S_X(f)\chi^2_1, & \text{for } f = 0 \text{ or } f_{\mathcal{N}}, \end{cases} \qquad \text{(269b)}$$

asymptotically as $N \to \infty$, where χ^2_η denotes an RV having a chi-square distribution with η degrees of freedom, as in Section 7.1. Note that, because $E\{\chi^2_\eta\} = \eta$, it follows from the above that $E\{\hat{S}^{(p)}_X(f)\} \approx S_X(f)$ for large N; i.e., asymptotically, the periodogram becomes an unbiased estimator of the SDF since its expected value comes into agreement with what it is intended to estimate. Furthermore, for $0 \le f' < f \le f_{\mathcal{N}}$, $\hat{S}^{(p)}_X(f)$ and $\hat{S}^{(p)}_X(f')$ are asymptotically independent. Using the result that $\text{var}\,\{\chi^2_\eta\} = 2\eta$, we thus have

$$\text{var}\,\{\hat{S}^{(p)}_X(f)\} = \begin{cases} S^2_X(f), & 0 < f < f_{\mathcal{N}}; \\ 2S^2_X(f), & f = 0 \text{ or } f_{\mathcal{N}}; \end{cases} \tag{270a}$$
$$\text{cov}\,\{\hat{S}^{(p)}_X(f), \hat{S}^{(p)}_X(f')\} = 0, \quad 0 \le f' < f \le f_{\mathcal{N}},$$

asymptotically as $N \to \infty$ (for finite sample sizes N, these asymptotic results are useful approximations if certain restrictions are observed – see Percival and Walden, 1993, p. 232).

The periodogram is thus seen to have one obvious major disadvantage, and one obvious major advantage. Since $S_X(f) > 0$ typically, Equation (270a) shows that the variance of $\hat{S}^{(p)}_X(f)$ does not decrease to 0 as $N \to \infty$, or, put another way, the probability that $\hat{S}^{(p)}_X(f)$ becomes arbitrarily close to its expected (mean) value of $S_X(f)$ is zero. Because of this, $\hat{S}^{(p)}_X(f)$ is said to be an inconsistent estimator of $S_X(f)$. On the other hand, the asymptotic independence of pairwise periodogram ordinates makes possible an understanding of the sampling properties of SDF estimators created by smoothing the periodogram across frequencies.

The periodogram has a second – less obvious – advantage in that we can easily transform the RV $\hat{S}^{(p)}_X(f)$ into an RV whose variance does not depend on the unknown SDF (Equation (270a) says that the periodogram itself does not enjoy this property). The 'variance stabilizing' transformation for the periodogram is just the logarithm, a fact that can be deduced from Equation (270a) via the following argument. Since $\hat{S}^{(p)}_X(f)$ has the same distribution as $S_X(f)\chi^2_2/2$ for $0 < f < f_{\mathcal{N}}$, it follows that $\log(\hat{S}^{(p)}_X(f))$ has the same distribution as $\log(S_X(f)\chi^2_2/2) = \log(S_X(f)/2) + \log(\chi^2_2)$. Since the variance of an RV plus a constant is just the variance of the RV itself, we have $\text{var}\,\{\log(\hat{S}^{(p)}_X(f))\} = \text{var}\,\{\log(\chi^2_2)\}$. Based on this, the results of Bartlett and Kendall (1946) show that, for $0 < f < f_{\mathcal{N}}$,

$$\text{var}\,\{\log(\hat{S}^{(p)}_X(f))\} = \frac{\pi^2}{6} \text{ and } E\{\log(\hat{S}^{(p)}_X(f))\} = \log(S_X(f)) - \gamma,$$

where $\gamma \doteq 0.57721$ is Euler's constant. Hence the log periodogram is 'variance stabilized.' For $0 < f < f_{\mathcal{N}}$, the RV

$$\epsilon(f) \equiv \log\left(\frac{\hat{S}^{(p)}_X(f)}{S_X(f)}\right) + \gamma \tag{270b}$$

has mean zero and variance $\sigma^2_\epsilon = \pi^2/6$; moreover,

$$\epsilon(f) \stackrel{\text{d}}{=} \log(\chi^2_2) + \gamma - \log(2) \tag{270c}$$

(Exercise [7.5] is to verify the above). If we let

$$Y^{(p)}(f) \equiv \log(\hat{S}_X^{(p)}(f)) + \gamma,$$

then, for $0 < f < f_{\mathcal{N}}$,

$$Y^{(p)}(f) = \log(S_X(f)) + \epsilon(f). \tag{271a}$$

Hence the log periodogram (plus a known constant γ) can be written as a 'signal' (the true log SDF) plus non-Gaussian noise with zero mean and known variance $\sigma_\epsilon^2 = \pi^2/6$.

- *Conventional smoothing of the periodogram*

Consistent estimators of the SDF can be developed by combining the periodogram ordinates in some way. For example, a traditional approach to this problem is to smooth across the periodogram evaluated at the Fourier frequencies, $f_j = j/(N\,\Delta t)$. The $\lfloor N/2 \rfloor + 1$ RVs

$$\hat{S}_X^{(p)}(f_0), \hat{S}_X^{(p)}(f_1), \ldots, \hat{S}_X^{(p)}(f_{\lfloor N/2 \rfloor})$$

are all approximately pairwise independent for N large enough; i.e.,

$$\operatorname{cov}\{\hat{S}_X^{(p)}(f_j), \hat{S}_X^{(p)}(f_k)\} \approx 0, \;\; j \neq k \text{ and } 0 \leq j, k \leq \lfloor N/2 \rfloor.$$

Thus

$$\hat{S}_X^{(p)}(f_j) \stackrel{d}{=} S_X(f_j) U_j \qquad j = 0, \ldots, \lfloor N/2 \rfloor, \tag{271b}$$

where U_j, $j = 1, \ldots, \lfloor (N-1)/2 \rfloor$, are asymptotically IID $\chi_2^2/2$ (in contrast, U_0 and $U_{N/2}$ – if N is even – are both distributed as χ_1^2). On the other hand, if we define $\epsilon(f_j)$ via Equation (270b), then $\{\epsilon(f_j) : 0 < f_j < f_{\mathcal{N}}\}$ constitutes a set of approximately IID RVs with mean zero and variance $\pi^2/6$ (note that neither U_j nor $\epsilon(f_j)$ are Gaussian).

Suppose that N is large enough so that the periodogram $\hat{S}_X^{(p)}(\cdot)$ is essentially an unbiased estimator of $S_X(\cdot)$ and is pairwise uncorrelated at the Fourier frequencies f_j. If $S_X(\cdot)$ is slowly varying in the neighborhood of, say, f_j, then

$$S_X(f_{j-M}) \approx \cdots \approx S_X(f_j) \approx \cdots \approx S_X(f_{j+M})$$

for some integer $M > 0$. Thus

$$\hat{S}_X^{(p)}(f_{j-M}), \ldots, \hat{S}_X^{(p)}(f_j), \ldots, \hat{S}_X^{(p)}(f_{j+M})$$

are a set of $2M+1$ unbiased and uncorrelated estimators of the same quantity, namely, $S_X(f_j)$. We can thus average them to produce the estimator

$$\bar{S}_X(f_j) \equiv \frac{1}{2M+1} \sum_{l=-M}^{M} \hat{S}_X^{(p)}(f_{j-l}). \tag{271c}$$

Under our assumptions we have

$$E\{\bar{S}_X(f_j)\} \approx S_X(f_j)$$

and

$$\operatorname{var}\{\bar{S}_X(f_j)\} \approx \frac{S_X^2(f_j)}{2M+1} \approx \frac{\operatorname{var}\{\hat{S}_X^{(p)}(f_j)\}}{2M+1}$$

	$c = 1$	$c = 2$	$c = 3$	$c = 4$
$r = 0$	2.5216281	−4.7715359	7.9199915	−11.9769211
$r = 1$	16.0778828	−20.6343346	25.0531521	−28.8738136
$r = 2$	31.8046265	−34.0071373	34.7700272	−34.3151321
$r = 3$	32.7861099	−30.2861233	26.7109356	−22.8838310
$r = 4$	18.7432098	−14.5717688	10.7177744	−7.5322194
$r = 5$	4.7226319	−2.6807923	1.3391306	−0.5167125

Table 272. Coefficients $\{\phi_{24,n} : n = 1, \ldots, 24\}$ for AR(24) process (Gao, 1997). The coefficient in row r and column c is $\phi_{24,4r+c}$. These coefficients are available on the Web site for this book (see page *xiv*).

using (270a) and assuming $f_{j-M} > 0$ and $f_{j+M} < f_{\mathcal{N}}$ for simplicity. If we now consider increasing both the sample size N and the index j in such a way that $j/(N\,\Delta t) = f_j$ is a constant, we can then let M get large also and claim that $\text{var}\,\{\bar{S}_X(f_j)\}$ can be made arbitrarily small so that $\bar{S}_X(f_j)$ is a consistent estimator of $S_X(f_j)$. An obvious generalization of the above is to use nonuniform weights in forming the summation in Equation (271c) (this and other variations are discussed in, for example, Cleveland and Parzen, 1975; Bloomfield, 1976; Walden, 1990; and Percival and Walden, 1993).

A different approach was taken in Wahba (1980), which investigates the smoothing of the log periodogram by a smoothing spline. This makes use of the signal plus noise model in Equation (271a) – recall, however, that this model is complicated by the fact that the noise has a $\log(\chi_2^2)$ distribution.

Methods such as periodogram smoothing and spline smoothing of the log periodogram are all characterized by a smoothing bandwidth that measures the width over which the underlying SDF estimate (such as the periodogram) is smoothed to produce the consistent estimator. For example, the uniform periodogram smoother averages over $2M + 1$ ordinates, each one themselves averaging over $\frac{1}{N\,\Delta t}$, and thus covers a bandwidth of $\frac{2M+1}{N\,\Delta t.}$ For a particular SDF, it is often possible to determine a bandwidth that trades off bias due to smoothing versus variance in a satisfactory way. Nevertheless, the bandwidth is fixed over the whole SDF and does not adapt in any manner to local features of the SDF; for example, sharp peaks or troughs require a narrow smoothing bandwidth in order to avoid a dominant bias due to smoothing, while slowly-varying areas of the SDF can benefit from variance reduction bestowed by a wide smoothing bandwidth. As we shall see in Sections 10.6 and 10.7, one approach to producing an adaptively smoothed SDF estimator is to use wavelet techniques to combine information about the SDF at different resolutions (scales).

- *Multitaper SDF estimation*

As we can deduce from Equation (269b), statistical theory says that, for large sample sizes N, the periodogram is approximately an unbiased estimator of the true SDF for a stationary process. It is important to realize, however, that, for finite sample sizes, this approximation can be quite poor due to a phenomenon known as *leakage*. Figure 273 illustrates the effect of leakage by comparing the periodogram (thin jagged curve) versus a true SDF (thick smooth curve) for a time series of length $N = 2048$ that is a realization of an AR(24) process with $\sigma_\varepsilon^2 = 1$ and with coefficients $\{\phi_{24,n}\}$ given in Table 272 (this process is also used as an example in Chapter 10 and by Gao, 1997; for details on how to generate realizations from AR processes, see Section 7.9).

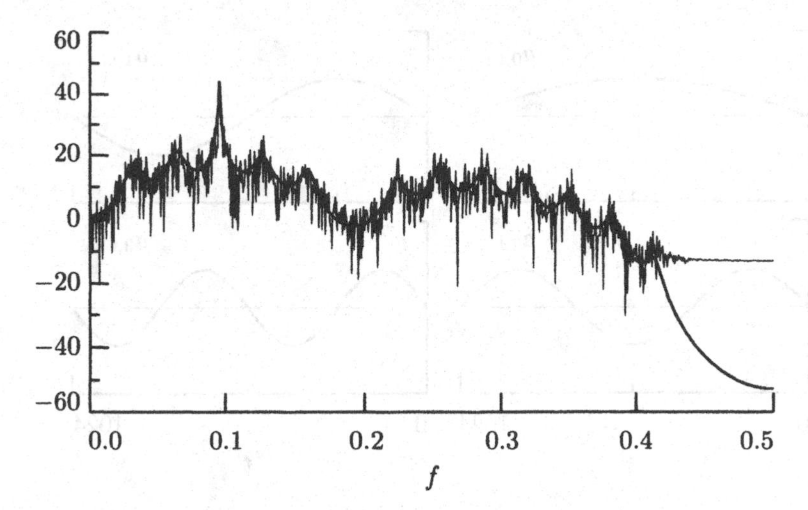

Figure 273. Periodogram (thin jagged curve) and true SDF (thick smooth) for a time series of length $N = 2048$ that is a realization of an AR(24) process (see Table 272 for the coefficients defining this process). Both the periodogram and true SDF are plotted on a decibel (dB) scale. Leakage is evident here in the periodogram at high frequencies, where the bias becomes as large as 40 dB (i.e., four orders of magnitude).

Both the periodogram and true SDF are plotted on a decibel (dB) scale; i.e., we plot $10 \cdot \log_{10}(\hat{S}_X^{(p)}(f))$ versus f. Leakage manifests itself here as the large discrepancy (bias) between the periodogram and the true SDF at high frequencies. In general, leakage is a concern in the periodogram for processes such as this AR(24) process whose SDFs cover a large dynamic range (i.e., the ratio of the largest SDF value to the smallest is more than, say, 40 or 50 dB – it is approximately 90 dB for the AR(24) SDF).

An accepted procedure for alleviating bias due to leakage in the periodogram is to apply a data taper (window) to the time series prior to computing the spectrum estimator, resulting in a so-called direct spectrum estimator. A direct spectrum estimator has better small sample bias properties than the periodogram, but, when subsequently smoothed across frequencies, has an asymptotic variance larger than that of a smoothed periodogram. The idea behind the multitaper spectrum estimator is to reduce this variance by computing a small number K of direct spectrum estimators, each with a different data taper, and then to average the K estimators together (Thomson, 1982; Percival and Walden, 1993, Chapter 7). If all K tapers are pairwise orthogonal and properly prevent leakage, the resulting multitaper estimator will be superior to the periodogram in terms of reduced bias and variance, particularly for spectra with high dynamic range and/or rapid variations. Multitapering has been successfully used to study, for example, the relationship between carbon dioxide and global temperature (Kuo *et al.*, 1990), turbulent plasma fluctuations (Riedel and Sidorenko, 1995) and heights of ocean waves (Walden *et al.*, 1995).

As before, let $\{X_t\}$ be a real-valued zero mean stationary process. Let $\{a_{n,t} : t = 0, \ldots, N-1\}$, $n = 0, \ldots, K-1$, denote the K different data tapers that are used to form the multitaper SDF estimator. These tapers are chosen to be orthonormal; i.e.,

$$\sum_{t=0}^{N-1} a_{n,t} a_{l,t} = \begin{cases} 1, & \text{if } n = l; \\ 0, & \text{otherwise.} \end{cases}$$

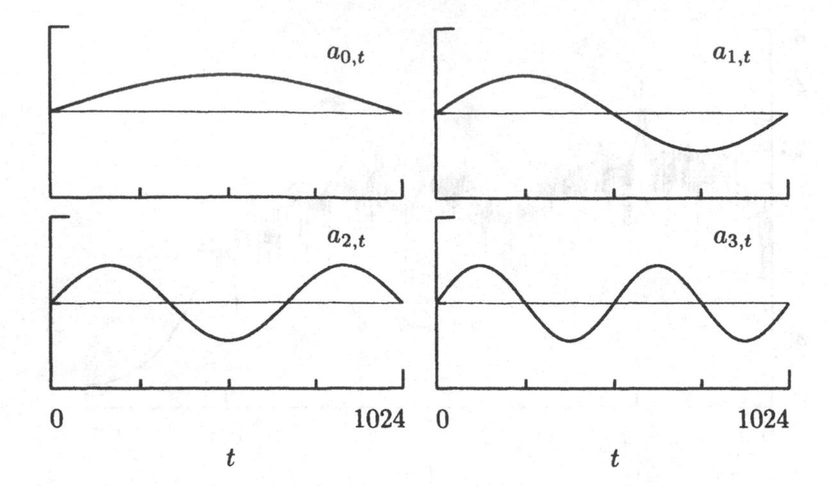

Figure 274. Sine tapers $\{a_{n,t}\}$ of orders $n = 0, 1, 2$ and 3 for $N = 1024$.

The primary role of each taper is to yield an SDF estimator that does not suffer from bias to the extent that the periodogram does. The simplest multitaper SDF estimator is the average of K direct spectral estimators (eigenspectra) and hence takes the form

$$\hat{S}_X^{(\mathrm{mt})}(f) \equiv \frac{1}{K} \sum_{n=0}^{K-1} \hat{S}_{X,n}^{(\mathrm{mt})}(f),$$

where

$$\hat{S}_{X,n}^{(\mathrm{mt})}(f) \equiv \Delta t \left| \sum_{t=0}^{N-1} a_{n,t} X_t e^{-i2\pi f t \, \Delta t} \right|^2 .$$

The primary role of the averaging operation is to recover 'information' that would be lost if but a single taper were to be used; a secondary role is to reduce the variability somewhat below what the periodogram gives.

A convenient set of easily computable orthonormal tapers is the set of sine tapers, the nth of which is given by

$$a_{n,t} = \left(\frac{2}{N+1}\right)^{1/2} \sin\left(\frac{(n+1)\pi(t+1)}{N+1}\right), \quad t = 0, \ldots, N-1. \tag{274a}$$

Figure 274 shows what these tapers look like when $N = 1024$ and $n = 0, 1, 2$ and 3. These simple tapers were introduced by Riedel and Sidorenko (1995) and are used in Sections 10.7. The estimator $\hat{S}_X^{(\mathrm{mt})}(\cdot)$ has an associated standardized bandwidth given by $\frac{K+1}{N+1}$ (Riedel and Sidorenko, 1995; Walden *et al.*, 1995). If the SDF is not rapidly varying over this bandwidth, the eigenspectra are approximately uncorrelated, which in turn yields the approximation (valid for large N)

$$\hat{S}_X^{(\mathrm{mt})}(f) \stackrel{\mathrm{d}}{=} \frac{S_X(f)\chi^2_{2K}}{2K}, \qquad 0 < f < f_{\mathcal{N}}. \tag{274b}$$

The multitaper estimator can be made consistent if we allow K to increase at a proper rate as $N \to \infty$, but typically K is kept fixed at a small number (10 or less). With K so

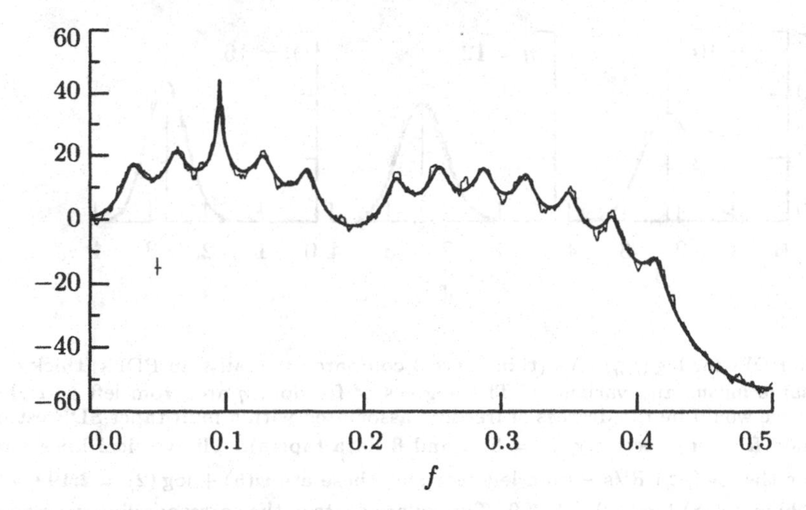

Figure 275. Multitaper SDF estimate $\hat{S}_X^{(mt)}(\cdot)$ (thin jagged curve) and true SDF (thick smooth) for a simulated AR(24) time series of length $N = 2048$ (the corresponding periodogram is shown in Figure 273). The multitaper estimate is based on $K = 10$ sine tapers. Both $\hat{S}_X^{(mt)}(\cdot)$ and the true SDF are plotted on a decibel scale. The width of the crisscross in the left-hand portion of the plot gives the bandwidth of $\hat{S}_X^{(mt)}(\cdot)$ (i.e., $\frac{K+1}{(N+1)} \doteq 0.0054$), while its height gives the length of a 95% confidence interval for a given $10 \cdot \log_{10}(S_X(f))$.

fixed, $\hat{S}_X^{(mt)}(\cdot)$ is an inconsistent SDF estimator; however, unlike the periodogram, the multitaper estimator now has small bias and $2K$ degrees of freedom, two for each of the K eigenspectra averaged together. Extensive discussion and background information on multitapering can be found in Thomson (1982) and Percival and Walden (1993, Chapter 7).

An example of a multitaper SDF estimate is shown in Figure 275 (thin curve) for the same AR(24) time series whose periodogram is shown in Figure 273 (in both figures, the thick smooth curve shows the true SDF). Here we use $K = 10$ sine tapers, yielding an estimate with $2K = 20$ degrees of freedom and a bandwidth of $\frac{K+1}{N+1} \doteq 0.0054$. This bandwidth is depicted as the width of the crisscross in the left-hand portion of the plot, while the height of the crisscross gives the length of a 95% confidence interval for the true SDF expressed in dBs (these are based on percentage points for the χ^2_{20} distribution, which can be approximated via Equation (264); for details on how to construct and interpret this confidence interval, see, for example, Percival and Walden, 1993, Section 6.10).

As we noted to be true for the periodogram, the logarithmic transform stabilizes the variance of a multitaper SDF estimator. The results of Bartlett and Kendall (1946) concerning the properties of a $\log(\chi^2_{2K})$ RV can again be called upon to show that, for $0 < f < f_{\mathcal{N}}$,

$$E\{\log(\hat{S}_X^{(mt)}(f))\} = \log(S_X(f)) + \psi(K) - \log(K)$$

and

$$\operatorname{var}\{\log(\hat{S}_X^{(mt)}(f))\} = \psi'(K),$$

where $\psi(\cdot)$ and $\psi'(\cdot)$ denote the digamma and trigamma functions, respectively:

$$\psi(z) \equiv \frac{d \log(\Gamma(z))}{dz} \text{ and } \psi'(z) \equiv \frac{d\psi(z)}{dz}. \tag{275}$$

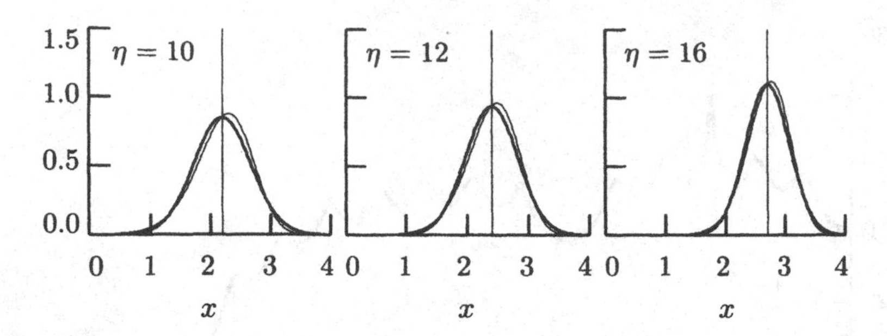

Figure 276. PDFs for $\log(\chi^2_\eta)$ RVs (thin curves) compared to Gaussian PDFs (thick curves) having the same means and variances. The degrees of freedom η are, from left to right, 10, 12 and 16 (these would be the degrees of freedom associated with a multitaper SDF estimator $\hat{S}^{(\text{mt})}_X(\cdot)$ formed from, respectively, $K = 5$, 6 and 8 data tapers). The vertical lines indicate the means for the $\log(\chi^2_\eta)$ RVs – from left to right, these are $\psi(5) + \log(2) \doteq 2.199$, $\psi(6) + \log(2) \doteq 2.399$ and $\psi(8) + \log(2) \doteq 2.709$. The square roots of the corresponding variances are, respectively, $\sqrt{\psi'(5)} \doteq 0.470$, $\sqrt{\psi'(6)} \doteq 0.426$ and $\sqrt{\psi'(8)} \doteq 0.365$. (Exercise [7.1] concerns the derivation of the $\log(\chi^2_\eta)$ PDF.)

As demonstrated in Figure 276, a comparison of a $\log(\chi^2_{2K})$ PDF with a Gaussian PDF having the same mean and variance shows very good agreement for $K \geq 5$, in line with Bartlett and Kendall (1946). For $0 < f < f_{\mathcal{N}}$, it thus follows that the RV

$$\eta(f) \equiv \log\left(\frac{\hat{S}^{(\text{mt})}_X(f)}{S_X(f)}\right) - \psi(K) + \log(K)$$

is approximately Gaussian distributed with mean zero and variance $\psi'(K)$.

If for $0 < f < f_{\mathcal{N}}$ we let

$$Y^{(\text{mt})}(f) \equiv \log(\hat{S}^{(\text{mt})}_X(f)) - \psi(K) + \log(K),$$

then

$$Y^{(\text{mt})}(f) = \log(S_X(f)) + \eta(f). \tag{276}$$

Hence the log multitaper SDF estimator (plus known constant $\log(K) - \psi(K)$) can be written as a signal (the true log SDF), plus approximately Gaussian noise with zero mean and known variance $\sigma^2_\eta = \psi'(K)$.

If we evaluate Equation (276) over the grid of Fourier frequencies $f_k = k/(N\,\Delta t)$, then we have the approximate result that, for $0 < f_k < f_{\mathcal{N}}$,

$$\eta(f_k) \overset{\text{d}}{=} \mathcal{N}(0, \psi'(K)),$$

but these RVs are correlated. Let us explore the covariance structure of $\eta(f)$ across frequencies. We take $\Delta t = 1$ for notational convenience (hence $f_{\mathcal{N}} = 1/2$). For a fixed f and ν such that $0 < f < 1/2$ and $0 < f + \nu < 1/2$, let us define

$$s_\eta(\nu) \equiv \text{cov}\{\eta(f), \eta(f+\nu)\} = \text{cov}\{\log(V(f)), \log(V(f+\nu))\},$$

where

$$V(f) \equiv \frac{\hat{S}^{(\text{mt})}_X(f)}{S_X(f)}.$$

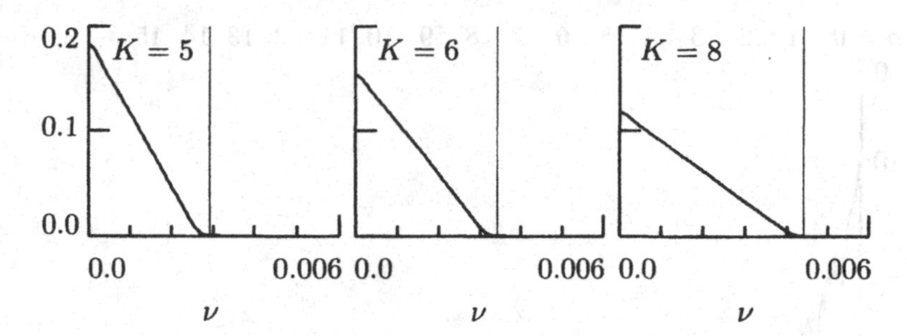

Figure 277. The autocovariance $\tilde{s}_\eta(\nu)$ versus ν for $N = 2048$ and $K = 5, 6$ and 8 sine tapers. Each vertical line shows the bandwidth $\frac{K+1}{N+1}$ of the associated multitaper SDF estimator.

If $\log(V(f))$ and $\log(V(f+\nu))$ were exactly jointly Gaussian, we would have

$$\mathrm{cov}\{\log(V(f)), \log(V(f+\nu))\} = \log(1 + \mathrm{cov}\{V(f), V(f+\nu)\})$$

(Granger and Hatanaka, 1964, Section 3.7); since they are approximately jointly Gaussian, we can take the above to be a reasonable approximation, and hence

$$s_\eta(\nu) \approx \log(1 + s_V(\nu)), \quad \text{where } s_V(\nu) \equiv \mathrm{cov}\{V(f), V(f+\nu)\}.$$

Under the assumption of a locally slowly varying SDF so that $S_X(f) \approx S_X(f+\nu)$ for small ν, we have

$$s_V(\nu) \approx \frac{1}{K^2 S_X^2(f)} \sum_{n=0}^{K-1} \sum_{l=0}^{K-1} \mathrm{cov}\left\{\hat{S}_{X,n}^{(\mathrm{mt})}(f), \hat{S}_{X,l}^{(\mathrm{mt})}(f+\nu)\right\}.$$

Under the same assumption on $S_X(\cdot)$, Thomson (1982, p. 1069) showed that

$$\mathrm{cov}\left\{\hat{S}_{X,n}^{(\mathrm{mt})}(f), \hat{S}_{X,l}^{(\mathrm{mt})}(f+\nu)\right\} \approx S_X^2(f) \left| \sum_{t=0}^{N-1} a_{n,t} a_{l,t} e^{i2\pi\nu t} \right|^2,$$

an approximation that neglects a frequency dependent term that is only significant for f close to 0 or 1/2. Hence, we have

$$s_V(\nu) \approx \tilde{s}_V(\nu) \equiv \frac{1}{K^2} \sum_{n=0}^{K-1} \sum_{l=0}^{K-1} \left| \sum_{t=0}^{N-1} a_{n,t} a_{l,t} e^{i2\pi\nu t} \right|^2.$$

The above development thus gives $s_\eta(\nu) \approx \tilde{s}_\eta(\nu) \equiv \log(1 + \tilde{s}_V(\nu))$.

Figure 277 shows a plot of $\tilde{s}_\eta(\nu)$ versus ν for $K = 5$, 6 and 8 sine tapers with $N = 2048$. We note that $\tilde{s}_\eta(\nu)$ is negligible for $\nu \geq \frac{K+1}{N+1}$, where the latter is the standardized bandwidth of a sine multitaper estimator and is indicated on the plots by vertical lines. The straight line shape of the autocovariance is maintained for other values of N and K. Recalling that $s_\eta(0) = \sigma_\eta^2$ and using $\frac{K+1}{N+1} \approx \frac{K+1}{N}$, we can thus formulate a very simple and convenient model, namely,

$$s_\eta(\nu) = \begin{cases} \sigma_\eta^2 \left(1 - \frac{|\nu| N}{K+1}\right), & \text{if } |\nu| \leq (K+1)/N; \\ 0, & \text{otherwise.} \end{cases} \tag{277}$$

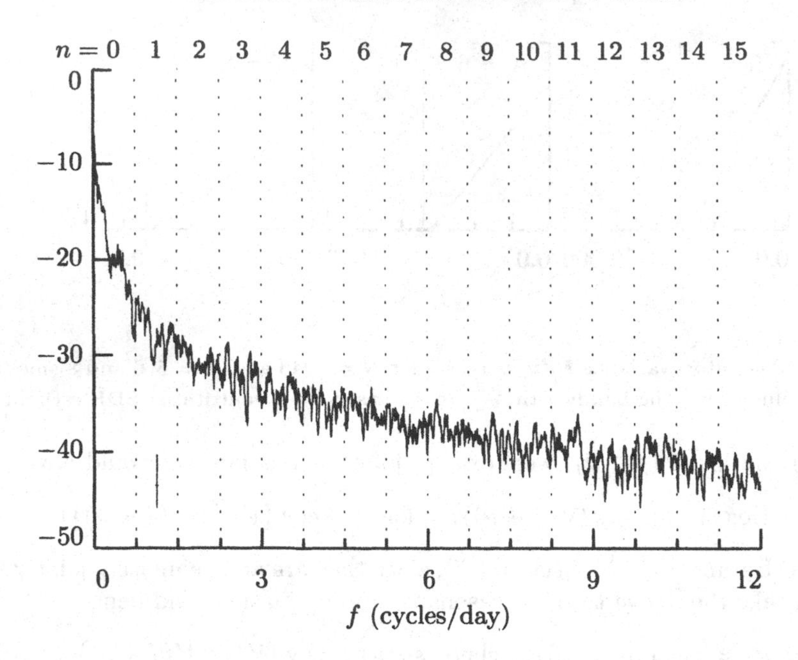

Figure 278. Multitaper SDF estimate $\hat{S}_X^{(\mathrm{mt})}(\cdot)$ (in decibels) of the solar physics time series using $K = 10$ sine tapers (this series of $N = 4096$ values is plotted in Figures 222 and 235). The vertical dotted lines partition the frequency interval [0, 12 cycles/day] into 16 subintervals, the same as would be achieved by a level $j = 4$ DWPT (see Figures 220) or MODWPT (Figure 236). The width of the crisscross in the lower left-hand corner of the plot gives the physical bandwidth of $\hat{S}_X^{(\mathrm{mt})}(\cdot)$ (i.e., $\frac{K+1}{(N+1)\,\Delta t} \doteq 0.0644$ cycles/day – here $\Delta t = 1/24$ days), while its height gives the length of a 95% confidence interval for a given $10 \cdot \log_{10}(S_X(f))$.

Comments and Extensions to Section 7.5

[1] As a second example of a multitaper SDF estimate, Figure 278 shows a $K = 10$ sine multitaper estimate for the solar physics series described in Section 6.2 (this series has $N = 4096$ data values and is plotted at the bottoms of Figures 222 and 235 – the vertical axis is larger in the latter). This estimate has $2K = 20$ degrees of freedom and an associated bandwidth of 0.0644 cycles/day (computed by multiplying the standardized bandwidth $\frac{K+1}{N+1}$ by $1/\Delta t = 24$, where $\Delta t = 1/24$ days). The (barely visible!) width of the crisscross in the lower left-hand corner of the plot depicts this bandwidth, while the height of the crisscross gives the length of a 95% confidence interval for a hypothesized true SDF for the time series. The vertical dotted lines show the partitioning of the frequency interval into the 16 subintervals corresponding to a level $j = 4$ DWPT (Figure 220) or MODWPT (Figure 236).

An obvious limitation of this (or any other) SDF estimate for this series is that, in keeping with the assumption of stationarity we use in interpreting it, we are declaring any observed time-dependent fluctuations in the solar physics series to be consistent with a realization of a stochastic process with time-independent properties (in particular, this implies that the observed bursts could have occurred equally likely at any other time); however, the physical considerations discussed in Section 6.2 would suggest using a time-dependent stochastic model. Thus, while the estimate $\hat{S}_X^{(\mathrm{mt})}(\cdot)$ tells us the relative importance of various frequencies in a global Fourier decomposition of this series, it is incapable of saying anything about localized time/frequency relationships (the DWPT/MODWPT decompositions achieve time localization at the expense of expanding the bandwidth up to 0.75 cycles/day).

7.6 Definition and Models for Long Memory Processes

Suppose that $\{X_t\}$ is a stationary process with an SDF denoted by $S_X(\cdot)$ (we assume for convenience that $\Delta t = 1$ so that $f_{\mathcal{N}} = 1/2$ throughout). We say that $\{X_t\}$ is a stationary long memory process if there exist constants α and C_S satisfying $-1 < \alpha < 0$ and $C_S > 0$ such that

$$\lim_{f \to 0} \frac{S_X(f)}{C_S |f|^\alpha} = 1 \tag{279a}$$

(Beran, 1994, p. 42). In other words, a stationary long memory process has an SDF $S_X(\cdot)$ such that $S_X(f) \approx C_S|f|^\alpha$, with the approximation improving as f approaches zero. An alternative definition can be stated in terms of the ACVS $\{s_{X,\tau}\}$ for $\{X_t\}$. We say that $\{X_t\}$ is a stationary long memory process if there exist constants β and C_s satisfying $-1 < \beta < 0$ and $C_s > 0$ such that

$$\lim_{\tau \to \infty} \frac{s_{X,\tau}}{C_s \tau^\beta} = 1,$$

where β is related to α in Equation (279a) via $\beta = -\alpha - 1$ (this definition and (279a) are equivalent in a sense made precise by Beran, 1994, Theorem 2.1). Standard time series models such as stationary autoregressive processes have ACVSs such that $s_{X,\tau} \approx C\phi^\tau$ for large τ, where $C \geq 0$ and $|\phi| < 1$. For a long memory process, we have $s_{X,\tau} \approx C_s\tau^\beta$ for large τ. In both cases, note that $s_{X,\tau} \to 0$ as $\tau \to \infty$, but the rate of decay toward zero is much slower for a long memory process, implying that observations that are widely separated in time can still have a nonnegligible covariance (in more colorful terms, current observations retain some 'memory' of the distant past).

Several different models for stationary long memory processes have been proposed and studied in the literature, of which the following three are particularly important.

- *Fractional Gaussian noise*

The first comprehensive model for a long memory process is due to Mandelbrot and van Ness (1968) and is known as fractional Gaussian noise (FGN). By definition, if $\{X_t\}$ is an FGN, then it is a stationary process whose ACVS is given by

$$s_{X,\tau} = \frac{\sigma_X^2}{2}\left(|\tau+1|^{2H} - 2|\tau|^{2H} + |\tau-1|^{2H}\right), \quad \tau = \ldots, -1, 0, 1, \ldots; \tag{279b}$$

here $\sigma_X^2 = \text{var}\{X_t\}$ is an arbitrary positive number, and H is the so-called Hurst (or self-similarity) parameter satisfying $0 < H < 1$. FGN can be regarded as increments of fractional Brownian motion (FBM) $\{B_H(t) : 0 \leq t < \infty\}$ with parameter H; i.e.,

$$X_t = B_H(t+1) - B_H(t), \quad t = 0, 1, 2, \ldots.$$

A precise definition of FBM is given in, for example, Beran (1994), from which we note that $B_H(0) \equiv 0$; that $B_H(t)$ for $t > 0$ is a zero mean Gaussian RV with variance $\sigma_X^2 t^{2H}$; and that $\text{cov}\{B_H(t), B_H(s) - B_H(t)\} = 0$ when $s \geq t \geq 0$. We can thus create samples of discrete fractional Brownian motion (DFBM) by cumulatively summing $\{X_t\}$:

$$B_t \equiv B_H(t) = \sum_{u=0}^{t-1} X_u, \quad t = 1, 2, \ldots$$

(see the Comments and Extensions to this section for a clarification on how 't' is to be interpreted in B_t and $B_H(t)$). The SDF for $\{B_H(t)\}$ can be taken to be

$$S_{B_H(t)}(f) = \frac{\sigma_X^2 C_H}{|f|^{2H+1}}, \quad -\infty < f < \infty, \tag{280a}$$

with $C_H \equiv \Gamma(2H+1)\sin(\pi H)/(2\pi)^{2H+1}$ (Mandelbrot and van Ness, 1968; Flandrin, 1989; Masry, 1993). Thus $S_{B(t)}(f) \propto |f|^{-1-2H}$. It follows from filtering and aliasing considerations (see, for example, Percival and Walden, 1993, Section 3.8) that the SDF for $\{B_t\}$ is given by

$$S_{B_t}(f) = \sigma_X^2 C_H \sum_{j=-\infty}^{\infty} \frac{1}{|f+j|^{2H+1}}, \quad -\frac{1}{2} \le f \le \frac{1}{2}, \tag{280b}$$

so that $S_{B_t}(f) \propto |f|^{-1-2H}$ approximately for small f, while the SDF for $\{X_t\}$ is given by

$$S_X(f) = 4\sigma_X^2 C_H \sin^2(\pi f) \sum_{j=-\infty}^{\infty} \frac{1}{|f+j|^{2H+1}}, \quad -\frac{1}{2} \le f \le \frac{1}{2}$$

(Sinai, 1976; Beran, 1994, p. 53, Equation (2.17)). For the latter result note that $X_t = B_{t+1} - B_t$, so that, except for an innocuous unit shift in the convention for indexing B_t, $\{X_t\}$ is the first difference of $\{B_t\}$. The squared gain function for a difference filter is given by $\mathcal{D}(f) \equiv 4\sin^2(\pi f)$, (see Exercise [105b]), and the result follows from Equation (268b). By using an Euler–Maclaurin summation (Dahlquist and Björck, 1974), we can approximate the infinite summation above using a finite one with $2M+3$ terms:

$$\begin{aligned} S_X(f) \approx 4\sigma_X^2 C_H \sin^2(\pi f) \Bigg(&\sum_{j=-M}^{M} \frac{1}{|f+j|^{2H+1}} \\ &+ \sum_{l=-1,1} \left[\frac{1}{2H(lf+M+1)^{2H}} - \frac{(2H+1)(2H+2)(2H+3)}{720(lf+M+1)^{2H+4}} \right. \\ &\left. + \frac{2H+1}{12(lf+M+1)^{2H+2}} + \frac{1}{2(lf+M+1)^{2H+1}} \right] \Bigg) \end{aligned}$$

(for details, see Percival *et al.*, 2000a). In practice, setting $M = 100$ yields sufficient accuracy. For small f we have $S_X(f) \propto |f|^{1-2H}$ approximately, so FGN satisfies the definition in Equation (279a) of a stationary long memory process when $-1 < 1-2H < 0$, i.e., when $\frac{1}{2} < H < 1$.

The top row of Figure 282 shows $S_X(\cdot)$ for $H = 0.55, 0.75, 0.90$ and 0.95 using both linear/log and log/log axes (left- and right-hand plots, respectively). Because $S_X(f) \propto f^{1-2H}$ for small f, processes with $1/2 < H < 1$ have increasingly prominent low frequency components as H increases, a pattern that is in agreement with the curves in the figure. For positive frequencies over which $S_X(f) \propto f^{1-2H}$ approximately, we have $\log_{10}(S_X(f)) \propto (1-2H)\log_{10}(f)$, so the linear appearance of the curves in the right-hand plot indicates that this approximation is quite good for frequencies below about $f = 0.2$.

The first and fourth plots of Figure 283 show realizations of FGNs for, respectively, $H = 0.55$ and $H = 0.95$ (these simulated series were created using the Davies–Harte method – see Section 7.8). Note that, in keeping with the properties of their SDFs, the series with $H = 0.95$ has proportionately stronger low frequency components than the $H = 0.55$ series.

When $H = 1/2$, an FGN reduces to a white noise process (i.e., Equation (279b) has the form of (268d) in this case); when $0 < H < \frac{1}{2}$, the process $\{X_t\}$ is characterized by high frequency fluctuations and has a deficiency in power at low frequencies (for these processes we in fact have $S_X(0) = 0$).

- *Pure power law process*

We say that the discrete parameter process $\{X_t\}$ is a pure power law (PPL) process if its SDF has the form

$$S_X(f) = C_S|f|^\alpha, \quad -\frac{1}{2} \le f \le \frac{1}{2},$$

where $C_S > 0$. These processes can be divided up into two categories corresponding to stationary and nonstationary processes. The first corresponds to the case $-1 < \alpha$. The process is stationary, and its SDF has the standard meaning. The corresponding ACVS can be obtained via the formula

$$s_{X,\tau} = \int_{-1/2}^{1/2} S_X(f) e^{i2\pi f\tau}\, df = 2C_S \int_0^{1/2} f^\alpha \cos(2\pi f\tau)\, df,$$

which in general does not have a simple closed form expression and hence must be evaluated via numerical integration. For $-1 < \alpha < 0$, a PPL process obviously obeys the definition in Equation (279a) for a stationary long memory process – the SDFs for four such processes ($\alpha = -0.1$, -0.5, -0.8 and -0.9) are shown in the middle row of Figure 282. Simulated series from PPL processes with $\alpha = -0.1$ and $\alpha = -0.9$ are shown in, respectively, the second and fifth plots of Figure 283. For $\alpha = 0$, a PPL process becomes a white noise process with variance C_S. All PPL processes with $\alpha > 0$ are stationary processes with a deficiency in low frequency components, which is reflected in the fact that $S_X(0) = 0$.

The second category applies when $\alpha \le -1$. Such a PPL process $\{X_t\}$ can be interpreted as a nonstationary process that can be turned into a stationary process through an appropriate differencing operation. Details are given in Section 7.7, where the meaning of the SDF in this case is explained. PPL processes are thus well defined for any α on the real axis.

Processes that satisfy Equation (279a) for some α on the real axis are sometimes called '$1/f$-type processes,' so stationary long memory processes are $1/f$-type processes with $-1 < \alpha < 0$, while $1/f$-type processes with $\alpha \le -1$ are sometimes called nonstationary long memory processes.

- *Fractionally differenced process*

A popular time series model $\{X_t\}$ for a stationary long memory process is the fractionally differenced (FD) process, which was introduced independently by Granger and Joyeux (1980) and Hosking (1981). Here $\{X_t\}$ is related to a (typically Gaussian)

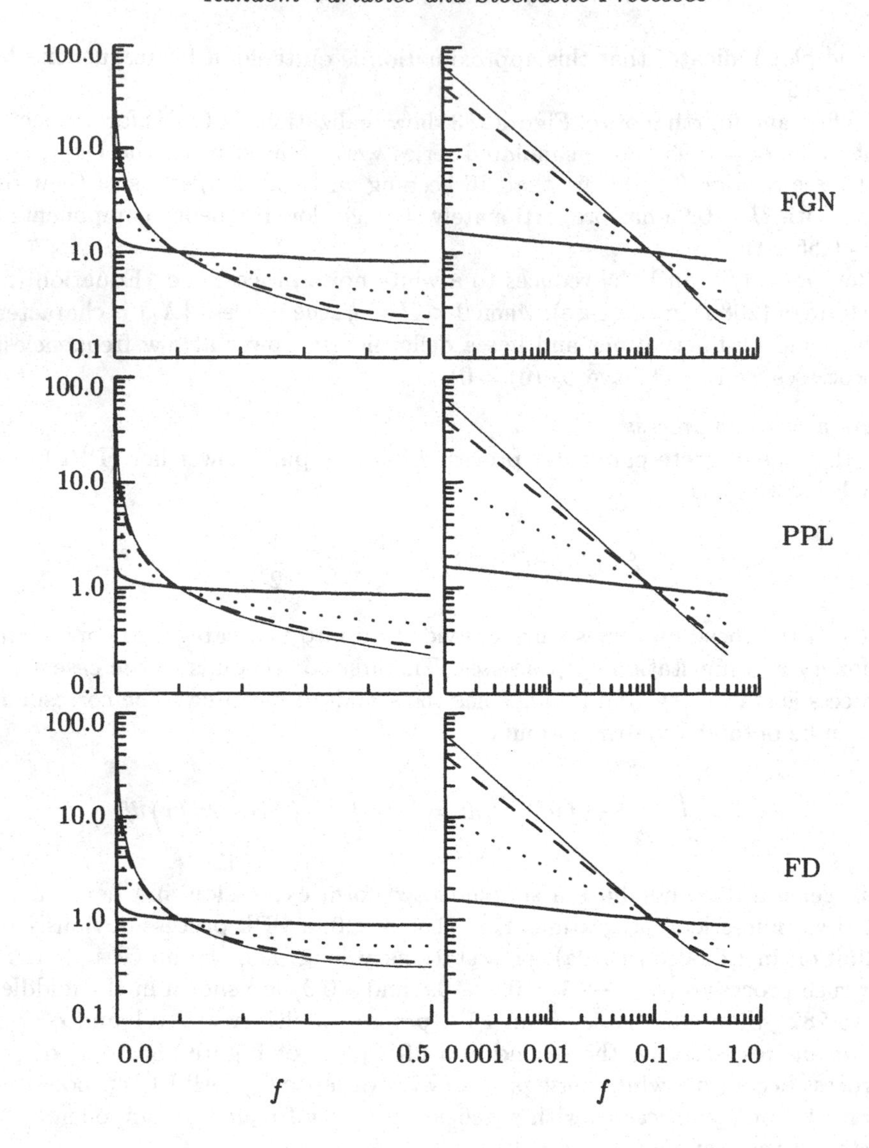

Figure 282. SDFs for FGN, PPL and FD processes (top to bottom rows, respectively) on both linear/log and log/log axes (left- and right-hand columns, respectively). Each SDF $S_X(\cdot)$ is normalized such that $S_X(0.1) = 1$. The table below gives the parameter values for the various plotted curves.

process	thick solid	dotted	dashed	thin solid
FGN	$H = 0.55$	$H = 0.75$	$H = 0.90$	$H = 0.95$
PPL	$\alpha = -0.1$	$\alpha = -0.5$	$\alpha = -0.8$	$\alpha = -0.9$
FD	$\delta = 0.05$	$\delta = 0.25$	$\delta = 0.40$	$\delta = 0.45$

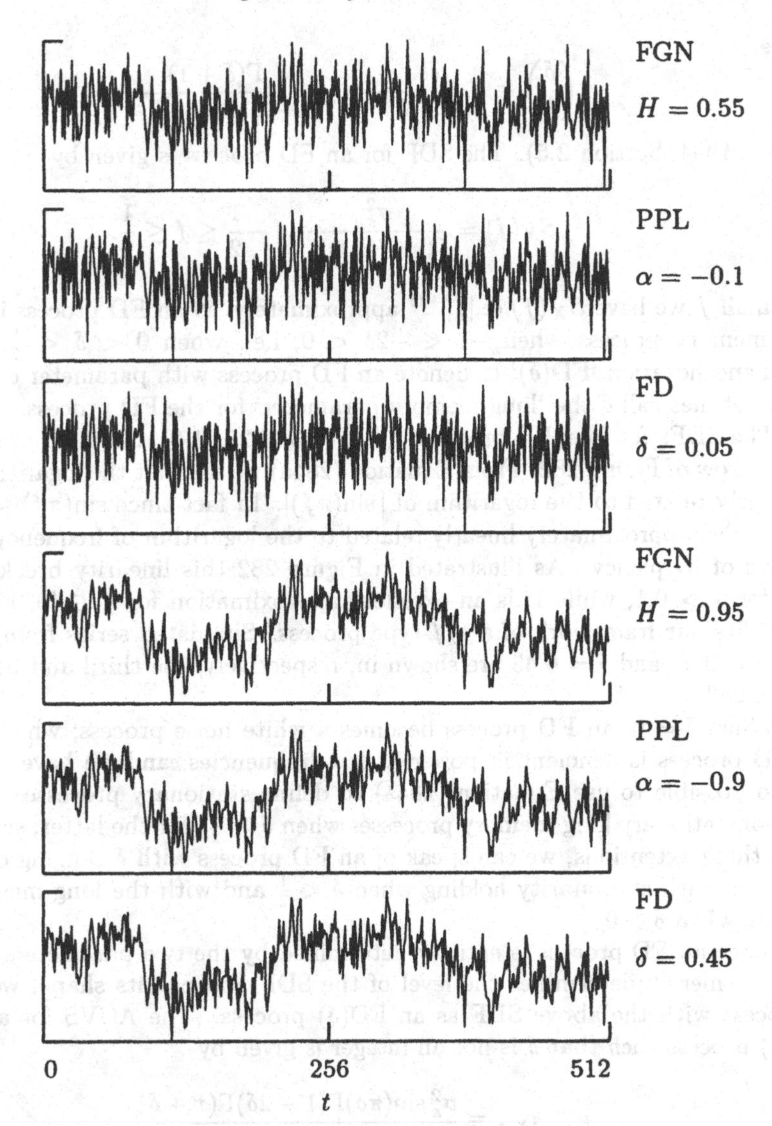

Figure 283. Simulated realizations of FGN, PPL and FD processes. The thick (thin) solid curves in Figure 282 show the SDFs for the top (bottom) three series – these SDFs differ markedly only at high frequencies. We formed each simulated $X_0, \ldots, X_{511}$ using the Davies–Harte method (see Section 7.8), which does so by transforming a realization of a portion $Z_0, \ldots, Z_{1023}$ of a white noise process (the Z_t values are on the Web site for this book – see page *xiv*). To illustrate the similarity of FGN, PPL and FD processes with comparable H, α and δ, we used the same Z_t to create all six series. Although the top (bottom) three series appear to be identical, estimates of their SDFs show high frequency differences consistent with their theoretical SDFs.

white noise process $\{\varepsilon_t\}$ with mean zero and variance σ_ε^2 through $(1 - B)^\delta X_t = \varepsilon_t$, where B is the backward shift operator (so that, for example, $(1-B)X_t = X_t - X_{t-1}$); $-\frac{1}{2} < \delta < \frac{1}{2}$; and $(1 - B)^\delta$ is interpreted as

$$(1 - B)^\delta = \sum_{k=0}^{\infty} \binom{\delta}{k} (-1)^k B^k \text{ so that } \sum_{k=0}^{\infty} \binom{\delta}{k} (-1)^k X_{t-k} = \varepsilon_t,$$

where

$$\binom{\delta}{k} \equiv \frac{\delta!}{k!(\delta-k)!} = \frac{\Gamma(\delta+1)}{\Gamma(k+1)\Gamma(\delta-k+1)}$$

(Beran, 1994, Section 2.5). The SDF for an FD process is given by

$$S_X(f) = \frac{\sigma_\varepsilon^2}{[4\sin^2(\pi f)]^\delta}, \quad -\frac{1}{2} \le f \le \frac{1}{2}. \tag{284a}$$

For small f we have $S_X(f) \propto |f|^{-2\delta}$ approximately, so an FD process is a stationary long memory process when $-1 < -2\delta < 0$, i.e., when $0 < \delta < \frac{1}{2}$. We use the shorthand notation 'FD(δ)' to denote an FD process with parameter δ. When $\delta > 0$, we sometimes call δ the 'long memory parameter' for the FD process.

The SDFs for four FD processes ($\delta = 0.05$, 0.25, 0.40 and 0.45) are shown in the bottom row of Figure 282. From Equation (284a) we see that the logarithm of the SDF is linearly related to the logarithm of $|\sin(\pi f)|$. In fact since $\sin(\pi f) \approx \pi f$ for small f, it is also approximately linearly related to the logarithm of frequency f over many octaves of frequency. As illustrated in Figure 282 this linearity breaks down badly only for $f > 0.1$, while it is an excellent approximation for $f \in (0, 0.1]$. Hence the model fits our framework of a $1/f$-type process. Simulated series from FD processes with $\delta = 0.05$ and $\delta = 0.45$ are shown in, respectively, the third and bottom plots of Figure 283.

When $\delta = 0$, an FD process becomes a white noise process; when $-\frac{1}{2} < \delta < 0$, an FD process is deficient in power at low frequencies, and we have $S_X(0) = 0$. It is also possible to use Equation (284a) to define stationary processes when $\delta \le -\frac{1}{2}$ and nonstationary long memory processes when $\delta \ge \frac{1}{2}$ (for the latter, see Section 7.7). With these extensions, we can speak of an FD process with δ ranging over the entire real axis, with stationarity holding when $\delta < \frac{1}{2}$ and with the long memory property holding when $\delta > 0$.

Since an FD process is entirely determined by the two parameters δ and σ_ε^2 and since σ_ε^2 merely determines the level of the SDF and not its shape, we will refer to a process with the above SDF as an FD(δ) process. The ACVS for any stationary FD(δ) process such that δ is not an integer is given by

$$s_{X,\tau} \equiv \frac{\sigma_\varepsilon^2 \sin(\pi\delta)\Gamma(1-2\delta)\Gamma(\tau+\delta)}{\pi\Gamma(\tau+1-\delta)}. \tag{284b}$$

Setting $\tau = 0$ gives us the process variance $s_{X,0} = \mathrm{var}\,\{X_t\}$, for which the above expression reduces somewhat to

$$s_{X,0} = \frac{\sigma_\varepsilon^2\Gamma(1-2\delta)}{\Gamma^2(1-\delta)}. \tag{284c}$$

This equation holds for all stationary FD processes (i.e., when $\delta < \frac{1}{2}$). Once $s_{X,0}$ has been determined, the remainder of the ACVS can be computed easily using the recursion

$$s_{X,\tau} = s_{X,\tau-1}\frac{\tau+\delta-1}{\tau-\delta}, \qquad \tau = 1, 2, \ldots \tag{284d}$$

(the above holds for all $\delta < \frac{1}{2}$; the ACVS at negative lags follows immediately from $s_{X,-\tau} = s_{X,\tau}$, which must be true for any real-valued stationary process).

FGN, PPL and FD processes are similar in that each depends upon just two parameters, one controlling the exponent of the approximating (or exact) power law as $f \to 0$, and the other, the variance of the process (or, equivalently, the level of the SDF). Proper choice of these parameters can yield FGN, PPL and FD processes whose SDFs are virtually indistinguishable for small f. In spite of these similarities, FD processes have several advantages:

[1] both the SDF and ACVS for an FD process are easy to compute;
[2] there is a natural extension to FD processes to cover nonstationary processes with stationary backward differences (see Section 7.7); and
[3] an FD process can be considered as a special case of an autoregressive, fractionally integrated, moving average (ARFIMA) process, which involves additional parameters that allow flexible modeling of the high frequency content of a time series.

Good discussions about the properties of FD processes and their extensions can be found in the seminal articles by Granger and Joyeux (1980) and Hosking (1981) and in the book by Beran (1994).

Table 286 summarizes parameter ranges for FGN, PPL and FD processes for which the form of the process is (a) stationary long memory, (b) white noise, (c) stationary (but not long memory) and (d) nonstationary long memory. Relationships amongst the Hurst coefficient H, the spectral slope α and the fractional difference parameter δ are illustrated in Figure 286, where, in order to distinguish between H as a parameter for DFBM and as a parameter for FGN, we have used H_B for the former and H_G for the latter. These relationships are obtained via Equation (279a), which allows us to connect H and δ to α and hence to each other.

Comments and Extensions to Section 7.6

[1] Fractional Gaussian noise $\{X_t\}$ is derived via sampling and differencing from the *continuous* parameter stochastic process known as fractional Brownian motion, which we denote as $\{B_H(t) : 0 \leq t < \infty\}$. Note that the '$t$' in X_t is a unitless index referring to the tth element of the process $\{X_t\}$, whereas the 't' in $B_H(t)$ has physically meaningful units (e.g., seconds, years, etc.). Hence, when we create the sampled process $B_t \equiv B_H(t)$ and difference it to form $X_t \equiv B_{t+1} - B_t$, the associated sampling interval Δt for both $\{B_t\}$ and $\{X_t\}$ is equal to unity, and the units in which Δt is expressed are the same as those for 't' in $B_H(t)$ (e.g., $\Delta t = 1$ second if 't' in $B_H(t)$ has units of seconds).

[2] It should be noted that the stationary PPL processes we define here are discrete parameter processes $\{X_t : t = \ldots, -1, 0, 1, \ldots\}$, which means in part that their SDFs are defined for $|f| \leq 1/2$ (the Nyquist frequency is one half because of our assumption that the sampling interval Δt is unity). Continuous parameter PPL processes $\{X(t) : -\infty < t < \infty\}$ are more commonly referred to in the literature, and these have SDFs that are defined over the entire real axis. These processes must be handled with some care. For example, note that, for $-1 < \alpha < 0$, we have

$$\int_{-\infty}^{\infty} |f|^{\alpha}\, df = \infty$$

because of the slow rate of decay of $|f|^{\alpha}$ to zero as $|f| \to \infty$. For this reason, it is necessary to introduce something like a high frequency cutoff in order to get an

process	nonstationary LMP	stationary LMP	white noise	stationary not LMP
FGN	—	$\frac{1}{2} < H < 1$	$H = \frac{1}{2}$	$0 < H \leq \frac{1}{2}$
PPL	$\alpha \leq -1$	$-1 < \alpha < 0$	$\alpha = 0$	$\alpha \geq 0$
FD	$\delta \geq \frac{1}{2}$	$0 < \delta < \frac{1}{2}$	$\delta = 0$	$\delta \leq 0$

Table 286. Parameter ranges for each named stochastic process for which the form of the process is (a) nonstationary long memory, (b) stationary long memory, (c) white noise or (d) stationary but not long memory.

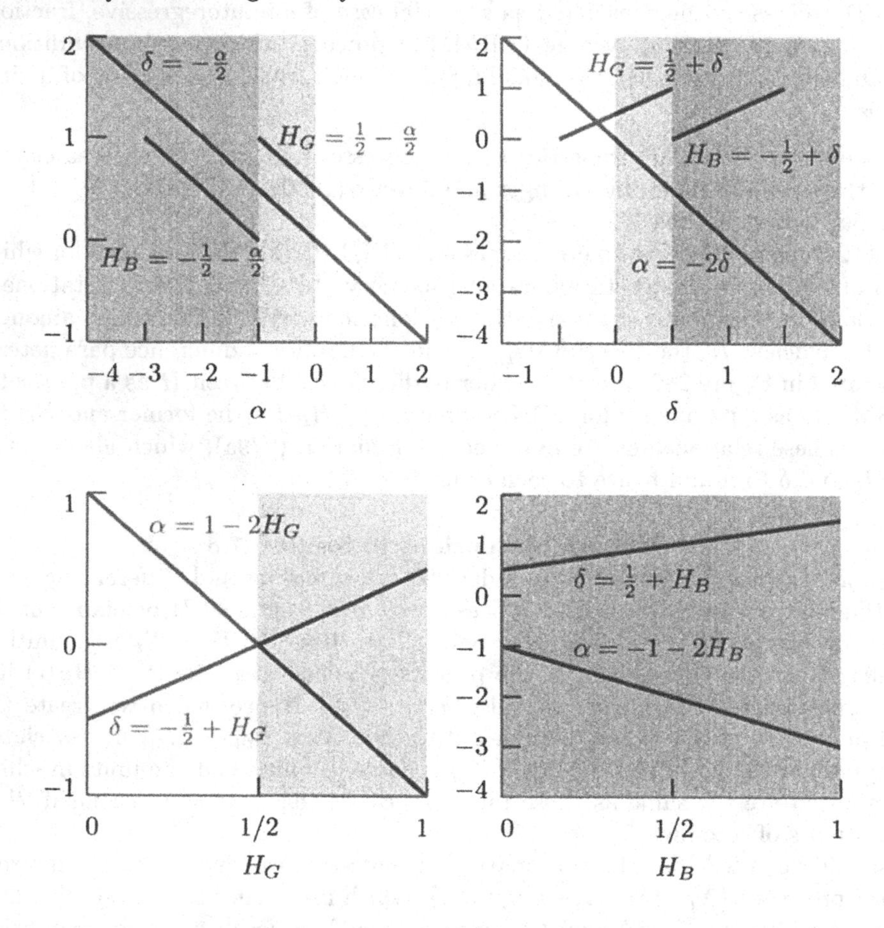

Figure 286. Relationships amongst the spectral slope α, the fractional difference parameter δ and the Hurst coefficient H (both for FGN and DFBM). The unshaded, lightly shaded and heavily shaded regions represent parameter values corresponding to, respectively, stationary processes without long memory, stationary long memory processes and nonstationary long memory processes (white noise processes occur when the boundary between the unshaded and lightly shaded regions crosses a thick line). For this plot only, we distinguish between H as a parameter for DFBM and for FGN by using H_B in the former case and H_G in the latter. Note that, while α and δ range over the entire real axis, we must have $0 < H < 1$ for both DFBM and FGN.

SDF that corresponds to some continuous parameter stationary process. Accordingly, suppose that we define a band-limited PPL process $\{X(t)\}$ to be one whose SDF is given by

$$S_{X(t)}(f) = \begin{cases} C_S|f|^{\alpha}, & |f| \leq f_c; \\ 0, & \text{otherwise,} \end{cases}$$

where $0 < f_c < \infty$ is the high frequency cutoff. If we now define $X_t = X(t)$, it is important to realize that we will *not* obtain a discrete parameter PPL process unless $f_c = 1/2$. Due to aliasing, the SDFs for $\{X_t\}$ and $\{X(t)\}$ are in fact related by

$$S_X(f) = \sum_{j=-\infty}^{\infty} S_{X(t)}(f + j), \quad |f| \leq \frac{1}{2},$$

a point that is sometimes overlooked in the literature (see also item [1] of the Comments and Extensions to the next section).

7.7 Nonstationary $1/f$-Type Processes

As mentioned in Section 7.4 any nonnegative and even function $S_X(\cdot)$ that integrates to a finite nonnegative value can be taken to be the SDF for some stationary real-valued process. The SDF $S_X(\cdot)$ of a stationary long memory process $\{X_t\}$ is such that $S_X(f) \to \infty$ as $f \to 0$ but this singularity at $f = 0$ still yields an integrable function because at low frequencies the SDF is of the form $S_X(f) \propto |f|^{\alpha}$ with $-1 < \alpha < 0$. It is possible to define an interesting class of nonstationary processes that have a well-defined SDF that in fact integrates to an infinite value.

Suppose that $\{X_t\}$ is a stochastic process whose dth order backward difference

$$Y_t \equiv (1 - B)^d X_t = \sum_{k=0}^{d} \binom{d}{k} (-1)^k X_{t-k}$$

is a stationary process with SDF $S_Y(\cdot)$ and mean μ_Y. Here d is a nonnegative *integer*, unlike δ for the FD process. Again B is the backward shift operator defined by $BX_t \equiv X_{t-1}$ so that $B^k X_t = X_{t-k}$. As examples, we have

$$Y_t = \begin{cases} X_t, & \text{if } d = 0; \\ X_t - X_{t-1}, & \text{if } d = 1; \\ X_t - 2X_{t-1} + X_{t-2}, & \text{if } d = 2; \text{ etc.} \end{cases}$$

If $\{X_t\}$ were itself stationary with SDF $S_X(\cdot)$, the result quoted in Equation (268b) says that $S_X(\cdot)$ and $S_Y(\cdot)$ would be related by $S_Y(f) = S_X(f)\mathcal{D}^d(f)$, where $\mathcal{D}(f) \equiv 4\sin^2(\pi f)$ is the squared gain function for a first order backward difference filter; if $\{X_t\}$ is not stationary, then $S_X(\cdot)$ can be *defined* via

$$S_X(f) \equiv \frac{S_Y(f)}{\mathcal{D}^d(f)}$$

(Yaglom, 1958). Note that, if $d = 0$, then $\{X_t\}$ is necessarily stationary, in which case the processes $\{X_t\}$ and $\{Y_t\}$ are identical. When $\{Y_t\}$ is an FD process with parameters $\sigma_\varepsilon^2 > 0$ and $-\frac{1}{2} \le \delta^{(s)} < \frac{1}{2}$ and hence has an SDF dictated by the right-hand side of Equation (284a), then $\{X_t\}$ has an SDF given by

$$S_X(f) = \frac{\sigma_\varepsilon^2}{[4\sin^2(\pi f)]^{d+\delta^{(s)}}}.$$

When $d > 0$, we can interpret the above as the SDF for a nonstationary FD process with long memory parameter $\delta \equiv d + \delta^{(s)}$.

Let us consider several examples of nonstationary processes with stationary backward differences of various orders. Our first example is a *random walk process,* defined in the following manner. Let $\{\varepsilon_t\}$ be a Gaussian white noise process with mean zero and variance σ_ε^2, and ACVS given by Equation (268d). Note that, since $\{\varepsilon_t\}$ is both Gaussian and uncorrelated, it is also an IID sequence. Using $\{\varepsilon_t\}$, a random walk process $\{X_t\}$ can be defined for $-\infty < t < \infty$ as

$$X_t \equiv \begin{cases} \sum_{u=1}^{t} \varepsilon_u, & \text{for } t \ge 1, \\ 0, & \text{for } t = 0, \\ -\sum_{u=0}^{|t|-1} \varepsilon_{-u}, & \text{for } t \le -1. \end{cases} \tag{288a}$$

▷ **Exercise [288a]:** Show that (a) $\{X_t\}$ is a nonstationary process; (b) the first order backward difference of $\{X_t\}$ is a stationary process with mean zero; and (c) at low frequencies the SDF for $\{X_t\}$ is approximately a power law process; i.e., $S_X(f) \propto |f|^\alpha$ approximately for f close to zero, where the power law exponent α is to be determined as part of the exercise. ◁

Note that, for $t \ge 0$, a random walk process is the same thing as DFBM $\{B_t\}$ with $H = 1/2$. A variation on the above is to let $\{\varepsilon_t\}$ have some nonzero mean μ_ε, in which case the first order backward difference is a stationary process with a nonzero mean, while the second order backward difference is a stationary process with zero mean (see Exercise [7.6]). See the top two plots in the first column of Figure 289 for examples of what a realization from a random walk – and this variation on it – looks like.

To form the random walk $\{X_t\}$ of Equation (288a), we cumulatively sum the zero mean process $\{\varepsilon_t\}$. Suppose now that we cumulatively sum a random walk process to form a *random run process* defined by:

$$X_t \equiv \begin{cases} \sum_{u=0}^{t} \sum_{u'=0}^{u} \varepsilon_{u'}, & \text{for } t \ge 0, \\ 0, & \text{for } t = -1, -2, \\ \sum_{u=0}^{|t|-3} \sum_{u'=0}^{u} \varepsilon_{-u'-1}, & \text{for } t \le -3. \end{cases} \tag{288b}$$

Exercise [7.7] is to show that the above is a nonstationary process whose first and second order backward differences are, respectively, a random walk process and a stationary process with mean zero. The third plot in the first column of Figure 289 shows what a realization of a random run process looks like.

As a final example, suppose that $\{U_t\}$ is a stationary process with mean zero and SDF $S_U(\cdot)$. Construct $X_t = a_0 + a_1 t + U_t$, where a_0 and $a_1 \ne 0$ are real-valued constants; i.e., $\{X_t\}$ is the sum of a linear trend and a stationary process.

▷ **Exercise [288b]:** Show that (a) $\{X_t\}$ is a nonstationary process; (b) the first order backward difference of $\{X_t\}$ is a stationary process with mean a_1; and (c) the second order backward difference of $\{X_t\}$ is a stationary process with mean zero. ◁

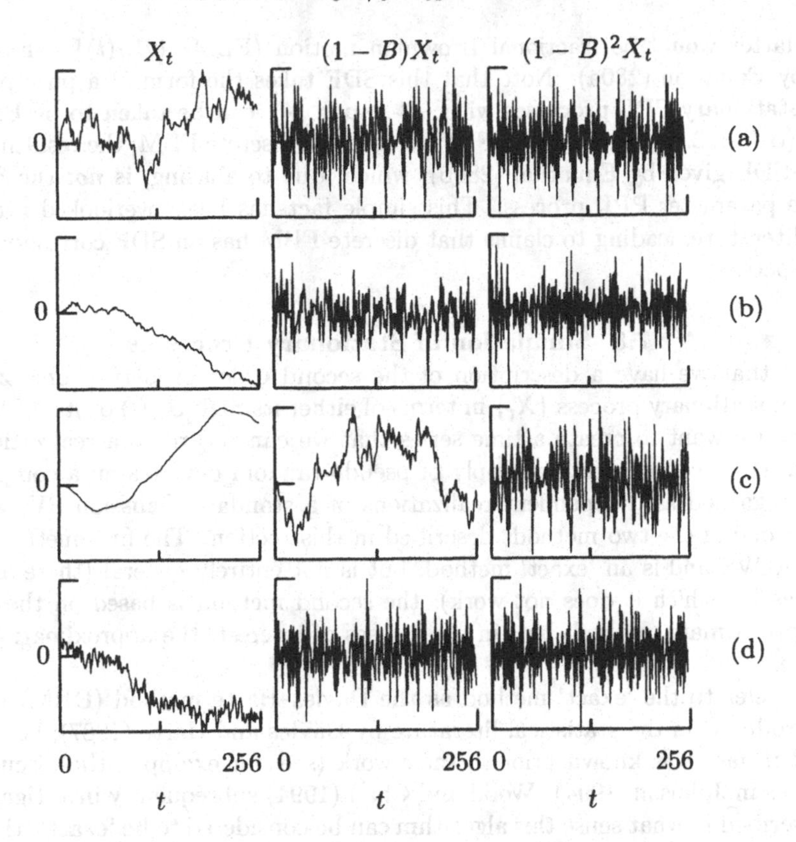

Figure 289. Simulated realizations of nonstationary processes $\{X_t\}$ with stationary backward differences of various orders (first column) along with their first backward differences $\{(1 - B)X_t\}$ (second column) and second backward differences $\{(1-B)^2X_t\}$ (final column). From top to bottom, the processes are (a) a random walk; (b) a modified random walk, formed using a white noise sequence with mean $\mu_\varepsilon = -0.2$; (c) a random run; and (d) a process formed by summing the line given by $-0.05t$ and a simulation of a stationary FD process with $\delta = 0.45$.

Using induction, the above exercise can be readily generalized to show that, if

$$X_t = \sum_{p=0}^{P} a_p t^p + U_t$$

with $a_P \neq 0$ (i.e., $\{X_t\}$ is the sum of a Pth order polynomial trend and a stationary process), then the Pth order backward difference of $\{X_t\}$ is a stationary process with a nonzero mean, while the $(P+1)$st order backward difference is stationary with mean zero (this is Exercise [7.8]). The bottom plot in the first column of Figure 289 gives an example of a realization of a process defined by $X_t = a_0 + a_1 t + U_t$, where here $\{U_t\}$ is a stationary FD process with $\delta = 0.45$.

Comments and Extensions to Section 7.7

[1] As in item [2] of the Comments and Extensions to the previous section, we emphasize that the nonstationary $1/f$-type processes we deal with here are discrete parameter processes $\{X_t\}$ rather than continuous parameter processes $\{X(t)\}$. An example

of the latter would be fractional Brownian motion (FBM) $\{B_H(t)\}$, whose SDF is given by Equation (280a). Note that this SDF takes the form of a pure power law, so nonstationary PPL processes with $-3 < \alpha < -1$ can be taken to be FBMs with $H = -(\alpha+1)/2$. If we sample an FBM to obtain a discrete FBM, the resulting process has an SDF given by Equation (280b), which, due to aliasing, is *not* the SDF for a discrete parameter PPL process. This simple fact has been overlooked rather often in the literature, leading to claims that discrete FBM has an SDF corresponding to a PPL process.

7.8 Simulation of Stationary Processes

Suppose that we have a description of the second order properties of a zero mean Gaussian stationary process $\{X_t\}$ in terms of either its SDF $S_X(\cdot)$ or its ACVS $\{s_{X,\tau}\}$ and that we want to create a time series that we can regard as a realization of this process. If we can generate a supply of pseudo-random deviates on a computer that can be regarded as independent realizations of a standard Gaussian RV, we can do so using one of the two methods described in this section. The first method is based on the ACVS and is an 'exact' method, but is not entirely general (there are certain processes for which it does not work); the second method is based on the SDF and is an approximate method, but, in most cases of interest, the approximation is quite accurate.

We refer to the 'exact' method as the Davies–Harte method (DHM) because it was introduced in the statistical literature by Davies and Harte (1987); however, this method in fact was known prior to their work (see, for example, the discussion and references in Johnson, 1994). Wood and Chan (1994) subsequently investigated DHM and described in what sense this algorithm can be considered to be 'exact' (the heart of their argument is the subject of Exercise [7.10]). The method assumes that the ACVS for $\{X_t\}$ is readily available (as is true for, for example, FGN and FD processes, but not for PPL processes without use of numerical integration). In order to simulate a realization of $X_0, X_1, \ldots, X_{N-1}$ from the process, we must take the following steps.

[1] Let $M = 2N$, and compute the real-valued sequence

$$\mathcal{S}_k \equiv \sum_{\tau=0}^{\frac{M}{2}} s_{X,\tau} e^{-i2\pi f_k \tau} + \sum_{\tau=\frac{M}{2}+1}^{M-1} s_{X,M-\tau} e^{-i2\pi f_k \tau},$$

where $f_k \equiv k/M$ for $k = 0, 1, \ldots, \frac{M}{2}$. Note that the $\mathcal{S}_k$'s can be obtained from the discrete Fourier transform (DFT) of the following sequence of length M:

$$s_{X,0}, s_{X,1}, \ldots, s_{X,\frac{M}{2}-1}, s_{X,\frac{M}{2}}, s_{X,\frac{M}{2}-1}, s_{X,\frac{M}{2}-2}, \ldots, s_{X,1}$$

(see Equation (36g)). The DFT can be efficiently computed using a fast Fourier transform (FFT) algorithm if the desired sample size N is a power of two (if we want, say, $N = 57$, we can always simulate a series whose length is 64 and then discard either the first or last 7 values).

[2] Check that $\mathcal{S}_k \geq 0$ for all k. Gneiting (2000) proves that FGNs with $\frac{1}{2} \leq H < 1$ always satisfy this condition, and his proof also works for FD processes with $0 \leq \delta < \frac{1}{2}$; for other processes, this nonnegativity condition might not be satisfied, in which case *this method will not work*!

[3] Let $Z_0, \ldots, Z_{M-1}$ be a set of M IID Gaussian RVs with zero mean and unit variance. Compute the complex-valued sequence

$$\mathcal{Y}(f_k) \equiv \begin{cases} Z_0\sqrt{MS_0}, & k = 0; \\ (Z_{2k-1} + iZ_{2k})\sqrt{\frac{M}{2}S_k}, & 1 \le k < \frac{M}{2}; \\ Z_{M-1}\sqrt{MS_{\frac{M}{2}}}, & k = \frac{M}{2}; \\ \mathcal{Y}^*(f_{M-k}), & \frac{M}{2} < k \le M-1. \end{cases}$$

(Here, as usual, the asterisk denotes complex conjugation).

[4] Finally use an inverse DFT algorithm to compute the real-valued sequence

$$Y_t \equiv \frac{1}{M}\sum_{k=0}^{M-1} \mathcal{Y}(f_k)e^{i2\pi f_k t}, \quad t = 0, \ldots, M-1. \tag{291}$$

The desired simulation of $X_0, \ldots, X_{N-1}$ is given by $Y_0, \ldots, Y_{N-1}$.

Hosking (1984) describes another exact method for simulating FD processes and other Gaussian stationary processes. This method again assumes the ready availability of the ACVS and has the advantage of being guaranteed to work for all Gaussian stationary processes, but it is an $O(N^2)$ algorithm (as compared to $O(N \log_2(N))$ for the FFT implementation of DHM) and hence can be very slow even for moderate sample sizes (say, $N = 1024$).

When either the nonnegativity condition [2] fails or the SDF for $\{X_t\}$ is readily available but its ACVS is not, it is possible to simulate a stationary Gaussian process using an approximate DFT-based simulation method (see Percival, 1992, and references therein for details). We refer to this as the Gaussian spectral synthesis method (GSSM). Assume the SDF $S_X(\cdot)$ is specified and is finite at all frequencies. To generate the sequences, let M be a positive even integer. From a Gaussian white noise (IID) sequence $\{Z_0, \ldots, Z_{M-1}\}$ construct

$$\mathcal{Z}(f_k) \equiv \begin{cases} Z_0\sqrt{MS_X(0)}, & k = 0; \\ (Z_{2k-1} + iZ_{2k})\sqrt{\frac{M}{2}S_X(f_k)}, & 1 \le k < \frac{M}{2}; \\ Z_{M-1}\sqrt{MS_X(\frac{1}{2})}, & k = \frac{M}{2}; \\ \mathcal{Z}^*(f_{M-k}), & \frac{M}{2} < k \le M-1. \end{cases}$$

Finally, use an inverse DFT algorithm to compute

$$Y_t \equiv \frac{1}{M}\sum_{k=0}^{M-1} \mathcal{Z}(f_k)e^{i2\pi f_k t}, \quad t = 0, \ldots, M-1.$$

By sampling $N \ll M$ values, with M large, this approximate frequency domain method gives quite accurate simulations (Percival, 1992). A good choice is to set M to $4N$. The desired simulation of $X_0, \ldots, X_{N-1}$ is given by $Y_0, \ldots, Y_{N-1}$.

To simulate a series from a stationary long memory process using GSSM, we face a problem because the SDF is infinite at $f = 0$. We can adjust for this by redefining $\mathcal{Z}(f_0)$ to be

$$Z_0\left[\frac{M^2}{2N-1}\left(\sum_{\tau=-(N-1)}^{N-1} s_{X,\tau} - \frac{1}{M}\sum_{k=1}^{M-1} S_X(f_k)\frac{\sin([2N-1]\pi f_k)}{\sin(\pi f_k)}\right)\right]^{1/2},$$

which presumes that the ACVS is also readily known. If this is not the case, the summation involving the ACVS can be replaced by an integral involving the SDF, namely,

$$\sum_{\tau=-(N-1)}^{N-1} s_{X,\tau} = 2\int_0^{1/2} \frac{\sin([2N-1]\pi f)}{\sin(\pi f)} S_X(f)\,df,$$

which – with sufficient care – can be evaluated numerically (for details on GSSM, see Percival *et al.*, 2000a).

Comments and Extensions to Section 7.8

[1] Processes whose dth order backward differences are stationary can be readily simulated by cumulatively summing a simulation of the underlying stationary process. For example, if $\{X_t\}$ is a nonstationary process whose first order backward difference $Y_t \equiv X_t - X_{t-1}$ is stationary, we can use one of the techniques described in this section to generate a simulation of $Y_1, \ldots, Y_{N-1}$, from which we would then form

$$X_t = \begin{cases} \sum_{u=1}^{t} Y_u, & t = 1, \ldots, N-1 \\ 0, & t = 0. \end{cases}$$

Similarly, if $\{X_t\}$ is nonstationary with stationary second order backward differences $Y_t \equiv X_t - 2X_{t-1} + X_{t-2}$, we would first simulate $Y_1, \ldots, Y_{N-1}$, then construct the intermediary series

$$U_t \equiv \sum_{u=1}^{t} Y_u, \quad t = 1, \ldots, N-1,$$

and finally form

$$X_t = \begin{cases} \sum_{u=1}^{t} U_u, & t = 1, \ldots, N-1 \\ 0, & t = 0. \end{cases}$$

7.9 Simulation of Stationary Autoregressive Processes

In the previous section we considered two methods of simulating a Gaussian stationary process that is specified in terms of its SDF or ACVS. These methods involved frequency domain techniques, one exact, and the other approximate. For certain classes of stationary processes, we can efficiently generate simulations using an exact time domain technique. Here we give a recipe for how to do this for a Gaussian stationary AR(p) process (for details, see Kay, 1981).

Suppose that $\{X_t\}$ is a stationary AR(p) process described as in Equation (268e). Let $Z_0, \ldots, Z_{N-1}$ be a set of N IID Gaussian RVs with zero mean and unit variance. To generate a realization of $X_0, X_1, \ldots, X_{N-1}$, we first calculate the $p-1$ sequences $\{\phi_{p-1,n} : n = 1, \ldots, p-1\}, \{\phi_{p-2,n} : n = 1, \ldots, p-2\}, \ldots, \{\phi_{2,n} : n = 1, 2\}$ and $\{\phi_{1,1}\}$ by computing the following for $k = p, p-1, \ldots, 2$:

$$\phi_{k-1,n} = \frac{\phi_{k,n} + \phi_{k,k}\phi_{k,k-n}}{1-\phi_{k,k}^2}, \qquad 1 \le n \le k-1.$$

Second, letting $\sigma_p^2 \equiv \sigma_\varepsilon^2$, we calculate

$$\sigma_{k-1}^2 = \frac{\sigma_k^2}{1-\phi_{k,k}^2}, \qquad k = p, p-1, \ldots, 1.$$

Third, we generate $X_0, X_1, \ldots, X_{p-1}$ via

$$\begin{aligned} X_0 &= \sigma_0 Z_0 \\ X_1 &= \phi_{1,1} X_0 + \sigma_1 Z_1 \\ X_2 &= \phi_{2,1} X_1 + \phi_{2,2} X_0 + \sigma_2 Z_2 \\ &\vdots \\ X_{p-1} &= \phi_{p-1,1} X_{p-2} + \phi_{p-1,2} X_{p-3} + \cdots + \phi_{p-1,p-1} X_0 + \sigma_{p-1} Z_{p-1}. \end{aligned}$$

Finally, the remaining $N - p$ values are generated using

$$X_t = \sum_{n=1}^{p} \phi_{p,n} X_{t-n} + \sigma_p Z_t, \qquad t = p, \ldots, N-1.$$

The generation of $X_0, \ldots, X_{p-1}$ as outlined above provides stationary start-up values, after which we generate the remaining values essentially via the definition of an AR(p) process in Equation (268e).

7.10 Exercises

[7.1] Suppose that X is a chi-square RV with η degrees of freedom; i.e., it has a PDF as stated in Equation (263b). Use Equation (258b) to determine the PDF $f_Y(\cdot;\eta)$ for $Y \equiv \log(X)$. Plot $f_Y(y;\eta)$ versus y for $0 < y \le 4$ and $\eta = 10, 12$ and 16 to verify that the thin curves plotted in Figure 276 are correct.

[7.2] Here we consider some basic properties of covariances.
(a) If X is an RV and c is a constant, show that $\operatorname{cov}\{X, c\} = 0$.
(b) Show that $|\operatorname{cov}\{X, Y\}|^2 \le \operatorname{var}\{X\} \operatorname{var}\{Y\}$.
(c) Suppose that X and Y are RVs, at least one of which has a zero mean. Show that $\operatorname{cov}\{X, Y\} = E\{XY\}$.
(d) Given two finite sets $\{X_j\}$ and $\{Y_k\}$ of RVs and corresponding sets $\{a_j\}$ and $\{b_k\}$ of constants, show that

$$\operatorname{cov}\left\{\sum_j a_j X_j, \sum_k b_k Y_k\right\} = \sum_j \sum_k a_j b_k \operatorname{cov}\{X_j, Y_k\}.$$

[7.3] Let $Z_1, Z_2, \ldots, Z_\eta$ be independent and identically distributed Gaussian RVs with zero mean and unit variance. Show that $\chi^2_\eta \equiv Z_1^2 + Z_2^2 + \cdots + Z_\eta^2$ has a PDF given by Equation (263b).

[7.4] Let $X_0, \ldots, X_{N-1}$ be a portion of a stationary process with ACVS $\{s_{X,\tau}\}$, and let $\overline{X} \equiv \frac{1}{N}\sum_{t=0}^{N-1} X_t$ be the sample mean. Show that

$$\operatorname{var}\{\overline{X}\} = \frac{1}{N} \sum_{\tau=-(N-1)}^{N-1} \left(1 - \frac{|\tau|}{N}\right) s_{X,\tau}. \tag{293}$$

[7.5] Based upon the distributional results given in Section 7.5 for the periodogram, verify Equation (270c).

[7.6] Let $\{\varepsilon_t\}$ be a white noise process such that $E\{\varepsilon_t\} = \mu_\varepsilon \ne 0$. Suppose that we use this nonzero process to construct the random walk process $\{X_t\}$ described by Equation (288a). Show that (a) the first order backward difference of $\{X_t\}$ is a stationary process with a nonzero mean but (b) the second order backward difference of $\{X_t\}$ is a stationary process with mean zero.

[7.7] Let $\{X_t\}$ be the random run process defined in Equation (288b). Show that (a) $\{X_t\}$ is a nonstationary process; (b) the second order backward difference of $\{X_t\}$ is a stationary process with mean zero; (c) the first order backward difference of $\{X_t\}$ is a random walk (Equation (288a)); and (d) at low frequencies the SDF for $\{X_t\}$ is approximately a power law process, where determination of the power law exponent α is part of the exercise.

[7.8] Suppose $\{U_t\}$ is a stationary process with mean zero. Show that, if

$$X_t = \sum_{p=0}^{P} a_p t^p + U_t,$$

with $a_P \neq 0$ then the Pth order backward difference of $\{X_t\}$ is a stationary process with a nonzero mean, while the $(P+1)$st order backward difference is stationary with mean zero.

[7.9] Using a random number generator that simulates uncorrelated deviates Z_t from a Gaussian (normal) distribution with zero mean and unit variance, generate 128 such deviates to construct a time series $\{Z_t : t = 0, \ldots, 127\}$ (this series can be regarded as a realization of length 128 from a Gaussian white noise process). Compute the D(4) MODWT wavelet coefficients $\widetilde{\mathbf{W}}_1$ of unit scale for $\{Z_t\}$, and then compute and plot the rotated cumulative variance series for $\widetilde{\mathbf{W}}_1$ (see Equation (189)). Generate a large number of similar white noise series; compute the rotated cumulative variance series for each series; and then plot the average of all the rotated cumulative variance series versus $t = 0, \ldots, 127$. If you repeated this process *ad infinitum*, what do your plots suggest the average rotated cumulative variance series would eventually converge to?

[7.10] Show that the process $\{Y_t\}$ of Equation (291) is a zero mean Gaussian stationary process whose ACVS at lags $\tau = 0, \ldots, N-1$ is given by $s_{X,0}, s_{X,1}, \ldots, s_{X,N-1}$ (we can conclude from this result that the RVs $Y_0, \ldots, Y_{N-1}$ have *exactly* the same statistical properties as $X_0, \ldots, X_{N-1}$).

[7.11] In terms of variables defined in Section 7.9, what is the variance of an AR(p) process?

8

The Wavelet Variance

8.0 Introduction

As we saw in Chapters 4 and 5, one important use for the discrete wavelet transform (DWT) and its variant, the maximal overlap DWT (MODWT), is to decompose the sample variance of a time series on a scale-by-scale basis. In this chapter we explore wavelet-based analysis of variance (ANOVA) in more depth by defining a theoretical quantity known as the *wavelet variance* (sometimes called the *wavelet spectrum*). This theoretical variance can be readily estimated based upon the DWT or MODWT and has been successfully used in a number of applications; see, for example, Gamage (1990), Bradshaw and Spies (1992), Flandrin (1992), Gao and Li (1993), Hudgins *et al.* (1993), Kumar and Foufoula–Georgiou (1993, 1997), Tewfik *et al.* (1993), Wornell (1993), Scargle (1997), Torrence and Compo (1998) and Carmona *et al.* (1998). The definition for the wavelet variance and rationales for considering it are given in Section 8.1, after which we discuss a few of its basic properties in Section 8.2. We consider in Section 8.3 how to estimate the wavelet variance given a time series that can be regarded as a realization of a portion of length N of a stochastic process with stationary backward differences. We investigate the large sample statistical properties of wavelet variance estimators and discuss methods for determining an approximate confidence interval for the true wavelet variance based upon the estimated wavelet variance (Section 8.4). We then describe in Section 8.5 how estimates of the wavelet variance can be turned into estimates of the spectral density function (SDF). Prior to concluding with a summary (Section 8.10), we consider what the wavelet variance can tell us about four time series related to the frequency stability of atomic clocks, subtidal sea level fluctuations, Nile River minima, and vertical shear in the ocean (Sections 8.6 to 8.9).

8.1 Definition and Rationale for the Wavelet Variance

As in Section 5.4, let $\{\tilde{h}_{j,l} : l = 0, \ldots, L_j - 1\}$ be a jth level MODWT wavelet filter associated with scale $\tau_j \equiv 2^{j-1}$, where $L_j \equiv (2^j - 1)(L - 1) + 1$ is the width of the filter (here $L = L_1$ is the width of the unit scale filter $\tilde{h}_l = \tilde{h}_{1,l}$ described in Section 5.2). Let $\{X_t : t = \ldots, -1, 0, 1, \ldots\}$ represent a discrete parameter real-valued stochastic process, and let

$$\overline{W}_{j,t} \equiv \sum_{l=0}^{L_j-1} \tilde{h}_{j,l} X_{t-l}, \quad t = \ldots, -1, 0, 1, \ldots, \tag{296a}$$

represent the stochastic process obtained by filtering $\{X_t\}$ with the MODWT wavelet filter $\{\tilde{h}_{j,l}\}$ (note carefully how the above compares to the definition for $\widetilde{W}_{j,t}$ in Equation (169a): the above involves filtering an *infinite* sequence, whereas $\widetilde{W}_{j,t}$ comes from *circularly* filtering a *finite* sequence). If it exists and is finite, the *time-dependent wavelet variance* for scale τ_j is defined to be the variance of $\overline{W}_{j,t}$; i.e.,

$$\nu^2_{X,t}(\tau_j) \equiv \text{var}\,\{\overline{W}_{j,t}\}.$$

In this chapter we will restrict ourselves to processes $\{X_t\}$ such that $\nu^2_{X,t}(\tau_j)$ exists, is finite and is independent of t (sufficient conditions for these statements to hold are given in the next section). These restrictions basically say that the statistical properties of $\{X_t\}$ at scale τ_j are invariant over time and hence can be usefully summarized by the *time-independent wavelet variance*

$$\nu^2_X(\tau_j) \equiv \text{var}\,\{\overline{W}_{j,t}\}. \tag{296b}$$

Although in developing the theory we restrict ourselves to processes yielding time-independent wavelet variances, we will demonstrate via examples in Sections 8.7 to 8.9 that the wavelet variance can be easily adapted to deal with certain processes having time-dependent wavelet variances.

We can offer three arguments as to why the wavelet variance is of interest. First, the wavelet variance decomposes (analyzes) the variance of certain stochastic processes on a scale-by-scale basis and hence has considerable appeal to scientists who think about physical phenomena in terms of variations operating over a range of different scales. Second, the wavelet variance is closely related to the concept of the SDF and offers a simple summary of the SDF in cases where this function has a fairly simple structure (as is often true in the physical sciences). Third, the wavelet variance is a useful substitute for the variance of a process for certain processes with infinite variance; similarly, estimators of the wavelet variance are a valuable substitute for the sample variance for processes for which the sample variance has poorly behaved sampling properties. Let us now consider each argument in more detail.

- *Decomposition of variance*

Suppose that $\{Y_t\}$ is a stationary process with SDF $S_Y(\cdot)$ defined over the frequency interval $[-1/2, 1/2]$. As noted by Equation (267d)(with Δt set to unity for convenience), a fundamental property of the SDF is that

$$\int_{-1/2}^{1/2} S_Y(f)\,df = \sigma^2_Y \equiv \text{var}\,\{Y_t\}; \tag{296c}$$

i.e., the SDF decomposes the process variance across frequencies. The analog of this fundamental result for the wavelet variance is

$$\sum_{j=1}^{\infty} \nu^2_Y(\tau_j) = \sigma^2_Y \tag{296d}$$

(see item [1] of the Comments and Extensions to this section). Thus, just as the SDF decomposes σ_Y^2 across frequencies, the wavelet variance decomposes σ_Y^2 with respect to a discrete independent variable τ_j known as scale (to simplify our discussion, we deal with standardized scales τ_j throughout most of this chapter, but, when dealing with practical applications, we must convert these to physical scales $\tau_j \, \Delta t$). Roughly speaking, $\nu_Y^2(\tau_j)$ compares two weighted averages of the process $\{Y_t\}$, the first involving values in an interval of width τ_j, and the second involving data surrounding this interval (or adjacent to it, in the case of the Haar wavelet) – the larger the difference is between these weighted averages, the larger is $\nu_Y^2(\tau_j)$. A plot of $\nu_Y^2(\tau_j)$ versus τ_j thus indicates which scales are important contributors to the process variance. (Note that the above is analogous to the discrete wavelet empirical power spectrum $P_{\mathcal{W}}(\tau_j)$, which decomposes the sample variance across scales – see Equation (62).)

Let us specialize to the simplest example of a wavelet variance, namely, one based upon the MODWT Haar wavelet filter $\{1/2, -1/2\}$. As discussed in Section 8.6, the wavelet variance is then proportional to the *Allan variance*, a well-known measure of the performance of atomic clocks. Plots of the square root of the Allan variance versus τ_j have been used routinely since the 1960s to characterize how well clocks keep time over various time periods (scales); however, as we shall see in Section 8.9, the Allan variance can be misleading for interpreting certain geophysical processes, for which wavelet variances based upon a wavelet filter with $L \geq 4$ are more appropriate (see also Percival and Guttorp, 1994).

- *Useful alternative to spectral density function*

The wavelet variance is also of interest because it provides a way of 'regularizing' the SDF. As we have noted in Chapters 4 and 5, the wavelet coefficients at scale τ_j are nominally associated with frequencies in the interval $[1/2^{j+1}, 1/2^j]$. Because $\nu_Y^2(\tau_j)$ is just the variance of the MODWT wavelet coefficients at scale τ_j, the close relationship between the notions of frequency and scale means that, under certain reasonable conditions,

$$\nu_Y^2(\tau_j) \approx 2 \int_{1/2^{j+1}}^{1/2^j} S_Y(f)\, df \tag{297a}$$

(see Equation (305) for the precise relationship – note also that the factor of 2 in front of the integral above is due to the fact that the SDF $S_Y(\cdot)$ is an even function of f over the interval $[-1/2, 1/2]$). The wavelet variance summarizes the information in the SDF using just one value per octave frequency band and is particularly useful when the SDF is relatively featureless within each octave band. Suppose, for example, that $\{Y_t\}$ is a pure power law process; i.e., its SDF is given by $S_Y(f) \propto |f|^\alpha$ (see Section 7.6). If we use the approximation in (297a), we see that

$$\nu_Y^2(\tau_j) \propto \tau_j^{-\alpha-1} \tag{297b}$$

approximately. Linear variation on a plot of $\log(\nu_Y^2(\tau_j))$ versus $\log(\tau_j)$ thus indicates the existence of a power law process, and the slope of the line can be used to deduce the exponent α of the power law (see Section 9.5). For this simple model, there is no information lost in using the summary given by the wavelet variance.

It is interesting to note that, if we again specialize to the Haar wavelet variance, the pilot spectrum of Blackman and Tukey (1958, Section 18) is identical to using (297a) with this wavelet variance. The pilot spectrum was recommended by

Blackman and Tukey as a 'simple and cheap' preliminary estimate of the SDF (see the Comments and Extensions to Section 8.5 for further discussion on the pilot spectrum). Wavelet variances based upon wavelets other than the Haar are a useful generalization because for certain processes the approximation in (297a) improves as the width of the wavelet filter increases.

Because the wavelet variance is a regularization of the SDF, estimation of the wavelet variance is more straightforward than nonparametric estimation of the SDF. Suppose for the moment that we have a time series of length N that can be regarded as a realization of a portion $Y_0, \ldots, Y_{N-1}$ of the stationary process $\{Y_t\}$ with unknown mean. As described in Section 7.5, the discrete Fourier transform produces a basic estimator of $S_Y(\cdot)$ called the periodogram, which is given by

$$\hat{S}_Y^{(\mathrm{p})}(f_k) \equiv \frac{1}{N}\left|\sum_{t=0}^{N-1}(Y_t - \overline{Y})e^{-i2\pi f_k t}\right|^2,$$

where $f_k \equiv k/N$, and $\overline{Y} \equiv \sum_{t=0}^{N-1} Y_t/N$ is the sample mean of the Y_t's. The periodogram satisfies a sampling version of Equation (296c), namely,

$$\frac{1}{N}\sum_{k=-\lfloor N/2\rfloor+1}^{\lfloor N/2\rfloor} \hat{S}_Y^{(\mathrm{p})}(f_k) = \frac{1}{N}\sum_{t=0}^{N-1}(Y_t - \overline{Y})^2$$

(here $\lfloor x\rfloor$ refers to the greatest integer less than or equal to x); however, as discussed in Section 7.5, the periodogram is limited in its usefulness because it is an inconsistent estimator of $S_Y(\cdot)$ and because it can be badly biased even for sample sizes that would normally be considered 'large' by statisticians (e.g., $N = 2048$, as demonstrated in Figure 273). We must go beyond the simplicity of the periodogram to get a useful SDF estimator.

By contrast, the MODWT of $Y_0, \ldots, Y_{N-1}$ described in Chapter 5 directly yields a useful estimator of the wavelet variance at scale τ_j, namely,

$$\tilde{\nu}_Y^2(\tau_j) \equiv \frac{1}{N}\sum_{t=0}^{N-1} \widetilde{W}_{j,t}^2.$$

As can be seen immediately from Equation (171a) and Exercise [171b], the above leads to the following sampling version of Equation (296d):

$$\sum_{j=1}^{J_0} \tilde{\nu}_Y^2(\tau_j) + \frac{1}{N}\sum_{t=0}^{N-1}(\widetilde{V}_{J_0,t} - \overline{Y})^2 = \frac{1}{N}\sum_{t=0}^{N-1}(Y_t - \overline{Y})^2.$$

In addition, whereas the periodogram can be badly biased, an unbiased estimator of $\nu_Y^2(\tau_j)$ can easily be constructed based upon the $\widetilde{W}_{j,t}$ terms uninfluenced by boundary conditions (see Section 8.3); for a caveat about this statement, see item [2] in the Comments and Extensions below. Moreover, whereas the periodogram is inconsistent, we establish consistency for this unbiased estimator in Section 8.3. For processes with relatively featureless spectra, the wavelet variance is an attractive alternative characterization that is easy to interpret and estimate.

• *Useful alternative to process variance*

A fundamental characteristic of a stationary process $\{Y_t\}$ is its variance σ_Y^2. If the process mean μ_Y for $\{Y_t\}$ is known, then

$$\tilde{\sigma}_Y^2 \equiv \frac{1}{N}\sum_{t=0}^{N-1}(Y_t - \mu_Y)^2$$

is an unbiased estimator for σ_Y^2 because

$$E\{\tilde{\sigma}_Y^2\} = \frac{1}{N}\sum_{t=0}^{N-1}E\{(Y_t - \mu_Y)^2\} = \sigma_Y^2.$$

On the other hand, if μ_Y is unknown and hence is estimated by the sample mean $\overline{Y}$, the usual estimator of the variance σ_Y^2 is

$$\hat{\sigma}_Y^2 \equiv \frac{1}{N}\sum_{t=0}^{N-1}(Y_t - \overline{Y})^2. \tag{299a}$$

▷ **Exercise [299]:** Show that

$$E\{\hat{\sigma}_Y^2\} = \sigma_Y^2 - \text{var}\,\{\overline{Y}\}. \tag{299b}$$ ◁

Since var $\{\overline{Y}\} \geq 0$ and since $E\{\hat{\sigma}_Y^2\}$ is necessarily nonnegative, we must have

$$0 \leq E\{\hat{\sigma}_Y^2\} \leq \sigma_Y^2 \tag{299c}$$

(this is a special case of an elegant result due to David, 1985). Thus, on the average, we cannot overestimate σ_Y^2 using the estimator $\hat{\sigma}_Y^2$. If $s_{Y,\tau} \to 0$ as $\tau \to \infty$ (as will always be true for stationary processes possessing an SDF), a standard result in time series analysis (Fuller, 1996, Corollary 6.1.1.1) says that var $\{\overline{Y}\} \to 0$ as $N \to \infty$, from which we can conclude that $E\{\hat{\sigma}_Y^2\} \to \sigma_Y^2$ as $N \to \infty$; i.e., $\hat{\sigma}_Y^2$ is an asymptotically unbiased estimator of σ_Y^2.

While bias in the sample variance is merely due to lack of knowledge about the process mean and while the asymptotic unbiasedness of $\hat{\sigma}_Y^2$ would seem to offer comfort to practitioners that all will be well if the sample size N is 'large' enough, in fact there are stationary processes that yield a sample variance $\hat{\sigma}_Y^2$ that can be a badly biased estimator of σ_Y^2 for quite large sample sizes. These processes are not pathological, but rather are quite reasonable models for many time series routinely collected in the physical sciences: in fact they are the stationary long memory processes of Section 7.6. We can use these processes to support the following (seemingly outrageous!) claim: for every sample size $N \geq 1$ and every $\epsilon > 0$, there exists a stationary process such that

$$\frac{E\{\hat{\sigma}_Y^2\}}{\sigma_Y^2} < \epsilon. \tag{299d}$$

In words, no matter how large a sample size N we pick, there is a stationary process for which the expected value of the sample variance is any arbitrarily small proportion of the true process variance. To establish this claim, suppose that $\{Y_t\}$ is fractional Gaussian noise (FGN), which by definition is a stationary process with an ACVS dictated by Equation (279b), namely,

$$s_{Y,\tau} = \frac{\sigma_Y^2}{2}\left(|\tau+1|^{2H} - 2|\tau|^{2H} + |\tau-1|^{2H}\right), \quad \tau = \ldots, -1, 0, 1, \ldots,$$

where H is the Hurst parameter satisfying $0 < H < 1$.

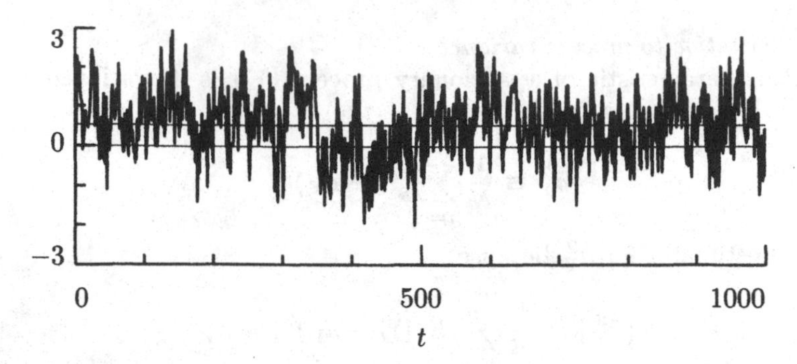

Figure 300. Realization of a fractional Gaussian noise (FGN) process with self-similarity parameter $H = 0.9$. The sample mean of approximately 0.53 and the true mean of zero are indicated by the thin horizontal lines.

▷ **Exercise [300]:** Show that

$$\text{var}\{\overline{Y}\} = \sigma_Y^2 N^{2H-2}. \tag{300}$$

Hint: use induction. ◁

From Equation (299b) we have

$$\frac{E\{\hat{\sigma}_Y^2\}}{\sigma_Y^2} = 1 - N^{2H-2}.$$

For $\epsilon < 1$ and $N > 1$, the claim follows by picking H such that

$$H > 1 + \frac{\log(1-\epsilon)}{2\log(N)}$$

(the case $\epsilon \geq 1$ follows immediately from Equation (299c), whereas the case $N = 1$ is trivial because then $\hat{\sigma}_Y^2 = 0$).

Note that, for small ϵ and large N, any FGN for which (299d) holds has a Hurst parameter just below unity. Writing $H = 1 - \frac{\delta}{2}$ for some small $\delta > 0$, we see that Equation (300) becomes $\text{var}\{\overline{Y}\} = \sigma_Y^2/N^\delta$. This can be compared to $\text{var}\{\overline{Y}\} = \sigma_Y^2/N$ when $\{Y_t\}$ is a white noise process (for FGNs, this is true when $H = 1/2$). Thus, for an FGN with H close to unity, the rate at which $\text{var}\{\overline{Y}\}$ decreases to 0 is much slower than the usual $1/N$ rate for uncorrelated observations.

As an example, Figure 300 shows a time series that is a realization of length $N = 1000$ from an FGN process with $\mu_Y = 0$, $\sigma_Y^2 = 1$ and $H = 0.9$. The sample mean for this time series is $\overline{Y} \doteq 0.53$ (indicated on the plot by a thin horizontal line). If we estimate the process variance with the knowledge that $\mu_Y = 0$, we obtain $\tilde{\sigma}_Y^2 \doteq 0.99$, which is quite close to the true variance; on the other hand, if we don't assume knowledge of μ_Y but instead make use of $\overline{Y}$, we obtain $\hat{\sigma}_Y^2 \doteq 0.71$, which underestimates the true σ_Y^2 and is close to what we would expect to get because $E\{\hat{\sigma}_Y^2\} \doteq 0.75$. In fact, for this particular FGN process, we would need $N = 10^{10}$ or more samples in order that $\sigma_Y^2 - E\{\hat{\sigma}_Y^2\} \leq 0.01$, i.e., for the bias to be 1% or less of the true variance.

The message we can draw from this discussion is that the sample variance has poor bias properties for certain stationary processes due to the necessity of estimating

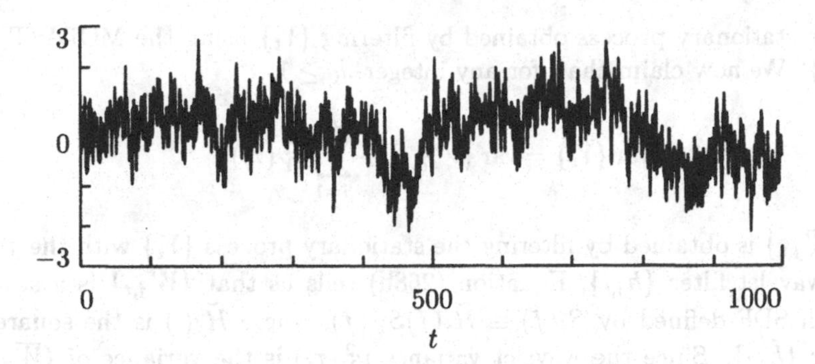

Figure 301. Realization of a stochastic process whose first order backward difference is an FGN process with self-similarity parameter $H = 0.1$. Depending upon its precise definition, the variance for this process is either infinite for all t or is time dependent and increases to infinity as $|t|$ goes to infinity.

the process mean μ_Y using the sample mean (μ_Y is rarely known *a priori* when a stationary process is used as a model for an observed time series). In addition, there are certain nonstationary processes with stationary differences (see Section 7.7) for which the sample variance is not a particularly useful statistic because a process variance cannot be defined that is both finite and time invariant. Nonetheless, finite portions of realizations from these processes can resemble many time series observed in the physical sciences and look disturbingly like realizations of stationary processes (see Figure 301)!

The wavelet variance is often a useful concept for stationary processes (and processes with stationary backward differences) for which the sample variance is of questionable utility. The basic idea is to substitute the notion of variability over certain scales for the 'global' measure of variability estimated by the sample variance. To see why the wavelet variance should be an easier quantity to estimate than the process variance, suppose again that $\{Y_t\}$ is a stationary process with unknown mean μ_Y. Since the stochastic process $\{\overline{W}_{j,t}\}$ is obtained by filtering $\{Y_t\}$ with the MODWT wavelet filter $\{\tilde{h}_{j,l}\}$, it follows that $\{\overline{W}_{j,t}\}$ is a stationary process whose process variance is the wavelet variance $\nu_Y^2(\tau_j)$ (see the discussion surrounding Equations (268a) and (267d)); however, the process mean for $\{\overline{W}_{j,t}\}$ is known because

$$E\{\overline{W}_{j,t}\} = \sum_{l=0}^{L_j-1} \tilde{h}_{j,l} E\{Y_{t-l}\} = \mu_Y \sum_{l=0}^{L_j-1} \tilde{h}_{j,l} = 0$$

(see Table 202). Because the process mean for $\{\overline{W}_{j,t}\}$ is known to be zero (essentially due to the differencing scheme embedded within wavelet filters), it is easy to obtain an unbiased estimate of its process variance, i.e., an unbiased estimate of the wavelet variance $\nu_Y^2(\tau_j)$.

Comments and Extensions to Section 8.1

[1] Here we show that Equation (296d) holds for any stationary process $\{Y_t\}$ with SDF $S_Y(\cdot)$. First, we define

$$\overline{V}_{J_0,t} \equiv \sum_{l=0}^{L_{J_0}-1} \tilde{g}_{J_0,l} Y_{t-l},$$

which is the stationary process obtained by filtering $\{Y_t\}$ using the MODWT scaling filter $\{\tilde{g}_{J_0,l}\}$. We now claim that, for any integer $J_0 \geq 1$,

$$\text{var}\,\{Y_t\} = \text{var}\,\{\overline{V}_{J_0,t}\} + \sum_{j=1}^{J_0} \nu_Y^2(\tau_j). \tag{302}$$

Because $\{\overline{W}_{j,t}\}$ is obtained by filtering the stationary process $\{Y_t\}$ with the jth level MODWT wavelet filter $\{\tilde{h}_{j,l}\}$, Equation (268b) tells us that $\{\overline{W}_{j,t}\}$ is a stationary process with SDF defined by $S_j(f) \equiv \widetilde{\mathcal{H}}_j(f)S_Y(f)$, where $\widetilde{\mathcal{H}}_j(\cdot)$ is the squared gain function for $\{\tilde{h}_{j,l}\}$. Since the wavelet variance $\nu_Y^2(\tau_j)$ is the variance of $\{\overline{W}_{j,t}\}$ and since the integral of the SDF $S_j(\cdot)$ is equal to this variance, we have

$$\nu_Y^2(\tau_j) = \int_{-1/2}^{1/2} \widetilde{\mathcal{H}}_j(f)S_Y(f)\,df;$$

similarly, since $\widetilde{\mathcal{G}}_{J_0}(\cdot)$ is the squared gain function for $\{\tilde{g}_{J_0,l}\}$,

$$\text{var}\,\{\overline{V}_{J_0,t}\} = \int_{-1/2}^{1/2} \widetilde{\mathcal{G}}_{J_0}(f)S_Y(f)\,df.$$

Using Equation (163c) we have $\widetilde{\mathcal{G}}_1(f) + \widetilde{\mathcal{H}}_1(f) \equiv \widetilde{\mathcal{G}}(f) + \widetilde{\mathcal{H}}(f) = 1$ for all f, from which we get

$$\text{var}\,\{Y_t\} = \int_{-1/2}^{1/2} S_Y(f)\,df = \int_{-1/2}^{1/2} [\widetilde{\mathcal{G}}_1(f) + \widetilde{\mathcal{H}}_1(f)]S_Y(f)\,df = \text{var}\,\{\overline{V}_{1,t}\} + \nu_Y^2(\tau_1),$$

so (302) holds for $J_0 = 1$.

▷ **Exercise [302]:** Complete the proof of the claim by showing that, if $J_0 \geq 2$ and if Equation (302) holds for $J_0 - 1$, it must also hold for J_0. Hint: study Table 202. ◁

A proof of Equation (296d) follows from (302) if we can show that $\text{var}\,\{\overline{V}_{J_0,t}\} \to 0$ as $J_0 \to \infty$. Intuitively, we can expect this to be true because $\{\overline{V}_{J_0,t}\}$ is a low-pass filtered version of $\{Y_t\}$ with a nominal pass-band given by $|f| \in [0, 1/2^{J_0+1}]$, the width of which decreases to 0 as $J_0 \to \infty$. To construct a formal proof, we must show that, for any $\epsilon > 0$, there exists a J_ϵ such that $\text{var}\,\{\overline{V}_{J_0,t}\} < \epsilon$ for all $J_0 > J_\epsilon$. First, if we note that $\sum_l \tilde{g}_{J_0,l}^2 = 1/2^{J_0}$ (see Table 202), Parseval's theorem then tells us that

$$\int_{-1/2}^{1/2} \widetilde{\mathcal{G}}_{J_0}(f)\,df = \sum_{l=0}^{L_{J_0}-1} \tilde{g}_{J_0,l}^2 = \frac{1}{2^{J_0}}.$$

If $S_Y(\cdot)$ is bounded above by a finite number C, we can argue that, for all $J_0 > J_\epsilon \equiv \log_2(C/\epsilon)$,

$$\text{var}\,\{\overline{V}_{J_0,t}\} = \int_{-1/2}^{1/2} \widetilde{\mathcal{G}}_{J_0}(f)S_Y(f)\,df \leq C\int_{-1/2}^{1/2} \widetilde{\mathcal{G}}_{J_0}(f)\,df = \frac{C}{2^{J_0}} < \epsilon;$$

if $S_Y(\cdot)$ cannot be bounded by any finite C, there exists a constant C_ϵ such that

$$\int_{S_Y(f) \geq C_\epsilon} S_Y(f)\,df < \frac{\epsilon}{2},$$

where now we must make use of a Lebesgue integral. Because $\widetilde{\mathcal{G}}_{J_0}(f) \leq 1$ for all f for all the wavelets discussed in Chapter 4, for all $J_0 > J_\epsilon \equiv \log_2(C_\epsilon/\epsilon) + 1$,

$$\begin{aligned}\int_{-1/2}^{1/2} \widetilde{\mathcal{G}}_{J_0}(f) S_Y(f)\, df &= \int_{S_Y(f)\geq C_\epsilon} + \int_{S_Y(f)<C_\epsilon} \widetilde{\mathcal{G}}_{J_0}(f) S_Y(f)\, df \\ &\leq \int_{S_Y(f)\geq C_\epsilon} S_Y(f)\, df + C_\epsilon \int_{S_Y(f)<C_\epsilon} \widetilde{\mathcal{G}}_{J_0}(f)\, df \\ &\leq \frac{\epsilon}{2} + C_\epsilon \int_{-1/2}^{1/2} \widetilde{\mathcal{G}}_{J_0}(f)\, df \leq \frac{\epsilon}{2} + \frac{C_\epsilon}{2^{J_0}} < \epsilon,\end{aligned}$$

which establishes Equation (296d).

[2] While it is true that the periodogram can be a badly biased estimator of the true SDF due to a phenomenon known as leakage (illustrated in Figure 273) and while it is easy to construct an unbiased estimator of the wavelet variance using the portion of the MODWT uninfluenced by boundary conditions (see Section 8.3), it is also true that the wavelet variance corresponding to a wavelet filter with too small a width (e.g., the Haar or D(4) filters) can give misleading or uninformative results that can be attributed to a form of leakage – see Section 8.9 for further discussion. Suffice it to say here that leakage in the wavelet variance essentially invalidates the approximate relationship (297a) between the wavelet variance and the SDF. While detection and elimination of leakage in the periodogram requires the use of techniques such as prewhitening or tapering, we just need to increase the width L of the associated wavelet filter to reduce leakage in the wavelet variance.

[3] Given two appropriate stochastic processes $\{X_t\}$ and $\{Y_t\}$ with wavelet coefficients $\{\overline{W}_{X,j,t}\}$ and $\{\overline{W}_{Y,j,t}\}$, we can define a wavelet covariance as

$$\nu_{XY}(\tau_j) \equiv \operatorname{cov}\{\overline{W}_{X,j,t}, \overline{W}_{Y,j,t}\},$$

which yields a scale-based decomposition of the covariance between $\{X_t\}$ and $\{Y_t\}$:

$$\sum_{j=1}^{\infty} \nu_{XY}(\tau_j) = \operatorname{cov}\{X_t, Y_t\}.$$

The wavelet covariance can be standardized to yield a wavelet correlation, namely,

$$\rho_{XY}(\tau_j) \equiv \frac{\operatorname{cov}\{\overline{W}_{X,j,t}, \overline{W}_{Y,j,t}\}}{(\operatorname{var}\{\overline{W}_{X,j,t}\} \operatorname{var}\{\overline{W}_{Y,j,t}\})^{1/2}} = \frac{\nu_{XY}(\tau_j)}{\nu_X(\tau_j)\nu_Y(\tau_j)},$$

where $\nu_X^2(\tau_j)$ and $\nu_Y^2(\tau_j)$ are the wavelet variances for $\{X_t\}$ and $\{Y_t\}$. In a manner similar to the wavelet variance, the wavelet covariance and correlation can be used to study how two processes are related on a scale-by-scale basis (for details on this methodology and examples of its application, see Hudgins, 1992; Hudgins *et al.*, 1993; Liu, 1994; Lindsay *et al.*, 1996; Torrence and Compo, 1998; Whitcher, 1998; Serroukh and Walden, 2000a and 2000b; Whitcher *et al.*, 2000b; and references therein).

8.2 Basic Properties of the Wavelet Variance

Suppose that $\{X_t\}$ is a stochastic process whose dth order backward difference

$$Y_t = (1-B)^d X_t \equiv \sum_{k=0}^{d} \binom{d}{k} (-1)^k X_{t-k}$$

is a stationary process with SDF $S_Y(\cdot)$ and mean μ_Y (for details, see Section 7.7). Recall that

$$\overline{W}_{j,t} \equiv \sum_{l=0}^{L_j-1} \tilde{h}_{j,l} X_{t-l}, \quad t = \ldots, -1, 0, 1, \ldots,$$

represents the output obtained from filtering $\{X_t\}$ using the MODWT wavelet filter $\{\tilde{h}_{j,l}\}$. We can then claim that, if $\{\tilde{h}_{j,l}\}$ is based on a Daubechies wavelet filter (as defined in Section 4.8) such that $L \geq 2d$, then $\{\overline{W}_{j,t}\}$ is a stationary process with SDF defined by

$$S_j(f) = \widetilde{\mathcal{H}}_j^{(\mathrm{D})}(f) S_X(f), \tag{304}$$

where $\widetilde{\mathcal{H}}_j^{(\mathrm{D})}(\cdot)$ is the squared gain function associated with $\{\tilde{h}_{j,l}\}$. To see that this is true for $j = 1$, first note that, because $\tilde{h}_{1,l} = \tilde{h}_l = h_l/\sqrt{2}$, we have from Equation (105b)

$$\widetilde{\mathcal{H}}_1^{(\mathrm{D})}(f) = \frac{1}{2}\mathcal{H}^{(\mathrm{D})}(f) = \sin^L(\pi f) \sum_{l=0}^{\frac{L}{2}-1} \binom{\frac{L}{2}-1+l}{l} \cos^{2l}(\pi f) = \mathcal{D}^{\frac{L}{2}}(f)\widetilde{\mathcal{A}}_L(f),$$

where

$$\mathcal{D}(f) \equiv 4\sin^2(\pi f) \text{ and } \widetilde{\mathcal{A}}_L(f) \equiv \frac{1}{2^L} \sum_{l=0}^{\frac{L}{2}-1} \binom{\frac{L}{2}-1+l}{l} \cos^{2l}(\pi f).$$

We can thus regard $\{\tilde{h}_{1,l}\}$ as equivalent to a two-stage cascade filter (cf. the discussion in Section 4.8). The first filter in the cascade can be taken to have a squared gain function given by $\mathcal{D}^{\frac{L}{2}}(\cdot)$, which corresponds to applying a backward difference filter of order $L/2$. Letting $\{O_t\}$ represent the output from this first filter, we have

$$O_t = (1-B)^{\frac{L}{2}} X_t = (1-B)^{\frac{L}{2}-d}\left[(1-B)^d X_t\right] = (1-B)^{\frac{L}{2}-d} Y_t.$$

Because $\{Y_t\}$ is by assumption a stationary process, it follows that filtering $\{Y_t\}$ repetitively with $\frac{L}{2}-d$ backward difference filters produces a process $\{O_t\}$ that must also be stationary (see the discussion concerning Equations (268a) and (268b)). The second stage filter has a squared gain function $\widetilde{\mathcal{A}}_L(\cdot)$, which by construction corresponds to a filter of finite width, so the output $\{\overline{W}_{1,t}\}$ obtained by filtering $\{O_t\}$ with a filter whose squared gain function is $\widetilde{\mathcal{A}}_L(\cdot)$ is a stationary process whose SDF is defined by $S_1(f) \equiv \widetilde{\mathcal{H}}_1^{(\mathrm{D})}(f) S_X(f)$. Figure 305 shows a flow diagram for the filter cascade that produces $\{\overline{W}_{1,t}\}$ from $\{X_t\}$, with $\{Y_t\}$ and $\{O_t\}$ being formed as intermediary processes.

▷ **Exercise [304]:** Establish the claim for $j \geq 2$. Hint: $\sin^2(2f) = 4\sin^2(f)\cos^2(f)$. ◁

$$X_t \longrightarrow \boxed{\mathcal{D}^d(\cdot)} \longrightarrow Y_t \longrightarrow \boxed{\mathcal{D}^{\frac{L}{2}-d}(\cdot)} \longrightarrow O_t \longrightarrow \boxed{\widetilde{\mathcal{A}}_L(\cdot)} \longrightarrow \overline{W}_{1,t}$$

Figure 305. Flow diagram depicting filtering of $\{X_t\}$ using the Daubechies MODWT wavelet filter $\{\tilde{h}_l\}$ to obtain $\overline{W}_{1,t}$. The actual filter can be decomposed into a cascade involving three filters (indicated above by their squared gain functions $\mathcal{D}^d(\cdot)$, $\mathcal{D}^{\frac{L}{2}-d}(\cdot)$ and $\widetilde{\mathcal{A}}_L(\cdot)$). By assumption the dth order backward difference of $\{X_t\}$ is a stationary process $\{Y_t\}$, while the wavelet filter has a width $L \geq 2d$ and hence implicitly involves $L/2$ backward difference filters $\mathcal{D}(\cdot)$. When $L = 2d$, the processes $\{Y_t\}$ and $\{O_t\}$ are identical (the filter $\mathcal{D}^0(\cdot)$ is taken to be a 'do nothing' filter); when $L > 2d$, we obtain $\{O_t\}$ by taking $\frac{L}{2} - d$ successive backward differences of $\{Y_t\}$. In either case, the process $\{O_t\}$ is then subjected to the smoothing filter $\widetilde{\mathcal{A}}_L(\cdot)$, yielding the first level wavelet coefficient process $\{\overline{W}_{1,t}\}$.

▷ **Exercise [305]:** Show that, on the one hand, if $L > 2d$ or if $L = 2d$ and $\mu_Y = 0$, then $E\{\overline{W}_{j,t}\} = 0$; and, on the other hand, if $\mu_Y \neq 0$ and $L = 2d$, then $E\{\overline{W}_{j,t}\} \neq 0$. ◁

Since the variance of a stationary process is equal to the integral of its SDF and since the variance of $\{\overline{W}_{j,t}\}$ is the wavelet variance $\nu_X^2(\tau_j)$, we have

$$\nu_X^2(\tau_j) = \int_{-1/2}^{1/2} S_j(f)\,df = \int_{-1/2}^{1/2} \widetilde{\mathcal{H}}_j^{(\mathrm{D})}(f) S_X(f)\,df \tag{305}$$

as long as the width L of the wavelet filter is at least as large as $2d$ (i.e., twice the number of backward differences we must apply in order to convert $\{X_t\}$ into a stationary process). In particular, for $j = 1$, we have

$$\nu_X^2(\tau_1) = \int_{-1/2}^{1/2} \widetilde{\mathcal{H}}_1^{(\mathrm{D})}(f) S_X(f)\,df = 2^{L-2d} \int_{-1/2}^{1/2} \widetilde{\mathcal{A}}_L(f) \sin^{L-2d}(\pi f) S_Y(f)\,df.$$

The condition $L \geq 2d$ ensures that the sinusoidal term in the integrand is bounded by unity and hence that the integral is finite.

Note that the wavelet variance is well defined for both stationary processes and nonstationary processes with stationary backward differences as long as the width L of the wavelet filter is large enough. For stationary processes, any L suffices; we have $E\{\overline{W}_{j,t}\} = 0$; and the sum of the wavelet variances over all scales is equal to the process variance (Equation (296d)). For nonstationary processes with dth order stationary backward differences, we need $L \geq 2d$; we might need the stronger condition $L > 2d$ in order to get $E\{\overline{W}_{j,t}\} = 0$; and the sum of the wavelet variances diverges to infinity. Typically, statistics that are designed to work with stationary processes (e.g., various SDF estimators) require specialized techniques in order to study their properties (e.g., bias and variance) when used on nonstationary processes with stationary backward differences. An advantage of the wavelet variance is that a single theory handles both types of processes equally well.

Comments and Extensions to Section 8.2

[1] In order to simplify our discussion here (and further on in this chapter), we made the assumption that $\{\tilde{h}_{j,l}\}$ is constructed using a Daubechies wavelet filter of width L. Because such filters contain an embedded backward difference filter of order $L/2$, we had to impose the condition $L \geq 2d$ to ensure that the filter could properly handle a process with stationary dth order backward differences. As discussed in Section 4.9, coiflet filters also contain an embedded backward difference filter, so they certainly can be used in place of the Daubechies filters; however, the relationship between the filter width L and the order of the difference filter is now $L/3$, so the results in this section need to be reformulated in terms of the condition $L \geq 3d$ (this amounts to simply replacing all occurrences of $2d$ and $L/2$ with, respectively, $3d$ and $L/3$).

8.3 Estimation of the Wavelet Variance

Suppose now that we are given a time series that can be regarded as a realization of one portion $X_0, \ldots, X_{N-1}$ of the process $\{X_t\}$ whose dth order backward differences form a stationary process. We want to estimate the wavelet variance $\nu_X^2(\tau_j)$ using a Daubechies wavelet filter of width $L \geq 2d$, where we assume that L is large enough so that $E\{\overline{W}_{j,t}\} = 0$ (see Exercise [305]). With this assumption we have

$$\nu_X^2(\tau_j) = \text{var}\,\{\overline{W}_{j,t}\} = E\{(\overline{W}_{j,t} - E\{\overline{W}_{j,t}\})^2\} = E\{\overline{W}_{j,t}^2\},$$

so we can base our estimator on the squared process $\{\overline{W}_{j,t}^2\}$.

We can obtain an unbiased estimator of $\nu_X^2(\tau_j)$ based upon the MODWT of $X_0, \ldots, X_{N-1}$ if we take care *not* to include the MODWT coefficients that involve use of the circularity assumption. To make this point explicit, recall that the MODWT of Chapter 5 is given by

$$\widetilde{W}_{j,t} \equiv \sum_{l=0}^{L_j-1} \tilde{h}_{j,l} X_{t-l \bmod N}, \quad t = 0, 1, \ldots, N-1, \tag{306a}$$

while in this chapter we have defined

$$\overline{W}_{j,t} \equiv \sum_{l=0}^{L_j-1} \tilde{h}_{j,l} X_{t-l}, \quad t = \ldots, -1, 0, 1, \ldots.$$

Now $\widetilde{W}_{j,t} = \overline{W}_{j,t}$ for those indices t such that construction of $\widetilde{W}_{j,t}$ remains the same if we were to eliminate the modulo operation – this is true as long as $t \geq L_j - 1$. Thus, if $N - L_j \geq 0$, we can construct an estimator of $\nu_X^2(\tau_j)$ based upon the MODWT using

$$\hat{\nu}_X^2(\tau_j) \equiv \frac{1}{M_j} \sum_{t=L_j-1}^{N-1} \widetilde{W}_{j,t}^2 = \frac{1}{M_j} \sum_{t=L_j-1}^{N-1} \overline{W}_{j,t}^2, \tag{306b}$$

where $M_j \equiv N - L_j + 1$. Since $E\{\hat{\nu}_X^2(\tau_j)\} = \nu_X^2(\tau_j)$, we refer to $\hat{\nu}_X^2(\tau_j)$ as the *unbiased* MODWT estimator of the wavelet variance.

By contrast, we refer to

$$\tilde{\nu}_X^2(\tau_j) \equiv \frac{1}{N} \sum_{t=0}^{N-1} \widetilde{W}_{j,t}^2 = \frac{1}{N} \left(\sum_{t=0}^{L_j-2} \widetilde{W}_{j,t}^2 + \sum_{t=L_j-1}^{N-1} \overline{W}_{j,t}^2 \right) \tag{306c}$$

as the *biased* MODWT estimator of the wavelet variance (even though there are some processes – an example being white noise – such that $E\{\tilde{\nu}_X^2(\tau_j)\} = \nu_X^2(\tau_j)$). Because of the ANOVA property of the MODWT (see Section 5.4), this biased estimator can lead to an exact decomposition of the sample variance, which certainly has some appeal (see also item [4] in the Comments and Extensions below). Note that, as $N \to \infty$, the ratio of the number of terms in the two estimators, i.e., M_j/N, goes to unity (see also Exercise [8.5]). In what follows, we will focus our attention on the unbiased estimator $\hat{\nu}_X^2(\tau_j)$.

Because the estimator $\hat{\nu}_X^2(\tau_j)$ is a random variable (RV), it is of interest to know how close it is likely to be to $\nu_X^2(\tau_j)$. To address this question, let us assume that $\{\overline{W}_{j,t}\}$ is a Gaussian stationary process with mean zero and SDF $S_j(\cdot)$; i.e., any collection of N RVs from $\{\overline{W}_{j,t}\}$ obeys a multivariate normal (or Gaussian) distribution. The following basic result holds (Percival, 1983, 1995): if $S_j(f) > 0$ almost everywhere and if

$$A_j \equiv \int_{-1/2}^{1/2} S_j^2(f)\,df < \infty, \tag{307a}$$

then the estimator $\hat{\nu}_X^2(\tau_j)$ of Equation (306b) is asymptotically normally distributed with mean $\nu_X^2(\tau_j)$ and large sample variance $2A_j/M_j$ (in practical terms, because $S_j(f) \geq 0$ for all f, the 'almost everywhere' qualifier means that, while $S_j(\cdot)$ is allowed to be zero at isolated single frequencies, it cannot be identically zero over an interval of frequencies). Thus, to a good approximation for large $M_j = N - L_j + 1$, we have

$$\frac{M_j^{1/2}(\hat{\nu}_X^2(\tau_j) - \nu_X^2(\tau_j))}{(2A_j)^{1/2}} \stackrel{\mathrm{d}}{=} \mathcal{N}(0,1) \tag{307b}$$

(see Equation (257a)).

To interpret the square integrability condition in (307a), we note that, since $\{\overline{W}_{j,t}\}$ is a stationary process, it has an ACVS, say, $\{s_{j,\tau}\}$. If the ACVS satisfies the square summability condition of Equation (267a), then $\{s_{j,\tau}\} \longleftrightarrow S_j(\cdot)$; i.e., the ACVS and the SDF form a Fourier transform pair. Parseval's theorem (Equation (35c)) then yields

$$\sum_{\tau=-\infty}^{\infty} s_{j,\tau}^2 = \int_{-1/2}^{1/2} S_j^2(f)\,df.$$

Thus, the square integrability condition holds as long as the ACVS for $\{\overline{W}_{j,t}\}$ dies down fast enough so that it is square summable. In practice, if $S_j(\cdot)$ is not square integrable, it is due to a singularity at $f = 0$, which can be cured by just increasing the width L of the wavelet filter.

To get some idea how good an approximation the above large sample result is for moderate sample sizes, suppose that $\{X_t\}$ is either a stationary fractionally differenced (FD) process or a nonstationary FD process whose backward differences of orders $d = 1$ or $d = 2$ form a stationary FD process (see Sections 7.6 and 7.7 for details). Collectively these processes are of $1/f$-type with a power law exponent α covering the range $-5 < \alpha \leq 1$. Simulation studies (Percival, 1995) indicate that the large sample variance quoted in Equation (307b) is – to within 2% – an accurate measure of the true variance of $\hat{\nu}_X^2(\tau_1)$ for sample sizes N such that $M_1 = 128$ for all possible combinations of

[1] FD processes with $\delta = -\frac{1}{2}, -\frac{1}{4}, -\frac{1}{8}, 0, \frac{1}{8}, \frac{1}{4}, \frac{1}{2}$, 1 and $\frac{3}{2}$; and
[2] Daubechies wavelet filters of widths $L = 2$, 4, 6 and 8

(with the exception of the combination $L = 2$ and $\delta = \frac{3}{2}$, for which the square integrability condition required by (307a) fails). Thus the large sample theory is a useful approximation even for relatively small sample sizes.

Comments and Extensions to Section 8.3

[1] We can also estimate the wavelet variance using the DWT instead of the MODWT. Under the assumption that the sample size N of $\{X_t\}$ is divisible by 2^j, we first compute the level j wavelet coefficients $W_{j,t}, t = 0, \ldots, N_j - 1$, where, as before, $N_j \equiv N/2^j$. We can then form an unbiased estimator of $\nu_X^2(\tau_j)$ by excluding all boundary coefficients (i.e., those that make explicit use of the circularity assumption) and then by averaging the squares of the remaining coefficients (after a renormalization to take into account the relationship between the DWT and MODWT coefficients, namely, $W_{j,t}/2^{j/2} = \widetilde{W}_{j,2^j(t+1)-1}$). As discussed in Section 4.11, the number of boundary coefficients is $\min\{L_j', N_j\}$, where L_j' is defined by Equation (146a) and is never more than $L - 2$. Under the assumption that $N_j > L_j'$, the unbiased DWT-based estimator is given by

$$\hat{\hat{\nu}}_X^2(\tau_j) \equiv \frac{1}{(N_j - L_j')2^j} \sum_{t=L_j'}^{N_j-1} W_{j,t}^2 = \frac{1}{N - 2^j L_j'} \sum_{t=L_j'}^{N_j-1} W_{j,t}^2. \tag{308a}$$

We can also define a biased DWT-based estimator via

$$\tilde{\tilde{\nu}}_X^2(\tau_j) \equiv \frac{1}{N_j 2^j} \sum_{t=0}^{N_j-1} W_{j,t}^2 = \frac{1}{N} \sum_{t=0}^{N_j-1} W_{j,t}^2. \tag{308b}$$

If the sample size N is not divisible by 2^j, we can still obtain an unbiased DWT-based estimator of the wavelet variance, as follows. Let $M > N$ be any power of two that is divisible by 2^j, and construct a new time series, say $\{X_t'\}$, of length M by appending $M - N$ zeros to the original series $\{X_t\}$:

$$X_t' \equiv \begin{cases} X_t, & 0 \le t \le N-1; \\ 0, & N \le t \le M-1. \end{cases} \tag{308c}$$

Let $W_{j,t}'$ denote the level j wavelet coefficients for $\{X_t'\}$, which can be readily computed since M is divisible by 2^j. The burden of Exercise [8.7] is to show that, upon division by $2^{j/2}$, a subsequence of these coefficients, namely,

$$W_{j,L_j'}', W_{j,L_j'+1}', \ldots, W_{j,\lfloor \frac{N}{2^j} - 1 \rfloor}', \tag{308d}$$

is equal to selected RVs from the infinite sequence $\{\overline{W}_{j,t}\}$. The desired unbiased estimator is thus

$$\hat{\hat{\nu}}_X^2(\tau_j) \equiv \frac{1}{(\lfloor \frac{N}{2^j} - 1 \rfloor - L_j' + 1)\, 2^j} \sum_{t=L_j'}^{\lfloor \frac{N}{2^j} - 1 \rfloor} [W_{j,t}']^2 = \frac{1}{M_j' 2^j} \sum_{t=L_j'}^{\lfloor \frac{N}{2^j} - 1 \rfloor} [W_{j,t}']^2,$$

where $M'_j \equiv \lfloor \frac{N}{2^j} - 1 \rfloor - L'_j + 1$ (note that this is in agreement with our previous definition when N is divisible by 2^j).

To get some idea about the relative merits of the unbiased MODWT and DWT estimators of the wavelet variance, let us specialize to $\nu_X^2(\tau_1)$, i.e., the unit scale variance. Under the assumptions that the ACVS $\{s_{1,\tau}\}$ for $\{\overline{W}_{1,t}\}$ is square summable and that its SDF satisfies $S_1(f) > 0$ almost everywhere, then $\hat{\hat{\nu}}_X^2(\tau_1)$ is asymptotically normally distributed with mean $\nu_X^2(\tau_j)$ and large sample variance

$$\frac{1}{2M'_1} \int_{-1/2}^{1/2} \left(S_1(\tfrac{f}{2}) + S_1(\tfrac{f}{2} + \tfrac{1}{2}) \right)^2 df$$

(Percival, 1995). This result parallels the one for the MODWT in that the integrand above is the square of the SDF for the subsampled stochastic process

$$2^{1/2}\overline{W}_{1,2t+1} = \sum_{l=0}^{L-1} h_l X_{2t+1-l}, \quad t = \ldots, -1, 0, 1, \ldots$$

(see Exercise [8.8]). We can now compare $\hat{\hat{\nu}}_X^2(\tau_1)$ to $\hat{\nu}_X^2(\tau_1)$ via the asymptotic relative efficiency (ARE) of the former to the latter:

$$e(\hat{\hat{\nu}}_X^2(\tau_1), \hat{\nu}_X^2(\tau_1)) \equiv \lim_{N\to\infty} \frac{\operatorname{var}\{\hat{\nu}_X^2(\tau_1)\}}{\operatorname{var}\{\hat{\hat{\nu}}_X^2(\tau_1)\}} = \frac{\int_{-1/2}^{1/2} S_1^2(f)\, df}{\int_{-1/2}^{1/2} S_1^2(f)\, df + \int_{-1/2}^{1/2} S_1(\frac{f}{2}) S_1(\frac{f}{2} + \frac{1}{2})\, df} \tag{309}$$

(verification of the above is Exercise [8.9]). Since any SDF is a nonnegative function, it is clear that $e(\hat{\hat{\nu}}_X^2(\tau_1), \hat{\nu}_X^2(\tau_1)) \le 1$, which means that, given the same large amount of data, the DWT-based estimator cannot be more efficient than the MODWT-based estimator in the sense that the former cannot have a smaller variance than the latter. It might seem, however, that the ARE should be quite close to unity because we might expect the second integral in the denominator to be close to zero, due to the following argument. Recall that

$$S_1(f) = \widetilde{\mathcal{H}}^{(\mathrm{D})}(f) S_X(f) \approx \begin{cases} S_X(f), & \frac{1}{4} \le |f| \le \frac{1}{2}; \\ 0, & 0 \le |f| \le \frac{1}{4}, \end{cases}$$

where the approximation is valid to the extent that the wavelet filter resembles a half-band high-pass filter. Because $S_1(\cdot)$ is periodic with unit period, it follows that

$$S_1(f + \tfrac{1}{2}) \approx \begin{cases} 0, & \frac{1}{4} \le |f| \le \frac{1}{2}; \\ S_X(f + \frac{1}{2}), & 0 \le |f| \le \frac{1}{4}. \end{cases}$$

This yields the approximation $S_1(\frac{f}{2}) S_1(\frac{f}{2} + \frac{1}{2}) \approx 0$ for all f, suggesting in turn that the ARE should be close to unity (with unity occurring if the wavelet filter were a perfect high-pass filter). By this argument, we really cannot lose much by using the (quicker to compute) DWT-based estimator in place of the MODWT-based estimator.

If, however, we actually compute the AREs for some specific processes $\{X_t\}$ in combination with wavelet filters of assorted widths, we find that in fact the DWT-based estimator can be rather inefficient – in the worst case, its large sample variance

δ	$L = 2$	$L = 4$	$L = 6$	$L = 8$
$-1/2$	0.85	0.89	0.91	0.92
$-1/4$	0.81	0.86	0.89	0.90
$-1/8$	0.78	0.84	0.87	0.89
0	0.75	0.82	0.85	0.87
1/8	0.72	0.80	0.83	0.86
1/4	0.68	0.77	0.81	0.84
1/2	0.61	0.72	0.77	0.80
1	0.50	0.61	0.67	0.71
3/2	—	0.52	0.58	0.62

Table 310. Asymptotic relative efficiencies (AREs) of the DWT-based estimator $\hat{\hat{\nu}}^2_X(\tau_1)$ with respect to the MODWT-based estimator $\hat{\nu}^2_X(\tau_1)$ for various combinations of FD processes and Daubechies wavelet filters. Note that an ARE of less than unity implies that $\hat{\nu}^2_X(\tau_1)$ has smaller large sample variance than $\hat{\hat{\nu}}^2_X(\tau_1)$.

is twice that of the MODWT-based estimator. This fact is demonstrated in Table 310, which gives the AREs for FD processes with δ ranging from $-1/2$ to $3/2$ and for $L = 2$, 4, 6 and 8. Since an FD process has an SDF proportional to $|\sin(\pi f)|^{-2\delta}$, we see that, with L fixed, the ARE decreases as the low frequency portion of the SDF becomes more prominent. Note also that, for a fixed δ, the ARE increases as L increases. This is reasonable because an ARE less than unity is attributable to imperfections in the wavelet filter as an ideal high-pass filter – as L increases, the wavelet filter becomes a better approximation to such a filter. Thus we can obtain a reduction in the large sample variance of up to a factor of two by using the MODWT-based estimator instead of the DWT-based estimator.

[2] We can easily compute the unbiased estimator $\hat{\nu}^2_X(\tau_j)$ by using the pyramid algorithm for the MODWT; however, because we discard all boundary coefficients in forming this estimator, we are computing more MODWT coefficients than we actually need. It is possible to adapt the MODWT pyramid algorithm so that it only computes the coefficients that are actually used by $\hat{\nu}^2_X(\tau_j)$ – for details, see Percival and Guttorp (1994).

[3] Suppose that $\{X_t : t = 0, 1, \ldots, N-1\}$ is a portion of length N of a power law process with $-2 \leq \alpha \leq 0$. Recently Greenhall *et al.* (1999) showed that, if $\{X_t\}$ is extended to a length $2N$ series by concatenating it with the 'reflected' series $\{X_t : t = N-1, N-2, \ldots, 0\}$ as per Equation (140), the biased MODWT estimator of the Haar wavelet variance based upon the length $2N$ series has significantly smaller mean square error at scales close to $N/2$ than the unbiased estimator based upon just the original series. This result suggests that dealing with the circularity assumption by creating an *ad hoc* circular series has some theoretical justification for long memory processes and that the biased MODWT wavelet variance estimator might have some important advantages for wavelets other than the Haar (both subjects are currently under study).

[4] Recently Serroukh *et al.* (2000) derived the asymptotic distribution of the unbiased MODWT wavelet variance estimator for a wide class of processes that are not necessarily Gaussian or linear and that are either (i) stationary, (ii) nonstationary but with stationary dth order differences or (iii) globally nonstationary but locally

stationary. They found that, under these (and additional) conditions, the estimator $\hat{\nu}^2_X(\tau_j)$ is again asymptotically Gaussian distributed with mean $\nu^2_X(\tau_j)$, but now the large sample variance is given by $S_{\overline{W}^2_j}(0)/M_j$, where $S_{\overline{W}^2_j}(\cdot)$ is the SDF for the process $\{\overline{W}^2_{j,t}\}$ (in practice, a good way to estimate $S_{\overline{W}^2_j}(\cdot)$ at $f = 0$ is via a multitaper estimator). At small scales, estimates of the variance of the MODWT estimator appropriate for non-Gaussian time series can be much larger than those computed under an (incorrect) Gaussian assumption. The Gaussian assumption on the stationary process $\{\overline{W}_{j,t}\}$ thus leads to a large sample result (Equation (307b)) that is not particularly robust against departures from Gaussianity.

8.4 Confidence Intervals for the Wavelet Variance

The chief use we will make of the large sample result of Equation (307b) is to compute a confidence interval for $\nu^2_X(\tau_j)$ based upon $\hat{\nu}^2_X(\tau_j)$. To do so, let $\Phi^{-1}(p)$ represent the $p \times 100\%$ percentage point for the standard Gaussian distribution (see Equation (257b)). The symmetric nature of the Gaussian distribution implies that, for $0 \leq p \leq 1/2$,

$$\mathbf{P}[-\Phi^{-1}(1-p) \leq Z \leq \Phi^{-1}(1-p)] = 1 - 2p.$$

For large N, Equation (307b) tells us that

$$\mathbf{P}\left[-\Phi^{-1}(1-p) \leq \frac{M_j^{1/2}(\hat{\nu}^2_X(\tau_j) - \nu^2_X(\tau_j))}{(2A_j)^{1/2}} \leq \Phi^{-1}(1-p)\right] = 1 - 2p$$

approximately. Now the event between the brackets occurs if and only if the events

$$-\Phi^{-1}(1-p) \leq \frac{M_j^{1/2}(\hat{\nu}^2_X(\tau_j) - \nu^2_X(\tau_j))}{(2A_j)^{1/2}}$$

and

$$\frac{M_j^{1/2}(\hat{\nu}^2_X(\tau_j) - \nu^2_X(\tau_j))}{(2A_j)^{1/2}} \leq \Phi^{-1}(1-p)$$

both occur; in turn, these two events are equivalent to, respectively,

$$\nu^2_X(\tau_j) \leq \hat{\nu}^2_X(\tau_j) + \Phi^{-1}(1-p)\left(\frac{2A_j}{M_j}\right)^{1/2}$$

and

$$\hat{\nu}^2_X(\tau_j) - \Phi^{-1}(1-p)\left(\frac{2A_j}{M_j}\right)^{1/2} \leq \nu^2_X(\tau_j).$$

Together these two events are equivalent to the single event

$$\hat{\nu}^2_X(\tau_j) - \Phi^{-1}(1-p)\left(\frac{2A_j}{M_j}\right)^{1/2} \leq \nu^2_X(\tau_j) \leq \hat{\nu}^2_X(\tau_j) + \Phi^{-1}(1-p)\left(\frac{2A_j}{M_j}\right)^{1/2},$$

which has probability $1 - 2p$ of occurring. Hence we can say that, with probability $1 - 2p$, the random interval

$$\left[\hat{\nu}^2_X(\tau_j) - \Phi^{-1}(1-p)\left(\frac{2A_j}{M_j}\right)^{1/2}, \hat{\nu}^2_X(\tau_j) + \Phi^{-1}(1-p)\left(\frac{2A_j}{M_j}\right)^{1/2}\right] \quad (311)$$

traps the true value of $\nu_X^2(\tau_j)$ and constitutes a $100 \times (1-2p)\%$ confidence interval for $\nu_X^2(\tau_j)$.

In order to use (311) in practical applications, we must estimate A_j, which is the integral of $S_j^2(\cdot)$. Since A_j will be determined for the most part by large values of $S_j(\cdot)$ and since a periodogram can estimate large values of an SDF with relatively little bias, we can just use the periodogram as our estimator of $S_j(\cdot)$:

$$\hat{S}_j^{(\mathrm{p})}(f) \equiv \frac{1}{M_j}\left|\sum_{t=L_j-1}^{N-1} \widetilde{W}_{j,t}e^{-i2\pi ft}\right|^2.$$

Equation (269b) says that, for $0 < |f| < 1/2$ and for large N, the ratio

$$\frac{2\hat{S}_j^{(\mathrm{p})}(f)}{S_j(f)} \stackrel{\mathrm{d}}{=} \chi_2^2,$$

where χ_2^2 is a χ^2 RV with two degrees of freedom (the properties of the χ^2 distribution are reviewed in Section 7.2).

▷ **Exercise [312]:** Use this large sample result to argue that

$$\hat{A}_j \equiv \frac{1}{2}\int_{-1/2}^{1/2} [\hat{S}_j^{(\mathrm{p})}(f)]^2\,df \tag{312a}$$

is an approximately unbiased estimator of A_j for large M_j. ◁

Equation (269a) says that $\{\hat{s}_{j,\tau}^{(\mathrm{p})}\} \longleftrightarrow \hat{S}_j^{(\mathrm{p})}(\cdot)$, where $\hat{s}_{j,\tau}^{(\mathrm{p})}$ is the usual biased estimator of the ACVS, i.e.,

$$\hat{s}_{j,\tau}^{(\mathrm{p})} \equiv \frac{1}{M_j}\sum_{t=L_j-1}^{N-1-|\tau|} \widetilde{W}_{j,t}\widetilde{W}_{j,t+|\tau|},\quad 0 \le |\tau| \le M_j - 1, \tag{312b}$$

and $\hat{s}_{j,\tau}^{(\mathrm{p})} \equiv 0$ for $|\tau| \ge M_j$. Evoking Parseval's theorem (Equation (35c)), we have

$$\int_{-1/2}^{1/2} [\hat{S}_j^{(\mathrm{p})}(f)]^2\,df = \sum_{\tau=-(M_j-1)}^{M_j-1} \left(\hat{s}_{j,\tau}^{(\mathrm{p})}\right)^2.$$

We thus obtain the convenient computational formula

$$\hat{A}_j = \frac{\left(\hat{s}_{j,0}^{(\mathrm{p})}\right)^2}{2} + \sum_{\tau=1}^{M_j-1}\left(\hat{s}_{j,\tau}^{(\mathrm{p})}\right)^2 = \frac{\hat{\nu}_X^4(\tau_j)}{2} + \sum_{\tau=1}^{M_j-1}\left(\hat{s}_{j,\tau}^{(\mathrm{p})}\right)^2.$$

Substitution of $\hat{A}_j$ for A_j in (311) leads to an approximate $100(1-2p)\%$ confidence interval for $\nu_X^2(\tau_j)$ (with the validity of the assumption being conditioned on the estimator $\hat{A}_j$ being close to A_j).

A difficulty with confidence intervals of the form given in Equation (311) is that there is nothing to prevent the lower confidence limit from being negative, which is a nuisance both because we know that $\nu_X^2(\tau_j)$ must be nonnegative and because we often want to plot the confidence intervals on a logarithmic scale. Now $\hat{\nu}_X^2(\tau_j)$ is proportional to the sum of squares of M_j zero mean Gaussian RVs, each having the same variance. If these RVs were also uncorrelated, Equation (263a) says that $\hat{\nu}_X^2(\tau_j)$ could be renormalized to obey a chi-square distribution with M_j degrees of freedom, from which a strictly positive confidence interval would then follow (its form would be similar to Equation (313c) below). We can approximate the distribution of the sum of squares of correlated Gaussian RVs with zero mean and a common variance using a scaled chi-square distribution with the degrees of freedom adjusted to account for the correlation in the RVs (Priestley, 1981, p. 466). In this approach we use the approximation

$$\frac{\eta \hat{\nu}_X^2(\tau_j)}{\nu_X^2(\tau_j)} \stackrel{\mathrm{d}}{=} \chi_\eta^2, \tag{313a}$$

where η is known as the 'equivalent degrees of freedom' (EDOF) and is determined as follows.

▷ **Exercise [313a]:** Under the assumption that $\hat{\nu}_X^2(\tau_j)$ is distributed as $a\chi_\eta^2$, determine a and η (the EDOF) by solving the two equations $E\{\hat{\nu}_X^2(\tau_j)\} = E\{a\chi_\eta^2\}$ and $\operatorname{var}\{\hat{\nu}_X^2(\tau_j)\} = \operatorname{var}\{a\chi_\eta^2\}$. In particular, show that

$$\eta = \frac{2\left(E\{\hat{\nu}_X^2(\tau_j)\}\right)^2}{\operatorname{var}\{\hat{\nu}_X^2(\tau_j)\}}.$$

Use this approach with the large sample approximations to the mean and variance of $\hat{\nu}_X^2(\tau_j)$ to show that the EDOF is given approximately by

$$\eta_1 \equiv \frac{M_j \nu_X^4(\tau_j)}{A_j}. \tag{313b}$$ ◁

The distribution assumed in Equation (313a) is in fact exact when $M_j = 1$ and asymptotically correct as $N \to \infty$ (because the χ^2 distribution approaches normality as the number of degrees of freedom goes to infinity).

▷ **Exercise [313b]:** Under the assumption that Equation (313a) holds, use an argument similar to the one leading up to Equation (311) to show that an approximate $100(1-2p)\%$ confidence interval for $\nu_X^2(\tau_j)$ is given by

$$\left[\frac{\eta \hat{\nu}_X^2(\tau_j)}{Q_\eta(1-p)}, \frac{\eta \hat{\nu}_X^2(\tau_j)}{Q_\eta(p)}\right], \tag{313c}$$

where $Q_\eta(p)$ is the $p \times 100\%$ percentage point for the χ_η^2 distribution, i.e., $\mathbf{P}[\chi_\eta^2 \le Q_\eta(p)] = p$. ◁

In practical applications, the degrees of freedom η required to compute the above confidence interval would be estimated using

$$\hat{\eta}_1 \equiv \frac{M_j \hat{\nu}_X^4(\tau_j)}{\hat{A}_j}. \tag{313d}$$

Another approach to obtaining a confidence interval for $\nu_X^2(\tau_j)$ is to assume that $S_X(\cdot)$, and hence $S_j(\cdot)$, is known to within a multiplicative constant; i.e., we suppose that, say, $S_j(f) = aC_j(f)$, where $C_j(\cdot)$ is a known function and a is an unknown constant. This assumption is used in, for example, Greenhall (1991) to obtain confidence intervals for the Allan variance (see Section 8.6). Because the periodogram is proportional to the squared modulus of the DFT of the $\widetilde{W}_{j,t}$'s, an application of Parseval's theorem yields

$$\sum_{t=L_j-1}^{N-1} \widetilde{W}_{j,t}^2 = \sum_{k=0}^{M_j-1} \hat{S}_j^{(\mathrm{p})}(f_k),$$

where $f_k \equiv k/M_j$. Hence we have

$$\hat{\nu}_X^2(\tau_j) = \frac{1}{M_j}\sum_{k=0}^{M_j-1} \hat{S}_j^{(\mathrm{p})}(f_k) \approx \frac{2}{M_j}\sum_{k=1}^{\lfloor (M_j-1)/2 \rfloor} \hat{S}_j^{(\mathrm{p})}(f_k), \tag{314a}$$

where $\lfloor (M_j-1)/2 \rfloor$ is the greatest integer less than or equal to $(M_j-1)/2$. The approximation above merely says that $\hat{S}_j^{(\mathrm{p})}(0)/M_j$ (the square of the sample mean of the $\widetilde{W}_{j,t}$'s) and $\hat{S}_j^{(\mathrm{p})}(\frac{1}{2})/M_j$ are negligible.

▷ **Exercise [314a]:** Under the usual large sample approximations that

$$\frac{2\hat{S}_j^{(\mathrm{p})}(f_k)}{S_j(f_k)} \stackrel{\mathrm{d}}{=} \chi_2^2 \text{ for } 0 < f_k < \frac{1}{2}$$

and that the RVs in the right-hand summation of Equation (314a) are independent of each other, use an EDOF argument to claim that

$$\frac{\eta_2 \hat{\nu}_X^2(\tau_j)}{\nu_X^2(\tau_j)} \stackrel{\mathrm{d}}{=} \chi_{\eta_2}^2$$

approximately, where

$$\eta_2 \equiv \frac{2\left(\sum_{k=1}^{\lfloor (M_j-1)/2 \rfloor} C_j(f_k)\right)^2}{\sum_{k=1}^{\lfloor (M_j-1)/2 \rfloor} C_j^2(f_k)}. \tag{314b}$$ ◁

An approximate $100(1-2p)\%$ confidence interval for $\nu_X^2(\tau_j)$ would be given by (313c) with η_2 replacing η.

A further simplification is to recall that the MODWT wavelet filter $\{\tilde{h}_{j,l}\}$ can be regarded as an approximate band-pass filter with pass-band defined by $1/2^{j+1} < |f| \leq 1/2^j$. This fact suggests that it might be reasonable to assume in certain practical problems that $S_j(\cdot)$ is band-limited and flat over its nominal pass-band (the validity of this assumption can readily be assessed by examining an estimate of $S_j(\cdot)$).

▷ **Exercise [314b]:** Show that an EDOF argument now leads to

$$\eta_3 = \max\{M_j/2^j, 1\}. \tag{314c}$$ ◁

If the sample size M_j is large enough (the simulation experiments reported in Percival, 1995, suggest that $M_j = 128$ is often sufficient), a confidence interval based upon Equation (313c) with η estimated by $\hat{\eta}_1$ of Equation (313d) is likely to be reasonably accurate, and we would recommend this as the method of choice. For smaller sample sizes, this method can yield overly optimistic confidence intervals in some instances, in which case a confidence interval based on η_2 from (314b) or η_3 from (314c) is a useful check and should be preferred if it is markedly wider than the one based on η_1. Use of η_2 requires a reasonable guess at the shape of $S_j(\cdot)$; if such a guess is not available, the EDOF should be based upon η_3.

Comments and Extensions to Section 8.4

[1] In practice, estimation of $S_j(\cdot)$ via the periodogram requires forming the DFT of $\{\widetilde{W}_{j,t} : t = L_j - 1, \ldots, N-1\}$, which can become problematic if the series length $M_j = N - L_j + 1$ is quite large. To get around this difficulty, we can use an alternative estimator of $S_j(\cdot)$ based upon breaking the series up into nonoverlapping blocks, each containing N_S contiguous wavelet coefficients. We would then have $N_B = \lfloor M_j/N_S \rfloor$ blocks in all. If we compute a periodogram for each block and average the N_B periodograms together, we would obtain an SDF estimator of the form

$$\hat{S}_j^{(\text{sa})}(f) = \frac{1}{N_B} \sum_{n=0}^{N_B-1} \left(\frac{1}{N_S} \left| \sum_{t=0}^{N_S-1} \widetilde{W}_{j,L_j-1+nN_S+t} e^{-i2\pi ft} \right|^2 \right).$$

This 'segment averaging' estimator of $S_j(\cdot)$ can be regarded as a special case of Welch's overlapped segment averaging (WOSA) SDF estimator (Welch, 1967). For details about WOSA (including guidance on choosing an appropriate block size N_S), see, for example, Percival and Walden (1993), Section 6.17.

We can now use $\hat{S}_j^{(\text{sa})}(\cdot)$ instead of $\hat{S}_j^{(\text{p})}(\cdot)$ to estimate A_j, but we need to adjust Equation (312a) slightly, as indicated by the following exercise.

▷ **Exercise [315]:** Under certain reasonable conditions, statistical theory suggests that $2N_B\hat{S}_j^{(\text{sa})}(f)/S_j(f)$ is approximately distributed as a chi-square RV with $2N_B$ degrees of freedom. Use this result to derive an approximately unbiased estimator for A_j based upon $\hat{S}_j^{(\text{sa})}(\cdot)$. What form does this estimator take as N_B gets large? ◁

8.5 Spectral Estimation via the Wavelet Variance

Let us now consider how estimates of the wavelet variance can be turned into SDF estimates. Suppose that $\{X_t\}$ is a process with SDF $S_X(\cdot)$ and sampling interval Δt – this implies that $S_X(\cdot)$ is defined over the interval $[-f_{\mathcal{N}}, f_{\mathcal{N}}]$, where $f_{\mathcal{N}} \equiv 1/(2\,\Delta t)$ is the Nyquist frequency. Let $\nu_X^2(\tau_j)$ be the wavelet variance at physical scale $\tau_j\,\Delta t \equiv 2^{j-1}\,\Delta t$. As noted in Section 8.1, the band-pass nature of the MODWT wavelet filter $\{\tilde{h}_{j,l}\}$ implies that we should have

$$\nu_X^2(\tau_j) \approx 2\int_{1/(2^{j+1}\Delta t)}^{1/(2^j\Delta t)} S_X(f)\,df \tag{315}$$

(this is Equation (297a) re-expressed to allow for a sampling interval other than unity). As the width L of the wavelet filter $\{\tilde{h}_l\}$ used to form $\{\tilde{h}_{j,l}\}$ increases, the above approximation improves because $\{\tilde{h}_{j,l}\}$ then becomes a better approximation to an ideal band-pass filter (most noticeably with regard to suppression of frequencies outside the nominal pass-band – see Figure 122).

Under the assumption that L is selected such that Equation (315) is a reasonable approximation, we can estimate $S_X(\cdot)$ using a function $\overline{S}_X(\cdot)$ that is piecewise constant over each interval $[\frac{1}{2^{j+1}\Delta t}, \frac{1}{2^j \Delta t}]$ for $j = 1, \ldots, J_0$; i.e., we assume

$$\overline{S}_X(f) = C_j \text{ when } \frac{1}{2^{j+1}\Delta t} < f \leq \frac{1}{2^j \Delta t},$$

where C_j is a constant defined such that

$$\int_{1/(2^{j+1}\Delta t)}^{1/(2^j \Delta t)} S_X(f)\,df = \int_{1/(2^{j+1}\Delta t)}^{1/(2^j \Delta t)} \overline{S}_X(f)\,df = \frac{C_j}{2^{j+1}\,\Delta t}.$$

Using Equation (315), we have

$$\nu_X^2(\tau_j) \approx \frac{C_j}{2^j\,\Delta t}, \text{ and hence we can use } \hat{C}_j \equiv 2^j \hat{\nu}_X^2(\tau_j)\,\Delta t \tag{316}$$

to estimate the SDF levels. Two examples of such wavelet-based SDF estimates are shown further on in Figure 330.

Given $\hat{\nu}_X^2(\tau_j)$ for $j = 1, \ldots, J_0$, the above scheme gives an estimate of the SDF over the interval $[1/(2^{J_0+1}\Delta t), f_{\mathcal{N}}]$, but not for low frequencies $f \in [0, 1/(2^{J_0+1}\Delta t)]$. Because the level J_0 MODWT scaling coefficients are associated with this frequency interval, we could fill out our estimate of $S_X(\cdot)$ based on an estimate of $\operatorname{var}\{\overline{V}_{J_0,t}\}$ if we can assume that $\{X_t\}$ is a stationary process (see Equation (302) and the discussion surrounding it). If we know that $E\{X_t\} = 0$, then we also have $E\{\overline{V}_{J_0,t}\} = 0$, so an appropriate estimator would be

$$\widehat{\operatorname{var}}\,\{\overline{V}_{J_0,t}\} \equiv \frac{1}{M_{J_0}} \sum_{t=L_{J_0}-1}^{N-1} \widetilde{V}_{J_0,t}^2;$$

however, if $E\{X_t\}$ is unknown so that $E\{\overline{V}_{J_0,t}\}$ is also unknown (as is usually the case in practice), then we must use

$$\widehat{\operatorname{var}}\,\{\overline{V}_{J_0,t}\} \equiv \frac{1}{M_{J_0}} \sum_{t=L_{J_0}-1}^{N-1} \left(\widetilde{V}_{J_0,t} - \overline{\widetilde{V}}_{J_0}\right)^2,$$

where $\overline{\widetilde{V}}_{J_0}$ is the sample mean of the observed $\widetilde{V}_{J_0,t}$. If $\{X_t\}$ is a long memory process, then so is $\{\overline{V}_{J_0,t}\}$, in which case we must be aware that, because the above estimator is of the form of Equation (299a), it can seriously underestimate $\operatorname{var}\{\overline{V}_{J_0,t}\}$. Thus, if the process mean is known or if $\{X_t\}$ does not exhibit long memory characteristics, we should be able to obtain a reasonable estimate for $S_X(\cdot)$ over low frequencies by taking the appropriate estimator $\widehat{\operatorname{var}}\,\{\overline{V}_{J_0,t}\}$ and using it to form the SDF level $2^{J_0}\widehat{\operatorname{var}}\,\{\overline{V}_{J_0,t}\}\,\Delta t$ to cover the interval $[0, 1/(2^{J_0+1}\Delta t)]$.

Comments and Extensions to Section 8.5

[1] As mentioned in Section 8.1, we can obtain the pilot spectrum estimator of Blackman and Tukey (1958) by using the Haar wavelet in conjunction with Equation (315) (see also Section 7.3.2, Jenkins and Watts, 1968). Blackman and Tukey describe a simple pilot estimator in Section 18 of their book and then propose several modifications in Section B.18. To within a constant of proportionality, their simple estimator of the spectral level C_j is given by $2^j \hat{\hat{\nu}}_X^2(\tau_j)\,\Delta t$, where $\hat{\hat{\nu}}_X^2(\tau_j)$ is the unbiased DWT-based estimator given in Equation (308a) but specialized to the Haar wavelet:

$$\hat{\hat{\nu}}_X^2(\tau_j) = \frac{1}{N}\sum_{t=0}^{N_j-1} W_{j,t}^2,$$

with $W_{j,t} = (V_{j-1,2t+1} - V_{j-1,2t})/\sqrt{2}$, $V_{j,t} = (V_{j-1,2t+1} + V_{j-1,2t})/\sqrt{2}$ and $V_{0,t} \equiv X_t$ (in comparing the above with (308a), recall that $L_j' = 0$ for the Haar wavelet). As indicated by a numerical example (Blackman and Tukey, 1958, Table III, p. 46), they actually compute the simple pilot estimate using the DWT pyramid algorithm for the Haar wavelet (again, ignoring a constant of proportionality).

In addition, one of the modifications to the simple pilot estimate is a 'complete version' (Blackman and Tukey, 1958, Table VI, p. 136), which is proportional to the unbiased Haar MODWT estimator of $\nu_X^2(\tau_j)$ and is implemented by an algorithm that is close to the MODWT pyramid algorithm for the Haar wavelet (the algorithms differ only by an interleaving operation and a constant of proportionality). They come to this estimator using an argument that closely parallels what we used in Section 5.3, i.e., elimination of an operation involving 'dropping every alternate value' (downsampling).

Finally, in Section 8.6 below, we discuss the fact that the Haar-based wavelet variance $\nu_X^2(\tau_j)$ can be adversely affected by linear trends, and, in situations where such a trend is of concern, we recommend using wavelet variances based upon Daubechies wavelets filters with $L \geq 4$. Blackman and Tukey (1958, p. 136) also recognize that pilot estimators are sensitive to linear trends and, as a remedy, suggest differencing the time series and then subjecting the differenced data to a pilot analysis. This remedy is similar in one aspect to using the D(4) wavelet in that, in comparison to the Haar, the D(4) filter has an additional embedded differencing operation.

8.6 Example: Atomic Clock Deviates

As a first example of the use of the wavelet variance, let us return to some data we exploited in Chapter 1 and consider the time series plotted at the top of Figure 318, which shows the differences in time $\{X_t\}$ as kept by a cesium beam atomic clock (labeled as 'clock 571') and by an official time scale known as UTC(USNO) maintained by the US Naval Observatory, Washington, DC (the 'UTC' portion of the acronym refers to 'coordinated universal time,' which is the international standard of time; the entire acronym refers to coordinated universal time as generated by clocks at the Naval Observatory). This time scale can be regarded as a surrogate for a 'perfect' clock and hence can be used to study imperfections in individual clocks such as clock 571 (in reality, UTC(USNO) itself is formed by averaging over a large ensemble of atomic clocks and – while certainly not perfect – is considered to be at least an order of magnitude better at keeping time than clock 571). The differences in time between clock 571 and UTC(USNO) were recorded at the same time each day for $N = 1026$

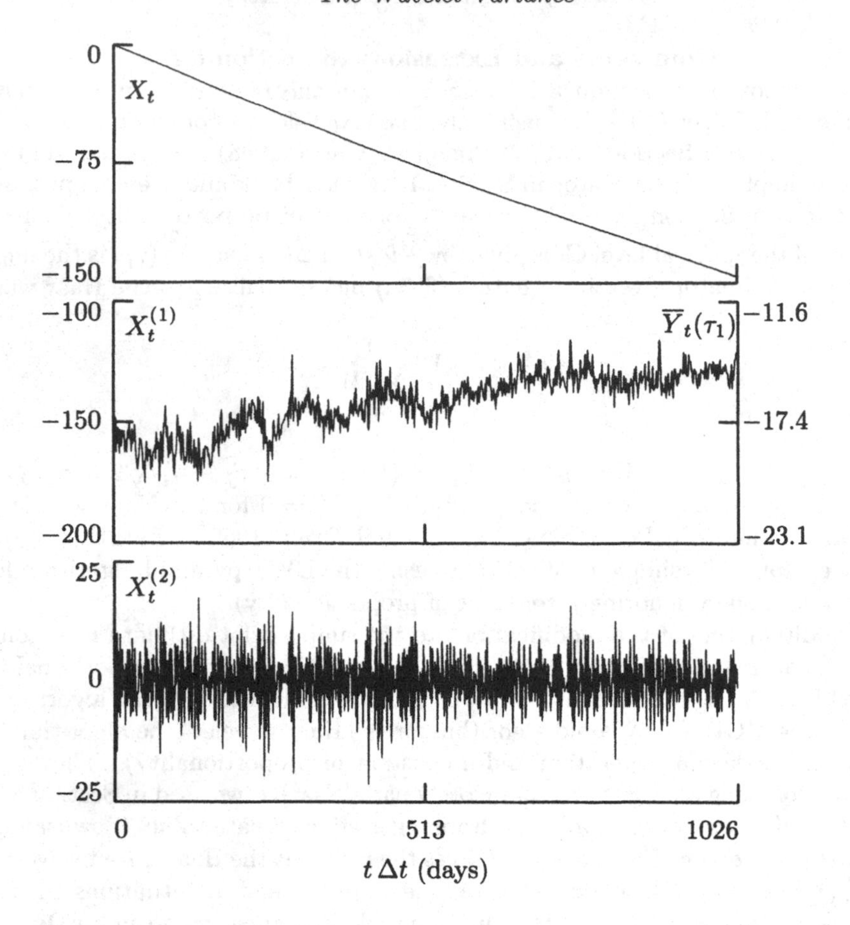

Figure 318. Plot of differences in time $\{X_t\}$ as kept by clock 571 (a cesium beam atomic clock) and as kept by the time scale UTC(USNO) maintained by the US Naval Observatory, Washington, DC (top plot); its first backward difference $\{X_t^{(1)}\}$ (middle); and its second backward difference $\{X_t^{(2)}\}$ (bottom). In the middle plot, $\overline{Y}_t(\tau_1)$ denotes the τ_1 average fractional frequency deviates (given in parts in 10^{13}) – these are defined by Equation (321c) and are proportional to $X_t^{(1)}$.

consecutive days in the 1970s, so the sampling interval is $\Delta t = 1$ day. Each value X_t was recorded in units of nanoseconds (there are 10^9 nanoseconds in one second), but for convenience the vertical axis of the top plot is expressed in units of microseconds (there are a thousand nanoseconds in one microsecond). If X_t were equal to zero on a particular day, the time kept by clock 571 and by UTC(USNO) would be in perfect agreement – this happens on the first day $t = 0$, but only because clock 571 was then set to be in agreement with UTC(USNO). If clock 571 is behind UTC(USNO) on the tth day (e.g., clock 571 declares it to be noon on a particular day some time after UTC(USNO) says it is noon), then $X_t < 0$ by convention. This is in fact true for all the values shown at the top of Figure 318, so clock 571 fell further and further behind UTC(USNO) over the almost three year period recorded here, ending up about 146 microseconds behind the time scale.

To estimate a wavelet variance for this time series using the estimator $\hat{\nu}_X^2(\tau_j)$ of

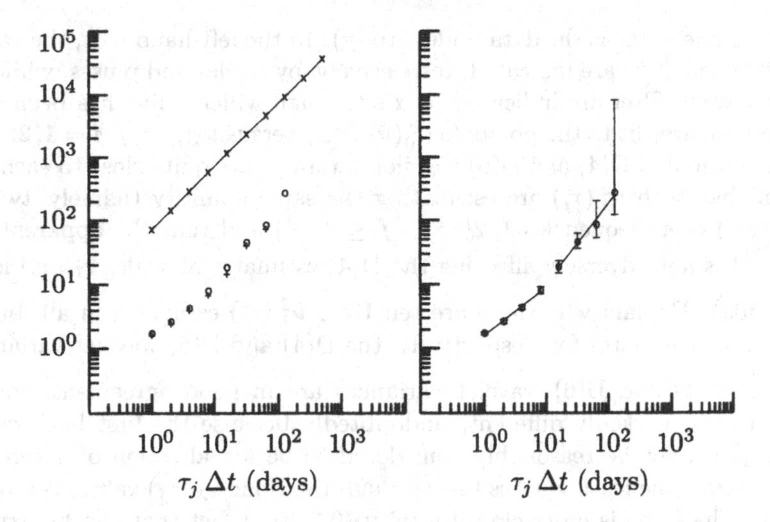

Figure 319. Square roots of wavelet variance estimates for atomic clock time differences $\{X_t\}$ based upon the unbiased MODWT estimator and the following wavelet filters: Haar (x's in left-hand plot, through which a least squares line has been fit), D(4) (circles in left- and right-hand plots) and D(6) (pluses in left-hand plot). The 95% confidence intervals in the second plot are the square roots of intervals computed using Equation (313c), with η given by $\hat{\eta}_1$ of Equation (313d) for $j = 1, \ldots, 6$ and by η_3 of Equation (314c) for $j = 7, 8$. The actual values for the various $\hat{\nu}_X(\tau_j)$ are listed on the Web site for this book – see page *xiv*.

Equation (306b), we first need to consider how well various backward differences of $\{X_t\}$ can be regarded as a realization of a stationary process with mean zero. The middle and bottom plots of Figure 318 show the first and second backward differences of $\{X_t\}$, i.e.,

$$X_t^{(1)} \equiv X_t - X_{t-1} \text{ and } X_t^{(2)} \equiv X_t^{(1)} - X_{t-1}^{(1)} = X_t - 2X_{t-1} + X_{t-2}$$

(note that now the units for the vertical axes are nanoseconds). The first difference series $\{X_t^{(1)}\}$ has a sample mean (-142.3) that is markedly different from zero and also has a tendency to increase linearly with time, so it is not reasonable to regard this series as a realization of a stationary process with mean zero. The second difference series $\{X_t^{(2)}\}$ has a very small sample mean (0.023) compared to its range of variation and also displays no obvious inconsistencies with being a realization of a stationary process; however, if indeed there is a linear trend in $\{X_t^{(1)}\}$ with slope a_1, Exercise [288b] says that the mean of $\{X_t^{(2)}\}$ will be equal to a_1. Since $\{X_t^{(1)}\}$ increases approximately 25 nanoseconds over a thousand days, a rough estimate of a_1 is 0.025, in good agreement with the sample mean. Since $d = 2$ is the smallest amount of differencing that appears to reduce $\{X_t\}$ to a process for which the wavelet variance makes sense, we should consider wavelet filters with widths $L \geq 2d = 4$. The D(4) wavelet should thus be adequate unless the linear trend in $\{X_t^{(1)}\}$ is indeed a problem, in which case we will need to use the D(6) wavelet (this is in keeping with the results of Exercise [305]).

The left-hand plot of Figure 319 shows the estimated wavelet standard deviation $\hat{\nu}_X(\tau_j)$ (i.e., the square root of $\hat{\nu}_X^2(\tau_j)$) versus τ_j for the D(4) and D(6) wavelet filters and – for illustrative purposes – the Haar wavelet filter also (it is common in the atomic clock literature to plot standard deviations rather than variances because the former has the same units as the data under study). In the left-hand plot, the values of $\hat{\nu}_X(\tau_j)$ for D(4) and D(6) are indicated, respectively, by circles and pluses, while those for the Haar wavelet filter are indicated by **x**'s through which a line has been drawn, namely, a least squares fit to the points $\log_{10}(\hat{\nu}_X(\tau_j))$ versus $\log_{10}(\tau_j)$, $j = 1, 2, \ldots, 10$. Note that the estimated D(4) and D(6) wavelet variances are quite close to each other, an indication that both $\hat{\nu}_X^2(\tau_j)$ are estimating the same quantity (namely, twice the integral of $S_X(\cdot)$ over frequencies $1/2^{j+1} < f \leq 1/2^j$) and that the apparent linear trend in $\{X_t^{(1)}\}$ is not adversely affecting the D(4) estimates at scales τ_7 and lower.

▷ **Exercise [320a]:** Explain why there are ten Haar $\hat{\nu}_X^2(\tau_j)$ estimates in all, but only eight and seven estimates for, respectively, the D(4) and D(6) wavelet variances. ◁

While the D(4) and D(6) wavelet variances are in good agreement, the Haar wavelet variance is markedly different, undoubtedly because the first backward difference of $\{X_t\}$ cannot be reasonably considered to be a realization of a zero mean stationary process. The least squares line through the Haar $\hat{\nu}_X(\tau_j)$ values is a remarkably good fit. The slope is quite close to unity (0.998), a fact that can be explained using the following exercise.

▷ **Exercise [320b]:** Let $\{X_t'\}$ be a process whose first backward differences are stationary with zero mean. Define $X_t \equiv X_t' + a_0 + a_1 t$; i.e., $\{X_t\}$ is the same as $\{X_t'\}$, but with the addition of a linear trend. Show that, if we use the Haar wavelet filter, then

$$E\{\hat{\nu}_X^2(\tau_j)\} = \nu_{X'}^2(\tau_j) + C\tau_j^2,$$

where the constant C depends on the slope a_1 and is to be determined as part of the exercise. Use the top plot of Figure 318 to get a rough estimate for a_1 and hence C, and explain the appearance of the Haar $\hat{\nu}_X(\tau_j)$ values in the left-hand plot of Figure 319. ◁

Note that, even without a careful consideration of the characteristics of $\{X_t\}$, it is easy to detect from Figure 319 that something is amiss with the Haar wavelet variance because it disagrees so markedly from the D(4) and D(6) variances. It is generally a good strategy to look at wavelet variances created using filters of different widths to guard against picking an L that is too small.

Since Figure 319 indicates little difference between the D(4) and D(6) wavelet variances, we can settle on the D(4) estimate as being adequate to assess the performance of clock 571 (however, we must reserve judgment about the D(4) estimate at scale τ_8, for which there is no corresponding D(6) estimate – as we shall see later, there is some concern that this D(4) estimate might be biased due to the apparent linear drift in $\{X_t^{(1)}\}$). The right-hand plot of Figure 319 shows the square root of the D(4) wavelet variance estimate along with the square root of a 95% confidence interval for the wavelet variance constructed using the chi-square approximation of Equation (313c) with degrees of freedom η estimated by $\hat{\eta}_1$ of Equation (313d) for scales τ_1 to τ_6 and by η_3 of Equation (314c) for τ_7 and τ_8 (the latter two scales have EDOFs $\eta_3 = 3$ and 1 as compared to $\hat{\eta}_1 = 13$ and 3, which yield confidence intervals that are undoubtedly too optimistic). This plot tells us, for example, that the differences in $\tau_1 = 1$ day averages of time deviations in clock 571 are close to two nanoseconds and that, at a confidence level of 95%, it is unreasonable that the true $\nu_X(\tau_1)$ is as large as, say, three nanoseconds or as small as one nanosecond. This analysis can be used to compare the performance of clock 571 with that of other clocks.

Since we have determined how close our estimate $\hat{\nu}_X(\tau_1)$ is likely to be to the true $\nu_X(\tau_1)$, we can use this assessment to tell if observed differences in the performance of clock 571 and any other clock are statistically significant or merely a statistical artifact due to the limited amount of data.

Figure 319 also indicates that changes in averaged time differences for clock 571 grow as τ_j increases. The manner in which $\hat{\nu}_X(\tau_j)$ increases with τ_j can be summarized somewhat by noting that there are two regions over which $\log_{10}(\hat{\nu}_X(\tau_j))$ versus $\log_{10}(\tau_j)$ appears to increase linearly, namely, for τ_1 up to τ_4 and for τ_5 up to τ_8. Figure 319 shows linear least squares fits over these two regions, yielding a slope of 0.74 for τ_1 up to τ_4 and a slope of 1.26 for τ_5 up to τ_8 (scale τ_4 seems to be the transition point since an extrapolation of the second regression line comes quite close to $\hat{\nu}_X(\tau_4)$). The above interpretation is consistent with the presence of two power laws over a very limited span of frequencies, one at small scales (8 days and less), and a second at large scales (16 days and greater). In terms of $\nu_X(\tau_j)$, we can summarize the performance of clock 571 as follows:

$$\hat{\nu}_X(\tau_j) \approx \begin{cases} 1.624\,\tau_j^{0.74}, & j = 1, 2, 3 \text{ and } 4; \\ 0.541\,\tau_j^{1.26}, & j = 5, 6, 7 \text{ and } 8. \end{cases} \tag{321a}$$

Note that, in terms of $\nu_X^2(\tau_j)$ rather than its square root, the above becomes $\nu_X^2(\tau_j) \propto \tau_j^{1.48}$, $j = 1, 2, 3$ and 4, and $\nu_X^2(\tau_j) \propto \tau_j^{2.52}$, $j = 5, 6, 7$ and 8. Since $\nu_X^2(\tau_j) \propto \tau_j^{-\alpha-1}$ translates into $S_X(f) \propto |f|^\alpha$ over $1/2^{j+1} < |f| \leq 1/2^j$, we have

$$S_X(f) \propto \begin{cases} |f|^{-2.48}, & 1/32 < |f| \leq 1/2; \\ |f|^{-3.52}, & 1/512 < |f| \leq 1/32, \end{cases} \tag{321b}$$

approximately.

Let us now discuss an interesting relationship between the Haar wavelet variance and the *Allan variance*, a well-established measure of clock performance that was introduced in Allan (1966) (see also Barnes *et al.*, 1971; Rutman, 1978; Percival, 1983, 1991; Greenhall, 1991; Flandrin, 1992; and Percival and Guttorp, 1994). To establish this relationship, we start with a definition for the τ_1 average fractional frequency deviates

$$\overline{Y}_t(\tau_1) \equiv \frac{X_t - X_{t-1}}{\tau_1\,\Delta t}, \tag{321c}$$

which are proportional to the first order backward difference of $\{X_t\}$. If X_t is measured in, say, nanoseconds and $\tau_1\,\Delta t$ is measured in, for example, days, then $\overline{Y}_t(\tau_1)$ can be turned into a dimensionless quantity if we re-express $\tau_1\,\Delta t$ in terms of nanoseconds. Since there are 86,400 seconds in one day and 10^9 nanoseconds in each second, in this example we would divide the first difference $X_t - X_{t-1}$ by 8.64×10^{13} to convert it into the unitless $\overline{Y}_t(\tau_1)$. A change of 8.64 nanoseconds in X_t over a one day period corresponds to a fractional frequency deviation of one part in 10^{13} (see the right-hand vertical axis on the middle plot of Figure 318).

Because a clock is just a device that counts the number of cycles produced by an oscillator operating at a particular nominal frequency f_0 (e.g., $f_0 = 5$ megahertz), the concept of an average fractional frequency deviation is useful because it tells how much the oscillator deviates on a relative basis from its nominal frequency on the average over a time period of $\tau_1\,\Delta t$. In particular, if we let $f(t)$ be the 'instantaneous'

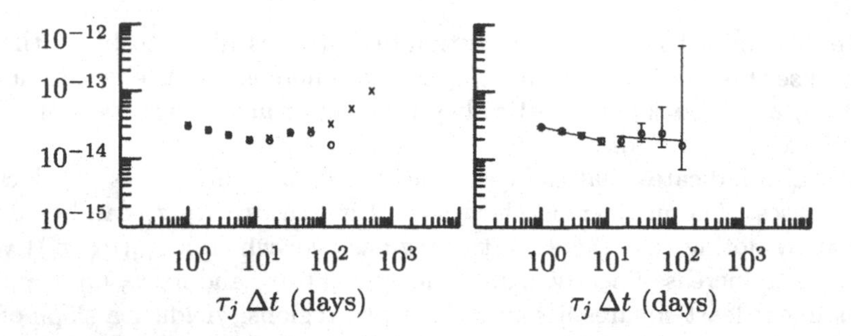

Figure 322. Square roots of wavelet variance estimates for atomic clock one day average fractional frequency deviates $\{\overline{Y}_t(\tau_1)\}$ based upon the unbiased MODWT estimator and the following wavelet filters: Haar (x's in left-hand plot) and D(4) (circles in left and right-hand plots). The actual values for $\hat{\nu}_{\overline{Y}}(\tau_j)$ are listed on the Web site for this book – see page *xiv*.

frequency of the oscillator at time t (this can be related to the time derivative of phase fluctuations), then we have

$$\overline{Y}_t(\tau_1) = \frac{1}{\tau_1\,\Delta t}\int_{t-\tau_1\,\Delta t}^{t} \frac{f(s) - f_0}{f_0}\,ds.$$

Likewise, the τ_j average fractional frequency deviates are given by

$$\overline{Y}_t(\tau_j) \equiv \frac{1}{\tau_j\,\Delta t}\int_{t-\tau_j\,\Delta t}^{t} \frac{f(s) - f_0}{f_0}\,ds = \frac{1}{\tau_j}\sum_{l=0}^{\tau_j-1}\overline{Y}_{t-l}(\tau_1) = \frac{X_t - X_{t-\tau_j}}{\tau_j\,\Delta t}.$$

Note in particular that $\overline{Y}_t(\tau_j)$ is an average of a subsequence of the τ_1 deviates and hence is a function of $\{\overline{Y}_t(\tau_1)\}$.

If we now regard $\{\overline{Y}_t(\tau_1)\}$ as a stochastic process whose first order backward difference is a stationary process, the Allan variance for these τ_1 deviates at scale τ_j is defined as

$$\sigma^2_{\overline{Y}}(2,\tau_j) = \frac{1}{2}E\left\{\left(\overline{Y}_t(\tau_j) - \overline{Y}_{t-\tau_j}(\tau_j)\right)^2\right\} \tag{322}$$

(Exercise [8.12] gives the rationale behind the notation for the Allan variance, while Exercise [8.13] explains why there is a factor of 1/2 in front of the expectation). We can now connect the Allan variance and the Haar wavelet variance for $\{\overline{Y}_t(\tau_1)\}$ as follows.

▷ **Exercise [322]:** Let $\{\overline{Y}_t(\tau_1)\}$ be a process with stationary first order backward differences. Let $\nu^2_{\overline{Y}}(\tau_j)$ be the Haar wavelet variance for this process. Show that, if the backward difference process has mean zero, then

$$\nu^2_{\overline{Y}}(\tau_j) = \frac{1}{2}\sigma^2_{\overline{Y}}(2,\tau_j).$$ ◁

Let us consider how our previous analysis of $\{X_t\}$ relates to an analysis of its rescaled first order backward difference $\{\overline{Y}_t(\tau_1)\}$. Figure 322 shows the square roots of the estimated Haar and D(4) wavelet variances for $\{\overline{Y}_t(\tau_1)\}$ (indicated on the plot by, respectively, x's and circles). Because we are now dealing with a process $\{\overline{Y}_t(\tau_1)\}$

whose first differences are reasonably close to a stationary process with mean zero, the Haar and D(4) wavelet variances are now quite comparable – the only apparent difference is at scale $\tau_8 = 128$ days, where the D(4) estimate is about three times smaller. If we use the Haar estimates to compute least squares fits comparable with Equation (321a), we obtain

$$\hat{\nu}_{\overline{Y}}(\tau_j) \approx \begin{cases} 3.093 \times 10^{-14}\, \tau_j^{-0.23}, & j = 1, 2, 3 \text{ and } 4; \\ 1.031 \times 10^{-14}\, \tau_j^{0.24}, & j = 5, 6, 7 \text{ and } 8. \end{cases} \tag{323}$$

These translate into

$$S_{\overline{Y}}(f) \propto \begin{cases} |f|^{-0.54}, & 1/32 < |f| \le 1/2; \\ |f|^{-1.48}, & 1/512 < |f| \le 1/32. \end{cases}$$

Because $\{\overline{Y}_t(\tau_1)\}$ is proportional to the first difference of $\{X_t\}$, we have

$$S_X(f) \propto \frac{S_{\overline{Y}}(f)}{\sin^2(\pi f)} \approx \frac{S_{\overline{Y}}(f)}{(\pi f)^2} \propto \begin{cases} |f|^{-2.54}, & 1/32 < |f| \le 1/2; \\ |f|^{-3.48}, & 1/512 < |f| \le 1/32. \end{cases}$$

The exponents above agree very well with those in Equation (321b).

Additionally, if we take the Haar estimates for the two largest scales and determine the line that passes through them (in log/log space), we find that $\hat{\nu}_{\overline{Y}}(\tau_j) \propto \tau_j^{0.88}$ for $j = 9$ and 10, so the exponent is rather close to unity. Exercise [320b] argues that a unit exponent can be due to a dominant linear trend in $\{\overline{Y}_t(\tau_1)\}$, which we have already noted as a concern based on the middle plot of Figure 318. This linear trend should not pose a problem for the D(4) wavelet variance since the D(4) filter has two embedded differencing operations. Since the Haar estimate $\hat{\nu}_{\overline{Y}}(\tau_8)$ lies quite close to an extrapolation based on the fit to scales τ_9 and τ_{10}, the difference between the observed Haar and D(4) estimates for $\nu_{\overline{Y}}(\tau_8)$ might be attributable to linear trend; i.e., the Haar estimate might be biased upwards due to this trend. The right-hand plot of Figure 322 shows the D(4) estimates by themselves with 95% confidence intervals constructed using Equation (313c) with η estimated by $\hat{\eta}_1$ of Equation (313d) for scales τ_1 to τ_6 and by η_3 of Equation (314c) for scales τ_7 to τ_8. The Haar estimate $\hat{\nu}_{\overline{Y}}(\tau_8)$ falls inside the confidence interval for $\nu_{\overline{Y}}(\tau_8)$ based upon the D(4) estimate, so sampling variability might also be the cause of the discrepancy.

Finally let us use this example to illustrate that wavelet variance plots can help to assess when a time series is best modeled as a stationary process as opposed to a nonstationary process with stationary backward differences. In terms of $1/f$-type processes, the former is characterized by an SDF varying at low frequencies approximately as $S_X(f) \propto |f|^\alpha$ with $\alpha > -1$, and the latter, with $\alpha \le -1$. For the wavelet variance, we have the approximation $\nu_X^2(\tau_j) \propto \tau_j^{-1-\alpha}$ at large scales, so stationarity is associated with negative exponents, and nonstationarity, with nonnegative exponents. When we plot $\log_{10}(\nu_X^2(\tau_j))$ versus $\log_{10}(\tau_j)$, the exponents become slopes, and a slope of zero at large scales becomes a convenient reference dividing stationary and nonstationary processes. Thus the markedly positive slopes in Figure 319 tell us immediately that the time differences $\{X_t\}$ are in keeping with a nonstationary model. The Haar estimates in the left-hand side of Figure 322 show a positive slope at large scales, so the fractional frequency deviates $\{\overline{Y}_t(\tau_1)\}$ are in keeping with a nonstationary model; however, the need for such a model might be just due to a linear trend – the D(4)

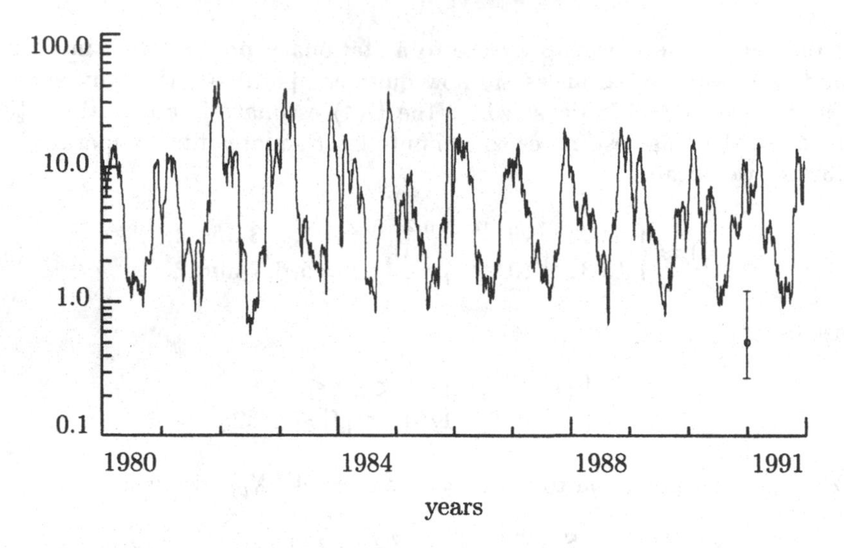

Figure 324. Estimated time-dependent LA(8) wavelet variances for physical scale $\tau_2 \, \Delta t = 1$ day for the Crescent City subtidal sea level variations of Figure 186, along with a representative 95% confidence interval based upon a hypothetical wavelet variance estimate of 1/2 and a chi-square distribution with $\nu = 15.25$ (see text for details).

estimates (which are impervious to such a trend) suggest a stationary model might be adequate, as is indicated by the negative slope of the linear least squares line depicted in the right-hand plot of Figure 322.

We examine the atomic clock data further in Section 9.7 as an example of a time series that can be modeled as a long memory process.

8.7 Example: Subtidal Sea Level Fluctuations

In Section 5.8 we saw that an MRA of the Crescent City subtidal sea level variations indicates a seasonally dependent variability at some scales (see Figure 186). Here we explore this variability by computing time-dependent wavelet variance estimates under the assumption that, within time segments encompassing approximately thirty days worth of data, the observed MODWT wavelet coefficients can be regarded as a realization of a zero mean Gaussian stationary process (past experience with this type of data suggests that this assumption is reasonable).

Figure 324 shows LA(8) MODWT wavelet variance estimates for scale $\tau_2 \, \Delta t = 1$ day based upon segments with 61 data values (since $\Delta t = 1/2$ day, each segment spans 30.5 days). Explicitly, we compute

$$\hat{\nu}^2_{X,t}(\tau_j) \equiv \frac{1}{N_S} \sum_{u=-(N_S-1)/2}^{(N_S-1)/2} \widetilde{W}^2_{j,t+|\nu_j^{(H)}|+u \bmod N} \tag{324}$$

for $j = 2$ and $N_S = 61$ and then plot this estimate versus the actual time associated with X_t (in the above, $|\nu_2^{(H)}| = 11$ is the circular shift required by the phase properties of the LA(8) filter to align $\{\widetilde{W}_{2,t}\}$ so that we can regard it as approximately the output from a zero phase filter – see Section 4.8 for details). Under our assumptions, we can compute a 95% confidence interval for the hypothesized true time-varying wavelet variance $\nu^2_{X,t}(\tau_2)$ based upon a chi-square approximation for the distribution of the RV $\hat{\nu}^2_{X,t}(\tau_2)$. We can set the EDOF for the chi-square distribution to be $\eta = 15.25$ by using an argument similar to the one yielding η_3 of Equation (314c): each estimate is formed using 61 MODWT wavelet coefficients, and $j = 2$. The confidence intervals thus take the form of

$$\left[\frac{\eta\hat{\nu}^2_{X,t}(\tau_2)}{Q_\eta(0.975)}, \frac{\eta\hat{\nu}^2_{X,t}(\tau_2)}{Q_\eta(0.025)}\right];$$

when plotted on a logarithmic scale, we effectively obtain a confidence interval for the log of $\nu^2_{X,t}(\tau_2)$ of the form

$$\left[\log\left(\hat{\nu}^2_{X,t}(\tau_2)\right) + \log\left(\frac{\eta}{Q_\eta(0.975)}\right), \log\left(\hat{\nu}^2_{X,t}(\tau_2)\right) + \log\left(\frac{\eta}{Q_\eta(0.025)}\right)\right].$$

The width of this interval is $\log\left(Q_\eta(0.975)/Q_\eta(0.025)\right)$, which is conveniently independent of the actual estimate. A confidence interval based upon a hypothetical estimate of $\hat{\nu}^2_{X,t}(\tau_2) = 1/2$ is depicted near the lower right-corner of Figure 324. Because this figure has a $\log_{10}$ vertical axis, we need only mentally shift this confidence interval so that the circle between the lower and upper confidence limits is centered around an actual estimate $\hat{\nu}^2_{X,t}(\tau_2)$ to obtain a 95% confidence interval for $\nu^2_{X,t}(\tau_2)$ (plotted on a logarithmic axis).

The confidence intervals for the time-varying wavelet variance indicate that, when we account for sampling variability, there are very few significant differences among the various $\hat{\nu}^2_{X,t}(\tau_2)$ that occur at the same dates in different years; i.e., we can model the curve in Figure 324 as a periodic function with an annual periodicity. A similar simplification can be justified for scales τ_3 up to τ_7. These results – along with a study of the MRA of Figure 186 – suggest that a reasonable way to summarize the annual variability in the scale dependent variance is by averaging – across all 12 years – all the squared circularly shifted MODWT wavelet coefficients that fall in a given calendar month. The results of this averaging are shown in Figure 326. Under the simplifying assumptions that the wavelet coefficients over each month can be regarded as one portion of a realization of a stationary process and that the coefficients from one year to the next are independent of each other, the methodology described in Section 8.4 can be readily adapted to develop approximate 95% confidence intervals for the underlying true variances. If we ignore the very small differences that arise because calendar months have different numbers of days, the widths of these intervals on a logarithmic scale again depend upon the particular scale, but – within a given scale – do not depend on the variance estimates themselves. Thus the confidence intervals based on the November estimates that are plotted for each scale in Figure 326 can be mentally translated to the other months to assess the sampling variability across a given scale.

We can draw the following conclusions from Figures 326 and 186. Although the variance estimates for a given month are different from scale to scale, the temporal patterns of variance are very similar for all six scales (the fact that this is true for scale $\tau_7\,\Delta t = 32$ days is not readily apparent in the MRA of Figure 186). The average variances are generally highest during the winter weather regime (November to March) and lowest in July and August, approximately 3 to 4 months after the lowest seasonal sea level of the year (cf. $\widetilde{\mathcal{S}}_7$ in Figure 186). The average winter variances are

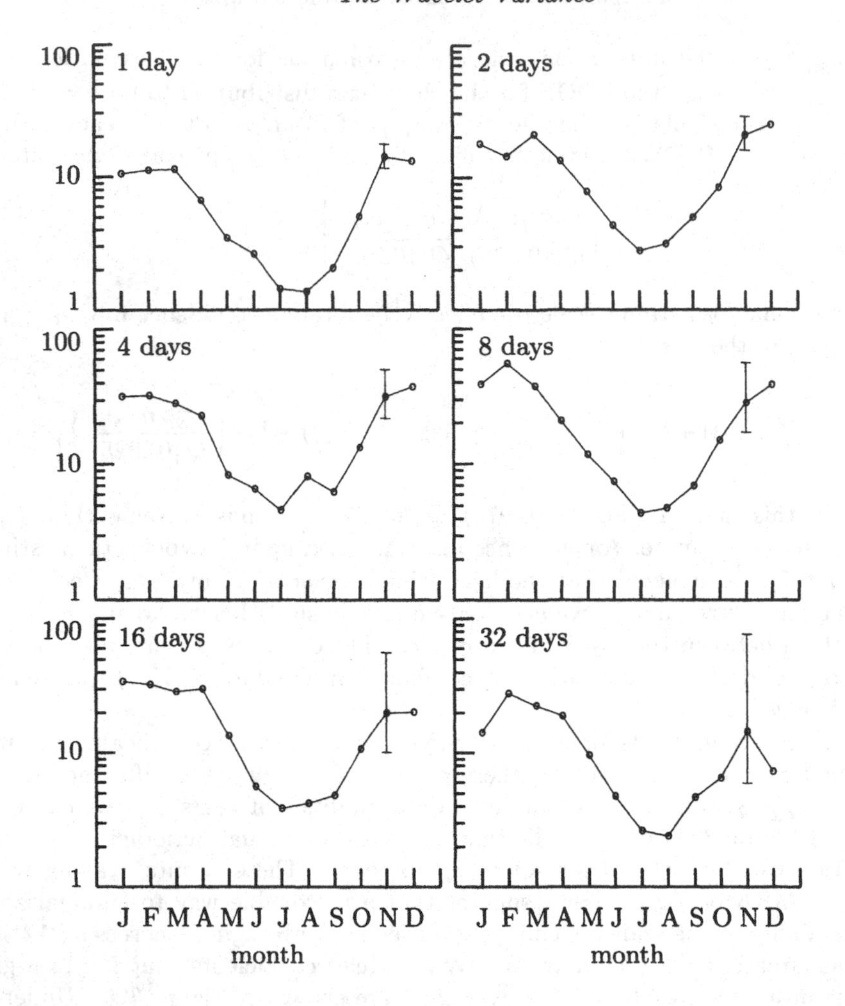

Figure 326. Estimated LA(8) wavelet variances for physical scales $\tau_j \Delta t = 2^{j-2}$ days, $j = 2, \ldots, 7$, grouped by calendar month for the subtidal sea level variations of Figure 186.

approximately an order of magnitude higher than the summer variances. This relatively simple scale-based description of the seasonal variability in subtidal sea level variations can be used, for example, to study their extreme value statistics via computer simulations – such statistics are an important component in forecasting 'worst case' effects of tsunamis (Percival and Mofjeld, 1997).

8.8 Example: Nile River Minima

In our discussion of the Nile River minima series in Section 5.9, we noted that a Haar-based MRA visually suggests a decrease in variability at scales $\tau_1 = 1$ and $\tau_2 = 2$ years after about year 715, which is around the time that historical records indicate a change in the way this series was measured. Because the human visual system has evolved in part so that we can rapidly pick out patterns, we can be easily tricked into seeing nonexistent patterns when presented with randomness. It is thus important to confirm our visual impression via a statistical procedure that takes into account the effects of sampling variability. To do so, we split the time series into two parts, consisting of the

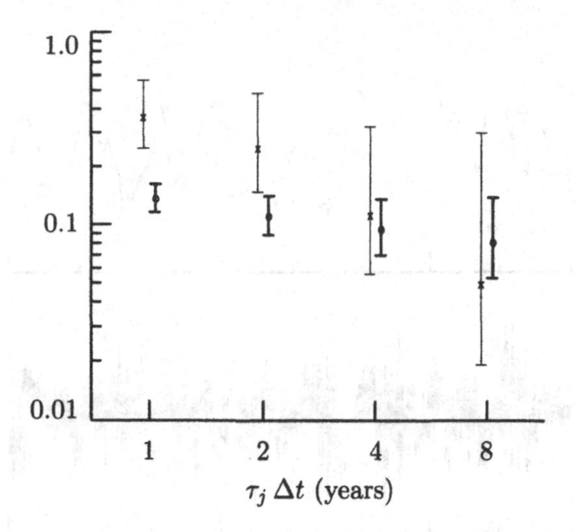

Figure 327. Estimated Haar wavelet variances for the Nile River minima time series before and after year 715.5 (x's and o's, respectively), along with 95% confidence intervals (thin and thick lines, respectively) based upon a chi-square approximation with EDOFs determined by η_3 of Equation (314c). The actual values for $\hat{\nu}_X^2(\tau_j)$ are listed on the Web site for this book – see page *xiv*.

first 94 observations of the series (i.e., 622 to 715) and the next 569 observations (716 to 1284). With this split, we then calculate the Haar wavelet variance at scales τ_1 to τ_4 separately for the two subseries. These estimated variances are plotted in Figure 327, along with 95% confidence intervals computed using the chi-square approximation with the EDOFs determined using η_3 of Equation (314c) (this method produced the widest confidence intervals amongst the ones discussed in Section 8.4 and hence are used so that we err on the conservative side). We see that the 95% confidence intervals for scales τ_1 and τ_2 indeed do not trap the same values, lending credence to our visual interpretation of the Haar MRA. By contrast, the 95% confidence intervals for scales τ_3 and τ_4 indicate little evidence that the wavelet variance has significantly changed at these scales.

This example illustrates the use of the wavelet variance on a series that is not globally stationary, but that can be considered to be locally stationary within two adjacent segments of time. The historical record allows us to split this series objectively at year 715, but the *a priori* identification of regions in a series over which stationarity is a reasonable assumption is not often possible. In Section 9.6 we consider a scaled-based statistical test for homogeneity of variance that does not require *a priori* partitioning, as was done here (when applied to the Nile River series in Section 9.8, we reach the same conclusions we came to here).

8.9 Example: Ocean Shear Measurements

Here we illustrate the use of the wavelet variance by applying it to the vertical ocean shear depth series, which is described in Section 5.10 and plotted at the bottom of Figure 194. There are two thin vertical lines marked on that plot, between which there are 4096 values ranging from 489.5 meters to 899.0 meters. In what follows, we will assume that this subseries (replotted at the top of Figure 328) can be regarded as a portion of one realization of a process whose first order backward difference is

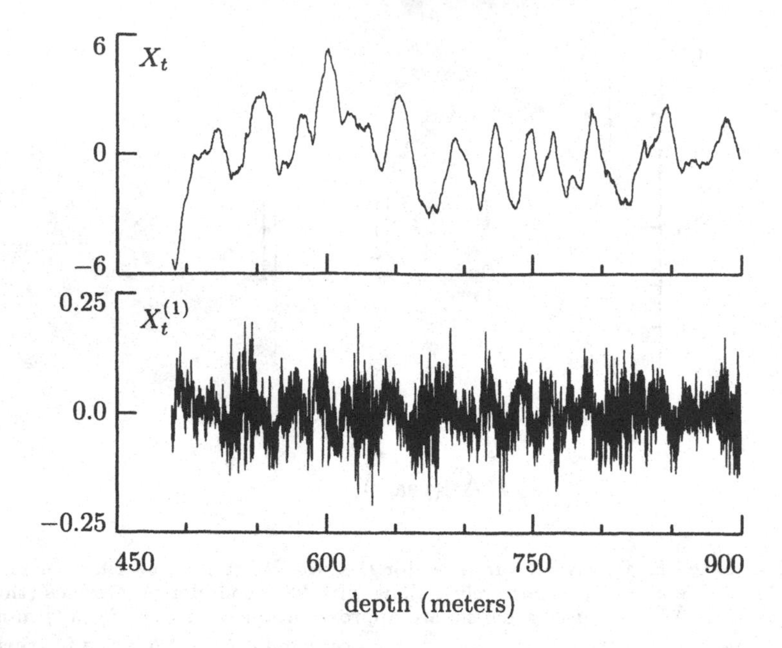

Figure 328. Selected portion $\{X_t\}$ of $N = 4096$ vertical shear measurements (top plot) and associated backward differences $\{X_t^{(1)}\}$ (bottom). The full series is plotted at the bottom of Figure 194, on which the subseries is delineated by two thin vertical lines.

a zero mean stationary process (we need this assumption to apply the methodology used below meaningfully). The bottom plot of Figure 328 shows these differences, and there is nothing obvious in the plot that would rule out a zero mean stationary process as a model.

The x's in the left-hand plot of Figure 329 show the unbiased MODWT Haar wavelet variance estimates for physical scales ranging from $\tau_1 \, \Delta t = 0.1$ meters up to $\tau_{12} \, \Delta t = 204.8$ meters. Theory suggests that regions of linearity in a plot of $\nu_X^2(\tau_j)$ versus τ_j on a log/log scale indicate the presence of a power law process over a particular region of frequencies, with the power law exponent being related to the slope of the line. The Haar wavelet variance estimates for the smallest seven scales fall on such a line almost perfectly. The line drawn through them on the plot was calculated via linear least squares and has a slope of $1.66 \doteq 5/3$. Since $\nu_X^2(\tau_j)$ varies approximately as $\tau_j^{-\alpha-1}$ for a power law process with exponent α (see Section 8.1), the Haar wavelet variance plot strongly suggests the presence of a power law process over scales of 0.1 to 6.4 meters with an exponent of $\alpha \doteq -8/3$.

In addition to the Haar estimates, the left-hand plot of Figure 329 also shows wavelet variance estimates based on the D(6) wavelet filter (the +'s). At small scales the D(6) estimates are systematically smaller than the Haar estimates. The D(6) estimates for the smallest seven scales are no longer aligned in an obvious straight line, and the slope that we found using the smallest seven scales of the Haar wavelet variance certainly does not look reasonable for the corresponding D(6) estimates. The right-hand plot replicates the D(6) estimates and also shows estimates based on the D(4) and LA(8) wavelet filters (small and large circles, respectively). Here we see good agreement between the D(6) and LA(8) estimates at all scales, and reasonably

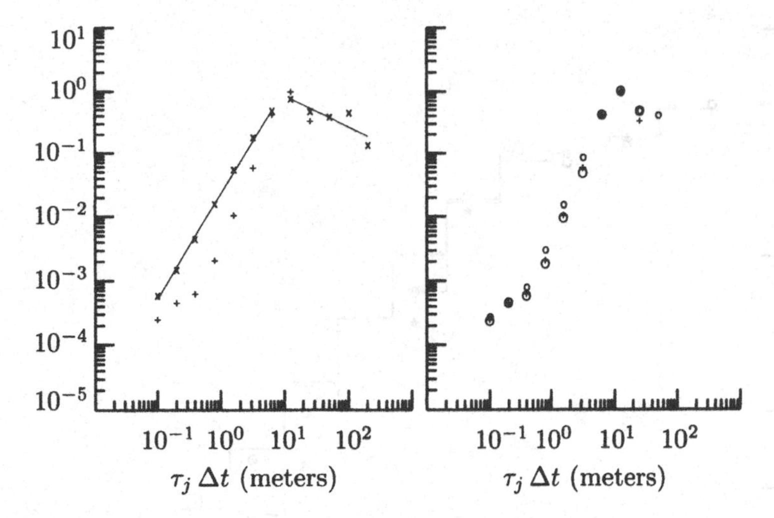

Figure 329. Wavelet variances estimated for vertical shear series using the unbiased MODWT estimator and the following wavelet filters: Haar (x's in left-hand plot, through which two regression lines have been fit), D(4) (small circles, right-hand plot), D(6) (+'s, both plots) and LA(8) (big circles, right-hand plot). The values for the various $\hat{\nu}_X^2(\tau_j)$ are listed on the Web site for this book – see page *xiv*.

good agreement between these and the D(4) estimates, with the exception of small discrepancies at intermediate scales. Thus, while the D(4), D(6) and LA(8) estimates are in reasonable agreement, the Haar wavelet variance estimates are systematically higher at small scales. In the atomic clock example of the previous section, we also noted discrepancies involving Haar estimates. We argued that these discrepancies were due to the fact that the backward differences for the atomic clock data cannot be reasonably modeled as a zero mean stationary process, as is quite evident from the middle plot of Figure 318 (the sample mean differs substantively from zero, and there is arguably a linear trend in the differenced data). For the vertical shear series, the discordance between the Haar and other wavelet variance estimates is not due to any obvious violation of stationarity, but rather can be explained by the concept of leakage.

Leakage is a well-known phenomenon in SDF estimation – see Section 7.5 (in particular the discussions pertaining to Figures 273 and 275). Leakage manifests itself as an upwards bias in the low power portion of the SDF. Since we can translate a wavelet variance estimate into an estimate of a region of the SDF, let us do so to illustrate that the elevated Haar estimates turn into SDF levels that match the typical pattern of leakage. Figure 330 shows the estimated SDF level $\hat{C}_j$ of Equation (316) plotted as a constant line over the octave band $[\frac{1}{2^{j+1}\,\Delta t}, \frac{1}{2^j\,\Delta t}]$, where $j = 1, \ldots, 12$ for the Haar-based estimates (thick 'staircase') and $j = 1, \ldots, 9$ for the D(6) estimates (thin staircase). While there is good agreement between the two SDF estimates at low frequencies, the Haar-based estimates are elevated upwards at high frequencies, which is the low power portion of the estimated SDFs. This pattern is thus consistent with leakage in the Haar-based estimates.

For comparison, we have also plotted SDF estimates based upon the leakage-prone periodogram $\hat{S}^{(\mathrm{p})}(\cdot)$ (circles) and a multitaper estimator $\hat{S}^{(\mathrm{mt})}(\cdot)$ (asterisks) that is based on $K = 7$ sine tapers and should be relatively free of leakage (for details on mul-

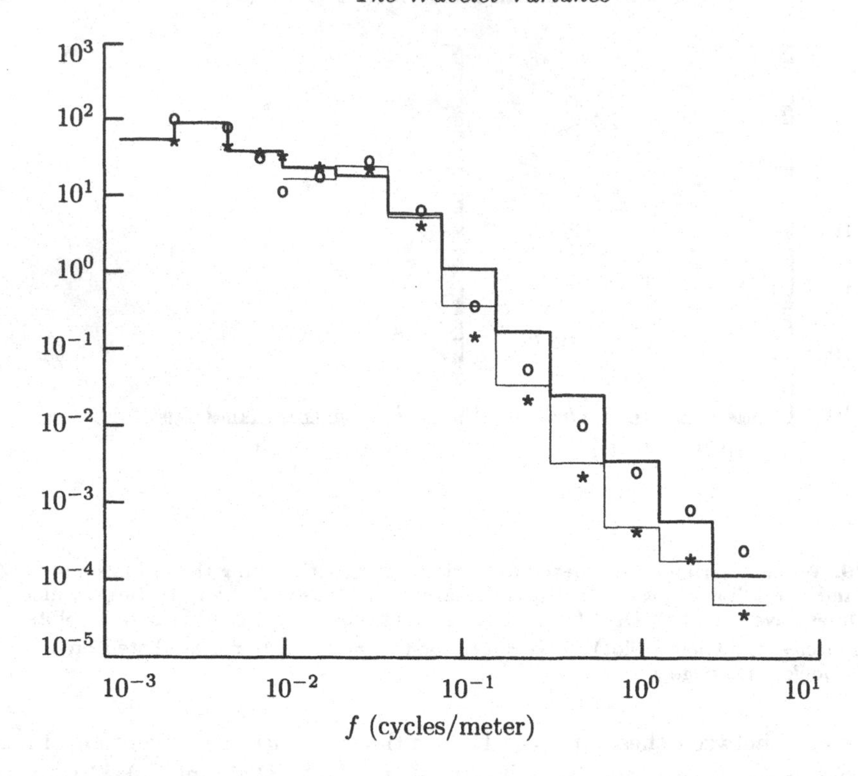

Figure 330. Comparison of 'octave band' SDF estimates for the vertical shear measurements based on the periodogram (o's), a multitaper SDF estimate formed using $K = 7$ sine tapers ($\star$'s) and Haar and D(6) wavelet variance estimates (thick and thin staircases, respectively). See text for details.

titaper SDF estimators, see Section 7.5). We computed the periodogram and the multitaper estimates at the nonzero Fourier frequencies $f_k \equiv \frac{k}{N\,\Delta t}$, $k = 1, \ldots, 2048$ (recall that $N = 4096$). The thirteen circles show, from left to right, $\hat{S}^{(\mathrm{p})}(f_1)$, $\hat{S}^{(\mathrm{p})}(f_2)$, $\hat{S}^{(\mathrm{p})}(f_3)$ and $\hat{S}^{(\mathrm{p})}(f_4)$; the average of the next four estimates, namely, $\hat{S}^{(\mathrm{p})}(f_5), \ldots, \hat{S}^{(\mathrm{p})}(f_8)$; the average of the next eight estimates $\hat{S}^{(\mathrm{p})}(f_9), \ldots, \hat{S}^{(\mathrm{p})}(f_{16})$; and so on, ending with an average of $\hat{S}^{(\mathrm{p})}(f_{1025}), \ldots, \hat{S}^{(\mathrm{p})}(f_{2048})$. Each of the 'octave band' averaged estimates is plotted versus the average of the Fourier frequencies associated with the estimates. The asterisks are obtained by treating the multitaper estimates in a similar manner. Note that, in the low power (high frequency) portion of the SDF, the periodogram-based estimates are consistently above the multitaper-based estimates, an indication of leakage in the periodogram. In general, the D(6) wavelet variance yields an SDF estimate that agrees reasonably well with the leakage-free multitaper estimate (there are some discrepancies in the mid-range octave bands, possibly attributable in part to sampling variability). By contrast, the Haar-based SDF estimate shows the same pattern as the periodogram; i.e., it is systematically higher at high frequencies than the D(6) or multitaper-based estimates. Again, the qualitative similarity between the Haar-based SDF estimate and the leakage-prone periodogram suggests a form of leakage in the Haar wavelet variance.

To see exactly why the Haar-based SDF estimates suffer from leakage, suppose that $\{X_t\}$ has the following theoretical SDF, which is roughly similar to the D(6) and

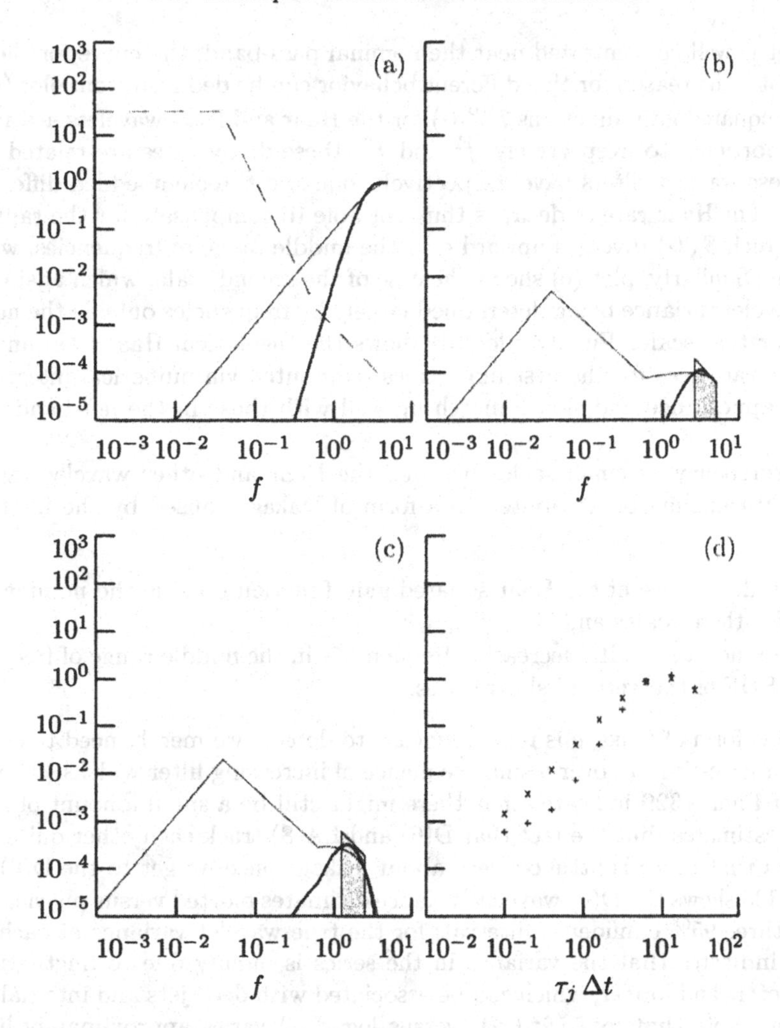

Figure 331. Leakage in Haar wavelet variance (see text for details).

multitaper SDF estimates in Figure 330:

$$S_X(f) = \begin{cases} 32.768, & |f| \le 5/128; \\ 32.768 \cdot |128f/5|^{-3.5}, & 5/128 < |f| \le 5/8; \text{ and} \\ |8f/5|^{-1.7}/500, & 5/8 < |f| \le 5. \end{cases} \tag{331}$$

Plot (a) of Figure 331 shows $S_X(\cdot)$ (dashed curve), along with the squared gain functions $\widetilde{\mathcal{H}}_1^{(\mathrm{D})}(\cdot)$ for the Haar (thin solid curve) and D(6) (thick) wavelet filters. The shaded area on plot (b) depicts the product of $S_X(\cdot)$ and the squared gain function for a perfect band-pass filter with pass-band $[5/2, 5]$, i.e., the nominal one associated with physical scale $\tau_1\,\Delta t$. This plot also shows the product of $S_X(\cdot)$ and $\widetilde{\mathcal{H}}_1^{(\mathrm{D})}(\cdot)$ for, respectively, the Haar (thin curve) and D(6) (thick) wavelet filters. If we recall that $\nu_X^2(\tau_1)$ is the integral of the product of $\widetilde{\mathcal{H}}_1^{(\mathrm{D})}(\cdot)$ and $S_X(\cdot)$ (see Equation (305)), we see that the integral under the thin and thick curves in plot (b) yields, respectively, the Haar and D(6) wavelet variances for the smallest scale. While the curve for the

D(6) wavelet is well concentrated near the nominal pass-band, the curve for the Haar wavelet is not. The reason for this different behavior can be deduced from plot (a): for small f, the squared gain functions $\widetilde{\mathcal{H}}_1^{(\mathrm{D})}(\cdot)$ for the Haar and D(6) wavelets are approximately proportional to, respectively, f^2 and f^6 (these decay rates are related to the fact that these wavelet filters have, respectively, one and three embedded differencing operations). The Haar rate of decay is thus not able to compensate for the rapid rate ($f^{-3.5}$) at which $S_X(\cdot)$ diverges upward over the middle range of frequencies, whereas the D(6) can. Similarly, plot (c) shows the case of the second scale, which again shows the Haar wavelet variance being determined largely by frequencies outside the nominal pass-band for that scale. Finally, plot (d) shows the theoretical Haar (x's) and D(6) (+'s) wavelet variances for the first nine scales (computed via numerical integration). The overall appearance and slopes match up well with those in the left-hand plot of Figure 329.

The discrepancy at small scales between the Haar and other wavelet variances in Figure 329 can thus be attributed to a form of leakage caused by the interaction between

[1] the slow decay rate of the Haar squared gain function outside the nominal pass-bands for these scales and

[2] the steep increase – with decreasing frequency – in the middle range of frequencies for the SDF of the vertical shear series.

Note that this form of leakage is relatively easy to detect: we merely need to compare wavelet variance estimates over a small sequence of increasing filter widths. The right-hand plot of Figure 329 indicates that there might still be a small amount of leakage in the D(4) estimates, but the fact that D(6) and LA(8) track each other quite well is an indication that there is little concern about leakage once we get to the D(6) filter.

Figure 333 shows the D(6) wavelet variance estimates plotted versus physical scale, along with three 95% confidence intervals for the true wavelet variance at each scale. This figure indicates that the variance in the series is mainly due to fluctuations at scales 6.4 meters and longer, which can be associated with deep jets and internal waves in the ocean. Note that $\log_{10}(\hat{\nu}_X^2(\tau_j))$ versus $\log_{10}(\tau_j)$ varies approximately linearly over small scales 0.1 to 0.4 meters and also over intermediate scales 0.8 to 6.4 meters (the extent of this approximate linearity, however, is quite limited, being less than an order of magnitude of scales in each case). The slopes associated with these two regions are, respectively, 0.7 and 2.5. The small scales are influenced mainly by turbulence, and the slope of $\log_{10}(\hat{\nu}_X^2(\tau_j))$ versus $\log_{10}(\tau_j)$ indicates that turbulence rolls off at a rate consistent with a power law of exponent $\alpha = -1.7$, a result that can be compared to physical models. The power law rolloff of $\alpha = -3.5$ at intermediate scales can be interpreted as a transition region between the internal wave and turbulent regions.

The three confidence intervals in Figure 333 are based upon the χ^2 approximation with EDOF η determined by $\hat{\eta}_1$, η_2 and η_3. The actual values for the EDOFs are listed in Table 333. For the seven smallest scales, the confidence intervals given by the three methods are interchangeable from a practitioner's point of view, but, not surprisingly, the agreement breaks down some at the two largest scales. The confidence intervals can be used to assess if fluctuations at, for example, scale 25.6 meters for this particular series agree with other sets of measurements taken at different locations in the ocean.

Finally let us again illustrate how the wavelet variance can easily be adapted to handle time series that are best modeled as being locally stationary within certain

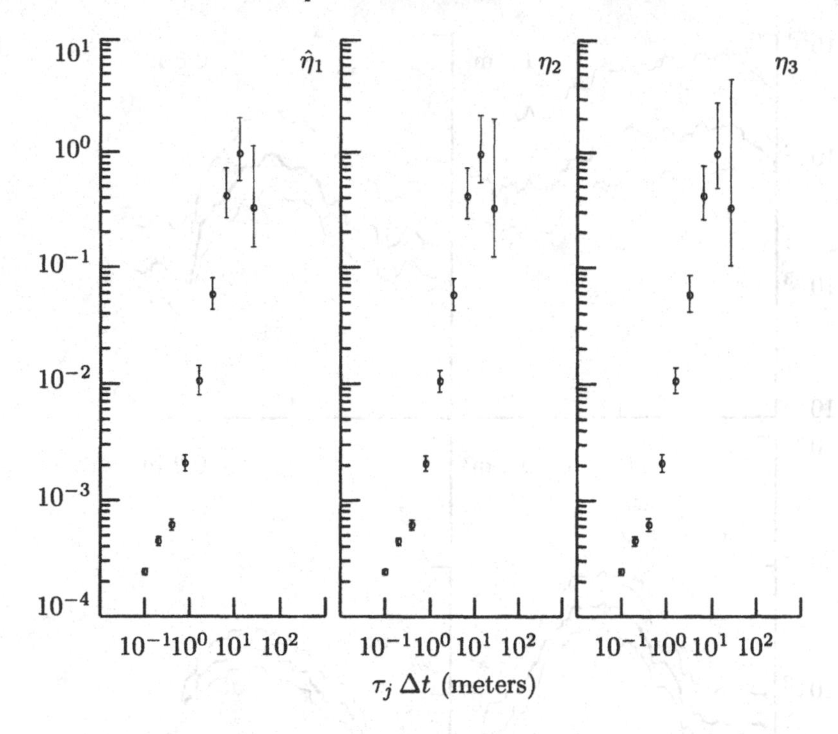

Figure 333. 95% confidence intervals for the D(6) wavelet variance for the vertical ocean shear series. The intervals are based upon the unbiased MODWT estimator (+'s in Figure 329 and o's above) and χ^2 approximations to its distribution with EDOFs determined by, from left to right, $\hat{\eta}_1$ of Equation (313d); η_2 of Equation (314b) using the nominal model for $S_X(\cdot)$ given by Equation (331); and η_3 of Equation (314c) (Table 333 lists the values for the EDOFs).

	j								
	1	2	3	4	5	6	7	8	9
$\hat{\eta}_1$	1890	1027	584	289	94	82	32	20	8
η_2	2850	1633	899	359	173	78	31	17	5
η_3	2046	1020	508	251	123	59	27	11	3
M_j	4091	4081	4061	4021	3941	3781	3461	2821	1541

Table 333. Equivalent degrees of freedom $\hat{\eta}_1$, η_2 and η_3 (rounded to the nearest integer) associated with the D(6) wavelet variance estimates $\hat{\nu}_X^2(\tau_j)$, $j = 1, \ldots, 9$, shown in Figure 333. The bottom row gives the number M_j of MODWT wavelet coefficients at each scale.

segments. In our discussion in Section 5.10 of the LA(8) MRA of the vertical shear series depicted in Figure 195, we noted that the prominent burst at shallow depths manifests itself first at a scale of $\tau_5\,\Delta t = 1.6$ meters, but then becomes more prominent at smaller scales as it increases in depth. Here we can confirm this pattern by computing the time-dependent LA(8) wavelet variance as per Equation (324) (note that we now use the LA(8) rather than D(6) wavelet because of the superior time alignment properties of the former). The size of the burst in Figure 195 is about 50 meters.

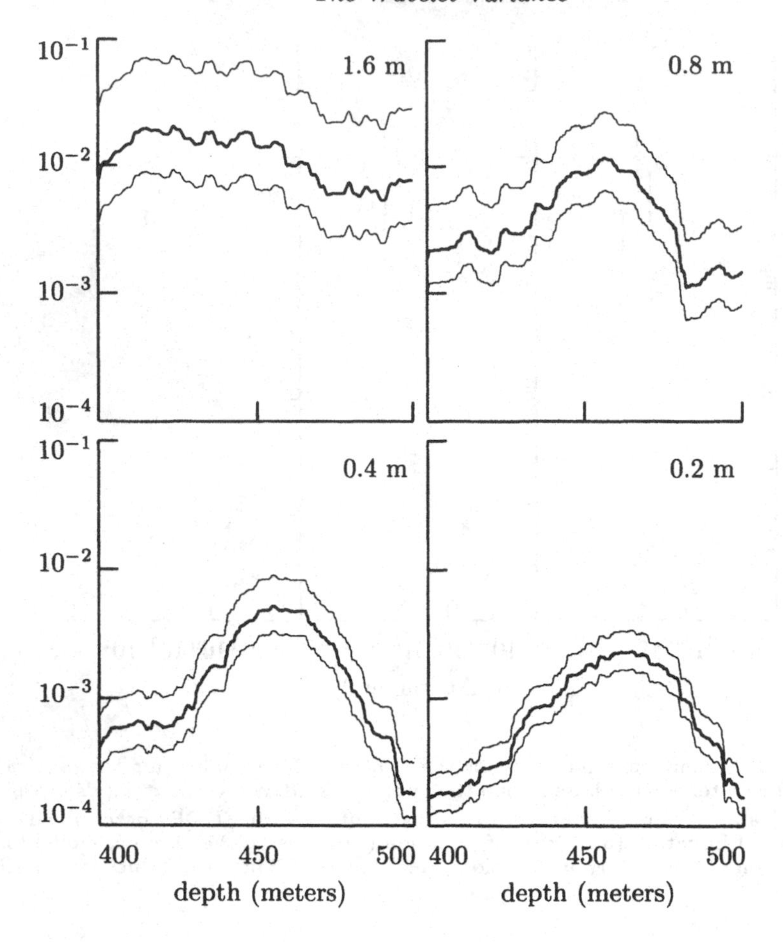

Figure 334. Estimated LA(8) wavelet variances (thick curves) for the shear data, computed using a running segment of 257 MODWT wavelet coefficients for depths surrounding 450 meters. As marked on the plots, the physical scales are 0.2, 0.4, 0.8 and 1.6 meters (i.e., $\tau_j\,\Delta t$, $j = 2, \ldots, 5$). The thin curves above and below the estimates are 95% confidence intervals as per Equation (313c) with $\eta = 257/2^j$ (this is in accordance with η_3 of Equation (314c)).

If we let $N_S = 257$, each segment spans 25.7 meters, which is a reasonable region over which to assume local stationarity. The resulting estimates for scales $\tau_5\,\Delta t = 1.6$ down to $\tau_2\,\Delta t = 0.2$ meters are plotted in Figure 334 as thick curves, along with upper and lower 95% confidence intervals (thin curves) computed via a chi-square approximation to the distribution of $\hat{\nu}^2_{X,t}(\tau_j)$ (the EDOFs are taken to be $\eta = N_S/2^j$ in accordance with η_3 of Equation (314c)). Note that, since the vertical axis of this figure is logarithmic, the heights of the confidence intervals appear to be independent of the estimated variances across each scale (in reality, the heights are proportional to the estimates). Scientifically, the important observation is a confirmation of the MRA of Figure 195, namely, that the peak in the wavelet variance corresponding to the burst appears increasingly deep with decreasing scale, suggesting that turbulence at a scale of 1.6 meters is driving shorter scale turbulence at greater depths (see Serroukh *et al.*, 2000, for a complementary treatment that drops the Gaussian assumption).

8.10 Summary

Let $\{X_t : t = \ldots, -1, 0, 1, \ldots\}$ be a discrete parameter real-valued stochastic process whose dth order backward difference

$$Y_t \equiv (1-B)^d X_t \equiv \sum_{k=0}^{d} \binom{d}{k} (-1)^k X_{t-k}$$

is a stationary process with SDF $S_Y(\cdot)$ and mean μ_Y (here d is a nonnegative integer, and μ_Y need not be zero). Let $S_X(\cdot)$ represent the SDF for $\{X_t\}$, for which we have $S_X(f) = S_Y(f)/\mathcal{D}^d(f)$, where $\mathcal{D}(f) \equiv 4\sin^2(\pi f)$ (if $\{X_t\}$ is in fact a nonstationary process, this relationship between $S_X(\cdot)$ and $S_Y(\cdot)$ provides a definition for $S_X(\cdot)$). Given a jth level MODWT wavelet filter $\{\tilde{h}_{j,l} : l = 0, \ldots, L_j - 1\}$ that is based upon a Daubechies wavelet filter $\{\tilde{h}_l\}$ of width L, define the jth level wavelet coefficient process as

$$\overline{W}_{j,t} \equiv \sum_{l=0}^{L_j-1} \tilde{h}_{j,l} X_{t-l}, \quad t = \ldots, -1, 0, 1, \ldots$$

(in the above, recall that $L_j \equiv (2^j - 1)(L-1) + 1$). If $L \geq 2d$, then $\{\overline{W}_{j,t}\}$ is a stationary process with SDF given by $S_j(f) \equiv \widetilde{\mathcal{H}}_j^{(\mathrm{D})}(f) S_X(f)$, where $\widetilde{\mathcal{H}}_j^{(\mathrm{D})}(\cdot)$ is the squared gain function for $\{\tilde{h}_{j,l}\}$. We can then define the wavelet variance for $\{X_t\}$ at scale $\tau_j \equiv 2^{j-1}$ as

$$\nu_X^2(\tau_j) \equiv \operatorname{var}\{\overline{W}_{j,t}\} = \int_{-1/2}^{1/2} \widetilde{\mathcal{H}}_j^{(\mathrm{D})}(f) S_X(f)\, df$$

(this variance is necessarily finite and independent of t). We have

$$\sum_{j=1}^{\infty} \nu_X^2(\tau_j) = \operatorname{var}\{X_t\},$$

where $\operatorname{var}\{X_t\}$ is taken to be infinite if $\{X_t\}$ is nonstationary. In words, $\nu_X^2(\tau_j)$ represents the contribution to the total variability in $\{X_t\}$ due to changes at scale τ_j. (Note that τ_j is in fact a unitless standardized scale, which – upon multiplication by the sampling interval Δt associated with $\{X_t\}$ – becomes the physically meaningful scale $\tau_j\,\Delta t$.)

Suppose that we are given a time series that can be regarded as a realization of one portion $X_0, \ldots, X_{N-1}$ of the process $\{X_t\}$. Provided that $M_j \equiv N - L_j + 1 > 0$ and that either (i) $L > 2d$ or (ii) $L = 2d$ and $\mu_Y = 0$ (either of these conditions implies that $E\{\overline{W}_{j,t}\} = 0$ and hence that $\nu_X^2(\tau_j) = E\{\overline{W}_{j,t}^2\}$), we can form an unbiased estimator of $\nu_X^2(\tau_j)$ via

$$\hat{\nu}_X^2(\tau_j) \equiv \frac{1}{M_j} \sum_{t=L_j-1}^{N-1} \widetilde{W}_{j,t}^2,$$

where $\{\widetilde{W}_{j,t}\}$ is the jth level MODWT wavelet coefficients for the time series:

$$\widetilde{W}_{j,t} \equiv \sum_{l=0}^{L_j-1} \tilde{h}_{j,l} X_{t-l \bmod N}, \quad t = 0, 1, \ldots, N-1.$$

Note that $\widetilde{W}_{j,t} = \overline{W}_{j,t}$ when $t = L_j - 1, \ldots, N-1$ so that $E\{\widetilde{W}_{j,t}^2\} = \nu_X^2(\tau_j)$ for this range of t (this relationship need not hold for $t < L_j - 1$, so taking the sample mean of all N possible $\widetilde{W}_{j,t}^2$ in general yields a biased estimator of $\nu_X^2(\tau_j)$). Under the additional assumption that $\{\overline{W}_{j,t}\}$ is a Gaussian process, then, for large M_j, the RV $\hat{\nu}_X^2(\tau_j)$ is approximately Gaussian distributed with mean $\nu_X^2(\tau_j)$ and variance $2A_j/M_j$, where

$$A_j \equiv \int_{-1/2}^{1/2} S_j^2(f)\,df$$

(this result holds provided that A_j is finite and that $S_j(f) > 0$ almost everywhere). We can use this result directly to construct a confidence interval for $\nu_X^2(\tau_j)$ based upon $\hat{\nu}_X^2(\tau_j)$ (see Equation (311)), but the resulting interval can have a negative lower limit, an inconvenience because in practice we often want to plot wavelet variance estimates – and their associated confidence intervals – on a log/log scale.

A more appealing (but asymptotically equivalent) approach to generating a confidence interval for $\nu_X^2(\tau_j)$ is to assume that $\eta\hat{\nu}_X^2(\tau_j)/\nu_X^2(\tau_j)$ has the same distribution as χ_η^2, i.e., a chi-square RV with η 'equivalent' degrees of freedom (EDOF). This assumption leads to a confidence interval of the form

$$\left[\frac{\eta\hat{\nu}_X^2(\tau_j)}{Q_\eta(1-p)}, \frac{\eta\hat{\nu}_X^2(\tau_j)}{Q_\eta(p)}\right],$$

where $Q_\eta(p)$ is the $p \times 100\%$ percentage point for the χ_η^2 distribution. The above requires a value for η, which can be determined using one of the following three methods.

[1] We can estimate η using

$$\hat{\eta}_1 \equiv \frac{M_j\hat{\nu}_X^4(\tau_j)}{\hat{A}_j},$$

where

$$\hat{A}_j \equiv \frac{1}{2}\int_{-1/2}^{1/2} [\hat{S}_j^{(\mathrm{p})}(f)]^2\,df = \frac{\hat{\nu}_X^4(\tau_j)}{2} + \sum_{\tau=1}^{M_j-1} \left(\hat{s}_{j,\tau}^{(\mathrm{p})}\right)^2;$$

here $\{\hat{s}_{j,\tau}^{(\mathrm{p})}\}$ is the biased estimator of the ACVS for $\{\widetilde{W}_{j,t}\}$, i.e.,

$$\hat{s}_{j,\tau}^{(\mathrm{p})} \equiv \frac{1}{M_j}\sum_{t=L_j-1}^{N-1-|\tau|} \widetilde{W}_{j,t}\widetilde{W}_{j,t+|\tau|}, \quad 0 \le |\tau| \le M_j - 1.$$

[2] If we assume that $S_X(\cdot) = aC_X(f)$, where $C_X(\cdot)$ is a known function and a is an unknown multiplicative constant, then we can express the SDF for $\{\overline{W}_{j,t}\}$ as $S_j(f) = a\widetilde{\mathcal{H}}_j^{(\mathrm{D})}(f)C_X(f) \equiv aC_j(f)$ and set η to

$$\eta_2 \equiv \frac{2\left(\sum_{k=1}^{\lfloor (M_j-1)/2\rfloor} C_j(f_k)\right)^2}{\sum_{k=1}^{\lfloor (M_j-1)/2\rfloor} C_j^2(f_k)}.$$

[3] Using an argument that appeals to the band-pass nature of the MODWT wavelet filter $\{\tilde{h}_{j,l}\}$, we can set η to

$$\eta_3 = \max\{M_j/2^j, 1\}.$$

The relative merits of these three ways of setting η are discussed at the end of Section 8.4.

Finally, we note that, due to the band-pass nature of the jth level MODWT wavelet filter, we can translate an estimate of the wavelet variance at scale τ_j into an estimate of the SDF $S_X(\cdot)$ over the pass-band of the filter. The approach uses $\hat{C}_j \equiv 2^j \hat{\nu}_X^2(\tau_j)\,\Delta t$ to estimate the SDF over the octave frequency band $[\frac{1}{2^{j+1}\Delta t}, \frac{1}{2^j \Delta t}]$. The resulting estimator is piecewise constant.

8.11 Exercises

[8.1] Let $\{\varepsilon_t\}$ be a white noise process with zero mean and variance σ_ε^2 (by definition, such a process has an ACVS given by Equation (268d)). Note that Equation (267b) can be used to argue that the SDF for a white noise process is given by $S_\varepsilon(f) = \sigma_\varepsilon^2$ for all f (assuming for convenience that Δt is unity). Derive an explicit expression for the wavelet variance $\nu_\varepsilon^2(\tau_j)$ of $\{\varepsilon_t\}$. Using this expression, verify that Equation (296d) holds for a white noise process (i.e., that the sum of the wavelet variance over all possible dyadic scales is equal to the process variance σ_ε^2).

[8.2] Using a Gaussian random number generator, create a realization of length $N = 1024$ of a Gaussian white noise process $\{\varepsilon_t\}$ with zero mean and unit variance. Use the unbiased MODWT estimator $\hat{\nu}_\varepsilon^2(\tau_j)$ of Equation (306b) to estimate the wavelet variance $\nu_\varepsilon^2(\tau_j)$, $\tau_j = 1, 2, 4, \ldots, 512$, for the generated series using any two of the following three wavelet filters: Haar, D(4) or LA(8). Repeat using the biased MODWT estimator of Equation (306c). Plot your estimates of $\nu_\varepsilon^2(\tau_j)$ versus τ_j on a log/log scale. Interpret your plots in terms of a power law process.

[8.3] Suppose that $\{X_t\}$ is a random walk process as defined in Equation (288a). Show that, for the Haar wavelet, we have

$$\nu_X^2(\tau_j) = \frac{\sigma_\varepsilon^2}{6}\left(\tau_j + \frac{1}{2\tau_j}\right),$$

where σ_ε^2 is the variance of the white noise process used to create $\{X_t\}$. Assuming that σ_ε^2 is unity, plot $\nu_X^2(\tau_j)$ versus τ_j for $j = 1, \ldots, 10$ on a log/log scale. How does your plot compare to the result suggested by the answer to the last part of Exercise [288a]?

[8.4] Using a Gaussian random number generator, create a realization of length $N = 1024$ of the random walk process $\{X_t\}$ (see the previous exercise). Use both the unbiased and biased MODWT estimators $\hat{\nu}_X^2(\tau_j)$ and $\tilde{\nu}_X^2(\tau_j)$ (Equations (306b) and (306c)) to estimate the wavelet variance $\nu_X^2(\tau_j)$, $\tau_j = 1, 2, 4, \ldots, 512$, for the generated series using any two of the following three wavelet filters: Haar, D(4) or LA(8). Plot your estimates of $\nu_X^2(\tau_j)$ versus τ_j on a log/log scale. Interpret your plots in terms of a power law process.

[8.5] Show that, if $\{X_t\}$ is a stationary process, the expected difference between the unbiased and biased MODWT estimators of the wavelet variance (Equations (306b) and (306c), respectively) goes to 0 as $N \to \infty$. Does the same result hold if $\{X_t\}$ is a nonstationary process with stationary first order backward differences?

[8.6] Suppose that $\{X_t\}$ is a first order stationary autoregressive process with zero mean; i.e., we can write $X_t = \phi X_{t-1} + \varepsilon_t$, where $|\phi| < 1$ and $\{\varepsilon_t\}$ is a zero mean white noise process. The autocovariance sequence $\{s_\tau\}$ for $\{X_t\}$ is given by $s_{X,\tau} = \phi^{|\tau|} s_{X,0}$, where $s_{X,0}$ is the variance of the process. As defined by Equations (296a) and (306a) and assuming use of the Haar wavelet, let $\{\overline{W}_{1,t}\}$ be the unit scale MODWT wavelet coefficient process, and let $\{\widetilde{W}_{1,t}\}$ be the unit scale MODWT wavelet coefficients based upon the time series $X_0, \ldots, X_{N-1}$. For what range of t do we have $\widetilde{W}_{1,t} = \overline{W}_{1,t}$? Determine $E\{\widetilde{W}_{1,t}^2\}$ for all t. Under what circumstances do we have $E\{\widetilde{W}_{1,0}^2\} = \nu_X^2(\tau_1)$?

[8.7] Let $X_0, \ldots, X_{N-1}$ be a portion of a process $\{X_t\}$ whose dth order backward differences form a stationary process. Let $M > N$ be any power of two divisible by 2^j, and construct X_t' as in Equation (308c). Let $W_{j,t}'$ denote the level j DWT wavelet coefficients for $\{X_t'\}$ based upon a Daubechies wavelet filter of width $L \geq 2d$, and let L_j' be as defined in Equation (146a). Under the assumption that $\lfloor \frac{N}{2^j} - 1 \rfloor \geq L_j'$, show that, upon division by $2^{j/2}$, the subsequence of wavelet coefficients displayed in Equation (308d) is equal to selected RVs from the infinite sequence $\{\overline{W}_{j,t}\}$ (this is defined in Equation (296a)). What would change if, in creating X_t', we padded X_t with the sample mean of $X_0, \ldots, X_{N-1}$ rather than zeros?

[8.8] Suppose that $\{X_t\}$ is such that its unit scale MODWT wavelet coefficient process $\{\overline{W}_{1,t}\}$ is stationary with SDF $S_1(\cdot)$ and square summable ACVS $\{s_{1,\tau}\}$. Define the corresponding unit scale DWT wavelet coefficient process as $\{2^{1/2}\overline{W}_{1,2t+1} : t = \ldots, -1, 0, 1, \ldots\}$ (cf. Equation (70c)). Show that this process is stationary with an SDF that is proportional to $S_1(\frac{f}{2}) + S_1(\frac{f}{2} + \frac{1}{2})$ for $|f| \leq 1/2$ (hint: recall Exercise [23b]).

[8.9] Verify Equation (309).

[8.10] Suppose that we rewrite the approximation in Equation (314a) to include the term at the Nyquist frequency when there is such:

$$\hat{\nu}_X^2(\tau_j) \approx \frac{2}{M_j} \sum_{k=1}^{\lfloor (M_j-1)/2 \rfloor} \hat{S}_j^{(\mathrm{p})}(f_k) + \frac{1}{M_j} \hat{S}_j^{(\mathrm{p})}(\tfrac{1}{2}) \mathcal{I}_{M_j},$$

where $\mathcal{I}_{M_j}$ is unity if M_j is even and zero if M_j is odd, an approximation that now just says that $\hat{S}_j^{(\mathrm{p})}(0)/M_j = \overline{W}_j^2$ is negligible. Large sample theory suggests that $\hat{S}_j^{(\mathrm{p})}(\frac{1}{2})/S_j(\frac{1}{2})$ has a χ_1^2 distribution and that this RV is independent of all the RVs in the summation above. Under the assumption that

$$\frac{\eta_2' \hat{\nu}_X^2(\tau_j)}{\nu_X^2(\tau_j)} \stackrel{\mathrm{d}}{=} \chi_{\eta_2'}^2,$$

use an EDOF argument to show that

$$\eta_2' = \frac{\left(2 \sum_{k=1}^{\lfloor (M_j-1)/2 \rfloor} C_j(f_k) + C_j(\frac{1}{2}) \mathcal{I}_{M_j}\right)^2}{2 \sum_{k=1}^{\lfloor (M_j-1)/2 \rfloor} C_j^2(f_k) + C_j^2(\frac{1}{2}) \mathcal{I}_{M_j}}$$

(cf. η_2 of Equation (314b)).

[8.11] Based upon Equation (313c) (with η estimated by $\hat{\eta}_1$ of Equation (313d)) and upon the D(4) MODWT unbiased estimator of the D(4) wavelet variance $\nu_X^2(\tau_1)$, compute a 95% confidence interval for $\nu_X^2(\tau_1)$ for the 16 point time series $\{X_{1,t}\}$

listed in the caption to Figure 42. Compute a second 95% confidence interval for $\nu_X^2(\tau_1)$, but this time using the EDOF η_3 of Equation (314c). Which measure of the EDOF would you regard as more appropriate to use here? (Note that Table 263 gives certain percentage points for the χ^2 distribution that might be of use here.)

[8.12] Show that, under the assumption that the process mean is unknown and hence is estimated using the sample mean, the Allan variance of Equation (322) for τ_1 is proportional to the sample variance for a sample size of two. The Allan variance is thus sometimes called the 'two sample' variance, which is the rationale for the '2' in the notation $\sigma_{\overline{Y}}^2(2, \tau_j)$.

[8.13] Show that, if $\{\overline{Y}_t(\tau_1)\}$ is a white noise process with variance $\sigma_{\overline{Y}}^2$, then we have $\sigma_{\overline{Y}}^2(2, \tau_1) = \sigma_{\overline{Y}}^2$, where $\sigma_{\overline{Y}}^2(2, \tau_1)$ is the Allan variance of Equation (322) (the desire to have the Allan variance reduce to the process variance in the case of white noise is the rationale for the factor 1/2 in Equation (322), without which we would not have this correspondence).

9

Analysis and Synthesis of Long Memory Processes

9.0 Introduction

In the previous chapter we noted that log/log plots of estimates of the wavelet variance $\nu_X^2(\tau_j)$ versus scale τ_j can help us to identify time series that might be well modeled as a power law process $\{X_t\}$; i.e., a process whose spectral density function (SDF) $S_X(\cdot)$ is such that $S_X(f) \propto |f|^\alpha$. When $\alpha < 0$ so that $S_X(f) \to \infty$ as $f \to 0$, we say that $\{X_t\}$ exhibits 'long memory' (see Section 7.6 for precise definitions and a review). Such processes are widely found and studied in scientific work on phenomena ranging from the microscopic to the cosmic:

- voltage fluctuations across cell membranes (Holden, 1976);
- density fluctuations in sand particles passing through an hour glass (Schick and Verveen, 1974);
- traffic fluctuations on an expressway (Musha and Higuchi, 1976);
- impedance fluctuations in a geophysical borehole (Walden, 1994);
- fluctuations in the Earth's rotation (Munk and MacDonald, 1975); and
- X-ray time variability of galaxies (McHardy and Czerny, 1987).

Plots of $\log(\nu_X^2(\tau_j))$ versus $\log(\tau_j)$ for long memory processes exhibit linear variation with a slope of $-\alpha - 1$ as τ_j gets large. This simple relationship suggests that wavelets might be particularly adept at handling such processes. That wavelet transforms are well suited for studying long memory and related processes has been recognized in recent years by, for example, Flandrin (1992), Masry (1993) and Wornell (1993) and others (the cited references mainly focus on wavelet series representations of continuous time nonstationary processes).

In this chapter we consider ways in which the discrete wavelet transform (DWT) can be used in the analysis and synthesis of both stationary long memory processes and related nonstationary processes, all of which have SDFs that plot approximately as straight lines with negative slopes on log-frequency/log-power axes, at least over several octaves of frequency f as f approaches 0. In Section 9.1 we argue that the DWT approximately decorrelates fractionally differenced (FD) processes (these give us a tractable set of simple models exhibiting long memory characteristics – see Section 7.6 for a discussion of the basic properties of FD processes). This decorrelation property is fundamental: whereas the random variables (RVs) in an FD process have a high degree of correlation, the DWT creates a new set of RVs, namely, the wavelet coefficients, that are approximately uncorrelated (both within and between scales) and hence are more amenable for statistical analysis. This basic result can be put to two uses. First, we can simulate a long memory process approximately by stochastically generating the DWT coefficients and then combining these via the inverse DWT to produce an output process (Section 9.2). Second, given a time series we want to model as an FD process, we can estimate the two unknown parameters for this process using approximate maximum likelihood estimators (MLEs) that are based directly on the decorrelation property of the DWT, leading to simple likelihood functions to be maximized (Sections 9.3 and 9.4). While MLEs are used extensively in statistics because of well-known optimality properties, we can also estimate the unknown parameters of an FD process via least squares fits (in log/log space) of estimated wavelet variances versus scales (Section 9.5). Least squares estimators are arguably suboptimal, but much easier to compute. In Section 9.6 we consider another use for the decorrelation property, yielding a simple test for homogeneity of variance that is applicable under the null hypothesis of a stationary FD process. Prior to a summary (Section 9.9), we use the atomic clock and Nile River series (9.7 and 9.8) to illustrate the methodology discussed in this chapter.

9.1 Discrete Wavelet Transform of a Long Memory Process

Let us begin our study of the statistical properties of the DWT of a long memory process with an example. The bottom left-hand plot of Figure 342 shows a realization $\mathbf{X}$ of length $N = 1024$ from a stationary FD process $\{X_t\}$ with zero mean and with parameters $\delta = 0.4$ and $\sigma_\varepsilon^2 = 1.0$. This simulated series was created using the Davies–Harte method (described in Section 7.8) and exhibits the large scale variations typical of a long memory process. To quantify these variations, we show the sample autocorrelation sequence (ACS) out to lag $\tau = 32$ in the bottom right-hand plot (the values in the ACS are indicated by vertical lines depicting deviations from zero). This sequence, defined via

$$\hat{\rho}_{X,\tau} \equiv \frac{\sum_{t=0}^{N-1-\tau} X_t X_{t+\tau}}{\sum_{t=0}^{N-1} X_t^2}, \quad \tau = 0, 1, \ldots, N-1,$$

is an estimator of $\rho_{X,\tau} \equiv s_{X,\tau}/s_{X,0}$, where $\{s_{X,\tau}\}$ is the autocovariance sequence (ACVS) for a stationary FD process (see Section 7.4 for a review of the ACVS). Note that the estimated ACS shows substantial positive autocorrelation at all displayed lags, which is also typical of long memory processes.

Above the time series, we show its partial DWT of level $J_0 = 7$ based on the LA(8) wavelet. The wavelet and scaling coefficients $\mathbf{W}_1, \ldots, \mathbf{W}_7$ and $\mathbf{V}_7$ are circularly shifted and then plotted versus times dictated by the phase properties of the LA(8) filters (see Section 4.8). The plots of the wavelet coefficient illustrate two important characteristics. First, the average magnitude of these coefficients increases with scale. Second, although the simulated series itself shows strong visual evidence of long term correlation, there are no apparent patterns amongst the wavelet coefficients within a given scale (an exception is arguably $\mathbf{W}_7$, whose coefficients alternate in sign, but it is problematic to deduce too much from just eight coefficients). We can quantify this lack of autocorrelation by computing the sample ACS for each vector of circularly shifted

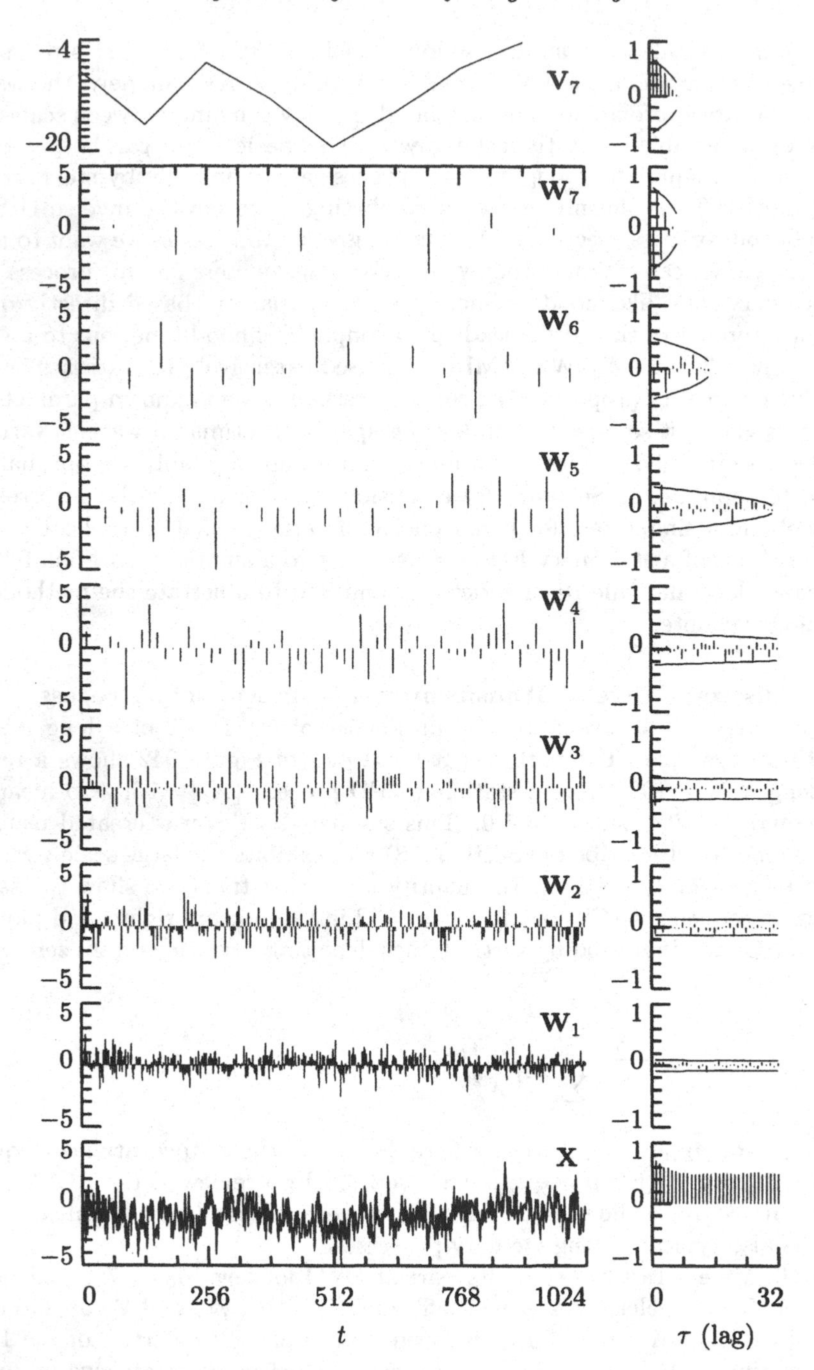

Figure 342. LA(8) DWT coefficients for simulated FD(0.4) time series and sample ACSs.

wavelet coefficients. These ACSs are plotted in the right-hand column of Figure 342 (again, as deviations from zero). The values in the sample ACSs are on the whole much smaller in magnitude than those for $\mathbf{X}$ itself, but we see that the average magnitudes do increase with scale. This increase can be attributed to sampling variability, as the following argument indicates. As we go up in scale, the number $N_j \equiv N/2^j$ of coefficients in $\mathbf{W}_j$ decreases. If the values in $\mathbf{W}_j$ were actually a realization of Gaussian white noise, then statistical theory suggests that approximately 95% of the sample ACS for the jth scale should fall between the limits $\pm\frac{2\sqrt{(N_j-\tau)}}{N_j}$ (indicated by the upper and lower curves in each plot; for details on the theory, see Fuller, 1996, Corollary 6.3.6.2). Note that these limits spread out as N_j decreases and that the total number of sample ACS values outside these limits is consistent with what can be attributed to sampling variability.

Let us verify that the rate of increase in the magnitude of the wavelet coefficients shown in Figure 342 is consistent with the known properties of an FD(0.4) process. In Section 4.6 we discussed the fact that the wavelet coefficients $\mathbf{W}_j$ for scale $\tau_j = 2^{j-1}$ are formed using an equivalent filter $\{h_{j,l}\}$ that is approximately a band-pass filter with nominal pass-band given by $[-1/2^j, -1/2^{j+1}] \cup [1/2^{j+1}, 1/2^j]$. If we recall that the squared gain function $\mathcal{H}_j(\cdot)$ for $\{h_{j,l}\}$ must integrate to unity because $\{h_{j,l}\}$ has unit energy, we can make the crude approximation

$$\mathcal{H}_j(f) \approx \begin{cases} 2^j, & 1/2^{j+1} \le |f| \le 1/2^j; \\ 0, & \text{otherwise.} \end{cases}$$

If $\{h_{j,l}\}$ is used to filter $\{X_t\}$, we obtain a process with an SDF given by $\mathcal{H}_j(f)S_X(f)$ (see Equation (268b)). This process is related to the 'circularity free' coefficients $W_{j,t}$ in $\mathbf{W}_j$, i.e., all except the boundary coefficients, of which there are no more than $L-2$ at a given scale (see Section 4.11; as usual, L is the width of the unit scale wavelet filter). If we ignore these boundary coefficients, the variance of $W_{j,t}$ follows from Equation (268c), namely,

$$\operatorname{var}\{W_{j,t}\} = \int_{-1/2}^{1/2} \mathcal{H}_j(f)S_X(f)\,df \tag{343a}$$

$$\approx 2\int_{1/2^{j+1}}^{1/2^j} 2^j S_X(f)\,df = \frac{1}{\frac{1}{2^j} - \frac{1}{2^{j+1}}} \int_{1/2^{j+1}}^{1/2^j} S_X(f)\,df \equiv C_j. \tag{343b}$$

If we recall that

$$\frac{1}{b-a}\int_a^b x(u)\,du$$

is the average value of the function $x(\cdot)$ over the interval $[a,b]$, we can interpret C_j as the average value of the SDF over the octave band $[\frac{1}{2^{j+1}}, \frac{1}{2^j}]$.

Let us now specialize to an FD process, which has an SDF $S_X(\cdot)$ given by Equation (284a), thus yielding

$$C_j = 2^{j+1}\int_{1/2^{j+1}}^{1/2^j} \frac{\sigma_\varepsilon^2}{[4\sin^2(\pi f)]^\delta}\,df. \tag{343c}$$

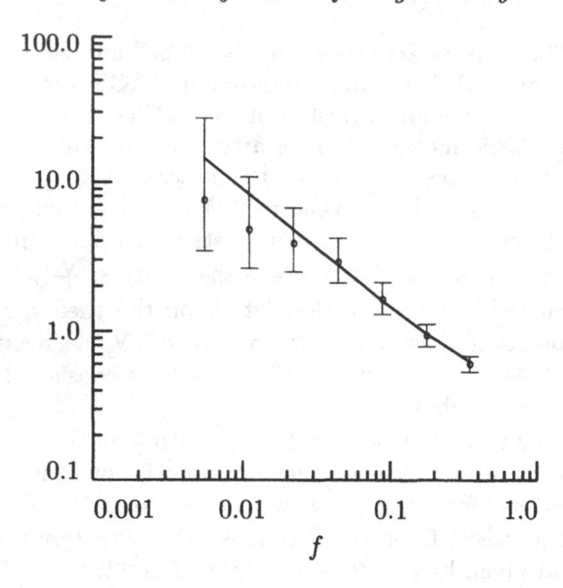

Figure 344. Sample variances of LA(8) wavelet coefficients from Figure 342 for – from right to left – levels $j = 1, \ldots, 7$ (circles) along with true FD(0.4) SDF evaluated at the center frequency $1/2^{j+\frac{1}{2}}$ of the octave bands $[\frac{1}{2^{j+1}}, \frac{1}{2^j}]$ (thick curve).

Since the approximation $\sin^{2\delta}(\pi f) \approx (\pi f)^{2\delta}$ is quite accurate when $f \leq 1/8$ (cf. Figure 282), we can use – for $j \geq 3$ –

$$C_j \approx 2^{j+1} \int_{1/2^{j+1}}^{1/2^j} \frac{\sigma_\varepsilon^2}{(2\pi f)^{2\delta}}\, df = \sigma_\varepsilon^2 \left(\frac{2^{j+\frac{1}{2}}}{2\pi}\right)^{2\delta} \frac{2^{2\delta}-2}{2^\delta(2\delta-1)} \equiv \widetilde{C}_j \tag{344}$$

when $\delta \neq 1/2$ and $\widetilde{C}_j = \sigma_\varepsilon^2 2^{j+\frac{1}{2}} \log(2)/(\pi\sqrt{2})$ when $\delta = 1/2$ (item [4] in the Comments and Extensions below considers a more accurate approximation, which can be used when $j = 1$ and 2). Note that $\widetilde{C}_j \propto [1/2^{j+\frac{1}{2}}]^{-2\delta}$, where $1/2^{j+\frac{1}{2}}$ is the center frequency (on a log scale) of the octave band $[\frac{1}{2^{j+1}}, \frac{1}{2^j}]$. Since the SDF at this frequency can be regarded as a surrogate for its average value C_j over this octave band, we should find that $\operatorname{var}\{W_{j,t}\}$ and $S_X(1/2^{j+\frac{1}{2}})$ are close to each other and, moreover, that a plot of either $\log(\operatorname{var}\{W_{j,t}\})$ or $\log(S_X(1/2^{j+\frac{1}{2}}))$ versus $\log(1/2^{j+\frac{1}{2}})$ should yield approximately a line with slope -2δ.

As an example, the circles in Figure 344 show the sample variances $\widehat{\operatorname{var}}\{W_{j,t}\} \equiv \|\mathbf{W}_j\|^2/N_j$ of the wavelet coefficients displayed in Figure 342 plotted versus the center frequencies $1/2^{j+\frac{1}{2}}$, $j = 1, \ldots, 7$, on log/log axes. The superimposed thick curve is a plot of $S_X(1/2^{j+\frac{1}{2}})$ versus these same frequencies for an FD(0.4) process with parameters $\delta = 0.4$ and $\sigma_\varepsilon^2 = 1.0$ (this curve appears to be very close to linear on this log/log plot and has a least squares slope of -0.79, which agrees well with the theoretical value $-2\delta = -0.8$). While $\widehat{\operatorname{var}}\{W_{j,t}\}$ is in good agreement with the SDF value for $j = 1$ to 4 (the four highest frequencies), there is noticeable disagreement for $j = 5, 6$ and 7. These sample variances are formed using just 32, 16 and 8 values, so we need to assess the effect of sampling variability. Under the assumption that elements of $\mathbf{W}_j$ constitute a random sample from a Gaussian distribution with zero mean and unknown variance $\operatorname{var}\{W_{j,t}\}$ (a reasonable approximation in view of the decorrelation property to be discussed shortly), it follows that the RV $N_j \widehat{\text{var}}\{W_{j,t}\}/\text{var}\{W_{j,t}\}$ obeys a chi-square distribution with N_j degrees of freedom (see Section 7.2). We can evoke the argument of Exercise [313b] to say that

$$\left[\frac{N_j \widehat{\text{var}}\{W_{j,t}\}}{Q_{N_j}(0.975)}, \frac{N_j \widehat{\text{var}}\{W_{j,t}\}}{Q_{N_j}(0.025)}\right]$$

is a 95% confidence interval for the unknown $\text{var}\{W_{j,t}\}$. These intervals are shown in Figure 344 and indicate that the discrepancies between $\widehat{\text{var}}\{W_{j,t}\}$ and $S_X(1/2^{j+\frac{1}{2}})$ are not significant when we take sampling variability into account. (The fact that $\log(\text{var}\{W_{j,t}\})$ versus $\log(1/2^{j+\frac{1}{2}})$ is approximately linear with a slope of -2δ suggests that we should be able to estimate δ from a least squares fit through these variables, a topic that we return to in Section 9.5.)

Let us now take a quantitative look at how well various DWTs decorrelate stationary FD processes.

▷ **Exercise [345]:** Let $W_{j,t}$ and $W_{j',t'}$ be any two nonboundary coefficients in, respectively, $\mathbf{W}_j$ and $\mathbf{W}_{j'}$ (i.e., coefficients not influenced by the circularity assumption). Show that

$$\text{cov}\{W_{j,t}, W_{j',t'}\} = \sum_{l=0}^{L_j-1} \sum_{l'=0}^{L_{j'}-1} h_{j,l} h_{j',l'} s_{X,2^j(t+1)-l-2^{j'}(t'+1)+l'}, \tag{345a}$$

where, as usual, L_j is the width of the filter $\{h_{j,l}\}$. In addition, show that, when $j = j'$ and $t' = t + \tau$, the above reduces to

$$\text{cov}\{W_{j,t}, W_{j,t+\tau}\} = \sum_{m=-(L_j-1)}^{L_j-1} s_{X,2^j\tau+m} \sum_{l=0}^{L_j-|m|-1} h_{j,l} h_{j,l+|m|}. \tag{345b}$$ ◁

We can use Equation (345b) to compute the theoretical ACS for the nonboundary wavelet coefficients within the jth level, i.e.,

$$\rho_{j,\tau} \equiv \frac{\text{cov}\{W_{j,t}, W_{j,t+\tau}\}}{\text{var}\{W_{j,t}\}}, \quad \tau = \ldots, -1, 0, 1, \ldots$$

(recall that $\text{var}\{W_{j,t+\tau}\} = \text{var}\{W_{j,t}\} = \text{cov}\{W_{j,t}, W_{j,t}\}$). Figure 346a shows the results of these computations out to level $j = 4$ and for lags $\tau = 1$ to 4 for an FD(0.4) process in combination with the Haar, D(4) and LA(8) wavelet filters (note that the vertical axis ranges just from -0.2 to 0.2). As in Figure 342, the values in each ACS are plotted as deviations from zero (some values are so close to zero that they are not visible on the plots). In each of the twelve plots, the largest autocorrelation in magnitude occurs at unit lag (the largest of these is $\rho_{4,1} \doteq -0.140$ for the D(4) DWT), and then the ACSs become very close to zero for $\tau \geq 2$. Note that, for a given DWT, $|\rho_{j,1}|$ increases with level j, but computations indicate that $|\rho_{j,1} - \rho_{4,1}| < 0.01$ for $j \geq 5$ for all three DWTs, so the unit lag correlation does not grow much beyond that for $j = 4$. For the sake of comparison, Figure 346b shows the ACS for an FD(0.4) process out to lag $\tau = 64$. All the autocorrelations given in that figure are greater in magnitude than the ones in Figure 346a. Thus, within a given scale, the decorrelation

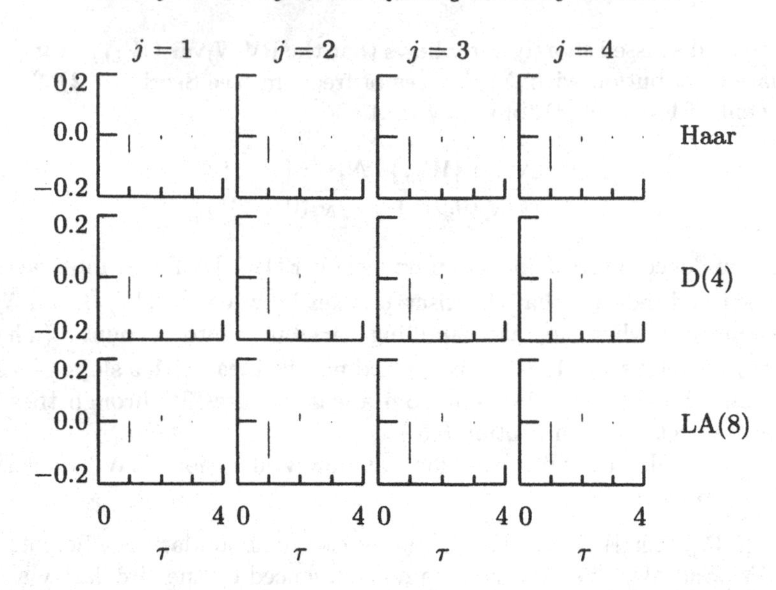

Figure 346a. ACSs at $\tau = 1, \ldots, 4$ for Haar, D(4) and LA(8) wavelet coefficients $W_{j,t}$, $j = 1, \ldots, 4$, of an FD(0.4) process. The ACS values are plotted as deviations from zero (some are not visible because they are so close to zero).

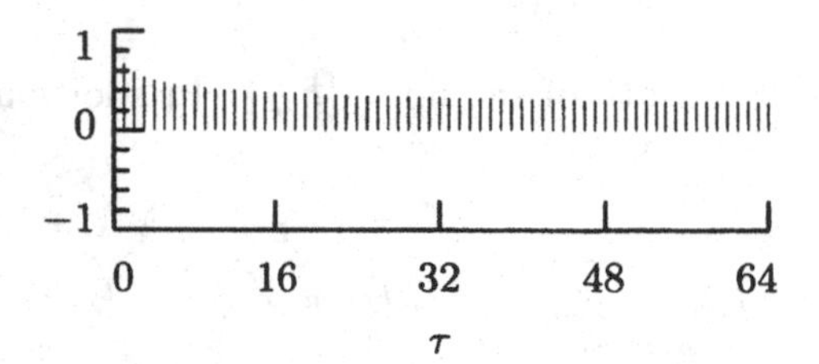

Figure 346b. ACS for FD process with $\delta = 0.4$ out to lag $\tau = 64$.

properties of all three DWTs are quite good (note also that the three DWTs perform virtually the same, although – interestingly enough – the Haar DWT does slightly better overall).

To examine the correlation between wavelet coefficients at different levels, we can use Equation (345a) to compute the correlation between $W_{j,t}$ and $W_{j',t'}$ over a grid of t and t' for a given j and j'. Figures 347a and 347b show the results of these computations for the Haar and LA(8) DWTs for $1 \le j < j' \le 4$ (Exercise [9.1] is to create a similar figure for the D(4) case). We set $t = 2^{|j'-j|-1}$ for the Haar DWT and

$$t = \begin{cases} 0, & \text{when } |j'-j| = 1; \\ 1, & \text{when } |j'-j| = 2; \text{ and} \\ 7, & \text{when } |j'-j| = 3, \end{cases} \tag{346}$$

for the LA(8) DWT, after which we let $t' = t + \tau$ for $\tau = -8, \ldots, 8$. These choices for t and t' ensure that each plot contains a correlation with the largest magnitude over all possible choices for these indices. The correlations across scales are small, but now

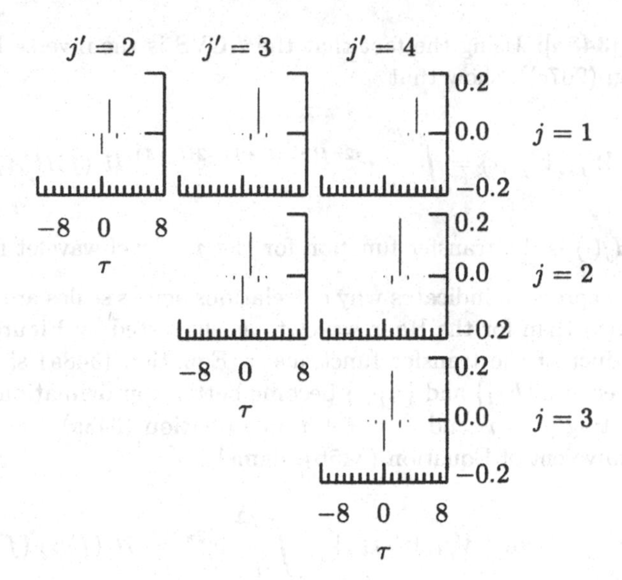

Figure 347a. Correlation between the Haar wavelet coefficients $W_{j,t}$ and $W_{j',t'}$ formed from an FD(0.4) process and for levels satisfying $1 \le j < j' \le 4$. By setting $t = 2^{|j'-j|-1}$ and $t' = t + \tau$ with $\tau = -8, \ldots, 8$, we capture two coefficients exhibiting the maximum absolute correlation over all possible combinations of t and t'.

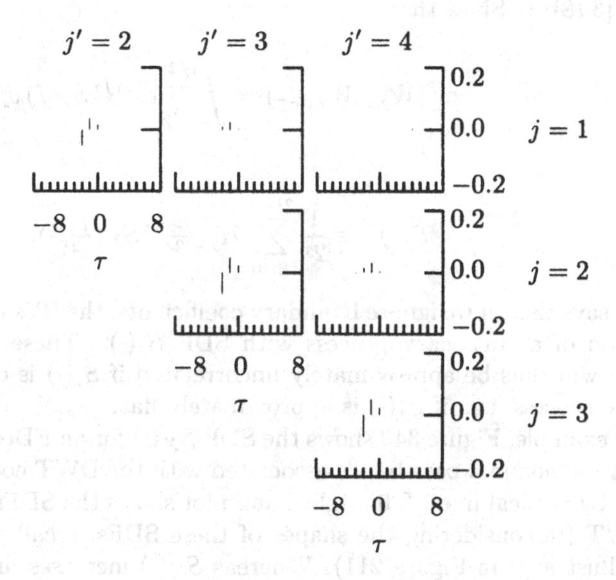

Figure 347b. As in Figure 347a, but now using the LA(8) DWT and with t set as per Equation (346).

we can see a benefit in using the LA(8) wavelet over the Haar wavelet: the maximum absolute correlations for the former are notably smaller than those for the latter.

We can gain some additional insight into the decorrelation properties of the DWT by formulating the frequency domain equivalent of Equation (345a).

▷ **Exercise [348a]:** Using the fact that the ACVS is the inverse DFT of the SDF (see Equation (267c)), show that

$$\operatorname{cov}\{W_{j,t}, W_{j',t'}\} = \int_{-1/2}^{1/2} e^{i2\pi f(2^{j'}(t'+1)-2^{j}(t+1))} H_j(f) H_{j'}^*(f) S_X(f)\, df, \qquad (348a)$$

where $H_j(\cdot)$ is the transfer function for the jth level wavelet filter. ◁

The above expression indicates why correlations across scales are smaller in magnitude for the LA(8) than for the Haar wavelet: as suggested by Figure 107, the magnitude of the product of the transfer functions in Equation (348a) should be smaller as L increases because $\{h_{j,l}\}$ and $\{h_{j',l}\}$ become better approximations to band-pass filters.

By setting $j = j'$ and $t' = t + \tau$ in Equation (348a), we obtain the frequency domain equivalent of Equation (345b), namely,

$$\operatorname{cov}\{W_{j,t}, W_{j,t+\tau}\} = \int_{-1/2}^{1/2} e^{i2\pi 2^j f\tau} \mathcal{H}_j(f) S_X(f)\, df. \qquad (348b)$$

Exercise [9.2] uses this expression to argue that $\operatorname{cov}\{W_{j,t}, W_{j,t+\tau}\} \approx 0$ for $\tau \neq 0$ when $S_X(\cdot)$ is approximately constant over the nominal pass-band $[1/2^{j+1}, 1/2^j]$. We can gain further insight based upon the following result.

▷ **Exercise [348b]:** Show that

$$\operatorname{cov}\{W_{j,t}, W_{j,t+\tau}\} = \int_{-1/2}^{1/2} e^{i2\pi f\tau} S_j(f)\, df,$$

where

$$S_j(f) \equiv \frac{1}{2^j} \sum_{k=0}^{2^j-1} \mathcal{H}_j(\tfrac{f+k}{2^j}) S_X(\tfrac{f+k}{2^j}).$$ ◁

The above says that, if we ignore boundary coefficients, the RVs in $\mathbf{W}_j$ can be regarded as a portion of a stationary process with SDF $S_j(\cdot)$. These nonboundary wavelet coefficients will thus be approximately uncorrelated if $S_j(\cdot)$ is close to the SDF for a white noise process, i.e., if $S_j(\cdot)$ is approximately flat.

As an example, Figure 349 shows the SDF $S_X(\cdot)$ for an FD(0.4) process (left-hand plot), with the nominal pass-bands associated with the DWT coefficients $\mathbf{W}_1, \ldots, \mathbf{W}_4$ delineated by vertical lines. The right-hand plot shows the SDFs $S_j(\cdot)$ when using the LA(8) DWT (in considering the shapes of these SDFs, recall the frequency reversal property illustrated in Figure 211). Whereas $S_X(\cdot)$ increases unboundedly as $f \to 0$, the SDFs for the nonboundary wavelet coefficients have minimum and maximum that differ by no more than 3 dB (i.e., a factor of two). The fact that these $S_j(\cdot)$ do not vary much can be traced to the fact that $S_X(\cdot)$ does not vary greatly within any given octave band: as $S_X(f) \to \infty$ as $f \to 0$, the lengths of the octave bands decrease, which explains why the DWT is well adapted to FD and other long memory processes (see item [3] in the Comments and Extensions for further discussion on this point).

As in Chapter 4, let us write the full DWT for a time series $\mathbf{X} = [X_0, \ldots, X_{N-1}]^T$ of length $N = 2^J$ in matrix form as $\mathbf{W} = \mathcal{W}\mathbf{X}$ (the rationale for considering a full –

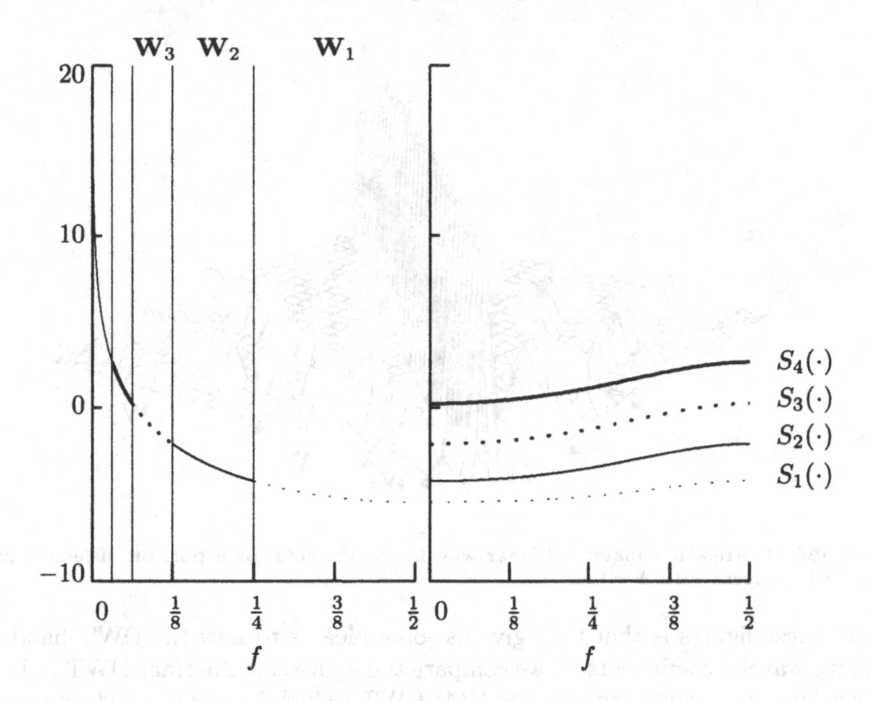

Figure 349. SDFs for an FD(0.4) process (left-hand plot) and for nonboundary LA(8) wavelet coefficients in $\mathbf{W}_1$, $\mathbf{W}_2$, $\mathbf{W}_3$ and $\mathbf{W}_4$ (right-hand). The vertical axis is in units of decibels (i.e., we plot $10 \cdot \log_{10}(S_X(f))$ versus f). The vertical lines in the left-hand plot denote the nominal pass-bands for the four $\mathbf{W}_j$.

rather than a partial – DWT is the subject of item [1] in the Comments and Extensions below). The covariance matrix of $\mathbf{W}$, say $\Sigma_{\mathbf{W}}$, is given by

$$\Sigma_{\mathbf{W}} = \mathcal{W}\Sigma_{\mathbf{X}}\mathcal{W}^T, \tag{349}$$

where $\Sigma_{\mathbf{X}}$ is the covariance matrix of $\mathbf{X}$ (see Equation (262c)). This result is exact and gives us the resulting covariance structure for whatever wavelet filter we happen to choose, with the periodic boundary conditions properly taken into account. Once we have formed the covariance matrix for $\mathbf{W}$, we can readily compute the corresponding correlation matrix: if we let $\Sigma_{\mathbf{W},m,n}$ be the (m,n)th element of $\Sigma_{\mathbf{W}}$, then the (m,n)th element of the correlation matrix is defined to be $\Sigma_{\mathbf{W},m,n}/\sqrt{(\Sigma_{\mathbf{W},m,m}\Sigma_{\mathbf{W},n,n})}$.

Figures 350, 351a and 351b show the correlation matrix for $\mathbf{W}$ created from a time series $\mathbf{X}$ that is a portion of length $N = 32$ from an FD(0.4) process in combination with, respectively, the Haar, D(4) and LA(8) DWTs. The line of peaks running through the middle of each plot shows the diagonal elements, which are all unity and can be used to gauge the magnitude of the off-diagonal correlations. The front- and back-most peaks are for, respectively, elements $(31, 31)$ and $(0, 0)$. Black dots along the right-hand edges of the figures indicate the boundaries between the subvectors $\mathbf{W}_j$ and $\mathbf{W}_{j+1}$, $j = 1, \ldots, 3$. These figures confirm our previous conclusions about the decorrelating properties of the various DWTs. For example, the peaks away from the main diagonal in Figure 350 for the Haar DWT are due to correlations between scales (see Figure 347a). In corresponding portions of Figure 351b for the LA(8) DWT, these correlations are smaller, in agreement with Figure 347b. The additional

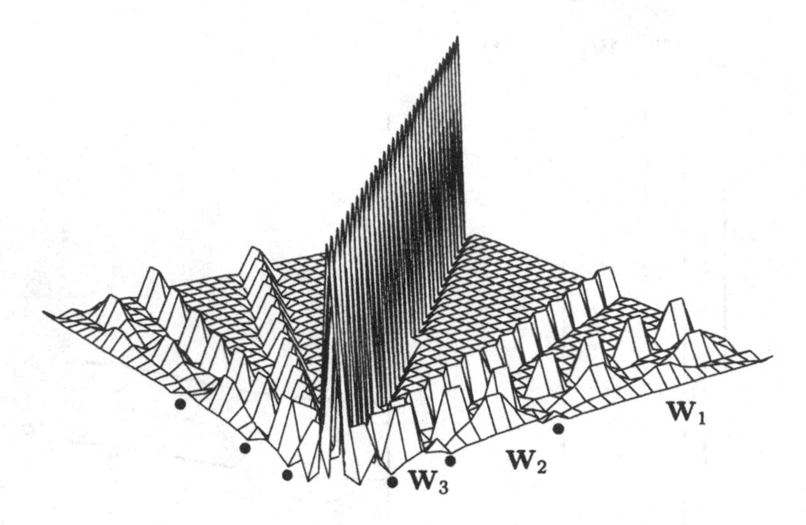

Figure 350. Correlation matrix of Haar wavelet coefficients for a portion of length $N = 32$ from an FD process with $\delta = 0.4$.

value of these figures is that they give us some idea as to how the DWT handles the boundary wavelet coefficients. If we compare the figures for the Haar DWT (which has no boundary coefficients) and for the D(4) DWT (which has one boundary coefficient in $\mathbf{W}_1$ and two in $\mathbf{W}_2$ to $\mathbf{W}_5$), we can see the correlations in the latter involving boundary coefficients: these are the peaks at the edges of the plot, along with some of like magnitude in the interior. A comparison of the D(4) and LA(8) figures shows that the LA(8) DWT reduces the magnitudes of these boundary influenced correlations, at the expense of spreading them out somewhat.

Finally, let us recall that in this example $\mathbf{X}$ has a correlation matrix whose (m, n)th element is given by $\rho_{X,|m-n|} \equiv s_{X,|m-n|}/s_{X,0}$, where $0 \leq |m-n| \leq 31$. This ACS is shown in Figure 346b out to lag 64, but, for lags less than or equal to 31, the smallest value is $\rho_{X,31} \doteq 0.338$, which is larger than the magnitudes of *any* of the off-diagonal elements for the three displayed correlation matrices for $\mathbf{W}$. Thus, even though the DWT does not do a perfect job of decorrelating an FD process, the DWT coefficients display dramatically less overall correlation than the original RVs in $\mathbf{X}$.

Comments and Extensions to Section 9.1

[1] Let us consider another aspect of the level $J_0 = 7$ LA(8) partial DWT displayed in Figure 342. In the upper right-hand corner, we show the sample ACS for the eight values in the circularly shifted vector of scaling coefficients. All eight values in $\mathbf{V}_7$ are negative, a reflection of the fact that the bulk of the values in this particular simulated series is also negative (even though the FD process $\{X_t\}$ has zero mean, one characteristic of realizations from a long memory process is a tendency to have quite long stretches that are either dominantly positive or dominantly negative). All seven values in the sample ACS are positive, with three being just outside the upper limits suggested as reasonable by sampling theory. It is admittedly dangerous to conclude too much from this example alone (particularly since the sampling theory upon which the upper limits are based presumes many more than just eight values); nevertheless, this example is in agreement with theory in that, whereas the wavelet coefficients for an FD process are approximately uncorrelated, the scaling coefficients in fact exhibit

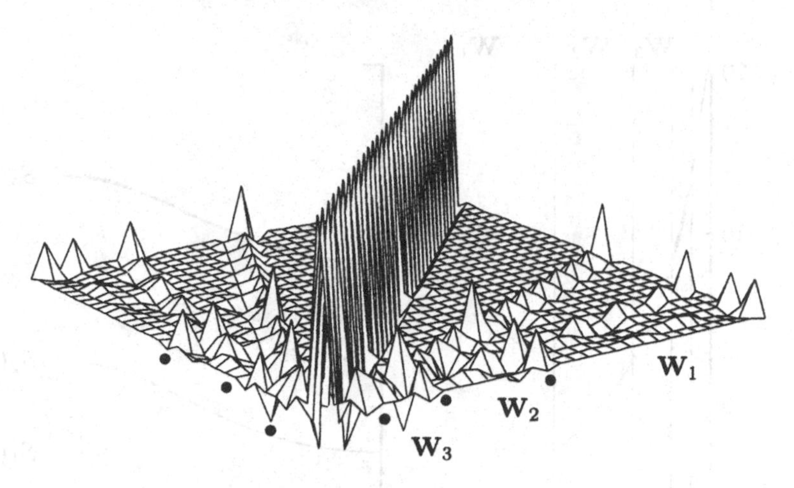

Figure 351a. As in Figure 350, but now using the D(4) DWT.

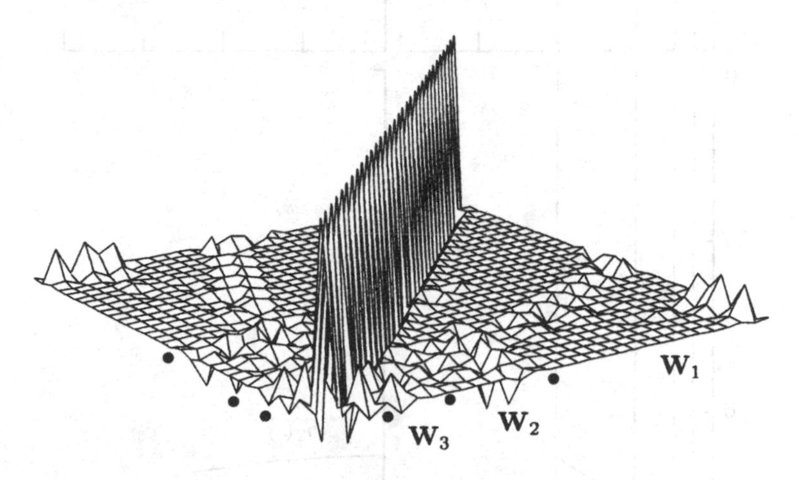

Figure 351b. As in Figure 350, but now using the LA(8) DWT.

a long memory structure themselves. It is thus judicious to deal with as few scaling coefficients as possible, which is why we use a full – rather than partial – DWT here.
[2] In the context of concepts discussed in Chapter 8, note that Figure 344 is an example of SDF estimation via the wavelet variance and is described by Equation (316) in terms of the MODWT-based unbiased estimator $\hat{\nu}_X^2(\tau_j)$ of Equation (306b); however, the procedure we use here amounts to using the DWT-based biased estimator $\tilde{\tilde{\nu}}_X^2(\tau_j)$ of Equation (308b) instead because $2^j\tilde{\tilde{\nu}}_X^2(\tau_j) = \|\mathbf{W}_j\|^2/N_j$.
[3] We have argued that the LA(8) DWT works well as a decorrelator of an FD(0.4) process (Whitcher, 1998, Section 4.1.1, shows that we get about the same degree of decorrelation when $-\frac{1}{2} < \delta < \frac{1}{2}$ and also with the Haar and D(4) wavelets). Here we explore briefly how well it works on other stationary processes. As a concrete example, let us consider an autoregressive process (AR) of order $p = 1$, which, per Equation (268e), can be written as $X_t = \phi X_{t-1} + \varepsilon_t$, where ϕ is the AR(1) parameter, while $\{\varepsilon_t\}$ is a white noise process with mean zero and variance σ_ε^2. The SDF for this process is given by Equation (268f) and is plotted in Figure 352 for $\sigma_\varepsilon^2 = 1$ and

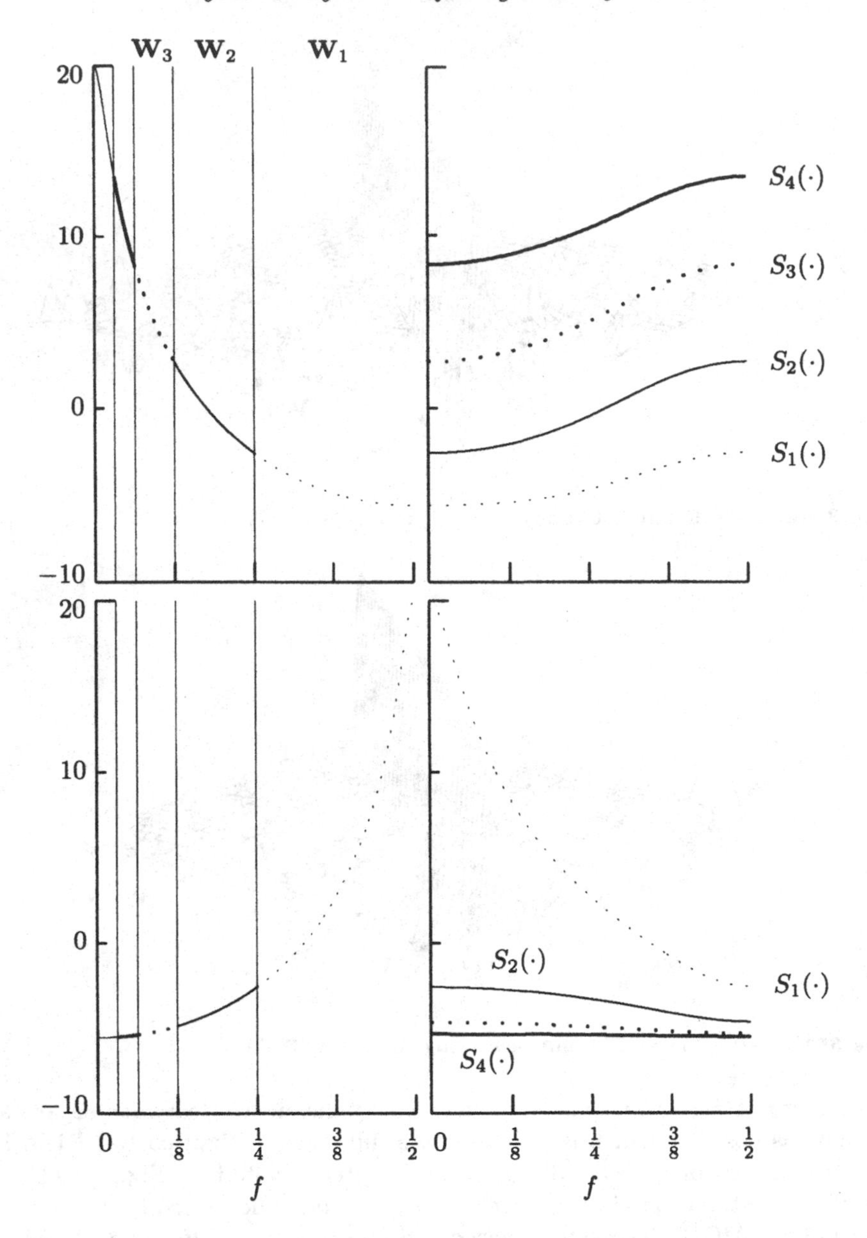

Figure 352. SDFs for AR(1) processes with $\phi = 0.9$ (top left-hand plot) and -0.9 (bottom left-hand) and for corresponding nonboundary LA(8) wavelet coefficients in $\mathbf{W}_1$ to $\mathbf{W}_4$ (right-hand plots). The vertical axes are in decibels, and the vertical lines in the left-hand plots delineate the nominal pass-bands for the four $\mathbf{W}_j$.

two settings for ϕ, namely, 0.9 (upper left-hand plot) and -0.9 (lower left-hand). The corresponding ACS for an AR(1) process is given by $\phi^{|\tau|}$. Because this ACS decays exponentially to zero as τ gets large, an AR(1) process is not considered to have long memory. The right-hand part of Figure 352 shows the SDFs $S_j(\cdot)$ for the nonboundary wavelet coefficients $\mathbf{W}_1, \ldots, \mathbf{W}_4$ (upper plot for $\phi = 0.9$, and lower for -0.9). Figure 353 shows portions of the corresponding ACSs ($\phi = 0.9$ in the top row,

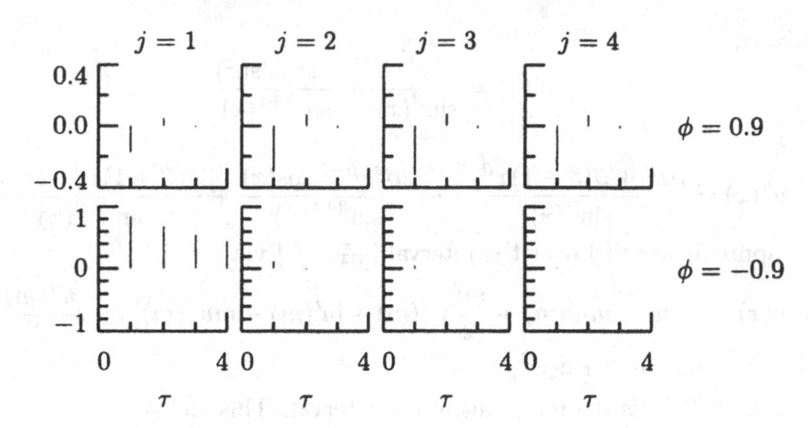

Figure 353. ACSs at $\tau = 1, \ldots, 4$ for LA(8) wavelet coefficients $W_{j,t}$, $j = 1, \ldots, 4$, of an AR(1) process with $\phi = 0.9$ and -0.9 (top and bottom rows, respectively).

and -0.9 in the bottom). We see that the DWT decorrelates the $\phi = 0.9$ process reasonably well (but not as well as for an FD(0.4) process: cf. the bottom row of plots in Figure 346a, and note also that the corresponding SDFs $S_j(\cdot)$ in Figure 352 show more variation than those in Figure 349); however, the DWT fails as a decorrelator for the $\phi = -0.9$ process in that the displayed ACS values for $\mathbf{W}_1$ are not close to zero. The reason why the DWT does poorly here is that the SDF for an AR(1) process with $\phi = -0.9$ has considerable variation over the nominal pass-band $[1/4, 1/2]$. associated with $\mathbf{W}_1$, which leads to a markedly non-flat $S_1(\cdot)$. Thus, while the DWT works as a decorrelator for many time series, it can fail whenever there is too much variation within a given octave band. (Section 4.1.2 of Whitcher, 1998, discusses AR processes covering the full range $-1 < \phi < 1$. Exercise [9.3] gives the reader a chance to create plots similar to Figures 352 and 353 for another type of stationary process.)

While the DWT fails here, it is interesting to note that the SDFs for $\phi = \pm 0.9$ are mirror images of each other so that, if we could do an octave-band splitting of $[0, 1/2]$ in the *opposite* direction from the usual way, then we would divide up the SDF for $\phi = -0.9$ into intervals having variations as small as when $\phi = 0.9$. This reversal can in fact be accomplished by picking an orthonormal transform that is the 'mirror image' of the DWT within a wavelet packet tree (see Section 6.1). More generally, in the context of wavelet-based bootstrapping of time series, Percival *et al.* (2000b) discuss a scheme for adaptively picking a transform from a wavelet packet tree for suitably decorrelating a given time series.

[4] Here we consider an approximation to C_1 and C_2 that is more accurate than what we would get by using Equation (344) (this approximation can also be used when $j \geq 3$ if so desired). The idea is to approximate the integrand of Equation (343c) via a Taylor series expansion about the midpoint of the interval of integration $[\frac{1}{2^{j+1}}, \frac{1}{2^j}]$. Letting $x = \pi f$ and $\beta = 2\delta$, we get

$$C_j = 2^{j+1} \int_{1/2^{j+1}}^{1/2^j} \frac{\sigma_\varepsilon^2}{[4 \sin^2(\pi f)]^\delta} \, df \equiv \frac{2^{j+1} \sigma_\varepsilon^2}{\pi 2^\beta} \int_{\pi/2^{j+1}}^{\pi/2^j} \frac{u(x)}{x^\beta} \, dx,$$

where $u(x) \equiv x^\beta / \sin^\beta(x)$. The first and second derivatives of $u(\cdot)$ are given by, respectively,

$$u'(x) = \frac{\beta x^{\beta-1}}{\sin^{\beta}(x)} - \frac{\beta x^{\beta}\cos(x)}{\sin^{\beta+1}(x)}$$

and

$$u''(x) = \frac{\beta x^{\beta} + \beta(\beta-1)x^{\beta-2}}{\sin^{\beta}(x)} - \frac{2\beta^2 x^{\beta-1}\cos(x)}{\sin^{\beta+1}(x)} + \frac{\beta(\beta+1)x^{\beta}\cos^2(x)}{\sin^{\beta+2}(x)}.$$

We can approximate $u(\cdot)$ over the interval $[\frac{\pi}{2^{j+1}}, \frac{\pi}{2^j}]$ via

$$u(x) \approx u(m) - mu'(m) + \frac{m^2}{2}u''(m) + [u'(m) - mu''(m)]\,x + \frac{u''(m)}{2}x^2$$
$$\equiv a_0 + a_1 x + a_2 x^2,$$

where $m \equiv 3\pi/2^{j+2}$ is the midpoint of the interval. This yields

$$C_j \approx \widetilde{C}_j \equiv \frac{\sigma_\varepsilon^2 2^{j+1}}{\pi 2^{\beta}} \int_{\pi/2^{j+1}}^{\pi/2^j} a_0 x^{-\beta} + a_1 x^{1-\beta} + a_2 x^{2-\beta}\,dx$$
$$\equiv \frac{\sigma_\varepsilon^2 2^{j+1}}{\pi 2^{\beta}} \left(A_{0,j} + A_{1,j} + A_{2,j}\right), \qquad (354)$$

where

$$A_{n,j} \equiv a_n \int_{\pi/2^{j+1}}^{\pi/2^j} x^{n-\beta}\,dx = \begin{cases} a_n \left(\frac{\pi}{2^{j+1}}\right)^{n+1-\beta} \frac{2^{n+1-\beta}-1}{n+1-\beta}, & \beta \neq n+1; \\ a_n \log(2), & \beta = n+1. \end{cases}$$

[5] The use of orthonormal wavelet decompositions as approximate whitening filters for fractional Brownian motion (FBM) is discussed in Flandrin (1992), who carried out discrete evaluations by taking the discrete sequence of interest to be a form of sampled FBM. Tewfik and Kim (1992) noted the effect of periodic boundary conditions when analyzing the correlation of wavelet coefficients generated from a form of sampled FBM; these were not predicted by their theory. Under certain conditions, including the idealized assumption that boundary coefficients do not exist, Dijkerman and Mazumdar (1994) showed that the discrete wavelet coefficients of FBM decay exponentially fast across scales and hyperbolically fast along time. Further results for FBM can be found in Masry (1993) and Wang (1996), which also discusses the 'derivative' of FBM (a continuous time analog of the discrete time fractional Gaussian noise (FGN) process).

[6] The idea of using wavelet-type transforms to analyze long memory time series dates back at least to Allan (1966) whose 'Allan variance' was used as a time domain measure of frequency stability in high-performance oscillators. As noted in Section 8.6, the Allan variance at a particular scale is directly related to the variance of wavelet coefficients at that scale when Haar wavelet filters are used. The Allan variance can be calculated for processes with stationary increments. A common process model utilized is FBM, which is a continuous time nonstationary process having stationary increments (see the discussion in Section 7.6). Such processes have a spectrum of power law form, with exponent α say, over a wide range of frequencies (Flandrin, 1989) and the exponent is simply related to the Allan variance at a particular scale, thus providing a way of estimating α, as investigated by Percival (1983). While the Allan variance is related to the variance of wavelet coefficients only when using Haar filters, if other suitable compact-support filters are used then the variance of wavelet coefficients at a scale can again be related to the spectral exponent – see Abry *et al.* (1993) for FBM examples, and also Percival (1995).

9.2 Simulation of a Long Memory Process

Let us now consider how we can use the fact that the DWT approximately decorrelates FD processes in order to generate simulations of such processes (the material in this and the next section builds upon Wornell, 1993 and 1996, and McCoy and Walden, 1996). Suppose that the vector $\mathbf{X} = [X_0, \ldots, X_{N-1}]^T$ contains a portion of length $N = 2^J$ from a stationary FD process $\{X_t\}$ with zero mean and with parameters δ and σ_ε^2. Let $\mathcal{W}_N$ be the $N \times N$ matrix defining a full DWT (to clarify our discussion in this section, we have added a subscript to our standard notation for this matrix). If we let $\Sigma_\mathbf{X}$ be the covariance matrix for $\mathbf{X}$, then, as noted previously, the covariance matrix for the DWT $\mathbf{W} = \mathcal{W}_N\mathbf{X}$ is given by $\Sigma_\mathbf{W} = \mathcal{W}_N\Sigma_\mathbf{X}\mathcal{W}_N^T$ (this is an application of Equation (262c)). The decorrelation property says that the off-diagonal elements of $\Sigma_\mathbf{W}$ are relatively small. As in McCoy and Walden (1996), let us approximate this matrix by an $N \times N$ diagonal matrix Λ_N with diagonal elements

$$\underbrace{C_1, \ldots, C_1}_{\frac{N}{2}\text{ of these}}, \underbrace{C_2, \ldots, C_2}_{\frac{N}{4}\text{ of these}}, \ldots, \underbrace{C_j, \ldots, C_j}_{\frac{N}{2^j}\text{ of these}}, \ldots, \underbrace{C_{J-1}, C_{J-1}}_{2\text{ of these}}, C_J, C_{J+1}. \tag{355a}$$

Here, as in Equation (343b), $C_j \approx \text{var}\{W_{j,t}\}$ for $j = 1, \ldots, J$, while C_{J+1} is a yet to be determined approximation to the variance of the final element of $\mathbf{W}$ (i.e., the single element of $\mathbf{V}_J$, namely, $V_{J,0}$). From Exercise [97] we know that, for any DWT based upon a Daubechies wavelet filter, the last row of $\mathcal{W}_N$ must have all of its elements equal to $1/\sqrt{N}$, from which it follows that $V_{J,0}$ must be equal to $\overline{X}\sqrt{N}$, where $\overline{X}$ is the sample mean of $\mathbf{X}$.

We can now obtain an approximate simulation of a Gaussian FD process via the following steps. First, we can use Equation (343c) to compute the approximate variances C_j, $j = 1, \ldots, J$, via numerical integration (alternatively, we could replace these approximate variances with either the exact variances $\text{var}\{W_{j,t}\}$, which we can compute from Equation (343a), or the simple approximations $\widetilde{C}_j$ to C_j given in Equation (344)). Second, we can determine C_{J+1} (which should be an approximation to $\text{var}\{V_{J,0}\}$) by making use of the following result.

▷ **Exercise [355]:** Show that

$$\text{var}\{X_t\} = \frac{\text{var}\{V_{J,0}\}}{N} + \sum_{j=1}^{J} \frac{\text{var}\{W_{j,t}\}}{2^j}.$$

Hint: start from Equation (302). ◁

If we combine the expression for $\text{var}\{X_t\}$ given in Equation (284c) with the already computed approximations C_j for $\text{var}\{W_{j,t}\}$, we obtain

$$C_{J+1} \equiv N\left(\frac{\sigma_\varepsilon^2\Gamma(1-2\delta)}{\Gamma^2(1-\delta)} - \sum_{j=1}^{J}\frac{C_j}{2^j}\right). \tag{355b}$$

Finally, if we let $\mathbf{Z}_N$ denote a vector of N deviates from a Gaussian white noise process with zero mean and unit variance, we can form

$$\mathbf{Y}_N \equiv \mathcal{W}_N^T\Lambda_N^{1/2}\mathbf{Z}_N,$$

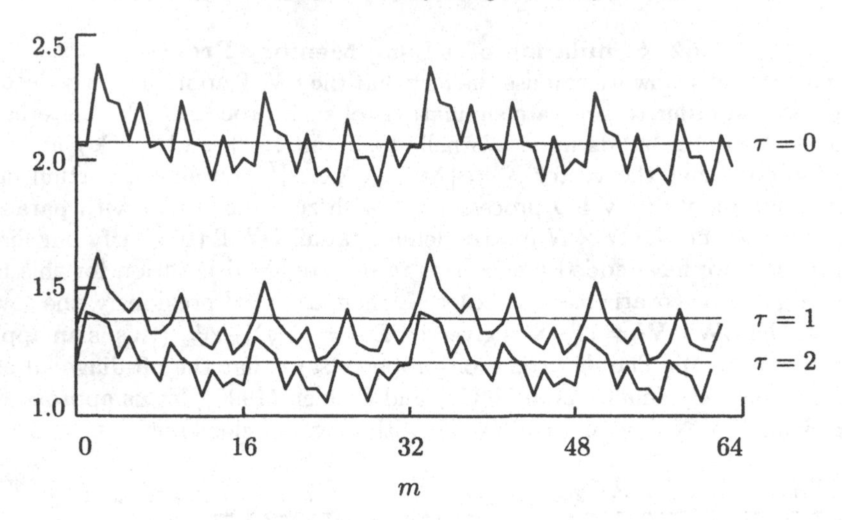

Figure 356. Diagonal elements $\Sigma_{\mathbf{Y},m,m+\tau}$ and $\Sigma_{\mathbf{X},m,m+\tau}$, $m = 0, \ldots, N-1-\tau$, of the covariance matrices $\Sigma_{\mathbf{Y}}$ and $\Sigma_{\mathbf{X}}$ (thick jagged curves and thin horizontal lines, respectively) for sample size $N = 64$ from an FD(0.4) process with $\sigma_\varepsilon^2 = 1$ and with $\Sigma_{\mathbf{Y}}$ constructed using an LA(8) DWT. Three diagonals are plotted for each covariance matrix, namely, the main diagonal ($\tau = 0$) and the first two off-diagonals ($\tau = 1$ and 2). Whereas $\Sigma_{\mathbf{X}}$ exhibits the Toeplitz structure required for a stationary process, its approximation $\Sigma_{\mathbf{Y}}$ does not.

which we take to be an approximation to $\mathbf{X}$ (in the above, $\Lambda_N^{1/2}$ is a diagonal matrix whose elements are the square roots of the elements of Λ_N; i.e., $\Lambda_N^{1/2}\Lambda_N^{1/2} = \Lambda_N$).

Let us now investigate how well the statistical properties of $\mathbf{Y}_N$ match up with those of $\mathbf{X}$. Because $\mathbf{Y}_N$ is a linear combination of Gaussian RVs, it obeys a multivariate Gaussian distribution, as is assumed to be true for $\mathbf{X}$ also. This distribution is completely determined by its vector of means and its covariance matrix. Because $E\{\mathbf{Z}_N\} = \mathbf{0}$ (a vector of N zeros), it follows that $E\{\mathbf{Y}_N\} = \mathbf{0}$ also, which agrees with the zero mean assumption on the FD process $\{X_t\}$. Since the covariance matrix for $\mathbf{Z}_N$ is just the identity matrix, an appeal to Equation (262c) says that the covariance matrix for $\mathbf{Y}_N$ is given by $\Sigma_{\mathbf{Y}} = \mathcal{W}_N^T \Lambda_N \mathcal{W}_N$, which we want to approximate $\Sigma_{\mathbf{X}}$ as closely as possible. Now $\{X_t\}$ is a stationary process, so the (m,n)th element of $\Sigma_{\mathbf{X}}$ is given by $s_{X,m-n}$, where $\{s_{X,\tau}\}$ is the ACVS for the process. The covariance matrix for any stationary process is said to have a Toeplitz structure because its values along any given diagonal are the same. Computations indicate that $\Sigma_{\mathbf{Y}}$ does not share this structure; i.e., the RVs in $\mathbf{Y}_N$ are not a portion of a stationary process. As an example (with $\Sigma_{\mathbf{Y},m,n}$ denoting the (m,n)th element of $\Sigma_{\mathbf{Y}}$), the thick jagged curves in Figure 356 show plots of $\Sigma_{\mathbf{Y},m,m+\tau}$ versus $m = 0, 1, \ldots, N-1-\tau$ for $\tau = 0, 1$ and 2 for the case of $N = 64$, the LA(8) DWT and an FD process with $\delta = 0.4$ and $\sigma_\varepsilon^2 = 1$. The superimposed thin horizontal lines show the corresponding values from $\Sigma_{\mathbf{X}}$, i.e., $s_{X,0}$, $s_{X,1}$ and $s_{X,2}$.

Although $\mathbf{Y}_N$ is not stationary, we can easily create a stationary process from it by applying a random circular shift (as noted in Chapter 4, the DWT naturally treats a time series as if it were circular). According, let κ be an RV that is uniformly distributed over the integers $0, 1, \ldots, N-1$, and let $\widetilde{\mathbf{Y}}_N \equiv \mathcal{T}^{-\kappa}\mathbf{Y}_N$, where $\mathcal{T}$ is the circular shift matrix defined in the discussion surrounding Equation (52a). The burden

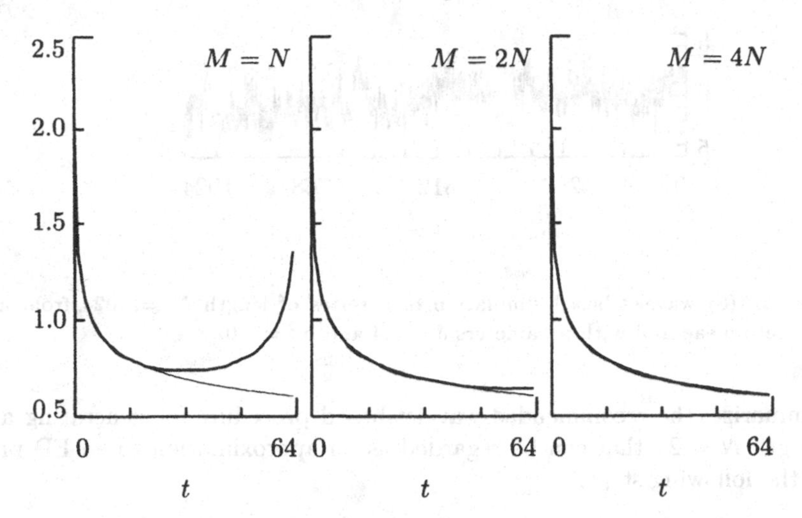

Figure 357. True ACVS (thin curves) and wavelet-based approximate ACVSs (thick) for an FD(0.4) process. The approximating ACVSs are based on an LA(8) DWT in which we generate a series of length M and then extract a series of length $N = 64$. As M goes from N to $4N$, the approximate ACVS gets closer to the true ACVS.

of Exercise [9.4] is to show that the RVs in $\widetilde{\mathbf{Y}}_N$ now form a portion of a stationary process whose ACVS is given by

$$s_{\widetilde{Y},\tau} \equiv \frac{1}{N} \sum_{m=0}^{N-1} \Sigma_{\mathbf{Y},m,m+\tau \bmod N}, \quad \tau = 0, 1, \ldots, N-1. \tag{357}$$

In terms of Figure 356, this ACVS is obtained by averaging the displayed diagonals (with one additional term for $\tau = 1$, namely, $\Sigma_{\mathbf{Y},N-1,0}$, and with two additional terms for $\tau = 2$, namely, $\Sigma_{\mathbf{Y},N-2,0}$ and $\Sigma_{\mathbf{Y},N-1,1}$). The left-hand plot of Figure 357 compares the ACVSs for the target FD(0.4) process (thin curve) and the approximating $\widetilde{\mathbf{Y}}_N$ process (thick) out to lag $\tau = 63$. While there is good correspondence for small lags, the same is not true at intermediate and large lags (one explanation for this discrepancy is indicated by the last part of Exercise [9.4]: as is evident from the plot, the ACVS for $\widetilde{\mathbf{Y}}_N$ is in fact constrained to be symmetric about $\tau = N/2$, i.e., $s_{\widetilde{Y},N-\tau} = s_{\widetilde{Y},\tau}$ for $\tau = 1, \ldots, N/2$).

We can obtain much better approximate simulations for $\mathbf{X}$ by resorting to the same strategy used by the DFT-based Gaussian spectral synthesis method (GSSM, described in Section 7.8), namely, to generate a simulated series of length $M > N$ and then just make use of the first N values. Essentially subsampling breaks the inherent circularity imposed by the DWT, hence leading to a better approximation to the desired ACVS out to lag $\tau = N - 1$. The middle and right-hand plots of Figure 357 compare the resulting desired and approximating ACVSs (thin and thick curves, respectively) for series of length $N = 64$ subsampled from simulated series of lengths, respectively, $M = 2N = 128$ and $M = 4N = 256$. While there is still some slight disagreement at the very largest lags for the $M = 2N$ case, there is quite good overall agreement in the $M = 4N$ case. This choice for M is the same one that works well for GSSM, so we adopt it here also.

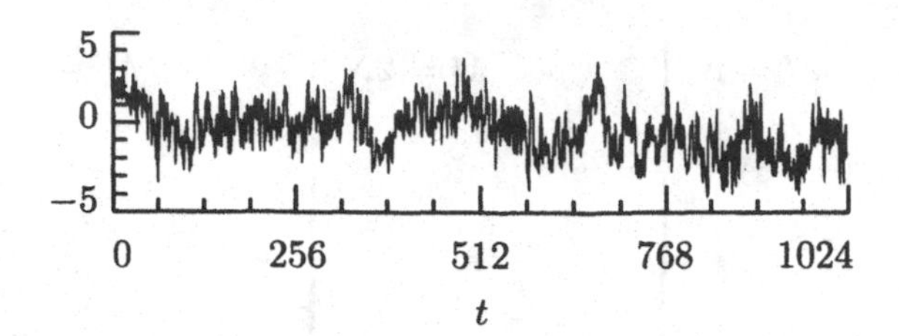

Figure 358. LA(8) wavelet-based simulation of a series of length $N = 1024$ from an FD process with zero mean and with parameters $\delta = 0.4$ and $\sigma_\varepsilon^2 = 1.0$.

To summarize, the recommended wavelet-based procedure for generating a time series of length $N = 2^J$ that can be regarded as an approximation to an FD process consists of the following steps.

[1] Using a pseudo-random number generator on a computer, generate a vector $\mathbf{Z}_M$ containing $M = 4N$ Gaussian white noise deviates with zero mean and unit variance.

[2] Using numerical integration, use Equation (343c) to compute the approximate variances C_j, $j = 1, \ldots, J+2$. Compute C_{J+3} via Equation (355b) by replacing J with $J+2$ on both sides of that equation.

[3] Multiply the first $M/2$ elements in $\mathbf{Z}_M$ by $\sqrt{C_1}$; the next $M/4$ values by $\sqrt{C_2}$; and so forth, until we come to the final four elements, which we multiply by, respectively, $\sqrt{C_{J+1}}, \sqrt{C_{J+1}}, \sqrt{C_{J+2}}$, and $\sqrt{C_{J+3}}$ (this is consistent with Equation (355a) when the sample size increases from N to $M = 4N = 2^{J+2}$). We can represent the result of this multiplication as $\Lambda_M^{1/2}\mathbf{Z}_M$.

[4] Compute $\mathbf{Y}_M \equiv \mathcal{W}_M^T \Lambda_M^{1/2}\mathbf{Z}_M$ (in practice, we do not generate $\mathcal{W}_M$ explicitly, but rather use the inverse DWT pyramid algorithm described in the Comments and Extensions to Section 4.6).

[5] Using a pseudo-random number generator, obtain a deviate κ from a uniform distribution over the integers $0, 1, \ldots, M-1$. The desired simulated series is then $Y_\kappa, Y_{\kappa+1 \bmod M}, \ldots, Y_{\kappa+N-1 \bmod M}$, where Y_t is the tth element of $\mathbf{Y}_M$.

As an example, Figure 358 shows a series of length $N = 1024$ simulated from an FD(0.4) process using the above procedure with the LA(8) wavelet. Qualitatively, this series is similar in character to the one shown at the bottom of Figure 342, which was generated using the DFT-based exact Davies–Harte method (DHM) described in Section 7.8.

The real test is how well the wavelet-based method does in simulating realizations with the correct properties. To study this, we generated 10 000 simulated FD(0.4) series of length $N = 64$ using both the LA(8) wavelet-based scheme and DHM. Let $\widetilde{X}_{k,t}, t = 0, \ldots, N-1$ represent the kth such series for either simulation method. For each series we computed an unbiased estimator of the ACVS using

$$\hat{s}_{k,\tau}^{(\mathrm{u})} \equiv \frac{1}{N-\tau} \sum_{t=0}^{N-1-\tau} \widetilde{X}_{k,t}\widetilde{X}_{k,t+\tau}, \quad \tau = 0, \ldots, N-1.$$

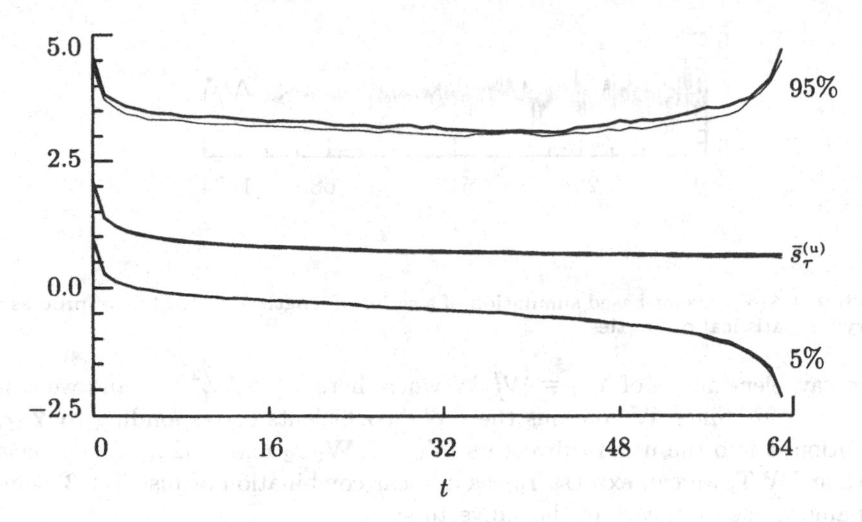

Figure 359. Estimated ACVSs averaged over 10 000 realizations generated via the Davies–Harte method (thin curve, middle of the plot) and the LA(8) wavelet-based method (thick curve) for an FD(0.4) process. The corresponding lower and upper pairs of curves indicate the 5% and 95% percentage points of the empirical distribution of the 10 000 simulations.

We then formed the sample mean of $\hat{s}_{k,\tau}^{(u)}$ over all 10 000 realizations:

$$\bar{s}_{\tau}^{(u)} \equiv \frac{1}{10\,000} \sum_{k=0}^{9999} \hat{s}_{k,\tau}^{(u)}.$$

The resulting $\bar{s}_{\tau}^{(u)}$ sequences for the two methods are shown in the middle portion of Figure 359 via thick and thin curves for, respectively, the wavelet-based method and DHM. These two curves are virtually the same (given the resolution of this figure, a plot of the true ACVS would be indistinguishable from the $\bar{s}_{\tau}^{(u)}$ for DHM). The lower and upper pairs of curves in the figure indicate the 5% and 95% percentage points of the empirical distributions for the $\hat{s}_{k,\tau}^{(u)}$ for both methods. While the lower percentage points are quite close, there is evidently a small upward bias in the upper percentage points for the wavelet-based method. The performance of the wavelet-based scheme is on the whole quite acceptable.

Because the DFT-based Davies–Harte method produces simulations of Gaussian FD processes with exactly the correct statistical properties, it certainly should be preferred over any approximate method, even one as accurate as the wavelet-based one; however, the wavelet-based method can be readily adapted to provide simulations in two scenarios for which it is problematic to use DHM: simulating very long time series (needed, for example, in certain real-time applications) and simulating time series whose statistical properties evolve over time. In the first scenario, it is problematic to use DHM to generate a very long time series because of the prohibitive amount of computer time and memory needed by a typical FFT algorithm to compute the DFT of, say, $N = 10^7$ values. The FFT algorithm and the DWT pyramid algorithm have computational complexities of $O(N \log_2(N))$ and $O(N)$, so the DWT has an important time advantage for very large N. Additionally, if a very long series is needed for a real-time application, the localized nature of the inverse DWT can be exploited to efficiently simulate the time series 'on the fly.' To get an idea how this works, let us

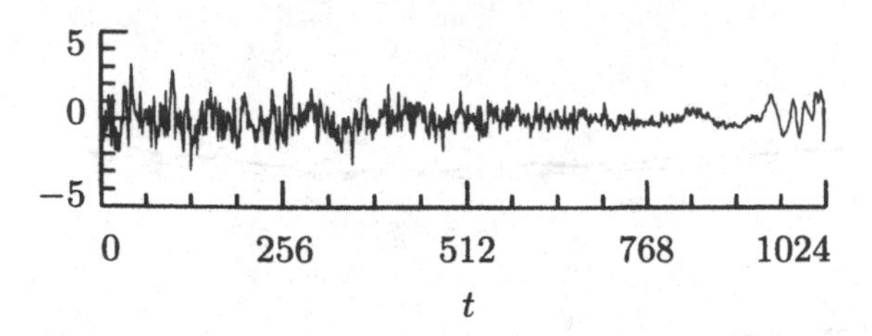

Figure 360. LA(8) wavelet-based simulation of a series of length $N = 1024$ from process with time varying statistical properties.

consider, say, element Y_{97} of $\mathbf{Y}_M \equiv \mathcal{W}_M^T \mathbf{W}$ where here $\mathbf{W} \equiv \Lambda_M^{1/2} \mathbf{Z}_M$, and we assume $M = 2^{J+2} \gg 97$. Since $\mathbf{W}$ contains the DWT coefficients corresponding to $\mathbf{Y}_M$, we can partition it into the usual subvectors $\mathbf{W}_1, \ldots, \mathbf{W}_{J+2}$ and $\mathbf{V}_{J+2}$. If we specialize to the Haar DWT, we can express Y_{97} as a linear combination of just $J+3$ elements in $\mathbf{W}$, namely, one from each of the subvectors:

$$W_{1,48}, W_{2,24}, W_{3,12}, W_{4,6}, W_{5,3}, W_{6,1}, W_{7,0}, \ldots, W_{J+2,0} \text{ and } V_{J+2,0}.$$

As we move from Y_{97} to Y_{98}, the linear combination involves almost the same set of elements: we just need to replace $W_{1,48}$ with $W_{1,49}$. In addition, all but the first two coefficients in the linear combinations forming Y_{97} and Y_{98} remain the same (these change from $\frac{1}{\sqrt{2}}$ and $-\frac{1}{4}$ to, respectively, $-\frac{1}{\sqrt{2}}$ and $\frac{1}{4}$). If so desired, we could thus compute Y_{98} in terms of Y_{97} using just

$$Y_{98} = Y_{97} - \frac{W_{1,48} + W_{1,49}}{\sqrt{2}} + \frac{W_{2,24}}{2}.$$

As we move along in time, we can continue to make direct use of just one element from each subvector at each time point. Memory requirements are quite modest, and this scheme is simple enough to satisfy requirements of real-time computations. Related schemes can be devised for, for example, the D(4) and LA(8) wavelets since Y_t still just depends on a small number of elements in each subvector (Percival and Constantine, 2000).

Let us now discuss briefly how the wavelet-based simulation method can be adapted to simulate time series whose statistical properties vary over time. As in the case of simulating very long time series, the key here is the localized influence of each element in $\mathbf{W}$ on a given element in $\mathbf{Y}_M$. As a simple example, suppose we set $M = N = 2^J$ and generate the elements of the diagonal matrix Λ_N in $J+1$ groups as per Equation (355a) so that the C_j values are determined by the LA(8) DWT applied to an FD(0.4) process, but with the modification that the tth member in the group for levels $j = 1, \ldots, 4$ is given by $(N_j - t)C_j/N_j$, $t = 0, \ldots, N_j - 1$. This mimics a process whose large scale properties are dictated by a long memory process, but whose small scale characteristics evolve with time. One realization of this process is shown in Figure 360. (In order to control precisely the evolution in time of the statistical properties of a process, we need to know which wavelet coefficients determine which portions of a time series. The relationship between time and each coefficient in $\mathbf{W}_j$ is simple to work out for the Haar wavelet and has been developed in Section 4.8 for the LA wavelets. To maintain the simplicity of this example, we have not made use of these relationships.)

Finally, we note that DHM requires knowledge of the ACVS, which – while readily available for FD processes – might not be easy to determine for other processes. Wavelet-based simulation works entirely from knowledge of average SDF levels or, equivalently, the wavelet variance: it is not necessary to know the covariance matrix of the process. Hence we can think of the technique described here as the dual to a common simulation technique for standard short-memory processes whereby a completely known covariance matrix is subjected to a Cholesky decomposition (see, for example, McLeod and Hipel, 1978, or Kay, 1981).

Comments and Extensions to Section 9.2

[1] The idea of synthesizing stochastic processes using wavelet expansions in terms of uncorrelated wavelet coefficients having a known variance progression was put forward by Wornell (1990, 1993). These papers concentrated on simulating continuous time nonstationary processes with power law properties as $f \to 0$. In practice, the simulation scheme is reformulated into an iterative 'upsample-filter-merge' discrete time algorithm followed by a single discrete-to-continuous conversion (Wornell, 1993, Appendix I). Masry (1993) criticizes Wornell's approach when applied to FBMs: he points out that FBM cannot be synthesized exactly via a wavelet series representation since the wavelet coefficients are not exactly uncorrelated for an FBM (it appears, however, that Wornell always intended his scheme to be an approximate one).

[2] It is interesting to note that the univariate distribution of any element $\widetilde{Y}_t$ of $\widetilde{\mathbf{Y}}_N$ is that of a Gaussian mixture model, a concept that we introduce and utilize in Section 10.4. The device of randomly circularly shifting a vector of multivariate Gaussian RVs is thus a viable way to construct a stationary process with a non-Gaussian univariate distribution.

9.3 MLEs for Stationary FD Processes

Suppose now that we have a time series that is regarded as a realization of a portion $\mathbf{X} = [X_0, \ldots, X_{N-1}]^T$ of a zero mean stationary FD process with unknown parameters $\delta \in [-\frac{1}{2}, \frac{1}{2})$ and $\sigma_\varepsilon^2 > 0$, where as before $N \equiv 2^J$ for some integer J (even though we are interested primarily in long memory FD processes, i.e., $\delta \in (0, \frac{1}{2})$, the theory we present here also holds for short memory FD processes with $\delta \in [-\frac{1}{2}, 0]$). We want to estimate these parameters based upon $\mathbf{X}$. Under the assumption that $\mathbf{X}$ obeys a multivariate Gaussian (normal) distribution, we can estimate δ and σ_ε^2 using the maximum likelihood method. Given the observations $\mathbf{X}$, the likelihood function for δ and σ_ε^2 is given by

$$L(\delta, \sigma_\varepsilon^2 \mid \mathbf{X}) \equiv \frac{1}{(2\pi)^{N/2} |\Sigma_{\mathbf{X}}|^{1/2}} e^{-\mathbf{X}^T \Sigma_{\mathbf{X}}^{-1} \mathbf{X}/2}, \tag{361}$$

where $\Sigma_{\mathbf{X}}$ is the covariance matrix of $\mathbf{X}$ (i.e., its (m, n)th element is given by $s_{X,m-n}$ of Equation (284b)), while $|\Sigma_{\mathbf{X}}|$ denotes the determinant of $\Sigma_{\mathbf{X}}$. Note that the dependence of the likelihood function on δ and σ_ε^2 is through $\Sigma_{\mathbf{X}}$ alone. We can in principle evaluate the above likelihood function for any particular parameter values. The maximum likelihood estimators (MLEs) of δ and σ_ε^2 are those values that maximize $L(\delta, \sigma_\varepsilon^2 \mid \mathbf{X})$ as a function of δ and σ_ε^2. There are, however, two practical problems with exact MLEs (Beran, 1994, Section 5.4). First, their determination can be very time consuming because $L(\delta, \sigma_\varepsilon^2 \mid \mathbf{X})$ is computationally expensive to evaluate, even for moderate N. Second, there can be potential numerical instabilities in computing the likelihood function when δ is close to 1/2. Efficient and numerically stable approximate estimators are thus of considerable interest.

Here we consider an approximate maximum likelihood scheme that exploits the fact that $\Sigma_{\mathbf{X}}$ can be approximated by $\widetilde{\Sigma}_{\mathbf{X}} \equiv \mathcal{W}^T \Lambda_N \mathcal{W}$, where $\mathcal{W}$ is a DWT matrix, and Λ_N is a diagonal matrix with diagonal elements C_j as depicted in Equation (355a). With this approximation, we have

$$L(\delta, \sigma_\varepsilon^2 \mid \mathbf{X}) \approx \tilde{L}(\delta, \sigma_\varepsilon^2 \mid \mathbf{X}) \equiv \frac{1}{(2\pi)^{N/2} |\widetilde{\Sigma}_{\mathbf{X}}|^{1/2}} e^{-\mathbf{X}^T \widetilde{\Sigma}_{\mathbf{X}}^{-1} \mathbf{X}/2}. \tag{362a}$$

Now maximization of $\tilde{L}(\delta, \sigma_\varepsilon^2 \mid \mathbf{X})$ is equivalent to minimization of the (rescaled and relocated) log likelihood function

$$\tilde{l}(\delta, \sigma_\varepsilon^2 \mid \mathbf{X}) \equiv -2\log\left(\tilde{L}(\delta, \sigma_\varepsilon^2 \mid \mathbf{X})\right) - N \log(2\pi) = \log\left(|\widetilde{\Sigma}_{\mathbf{X}}|\right) + \mathbf{X}^T \widetilde{\Sigma}_{\mathbf{X}}^{-1} \mathbf{X}. \tag{362b}$$

We can simplify the first term using the following result.

▷ **Exercise [362a]:** With $N_j \equiv N/2^j$ as before, show that

$$\log\left(|\widetilde{\Sigma}_{\mathbf{X}}|\right) = \log\left(C_{J+1}\right) + \sum_{j=1}^{J} N_j \log\left(C_j\right). \quad ◁$$

It is convenient at this point to note exactly how C_j depends on δ and σ_ε^2. For $j = 1, \ldots, J$, Equation (343c) says that we can write

$$C_j = \sigma_\varepsilon^2 C_j'(\delta), \text{ where } C_j'(\delta) \equiv 2^{j+1} \int_{1/2^{j+1}}^{1/2^j} \frac{1}{[4\sin^2(\pi f)]^\delta}\, df; \tag{362c}$$

on the other hand, for $j = J+1$, Equation (355b) yields

$$C_{J+1} = \sigma_\varepsilon^2 C_{J+1}'(\delta), \text{ where } C_{J+1}'(\delta) \equiv N \frac{\Gamma(1-2\delta)}{\Gamma^2(1-\delta)} - \sum_{j=1}^{J} N_j C_j'(\delta). \tag{362d}$$

We can now write

$$\log\left(|\widetilde{\Sigma}_{\mathbf{X}}|\right) = N \log\left(\sigma_\varepsilon^2\right) + \log\left(C_{J+1}'(\delta)\right) + \sum_{j=1}^{J} N_j \log\left(C_j'(\delta)\right).$$

The second term of Equation (362b) can be reduced as follows.

▷ **Exercise [362b]:** Show that

$$\mathbf{X}^T \widetilde{\Sigma}_{\mathbf{X}}^{-1} \mathbf{X} = \frac{1}{\sigma_\varepsilon^2} \left(\frac{V_{J,0}^2}{C_{J+1}'(\delta)} + \sum_{j=1}^{J} \frac{1}{C_j'(\delta)} \sum_{t=0}^{N_j - 1} W_{j,t}^2 \right). \quad ◁$$

		MLE			
δ		Haar	D(4)	LA(8)	exact
0.25	mean	0.2184	0.2293	0.2328	0.2374
	bias	−0.0316	−0.0207	−0.0172	−0.0126
	SD	0.0713	0.0705	0.0710	0.0673
	RMSE	0.0780	0.0735	0.0731	0.0685
0.4	mean	0.3614	0.3727	0.3768	0.3797
	bias	−0.0386	−0.0273	−0.0232	−0.0203
	SD	0.0675	0.0652	0.0640	0.0604
	RMSE	0.0778	0.0707	0.0681	0.0637

Table 363. Sample mean, bias, standard deviation and root mean square error of 1024 wavelet-based approximate MLEs $\tilde{\delta}^{(s)}$ of the parameter δ based on the likelihood function of Equation (362a) using Haar, D(4) and LA(8) wavelet filters. All 1024 time series were of length $N = 128$ and were simulated using the Davies–Harte method. Corresponding statistics for exact MLEs $\hat{\delta}$ are given in the final column.

With the above expressions for $\log(|\widetilde{\Sigma}_{\mathbf{X}}|)$ and $\mathbf{X}^T\widetilde{\Sigma}_{\mathbf{X}}^{-1}\mathbf{X}$, we can re-express Equation (362b) as

$$\begin{aligned}\tilde{l}(\delta, \sigma_\varepsilon^2 \mid \mathbf{X}) = N \log(\sigma_\varepsilon^2) + \log(C'_{J+1}(\delta)) + \sum_{j=1}^{J} N_j \log(C'_j(\delta)) \\ + \frac{1}{\sigma_\varepsilon^2}\left(\frac{V_{J,0}^2}{C'_{J+1}(\delta)} + \sum_{j=1}^{J}\frac{1}{C'_j(\delta)}\sum_{t=0}^{N_j-1} W_{j,t}^2\right).\end{aligned} \tag{363a}$$

We can obtain an expression for the approximate MLE $\tilde{\sigma}_\varepsilon^2$ of σ_ε^2 by differentiating the right-hand side of Equation (363a) with respect to σ_ε^2 and setting the resulting expression to zero. Solving for σ_ε^2 yields the estimator $\tilde{\sigma}_\varepsilon^2$, which can be regarded as a function of δ given by

$$\tilde{\sigma}_\varepsilon^2(\delta) \equiv \frac{1}{N}\left(\frac{V_{J,0}^2}{C'_{J+1}(\delta)} + \sum_{j=1}^{J}\frac{1}{C'_j(\delta)}\sum_{t=0}^{N_j-1} W_{j,t}^2\right). \tag{363b}$$

Using the above expression, we can eliminate the parameter σ_ε^2 from Equation (363a), yielding what Brockwell and Davis (1991) refer to as the *reduced log likelihood*:

$$\begin{aligned}\tilde{l}(\delta \mid \mathbf{X}) &\equiv \tilde{l}(\delta, \tilde{\sigma}_\varepsilon^2(\delta) \mid \mathbf{X}) - N \\ &= N \log(\tilde{\sigma}_\varepsilon^2(\delta)) + \log(C'_{J+1}(\delta)) + \sum_{j=1}^{J} N_j \log(C'_j(\delta)).\end{aligned} \tag{363c}$$

Once we have substituted Equation (363b) into the above, the reduced likelihood depends on just the single parameter δ, so it is an easy matter to search numerically over the interval $[-\frac{1}{2}, \frac{1}{2})$ for the value of δ, say $\tilde{\delta}^{(s)}$, minimizing $\tilde{l}(\delta \mid \mathbf{X})$. This minimizing value is the approximate MLE for δ. After we have determined $\tilde{\delta}^{(s)}$ numerically, we get the corresponding approximate MLE $\tilde{\sigma}_\varepsilon^2$ for σ_ε^2 by plugging $\tilde{\delta}^{(s)}$ into Equation (363b).

δ		MLE Haar	D(4)	LA(8)	exact
0.25	mean	0.2256	0.2363	0.2402	0.2443
	bias	−0.0244	−0.0137	−0.0098	−0.0057
	SD	0.0505	0.0495	0.0502	0.0479
	RMSE	0.0561	0.0514	0.0511	0.0483
0.4	mean	0.3710	0.3832	0.3886	0.3900
	bias	−0.0290	−0.0168	−0.0114	−0.0100
	SD	0.0488	0.0478	0.0465	0.0437
	RMSE	0.0567	0.0506	0.0479	0.0448

Table 364. As in Table 363, but now with $N = 256$.

To see how well these approximate MLEs work, we simulated 1024 Gaussian FD(0.25) and FD(0.4) time series of lengths $N = 128$ and $N = 256$ using the Davies–Harte method (see Section 7.8) and then estimated δ using the value $\tilde{\delta}^{(s)}$ minimizing Equation (363c). We considered three different wavelet filters: the Haar, D(4) and LA(8). Let $\tilde{\delta}^{(s),n}$ be the MLE for the nth simulated series for a given choice of δ and N. For the two values of δ and $N = 128$, Table 363 gives the sample mean $\bar{\delta}^{(s)}$ of the $\tilde{\delta}^{(s),n}$ and the sample bias (i.e., $\bar{\delta}^{(s)} - \delta$), along with the sample standard deviation (SD) and root mean square error (RMSE), i.e.,

$$\left(\frac{1}{1024}\sum_{n=0}^{1023}(\tilde{\delta}^{(s),n} - \bar{\delta}^{(s)})^2\right)^{1/2} \text{ and } \left(\frac{1}{1024}\sum_{n=0}^{1023}(\tilde{\delta}^{(s),n} - \delta)^2\right)^{1/2}.$$

Table 364 shows corresponding results when $N = 256$. For both choices of δ, the Haar wavelet leads to larger biases and RMSEs than the other two wavelets. When $\delta = 0.25$ there is virtually nothing to choose between the D(4) and LA(8) results: the former has larger bias, but the RMSEs are almost identical. The LA(8) does somewhat better than the D(4) when $\delta = 0.4$. As we would expect, both the biases and the RMSEs decrease as we increase the sample size from length 128 to 256.

The final columns in both tables give corresponding results for the exact MLE $\hat{\delta}$ of δ (see item [4] of the Comments and Extensions). We see that the RMSEs for the LA(8) wavelet are about 10% larger than for the exact method, so the wavelet-based approximation is quite good. Kashyap and Eom (1988) derived the asymptotic Cramér–Rao lower bound for the MSE of unbiased estimators of δ as $\frac{6}{N\pi^2}$, which yields RMSEs of 0.0689 and 0.0487 for $N = 128$ and 256, respectively. These values are in reasonable agreement with the RMSEs for the D(4), LA(8) and exact methods.

Our development so far assumes that $\mathbf{X}$ is a portion of a realization of a stationary FD process with zero mean. In practical applications, we rarely know the process mean $\mu \equiv E\{X_t\}$ for a time series *a priori*, so it is of interest to modify our scheme so that μ becomes an additional parameter to be estimated. Accordingly, given the observations $\mathbf{X}$, we now regard the likelihood as a function of δ, σ_ε^2 and μ so that Equation (361) becomes

$$L(\delta, \sigma_\varepsilon^2, \mu \mid \mathbf{X}) \equiv \frac{1}{(2\pi)^{N/2}|\Sigma_{\mathbf{X}}|^{1/2}} e^{-(\mathbf{X}-\mu\mathbf{1})^T\Sigma_{\mathbf{X}}^{-1}(\mathbf{X}-\mu\mathbf{1})/2},$$

where $\mathbf{1}$ is an N dimensional vector, all of whose elements are unity. The log likelihood function of Equation (362b) now becomes

$$\tilde{l}(\delta, \sigma_\varepsilon^2, \mu \mid \mathbf{X}) = \log(|\widetilde{\Sigma}_\mathbf{X}|) + (\mathbf{X} - \mu\mathbf{1})^T \widetilde{\Sigma}_\mathbf{X}^{-1}(\mathbf{X} - \mu\mathbf{1}).$$

The first term can be simplified again as per Exercise [362a], whereas the second term can be reduced as follows (cf. Exercise [362b]).

▷ **Exercise [365a]:** Show that

$$(\mathbf{X} - \mu\mathbf{1})^T \widetilde{\Sigma}_\mathbf{X}^{-1}(\mathbf{X} - \mu\mathbf{1}) = \frac{1}{\sigma_\varepsilon^2}\left(\frac{N^2(\overline{X} - \mu)^2}{C'_{J+1}(\delta)} + \sum_{j=1}^{J} \frac{1}{C'_j(\delta)} \sum_{t=0}^{N_j-1} W_{j,t}^2\right). \quad ◁$$

Equation (363a) now becomes

$$\tilde{l}(\delta, \sigma_\varepsilon^2, \mu \mid \mathbf{X}) = N \log(\sigma_\varepsilon^2) + \log(C'_{J+1}(\delta)) + \sum_{j=1}^{J} N_j \log(C'_j(\delta)) + \frac{1}{\sigma_\varepsilon^2}\left(\frac{N^2(\overline{X} - \mu)^2}{C'_{J+1}(\delta)} + \sum_{j=1}^{J} \frac{1}{C'_j(\delta)} \sum_{t=0}^{N_j-1} W_{j,t}^2\right),$$

from which it is clear that the value of μ minimizing the above yields the MLE $\tilde{\mu} \equiv \overline{X}$, which is the same no matter what the minimizing values for δ and σ_ε^2 are. With μ so determined, the MLE for σ_ε^2 now takes the form

$$\tilde{\sigma}_\varepsilon^2(\delta) = \frac{1}{N} \sum_{j=1}^{J} \frac{1}{C'_j(\delta)} \sum_{t=0}^{N_j-1} W_{j,t}^2 \tag{365}$$

(cf. Equation (363b)). With this new definition for $\tilde{\sigma}_\varepsilon^2(\delta)$, the reduced log likelihood for δ has the same form as before (Equation (363c)).

▷ **Exercise [365b]:** Suppose that we have a computer routine that takes a time series $\mathbf{X}$ and computes the MLEs $\tilde{\delta}^{(s)}$ and $\tilde{\sigma}_\varepsilon^2$ under the assumption that the process mean is known to be zero. Argue that, if $\mathbf{X}$ has an unknown process mean of μ instead, we can obtain the MLEs for δ and σ_ε^2 by using the routine with $\mathbf{X} - \overline{X}\mathbf{1}$ in place of $\mathbf{X}$ (i.e., we need only subtract off the sample mean from each observation prior to evoking the MLE routine). ◁

To see the effect of not knowing the process mean, we redid the calculations that led to Table 364, with the sole difference being that each simulated $\mathbf{X}$ was replaced by $\mathbf{X} - \overline{X}\mathbf{1}$. Table 366 gives summary statistics for the various MLEs of δ over the 1024 simulated series. The biases, SDs and RMSEs are all markedly worse here than for the corresponding entries in Table 364. We thus pay a substantial price for lack of knowledge about the process mean when trying to estimate δ (we noted a similar result in Section 8.1 in our discussion of the sample variance for a long memory process).

For an example using an actual time series with unknown mean, see Section 9.8, where we compute $\tilde{\delta}^{(s)}$ for the Nile River minima.

		MLE			
δ		Haar	D(4)	LA(8)	exact
0.25	mean	0.2058	0.2182	0.2227	0.2274
	bias	−0.0442	−0.0318	−0.0273	−0.0226
	SD	0.0559	0.0551	0.0557	0.0528
	RMSE	0.0712	0.0636	0.0620	0.0575
0.4	mean	0.3449	0.3602	0.3672	0.3687
	bias	−0.0551	−0.0398	−0.0328	−0.0313
	SD	0.0550	0.0538	0.0525	0.0494
	RMSE	0.0778	0.0669	0.0619	0.0585

Table 366. As in Table 364, but now with the process mean assumed unknown and hence estimated using the sample mean $\overline{X}$.

Comments and Extensions to Section 9.3

[1] Wornell and Oppenheim (1992), Wornell (1993) and Wornell (1996) consider estimation of the equivalent of δ and σ_ε^2 for continuous time Gaussian $1/f$-type processes. Their approach formulates the likelihood function purely in terms of wavelet coefficients. Our approach includes the scaling coefficient $V_{J,0}$, which is justified under the assumption that $\{X_t\}$ is known to have zero mean. When this is true, the coefficient $V_{J,0}$ contains information regarding the band-pass variance in the lowest frequency band $[-1/2^{J+1}, 1/2^{J+1}]$, without which the estimation scheme is poorer (cf. Tables 364 and 366). When $\{X_t\}$ is not known to have zero mean (the usual case in practical applications), the coefficient $V_{J,0}$ can be a badly biased estimator of the band-pass variance, which is why this coefficient is dropped in forming the reduced log likelihood function for δ after μ has been estimated by $\overline{X}$; however, it is interesting to note that this function still depends on $\operatorname{var}\{V_{J,0}\} = C_{J+1} = \sigma_\varepsilon^2 C'_{J+1}(\delta)$ through the term $\log(C'_{J+1}(\delta))$. This dependence is intuitively reasonable because C_{J+1} is an important component in constructing the wavelet-based approximation $\widetilde{\Sigma}_{\mathbf{X}}$ to $\Sigma_{\mathbf{X}}$ (here is one measure of the importance of the scaling coefficient variance: if we replace it with zero in constructing the $M = 4N$ wavelet-based approximate ACVS of Figure 357, the resulting ACVS sweeps from 1.41 at lag zero down to −0.05 at lag 63, whereas the true ACVS goes from 2.07 to 0.61).

Kaplan and Kuo (1993) discuss a modification to the approach of Wornell and Oppenheim (1992). They point out that, in the latter reference, the DWT is actually applied to DFBM (i.e., the discretely sampled continuous time FBM process), so that the relied-on variance progression with scale is biased, leading to potential problems with the estimate of the spectral exponent. They show that use of FGN (the first difference of DFBM) along with the Haar wavelet leads to an unbiased variance progression.

[2] Jensen (1999b, 2000) extends wavelet-based MLEs to handle autoregressive, fractionally integrated, moving average (ARFIMA) processes. These extensions to FD processes entertain p autoregressive and q moving average parameters in order to model the small scale (high frequency) properties of a time series. A valid critique of FD processes is that they only depend on the two parameters δ and σ_ε^2. Essentially δ controls the approximate power law of the SDF as $f \to 0$, while σ_ε^2 merely sets the spectral level and hence has no effect on the shape of the SDF. Using δ and σ_ε^2 to

capture large scale (low frequency) characteristics and spectral levels leaves us with no way to deal with small scale properties – these are imputed in a manner that need not be a good match for a particular time series. Thus, if we try to estimate δ and σ_ε^2 from a time series that does not obey an FD model at small scales, the MLE of δ can be seriously biased because the ML scheme requires that the spectral levels in each octave band follow an FD model. The additional $p + q$ parameters in an ARFIMA model allow the small and large scale properties to be decoupled, and hence the corresponding estimator of δ is freed to reflect accurately the large scale properties of a time series. (It should be noted that the estimators of δ to be discussed in the next two sections can be easily adjusted to reduce adverse dependency on discordant small scales – see the examples in Section 9.7.)

The above critique of FD processes also applies to FGN and PPL processes, both of which depend on just two parameters (Figure 282 illustrates that proper choice of parameter pairs for FGN, PPL and FD processes leads to SDFs with similar low frequency rolloffs and heights, but with high frequency characteristics that are then uncontrollable and hence arbitrary). In response to this limitation, Kaplan and Kuo (1994) propose an extension to FGN that allows for different correlation effects at small scales.

[3] In contrast to the estimators of δ presented in the next two sections, the wavelet-based MLE discussed here is admittedly limited to time series whose sample sizes N are powers of two. If N is not such, a simple scheme to get around this limitation is to average the two estimators of δ computed for the two subseries $X_0, \ldots, X_{N'-1}$ and $X_{N-N'}, \ldots, X_{N-1}$, where N' is the largest power of two satisfying $N' < N$.

[4] For completeness, here we note the equations needed to find the exact MLEs for a stationary Gaussian FD process with zero mean (for details, see Beran, 1994, Sections 5.3 and 2.5, or Brockwell and Davis, 1991, Section 13.2). Given a particular δ we first compute the partial autocorrelation sequence (PACS) $\phi_{t,t}$, $t = 1, \ldots, N-1$, which takes the form $\phi_{t,t} = \frac{\delta}{t-\delta}$ for an FD process (Hosking, 1981). We put this PACS to two uses. First, we use it recursively to compute the remaining coefficients of the best linear predictor of X_t given $X_{t-1}, \ldots, X_0$ for $t = 2, \ldots, N-1$ via

$$\phi_{t,k} = \phi_{t-1,k} - \phi_{t,t}\phi_{t-1,t-k}, \quad k = 1, \ldots, t-1$$

(these are the key equations in what are known as the Levinson–Durbin recursions). We use these coefficients to form the observed prediction errors:

$$e_t \equiv X_t - \sum_{k=1}^{t} \phi_{t,k} X_{t-k}, \quad t = 1, \ldots, N-1$$

(for convenience, let $e_0 \equiv X_0$). Our second use for the PACS is to compute a sequence $\{v_t\}$ relating the variances of e_t and e_0:

$$v_t = \operatorname{var}\{e_0\} \prod_{n=1}^{t} (1 - \phi_{n,n}^2), \quad t = 0, \ldots, N-1, \quad \text{where} \quad \operatorname{var}\{e_0\} = \frac{\Gamma(1-2\delta)}{\Gamma^2(1-\delta)}$$

(see Equation (284c); here $v_0 = \operatorname{var}\{e_0\} = \operatorname{var}\{X_t\}$). Given $\mathbf{X}$, the sequences $\{\phi_{t,k}\}$, $\{e_t\}$ and $\{v_t\}$ are all implicit functions of δ and are entirely determined by it.

The equations that we need to compute the exact MLEs closely parallel those for the wavelet-based approximate MLEs. In place of the approximate ML estimator $\tilde{\sigma}_\varepsilon^2(\delta)$ of Equation (363b), we have the exact estimator

$$\hat{\sigma}_\varepsilon^2(\delta) \equiv \frac{1}{N}\sum_{t=0}^{N-1}\frac{e_t^2}{v_t}. \tag{368}$$

In lieu of the reduced log likelihood of Equation (363c), we have

$$l(\delta \mid \mathbf{X}) \equiv N \log(\hat{\sigma}_\varepsilon^2(\delta)) + N \log\left(\frac{\Gamma(1-2\delta)}{\Gamma^2(1-\delta)}\right) + \sum_{t=1}^{N-1}(N-t)\log(1-\phi_{t,t}^2).$$

Upon substitution of Equation (368) into the above, we obtain a function of δ only, which we can numerically minimize to find the MLE $\hat{\delta}$. After we have $\hat{\delta}$, we can obtain the MLE $\hat{\sigma}_\varepsilon^2$ for σ_ε^2 by substituting $\hat{\delta}$ into Equation (368).

9.4 MLEs for Stationary or Nonstationary FD Processes

In this section we consider an alternative way of formulating MLEs for the parameters δ and σ_ε^2 for an FD process. The idea we explore here is to formulate the likelihood function directly in terms of the wavelet coefficients rather than through the time series itself. This approach has the benefit of allowing us to handle general FD processes observed in the presence of deterministic trends; i.e., we no longer need to assume a stationary model for our time series (for additional details, see Craigmile *et al.*, 2000a and 2000b).

Accordingly, for a given nonnegative integer d, suppose that $\{U_t\}$ is a stochastic process whose dth order backward difference $\{Y_t\}$ is a zero mean Gaussian stationary FD process with parameters $\delta^{(s)}$ and σ_ε^2, where $-\frac{1}{2} \le \delta^{(s)} < \frac{1}{2}$ and $\sigma_\varepsilon^2 > 0$. With this setup, $\{U_t\}$ is an FD process with parameters $\delta \equiv d + \delta^{(s)}$ and σ_ε^2 and is stationary (nonstationary) when $d = 0$ ($d > 0$). We consider the problem of estimating δ and σ_ε^2 under the assumption that what we actually observe is a realization of

$$X_t \equiv T_t + U_t, \quad t = 0, 1, \ldots, N-1,$$

where T_t is an unknown deterministic trend (Brillinger, 1994 and 1996, considers a model similar to this, but proposes a different wavelet-based scheme for trend estimation than the one presented here). For simplicity, we assume the trend to be a polynomial of order r:

$$T_t \equiv \sum_{j=0}^{r} a_j t^j$$

(as discussed in Craigmile *et al.*, 2000b, the scheme outlined below can in fact handle certain departures from a pure polynomial trend). In vector notation, we write the above as $\mathbf{X} = \mathbf{T} + \mathbf{U}$.

Let $\mathcal{W}$ be the DWT matrix for a J_0th level partial DWT based upon a Daubechies wavelet filter of width L, and let $\mathbf{W} = \mathcal{W}\mathbf{X}$ be the corresponding DWT coefficients. Recall that such a filter involves an implicit differencing operation of order $L/2$ and hence has the important property of reducing an rth order polynomial trend to zero

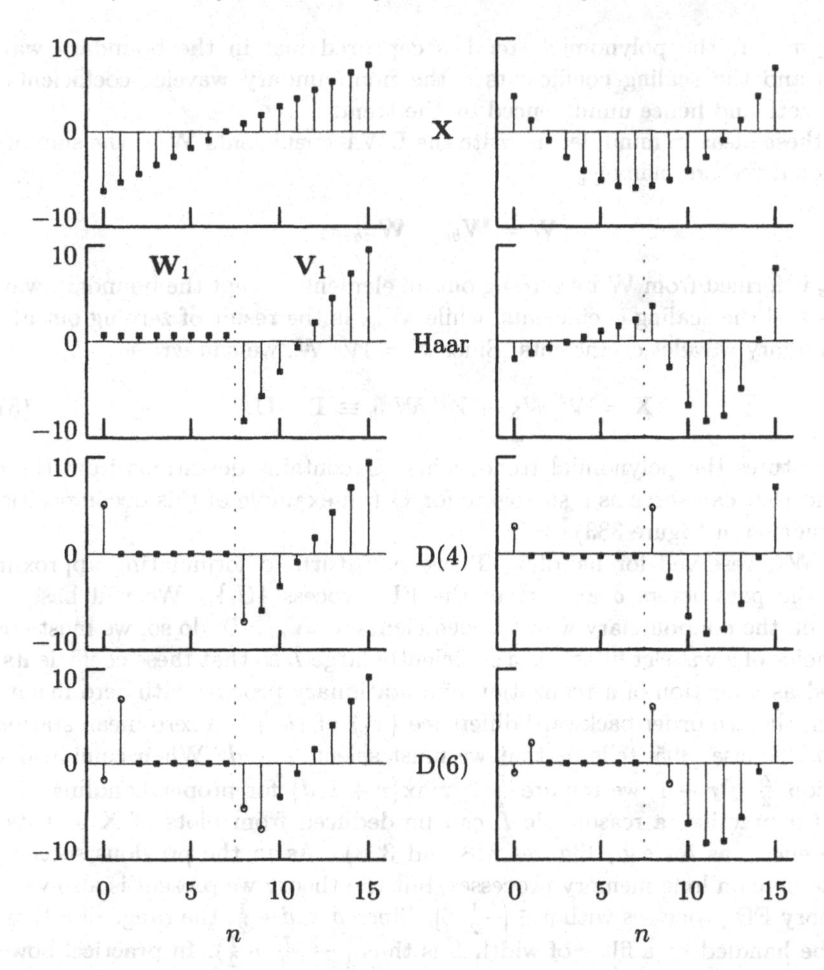

Figure 369. Linear and quadratic trends **T** (top plots), below which are shown their partial DWT coefficients $\mathbf{W} = [\mathbf{W}_1^T, \mathbf{V}_1^T]^T$ based on the Haar, D(4) and D(6) wavelet filters (second to fourth rows, respectively). The vertical dotted lines delineate the subvectors $\mathbf{W}_1$ and $\mathbf{V}_1$. Boundary wavelet and scaling coefficients are indicated by circles (there are none for the Haar; one each in $\mathbf{W}_1$ and $\mathbf{V}_1$ for the D(4); and two in each of the subvectors for the D(6)). Because the Haar wavelet does not reduce either linear or quadratic polynomials to zero, its wavelet coefficients are all nonzero; on the other hand, the D(6) wavelet reduces both polynomials to zero, so its six nonboundary wavelet coefficients are zero in both $\mathbf{W}_1$. The D(4) wavelet can handle a linear polynomial, but not a quadratic, which is why its seven nonboundary wavelet coefficients are zero for the former and nonzero for the latter. (For the record, the linear and quadratic trends are defined by $T_t = 0.9 \cdot (t - 7)$ and $T_t = 0.2 \cdot (t - 7)^2 - 6$. This illustration is due to W. Constantine, MathSoft, Seattle.)

if $\frac{L}{2} \geq r + 1$. The implications of this property are explored in Figure 369, which shows what the unit level DWT coefficients $\mathbf{W}_1$ and $\mathbf{V}_1$ look like when computed for linear ($r = 1$) and quadratic ($r = 2$) trends using the Haar, D(4) and D(6) filters. The highest order polynomial that can be reduced to zero by a wavelet filter is $r = \frac{L}{2} - 1$, which is 0, 1 and 2, respectively, for filters of widths 2, 4 and 6. If we recall that the first $L_1' = \frac{L}{2} - 1$ coefficients in $\mathbf{W}_1$ (and $\mathbf{V}_1$) are influenced by the periodic boundary conditions of the DWT (see Equation (146b)), this figure illustrates that, when $\frac{L}{2} \geq r + 1$, the polynomial trend is captured just in the boundary wavelet coefficients and the scaling coefficients – the nonboundary wavelet coefficients are necessarily zero and hence uninfluenced by the trend.

With these ideas in mind, let us write the DWT coefficients $\mathbf{W}$ as the sum of two N dimensional vectors, namely,

$$\mathbf{W} = \mathbf{W}_{bs} + \mathbf{W}_{nb},$$

where $\mathbf{W}_{bs}$ is formed from $\mathbf{W}$ by zeroing out all elements except the boundary wavelet coefficients and the scaling coefficients, while $\mathbf{W}_{nb}$ is the result of zeroing out all but the nonboundary wavelet coefficients. Since $\mathbf{X} = \mathcal{W}^T\mathbf{W}$, we can write

$$\mathbf{X} = \mathcal{W}^T\mathbf{W}_{bs} + \mathcal{W}^T\mathbf{W}_{nb} \equiv \widehat{\mathbf{T}} + \widehat{\mathbf{U}}, \tag{370a}$$

where $\widehat{\mathbf{T}}$ captures the polynomial trend, while $\widehat{\mathbf{U}}$ contains deviations from the estimated trend that can serve as a surrogate for $\mathbf{U}$ (an example of this decomposition is shown further on in Figure 383).

With $\mathbf{W}_{bs}$ reserved for handling $\mathbf{T}$, we now turn to formulating approximate MLEs for the parameters δ and σ_ε^2 of the FD process $\{U_t\}$. We will base these estimators on the nonboundary wavelet coefficients in $\mathbf{W}_{nb}$. To do so, we must ensure that our choice of a wavelet filter has a sufficiently large L so that these coefficients can be regarded as a portion of a realization of a stationary process with zero mean. By assumption, the dth order backward difference $\{Y_t\}$ of $\{U_t\}$ is a zero mean stationary process, so Exercise [305] tells us that we must select $\frac{L}{2} \geq d$. When combined with the condition $\frac{L}{2} \geq r + 1$, we require $\frac{L}{2} \geq \max\{r + 1, d\}$ for proper handling of both $\mathbf{T}$ and $\mathbf{U}$ (in practice, a reasonable L can be deduced from plots of $\mathbf{X}$ and its low order differences, as in, e.g., Figures 318 and 328). As in the previous section, we concentrate here on long memory processes, but the theory we present is also valid for short memory FD processes with $\delta \in [-\frac{1}{2}, 0]$. Since $\delta < d + \frac{1}{2}$, the range of δ that can in theory be handled by a filter of width L is thus $[-\frac{1}{2}, \frac{L}{2} + \frac{1}{2})$. In practice, however, there can be substantial biases in estimating δ at the extremes of this interval: the Haar and – to a lesser extent – the D(4) filters are problematic when $\delta \in [-\frac{1}{2}, -\frac{1}{4}]$, and a filter of width L can have difficulties at the upper range of the interval, say, $\delta \in [\frac{L}{2} + \frac{1}{4}, \frac{L}{2} + \frac{1}{2})$. Hence our preferred interval for δ is

$$\delta \in [-\tfrac{1}{4}, \tfrac{L}{2} + \tfrac{1}{4}]. \tag{370b}$$

If the estimate of δ that we get from the procedure described below is greater than, say, $\frac{L}{2}$, it is a good idea to compute an estimate using a wavelet filter of width $L+2$ for comparison (similarly, if we get an estimate below zero using a Haar or D(4) wavelet, we should compare it to a D(6) estimate).

Under the assumption that $\frac{L}{2} \geq \max\{r+1, d\}$, the jth level nonboundary wavelet coefficients are given by

$$W_{j,t}, \quad t = L_j', \ldots, N_j - 1, \text{ where } L_j' \equiv \left\lceil (L-2)\left(1 - \frac{1}{2^j}\right) \right\rceil, \tag{370c}$$

and we require that $N_j' \equiv N_j - L_j' > 0$ (this expression for L_j' is from Equation (146a) and is evaluated in Table 136 for filter widths $L = 2, 4, \ldots, 20$). The above coefficients are a portion of a zero mean Gaussian stationary process with SDF given by

$\mathcal{H}_j(f)S_X(f)$ and dependent on the unknown parameters δ and σ_ε^2. Let J_0 be a level such that $N'_{J_0} > 0$ (in practice, we usually set J_0 to be as large as possible; i.e., we set it such that $N'_{J_0} > 0$ but $N'_{J_0+1} \le 0$). Define $\mathbf{W}'_{nb}$ to be a vector containing all of the nonboundary wavelet coefficients (recall that some of the elements of $\mathbf{W}_{nb}$ are forced to be zero – removal of these zeros yields $\mathbf{W}'_{nb}$). The length of $\mathbf{W}'_{nb}$ is $N' \equiv \sum_{j=1}^{J_0} N'_j$. Given $\mathbf{W}'_{nb}$, we write the exact likelihood function for δ and σ_ε^2 as

$$L(\delta, \sigma_\varepsilon^2 \mid \mathbf{W}'_{nb}) \equiv \frac{1}{(2\pi)^{N'/2}|\Sigma_{\mathbf{W}'_{nb}}|^{1/2}} e^{-[\mathbf{W}'_{nb}]^T \Sigma^{-1}_{\mathbf{W}'_{nb}} \mathbf{W}'_{nb}/2}.$$

We now appeal to the decorrelating property of the DWT to obtain a convenient approximation to the above by treating the RVs in $\mathbf{W}_{nb}$ as if they were uncorrelated. Because of the Gaussian assumption, we can now approximate the likelihood function as the product of univariate Gaussian densities, which differ only in their associated variances. Equation (343a) gives an exact expression for var $\{W_{j,t}\}$, but we can approximate this by either C_j of Equation (343c) or $\widetilde{C}_j$ of Equations (344) and (354) (the former is valid for $j \ge 3$, while the latter is needed for $j = 1$ or 2, but can be used for all j if so desired). For notational convenience, we adopt the C_j approximation, but it is an easy matter to replace C_j with either var $\{W_{j,t}\}$ or $\widetilde{C}_j$ in what follows. We thus obtain

$$L(\delta, \sigma_\varepsilon^2 \mid \mathbf{W}'_{nb}) \approx \tilde{L}(\delta, \sigma_\varepsilon^2 \mid \mathbf{W}'_{nb}) \equiv \prod_{j=1}^{J_0} \prod_{t=0}^{N'_j-1} \frac{1}{(2\pi C_j)^{1/2}} e^{-W^2_{j,t+L'_j}/(2C_j)}. \tag{371a}$$

▷ **Exercise [371]:** Using an approach similar to what led to up Equations (363b) and (363c), show that the approximate MLE of σ_ε^2 for a given δ is

$$\tilde{\sigma}_\varepsilon^2(\delta) = \frac{1}{N'} \sum_{j=1}^{J_0} \frac{1}{C'_j(\delta)} \sum_{t=0}^{N'_j-1} W^2_{j,t+L'_j}. \tag{371b}$$

Show also that the reduced log likelihood function is

$$\tilde{l}(\delta \mid \mathbf{W}'_{nb}) \equiv \tilde{l}(\delta, \tilde{\sigma}_\varepsilon^2 \mid \mathbf{W}'_{nb}) - N' = N' \log(\tilde{\sigma}_\varepsilon^2(\delta)) + \sum_{j=1}^{J_0} N'_j \log(C'_j(\delta)), \tag{371c}$$

which, upon substituting in the expression for $\tilde{\sigma}_\varepsilon^2(\delta)$, is seen to be a function of δ alone. ◁

The scheme to get the approximate MLEs for δ and σ_ε^2 is the same as before: we numerically determine the value, say $\tilde{\delta}^{(s/ns)}$, minimizing the above as a function δ, after which we evaluate $\tilde{\sigma}_\varepsilon^2(\delta)$ at $\delta = \tilde{\delta}^{(s/ns)}$ to obtain the MLE $\tilde{\sigma}_\varepsilon^2 \equiv \tilde{\sigma}_\varepsilon^2(\tilde{\delta}^{(s/ns)})$ for σ_ε^2. Craigmile *et al.* (2000a) show that, under the assumption that the wavelet coefficients in $\mathbf{W}'_{nb}$ are indeed uncorrelated, then, for large N and for $\delta \in [-\frac{1}{2}, \frac{L}{2}]$, the estimator $\tilde{\delta}^{(s/ns)}$ is approximately Gaussian distributed with mean δ and variance

$$\sigma_{\tilde{\delta}}^2 \equiv 2\left[\sum_{j=1}^{J_0} N'_j \gamma_j^2 - \frac{1}{N'}\left(\sum_{j=1}^{J_0} N'_j \gamma_j\right)^2\right]^{-1}, \tag{371d}$$

δ		MLE Haar	D(4)	LA(8)	exact
0.4	mean	0.3670	0.3762	0.3792	0.3900
	bias	−0.0330	−0.0238	−0.0208	−0.0100
	SD	0.0588	0.0732	0.0943	0.0437
	RMSE	0.0674	0.0769	0.0966	0.0448
	σ_δ	0.0530	0.0673	0.0869	
0.75	mean	0.7230	0.7277	0.7346	0.7677
	bias	−0.0270	−0.0223	−0.0154	0.0177
	SD	0.0783	0.0878	0.0863	0.0272
	RMSE	0.0829	0.0906	0.0877	0.0325
	σ_δ	0.0526	0.0665	0.0857	

Table 372. As in Table 364, but now using the likelihood function of Equation (371a) to define the wavelet-based approximate MLE $\tilde{\delta}^{(\mathrm{s/ns})}$ for δ (the $\delta = 0.4$ results for the exact MLE in the final column are replicated from Table 364). The nature of the term σ_δ is explained in the text.

where

$$\gamma_j \equiv \frac{\frac{d\,\mathrm{var}\,\{W_{j,t}\}}{d\delta}}{\mathrm{var}\,\{W_{j,t}\}} = -\frac{4\sigma_\varepsilon^2}{\mathrm{var}\,\{W_{j,t}\}} \int_0^{1/2} \mathcal{H}_j(f) \frac{\log\left(2\sin(\pi f)\right)}{[2\sin(\pi f)]^{2\delta}}\,df. \tag{372}$$

To study the statistical properties of $\tilde{\delta}^{(\mathrm{s/ns})}$, we carried out a simulation study similar to the one described in the previous section for a sample size of $N = 256$ (see Table 364), with the exception that we replaced $\delta = 0.25$ with $\delta = 0.75$ (as discussed in the Comments and Extensions to Section 7.8, we can simulate an FD(0.75) process by cumulatively summing a simulated FD(−0.25) process). The first three columns of Table 372 summarize the results of the study for the Haar, D(4) and LA(8) wavelets. As a benchmark, the last column gives corresponding statistics for the exact MLE $\hat{\delta}$ (the $\delta = 0.4$ case is copied from Table 364, while the $\delta = 0.75$ case is based on adding one to the MLEs of δ obtained after differencing the simulated FD(0.75) series). While the wavelet-based approximate MLEs have RMSEs that are two or three times the RMSEs for the exact MLEs, we must recall that the underlying model allows for the presence of a trend even though we set $\mathbf{T} = \mathbf{0}$ in the simulation study – if such a trend were actually present, we can expect the performance of the exact MLE to deteriorate considerably. In addition, use of the exact MLE with $\delta \geq \frac{1}{2}$ requires differencing of the original time series, whereas there is no need for an *a priori* decision about differencing when using $\tilde{\delta}^{(\mathrm{s/ns})}$.

Table 372 also shows σ_δ, i.e., the square root of the large sample variance for $\tilde{\delta}^{(\mathrm{s/ns})}$ in Equation (371d). We computed σ_δ using the band-pass approximation C_j of Equation (343c) for $\mathrm{var}\,\{W_{j,t}\}$ and a similar approximation for the integral in Equation (372). While σ_δ underestimates the observed SDs and RMSEs (particularly for the Haar and D(4) wavelets when $\delta = 0.75$), we can use it in practice – with some caution – as a measure of the variability in $\tilde{\delta}^{(\mathrm{s/ns})}$ (additional computer simulations show σ_δ and the observed RMSEs converging with increasing N).

Suppose now that we have obtained an estimate $\tilde{\delta}^{(\mathrm{s/ns})}$ for our time series and that we want to assess the hypothesis that there is a significant trend component in the time series versus the null hypothesis that there is no trend (i.e., $\mathbf{T} = \mathbf{0}$). Let us consider $P \equiv \|\mathbf{X}\|^2/\|\mathbf{X} - \widehat{\mathbf{T}}\|^2$ as a test statistic. This statistic compares the energy in the entire time series to the energy in the detrended series, or, equivalently, in the nonboundary wavelet coefficients, as the following exercise indicates.

▷ **Exercise [373]:** Show that

$$P = \frac{\|\mathbf{X}\|^2}{\|\mathbf{W}_{nb}\|^2}. \quad \triangleleft$$

Note that, because P is a ratio of energies, it does not depend on the value of σ_ε^2. Under the null hypothesis and conditional on $\tilde{\delta}^{(\mathrm{s/ns})}$, we can determine the distribution of P empirically by repetitively simulating from a zero mean FD($\tilde{\delta}^{(\mathrm{s/ns})}$) process with, say, $\sigma_\varepsilon^2 = 1$ and computing P for each simulation. Let P_α be the upper $100\alpha\%$ percentage point of the simulated P values, where typically we set $\alpha = 0.05$ or smaller. When $\mathbf{T} \neq \mathbf{0}$, the statistic P should be large, so we reject the null hypothesis at an α level of significance if the observed P exceeds P_α.

We note two refinements to this test. First, although the intuitive notion of trend is one of smooth nonstochastic variations over large scales, the test statistic P will also be large when the time series has a prominent nonzero sample mean. To prevent rejection of the null hypothesis due merely to a nonzero mean, we adjust the test statistic to be

$$\bar{P} \equiv \frac{\sum_{t=0}^{N-1}(X_t - \overline{X})^2}{\|\mathbf{W}_{nb}\|^2}, \qquad (373)$$

which we then use both on the time series under study and on the simulated series that empirically determine the distribution of $\bar{P}$ under the null hypothesis (the denominators of P and $\bar{P}$ are the same because wavelet coefficients are impervious to the sample mean). Second, we can compensate for the effect of conditioning on the observed estimate $\tilde{\delta}^{(\mathrm{s/ns})}$ by simulating from its large sample distribution. Thus, prior to generating each simulated series to determine the distribution of P or $\bar{P}$, we generate a deviate from a Gaussian distribution with mean $\tilde{\delta}^{(\mathrm{s/ns})}$ and variance dictated by Equation (371d) with $\delta = \tilde{\delta}^{(\mathrm{s/ns})}$. We then use this deviate as the δ parameter when simulating from the FD process.

An example of the use of the test statistic $\bar{P}$ on the atomic clock data is given in Section 9.7.

Comments and Extensions to Section 9.4

[1] The likelihood function of Equation (371a) can be readily adapted to handle time series that are well modeled at large scales by an FD process, . but not at small scales: if the model is deemed to hold over scales $j = J_1, \ldots, J_0$ for some $J_1 > 1$, we merely replace the lower limit of the product in (371a) with $j = J_1$. The test for trend can likewise be easily adjusted by (i) replacing $\mathbf{X}$ with the scaling coefficients $\mathbf{V}_{J_1-1}$ and (ii) eliminating scales of levels $j = 1, \ldots, J_1 - 1$ from $\mathbf{W}_{nb}$.

Additionally, if trend is not a concern so that we are only interested in estimating δ and σ_ε^2, it is easy to adjust the likelihood function of Equation (371a) so that it handles sample sizes N that are not powers of two. To do so, let M be the smallest power of two such that $M > N$, and construct $\{X'_t\}$ as per Equation (308c) by appending $M - N$ zeros to the end of $\{X_t\}$. Let $\mathbf{W}'_j$ be the jth level wavelet coefficients of $\{X'_t\}$. The idea is to form the likelihood function using just those elements of $\mathbf{W}'_j$ whose construction did not involve any of the $M - N$ zeros at the end of $\mathbf{X}'$ – these are precisely the M'_j

coefficients displayed in Equation (308d), where $M'_j \equiv \lfloor \frac{N}{2^j} - 1 \rfloor - L'_j + 1$. If we let J_0 be the largest integer such that $M'_{J_0} > 0$, then the likelihood function of Equation (371a) now takes the form

$$\prod_{j=1}^{J_0} \prod_{t=0}^{M'_j - 1} \frac{1}{(2\pi C_j)^{1/2}} e^{-[W'_{j,t+L'_j}]^2/(2C_j)},$$

where $W'_{j,t}$ is the tth element of $\mathbf{W}'_j$. (A refinement to this procedure is to use the notion of 'cycle spinning' discussed in Coifman and Donoho, 1995, in which we circularly shift $\{X'_t\}$, form a likelihood function that uses only those wavelet coefficients that do not involve the extra $M - N$ zeros, determine the MLE of δ from the likelihood equation, and then average the MLEs obtained over all possible circular shifts.)

[2] We can adapt the above scheme to obtain an 'instantaneous' estimator of δ, which can be useful for tracking a nonstationary long memory process whose long memory parameter is believed to evolve over time. The basic idea is to use just wavelet coefficients that are colocated in time. To avoid ambiguities in time due to the subsampling inherent in the DWT, we can pick off one wavelet coefficient per scale from a MODWT, multiply it by $2^{j/2}$, and then form a likelihood function consisting of these renormalized MODWT coefficients. To ensure proper colocation in time, it is imperative to use one of the LA or coiflet filters since – with proper shifts – these filters have approximately zero phase and hence have outputs that can be meaningfully aligned in time with respect to the original time series (see Sections 4.8 and 4.9 for details). The resulting reduced log likelihood function for an instantaneous estimator of δ at time t takes the form

$$J_0 \log \left(\frac{1}{J_0} \sum_{j=1}^{J_0} \frac{2^j}{C'_j(\delta)} \widetilde{W}^2_{j,t_j} \right) + \sum_{j=1}^{J_0} \log \left(C'_j(\delta) \right), \tag{374}$$

where t_j is the index of the jth level MODWT coefficient that is associated with time t (verification of the above is part of Exercise [9.6]).

9.5 Least Squares Estimation for FD Processes

We noted in the introduction to this chapter that, for a long memory process $\{X_t\}$, log/log plots of the wavelet variance $\nu_X^2(\tau_j)$ versus scale τ_j show an approximate linear variation as τ_j gets large; i.e., we have $\log(\nu_X^2(\tau_j)) \approx \zeta + \beta \log(\tau_j)$ for some ζ and β. The slope β is related to the exponent α of the approximating power law SDF at low frequencies via $\beta = -\alpha - 1$. Now $\alpha = -2\delta$ for an FD(δ) process (see the upper right-hand plot of Figure 286), so we should have $\beta = 2\delta - 1$ approximately. This simple relationship suggests that we can formulate an estimator of δ via an estimator of β by regressing log wavelet variance estimates versus $\log(\tau_j)$ over a range of scales (see, for example, Abry *et al.*, 1993 and 1995, Abry and Veitch, 1998, and Jensen, 1999a). In this section we discuss how to create a weighted least squares estimator (WLSE) of β – and hence δ – that takes into account the large sample properties of wavelet variance estimators (Abry *et al.*, 1995, and Abry and Veitch, 1998, adopt similar approaches).

Let us begin by relating the wavelet variance to the average SDF levels C_j of Equation (343b), from which we can get some idea as to how large τ_j must be for the linear approximation to be reasonably good. By definition the wavelet variance $\nu_X^2(\tau_j)$ is the variance of the MODWT wavelet coefficient $\widetilde{W}_{j,t}$. The relationship between

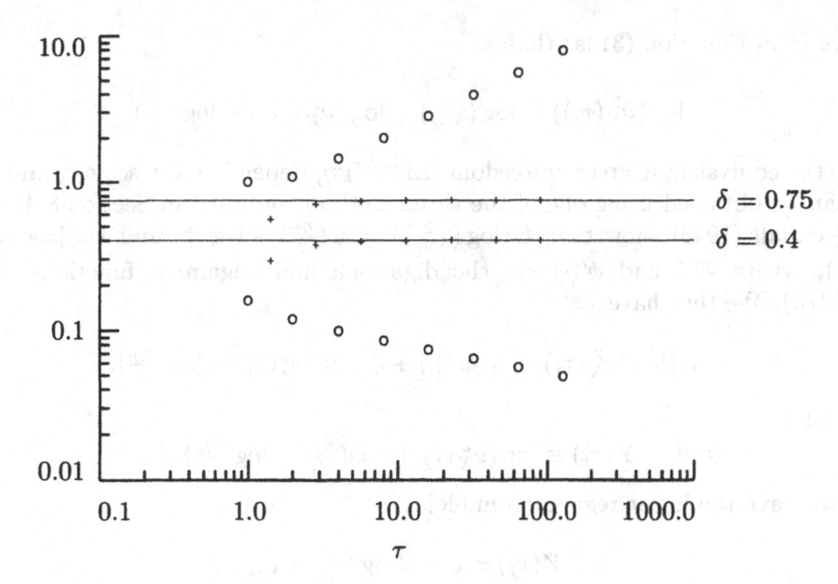

Figure 375. Approximations $C_j/2^j$ (computed from Equation (343b) via numerical integration) to wavelet variances $\nu_X^2(\tau_j)$ versus scales τ_j, $j = 1, \ldots, 8$, for FD processes with $\delta = 0.4$ (lower circles) and 0.75 (upper). See the text for an explanation of the portion of the plot between these sets of circles.

the DWT and MODWT coefficients, namely, $W_{j,t}/2^{j/2} = \widetilde{W}_{j,2^j(t+1)-1}$, tells us that $\nu_X^2(\tau_j) = \text{var}\,\{W_{j,t}\}/2^j$. Equation (343b) says that the average SDF level C_j is an approximation to var $\{W_{j,t}\}$. Thus we have $\nu_X^2(\tau_j) \approx C_j/2^j$, where we can determine C_j for FD processes by evaluating the integral in Equation (343c) numerically.

Figure 375 shows $C_j/2^j$ versus τ_j, $j = 1, \ldots, 8$, on log/log axes for FD processes with $\delta = 0.4$ (lower circles) and 0.75 (upper). For either δ, the circles roughly fall along a line, but the left-most ($j = 1$) circle deviates noticeably upwards. Let us determine how well the slope between adjacent circles maps onto δ with increasing τ_j by making use of the relationship $\delta = \frac{1}{2}(\beta + 1)$. Let δ_j be determined by the slope (in log/log space) between circles associated with indices j and $j + 1$:

$$\delta_j \equiv \frac{1}{2}\left(\frac{\log_{10}(C_{j+1}/2^{j+1}) - \log_{10}(C_j/2^j)}{\log_{10}(\tau_{j+1}) - \log_{10}(\tau_j)} + 1\right).$$

Figure 375 shows plots of δ_j versus $(\tau_{j+1} + \tau_j)/2$, $j = 1, \ldots, 7$, when $\delta = 0.4$ (lower pluses) and 0.75 (upper). We see that δ_j quickly converges to δ (indicated by a horizontal line) as j increases. These plots suggest that the linear approximation is reasonable when $j \geq 2$ or 3 (see Jensen, 1999a, for a formal proof that $\nu_X^2(\tau_j) \to C\tau_j^{2\delta-1}$ as $j \to \infty$ for FD processes with $|\delta| < 1/2$, where $C = e^{\zeta}$).

Accordingly, let us now take $\zeta + \beta \log(\tau_j)$ as an approximation for $\log(\nu_X^2(\tau_j))$ when $j \geq J_1$, where ζ and β are unknown parameters to be estimated, and typically we set J_1 to be two or three. Suppose that we use a wavelet filter of width L and that we regard our time series as a realization of a portion $X_0, \ldots, X_{N-1}$ of an FD process with $\delta \in [-\frac{1}{4}, \frac{L}{2} + \frac{1}{4}]$ (as per Equation (370b)). We can then use the unbiased MODWT estimator $\hat{\nu}_X^2(\tau_j)$ of Equation (306b) to estimate $\nu_X^2(\tau_j)$ over scales $J_1 \leq j \leq J_0$, where J_0 is as large as possible given the restrictions of sample size N and filter width L. It follows from Equation (313a) that

$$\log(\hat{\nu}_X^2(\tau_j)) \stackrel{d}{=} \log(\chi^2_{\eta_j}) + \log(\nu_X^2(\tau_j)) - \log(\eta_j),$$

where the equivalent degrees of freedom (EDOF) η_j depend on the scale τ_j and in practice can be obtained using one of the three methods outlined in Section 8.4. Bartlett and Kendall (1946) show that $E\{\log(\chi^2_{\eta_j})\} = \psi(\frac{\eta_j}{2}) + \log(2)$ and $\operatorname{var}\{\log(\chi^2_{\eta_j})\} = \psi'(\frac{\eta_j}{2})$, where $\psi(\cdot)$ and $\psi'(\cdot)$ are the digamma and trigamma functions of Equation (275). We thus have

$$E\{\log(\hat{\nu}_X^2(\tau_j))\} = \psi(\tfrac{\eta_j}{2}) + \zeta + \beta\log(\tau_j) - \log(\tfrac{\eta_j}{2}).$$

If we let

$$Y(\tau_j) \equiv \log(\hat{\nu}_X^2(\tau_j)) - \psi(\tfrac{\eta_j}{2}) + \log(\tfrac{\eta_j}{2}), \tag{376a}$$

then we have the linear regression model

$$Y(\tau_j) = \zeta + \beta\log(\tau_j) + e_j,$$

for which the error term

$$e_j \equiv \log\left(\frac{\hat{\nu}_X^2(\tau_j)}{\nu_X^2(\tau_j)}\right) - \psi(\tfrac{\eta_j}{2}) + \log(\tfrac{\eta_j}{2})$$

is equal in distribution to the RV $\log(\chi^2_{\eta_j}) - \psi(\frac{\eta_j}{2}) - \log(2)$. Note that e_j has zero mean and variance $\psi'(\frac{\eta_j}{2})$ and is approximately Gaussian distributed as long as η_j is around ten or greater (see Figure 276). We further assume that the e_j are pairwise uncorrelated, which is reasonable in view of our discussion in Section 9.1. In vector notation, we can write

$$\mathbf{Y} = A\mathbf{b} + \mathbf{e}, \tag{376b}$$

where $\mathbf{Y} \equiv [Y(\tau_{J_1}), \ldots, Y(\tau_{J_0})]^T$; A is a $(J_0 - J_1 + 1) \times 2$ matrix, the first column of which has all ones, while the second has $\log(\tau_{J_1}), \ldots, \log(\tau_{J_0})$; $\mathbf{b} \equiv [\zeta, \beta]^T$; and $\mathbf{e} \equiv [e_{J_1}, \ldots, e_{J_0}]^T$ is a random vector with mean $\mathbf{0}$ and covariance matrix $\Sigma_{\mathbf{e}}$ that is diagonal with diagonal elements $\psi'(\frac{\eta_{J_1}}{2}), \ldots, \psi'(\frac{\eta_{J_0}}{2})$.

We can now appeal to the theory of linear least squares to obtain the WLSE of $\mathbf{b}$, namely,

$$\hat{\mathbf{b}}^{(\mathrm{wls})} \equiv (A^T \Sigma_{\mathbf{e}}^{-1} A)^{-1} A^T \Sigma_{\mathbf{e}}^{-1} \mathbf{Y}$$

(see, for example, Draper and Smith, 1998). The second element of $\hat{\mathbf{b}}^{(\mathrm{wls})}$ is the desired WLSE of β, namely,

$$\hat{\beta}^{(\mathrm{wls})} = \frac{\sum w_j \sum w_j \log(\tau_j) Y(\tau_j) - \sum w_j \log(\tau_j) \sum w_j Y(\tau_j)}{\sum w_j \sum w_j \log^2(\tau_j) - \left(\sum w_j \log(\tau_j)\right)^2}, \tag{376c}$$

where $w_j \equiv 1/\psi'(\frac{\eta_j}{2})$, and all summations are over $j = J_1, \ldots, J_0$ (proof of the above is the first part of Exercise [9.5]). Under the assumptions behind our model, the estimator $\hat{\mathbf{b}}^{(\mathrm{wls})}$ has mean value $\mathbf{b}$ and covariance

$$\Sigma_{\hat{\mathbf{b}}^{(\mathrm{wls})}} = (A^T \Sigma_{\mathbf{e}}^{-1} A)^{-1}.$$

δ		WLSE			MLE
		Haar	D(4)	LA(8)	exact
0.4	mean	0.3925	0.4006	0.4044	0.3900
	bias	−0.0075	0.0006	0.0044	−0.0100
	SD	0.0715	0.0886	0.1185	0.0437
	RMSE	0.0719	0.0886	0.1186	0.0448
0.75	mean	0.7398	0.7443	0.7435	0.7677
	bias	−0.0102	−0.0057	−0.0065	0.0177
	SD	0.0779	0.0877	0.1196	0.0272
	RMSE	0.0786	0.0879	0.1198	0.0325
	$\sqrt{\operatorname{var}\{\hat{\beta}^{(\mathrm{wls})}\}}$	0.0891	0.1145	0.1552	

Table 377. As in Table 372, but now using the wavelet-based WLSE $\hat{\delta}^{(\mathrm{wls})}$ (the final column is replicated from Table 372). The WLSEs are based on the unbiased MODWT estimators $\hat{\nu}_X^2(\tau_j)$ of the wavelet variance, which are presumed to have EDOFs $\eta_j = \max\{M_j/2^j, 1\}$. We set $J_1 = 2$ for all three wavelets. The results reported in the table are based on 1024 simulated time series of sample size $N = 256$.

In particular, the lower right-hand element of this covariance matrix tells us that

$$\operatorname{var}\{\hat{\beta}^{(\mathrm{wls})}\} = \frac{\sum w_j}{\sum w_j \sum w_j \log^2(\tau_j) - \left(\sum w_j \log(\tau_j)\right)^2} \tag{377}$$

(this is the second part of Exercise [9.5]). Since $\delta = \frac{1}{2}(\beta + 1)$, we can let $\hat{\delta}^{(\mathrm{wls})} \equiv \frac{1}{2}(\hat{\beta}^{(\mathrm{wls})} + 1)$ be the corresponding estimator of the FD parameter. We note that $\operatorname{var}\{\hat{\delta}^{(\mathrm{wls})}\} = \frac{1}{4}\operatorname{var}\{\hat{\beta}^{(\mathrm{wls})}\}$.

To get an idea as to how well the WLSE performs, we carried out a simulation study similar to the one in the previous section (see Table 372), in which $N = 256$ and $\delta = 0.4$ and 0.75. For simplicity, we determined the unknown EDOFs in keeping with Equation (314c); i.e., we set $\eta_j = \max\{M_j/2^j, 1\}$, where, as usual, $M_j \equiv N - L_j + 1$ is the number of MODWT coefficients used to construct the estimator $\hat{\nu}_X^2(\tau_j)$. We looked at three choices for J_1 namely, 1, 2 and 3, and found that use of $J_1 = 2$ gave the best RMSEs for both sample sizes and all three wavelets (this is reasonable in view of Figure 375). The sample size and filter widths dictate that $J_0 = 8$, 6 and 5 for, respectively, the Haar, D(4) and LA(8) wavelets. Table 377 summarizes the results of the simulation study. We see that the RMSEs for the Haar wavelet are smaller than for the other two wavelets, undoubtedly because the shorter width of the Haar wavelet filter give us more scales to use. Note that biases are quite small in most cases, which means that the RMSEs are about the same as the reported SDs. We also see that the RMSEs for a given wavelet are quite similar for both $\delta = 0.4$ and 0.75 (this result is consistent with the expression for $\operatorname{var}\{\hat{\beta}^{(\mathrm{wls})}\}$ in Equation (377), which is seen to have no dependence on δ with the scheme for determining the EDOFs adopted here). For the sake of comparison, the bottom row of Table 377 shows the square root of the theoretical variances $\operatorname{var}\{\hat{\beta}^{(\mathrm{wls})}\}$ for the three wavelets. We see that the actual RMSEs are somewhat smaller than what theory would suggest (possible explanations are a mismatch between the actual and assumed EDOFs or small nonzero correlations in the error terms e_j). From a practical point of view, $\operatorname{var}\{\hat{\beta}^{(\mathrm{wls})}\}$ seems to provide

		WLSE			MLE
δ		Haar	D(4)	LA(8)	exact
0.4	mean	0.3881	0.3994	0.4000	0.3900
	bias	−0.0119	−0.0006	−0.0000	−0.0100
	SD	0.0669	0.0700	0.0728	0.0437
	RMSE	0.0680	0.0700	0.0728	0.0448
0.75	mean	0.7278	0.7432	0.7442	0.7677
	bias	−0.0222	−0.0068	−0.0058	0.0177
	SD	0.0739	0.0718	0.0726	0.0272
	RMSE	0.0772	0.0721	0.0728	0.0325

Table 378. As in Table 377, but now using an WLSE $\tilde{\delta}^{(\mathrm{wls})}$ based upon a biased MODWT estimator of the wavelet variance along with reflection boundary conditions.

a conservative upper bound on the variability in $\hat{\beta}^{(\mathrm{wls})}$. Finally, if we compare the RMSEs reported in Table 377 with those for the MLEs in Table 372, we see that the approximate MLEs outperform the WLSEs for $\delta = 0.4$, but, except for the LA(8) wavelet, the opposite is true for $\delta = 0.75$.

Comments and Extensions to Section 9.5

[1] Jensen (1999a) considers estimation of δ using an ordinary least squares estimator (OLSE). In terms of the regression model of Equation (376b) (which differs somewhat from the model used by Jensen), this estimator is given by

$$\hat{\mathbf{b}}^{(\mathrm{ols})} \equiv (A^T A)^{-1} A^T \mathbf{Y}$$

and has mean $\mathbf{b}$ and covariance $\Sigma_{\hat{\mathbf{b}}^{(\mathrm{ols})}} = (A^T A)^{-1}(A^T \Sigma_{\mathbf{e}} A)(A^T A)^{-1}$, from which we can extract an expression for $\mathrm{var}\,\{\hat{\beta}^{(\mathrm{ols})}\}$ (for details, see, for example, Draper and Smith, 1998). While the OLSE and WLSE are both unbiased estimators of $\mathbf{b}$, the WLSE is theoretically superior to the OLSE in the sense that we should have $\mathrm{var}\,\{\hat{\beta}^{(\mathrm{wls})}\} < \mathrm{var}\,\{\hat{\beta}^{(\mathrm{ols})}\}$ if all our modeling assumptions are correct. If we repeat the simulation study described in this section using OLSEs instead of WLSEs, we find that, for all combinations of δ and wavelets in Table 377, the RMSEs for the OLSEs are about twice as large as those reported for the WLSEs. We can thus recommend $\hat{\beta}^{(\mathrm{wls})}$ as the LSE of choice.

[2] We noted in item [3] of the Comments and Extensions to Section 8.3 that Greenhall *et al.* (1999) consider a biased MODWT estimator of the wavelet variance defined by

$$\tilde{\nu}_X^2(\tau_j) = \frac{1}{2N} \sum_{t=0}^{2N-1} \widetilde{W}_{j,t}^2$$

(cf. Equation (306c)), where the $\widetilde{W}_{j,t}^2$ variables are now taken to be the MODWT coefficients for

$$X_0, X_1, \ldots, X_{N-2}, X_{N-1}, X_{N-1}, X_{N-2}, \ldots, X_1, X_0$$

(see the discussion concerning Equation (140)). Their results suggest that, at large scales, this biased estimator has statistical properties superior to those of the unbiased estimator $\hat{\nu}_X^2(\tau_j)$. To see if the biased estimator can lead to an improved WLSE of δ, let us construct an estimator, say $\tilde{\delta}^{(\mathrm{wls})}$, that is defined as in the right-hand side of Equation (376c), but now with $Y(\tau_j)$ formed using $\tilde{\nu}_X^2(\tau_j)$ instead of $\hat{\nu}_X^2(\tau_j)$. An immediate advantage of the biased estimator is that we can let $J_0 = \log_2(N)$ for any wavelet. Pending further study of the distribution of $\tilde{\delta}^{(\mathrm{wls})}$, let us assume weights w_j based on $\eta_j = N/2^j$. With this setup, we redid the simulation study and obtained the results in Table 378, which indicates an improvement in RMSEs for all combinations of δ and wavelets over what is reported in Table 377. The D(4) and LA(8) wavelets now perform much better and seem to offer some improvement over the Haar for $\delta = 0.75$. More study is called for, however, to ascertain the relative merits of $\hat{\delta}^{(\mathrm{wls})}$ and $\tilde{\delta}^{(\mathrm{wls})}$ (in particular, whereas the former works well in the presence of certain types of trends, the latter might not).

[3] As is true for the MLE discussed in the previous section (see item [2] of the Comments and Extensions there), we can adapt the estimator described in this section to obtain an 'instantaneous' estimator of δ. We do so by forming wavelet variance estimators using a single squared MODWT coefficient from each level j, with the coefficients being chosen so that they are colocated in time. The resulting estimator of δ at time t is given by

$$\frac{1}{2}\left(\frac{(J_0 - J_1 + 1)\sum \log(\tau_j) Y_t(\tau_j) - \sum \log(\tau_j) \sum Y_t(\tau_j)}{(J_0 - J_1 + 1)\sum \log^2(\tau_j) - \left(\sum \log(\tau_j)\right)^2} + 1\right), \tag{379}$$

where $Y_t(\tau_j) \equiv \log(\widetilde{W}_{j,t_j}^2) - \psi(\frac{1}{2}) - \log(2)$; t_j is the same as in Equation (374); and, as before, all summations are from $j = J_1$ up to J_0 (verification of the above is part of Exercise [9.6]). The relative merits of the LS and ML estimators are currently unknown.

[4] Geweke and Porter-Hudak (1983), Robinson (1995) and Hurvich *et al.* (1998) discuss the statistical properties of LSEs of the long memory parameter based upon regressing the log periodogram versus log frequency. McCoy *et al.* (1998) consider a WLSE based upon a log multitaper SDF estimator (this was found to be superior to log periodogram-based estimators). While a definitive comparison between Fourier- and wavelet-based LSEs has yet to be undertaken, Jensen (1999a) finds the periodogram-based estimator to be inferior to the OLSEs he formulated, while McCoy and Walden (1996) find that the wavelet-based $\tilde{\delta}^{(\mathrm{s})}$ outperforms both periodogram- and multitaper-based estimators of δ.

9.6 Testing for Homogeneity of Variance

In this section we consider the problem of testing for homogeneity of variance in a time series that exhibits long memory structure. In Section 7.6 we reviewed stochastic models with long memory properties, all of which make the assumption of stationarity. One implication of this assumption is that the variance of the process is independent of time. Because realizations of long memory processes can appear to have bursts of increased variability or periods of relative quiescence, it is important to have a statistical test to help assess the reasonableness of the homogeneity assumption when we come across a time series whose appearance makes this assumption questionable. An obvious test would be to partition the time series into subseries, compute the sample variance for each subseries and then determine whether or not the observed differences among these sample variances are in keeping with what we can reasonably expect to observe from a stationary long memory process; however, as we have noted in Section 8.1, the sampling properties of the sample variance are problematic for long memory processes, and we would also need to study carefully how the scheme for choosing the subseries influences the results of the test. As we demonstrate in this section, we can use the fact that the DWT decorrelates long memory processes in terms of time localized coefficients to produce an attractive test for homogeneity of variance (for details, see Whitcher, 1998, and Whitcher *et al.*, 2000a).

Let $\mathbf{X} \equiv [X_0, \ldots, X_{N-1}]^T$ be a time series that we propose to model using a stationary Gaussian FD process, but for which the hypothesis of homogeneity of variance is in question. For convenience, we assume that N is an integer multiple of 2^{J_0} for some J_0 (this restriction can easily be relaxed – see item [1] in the Comments and Extensions below). The key idea is to postulate that the inhomogeneity of variance in $\mathbf{X}$ is such that it also manifests itself at certain selected scales. Accordingly, let $\mathbf{W}_j$ be the jth level wavelet coefficients of the DWT of $\mathbf{X}$. We assume that the level j is such that, as displayed in Equation (370c), there are $N_j' > 0$ nonboundary wavelet coefficients $W_{j,L_j'}, \ldots, W_{j,N_j-1}$, where, as before, $N_j' \equiv N_j - L_j'$. We will use these coefficients to construct a test statistic for the null hypothesis

$$H_0 : \operatorname{var}\{W_{j,L_j'}\} = \operatorname{var}\{W_{j,L_j'+1}\} = \cdots = \operatorname{var}\{W_{j,N_j-1}\}.$$

We assume that the decorrelating property of the DWT is effective here so that, under H_0, these nonboundary coefficients can be regarded as a sample from a Gaussian white noise process with mean zero (simulation studies in Whitcher, 1998, and Whitcher *et al.*, 2000a, demonstrate that the decorrelation is quite good in terms of the test statistic presented here). Brown *et al.* (1975), Hsu (1977), Inclán and Tiao (1994) and others investigate a test statistic D that can discriminate between H_0 and a variety of alternative hypotheses, including

$$\operatorname{var}\{W_{j,L_j'}\} = \cdots = \operatorname{var}\{W_{j,t'}\} \neq \operatorname{var}\{W_{j,t'+1}\} = \cdots = \operatorname{var}\{W_{j,N_j-1}\}, \qquad (380a)$$

where t' is an unknown change point. This test statistic is based on normalized cumulative sums of squares and can be defined in our context as follows. Let

$$\mathcal{P}_k \equiv \frac{\sum_{t=L_j'}^{k} W_{j,t}^2}{\sum_{t=L_j'}^{N_j-1} W_{j,t}^2}, \quad k = L_j', \ldots, N_j - 2, \qquad (380b)$$

and define $D \equiv \max\{D^+, D^-\}$, where

$$D^+ \equiv \max_k \left(\frac{k - L_j' + 1}{N_j' - 1} - \mathcal{P}_k \right) \text{ and } D^- \equiv \max_k \left(\mathcal{P}_k - \frac{k - L_j'}{N_j' - 1} \right).$$

Note that, under H_0, the ratio of the expected value of the numerator of $\mathcal{P}_k$ to the expected value of its denominator is just $(k - L_j' + 1)/N_j'$, which varies linearly with k. In essence the test statistic is measuring the magnitude of the largest deviation (positive or negative) from the expected linear increase under H_0. When H_0 is false so that there are departures from the expected linear increase, we can expect D to be large for certain alternative hypotheses, including the one displayed in Equation (380a). We thus will reject H_0 when D is 'too large.'

In order to quantify when D is 'too large' and hence to evaluate the evidence that our test statistic provides for or against the null hypothesis for a given time series, we need to know the distribution of D under H_0. Let $d(p)$ be the $p \times 100\%$ percentage point for this distribution; i.e., $\mathbf{P}[D \leq d(p)] = p$. There is no known tractable analytic expression for determining $d(p)$ for arbitrary N'_j, but fortunately there are two good approaches for approximating it. First, when N'_j is about 128 or larger, we can use a large sample result stated in Inclán and Tiao (1994), namely, that

$$\mathbf{P}[D > d(p)] = 1 - p \approx 2 \sum_{l=1}^{\infty} (-1)^{l+1} e^{-l^2 N'_j d^2(p)} \tag{381}$$

(the right-hand side is based on Equation (11.39) of Billingsley, 1968). To see how to use this expression in practice, suppose that we want our test to have a significance level of α; i.e., α is the proportion of times that we would (incorrectly) reject H_0 if we were to repeatedly apply this test under a sampling setup in which H_0 is in fact true (typically α is chosen to be a small value such as 0.05 or 0.01). We would then replace $d(p)$ in the right-hand side of (381) with the observed value of our test statistic and evaluate the resulting expression – if this expression is less than (greater than) α, then we reject (fail to reject) the null hypothesis at significance level α.

The second approach for approximating the distribution of D is to conduct a simulation study in which we obtain N'_j deviates from a Gaussian random number generator, compute D, and then repeat this process over and over again until we build up an empirical approximation to $\mathbf{P}[D \leq d(p)]$. In particular, to evaluate an observed D for a test of level α, suppose that we generate M Gaussian white noise series and that M' of these series yield a D exceeding the D for the actual wavelet coefficients. We would then reject the null hypothesis if $M'/M < \alpha$. Here we need to be careful that M is large enough so that the variability in the observed proportion M'/M is small compared to its distance from α. A rough rule of thumb is to check that

$$\left(\frac{M'}{M} - \alpha\right)^2 > \frac{4M'(M - M')}{M^3};$$

if this inequality is not satisfied, we need to increase M until it is (if increasing M beyond, say, 100 000 when $\alpha = 0.05$ or 0.01 still does not result in the inequality being satisfied, then the observed significance of the test is evidently quite close to α, so we cannot formally reject or fail to reject H_0).

We can thus use the test statistic D to give us a scale-by-scale evaluation of the hypothesis of homogeneity of variance (an example is given in Section 9.8). Suppose that, at some level j, we can reject H_0 using D and that it is reasonable to entertain the alternative hypothesis displayed in Equation (380a); i.e., that there is a single change in variability at an unknown change point. By considering the ratio of the expected value of the numerator of $\mathcal{P}_k$ to the expected value of its denominator under this alternative hypothesis, we can argue that the index k actually yielding D should be quite close to the change point t'. This fact suggests letting the maximizing index be our estimator of t'; however, because it is usually of more interest to identify the change point in the time series itself and because the subsampling inherent in the DWT introduces ambiguities in time, let us consider using the MODWT instead. Under this scheme, we form a modified version of the D statistic, with the nonboundary MODWT wavelet coefficients replacing their DWT analogs in Equation (380b). Thus, with

$$\widetilde{\mathcal{P}}_k \equiv \frac{\sum_{t=L_j-1}^{k} \widetilde{W}_{j,t}^2}{\sum_{t=L_j-1}^{N-1} \widetilde{W}_{j,t}^2}, \quad k = L_j - 1, \ldots, N-2,$$

and with

$$\Delta_k^+ \equiv \frac{k - L_j + 2}{N - L_j} - \widetilde{\mathcal{P}}_k \text{ and } \Delta_k^- \equiv \widetilde{\mathcal{P}}_k - \frac{k - L_j + 1}{N - L_j}, \tag{382}$$

define $\widetilde{D}^+ \equiv \max_k \{\Delta_k^+\}$, $\widetilde{D}^- \equiv \max_k \{\Delta_k^-\}$ and $\widetilde{D} \equiv \max\{\widetilde{D}^+, \widetilde{D}^-\}$. Our estimate of the change point is the index, say $\hat{k}$, that yields the auxiliary statistic $\widetilde{D}$; i.e., we have either $\widetilde{D} = \Delta_{\hat{k}}^+$ or $\widetilde{D} = \Delta_{\hat{k}}^-$. If we use a Daubechies LA or coiflet wavelet filter to form $\widetilde{D}$, we can then map $\hat{k}$ onto the physically meaningful time $t_0 + (\hat{k} + |\nu_j^{(H)}| \bmod N)\,\Delta t$, where t_0 is the actual time associated with X_0; $\nu_j^{(H)}$ is given in Equations (114c) and (124) for, respectively, the LA and coiflet filters; and Δt is the sampling time between observations. Simulation studies in Whitcher (1998) and Whitcher *et al.* (2000a) indicate that $\hat{k}$ is an effective estimator of the change point in the time series itself, with the distribution of $\hat{k}$ about the actual change point becoming tighter as the ratio of the variances before and after the change progresses away from unity. These studies also indicate that, if we reject the null hypothesis H_0 over several scales using the DWT-based statistic D, the smallest such scale should be used to estimate the change point via the corresponding $\widetilde{D}$ (an example of change point estimation is given in Section 9.8 for the Nile River series).

Comments and Extensions to Section 9.6

[1] Although our formulation of the DWT-based test statistic D in this section presumes that the sample size N is a multiple of a power of two, in fact we can use the same strategy for handling arbitrary sample sizes as described in the last part of item [1] in the Comments and Extensions to Section 9.4; i.e., we can add zeros to the end of the time series so that the padded series has a length that is easily handled by a standard DWT, and then we can form D using just the M_j' wavelet coefficients that do not involve any of these zeros (there is no need to make a similar adjustment when forming the MODWT-based auxiliary statistic $\widetilde{D}$ since the MODWT naturally handles any sample size).

[2] Let us note a connection between the auxiliary statistic $\widetilde{D}$ and the rotated cumulative variance plot defined by Equation (189) (an example of such a plot in shown in Figure 190 for the subtidal sea level variations). Since $\|\widetilde{\mathbf{W}}_j\|^2$ is the contribution to $\|\mathbf{X}\|^2$ due to changes at scale τ_j and since $\|\widetilde{\mathbf{W}}_j\|^2$ remains the same if we circularly shift $\widetilde{\mathbf{W}}_j$ to achieve proper time alignment, this plot is essentially proportional to the difference between the accumulated circularly shifted $\widetilde{W}_{j,t}^2$ and a uniform build up to $\|\widetilde{\mathbf{W}}_j\|^2$. The statistic $\widetilde{D}$ is closely related to a normalized version of the maximum of these absolute differences. In particular, if we plot Δ_k^- versus the physically meaningful time corresponding to k, we get a plot that closely resembles a rotated cumulative variance plot (see Figure 388 below for an example); however, note that Δ_k^- just involves the nonboundary MODWT coefficients, whereas a rotated cumulative variance plot uses all the MODWT coefficients (the latter is arguably reasonable for series such as the sea level variations, which – with a carefully chosen sample size – can be treated as if it were actually circular).

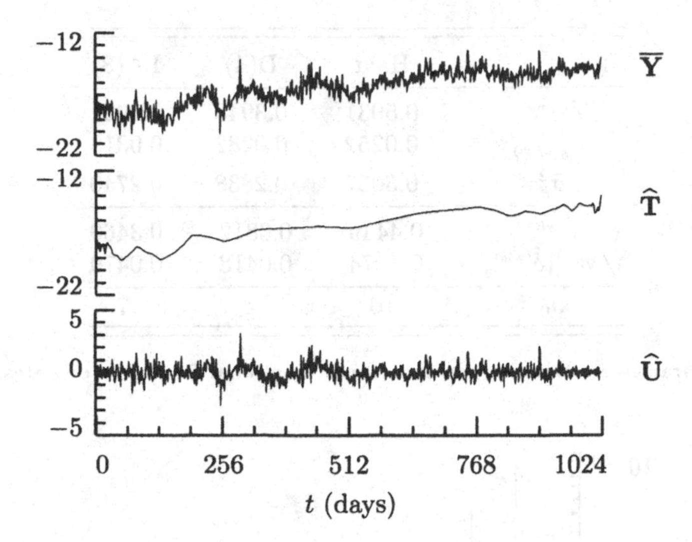

Figure 383. Wavelet-based decomposition of atomic clock fractional frequency deviates $\overline{\mathbf{Y}}$ into an estimated trend $\widehat{\mathbf{T}}$ and residuals $\widehat{\mathbf{U}}$ about the trend. Here we used an LA(8) partial DWT of level $J_0 = 7$. Note that $\widehat{\mathbf{T}}$ has much more structure than a low order polynomial and in fact resembles the output from a variable bandwidth smoother: it is quite smooth near the middle of the series, but then becomes rougher in appearance toward the end points.

9.7 Example: Atomic Clock Deviates

Let us now illustrate some of the ideas in this chapter using the atomic clock fractional frequency deviates $\{\overline{Y}_t(\tau_1)\}$, which we looked at previously in Sections 1.2, 1.3 and 8.6. These deviates are defined in Equation (321c) in terms of the measured time differences $\{X_t\}$ and are plotted in the middle of Figure 318 (this plot has two vertical axes – the right-hand axis is the appropriate one for $\{\overline{Y}_t(\tau_1)\}$).

As noted toward the end of Section 8.6, the square roots of the wavelet variance estimates for $\{\overline{Y}_t(\tau_1)\}$ in Figure 322 are consistent with a long memory interpretation, but the question of whether or not these deviates are best modeled by a stationary or nonstationary process is confounded by the possible presence of a linear trend. Because of this, let us first entertain a model of $\overline{\mathbf{Y}} = \mathbf{T} + \mathbf{U}$ (as per Section 9.4), where $\overline{\mathbf{Y}}$ is a vector containing $\overline{Y}_0(\tau_1), \ldots, \overline{Y}_{1023}(\tau_1)$, and $\mathbf{T}$ and $\mathbf{U}$ are the components due to, respectively, a polynomial trend and a zero mean FD process with unknown parameters δ and σ_ε^2. An example of the estimated additive decomposition $\overline{\mathbf{Y}} = \widehat{\mathbf{T}} + \widehat{\mathbf{U}}$ of Equation (370a) is shown in Figure 383 for a level $J_0 = 7$ LA(8) DWT (Exercise [9.7] is to create similar decompositions using the D(4) and C(6) wavelets).

Upon minimizing the reduced log likelihood function of Equation (371c), we obtain the MLE $\tilde{\delta}^{(\text{s/ns})}$ for δ, after which we can use Equation (371b) to get the corresponding estimator for σ_ε^2, namely, $\tilde{\sigma}_\varepsilon^2 \equiv \tilde{\sigma}_\varepsilon^2(\tilde{\delta}^{(\text{s/ns})})$. These estimates are listed in the upper portion of Table 384 for the Haar, D(4) and LA(8) wavelets (we also give the large sample standard deviation $\sigma_{\tilde{\delta}^{(\text{s/ns})}}$ for $\tilde{\delta}^{(\text{s/ns})}$ based on Equation (371d), and at the bottom of the table we give the value of J_0 appropriate for each wavelet). While the D(4) and LA(8) MLEs for δ agree quite nicely and indicate that a stationary FD model is appropriate, the Haar estimate is considerably higher and is just barely above the stationary region; however, this estimate might well be inappropriate because the Haar wavelet cannot handle even a linear trend.

	Haar	D(4)	LA(8)
$\tilde{\delta}^{(\mathrm{s/ns})}$	0.5031	0.3943	0.3921
$\sigma_{\tilde{\delta}^{(\mathrm{s/ns})}}$	0.0252	0.0282	0.0318
$\tilde{\sigma}_{\varepsilon}^2$	0.3057	0.2838	0.2740
$\hat{\delta}^{(\mathrm{wls})}$	0.4449	0.3812	0.3460
$\sqrt{\mathrm{var}\,\{\hat{\delta}^{(\mathrm{wls})}\}}$	0.0374	0.0418	0.0479
J_0	10	8	7

Table 384. Parameter estimation for the atomic fractional frequency deviates (see text for details).

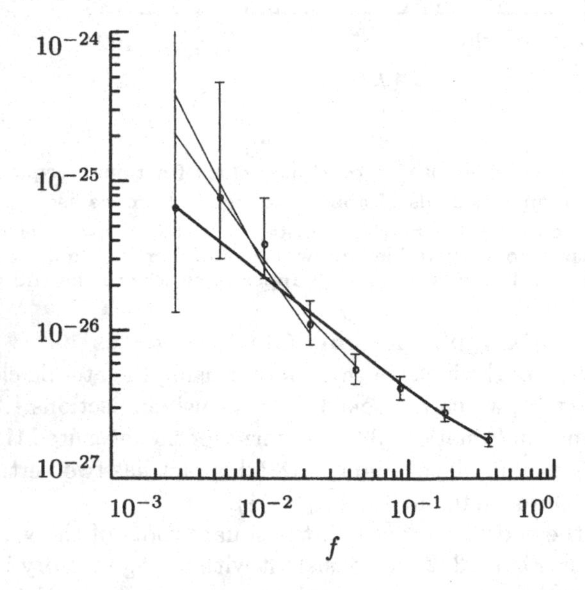

Figure 384. D(4) MODWT wavelet variance estimates for atomic clock fractional frequency deviates (re-expressed as spectral levels C_j), along with spectral levels deduced from MLEs of δ and σ_{ε}^2 (thick curve). The two shorter thin curves are discussed in the text.

Conditional on the D(4) MLE $\tilde{\delta}^{(\mathrm{s/ns})}$, we can now test the hypothesis of no trend, i.e., $\mathbf{T} = \mathbf{0}$, in the model $\overline{\mathbf{Y}} = \mathbf{T} + \mathbf{U}$. As is obvious from Figure 318, the fractional frequency deviates have a substantial nonzero mean. We thus use the test statistic $\overline{P}$ of Equation (373) since we do not wish this nonzero mean to be considered as part of the trend. We obtain $\overline{P} \doteq 4.581$ for the observed $\overline{\mathbf{Y}}$. To evaluate $\overline{P}$ under the null hypothesis, we simulated 10 000 FD(0.3943) time series of length $N = 1024$ using the Davies–Harte algorithm (see Section 7.8). The smallest, median and largest values of the $\overline{P}$ statistics for the 10 000 simulated series were, respectively, 1.019, 1.182 and 2.047, so we can soundly reject the null hypothesis!

Let us, however, back up and consider if the FD model obtained from the D(4) wavelet is really appropriate for $\overline{\mathbf{Y}}$. To do so, we took the D(4) estimates $\hat{\nu}_{\overline{Y}}(\tau_j)$ (plotted as circles in Figure 322), converted them (along with their associated 95% confidence intervals) into spectral levels using the relationship $C_j \approx 2^j \nu_{\overline{Y}}^2(\tau_j)$ and then plotted them versus $1/2^{j+\frac{1}{2}}$ cycles/day (this is the center frequency of the octave band $[\frac{1}{2^{j+1}\Delta t}, \frac{1}{2^j \Delta t}]$ on a log scale). The results are shown in Figure 384 as the circles, through which pass vertical lines indicating the confidence intervals. There is also a thick curve on this figure, and this shows the spectral levels C_j corresponding to the D(4) MLEs $\tilde{\delta}^{(\text{s/ns})}$ and $\tilde{\sigma}_\varepsilon^2$ (these levels were calculated from Equation (343c) using numerical integration). While all but one of the 95% confidence intervals trap the D(4) spectral levels, the overall agreement is questionable and suggests that there might be more structure in $\overline{\mathbf{Y}}$ than can be reasonably accounted for with our simple two parameter FD model.

In the spirit of Jensen (1999b, 2000), we could adapt our estimation scheme to handle ARFIMA models and hence entertain a more complicated model for $\overline{\mathbf{Y}}$; however, for the purpose of reassessing the trend in view of our concerns about the adequacy of the FD model, we can consider a simpler approach. As indicated by the circles in Figure 384, the three or four smallest scales (i.e., high frequencies) seem to obey a linear relationship (in log/log space) that is arguably inconsistent with some of the larger scales. We can easily adapt our scheme so that the MLE of δ just uses the nonboundary wavelet coefficients for the four or five largest scales and so that the subsequent evaluation of the trend is done using this scale-restricted MLE. If we let J_1 be the index for the smallest scale we decide to retain, our modified scheme is to take a level $J_1 - 1$ partial DWT of $\overline{\mathbf{Y}}$ and use the scaling coefficients $\mathbf{V}_{J_1-1}$ as the series to which we fit an FD model. The rationale for this scheme is that the only difference between $\mathbf{V}_{J_1-1}$ and $\overline{\mathbf{Y}}$ is in the wavelet coefficients of levels $j = 1, \ldots, J_1 - 1$, which we now want to ignore when fitting the FD model. Except for a limited number of small scale boundary wavelet coefficients, the trend in $\overline{\mathbf{Y}}$ is also captured in $\mathbf{V}_{J_1-1}$, so we can assess the trend by modeling $\mathbf{V}_{J_1-1}$ as an FD process.

Let us now apply this modified scheme to the fractional frequency deviates, first with $J_1 = 4$ and then with $J_1 = 5$. When $J_1 = 4$, we use $\mathbf{V}_3$ to obtain a D(4) MLE of $\tilde{\delta}^{(\text{s/ns})} \doteq 0.7025$. The spectral levels that are given by this fitted model are shown in Figure 384 as the longer of the two thin curves. The corresponding observed test statistic is $\overline{P} \doteq 8.113$. When the latter is compared to similar statistics for 10 000 simulated FD(0.7025) series (each of length $N_3 = 128$), we find that 1.74% of the simulated $\overline{P}$ exceed the observed $\overline{P}$, so we would reject the null hypothesis at a significance level of $\alpha = 0.05$, but not at $\alpha = 0.01$. On the other hand, when $J_1 = 5$ so that we use $\mathbf{V}_4$ and hence retain only the largest four scales in $\overline{\mathbf{Y}}$ (in keeping with the partitioning shown in Figure 322), we get $\tilde{\delta}^{(\text{s/ns})} \doteq 0.9978$ (i.e., very close to a random walk model). The corresponding spectral levels are shown in Figure 384 as the shorter thin curve. We now obtain $\overline{P} \doteq 9.697$, which is exceeded by the corresponding statistics for the 10 000 simulated series fully 32% of the time – here we cannot reject the null hypothesis of no trend at *any* reasonable level of significance! This example points out that the assessment of the significance of trends in time series can be highly dependent on the assumed statistical model, particularly when dealing with nonstationary long memory processes.

For the sake of comparison with the MLEs presented in Table 384, let us also compute corresponding WLSEs $\hat{\delta}^{(\text{wls})}$. As discussed in Section 9.5, we now make use of the nonboundary MODWT coefficients for scales $j = J_1, \ldots, J_0$ with $J_1 = 2$. The resulting WLS estimates are given in the middle portion of Table 384, along with an estimate of their standard error based on Equation (377) and the relationship $\text{var}\{\hat{\delta}^{(\text{wls})}\} = \frac{1}{4}\text{var}\{\hat{\beta}^{(\text{wls})}\}$. We see that the WLSEs are comparable to – but all some-

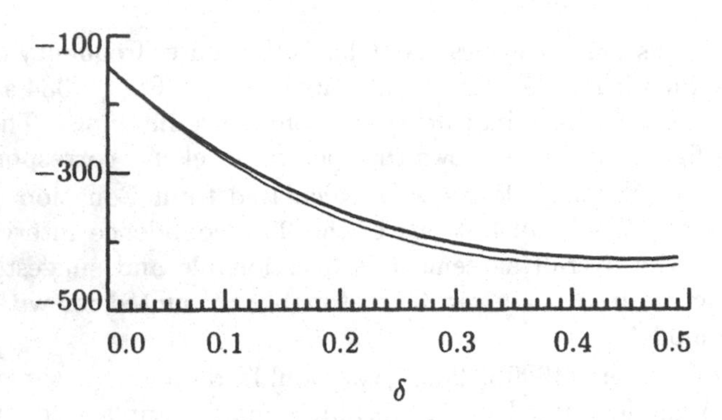

Figure 386. Reduced log likelihood functions for the FD parameter δ computed using the final 512 observations of the Nile River time series (see Figure 192). The thick curve is based on the LA(8) wavelet and attains its minimum at the estimate $\tilde{\delta}^{(s)} \doteq 0.4532$, while the thin curve is for the exact method and reaches its minimum at $\hat{\delta} \doteq 0.4452$.

what less than – the corresponding MLEs and that the standard errors are somewhat larger.

9.8 Example: Nile River Minima

Let us now illustrate the methodology of Sections 9.3 and 9.6 using the Nile River minima (this series is replotted at the top of Figure 388). As discussed in Sections 5.9 and 8.8, approximately the first 100 years of this time series appear to have different small scale variability than the rest of the series (possibly due to a change in the measuring instrument around the year 715). Accordingly, we follow Beran (1994) and model all but the first part of the Nile River time series as a stationary FD process with unknown process mean. Since there are 663 values in all and since the methodology of Section 9.3 presumes a sample size with a power of two, it is convenient to use just the final $N = 512$ years for this example (i.e., years 773 to 1284). We estimate the process mean using the sample mean $\overline{X} \doteq 11.553$ of the final 512 data values. Using the LA(8) wavelet, the MLEs for δ and σ_ε^2 are $\tilde{\delta}^{(s)} \doteq 0.4532$ and $\tilde{\sigma}_\varepsilon^2(\tilde{\delta}^{(s)}) \doteq 0.4293$, while the corresponding exact MLEs are $\hat{\delta} \doteq 0.4452$ and $\hat{\sigma}_\varepsilon^2(\hat{\delta}) \doteq 0.4230$ (computation of Haar and D(4) estimates is the subject of Exercise [9.8]). The reduced log likelihood functions from which each MLE of δ is determined are plotted in Figure 386 (the thick and thin curves are based on, respectively, the wavelet approximation and the exact method). The asymptotic Cramér–Rao lower bound for unbiased estimators of δ yields an RMSE of 0.0344 for $N = 512$, using which we can form approximate 95% confidence intervals for δ by subtracting and adding 0.0689 from a given estimate (this assessment of the variability in the MLEs also tells us that the observed difference between $\tilde{\delta}^{(s)}$ and $\hat{\delta}$ is not of much concern).

Figure 387 shows $\tilde{\delta}^{(s)}$ and $\tilde{\sigma}_\varepsilon^2(\tilde{\delta}^{(s)})$ converted into estimates of C_j via Equations (344) for $j \geq 3$ and (354) for $j = 1$ and 2 (thick curve), along with sample estimates of C_j based on $\widehat{\text{var}}\,\{W_{j,t}\} \equiv \|\mathbf{W}_j\|^2/N_j$ (circles). Both estimates of C_j are plotted versus the center frequency of the octave bands (cf. Figure 344). The agreement is remarkably good at all scales.

Let us now turn to the question of assessing the hypothesis of homogeneity of variance using the test statistic D discussed in Section 8.8. For simplicity we use the Haar wavelet (results using the D(4) and LA(8) are virtually the same). As noted

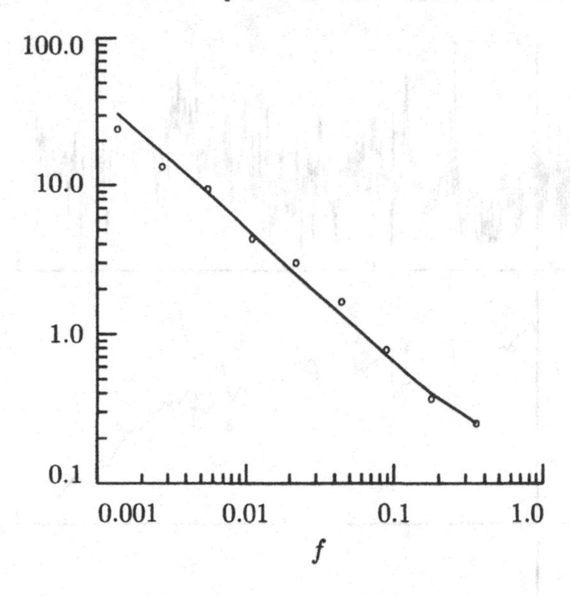

Figure 387. Variance of wavelet coefficients computed via LA(8) MLEs $\tilde{\delta}^{(s)}$ and $\tilde{\sigma}_{\varepsilon}^2(\tilde{\delta}^{(s)})$ (solid curve) as compared to sample variances of LA(8) wavelet coefficients (circles).

τ_j	M'_j	D	critical levels 10%	5%	1%
1 year	331	0.1559	0.0945	0.1051	0.1262
2 years	165	0.1754	0.1320	0.1469	0.1765
4 years	82	0.1000	0.1855	0.2068	0.2474
8 years	41	0.2313	0.2572	0.2864	0.3436

Table 387. Results of testing Nile River minima for homogeneity of variance using the Haar wavelet filter with critical values determined by computer simulations (Whitcher *et al.*, 2000a).

in item [1] of the Comments and Extensions to Section 9.6, we can easily adapt this test to work with time series that are not multiples of powers of two, as is true here ($N = 663$). Here we do so by padding the series with $1024 - 663 = 361$ zeros and then picking off the first $M'_j = \lfloor 663/2^j \rfloor$ wavelet coefficients in $\mathbf{W}_j$ (M'_j has such a simple formula because we are using the Haar wavelet). The values of the D statistic for the first four scales are given in Table 387, along with critical values determined by computer simulations. Because D for $\tau_1 = 1$ year is greater than any of the tabulated critical values, we can reject the null hypothesis of homogeneity of variance at all three critical levels. For $\tau_2 = 2$ years we can reject at level $\alpha = 0.05$, but we (just barely) fail to reject at level $\alpha = 0.01$. For the other scales, we cannot reject the null hypothesis at any reasonable level of significance (these results are in good agreement with the conclusions that we drew from Figure 327).

Let us now turn to the question of identifying the change point using the auxiliary statistic $\widetilde{D}$. The middle and lower plots of Figure 388 are Δ_k^- versus k – for both scales of interest, this (rather than Δ_k^+) determines $\widetilde{D}$. The estimated change point occurs quite close to the time of the construction of the new measuring device in the year 715.

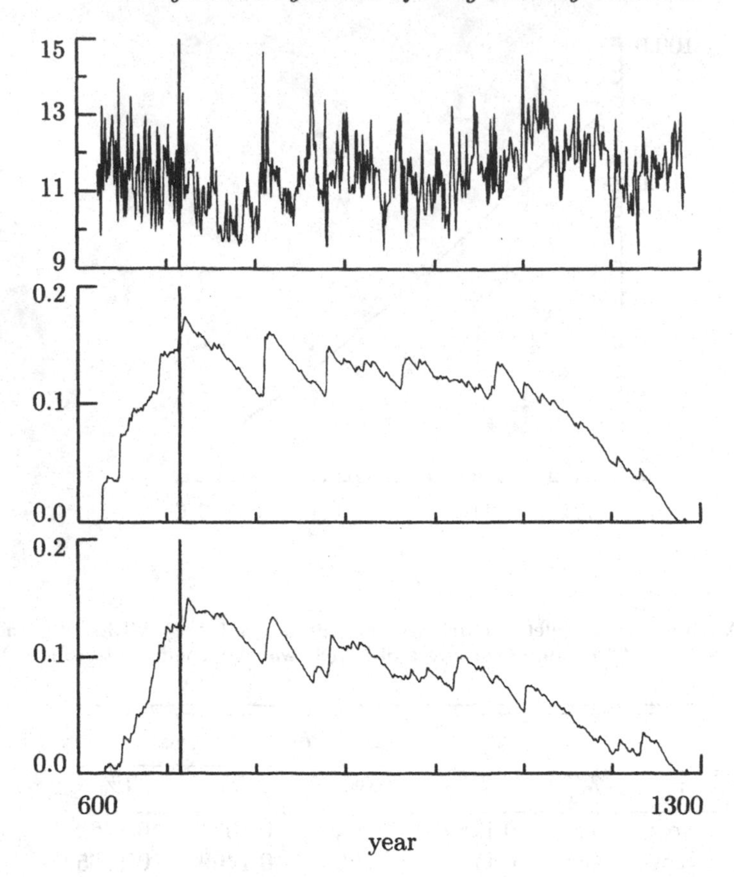

Figure 388. The Nile River minima (top plot) along with Δ_k^- versus k for scales τ_1 and τ_2 (middle and bottom plots, respectively). The thick vertical lines denote the year 715.

Beran and Terrin (1996) also looked at the Nile River minima and used a test statistic to argue for a change in the long memory parameter in the time series. The results from this analysis, in conjunction with the historical record, suggest an alternative interpretation, namely, that the decrease is due to a new measurement instrument rather than a change in the long term characteristics of the Nile River.

9.9 Summary

Let $\{X_t\}$ be a zero mean FD process with parameters δ and σ_ε^2 (see Section 7.6 for a precise definition). When $0 < \delta < \frac{1}{2}$, this process is stationary and exhibits long memory in the sense that its ACVS $s_{X,\tau} \equiv \text{cov}\{X_t, X_{t+\tau}\}$ dies down slowly to zero, particularly for δ close to 1/2. Let $\mathbf{X} \equiv [X_0, X_1, \ldots, X_{N-1}]^T$ for some $N = 2^J$, and consider its level J (i.e., full) DWT $\mathbf{W} \equiv \mathcal{W}\mathbf{X}$. Whereas the covariance matrix $\Sigma_\mathbf{X}$ for $\mathbf{X}$ can have significant nonzero off-diagonal elements, the corresponding elements in the covariance matrix $\Sigma_\mathbf{W} = \mathcal{W}\Sigma_\mathbf{X}\mathcal{W}^T$ for $\mathbf{W}$ are relatively small. We can thus claim that the DWT acts as a decorrelating transform for FD and related stationary long memory processes.

Let C_j, $j = 1, \ldots, J$, be the average level of the SDF for the FD process over the octave band $[\frac{1}{2^{j+1}}, \frac{1}{2^j}]$ associated with the jth level wavelet coefficients:

$$C_j \equiv 2^{j+1} \int_{1/2^{j+1}}^{1/2^j} S_X(f)\, df.$$

A convenient approximation for $\Sigma_{\mathbf{W}}$ is the diagonal matrix Λ_N, whose diagonal elements are

$$\underbrace{C_1, \ldots, C_1}_{\frac{N}{2} \text{ of these}}, \underbrace{C_2, \ldots, C_2}_{\frac{N}{4} \text{ of these}}, \ldots, \underbrace{C_j, \ldots, C_j}_{\frac{N}{2^j} \text{ of these}}, \ldots, \underbrace{C_{J-1}, C_{J-1}}_{2 \text{ of these}}, C_J, C_{J+1},$$

where the final element C_{J+1} is given by Equation (355b) and is an approximation to the variance of the single scaling coefficient in $\mathbf{W}$. When $\mathbf{X}$ is multivariate Gaussian, we can thus claim that its DWT $\mathbf{W}$ is also multivariate Gaussian with mean vector $\mathbf{0}$ and covariance that is well approximated by Λ_N.

Two uses for this approximation to the statistical properties of $\mathbf{W}$ are to synthesize stationary FD processes with long memory and to estimate their parameters. To synthesize (i.e., simulate) a time series of length $N/4$ from an FD process, we first form the diagonal matrix $\Lambda_N^{1/2}$ by filling it with the square roots of the spectral levels C_j corresponding to the desired parameters δ and σ_ε^2; i.e., we have $\Lambda_N^{1/2}\Lambda_N^{1/2} = \Lambda_N$. We next take a vector $\mathbf{Z}_N$ of N deviates from a Gaussian white noise process with zero mean and unit variance and create

$$\mathbf{Y}_N \equiv \mathcal{W}^T \Lambda_N^{1/2} \mathbf{Z}_N.$$

We then obtain a realization of an RV κ that is uniformly distributed over the integers $0, \ldots, N-1$. We use this realization to circularly shift $\mathbf{Y}_N$; i.e., we form $\widetilde{\mathbf{Y}}_N \equiv \mathcal{T}^{-\kappa}\mathbf{Y}_N$, where $\mathcal{T}$ is the circular shift matrix (see the discussion surrounding Equation (52a)). The first $N/4$ elements of $\widetilde{\mathbf{Y}}_N$ are a realization of a stationary process whose ACVS closely matches that of the desired FD process at lags $\tau = 0, \ldots, \frac{N}{4} - 1$.

Suppose now that we have a time series $\mathbf{X}$ that we want to model as a zero mean FD process. We can use the approximate statistical properties of $\mathbf{W}$ to formulate simple maximum likelihood estimators (MLEs) of δ and σ_ε^2. Given $\mathbf{X}$, the exact likelihood function for these two parameters is given by

$$L(\delta, \sigma_\varepsilon^2 \mid \mathbf{X}) \equiv \frac{1}{(2\pi)^{N/2}|\Sigma_{\mathbf{X}}|^{1/2}} e^{-\mathbf{X}^T\Sigma_{\mathbf{X}}^{-1}\mathbf{X}/2}.$$

The approximation $\Sigma_{\mathbf{W}} = \mathcal{W}\Sigma_{\mathbf{X}}\mathcal{W}^T \approx \Lambda_N$ allows us to simplify two key parts of the above. First, we have $|\Sigma_{\mathbf{X}}| \approx |\Lambda_N|$, which in turn is equal to the product of all the diagonal elements in Λ_N. Second, we have $\Sigma_{\mathbf{X}}^{-1} \approx \mathcal{W}^T\Lambda_N^{-1}\mathcal{W}$. With these substitutions, we are led to a reduced log likelihood function for δ, namely,

$$\tilde{l}(\delta \mid \mathbf{X}) \equiv N \log(\tilde{\sigma}_\varepsilon^2(\delta)) + \log(C'_{J+1}(\delta)) + \sum_{j=1}^{J} N_j \log(C'_j(\delta)),$$

where

$$C'_j(\delta) \equiv 2^{j+1} \int_{1/2^{j+1}}^{1/2^j} \frac{1}{[4\sin^2(\pi f)]^\delta}\, df, \quad j = 1, \ldots, J$$

(the remaining variable C_{J+1} is given by Equation (362d)). The MLE of δ is the value $\tilde{\delta}^{(s)}$ that minimizes $\tilde{l}(\delta \mid \mathbf{X})$, and the corresponding MLE for σ_ε^2 is obtained by plugging $\tilde{\delta}^{(s)}$ into the right-hand side of Equation (363b). The wavelet-based approximate MLE $\tilde{\delta}^{(s)}$ compares quite favorably with the exact MLE $\hat{\delta}$ (see Tables 363 and 364). As indicated by Exercise [365b], we can easily adjust this scheme to handle time series that cannot be assumed *a priori* to have a zero mean.

The MLE $\tilde{\delta}^{(s)}$ is based on formulating an approximation to the likelihood function for δ given the time series $\mathbf{X}$; however, what we end up with can also be interpreted as an approximate likelihood function for δ given the DWT $\mathbf{W}$ of $\mathbf{X}$. By limiting ourselves to just the nonboundary wavelet coefficients in $\mathbf{W}$, we can construct a likelihood function for δ that is appropriate for (i) nonstationary FD processes and (ii) stationary or nonstationary FD processes observed in the presence of a deterministic polynomial trend. Minimization of the corresponding reduced log likelihood leads to the wavelet-based approximate MLE $\tilde{\delta}^{(s/ns)}$. This estimator has a noticeably larger RMSE than either the exact MLE $\hat{\delta}$ or approximate MLE $\tilde{\delta}^{(s)}$ in circumstances where these latter two are applicable (i.e., a stationary FD process with no trend), but $\tilde{\delta}^{(s/ns)}$ is directly applicable over a much wider class of processes than these other two MLEs. For example, we can make use of $\tilde{\delta}^{(s/ns)}$ in testing the hypothesis that $\mathbf{T} = \mathbf{0}$ in the model $\mathbf{X} = \mathbf{T} + \mathbf{U}$, where $\mathbf{T}$ is a polynomial trend, and $\mathbf{U}$ is a zero mean FD process. The test is based on using the scaling and boundary wavelet coefficients in $\mathbf{W}$ to form an estimate $\widehat{\mathbf{T}}$ of the trend and then computing the test statistic $P \equiv \|\mathbf{X}\|^2 / \|\mathbf{X} - \widehat{\mathbf{T}}\|^2$. Conditional on the observed $\tilde{\delta}^{(s/ns)}$, we can determine the distribution for P under the null hypothesis empirically via a simulation study and hence evaluate the evidence for or against the null hypothesis.

As an alternative to MLEs, we can also estimate δ using least squares (LS) estimators. This idea is based on the fact that log/log plots of the wavelet variance $\nu_X^2(\tau_j)$ versus scale τ_j for an FD process are approximately linear. This approximation becomes increasingly accurate as $\tau_j \to \infty$, but is quite acceptable over all scales except τ_1. We thus can write $\log(\nu_X^2(\tau_j)) \approx \zeta + \beta \log(\tau_j)$ for $j \geq 2$, where the slope β and δ are related via $\delta = \frac{1}{2}(\beta + 1)$. Given a wavelet filter of length L and a time series that can be regarded as a realization of an FD process with $\delta \in [-\frac{1}{4}, \frac{L}{2} + \frac{1}{4}]$ (see the discussion surrounding Equation (370b)), we form the unbiased estimator $\hat{\nu}_X^2(\tau_j)$ of the wavelet variance using the M_j nonboundary MODWT wavelet coefficients (see Equation (306b)). The distributional result of Equation (313a) says that

$$\hat{\nu}_X^2(\tau_j) \stackrel{\mathrm{d}}{=} \nu_X^2(\tau_j)\chi_{\eta_j}^2/\eta_j$$

approximately, where, for simplicity, we can take the equivalent degrees of freedom (EDOFs) η_j to be $\max\{M_j/2^j, 1\}$ (this is in keeping with Equation (314c)). We can now formulate the linear regression model

$$Y(\tau_j) = \zeta + \beta \log(\tau_j) + e_j,$$

where

$$Y(\tau_j) \equiv \log(\hat{\nu}_X^2(\tau_j)) - \psi(\tfrac{\eta_j}{2}) + \log(\tfrac{\eta_j}{2}),$$

and $\psi(\cdot)$ is the digamma function. Here the error terms e_j are approximately uncorrelated and equal in distribution to the RV $\log(\chi_{\eta_j}^2) - \psi(\frac{\eta_j}{2}) - \log(2)$; moreover, e_j has zero mean and variance $\psi'(\frac{\eta_j}{2})$ and is approximately Gaussian distributed as long as the EDOFs is around ten or greater, where $\psi'(\cdot)$ is the trigamma function. Standard least squares theory now leads us to the weighted least squares estimator (WLSE) $\hat{\beta}^{(\text{wls})}$ displayed in Equation (376c) (the theoretical variance for this WLSE is given by Equation (377)). The corresponding estimator for δ is given by $\hat{\delta}^{(\text{wls})} \equiv \frac{1}{2}(\hat{\beta}^{(\text{wls})} + 1)$. While Table 377 indicates that the WLSEs have considerably larger RMSEs than the exact MLEs, the former is much easier to compute and – similar to the wavelet-based approximate MLE $\tilde{\delta}^{(\text{s/ns})}$ – enjoys the advantage of being directly applicable with (i) nonstationary FD processes and (ii) stationary and nonstationary FD processes observed in the presence of polynomial trends.

The fact that the DWT approximately decorrelates an FD process can also be used to devise a test for the homogeneity of variance of a time series exhibiting a long memory structure. The idea is that, under the null hypothesis that $\mathbf{X}$ is a sample from a Gaussian FD process, the nonboundary wavelet coefficients in $\mathbf{W}_j$ should be approximately a sample of a Gaussian white noise process with homogeneous variance. We can test the null hypothesis of homogeneous variance on a scale-by-scale basis using a test statistic D defined via Equation (380b). If we reject the null hypothesis for the jth level wavelet coefficients and if it is reasonable to entertain an alternative hypothesis of a single change of variance in the time series itself, then we can estimate location of the change point using an auxiliary statistic $\widetilde{D}$ based upon the MODWT wavelet coefficients (see Equation (382) and surrounding discussion).

9.10 Exercises

[9.1] Create plots similar to Figures 347a and 347b for the D(4) DWT.

[9.2] Show that, if we approximate $\mathcal{H}_j(\cdot)$ as the squared gain function for an ideal band-pass filter and if we assume that $S_X(\cdot)$ is constant over this pass-band, Equation (348b) yields $\text{cov}\{W_{j,t}, W_{j,t+\tau}\} \approx 0$.

[9.3] Let $\{X_t\}$ be a first order moving average (MA) process; i.e., we can write

$$X_t = \varepsilon_t + \theta\varepsilon_{t-1},$$

where $\{\varepsilon_t\}$ is a white noise process, and θ is an arbitrary constant. Derive the ACVS and SDF for this process, and create a plot similar to Figure 349 for the cases $\theta = 0.5$ and $\theta = 0.98$. How well does the DWT decorrelate these MA processes? (For a discussion of MA processes with $-1 < \theta < 1$, see Whitcher, 1998, Section 4.1.2.)

[9.4] Suppose that the N dimensional vector $\mathbf{Y}$ of RVs obeys a multivariate Gaussian distribution with a mean vector containing all zeros and with a covariance matrix $\Sigma_{\mathbf{Y}}$, whose (m,n)th element is denoted as $\Sigma_{\mathbf{Y},m,n}$. Let κ be an RV that is uniformly distributed over the integers $0, 1, \ldots, N-1$, and consider $\widetilde{\mathbf{Y}} \equiv \mathcal{T}^{-\kappa}\mathbf{Y}$, where $\mathcal{T}$ is the circular shift matrix of Equation (52a) (recall that $\mathcal{T}^2 = \mathcal{T}\mathcal{T}$ and so forth). Show that the RVs in $\widetilde{\mathbf{Y}}$ constitute a segment of a stationary process whose ACVS is given by Equation (357). Show also that this ACVS is symmetric about $\tau = N/2$ in the sense that $s_{\widetilde{Y},N-\tau} = s_{\widetilde{Y},\tau}$ for $\tau = 1, \ldots, N/2$.

[9.5] Verify Equations (376c) and (377).

[9.6] Verify Equations (374) and (379), which can be used to obtain, respectively, instantaneous ML and LS estimators of δ. Determine an expression for t_j for the case of the LA(8) wavelet. Finally, argue that an expression for the variance of the instantaneous LSE of δ is

$$\frac{(J_0 - J_1 + 1)\pi^2}{8\left[(J_0 - J_1 + 1)\sum \log^2(\tau_j) - \left(\sum \log(\tau_j)\right)^2\right]}.$$

[9.7] Create plots similar to Figure 383, but now using the D(4) and C(6) wavelets. Compare the estimated trends $\widehat{\mathbf{T}}$ with the one in Figure 383, and comment on how each wavelet influences the resulting trend estimate.

[9.8] Based upon the last $N = 512$ data values of the Nile River minima, compute approximate wavelet-based MLEs for the parameters δ and σ_ε^2 of a stationary FD process using the Haar and D(4) wavelets. Use these MLEs to create plots similar to Figure 387. What are the relative advantages and disadvantages of the Haar and D(4) MLEs as compared to the LA(8) MLEs given in Section 9.8? (The Nile River series is available via the Web site for this book– see page *xiv*.)

10

Wavelet-Based Signal Estimation

10.0 Introduction

As discussed in Chapter 4, the discrete wavelet transform (DWT) allows us to analyze (decompose) a time series $\mathbf{X}$ into DWT coefficients $\mathbf{W}$, from which we can then synthesize (reconstruct) our original series. We have already noted that the synthesis phase can be used, for example, to construct a multiresolution analysis of a time series (see Equation (64) or (104a)) and to simulate long memory processes (see Section 9.2). In this chapter we study another important use for the synthesis phase that provides an answer to the *signal estimation* (or *function estimation*, or *denoising*) problem, in which we want to estimate a signal hidden by noise within an observed time series. The basic idea here is to modify the elements of $\mathbf{W}$ to produce, say, $\mathbf{W}'$, from which an estimate of the signal can be synthesized. With the exception of methods briefly discussed in Section 10.8, once certain parameters have been estimated, the elements W_n of $\mathbf{W}$ are treated one at a time; i.e., how we modify W_n is not directly influenced by the remaining DWT coefficients. The wavelet-based techniques that we concentrate on here are thus conceptually very simple, yet they are remarkably adaptive to a wide variety of signals.

We will consider both deterministic and stochastic signals, and uncorrelated and correlated noise. We adopt the following notational conventions in order to emphasize which models are under discussion:

[1] $\mathbf{D}$ represents an N dimensional deterministic signal vector;

[2] $\mathbf{C}$ represents an N dimensional stochastic (random) signal vector; i.e., $\mathbf{C}$ is a multivariate random variable (RV) with covariance matrix $\Sigma_{\mathbf{C}}$;

[3] $\boldsymbol{\epsilon}$ represents an N dimensional vector of independent and identically distributed (IID) noise with mean zero; i.e., $\boldsymbol{\epsilon}$ is a multivariate RV with each component having the same univariate distribution, with $E\{\boldsymbol{\epsilon}\} = \mathbf{0}$ (an N dimensional vector of zeros) and with covariance matrix $\Sigma_{\boldsymbol{\epsilon}} = \sigma_\epsilon^2 I_N$, where σ_ϵ^2 is the common variance; and

[4] $\boldsymbol{\eta}$ represents an N dimensional vector of (in general) non-IID noise with mean zero; i.e., $\boldsymbol{\eta}$ is a multivariate RV with $E\{\boldsymbol{\eta}\} = \mathbf{0}$ and covariance matrix $\Sigma_{\boldsymbol{\eta}}$ (depending upon specific assumptions made in the text, the RVs in $\boldsymbol{\eta}$ might deviate from being IID by having different variances or by being correlated).

Any assumptions about the distributional properties of the RVs will be made in the appropriate places in the text.

This chapter is organized as follows. We first consider a general orthonormal transform $\mathcal{O}$ applied to a deterministic signal. In Section 10.1 we show that some orthonormal transforms can require substantially fewer coefficients than others in order to succinctly summarize a signal (in particular, the DWT of a wide variety of deterministic signals $\mathbf{D}$ can often isolate the key features of $\mathbf{D}$ into a small number of coefficients). This supposition leads to the idea – explored in Section 10.2 – of applying thresholding to an orthonormal transform of a deterministic signal plus noise. We discuss different types of threshold functions and explain techniques for choosing threshold levels. In Section 10.3 we show that minimum mean square estimation of a *stochastic* signal $\mathbf{C}$ in the presence of noise can be accomplished by a simple scaling of the observed coefficients, leading to an estimator (similar to the well-known Wiener filter) that shrinks observed transform coefficients toward zero. We discuss shrinkage rules that fully exploit the statistical structure of the signal transform coefficients and the noise transform coefficients in Section 10.4 – these include conditional mean and median methods, and Bayesian formulations. What we get by applying thresholding or shrinkage to observed DWT coefficients depends critically on the structure of the noise transform coefficients. When the additive noise terms in a time series are IID and Gaussian, then so are the noise transform coefficients. We examine this case in Section 10.5, while in Section 10.6 we look at the situation where the additive noise terms are uncorrelated, but non-Gaussian, so that the DWT noise coefficients are not identically distributed, and thresholding becomes problematic. An interesting application with this structure is spectrum estimation using the periodogram. In Section 10.7 we consider the case of correlated Gaussian DWT noise coefficients in the context of a specific example for which this model is a good approximation, namely, spectrum estimation based upon the multitaper spectrum estimator. While Sections 10.5 to 10.7 focus on dealing with the structure of the noise process $\boldsymbol{\epsilon}$ or $\boldsymbol{\eta}$, we discuss an approach to modeling a stochastic signal $\mathbf{C}$ in Section 10.8. We conclude with a summary in Section 10.9.

10.1 Signal Representation via Wavelets

Let $\mathbf{D}$ be an N dimensional vector whose elements constitute a deterministic signal of interest. The central argument of this section is that the DWT of $\mathbf{D}$ can be a useful way of representing (re-expressing) the signal in a manner that will facilitate the signal estimation problem. As we shall see in Section 10.2, representations that enable signal estimation are those that essentially summarize the important features in $\mathbf{D}$ in a small number, say M, of terms. If M is considerably smaller than N, our estimation problem is simplified in that we need only worry about estimating these M terms rather than all N elements of $\mathbf{D}$. What we are seeking, then, is an efficient way of representing a signal $\mathbf{D}$ using a small number of terms, and what we are claiming is that the DWT can do this.

To quantify how well the DWT does at representing a signal $\mathbf{D}$ as compared to other orthonormal transforms, let us make the trivial assumption that $\|\mathbf{D}\|^2 > 0$ (i.e., we rule out a signal that is identically zero). We can then use the notion of a normalized partial energy sequence (NPES) to see how well a particular orthonormal transform does in capturing the key features in $\mathbf{D}$ in a small number of terms. As described previously in Section 4.10, if $\mathbf{O} = \mathcal{O}\mathbf{D}$ is the vector of transform coefficients resulting from a real- or complex-valued orthonormal transform $\mathcal{O}$ of the signal $\mathbf{D}$, the NPES for $\mathbf{O}$ is defined as follows. Let $O_{(l)}$ be the element $O_{l'}$ of $\mathbf{O}$ whose squared magnitude $|O_{l'}|^2$ is the lth largest of the N such possible squared magnitudes. We then have

$$|O_{(0)}|^2 \geq |O_{(1)}|^2 \geq \cdots \geq |O_{(N-1)}|^2.$$

We next cumulatively sum the $|O_{(l)}|^2$ terms and normalize this sum by the energy $\|\mathbf{O}\|^2 = \|\mathbf{D}\|^2$ to define the NPES $\{C_n\}$:

$$C_n \equiv \frac{\sum_{l=0}^{n} |O_{(l)}|^2}{\|\mathbf{O}\|^2} = \frac{\sum_{l=0}^{n} |O_{(l)}|^2}{\sum_{l=0}^{N-1} |O_{(l)}|^2}, \qquad n = 0, \ldots, N-1.$$

By construction, $0 < C_n \leq 1$ and $C_n \leq C_{n+1}$. To see how well the M coefficients in $\mathbf{O}$ with the largest squared magnitudes do at capturing the features in $\mathbf{D}$, let $\mathcal{I}_M$ be an $N \times N$ diagonal matrix, whose diagonal elements consist of M ones (indicating the positions of the M largest $|O_{l'}|^2$) and $N - M$ zeros. Let $\widehat{\mathbf{D}}_M \equiv \mathcal{O}^T \mathcal{I}_M \mathbf{O}$; i.e., $\widehat{\mathbf{D}}_M$ is an approximation to $\mathbf{D}$ formed using just the M largest transform coefficients.

▷ **Exercise [395]:** Show that

$$C_{M-1} = 1 - \frac{\|\mathbf{D} - \widehat{\mathbf{D}}_M\|^2}{\|\mathbf{D}\|^2}.$$ ◁

Note that $1 - C_{M-1}$ represents the *relative approximation error* we would encounter in approximating $\mathbf{D}$ using $\widehat{\mathbf{D}}_M$, i.e., the ratio of the sum of the magnitude squared approximation errors to the sum of magnitude squared signal values. Thus, the smaller $1 - C_{M-1}$ is (or, equivalently, the larger C_{M-1} is), the better the approximation. If $\{C_n\}$ and $\{C'_n\}$ are the NPESs for two orthonormal transforms and if $C_{M-1} > C'_{M-1}$, we would prefer the first transform over the second for a representation scheme that uses exactly M transform coefficients.

To see why the DWT might be a good orthonormal transform to use for signal representation, let us consider the three signals shown in the left-hand column of plots in Figure 396. Each signal $\mathbf{D}_j$ consists of $N = 128$ elements, denoted as $D_{j,t}, t = 0, \ldots, 127$. From top to bottom, the three signals are

[1] a sampled sinusoid $D_{1,t} = \frac{1}{2}\cos(\frac{3\pi t}{32} + 0.08)$;

[2] a 'bump' defined by $D_{2,t} = 0$ for all t except $t = 59, \ldots, 69$ at which $D_{2,t} = D_{1,t}$; and

[3] $D_{3,t} = \frac{1}{\sqrt{120}} D_{1,t} + D_{2,t}$.

Qualitatively speaking, $\mathbf{D}_1$ is a 'frequency domain' signal (because its orthonormal discrete Fourier transform (odft) consists of just two nonzero coefficients), while $\mathbf{D}_2$ is a 'time domain' signal (because it has only 11 nonzero values). The third signal $\mathbf{D}_3$ is a mixture of the two domains. The right-hand column of plots shows the NPESs for three different orthonormal transforms of each signal, namely, the identity transform I_N (curves broken by intermittent dots), the ODFT $\mathcal{F}$ (thin solid curves) and the DWT $\mathcal{W}$ based upon the LA(8) Daubechies wavelet filter (dashed curves). Table 396 indicates the smallest value of M that yields a relative approximation error of no more than 1% for various combinations of signals and transforms. While the DWT is surpassed by the identity and ODFT transforms when applied to signals ideally

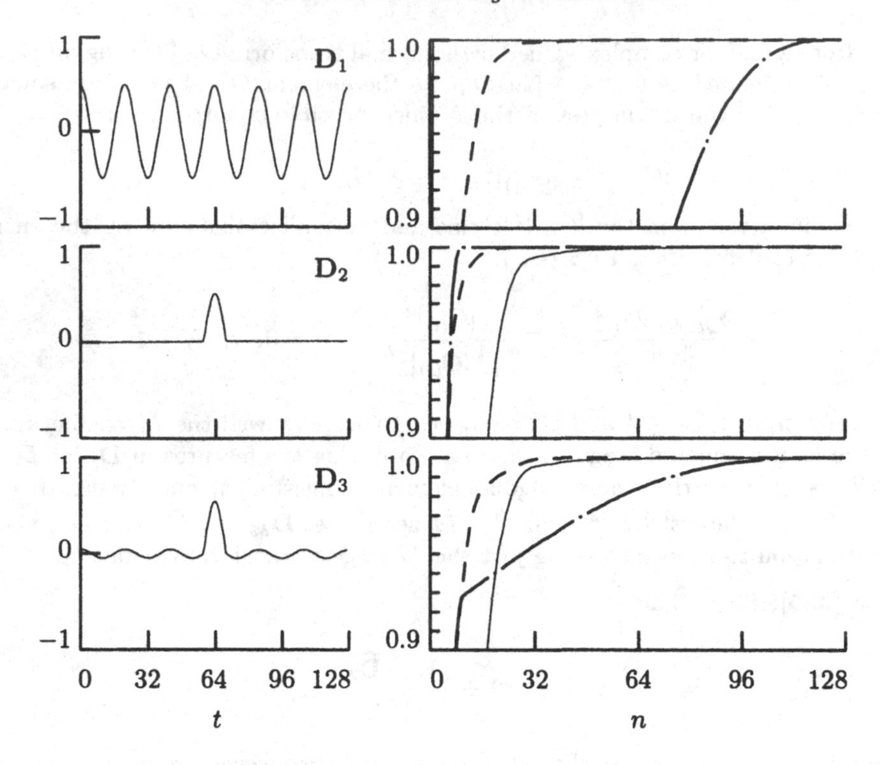

Figure 396. Three signals (left-hand column) and their corresponding normalized partial energy sequences (right-hand column) for the signals themselves (curves broken by intermittent dots), their ODFT coefficients (thin solid curves) and their LA(8) DWT coefficients (dashed curves).

	domain of signal		
	frequency	time	mixture
$\mathcal{F}$	2	29	28
I_N	105	9	75
$\mathcal{W}$	22	14	21

Table 396. Number of coefficients required to obtain no more than a 1% relative approximation error for three signals using an orthonormal discrete Fourier transform $\mathcal{F}$, an identity transform I_N and an LA(8) DWT transform $\mathcal{W}$ (see text for details).

suited to them, it comes in second in both cases, and it is in fact superior for the mixture signal. This simple example suggests that the DWT should work well for the many signals occurring in practice that exhibit a combination of time domain characteristics (transients) and frequency domain characteristics (narrow-band and broad-band features).

As an example of how well the DWT can represent an actual fairly complicated signal, let us consider a portion of the vertical shear ocean measurements that are shown in the bottom panel of Figure 194 and described in Section 5.10 (this set of data cannot really be regarded as a deterministic signal, but we do so here anyway to illustrate a point about the DWT). In order to make use of a partial LA(8) DWT of

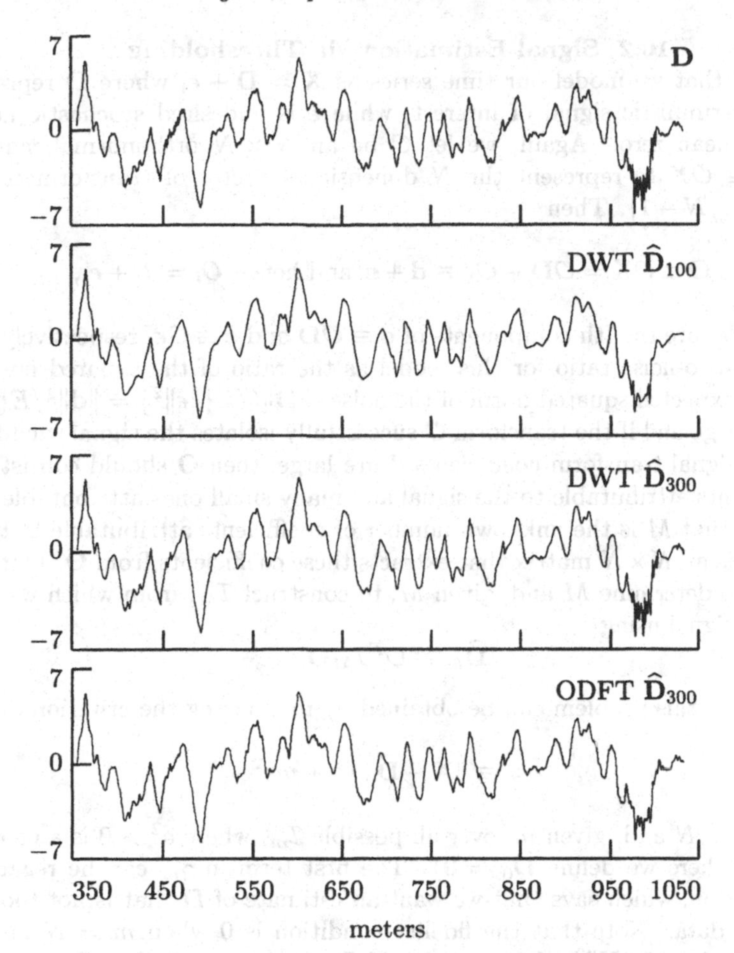

Figure 397. Plot of $N = 6784$ values of a hypothesized signal $\mathbf{D}$ related to vertical shear in the ocean versus depth in meters (top plot), along with reconstructions using 100 LA(8) DWT coefficients, 300 LA(8) DWT coefficients and 300 ODFT coefficients. See Section 5.10 for more discussion about these data.

level $J_0 = 7$, we need to have a sample size that is a multiple of $2^{J_0} = 128$, so we have shortened the original series by 91 observations to obtain a subseries of length $N = 6784 = 53 \cdot 128$ – this subseries is shown in the top panel of Figure 397 (if we denote the original series by $X_0, \ldots, X_{6874}$, the subseries is given by $X_{54}, \ldots, X_{6837}$). The three plots below this one show $\widehat{\mathbf{D}}_M$ based upon, respectively, 100 LA(8) DWT coefficients, 300 such coefficients and 300 ODFT coefficients. Visually the approximation using 300 DWT coefficients does a more reasonable job of preserving the transient events in the original data than does the ODFT with 300 terms (note that 300 coefficients is less than 5% of N). In terms of the NPESs, we have $C_{299}^{(\mathrm{dwt})} \doteq 0.9983$ and $C_{299}^{(\mathrm{odft})} \doteq 0.9973$, so the DWT outperforms the ODFT by this measure (in fact we have $C_{422}^{(\mathrm{odft})} < C_{299}^{(\mathrm{dwt})} < C_{423}^{(\mathrm{odft})}$, so we need about 123 additional ODFT coefficients in order to achieve the same relative approximation error as the LA(8) $\widehat{\mathbf{D}}_{300}$).

10.2 Signal Estimation via Thresholding

Suppose now that we model our time series as $\mathbf{X} = \mathbf{D} + \boldsymbol{\epsilon}$, where $\mathbf{D}$ represents an unknown deterministic signal of interest, while $\boldsymbol{\epsilon}$ is undesired stochastic noise that is IID with mean zero. Again, we let $\mathcal{O}$ be an $N \times N$ orthonormal matrix, and we take $\mathbf{O} \equiv \mathcal{O}\mathbf{X}$ to represent the N dimensional vector of transform coefficients $\{O_l : l = 0, \ldots, N-1\}$. Then

$$\mathbf{O} \equiv \mathcal{O}\mathbf{X} = \mathcal{O}\mathbf{D} + \mathcal{O}\boldsymbol{\epsilon} \equiv \mathbf{d} + \mathbf{e} \text{ and hence } O_l = d_l + e_l, \tag{398}$$

where d_l and e_l are the lth components of $\mathbf{d} \equiv \mathcal{O}\mathbf{D}$ and $\mathbf{e} \equiv \mathcal{O}\boldsymbol{\epsilon}$, respectively. We can define a signal-to-noise ratio for this model as the ratio of the squared norm of the signal to the expected squared norm of the noise: $\|\mathbf{D}\|^2/E\{\|\boldsymbol{\epsilon}\|^2\} = \|\mathbf{d}\|^2/E\{\|\mathbf{e}\|^2\}$. If this ratio is large and if the transform $\mathcal{O}$ successfully isolates the signal such that only a few of the signal transform coefficients $\mathbf{d}$ are large, then $\mathbf{O}$ should consist of a few large coefficients attributable to the signal and many small ones attributable to noise.

Suppose that M is the unknown number of coefficients attributable to the signal and that $\mathcal{I}_M$ is an $N \times N$ matrix that extracts these coefficients from $\mathbf{O}$. Our problem then is how to determine M and, given M, to construct $\mathcal{I}_M$, from which we will then estimate the signal using

$$\widehat{\mathbf{D}}_M \equiv \mathcal{O}^T \mathcal{I}_M \mathbf{O}.$$

One solution to this problem can be obtained by minimizing the criterion

$$\gamma_m \equiv \|\mathbf{X} - \widehat{\mathbf{D}}_m\|^2 + m\delta^2$$

over $m = 0, \ldots, N$ and, given m, over all possible $\mathcal{I}_m$, where $\delta^2 > 0$ is a constant yet to be chosen (here we define $\widehat{\mathbf{D}}_0 = \mathbf{0}$). The first term in γ_m can be regarded as a *fidelity* condition, which says that we want an estimate of $\mathbf{D}$ that is not too far from the observed data. Note that the fidelity condition is 0 when $m = N$ and attains a maximum value of $\|\mathbf{X}\|^2$ when $m = 0$. If $\mathcal{I}_{m+1} - \mathcal{I}_m = \mathcal{I}_1$ for all m, the fidelity condition will be a nonincreasing function of m. The second term in γ_m is a *penalty* term which mitigates against the use of a large number of coefficients in constructing $\widehat{\mathbf{D}}_m$ and is a strictly increasing function of m.

We claim that γ_m is minimized at $m = M$ if we define M to be the number of transform coefficients such that $|O_l|^2 > \delta^2$. To see this, let $\mathcal{J}_m$ be the set of m indices l such that the lth diagonal element of $\mathcal{I}_m$ is equal to 1, and note that

$$\begin{aligned}\gamma_m = \|\mathbf{X} - \widehat{\mathbf{D}}_m\|^2 + m\delta^2 &= \|\mathcal{O}^T\mathbf{O} - \mathcal{O}^T\mathcal{I}_m\mathbf{O}\|^2 + m\delta^2 \\ &= \|(I_N - \mathcal{I}_m)\mathbf{O}\|^2 + m\delta^2 = \sum_{l \notin \mathcal{J}_m} |O_l|^2 + \sum_{l \in \mathcal{J}_m} \delta^2.\end{aligned}$$

Note that, if $l \in \mathcal{J}_m$, we contribute δ^2 to the summations forming γ_m, whereas, if $l \notin \mathcal{J}_m$, we contribute $|O_l|^2$. We can thus minimize γ_m by putting into $\mathcal{J}_m$ all the l's such that $|O_l|^2 > \delta^2$. This result shows that minimizing the criterion γ_m over m naturally leads to an estimator $\widehat{\mathbf{D}}_M$ based upon thresholding the transform coefficients (Moulin, 1995).

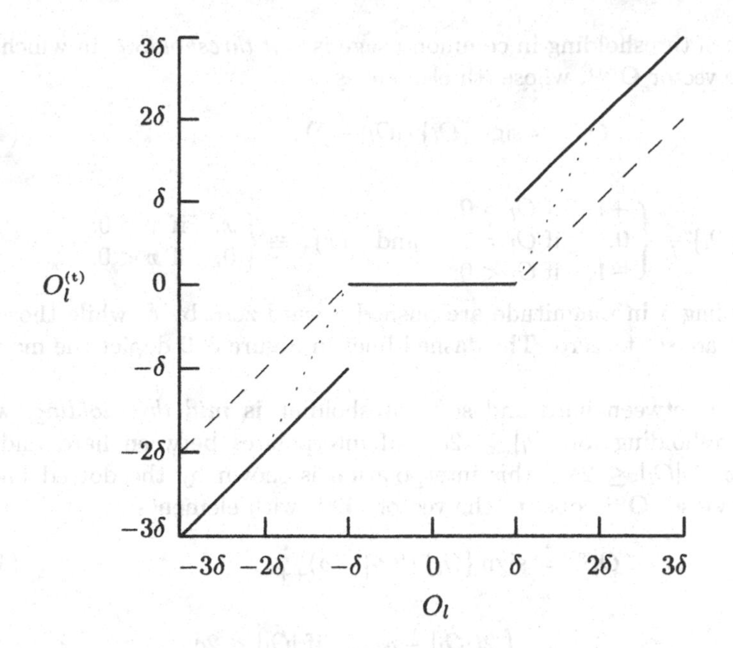

Figure 399. Mappings from O_l to $O_l^{(t)}$, where $O_l^{(t)}$ is either $O_l^{(ht)}$ for hard thresholding (solid lines), $O_l^{(st)}$ for soft thresholding (dashed lines), or $O_l^{(mt)}$ for mid thresholding (dotted lines). Note that the effect of all three thresholding schemes is the same when $-\delta \le O_l \le \delta$.

- *Threshold functions*

A common thread in the formulation of threshold functions is that, if the magnitude of the observed transform coefficient O_l is less than the threshold $\delta > 0$, it is set to zero. A *thresholding* scheme for estimating $\mathbf{D}$ thus consists of three basic steps.

[1] Compute the transform coefficients $\mathbf{O} \equiv \mathcal{O}\mathbf{X}$.

[2] Define the thresholded coefficients $\mathbf{O}^{(t)}$ to be the vector whose lth element $O_l^{(t)}$ is such that

$$O_l^{(t)} = \begin{cases} 0, & \text{if } |O_l| \le \delta; \\ \text{some nonzero value}, & \text{otherwise}, \end{cases}$$

where the nonzero values are yet to be defined.

[3] Estimate $\mathbf{D}$ via $\widehat{\mathbf{D}}^{(t)} \equiv \mathcal{O}^T\mathbf{O}^{(t)}$.

Note that the same threshold applies to every term O_l to be thresholded. This type of thresholding is sometimes known as *global thresholding.*

For coefficients exceeding δ in magnitude, there are several possibilities. The simplest scheme is *hard thresholding,* for which we define the thresholded coefficients $\mathbf{O}^{(t)}$ to be the vector $\mathbf{O}^{(ht)}$ with elements

$$O_l^{(ht)} = \begin{cases} 0, & \text{if } |O_l| \le \delta; \\ O_l, & \text{otherwise}. \end{cases} \tag{399}$$

Here coefficients whose magnitudes exceed δ are left untouched, while those less than or equal to δ are 'killed' or set to zero. The mapping from O_l to $O_l^{(ht)}$ is shown by the solid lines in Figure 399.

Another type of thresholding in common usage is *soft thresholding,* in which $\mathbf{O}^{(t)}$ is set equal to the vector $\mathbf{O}^{(st)}$, whose lth element is

$$O_l^{(st)} = \text{sign}\,\{O_l\}\,(|O_l| - \delta)_+, \tag{400a}$$

where

$$\text{sign}\,\{O_l\} \equiv \begin{cases} +1, & \text{if } O_l > 0; \\ 0, & \text{if } O_l = 0; \\ -1, & \text{if } O_l < 0; \end{cases} \quad \text{and} \quad (x)_+ \equiv \begin{cases} x, & \text{if } x \geq 0; \\ 0, & \text{if } x < 0. \end{cases}$$

Coefficients exceeding δ in magnitude are pushed toward zero by δ, while those less than or equal to δ are set to zero. The dashed lines in Figure 399 depict the mapping from O_l to $O_l^{(st)}$.

A compromise between hard and soft thresholding is *mid thresholding,* which acts like hard thresholding for $|O_l| \geq 2\delta$ and interpolates between hard and soft thresholding for $\delta \leq |O_l| \leq 2\delta$ – this interpolation is shown by the dotted lines in Figure 399. Here we set $\mathbf{O}^{(t)}$ equal to the vector $\mathbf{O}^{(mt)}$ with elements

$$O_l^{(mt)} = \text{sign}\,\{O_l\}\,(|O_l| - \delta)_{++}, \tag{400b}$$

where

$$(|O_l| - \delta)_{++} \equiv \begin{cases} 2(|O_l| - \delta)_+, & \text{if } |O_l| < 2\delta; \\ |O_l|, & \text{otherwise.} \end{cases}$$

Here large coefficients, i.e., those exceeding 2δ in magnitude, are left untouched, those between δ and 2δ are shrunk, while those less than δ are set to zero.

Mid thresholding is a particular case of the more general *firm thresholding* (Bruce and Gao, 1996b), in which step [2] is replaced by a thresholding procedure defined in terms of two threshold parameters, δ and δ'. Here the vector of thresholded coefficients is given by $\mathbf{O}^{(ft)}$ with elements

$$O_l^{(ft)} = \begin{cases} 0, & \text{if } |O_l| \leq \delta; \\ \text{sign}\{O_l\}\frac{\delta'(|O_l|-\delta)}{\delta'-\delta}, & \text{if } \delta < |O_l| \leq \delta'; \\ O_l, & \text{if } |O_l| > \delta'. \end{cases} \tag{400c}$$

We will henceforth concentrate on threshold rules that depend on just the threshold level δ.

A number of ways for choosing δ have been proposed in the literature.

- *Universal threshold*

A particularly interesting way to derive a threshold level was given by Donoho and Johnstone (1994) for the case of IID noise $\boldsymbol{\epsilon}$ that is Gaussian distributed; i.e., $\boldsymbol{\epsilon}$ is a multivariate Gaussian (normal) RV with zero mean and covariance $\sigma_\epsilon^2 I_N$. A key property about an orthonormal transform of IID Gaussian noise is that the transformed noise has the same statistical properties as the untransformed noise (see Exercise [263]), which tells us that $\mathbf{e}$ in Equation (398) is a vector of IID Gaussian RVs with mean zero and variance σ_ϵ^2; i.e.,

$$e_l \stackrel{\text{d}}{=} \mathcal{N}(0, \sigma_\epsilon^2) \text{ and } \text{cov}\,\{e_l, e_{l'}\} = 0 \text{ when } l \neq l'.$$

The form for δ given by Donoho and Johnstone (1994) is

$$\delta^{(u)} \equiv \sqrt{[2\sigma_\epsilon^2 \log(N)]}, \tag{400d}$$

which is known as the *universal threshold* (in the above, the base of the logarithm is e). The rationale for this threshold is the following. Suppose that the signal $\mathbf{D}$ is in fact a vector of zeros so that the transform coefficients $\{O_l\}$ are a portion of an IID Gaussian sequence $\{e_l\}$ with zero mean and variance σ_ϵ^2; i.e., $\mathbf{O} \equiv \mathbf{e}$. Then, as $N \to \infty$, we have

$$\mathbf{P}\big[\max_l\{|O_l|\} \le \delta^{(\mathrm{u})}\big] \equiv \mathbf{P}\big[\max_l\{|e_l|\} \le \delta^{(\mathrm{u})}\big] \to 1,$$

so that asymptotically we will correctly estimate the signal vector. This result must be interpreted with caution for two reasons. First, this threshold is slanted toward eliminating the vast majority of the noise. For IID Gaussian $\{e_l\}$, as we have here,

$$\mathbf{P}\big[\max_l\{|e_l|\} > \delta^{(\mathrm{u})}\big] \le \frac{1}{\sqrt{[4\pi \log{(N)}]}},$$

so that in the limit no noise will be let through the threshold. For finite samples, this probability bound says that in no more than 13% of the realizations of length $N = 128$ or greater will *any* pure noise variables exceed the threshold (Johnstone and Silverman, 1997, p. 325). Universal thresholding thus typically removes all the noise, but, in doing so, it can inadvertently set some small signal transform coefficients to zero. Since this tends to smooth out the signal, universal thresholding thus ensures, with high probability, that the reconstruction is as smooth as – or smoother than – the true deterministic signal.

Second, if $\mathbf{D}$ is a nonzero signal such that the maximum element of $\mathbf{d} = \mathcal{O}\mathbf{D}$ is bounded as $N \to \infty$, then asymptotically we would estimate $\mathbf{D}$ as a vector of zeros (this is because $\delta^{(\mathrm{u})}$ will eventually get so large that it is highly unlikely that $|d_l + e_l|$ will ever exceed it). In developing asymptotic arguments to justify the use of universal thresholding, it is thus imperative that all nonzero signal transform coefficients keep ahead of the $\delta^{(\mathrm{u})}$ threshold as $N \to \infty$ if we are to extract $\mathbf{D}$ successfully. This requires 'playing the asymptotic game' in a manner that might seem peculiar to time series analysts. As an example (the details of which are the subject of Exercise [10.2]), suppose that we have a time series $\mathbf{X}$ of length $N = 16$ containing a signal $\mathbf{D}$ whose values are proportional to row $n = 14$ of the Haar DWT matrix (see Figure 57):

$$D_t \equiv \begin{cases} -1, & 0 \le t \le 7; \\ 1, & 8 \le t \le 15. \end{cases} \tag{401a}$$

The Haar DWT of this signal consists of $d_{14} = 4 = \sqrt{N}$ and $d_l = 0$ for $l \ne 14$, so the signal is isolated into one signal transform coefficient. Suppose that we increase N to infinity through powers of two; i.e., we let $N = 2^J$ for $J = 5, 6, \ldots$. If we decide to extend our original signal periodically, then the Haar DWT of the extended $\mathbf{D}$ has 2^{J-4} nonzero coefficients, each of which is equal to 4 (the proportion of nonzero coefficients is thus fixed at 1/16). This extension mimics the situation in which we observe more of a signal with the passage of time, as is common in signal processing (i.e., the sampling interval Δt between observations remains fixed at, say, unity, and we observe a total time span of $N\,\Delta t$ when N increases). We can argue that $\mathbf{P}\,[|O_l| \le \delta^{(\mathrm{u})}] \to 1$ as $N \to \infty$ for all l, so that the hard thresholding signal estimator $\widehat{\mathbf{D}}^{(\mathrm{ht})}$ becomes a vector of zeros with high probability and hence does not recover $\mathbf{D}$. On the other hand, suppose that we decide to extend our original signal by 'filling it in:'

$$D_t \equiv \begin{cases} -1, & 0 \le t \le \frac{N}{2} - 1; \\ 1, & \frac{N}{2} \le t \le N - 1. \end{cases} \tag{401b}$$

This extension mimics what we would get by more finely sampling a signal over a *fixed* time span (i.e., as N increases, we decrease Δt so that the product $N\,\Delta t$ remains fixed at, say, unity). This setup is not common in signal processing, where Δt is usually set so that the resulting Nyquist frequency $f_{\mathcal{N}} \equiv 1/(2\,\Delta t)$ is greater than the dominant frequency content in the signal (see, however, item [2] of the Comments and Extensions to Section 10.6 for an application in which the scheme of sampling more finely is quite natural). The Haar DWT of this signal has one nonzero coefficient, namely, $d_{N-2} = \sqrt{N}$. We can now argue that $\mathbf{P}\left[|O_l| \leq \delta^{(\mathrm{u})}\right] \to 1$ for all $l \neq N-2$ as $N \to \infty$, while $\mathbf{P}\left[|O_{N-2}| \leq \delta^{(\mathrm{u})}\right] \to 0$ so that $\widehat{\mathbf{D}}^{(\mathrm{ht})}$ becomes arbitrarily close to $\mathbf{D}$ with high probability. A practitioner who has just the original sixteen point time series could thus glean from these asymptotic arguments two quite conflicting opinions about the validity of using $\widehat{\mathbf{D}}^{(\mathrm{ht})}$!

In spite of these concerns about the interpretation of $\delta^{(\mathrm{u})}$, signal extraction based upon this thresholding level seems to work remarkably well in certain practical applications, particularly in view of its inherent simplicity. We give some examples in Section 10.5 and 10.7.

- *Cross-validation*

Another approach to choosing δ in the presence of IID Gaussian noise is known as cross-validation. This procedure has been extensively applied in the statistical literature, in particular as an automatic way of choosing a smoothing parameter in nonparametric regression (e.g., Green and Silverman, 1994). Adapted to the current context, we can formulate cross-validation in two ways (Nason, 1996). The first way, called 'two-fold' cross-validation, consists of splitting $X_0, \ldots, X_{N-1}$ into even and odd indexed downsamples. Given a threshold, we use the odd-indexed (even-indexed, respectively) downsamples to estimate a signal, which we then interpolate and compare – using squared differences – with the even-indexed (odd-indexed) downsamples. We select the threshold – call it $\hat{\delta}^{(\mathrm{tfcv})}$ – as the one minimizing the sum of all the squared differences. The second way, called 'leave-one-out' cross-validation, starts with $X_0, \ldots, X_{N-1}$ and removes X_t to obtain two subsequences, namely, the variables before X_t, and those after it. Given a threshold δ, we use these subsequences to produce an estimate $X_t^{(\delta)}$ of X_t. This is repeated for each t. We then choose the threshold – call it $\hat{\delta}^{(\mathrm{loocv})}$ – to be the value of $\delta > 0$ minimizing

$$\sum_{t=0}^{N-1} (X_t^{(\delta)} - X_t)^2. \qquad (402)$$

There are, however, two problems with using this procedure with transforms like the DWT. First, the partial DWT of level J_0 is only naturally defined for sample sizes that are multiples of 2^{J_0}, whereas now we have two subsequences of total length $N-1$. Second, we need some special rule for defining $X_t^{(\delta)}$. As we note at the end of Section 10.5, these difficulties can be overcome.

- *Minimum unbiased risk*

So far we have assumed the noise transform coefficients to be an IID Gaussian vector $\mathbf{e}$ that arises by transforming a similar vector $\boldsymbol{\epsilon}$ forming one part of the 'signal plus noise' model for $\mathbf{X}$. When the additive noise in $\mathbf{X}$ is instead a non-IID vector $\boldsymbol{\eta}$, we must consider noise transform coefficients that are also non-IID. We now consider the choice of threshold level for a simple non-IID model that we formulate directly in terms of the noise transform coefficients. Accordingly, let us assume that our transform model is

$$\mathbf{O} \equiv \mathcal{O}\mathbf{X} = \mathcal{O}\mathbf{D} + \mathcal{O}\boldsymbol{\eta} \equiv \mathbf{d} + \mathbf{n} \text{ so that } O_l = d_l + n_l,$$

where d_l and n_l are the lth components of $\mathbf{d} \equiv \mathcal{O}\mathbf{D}$ and $\mathbf{n} \equiv \mathcal{O}\boldsymbol{\eta}$, respectively. We further assume that

$$n_l \stackrel{\mathrm{d}}{=} \mathcal{N}(0, \sigma_{n_l}^2).$$

Thus each component n_l has a Gaussian distribution with mean zero and variance $\sigma_{n_l}^2$, but the model for the noise transform coefficients is no longer IID because the RVs n_l are allowed to have different variances and to be correlated.

Now let $O_l^{(\delta)}$ be an estimator of d_l that will use a yet to be determined threshold δ. Let $\mathbf{O}^{(\delta)}$ be the vector with lth component $O_l^{(\delta)}$. Define $\widehat{\mathbf{D}}^{(\delta)} \equiv \mathcal{O}^T\mathbf{O}^{(\delta)}$ and note that $\mathcal{O}\widehat{\mathbf{D}}^{(\delta)} = \mathcal{O}\mathcal{O}^T\mathbf{O}^{(\delta)} = \mathbf{O}^{(\delta)}$. Define the risk $R(\cdot,\cdot)$ involved in the estimation of $\mathbf{D}$ by $\widehat{\mathbf{D}}^{(\delta)}$ as

$$\begin{aligned} R(\widehat{\mathbf{D}}^{(\delta)}, \mathbf{D}) \equiv E\{\|\widehat{\mathbf{D}}^{(\delta)} - \mathbf{D}\|^2\} = E\{\|\mathcal{O}(\widehat{\mathbf{D}}^{(\delta)} - \mathbf{D})\|^2)\} &= E\{\|\mathbf{O}^{(\delta)} - \mathbf{d}\|^2)\} \\ &= E\Big\{\sum_{l=0}^{N-1} (O_l^{(\delta)} - d_l)^2\Big\}, \end{aligned}$$

where we have used the fact that the orthonormal transform $\mathcal{O}$ does not affect the squared norm. Following Stein (1981), let us consider a restricted class of estimators, namely, those that can be written as

$$O_l^{(\delta)} = O_l + A^{(\delta)}(O_l), \qquad l = 0, \ldots, N-1, \tag{403}$$

where formally $A^{(\delta)}(\cdot)$ must be a 'weakly differentiable' real-valued function (for a precise definition of this concept, see, for example, Härdle *et al.*, 1998, p. 72; roughly speaking, we should expect $A^{(\delta)}(\cdot)$ to be at least piecewise continuous). Then we have

$$O_l^{(\delta)} - d_l = n_l + A^{(\delta)}(O_l), \qquad l = 0, \ldots, N-1,$$

so that

$$E\{(O_l^{(\delta)} - d_l)^2\} = \sigma_{n_l}^2 + 2E\{n_l A^{(\delta)}(O_l)\} + E\{[A^{(\delta)}(O_l)]^2\}.$$

Since n_l is Gaussian with mean zero and variance $\sigma_{n_l}^2$, we have (making the change of variable $\lambda \equiv d_l + n_l$)

$$\begin{aligned} E\{n_l A^{(\delta)}(O_l)\} &= E\{n_l A^{(\delta)}(d_l + n_l)\} \\ &= \frac{1}{\sqrt{(2\pi\sigma_{n_l}^2)}} \int_{-\infty}^{\infty} n_l A^{(\delta)}(d_l + n_l) e^{-n_l^2/(2\sigma_{n_l}^2)}\, dn_l \\ &= \frac{1}{\sqrt{(2\pi\sigma_{n_l}^2)}} \int_{-\infty}^{\infty} (\lambda - d_l) A^{(\delta)}(\lambda) e^{-(\lambda - d_l)^2/(2\sigma_{n_l}^2)}\, d\lambda \\ &= -\frac{1}{\sqrt{(2\pi\sigma_{n_l}^2)}} \int_{-\infty}^{\infty} \sigma_{n_l}^2 A^{(\delta)}(\lambda) \frac{d}{d\lambda} e^{-(\lambda - d_l)^2/(2\sigma_{n_l}^2)}\, d\lambda. \end{aligned}$$

Using integration by parts,

$$E\{n_l A^{(\delta)}(O_l)\} = \frac{1}{\sqrt{(2\pi\sigma_{n_l}^2)}} \int_{-\infty}^{\infty} \left[\sigma_{n_l}^2 \frac{d}{d\lambda} A^{(\delta)}(\lambda)\right] e^{-(\lambda - d_l)^2/(2\sigma_{n_l}^2)}\, d\lambda,$$

which follows from the weak differentiability property. Hence,

$$E\{n_l A^{(\delta)}(O_l)\} = \sigma_{n_l}^2 E\left\{ \frac{d}{d\lambda} A^{(\delta)}(\lambda)\Big|_{\lambda=O_l} \right\}.$$

We can now write

$$E\{(O_l^{(\delta)} - d_l)^2\} = E\{\mathcal{R}(\sigma_{n_l}, O_l, \delta)\}, \tag{404a}$$

where the functional form of $\mathcal{R}(\cdot,\cdot,\cdot)$ is given by

$$\mathcal{R}(\sigma_{n_l}, x, \delta) \equiv \sigma_{n_l}^2 + 2\sigma_{n_l}^2 \frac{d}{dx} A^{(\delta)}(x) + [A^{(\delta)}(x)]^2.$$

Using Equation (404a), we now have

$$R(\widehat{\mathbf{D}}^{(\delta)}, \mathbf{D}) = E\left\{ \sum_{l=0}^{N-1} (O_l^{(\delta)} - d_l)^2 \right\} = E\left\{ \sum_{l=0}^{N-1} \mathcal{R}(\sigma_{n_l}, O_l, \delta) \right\},$$

where the quantity inside of the right-most braces is called Stein's unbiased risk estimator (SURE) for $R(\widehat{\mathbf{D}}^{(\delta)}, \mathbf{D})$ (Stein, 1981). In practice, given realizations o_l of the RVs O_l, $l = 0, \ldots, N-1$, we seek the value of δ minimizing

$$\sum_{l=0}^{N-1} \mathcal{R}(\sigma_{n_l}, o_l, \delta).$$

With δ based on SURE so chosen, we can then form the estimator $O_l^{(\delta)}$ of the signal transform coefficient d_l via Equation (403).

As an example of an estimator that fits into the framework of Equation (403), let us consider the soft threshold function defined in Equation (400a). If we take $A^{(\delta)}(\cdot)$ to be the piecewise continuous function

$$A^{(\delta)}(O_l) = \begin{cases} -O_l, & \text{if } |O_l| < \delta; \\ -\delta \operatorname{sign}\{O_l\}, & \text{if } |O_l| \geq \delta, \end{cases}$$

then we obtain $O_l^{(\delta)} = O_l + A^{(\delta)}(O_l) = O_l^{(\mathrm{st})}$. We can rewrite this $O_l^{(\delta)}$ as

$$O_l^{(\delta)} = O_l - O_l 1_{[0,\delta^2)}(O_l^2) - \delta \operatorname{sign}\{O_l\} 1_{[\delta^2,\infty)}(O_l^2),$$

where $1_{\mathcal{J}}(x)$ denotes the indicator function:

$$1_{\mathcal{J}}(x) \equiv \begin{cases} 1, & \text{if } x \in \mathcal{J}; \\ 0, & \text{otherwise.} \end{cases}$$

▷ **Exercise [404]:** Show that

$$\mathcal{R}(\sigma_{n_l}, O_l, \delta) = O_l^2 - \sigma_{n_l}^2 + (2\sigma_{n_l}^2 - O_l^2 + \delta^2) 1_{[\delta^2,\infty)}(O_l^2) \tag{404b}$$

for this choice of $A^{(\delta)}(O_l)$. ◁

Now $O_l^2 - \sigma_{n_l}^2$ in Equation (404b) does not depend on δ, so, given $O_l = o_l, l = 0, \ldots, N-1$, the SURE of δ, say $\delta^{(S)}$, is given by the value of δ minimizing

$$\sum_{l=0}^{N-1} (2\sigma_{n_l}^2 - o_l^2 + \delta^2) 1_{[\delta^2,\infty)}(o_l^2). \tag{405a}$$

Suppose we order $\{o_l\}$ to give $\{o_{(l)}\}$ such that

$$o_{(0)}^2 \le o_{(1)}^2 \le \cdots \le o_{(N-1)}^2,$$

and we then order $\{\sigma_{n_l}^2\}$ according to the ordering of the corresponding $\{o_l\}$ to give $\{\sigma_{n_{(l)}}^2\}$. To find $\delta^{(S)}$, we need to find the square root of the δ^2 that minimizes the function

$$\Upsilon(\delta^2) \equiv \sum_{l=0}^{N-1} \left(2\sigma_{n_{(l)}}^2 - o_{(l)}^2 + \delta^2\right) 1_{[\delta^2,\infty)}(o_{(l)}^2).$$

Define $o_{(-1)} = 0$ so that $o_{(k-1)}^2 \le o_{(k)}^2$ for $k = 0, \ldots, N-1$. If δ satisfies $o_{(k-1)}^2 < \delta^2 < o_{(k)}^2$ for some k, then we have

$$\begin{aligned}\Upsilon(\delta^2) &= (N-k)\delta^2 + \sum_{l=k}^{N-1} \left(2\sigma_{n_{(l)}}^2 - o_{(l)}^2\right) \\ &> (N-k)o_{(k-1)}^2 + \sum_{l=k}^{N-1} \left(2\sigma_{n_{(l)}}^2 - o_{(l)}^2\right) = \Upsilon(o_{(k-1)}^2)\end{aligned}$$

and hence the minimum value of $\Upsilon(\cdot)$ over $o_{(k-1)}^2 \le \delta^2 < o_{(k)}^2$ must be at $\delta^2 = o_{(k-1)}^2$. On the other hand, if $\delta^2 \ge o_{(N-1)}^2$, then $\Upsilon(\delta^2) = 0$ always, so the choice $\delta^2 = o_{(N-1)}^2$ achieves the minimum of $\Upsilon(\cdot)$ for this range of δ^2. We can thus find the minimum value of $\Upsilon(\cdot)$ over δ^2 by just computing the values $\Upsilon(o_{(k)}^2)$, $k = -1, 0, \ldots, N-1$. If we compare

$$\Upsilon(o_{(k)}^2) = (N-k-1)o_{(k)}^2 + \sum_{l=k+1}^{N-1} \left(2\sigma_{n_{(l)}}^2 - o_{(l)}^2\right)$$

with

$$\Upsilon(o_{(k+1)}^2) = (N-k-2)o_{(k+1)}^2 + \sum_{l=k+2}^{N-1} \left(2\sigma_{n_{(l)}}^2 - o_{(l)}^2\right),$$

we can see a convenient recursion that starts with $\Upsilon(o_{(N-1)}^2) = 0$ and then computes, for $k = N-2, N-3, \ldots, 0, -1$,

$$\begin{aligned}\Upsilon(o_{(k)}^2) &= \Upsilon(o_{(k+1)}^2) + \left(2\sigma_{n_{(k+1)}}^2 - o_{(k+1)}^2\right) + (N-k-1)o_{(k)}^2 - (N-k-2)o_{(k+1)}^2 \\ &= \Upsilon(o_{(k+1)}^2) + 2\sigma_{n_{(k+1)}}^2 + (N-k-1)\left(o_{(k)}^2 - o_{(k+1)}^2\right).\end{aligned} \tag{405b}$$

The approach just described results in a single threshold level δ to apply to all components O_l, even though their associated noise terms n_l have possibly differing variances. If in fact each n_l has a common variance so that $\sigma^2_{n_l} = \sigma^2_0$, say, for all l, the single SURE-based threshold level would be the (nonnegative) value of δ minimizing

$$\sum_{l=0}^{N-1} (2\sigma_0^2 - o_l^2 + \delta^2) 1_{[\delta^2,\infty)}(o_l^2). \tag{406a}$$

- *Groups of coefficients with common variance*

For some signal estimation problems (see, for example, Section 10.7) certain groups of noise transform coefficients will have a common variance. Suppose we relabel the components $\{O_l\}$ so that $O_{j,t}$ denotes the tth member of the jth group of coefficients, C_j, say, and that

$$O_{j,t} = d_{j,t} + n_{j,t} \text{ with } n_{j,t} \stackrel{\text{d}}{=} \mathcal{N}(0, \sigma_j^2);$$

i.e., members of the jth group each have the same variance σ_j^2. Then, for each group j and given $O_{j,t} = o_{j,t}$, it follows from Equation (406a) that a threshold value $\delta_j^{(\text{S})}$ could be found as the value of δ minimizing

$$\sum_{t \in C_j} (2\sigma_j^2 - o_{j,t}^2 + \delta^2) 1_{[\delta^2,\infty)}(o_{j,t}^2), \tag{406b}$$

where, for each j, we can use the same algorithm as before. This defines a level-dependent SURE-based threshold, valid for soft thresholding.

The universal thresholding approach can also be readily modified to this situation. Under the assumption that the signal $\mathbf{D}$ is a vector of zeros, the transform coefficients are identical to the noise coefficients, i.e., $O_{j,t} = n_{j,t}$. The $n_{j,t}$ RVs are correlated, but Gaussian, and

$$\mathbf{P}\Big[\max_{j,t} \{|n_{j,t}|/\sigma_j\} > \sqrt{(2\log{(N)})}\Big] \le B_c \le \frac{1}{\sqrt{[4\pi \log{(N)}]}},$$

(Johnstone and Silverman, 1997), where B_c denotes a bound applicable to the correlated case that is lower than the bound in the independence case. Furthermore, we can write

$$\mathbf{P}\Big[\max_j \big\{ \max_t \{|n_{j,t}|\} > \sqrt{(2\sigma_j^2 \log{(N)})}\big\}\Big] \le B_c,$$

so that

$$\mathbf{P}\Big[\max_j \big\{ \max_t \{|n_{j,t}|\} \le \delta_j^{(\text{u})}\big\}\Big] \to 1,$$

where

$$\delta_j^{(\text{u})} \equiv \sqrt{[2\sigma_j^2 \log{(N)}]} \tag{406c}$$

provides a conservative threshold level for coefficients in group j, thus defining a level-dependent 'universal' threshold. With $\delta_j^{(\text{u})}$ being thus set and with realizations $o_{j,t}$ of $O_{j,t}$, we would now implement, for example, hard thresholding (see Equation (399)) in each group j via

$$o_{j,t}^{(\text{ht})} = \begin{cases} 0, & \text{if } |o_{j,t}| \le \delta_j^{(\text{u})}; \\ o_{j,t}, & \text{otherwise,} \end{cases} \quad \text{for } t \in C_j.$$

10.3 Stochastic Signal Estimation via Scaling

Suppose that $\mathbf{X} = \mathbf{C} + \boldsymbol{\eta}$, where $\mathbf{C}$ represents an unknown Gaussian signal of interest, while $\boldsymbol{\eta}$ is undesired zero mean Gaussian noise; i.e., $\mathbf{C}$ and $\boldsymbol{\eta}$ are both multivariate Gaussian RVs, $E\{\boldsymbol{\eta}\} = \mathbf{0}$, and the covariance matrices are given by, respectively, $\Sigma_{\mathbf{C}}$ and $\Sigma_{\boldsymbol{\eta}}$. We assume that the random signal $\mathbf{C}$ is uncorrelated with the noise $\boldsymbol{\eta}$ (i.e., the covariance between any RV in $\mathbf{C}$ and any RV in $\boldsymbol{\eta}$ is zero – recall also that two uncorrelated Gaussian RVs are in fact independent).

Now let $\mathcal{O}$ be an $N \times N$ orthonormal matrix, and let $\mathbf{O} \equiv \mathcal{O}\mathbf{X}$ represent the N dimensional vector of transform coefficients $\{O_l : l = 0, \ldots, N-1\}$ corresponding to $\mathbf{X}$. Then

$$\mathbf{O} \equiv \mathcal{O}\mathbf{X} = \mathcal{O}\mathbf{C} + \mathcal{O}\boldsymbol{\eta} \equiv \mathbf{R} + \mathbf{n} \text{ and hence } O_l = R_l + n_l,$$

where R_l and n_l are the lth components of $\mathbf{R} \equiv \mathcal{O}\mathbf{C}$ and $\mathbf{n} \equiv \mathcal{O}\boldsymbol{\eta}$, respectively. A standard result from the theory of multivariate Gaussian RVs says that $\mathbf{R}$ and $\mathbf{n}$ are Gaussian RVs with covariance matrices given by, respectively, $\mathcal{O}\Sigma_{\mathbf{C}}\mathcal{O}^T$ and $\mathcal{O}\Sigma_{\boldsymbol{\eta}}\mathcal{O}^T$ (see Equation (262c)); moreover, because $\mathbf{C}$ and $\boldsymbol{\eta}$ are uncorrelated, their transforms $\mathbf{R}$ and $\mathbf{n}$ are also such, implying that R_l and $n_{l'}$ are uncorrelated for all l and l'.

Since $\boldsymbol{\eta}$ has zero mean, we know from Equation (262b) that $E\{\mathbf{n}\} = 0$, so in particular n_l has a zero mean. We assume here that any component l of interest to us is such that $E\{R_l\} = 0$. Note this does not require that $E\{\mathbf{C}\} = 0$, since the transform $\mathcal{O}$ can achieve this for many, if not all, of its coefficients (recall, for example, that DWTs trap the sample mean of a time series in the scaling coefficients, which are usually only a small portion of the total number of DWT coefficients). Henceforth in this chapter, when looking at stochastic signal estimation, we assume that we are dealing with components for which $E\{R_l\} = 0$. As shown in Section 10.5, this is easily achieved for the DWT.

Suppose that we wish to estimate the signal coefficient R_l in terms of the observed signal-plus-noise coefficient O_l using a simple weighting a_l, say,

$$\widehat{R}_l \equiv a_l O_l. \tag{407}$$

Suppose additionally that we decide to select a_l such that $E\{(R_l - \widehat{R}_l)^2\}$ is as small as possible.

▷ **Exercise [407]:** Show that the minimization of $E\{(R_l - a_l O_l)^2\}$ requires that $E\{(R_l - a_l O_l)O_l\} = 0$, a result that is known as the orthogonality principle of linear mean square estimation. ◁

In words, the estimation error and quantity being estimated are uncorrelated (i.e., 'orthogonal') to each other. Hence

$$E\{R_l O_l\} - a_l E\{O_l^2\} = 0, \text{ so } a_l = \frac{E\{R_l O_l\}}{E\{O_l^2\}}.$$

From the fact that $\{R_l\}$ and $\{n_l\}$ are uncorrelated and Gaussian, we have

$$E\{R_l O_l\} = E\{R_l^2\} \text{ and } E\{O_l^2\} = E\{R_l^2\} + E\{n_l^2\}.$$

Hence,

$$\widehat{R}_l = \frac{E\{R_l^2\}}{E\{R_l^2\} + E\{n_l^2\}} O_l.$$

Because the RVs R_l and n_l have zero mean, the quantities $\sigma^2_{R_l} \equiv E\{R_l^2\}$ and $\sigma^2_{n_l} \equiv E\{n_l^2\}$ are, respectively, signal and noise variances. Thus our estimate $\widehat{R}_l$ of the signal component R_l can be written in terms of the 'observed' coefficient O_l weighted by a term reflecting the signal variance as a proportion of the total variance for the lth component:

$$\widehat{R}_l = \frac{\sigma^2_{R_l}}{\sigma^2_{R_l} + \sigma^2_{n_l}} O_l. \tag{408}$$

In the presence of noise in the lth component, this weight effectively 'shrinks' O_l toward zero, and indeed, if the noise level is very high, can almost 'kill' O_l, delivering an estimate $\widehat{R}_l$ that is nearly zero. In the absence of noise in the lth component, the coefficient O_l is returned as the estimate of R_l, as is intuitively reasonable.

10.4 Stochastic Signal Estimation via Shrinkage

In Section 10.2 we looked at thresholding rules, which set some transform coefficients exactly to zero and handle other coefficients by leaving them unaltered or by reducing their magnitudes. In this section we concentrate on shrinkage rules. These differ from thresholding rules in that all nonzero transform coefficients are mapped to nonzero values – in fact they are either reduced in magnitude in some smooth nonlinear manner or left unaltered. In practice, however, shrinkage rules often resemble threholding rules in that there is a substantive range of nonzero coefficients that are shrunk to values very close to zero.

- *Conditional mean and median*

We continue here with the idea (introduced in Section 10.3) of treating the signal and noise as independent stochastic components, but note that we do not assume *a priori* that the signal and noise are Gaussian RVs. The orthonormal transform takes the form

$$\mathbf{O} \equiv \mathcal{O}\mathbf{X} = \mathcal{O}\mathbf{C} + \mathcal{O}\boldsymbol{\eta} \equiv \mathbf{R} + \mathbf{n} \text{ and hence } O_l = R_l + n_l,$$

where R_l is the lth component of the transformed stochastic signal $\mathbf{R}$, n_l is the lth component of the transformed noise $\mathbf{n}$, and we assume $E\{R_l\} = E\{n_l\} = 0$. We saw in Section 10.3 that, if we wish to produce a linear estimator of R_l based on scaling O_l and if we use a mean square error criterion, then the scaling factor is the signal variance as a proportion of the total variance for component l. Suppose that we still use the mean square error criterion, but now we consider a more general (nonlinear) estimator, say, $U_2(O_l)$ of R_l; i.e., $\widehat{R}_l = U_2(O_l)$, and $U_2(O_l)$ defines a shrinkage rule operating on O_l. As will be shown, the estimator $U_2(O_l)$ of R_l that minimizes $E\{(R_l - U_2(O_l))^2\}$ is the conditional mean, $E\{R_l|O_l\}$ (the subscript '2' emphasizes this is a squared norm minimizer).

Accordingly, let $\mathbf{U}_2$ be the vector whose lth component is $U_2(O_l)$. Since $U_2(O_l)$ is an estimator of R_l, the vector $\mathbf{U}_2$ is an estimator of $\mathbf{R} = \mathcal{O}\mathbf{C}$, i.e., the transformed stochastic signal. Define $\widehat{\mathbf{C}} \equiv \mathcal{O}^T\mathbf{U}_2$, and note that $\mathcal{O}\widehat{\mathbf{C}} = \mathbf{U}_2$. The risk $R(\cdot,\cdot)$ involved in the estimation of $\mathbf{C}$ by $\widehat{\mathbf{C}}$ is given by

$$\begin{aligned} R(\widehat{\mathbf{C}}, \mathbf{C}) \equiv E\{\|\widehat{\mathbf{C}} - \mathbf{C}\|^2\} &= E\{\|\mathcal{O}(\widehat{\mathbf{C}} - \mathbf{C})\|^2\} \\ &= E\{\|\mathbf{U}_2 - \mathbf{R}\|^2\} \\ &= E\left\{\sum_{l=0}^{N-1} (U_2(O_l) - R_l)^2\right\} = \sum_{l=0}^{N-1} E\left\{(U_2(O_l) - R_l)^2\right\}, \end{aligned}$$

where we have again used the fact that the orthonormal transform $\mathcal{O}$ does not affect the squared norm. This risk is thus clearly minimized by minimizing $E\{(U_2(O_l) - R_l)^2\}$ individually for each l. We claim that the estimator $U_2(O_l)$ of R_l that minimizes this mean square error criterion is the conditional mean $E\{R_l|O_l\}$. To see this, note that

$$E\{(U_2(O_l) - R_l)^2\} = \int_{-\infty}^{\infty}\int_{-\infty}^{\infty} (U_2(o_l) - r_l)^2 f_{R_l,O_l}(r_l, o_l)\, dr_l\, do_l,$$

where $f_{R_l,O_l}(\cdot,\cdot)$ is the joint probability density function (PDF) of the RVs R_l and O_l. Conditioning on the event that O_l assumes the value o_l, we can write $f_{R_l,O_l}(r_l, o_l) = f_{R_l|O_l=o_l}(r_l) f_{O_l}(o_l)$, yielding

$$E\{(U_2(O_l) - R_l)^2\} = \int_{-\infty}^{\infty} f_{O_l}(o_l) \int_{-\infty}^{\infty} (U_2(o_l) - r_l)^2 f_{R_l|O_l=o_l}(r_l)\, dr_l\, do_l.$$

The inner integrand is nonnegative, as is $f_{O_l}(o_l)$, so the double integral is minimized by minimizing

$$\int_{-\infty}^{\infty} (U_2(o_l) - r_l)^2 f_{R_l|O_l=o_l}(r_l)\, dr_l$$

for every outcome o_l. If we replace X_0 and X_1 in Equation (261a) by O_l and R_l, respectively, then we see that the integral above is identical to the one in Equation (261a), and hence

$$E\{R_l|O_l = o_l\} = \arg\min_{U_2(o_l)} \int (U_2(o_l) - r_l)^2 f_{R_l|O_l=o_l}(r_l)\, dr_l,$$

as claimed.

Now the conditional PDF $f_{R_l|O_l=o_l}(r_l)$ can be written

$$f_{R_l|O_l=o_l}(r_l) = \frac{f_{R_l,O_l}(r_l, o_l)}{f_{O_l}(o_l)}.$$

From Exercise [262a] we know that

$$f_{R_l,O_l}(r_l, o_l) = f_{R_l,n_l}(r_l, o_l - r_l),$$

where $f_{R_l,n_l}(\cdot,\cdot)$ is the joint PDF of the RVs R_l and n_l. Because we have assumed $\mathbf{C}$ and $\boldsymbol{\eta}$ to be independent, their transforms $\mathbf{R} = \mathcal{O}\mathbf{C}$ and $\mathbf{n} = \mathcal{O}\boldsymbol{\eta}$ must be also. Since R_l and n_l are thus independent, we have $f_{R_l,n_l}(r_l, o_l - r_l) = f_{R_l}(r_l) f_{n_l}(o_l - r_l)$. Hence, for additive noise, independent of the signal, we obtain

$$f_{R_l,O_l}(r_l, o_l) = f_{R_l}(r_l) f_{n_l}(o_l - r_l).$$

Furthermore, since $O_l = R_l + n_l$, the PDF of O_l is that of a sum, and we can use the result (see Equation (262a)) that the PDF of a sum of independent RVs is the convolution of their densities to get

$$f_{O_l}(o_l) = f_{R_l} * f_{n_l}(o_l) = \int_{-\infty}^{\infty} f_{R_l}(r_l) f_{n_l}(o_l - r_l)\, dr_l$$

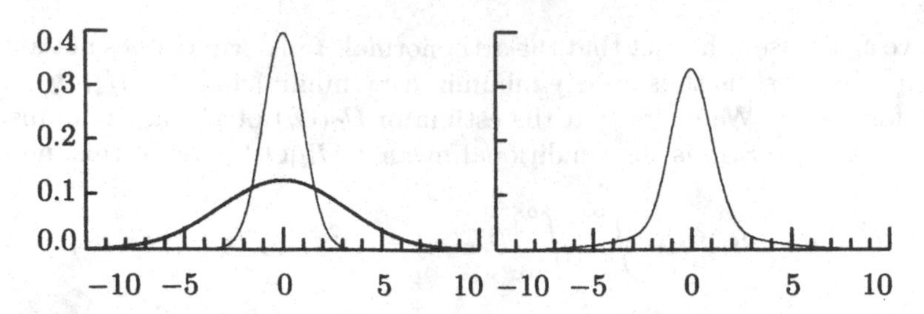

Figure 410. PDFs for $\mathcal{N}(0,1)$ and $\mathcal{N}(0,10)$ RVs (left-hand plot, thin and thick curves, respectively) and for an RV obeying a Gaussian mixture model (right-hand plot). The mixture PDF is non-Gaussian and is formed by adding the $\mathcal{N}(0,1)$ and $\mathcal{N}(0,10)$ PDFs, weighted by $p_l = 0.75$ and $1 - p_l = 0.25$, respectively (adapted from Figure 1 of Chipman *et al.*, 1997).

and hence

$$f_{R_l|O_l=o_l}(r_l) = \frac{f_{R_l}(r_l) f_{n_l}(o_l - r_l)}{\int_{-\infty}^{\infty} f_{R_l}(r_l) f_{n_l}(o_l - r_l)\, dr_l}. \tag{410a}$$

Given the realization o_l of O_l, we can write the conditional mean as

$$E\{R_l|O_l = o_l\} = \int_{-\infty}^{\infty} r_l f_{R_l|O_l=o_l}(r_l)\, dr_l = \frac{\int_{-\infty}^{\infty} r_l f_{R_l}(r_l) f_{n_l}(o_l - r_l) dr_l}{\int_{-\infty}^{\infty} f_{R_l}(r_l) f_{n_l}(o_l - r_l)\, dr_l}. \tag{410b}$$

Let us now specialize to the case in which the stochastic signal obeys a *Gaussian mixture model*, defined as follows. Let $\mathcal{I}_l$ be a binary-valued RV such that

$$\mathbf{P}[\mathcal{I}_l = 1] = p_l \text{ and } \mathbf{P}[\mathcal{I}_l = 0] = 1 - p_l, \tag{410c}$$

where $0 \le p_l \le 1$. We say that R_l follows a Gaussian mixture model if

$$R_l \stackrel{\text{d}}{=} \mathcal{I}_l \mathcal{N}(0, \gamma_l^2 \sigma_{G_l}^2) + (1 - \mathcal{I}_l)\mathcal{N}(0, \sigma_{G_l}^2), \tag{410d}$$

where the Gaussian RVs $\mathcal{N}(0, \gamma_l^2 \sigma_{G_l}^2)$ and $\mathcal{N}(0, \sigma_{G_l}^2)$ are independent of each other and of $\mathcal{I}_l$. We can obtain a realization of R_l by first generating a realization of $\mathcal{I}_l$ and then generating a realization of either $\mathcal{N}(0, \gamma_l^2 \sigma_{G_l}^2)$ if $\mathcal{I}_l = 1$ or $\mathcal{N}(0, \sigma_{G_l}^2)$ if $\mathcal{I}_l = 0$. The stochastic signal is thus a mixture of two Gaussian distributions with zero means and with variances $\gamma_l^2 \sigma_{G_l}^2$ and $\sigma_{G_l}^2$, respectively, in the proportion p_l to $1 - p_l$. The Gaussian mixture model is a way of modeling the overall signal as non-Gaussian. For example, if $\gamma_l \ll 1$ and $p_l = 0.95$, then on average we allow 5% of the signal terms to have a much larger variance $\sigma_{G_l}^2$ than the variance $\gamma_l^2 \sigma_{G_l}^2$ for the remaining 95%. The PDF for R_l is formed by adding the PDFs for $\mathcal{N}(0, \gamma_l^2 \sigma_{G_l}^2)$ and $\mathcal{N}(0, \sigma_{G_l}^2)$, weighted by, respectively, p_l and $1 - p_l$. Figure 410 shows an example for the case $p_l = 0.75$, $\gamma_l^2 \sigma_{G_l}^2 = 1$ and $\sigma_{G_l}^2 = 10$.

To complete our model for this special case, we assume that the noise is Gaussian with mean zero and variance $\sigma_{n_l}^2$:

$$n_l \stackrel{\text{d}}{=} \mathcal{N}(0, \sigma_{n_l}^2). \tag{410e}$$

Exercise [10.5] shows that, given this setup, we have

$$E\{R_l|O_l = o_l\} = \frac{a_l A_l(o_l) + b_l B_l(o_l)}{A_l(o_l) + B_l(o_l)} o_l, \tag{410f}$$

where

$$a_l \equiv \frac{\gamma_l^2 \sigma_{G_l}^2}{\gamma_l^2 \sigma_{G_l}^2 + \sigma_{n_l}^2}; \quad A_l(o_l) \equiv \frac{p_l}{\sqrt{(2\pi[\gamma_l^2 \sigma_{G_l}^2 + \sigma_{n_l}^2])}} e^{-o_l^2/[2(\gamma_l^2 \sigma_{G_l}^2 + \sigma_{n_l}^2)]};$$
$$b_l \equiv \frac{\sigma_{G_l}^2}{\sigma_{G_l}^2 + \sigma_{n_l}^2}; \quad \text{and } B_l(o_l) \equiv \frac{1 - p_l}{\sqrt{(2\pi[\sigma_{G_l}^2 + \sigma_{n_l}^2])}} e^{-o_l^2/[2(\sigma_{G_l}^2 + \sigma_{n_l}^2)]}. \tag{411a}$$

Since $U_2(O_l) = E\{R_l|O_l\}$, the conditional mean estimator $U_2(O_l)$ of R_l is now nonlinear in O_l because O_l^2 appears in the exponents of the expressions for $A_l(\cdot)$ and $B_l(\cdot)$ when these are regarded as RVs. The form of the conditional mean in Equation (410f) is discussed further in the Comments and Extensions to this section.

To gain some insight into how Equation (410f) shrinks a given o_l, let us consider an important special case in which $\gamma_l = 0$ (we can interpret $\mathcal{N}(0,0)$ as a degenerate RV whose realizations are always zero). The model for the signal becomes

$$R_l \stackrel{\text{d}}{=} (1 - \mathcal{I}_l)\mathcal{N}(0, \sigma_{G_l}^2), \tag{411b}$$

so that $\mathbf{P}[R_l = 0] = \mathbf{P}[\mathcal{I}_l = 1] = p_l$. For values of p_l approaching unity, this gives the so-called *sparse signal model*; i.e., on most occasions the signal is zero, but every once in a while it is nonzero. This sparse signal model leads to a particularly simple form for the conditional mean:

$$E\{R_l|O_l = o_l\} = \frac{b_l}{1 + c_l} o_l, \tag{411c}$$

where

$$c_l = \frac{p_l \sqrt{(\sigma_{G_l}^2 + \sigma_{n_l}^2)}}{(1 - p_l)\sigma_{n_l}} e^{-o_l^2 b_l/(2\sigma_{n_l}^2)}. \tag{411d}$$

We note that $E\{R_l|O_l = o_l\} \approx o_l b_l$ for large o_l, i.e., the linear form familiar from Equation (407). This form of conditional mean has been successfully employed in geophysical applications (Godfrey and Rocca, 1981; Walden, 1985). In order to compute the shrinkage rule $U_2(o_l) = E\{R_l|O_l = o_l\}$ in practice for this sparse signal model, three parameter values are required, namely, $p_l, \sigma_{G_l}^2$ and $\sigma_{n_l}^2$. On the other hand, if we were to use SURE with soft thresholding for this model, we would only require the noise variance $\sigma_{n_l}^2$ – essentially the need to deal with three parameters to determine the shrinkage rule in Equation (411c) is due to the fact that we are not preselecting the shape of the threshold function.

Figure 412 illustrates the conditional mean shrinkage rule of Equation (411c) when $p_l = 0.95, \sigma_{n_l}^2 = 1$ and $\sigma_{G_l}^2 = 5, 10$ and 25 (thick to thin curves, respectively). Note that, for $\sigma_{G_l}^2 = 25$, the shrinkage rule resembles a smoothed version of a firm thresholding rule (see Equation (400c)).

The conditional mean estimator $U_2(\cdot)$ arises from minimizing a mean square error. The function $U_2(\cdot)$ is not a thresholding rule, but rather a (nonlinear) shrinkage rule. No coefficients are completely killed by this procedure. We could equally well choose to minimize the mean absolute error. In this case the estimator, $U_1(\cdot)$ say, is the conditional median (see Equation (261b)). When O_l takes on the value o_l, we need to solve the following for $U_1(o_l)$:

$$\int_{-\infty}^{U_1(o_l)} f_{R_l|O_l=o_l}(r_l)\, dr_l = \frac{\int_{-\infty}^{U_1(o_l)} f_{R_l}(r_l) f_{n_l}(o_l - r_l)\, dr_l}{\int_{-\infty}^{\infty} f_{R_l}(r_l) f_{n_l}(o_l - r_l)\, dr_l} = 0.5, \tag{411e}$$

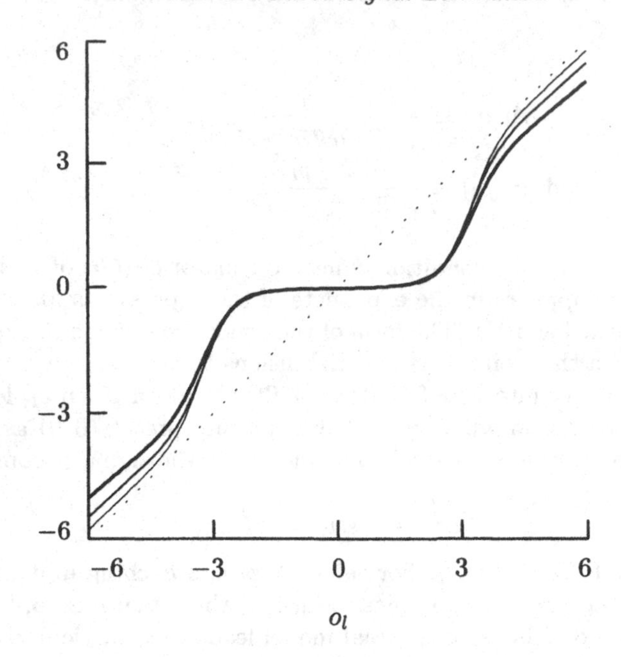

Figure 412. The conditional mean shrinkage rule of Equation (411c) for $p_l = 0.95, \sigma^2_{n_l} = 1$ and $\sigma^2_{G_l} = 5$ (thickest curve, furthest from dotted diagonal), 10 and 25 (thinnest curve, nearest to diagonal). Because of the correspondence between conditional mean shrinkage rules and the Bayes rule estimators of Chipman *et al.* (1997) with respect to squared error loss, the above also illustrates $B_2(\cdot)$ of Equation (414a).

where we have used Equation (410a). Godfrey and Rocca (1981) show that, for the sparse signal model, the shrinkage rule $U_1(\cdot)$ is approximately equal to a hard thresholding rule, namely,

$$U_1(O_l) \approx \begin{cases} 0, & \text{if } |O_l| \leq \delta; \\ b_l O_l, & \text{otherwise,} \end{cases} \tag{412a}$$

where the threshold is given by

$$\delta = \sigma_{n_l} \left[2 \log \left(\frac{p_l \sigma_{G_l}}{(1 - p_l)\sigma_{n_l}} \right) \right]^{1/2} \tag{412b}$$

(this approximation is valid provided that $p_l/(1-p_l) \gg \sigma^2_{n_l}/(\sigma_{G_l}\delta)$ and $\sigma^2_{G_l} \gg \sigma^2_{n_l}$).

- *Bayesian methods*

Much of the recent work on shrinkage has been done in the context of wavelet transforms and Bayesian methodology (the latter is briefly reviewed in Section 7.3). Some of the resultant shrinkage rules will be recognizable from our discussion of conditional means and medians. The setup that we consider for this discussion assumes that the signal is stochastic and that the noise is not necessarily IID, so we write $\mathbf{X} = \mathbf{C} + \boldsymbol{\eta}$. Applying the orthonormal transform $\mathcal{O}$, we obtain the model $\mathbf{O} = \mathbf{R} + \mathbf{n}$. We write the lth component of this model as $O_l = R_l + n_l$. For any given l, we make the assumption that the RVs R_l and n_l are independent of each other. Note that we have not made any assumptions at this point as to how the RVs in $\mathbf{n}$ are related to each other (the same holds for the RVs in $\mathbf{R}$). We need not do so because the signal extraction schemes we discuss below work on each signal component separately; i.e., we only use O_l to estimate the signal component R_l. Practical implementation of these schemes, however, requires certain parameters to be estimated, and it is at this point that we need to specify a model structure for $\mathbf{n}$ as a whole (we return to this point in Section 10.5).

Chipman *et al.* (1997) assume that the noise transform coefficient n_l is Gaussian with mean zero and variance $\sigma^2_{n_l}$ as in Equation (410e) and that the signal transform coefficient R_l has a Gaussian mixture PDF as dictated by Equation (410d). In the Bayesian formulation, we write this mixture model as

$$R_l|\mathcal{I}_l \stackrel{\mathrm{d}}{=} \mathcal{I}_l\mathcal{N}(0,\gamma_l^2\sigma^2_{G_l}) + (1-\mathcal{I}_l)\mathcal{N}(0,\sigma^2_{G_l}), \tag{413a}$$

where now $\mathcal{I}_l$ is regarded as a 'mixing parameter' with a prior distribution governed by Equation (410c). The distribution of $\mathcal{I}_l$ depends on p_l, which is now called a 'hyperparameter;' in addition, the RV $R_l|\mathcal{I}_l$ depends on this and two other hyperparameters, namely, γ_l^2 and $\sigma^2_{G_l}$. Note that the PDF for R_l itself (i.e., without the conditioning RV $\mathcal{I}_l$) is formed by adding the PDFs for $\mathcal{N}(0,\gamma_l^2\sigma^2_{G_l})$ and $\mathcal{N}(0,\sigma^2_{G_l})$, weighted by p_l and $1-p_l$. As we can see by comparing the right-hand sides of Equations (410d) and (413a), there is no real distinction between the RVs R_l and $R_l|\mathcal{I}_l$ when we don't have access to the realized values of $\mathcal{I}_l$ (when these are available, then R_l given $\mathcal{I}_l = 1$ is equal in distribution to the RV $\mathcal{N}(0,\gamma_l^2\sigma^2_{G_l})$, whereas R_l given $\mathcal{I}_l = 0$ is equal in distribution to $\mathcal{N}(0,\sigma^2_{G_l})$). Because $O_l = R_l + n_l$ and because n_l is Gaussian with mean zero and variance $\sigma^2_{n_l}$, it follows that

$$O_l|R_l \stackrel{\mathrm{d}}{=} \mathcal{N}(R_l,\sigma^2_{n_l}). \tag{413b}$$

In practice, $\sigma^2_{n_l}$ and the three hyperparameters all have to be estimated in order to obtain what is now called the 'Bayesian shrinkage rule.'

Vidakovic (1998) also looks at Bayesian shrinkage rules. Similar to Chipman *et al.* (1997), he assumes that n_l is Gaussian with mean zero and variance $\sigma^2_{n_l}$, but now only conditionally so: he does not regard the variance $\sigma^2_{n_l}$ as known, but rather gives it a prior distribution, namely, the exponential with hyperparameter v_l. Thus, in place of Equation (410e), we now have

$$n_l|\sigma^2_{n_l} \stackrel{\mathrm{d}}{=} \mathcal{N}(0,\sigma^2_{n_l}),$$

which in turn implies that

$$O_l|R_l,\sigma^2_{n_l} \stackrel{\mathrm{d}}{=} \mathcal{N}(R_l,\sigma^2_{n_l}). \tag{413c}$$

As discussed in Section 7.3, a variance mixture of the Gaussian distribution for which the variance has an exponential distribution produces a Laplace (two-sided exponential) distribution so that the marginal PDF for n_l is

$$f_{n_l}(n_l) = \frac{1}{\sqrt{(2/v_l)}}e^{-(2v_l)^{1/2}|n_l|}$$

(see Equation (257c)). The same argument holds for $O_l|R_l$, so we also have

$$f_{O_l|R_l=r_l}(o_l) = \frac{1}{\sqrt{(2/v_l)}}e^{-(2v_l)^{1/2}|o_l-r_l|}. \tag{413d}$$

In place of Equation (413a), Vidakovic (1998) completes the model specification by giving R_l a t distribution (Equation (258a)) with ϑ_l degrees of freedom and scale parameter κ_l (the location parameter is taken to be zero). The hyperparameters for this model are υ_l, ϑ_l and κ_l.

Let $B_2(O_l)$ be the Bayes rule estimator of R_l with respect to squared error loss; i.e., $B_2(O_l)$ is the minimizer of $E_{R_l}\{E_{O_l|R_l}\{(B_2(O_l) - R_l)^2\}\}$. As stated in Equation (265b), the Bayes rule in this case is given by the posterior mean. Given a realization o_l of O_l, the posterior mean is the same as the conditional mean of Equation (410b), namely

$$B_2(o_l) = \int_{-\infty}^{\infty} r_l f_{R_l|O_l=o_l}(r_l)\, dr_l,$$

and $B_2(o_l) \equiv U_2(o_l)$. Now

$$f_{R_l|O_l=o_l}(r_l) = \frac{f_{R_l,O_l}(r_l,o_l)}{f_{O_l}(o_l)} = \frac{f_{R_l}(r_l)f_{O_l|R_l=r_l}(o_l)}{f_{O_l}(o_l)} = \frac{f_{R_l}(r_l)f_{O_l|R_l=r_l}(o_l)}{\int_{-\infty}^{\infty} f_{R_l}(r_l)f_{O_l|R_l=r_l}(o_l)\, dr_l},$$

and hence we can write $B_2(o_l)$ as

$$B_2(o_l) = \frac{\int_{-\infty}^{\infty} r_l f_{R_l}(r_l)f_{O_l|R_l=r_l}(o_l)\, dr_l}{\int_{-\infty}^{\infty} f_{R_l}(r_l)f_{O_l|R_l=r_l}(o_l)\, dr_l}. \tag{414a}$$

For the model of Chipman *et al.* (1997), the resulting posterior mean (i.e., Bayes rule) is identical to the conditional mean given by Equation (410f) (and hence is illustrated in Figure 412). This follows since $f_{O_l|R_l=r_l}(o_l)$ specified by (413b) is

$$f_{O_l|R_l=r_l}(o_l) = \frac{1}{\sqrt{(2\pi\sigma_{n_l}^2)}} e^{-(o_l-r_l)^2/(2\sigma_{n_l}^2)},$$

which is identical to $f_{n_l}(o_l - r_l)$ in (410b) (the PDF of n_l is specified in Equation (410e)). The PDF of R_l is the same in both cases and is specified by Equation (410d).

For the model of Vidakovic (1998), we obtain the Bayes rule of Equation (414a) by using the t distribution for the signal PDF $f_{R_l}(\cdot)$ and by using the Laplace distribution for $f_{O_l|R_l=r_l}(o_l)$ (as per Equation (413d)). The resulting $B_2(\cdot)$ cannot be reduced to a tractable form, so we must resort to numerical integration. To compute the required integrals, we note that, for $o_l \geq 0$,

$$B_2(o_l) = \kappa_l\sqrt{\vartheta_l}\frac{A_1^-(o_l', o_l', \infty) + A_1^+(o_l', 0, o_l') - e^{-o_l\sqrt{(2\upsilon_l)}}A_1^-(0,0,\infty)}{A_0^-(o_l', o_l', \infty) + A_0^+(o_l', 0, o_l') + e^{-o_l\sqrt{(2\upsilon_l)}}A_0^-(0,0,\infty)}, \tag{414b}$$

where $o_l' \equiv o_l/(\kappa_l\sqrt{\vartheta_l})$ and

$$A_n^{\pm}(o_l', a, b) \equiv \int_a^b \frac{x^n}{(1+x^2)^{(\vartheta_l+1)/2}} e^{\pm(x-o_l')\kappa_l\sqrt{(2\upsilon_l\vartheta_l)}}\, dx$$

(proof of the above is Exercise [10.6]; note that we can obtain $B_2(o_l)$ for $o_l < 0$ using the relationship $B_2(o_l) = -B_2(-o_l)$). Figure 427 below shows two examples of $B_2(\cdot)$ in the context of wavelet coefficients.

If we let $B_1(O_l)$ be the Bayes rule estimator of R_l with respect to absolute error loss, i.e., $B_1(O_l)$ is the minimizer of $E_{R_l}\{E_{O_l|R_l}\{|B_1(O_l) - R_l|\}\}$, then the Bayes rule is the same as the posterior median (Equation (266)). Again, given a realization o_l of O_l, the posterior median is the same as the conditional median of Equation (411e), namely, the solution to

$$\int_{-\infty}^{B_1(o_l)} f_{R_l|O_l=o_l}(r_l)\, dr_l = 0.5.$$

Thus $B_1(o_l) \equiv U_1(o_l)$, and a computable form is given by the right side of Equation (411e) under our assumption of additive noise, independent of the signal (Equation (412a) gives a convenient approximation). Abramovich *et al.* (1998) examine the posterior median for the sparse signal model of Equation (411b). They find that

$$B_1(o_l) = \text{sign}\{o_l\} \max\{0, \zeta_l\}, \qquad (415)$$

where

$$\zeta_l = |o_l|b_l - (b_l\sigma_{n_l}^2)^{1/2}\Phi^{-1}\left(\frac{1 + \min\{c_l, 1\}}{2}\right),$$

and c_l is defined in Equation (411d) (here $\Phi^{-1}(\cdot)$ is the inverse of the standard Gaussian cumulative distribution function – see Equation (257b)). The quantity ζ_l is negative for all o_l in some implicitly defined interval $[-\delta, \delta]$. Hence $B_1(o_l)$ is set to zero whenever $|o_l|$ falls below the threshold δ, and we obtain a thresholding rule rather than a shrinkage rule.

Figure 416 compares the posterior median computed from Equation (415) (thin curve) to the *approximate* conditional median given in Equation (412a) (thick) for the case $\sigma_{G_l}^2 = 25, p_l = 0.95$ and $\sigma_{n_l}^2 = 1$, as used by Abramovich *et al.* (1998). It is clear that there is very little practical difference between these two thresholding rules, so the approximation of Godfrey and Rocca (1981) is adequate for this case. Note that the two conditions for this approximation to be valid (given after Equation (412b)) are indeed satisfied for the parameters chosen here.

Comments and Extensions to Section 10.4

[1] Here we develop an interpretation for the form of the conditional mean in Equation (410f). In view of Equation (410d), we can express R_l as

$$R_l = \mathcal{I}_l R_l^{(1)} + (1 - \mathcal{I}_l) R_l^{(2)},$$

where $R_l^{(1)}$ and $R_l^{(2)}$ are independent Gaussian RVs with zero means and variances given by, respectively, $\gamma_l^2\sigma_{G_l}^2$ and $\sigma_{G_l}^2$. Let $f_{R_l}(\cdot)$, $f_{R_l^{(1)}}(\cdot)$ and $f_{R_l^{(2)}}(\cdot)$ be the PDFs for, respectively, R_l, $R_l^{(1)}$ and $R_l^{(2)}$. Because R_l obeys a Gaussian mixture model, we have, as previously noted,

$$f_{R_l}(r_l) = p_l f_{R_l^{(1)}}(r_l) + (1 - p_l) f_{R_l^{(2)}}(r_l).$$

Recall that the transform coefficients O_l for the time series can be written as $O_l = R_l + n_l$, where n_l is a Gaussian RV with zero mean and variance $\sigma_{n_l}^2$ that is independent

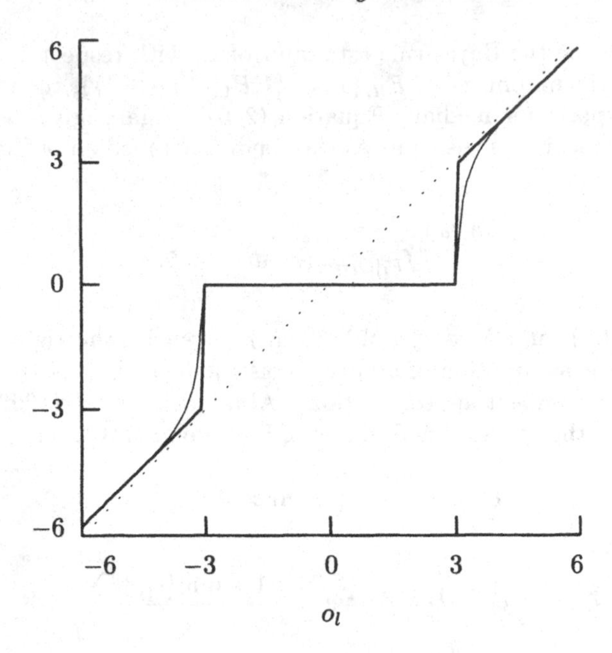

Figure 416. Comparison of the posterior median $B_1(o_l)$ (thin curve) to the *approximate* conditional median $U_1(o_l)$ (thick) when $\sigma^2_{G_l} = 25, p_l = 0.95$ and $\sigma^2_{n_l} = 1$. The dotted line marks the diagonal.

of R_l. Let $f_{O_l}(\cdot)$ and $f_{n_l}(\cdot)$ represent the PDFs for O_l and n_l. Now the PDF of a sum of independent RVs is the convolution of their densities (see Equation (262a)), so the PDF for O_l is given by the convolution of the PDFs for R_l and n_l:

$$f_{O_l}(o_l) = f_{R_l} * f_{n_l}(o_l) = p_l f_{R_l^{(1)}} * f_{n_l}(o_l) + (1 - p_l) f_{R_l^{(2)}} * f_{n_l}(o_l).$$

Since $R_l^{(1)}$ and n_l are independent Gaussian RVs with zero mean, we can interpret $f_{R_l^{(1)}} * f_{n_l}(\cdot)$ as the PDF $f_{O_l^{(1)}}(\cdot)$ of the sum $O_l^{(1)} \equiv R_l^{(1)} + n_l$. If we use the fact that the sum of two independent Gaussian RVs with zero means is also Gaussian with zero mean and a variance that is the sum of the individual variances, then the PDF $f_{O_l^{(1)}}(\cdot)$ is seen to be Gaussian with zero mean and variance $\gamma_l^2 \sigma^2_{G_l} + \sigma^2_{n_l}$. If we set $\mu = 0$ and $\sigma^2 = \gamma_l^2 \sigma^2_{G_l} + \sigma^2_{n_l}$ in the expression for the Gaussian PDF in Equation (256) and if we compare the resulting expression with $A_l(\cdot)$ of Equation (411a), we can conclude that $A_l(o_l) = p_l f_{O_l^{(1)}}(o_l)$. A similar argument tells us that $B_l(o_l) = (1 - p_l) f_{O_l^{(2)}}(o_l)$, where $B_l(o_l)$ is also given in Equation (411a), and $f_{O_l^{(2)}}(\cdot)$ is the PDF for $O_l^{(2)} \equiv R_l^{(2)} + n_l$. Hence, given the realization o_l of O_l, we can rewrite the conditional mean of Equation (410f) in the form

$$E\{R_l | O_l = o_l\} = \frac{a_l o_l p_l f_{O_l^{(1)}}(o_l) + b_l o_l (1 - p_l) f_{O_l^{(2)}}(o_l)}{p_l f_{O_l^{(1)}}(o_l) + (1 - p_l) f_{O_l^{(2)}}(o_l)}. \tag{416}$$

By comparing the definition of a_l in Equation (411a) with the simple shrinkage estimator of Equation (408), we see that the term $a_l o_l$ corresponds to the estimator that minimizes $E\{(R_l^{(1)} - a_l O_l)^2\}$, evaluated at $O_l = o_l$. Similarly, $b_l o_l$ corresponds to the estimator that minimizes $E\{(R_l^{(2)} - b_l O_l)^2\}$, evaluated at $O_l = o_l$. Thus the conditional mean can be thought of as minimum mean square error estimates for signal components $R_l^{(1)}$ and $R_l^{(2)}$, weighted by the likelihoods of o_l coming from the two signal components.

To understand this further, note that, in minimum mean square estimation, we require that $E\{(R_l^{(1)} - a_l O_l) O_l\} = 0$, as shown in Exercise [407]. This means that the RVs $R_l^{(1)} - a_l O_l$ and O_l are uncorrelated, but, since they are here jointly Gaussian, they are also independent. Thus

$$E\{(R_l^{(1)} - a_l O_l)|O_l\} = E\{R_l^{(1)} - a_l O_l\} = E\{R_l^{(1)}\} - a_l E\{O_l\} = 0;$$

but we also have

$$E\{(R_l^{(1)} - a_l O_l)|O_l\} = E\{R_l^{(1)}|O_l\} - E\{a_l O_l|O_l\} = E\{R_l^{(1)}|O_l\} - a_l O_l.$$

Given the realization o_l of O_l, the above tells us that $E\{R_l^{(1)}|O_l = o_l\} = a_l o_l$, so that the nonlinear conditional mean approach and the linear minimum mean square approach give the same estimator in the jointly Gaussian case. Clearly the same is true for the second component of the signal, and this lends some insight into why, for this model, the conditional mean takes the form of Equation (416).

[2] The Gaussian mixture model is certainly not the only possible choice for the distribution of the stochastic signal component in Equation (410b). Another flexible model is the generalized Gaussian distribution with PDF given in Equation (257d), for which it is easy to find the conditional mean using standard numerical integration techniques (Walden, 1985).

10.5 IID Gaussian Wavelet Coefficients

In Sections 10.2 and 10.4 we looked at the design of thresholding and shrinkage rules, which in principle can be applied to any collection of orthonormal transform coefficients. How well thresholding or shrinkage does in extracting a signal depends critically on the properties of the transform coefficients. We have already argued in Section 10.1 that the DWT can succinctly represent signals with a combination of time/frequency characteristics. We now specialize the application of thresholding and shrinkage rules to wavelet coefficients, and in so doing we discuss the estimation of unknown parameters. When the orthonormal matrix $\mathcal{O}$ is replaced by the DWT matrix $\mathcal{W}$ and thresholding or shrinkage is carried out, we talk about *wavelet thresholding* or *wavelet shrinkage* (sometimes abbreviated as *waveshrink*).

We look at IID Gaussian noise in this section, after which we consider uncorrelated non-Gaussian noise and then correlated Gaussian noise in, respectively, Sections 10.6 and 10.7.

• *Wavelet-based thresholding*

Let the model under consideration be $\mathbf{X} = \mathbf{D} + \boldsymbol{\epsilon}$, where $\mathbf{D}$ contains a deterministic signal. Upon application of the DWT $\mathcal{W}$, we obtain

$$\mathbf{W} = \mathcal{W}\mathbf{D} + \mathcal{W}\boldsymbol{\epsilon} = \mathbf{d} + \mathbf{e},$$

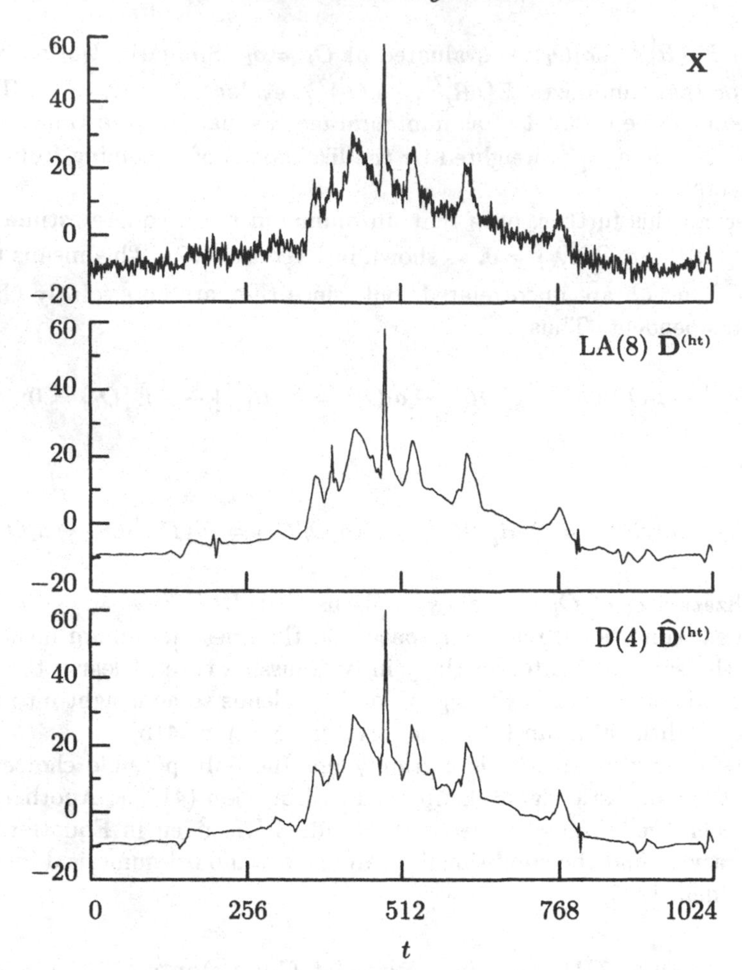

Figure 418. Nuclear magnetic resonance (NMR) spectrum (top plot), along with wavelet-based hard threshold signal estimates using the level $J_0 = 6$ partial LA(8) DWT (middle) and a similar D(4) DWT (bottom). In both cases, we determine the noise variance σ_ϵ^2 using the MAD standard deviation estimate $\hat{\sigma}_{(\mathrm{mad})}$, after which we set the universal threshold level $\hat{\delta}^{(\mathrm{u})} \equiv \sqrt{[2\hat{\sigma}^2_{(\mathrm{mad})} \log(N)]}$. This NMR spectrum was extracted from the public domain software package WaveLab, to which it was provided by Andrew Maudsley, Department of Radiology, University of California, San Francisco (the data can be accessed via the Web site for this book – see page *xiv*).

which we can regard as a special case of Equation (398). As discussed in Section 10.2, the universal threshold is suitable for use when the noise is IID Gaussian with a common variance σ_ϵ^2 so that the noise transform coefficients $\mathbf{e}$ are IID Gaussian, also with the common variance σ_ϵ^2. Donoho and Johnstone (1994) recommend using a level J_0 partial DWT rather than a full DWT, where J_0 must be specified by the user (see Section 4.11). In this scheme, the transform coefficients are contained in the vectors $\mathbf{W}_1, \ldots, \mathbf{W}_{J_0}$ and $\mathbf{V}_{J_0}$ making up $\mathbf{W}$, but only the coefficients in the $\mathbf{W}_k$ vectors are subjected to thresholding; i.e., the elements of $\mathbf{V}_{J_0}$ are untouched, so the portion of $\mathbf{X}$ attributable to the 'coarse' scale λ_{J_0} is automatically assigned to the signal $\mathbf{D}$.

The thresholding algorithm consists of the following steps.

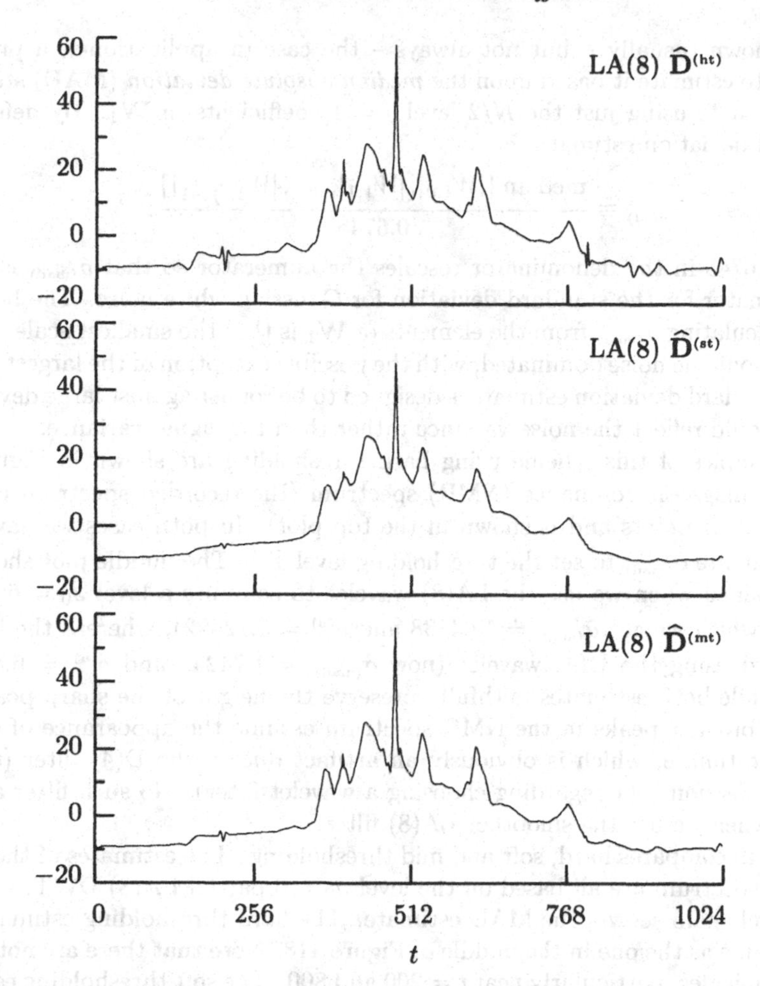

Figure 419. Thresholding signal estimates of the NMR spectrum based upon the level $J_0 = 6$ partial LA(8) DWT with – from top to bottom – hard, soft and mid thresholding (the top plot is a repeat of the middle of Figure 418). For all three estimates, we use the universal threshold level $\hat{\delta}^{(\mathrm{u})} \equiv \sqrt{[2\hat{\sigma}^2_{(\mathrm{mad})} \log(N)]} \doteq 6.12622$ based upon the MAD standard deviation estimate $\hat{\sigma}_{(\mathrm{mad})} \doteq 1.64538$.

[1] Compute a level J_0 partial DWT to obtain the coefficient vectors $\mathbf{W}_1, \ldots, \mathbf{W}_{J_0}$ and $\mathbf{V}_{J_0}$. Component-wise, we have

$$W_{j,t} = d_{j,t} + e_{j,t} \quad j = 1, \ldots, J_0; \ t = 0, \ldots, N_j - 1.$$

[2] Next specify the threshold level $\delta = \delta^{(\mathrm{u})}$. If σ_ϵ^2 is known, compute $\delta^{(\mathrm{u})}$ in Equation (400d). If σ_ϵ^2 is unknown, calculate $\hat{\sigma}_{(\mathrm{mad})}$ as shown in Equation (420) below, and replace $\delta^{(\mathrm{u})}$ by $\hat{\delta}^{(\mathrm{u})} \equiv \sqrt{[2\hat{\sigma}^2_{(\mathrm{mad})} \log(N)]}$.

[3] For $W_{j,t}, j = 1, \ldots, J_0$ and $t = 0, \ldots, N_j - 1$, apply one of the thresholding rules (hard, mid, soft or firm), with $W_{j,t}$ replacing O_l, to obtain the thresholded coefficients $W^{(\mathrm{t})}_{j,t}$, which are then used to form $\mathbf{W}^{(\mathrm{t})}_j, j = 1, \ldots, J_0$.

[4] Estimate $\mathbf{D}$ via $\widehat{\mathbf{D}}^{(\mathrm{t})}$ obtained by inverse transforming $\mathbf{W}^{(\mathrm{t})}_1, \ldots, \mathbf{W}^{(\mathrm{t})}_{J_0}$ and $\mathbf{V}_{J_0}$.

If σ_ϵ^2 is unknown (usually – but not always – the case in applications), a practical procedure is to estimate it based upon the *median absolute deviation* (MAD) standard deviation estimate using just the $N/2$ level $j = 1$ coefficients in $\mathbf{W}_1$. By definition, this standard deviation estimator is

$$\hat{\sigma}_{(\text{mad})} \equiv \frac{\text{median}\,\{|W_{1,0}|, |W_{1,1}|, \ldots, |W_{1,\frac{N}{2}-1}|\}}{0.6745}. \tag{420}$$

The factor 0.6745 in the denominator rescales the numerator so that $\hat{\sigma}_{(\text{mad})}$ is also a suitable estimator for the standard deviation for Gaussian white noise. The heuristic reason for calculating $\hat{\sigma}_{(\text{mad})}$ from the elements of $\mathbf{W}_1$ is that the smallest scale wavelet coefficients should be noise dominated, with the possible exception of the largest values. The MAD standard deviation estimate is designed to be robust against large deviations and hence should reflect the noise variance rather than the signal variance.

Two examples of this scheme using hard thresholding are shown in Figure 418 for a nuclear magnetic resonance (NMR) spectrum (the recorded spectrum consists of $N = 1024$ data points and is shown in the top plot). In both cases we have used the MAD estimate $\hat{\sigma}_{(\text{mad})}$ to set the thresholding level $\hat{\delta}^{(\text{u})}$. The middle plot shows the estimate obtained when we use the LA(8) wavelet to compute a level $J_0 = 6$ partial DWT (from which we get $\hat{\sigma}_{(\text{mad})} \doteq 1.64538$ and $\hat{\delta}^{(\text{u})} \doteq 6.12622$), whereas the bottom plot is formed using the D(4) wavelet (now $\hat{\sigma}_{(\text{mad})} \doteq 1.74245$ and $\hat{\delta}^{(\text{u})} \doteq 6.48766$). Note that, while both estimates faithfully preserve the height of the sharp peak near $t = 500$, the broader peaks in the NMR spectrum assume the appearance of sharks' fins in D(4) estimate, which is obviously an artifact due to the D(4) filter (see the discussion in Section 4.11 regarding choosing a wavelet filter). No such filter artifact is apparent when we use the smoother LA(8) filter.

Figure 419 compares hard, soft and mid thresholding. The estimates of the signal in the NMR spectrum are all based on the level $J_0 = 6$ partial LA(8) DWT, with the threshold level again set via the MAD estimate. The hard thresholding estimate (top plot) is the same as the one in the middle of Figure 418. Note that there are noticeable small scale squiggles, particularly near $t = 200$ and 800. The soft thresholding estimate (middle) plot suppresses these squiggles almost completely and is quite a bit smoother than the hard thresholding estimate, but a critique is that it suppresses the peaks somewhat. In particular, the height of the sharp peak is not faithfully rendered as it is in the case of hard thresholding. A general critique of soft thresholding is that energy levels are materially altered for large coefficients. The bottom plot of Figure 419 indicates that mid thresholding alleviates problems with energy levels, while offering adequate suppression of squiggles. Visually this last estimate is the most pleasing.

When the scheme outlined in steps [1]–[4] is implemented using soft thresholding, Donoho and Johnstone (1994) refer to this procedure as *VisuShrink*, but note that we would classify this as a thresholding rule since some wavelet coefficients will be set to zero by soft thresholding. Theorem 1 of Donoho and Johnstone (1994) states an important justification for VisuShrink, namely, that, for *all* possible signals $\mathbf{D}$, the risk achieved using it is within a logarithmic factor of the ideal risk $R(\widehat{\mathbf{D}}^{(\text{i})}, \mathbf{D})$ that can be obtained via an 'oracle' telling us which of the $W_{j,t}$ coefficients are noise dominated:

$$R(\widehat{\mathbf{D}}^{(\text{st})}, \mathbf{D}) \equiv E\{\|\widehat{\mathbf{D}}^{(\text{st})} - \mathbf{D}\|^2\} \leq [2\log(N) + 1][\sigma_\epsilon^2 + R(\widehat{\mathbf{D}}^{(\text{i})}, \mathbf{D})].$$

Instead of using the universal threshold level $\delta = \delta^{(\text{u})}$, we can use the SURE approach to define the threshold level. For the current assumption of IID noise wavelet

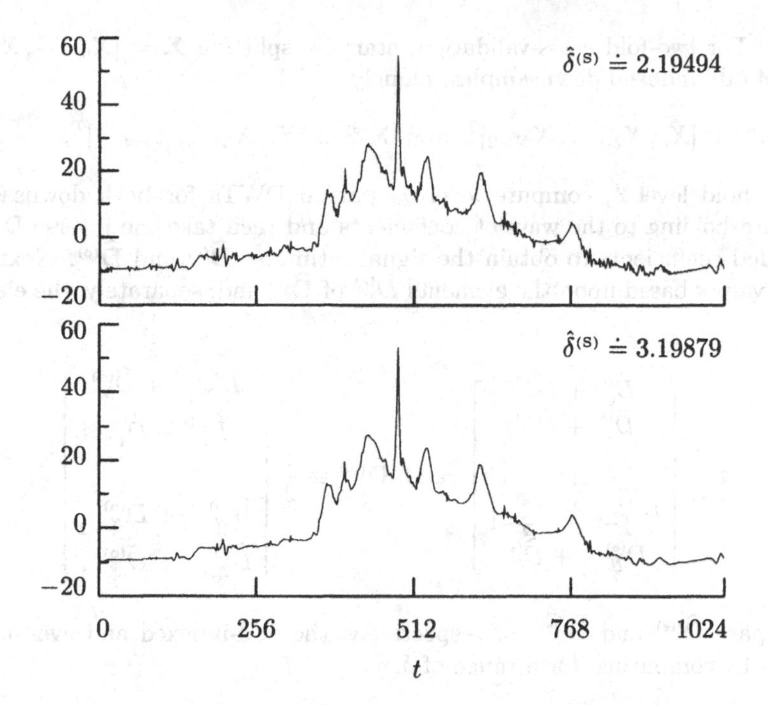

Figure 421. Thresholding signal estimates of the NMR spectrum based upon the level $J_0 = 6$ partial LA(8) DWT with soft thresholding and SURE threshold levels $\hat{\delta}^{(S)}$, which are computed using MAD scale estimates based on, respectively, just the unit scale wavelet coefficients (top plot) and wavelet coefficients from all six scales (bottom).

coefficients with variance σ_ϵ^2, we can use $\hat{\sigma}^2_{(\text{mad})}$ to estimate σ_ϵ^2, and, given realizations $w_{j,t}$ of $W_{j,t}$, we can then determine $\hat{\delta}^{(S)}$ by setting it equal to the square root δ of the δ^2 minimizing

$$\Upsilon(\delta^2) \equiv \sum_{j=1}^{J_0} \sum_{t=0}^{N_j-1} (2\hat{\sigma}^2_{(\text{mad})} - w_{j,t}^2 + \delta^2) 1_{[\delta^2,\infty)}(w_{j,t}^2) \tag{421}$$

(cf. Equation (406a)). Using the recursion given by Equation (405b) to compute $\Upsilon(\delta^2)$ at all possible minimizing values (i.e., $w_{j,t}^2$, $j = 1, \ldots, J_0 = 6$, $t = 0, \ldots, N_j - 1$), we obtain $\hat{\delta}^{(S)} \doteq 2.19494$ for the NMR spectrum, which is considerably smaller than the corresponding universal threshold level $\hat{\delta}^{(u)} \doteq 6.48766$. The resulting soft thresholding signal estimate is shown in the top plot of Figure 421 and is considerably noisier in appearance than the estimates in Figures 418 and 419. Since $\hat{\sigma}^2_{(\text{mad})}$ in the above is intended to be an estimate of the assumed common variance for all the wavelet coefficients, we can construct an alternative estimate based on the MAD of all such coefficients rather than just the unit scale coefficients. This procedure yields a noise variance estimate of 2.05270, a threshold of $\hat{\delta}^{(u)} \doteq 3.19879$ and the less noisy signal estimate shown on the bottom plot of Figure 421.

Finally, as noted in Section 10.2, an alternative to specifying the threshold level via the universal threshold or SURE is to use either two-fold or leave-one-out cross-validation, both of which Nason (1996) formulates in the wavelet setting using soft thresholding. For two-fold cross-validation, start by splitting $\mathbf{X} = [X_0, \ldots, X_{N-1}]^T$ into even and odd indexed downsamples, namely

$$\mathbf{X}^{(e)} \equiv [X_0, X_2, \ldots, X_{N-2}]^T \text{ and } \mathbf{X}^{(o)} \equiv [X_1, X_3, \ldots, X_{N-1}]^T.$$

Using a threshold level δ, compute level J_0 partial DWTs for both downsamples, apply soft thresholding to the wavelet coefficients and then take the inverse DWT of the thresholded coefficients to obtain the signal estimates $\widehat{\mathbf{D}}^{(e)}$ and $\widehat{\mathbf{D}}^{(o)}$. Next, form interpolated values based upon the elements $\widehat{D}_t^{(e)}$ of $\widehat{\mathbf{D}}^{(e)}$ and, separately, the elements $\widehat{D}_t^{(o)}$ of $\widehat{\mathbf{D}}^{(o)}$:

$$\widehat{\mathbf{D}}^{(ie)} \equiv \frac{1}{2} \begin{bmatrix} \widehat{D}_0^{(e)} + \widehat{D}_1^{(e)} \\ \widehat{D}_1^{(e)} + \widehat{D}_2^{(e)} \\ \vdots \\ \widehat{D}_{\frac{N}{2}-2}^{(e)} + \widehat{D}_{\frac{N}{2}-1}^{(e)} \\ \widehat{D}_{\frac{N}{2}-1}^{(e)} + \widehat{D}_0^{(e)} \end{bmatrix} \text{ and } \widehat{\mathbf{D}}^{(io)} \equiv \frac{1}{2} \begin{bmatrix} \widehat{D}_{\frac{N}{2}-1}^{(o)} + \widehat{D}_0^{(o)} \\ \widehat{D}_0^{(o)} + \widehat{D}_1^{(o)} \\ \vdots \\ \widehat{D}_{\frac{N}{2}-3}^{(o)} + \widehat{D}_{\frac{N}{2}-2}^{(o)} \\ \widehat{D}_{\frac{N}{2}-2}^{(o)} + \widehat{D}_{\frac{N}{2}-1}^{(o)} \end{bmatrix}.$$

Finally, compare $\widehat{\mathbf{D}}^{(ie)}$ and $\widehat{\mathbf{D}}^{(io)}$ to, respectively, the odd-indexed and even-indexed downsamples by computing, for a range of δ,

$$\|\mathbf{X}^{(o)} - \widehat{\mathbf{D}}^{(ie)}\|^2 + \|\mathbf{X}^{(e)} - \widehat{\mathbf{D}}^{(io)}\|^2. \tag{422}$$

Denote the threshold level δ minimizing the above as $\hat{\delta}$. Because this level is based on downsamples of size $N/2$ whereas the time series $\mathbf{X}$ is of size N, Nason (1996) recommends increasing $\hat{\delta}$ commensurate with the increase in the universal threshold level in going from sample size $N/2$ to N; i.e., the threshold level we actually use is $\hat{\delta}^{(tfcv)} \equiv \hat{\delta}\sqrt{\frac{\log(N)}{\log(N/2)}}$.

To implement leave-one-out cross-validation, fix a threshold level δ, and repeat the following steps for $1 \le l \le N-2$.

[1] Remove X_l from $\mathbf{X} = [X_0, \ldots, X_{N-1}]^T$, and split the remaining time series values into 'left' and 'right' vectors, namely,

$$\mathbf{X}^{(l)} \equiv [X_0, \ldots, X_{l-1}]^T \text{ and } \mathbf{X}^{(r)} \equiv [X_{l+1}, \ldots, X_{N-1}]^T.$$

[2] Reflect $\mathbf{X}^{(l)}$ and $\mathbf{X}^{(r)}$ at the top and bottom, respectively, and extend each vector so that its length is the next largest power of two by padding with X_{l-1} and X_{l+1}, respectively, to obtain the column vectors

$$\mathbf{X}^{(le)} \equiv [X_{l-1}, \ldots, X_{l-1}, X_{l-1}, \ldots, X_0, X_0, \ldots, X_{l-1}]^T \text{ and}$$
$$\mathbf{X}^{(re)} \equiv [X_{l+1}, \ldots, X_{N-1}, X_{N-1}, \ldots, X_{l+1}, X_{l+1}, \ldots, X_{l+1}]^T.$$

Note that the number of terms N' in $\mathbf{X}^{(le)}$ is the smallest power of two greater than $2l$, while the number of terms in $\mathbf{X}^{(re)}$ is the smallest power of two greater than $2(N-l-1)$.

[3] Apply the *full* DWT to the vectors $\mathbf{X}^{(le)}$ and $\mathbf{X}^{(re)}$, and then apply soft thresholding to the wavelet coefficients. Apply the inverse DWT to the thresholded vectors to obtain the signal estimates $\widehat{\mathbf{D}}^{(le)}$ and $\widehat{\mathbf{D}}^{(re)}$.

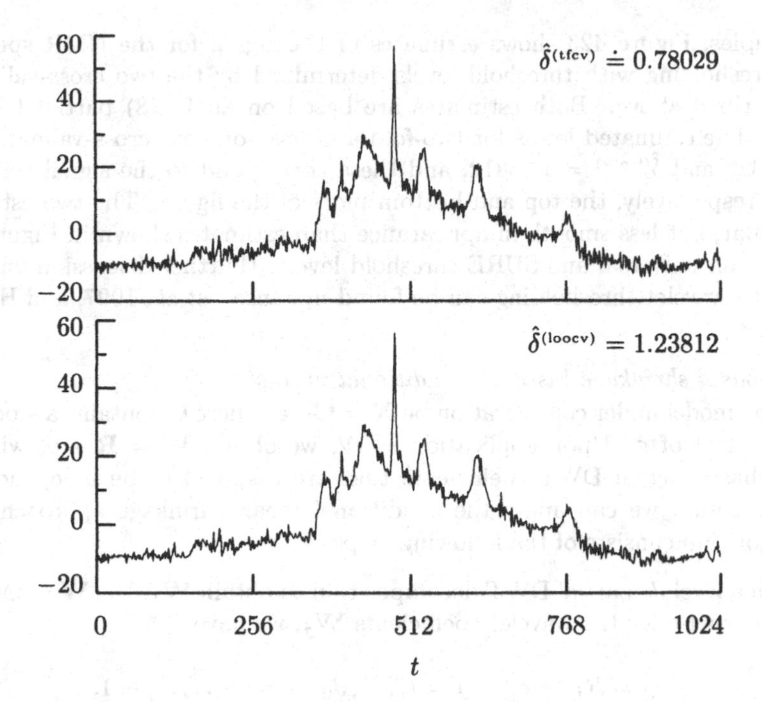

Figure 423. Thresholding signal estimates of the NMR spectrum based upon the level $J_0 = 6$ partial LA(8) DWT with soft thresholding and threshold levels determined by two-fold cross-validation (top plot) and leave-one-out cross-validation (bottom).

[4] Form $X_l^{(\delta)} \equiv (\widehat{D}_{N'-1}^{(\text{le})} + \widehat{D}_0^{(\text{re})})/2$, which is the resulting estimate of the removed point X_l based upon the bottom point $\widehat{D}_{N'-1}^{(\text{le})}$ in $\widehat{\mathbf{D}}^{(\text{le})}$ and the top point $\widehat{D}_0^{(\text{re})}$ in $\widehat{\mathbf{D}}^{(\text{re})}$.

The sum of squared errors $\sum_{l=1}^{N-2}(X_l^{(\delta)} - X_l)^2$ can then be constructed. This is calculated for a range of δ values, and then $\hat{\delta}^{(\text{loocv})}$ is chosen accordingly by minimizing Equation (402).

Three comments are in order regarding selecting δ by one of these cross-validation schemes. First, while we are assuming IID Gaussian noise $\boldsymbol{\epsilon}$, in fact Gaussianity is not a requirement, but Nason (1996) notes that neither scheme works well if we were to replace $\boldsymbol{\epsilon}$ in the model $\mathbf{X} = \mathbf{D} + \boldsymbol{\epsilon}$ with correlated noise $\boldsymbol{\eta}$ (this is a well-known problem with cross-validation methods). Second, for leave-one-out cross-validation, the method of extending the data length to powers of two by padding with X_{l-1} or X_{l+1} ensures circular vectors $\mathbf{X}^{(\text{le})}$ and $\mathbf{X}^{(\text{re})}$ for input to the DWT. This method of padding is motivated by the procedure at hand and differs from what we discuss in Section 4.11, where we advocate padding with the sample mean because this has the desirable property of preserving the sample variance. Third, while leave-one-out cross-validation does have an advantage over two-fold cross-validation in that the former is applicable to any sample size, the latter is much easier to compute; moreover, while the latter makes use of a single sample size ($N/2$) throughout (with an adjustment at the end to produce a level appropriate for a time series with N values), leave-one-out cross-validation involves a mixture of sample sizes ranging from two up to at least $2N - 2$, so it is not clear how relevant the resulting level is for sample size N.

As examples, Figure 423 shows estimates of the signal for the NMR spectrum using soft thresholding with threshold levels determined by the two cross-validation methods described above. Both estimates are based on an LA(8) partial DWT of level $J_0 = 6$. The estimated levels for two-fold and level-one-out cross-validation are $\hat{\delta}^{(\text{tfcv})} \doteq 0.78029$ and $\hat{\delta}^{(\text{loocv})} = 1.23812$, and these correspond to the signal estimates displayed in, respectively, the top and bottom plots of the figure. The two estimates are quite similar, but less smooth in appearance than estimates shown in Figures 419 and 421 based on universal and SURE threshold levels. (Further discussion on cross-validation with wavelet thresholding can be found in Jansen *et al.*, 1997, and Hurvich and Tsai, 1998.)

- *Wavelet-based shrinkage based on conditional means*

We now let the model under consideration be $\mathbf{X} = \mathbf{C} + \boldsymbol{\epsilon}$, where $\mathbf{C}$ contains a stochastic signal independent of $\boldsymbol{\epsilon}$. Upon application of $\mathcal{W}$, we obtain $\mathbf{W} = \mathbf{R} + \mathbf{e}$, where $\mathbf{R}$ contains stochastic signal DWT coefficients that are assumed to be independent of $\mathbf{e}$. With this setup, we can apply the conditional mean shrinkage approach. The shrinkage algorithm consists of the following steps.

[1] Compute a level J_0 partial DWT decomposition to obtain $\mathbf{W}_1, \ldots, \mathbf{W}_{J_0}$ and $\mathbf{V}_{J_0}$. Component-wise for the wavelet coefficients $\mathbf{W}_j$, we have

$$W_{j,t} = R_{j,t} + e_{j,t} \quad j = 1, \ldots, J_0; \;\; t = 0, \ldots, N_j - 1.$$

Recall our assumption that the signal coefficients $R_{j,t}$ have zero mean; as shown in Section 8.2, this is guaranteed for wavelet coefficients provided that the stochastic signal $\mathbf{C}$ is from a stochastic process whose dth order backward difference is stationary, and we take the filter length L such that $L > 2d$ (note that it is *not* reasonable to assume that the scaling coefficients in $\mathbf{V}_{J_0}$ have zero mean, but this is not of concern since the scheme is to leave $\mathbf{V}_{J_0}$ untouched).

[2] Suppose that we use the sparse signal model (Equation (411b)) and assume the parameters do not vary with level j or position t; i.e.,

$$R_{j,t} \stackrel{\text{d}}{=} (1 - \mathcal{I}_{j,t})\mathcal{N}(0, \sigma_G^2) \text{ and } e_{j,t} \stackrel{\text{d}}{=} \mathcal{N}(0, \sigma_\epsilon^2), \tag{424}$$

where, for $0 \le p \le 1$,

$$\mathbf{P}[\mathcal{I}_{j,t} = 1] = p \text{ and } \mathbf{P}[\mathcal{I}_{j,t} = 0] = 1 - p.$$

Hence the signal and noise components are identically distributed for all j and t. If we let σ_R^2 and σ_W^2 stand for the common variances of, respectively, the RVs $R_{j,t}$ and $W_{j,t}$, we have $\sigma_R^2 = (1-p)\sigma_G^2$ and $\sigma_W^2 = \sigma_R^2 + \sigma_\epsilon^2$. To estimate the parameters needed to implement the shrinkage rule, we can

(a) take $\hat{\sigma}_\epsilon = \hat{\sigma}_{(\text{mad})}$;

(b) form $\hat{\sigma}_W^2$ by taking the sample mean of the squared coefficients $W_{j,t}^2$ over $j = 1, \ldots, J_0$ and $t = 0, \ldots, N_j - 1$; and

(c) use $\hat{\sigma}_G^2 = (\hat{\sigma}_W^2 - \hat{\sigma}_\epsilon^2)/(1-p)$ to estimate σ_G^2 for a given value of p (typically p itself is set by trying out several different values and making a subjective judgment as to what is most appropriate for a particular application – this is illustrated in Figure 425).

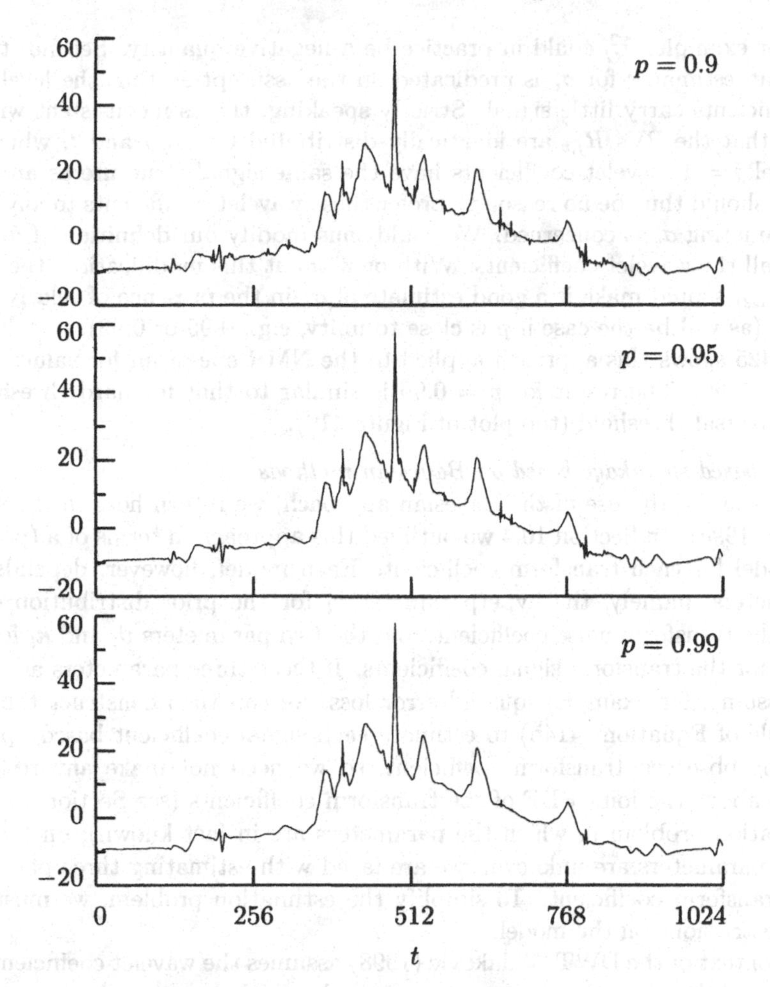

Figure 425. Shrinkage signal estimates of the NMR spectrum based upon the level $J_0 = 6$ partial LA(8) wavelet transform and the conditional mean with $p = 0.9$ (top plot), 0.95 (middle) and 0.99 (bottom). The remaining parameters (namely, $\sigma_\epsilon^2, \sigma_W^2$ and σ_G^2) are estimated as explained in the text.

[3] Given realizations $w_{j,t}$ of $W_{j,t}$, we then apply Equation (411c) to obtain the shrinkage estimates of the signal coefficients:

$$w_{j,t}^{(\mathrm{shr})} = \frac{\hat{b}}{1+\hat{c}} w_{j,t}, \tag{425}$$

where, for a given value of p,

$$\hat{b} = \frac{\hat{\sigma}_G^2}{\hat{\sigma}_G^2 + \hat{\sigma}_\epsilon^2} \text{ and } \hat{c} = \frac{p\sqrt{(\hat{\sigma}_G^2 + \hat{\sigma}_\epsilon^2)}}{(1-p)\hat{\sigma}_\epsilon} e^{-w_{j,t}^2 \hat{b}/(2\hat{\sigma}_\epsilon^2)}.$$

[4] Estimate the realization, say $\mathbf{c}$, of $\mathbf{C}$ via $\widehat{\mathbf{c}}^{(\mathrm{shr})}$ obtained by inverse transforming $\mathbf{w}_1^{(\mathrm{shr})}, \ldots, \mathbf{w}_{J_0}^{(\mathrm{shr})}$ and $\mathbf{V}_{J_0}$, where $\mathbf{w}_j^{(\mathrm{shr})}$ has components $w_{j,t}^{(\mathrm{shr})}, t = 0, \ldots, N_j - 1$.

We must make two comments about the approach to parameter estimation given in step [2]. First, the subtraction of estimated variances in general is fraught with problems; for example, $\hat{\sigma}_G^2$ could in practice be a negative quantity. Second, the use of $\hat{\sigma}_{(\text{mad})}$ as an estimator for σ_ϵ is predicated on the assumption that the level $j = 1$ wavelet coefficients carry little signal. Strictly speaking, this is inconsistent with our assumption that the RVs $R_{j,t}$ are identically distributed for all j and t, which says that the level $j = 1$ wavelet coefficients have the same signal structure as any other level – there should thus be no reason to prefer these wavelet coefficients to any others as far as estimating σ_ϵ is concerned. We could thus modify our definition of $\hat{\sigma}_{(\text{mad})}$ so that it uses all the wavelet coefficients. With or without this modification, the robust nature of $\hat{\sigma}_{(\text{mad})}$ should make it a good estimate of σ_ϵ in the presence of relatively few signal terms (as will be the case if p is close to unity, e.g., 0.95 or 0.99).

Figure 425 shows this approach applied to the NMR spectrum for values of p of 0.9, 0.95 and 0.99. The result for $p = 0.99$ is similar to that for hard thresholding using the universal threshold (top plot of Figure 419).

- *Wavelet-based shrinkage based on Bayesian methods*

As an illustration of the use of the Bayesian approach, we return here to the scheme of Vidakovic (1998). In Section 10.4 we outlined this approach in terms of a (possibly) separate model for each transform coefficient. Each model, however, depends upon three parameters, namely, the hyperparameter υ_l for the prior distribution on the variance of the transform noise coefficient, and the two parameters ϑ_l and κ_l for the t distribution for the transform signal coefficients. If these three parameters are known and if we assume, for example, squared error loss, we can then construct the Bayes shrinkage rule of Equation (414b) to estimate each signal coefficient based upon the corresponding observed transform coefficient, so we need not make any restrictive assumptions about the joint PDF of the transform coefficients (see Section 10.7 for a signal estimation problem in which the parameters are in fact known); on the other hand, if the parameters are unknown, we are faced with estimating three parameters from each transform coefficient. To simplify the estimation problem, we must place additional restrictions on the model.

In the context of the DWT, Vidakovic (1998) assumes the wavelet coefficients $W_{j,t}$ all depend upon the same three parameters, namely, υ (the common hyperparameter for the exponential prior distribution on the variance of the noise wavelet coefficients) and ϑ and κ (the common parameters for the PDF of the signal wavelet coefficients). Additionally, he assumes that each noise wavelet coefficient is conditioned on a single RV σ_ϵ^2. Thus, conditional on the outcome of this one RV, the noise DWT coefficients $\mathbf{e} = \mathcal{W}\boldsymbol{\epsilon}$ are in fact IID Gaussian (Chipman *et al.*, 1997, and Abramovich *et al.*, 1998, also arrive at this noise model, but from different considerations). Component-wise, for a level J_0 partial DWT, Vidakovic's approach uses

$$W_{j,t} = R_{j,t} + e_{j,t}.$$

The model is then

$$e_{j,t}|\sigma_\epsilon^2 \stackrel{\text{d}}{=} \mathcal{N}(0, \sigma_\epsilon^2) \text{ and hence } W_{j,t}|R_{j,t}, \sigma_\epsilon^2 \stackrel{\text{d}}{=} \mathcal{N}(R_{j,t}, \sigma_\epsilon^2),$$

where σ_ϵ^2 has a prior (exponential) distribution (cf. Equation (413c)). Given observed wavelet coefficients, this model has three unknown parameters (υ, ϑ and κ) and the unknown realization of the RV σ_ϵ^2. Since the signal and noise terms are identically distributed for all j and t, we can again let σ_R^2, σ_ϵ^2 and σ_W^2 represent the variances

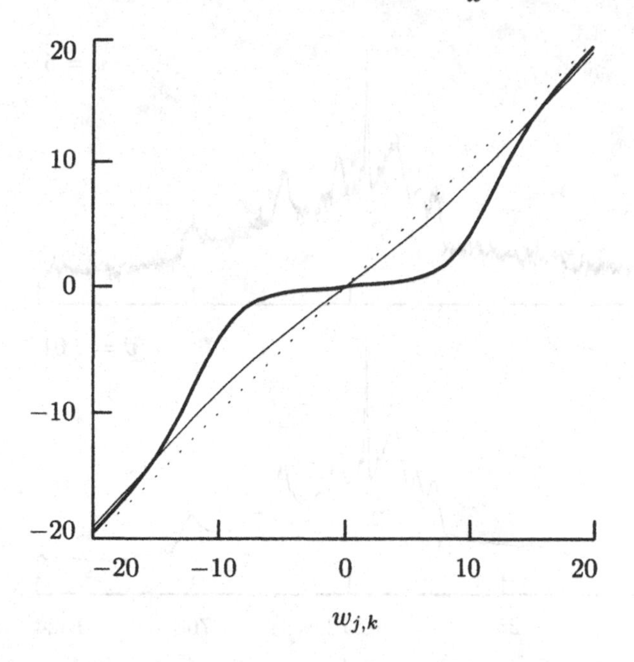

Figure 427. Bayes rules $B_2(w_{j,k})$ versus $w_{j,k}$ for the Vidakovic (1998) scheme as applied to the NMR spectrum. The thin and thick curves depict $B_2(\cdot)$ assuming degrees of freedom ϑ of, respectively, 5 and 2.01. The dotted line marks the diagonal. The corresponding signal estimates are shown in Figure 428.

common to, respectively, the RVs $R_{j,t}$, $e_{j,t}$ and $W_{j,t}$. We can thus obtain the required parameters as follows.

[1] We can estimate the realization of σ_ϵ using $\hat{\sigma}_{(\text{mad})}$ (Equation (420)), since, conditional on this realization, the IID assumption is in place.

[2] For the hyperparameter υ, we know that $E\{\sigma_\epsilon^2\} = 1/\upsilon$ because σ_ϵ^2 has an exponential distribution. Thus Vidakovic suggests taking $\hat{\upsilon} = 1/\hat{\sigma}_\epsilon^2$.

[3] Since the zero mean RVs $R_{j,t}$ have a t distribution with ϑ degrees of freedom and scale parameter κ, we know that their common variance is

$$\sigma_R^2 = \frac{\kappa^2 \vartheta}{\vartheta - 2}.$$

A value for κ can be specified via

$$\hat{\kappa} = \left(\hat{\sigma}_R^2 \frac{\vartheta - 2}{\vartheta} \right)^{1/2},$$

where $\hat{\sigma}_R^2$ is found from $\hat{\sigma}_R^2 = \hat{\sigma}_W^2 - \hat{\sigma}_\epsilon^2$, while $\hat{\sigma}_W^2$ is formed by averaging $W_{j,t}^2$ over $j = 1, \ldots, J_0$ and $t = 0, \ldots, N_j - 1$; however, as mentioned previously, the subtraction of estimated variances can be problematic.

[4] Vidakovic (1998) finds that setting the number of degrees of freedom ϑ to five works well.

We note that, unlike Vidakovic (1998), Chipman *et al.* (1997) and Abramovich *et al.* (1998) allow hyperparameters to vary with the level j.

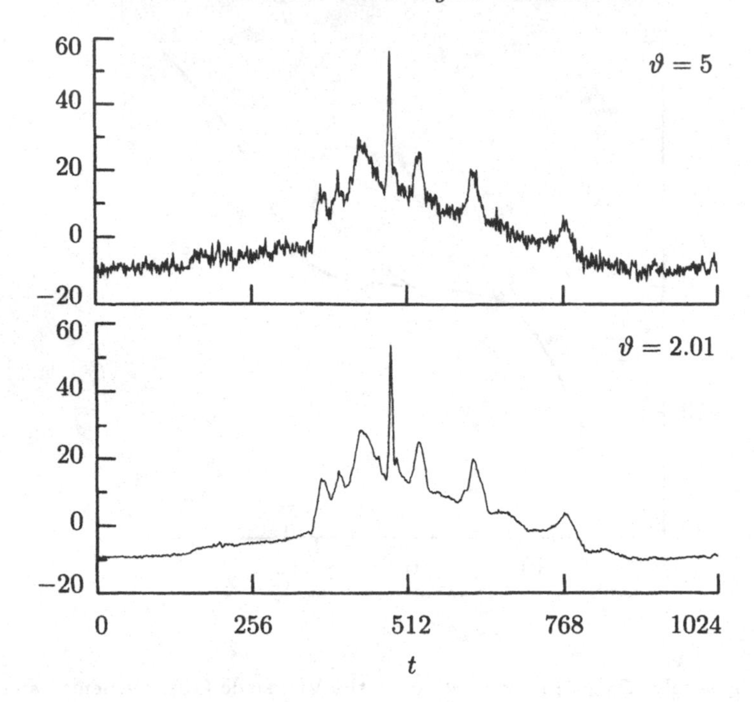

Figure 428. Shrinkage signal estimates of the NMR spectrum based upon the level $J_0 = 6$ partial LA(8) wavelet transform and the Bayes rule $B_2(\cdot)$ as formulated by Vidakovic (1998) and given in Equation (414b) (the specific rule for each estimate is plotted in Figure 427). The difference between the two estimates is solely due to the choice of the degrees of freedom ϑ for the signal PDF $f_R(\cdot)$, with the remaining two parameters (κ and υ) estimated as described in the text.

As an example, let us apply the Vidakovic procedure to the NMR data, again using a level $J_0 = 6$ LA(8) partial DWT. To estimate the required parameter υ, we compute $\hat{\sigma}_{(\text{mad})} \doteq 1.64538$, yielding $\hat{\upsilon} \doteq 0.36938$; to estimate κ, we compute $\hat{\sigma}^2_W \doteq 14.36755$ and $\hat{\sigma}^2_R \doteq 11.66029$ and set $\vartheta = 5$, from which we then get $\hat{\kappa} \doteq 2.64503$. The resulting Bayes rule $B_2(\cdot)$ can be computed using numerical integration via Equation (414b) and is plotted as the thin curve in Figure 427. The corresponding estimate of the signal is shown in the top plot of Figure 428. Here the Bayes rule dictates a very small amount of shrinkage, so the signal estimate and the original data (top plot of Figure 418) are quite similar. Suppose, however, we assume that the transform signal PDF $f_R(\cdot)$ is heavier tailed than what would be dictated by setting the degrees of freedom ϑ to five; i.e., we allow for the possibility that a small number of signal wavelet coefficients can be quite large, as might be reasonable given the sharp peaks in the observed NMR spectrum. If we hold $\text{var}\{R\}$ fixed at its estimate $\hat{\sigma}^2_R \doteq 11.66029$, we can make the tails of $f_R(\cdot)$ more prominent by decreasing ϑ toward two (doing this also has the effect of shortening the symmetric interval about zero that traps a given transform signal coefficient with, say, 95% probability). The subjective choice $\vartheta = 2.01$ yields the Bayes rule shown by the thick curve in Figure 427 and the estimate shown in the bottom plot of Figure 428. The signal estimate is now more in keeping with those obtained using a conditional mean estimator (Figure 425), although the former does show very small squiggles that are less evident in the latter.

Comments and Extensions to Section 10.5

[1] As noted in Chapter 5, one potential problem with the DWT is its sensitivity to where we 'break into' a time series. Coifman and Donoho (1995), Lang *et al.* (1996) and Bruce *et al.* (1999) seek to alleviate this sensitivity by using the notion of 'cycle spinning' (we mentioned this concept briefly in Exercise [5.1]). When used with a partial DWT of level J_0 on a time series whose length N is an integer multiple of 2^{J_0}, the basic idea is to apply a given signal extraction procedure not only to the original series $\mathbf{X}$, but also to all possible circularly shifted series of interest, namely, $\mathcal{T}^n\mathbf{X}$. For simplicity, let us focus on a wavelet-based thresholding scheme as applied to the model $\mathbf{X} = \mathbf{D} + \boldsymbol{\epsilon}$, where $\mathbf{D}$ is a deterministic signal, and $\boldsymbol{\epsilon}$ is a vector of IID Gaussian RVs, each with variance σ_ϵ^2. If $\mathcal{T}^n\mathbf{X}$ leads to the signal estimate $\widehat{\mathbf{D}}_n^{(\mathrm{t})}$, then we can obtain an estimator that is less influenced by the choice of starting time by averaging these individual estimates together, after each has been shifted to bring it back into alignment with the original series $\mathbf{X}$; i.e., cycle spinning yields the signal estimator

$$\widetilde{\mathbf{D}}^{(\mathrm{t})} \equiv \frac{1}{2^{J_0}} \sum_{n=0}^{2^{J_0}-1} \mathcal{T}^{-n}\widehat{\mathbf{D}}_n^{(\mathrm{t})}$$

(we need not consider $n \geq 2^{J_0}$: such a shift yields a $\mathcal{T}^n\mathbf{X}$ whose DWT is redundant because $\mathcal{T}^{-n}\widehat{\mathbf{D}}_n^{(\mathrm{t})} = \mathcal{T}^{-(n \bmod 2^{J_0})}\widehat{\mathbf{D}}^{(\mathrm{t})}_{n \bmod 2^{J_0}}$). Simulation experiments in the references cited above indicate that $\widetilde{\mathbf{D}}^{(\mathrm{t})}$ outperforms $\widehat{\mathbf{D}}_0^{(\mathrm{t})}$ in terms of mean square error.

Because the DWTs for the various $\mathcal{T}^n\mathbf{X}$ can be extracted (after appropriate renormalization) from the MODWT for $\mathbf{X}$, cycle spinning in fact can be implemented efficiently in terms of the MODWT, as follows (for simplicity, we assume that, in the corresponding DWTs, we use the universal threshold level of Equation (400d), i.e., $\delta^{(\mathrm{u})} \equiv \sqrt{[2\sigma_\epsilon^2 \log(N)]}$, with σ_ϵ^2 taken to be known). We first compute a level J_0 MODWT to obtain the coefficient vectors $\widetilde{\mathbf{W}}_j$, $j = 1, \ldots, J_0$ and $\widetilde{\mathbf{V}}_{J_0}$. We then apply one of the thresholding rules to each element of $\widetilde{\mathbf{W}}_j$ using the level-dependent threshold $\tilde{\delta}_j^{(\mathrm{u})} \equiv \sqrt{[2\sigma_\epsilon^2 \log(N)/2^j]}$ (note that $\tilde{\delta}_j^{(\mathrm{u})} = \delta^{(\mathrm{u})}/2^{j/2}$, which mimics the relationship between the jth level DWT and MODWT equivalent wavelet filters given in Section 5.4; i.e., $\tilde{h}_{j,l} = h_{j,l}/2^{j/2}$). Letting $\widetilde{\mathbf{W}}_j^{(\mathrm{t})}$ represent the thresholded version of $\widetilde{\mathbf{W}}_j$, we can obtain $\widetilde{\mathbf{D}}^{(\mathrm{t})}$ by taking the inverse MODWT of $\widetilde{\mathbf{W}}_j^{(\mathrm{t})}$, $j = 1, \ldots, J_0$ and $\widetilde{\mathbf{V}}_{J_0}$.

Two comments are in order. First, whereas the argument behind cycle spinning assumes that the sample size N of $\mathbf{X}$ is an integer multiple of 2^{J_0}, we can in fact apply the above procedure to any desired sample size, thus leading to a definition for $\widetilde{\mathbf{D}}^{(\mathrm{t})}$ valid for general N. Second, if σ_ϵ^2 is unknown, we can adapt the MAD scale estimate to work with the MODWT. To do so, we define the MODWT-based MAD estimator of the standard deviation as

$$\tilde{\sigma}_{(\mathrm{mad})} \equiv \frac{2^{1/2}\,\mathrm{median}\,\{|\widetilde{W}_{1,0}|, |\widetilde{W}_{1,1}|, \ldots, |\widetilde{W}_{1,N-1}|\}}{0.6745}.$$

As can be seen by comparing the above with Equation (420), the main difference between the DWT-based and MODWT-based estimators is the extra '$2^{1/2}$' in the numerator above, which is in keeping with the relationship between the unit scale DWT and MODWT coefficients, namely, $W_{1,t} = 2^{1/2}\widetilde{W}_{1,2t+1}$.

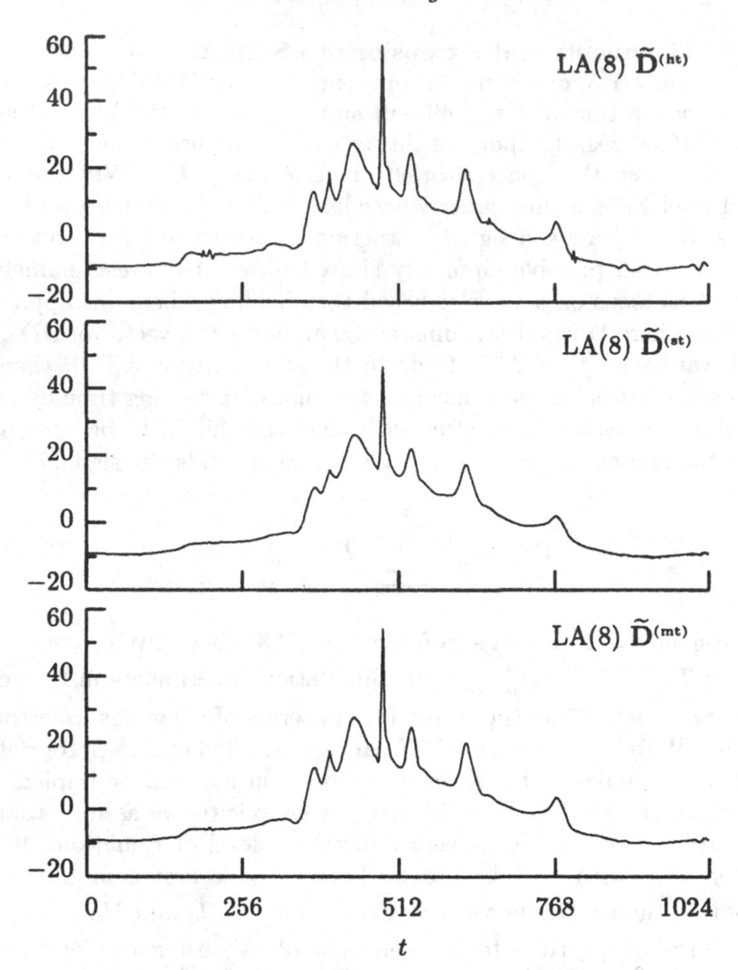

Figure 430. Thresholding signal estimates of the NMR spectrum based upon the level $J_0 = 6$ LA(8) MODWT with – from top to bottom – hard, soft and mid thresholding (Figure 419 has corresponding plots for the DWT). Each estimate uses the universal threshold levels $\tilde{\delta}_j^{(\mathrm{u})} \equiv \sqrt{[2\tilde{\sigma}^2_{(\mathrm{mad})} \log(N)/2^j]} \doteq 6.49673/2^{j/2}$ computed via the MODWT-based MAD standard deviation estimate $\tilde{\sigma}_{(\mathrm{mad})} \doteq 1.74489$.

As an example, Figure 430 shows hard, soft and mid thresholding estimates of the signal in the NMR spectrum based on the level $J_0 = 6$ LA(8) MODWT. Each estimate uses the level-dependent universal threshold $\tilde{\delta}_j^{(\mathrm{u})} = \sqrt{[2\tilde{\sigma}^2_{(\mathrm{mad})} \log(N)/2^j]} \doteq 6.49673/2^{j/2}$, for which we computed $\tilde{\sigma}_{(\mathrm{mad})} \doteq 1.74489$. In comparison to the corresponding DWT-based estimate in Figure 419, the small scale squiggles in the hard thresholding estimate near $t = 200$ and 800 are noticeably attenuated in the MODWT-based estimate, whereas the peaks are generally comparable in the two estimates. The MODWT-based soft thresholding and mid thresholding estimates are smoother in appearance than their DWT counterparts. The height of the most prominent peak in the MODWT-based mid thresholding estimate is slightly lower than the DWT-based estimate, which arguably might be a point in favor of the latter estimate (in the bias-variance trade-off, the extra averaging inherent in the MODWT-based estimator

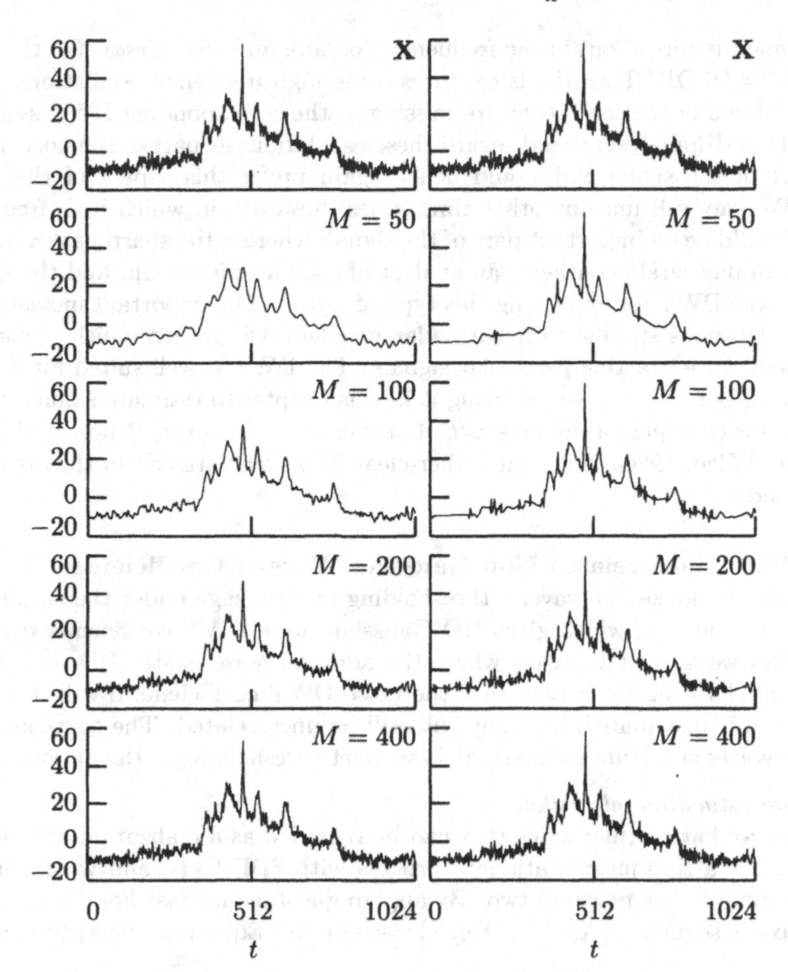

Figure 431. Denoising of NMR spectrum (top plot) using hard thresholding based upon keeping the M coefficients with the largest magnitudes in the ODFT (left-hand column) and the level $J_0 = 6$ partial LA(8) DWT (right-hand) for $M = 50, 100, 200$ and 400 (second to fifth rows, respectively).

tends to decrease variance at the expense of greater bias).

[2] It is instructive to use the NMR spectrum $\mathbf{X}$ to compare further what kinds of features in a time series are well-preserved by the ODFT and the DWT (this is a follow-up to the discussion in Section 10.1). This comparison is done in Figure 431, the first row of which shows plots of $\mathbf{X}$ itself. The bottom four rows show estimates of the signal in $\mathbf{X}$ constructed using, respectively, just the $M = 50, 100, 200$ and 400 coefficients with the largest magnitudes in the ODFT (left-hand column) and in the level $J_0 = 6$ partial LA(8) DWT (right-hand). In particular, note the sharp peak in $\mathbf{X}$ near $t = 500$: while this is captured faithfully using just 50 DWT coefficients, it is virtually missing from the $M = 50$ ODFT synthesis and is still visibly underestimated in the $M = 400$ synthesis. For data such as the NMR spectrum, it is important to preserve such peaks, and the DWT does so because the peak is well isolated in a few coefficients, whereas the ODFT spreads it out across many coefficients (a sharp peak in

the time domain is spread out in the frequency domain, and *vice versa*). On the other hand, the $M = 50$ ODFT synthesis captures some high frequency oscillations at the beginning and end of the series that are missing in the corresponding DWT synthesis – presumably NMR analysts would regard these oscillations as part of the noise rather than as part of the signal and would hence again prefer this aspect of the DWT synthesis. We can well imagine other time series, however, in which high frequency oscillations would be an important part of the signal, whereas the sharp peak would be regarded as an undesirable outlier – an analyst of such a series would find the ODFT preferable to the DWT for extracting this type of signal. The important message here is that considerations specific to a particular problem will govern which transforms are most useful for extracting particular signals. The DWT is well suited for a broad class of signals, but the way we are using it in this chapter to estimate signals can be problematic, for example, in the presence of outliers (see, however, Bruce *et al.*, 1994, and Bruce and Gao, 1996a, for a 'smoother-cleaner' wavelet transform that is robust against outliers).

10.6 Uncorrelated Non-Gaussian Wavelet Coefficients

In Section 10.5 we looked at wavelet thresholding or shrinkage under the assumption of IID Gaussian noise $\boldsymbol{\epsilon}$, which gives IID Gaussian noise DWT coefficients $\mathbf{e} = \mathcal{W}\boldsymbol{\epsilon}$. In this section we look at the case where the additive noise is still IID, but now is non-Gaussian. This means in turn that the noise DWT coefficients $\mathbf{e}$ will in general *not* be identically distributed, but they will still be uncorrelated. The particular case of interest involves spectrum estimation via wavelet thresholding of the periodogram.

- *Spectrum estimation algorithm*

Suppose that we have a time series that can be regarded as a realization of a portion $X_0, \ldots, X_{N-1}$ of a zero mean stationary process with SDF $S_X(\cdot)$ and with sampling time Δt. Assume N is a power of two. By applying a standard fast Fourier transform algorithm to the sequence $\{X_0, \ldots, X_{N-1}\}$, we can use Equation (269a) to compute

$$Y^{(\mathrm{p})}(f_k) = \log(\hat{S}_X^{(\mathrm{p})}(f_k)) + \gamma = \log(S_X(f_k)) + \epsilon(f_k)$$

at the Fourier frequencies $f_k = k/(N\,\Delta t), k = 0, \ldots, M$, where $M \equiv N/2$ (see Equation (271a)). The above says that the log periodogram – after addition of Euler's constant $\gamma \doteq 0.57721$ – can be written as a signal (the true log SDF) plus noise with zero mean. Furthermore, $\{\epsilon(f_k)\}$ for $0 < f_k < f_{\mathcal{N}}$ constitutes a set of approximately IID RVs such that

$$\epsilon(f_k) \stackrel{\mathrm{d}}{=} \log(\chi_2^2) + \gamma - \log(2)$$

(see Equation (270c)). The above implies that $\sigma_\epsilon^2 \equiv \mathrm{var}\{\epsilon(f_k)\} = \pi^2/6$. How can wavelet thresholding methods be applied to the problem of estimating this signal?

The basic algorithm for using wavelet thresholding to produce a smooth estimate of the logarithm of an SDF of a stationary process is very simple and, as implemented by Moulin (1994) or Gao (1993, 1997), consists of four basic steps.

[1] Calculate the logarithm of the periodogram at the Fourier frequencies, and form $Y^{(\mathrm{p})}(f_k) = \log(\hat{S}_X^{(\mathrm{p})}(f_k)) + \gamma$.

[2] Apply a level J_0 partial DWT to $\{Y^{(\mathrm{p})}(f_k)\}$.

[3] Apply a thresholding procedure to the empirical wavelet coefficients, thus producing a set of thresholded coefficients.

[4] Apply the inverse DWT to the thresholded empirical wavelet coefficients, producing a smoothed estimate of the log SDF at the Fourier frequencies.

Let us look at the algorithm, step by step, in more detail.

[1] If we want to pose the log SDF estimation problem in terms of a model with IID additive noise, then we must restrict ourselves to using $Y^{(p)}(f_k)$ such that $0 < f_k < f_{\mathcal{N}}$, i.e., $k = 1, \ldots, M-1$ (the RVs $Y^{(p)}(f_0)$ and $Y^{(p)}(f_M)$ have distributions related to a $\log(\chi_1^2)$ RV rather than a $\log(\chi_2^2)$ RV). Unfortunately, this yields a sample size of $M-1$, which is not a power of two and hence does not match the usual restriction imposed by the DWT. Accordingly, we consider including the $Y^{(p)}(f_0)$ term to obtain, in vector notation,

$$\begin{bmatrix} Y^{(p)}(f_0) \\ \vdots \\ Y^{(p)}(f_{M-1}) \end{bmatrix} = \begin{bmatrix} \log(S_X(f_0)) \\ \vdots \\ \log(S_X(f_{M-1})) \end{bmatrix} + \begin{bmatrix} \epsilon(f_0) \\ \vdots \\ \epsilon(f_{M-1}) \end{bmatrix}, \tag{433a}$$

which gives us an observable vector whose length is a power of two at the expense of including a single term with an anomalous distribution (its influence should be negligible for large $M = N/2$). The above formulation is still not entirely satisfactory because the DWT makes use of circular filtering. Since there is no requirement that $S_X(0)$ and $S_X(f_{\mathcal{N}})$ be at all close to each other, we could end up with distortions in our estimates of $\log(S_X(f))$ at frequencies close to zero and $f_{\mathcal{N}}$. A natural way to eliminate these distortions is to recall that an SDF for a real-valued process is by definition an even periodic function with a period of $2f_{\mathcal{N}}$. Accordingly, let us apply the DWT to an expanded version of Equation (433a), namely,

$$\mathbf{Y}^{(p)} \equiv \begin{bmatrix} Y^{(p)}(f_0) \\ \vdots \\ Y^{(p)}(f_{2M-1}) \end{bmatrix} = \begin{bmatrix} \log(S_X(f_0)) \\ \vdots \\ \log(S_X(f_{2M-1})) \end{bmatrix} + \begin{bmatrix} \epsilon(f_0) \\ \vdots \\ \epsilon(f_{2M-1}) \end{bmatrix} \equiv \mathbf{D} + \boldsymbol{\epsilon}. \tag{433b}$$

Since $Y^{(p)}(f_{2M-k}) = Y^{(p)}(f_k)$ for $k = 1, \ldots, M-1$, the above is quite close to what we would need to analyze the left-hand side of Equation (433a) using the reflection boundary condition scheme of Equation (140) (we only deviate from this scheme by using both anomalous RVs $Y^{(p)}(f_0)$ and $Y^{(p)}(f_M)$ once rather than $Y^{(p)}(f_0)$ twice). Moulin (1994) makes a similar argument, and in fact a standard way of forming a spectral estimator is to smooth the periodogram circularly over a complete period (see Equation (271c) for an example).

[2] Compute a level J_0 partial DWT decomposition of $\mathbf{Y}^{(p)}$ to obtain the vectors $\mathbf{W}_1^{(p)}, \ldots, \mathbf{W}_{J_0}^{(p)}$ and $\mathbf{V}_{J_0}^{(p)}$. Component-wise,

$$W_{j,t}^{(p)} = d_{j,t} + e_{j,t}, \quad j = 1, \ldots, J_0; \; t = 0, \ldots, 2M_j - 1, \tag{433c}$$

where $M_j \equiv M/2^j = N/2^{j+1}$. Here $\{d_{j,t}\}$ and $\{e_{j,t}\}$ are the level j wavelet coefficients of the signal $\mathbf{D}$ and of the noise $\boldsymbol{\epsilon}$, respectively.

[3] Apply a thresholding scheme to $\mathbf{W}_1^{(p)}, \ldots, \mathbf{W}_{J_0}^{(p)}$ to obtain $\mathbf{W}_1^{(t)}, \ldots, \mathbf{W}_{J_0}^{(t)}$. Note that, even though the DWT is applied to a vector of length $2M$, we must compute

threshold levels appropriate for a sample size of M (the number of independent noise terms in Equation (433a) is M, and the M additional terms in $\mathbf{Y}^{(\mathrm{p})}$ arise merely to handle boundary conditions in a sensible manner).

[4] Estimate $\mathbf{D}$ via $\widehat{\mathbf{D}}^{(\mathrm{t})}$ obtained by inverse transforming $\mathbf{W}_1^{(\mathrm{t})}, \ldots, \mathbf{W}_{J_0}^{(\mathrm{t})}$ and $\mathbf{V}_{J_0}^{(\mathrm{p})}$. Let $\widehat{D}_k^{(\mathrm{t})}$ denote the kth element of $\widehat{\mathbf{D}}^{(\mathrm{t})}$. We can now form the spectrum estimate via

$$\hat{S}_X^{(\mathrm{pt})}(f_k) \equiv \begin{cases} e^{\widehat{D}_0^{(\mathrm{t})}}, & k = 0; \\ \frac{1}{2}\left(e^{\widehat{D}_k^{(\mathrm{t})}} + e^{\widehat{D}_{2M-k}^{(\mathrm{t})}}\right), & k = 1, \ldots, M-1; \text{ and} \\ e^{\widehat{D}_M^{(\mathrm{t})}}, & k = M. \end{cases}$$

We note that, in contrast to typical SDF estimates, the thresholded estimates $\exp(\widehat{D}_k^{(\mathrm{t})})$ need *not* be symmetric about $f = f_{\mathcal{N}}$, which is the reason we have chosen to average estimates at frequencies f_k and f_{2M-k} to form $\hat{S}_X^{(\mathrm{pt})}(f_k)$ in step [4] (the source of this asymmetry is the subject of Exercise [10.8]).

- *Statistical properties of coefficients and threshold specification*

Because of the localized nature of the DWT, because $Y^{(\mathrm{p})}(f_{2M-k}) = Y^{(\mathrm{p})}(f_k)$ for $k = 1, \ldots, M-1$, and because $Y^{(\mathrm{p})}(f_0)$ and $Y^{(\mathrm{p})}(f_M)$ become negligible contributors as M gets large, we can argue that, for large sample sizes, the statistical properties of the DWT of $\mathbf{Y}^{(\mathrm{p})}$ that are important for formulating threshold levels are the same as those of the DWT of just $Y^{(\mathrm{p})}(f_1), \ldots, Y^{(\mathrm{p})}(f_{M-1})$ (as can be seen from what follows, we can ignore the complication that $M-1$ is not a power of two). With this simplified approach, we can follow Moulin (1994) to obtain the statistical properties of the wavelet coefficients $\{e_{j,t}\}$. Because the $\{\epsilon(f_k)\}$ are IID by assumption and the DWT is orthonormal, the $\{e_{j,t}\}$ are uncorrelated, but, since the $\{\epsilon(f_k)\}$ are related to a $\log(\chi_2^2)$ distribution rather than a Gaussian distribution, the $\{e_{j,t}\}$ are not identically distributed RVs – their distribution will now depend on the effect of the level j wavelet filter.

The number of terms in the linear combination of IID RVs $\{\epsilon(f_k)\}$ from which $e_{j,t}$ is formed increases rapidly with j (recall that the width of the level j wavelet filter is given as $L_j \equiv (2^j - 1)(L-1) + 1$ by Equation (96a)). We can thus appeal to an appropriate version of the Central Limit Theorem to argue that the distribution of $e_{j,t}$ converges to Gaussian as j increases. Since the $\{e_{j,t}\}$ are uncorrelated, Moulin (1994) points out that the $\{e_{j,t}\}$ can thus be treated as independent RVs for large j. In this case it would be legitimate to use a version of wavelet thresholding or shrinkage that assumes IID Gaussian wavelet coefficients (we discussed such methodology in Section 10.5). At small and moderate scales, however, the tails of the distribution of $e_{j,t}$ are not close to Gaussian, but merely become closer to Gaussian with increasing j. For these scales, Moulin (1994) uses the saddlepoint method to approximate the tail behavior of the distribution of $e_{j,t}$ and hence to compute appropriate thresholds (for background on this method, see Lugannani and Rice, 1980; Daniels, 1987; Reid, 1988; and Davison and Hinkley, 1997).

In accordance with these ideas, Moulin (1994) outlines the following approach to thresholding. Consider the null hypothesis $H_0 : d_{j,t} = 0$ against the alternative $H_1 : d_{j,t} \neq 0$ for each wavelet coefficient. Under the null hypothesis, $W_{j,t}^{(\mathrm{p})}$ has the same distribution as the noise term $e_{j,t}$, which – since it arises by filtering an RV with a $\log(\chi_2^2)$ distribution – will be asymmetric for small and moderate scales. We fail to reject the null hypothesis at significance level α when $W_{j,t}^{(\mathrm{p})}$ belongs to an acceptance region $[\delta^{(\mathrm{l})}_{j,\frac{\alpha}{2}}, \delta^{(\mathrm{u})}_{j,\frac{\alpha}{2}}]$, where $\delta^{(\mathrm{l})}_{j,\frac{\alpha}{2}}$ and $\delta^{(\mathrm{u})}_{j,\frac{\alpha}{2}}$ are, respectively, lower and upper thresholds such that

$$\mathbf{P}[e_{j,t} < \delta^{(\mathrm{l})}_{j,\frac{\alpha}{2}}] = \frac{\alpha}{2} \text{ and } \mathbf{P}[e_{j,t} > \delta^{(\mathrm{u})}_{j,\frac{\alpha}{2}}] = \frac{\alpha}{2}$$

and hence, when H_0 is true,

$$\mathbf{P}[\delta^{(\mathrm{l})}_{j,\frac{\alpha}{2}} \le W^{(\mathrm{p})}_{j,t} \le \delta^{(\mathrm{u})}_{j,\frac{\alpha}{2}}] = \mathbf{P}[\delta^{(\mathrm{l})}_{j,\frac{\alpha}{2}} \le e_{j,t} \le \delta^{(\mathrm{u})}_{j,\frac{\alpha}{2}}] = 1 - \alpha. \tag{435}$$

The lower and upper thresholds depend on the level j of $W^{(\mathrm{p})}_{j,t}$ since different wavelet filters are used for each j. This hypothesis test is equivalent to applying the following thresholding rule to the wavelet coefficients:

$$W^{(\mathrm{t})}_{j,t} = \begin{cases} 0, & \text{if } \delta^{(\mathrm{l})}_{j,\frac{\alpha}{2}} \le W^{(\mathrm{p})}_{j,t} \le \delta^{(\mathrm{u})}_{j,\frac{\alpha}{2}}; \\ W^{(\mathrm{p})}_{j,t}, & \text{otherwise.} \end{cases}$$

The value of α can be chosen by requiring that the probability of a false alarm P_F takes a specified small value, where by definition P_F is the probability that at least one noise wavelet coefficient is outside the acceptance region. Hence,

$$P_F = \mathbf{P}\left[\bigcup_{j,t}\left(e_{j,t} > \delta^{(\mathrm{u})}_{j,\frac{\alpha}{2}} \text{ or } e_{j,t} < \delta^{(\mathrm{l})}_{j,\frac{\alpha}{2}}\right)\right] = 1 - \mathbf{P}\left[\bigcap_{j,t}\left(\delta^{(\mathrm{l})}_{j,\frac{\alpha}{2}} \le e_{j,t} \le \delta^{(\mathrm{u})}_{j,\frac{\alpha}{2}}\right)\right].$$

The $\{e_{j,t}\}$ are uncorrelated, but, if we make an approximation by treating them as independent coefficients, the above becomes

$$P_F = 1 - \prod_{j,t} \mathbf{P}[\delta^{(\mathrm{l})}_{j,\frac{\alpha}{2}} \le e_{j,t} \le \delta^{(\mathrm{u})}_{j,\frac{\alpha}{2}}] = 1 - \prod_{j,t}(1-\alpha).$$

With the simplification that we take close to a full DWT of $Y^{(\mathrm{p})}(f_1), \ldots, Y^{(\mathrm{p})}(f_{M-1})$ (i.e., there are relatively few scaling coefficients in the partial DWT), a level J_0 transform has approximately M wavelet coefficients, which gives the approximate number of terms in the product above. We can thus obtain a desired false alarm probability (e.g., $P_F = 0.1$) by setting

$$\alpha = 1 - (1 - P_F)^{1/M} \approx \frac{P_F}{M},$$

where the approximation (based on $\log(1-x) \approx -x$ for small x) is convenient to use when P_F is taken to be small.

- *Examples*

Let us apply the periodogram-based scheme to three Gaussian stationary processes that Gao (1993, 1997) and Moulin (1994) use as examples.

[1] Recall that by definition a pth order autoregressive process (AR(p)) satisfies Equation (268e). Gao (1993, 1997) considers the AR(24) process whose coefficients $\{\phi_{24,n}\}$ are given in Table 272. The corresponding SDF is given by Equation (268f) with $\sigma_\varepsilon^2 = 1$ and $\Delta t = 1$ and is shown as the thin curve in the top plot of Figure 438 (and also in Figure 273). We can form realizations of a portion $X_0, \ldots, X_{N-1}$ of this process using the procedure described in Section 7.9.

	j							
	1	2	3	4	5	6	7	8
$-\delta^{(l)}_{j,\frac{\alpha}{2}}$	7.825	7.031	6.228	5.750	5.460	5.287	5.182	5.118
$\delta^{(u)}_{j,\frac{\alpha}{2}}$	5.556	5.601	5.142	4.976	4.913	4.901	4.910	4.925

Table 436. **Lower and upper thresholds $\delta^{(l)}_{j,\frac{\alpha}{2}}$ and $\delta^{(u)}_{j,\frac{\alpha}{2}}$, $j = 1, \ldots, 8$, for wavelet-based thresholding of the log periodogram using the LA(8) DWT. Here we use the approximation $\alpha = P_F/M$ with $P_F = 0.1$ and $M = 1024$. For convenience, we have tabulated $-\delta^{(l)}_{j,\frac{\alpha}{2}}$ instead of $\delta^{(l)}_{j,\frac{\alpha}{2}}$. Note that, as j increases, the lower and upper thresholds come closer to each other in magnitude, as would be expected due to convergence to Gaussianity.**

[2] Moulin (1994) defines an AR(2) process specified by

$$X_t = \phi_{2,1} X_{t-1} + \phi_{2,2} X_{t-2} + \varepsilon_t, \tag{436a}$$

where $\phi_{2,1} \equiv 0.97\sqrt{2}$ and $\phi_{2,2} \equiv -(0.97)^2$, and $\{\varepsilon_t\}$ is Gaussian white noise with zero mean and unit variance. The SDF for this process can be computed via Equation (268f) and is shown as the thin curve in the middle plot of Figure 438. Again, we can generate realizations as described in Section 7.9.

[3] Moulin (1994) also defines a 'typical mobile radio communications' (MRC) SDF, which is a superposition of two band-limited, fading, mobile radio signals, a white background noise, and a narrow-band interference term with a Gaussian-shaped SDF. The overall SDF is, for $0 \leq f \leq 1/2$,

$$\begin{aligned} S_X(f) =& 10^{-3} + 0.2e^{-(f-0.45)^2/(2\cdot 10^{-6})} + \left[1 - \left(\tfrac{f}{B_0}\right)^2\right]^{1/2} 1_{[0,B_0)}(f) \\ &+ \left[1 - \left(\tfrac{f-f_0}{B_0}\right)^2\right]^{1/2} 1_{(f_0-B_0, f_0+B_0)}(f), \end{aligned}$$

with $f_0 = 0.3$, $B_0 = 0.1$ (as usual, $S_X(f) = S_X(-f)$ for $f < 0$). This SDF is shown as the thin curve in the bottom plot of Figure 438. Here we can use the method for simulating from a Gaussian process with a known spectrum described in Section 7.8.

For each of these models, we generated a thousand simulated time series of length $N = 2048$ (so that $M = 1024$) and formed a periodogram for each series. With $P_F = 0.1$, we set $\alpha = P_F/M \doteq 9.7656 \times 10^{-5}$ and then found, via the saddlepoint method, the lower and upper thresholds $\delta^{(l)}_{j,\frac{\alpha}{2}}$ and $\delta^{(u)}_{j,\frac{\alpha}{2}}$, $j = 1, \ldots, 8$, for use with an LA(8) DWT (the thresholds are listed in Table 436). Using these thresholds in conjunction with soft thresholding, we then computed $\hat{S}^{(\text{pt})}_X(\cdot)$ based upon partial LA(8) DWTs of levels $J_0 = 5, 6$ and 8 (these settings for J_0 result in, respectively, 64, 32 and 8 scaling coefficients left untouched by thresholding). We judged the quality of each SDF estimate by computing its root mean square error (RMSE), namely,

$$\left[\frac{1}{M+1}\sum_{k=0}^{M}\left[10\log_{10}(\hat{S}^{(\text{pt})}_X(f_k)) - 10\log_{10}(S_X(f_k))\right]^2\right]^{1/2} \tag{436b}$$

(note that this has units of decibels). For each level J_0, we computed an average RMSE over the thousand simulated series (these are plotted as asterisks in Figure 446 below). In addition, for the level yielding the smallest average RMSE, we have plotted (thick curves in Figure 438) the SDF estimate – from amongst the thousand such estimates – whose RMSE was closest to the average RMSE (the selected levels are $J_0 = 5, 8$ and 5 for, respectively, the AR(24), AR(2) and MRC models).

Let us now comment on the quality of the spectrum estimates in Figure 438 for each of the three models.

[1] The SDF of the AR(24) process has a dynamic range close to 90 dB (by definition, this range is the ratio of the largest to smallest SDF value). Although Gao (1997) uses this process in his study of wavelet thresholding of the log periodogram, in fact the periodogram is badly biased due to leakage at least for sample sizes $N \leq 2048$ and $f \geq 0.4$ (see Figure 273 or Figure 1 of Gao, 1997). This leakage is evident at high frequencies in the SDF estimate shown at the top of Figure 438. In addition, for $f < 0.4$, the estimate misrepresents several of the peaks and, in particular, underestimates the dominant peak close to $f = 0.1$ by about 8 dB. Due to the severe leakage, the average RMSEs here are much larger than for the other two models – see the asterisks in Figure 446.

[2] The middle plot of Figure 438 shows the representative estimate for the AR(2) process. The SDF for this process has a much simpler structure than the AR(24) process, and the resulting SDF estimate is quite good overall (there are only slight distortions near the dominant peak and at high frequencies).

[3] The bottom plot shows the representative estimate for the MRC process. The true SDF has sharper features than either of the AR SDFs, but it also has a smaller dynamic range. The estimate is less smooth than the one for the AR(2) model, but is generally quite acceptable; however, there is noticeable distortion just before the rapid 30 dB rise at $f = 0.2$.

- *Critique of spectrum estimation method*

There are three main disadvantages with using the log periodogram as the starting point for a log SDF estimator.

[1] First, the 'noise' in $\mathbf{Y}^{(p)}$ is $\log(\chi_2^2)$ distributed and hence has tails that differ significantly enough from the Gaussian distribution so that one cannot use simple Gaussian-based thresholds for all levels of wavelet coefficients. The threshold levels used by Moulin (1994) are sample size, level (scale) and wavelet dependent and hence – without readily available software – require a fair amount of effort to compute.

[2] Second, irrespective of wavelet thresholding, the periodogram can give very poor SDF estimates whenever the true SDF has a high dynamic range and/or is rapidly varying (even for sample sizes that would normally be regarded as large). This deficiency stems from the fact that the periodogram offers poor protection against sidelobe leakage. This point is discussed in detail in, for example, Percival and Walden (1993, Section 6.3).

[3] In order to preserve the uncorrelated nature of the periodogram, we must sample it at the Fourier frequencies, which dictates that the number of observations in the time series must be a power of two. If this assumption is not met, a simple – but unappealing – alternative is to sample the periodogram over a grid of frequencies coarser than the Fourier frequencies, which effectively amounts to

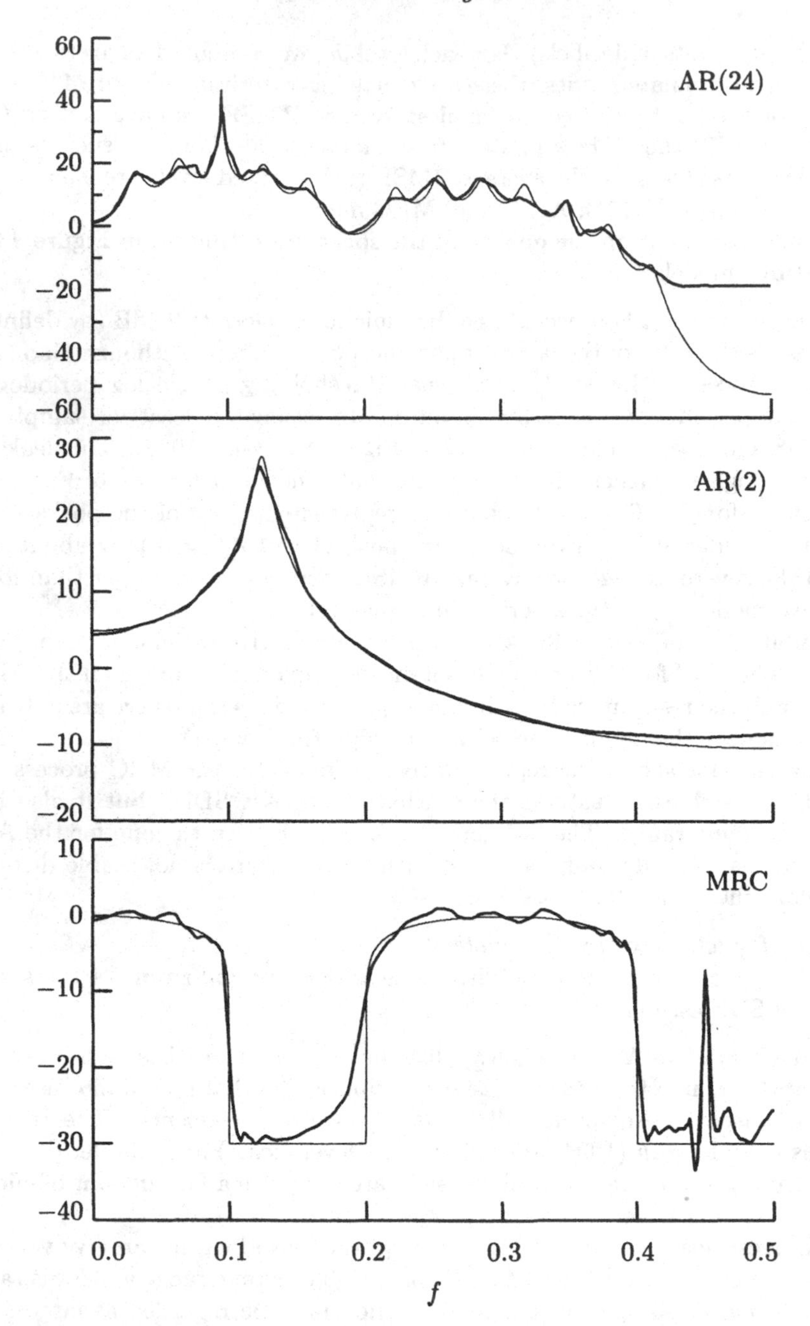

Figure 438. Periodogram-based estimated SDFs (thick curves) and true SDFs (thin) for the AR(24), AR(2) and MRC processes (see the text for details).

discarding some time series values (this alternative could be made more attractive by using an 'overlapped segment averaging' approach – see, for example, Percival and Walden, 1993, Section 6.17).

In Section 10.7 we give an alternative scheme based on multitaper SDF estimation that gets around these three practical disadvantages.

Comments and Extensions to Section 10.6

[1] Gao (1993, 1997) develops a slightly less complicated approach to thresholding the log periodogram. He derives a single threshold level $\delta_{j,M}$ such that, as $M \to \infty$ (i.e., $N \to \infty$),

$$P_F = \mathbf{P}\Big[\bigcup_{j,t}(|e_{j,t}| > \delta_{j,M})\Big] \to 0.$$

An alternative way of writing this is

$$P_F = \mathbf{P}\Big[\bigcup_{j}\Big(\max_t |e_{j,t}| > \delta_{j,M}\Big)\Big] \to 0.$$

Hence the probability that at least one noise wavelet coefficient exceeds the threshold $\delta_{j,M}$ tends to zero with increasing sample size. The threshold is defined by

$$\delta_j^{(\mathrm{G})} = \max\{\alpha_j \log(M), \pi\sqrt{[\log(M)/3]}\},$$

where the coefficients $\{\alpha_j\}$ were tabulated by Gao (1993, 1997) for $j = 1, \ldots, 10$ as

$$\{\alpha_1, \ldots, \alpha_{10}\} = \{1.29, 1.09, 0.92, 0.77, 0.65, 0.54, 0.46, 0.39, 0.32, 0.27\}.$$

These are valid for the commonly used compactly supported orthonormal wavelet designs. Now $\delta_j^{(\mathrm{G})} = \alpha_j \log(M)$ when $\alpha_j \geq \pi/\sqrt{[3\log(M)]}$. For example, if $M = 512$, this condition becomes $\alpha_j \geq 0.73$. The threshold for this sample size takes the form $\delta_j^{(\mathrm{G})} = \alpha_j \log(M)$ at the four smallest scales, and then $\delta_j^{(\mathrm{G})} = \pi\sqrt{[\log(M)/3]}$ at coarse levels $j \geq 5$. This is undoubtedly a simpler approach to thresholding, but nevertheless still uses the log periodogram as the basic SDF estimate, with all its inherent inability to suppress sidelobe leakage.

[2] In Section 10.2 we noted that the asymptotic argument that is used to justify, for example, universal thresholding requires that the signal be more finely sampled over a fixed interval as we collect more data. For the spectrum estimation problem considered here (and in the next section), this sampling scheme is in fact quite natural. We regard the log spectrum as a signal that we can observe only over the fixed interval of frequencies $[0, f_{\mathcal{N}}]$. As the sample size N for our time series grows, we can sample our signal via the log periodogram over a grid of frequencies whose spacing is $\Delta f \equiv 1/(N\,\Delta t)$. The sampling interval Δf for the log periodogram hence decreases to zero as $N \to \infty$, so our sampling of the signal becomes finer and finer, as required for the asymptotic argument.

[3] In formulating the theory in this section, we assumed that $E\{X_t\} = 0$. In practical applications, we do not know the mean of $\{X_t\}$ *a priori*. The usual procedure is to estimate the unknown process mean using the sample mean $\overline{X}$ of $X_0, \ldots, X_{N-1}$. We would then compute the periodogram using $X_t - \overline{X}$ in place of X_t, a procedure that renders $\hat{S}_X^{(\mathrm{p})}(f_0) = 0$ and hence $Y^{(\mathrm{p})}(f_0) = -\infty$. To overcome this difficulty, we can redefine $\mathbf{Y}^{(\mathrm{p})}$ of Equation (433b) so that it becomes

$$Y^{(\mathrm{p})}(f_1), Y^{(\mathrm{p})}(f_2), \ldots, Y^{(\mathrm{p})}(f_M), Y^{(\mathrm{p})}(f_M), \ldots, Y^{(\mathrm{p})}(f_2), Y^{(\mathrm{p})}(f_1),$$

in keeping with the reflection boundary conditions of Equation (140). Note that, because $Y^{(\mathrm{p})}(f_{2M-k}) = Y^{(\mathrm{p})}(f_k)$ for $k = 1, \ldots M-1$, the above is equivalent to removing $Y^{(\mathrm{p})}(f_0)$ from $\mathbf{Y}^{(\mathrm{p})}$ and then replicating $Y^{(\mathrm{p})}(f_M)$ to get the size of the vector back up to a power of two.

10.7 Correlated Gaussian Wavelet Coefficients

In Section 10.6 we looked at spectrum estimation via wavelet thresholding of the periodogram. While the noise wavelet coefficients $\{e_{j,t}\}$ were uncorrelated, they were not identically distributed, leading to much more complicated thresholding rules. In this section we examine the situation where the noise wavelet coefficients $\{n_{j,t}\}$ are correlated, but approximately Gaussian distributed. The application of interest here again involves spectrum estimation, but this time via wavelet thresholding of a multitaper spectrum estimator utilizing the K sine tapers $\{a_{n,t}, t = 0, \ldots, N-1\}$, $n = 0, \ldots, K-1$ (see Section 7.5 and Equation (274a)).

- *Outline of spectrum estimation algorithm*

As in Section 10.6, let $X_0, \ldots, X_{N-1}$ be a portion of a zero mean stationary process with SDF $S_X(\cdot)$ and sampling interval Δt. Let $2M = 2^{J+1}$ be a power of two greater than or equal to the sample size N. We extend the tapered time series $\{a_{n,t}X_t,\ t = 0, \ldots, N-1\}$ to length $2M$ by 'padding with zeros,' yielding the K sequences

$$\{a_{n,0}X_0, \ldots, a_{n,N-1}X_{N-1}, \underbrace{0, \ldots, 0}_{2M-N \text{ zeros}}\}, \quad n = 0, \ldots, K-1.$$

By applying a standard fast Fourier transform algorithm to these sequences, we can readily compute the log multitaper SDF ordinates at the $M = 2^J$ frequencies $f_k \equiv k/(2M\,\Delta t), k = 0, \ldots, M-1$:

$$Y^{(\mathrm{mt})}(f_k) = \log(\hat{S}_X^{(\mathrm{mt})}(f_k)) - \psi(K) + \log(K) = \log(S_X(f_k)) + \eta(f_k),$$

where $\psi(\cdot)$ is the digamma function. The above says that the log multitaper estimator (plus known constants) can be written as a signal (the true log spectrum) plus noise, where – assuming we use a moderate number of tapers ($K \geq 5$) – the noise is approximately Gaussian with zero mean and known variance $\sigma_\eta^2 = \psi'(K)$ (see Section 7.5; as before, $\psi'(\cdot)$ is the trigamma function). Since there is no necessity to estimate the spectrum at the Fourier frequencies $k/(N\,\Delta t)$, the number of observations need not be a power of two. Note that padding with *zeros*, as done here, is justified in that we have assumed that our process $\{X_t\}$ has a mean of zero.

We can express these results in vector notation as

$$\begin{bmatrix} Y^{(\mathrm{mt})}(f_0) \\ \vdots \\ Y^{(\mathrm{mt})}(f_{M-1}) \end{bmatrix} = \begin{bmatrix} \log(S_X(f_0)) \\ \vdots \\ \log(S_X(f_{M-1})) \end{bmatrix} + \begin{bmatrix} \eta(f_0) \\ \vdots \\ \eta(f_{M-1}) \end{bmatrix} \equiv \mathbf{D} + \boldsymbol{\eta} \tag{440}$$

(cf. Equation (433a)). As for the log periodogram, this implicitly extends the model of Equation (276) to include the case $f = f_0 = 0$, even though this is not strictly correct; again, the influence of this one additional term should be negligible for large N. Application of a full DWT to the above gives

$$W_{j,t}^{(\mathrm{mt})} = d_{j,t} + n_{j,t} \text{ and } V_{J,0}^{(\mathrm{mt})} = d_{J+1,0} + n_{J+1,0}$$

for $j = 1, \ldots, J$ and $t = 0, \ldots, M_j - 1$, where $V_{J,0}^{(\mathrm{mt})}$ is the single scaling coefficient with signal and noise components $d_{J+1,0}$ and $n_{J+1,0}$, and $M_j \equiv M/2^j$.

• *Statistical properties of coefficients and threshold specification*
We model $\boldsymbol{\eta}$ as a multivariate Gaussian vector with zero mean and covariance matrix

$$\Sigma_{\boldsymbol{\eta}} \equiv \begin{bmatrix} s_\eta(f_0) & \cdots & s_\eta(f_{\frac{M}{2}-1}) & s_\eta(f_{\frac{M}{2}}) & s_\eta(f_{\frac{M}{2}-1}) & \cdots & s_\eta(f_1) \\ s_\eta(f_1) & \cdots & s_\eta(f_{\frac{M}{2}-2}) & s_\eta(f_{\frac{M}{2}-1}) & s_\eta(f_{\frac{M}{2}}) & \cdots & s_\eta(f_2) \\ s_\eta(f_2) & \cdots & s_\eta(f_{\frac{M}{2}-3}) & s_\eta(f_{\frac{M}{2}-2}) & s_\eta(f_{\frac{M}{2}-1}) & \cdots & s_\eta(f_3) \\ \vdots & & \vdots & \vdots & \vdots & & \vdots \\ s_\eta(f_1) & \cdots & s_\eta(f_{\frac{M}{2}}) & s_\eta(f_{\frac{M}{2}-1}) & s_\eta(f_{\frac{M}{2}-2}) & \cdots & s_\eta(f_0) \end{bmatrix} \tag{441a}$$

with $s_\eta(f_k)$ defined in Equation (277). This circular covariance matrix is symmetric and positive semidefinite. Its elements are in accordance with the discussion in Section 7.5 for $\eta(f_k)$ such that f_k is sufficiently far from 0 and $f_{\mathcal{N}}$; for other frequencies (these are relatively few in number and decreasing in importance as N increases), it provides a convenient mathematical structure that is put to good use below. The matrix $\Sigma_{\boldsymbol{\eta}}$ can be regarded as an approximation to what would be a more natural assumption for the covariance structure of $\boldsymbol{\eta}$, namely, a symmetric Toeplitz covariance matrix $\Sigma'_{\boldsymbol{\eta}}$ whose first row is given by $s_\eta(f_k)$, $k = 0, \ldots, M-1$ (a Toeplitz matrix by definition has elements $\Sigma'_{\boldsymbol{\eta},l,m} = s_\eta(f_{l-m})$, i.e., a single value along a given diagonal, and hence the first row determines the entire matrix). Because of Equation (277), the matrix $\Sigma_{\boldsymbol{\eta}}$ is almost a banded Toeplitz matrix: it has $2K+1$ nonzero diagonals, but it also has $K(K+1)/2$ nonzero 'off-band' elements in both the upper right- and lower left-hand corners. These relatively few off-band elements are what distinguish $\Sigma_{\boldsymbol{\eta}}$ and $\Sigma'_{\boldsymbol{\eta}}$.

Because the covariance matrix $\Sigma_{\boldsymbol{\eta}}$ in Equation (441a) is circular, we can evoke a well-known result in time series analysis (see, for example, Fuller, 1996, p. 151) to write $\Sigma_{\boldsymbol{\eta}} = \mathcal{F}^H D \mathcal{F}$, where $\mathcal{F}$ is the $M \times M$ ODFT matrix (see Section 3.4); 'H' denotes Hermitian transpose; and D is a diagonal matrix with diagonal elements $\{S_k : k = 0, \ldots, M-1\}$, which is the DFT of the first row of $\Sigma_{\boldsymbol{\eta}}$. This following result gives us a way to compute the variance for $n_{j,t}$ easily.

▷ **Exercise [441]:** For $M = 2^J$, let $\mathbf{n} = \mathcal{W}\boldsymbol{\eta}$ be the DWT coefficients for a Gaussian stationary process with $M \times M$ covariance matrix $\Sigma_{\boldsymbol{\eta}}$ given as in Equation (441a), and let $\{n_{j,t} : t = 0, \ldots, M_j - 1\}$ be the elements of $\mathbf{n}$ associated with level j (recall that $M_j \equiv M/2^j$). Assuming that the DWT is based on a wavelet filter with width $L \leq M$, show that

$$\operatorname{var}\{n_{j,t}\} = \frac{1}{M} \sum_{k=0}^{M-1} S_k \mathcal{H}_j(\tfrac{k}{M}) \equiv \sigma_j^2, \tag{441b}$$

where, as usual, $\mathcal{H}_j(\cdot)$ is the squared gain function for the jth level wavelet filter $\{h_{j,l}\}$ (note in particular that $\operatorname{var}\{n_{j,t}\}$ depends on j but not on t). ◁

Equation (441b) can be readily computed, giving us *known* values for the level-dependent variances σ_j^2.

How do these variances change with level j? For the special case of the Haar DWT with $M = 2^J = N/2$ and under the mild condition $K < M/2$, it can be shown that

$$\sigma_1^2 < \sigma_2^2 < \cdots < \sigma_J^2 < \bar{\sigma}_{J+1}^2, \tag{441c}$$

where $\bar{\sigma}^2_{J+1}$ is the variance of the noise scaling coefficient $n_{J+1,0}$ (for details, see Walden *et al.*, 1998). Equation (441b) can be used to demonstrate numerically that the above still holds for DWTs based on other Daubechies wavelets (see Exercise [10.9]).

A final question to answer about the variances of the noise wavelet coefficients is how they are related to $\sigma^2_\eta = \psi'(K)$. Since $\mathcal{W}$ is an orthonormal transform, $\text{tr}\,\{\Sigma_{\boldsymbol{\eta}}\} = \text{tr}\,\{\mathcal{W}\Sigma_{\boldsymbol{\eta}}\mathcal{W}^T\} = \text{tr}\,\{\Sigma_{\mathbf{n}}\}$, where $\text{tr}\{\cdot\}$ denotes the trace of a matrix (i.e., the sum of its diagonal elements). The diagonal of $\Sigma_{\boldsymbol{\eta}}$ consists of the common element $s_\eta(f_0) = \sigma^2_\eta$, so that $\text{tr}\,\{\Sigma_{\boldsymbol{\eta}}\} = M\sigma^2_\eta = 2^J\sigma^2_\eta$. From Equation (441b) we know that the diagonal of $\mathcal{W}\Sigma_{\boldsymbol{\eta}}\mathcal{W}^T$ has the form

$$\underbrace{\sigma_1^2,\ldots,\sigma_1^2}_{\frac{M}{2}\text{ of these}},\underbrace{\sigma_2^2,\ldots,\sigma_2^2}_{\frac{M}{4}\text{ of these}},\ldots,\sigma_J^2,\bar{\sigma}^2_{J+1}.$$

Hence,

$$\sigma^2_\eta = \frac{\bar{\sigma}^2_{J+1}}{2^J} + \sum_{j=1}^{J}\frac{\sigma^2_j}{2^j},$$

which, combined with Equation (441c), shows that

$$\sigma^2_1 < \sigma^2_\eta < \bar{\sigma}^2_{J+1}. \tag{442}$$

Computations indicate that, under the special case of the Haar DWT and $M = N/2$, in fact $\sigma^2_2 < \sigma^2_\eta < \sigma^2_3$ when $K = 5,\ldots,9$; $\sigma^2_\eta \approx \sigma^2_3$ when $K = 10$; and $\sigma^2_3 < \sigma^2_\eta < \sigma^2_4$ when $K = 11,\ldots,41$ (this covers the number of tapers K likely to be used in practice).

To confirm the increase in σ^2_j with j and, more importantly, to validate the statistical model defined by Equations (277), (440) and (441a), we carried out a simulation study in which the following steps were repeated a thousand times.

[1] A sample $X_0,\ldots,X_{N-1}$ from a specified Gaussian stationary process was generated, where $N = 2048$ and $M = N/2$.

[2] The multitaper estimator was calculated using $K = 10$ sinusoidal tapers.

[3] The DWT of $\boldsymbol{\eta}$ of Equation (440) was computed based upon wavelet filters of lengths $L = 2, 4, 8$ and 16, yielding the wavelet coefficients $\{n_{j,t}\}$.

[4] Finally, the standard deviation of the wavelet coefficients $\{n_{j,t}\}$ at levels $j = 1$, 2, 3 and 4 was estimated from the square root of $\hat{\sigma}^2_j \equiv \frac{1}{M_j}\sum_{t=0}^{M_j-1} n^2_{j,t}$.

Two different processes were used, a white noise process and the AR(2) process defined in Equation (436a), but the results proved to be virtually identical, so we only report on the AR(2) case here. Box plots for the thousand standard deviations for each wavelet filter and each level are plotted in Figure 443. The box plots shown there are defined as follows: a thin horizontal line is drawn through the box at the median of the data, the upper and lower ends of the box are at the upper and lower quartiles, and the vertical lines extend from the box to points within a standard range of the data, defined as 1.5 times the inter-quartile range. The variability of the estimates $\hat{\sigma}_j$ increases with level j. The thick horizontal lines extending beyond each box plot indicate the value of σ_j derived from Equation (441b). The 'nominal' standard deviation $\sigma_\eta = \sqrt{[\psi'(K)]} = \sqrt{[\psi'(10)]} \doteq 0.32$ is marked as a thin horizontal line spanning the width of each plot. This exceeds the sample median of the $\hat{\sigma}_j$'s for levels 1, 2 and 3 when we use the D(4),

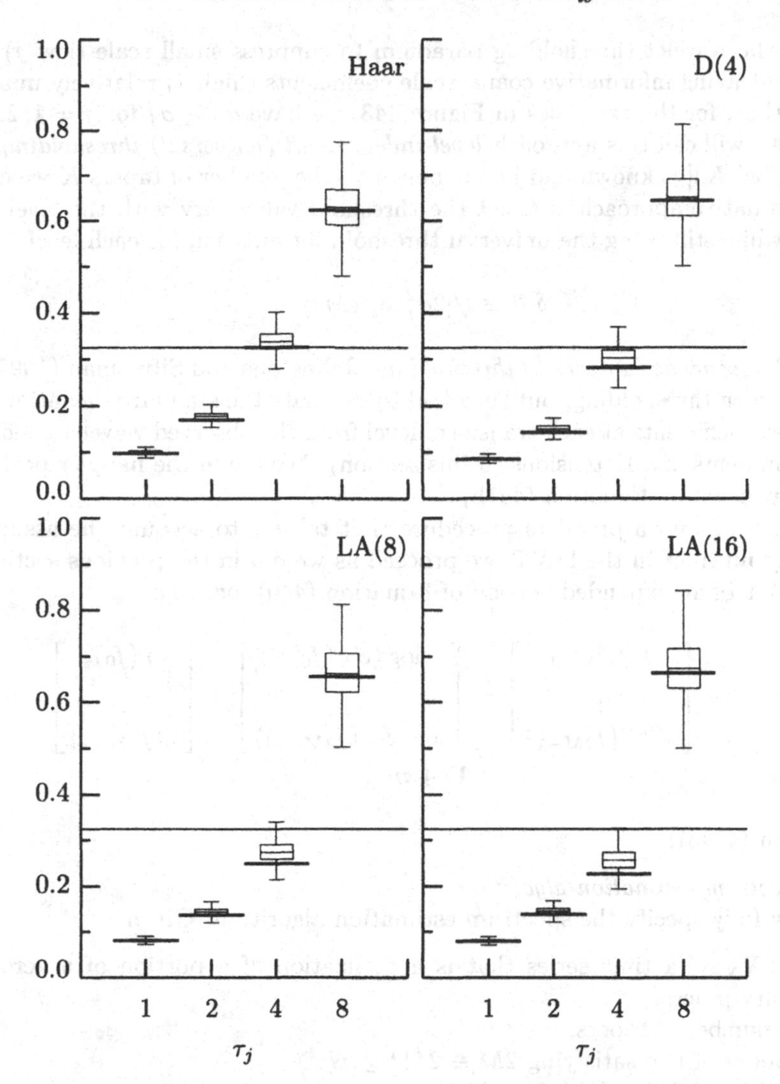

Figure 443. Box plots of the estimated standard deviations $\hat{\sigma}_j$ of wavelet coefficients $n_{j,t}$ at levels $j = 1, 2, 3$ and 4 derived from the AR(2) process using different wavelet filters, for $N =$ 2048 and $K = 10$. The horizontal solid lines extending beyond each box plot indicate the value of σ_j derived from Equation (441b). The 'nominal' standard deviation $\sigma_\eta = \sqrt{[\psi'(10)]} \doteq 0.32$ is marked as a solid horizontal line right across each of the plots.

LA(8) or LA(16) wavelet filters. For the Haar wavelet filter, the nominal standard deviation exceeds the sample median of the $\hat{\sigma}_j$'s for levels 1 and 2. Figure 443 thus illustrates Equation (441c) and lends credence to the approximations leading up to Equations (277), (440) and (441a).

Let us now turn our attention to thresholding the wavelet coefficients $\{W_{j,t}^{(\text{mt})}\}$. Suppose we choose the threshold value to be fixed for all transform levels j according to the universal threshold

$$\delta^{(\text{u})} \equiv \sqrt{[2\sigma_\eta^2 \log{(M)}]}. \tag{443}$$

From Figure 443 and Equation (442), it is clear that the correlation will act in accordance with the wavelet thresholding paradigm to suppress small scale (low j) 'noise spikes' while leaving informative coarse scale coefficients (high j) relatively unattenuated (note that, for the examples in Figure 443, we have $\sigma_\eta \geq \sigma_j$ for $j = 1, 2, 3$ and $\sigma_4 > \sigma_\eta$). We will call this approach *level-independent (universal) thresholding*. Note that $\sigma_\eta = \sqrt{[\psi'(K)]}$ is known and just depends on the number of tapers K we choose.

An alternative approach is to let the threshold value vary with the level of the transform, while still using the universal threshold formulation for each level:

$$\delta_j^{(\mathrm{u})} \equiv \sqrt{[2\sigma_j^2 \log(M)]}. \tag{444a}$$

This is *level-dependent (universal) thresholding*. Johnstone and Silverman (1997) gave examples of such thresholding, but they had to estimate the standard deviation of the noise wavelet coefficients at each transform level from the observed wavelet coefficients (see the Comments and Extensions to this section). We are in the happier position of having σ_j^2 specified by Equation (441b).

Finally, to obtain a practical procedure that takes into account the assumption of circularity inherent in the DWT, we proceed as we did in the previous section and take the DWT of an expanded version of Equation (440), namely,

$$\mathbf{Y}^{(\mathrm{mt})} \equiv \begin{bmatrix} Y^{(\mathrm{mt})}(f_0) \\ \vdots \\ Y^{(\mathrm{mt})}(f_{2M-1}) \end{bmatrix} = \begin{bmatrix} \log(S_X(f_0)) \\ \vdots \\ \log(S_X(f_{2M-1})) \end{bmatrix} + \begin{bmatrix} \eta(f_0) \\ \vdots \\ \eta(f_{2M-1}) \end{bmatrix}$$
$$\equiv \mathbf{D} + \boldsymbol{\eta} \tag{444b}$$

(cf. Equation (433b)).

- *Full spectrum estimation algorithm*

We can now fully specify the spectrum estimation algorithm. Given

[1] $X_0, \ldots, X_{N-1}$, a time series that is a realization of a portion of a zero mean stationary process;
[2] K, the number of tapers;
[3] M, a power of two satisfying $2M \equiv 2^{J+1} \geq N$;
[4] J_0, a level satisfying $J_0 \leq J$; and
[5] a choice of thresholding method (e.g., hard, mid, soft),

we construct our spectrum estimator $\hat{S}_X^{(\mathrm{mthr})}$ in the following manner.

[1] We begin by calculating the logarithm of the multitaper estimate. For each sine taper $\{a_{n,t}\}$ defined as in Equation (274a) for $n = 0, \ldots, K-1$, we form the tapered series $a_{n,0}X_0, \ldots, a_{n,N-1}X_{N-1}$, append $2M - N$ zeros to this series and use a standard 'power of two' fast Fourier transform algorithm to form the eigenspectrum

$$\hat{S}_{X,n}^{(\mathrm{mt})}(f_k) = \Delta t \left| \sum_{t=0}^{2M-1} a_{n,t} X_t e^{-i2\pi f_k t\,\Delta t} \right|^2, \quad k = 0, \ldots, 2M-1,$$

where $f_k = k/(2M\,\Delta t)$ and $a_{n,t}X_t \equiv 0$ for $t \geq N$. The K eigenspectra are then averaged to form the multitaper estimate $\hat{S}_X^{(\mathrm{mt})}(f_k)$ as in Section 7.5. Then we form $Y^{(\mathrm{mt})}(f_k) \equiv \log(\hat{S}_X^{(\mathrm{mt})}(f_k)) - \psi(K) + \log(K)$ and place these $2M$ values in the column vector $\mathbf{Y}^{(\mathrm{mt})}$, as per Equation (444b).

[2] Next we apply a partial DWT out to level J_0. We obtain the partial DWT transform coefficients $\mathbf{W}_1^{(\mathrm{mt})}, \ldots, \mathbf{W}_{J_0}^{(\mathrm{mt})}$ and $\mathbf{V}_{J_0}^{(\mathrm{mt})}$. Component-wise the wavelet coefficients can be written

$$W_{j,t}^{(\mathrm{mt})} = d_{j,t} + n_{j,t} \quad j = 1, \ldots, J_0 \text{ and } t = 0, \ldots, M_{j-1} - 1.$$

The M_{J_0} scaling coefficients in $\mathbf{V}_{J_0}^{(\mathrm{mt})}$ are left entirely alone.

[3] We threshold $\mathbf{W}_1^{(\mathrm{mt})}, \ldots, \mathbf{W}_{J_0}^{(\mathrm{mt})}$ to obtain $\mathbf{W}_1^{(\mathrm{t})}, \ldots, \mathbf{W}_{J_0}^{(\mathrm{t})}$ via either
   (a) level-independent thresholding using Equation (443) applied to wavelet coefficients for all levels $j = 1, \ldots, J_0$ or
   (b) level-dependent thresholding using Equation (444a) applied separately to each level $j = 1, \ldots, J_0$.

[4] We estimate $\mathbf{D}$ of Equation (444b) via $\widehat{\mathbf{D}}^{(\mathrm{t})}$ obtained by inverse transforming $\mathbf{W}_1^{(\mathrm{t})}, \ldots, \mathbf{W}_{J_0}^{(\mathrm{t})}$ and $\mathbf{V}_{J_0}^{(\mathrm{mt})}$. Let $\widehat{D}_k^{(\mathrm{t})}$ denote the kth element of $\widehat{\mathbf{D}}^{(\mathrm{t})}$, and note that $\exp(\widehat{D}_k^{(\mathrm{t})})$ gives us an estimate of $S_X(f_k)$. We can now form the spectrum estimate based on multitapering with thresholding via

$$\hat{S}_X^{(\mathrm{mthr})}(f_k) \equiv \begin{cases} e^{\widehat{D}_0^{(\mathrm{t})}}, & k = 0; \\ \frac{1}{2}(e^{\widehat{D}_k^{(\mathrm{t})}} + e^{\widehat{D}_{2M-k}^{(\mathrm{t})}}), & k = 1, \ldots, M-1; \text{ and} \\ e^{\widehat{D}_M^{(\mathrm{t})}}, & k = M. \end{cases}$$

If $\{X_t\}$ cannot be assumed to have zero mean (usually the case in practical applications), we replace X_t in step [1] above with $X_t - \overline{X}$, where $\overline{X}$ is the sample mean. (As noted in item [3] of the Comments and Extensions to the previous section, recentering the X_t in this manner has the effect of rendering the periodogram equal to zero at zero frequency; in contrast, a multitaper estimate is usually *not* similarly affected, so we can usually get along without having to readjust the contents of $\mathbf{Y}^{(\mathrm{mt})}$.)

- *Examples*

Let us apply the level-independent and level-dependent thresholding schemes to the same AR(24), AR(2) and MRC models used to study the periodogram-based scheme. For each of these models, we generated a thousand simulated time series of length $N = 2048$ and then formed multitaper SDF estimates based on $K = 10$ sine tapers. The bandwidth for these estimates is $(K+1)/(N+1) \approx 0.0054$, which is small relative to the widths of the peaks in the three models. We then computed $\hat{S}_X^{(\mathrm{mthr})}(\cdot)$ based upon the LA(8) wavelet filter in combination with level-independent and level-dependent hard, mid and soft thresholds and $J_0 = 5$, 6 and 8. We judged the quality of each SDF estimate by computing the RMSE as in Equation (436b), but with $\hat{S}_X^{(\mathrm{pt})}(f_k)$ replaced by $\hat{S}_X^{(\mathrm{mthr})}(f_k)$. The average of these RMSEs for each type of thresholding and choice of J_0 over the thousand simulations is plotted in Figure 446 using various types of connected lines (the asterisks show similar RMSEs for the periodogram-based method). In addition, the SDF estimate whose RMSE was closest to the average RMSE is plotted in Figures 447, 448 and 449 as a representative spectrum estimate for the stated choice of model, threshold type and J_0.

Let us now comment on the quality of the spectrum estimates for each of the three models.

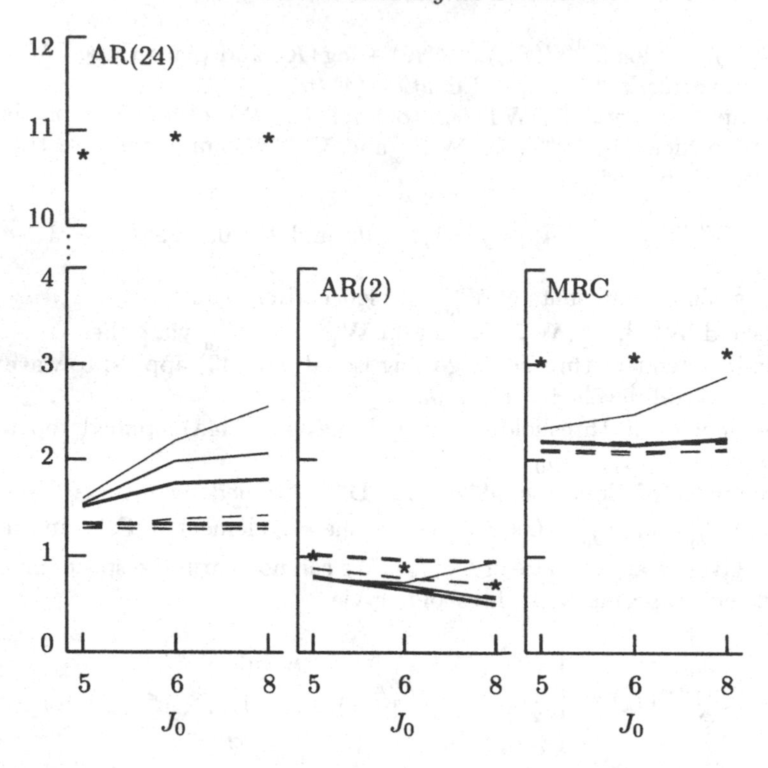

Figure 446. Average value over a thousand simulations of the RMSEs for the AR(24) (left-hand plot), AR(2) (center) and MRC models (right-hand). In each plot, the average RMSE (in dBs) is plotted for the level-dependent multitaper-based method with hard (solid thick curve), mid (solid medium) and soft (solid thin) thresholding and also for the level-independent method with hard (dashed thick curve), mid (dashed medium) and soft (dashed thin) thresholding. Three values of level J_0 are considered, namely 5, 6 and 8, corresponding to 64, 32 and 8 scaling coefficients left untouched by the thresholding. The asterisks show average RMSEs for the periodogram-based method. In all cases, the series length is $N = 2048$, and we use the LA(8) wavelet to compute the DWT.

[1] In contrast to the periodogram, the multitaper scheme suppresses leakage and, when combined with wavelet thresholding, produces a smoothed log SDF estimate with no leakage, as shown in Figure 447. This figure shows the representative estimates for level-independent soft thresholding (upper plot) and level-dependent hard thresholding (lower plot), both with $J_0 = 5$ (these choices give the minimum average RMSE across parameter combinations for both level-independent and level-dependent thresholding; see the left-hand plot of Figure 446). Both estimates are good and tend to capture the peaks better than the corresponding periodogram-based estimate (top plot of Figure 438); however, the level-independent estimate would probably be judged superior to the level-dependent one in the estimation of the three highest-frequency peaks in the SDF. The level-dependent estimate shows evidence of small scale noise coefficients unattenuated by the thresholding.

[2] Figure 448 shows the representative estimates for the AR(2) process for level-independent soft thresholding and level-dependent hard thresholding, both with $J_0 = 8$ (these choices again giving the minimum average RMSE across parameter combinations for both level-independent and level-dependent thresholding; see the

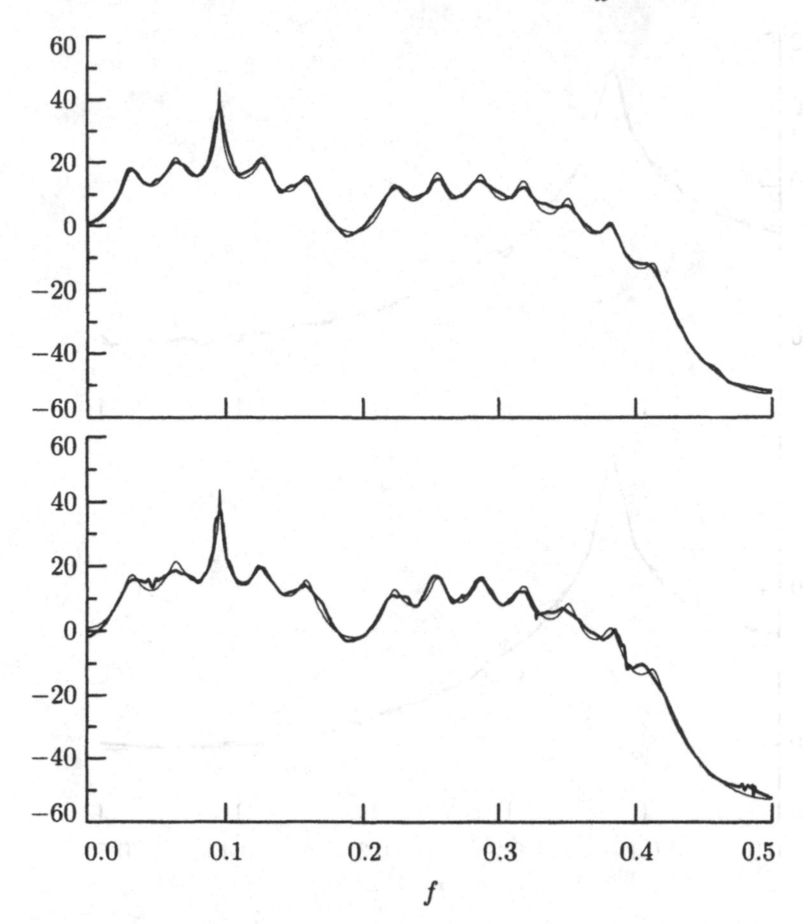

Figure 447. Estimated SDFs (thick curves) and true SDF (thin) for the AR(24) process. The SDF estimates are representative in that they have RMSEs closest to the average RMSE over a thousand simulations. The upper plot is for level-independent soft thresholding, and the lower plot, level-dependent hard thresholding, with $J_0 = 5$ in both cases. The simulated series are of length $N = 2048$, and we use an LA(8) DWT.

middle plot of Figure 446). Both estimates are quite good. The level-dependent method (lower plot) again shows evidence of small scale noise coefficients unattenuated by the thresholding, but also does a little better in estimating the height of the peak in the SDF. Even though the periodogram for this process is essentially leakage free even for moderate sample sizes, the multitaper-based estimates in fact have somewhat smaller average RMSEs than those obtained using the periodogram-based scheme.

[3] Figure 449 shows the representative estimates for the MRC model for level-independent soft thresholding and level-dependent hard thresholding, both with $J_0 = 6$ (chosen as before based upon a study of the right-hand plot of Figure 446). Both the 30 dB rises and falls of the band-limited part of the SDF and the narrow-band interference at $f = 0.45$ are well estimated by both methods, with the level-independent estimate (upper plot) being somewhat better at capturing the height of the peak (as Figure 438 shows, the periodogram-based

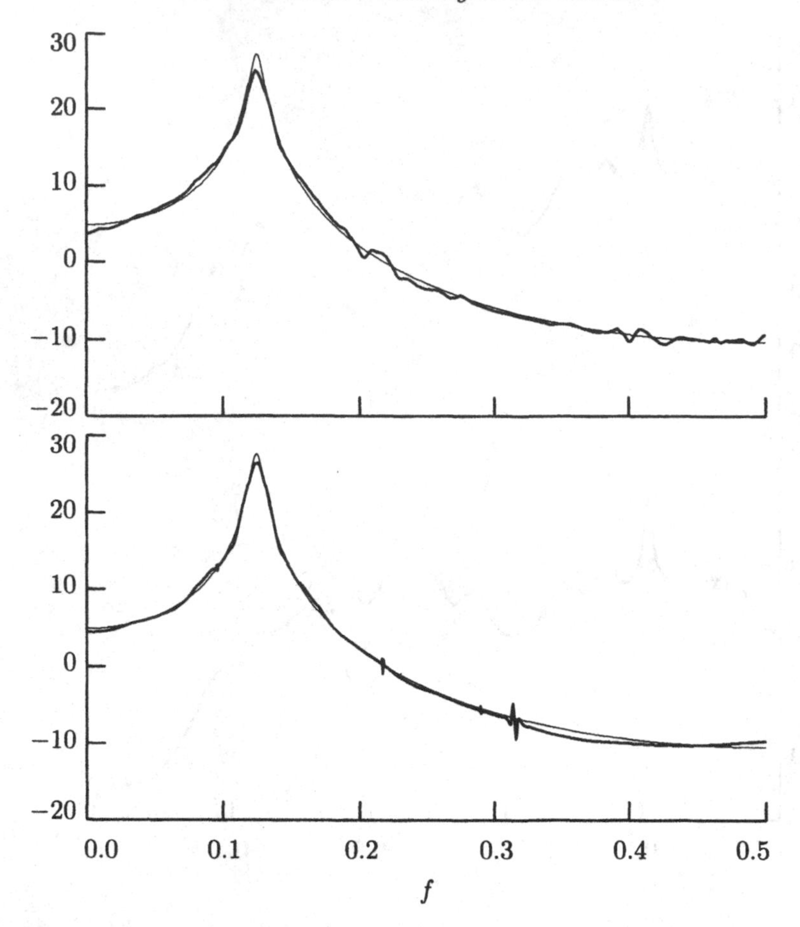

Figure 448. Estimated SDFs (thick curves) and true SDF (thin) for the AR(2) process. Layout and parameters are as for Figure 447.

estimate does not capture the sharp transitions in the SDF as well). Again, the level-dependent estimate (lower plot) shows evidence of small scale noise coefficients unattenuated by the thresholding. The multitaper-based estimates show a significant decrease in RMSE when compared to the periodogram-based estimates (see the right-hand plot of Figure 446).

Note that the best level J_0 is different for the three models and that the RMSE of the spectrum estimators does vary somewhat with the choice of J_0, as shown in Figure 446; however, provided that soft thresholding is used with level-independent universal thresholds, and hard thresholding, with level-dependent universal thresholds, the variation of RMSE with J_0 does not cause the corresponding spectrum estimates to be markedly different in their visual appearances.

Finally let us comment on level-dependent versus level-independent thresholding. In Figure 446 the solid curves show the performance of the former, and the dashed curves, the latter. Level-independent thresholding does better overall for the AR(24) and model radio communications models and, when used with soft thresholding for the AR(2) model, is comparable to the best that level-dependent thresholding can offer.

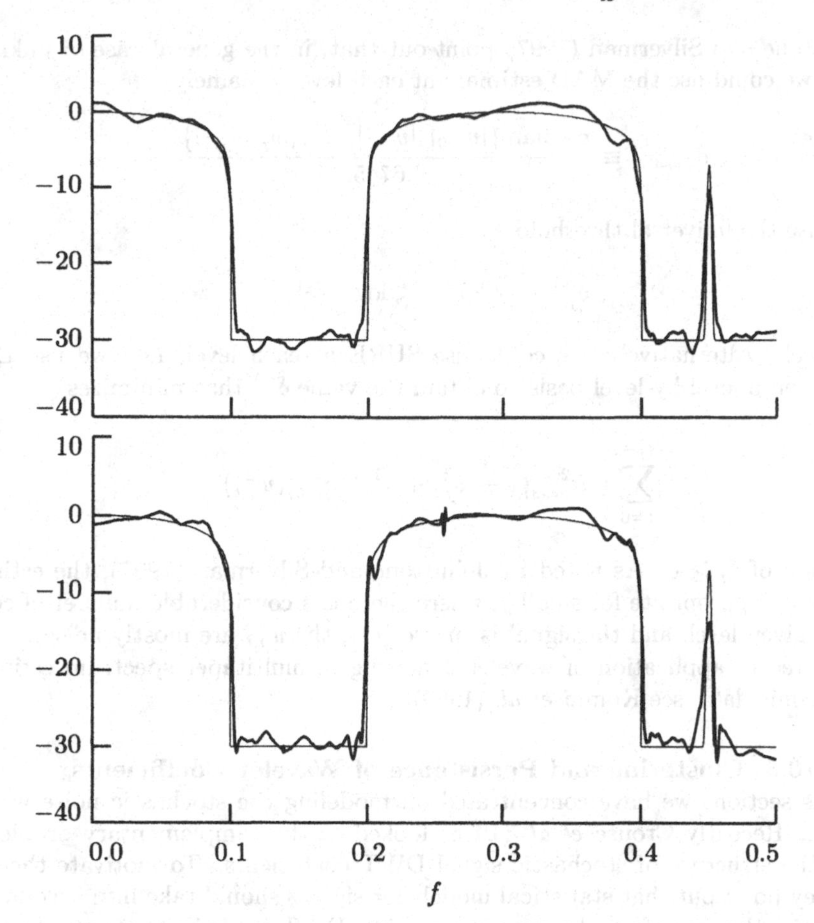

Figure 449. Estimated SDF (thick curve) and true SDF (thin curve) for the mobile radio communications process. Layout and parameters are as for Figure 447.

Some insight into why level-independent thresholding works so well can be gleaned from Figure 443. The thin horizontal lines spanning the plots in this figure indicate the level-independent noise variance σ_η^2. Because this variance falls above just the lowest two or three level-dependent variances σ_j^2, the correlation that multitapering induces acts in accordance with the underlying paradigm for wavelet thresholding in that it actually helps suppress small scale 'noise spikes' while leaving informative large scale wavelet coefficients relatively unattenuated.

Comments and Extensions to Section 10.7

[1] In the multitaper SDF estimation problem, we know the variance σ_j^2 of the noise wavelet coefficients at each level. Hence, given realizations $w_{j,t}$ of $W_{j,t}$, the minimizer of

$$\sum_{t=0}^{N_j-1} (2\sigma_j^2 - w_{j,t}^2 + \delta_j^2) 1_{[\delta_j^2,\infty)}(w_{j,t}^2) \tag{449}$$

as a function of $\delta > 0$ gives us the SURE-based threshold value $\delta_j^{(\mathrm{S})}$ for the level j coefficients (cf. Equation (406b)).

[2] Johnstone and Silverman (1997) point out that, in the general case of unknown variances, we could use the MAD estimate at each level j, namely,

$$\hat{\sigma}_{(\text{mad}),j} \equiv \frac{\text{median}\,\{|w_{j,0}|, |w_{j,1}|, \ldots, |w_{j,N_j-1}|\}}{0.6745},$$

and then use the universal threshold

$$\hat{\delta}_j^{(\text{u})} \equiv \sqrt{[2\hat{\sigma}^2_{(\text{mad}),j} \log(N)]}$$

at each level. Alternatively, we could use SURE at each level; i.e., we use Equation (405a) on a level-by-level basis to obtain the value $\hat{\delta}_j^{(\text{S})}$ that minimizes

$$\sum_{t=0}^{N_j-1} (2\hat{\sigma}^2_{(\text{mad}),j} - w_{j,t}^2 + \delta_j^2) 1_{[\delta_j^2,\infty)}(w_{j,t}^2)$$

as a function of $\delta_j \geq 0$. As noted by Johnstone and Silverman (1997), the estimate $\hat{\sigma}_{(\text{mad}),j}$ is only appropriate for small j, where there is a considerable number of coefficients in a given level, and the signal is sparse (i.e., the $w_{j,t}$ are mostly noise).
[3] For a recent application of wavelet denoising of multitaper spectrum estimates for helioseismic data, see Komm *et al.* (1999).

10.8 Clustering and Persistence of Wavelet Coefficients

In previous sections we have concentrated on modeling the stochastic noise wavelet coefficients. Recently Crouse *et al.* (1998) looked at the complementary problem of modeling the structure of stochastic signal DWT coefficients. To motivate their approach, they point out that statistical models for signals should take into account two key properties that are often observed in empirical DWT coefficients of actual signals. The first property is called *clustering*, which says that, if a particular DWT coefficient within a level j is large or small, then adjacent coefficients within that same level are likely also to be large or small. The second property is called *persistence*, which says that large or small values of wavelet coefficients tend to propagate across different levels. These ideas are illustrated by the LA(8) DWT of the ECG time series shown in Figure 127 (it is a stretch of the imagination to regard this series as a noise-free signal, but we do so anyway in this section for the sake of pedagogy). Clustering is particularly evident in the level $j = 3$ coefficients $\mathbf{W}_3$, where there tend to be two or more large coefficients adjacent to each other when an R wave event occurs. Persistence is also connected with R wave events, as can be seen from the coaligned large coefficients at, for example, levels $j = 2$ and 3. Taken together, clustering and persistence argue that the DWT cannot completely decorrelate actual signals, so we must move beyond simple IID models for stochastic signals.

With these ideas in mind, Crouse *et al.* (1998) develop a new framework for modeling wavelet coefficient dependencies and non-Gaussianity based around hidden Markov models (HMMs) – see Rabiner (1989) for a good review of HMMs. In the context of DWTs, the HMM idea is to associate a *hidden state variable* to each signal DWT coefficient (note in particular that the state variables are *not* directly associated with the original time series or signal, but rather with the transform of the signal). To match the non-Gaussian nature of signal DWT coefficients, they model the PDF of each coefficient as a Gaussian mixture density involving two or more components (the right-hand plot of Figure 410 depicts a PDF for a mixture density with two components). The hidden state variable can take one of a finite set of state values, and the state value will say something statistically about the corresponding signal DWT coefficient, namely, which component of the Gaussian mixture corresponds to its PDF. Although each signal DWT coefficient is conditionally Gaussian given the value of its hidden state variable, the coefficient has an overall (unconditional) non-Gaussian density (again, see Figure 410).

To allow for dependencies between signal DWT coefficients, Crouse *et al.* (1998) introduce Markovian dependencies between hidden state variables based on a natural 'tree' structure for the wavelet and scaling coefficients. To see how this tree structure arises, suppose we compute a level J_0 partial DWT for a signal, yielding the vectors $\mathbf{W}_1, \ldots, \mathbf{W}_{J_0}$ and $\mathbf{V}_{J_0}$. We can plot the vectors $\mathbf{V}_{J_0}, \mathbf{W}_{J_0}, \ldots, \mathbf{W}_1$ as rows stacked on top of each other, as in Figure 127, which depicts a level $J_0 = 6$ partial DWT of the ECG series (again, we are pretending that this series is a noise-free signal). Here each coefficient vector has been circularly shifted so that it is plotted against a physically meaningful time. If we ignore the complication introduced by the small shift in the times associated with $\mathbf{V}_6$ and $\mathbf{W}_6$ in that figure, then the 'node' $V_{6,2}$ is above $W_{6,3}$, which in turn is above $W_{5,3}$ and $W_{5,4}$. Continuing on down, $W_{5,3}$ is above $W_{4,3}$ and $W_{4,4}$, while $W_{5,4}$ is above $W_{4,5}$ and $W_{4,6}$ and so forth. In other words, there is a tree structure of nodes under each of the level J_0 coefficients. Formally we can say that each of the N_{J_0} scaling coefficients in $\mathbf{V}_{J_0}$ is above one coefficient in $\mathbf{W}_{J_0}$, and each coefficient in $\mathbf{W}_j$ is above two coefficients in $\mathbf{W}_{j-1}$, $j = J_0, \ldots, 2$. This structure thus defines N_{J_0} trees. Since a state variable is associated with each wavelet or scaling coefficient, it follows that the state variables could also be represented by a set of N_{J_0} trees.

With the tree structure so defined, Crouse *et al.* (1998) consider the following three models for the relationships between state variables in the tree.

[1] There is no connection between state variables within the same tree or between different trees. This treats DWT state variables (and hence signal DWT coefficients) as independent and does not account for either clustering or persistence.

[2] The state variables have a first order Markov chain dependency (Rabiner, 1989) within a level j (connected at the same level across trees), but are independent from level to level (unconnected down levels within a tree). This forms the *hidden Markov chain model* and can account for clustering, but not persistence.

[3] The state variables are connected vertically (down levels within a tree); i.e., there are N_{J_0} trees, each consisting of connected state variables. This is the *hidden Markov tree model* and provides for both clustering and persistence.

Crouse *et al.* (1998) emphasize that, for cases [2] and [3], the Markov structure is on the hidden state variables, and not directly on the DWT coefficients for $\mathbf{X}$ itself. They then consider the DWT of stochastic signal plus stochastic IID Gaussian noise, with the signal and noise assumed independent. With this additional structure, because the signal DWT coefficients obey an HMM, then so do the DWT coefficients for $\mathbf{X}$ itself. To see this, note that, given the value of the hidden state variable, the signal DWT coefficient is Gaussian, and the sum of two independent Gaussian RVs (signal plus noise) is Gaussian, with the variance given by the sum of the variances. Hence the signal DWT model induces a parallel structure on the DWT for $\mathbf{X}$, and we can estimate the signal by fitting an HMM to the observed DWT coefficients (i.e., those for $\mathbf{X}$). Crouse *et al.* (1998) take the variance of the signal DWT coefficient given a particular state value to be that of the observed DWT coefficient given the same state value, less the variance of the noise DWT coefficients (we can estimate the latter by $\hat{\sigma}^2_{(\text{mad})}$, defined in Equation (420)). For further details on this estimation method, see Crouse *et al.* (1998).

10.9 Summary

Let $\mathbf{X}$ be an N dimensional vector whose elements are the real-valued time series $\{X_t : t = 0, \ldots, N-1\}$, and let O_l be the lth component of $\mathbf{O}$, the vector of coefficients resulting from applying an orthonormal transform $\mathcal{O}$ to $\mathbf{X}$. We consider $\mathbf{X}$ to be an additive sum of a signal vector and a noise vector. We can think of the signal vector as either deterministic (fixed) or stochastic (random) and the stochastic noise vector as always having mean zero but containing either IID RVs or non-IID RVs (the departure from the IID assumption typically takes one of two forms, namely, either the RVs are independent but have different variances or the RVs are correlated). Hence, we can think of $\mathbf{X}$ as obeying one of the following four models:

[i] $\mathbf{X} = \mathbf{D} + \boldsymbol{\epsilon}$, where $\mathbf{D}$ is a deterministic signal and $\boldsymbol{\epsilon}$ is IID noise, so that application of the orthonormal transform gives

$$\mathbf{O} \equiv \mathcal{O}\mathbf{X} = \mathcal{O}\mathbf{D} + \mathcal{O}\boldsymbol{\epsilon} \equiv \mathbf{d} + \mathbf{e} \text{ and hence } O_l = d_l + e_l;$$

[ii] $\mathbf{X} = \mathbf{D} + \boldsymbol{\eta}$, where $\mathbf{D}$ is a deterministic signal and $\boldsymbol{\eta}$ is non-IID noise, so that

$$\mathbf{O} \equiv \mathcal{O}\mathbf{X} = \mathcal{O}\mathbf{D} + \mathcal{O}\boldsymbol{\eta} \equiv \mathbf{d} + \mathbf{n} \text{ and hence } O_l = d_l + n_l;$$

[iii] $\mathbf{X} = \mathbf{C} + \boldsymbol{\epsilon}$, where $\mathbf{C}$ is a stochastic signal and $\boldsymbol{\epsilon}$ is IID noise (with signal and noise independent of each other), so that

$$\mathbf{O} \equiv \mathcal{O}\mathbf{X} = \mathcal{O}\mathbf{C} + \mathcal{O}\boldsymbol{\epsilon} \equiv \mathbf{R} + \mathbf{e} \text{ and hence } O_l = R_l + e_l; \text{ or}$$

[iv] $\mathbf{X} = \mathbf{C} + \boldsymbol{\eta}$, where $\mathbf{C}$ is a stochastic signal and $\boldsymbol{\eta}$ is non-IID noise, so that

$$\mathbf{O} \equiv \mathcal{O}\mathbf{X} = \mathcal{O}\mathbf{C} + \mathcal{O}\boldsymbol{\eta} \equiv \mathbf{R} + \mathbf{n} \text{ and hence } O_l = R_l + n_l$$

(typically the signal and noise are independent of each other, but this is not true in, for example, the model studied in Vidakovic, 1998).

It is assumed that, for any component l of interest in [iii] or [iv], the orthonormal transform gives $E\{R_l\} = 0$. (If we specialize to the DWT, this zero mean assumption is not a serious restriction: it is guaranteed if the stochastic signal $\mathbf{C}$ is from a stochastic process whose dth order backward difference is stationary, and we take the filter length L such that $L > 2d$.)

In the case of IID input noise (cases [i] and [iii]), the noise transform coefficients will have identical variances (no matter how $\boldsymbol{\epsilon}$ is distributed) and will have identical distributions if $\boldsymbol{\epsilon}$ is Gaussian. In symbols we have

$$\Sigma_{\boldsymbol{\epsilon}} = \sigma_\epsilon^2 I_N \text{ and } \Sigma_{\mathbf{e}} = \mathcal{O}\Sigma_{\boldsymbol{\epsilon}}\mathcal{O}^T = \sigma_\epsilon^2 I_N, \text{ so } \operatorname{var}\{e_l\} = \operatorname{var}\{\epsilon_l\} = \sigma_\epsilon^2,$$

and, if $\epsilon_l \stackrel{d}{=} \mathcal{N}(0, \sigma_\epsilon^2)$, then $e_l \stackrel{d}{=} \mathcal{N}(0, \sigma_\epsilon^2)$. In the case of non-IID noise (cases [ii] and [iv]), we can only say

$$\Sigma_{\mathbf{n}} = \mathcal{O}\Sigma_{\boldsymbol{\eta}}\mathcal{O}^T;$$

however, if $\eta_l \stackrel{d}{=} \mathcal{N}(0, \sigma_{\eta_l}^2)$, then $n_l \stackrel{d}{=} \mathcal{N}(0, \sigma_{n_l}^2)$, but we need not have $\sigma_{n_l}^2 = \sigma_{\eta_l}^2$, and the variances can depend on the index l.

We carry out our signal estimation using thresholding or shrinkage approaches, both of which employ the following steps.

[1] Compute the transform coefficients $\mathbf{O} \equiv \mathcal{O}\mathbf{X}$.
[2] Define $\mathbf{O}^{(\cdot)}$ to be the vector whose lth element $O_l^{(\cdot)}$ is the thresholded or shrunken version of the coefficient O_l, where the generic superscript '(.)' is replaced by, for example, '(t)' or '(shr)' to denote thresholding or shrinkage (or by a more specific designation such as '(ht)' for hard thresholding).
[3] Estimate the signal via $\mathcal{O}^T\mathbf{O}^{(\cdot)}$. Here we do not distinguish between the estimation of a deterministic signal and the estimation of a realization of a stochastic one.

We emphasize that this scheme carries out the thresholding or shrinkage component-wise, i.e., on a term by term basis. Thresholding is appropriate for cases [i] and [ii]; i.e., the signal is modeled as deterministic. Rules include hard, mid and soft thresholding (Equations (399), (400b) and (400a), respectively), all specified by a single threshold parameter $\delta > 0$, and firm thresholding (Equation (400c)), specified by two. The threshold level δ for O_l must be defined, and an appropriate definition depends on the statistical properties of the noise transform coefficients. If these coefficients are IID, then a single threshold level will apply to all O_l. For IID Gaussian noise, a popular threshold choice is the universal threshold

$$\delta^{(\mathrm{u})} \equiv \sqrt{[2\sigma_\epsilon^2 \log(N)]}$$

(this is Equation (400d)). Other ways for choosing the threshold level are through minimizing criteria using cross-validation (Equations (422) and (402)) and via an approach based on Stein's unbiased risk estimator (SURE), in which, given observed values o_l of the RV O_l and assuming soft thresholding, we set the level $\delta^{(\mathrm{S})}$ to the value of δ minimizing

$$\sum_{l=0}^{N-1} (2\sigma_\epsilon^2 - o_l^2 + \delta^2) 1_{[\delta^2,\infty)}(o_l^2)$$

(this is Equation (406a), with the common variance σ_0^2 set to σ_ϵ^2). If the noise transform coefficients come from non-IID input noise $\boldsymbol{\eta}$, then the threshold levels need to be varied as appropriate. For example, when coefficients divide into groups $\mathcal{C}_j$ that are IID within a group with common variance σ_j^2, suitable threshold levels for coefficients in group j can be defined via the level-dependent 'universal' threshold

$$\delta_j^{(\mathrm{u})} \equiv \sqrt{[2\sigma_j^2 \log(N)]}$$

(this is Equation (406c)) or via a SURE-based approach in which we define $\delta_j^{(\mathrm{S})}$ as the value of δ minimizing

$$\sum_{t \in \mathcal{C}_j} (2\sigma_j^2 - o_{j,t}^2 + \delta^2) 1_{[\delta^2,\infty)}(o_{j,t}^2)$$

(this is Equation (406b)).

Shrinkage is appropriate for cases [iii] and [iv]; i.e., the signal is modeled as stochastic. The simplest shrinkage rule is a rescaling of the observation by the ratio of the signal variance $\sigma^2_{R_l}$ to the total variance $\sigma^2_{R_l} + \sigma^2_{n_l}$ (see Section 10.3). Well-known formulations for nonlinear shrinkage functions include the conditional mean $E\{R_l | O_l = o_l\}$ of the signal, given the observation (Equation (410b)). For example, the sparse signal model

$$R_l \stackrel{\mathrm{d}}{=} (1 - \mathcal{I}_l)\mathcal{N}(0, \sigma^2_{G_l}) \quad \text{with} \quad n_l \stackrel{\mathrm{d}}{=} \mathcal{N}(0, \sigma^2_{n_l})$$

leads to the conditional mean

$$E\{R_l | O_l = o_l\} = \frac{b_l}{1 + c_l} o_l,$$

where, with p_l being the probability that the binary-valued RV $\mathcal{I}_l$ is unity, we have

$$b_l = \frac{\sigma^2_{G_l}}{\sigma^2_{G_l} + \sigma^2_{n_l}} \quad \text{and} \quad c_l = \frac{p_l \sqrt{(\sigma^2_{G_l} + \sigma^2_{n_l})}}{(1 - p_l)\sigma_{n_l}} e^{-o_l^2 b_l / (2\sigma^2_{n_l})}$$

(see Equations (411b), (410e), (411c), (411a) and (411d)). We could also use the conditional median instead (Equation (411e)). Shrinkage functions depend on the statistical properties of the signal transform coefficients $\mathbf{R}$ and noise transform coefficients ($\mathbf{e}$ or $\mathbf{n}$). The specification of a shrinkage rule defines both the global shape of the shrinkage function and the region over which coefficients are shrunk to very small values. Determination of the shrinkage rule is thus equivalent in spirit to specifying both a threshold value and threshold function (compare Figure 412 with 399). Bayesian methods define the shrinkage function via Bayes rule. Some of the posterior mean and posterior median Bayes rules defined in the literature turn out to be the same as previously studied conditional mean and median shrinkage functions.

These results on signal estimation are valid for general orthonormal transform coefficients O_l but they can be readily specialized and applied to the DWT case. The thresholding and shrinkage algorithms consist of the following steps.

[1] Compute a level J_0 partial DWT decomposition to obtain $\mathbf{W}_1, \ldots, \mathbf{W}_{J_0}$ and $\mathbf{V}_{J_0}$.

[2] For each of these wavelet coefficients $W_{j,t}$, determine either
  (a) the threshold level to use with a preselected threshold function or
  (b) the shrinkage rule to apply.

[3] For $j = 1, \ldots, J_0$ and $t = 0, \ldots, N_j - 1$, let $W^{(\cdot)}_{j,t}$ be the thresholded or shrunken version of the wavelet coefficient $W_{j,t}$, where again the generic superscript '(.)' would be replaced by, e.g., '(t)' or '(shr)' to indicate exactly how $W_{j,t}$, is manipulated.

[4] Estimate the signal by inverse transforming $\mathbf{W}^{(\cdot)}_1, \ldots, \mathbf{W}^{(\cdot)}_{J_0}$ and $\mathbf{V}_{J_0}$.

For IID Gaussian input noise $\boldsymbol{\epsilon}$ (Section 10.5), the single threshold level can be estimated via

$$\hat{\delta}^{(\mathrm{u})} = \sqrt{[2\hat{\sigma}^2_{(\mathrm{mad})} \log(N)]}$$

or, given realizations $w_{j,t}$ of the RVs $W_{j,t}$, via the SURE-based estimator $\hat{\delta}^{(\mathrm{S})}$, which minimizes

$$\sum_{j=1}^{J_0} \sum_{t=0}^{N_j - 1} (2\hat{\sigma}^2_{(\mathrm{mad})} - w^2_{j,t} + \delta^2) 1_{[\delta^2, \infty)}(w^2_{j,t})$$

as a function of δ, where

$$\hat{\sigma}_{(\text{mad})} \equiv \frac{\text{median}\,\{|w_{1,0}|, |w_{1,1}|, \ldots, |w_{1,\frac{N}{2}-1}|\}}{0.6745}$$

(the above are, respectively, Equations (421) and (420)).

Shrinkage rules can readily be implemented given a statistical model structure. When the signal and noise variables composing $W_{j,t}$ were assumed to have PDFs with parameters invariant to level j or position t, the shrinkage rule for the sparse signal model takes the form of Equation (425), namely,

$$w_{j,t}^{(\text{shr})} = \frac{\hat{b}}{1+\hat{c}} w_{j,t},$$

where $\hat{b}$ and $\hat{c}$ are estimators of the parameters b and c (Figure 425 shows examples of signal estimates formed via $w_{j,t}^{(\text{shr})}$).

When the noise wavelet coefficients have PDFs that vary in shape with level j (this is true for the periodogram-based spectrum estimator discussed in Section 10.6), the formulation of thresholding becomes problematic. Correlated – but Gaussian – noise wavelet coefficients arise in the spectrum estimation scheme of Section 10.7, where the variances σ_j^2 for each transform level j are known. Because of this rather fortuitous *a priori* knowledge, it is easy to carry out thresholding using the level-dependent universal threshold of Equation (406c) after we adapt it for use with log multitaper estimates evaluated at M nonnegative Fourier frequencies:

$$\delta_j^{(\text{u})} \equiv \sqrt{[2\sigma_j^2 \log{(M)}]}$$

(this is Equation (444a)). Perhaps surprisingly, application of the level-independent universal threshold of Equation (443) produces good results also, but, in retrospect, this is reasonable: the level-independent universal threshold acts in accordance with the wavelet thresholding paradigm in that it suppresses small scale noise spikes while leaving informative large scale coefficients relatively unaffected.

Finally, in Section 10.8 we summarize recent work by Crouse *et al.* (1998), who point out that the DWT coefficients for 'real world' signals tend to have large values that cluster within a level and persist between levels. They propose to account for these observed properties of clustering and persistence by modeling DWT signal coefficients using a hidden Markov tree, in which a state variable is associated with each signal DWT coefficient, and the state variables are connected in a tree structure. Such an approach opens the way for a more realistic modeling methodology.

10.10 Exercises

[10.1] Verify that the dashed curves in the right-hand column of Figure 396 are correct by computing and plotting the NPESs for the signals $\mathbf{D}_1$, $\mathbf{D}_2$ and $\mathbf{D}_3$ based upon the LA(8) DWT. Compute similar NPESs for each signal using the Haar and D(4) DWTs. How well do the NPESs for these latter two DWTs compare to that for the LA(8) case and to the NPESs for the identity and ODFT transforms?

[10.2] Suppose that $\mathbf{X} \equiv \mathbf{D} + \boldsymbol{\epsilon}$ is a time series of length $N = 2^J$ for some integer $J \geq 4$, where $\mathbf{D}$ is a deterministic signal, while $\boldsymbol{\epsilon}$ is an N dimensional vector of IID Gaussian RVs, each with mean zero and variance σ_ϵ^2. If we estimate $\mathbf{D}$

using a hard thresholding estimator $\widehat{\mathbf{D}}^{(\text{ht})}$ based upon the Haar DWT (as defined in Section 4.1) in combination with the universal threshold $\delta^{(\text{u})} \equiv \sqrt{[2\sigma_\epsilon^2 \log(N)]}$ (Donoho and Johnstone, 1994), show that

(a) if $\mathbf{D}$ is constructed by periodically extending the sixteen point sequence defined by Equation (401a), then, for some small $\alpha > 0$, we have $\mathbf{P}\left[\frac{1}{N}\|\widehat{\mathbf{D}}^{(\text{ht})} - \mathbf{D}\|^2 < \alpha\right] \to 0$ as $J \to \infty$; i.e., $\widehat{\mathbf{D}}^{(\text{ht})}$ is *not* a consistent estimator of $\mathbf{D}$;

(b) on the other hand, if $\mathbf{D}$ is constructed via Equation (401b), then, for any $\alpha > 0$, we have $\mathbf{P}\left[\frac{1}{N}\|\widehat{\mathbf{D}}^{(\text{ht})} - \mathbf{D}\|^2 < \alpha\right] \to 1$ as $J \to \infty$; i.e., $\widehat{\mathbf{D}}^{(\text{ht})}$ is a consistent estimator of $\mathbf{D}$.

[10.3] Show that, for soft thresholding, the SURE-based function $\mathcal{R}(\cdot,\cdot,\cdot)$ in Equation (404b) can be written in the alternative form

$$\mathcal{R}(\sigma_{n_l}, O_l, \delta) = \sigma_{n_l}^2 - 2\sigma_{n_l}^2 1_{[0,\delta^2)}(O_l^2) + \min\{O_l^2, \delta^2\}.$$

[10.4] Gao (1998) proposes an alternative thresholding rule to soft, mid or hard thresholding called the nonnegative garrote. This mapping is defined by

$$O_l^{(\text{g})} = O_l\left(1 - \frac{\delta^2}{O_l^2}\right)_+ = \begin{cases} 0, & \text{if } |O_l| \le \delta; \\ O_l - \frac{\delta^2}{O_l}, & \text{otherwise.} \end{cases}$$

Show that, if we use SURE with this thresholding rule, the function defined by $\mathcal{R}(\sigma_{n_l}, O_l, \delta)$ in Equation (404b) is replaced by

$$\mathcal{R}(\sigma_{n_l}, O_l, \delta) = O_l^2 - \sigma_{n_l}^2 + \left(2\sigma_{n_l}^2 - O_l^2 + \frac{\delta^4 + 2\sigma_{n_l}^2\delta^2}{O_l^2}\right) 1_{[\delta^2,\infty)}(O_l^2).$$

(It can be shown that the nonnegative garrote makes $A^{(\delta)}(\cdot)$ weakly differentiable, as required for SURE.)

[10.5] Suppose that R_l and n_l are independent and distributed as in Equations (410d) and (410e). Show that the conditional mean, defined by Equation (410b) can be written in the form of Equation (410f). Hint: use integration by parts to express the numerator of (410f) in terms of the two integrals involved in the denominator, and complete the squares of the integrands to obtain integrals in recognizable forms (i.e., Gaussian PDFs).

[10.6] Verify Equation (414b).

[10.7] Using an LA(8) partial DWT of level $J_0 = 6$ along with the universal threshold and the hard, mid and soft thresholding rules, apply the wavelet-based thresholding procedure described in Section 10.5 to the ECG data, but, instead of retaining the scaling coefficients untouched, set all the scaling coefficients to zero. Plot the three signal estimates and comment on any noticeable differences among the different estimates. (The ECG time series can be obtained via the Web site for this book – see page *xiv*.)

[10.8] As discussed in Section 10.6, we apply wavelet-based thresholding to the partial DWT of the (recentered) log periodogram $\mathbf{Y}^{(\text{p})}$ in order to obtain the log spectrum estimator $\widehat{\mathbf{D}}^{(\text{t})}$. By construction, for $k = 1, \ldots, M-1$, elements k and $2M - k$ of $\mathbf{Y}^{(\text{p})}$ are identical and associated with frequency f_k. Explain why the corresponding elements of $\widehat{\mathbf{D}}^{(\text{t})}$ need not be identical even though both are still associated with frequency f_k.

[10.9] Verify the progression of level-dependent variances stated in Equation (441c) by computing $\sigma_1^2, \ldots, \sigma_J^2$ using the D(4) and LA(8) wavelet filters for the case $M = 2N = 1024 = 2^{10} = 2^J$ and $K = 5$.

11

Wavelet Analysis of Finite Energy Signals

11.0 Introduction

The continuous time wavelet transform is becoming a well-established tool for multiple scale representation of a continuous time 'signal,' which by definition is a finite energy function defined over the entire real axis. This transform essentially correlates a signal with 'stretched' versions of a wavelet function (in essence a continuous time band-pass filter) and yields a multiresolution representation of the signal. In this chapter we summarize the important ideas and results for the multiresolution view of the continuous time wavelet transform. Our primary intent is to demonstrate the close relationship between continuous time wavelet analysis and the discrete time wavelet analysis presented in Chapter 4. To make this connection, we adopt a formalism that allows us to bridge the gap between the inner product convention used in mathematical discussions on wavelets and the filtering convention favored by engineers. For simplicity we deal only with signals, scaling functions and wavelet functions that are all taken to be *real-valued.* Only the case of dyadic wavelet analysis (where the scaling factor in the dilation of the basis function takes the value of two) is considered here.

The summary given in this chapter is primarily a distillation of Vetterli and Herley (1992), Mallat (1989a, b, c) and Daubechies (1988, 1992). In order to make progress we initially assume the existence of scaling functions and wavelet functions with the properties discussed. In Sections 11.9 and 11.10 we discuss the construction of wavelet filters using the approach of Daubechies (1988), principally in order to understand the ideas of vanishing moments and regularity. A well-organized account concentrating on the construction of wavelet functions is given in Strichartz (1993).

11.1 Translation and Dilation

Multiresolution analysis involves the operations of translation and dilation, both of which can be easily defined for a real-valued function $\gamma(\cdot)$ whose domain of definition is the real axis $\mathbb{R}$. For the unit translation, or delay, operator we again use the notation $\mathcal{T}$ (as in Equation (52a) of Section 3.4 where it represented a discrete time (circular) unit delay matrix). The application of $\mathcal{T}$ to the function $\gamma(\cdot)$ yields a new function to be denoted as $\gamma_{0,1}(\cdot)$. The value of $\gamma_{0,1}(\cdot)$ at t is given by

$$\gamma_{0,1}(t) \equiv \gamma(t-1), \quad -\infty < t < \infty.$$

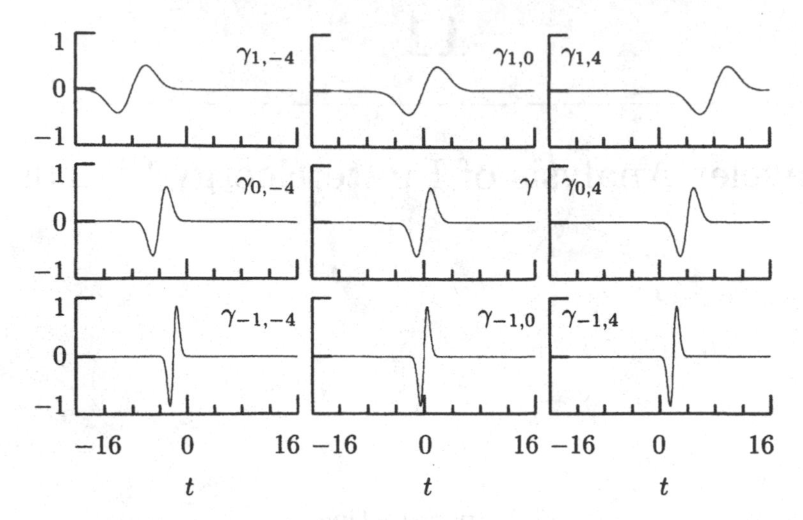

Figure 458. Translation and dilation of the function defined by $\gamma(t) = t\exp(-t^2/2)$.

The application of $\mathcal{T}$ to $\gamma_{0,1}(\cdot)$ yields the function $\gamma_{0,2}(\cdot)$, whose value at t is

$$\gamma_{0,2}(t) \equiv \gamma(t-2), \quad -\infty < t < \infty;$$

likewise, we can define $\gamma_{0,k}(t)$ via

$$\gamma_{0,k}(t) \equiv \gamma(t-k), \quad -\infty < t < \infty.$$

This definition works for all $k \in \mathbb{Z}$ (the set of all integers). In particular, the choice $k = -1$ defines the inverse unit translation operator $\mathcal{T}^{-1}$, so that application of $\mathcal{T}^{-1}$ to the function $\gamma_{0,k}(\cdot)$ yields the function $\gamma_{0,k-1}(\cdot)$ (with $\gamma_{0,0}(\cdot)$ defined to be $\gamma(\cdot)$).

As an example, the middle row of plots in Figure 458 shows, from left to right, $\gamma_{0,-4}(\cdot)$, $\gamma(\cdot)$ and $\gamma_{0,4}(\cdot)$ for $\gamma(t) = t\exp(-t^2/2)$. Note that $k > 0$ translates a function to the right on its graph (delaying it, relative to time zero), whereas $k < 0$ does the opposite (to the left, corresponding to an advance, relative to time zero).

In a similar manner, we can define a dyadic dilation (or scaling) operator $\mathcal{K}$ (the term 'dyadic' pertains to the number two in the definition below). The application of $\mathcal{K}$ to the function $\gamma(\cdot)$ yields a new function to be denoted as $\gamma_{1,0}(\cdot)$. The value of $\gamma_{1,0}(\cdot)$ at t is given by

$$\gamma_{1,0}(t) \equiv \frac{\gamma(\frac{t}{2})}{\sqrt{2}}, \quad -\infty < t < \infty.$$

The factor $1/\sqrt{2}$ is included so that, if $\gamma(\cdot)$ is a member of $L^2(\mathbb{R})$ (the set of square integrable real-valued functions defined over $\mathbb{R}$) with a squared norm given by

$$\|\gamma(\cdot)\|^2 \equiv \int_{-\infty}^{\infty} \gamma^2(t)\,dt < \infty,$$

then $\|\gamma_{1,0}(\cdot)\|^2 = \|\gamma(\cdot)\|^2$ (i.e., the dilation operator preserves the norm, and – in fact – so does the translation operator). The application of $\mathcal{K}$ to $\gamma_{1,0}(\cdot)$ yields the function $\gamma_{2,0}(\cdot)$, whose value at t is

$$\gamma_{2,0}(t) \equiv \frac{\gamma(\frac{t}{4})}{2}, \quad -\infty < t < \infty;$$

likewise, we can define $\gamma_{j,0}(\cdot)$ via

$$\gamma_{j,0}(\cdot) \equiv \frac{\gamma(\frac{t}{2^j})}{\sqrt{2^j}}, \quad -\infty < t < \infty.$$

Again, this definition holds for all $j \in \mathbb{Z}$, and the choice $j = -1$ defines the inverse dyadic dilation operator $\mathcal{K}^{-1}$, so that application of $\mathcal{K}^{-1}$ to the function $\gamma_{j,0}(\cdot)$ yields $\gamma_{j-1,0}(\cdot)$.

As an example of dyadic dilation, the middle column of plots in Figure 458 shows, from top to bottom, $\gamma_{1,0}(\cdot)$, $\gamma(\cdot)$ and $\gamma_{-1,0}(\cdot)$. Note that $j > 0$ visually expands the graph of a function, whereas $j < 0$ contracts it.

Finally we note that the (j, k)th translation followed by dilation of $\gamma(\cdot)$ is defined via

$$\gamma_{j,k}(t) \equiv \frac{\gamma(\frac{t}{2^j} - k)}{\sqrt{2^j}}, \quad -\infty < t < \infty.$$

As examples, the left- and right-hand plots on the top row of Figure 458 show $\gamma_{1,-4}(\cdot)$ and $\gamma_{1,4}(\cdot)$, while the corresponding plots on the bottom row show $\gamma_{-1,-4}(\cdot)$ and $\gamma_{-1,4}(\cdot)$. Note that $\gamma_{j,k}(\cdot)$ is the jth level dilation of $\gamma_{0,k}(\cdot)$, but that in general $\gamma_{j,k}(\cdot)$ is *not* the kth order translation of $\gamma_{j,0}(\cdot)$ (i.e., the order in which we apply the translation and dilation operators matters – in mathematical terminology, these operators do not commute).

11.2 Scaling Functions and Approximation Spaces

The wavelet transform was originally developed as an analysis and synthesis tool for a finite energy signal $x(\cdot)$, where by definition the finite energy condition is (with $x(\cdot)$ regarded as real-valued)

$$\|x\|^2 \equiv \int_{-\infty}^{\infty} x^2(t)\,dt < \infty;$$

i.e., $x(\cdot)$ belongs to $L^2(\mathbb{R})$. Let $\phi(\cdot) \in L^2(\mathbb{R})$ be such that its integer translates $\{\phi_{0,k}(\cdot) : k \in \mathbb{Z}\}$ form an orthonormal basis for some closed subspace $V_0 \subset L^2(\mathbb{R})$, where

$$V_0 \equiv \overline{\text{span}\,\{\phi_{0,k}(\cdot) : k \in \mathbb{Z}\}}.$$

Here the notation span $\{S\}$ refers to the linear span, which is the subspace of linear combinations of elements of S, while $\overline{\text{span}\,\{S\}}$ is the closed subspace generated by S (the distinction between span $\{S\}$ and $\overline{\text{span}\,\{S\}}$ is that the latter includes the limits of linear combinations). We thus have

$$\int_{-\infty}^{\infty} \phi_{0,k}(t)\phi_{0,m}(t)\,dt = \int_{-\infty}^{\infty} \phi(t-k)\phi(t-m)\,dt = \begin{cases} 1, & k = m; \\ 0, & \text{otherwise.} \end{cases} \tag{459}$$

The function $\phi(\cdot)$ is known as the *scaling function*, while the space V_0 is called the *approximation space* with scale unity. We can analyze any function $x(\cdot)$ with respect to V_0 by projecting it against the basis functions for V_0. If we let

$$\langle x(\cdot), y(\cdot)\rangle \equiv \int_{-\infty}^{\infty} x(t)y(t)\,dt$$

represent the inner product between two finite energy real-valued signals $x(\cdot)$ and $y(\cdot)$, the analysis of $x(\cdot)$ yields

$$v_{0,k} \equiv \langle x(\cdot), \phi_{0,k}(\cdot)\rangle = \int_{-\infty}^{\infty} x(t)\phi_{0,k}(t)\,dt = \int_{-\infty}^{\infty} x(t)\phi(t-k)\,dt, \tag{460a}$$

where the $v_{0,k}$'s are called the *scaling coefficients* for the function $x(\cdot)$. If $x(\cdot) \in V_0$, we can synthesize $x(\cdot)$ from its scaling coefficients:

$$x(t) = \sum_{k=-\infty}^{\infty} v_{0,k}\phi_{0,k}(t) = \sum_{k=-\infty}^{\infty} v_{0,k}\phi(t-k). \tag{460b}$$

If $x(\cdot) \notin V_0$, the infinite summations above are an approximation to $x(\cdot)$ obtained by projecting $x(\cdot)$ onto the space V_0.

As a simple example, consider the *Haar scaling function*, namely,

$$\phi^{(\mathrm{H})}(t) \equiv \begin{cases} 1, & -1 < t \le 0; \\ 0, & \text{otherwise} \end{cases} \tag{460c}$$

(see item [1] of the Comments and Extensions below). Hence we have

$$\phi^{(\mathrm{H})}_{0,k}(t) = \phi^{(\mathrm{H})}(t-k) = \begin{cases} 1, & k-1 < t \le k; \\ 0, & \text{otherwise.} \end{cases}$$

The orthonormality of the $\phi^{(\mathrm{H})}_{0,k}(\cdot)$'s is easily established. The Haar approximation space $V_0^{(\mathrm{H})}$ consists of all $L^2(\mathbb{R})$ functions $x(\cdot)$ that are constant over each interval defined by $k-1 < t \le k$ for $k \in \mathbb{Z}$; i.e., for some real-valued constant $v_{0,k}$, we have $x(t) = v_{0,k}$ for $k-1 < t \le k$, where $v_{0,k}$ can be obtained via Equation (460a). The middle row of plots in Figure 461 shows, from left to right, $\phi^{(\mathrm{H})}_{0,-1}(\cdot)$, $\phi^{(\mathrm{H})}(\cdot)$ and $\phi^{(\mathrm{H})}_{0,1}(\cdot)$, along with an example of a function contained in $V_0^{(\mathrm{H})}$.

A change of variable in Equation (459) shows that

$$\int_{-\infty}^{\infty} \phi_{0,k}(t)\phi_{0,m}(t)\,dt = \int_{-\infty}^{\infty} \frac{\phi_{0,k}(\frac{t}{2^j})}{\sqrt{2^j}}\frac{\phi_{0,m}(\frac{t}{2^j})}{\sqrt{2^j}}\,dt = \int_{-\infty}^{\infty} \phi_{j,k}(t)\phi_{j,m}(t)\,dt,$$

where

$$\phi_{j,k}(t) \equiv \frac{\phi_{0,k}(\frac{t}{2^j})}{\sqrt{2^j}} = \frac{\phi(\frac{t}{2^j}-k)}{\sqrt{2^j}}.$$

Thus, given that $\{\phi_{0,k}(\cdot) : k \in \mathbb{Z}\}$ constitutes an orthonormal basis for V_0, it follows that $\{\phi_{j,k}(\cdot) : k \in \mathbb{Z}\}$ constitutes an orthonormal basis for V_j, where

$$V_j \equiv \overline{\operatorname{span}\{\phi_{j,k}(\cdot) : k \in \mathbb{Z}\}}.$$

The approximation space V_j with scale 2^j is hence the closed linear span of $\phi_{j,k}(\cdot)$ with $k \in \mathbb{Z}$.

Suppose now that the following condition holds:

$$\cdots \subset V_3 \subset V_2 \subset V_1 \subset V_0 \subset V_{-1} \subset \cdots.$$

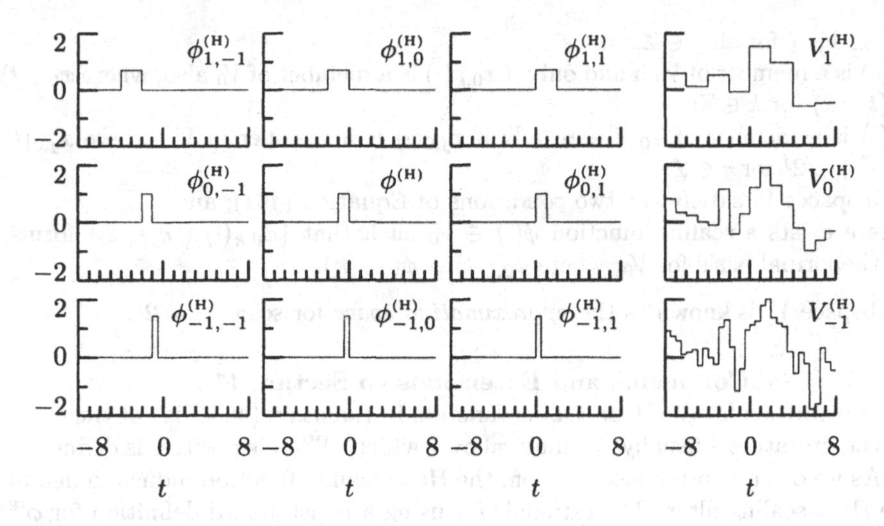

Figure 461. The Haar scaling function $\phi^{(\mathrm{H})}(\cdot)$ and corresponding approximation spaces. The first three plots on the middle row show three of the basis functions for the Haar approximation space $V_0^{(\mathrm{H})}$, namely, from left to right, $\phi^{(\mathrm{H})}_{0,-1}(\cdot)$, $\phi^{(\mathrm{H})}(\cdot)$ and $\phi^{(\mathrm{H})}_{0,1}(\cdot)$. The right-most plot on this row is an example of a function contained in $V_0^{(\mathrm{H})}$. The top and bottom rows show, respectively, corresponding plots for the Haar approximation spaces $V_1^{(\mathrm{H})}$ (a coarser approximation than $V_0^{(\mathrm{H})}$) and $V_{-1}^{(\mathrm{H})}$ (a finer approximation than $V_0^{(\mathrm{H})}$). The right-most column of plots can be regarded as three Haar approximations of a single $L^2(\mathbb{R})$ function, with the associated scales of 2, 1 and 1/2 (top to bottom).

The above conditions says that projections of any finite energy signal onto the V_j will give successive approximations. If we want this succession of approximations to be useful for any function in $L^2(\mathbb{R})$, we must further assume

$$\overline{\bigcup_{j\in\mathbb{Z}} V_j} = L^2(\mathbb{R}) \text{ and } \bigcap_{j\in\mathbb{Z}} V_j = \{0\}, \tag{461}$$

where here 0 stands for the null function (i.e., the function that is identically zero over the entire real axis). The above ensures that, for all $y(\cdot) \in L^2(\mathbb{R})$, the projection of $y(\cdot)$ onto V_j is equal to $y(\cdot)$ as $j \to -\infty$. In the nesting of the subspaces, note that $l < j$ implies that V_l is larger than V_j (i.e., $V_j \subset V_l$). This is the same convention as Daubechies (1988, 1992) but is opposite to that used by Mallat (1989b).

As an example, the top row of Figure 461 shows, from left to right, three of the basis functions for the Haar approximation space $V_1^{(\mathrm{H})}$ of scale two along with an example of a function contained in this space (the three basis functions are, from left to right, $\phi^{(\mathrm{H})}_{1,-1}(\cdot)$, $\phi^{(\mathrm{H})}_{1,0}(\cdot)$ and $\phi^{(\mathrm{H})}_{1,1}(\cdot)$). The bottom row shows corresponding plots for the Haar approximation space $V_{-1}^{(\mathrm{H})}$ of scale one half. The right-most column of plots can be regarded as, from top to bottom, successively finer approximations to some $L^2(\mathbb{R})$ function $x(\cdot)$, where the approximations are obtained by projecting $x(\cdot)$ onto the subspaces $V_1^{(\mathrm{H})}$, $V_0^{(\mathrm{H})}$ and $V_{-1}^{(\mathrm{H})}$ (see Equation (462b)).

The key points of this section can be summarized in the following definition (Jawerth and Sweldens, 1994; Daubechies, 1988; Mallat, 1989b). A *multiresolution analysis* (MRA) is by definition a sequence of closed subspaces V_j of $L^2(\mathbb{R})$, $j \in \mathbb{Z}$, such that

[1] $V_j \subset V_{j-1}$ for all $j \in \mathbb{Z}$;
[2] $x(\cdot)$ is a member of V_0 if and only if $x_{0,k}(\cdot)$ is a member of V_0 also, where $x_{0,k}(t) \equiv x(t-k)$ for $k \in \mathbb{Z}$;
[3] $x(\cdot)$ is a member of V_0 if and only if $x_{j,0}(\cdot)$ is a member of V_j, where $x_{j,0}(t) = x(\frac{t}{2^j})/\sqrt{2^j}$ for $j \in \mathbb{Z}$;
[4] the spaces V_j satisfy the two conditions of Equation (461); and
[5] there exists a scaling function $\phi(\cdot) \in V_0$ such that $\{\phi_{0,k}(\cdot) : k \in \mathbb{Z}\}$ forms an orthonormal basis for V_0, where $\phi_{0,k}(t) \equiv \phi(t-k)$.

The subspace V_j is known as the *approximation space* for scale $\lambda_j = 2^j$.

Comments and Extensions to Section 11.2

[1] Our definition for the Haar scaling function in Equation (460c) is not the common one in the literature – usually the interval over which $\phi^{(\mathrm{H})}(\cdot)$ is positive is defined to be $(0, 1]$. As we discuss in the next section, the Haar scaling function induces a definition for the Haar scaling filter. The rationale for using a nonstandard definition for $\phi^{(\mathrm{H})}(\cdot)$ is to ensure that the Haar scaling function yields the same Haar scaling filter we introduced in Section 4.3; that it does is shown by Equation (463d). The need for this nonstandard definition merely reflects the fact that the operations of inner products and convolutions differ by a time reversal.

11.3 Approximation of Finite Energy Signals

As before, suppose that $\{V_j : j \in \mathbb{Z}\}$ constitutes an MRA of $L^2(\mathbb{R})$. As indicated by Equations (460b) and (460a), a signal $x(\cdot) \in V_0$ can be written as

$$x(t) \equiv \sum_{k=-\infty}^{\infty} \langle x(\cdot), \phi_{0,k}(\cdot)\rangle \phi_{0,k}(t) = \sum_{k=-\infty}^{\infty} v_{0,k}\phi_{0,k}(t), \tag{462a}$$

where $v_{0,k} \equiv \langle x(\cdot), \phi_{0,k}(\cdot)\rangle$. Because $x(\cdot) \in V_0$, we say that $x(\cdot)$ is a signal with scale $\lambda_0 = 2^0 = 1$. We can obtain an approximation to $x(\cdot)$ by projecting it onto the subspace $V_1 \subset V_0$ to obtain

$$s_1(t) \equiv \sum_{k=-\infty}^{\infty} \langle x(\cdot), \phi_{1,k}(\cdot)\rangle \phi_{1,k}(t) = \sum_{k=-\infty}^{\infty} v_{1,k}\phi_{1,k}(t), \tag{462b}$$

where

$$v_{1,k} \equiv \langle x(\cdot), \phi_{1,k}(\cdot)\rangle = \int_{-\infty}^{\infty} x(t)\phi_{1,k}(t)\,dt = \int_{-\infty}^{\infty} x(t)\frac{\phi(\frac{t}{2}-k)}{\sqrt{2}}\,dt. \tag{462c}$$

Because $s_1(\cdot) \in V_1$, the signal $s_1(\cdot)$ has scale $\lambda_1 = 2^1 = 2$ and represents a *coarse approximation* to the unit scale signal $x(\cdot)$. The $v_{1,k}$'s are called the *scaling coefficients* of scale two.

There is an interesting relationship between the $v_{0,k}$'s and the $v_{1,k}$'s that we can obtain by using (462a) in (462c) to write

$$v_{1,k} = \int_{-\infty}^{\infty} \left[\sum_{m=-\infty}^{\infty} v_{0,m}\phi_{0,m}(t)\right] \frac{\phi(\frac{t}{2}-k)}{\sqrt{2}}\,dt$$

$$= \sum_{m=-\infty}^{\infty} v_{0,m} \int_{-\infty}^{\infty} \phi(t-m) \frac{\phi(\frac{t}{2}-k)}{\sqrt{2}}\, dt$$

$$= \sum_{m=-\infty}^{\infty} v_{0,m} \int_{-\infty}^{\infty} \phi(t-[m-2k]) \frac{\phi(\frac{t}{2})}{\sqrt{2}}\, dt$$

$$= \sum_{m=-\infty}^{\infty} \bar{g}_{m-2k} v_{0,m}, \tag{463a}$$

where

$$\bar{g}_l \equiv \int_{-\infty}^{\infty} \phi(t-l) \frac{\phi(\frac{t}{2})}{\sqrt{2}}\, dt = \langle \phi_{0,l}(\cdot), \phi_{1,0}(\cdot) \rangle. \tag{463b}$$

If we define $g_l \equiv \bar{g}_{-l}$ and let $l = 2k - m$ in Equation (463a), we can write

$$v_{1,k} = \sum_{l=-\infty}^{\infty} g_l v_{0,2k-l}.$$

Thus $\{v_{1,k}\}$ corresponds to every other value of the output (i.e., a *downsampling* by a factor of two) produced by filtering the sequence $\{v_{0,k}\}$ with $\{g_l\}$ (the filter $\{g_l\}$ is the same as the scaling filter we introduced in Section 4.3).

As an example, let us compute $\{\bar{g}_l\}$ for the Haar scaling function $\phi^{(H)}(\cdot)$. Since

$$\phi^{(H)}(t-l) = \begin{cases} 1, & l-1 < t \le l; \\ 0, & \text{otherwise;} \end{cases} \quad \phi^{(H)}(\tfrac{t}{2}) = \begin{cases} 1, & -2 < t \le 0; \\ 0, & \text{otherwise;} \end{cases}$$

Equation (463b) yields

$$\bar{g}_l = \frac{1}{\sqrt{2}} \int_{-2}^{0} \phi^{(H)}(t-l)\, dt = \begin{cases} 1/\sqrt{2}, & l = 0 \text{ or } -1; \\ 0, & \text{otherwise.} \end{cases} \tag{463c}$$

Since $g_l \equiv \bar{g}_{-l}$, we have

$$g_l = \begin{cases} 1/\sqrt{2}, & l = 0 \text{ or } 1; \\ 0, & \text{otherwise.} \end{cases} \tag{463d}$$

This indeed agrees with the definition of the Haar scaling filter $\{g_l\}$ given in Equation (75c).

Since $s_1(\cdot) \in V_1$, we can approximate $s_1(\cdot)$ by projecting it onto the subspace $V_2 \subset V_1$ to obtain

$$s_2(t) \equiv \sum_{k=-\infty}^{\infty} v_{2,k} \phi_{2,k}(t) \text{ with } v_{2,k} \equiv \langle s_1(\cdot), \phi_{2,k}(\cdot) \rangle,$$

which also represents an approximation to $x(\cdot)$ of scale $\lambda_2 = 2^2 = 4$. Using the expression for $s_1(\cdot)$ in Equation (462b), we have

$$v_{2,k} = \left\langle \sum_{m=-\infty}^{\infty} v_{1,m} \phi_{1,m}(\cdot), \phi_{2,k}(\cdot) \right\rangle$$

$$= \sum_{m=-\infty}^{\infty} v_{1,m} \langle \phi_{1,m}(\cdot), \phi_{2,k}(\cdot) \rangle$$

$$= \sum_{m=-\infty}^{\infty} \bar{g}_{m-2k} v_{1,m} = \sum_{l=-\infty}^{\infty} g_l v_{1,2k-l}, \tag{463e}$$

where we have made use of the following:

▷ **Exercise [464]:** Show that

$$\langle \phi_{j,m}(\cdot), \phi_{j+1,k}(\cdot)\rangle = \langle \phi_{0,m-2k}(\cdot), \phi_{1,0}(\cdot)\rangle = \bar{g}_{m-2k}$$

for all j, where $\bar{g}_l$ is defined in Equation (463b) (letting $j = 1$ establishes Equation (463e)). As a corollary, show also that

$$\langle \phi_{j,m}(\cdot), \phi_{j-1,k}(\cdot)\rangle = \bar{g}_{k-2m} \tag{464a}$$

for all j. ◁

In general, given $s_{j-1}(\cdot) \in V_{j-1}$, we can approximate $s_{j-1}(\cdot)$ by projecting it onto the subspace $V_j \subset V_{j-1}$ to obtain

$$s_j(t) \equiv \sum_{k=-\infty}^{\infty} v_{j,k}\phi_{j,k}(t)$$

with $v_{j,k} \equiv \langle s_{j-1}(\cdot), \phi_{j,k}(\cdot)\rangle = \langle x(\cdot), \phi_{j,k}(\cdot)\rangle$. The signal $s_j(\cdot)$ represents an approximation of scale $\lambda_j = 2^j$ to $x(\cdot)$. A recursive computational expression for $v_{j,k}$ is given by

$$v_{j,k} = \sum_{m=-\infty}^{\infty} \bar{g}_{m-2k} v_{j-1,m} = \sum_{l=-\infty}^{\infty} g_l v_{j-1,2k-l}. \tag{464b}$$

Comments and Extensions to Section 11.3

[1] Suppose that we have a time series $\{x_t : t \in \mathbb{Z}\}$ that can be regarded as samples of the signal $x(\cdot)$; i.e., $x_t = x(t)$. From Equation (460a) we know that each element $v_{0,k}$ of the initial scaling sequence is the inner product of $x(\cdot)$ with $\phi_{0,k}(\cdot)$. If our knowledge of $x(\cdot)$ is limited to its values on $\mathbb{Z}$, i.e., to $\{x_t\}$, then strictly speaking we cannot obtain an MRA of $x(\cdot)$ because we do not have enough information to construct the $v_{0,k}$'s. We can, however, approximate $v_{0,k}$ using a Riemann sum to obtain

$$v_{0,k} = \int_{-\infty}^{\infty} x(t)\phi_{0,k}(t)\,dt \approx \sum_{t=-\infty}^{\infty} x(t)\phi_{0,k}(t) = \sum_{t=-\infty}^{\infty} x_t\phi_{0,k}(t).$$

In practice, the initialization of the algorithm often takes the naïve approximation $v_{0,k} = x_k$. Jawerth and Sweldens (1994) point out that other sampling procedures, including quadrature formulae, have been proposed and give references. Abry and Flandrin (1994) also give examples of initialization effects.

11.4 Two-Scale Relationships for Scaling Functions

In this section we explore the so-called *two-scale relationship* for the scaling function. This relationship arises by noting that, since $V_0 \subset V_{-1}$, $\phi(\cdot) \in V_0$ implies that $\phi(\cdot) \in V_{-1}$. We can thus write

$$\phi(t) = \sum_{l=-\infty}^{\infty} \langle \phi(\cdot), \phi_{-1,l}(\cdot)\rangle \phi_{-1,l}(t) = \sum_{l=-\infty}^{\infty} \bar{g}_l \phi_{-1,l}(t)$$

where we have made use of Equation (464a) to get $\langle \phi(\cdot), \phi_{-1,l}(\cdot)\rangle = \bar{g}_l$ (in using this equation, note that $\phi(\cdot) \equiv \phi_{0,0}(\cdot)$). Since $\phi_{-1,l}(t) = \sqrt{2}\phi(2t-l)$, the above yields the two-scale relationship

$$\phi(t) = \sqrt{2} \sum_{l=-\infty}^{\infty} \bar{g}_l \phi(2t-l), \tag{465a}$$

which is also sometimes called a *two-scale difference equation.*

Let us explore some of the implications of the two-scale relationship. Suppose that $\phi(\cdot)$ is zero outside an interval $a < t \leq b$, in which case we say that $\phi(\cdot)$ has *finite support* on $(a, b]$. We can express this fact by writing support $\{\phi(\cdot)\} = (a, b]$. Since the function defined by $\phi(2t-l)$ is then supported on $(\frac{a+l}{2}, \frac{b+l}{2}]$, it follows from Equation (465a) that only finitely many $\bar{g}_l$ are nonzero. Hence the right-hand side of the two-scale equation is supported between $\frac{a-(L-1)}{2}$ and $\frac{b}{2}$. For equality with the left-hand side of the two-scale equation, we must have $(a,b] = (\frac{a-(L-1)}{2}, \frac{b}{2}]$, which implies that $a = -(L-1)$ and $b = 0$; i.e., $\phi(\cdot)$ is zero outside $(-(L-1), 0]$. Note that, if the filter $\{g_l\}$ is zero for l outside $0, \ldots, L-1$, then $\{\bar{g}_l\} \equiv \{g_{-l}\}$ will be zero for l outside $-(L-1), \ldots, 0$. Thus a scaling filter $\{g_l\}$ of width L has an associated scaling function $\phi(\cdot)$ with support on $(-(L-1), 0]$. We have already seen an example of this relationship: the Haar scaling filter has width $L = 2$, and the Haar scaling function of Equation (460c) has support on $(-1, 0]$. (For an argument justifying the assumption of finite support, see, for example, Strang and Nguyen, 1996, p. 186).

We can obtain a solution for the scaling function $\phi(\cdot)$ in terms of $\{\bar{g}_l\}$ by using the Fourier transform. Let

$$\Phi(f) \equiv \int_{-\infty}^{\infty} \phi(t) e^{-i2\pi ft}\, dt$$

be the Fourier transform of $\phi(\cdot)$. Since

$$\int_{-\infty}^{\infty} \phi(2t-l) e^{-i2\pi ft}\, dt = \frac{e^{-i2\pi\frac{f}{2}l}}{2} \int_{-\infty}^{\infty} \phi(t) e^{-i2\pi \frac{f}{2} t}\, dt = \frac{e^{-i2\pi\frac{f}{2}l}}{2} \Phi(\tfrac{f}{2}),$$

it follows from applying the Fourier transform to both sides of Equation (465a) that

$$\Phi(f) = \sqrt{2} \sum_{l=-\infty}^{\infty} \bar{g}_l \frac{e^{-i2\pi\frac{f}{2}l}}{2} \Phi(\tfrac{f}{2}) = \frac{\Phi(\frac{f}{2})}{\sqrt{2}} \sum_{l=-\infty}^{\infty} \bar{g}_l e^{-i2\pi\frac{f}{2}l} = \Phi(\tfrac{f}{2}) \frac{\overline{G}(\frac{f}{2})}{\sqrt{2}}, \tag{465b}$$

where $\overline{G}(\cdot)$ is the Fourier transform of $\{\bar{g}_l\}$, i.e.,

$$\overline{G}(f) \equiv \sum_{l=-\infty}^{\infty} \bar{g}_l e^{-i2\pi fl}.$$

Iterating this result n times yields

$$\Phi(f) = \Phi(\tfrac{f}{2^n}) \prod_{m=1}^{n} \frac{\overline{G}(\frac{f}{2^m})}{\sqrt{2}}. \tag{465c}$$

In the limit as $n \to \infty$, we have

$$\Phi(f) = \Phi(0) \prod_{m=1}^{\infty} \frac{\overline{G}(\frac{f}{2^m})}{\sqrt{2}}, \tag{466a}$$

from which we can obtain the scaling function from the inverse Fourier transform of $\Phi(\cdot)$.

If we let $f = 0$ in the top line of Equation (465b) and if we recall that $\Phi(0)$ is just the integral of $\phi(\cdot)$, we obtain

$$\int_{-\infty}^{\infty} \phi(t)\, dt = \sqrt{2} \sum_{l=-\infty}^{\infty} \bar{g}_l \int_{-\infty}^{\infty} \phi(2t - l)\, dt = \frac{1}{\sqrt{2}} \sum_{l=-\infty}^{\infty} \bar{g}_l \int_{-\infty}^{\infty} \phi(t)\, dt.$$

Under the mild assumption that $\phi(\cdot)$ integrates to a nonzero value, we thus have the normalization

$$\sum_{l=-\infty}^{\infty} \bar{g}_l = \sqrt{2}, \text{ which implies that } \overline{G}(0) = \sum_{l=-\infty}^{\infty} \bar{g}_l = \sqrt{2}. \tag{466b}$$

The two-scale difference equation (465a) can be rewritten as

$$\phi(t) = \sum_{l=-\infty}^{\infty} \bar{g}_l \phi_{-1,0}(t - \tfrac{l}{2})$$

because

$$\phi_{-1,0}(t - \tfrac{l}{2}) = \phi_{-1,0}(\tfrac{2t-l}{2}) = \sqrt{2}\phi(2t - l).$$

Similarly, the difference equation can be written as either

$$\phi_{1,0}(t) = \sum_{l=-\infty}^{\infty} \bar{g}_l \phi_{0,0}(t - l) \text{ or } \phi_{2,0}(t) = \sum_{l=-\infty}^{\infty} \bar{g}_l \phi_{1,0}(t - 2l),$$

and in general as

$$\phi_{j+1,0}(t) = \sum_{l=-\infty}^{\infty} \bar{g}_l \phi_{j,0}(t - 2^j l). \tag{466c}$$

▷ **Exercise [466]:** Let $\Phi_{j+1}(\cdot)$ denote the Fourier transform of $\phi_{j+1,0}(\cdot)$. Show that

$$\Phi_{j+1}(f) = \overline{G}(2^j f)\Phi_j(f).$$ ◁

For example, starting with $j = 2$ and noting that $\Phi_0(\cdot) \equiv \Phi(\cdot)$, we have (after three applications of the above)

$$\Phi_3(f) = \overline{G}(4f)\overline{G}(2f)\overline{G}(f)\Phi(f) = \overline{G}(4f)\overline{G}(2f)\overline{G}(f)\Phi(0) \prod_{m=1}^{\infty} \frac{\overline{G}(\frac{f}{2^m})}{\sqrt{2}}.$$

The fact that $\{\phi_{0,k}(\cdot) : k \in \mathbb{Z}\}$ constitutes an orthonormal set of functions places an important constraint on the form of $\overline{G}(\cdot)$. Since $\phi(\cdot) \longleftrightarrow \Phi(\cdot)$ implies that

$$\int_{-\infty}^{\infty} \phi_{0,k}(t) e^{-i2\pi f t}\, dt = e^{-i2\pi f k}\Phi(f),$$

the two function version of Parseval's theorem (Equation (38b)) tells us that

$$\delta_{k,0} = \int_{-\infty}^{\infty} \phi_{0,k}(t)\phi(t)\, dt = \int_{-\infty}^{\infty} |\Phi(f)|^2 e^{i2\pi f k}\, df,$$

where $\delta_{k,0}$ is unity if $k = 0$ and is zero otherwise. Hence we have

$$\begin{aligned}\delta_{k,0} = \sum_{m=-\infty}^{\infty} \int_{(2m-1)/2}^{(2m+1)/2} |\Phi(f)|^2 e^{i2\pi f k}\, df &= \sum_{m=-\infty}^{\infty} \int_{-1/2}^{1/2} |\Phi(f+m)|^2 e^{i2\pi (f+m) k}\, df \\ &= \int_{-1/2}^{1/2} \left(\sum_{m=-\infty}^{\infty} |\Phi(f+m)|^2 \right) e^{i2\pi f k}\, df.\end{aligned}$$

The above says that the inverse Fourier transform of the sum in parentheses is the sequence $\{\delta_{k,0}\}$; however, the Fourier transform of $\{\delta_{k,0}\}$ is equal to unity at all frequencies, so we obtain

$$\sum_{m=-\infty}^{\infty} |\Phi(f+m)|^2 = 1. \tag{467a}$$

Using Equation (465b), we have

$$\sum_{m=-\infty}^{\infty} |\Phi(f+m)|^2 = \frac{1}{2} \sum_{m=-\infty}^{\infty} |\overline{G}(\tfrac{f}{2} + \tfrac{m}{2})|^2 |\Phi(\tfrac{f}{2} + \tfrac{m}{2})|^2,$$

which implies that

$$\sum_{m=-\infty}^{\infty} |\overline{G}(f' + \tfrac{m}{2})|^2 |\Phi(f' + \tfrac{m}{2})|^2 = 2,$$

where $f' = f/2$. The portion of the above sum indexed by even m can be written as

$$\sum_{n=-\infty}^{\infty} |\overline{G}(f'+n)|^2 |\Phi(f'+n)|^2 = |\overline{G}(f')|^2 \sum_{n=-\infty}^{\infty} |\Phi(f'+n)|^2 = |\overline{G}(f')|^2,$$

where we have used Equation (467a) again along with the fact that $\overline{G}(\cdot)$ has unit periodicity (i.e., $\overline{G}(f'+n) = \overline{G}(f')$ for all integers n); likewise, the portion indexed by odd m gives

$$\sum_{n=-\infty}^{\infty} |\overline{G}(f'+n+\tfrac{1}{2})|^2 |\Phi(f'+n+\tfrac{1}{2})|^2 = |\overline{G}(f'+\tfrac{1}{2})|^2 \sum_{n=-\infty}^{\infty} |\Phi(f'+\tfrac{1}{2}+n)|^2 = |\overline{G}(f'+\tfrac{1}{2})|^2.$$

Combining yields the constraint

$$|\overline{G}(f)|^2 + |\overline{G}(f+\tfrac{1}{2})|^2 = 2. \tag{467b}$$

In particular, if we let $f = 0$ in the equation, we obtain

$$|\overline{G}(0)|^2 + |\overline{G}(\tfrac{1}{2})|^2 = 2;$$

on the other hand, Equation (466b) tells us that $\overline{G}(0) = \sqrt{2}$, which allows us to conclude that

$$\overline{G}(\tfrac{1}{2}) = 0, \tag{468a}$$

a result that will prove useful in Section 11.6.

This result also enables us to fix the normalization of $\phi(\cdot)$. Returning first to Equation (467a), we can write

$$|\Phi(f)|^2 + \sum_{\substack{m=-\infty \\ m\neq 0}}^{\infty} |\Phi(f+m)|^2 = 1 \text{ and hence } |\Phi(0)|^2 + \sum_{\substack{m=-\infty \\ m\neq 0}}^{\infty} |\Phi(m)|^2 = 1$$

when $f = 0$; however, from the following, we see that $\Phi(m) = 0$ for all terms in the last summation.

▷ **Exercise [468a]:** Let m be any nonzero integer. By noting that $m = 2^l[2j+1]$ for some $l \geq 0$ and $j \in \mathbb{Z}$ and by using Equation (465c), show that

$$\Phi(m) = \Phi(\tfrac{2j+1}{2}) \prod_{k=1}^{l+1} \frac{\overline{G}(2^{l-k}[2j+1])}{\sqrt{2}}.$$

Use the above to argue that $\Phi(m) = 0$. ◁

We can conclude that $|\Phi(0)|^2 = 1$, so we must have $\Phi(0) = \pm 1$. It is convenient to choose $\Phi(0) = 1$, yielding the normalization

$$\int_{-\infty}^{\infty} \phi(t)\, dt = \Phi(0) = 1 \tag{468b}$$

(Daubechies, 1992, p. 175). Hence Equation (466a) simplifies to

$$\Phi(f) = \prod_{m=1}^{\infty} \frac{\overline{G}(\frac{f}{2^m})}{\sqrt{2}}. \tag{468c}$$

The two-scale relationship and orthonormality of the $\phi_{0,k}$ can also be used to show that the filter $\{\bar{g}_l\}$ has unit energy and is orthogonal to its even shifts, as indicated by the following exercise (cf. Exercise [76b]).

▷ **Exercise [468b]:** Show that

$$\sum_{l=-\infty}^{\infty} \bar{g}_l \bar{g}_{l+2k} = \begin{cases} 1, & k = 0; \\ 0, & \text{otherwise.} \end{cases} \tag{468d}$$ ◁

11.5 Scaling Functions and Scaling Filters

In this section we explore an interesting connection between the scaling function $\phi(\cdot)$ and jth level equivalent scaling filter $\{g_{j,l}\}$ (this filter is discussed in Section 4.6, and some of its key properties are summarized in Table 154). Because $g_l \equiv \bar{g}_{-l}$, the scaling filter $\{g_l\}$ of unit level has a transfer function given by

$$G(f) \equiv \sum_{l=-\infty}^{\infty} g_l e^{-i2\pi fl} = \sum_{l=-\infty}^{\infty} \bar{g}_{-l} e^{-i2\pi fl} = \overline{G}(-f).$$

In Section 5.2, we defined the MODWT scaling filter as $\tilde{g}_l \equiv g_l/\sqrt{2}$. This filter has a transfer function given by $\widetilde{G}(f) \equiv G(f)/\sqrt{2}$. In terms of this transfer function, we can rewrite Equation (468c) as

$$\Phi(f) = \prod_{m=1}^{\infty} \widetilde{G}(-\tfrac{f}{2^m}).$$

This result enables us to make a connection between $\phi(\cdot)$ and $\{g_l\}$. We note first that

$$\bar{\Phi}(f) \equiv \Phi(-f) = \prod_{m=1}^{\infty} \widetilde{G}(\tfrac{f}{2^m}) \approx \prod_{m=1}^{j} \widetilde{G}(\tfrac{f}{2^m})$$

for a large value of j, where the approximation can be justified by appealing to properties of, for example, the Daubechies scaling filters (intuitively, the above is reasonable because $\widetilde{G}(0) = 1$ and $\widetilde{G}(\cdot)$ is a continuous function with derivatives of all orders for those filters). Thus, from Equation (169b),

$$\bar{\Phi}(2^j f) \approx \prod_{m=1}^{j} \widetilde{G}(2^{j-m} f) = \prod_{l=0}^{j-1} \widetilde{G}(2^l f) = \widetilde{G}_j(f),$$

where $\widetilde{G}_j(f)$ is the transfer function for $\{\tilde{g}_{j,l}\}$. The inverse DFT (Equation (35a)) of $\widetilde{G}_j(f)$ is $\{\tilde{g}_{j,l}\}$, and hence if we take the inverse DFT of both sides and consider the case of large j we obtain

$$\begin{aligned}\tilde{g}_{j,l} &\approx \int_{-1/2}^{1/2} \bar{\Phi}(2^j f) e^{i2\pi fl}\, df = \tfrac{1}{2^j} \int_{-2^{j-1}}^{2^{j-1}} \bar{\Phi}(f') e^{i2\pi f'(l/2^j)}\, df' \\ &\approx \tfrac{1}{2^j} \int_{-\infty}^{\infty} \bar{\Phi}(f') e^{i2\pi f'(l/2^j)}\, df' = \tfrac{1}{2^j} \phi(-\tfrac{l}{2^j}).\end{aligned}$$

Hence we conclude that

$$2^j \tilde{g}_{j,-l} = 2^{j/2} g_{j,-l} \approx \phi(\tfrac{l}{2^j}), \tag{469}$$

so that the time reverse of the jth level equivalent scaling filter for large j should have the same shape as $\phi(\cdot)$.

To verify the approximation of Equation (469), we need some way of computing $\phi(\cdot)$ over the grid defined by $l/2^j$. In fact, we can use the two-scale equation to compute the scaling function – to within a constant of proportionality – over this grid by a recursion. To see how this is done, let us use the D(4) scaling filter as an example, for which we have (via Equation (75d))

$$\bar{g}_0 = g_0 = \frac{1+\sqrt{3}}{4\sqrt{2}}, \ \bar{g}_{-1} = g_1 = \frac{3+\sqrt{3}}{4\sqrt{2}},$$
$$\bar{g}_{-2} = g_2 = \frac{3-\sqrt{3}}{4\sqrt{2}}, \ \bar{g}_{-3} = g_3 = \frac{1-\sqrt{3}}{4\sqrt{2}}.$$

When $L = 4$, the support of $\phi(\cdot)$ is over $(-(L-1), 0] = (-3, 0]$. From the two-scale difference equation (Equation (465a)), we obtain

$$\begin{bmatrix} \phi(-3) \\ \phi(-2) \\ \phi(-1) \\ \phi(0) \end{bmatrix} = \sqrt{2} \begin{bmatrix} \bar{g}_{-3} & 0 & 0 & 0 \\ \bar{g}_{-1} & \bar{g}_{-2} & \bar{g}_{-3} & 0 \\ 0 & \bar{g}_0 & \bar{g}_{-1} & \bar{g}_{-2} \\ 0 & 0 & 0 & \bar{g}_0 \end{bmatrix} \begin{bmatrix} \phi(-3) \\ \phi(-2) \\ \phi(-1) \\ \phi(0) \end{bmatrix}.$$

From the bottom row, we obtain $\phi(0) = 2^{1/2}\bar{g}_0\phi(0)$, which tells us that $\phi(0) = 0$ since $2^{1/2}\bar{g}_0 \neq 1$. We can thus delete the last row and column and use $\bar{g}_{-l} = g_l$ to write

$$\begin{bmatrix} \phi(-3) \\ \phi(-2) \\ \phi(-1) \end{bmatrix} = \sqrt{2} \begin{bmatrix} g_3 & 0 & 0 \\ g_1 & g_2 & g_3 \\ 0 & g_0 & g_1 \end{bmatrix} \begin{bmatrix} \phi(-3) \\ \phi(-2) \\ \phi(-1) \end{bmatrix} = \mathbf{G}_0 \begin{bmatrix} \phi(-3) \\ \phi(-2) \\ \phi(-1) \end{bmatrix}, \text{ say.} \tag{470a}$$

A similar argument says $\phi(-3) = 0$ also (this agrees with the assumption that $t = -3$ is outside the interval of support for $\phi(\cdot)$). We can thus delete the first row and column to obtain

$$\begin{bmatrix} \phi(-2) \\ \phi(-1) \end{bmatrix} = \sqrt{2} \begin{bmatrix} g_2 & g_3 \\ g_0 & g_1 \end{bmatrix} \begin{bmatrix} \phi(-2) \\ \phi(-1) \end{bmatrix}.$$

The above says that the vector $[\phi(-2), \phi(-1)]^T$ is an eigenvector for the 2×2 matrix (the associated eigenvalue is unity). Since $c[\phi(-2), \phi(-1)]^T$ is also an eigenvector for any constant c, we can only obtain $\phi(-2)$ and $\phi(-1)$ to within a constant of proportionality (in practice, we can compute this constant quite accurately later on by making use of the fact that $\phi(\cdot)$ must integrate to unity). The first row of the above gives us

$$\frac{\phi(-2)}{\phi(-1)} = \frac{g_3\sqrt{2}}{1 - g_2\sqrt{2}} = \frac{1-\sqrt{3}}{1+\sqrt{3}}, \tag{470b}$$

which – since $\phi(-3) = 0$ – fixes the relative element sizes of $[\phi(-3), \phi(-2), \phi(-1)]^T$. Suppose we now add 0.5 (binary representation .1) to each of the integers $-3, -2$ and -1 and evaluate $\phi(\cdot)$ at these points. If we again use the two-scale equation, we obtain

$$\begin{bmatrix} \phi(-2.5) \\ \phi(-1.5) \\ \phi(-0.5) \end{bmatrix} = \sqrt{2} \begin{bmatrix} g_2 & g_3 & 0 \\ g_0 & g_1 & g_2 \\ 0 & 0 & g_0 \end{bmatrix} \begin{bmatrix} \phi(-3) \\ \phi(-2) \\ \phi(-1) \end{bmatrix} = \mathbf{G}_1 \begin{bmatrix} \phi(-3) \\ \phi(-2) \\ \phi(-1) \end{bmatrix}, \text{ say;}$$

i.e., we obtain values half way between the integers from values at the integers. If we add 0.25 (binary representation .01) to each of the integers $-3, -2$ and -1 and evaluate $\phi(\cdot)$ at these points, the two-scale equation gives

$$\begin{bmatrix} \phi(-2.75) \\ \phi(-1.75) \\ \phi(-0.75) \end{bmatrix} = \mathbf{G}_0 \begin{bmatrix} \phi(-2.5) \\ \phi(-1.5) \\ \phi(-0.5) \end{bmatrix} = \mathbf{G}_0\mathbf{G}_1 \begin{bmatrix} \phi(-3) \\ \phi(-2) \\ \phi(-1) \end{bmatrix};$$

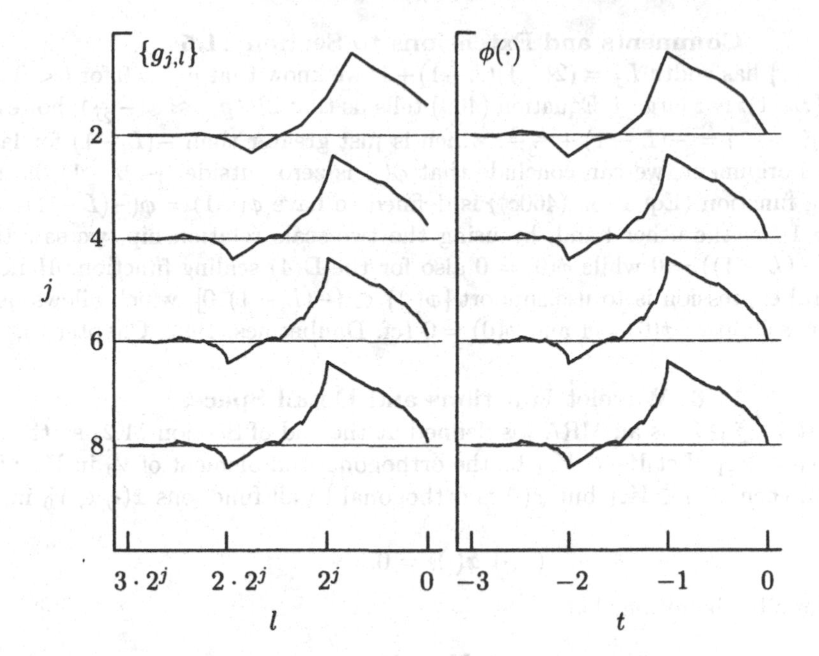

Figure 471. Level j equivalent D(4) scaling filters $\{g_{j,l}\}$ (left-hand column) and the D(4) scaling function $\phi(\cdot)$ evaluated over the grid defined by $\frac{l}{2^j}$, $l = -3\cdot 2^j, \ldots, -1, 0$ (right-hand) for $j = 2, 4, 6$ and 8 (top to bottom). For a given j, the two plotted sequences consist of $3\cdot 2^j + 1$ values connected by line segments. In the right-hand column, the function $\phi(\cdot)$ is plotted at values $t = -3.0$ to $t = 0$ in steps of, from top to bottom, 0.25, 0.0625, 0.015625 and 0.00390625. The filters in the left-hand column are plotted in such a manner as to illustrate the approximation of Equation (469), whose validity increases with increasing j.

similarly, if we add 0.75 (binary representation .11) to each of the integers, the two-scale equation now gives

$$\begin{bmatrix}\phi(-2.25)\\ \phi(-1.25)\\ \phi(-0.25)\end{bmatrix} = \mathbf{G}_1\begin{bmatrix}\phi(-2.5)\\ \phi(-1.5)\\ \phi(-0.5)\end{bmatrix} = \mathbf{G}_1\mathbf{G}_1\begin{bmatrix}\phi(-3)\\ \phi(-2)\\ \phi(-1)\end{bmatrix}.$$

We note that the binary representation of the quantity being added to the integers tells us the sequence of subscripts of the matrices $\mathbf{G}_0$ and $\mathbf{G}_1$ to use. For example, if we add 0.625 (binary representation .101), we use $\mathbf{G}_1\mathbf{G}_0\mathbf{G}_1$ as a premultiplier of $[\phi(-3), \phi(-2), \phi(-1)]^T$ to obtain $[\phi(-2.375), \phi(-1.375), \phi(-0.375)]^T$. (For a more formal discussion of this recursive scheme, see Strang and Nguyen, 1996, p. 196.)

As an example, the right-hand column of Figure 471 shows the sequences $\{\phi(\frac{l}{2^j}) : l = -3\cdot 2^j, \ldots, -1, 0\}$ for (from top to bottom) $j = 2, 4, 6$ and 8. The top-most plotted curve is thus an evaluation of $\phi(\cdot)$ from −3 to 0 in steps of 0.25; the bottom-most, in steps of 0.00390625. As the step size decreases, we see the plotted curve converging to the characteristic 'shark's fin' shape of the D(4) scaling function. The left-hand column shows corresponding plots for $g_{j,l}$ versus $l = 3\cdot 2^j, \ldots, 1, 0$ (the curves for $j = 2, 4$ and 6 are also shown – in reversed order – in the left-hand column of Figure 98a). Note that, while there are visible differences between the left- and right-hand curves for $j = 2$ and 4, they are largely gone for $j = 6$ and 8, thus verifying the approximation of Equation (469).

Comments and Extensions to Section 11.5

[1] Since $\{g_{j,l}\}$ has width $L_j = (2^j - 1)(L-1)+1$, we know that $g_{j,l} = 0$ for $l < 0$ and $l > (2^j - 1)(L-1)$. For large j, Equation (469) tells us that $2^{j/2} g_{j,l} \approx \phi(-\frac{l}{2^j})$; however, $-\frac{1}{2^j}(2^j - 1)(L-1) = -(L-1) + \frac{L-1}{2^j}$, which is just greater than $-(L-1)$ for large j. From this argument, we can conclude that $\phi(\cdot)$ is zero outside $(-(L-1), 0]$. The Haar scaling function (Equation (460c)) is defined to have $\phi(-1) = \phi(-(L-1)) = 0$ and $\phi(0) = 1$; on the other hand, by using the two-scale relationship, we saw that $\phi(-3) = \phi(-(L-1)) = 0$ while $\phi(0) = 0$ also for the D(4) scaling function. Hence a useful general expression is to use support $\{\phi(\cdot)\} \subset (-(L-1), 0]$, which allows us to handle the possibilities $\phi(0) = 1$ and $\phi(0) = 0$ (cf. Daubechies, 1992, Chapter 6).

11.6 Wavelet Functions and Detail Spaces

Suppose that V_j, $j \in \mathbb{Z}$, is an MRA (as defined at the end of Section 11.2) so that, in particular, $V_0 \subset V_{-1}$. Let $W_0 \subset V_{-1}$ be the orthogonal complement of V_0 in V_{-1}; i.e., if $\varphi(\cdot) \in W_0$, then $\varphi(\cdot) \in V_{-1}$ but $\varphi(\cdot)$ is orthogonal to all functions $x(\cdot) \in V_0$ in the sense that

$$\langle \varphi(\cdot), x(\cdot) \rangle = 0.$$

Then it follows by definition that

$$V_{-1} = V_0 \oplus W_0;$$

i.e., the space V_{-1} is the so-called *direct sum* of V_0 and W_0, which means that any element in V_{-1} can be expressed as the sum of two orthogonal elements, one from V_0 and the other from W_0. For the general case, $V_j \subset V_{j-1}$, and $W_j \subset V_{j-1}$ is the orthogonal complement in V_{j-1} of V_j, so that

$$V_j = V_{j+1} \oplus W_{j+1}.$$

By iteration of the above we obtain

$$V_j = W_{j+1} \oplus W_{j+2} \oplus \cdots.$$

The subspace W_j is called the *detail space* for scale $\tau_j = 2^{j-1}$. The nesting of the V_j's and W_j's is illustrated in Figure 473a.

As an example, Figure 473b shows, from left to right, a function in each of the subspaces $V_0^{(\mathrm{H})}$, $W_0^{(\mathrm{H})}$ and $V_{-1}^{(\mathrm{H})}$ based upon the Haar scaling function. Recall that all functions in $V_0^{(\mathrm{H})}$ must be constant over intervals defined by $k - 1 < t \le k$ for $k \in \mathbb{Z}$; on the other hand, all functions in $V_{-1}^{(\mathrm{H})}$ must be constant over intervals $\frac{k-1}{2} < t \le \frac{k}{2}$. In order for a function $\varphi(\cdot)$ to be in $V_{-1}^{(\mathrm{H})}$ and yet be orthogonal to all functions in $V_0^{(\mathrm{H})}$, the two values it assumes over the intervals given by $\frac{k-2}{2} < t \le \frac{k-1}{2}$ and $\frac{k-1}{2} < t \le \frac{k}{2}$ for even k must sum to zero. This is illustrated by the middle plot of Figure 473b, where the values of $\varphi(\cdot)$ are represented as shaded deviations from zero, and the intervals $k - 1 < t \le k$ are indicated by vertical dotted lines. In fact, if we sum the functions shown in the left-hand and middle plots, we obtain the function shown on the right-hand plot.

Let us now define the *wavelet function* $\psi(\cdot)$ as

$$\psi(t) \equiv \sqrt{2} \sum_{l=-\infty}^{\infty} \bar{h}_l \phi(2t - l), \quad \text{where } \bar{h}_l \equiv (-1)^l \bar{g}_{1-l-L} \tag{472}$$

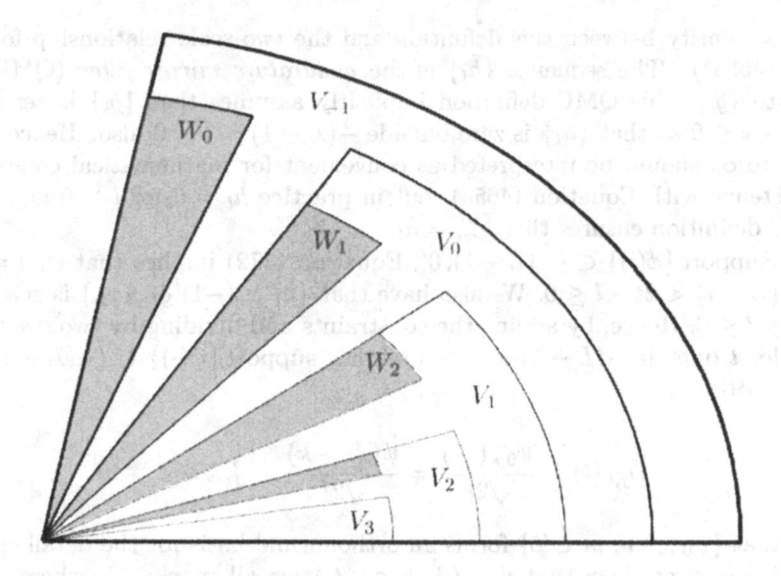

Figure 473a. Venn diagram illustrating the nesting of subspaces V_j and W_j. There are five arcs emanating from the baseline. Starting at each end of an arc, there is a line segment that continues to the lower left-hand corner of the plot. The area enclosed by a given arc and the two line segments emanating from its ends represents an approximation space V_j. The largest such area outlines the entire figure and represents V_{-1}, while the smallest area represents V_3 (note that $V_3 \subset V_2 \subset V_1 \subset V_0 \subset V_{-1}$, as required). The shaded areas represent the detail spaces W_j. Note that $W_0 \subset V_{-1}$, $W_1 \subset V_0$, $W_2 \subset V_1$ and $W_3 \subset V_2$ (there is no label for W_3 due to lack of space). Note also that, while $V_0 \subset V_{-1}$ and $W_0 \subset V_{-1}$, it is the case that $V_0 \cup W_0 \neq V_{-1}$ because V_{-1} also contains linear combinations of functions that are in both V_0 and W_0 – such linear combinations need not be in either V_0 or W_0, but rather can be in the space disjoint to V_0 and W_0 and represented by the scythe-like shape bearing the label V_{-1}. Finally, note that all the V_j and W_j intersect at a single point (represented by the lower left-hand corner of the plot) because all these spaces must contain the null function.

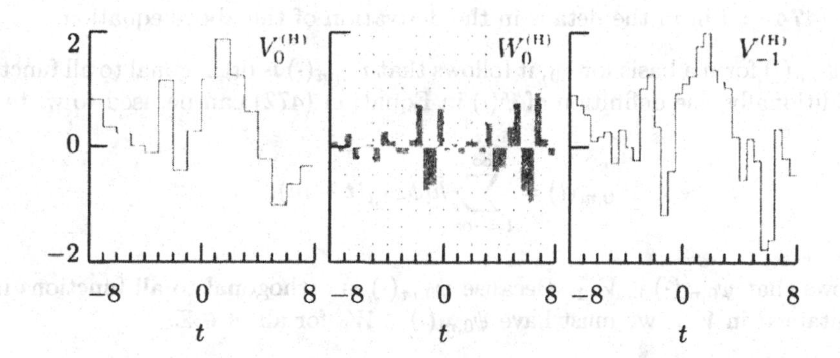

Figure 473b. Examples of functions in $V_0^{(H)}$, $W_0^{(H)}$ and $V_{-1}^{(H)}$. Note that, while the function in $V_0^{(H)}$ is constant over intervals of the form $(k-1, k]$ for $k \in \mathbb{Z}$, the function in $W_0^{(H)}$ integrates to zero over such intervals (the ends of these intervals are indicated in the middle plot by the vertical dotted lines). The function in $V_{-1}^{(H)}$ is in fact formed by point-wise addition of the functions in the other two spaces.

(note the similarity between this definition and the two-scale relationship for $\phi(\cdot)$ in Equation (465a)). The sequence $\{\bar{h}_l\}$ is the *quadrature mirror filter* (QMF) corresponding to $\{\bar{g}_l\}$; this QMF definition implicitly assumes that $\{\bar{g}_l\}$ is zero outside $-(L-1) \le l \le 0$ so that $\{\bar{h}_l\}$ is zero outside $-(L-1) \le l \le 0$ also. Hence the sum from $-\infty$ to ∞ should be interpreted as convenient for mathematical computations and consistency with Equation (465a), but in practice $\bar{h}_l = 0$ for $l > 0$ and $l \le -L$. This QMF definition ensures that $\bar{h}_{-l} = h_l$.

With support $\{\phi(\cdot)\} \subset (-(L-1), 0]$, Equation (472) implies that $\psi(t) = 0$ for t outside $-(L-1) < 2t - l \le 0$. We also have that $\{\bar{h}_l \equiv (-1)^l \bar{g}_{1-l-L}\}$ is zero outside $-(L-1) \le l \le 0$. Hence, by adding the constraints and dividing by two, we find that $\psi(t) = 0$ for t outside $-(L-1) < t \le 0$ so that support $\{\psi(\cdot)\} \subset (-(L-1), 0]$, the same as for $\phi(\cdot)$.

With

$$\psi_{j,k}(t) \equiv \frac{\psi_{0,k}(\frac{t}{2^j})}{\sqrt{2^j}} = \frac{\psi(\frac{t}{2^j} - k)}{\sqrt{2^j}}, \quad j, k \in \mathbb{Z},$$

we claim that $\{\psi_{0,m}(\cdot) : m \in \mathbb{Z}\}$ forms an orthonormal basis for the detail space W_0. To see this, we first show that $\psi_{0,m}(\cdot) \perp \phi_{0,n}(\cdot)$ for all $m, n \in \mathbb{Z}$, where $\perp$ means 'orthogonal to;' i.e.,

$$\langle \psi_{0,m}(\cdot), \phi_{0,n}(\cdot) \rangle \equiv \int_{-\infty}^{\infty} \psi_{0,m}(t)\phi_{0,n}(t)\, dt = 0.$$

This result follows from the orthonormality of the $\phi_{0,n}$ along with Equations (465a) and (472) because we have

$$\int_{-\infty}^{\infty} \psi_{0,m}(t)\phi_{0,n}(t)\, dt = \sum_{l=-\infty}^{\infty} \bar{h}_l \bar{g}_{l+2m-2n} = 0.$$

▷ **Exercise [474a]:** Fill in the details in the derivation of the above equation. ◁

Since the $\phi_{0,n}(\cdot)$ form a basis for V_0, it follows that $\psi_{0,m}(\cdot)$ is orthogonal to all functions in V_0. Additionally, the definition of $\psi(\cdot)$ in Equation (472) can be used to write

$$\psi_{0,m}(t) = \sum_{l=-\infty}^{\infty} \bar{h}_l \phi_{-1,l}(t - m),$$

which shows that $\psi_{0,m}(\cdot) \in V_{-1}$. Because $\psi_{0,m}(\cdot)$ is orthogonal to all functions in V_0 and is contained in V_{-1}, we must have $\psi_{0,m}(\cdot) \in W_0$ for all $m \in \mathbb{Z}$.

▷ **Exercise [474b]:** Using the definition of $\psi(\cdot)$, the orthonormality of the $\phi_{0,m}(\cdot)$ and the orthogonality of $\{\bar{g}_l\}$ to its even shifts given by Equation (468d), show that the $\psi_{0,m}(\cdot)$ are orthonormal. ◁

The final step needed to establish that the $\psi_{0,m}(\cdot)$ constitute an orthonormal basis for W_0 is the subject of Exercise [11.2]. Thus, just as the basis functions for V_j are dilations and translations of the scaling function $\phi(\cdot)$, the basis functions for W_j are dilations and translations of the wavelet function $\psi(\cdot)$.

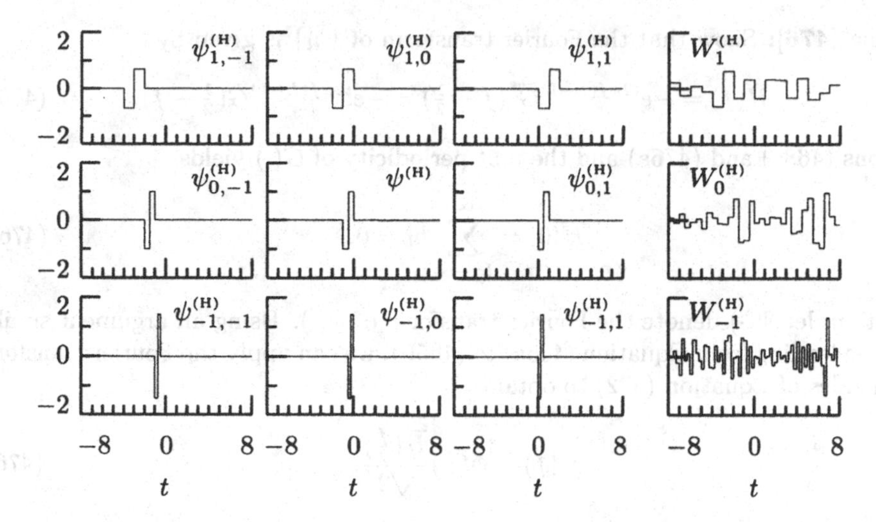

Figure 475. The Haar wavelet function $\psi^{(H)}(\cdot)$ and corresponding detail spaces. The first three plots on the middle row shows three of the basis functions for the Haar detail space $W_0^{(H)}$, namely, from left to right, $\psi_{0,-1}^{(H)}(\cdot)$, $\psi^{(H)}(\cdot)$ and $\psi_{0,1}^{(H)}(\cdot)$. The right-most plot on this row is an example of a function contained in $W_0^{(H)}$. The top and bottom rows show, respectively, corresponding plots for the Haar detail spaces $W_1^{(H)}$ and $W_{-1}^{(H)}$ (Figure 461 shows the corresponding Haar scaling function and approximation spaces).

The simplest example of a wavelet function is the *Haar wavelet function*, defined as

$$\psi^{(H)}(t) \equiv \sqrt{2}\sum_{l=-\infty}^{\infty} \bar{h}_l \phi^{(H)}(2t-l) = \sqrt{2}\left[-\frac{\phi^{(H)}(2t+1)}{\sqrt{2}} + \frac{\phi^{(H)}(2t)}{\sqrt{2}}\right]$$

$$= \begin{cases} -1, & -1 < t \le -1/2; \\ 1, & -1/2 < t \le 0; \\ 0, & \text{otherwise}, \end{cases} \tag{475}$$

since the QMF relationship yields $\bar{h}_{-1} = -\bar{g}_0 = -1/\sqrt{2}$, $\bar{h}_0 = \bar{g}_{-1} = 1/\sqrt{2}$ and $\bar{h}_l = 0$ for $l \neq -1$ or 0 (note that this definition differs from the one given in Equation (2c)). Figure 475 shows Haar wavelet basis functions for, from top to bottom rows, the subspaces $W_1^{(H)}$, $W_0^{(H)}$ and $W_{-1}^{(H)}$, along with an example of a function in each subspace (cf. Figure 461 for the Haar scaling function).

Since $\psi_{1,k} \in W_1$ and $\phi_{1,m}(\cdot) \in V_1$ and since all functions in W_1 are orthogonal to all functions in V_1, we have $\psi_{1,k}(\cdot) \perp \phi_{1,m}(\cdot)$ for all $k, m \in \mathbb{Z}$. Further, since

$$V_0 = V_2 \oplus W_2 \oplus W_1,$$

we have that $\phi_{2,k}(\cdot) \perp \psi_{2,m}(\cdot)$, $\phi_{2,k}(\cdot) \perp \psi_{1,m}(\cdot)$ and $\psi_{2,k}(\cdot) \perp \psi_{1,m}(\cdot)$ for $k, m \in \mathbb{Z}$. In general $\phi_{j,k}(\cdot) \perp \psi_{l,m}(\cdot)$ for $j,k,l,m \in \mathbb{Z}$ such that $l \le j$. Also, $\psi_{j,k}(\cdot) \perp \psi_{l,m}(\cdot)$ for all $j,k,l,m \in \mathbb{Z}$ such that $j \neq l$, and $\psi_{j,k}(\cdot) \perp \psi_{j,m}(\cdot)$ for all $j,k,m \in \mathbb{Z}$ such that $k \neq m$.

The QMF relationship between $\{\bar{g}_l\}$ and $\{\bar{h}_l\}$ can be used to establish the following result:

▷ **Exercise [476]:** Show that the Fourier transform of $\{\bar{h}_l\}$ is given by

$$\overline{H}(f) = -e^{i2\pi f(L-1)}\overline{G}^*(f - \tfrac{1}{2}) = -e^{i2\pi f(L-1)}\overline{G}(\tfrac{1}{2} - f). \quad (476a) ◁$$

Equations (468a) and (476a) and the unit periodicity of $\overline{G}(\cdot)$ yields

$$\overline{H}(0) = \sum_{l=-\infty}^{\infty} \bar{h}_l = 0. \quad (476b)$$

In addition, let $\Psi(\cdot)$ denote the Fourier transform of $\psi(\cdot)$. Using an argument similar to the one leading from Equation (465a) to (465b), we can apply the Fourier transform to both sides of Equation (472) to obtain

$$\Psi(f) = \Phi(\tfrac{f}{2})\frac{\overline{H}(\frac{f}{2})}{\sqrt{2}}. \quad (476c)$$

Hence $\overline{H}(0) = 0$ implies that

$$\Psi(0) = \int_{-\infty}^{\infty} \psi(t)\,dt = 0.$$

(Daubechies, 1992, Section 2.4). Put simply, the above means that the wavelet function looks visually like a 'small wave;' since $\Psi(0) = 0$, the wavelet function has no energy at zero frequency. Thus, like the scaling function, the wavelet function $\psi(\cdot)$ has unit norm, but, unlike the scaling function, it integrates to zero. By using Equation (465b) iteratively in Equation (476c), and recalling that $\Phi(0) = 1$, we get

$$\Psi(f) = \frac{\overline{H}(\frac{f}{2})}{\sqrt{2}} \prod_{m=2}^{\infty} \frac{\overline{G}(\frac{f}{2^m})}{\sqrt{2}}. \quad (476d)$$

11.7 Wavelet Functions and Wavelet Filters

Explicit connections exist between the wavelet function $\psi(\cdot)$ and the jth level equivalent wavelet filters $\{h_{j,l}\}$ that are analogous to those we found in Section 11.5 for the scaling function and filters. Since $h_l \equiv \bar{h}_{-l}$, we know that $H(f) = \overline{H}(-f)$. Because the MODWT wavelet filter is $\tilde{h}_l \equiv h_l/\sqrt{2}$ (see Section 5.2), its transfer function is given by $\widetilde{H}(f) \equiv H(f)/\sqrt{2}$. Hence Equation (476d) gives

$$\Psi(f) = \widetilde{H}(-\tfrac{f}{2}) \prod_{m=2}^{\infty} \widetilde{G}(-\tfrac{f}{2^m}),$$

so that

$$\bar{\Psi}(f) \equiv \Psi(-f) = \widetilde{H}(\tfrac{f}{2}) \prod_{m=2}^{\infty} \widetilde{G}(\tfrac{f}{2^m}) \approx H(\tfrac{f}{2}) \prod_{m=2}^{j} \widetilde{G}(\tfrac{f}{2^m})$$

for large j. Thus, from Equation (169b),

$$\bar{\Psi}(2^j f) \approx \widetilde{H}(2^{j-1} f) \prod_{m=2}^{j} \widetilde{G}(2^{j-m} f) = \widetilde{H}(2^{j-1} f) \prod_{l=0}^{j-2} \widetilde{G}(2^l f) = \widetilde{H}_j(f),$$

where $\widetilde{H}_j(\cdot)$ is the transfer function for $\{\tilde{h}_{j,l}\}$. For large j, the inverse DFT yields

$$\tilde{h}_{j,l} \approx \int_{-1/2}^{1/2} \bar{\Psi}(2^j f) e^{i2\pi fl}\, df \approx \tfrac{1}{2^j}\psi(-\tfrac{l}{2^j}).$$

Hence

$$2^j \tilde{h}_{j,-l} = 2^{j/2} h_{j,-l} \approx \psi(\tfrac{l}{2^j}) \qquad (477a)$$

for large j, so the time reverse of the jth level equivalent wavelet filter should have approximately the same shape as $\psi(\cdot)$.

Let us now turn to the calculation of $\psi(\cdot)$, which can be accomplished by recursion using Equation (472). As a specific example, consider again the D(4) design, with wavelet filter given by Equation (59a). From Equation (472) and using $h_l = \bar{h}_{-l}$ we get

$$\begin{bmatrix} \psi(-3) \\ \psi(-2) \\ \psi(-1) \\ \psi(0) \end{bmatrix} = \sqrt{2} \begin{bmatrix} h_3 & 0 & 0 & 0 \\ h_1 & h_2 & h_3 & 0 \\ 0 & h_0 & h_1 & h_2 \\ 0 & 0 & 0 & h_0 \end{bmatrix} \begin{bmatrix} \phi(-3) \\ \phi(-2) \\ \phi(-1) \\ \phi(0) \end{bmatrix}.$$

Since we know that $\phi(-3) = \phi(0) = 0$, it follows that $\psi(-3) = \psi(0) = 0$ also. Deleting the first and last rows and columns, we obtain

$$\begin{bmatrix} \psi(-2) \\ \psi(-1) \end{bmatrix} = \sqrt{2} \begin{bmatrix} h_2 & h_3 \\ h_0 & h_1 \end{bmatrix} \begin{bmatrix} \phi(-2) \\ \phi(-1) \end{bmatrix}, \qquad (477b)$$

from which we can determine the ratio

$$\frac{\psi(-2)}{\psi(-1)} = -\frac{1+\sqrt{3}}{1-\sqrt{3}} \qquad (477c)$$

and hence the relative sizes of the elements of $[\psi(-3), \psi(-2), \psi(-1)]^T$ (verification of this ratio is the subject of Exercise [11.3]). In analogy to Equation (470a), we can write

$$\begin{bmatrix} \psi(-3) \\ \psi(-2) \\ \psi(-1) \end{bmatrix} = \sqrt{2} \begin{bmatrix} h_3 & 0 & 0 \\ h_1 & h_2 & h_3 \\ 0 & h_0 & h_1 \end{bmatrix} \begin{bmatrix} \phi(-3) \\ \phi(-2) \\ \phi(-1) \end{bmatrix} = \mathbf{H}_0 \begin{bmatrix} \phi(-3) \\ \phi(-2) \\ \phi(-1) \end{bmatrix}, \text{ say.}$$

Suppose we now add 0.5 (binary representation .1) to each of the integers −3, −2 and −1 and evaluate $\psi(\cdot)$ at these points. Using Equation (472) again, we can write

$$\begin{bmatrix} \psi(-2.5) \\ \psi(-1.5) \\ \psi(-0.5) \end{bmatrix} = \sqrt{2} \begin{bmatrix} h_2 & h_3 & 0 \\ h_0 & h_1 & h_2 \\ 0 & 0 & h_0 \end{bmatrix} \begin{bmatrix} \phi(-3) \\ \phi(-2) \\ \phi(-1) \end{bmatrix} = \mathbf{H}_1 \begin{bmatrix} \phi(-3) \\ \phi(-2) \\ \phi(-1) \end{bmatrix}, \text{ say.}$$

If we add 0.25 (binary representation .01) to the integers −3, −2 and −1 and evaluate $\psi(\cdot)$ at these points, Equation (472) gives

$$\begin{bmatrix} \psi(-2.75) \\ \psi(-1.75) \\ \psi(-0.75) \end{bmatrix} = \mathbf{H}_0 \begin{bmatrix} \phi(-2.5) \\ \phi(-1.5) \\ \phi(-0.5) \end{bmatrix} = \mathbf{H}_0 \mathbf{G}_1 \begin{bmatrix} \phi(-3) \\ \phi(-2) \\ \phi(-1) \end{bmatrix}.$$

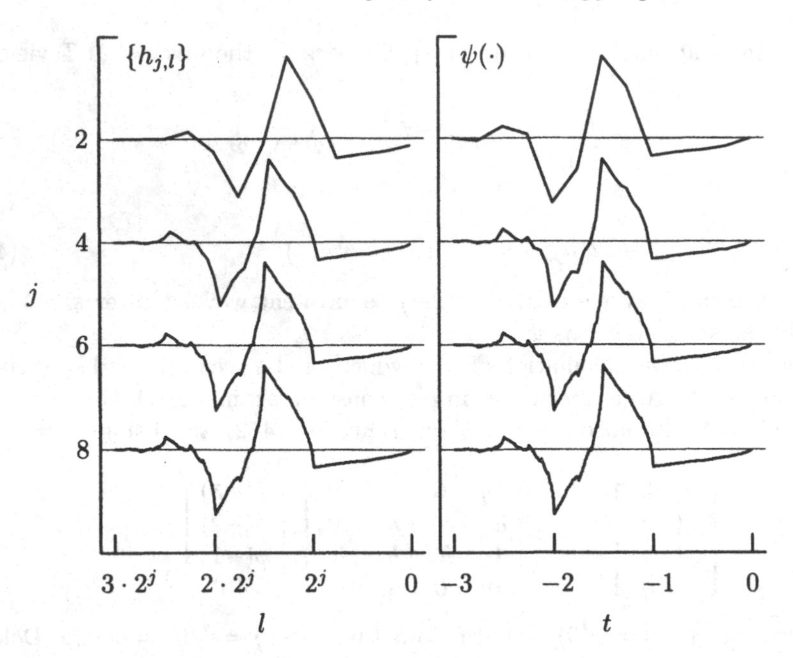

Figure 478. Level j equivalent D(4) wavelet filters $\{h_{j,l}\}$ (left-hand column) and the D(4) wavelet function $\psi(\cdot)$ evaluated over the grid defined by $\frac{l}{2^j}$, $l = -3 \cdot 2^j, \ldots, -1, 0$ (right-hand) for $j = 2, 4, 6$ and 8 (top to bottom). For details on the layout, see the caption for the analogous Figure 471.

Similarly, if we add 0.75 (binary representation .11) to the integers $-3, -2$ and -1 and evaluate $\psi(\cdot)$ at these points, we obtain

$$\begin{bmatrix} \psi(-2.25) \\ \psi(-1.25) \\ \psi(-0.25) \end{bmatrix} = \mathbf{H}_1 \begin{bmatrix} \phi(-2.5) \\ \phi(-1.5) \\ \phi(-0.5) \end{bmatrix} = \mathbf{H}_1\mathbf{G}_1 \begin{bmatrix} \phi(-3) \\ \phi(-2) \\ \phi(-1) \end{bmatrix}.$$

Again the binary representation of the quantity being added to the integers tells us the sequence of matrices to use. For example, if we add 0.625 (binary representation .101), we multiply $\mathbf{H}_1\mathbf{G}_0\mathbf{G}_1$ times the vector $[\phi(-3), \phi(-2), \phi(-1)]^T$. The last matrix multiplication will be by an $\mathbf{H}_n$ matrix, while all previous will involve $\mathbf{G}_n$ matrices.

As an example, the right-hand column of Figure 478 shows the sequences $\{\psi(\frac{l}{2^j}) : l = -3 \cdot 2^j, \ldots, -1, 0\}$ for (from top to bottom) $j = 2, 4, 6$ and 8, while the left-hand column shows corresponding plots for $h_{j,l}$ versus $l = 3 \cdot 2^j, \ldots, 1, 0$ (the curves for $j = 2, 4$ and 6 are shown reversed in the right-hand column of Figure 98a). As j increases, the visible discrepancy between plots in the same row decreases, thus verifying the approximation of Equation (477a).

11.8 Multiresolution Analysis of Finite Energy Signals

In Section 11.3 we considered approximating a signal $x(\cdot) \in V_0$ via $s_j(\cdot)$, which is obtained by projecting $s_{j-1}(\cdot)$ onto the subspace $V_j \subset V_0$. Here we expand upon this discussion by showing that the difference between $x(\cdot)$ and its approximation $s_j(\cdot)$ can be expressed in terms of projections onto the detail subspaces $W_j, W_{j-1}, \ldots, W_1$. A signal $x(\cdot) \in V_0 = V_1 \oplus W_1$ can be decomposed into its components along V_1 and W_1 to give the additive form (MRA)

$$x(t) = s_1(t) + d_1(t),$$

where $s_1(\cdot)$ and $d_1(\cdot)$ are signals contained in, respectively, V_1 and W_1. The signal $s_1(\cdot)$ is the projection of $x(\cdot)$ onto the subspace V_1 and is given by Equation (462b). Similarly, $d_1(\cdot)$ is the projection of $x(\cdot)$ onto the subspace W_1. Since $\{\psi_{1,k}(\cdot)\}$ forms an orthonormal basis for W_1, we have

$$d_1(t) = \sum_{k=-\infty}^{\infty} w_{1,k}\psi_{1,k}(t), \text{ where } w_{1,k} \equiv \langle x(\cdot), \psi_{1,k}(\cdot)\rangle. \tag{479a}$$

Here $d_1(\cdot)$ represents the *detail* in $x(\cdot)$ missing from the coarse approximation $s_1(\cdot)$. The $w_{1,k}$'s are the so-called *wavelet coefficients* for $d_1(\cdot)$. Using a development paralleling that of Equation (463a), we have

$$\begin{aligned} w_{1,k} &= \int_{-\infty}^{\infty} \left[\sum_{m=-\infty}^{\infty} v_{0,m}\phi_{0,m}(t) \right] \frac{\psi(\frac{t}{2} - k)}{\sqrt{2}}\, dt \\ &= \sum_{m=-\infty}^{\infty} v_{0,m} \int_{-\infty}^{\infty} \phi(t - [m - 2k]) \frac{\psi(\frac{t}{2})}{\sqrt{2}}\, dt \\ &= \sum_{m=-\infty}^{\infty} v_{0,m} \langle \phi_{0,m-2k}(\cdot), \psi_{1,0}(\cdot)\rangle. \end{aligned} \tag{479b}$$

▷ **Exercise [479]**: Show that $\langle \phi_{0,m-2k}(\cdot), \psi_{1,0}(\cdot)\rangle = \bar{h}_{m-2k}$, where $\{\bar{h}_l\}$ is the QMF corresponding to $\{\bar{g}_l\}$. Additionally, show that

$$\langle \phi_{j,m}(\cdot), \psi_{j+1,k}(\cdot)\rangle = \langle \phi_{0,m-2k}(\cdot), \psi_{1,0}(\cdot)\rangle = \bar{h}_{m-2k} \tag{479c}$$

for all j. ◁

We note that

$$\bar{h}_l = \langle \phi_{0,l}(\cdot), \psi_{1,0}(\cdot)\rangle = \int_{-\infty}^{\infty} \phi(t - l) \frac{\psi(\frac{t}{2})}{\sqrt{2}}\, dt. \tag{479d}$$

As an example, let us compute $\{\bar{h}_l\}$ for $\phi^{(H)}(\cdot)$ and $\psi^{(H)}(\cdot)$, the the Haar scaling and wavelet functions of Equations (460c) and (475). Since

$$\phi^{(H)}(t - l) = \begin{cases} 1, & l - 1 < t \le l; \\ 0, & \text{otherwise}; \end{cases} \quad \psi^{(H)}(\tfrac{t}{2}) = \begin{cases} -1, & -2 < t \le -1; \\ 1, & -1 < t \le 0; \\ 0, & \text{otherwise}, \end{cases}$$

Equation (479d) yields

$$\bar{h}_l = -\frac{1}{\sqrt{2}} \int_{-2}^{-1} \phi^{(H)}(t - l)\, dt + \frac{1}{\sqrt{2}} \int_{-1}^{0} \phi^{(H)}(t - l)\, dt = \begin{cases} -1/\sqrt{2}, & l = -1; \\ 1/\sqrt{2}, & l = 0; \\ 0, & \text{otherwise}. \end{cases}$$

Since $h_l \equiv \bar{h}_{-l}$, we get,

$$h_l = \begin{cases} 1/\sqrt{2}, & l = 0; \\ -1/\sqrt{2}, & l = 1; \\ 0, & \text{otherwise.} \end{cases}$$

This agrees with the definition of the Haar wavelet filter $\{h_l\}$ introduced in Section 4.2.

From Equation (479b) and the first part of Exercise [479], we have

$$w_{1,k} = \sum_{m=-\infty}^{\infty} \bar{h}_{m-2k} v_{0,m} = \sum_{l=-\infty}^{\infty} h_l v_{0,2k-l}, \tag{480a}$$

where we have used $h_l \equiv \bar{h}_{-l}$ and let $l = 2k - m$. This equation closely parallels (463a) for $v_{1,k}$. As was true of $\{v_{1,k}\}$, the sequence $\{w_{1,k}\}$ is produced by downsampling (by a factor of 2) the output of a filter, in this case $\{h_l\}$.

Since $s_1(\cdot) \in V_1 = V_2 \oplus W_2$, we can decompose it into components in V_2 and W_2 to obtain

$$s_1(t) = s_2(t) + d_2(t) \text{ with } d_2(t) \equiv \sum_{k=-\infty}^{\infty} w_{2,k} \psi_{2,k}(t),$$

where the wavelet coefficients for level $j = 2$ are given by

$$w_{2,k} \equiv \langle s_1(\cdot), \psi_{2,k}(\cdot) \rangle.$$

Using the expression for $s_1(\cdot)$ in Equation (462b), we have

$$\begin{aligned} w_{2,k} &= \left\langle \sum_{m=-\infty}^{\infty} v_{1,m} \phi_{1,m}(\cdot), \psi_{2,k}(\cdot) \right\rangle \\ &= \sum_{m=-\infty}^{\infty} v_{1,m} \langle \phi_{1,m}(\cdot), \psi_{2,k}(\cdot) \rangle \\ &= \sum_{m=-\infty}^{\infty} \bar{h}_{m-2k} v_{1,m} = \sum_{l=-\infty}^{\infty} h_l v_{1,2k-l}, \end{aligned} \tag{480b}$$

where we have used Equation (479c) with $j = 1$.

In a similar manner, we can decompose $s_{j-1}(\cdot) \in V_{j-1} = V_j \oplus W_j$ into components in V_j and W_j to obtain

$$s_{j-1}(t) = s_j(t) + d_j(t) \text{ with } d_j(t) \equiv \sum_{k=-\infty}^{\infty} w_{j,k} \psi_{j,k}(t), \tag{480c}$$

where the wavelet coefficients for level j are given by

$$w_{j,k} \equiv \langle s_{j-1}(\cdot), \psi_{j,k}(\cdot) \rangle.$$

A recursive computational expression for $w_{j,k}$ is given by

$$w_{j,k} = \sum_{m=-\infty}^{\infty} \bar{h}_{m-2k} v_{j-1,m} = \sum_{l=-\infty}^{\infty} h_l v_{j-1,2k-l} \tag{480d}$$

(cf. Equation (464b) for $v_{j,k}$).

At step j, we can write

$$\begin{aligned} x(t) &= s_1(t) + d_1(t) \\ &= s_2(t) + d_2(t) + d_1(t) \\ &\ \ \vdots \\ &= s_j(t) + d_j(t) + \cdots + d_1(t), \end{aligned} \tag{481a}$$

an MRA reflecting the fact that

$$V_0 = V_j \oplus W_j \oplus \cdots \oplus W_1.$$

This result leads naturally to noting the important result that projecting $s_{j-1}(\cdot)$ onto V_j is equivalent to projecting $x(\cdot)$ onto V_j and projecting $s_{j-1}(\cdot)$ onto W_j is equivalent to projecting $x(\cdot)$ onto W_j, as the following exercise makes explicit.

▷ **Exercise [481]:** Show that alternative expressions for $v_{2,k}$ and $w_{2,k}$ are given by

$$v_{2,k} = \langle x(\cdot), \phi_{2,k}(\cdot)\rangle \text{ and } w_{2,k} = \langle x(\cdot), \psi_{2,k}(\cdot)\rangle.$$

Generalize this result to show that

$$v_{j,k} = \langle x(\cdot), \phi_{j,k}(\cdot)\rangle \text{ and } w_{j,k} = \langle x(\cdot), \psi_{j,k}(\cdot)\rangle$$ ◁

In terms of the scaling and wavelet coefficients, the expression for $x(\cdot)$ in Equation (481a) can be rewritten explicitly as

$$x(t) = \sum_{k=-\infty}^{\infty} v_{j,k}\phi_{j,k}(t) + \sum_{l=1}^{j}\sum_{k=-\infty}^{\infty} w_{l,k}\psi_{l,k}(t). \tag{481b}$$

The orthonormality of the $\phi_{j,k}(\cdot)$'s and the $\psi_{l,k}(\cdot)$'s in the above expression readily yields the following form of Parseval's theorem:

$$\int_{-\infty}^{\infty} x^2(t)\,dt = \sum_{k=-\infty}^{\infty} v_{j,k}^2 + \sum_{l=1}^{j}\sum_{k=-\infty}^{\infty} w_{l,k}^2. \tag{481c}$$

Suppose now that $x(\cdot) \in V_{j_0} = V_{j_0+1} \oplus W_{j_0+1}$ for some $j_0 \in \mathbb{Z}$; i.e., we do not necessarily assume that $x(\cdot) \in V_0$ as before, so that $x(\cdot)$ is now considered to have scale $\lambda_{j_0} = 2^{j_0}$ rather than unit scale. A parallel development to the above shows that, for any $j > j_0$, we have

$$x(t) = s_j(t)+d_j(t)+\cdots+d_{j_0+1}(t) = \sum_{k=-\infty}^{\infty} v_{j,k}\phi_{j,k}(t) + \sum_{l=j_0+1}^{j}\sum_{k=-\infty}^{\infty} w_{l,k}\psi_{l,k}(t), \tag{481d}$$

while Equation (481c) becomes

$$\int_{-\infty}^{\infty} x^2(t)\,dt = \sum_{k=-\infty}^{\infty} v_{j,k}^2 + \sum_{l=j_0+1}^{j}\sum_{k=-\infty}^{\infty} w_{l,k}^2.$$

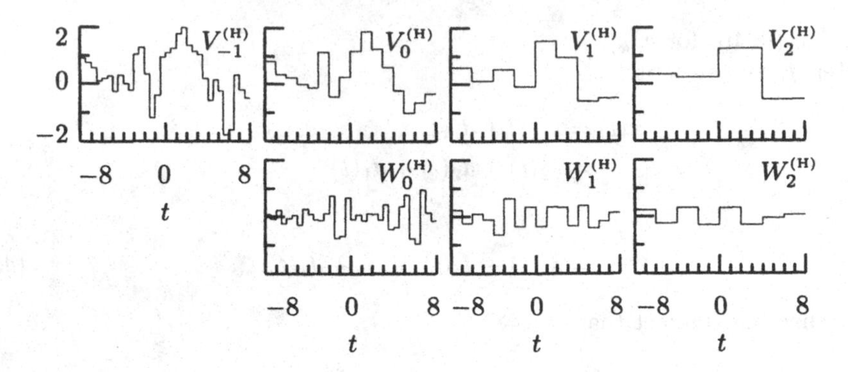

Figure 482. Multiresolution analysis of a function $x(\cdot) \in V_{-1}^{(\mathrm{H})}$ (upper left-hand plot), yielding three approximations, namely, $s_0(\cdot) \in V_0^{(\mathrm{H})}$, $s_1(\cdot) \in V_1^{(\mathrm{H})}$ and $s_2(\cdot) \in V_2^{(\mathrm{H})}$ (remaining plots on top row, from left to right), along with corresponding details $d_0(\cdot) \in W_0^{(\mathrm{H})}$, $d_1(\cdot) \in W_1^{(\mathrm{H})}$ and $d_2(\cdot) \in W_2^{(\mathrm{H})}$ (bottom row, from left to right).

As an example, Figure 482 shows an MRA of a 'half scale' function $x(\cdot) \in V_{-1}^{(\mathrm{H})}$ up to scale $\lambda_2 = 4$.

The sum over l in Equation (481d) is finite because we are considering $x(\cdot) \in V_{j_0} \subset L^2(\mathbb{R})$. Suppose, however, that we know merely that $x(\cdot) \in L^2(\mathbb{R})$.

▷ **Exercise [482]:** Argue that, for any given j,

$$\overline{V_j \oplus W_j \oplus W_{j-1} \cdots} = \overline{\bigcup_{l \in \mathbb{Z}} V_l} = L^2(\mathbb{R}). \qquad \triangleleft$$

From the above we obtain the *inhomogeneous wavelet expansion*

$$x(t) = \sum_{k=-\infty}^{\infty} v_{j,k}\phi_{j,k}(t) + \sum_{l=-\infty}^{j} \sum_{k=-\infty}^{\infty} w_{l,k}\psi_{l,k}(t).$$

We can also argue that

$$\overline{\bigoplus_{l \in \mathbb{Z}} W_l} = L^2(\mathbb{R}),$$

yielding a *homogeneous wavelet expansion* in terms of the wavelet function only:

$$x(t) = \sum_{l=-\infty}^{\infty} \sum_{k=-\infty}^{\infty} w_{l,k}\psi_{l,k}(t).$$

Finally, we note that, given the scaling and wavelet coefficients for level j, we can reconstruct the scaling coefficients for level $j-1$ by making use of Equation (480c) to write

$$\begin{aligned} v_{j-1,k} &= \langle s_{j-1}(\cdot), \phi_{j-1,k}(\cdot)\rangle \\ &= \langle s_j(\cdot) + d_j(\cdot), \phi_{j-1,k}(\cdot)\rangle \end{aligned}$$

$$= \langle s_j(\cdot), \phi_{j-1,k}(\cdot)\rangle + \langle d_j(\cdot), \phi_{j-1,k}(\cdot)\rangle$$
$$= \sum_{m=-\infty}^{\infty} v_{j,m}\langle \phi_{j,m}(\cdot), \phi_{j-1,k}(\cdot)\rangle + \sum_{m=-\infty}^{\infty} w_{j,m}\langle \psi_{j,m}(\cdot), \phi_{j-1,k}(\cdot)\rangle$$
$$= \sum_{m=-\infty}^{\infty} \bar{g}_{k-2m} v_{j,m} + \sum_{m=-\infty}^{\infty} \bar{h}_{k-2m} w_{j,m}, \qquad (483a)$$

where we have evoked the results of Exercises [464] and [479] to obtain the last equality. Note that the first sum in the last line does *not* represent filtering $\{v_{j,m}\}$ with $\{\bar{g}_m\}$ because, when k is even, the sum makes use of just those values of $\{\bar{g}_m\}$ with even indices, whereas, when k is odd, the sum only uses those values with odd indices (a similar comment holds for the sum involving $\{\bar{h}_m\}$).

11.9 Vanishing Moments

Let $\mathcal{N}_m$ be the mth moment of the wavelet function $\psi(\cdot)$:

$$\mathcal{N}_m \equiv \int_{-\infty}^{\infty} t^m \psi(t)\, dt. \qquad (483b)$$

In this section we explore the impact on $\psi(\cdot)$ when r of its moments vanish, by which we mean that $\mathcal{N}_m = 0$ for $m = 0, \ldots, r-1$, where r is a positive integer. The significance of vanishing moments stems from the fact (discussed briefly in item [1] in the Comments and Extensions) that, if $\psi(\cdot)$ and $r-1$ of its derivatives are continuous everywhere and satisfy certain boundedness conditions, then $\psi(\cdot)$ has r vanishing moments. Continuity of $\psi(\cdot)$ and enough of its derivatives helps eliminate artifacts in an analysis of a signal due to the wavelet function itself - loosely speaking, the more vanishing moments that $\psi(\cdot)$ has, the smoother is the transition from one approximation space to the next. Now, only if $\psi(\cdot)$ has r vanishing moments can $\psi(\cdot)$ and $r-1$ of its derivatives be continuous. This was the motivation for Daubechies (1993) to incorporate the vanishing moments requirement into her filter design, as discussed in this section. It is important to note that a wavelet $\psi(\cdot)$ with r vanishing moments is *not* guaranteed to be continuous with $r-1$ continuous derivatives; i.e., 'if and only if' does not hold, just 'only if' (we return to this point in item [2] of the Comments and Extensions).

A vanishing moment condition on $\psi(\cdot)$ imposes conditions on its Fourier transform $\Psi(\cdot)$. If we let $\Psi^{(m)}(\cdot)$ represent the mth derivative of $\Psi(\cdot)$, then we have

$$\Psi^{(m)}(f) = (-i2\pi)^m \int_{-\infty}^{\infty} t^m \psi(t) e^{-i2\pi ft}\, dt,$$

from which we see that $\mathcal{N}_m = 0$ implies that $\Psi^{(m)}(0) = 0$; i.e., a vanishing moment condition on $\psi(\cdot)$ implies that the Fourier transform of $\psi(\cdot)$ and a certain number of its derivatives are zero at zero frequency.

The vanishing moment condition also implies certain conditions on $\overline{H}(\cdot)$, which is the Fourier transform for the sequence $\{\bar{h}_l\}$ used to define $\psi(\cdot)$ in Equation (472). To see this, let us differentiate Equation (476c) (the frequency domain version of Equation (472)) to obtain

$$\Psi^{(1)}(f) = \frac{\Phi^{(1)}(\frac{f}{2})}{2^{3/2}} \overline{H}(\tfrac{f}{2}) + \frac{\Phi(\frac{f}{2})}{2^{3/2}} \overline{H}^{(1)}(\tfrac{f}{2}).$$

Since $\overline{H}(0) = 0$ (see Equation (476b)) and since $\Phi(0) = 1$ (see Equation (468b)), we can see that $\Psi^{(1)}(0) = 0$ implies that $\overline{H}^{(1)}(0) = 0$.

▷ **Exercise [484]:** Generalize the above by showing that $\Psi^{(m)}(0) = 0$ implies $\overline{H}^{(m)}(0) = 0$ also for $m = 0, \ldots, r-1$. ◁

Thus, if $\psi(\cdot)$ has r vanishing moments, then $\overline{H}^{(m)}(0) = 0$ for $m = 0, \ldots, r-1$. Hence r vanishing moments for $\psi(\cdot)$ means that the sequence $\{\bar{h}_l\}$ also has r vanishing discrete moments since

$$\overline{H}^{(m)}(0) = (-i2\pi)^m \sum_{l=-\infty}^{\infty} l^m \bar{h}_l.$$

Now $\{\bar{h}_l\}$ is the QMF corresponding to $\{\bar{g}_l\}$, so conditions on $\overline{H}(\cdot)$ translate into conditions on $\overline{G}(\cdot)$ via Equation (476a). First note that

$$\overline{H}(\tfrac{1}{2}) = -e^{i\pi(L-1)}\overline{G}(0) = \overline{G}(0) = \sqrt{2} \text{ and } \overline{H}(0) = -\overline{G}(\tfrac{1}{2}) = 0$$

(using Equations (466b) and (476b), and recalling that L is even). Differentiating the left- and right-hand sides of Equation (476a) and setting $f = 0$ yields

$$\overline{H}^{(1)}(0) = -i2\pi(L-1)\overline{G}(\tfrac{1}{2}) + \overline{G}^{(1)}(\tfrac{1}{2}) = \overline{G}^{(1)}(\tfrac{1}{2}),$$

so the condition $\overline{H}^{(1)}(0) = 0$ implies that $\overline{G}^{(1)}(\frac{1}{2}) = 0$. Further differentiation reveals that $\overline{H}^{(m)}(0) = 0$ for $m = 0, \ldots, r-1$ implies that $\overline{G}^{(m)}(\frac{1}{2}) = 0$ for $m = 0, \ldots, r-1$ also. Hence, r vanishing moments for $\psi(\cdot)$ also implies $\overline{G}^{(m)}(\frac{1}{2}) = 0$ for $m = 0, \ldots, r-1$.

For a complex-valued variable z, let us define

$$\overline{\overline{G}}(z) \equiv \sum_{l=-\infty}^{\infty} \bar{g}_l z^{-l},$$

which is the so-called z transform of the sequence $\{\bar{g}_l\}$ (Oppenheim and Schafer, 1989, p. 149). Note that, if we let $z = e^{i2\pi f}$, we obtain

$$\overline{\overline{G}}(e^{i2\pi f}) = \sum_{l=-\infty}^{\infty} \bar{g}_l e^{-i2\pi fl} \equiv \overline{G}(f),$$

i.e., the Fourier transform for $\{\bar{g}_l\}$ (a function of the real-valued variable f). Recall that finite support for both $\phi(\cdot)$ and $\psi(\cdot)$ within the interval $(-(L-1), 0]$ corresponds to both $\{\bar{h}_l\}$ and $\{\bar{g}_l\}$ being zero outside of $-(L-1) \leq l \leq 0$. Hence $\overline{\overline{G}}(\cdot)$ is a polynomial of order $L-1$ in nonnegative powers of z, and we can write

$$\overline{\overline{G}}(z) = \sum_{l=-(L-1)}^{0} \bar{g}_l z^{-l} = \bar{g}_{-(L-1)} \prod_{l=1}^{L-1} (z - z_l), \tag{484}$$

where z_l is the lth root of the polynomial $\overline{\overline{G}}(\cdot)$ after dividing by $\bar{g}_{-(L-1)}$ (we can take $\bar{g}_{-(L-1)} \neq 0$ to be an assumption). Since

$$\overline{G}(f) = \overline{\overline{G}}(e^{i2\pi f}) = \bar{g}_{-(L-1)} \prod_{l=1}^{L-1} (e^{i2\pi f} - z_l),$$

we can refer to $\overline{\overline{G}}(\cdot)$ as a *trigonometric polynomial* of order $L-1$ in $e^{i2\pi f}$.

Suppose that $\psi(\cdot)$ has exactly r vanishing moments so that $\overline{G}^{(m)}(\frac{1}{2}) = 0$ for $m = 0, \ldots, r-1$. Since $\overline{G}^{(m)}(\frac{1}{2}) = \overline{\overline{G}}^{(m)}(e^{i\pi}) = \overline{\overline{G}}^{(m)}(-1)$, it follows that $\overline{\overline{G}}^{(m)}(-1) = 0$ also. We claim that this condition on the derivatives of $\overline{\overline{G}}(\cdot)$ implies that r roots of $\overline{\overline{G}}(\cdot)$, say $z_1, \ldots, z_r$, are equal to -1. To see this, we expand $\overline{\overline{G}}(\cdot)$ in a Taylor series about $z = -1$ to obtain (using the fact that $\overline{\overline{G}}^{(l)}(z) = 0$ for all $l \geq L$ because $\overline{\overline{G}}(\cdot)$ is a polynomial of order $L-1$)

$$\overline{\overline{G}}(z) = \sum_{l=0}^{L-1} \frac{(z+1)^l}{l!} \overline{\overline{G}}^{(l)}(-1) = \sum_{l=r}^{L-1} \frac{(z+1)^l}{l!} \overline{\overline{G}}^{(l)}(-1)$$

$$= (z+1)^r \sum_{l=r}^{L-1} \frac{(z+1)^{l-r}}{l!} \overline{\overline{G}}^{(l)}(-1).$$

Equating the above with Equation (484) yields

$$\overline{\overline{G}}(z) = \bar{g}_{-(L-1)}(z+1)^r \prod_{l=r+1}^{L-1} (z - z_l). \tag{485a}$$

Letting $z = e^{i2\pi f}$ gives us the frequency domain equivalent of the above:

$$\overline{G}(f) = \overline{\overline{G}}(e^{i2\pi f}) = \bar{g}_{-(L-1)}(e^{i2\pi f}+1)^r \prod_{l=r+1}^{L-1} (e^{i2\pi f} - z_l) = \left(\frac{e^{i2\pi f}+1}{2}\right)^r Q(f), \tag{485b}$$

where

$$Q(f) \equiv 2^r \bar{g}_{-(L-1)} \prod_{l=r+1}^{L-1} (e^{i2\pi f} - z_l).$$

Note that, since $\overline{G}(0) = \sqrt{2}$, we must also have $Q(0) = \sqrt{2}$. Additionally, since $z_l \neq -1$ for $l = r+1, \ldots, L-1$, we know that $Q(\frac{1}{2}) \neq 0$. It is also easy to see that $Q(\cdot)$ is a trigonometric polynomial of order $L-r-1$ with period unity.

The key to constructing wavelets with finite support is Equation (485b). The squared gain function is defined by $\overline{\mathcal{G}}(f) \equiv |\overline{G}(f)|^2$ so that

$$\overline{\mathcal{G}}(f) = \left|\frac{e^{i2\pi f}+1}{2}\right|^{2r} |Q(f)|^2 = \cos^{2r}(\pi f)\mathcal{Q}(f),$$

where $\mathcal{Q}(f) \equiv |Q(f)|^2$.

▷ **Exercise [485]:** Show that, because the sequence $\{\bar{g}_l\}$ is real-valued, $\overline{\mathcal{G}}(f)$ is a polynomial in $\cos(2\pi f)$, and hence so is $\mathcal{Q}(f)$. ◁

Since $\cos(2\pi f) = 1 - 2\sin^2(\pi f)$, we can write $\mathcal{Q}(f)$ as a polynomial in $\sin^2(\pi f)$, so that

$$\overline{\mathcal{G}}(f) = \cos^{2r}(\pi f)\mathcal{P}(\sin^2(\pi f)),$$

where $\mathcal{P}(\cdot)$ is the polynomial in question. In terms of $\overline{\mathcal{G}}(\cdot)$, the orthogonality condition of Equation (467b) can be written

$$\overline{\mathcal{G}}(f) + \overline{\mathcal{G}}(f + \tfrac{1}{2}) = 2,$$

i.e.,

$$\cos^{2r}(\pi f)\mathcal{P}(\sin^2(\pi f)) + \sin^{2r}(\pi f)\mathcal{P}(\cos^2(\pi f)) = 2$$

or

$$(1-y)^r \mathcal{P}(y) + y^r \mathcal{P}(1-y) = 2,$$

which should hold for all $y \in [0,1]$. Daubechies (1992, Proposition 6.1.2, p. 171) shows that $\mathcal{P}(\cdot)$ must be of the form

$$\mathcal{P}(y) = 2[\mathcal{P}_r(y) + y^r \mathcal{R}(\tfrac{1}{2} - y)],$$

where

$$\mathcal{P}_r(y) \equiv \sum_{k=0}^{r-1} \binom{r-1+k}{k} y^k,$$

and $\mathcal{R}(\cdot)$ is an odd polynomial, chosen so that $\mathcal{P}(y) \geq 0$ for $y \in [0,1]$. With this result, the form of $\mathcal{Q}(\cdot)$ is completely known, but, in order to use Equation (485b) to obtain the filter coefficients $\{\bar{g}_l\}$, we actually need to know $Q(f)$, which is the 'square root' of $\mathcal{Q}(f)$. Extracting the square root of $\mathcal{Q}(f) = \mathcal{P}(\sin^2(\pi f))$ is known in the mathematical literature as *spectral factorization.*

Daubechies (1988) constructed an appealing class of scaling filters by choosing $\mathcal{R}(y) \equiv 0$, in which case $\mathcal{P}(\cdot) \equiv 2\mathcal{P}_r(\cdot)$ and the squared gain function is given by

$$\overline{\mathcal{G}}(f) = 2\cos^{2r}(\pi f) \sum_{k=0}^{r-1} \binom{r-1+k}{k} \sin^{2k}(\pi f), \tag{486}$$

which is identical in form to Equation (105a) with $r = L/2$. In the next section we consider the spectral factorization of the above, with specific details concerning the cases $L = 2, 4, 6$ and 8.

Comments and Extensions to Section 11.9

[1] Let C^n denote the collection of functions $\gamma(\cdot)$ such that $\gamma^{(0)}(\cdot), \gamma^{(1)}(\cdot), \ldots, \gamma^{(n)}(\cdot)$ are continuous everywhere, where $\gamma^{(k)}(\cdot)$ is the kth derivative of $\gamma(\cdot)$ (by definition, $\gamma^{(0)}(\cdot) = \gamma(\cdot)$). Let $\psi(\cdot)$ be a function such that $\{\psi_{j,k}(\cdot) : k \in \mathbb{Z}\}$ constitutes an orthonormal set of function in $L^2(\mathbb{R})$ (this holds if $\psi(\cdot)$ is a wavelet function). Daubechies (1992, Corollary 5.5.2, p. 154) shows that, if $\psi(\cdot) \in C^{r-1}$, if $\psi^{(k)}(\cdot)$ is bounded for $k \leq r-1$ and if

$$|\psi(t)| \leq \frac{c}{(1+|t|)^{r+\epsilon}},$$

where c is a constant and $\epsilon > 0$, then $\psi(\cdot)$ has r vanishing moments (i.e., $\mathcal{N}_m$ of Equation (483b) is zero for $m = 0, \ldots, r-1$).

[2] One measure of smoothness of the scaling and wavelet functions is given by their so-called *regularity*, which we can define in terms of Lipschitz (or Hölder) exponents.

A function $\gamma(\cdot)$ is called Lipschitz of order β, where $0 < \beta \le 1$, if, for any $t, t' \in \mathbb{R}$, we have

$$|\gamma(t) - \gamma(t + t')| < c|t'|^{\beta},$$

where c is a constant (in practice, t' is regarded as being small). Higher orders $\alpha = n+\beta$ are obtained by replacing $\gamma(\cdot)$ by its nth derivative, and this defines Hölder spaces of order α that interpolate between C^n and C^{n+1} (recall that C^n denotes the collection of functions $\gamma(\cdot)$ such that $\gamma(\cdot)$ and its first n derivatives are continuous everywhere). A function in a Hölder spaces of order α is said to be (Hölder) regular of order α.

Rioul (1992) estimated the Hölder regularity of the D(L) $\phi(\cdot)$ and $\psi(\cdot)$ functions for $L = 4$, 6 and 8 as, respectively,

$$\begin{aligned}
\phi(\cdot), \psi(\cdot) &\in C^{0.5500}, \quad \text{i.e., } \phi(\cdot), \psi(\cdot) \in C^0, \notin C^1; \\
\phi(\cdot), \psi(\cdot) &\in C^{1.0878}, \quad \text{i.e., } \phi(\cdot), \psi(\cdot) \in C^1, \notin C^2; \\
\phi(\cdot), \psi(\cdot) &\in C^{1.6179}, \quad \text{i.e., } \phi(\cdot), \psi(\cdot) \in C^1, \notin C^2,
\end{aligned}$$

where, for example, the D(4) $\phi(\cdot)$ and $\psi(\cdot)$ functions are continuous and hence in C^0, but they are not in C^1 (i.e., they do not have continuous derivatives). Note that, since $\psi(\cdot)$ is a finite linear combination of translates of the function defined by $\phi(2t)$, it is reasonable that both $\phi(\cdot)$ and $\psi(\cdot)$ have the same regularity properties.

For the D(L) class with large values of L, Daubechies (1992) finds that, approximately,

$$\phi(\cdot), \psi(\cdot) \in C^{L/10}, \quad \text{i.e., } \phi(\cdot), \psi(\cdot) \in C^{r/5},$$

so that approximately 80% of the zero moments are 'wasted;' i.e., the same regularity could in theory be achieved with only $L/10$ vanishing moments. This illustrates that $\psi(\cdot)$ having r vanishing moments does not guarantee that $\psi(\cdot) \in C^{r-1}$.

[3] Recall that r vanishing moments for $\psi(\cdot)$ implies that

$$\overline{G}^{(m)}(\tfrac{1}{2}) = \overline{G}^{(m)}(-\tfrac{1}{2}) = 0, \quad m = 0, \ldots, r-1;$$

since $\overline{G}(f) = G(-f)$, we also have $G^{(m)}(\frac{1}{2}) = 0$. The gain function $|G(\cdot)|$ for $\{g_l\}$ thus gets flatter $f = 1/2$ as r increases. Daubechies (1992, p. 245–7) points out that, in a cascade of filters involving multiple uses of $G(\cdot)$ interleaved with downsampling (such as is used in the DWT), the gain function for the equivalent filter can have ripples at high frequencies unless $|G(\cdot)|$ is sufficiently flat at $f = 1/2$.

11.10 Spectral Factorization and Filter Coefficients

In preparation for the factorization of $\overline{\mathcal{G}}(\cdot)$ in Equation (486), we first need to review some results on the delay, or phase, properties of filters with real-valued coefficients (i.e., a real-valued impulse response sequence). Our starting point is Equation (485a) with $r = L/2$. Let

$$\overline{\overline{G}}(z) = \overline{g}_{-(L-1)}(z+1)^{L/2}\overline{\overline{Q}}_0(z), \tag{487}$$

for which we define

$$\overline{\overline{Q}}_0(z) \equiv \prod_{l'=\frac{L}{2}+1}^{L-1} (z - z_{l'}) = \prod_{l=1}^{\frac{L}{2}-1} (z - z_l);$$

here the right-hand equality means nothing more than that we are reindexing the roots for convenience. Note that $\overline{\overline{Q}}_0(z)$ is a polynomial of order $\frac{L}{2}-1$ in nonnegative powers of z; that the coefficient of $z^{\frac{L}{2}-1}$ is unity; and that all of its $\frac{L}{2}$ coefficients must be real-valued (since $\overline{\overline{G}}(z)$ is the z transform of a real-valued filter). With $z = e^{i2\pi f}$, the above becomes a transfer function, so we can regard the above as the z transform domain representation for a filter of width $L/2$. Since convolving filters of width two is equivalent to multiplying their z transforms, an interpretation of the filter with the above z transform is that it is the result of convolving $\frac{L}{2}-1$ filters, each of width two. A filter of width two is often called a 'dipole,' a terminology we adopt henceforth. The lth such dipole has a z transform of $z - z_l \equiv z + b_l$, a transfer function of $e^{i2\pi f} + b_l$ and an impulse response sequence that is equal to unity at $t = -1$, to $b_l \equiv -z_l$ at $t = 0$, and to zero otherwise. Let us write the impulse response sequence for this dipole as $\{1, \boldsymbol{b_l}\}$, where we indicate the coefficient with index $t = 0$ using a bold type. Note that, in order to get a filter of width $L/2$ with real-valued coefficients, we can have either dipoles $\{1, \boldsymbol{b_l}\}$ with b_l real or pairs of dipoles with coefficients that are complex conjugates, say, $\{1, \boldsymbol{b_m}\}$ and $\{1, \boldsymbol{b_n}\}$ with $b_n = b_m^*$.

Now let us compare the magnitudes of the two nonzero coefficients in the lth dipole; i.e., we want to compare $|b_l| = |z_l|$ to unity. Suppose that $|b_l| > 1$, in which case the root z_l lies outside the unit circle in the complex plane. Such a dipole is called a *maximum delay* or *maximum phase* dipole because, of the two nonzero coefficients, the one having the larger magnitude occurs later in time. If all $\frac{L}{2} - 1$ roots z_l are outside the unit circle, then the width $L/2$ real-valued filter is the convolution of $\frac{L}{2}-1$ maximum delay dipoles, and the resulting filter is described as a maximum delay or maximum phase filter (see e.g., Robinson, 1980; Robinson and Treitel, 1980). The build-up of its partial energy sequence (discussed in Section 4.8) is the slowest of any filter of the same width with the same squared gain function.

Let us next consider the dipole $\{\boldsymbol{b_l}, 1\}$; i.e., its impulse response sequence is equal to b_l at $t = 0$, to unity at $t = 1$, and to zero otherwise (this amounts to time-reversing $\{1, \boldsymbol{b_l}\}$). If we still have $|b_l| > 1$, then, of the two nonzero coefficients, the one having the larger magnitude now occurs earlier in time. The z transform for $\{\boldsymbol{b_l}, 1\}$ is $b_l + z^{-1}$, which has a root $-1/b_1 = 1/z_l$ that lies inside the unit circle. Such a dipole is called a *minimum delay* or *minimum phase* dipole. If $\{\boldsymbol{b_l}, 1\}$, $l = 1, \ldots, \frac{L}{2}-1$, are all minimum delay dipoles, then the width $L/2$ real-valued filter that we obtain by convolving them together is described as a minimum delay or minimum phase filter. The build-up of its partial energy sequence is the fastest of any filter of width $L/2$ with the same squared gain function.

Now the squared gain function for the filter whose z transform is $\overline{\overline{Q}}_0(z)$ is obtained by evaluating $|\overline{\overline{Q}}_0(z)|^2$ at $z = e^{i2\pi f}$.

▷ **Exercise [488]:** Show that, when $z = e^{i2\pi f}$, we have

$$\left[\overline{\overline{Q}}_0(z)\right]^* = \overline{\overline{Q}}_0(z^{-1}) \text{ and hence } |\overline{\overline{Q}}_0(z)|^2 = \overline{\overline{Q}}_0(z)\overline{\overline{Q}}_0(z^{-1}).$$ ◁

Now $\overline{\overline{Q}}_0(z)$ is a product of dipole z transform terms. A dipole term $\{1, \boldsymbol{b_l}\}$ with real-valued coefficients will contribute $(z + b_l)(z^{-1} + b_l)$ to $|\overline{\overline{Q}}_0(z)|^2$. Note that with $|b_l| > 1$ this corresponds to a maximum delay dipole $\{1, \boldsymbol{b_l}\}$ convolved with a minimum delay dipole $\{\boldsymbol{b_l}, 1\}$: the result is simply the autocorrelation of the maximum delay dipole, since it is the convolution of the filter with its time reverse. A pair of dipoles with coefficients that are complex conjugates, $\{1, b_m\}$ and $\{1, b_m^*\}$ will contribute $(z + b_m)(z + b_m^*)(z^{-1} + b_m)(z^{-1} + b_m^*)$ to $|\overline{\overline{Q}}_0(z)|^2$. Here $(z + b_m^*)(z^{-1} + b_m)$ is the z transform of the autocorrelation of the dipole $\{1, b_m\}$ since the autocorrelation is the convolution of the complex conjugate of $\{1, b_m\}$ with its time reverse. With $|b_m| > 1$ this corresponds to a maximum delay dipole convolved with a minimum delay dipole; also, $(z + b_m)(z^{-1} + b_m^*)$ is the z transform of the autocorrelation of $\{1, b_m^*\}$, and again the autocorrelation corresponds to a maximum delay dipole convolved with a minimum delay dipole.

As will be made explicit in the examples that follow, starting with $|\overline{\overline{Q}}_0(z)|^2$, there are many ways to derive $\overline{\overline{Q}}_0(z)$: we could take all maximum delay dipoles, all minimum delay dipoles, or we could take mixtures of minimum and maximum delay dipoles having real-valued coefficients, along with pairs of dipoles with complex-valued coefficients where both terms are either minimum delay or maximum delay. Starting with a set of $\frac{L}{2} - 1$ dipoles with real-valued coefficients, there are $2^{\frac{L}{2}-1}$ real-valued filters of width $L/2$ having the same magnitude squared z transform (or squared gain function), but different phase functions. If K_0 dipoles with real-valued coefficients and K_1 complex conjugate pairs of dipoles are convolved, with $K_0 + 2K_1 = \frac{L}{2} - 1$, there are $2^{K_0+K_1}$ real-valued filters of width $L/2$ having the same magnitude squared z transform (or squared gain function), but different phase functions.

- *One vanishing moment and the Haar scaling filter*

When $L = 2$ so that there is $r = L/2 = 1$ vanishing moment, we have $\mathcal{P}(\sin^2(\pi f)) \equiv 2\mathcal{P}_1(\sin^2(\pi f)) = 2$. Hence,

$$\mathcal{G}(f) = |G(f)|^2 = 2\cos^2(\pi f) = 2[\tfrac{1}{2} + \tfrac{1}{2}\cos(2\pi f)] = 2[\tfrac{1}{2} + \tfrac{1}{4}(e^{-i2\pi f} + e^{i2\pi f})],$$

so that

$$|\overline{\overline{G}}(z)|^2 = 2[\tfrac{1}{4}(z^{-1} + 2 + z)] = \tfrac{1}{2}(z+1)(z^{-1}+1).$$

Taking $\overline{\overline{G}}(z) = \sum_{l=-1}^{0} \bar{g}_l z^{-l} = \bar{g}_{-1} z + \bar{g}_0 = \pm(z+1)/\sqrt{2}$, we note that, as expected, this has a single ($r = 1$) root at $z = -1$ and that $\overline{\overline{G}}(z)\overline{\overline{G}}(z^{-1}) = |\overline{\overline{G}}(z)|^2$. We choose the plus sign in order that $\overline{\overline{G}}(1) = \sqrt{2}$ (i.e., $\overline{G}(0) = \sqrt{2}$). The corresponding filter coefficients are $\bar{g}_{-1} = \bar{g}_0 = 1/\sqrt{2}$. The Haar scaling filter is the filter corresponding to a time reversal, i.e., $\{g_l = \bar{g}_{-l}\}$, so that $g_0 = g_1 = 1/\sqrt{2}$, as in Equation (463d).

- *Two vanishing moments and the D(4) scaling filter*

When $L = 4$ (i.e., $r = 2$ vanishing moments), we have $\mathcal{P}(\sin^2(\pi f)) \equiv 2\mathcal{P}_2(\sin^2(\pi f)) = 2 + 4\sin^2(\pi f)$. Hence,

$$\mathcal{G}(f) = \cos^4(\pi f)[2 + 4\sin^2(\pi f)] = [\tfrac{1}{2} + \tfrac{1}{2}\cos(2\pi f)]^2[4 - 2\cos(2\pi f)],$$

so that

$$|\overline{\overline{G}}(z)|^2 = \tfrac{1}{16}(z^{-1} + 2 + z)^2(-z^{-1} + 4 - z).$$

We know from the case of one vanishing moment that $(z^{-1} + 2 + z)^2$ can be factored as $(z+1)^2(z^{-1}+1)^2$. We need only additionally to factor $-z^{-1} + 4 - z$. If we multiply through by $-z$, we get $z^2 - 4z + 1$, which has real-valued reciprocal roots $2 + \sqrt{3} \equiv z_1$ and $2 - \sqrt{3} = z_1^{-1}$; i.e.,

$$-z^{-1} + 4 - z = \frac{z^2 - 4z + 1}{-z} = \frac{(z - z_1)(z - z_1^{-1})}{-z} = z_1^{-1}(z - z_1)(z^{-1} - z_1).$$

Letting $b_1 = -z_1$ and noting that $|b_1| > 1$, we thus have

$$|\overline{\overline{G}}(z)|^2 = C^2(z+1)^2(z^{-1}+1)^2(z+b_1)(z^{-1}+b_1),$$

where $C^2 \equiv z_1^{-1}/16$ so that $C = \pm\frac{1-\sqrt{3}}{4\sqrt{2}}$. If we proceed as in the $r = 1$ case and use the term involving $z+1$ in the construction of $\overline{\overline{G}}(z)$ and if we pick $C < 0$, we have two possible definitions for $\overline{\overline{G}}(z)$, namely, either

$$\overline{\overline{G}}(z) \equiv C(z+1)^2(z+b_1)$$

so that

$$\overline{\overline{G}}(z^{-1}) = C(z^{-1}+1)^2(z^{-1}+b_1)$$

or

$$\overline{\overline{G}}(z) \equiv Cz(z+1)^2(z^{-1}+b_1) = b_1C(z+1)^2(z+b_1^{-1}) \tag{490a}$$

so that

$$\overline{\overline{G}}(z^{-1}) = Cz^{-1}(z^{-1}+1)^2(z+b_1),$$

both of which lead to the required properties that $\overline{\overline{G}}(z)$ is a polynomial in nonnegative powers of z, that $\overline{\overline{G}}(z)\overline{\overline{G}}(z^{-1}) = |\overline{\overline{G}}(z)|^2$ and that $\overline{\overline{G}}(1) = \sqrt{2}$ (note also that either definition for $\overline{\overline{G}}(z)$ has a double root at $z = -1$, as expected). The first definition makes use of the maximum delay dipole $\{1, b_1\}$, and the second, the minimum delay dipole $\{b_1, 1\}$. If we compare the first definition with Equation (487), we note that $\overline{Q}_0(z) = z + b_1 = z - 2 - \sqrt{3}$. Expanding this first definition, we obtain

$$\begin{aligned}\overline{\overline{G}}(z) &= \frac{1-\sqrt{3}}{4\sqrt{2}}\left(z^3 + [2+b_1]z^2 + (1+2b_1)z + b_1\right) \\ &= \frac{1-\sqrt{3}}{4\sqrt{2}}z^3 + \frac{3-\sqrt{3}}{4\sqrt{2}}z^2 + \frac{3+\sqrt{3}}{4\sqrt{2}}z + \frac{1+\sqrt{3}}{4\sqrt{2}},\end{aligned} \tag{490b}$$

so that the coefficients are given by

$$\{\overline{g}_{-3}, \overline{g}_{-2}, \overline{g}_{-1}, \overline{g}_0\} = \left\{\frac{1-\sqrt{3}}{4\sqrt{2}}, \frac{3-\sqrt{3}}{4\sqrt{2}}, \frac{3+\sqrt{3}}{4\sqrt{2}}, \frac{1+\sqrt{3}}{4\sqrt{2}}\right\}.$$

The coefficients of the filter $\{g_l\}$ follow by a time reversal, i.e.,

$$\{g_0, g_1, g_2, g_3\} = \left\{\frac{1+\sqrt{3}}{4\sqrt{2}}, \frac{3+\sqrt{3}}{4\sqrt{2}}, \frac{3-\sqrt{3}}{4\sqrt{2}}, \frac{1-\sqrt{3}}{4\sqrt{2}}\right\},$$

which is the D(4) scaling filter as stated in Equation (75d).

▷ **Exercise [490]:** Show that the second definition for $\overline{\overline{G}}(z)$ (i.e., the one in Equation (490a) involving a minimum delay dipole) leads to

$$\{g_0, g_1, g_2, g_3\} = \left\{\frac{1-\sqrt{3}}{4\sqrt{2}}, \frac{3-\sqrt{3}}{4\sqrt{2}}, \frac{3+\sqrt{3}}{4\sqrt{2}}, \frac{1+\sqrt{3}}{4\sqrt{2}}\right\},$$

the D(4) filter with coefficients in reversed order. ◁

• *Three vanishing moments and the D(6) scaling filter*
When $L = 6$ so that $r = 3$, we have

$$\mathcal{P}(\sin^2(\pi f)) \equiv 2\mathcal{P}_3(\sin^2(\pi f)) = 2 + 6\sin^2(\pi f) + 12\sin^4(\pi f).$$

Hence,

$$\begin{aligned}\mathcal{G}(f) &= \cos^6(\pi f)\left[2 + 6\sin^2(\pi f) + 12\sin^4(\pi f)\right]\\ &= \left[\tfrac{1}{2} + \tfrac{1}{2}\cos(2\pi f)\right]^3\left[8 - 9\cos(2\pi f) + 3\cos^2(2\pi f)\right].\end{aligned}$$

Substituting $(z + z^{-1})/2$ for $\cos(2\pi f)$ in the above yields

$$|\overline{\overline{G}}(z)|^2 = \tfrac{1}{64}(z+1)^3(z^{-1}+1)^3\tfrac{1}{4}(3z^{-2} - 18z^{-1} + 38 - 18z + 3z^2).$$

We can use the S-PLUS function **polyroot** to find that

$$1 - 6z + \tfrac{38}{3}z^2 - 6z^3 + z^4 = (z - z_1)(z - z_1^*)(z - z_2)(z - z_2^*),$$

where

$$z_1 \doteq 2.7127486219559795 + 1.4438867826180040i \tag{491}$$

and

$$z_2 \doteq 0.2872513780440209 + 0.1528923338821992i \doteq \frac{1}{z_1^*} = \frac{z_1}{|z_1|^2}.$$

We can thus write

$$|\overline{\overline{G}}(z)|^2 = \frac{3}{256|z_1|^2}(z+1)^3(z^{-1}+1)^3(z - z_1)(z - z_1^*)(z^{-1} - z_1)(z^{-1} - z_1^*),$$

where we note that, because $|z_1| > 1$, the terms $z - z_1$ and $z - z_1^*$ correspond to the maximum delay dipoles $\{1, -z_1\}$ and $\{1, -z_1^*\}$. If we again use the terms involving $z + 1$ and the maximum delay dipoles to construct $\overline{\overline{G}}(z)$, we obtain

$$\overline{\overline{G}}(z) = \frac{\sqrt{3}}{16|z_1|}(z+1)^3(z - z_1)(z - z_1^*),$$

which is a polynomial in nonnegative powers of z satisfying the required conditions $\overline{\overline{G}}(z)\overline{\overline{G}}(z^{-1}) = |\overline{\overline{G}}(z)|^2$ and $\overline{\overline{G}}(1) = \sqrt{2}$. If we compare the above with Equation (487), we note that $\overline{\overline{Q}}_0(z) = (z - z_1)(z - z_1^*)$. Expanding the above and equating the result to $\bar{g}_{-5}z^5 + \bar{g}_{-4}z^4 + \cdots + \bar{g}_0$ yields

$$\begin{bmatrix}\bar{g}_{-5}\\ \bar{g}_{-4}\\ \bar{g}_{-3}\\ \bar{g}_{-2}\\ \bar{g}_{-1}\\ \bar{g}_0\end{bmatrix} = \frac{\sqrt{3}}{16|z_1|}\begin{bmatrix}1\\ 3 - 2r_1\\ 3 - 6r_1 + |z_1|^2\\ 1 - 6r_1 + 3|z_1|^2\\ -2r_1 + 3|z_1|^2\\ |z_1|^2\end{bmatrix},$$

where r_1 is the real component of z_1. Recalling that $g_l = \bar{g}_{-l}$, computations give us the values for the D(6) scaling filter as set out in Table 109.

- *General form for D(L) filters*

For general $L \geq 2$, we can use Equation (487) and Exercise [488] to write (assuming $z = e^{i2\pi f}$)

$$|\overline{\overline{G}}(z)|^2 = \bar{g}^2_{-(L-1)}|z+1|^L|\overline{\overline{Q}}_0(z)|^2 = \bar{g}^2_{-(L-1)}|z+1|^L\overline{\overline{Q}}_0(z)\overline{\overline{Q}}_0(z^{-1}),$$

where $\overline{\overline{Q}}_0(z)$ is a polynomial of order $\frac{L}{2}-1$ in nonnegative powers of z with real-valued coefficients. For the case $L=4$, we have $\overline{\overline{Q}}_0(z) = z - 2 - \sqrt{3} \equiv z - r_1$, where r_1 is a real-valued root; for the case $L=6$, we have $\overline{\overline{Q}}_0(z) = (z-z_1)(z-z_1^*)$, where z_1 is the complex-valued root of Equation (491). For general $L \geq 4$, we can always write

$$|\overline{\overline{Q}}_0(z)|^2 = \prod_{l=1}^{K_0}(z-r_l)(z^{-1}-r_l)\prod_{l=1}^{K_1}(z-z_l)(z-z_l^*)(z^{-1}-z_l)(z^{-1}-z_l^*),$$

where $K_0+2K_1 = \frac{L}{2}-1$; $r_1,\ldots,r_{K_0}$ are real-valued roots (none of these can be zero or minus one); and $z_1,\ldots,z_{K_1}$ are complex-valued roots (Daubechies, 1992, p. 173; Chui, 1997, p. 81). Whereas the representation for $|\overline{\overline{G}}(z)|^2$ is unique because $|\overline{\overline{G}}(e^{i2\pi f})|^2 = |\overline{G}(f)|^2 = \overline{\mathcal{G}}(f)$, where $\overline{\mathcal{G}}(f)$ is given by Equation (486), the representation for $|\overline{\overline{Q}}_0(z)|^2$ is not unique. In particular, if we invert any of the real-valued roots r_l or the complex-valued pair of roots z_l, z_l^*, then $|\overline{\overline{Q}}_0(z)|^2$ changes. For example, since

$$(z-r_l)(z^{-1}-r_l) = r_l^2(z-r_l^{-1})(z^{-1}-r_l^{-1}),$$

we can write

$$|\overline{\overline{Q}}_0(z)|^2 = |\overline{\overline{Q}}_1(z)|^2\left(\prod_{l=1}^{K_0} r_l^2\right),$$

where

$$|\overline{\overline{Q}}_1(z)|^2 \equiv \prod_{l=1}^{K_0}(z-r_l^{-1})(z^{-1}-r_l^{-1})\prod_{l=1}^{K_1}(z-z_l)(z-z_l^*)(z^{-1}-z_l)(z^{-1}-z_l^*),$$

from which we obtain

$$|\overline{\overline{G}}(z)|^2 = \left(\bar{g}^2_{-(L-1)}\prod_{l=1}^{K_0} r_l^2\right)|z+1|^L|\overline{\overline{Q}}_1(z)|^2,$$

where now $|\overline{\overline{Q}}_1(z)|^2$ plays the role of the original $|\overline{\overline{Q}}_0(z)|^2$. Hence there are $2^{K_0+K_1}$ possible versions of $\overline{\overline{Q}}_0(z)$ yielding the same squared gain function, but different phase functions.

As is by now clear, the choice for $\{\bar{g}_l\}$ that gives the Daubechies D(L) filters $\{g_l\}$ is to take

$$\overline{\overline{Q}}_0(z) = \prod_{l=1}^{K_0}(z-r_l)\prod_{l=1}^{K_1}(z-z_l)(z-z_l^*),$$

where $|r_1|, \ldots, |r_{K_0}|, |z_1|, \ldots, |z_{K_1}|$ are all greater than unity; in other words, $\overline{\overline{Q}}_0(z)$ is constructed from maximum delay dipoles, as was done for the $L = 4$ and 6 examples. The filter $\{\bar{g}_l, l = -(L-1), \ldots, 0\}$ is thus made up of the maximum delay dipoles in $\overline{\overline{Q}}_0(z)$ convolved with $L/2$ dipoles of the form $\{1, \mathbf{1}\}$. Hence the filter of width L (with $L/2$ dipoles of the form $\{1, \mathbf{1}\}$) and gain function $|\overline{G}(\cdot)|$, namely $\{\bar{g}_l\}$, has the slowest possible build-up of its partial energy sequence, so $\{\bar{g}_l\}$ is a maximum delay filter. On the other hand, the filter of width L (again with $L/2$ dipoles of the form $\{1, \mathbf{1}\}$) and gain function $|\overline{G}(\cdot)|$ that is the time reverse of $\{\bar{g}_l\}$, i.e., a Daubechies D(L) filter $\{g_l, l = 0, \ldots, L-1\}$, has the fastest possible build-up of its partial energy sequence and is hence a minimum delay filter.

The phase of a minimum delay filter departs maximally from zero, so that the minimum delay or minimum phase design is sometimes called *extremal phase* in the wavelet literature (e.g., Daubechies, 1992, p. 255). Since a minimum phase filter is necessarily 'front-loaded,' the resulting scaling function is very asymmetric (the only exception is the $L = 2$ (Haar) case, for which the scaling function is symmetric about $t = -\frac{1}{2}$).

- *Four vanishing moments and the LA(8) scaling filter*

It was pointed out in Section 4.8 that the least asymmetric (LA) family of Daubechies scaling filters arises by choosing a different spectral factorization; i.e., rather than choosing terms for $\overline{\overline{Q}}_0(z)$ corresponding to maximum delay dipoles as is done for the D(L) filters, the roots are chosen so that the expression in Equation (112a) is as small as possible. For $L = 8$,

$$\mathcal{P}(\sin^2(\pi f)) = 2\mathcal{P}_4(\sin^2(\pi f)) = 2 + 8\sin^2(\pi f) + 20\sin^4(\pi f) + 40\sin^6(\pi f),$$

from which we obtain

$$|\overline{\overline{Q}}_0(z)|^2 \propto z^{-3} - 8z^{-2} + 26.2z^{-1} - 41.6 + 26.2z - 8z^2 + z^3 \tag{493a}$$

(verification of the above is the first part of Exercise [11.5]). Using the S-PLUS routine **polyroot**, we find that we can write

$$|\overline{\overline{Q}}_0(z)|^2 = (z - r_1)(z^{-1} - r_1)(z - z_1)(z - z_1^*)(z^{-1} - z_1)(z^{-1} - z_1^*)$$

where

$$r_1 \doteq 3.0406604616474480 \text{ and } z_1 \doteq 2.0311355120914390 + 1.738950807644819i.$$

Because $|r_1| > 1$ and $|z_1| > 1$, the terms $z - r_1$, $z - z_1$ and $z - z_1^*$ in Equation (493b) correspond to maximum delay dipoles. Taking

$$\overline{G}(z) = -\left[\frac{5}{2^{11}} r_1^{-1} |z_1|^{-2}\right]^{1/2} (z+1)^4 (z - r_1)(z - z_1)(z - z_1^*) \tag{493b}$$

yields a filter $\{\bar{g}_l\}$ that, after time reversal, gives the Daubechies 'minimum delay' D(8) filter $\{g_l\}$ in Table 109 (see the second part of Exercise [11.5]). For this choice,

$$\overline{\overline{Q}}_0(z) = (z - r_1)(z - z_1)(z - z_1^*).$$

However, there are four possible ways of constructing $\overline{\overline{G}}(z)$, corresponding to roots $\{r_1, z_1, z_1^*\}$ as above, $\{1/r_1, z_1, z_1^*\}$, $\{r_1, 1/z_1, 1/z_1^*\}$ or $\{1/r_1, 1/z_1, 1/z_1^*\}$. The second of these gives

$$\overline{\overline{G}}(z) = \left[\frac{5}{2^{11}} r_1 |z_1|^{-2}\right]^{1/2} (z+1)^4 (z - r_1^{-1})(z - z_1)(z - z_1^*),$$

and the resulting filter $\{\bar{g}_l\}$, after time reversal, gives the Daubechies 'mixed delay' LA(8) filter $\{g_l\}$ in Table 109. In this case,

$$\overline{\overline{Q}}_0(z) = (z - r_1^{-1})(z - z_1)(z - z_1^*).$$

(The third and fourth representations give larger values for the expression in Equation (112a) and so are not chosen for the LA(8) filter.)

Comments and Extensions to Section 11.10

[1] The dipoles that we have considered here all have one coefficient that is equal to unity, so we could state the maximum/minimum delay conditions in terms of comparing the magnitude of the other coefficient to unity. The most general form for a dipole is $\{a_l, b_l\}$, where a_l has an index occurring before the index for b_l, and both a_l and b_l can be complex-valued. In this case, the dipole is said to have maximum delay if $|a_l| < |b_l|$, and minimum delay if $|a_l| > |b_l|$.

[2] Hölder regularity for Daubechies' least asymmetric, LA(L), scaling and wavelet functions is lower than for the minimum delay, D(L), scaling and wavelet functions; see Rioul (1992, Table 1) for details. Thus, perhaps surprisingly, given the same gain function, a greater degree of symmetry decreases regularity.

11.11 Summary

In this chapter we investigate connections between continuous time and discrete time wavelet analyses. Continuous time analysis deals with functions $\gamma(\cdot)$ (also called signals) whose domain of definition is the entire real axis $\mathbb{R}$. For simplicity we assume that $\gamma(\cdot)$ is real-valued and that it belongs to $L^2(\mathbb{R})$, the space of all square integrable functions defined over $\mathbb{R}$. We define the jth level dilation of the kth level translation of $\gamma(\cdot)$ by

$$\gamma_{j,k}(t) \equiv \frac{\gamma(\frac{t}{2^j} - k)}{\sqrt{2^j}}, \quad -\infty < t < \infty,$$

where $j, k \in \mathbb{Z}$ (the set of all integers).

The key concept in the wavelet analysis of a signal is that of a multiresolution analysis (MRA), which by definition is a sequence of closed subspaces $V_j \subset L^2(\mathbb{R}), j \in \mathbb{Z}$, obeying the following properties. The subspaces must be nested such that

$$\cdots \subset V_3 \subset V_2 \subset V_1 \subset V_0 \subset V_{-1} \subset \cdots,$$

and their closed union and intersection must satisfy

$$\overline{\bigcup_{j\in\mathbb{Z}} V_j} = L^2(\mathbb{R}) \text{ and } \bigcap_{j\in\mathbb{Z}} V_j = \{0\},$$

where $\{0\}$ is the set containing just the null function (i.e., the function that is zero for all $t \in \mathbb{R}$). The space V_0 is such that $\gamma(\cdot) \in V_0$ if and only if $\gamma_{0,k}(\cdot) = \gamma(t-k) \in V_0$, $k \in \mathbb{Z}$; additionally, $\gamma(\cdot) \in V_0$ if and only if $\gamma_{j,0}(\cdot) \in V_j$, $j \in \mathbb{Z}$. Finally, there must exist a function $\phi(\cdot) \in L^2(\mathbb{R})$ such that its integer translates $\{\phi_{0,k}(\cdot) : k \in \mathbb{Z}\}$ form an orthonormal basis for the closed subspace V_0 (this implies that $\{\phi_{j,k}(\cdot) : k \in \mathbb{Z}\}$ forms an orthonormal basis for V_j). The function $\phi(\cdot)$ is called a scaling function.

The subspace V_j in an MRA is called an approximation space, and it is associated with scale $\lambda_j = 2^j$. Given an MRA, a signal $x(\cdot) \in V_0$ can be represented as

$$x(t) = \sum_{k=-\infty}^{\infty} \langle x(\cdot), \phi_{0,k}(\cdot)\rangle \phi_{0,k}(t) = \sum_{k=-\infty}^{\infty} v_{0,k}\phi_{0,k}(t),$$

where

$$v_{0,k} \equiv \langle x(\cdot), \phi_{0,k}(\cdot)\rangle = \int_{-\infty}^{\infty} x(t)\phi_{0,k}(t)\,dt.$$

We can obtain a coarser scale approximation to $x(\cdot)$ by projecting it onto the subspace $V_1 \subset V_0$ to obtain

$$s_1(t) = \sum_{k=-\infty}^{\infty} v_{1,k}\phi_{1,k}(t),$$

where $v_{1,k} \equiv \langle x(\cdot), \phi_{1,k}(\cdot)\rangle$. The function $s_1(\cdot) \in V_1$ is a scale two coarse approximation to the unit scale signal $x(\cdot)$. The $v_{1,k}$'s are scale two scaling coefficients. The relationship between the $v_{0,k}$'s and the $v_{1,k}$'s is given by

$$v_{1,k} = \sum_{m=-\infty}^{\infty} \bar{g}_{m-2k} v_{0,m} = \sum_{l=-\infty}^{\infty} g_l v_{0,2k-l},$$

where $\bar{g}_l \equiv \langle \phi_{0,l}(\cdot), \phi_{1,0}(\cdot)\rangle$, and $g_l \equiv \bar{g}_{-l}$ (here $\{g_l\}$ is the same as the scaling filter introduced in Section 4.3, so $\{\bar{g}_l\}$ is just its time reverse). This scheme can be generalized: the function $s_{j-1}(\cdot) \in V_{j-1}$, can be projected onto the subspace $V_j \subset V_{j-1}$ to obtain

$$s_j(t) \equiv \sum_{k=-\infty}^{\infty} v_{j,k}\phi_{j,k}(t)$$

with $v_{j,k} \equiv \langle s_{j-1}(\cdot), \phi_{j,k}(\cdot)\rangle = \langle x(\cdot), \phi_{j,k}(\cdot)\rangle$ being scaling coefficients of scale $\lambda_j = 2^j$. Here $s_j(\cdot)$ is a scale λ_j approximation to $x(\cdot)$ (it can also be regarded as an approximation to $s_{j-1}(\cdot)$). The $v_{j,k}$'s may be computed recursively via

$$v_{j,k} = \sum_{m=-\infty}^{\infty} \bar{g}_{m-2k} v_{j-1,m} = \sum_{l=-\infty}^{\infty} g_l v_{j-1,2k-l}.$$

The nesting of the multiresolution subspaces gives rise to the two-scale difference equation of Equation (465a), namely,

$$\phi(t) = \sqrt{2} \sum_{l=-\infty}^{\infty} \bar{g}_l \phi(2t-l).$$

If we assume that $\{\bar{g}_l\} \equiv \{g_{-l}\}$ is zero for l outside the range $-(L-1), \ldots, 0$, then $\phi(\cdot)$ is zero outside the interval $(-(L-1), 0]$, in which case $\phi(\cdot)$ is said to have finite support. We write this fact as support$\{\phi(\cdot)\} \subset (-(L-1), 0]$ (we use '$\subset$' here rather than '=' merely because $\phi(\cdot)$ might or might not be equal to zero at $t = 0$). The two-scale difference equation can be used to compute values of the scaling function $\phi(\cdot)$ using a recursive approach (this is detailed in Section 11.5). Application of the Fourier transform to both sides of the two-scale difference equation yields relationships between $\Phi(\cdot)$ and $\overline{G}(\cdot)$, namely,

$$\Phi(f) = \Phi(\tfrac{f}{2^n}) \prod_{m=1}^{n} \frac{\overline{G}(\frac{f}{2^m})}{\sqrt{2}}, \quad n = 1, 2, \ldots.$$

where $\Phi(\cdot)$ is the Fourier transform of $\phi(\cdot)$, while $\overline{G}(\cdot)$ is the Fourier transform of $\{\bar{g}_l\}$ (when the latter is regarded as an impulse response sequence, $\overline{G}(\cdot)$ becomes its transfer function). Integration of both sides of the two-scale difference equation (combined with the mild assumption that $\phi(\cdot)$ integrates to a nonzero value) gives the normalization $\overline{G}(0) = \sum_l \bar{g}_l = \sqrt{2}$.

The orthonormality of the $\{\phi_{0,k}(\cdot) : k \in \mathbb{Z}\}$ means that $\Phi(\cdot)$ must satisfy Equation (467a), namely,

$$\sum_{m=-\infty}^{\infty} |\Phi(f+m)|^2 = 1.$$

Three useful results now follow. First, we find that $\phi(\cdot)$ must integrate to either plus or minus one. Choosing the former leads to the result $\Phi(0) = 1$ and thence to

$$\Phi(f) = \prod_{m=1}^{\infty} \frac{\overline{G}(\frac{f}{2^m})}{\sqrt{2}}.$$

Second, we have $|\overline{G}(f)|^2 + |\overline{G}(f+\frac{1}{2})|^2 = 2$ (this is Equation (467b)) so that $\overline{G}(\frac{1}{2}) = 0$. Third, we can argue that, for large j, the time reverse of the jth level scaling filter should have approximately the same shape as $\phi(\cdot)$; i.e.,

$$2^j \tilde{g}_{j,-l} = 2^{j/2} g_{j,-l} \approx \phi(\tfrac{l}{2^j})$$

(this approximation is illustrated in Figure 471).

Let $W_0 \subset V_{-1}$ be the orthogonal complement in V_{-1} of V_0; i.e., if $\psi(\cdot) \in W_0$, then $\psi(\cdot) \in V_{-1}$, but $\psi(\cdot)$ is orthogonal to all functions in V_0 in the sense that $\langle \psi(\cdot), x(\cdot) \rangle = 0$ for all $x(\cdot) \in V_0$. The space W_0 is called a detail space. If $\psi(\cdot) \in L^2(\mathbb{R})$ is such that its integer translates $\{\psi_{0,k}(\cdot) : k \in \mathbb{Z}\}$ form an orthonormal basis for W_0, then $\psi(\cdot)$ is called a wavelet function. More generally, the subspace $W_j \subset V_{j-1}$ is the orthogonal complement in V_{j-1} of V_j and is known as the detail space for scale $\tau_j = 2^{j-1}$. Hence

$$V_j = V_{j+1} \oplus W_{j+1} \text{ and } V_j = W_{j+1} \oplus W_{j+2} \oplus \cdots.$$

A signal $x(\cdot) \in V_0 = V_1 \oplus W_1$ can be decomposed into components from V_1 and W_1 as

$$x(t) = s_1(t) + d_1(t),$$

where

$$d_1(t) = \sum_{k=-\infty}^{\infty} w_{1,k}\psi_{1,k}(t) \text{ with } w_{1,k} \equiv \langle x(\cdot), \psi_{1,k}(\cdot)\rangle.$$

Here $d_1(\cdot) \in W_1$ is the detail in $x(\cdot)$ missing from the coarse approximation $s_1(\cdot)$. The $w_{1,k}$'s are known as the unit scale wavelet coefficients. The relationship between the $v_{0,k}$'s and the $w_{1,k}$'s is given by

$$w_{1,k} = \sum_{m=-\infty}^{\infty} \bar{h}_{m-2k}v_{0,m} = \sum_{l=-\infty}^{\infty} h_l v_{0,2k-l},$$

where $\bar{h}_l \equiv \langle \phi_{0,l}(\cdot), \psi_{1,0}(\cdot)\rangle$, and $h_l = \bar{h}_{-l}$ (the filter $\{h_l\}$ is the wavelet filter introduced in Section 4.2, so $\{\bar{h}_l\}$ is its time reverse). We can decompose $s_{j-1}(\cdot) \in V_{j-1} = V_j \oplus W_j$ into components in V_j and W_j to obtain

$$s_{j-1}(t) = s_j(t) + d_j(t) \text{ with } d_j(t) \equiv \sum_{k=-\infty}^{\infty} w_{j,k}\psi_{j,k}(t),$$

where $w_{j,k} \equiv \langle s_{j-1}(\cdot), \psi_{j,k}(\cdot)\rangle$. The $w_{j,k}$ coefficients can be computed recursively using Equation (480d), namely,

$$w_{j,k} = \sum_{m=-\infty}^{\infty} \bar{h}_{m-2k}v_{j-1,m} = \sum_{l=-\infty}^{\infty} h_l v_{j-1,2k-l}.$$

Since

$$V_{j_0} = V_j \oplus W_j \oplus \cdots \oplus W_{j_0+1},$$

we know that, if $x(\cdot) \in V_{j_0}$ for some $j_0 \in \mathbb{Z}$, then, for any $j > j_0$,

$$x(t) = \sum_{k=-\infty}^{\infty} v_{j,k}\phi_{j,k}(t) + \sum_{l=j_0+1}^{j} \sum_{k=-\infty}^{\infty} w_{l,k}\psi_{l,k}(t)$$

and

$$\int_{-\infty}^{\infty} x^2(t)\, dt = \sum_{k=-\infty}^{\infty} v_{j,k}^2 + \sum_{l=j_0+1}^{j} \sum_{k=-\infty}^{\infty} w_{l,k}^2.$$

The scaling coefficients for level $j-1$ can be reconstructed from the scaling and wavelet coefficients for level j using

$$v_{j-1,k} = \sum_{m=-\infty}^{\infty} \bar{g}_{k-2m}v_{j,m} + \sum_{m=-\infty}^{\infty} \bar{h}_{k-2m}w_{j,m}.$$

The wavelet function $\psi(\cdot)$ is related to the scaling function $\phi(\cdot)$ through Equation (472), namely,

$$\psi(t) \equiv \sqrt{2} \sum_{l=-\infty}^{\infty} \bar{h}_l \phi(2t-l), \text{ where } \bar{h}_l \equiv (-1)^l \bar{g}_{1-l-L}.$$

The above is analogous to the two-scale difference equation. The wavelet filter $\{\bar{h}_l\}$, which is zero outside $-(L-1) \leq l \leq 0$, is the quadrature mirror filter (QMF) corresponding to the scaling filter $\{\bar{g}_l\}$, which is also zero outside $-(L-1) \leq l \leq 0$. This QMF definition ensures that $\bar{h}_{-l} = h_l$. The above expression for $\psi(\cdot)$ in terms of $\phi_{-1,l}(\cdot)$ along with the fact that support $\{\phi(\cdot)\} \subset (-(L-1), 0]$ implies that the same statement can be made about the support for $\psi(\cdot)$, i.e., support $\{\psi(\cdot)\} \subset (-(L-1), 0]$. The above expression can be used to compute values of the wavelet function $\psi(\cdot)$ using the recursive method described in Section 11.7. The Fourier transform $\Psi(\cdot)$ of $\psi(\cdot)$ follows as

$$\Psi(f) = \frac{\overline{H}(\frac{f}{2})}{\sqrt{2}} \prod_{m=2}^{\infty} \frac{\overline{G}(\frac{f}{2^m})}{\sqrt{2}},$$

where $\overline{H}(\cdot)$ is the Fourier transform of $\{\bar{h}_l\}$. The QMF relationship implies that

$$\overline{H}(f) = -e^{i2\pi f(L-1)}\overline{G}(\tfrac{1}{2} - f)$$

and, since $\overline{G}(\frac{1}{2}) = 0$, it follows that $\overline{H}(0) = 0$ and hence that $\Psi(0) = \int_{-\infty}^{\infty} \psi(t)\, dt = 0$. For large j, the time reverse of the jth level wavelet filter should have approximately the same shape as $\psi(\cdot)$; i.e.,

$$2^j \tilde{h}_{j,-l} = 2^{j/2} h_{j,-l} \approx \psi(\tfrac{l}{2^j})$$

(this approximation is illustrated in Figure 478).

If $\psi(\cdot)$ has r vanishing moments, i.e.,

$$\int_{-\infty}^{\infty} t^m \psi(t)\, dt = 0, \quad \text{then } \overline{H}^{(m)}(0) = 0 \text{ and } \sum_{l=-\infty}^{\infty} l^m \bar{h}_l = 0$$

for $m = 0, \ldots, r-1$. The filter sequence $\{\bar{h}_l\}$ can thus also be said to have r vanishing discrete moments. Furthermore, $\overline{H}^{(m)}(0) = 0$ for $m = 0, \ldots, r-1$ implies that $\overline{G}^{(m)}(\frac{1}{2}) = 0$ for $m = 0, \ldots, r-1$ also, and, as a consequence, r roots of $\overline{\overline{G}}(z)$ are equal to -1, where $\overline{\overline{G}}(z)$ is the z transform of $\{\bar{g}_l\}$. When $|z| = 1$, i.e., $z = e^{i2\pi f}$, the magnitude squared of the z transform takes the form

$$|\overline{\overline{G}}(z)|^2 = \overline{\overline{G}}(z)\overline{\overline{G}}(z^{-1}) = g^2_{-(L-1)}(z^{-1} + 2 + z)^r \prod_{l=r+1}^{L-1} (z - z_l)\,(z^{-1} - z_l)\,.$$

If we let $\overline{\mathcal{G}}(f) \equiv |\overline{G}(f)|^2$, where $\overline{G}(f) = \overline{\overline{G}}(e^{i2\pi f})$, then the corresponding squared gain function of the filter coefficients $\{\bar{g}_l\}$ is given by

$$\overline{\mathcal{G}}(f) = \left|\frac{1 + e^{i2\pi f}}{2}\right|^{2r} |Q(f)|^2 = \cos^{2r}(\pi f)\mathcal{Q}(f),$$

where $\mathcal{Q}(f)$ is a polynomial in $\cos(2\pi f)$. Daubechies (1988) constructed an appealing class of scaling filters with $L = 2r$ coefficients by specifying the squared gain function to be of the form

$$\overline{\mathcal{G}}(f) = 2\cos^{2r}(\pi f) \sum_{k=0}^{r-1} \binom{r-1+k}{k} \sin^{2k}(\pi f).$$

$h_l \equiv \bar{h}_{-l},\ l = 0, \ldots, L-1$	$g_l \equiv \bar{g}_{-l},\ l = 0, \ldots, L-1$
$\bar{h}_l = (-1)^l \bar{g}_{1-l-L}$	$\bar{g}_l \equiv (-1)^{l+1} \bar{h}_{1-l-L}$
$\{\bar{h}_l\} \longleftrightarrow \overline{H}(\cdot)$	$\{\bar{g}_l\} \longleftrightarrow \overline{G}(\cdot)$
$\{h_l\} \longleftrightarrow H(\cdot)$	$\{g_l\} \longleftrightarrow G(\cdot)$
$H(f) = \overline{H}(-f)$	$G(f) = \overline{G}(-f)$
$\overline{H}(0) = 0$	$\overline{G}(0) = \sqrt{2}$
$\overline{H}(f) = -e^{i2\pi f(L-1)}\overline{G}(\frac{1}{2} - f)$	$\overline{G}(f) = e^{i2\pi f(L-1)}\overline{H}(\frac{1}{2} - f)$
$\overline{H}^{(m)}(0) = 0,\ m = 0, \ldots, r-1$	$\overline{G}^{(m)}(\frac{1}{2}) = 0,\ m = 0, \ldots, r-1$
$\bar{h}_l = \int \phi(t-l)\frac{\psi(\frac{t}{2})}{\sqrt{2}}\, dt$	$\bar{g}_l = \int \phi(t-l)\frac{\phi(\frac{t}{2})}{\sqrt{2}}\, dt$
$\int \psi(t) dt = 0$	$\int \phi(t)\, dt = 1$
$\text{support}\,\{\psi(\cdot)\} \subset (-(L-1), 0]$	$\text{support}\,\{\phi(\cdot)\} \subset (-(L-1), 0]$
$\psi(\cdot) \longleftrightarrow \Psi(\cdot)$	$\phi(\cdot) \longleftrightarrow \Phi(\cdot)$
$\Psi(-2^j f) \approx \widetilde{H}_j(f)$	$\Phi(-2^j f) \approx \widetilde{G}_j(f)$
$\psi(-\frac{l}{2^j}) \approx 2^j \tilde{h}_{j,l} = 2^{j/2} h_{j,l}$	$\phi(-\frac{l}{2^j}) \approx 2^j \tilde{g}_{j,l} = 2^{j/2} g_{j,l}$
$\Psi(f) = \Phi(\frac{f}{2})\frac{\overline{H}(\frac{f}{2})}{\sqrt{2}}$	$\Phi(f) = \Phi(\frac{f}{2})\frac{\overline{G}(\frac{f}{2})}{\sqrt{2}}$
$\Psi(f) = \frac{\overline{H}(\frac{f}{2})}{\sqrt{2}} \prod_{m=2}^{\infty} \frac{\overline{G}(\frac{f}{2^m})}{\sqrt{2}}$	$\Phi(f) = \prod_{m=1}^{\infty} \frac{\overline{G}(\frac{f}{2^m})}{\sqrt{2}}$
$\psi_{j,k}(t) \equiv \psi(\frac{t}{2^j} - k)/\sqrt{2^j}$	$\phi_{j,k}(t) \equiv \phi(\frac{t}{2^j} - k)/\sqrt{2^j}$
$\psi(t) = \sqrt{2}\sum_l \bar{h}_l \phi(2t - l)$	$\phi(t) = \sqrt{2}\sum_l \bar{g}_l \phi(2t - l)$
$w_{j,k} = \int x(t)\psi_{j,k}(t)\, dt$	$v_{j,k} = \int x(t)\phi_{j,k}(t)\, dt$
$w_{j,k} = \sum_l h_l v_{j-1,2k-l}$	$v_{j,k} = \sum_l g_l v_{j-1,2k-l}$

Table 499. Key relationships involving (i) wavelet and scaling filters $\{h_l\}$ and $\{g_l\}$ and their time reverses $\{\bar{h}_l\}$ and $\{\bar{g}_l\}$ and (ii) wavelet functions $\psi(\cdot)$ and scaling functions $\phi(\cdot)$.

For a given r, this expression can be rewritten in the form of

$$|\overline{\overline{G}}(z)|^2 = \bar{g}^2_{-(L-1)}|z+1|^L \overline{\overline{Q}}_0(z)\overline{\overline{Q}}_0(z^{-1}),$$

where $\overline{\overline{Q}}_0(z)$ is a polynomial of order $\frac{L}{2} - 1$ in nonnegative powers of z that has real-valued coefficients and can be written as

$$\overline{\overline{Q}}_0(z) = \prod_{l=1}^{K_0}(z - r_l)\prod_{l=1}^{K_1}(z - z_l)(z - z_l^*);$$

here r_l and z_l are, respectively real- and complex-valued roots, and $K_0 + 2K_1 = \frac{L}{2} - 1$. The z transform $\overline{\overline{G}}(z)$ of $\{\bar{g}_l\}$ is constructed by 'spectral factorization,' in which we select $r = L/2$ roots of value -1 and then specify $\overline{\overline{Q}}_0(z)$ by using the $\frac{L}{2} - 1$ roots outside the unit circle (corresponding to 'maximum delay' dipoles) obtained from a factorization of $\overline{\overline{Q}}_0(z)\overline{\overline{Q}}_0(z^{-1})$. The resulting filter $\{g_l = \bar{g}_{-l}\}$ is the D(L) scaling filter and has a z transform with $L/2$ roots of the form -1 and $\frac{L}{2} - 1$ roots inside the unit circle (corresponding to 'minimum delay' dipoles). This approach can be modified to produce the LA(L) scaling filters: rather than selecting $\frac{L}{2} - 1$ maximum delay dipoles for $\overline{\overline{G}}(z)$, we choose that combination of minimum and maximum delay dipoles such that the resulting filter $\{\bar{g}_l\}$ gives rise to $\{g_l\}$ satisfying the optimality criterion of Equation (112a).

Finally we note that Table 499 summarizes key relationships involving the wavelet and scaling filters, the wavelet and scaling functions and related quantities.

11.12 Exercises

[11.1] Using a generalization of the scheme outlined in Section 11.6, create a figure similar to Figure 471 for the D(6) case (the coefficients for the D(6) scaling filter are given in Table 109).

[11.2] Show that $\psi_{0,k}(\cdot)$, $k \in \mathbb{Z}$, constitute an orthonormal basis for W_0. Hint: Given that $\int_{-\infty}^{\infty} \psi_{0,k}(t)\phi_{0,l}(t)\,dt = 0$ and the result of Exercise [474b], it is necessary to show that any $x(\cdot) \in V_{-1} \subset L^2(\mathbb{R})$ can be written as $x(t) = \sum_k c_k^\phi \phi_{0,k}(t) + \sum_k c_k^\psi \psi_{0,k}(t)$, where the coefficients $\{c_k^\phi\}$ and $\{c_k^\psi\}$ must satisfy $\sum_k (c_k^\phi)^2 < \infty$ and $\sum_k (c_k^\psi)^2 < \infty$. This is most easily proven using the special case of Haar scaling and wavelet functions.

[11.3] Verify Equation (477c).

[11.4] Suppose that $x(\cdot) \in L^2(\mathbb{R})$ and that $V_j, j \in \mathbb{Z}$, is an MRA with associated scaling function $\phi(\cdot)$. Consider the approximation to $x(\cdot)$ formed by projecting it onto the subspace V_j:

$$s_j(t) = \sum_{k=-\infty}^{\infty} v_{j,k}\phi_{j,k}(t), \text{ where, as usual, } v_{j,k} \equiv \langle x(\cdot), \phi_{j,k}(\cdot)\rangle.$$

For any $j_0 < j$, show that, for this $x(\cdot)$, the right-hand side of Equation (481d) can be interpreted as $s_{j_0}(t)$, i.e., the projection of $x(\cdot)$ on the subspace V_{j_0}.

[11.5] Verify the form of Equations (493a) and (493b).

Appendix

Answers to Embedded Exercises

Here we give solutions to all exercises that are embedded in the main text of each chapter (instructors who want solutions to the exercises at the end of each chapter should contact us). We believe that the reader can develop a basic understanding of wavelets by making an honest effort to do these exercises, so we encourage him or her to do this before looking at our solutions. Working these exercises will also give the reader the opportunity to find better solutions than the ones given here, and we would certainly appreciate hearing from readers about improvements or any errors (our e-mail and 'paper mail' addresses are given in the front of the book).

Answer to Exercise [22a]: Note first that, because t and k are both integers, $e^{-i2\pi kt} = \cos(2\pi kt) - i\sin(2\pi kt) = 1$. Hence

$$A(f+k) = \sum_{t=-\infty}^{\infty} a_t e^{-i2\pi(f+k)t} = \sum_{t=-\infty}^{\infty} a_t e^{-i2\pi ft} e^{-i2\pi kt} = \sum_{t=-\infty}^{\infty} a_t e^{-i2\pi ft} = A(f),$$

as required.

Answer to Exercise [22b]: On the one hand, we have

$$A(-f) = \sum_{t=-\infty}^{\infty} a_t e^{-i2\pi(-f)t} = \sum_{t=-\infty}^{\infty} a_t e^{i2\pi ft};$$

on the other hand, since $a_t^* = a_t$ for a real-valued variable,

$$A^*(f) = \left(\sum_{t=-\infty}^{\infty} a_t e^{-i2\pi ft}\right)^* = \sum_{t=-\infty}^{\infty} a_t^* \left(e^{-i2\pi ft}\right)^* = \sum_{t=-\infty}^{\infty} a_t e^{i2\pi ft}$$

also.

Answer to Exercise [22c]: Using the definition for $A(f)$, we have

$$\begin{aligned}\int_{-1/2}^{1/2} A(f)e^{i2\pi ft}\, df &= \int_{-1/2}^{1/2} \left(\sum_{t'=-\infty}^{\infty} a_{t'} e^{-i2\pi ft'}\right) e^{i2\pi ft}\, df \\ &= \sum_{t'=-\infty}^{\infty} a_{t'} \int_{-1/2}^{1/2} e^{i2\pi f(t-t')}\, df.\end{aligned}$$

If $t = t'$, the integral above is unity; on the other hand, if $t \neq t'$, then

$$\int_{-1/2}^{1/2} e^{i2\pi f(t-t')}\,df = \frac{1}{i2\pi(t-t')} e^{i2\pi f(t-t')}\Big|_{-1/2}^{1/2}$$
$$= \frac{1}{i2\pi(t-t')}\left(e^{i\pi(t-t')} - e^{-i\pi(t-t')}\right) = \frac{\sin(\pi(t-t'))}{\pi(t-t')} = 0,$$

since $t - t'$ is always an integer. Hence all the terms in the infinite summation above are zero except when $t' = t$, in which case it is a_t.

Answer to Exercise [23a]: Since

$$A(f) = \sum_{t=-\infty}^{\infty} a_t e^{-i2\pi ft} \text{ and } B^*(f) = \sum_{t'=-\infty}^{\infty} b_{t'}^* e^{i2\pi ft'},$$

we have

$$\int_{-1/2}^{1/2} A(f)B^*(f)\,df = \int_{-1/2}^{1/2} \left(\sum_{t=-\infty}^{\infty} a_t e^{-i2\pi ft}\right)\left(\sum_{t'=-\infty}^{\infty} b_{t'}^* e^{i2\pi ft'}\right) df$$
$$= \sum_{t=-\infty}^{\infty} a_t \sum_{t'=-\infty}^{\infty} b_{t'}^* \int_{-1/2}^{1/2} e^{i2\pi f(t'-t)}\,df.$$

By the same argument used in Exercise [22c], the integral above is unity if $t' = t$ and is zero otherwise, so the inner summation reduces to just b_t^*.

Answer to Exercise [23b]: Since

$$a_t = \int_{-1/2}^{1/2} A(f)e^{i2\pi ft}\,df,$$

we have (letting $f' = 2f$)

$$a_{2n} = \int_{-1/2}^{1/2} A(f)e^{i2\pi f(2n)}\,df$$
$$= \frac{1}{2}\int_{-1}^{1} A(\tfrac{f'}{2})e^{i2\pi f'n}\,df' = \frac{1}{2}\left[\int_{-1}^{-1/2} + \int_{-1/2}^{1/2} + \int_{1/2}^{1} A(\tfrac{f'}{2})e^{i2\pi f'n}\,df'\right].$$

In the first and third integrals within the brackets, we change the variables of integration to, respectively, $f = f' + 1$ and $f = f' - 1$ and then group the results to obtain

$$a_{2n} = \int_{-1/2}^{1/2} \frac{1}{2}\left[A(\tfrac{f}{2}) + A(\tfrac{f}{2} + \tfrac{1}{2})\right] e^{i2\pi fn}\,df$$

(we also make use of the facts that $A(\frac{f}{2} - \frac{1}{2}) = A(\frac{f}{2} + \frac{1}{2})$ because of unit periodicity and that $\exp(\pm i2\pi n) = 1$ because n is an integer). The integral above has the form of an inverse DFT, which tells us that the DFT of $\{a_{2n}\}$ is as stated in the exercise.

Answer to Exercise [24]: Using the definition of $\{a * b_t\}$, we have (letting $v \equiv t - u$)

$$\begin{aligned}
\sum_{t=-\infty}^{\infty} a * b_t e^{-i2\pi ft} &= \sum_{t=-\infty}^{\infty} \left(\sum_{u=-\infty}^{\infty} a_u b_{t-u} \right) e^{-i2\pi ft} \\
&= \sum_{u=-\infty}^{\infty} a_u \left(\sum_{t=-\infty}^{\infty} b_{t-u} e^{-i2\pi ft} \right) \\
&= \sum_{u=-\infty}^{\infty} a_u \left(\sum_{v=-\infty}^{\infty} b_v e^{-i2\pi f(v+u)} \right) \\
&= \sum_{u=-\infty}^{\infty} a_u e^{-i2\pi fu} \left(\sum_{v=-\infty}^{\infty} b_v e^{-i2\pi fv} \right) = A(f)B(f),
\end{aligned}$$

as required.

Answer to Exercise [27]: Let $B(\cdot)$ be the Fourier transform of the input to the filter cascade. It follows from Exercise [24] that the output from the first filter has a DFT defined by $B(f)A_1(f)$. Since the DFT of the input to the second filter is also defined by $B(f)A_1(f)$, a second application of [24] says that the output from the second filter has a DFT defined by $B(f)A_1(f)A_2(f)$. After M repetitions of this argument, we have that the output $\{c_t\}$ from the filter cascade has DFT $C(f) \equiv B(f)A_1(f)A_2(f) \cdots A_M(f)$. With

$$A(f) \equiv \prod_{m=1}^{M} A_m(f),$$

we have $C(f) = B(f)A(f)$. Letting $\{a_t\}$ be the inverse Fourier transform of $A(\cdot)$, Exercise [24] tells us that $\{c_t\}$ is the convolution of $\{b_t\}$ with $\{a_t\}$. The statement that $\{a_t\}$ is the successive convolution of $\{a_{m,t}\}$, $m = 1, \ldots, M$, follows easily from [24] and an induction argument.

Answer to Exercise [28a]: First, we show that the result holds for $M = 2$. From Exercise [27], the transfer function for the filter cascade is given by

$$\begin{aligned}
A(f) = A_1(f)A_2(f) &= \left(\sum_{t=K_1}^{K_1+L_1-1} a_{1,t} e^{-i2\pi ft} \right) \left(\sum_{u=K_2}^{K_2+L_2-1} a_{2,u} e^{-i2\pi fu} \right) \\
&= \sum_{t=K_1}^{K_1+L_1-1} \sum_{u=K_2}^{K_2+L_2-1} a_{1,t} a_{2,u} e^{-i2\pi f(t+u)}.
\end{aligned}$$

Now the maximum value that $t + u$ can assume within the double summation occurs when $t = K_1 + L_1 - 1$ and $u = K_2 + L_2 - 1$; likewise, the minimum value that $t + u$ can assume occurs when $t = K_1$ and $u = K_2$. Letting $\{a_t\}$ be the inverse DFT of $A(\cdot)$, it follows that we can write

$$A(f) = \sum_{v=K_1+K_2}^{K_1+L_1-1+K_2+L_2-1} a_v e^{-i2\pi fv},$$

so $\{a_t\}$ has a width of

$$K_1 + L_1 - 1 + K_2 + L_2 - 1 - (K_1 + K_2) + 1 = L_1 + L_2 - 1,$$

Hence the result is established for $M = 2$.

Next we argue that, if the result holds for $M-1$ filters, it must hold for M filters. From Exercise [27], the transfer function of the equivalent filter for the filter cascade of M filters is given by

$$A(f) = \prod_{m=1}^{M} A_m(f) = A_M(f) \prod_{m=1}^{M-1} A_m(f) \equiv A_M(f)\overline{A}(f),$$

where $\overline{A}(\cdot)$ is the transfer function of the equivalent filter for a filter cascade of $M-1$ filters with transfer functions $A_1(\cdot), \ldots, A_{M-1}(\cdot)$. The induction hypothesis says that $\{\overline{a}_t\} \longleftrightarrow \overline{A}(\cdot)$ has a width of $L' = \sum_{m=1}^{M-1} L_m - M + 2$. Since $A_M(f)\overline{A}(f)$ defines the transfer function for a filter cascade of two filters, the $M = 2$ case says that the width for $\{a_t\} \longleftrightarrow A(\cdot)$ is given by

$$L' + L_M - 1 = \sum_{m=1}^{M-1} L_m - M + 2 + L_M - 1 = \sum_{m=1}^{M} L_m - M + 1,$$

as required.

Answer to Exercise [28b]: Since $\exp(i2\pi m) = 1$ for all integers m and since the product tn is an integer, we have

$$A_{k+nN} = \sum_{t=0}^{N-1} a_t e^{-i2\pi t(k+nN)/N} = \sum_{t=0}^{N-1} a_t e^{-i2\pi tk/N} e^{-i2\pi tn} = \sum_{t=0}^{N-1} a_t e^{-i2\pi tk/N} = A_k,$$

as required.

Answer to Exercise [29a]: To prove the hint, note that

$$(1-z)\sum_{u=0}^{N-1} z^u = \sum_{u=0}^{N-1} z^u - \sum_{u=0}^{N-1} z^{u+1} = 1 - z^N,$$

from which we can conclude that, if $z \neq 1$,

$$\sum_{u=0}^{N-1} z^u = \frac{1-z^N}{1-z}.$$

On the other hand, if $z = 1$, then $\sum_{u=0}^{N-1} z^u = N$. In particular, if $z = e^{i\theta}$, then $z = 1$ if and only if $\theta = 0, \pm 2\pi, \pm 4\pi, \ldots$. Using the definition for A_k, we have

$$\begin{aligned}\frac{1}{N}\sum_{k=0}^{N-1} A_k e^{i2\pi tk/N} &= \frac{1}{N}\sum_{k=0}^{N-1}\left(\sum_{u=0}^{N-1} a_u e^{-i2\pi uk/N}\right) e^{i2\pi tk/N} \\ &= \frac{1}{N}\sum_{u=0}^{N-1} a_u \sum_{k=0}^{N-1} e^{i2\pi(t-u)k/N}.\end{aligned}$$

If $t = u$, the inner summation reduces to N; if $t \neq u$, note that $z \equiv \exp(i2\pi(t-u)/N) \neq 1$ (because $0 < 2\pi|t-u|/N \leq 2\pi(N-1)/N < 2\pi$), so the inner summation is

$$\sum_{k=0}^{N-1} e^{i2\pi(t-u)k/N} = \frac{1 - e^{i2\pi(t-u)}}{1 - e^{i2\pi(t-u)/N}} = 0$$

since $\exp(i2\pi(t-u)) = 1$ because $t - u$ is always an integer. The double summation thus collapses to just a_t, yielding the desired result.

Answer to Exercise [29b]: Since

$$A_k = \sum_{t=0}^{N-1} a_t e^{-i2\pi tk/N} \text{ and } B_k^* = \sum_{u=0}^{N-1} b_u^* e^{i2\pi uk/N},$$

we have

$$\frac{1}{N}\sum_{k=0}^{N-1} A_k B_k^* = \frac{1}{N}\sum_{k=0}^{N-1}\left(\sum_{t=0}^{N-1} a_t e^{-i2\pi tk/N}\right)\left(\sum_{u=0}^{N-1} b_u^* e^{i2\pi uk/N}\right)$$
$$= \frac{1}{N}\sum_{t=0}^{N-1} a_t \sum_{u=0}^{N-1} b_u^* \sum_{k=0}^{N-1} e^{i2\pi(u-t)k/N}.$$

In Exercise [29a] we established that

$$\sum_{k=0}^{N-1} e^{i2\pi(u-t)k/N} = \begin{cases} N, & \text{if } u = t; \\ 0, & \text{otherwise.} \end{cases}$$

Hence the triple summation reduces to a single summation, yielding the desired result.

Answer to Exercise [30]: The definition of $a * b_t$ yields

$$\sum_{t=0}^{N-1} a * b_t e^{-i2\pi tk/N} = \sum_{t=0}^{N-1}\left(\sum_{u=0}^{N-1} a_u b_{t-u \bmod N}\right) e^{-i2\pi tk/N}$$
$$= \sum_{u=0}^{N-1} a_u \sum_{t=0}^{N-1} b_{t-u \bmod N} e^{-i2\pi tk/N}.$$

Now the infinite sequence $\{b_{t-u \bmod N} e^{-i2\pi tk/N} : t = \ldots, -1, 0, 1, \ldots\}$ is periodic with a period of N, so the sum over any N consecutive variables in the sequence is the same. Using the additional fact that $b_{t \bmod N} = b_t$ for $0 \le t \le N-1$, we have

$$\sum_{t=0}^{N-1} a * b_t e^{-i2\pi tk/N} = \sum_{u=0}^{N-1} a_u \sum_{t=0}^{N-1} b_t e^{-i2\pi(t+u)k/N}$$
$$= \sum_{u=0}^{N-1} a_u e^{-i2\pi uk/N} \sum_{t=0}^{N-1} b_t e^{-i2\pi tk/N} = A_k B_k,$$

as required.

Answer to Exercise [31]: Letting $\tilde{a}_0 = a_0^*$ and $\tilde{a}_t = a_{N-t}^*$ for $t = 1, \ldots, N-1$ does the trick. This can be seen either from an element by element comparison of the labels outside the circles in Figures 31a and 31b, or from the following formal argument. Letting $v = -u$, we have

$$a^* \star b_t \equiv \sum_{u=0}^{N-1} a_u^* b_{u+t \bmod N}$$
$$= \sum_{v=-(N-1)}^{0} a_{-v}^* b_{t-v \bmod N}$$
$$= \sum_{v=1}^{N} a_{N-v}^* b_{t-v \bmod N} \qquad (\text{since } b_{t-v+N \bmod N} = b_{t-v \bmod N})$$
$$= \sum_{v=0}^{N-1} a_{N-v \bmod N}^* b_{t-v \bmod N} \equiv \sum_{v=0}^{N-1} \tilde{a}_v b_{t-v \bmod N},$$

where $\tilde{a}_t \equiv a^*_{N-t \bmod N}$ (recall that $\tilde{a}_0 = a^*_{N \bmod N} = a^*_0$, while $\tilde{a}_t = a^*_{N-t}$ for $t = 1, \ldots, N-1$). Finally, the DFT of $\{\tilde{a}_t\}$ is given by

$$\begin{aligned}\sum_{t=0}^{N-1} \tilde{a}_t e^{-i2\pi tk/N} &= a_0^* + \sum_{t=1}^{N-1} a^*_{N-t} e^{-i2\pi tk/N} \\ &= a_0^* + \sum_{u=1}^{N-1} a^*_u e^{-i2\pi(N-u)k/N} \qquad \text{(letting } u = N-t\text{)} \\ &= \sum_{u=0}^{N-1} a^*_u e^{i2\pi uk/N} = \left(\sum_{u=0}^{N-1} a_u e^{-i2\pi uk/N}\right)^* = A^*_k,\end{aligned}$$

as required.

Answer to Exercise [33]: Using the fact that

$$e^{-i2\pi(u+nN)k/N} = e^{-i2\pi uk/N} e^{-i2\pi nk} = e^{-i2\pi uk/N}$$

for all integers n and k, we have

$$\begin{aligned}A^\circ_k \equiv \sum_{u=0}^{N-1} a^\circ_u e^{-i2\pi uk/N} &= \sum_{u=0}^{N-1}\left(\sum_{n=-\infty}^{\infty} a_{u+nN}\right) e^{-i2\pi uk/N} \\ &= \sum_{n=-\infty}^{\infty}\sum_{u=0}^{N-1} a_{u+nN} e^{-i2\pi(u+nN)k/N} \\ &= \sum_{t=-\infty}^{\infty} a_t e^{-i2\pi tk/N} = A(\tfrac{k}{N}),\end{aligned}$$

where the double summation reduces to the single summation because in both expressions each term of the infinite sequence $\{a_t e^{-i2\pi tk/N}\}$ occurs exactly once (the double summation merely adds groups of N consecutive variables together first and then adds the group sums together).

Answer to Exercise [34]: As a simple counter-example, consider the following three filters, for each of which $a_t \equiv 0$ when $t = -1, -2, \ldots$ or $t = 4, 5, \ldots$:

$$a_t = \begin{cases} \frac{1}{2}, & t = 0, 1; \\ -\frac{1}{2}, & t = 2, 3; \end{cases} \quad a_t = \begin{cases} \frac{1-\sqrt{3}}{4\sqrt{2}}, & t = 0; \\ \frac{-3+\sqrt{3}}{4\sqrt{2}}, & t = 1; \\ \frac{3+\sqrt{3}}{4\sqrt{2}}, & t = 2; \\ \frac{-1-\sqrt{3}}{4\sqrt{2}}, & t = 3; \end{cases} \quad \text{and } a_t = \tfrac{1}{2},\ 0 \le t \le 3$$

(these filters are all special cases of first and second level wavelet and scaling filters discussed in Chapter 4). Each of the above satisfies $\sum a_t^2 = 1$. Periodization of the above to length $N = 2$ yields the filters $\{a^\circ_0 = 0, a^\circ_1 = 0\}$, $\{a^\circ_0 = 1/\sqrt{2}, a^\circ_1 = -1/\sqrt{2}\}$ and $\{a^\circ_0 = 1, a^\circ_1 = 1\}$, for which we have, respectively,

$$\sum_{t=0}^{1} |a^\circ_t|^2 = 0; \ \sum_{t=0}^{1} |a^\circ_t|^2 = 1; \ \text{and} \ \sum_{t=0}^{1} |a^\circ_t|^2 = 2.$$

This example demonstrates that the sum of squares of a periodized filter can be less than, equal to, or greater than the sum of squares for the original filter.

Here is a more formal proof. Since

$$\{a_t^\circ : t = 0, \ldots, N-1\} \longleftrightarrow \{A(\tfrac{k}{N}) : k = 0, \ldots, N-1\},$$

Parseval's theorem states that

$$\sum_{t=0}^{N-1} |a_t^\circ|^2 = \frac{1}{N} \sum_{k=0}^{N-1} |A(\tfrac{k}{N})|^2$$

(see Equation (36h)). Since $A(\cdot)$ is the DFT of the nonperiodized filter $\{a_t\}$, we can write

$$\begin{aligned}\sum_{t=0}^{N-1} |a_t^\circ|^2 &= \frac{1}{N} \sum_{k=0}^{N-1} \left(\sum_{t=-\infty}^{\infty} a_t e^{-i2\pi tk/N} \right) \left(\sum_{u=-\infty}^{\infty} a_u^* e^{i2\pi uk/N} \right) \\ &= \frac{1}{N} \sum_{t=-\infty}^{\infty} \sum_{u=-\infty}^{\infty} a_t a_u^* \sum_{k=0}^{N-1} e^{i2\pi(u-t)k/N}.\end{aligned}$$

Using an argument similar to that given in the solution to Exercise [29a], the innermost summation is nonzero only when $u-t$ is an integer multiple of N, in which case this summation is equal to N. For a given t, this happens when $u = t, t \pm N, t \pm 2N, \ldots$, so we have

$$\begin{aligned}\sum_{t=0}^{N-1} |a_t^\circ|^2 &= \sum_{t=-\infty}^{\infty} a_t \sum_{n=-\infty}^{\infty} a_{t+nN}^* \\ &= \sum_{t=-\infty}^{\infty} |a_t|^2 + \sum_{t=-\infty}^{\infty} a_t \sum_{n=1}^{\infty} (a_{t+nN}^* + a_{t-nN}^*) \neq \sum_{t=-\infty}^{\infty} |a_t|^2\end{aligned}$$

in general.

Answer to Exercise [43]: Note that $\widetilde{\mathcal{O}} \equiv \mathcal{O}^T$ is an orthonormal matrix, and hence Equation (43c) tells us that

$$\mathbf{X} = \sum_{k=0}^{N-1} \langle \mathbf{X}, \widetilde{\mathcal{O}}_{k\bullet} \rangle \widetilde{\mathcal{O}}_{k\bullet}; \text{ however } \widetilde{\mathcal{O}}_{k\bullet} = \mathcal{O}_{\bullet k}, \text{ so } \mathbf{X} = \sum_{k=0}^{N-1} \langle \mathbf{X}, \mathcal{O}_{\bullet k} \rangle \mathcal{O}_{\bullet k},$$

as required.

Answer to Exercise [44a]: Since $\mathcal{O}_{j\bullet}^T \mathcal{O}_{j'\bullet} = 0$ when $j \neq j'$, we have

$$\begin{aligned}\|\mathbf{X}\|^2 = \left\| \sum_{j=0}^{N-1} O_j \mathcal{O}_{j\bullet} \right\|^2 &= \left(\sum_{j=0}^{N-1} O_j \mathcal{O}_{j\bullet} \right)^T \left(\sum_{j'=0}^{N-1} O_{j'} \mathcal{O}_{j'\bullet} \right) \\ &= \sum_{j=0}^{N-1} \sum_{j'=0}^{N-1} O_j O_{j'} \mathcal{O}_{j\bullet}^T \mathcal{O}_{j'\bullet} = \sum_{j=0}^{N-1} O_j^2 \mathcal{O}_{j\bullet}^T \mathcal{O}_{j\bullet} \\ &= \sum_{j=0}^{N-1} (O_j \mathcal{O}_{j\bullet})^T (O_j \mathcal{O}_{j\bullet}) = \sum_{j=0}^{N-1} \|O_j \mathcal{O}_{j\bullet}\|^2,\end{aligned}$$

as required.

Answer to Exercise [44b]: Since by definition $\mathbf{e} = \mathbf{X} - \widehat{\mathbf{X}}$, we have

$$\|\mathbf{e}\|^2 = \|\mathbf{X} - \widehat{\mathbf{X}}\|^2 = \left\| \sum_{j=0}^{N-1} O_j \mathcal{O}_{j\bullet} - \sum_{j=0}^{N'-1} \alpha_j \mathcal{O}_{j\bullet} \right\|^2$$

$$= \left\| \sum_{j=0}^{N'-1} (O_j - \alpha_j)\mathcal{O}_{j\bullet} + \sum_{j=N'}^{N-1} O_j \mathcal{O}_{j\bullet} \right\|^2 = \left\| \sum_{j=0}^{N-1} \beta_j \mathcal{O}_{j\bullet} \right\|^2,$$

where

$$\beta_j \equiv \begin{cases} O_j - \alpha_j, & \text{for } j = 0, \ldots, N' - 1; \\ O_j, & \text{for } j = N', \ldots, N - 1. \end{cases}$$

Now

$$\left\| \sum_{j=0}^{N-1} \beta_j \mathcal{O}_{j\bullet} \right\|^2 = \left(\sum_{j=0}^{N-1} \beta_j \mathcal{O}_{j\bullet} \right)^T \left(\sum_{k=0}^{N-1} \beta_k \mathcal{O}_{k\bullet} \right) = \sum_{j=0}^{N-1} \sum_{k=0}^{N-1} \beta_j \beta_k \mathcal{O}_{j\bullet}^T \mathcal{O}_{k\bullet}.$$

Orthonormality of the $\mathcal{O}_{j\bullet}$'s means that $\mathcal{O}_{j\bullet}^T \mathcal{O}_{k\bullet} = 0$ when $j \neq k$. Hence the double summation above reduces to

$$\sum_{j=0}^{N-1} \beta_j^2 = \sum_{j=0}^{N'-1} (O_j - \alpha_j)^2 + \sum_{j=N'}^{N-1} O_j^2,$$

as required.

Answer to Exercise [46a]: By definition, the jth element of $\mathbf{O}$ is $O_j \equiv \langle \mathbf{X}, \mathcal{O}_{j,\bullet} \rangle = \mathcal{O}_{j,\bullet}^H \mathbf{X}$. To see that this yields $\mathbf{O} = \mathcal{O}\mathbf{X}$, note that

$$\mathcal{O}\mathbf{X} = \begin{bmatrix} \mathcal{O}_{0\bullet}^H \\ \mathcal{O}_{1\bullet}^H \\ \vdots \\ \mathcal{O}_{N-1\bullet}^H \end{bmatrix} \mathbf{X} = \begin{bmatrix} \mathcal{O}_{0\bullet}^H \mathbf{X} \\ \mathcal{O}_{1\bullet}^H \mathbf{X} \\ \vdots \\ \mathcal{O}_{N-1\bullet}^H \mathbf{X} \end{bmatrix} = \begin{bmatrix} \langle \mathbf{X}, \mathcal{O}_{0\bullet} \rangle \\ \langle \mathbf{X}, \mathcal{O}_{1\bullet} \rangle \\ \vdots \\ \langle \mathbf{X}, \mathcal{O}_{N-1\bullet} \rangle \end{bmatrix} = \begin{bmatrix} O_0 \\ O_1 \\ \vdots \\ O_{N-1} \end{bmatrix},$$

which establishes the analysis equation. Because by definition $\mathcal{O}^H \mathcal{O} = I_N$, the synthesis equation follows easily by premultiplying both sizes of the analysis equation by $\mathcal{O}^H$.

Answer to Exercise [46b]: First, note that, because $e^0 = 1$,

$$F_0 = \frac{1}{\sqrt{N}} \sum_{t=0}^{N-1} X_t,$$

which is real-valued because the X_t's are so. Next, note that, because $e^{-i2\pi t} = 1$ for all integers t,

$$F_{N-k} = \frac{1}{\sqrt{N}} \sum_{t=0}^{N-1} X_t e^{-i2\pi t(N-k)/N} = \frac{1}{\sqrt{N}} \sum_{t=0}^{N-1} X_t e^{-i2\pi t} e^{i2\pi tk/N}$$

$$= \frac{1}{\sqrt{N}} \sum_{t=0}^{N-1} X_t e^{i2\pi tk/N} = \left(\frac{1}{\sqrt{N}} \sum_{t=0}^{N-1} X_t e^{-i2\pi tk/N} \right)^* = F_k^*.$$

Finally, note that, because $e^{-i\pi t} = \left(e^{-i\pi}\right)^t = (-1)^t$,

$$F_{\frac{N}{2}} = \frac{1}{\sqrt{N}} \sum_{t=0}^{N-1} X_t e^{-i2\pi t(\frac{N}{2})/N} = \frac{1}{\sqrt{N}} \sum_{t=0}^{N-1} X_t (-1)^t,$$

which is necessarily real-valued.

Answer to Exercise [47]: The elements of the jth row of $\mathcal{F}^H$ are $\exp(i2\pi tj/N)/\sqrt{N}$, $t = 0, \ldots, N-1$, while the elements of the kth column of $\mathcal{F}$ are given by $\exp(-i2\pi tk/N)/\sqrt{N}$, $t = 0, \ldots, N-1$. Hence the (j,k)th element of $\mathcal{F}^H\mathcal{F}$ is given by

$$A_{j,k} \equiv \frac{1}{N}\sum_{t=0}^{N-1} e^{i2\pi tj/N}e^{-i2\pi tk/N} = \frac{1}{N}\sum_{t=0}^{N-1} e^{i2\pi(j-k)t/N}$$
$$= \frac{1}{N}\sum_{t=0}^{N-1} z^t \text{ with } z \equiv e^{i2\pi(j-k)/N}$$

The complex exponential $\exp(i2\pi x) \equiv \cos(2\pi x) + i\sin(2\pi x)$ is equal to unity if and only if x is an integer; because $-(N-1) \le j-k \le N-1$, we have $z = 1$ only when $j = k$, for which the last summation above is equal to N. Hence, when $j = k$, we have $A_{k,k} = 1$, while for $j \ne k$, we have (using the hint for Exercise [29a])

$$A_{j,k} = \frac{1}{N}\left(\frac{1-z^N}{1-z}\right) = \frac{1}{N}\left(\frac{1-e^{i2\pi(j-k)}}{1-e^{i2\pi(j-k)/N}}\right) = 0$$

because $1 - e^{i2\pi(j-k)} = 1 - \cos[2\pi(j-k)] - i\sin[2\pi(j-k)] = 0$.

Answer to Exercise [50a]: We need to show that, for $0 < k < \frac{N}{2}$,

$$F_k\mathcal{F}_{k\bullet} + F_{N-k}\mathcal{F}_{N-k\bullet} = 2\Re(F_k\mathcal{F}_{k\bullet}).$$

First, we note that

$$F_{N-k} \equiv \frac{1}{\sqrt{N}}\sum_{t=0}^{N-1} X_t e^{-i2\pi t(N-k)/N} = \frac{1}{\sqrt{N}}\sum_{t=0}^{N-1} X_t e^{i2\pi t}e^{i2\pi tk/N}$$
$$= \frac{1}{\sqrt{N}}\sum_{t=0}^{N-1} X_t e^{i2\pi tk/N} = F_k^*,$$

where we have used the fact that $\exp(i2\pi t) = 1$ because t is an integer. Second, note that the $(N-k)$th row of $\mathcal{F}$ is $\mathcal{F}_{N-k\bullet}^H$, so the tth element of $\mathcal{F}_{N-k\bullet}$ is given by

$$\left(\frac{1}{\sqrt{N}}e^{-i2\pi t(N-k)/N}\right)^* = \frac{1}{\sqrt{N}}e^{-i2\pi tk/N},$$

which is equal to the tth element of $\mathcal{F}_{k\bullet}^*$; i.e., $\mathcal{F}_{N-k\bullet} = \mathcal{F}_{k\bullet}^*$. Hence

$$F_k\mathcal{F}_{k\bullet} + F_{N-k}\mathcal{F}_{N-k\bullet} = F_k\mathcal{F}_{k\bullet} + F_k^*\mathcal{F}_{k\bullet}^* = F_k\mathcal{F}_{k\bullet} + (F_k\mathcal{F}_{k\bullet})^*.$$

Now, if $z \equiv x+iy$ is any complex-valued variable, then $z + z^* = x + iy + x - iy = 2x = 2\Re(z)$, so we can conclude that $F_k\mathcal{F}_{k\bullet} + F_{N-k}\mathcal{F}_{N-k\bullet} = 2\Re(F_k\mathcal{F}_{k\bullet})$.

Answer to Exercise [50b]: We have $\mathcal{D}_{\mathcal{F},k} = 2\Re(F_k\mathcal{F}_{k\bullet})$ from the result of Exercise [50a]. The kth row of $\mathcal{F}$ is $\mathcal{F}_{k\bullet}^H$, so its tth component is given by

$$\frac{e^{i2\pi tk/N}}{\sqrt{N}} = \frac{e^{i2\pi f_k t}}{\sqrt{N}},$$

and the tth component of $F_k \mathcal{F}_{k\bullet}$ is given by

$$\begin{aligned}\frac{1}{\sqrt{N}} F_k e^{i2\pi f_k t} &= \frac{1}{\sqrt{N}} (A_k - iB_k)(\cos(2\pi f_k t) + i \sin(2\pi f_k t)) \\ &= \frac{1}{\sqrt{N}} \big[A_k \cos(2\pi f_k t) + B_k \sin(2\pi f_k t) \\ &\qquad + i \left(A_k \sin(2\pi f_k t) - B_k \cos(2\pi f_k t) \right) \big],\end{aligned}$$

from which it follows that the tth component of $2\Re(F_k \mathcal{F}_{k\bullet})$, i.e., $\mathcal{D}_{\mathcal{F},k,t}$, is given by

$$\frac{2}{\sqrt{N}} \left[A_k \cos(2\pi f_k t) + B_k \sin(2\pi f_k t) \right],$$

as required.

Answer to Exercise [52]: Note that, for any integer m, the elements of $\mathcal{T}^m \mathbf{X}$ can be expressed as

$$X_{-m \bmod N}, X_{1-m \bmod N}, X_{2-m \bmod N}, \ldots, X_{N-1-m \bmod N},$$

so the kth Fourier coefficient for $\mathcal{T}^m \mathbf{X}$ is given by

$$\begin{aligned}\frac{1}{\sqrt{N}} \sum_{t=0}^{N-1} X_{t-m \bmod N} e^{-i2\pi tk/N} &= \frac{1}{\sqrt{N}} \sum_{t=-m}^{N-1-m} X_{t \bmod N} e^{-i2\pi(t+m)k/N} \\ &= \frac{e^{-i2\pi mk/N}}{\sqrt{N}} \sum_{t=-m}^{N-1-m} X_{t \bmod N} e^{-i2\pi tk/N} \\ &= \frac{e^{-i2\pi mk/N}}{\sqrt{N}} \sum_{t=0}^{N-1} X_{t \bmod N} e^{-i2\pi tk/N} \\ &= \frac{e^{-i2\pi mk/N}}{\sqrt{N}} \sum_{t=0}^{N-1} X_t e^{-i2\pi tk/N} = e^{-i2\pi mk/N} F_k,\end{aligned}$$

where we have made use of the following facts: (i) the infinite sequence $\{X_{t \bmod N} e^{-i2\pi tk/N} : t = \ldots, -1, 0, 1, \ldots\}$ is periodic with period N, and hence a sum over any N consecutive elements is equal to the sum over any other N consecutive elements; and (ii) $X_{t \bmod N} = X_t$ for $t = 0, \ldots, N-1$.

Answer to Exercise [58]: Let $r_1 \equiv 0$, and define $r_j = r_{j-1} + N/2^{j-1}$ for $j = 2, 3, \ldots, J$, where r_j will be used to index specific rows of $\mathcal{W}$ (for example, when $N = 16$ so that $J = 4$, the r_j's index rows $r_1 = 0$, $r_2 = 8$, $r_3 = 12$ and $r_4 = 14$). Define row r_j, $j = 1, 2, \ldots, J$, of $\mathcal{W}$ as

$$\mathcal{W}_{r_j\bullet}^T = \Big[\underbrace{-\tfrac{1}{2^{j/2}}, \ldots, -\tfrac{1}{2^{j/2}}}_{2^{j-1} \text{ of these}}, \underbrace{\tfrac{1}{2^{j/2}}, \ldots, \tfrac{1}{2^{j/2}}}_{2^{j-1} \text{ of these}}, \underbrace{0, \ldots, 0}_{N-2^j \text{ zeros}} \Big].$$

Since

$$\|\mathcal{W}_{r_j\bullet}\|^2 = 2^{j-1} \left(-\frac{1}{2^{j/2}} \right)^2 + 2^{j-1} \left(\frac{1}{2^{j/2}} \right)^2 = \frac{2^{j-1}}{2^j} + \frac{2^{j-1}}{2^j} = 1,$$

all of the $\mathcal{W}_{r_j\bullet}$ have unit norm; in addition, if $j < j'$, we have

$$\begin{aligned}\langle \mathcal{W}_{r_j\bullet}, \mathcal{W}_{r_{j'}\bullet} \rangle &= 2^{j-1} \left(-\frac{1}{2^{j/2}} \right) \left(-\frac{1}{2^{j'/2}} \right) + 2^{j-1} \left(\frac{1}{2^{j/2}} \right) \left(-\frac{1}{2^{j'/2}} \right) \\ &= \frac{2^{j-1}}{2^{(j+j')/2}} - \frac{2^{j-1}}{2^{(j+j')/2}} = 0,\end{aligned}$$

so the $\mathcal{W}_{r_j \bullet}$'s are orthonormal. Next define the remaining rows of $\mathcal{W}$ via the following:

$$\mathcal{W}_{r_j+k\bullet} = \mathcal{T}^{2^j k}\mathcal{W}_{r_j\bullet}, \quad j = 1, \ldots, J-1 \text{ and } k = 1, \ldots, \frac{N}{2^j} - 1;$$

$$\mathcal{W}_{N-1\bullet} = \left[\tfrac{1}{2^{J/2}}, \ldots, \tfrac{1}{2^{J/2}}\right]^T.$$

Since

$$\|\mathcal{W}_{r_j+k\bullet}\|^2 = \|\mathcal{T}^{2^j k}\mathcal{W}_{r_j\bullet}\|^2 = \|\mathcal{W}_{r_j\bullet}\|^2 = 1$$

and

$$\|\mathcal{W}_{N-1\bullet}\|^2 = N\left(\frac{1}{2^{J/2}}\right)^2 = \frac{2^J}{2^J} = 1,$$

each row of $\mathcal{W}$ has unit norm. Since

$$\langle \mathcal{W}_{N-1\bullet}, \mathcal{W}_{r_j+k\bullet}\rangle = \frac{1}{2^{J/2}}\left[2^{j-1}\left(-\frac{1}{2^{j/2}}\right) + 2^{j-1}\left(\frac{1}{2^{j/2}}\right)\right] = 0$$

for $j = 1, \ldots, J$ and $k = 0, \ldots, \frac{N}{2^j} - 1$, it follows that $\mathcal{W}_{N-1\bullet}$ is orthogonal to the remaining $N-1$ rows of $\mathcal{W}$. Finally, we must show that any two other arbitrarily selected rows are orthogonal, i.e.,

$$\langle \mathcal{W}_{r_j+k\bullet}, \mathcal{W}_{r_{j'}+k'\bullet}\rangle = 0.$$

Now

$$\begin{aligned}\langle \mathcal{W}_{r_j+k\bullet}, \mathcal{W}_{r_{j'}+k'\bullet}\rangle &= \langle \mathcal{T}^{2^j k}\mathcal{W}_{r_j\bullet}, \mathcal{T}^{2^{j'} k'}\mathcal{W}_{r_{j'}\bullet}\rangle \\ &= \mathcal{W}_{r_j\bullet}^T \mathcal{T}^{-2^j k}\mathcal{T}^{2^{j'} k'}\mathcal{W}_{r_{j'}\bullet} = \mathcal{W}_{r_j\bullet}^T \mathcal{T}^{2^{j'} k' - 2^j k}\mathcal{W}_{r_{j'}\bullet}.\end{aligned}$$

We can choose $j < j'$ such that $j' = j + \delta$, say, with $\delta > 0$. Then $2^{j'}k' - 2^j k = 2^j(2^\delta k' - k)$, which shows that the shift is an integer multiple of 2^j; however, a forward or backward shift of that amount applied to $\mathcal{W}_{r_{j'}\bullet}$ will always result in aligning $\mathcal{W}_{r_j\bullet}^T$ with a string of equal values (the common value can be either zero, $1/2^{j'/2}$ or $-1/2^{j'/2}$). The inner product will thus always yield zero, as required.

Answer to Exercise [66]: Recall that $\mathcal{V}_J$ is the final row of the orthonormal matrix $\mathcal{W}$ and that $\mathcal{W}_j$ is an $N/2^j \times N$ matrix containing $N/2^j$ other rows of $\mathcal{W}$. Orthonormality thus implies that $\mathcal{W}_j\mathcal{V}_J^T = \mathbf{0}$, where $\mathbf{0}$ is equal to an $N/2^j$ dimensional vector of zeros. Hence, for $1 \le j \le J$,

$$\mathcal{D}_j^T\mathcal{V}_J^T = \mathbf{W}_j^T\mathcal{W}_j\mathcal{V}_J^T = \mathbf{0}.$$

Exercise [97] says that all the elements of $\mathcal{V}_J$ are equal to $1/\sqrt{N}$, so the left-hand side above is proportional to the sum of the elements of $\mathcal{D}_j$, and hence we can conclude that the sample mean of the elements of $\mathcal{D}_j$ is always zero.

Answer to Exercise [69]: Suppose that $\{h'_l : l = 0, \ldots, L-1\}$ is a wavelet filter with odd width L. The orthogonality condition (Equation (69c)) states that we must have

$$\sum_{l=0}^{L-1} h'_l h'_{l+2n} = 0$$

for all nonzero integers n (here $h'_l \equiv 0$ for $l < 0$ and $l \ge L$). If L is odd, then $L-1$ is even, so let $n = (L-1)/2$ to obtain (because $h'_0 \ne 0$ and $h'_{L-1} \ne 0$)

$$\sum_{l=0}^{L-1} h'_l h'_{l+L-1} = h'_0 h'_{L-1} \ne 0,$$

which is in violation of Equation (69c).

Answer to Exercise [70]: Because $\{h \star h_j\}$ is the autocorrelation of $\{h_j\}$, the DFT of $\{h \star h_j\}$ is given by the squared modulus of the DFT of $\{h_j\}$, namely, $\mathcal{H}(f) = |H(f)|^2$ (see Equation (36e)). Using the fact that $\exp(-i\pi) = -1$ and the assumption $h \star h_j = 0$ for nonzero even j, we have

$$\mathcal{H}(f) = \sum_{j=-\infty}^{\infty} h \star h_j e^{-i2\pi fj} = 1 + \sum_{j=\pm1,\pm3,\ldots}^{\infty} h \star h_j e^{-i2\pi fj}$$

$$\begin{aligned}\mathcal{H}(f+\tfrac{1}{2}) = \sum_{j=-\infty}^{\infty} h \star h_j e^{-i2\pi(f+\frac{1}{2})j} &= 1 + \sum_{j=\pm1,\pm3,\ldots}^{\infty} \left[e^{-i\pi}\right]^j h \star h_j e^{-i2\pi fj} \\ &= 1 + \sum_{j=\pm1,\pm3,\ldots}^{\infty} [-1]^j\, h \star h_j e^{-i2\pi fj} \\ &= 1 - \sum_{j=\pm1,\pm3,\ldots}^{\infty} h \star h_j e^{-i2\pi fj}\end{aligned}$$

since $[-1]^j = -1$ for all odd integers j. Addition yields $\mathcal{H}(f) + \mathcal{H}(f+\frac{1}{2}) = 2$.

Alternatively, since $\{h \star h_j\} \longleftrightarrow \mathcal{H}(\cdot)$, we can use Exercise [23b] to write

$$h \star h_{2n} = \frac{1}{2}\int_{-1/2}^{1/2} \left[\mathcal{H}(\tfrac{f}{2}) + \mathcal{H}(\tfrac{f}{2}+\tfrac{1}{2})\right] e^{i2\pi fn}\, df.$$

The above says that $\mathcal{H}(\frac{f}{2}) + \mathcal{H}(\frac{f}{2}+\frac{1}{2})$ defines the function whose inverse DFT is the sequence that is 2 at index $n = 0$ and is 0 otherwise; however, the DFT of this sequence is the function that is equal to 2 everywhere, i.e., $\mathcal{H}(\frac{f}{2}) + \mathcal{H}(\frac{f}{2}+\frac{1}{2}) = 2$.

Answer to Exercise [72]: Because $e^{i2\pi t} = 1$ for all integers t and $\mathcal{H}(f) + \mathcal{H}(f+\frac{1}{2}) = 2$ for all f, we have

$$\begin{aligned}h^\circ \star h^\circ_{2t} &= \frac{1}{N}\left(\sum_{k=0}^{\frac{N}{2}-1} \mathcal{H}(\tfrac{k}{N}) e^{i2\pi(2t)k/N} + \sum_{k=0}^{\frac{N}{2}-1} \mathcal{H}(\tfrac{k}{N}+\tfrac{1}{2}) e^{i2\pi(2t)(\frac{k}{N}+\frac{1}{2})}\right) \\ &= \frac{1}{N}\sum_{k=0}^{\frac{N}{2}-1} \left(\mathcal{H}(\tfrac{k}{N}) + \mathcal{H}(\tfrac{k}{N}+\tfrac{1}{2})\right) e^{i2\pi(2t)k/N} = \frac{2}{N}\sum_{k=0}^{\frac{N}{2}-1} e^{i2\pi(2t)k/N}.\end{aligned}$$

To obtain the desired result, note that

$$\sum_{k=0}^{\frac{N}{2}-1} e^{i2\pi(2t)k/N} = \begin{cases} N/2, & \text{if } t = 0; \\ 0, & \text{if } t = 1, \ldots, \frac{N}{2}-1, \end{cases}$$

which is obvious for $t = 0$ and which is an application of

$$\sum_{k=0}^{\frac{N}{2}-1} z^k = \frac{1 - z^{N/2}}{1 - z} \text{ with } z = e^{i2\pi(2t)/N}, \text{ so } z^{N/2} = e^{i2\pi t} = 1$$

for other t.

Answer to Exercise [76a]: Since $g_l = (-1)^{l+1} h_{L-l-1}$, we have

$$G(f) \equiv \sum_{l=0}^{L-1} g_l e^{-i2\pi fl} = \sum_{l=0}^{L-1} (-1)^{l+1} h_{L-l-1} e^{-i2\pi fl}$$
$$= \sum_{l=0}^{L-1} (-1)^{L-l} h_l e^{-i2\pi f(L-1-l)} = \sum_{l=0}^{L-1} (-1)^l h_l e^{-i2\pi f(L-1-l)},$$

where we have used the fact that $(-1)^{L-l} = (-1)^L (-1)^{-l} = (-1)^l$ because L is even (so that $(-1)^L = 1$) and because $(-1)^{-l} = (-1)^l$. Noting that $e^{-i\pi} = -1$ and hence that $(-1)^l = e^{-i\pi l}$, we have

$$G(f) = \sum_{l=0}^{L-1} e^{-i\pi l} h_l e^{-i2\pi f(L-1-l)} = e^{-i2\pi f(L-1)} \sum_{l=0}^{L-1} h_l e^{-i2\pi(\frac{1}{2}-f)l}$$
$$= e^{-i2\pi f(L-1)} H(\tfrac{1}{2} - f).$$

Since

$$\mathcal{G}(f) \equiv |G(f)|^2 = \left| e^{-i2\pi f(L-1)} H(\tfrac{1}{2} - f) \right|^2 = \left| H(\tfrac{1}{2} - f) \right|^2 = \mathcal{H}(\tfrac{1}{2} - f),$$

the second part follows easily.

Answer to Exercise [76b]: From Exercise [76a] we have

$$\sum_{l=0}^{L-1} g_l = G(0) = H(\tfrac{1}{2}).$$

In Equation (69d), set $f = 0$ to get $|H(0)|^2 + |H(\frac{1}{2})|^2 = 2$; however, from Equation (69a), we have

$$H(0) = \sum_{l=0}^{L-1} h_l = 0,$$

so $|H(\frac{1}{2})|^2 = \mathcal{H}(\frac{1}{2}) = 2$, and hence $H(\frac{1}{2}) = \pm\sqrt{2}$, establishing the first part of the exercise.

Because the squared gain function $\mathcal{G}(\cdot)$ for $\{g_l\}$ satisfies $\mathcal{G}(f) + \mathcal{G}(f + \frac{1}{2}) = 2$ (this is the first part of Equation (76b)), we know from the discussion following Equation (69d) that $\{g_l\}$ must satisfy the orthonormality property, thus establishing the second part of the exercise. Alternatively, we can establish the two parts of the orthonormality property directly as follows. Using $g_l \equiv (-1)^{l+1} h_{L-1-l}$, we have

$$\sum_{l=0}^{L-1} g_l^2 = \sum_{l=0}^{L-1} (-1)^{2l+2} h_{L-1-l}^2 = \sum_{l=0}^{L-1} h_l^2 = 1$$

because a wavelet filter has unit energy (Equation (69a)). Second, for all nonzero integers n we have

$$\sum_{l=0}^{L-1} g_l g_{l+2n} = \sum_{l=0}^{L-1} (-1)^{l+1} h_{L-1-l} (-1)^{l+2n+1} h_{L-1-(l+2n)}$$
$$= \sum_{l=0}^{L-1} (-1)^{2(l+n+1)} h_{L-1-l} h_{L-1-(l+2n)} = \sum_{l=0}^{L-1} h_l h_{l-2n} = 0$$

because a wavelet filter is orthogonal to its even shifts (Equation (69c)).

Answer to Exercise [77]: For all integers n we have

$$\sum_{l=-\infty}^{\infty} g_l h_{l+2n} = \sum_{l=-\infty}^{\infty} (-1)^{l+1} h_{L-1-l} h_{l+2n} \equiv S,$$

say. If we change variable from l to $m \equiv L-1-l-2n$ (so that $l = L-1-m-2n$), we have

$$\begin{aligned}
S &= \sum_{m=-\infty}^{\infty} (-1)^{(L-1-m-2n)+1} h_{L-1-(L-1-m-2n)} h_{(L-1-m-2n)+2n} \\
&= \sum_{m=-\infty}^{\infty} (-1)^{L-m-2n} h_{m+2n} h_{L-1-m} \\
&= \sum_{m=-\infty}^{\infty} (-1)^{-m} h_{m+2n} h_{L-1-m} \quad \text{(because } L-2n \text{ is even)} \\
&= \sum_{m=-\infty}^{\infty} (-1)^{m} h_{m+2n} h_{L-1-m} = -\sum_{m=-\infty}^{\infty} (-1)^{m+1} h_{L-1-m} h_{m+2n} = -S,
\end{aligned}$$

from which we obtain $S = 0$, as required.

Alternatively, here is a frequency domain approach that is closely related to the solution of Exercise [78]. Because g_l is real-valued, we know from Equation (36d) that the DFT of the cross-correlation

$$g \star h_n \equiv \sum_{l=-\infty}^{\infty} g_l h_{l+n}$$

is given by $G^*(f)H(f)$. Let $b_n \equiv g \star h_{2n}$. We need to show that $b_n = 0$ for all n. From Exercise [23b], it follows that the DFT of $\{b_n\}$ is given by

$$B(f) = \frac{1}{2}\left[G^*(\tfrac{f}{2})H(\tfrac{f}{2}) + G^*(\tfrac{f}{2}+\tfrac{1}{2})H(\tfrac{f}{2}+\tfrac{1}{2})\right].$$

Equation (76a) says that $G(f) = e^{-i2\pi f(L-1)}H(\tfrac{1}{2}-f)$, so

$$G^*(f) = e^{i2\pi f(L-1)}H^*(\tfrac{1}{2}-f) = e^{i2\pi f(L-1)}H(f-\tfrac{1}{2}) = e^{i2\pi f(L-1)}H(f+\tfrac{1}{2})$$

because $H^*(f) = H(-f)$ for a real-valued sequence (see Exercise [22b]) and $H(\cdot)$ is of unit periodicity (see Exercise [22a]). Since $G(\cdot)$ also has unit periodicity, the terms in the brackets above thus become

$$\begin{aligned}
&e^{i\pi f(L-1)}H(\tfrac{f}{2}+\tfrac{1}{2})H(\tfrac{f}{2}) + e^{i\pi(f+1)(L-1)}H(\tfrac{f}{2})H(\tfrac{f}{2}+\tfrac{1}{2}) \\
&= e^{i\pi f(L-1)}\left[H(\tfrac{f}{2}+\tfrac{1}{2})H(\tfrac{f}{2}) + e^{i\pi(L-1)}H(\tfrac{f}{2})H(\tfrac{f}{2}+\tfrac{1}{2})\right] \\
&= e^{i\pi f(L-1)}\left[H(\tfrac{f}{2}+\tfrac{1}{2})H(\tfrac{f}{2}) - H(\tfrac{f}{2})H(\tfrac{f}{2}+\tfrac{1}{2})\right] = 0
\end{aligned}$$

because $e^{i\pi(L-1)} = [e^{i\pi}]^{L-1} = [-1]^{L-1} = -1$ (recall that $L-1$ must be an odd integer). Thus $B(f) = 0$ for all f, which tells us that $b_n = 0$ for all n.

Answer to Exercise [78]: The first displayed equation follows from

$$\begin{aligned} g^\circ \star h^\circ_{2n} &= \frac{1}{N}\sum_{k=0}^{N-1} G^*(\tfrac{k}{N})H(\tfrac{k}{N})e^{i4\pi nk/N} \\ &= \frac{1}{N}\Bigg(\sum_{k=0}^{\frac{N}{2}-1} G^*(\tfrac{k}{N})H(\tfrac{k}{N})e^{i4\pi nk/N} \\ &\qquad + \sum_{k=0}^{\frac{N}{2}-1} G^*(\tfrac{k}{N}+\tfrac{1}{2})H(\tfrac{k}{N}+\tfrac{1}{2})e^{i4\pi n(\frac{k}{N}+\frac{1}{2})}\Bigg) \\ &= \frac{1}{N}\sum_{k=0}^{\frac{N}{2}-1} \left[G^*(\tfrac{k}{N})H(\tfrac{k}{N}) + G^*(\tfrac{k}{N}+\tfrac{1}{2})H(\tfrac{k}{N}+\tfrac{1}{2})\right] e^{i4\pi nk/N} \end{aligned}$$

To obtain the second displayed equation, first note the following facts:

$$\begin{aligned} G(f) &= e^{-i2\pi f(L-1)}H(\tfrac{1}{2}-f) \text{ from Equation (76a)} \\ G^*(f) &= e^{i2\pi f(L-1)}H^*(\tfrac{1}{2}-f) = e^{i2\pi f(L-1)}H(f-\tfrac{1}{2}) \\ G^*(f+\tfrac{1}{2}) &= e^{i2\pi(f+\frac{1}{2})(L-1)}H(f) = -e^{i2\pi f(L-1)}H(f), \end{aligned}$$

where we have used the fact that $e^{i\pi(L-1)} = (e^{i\pi})^{L-1} = (-1)^{L-1} = -1$ because L is an even integer. Hence

$$\begin{aligned} &G^*(\tfrac{k}{N})H(\tfrac{k}{N}) + G^*(\tfrac{k}{N}+\tfrac{1}{2})H(\tfrac{k}{N}+\tfrac{1}{2}) \\ &\quad = e^{i2\pi\frac{k}{N}(L-1)}H(\tfrac{k}{N}-\tfrac{1}{2})H(\tfrac{k}{N}) - e^{i2\pi\frac{k}{N}(L-1)}H(\tfrac{k}{N})H(\tfrac{k}{N}+\tfrac{1}{2}) = 0 \end{aligned}$$

because $H(\cdot)$ is periodic with unit period so that $H(\frac{k}{N}-\frac{1}{2}) = H(\frac{k}{N}+\frac{1}{2})$.

Answer to Exercise [91]: Let $\{a_l\} \longleftrightarrow A(\cdot)$ denote the new filter. Since

$$a_l = \begin{cases} h_{l/(m+1)}, & \text{if } l \text{ is an integer multiple of } m+1; \\ 0, & \text{otherwise,} \end{cases}$$

we have

$$\begin{aligned} A(f) = \sum_{l=0}^{(m+1)(L-1)} a_l e^{-i2\pi fl} &= \sum_{l=0}^{L-1} a_{(m+1)l}e^{-i2\pi f(m+1)l} \\ &= \sum_{l=0}^{L-1} h_l e^{-i2\pi([m+1]f)l} = H([m+1]f), \end{aligned}$$

as required.

Answer to Exercise [92]: Since $\{h_{2,l}\} \longleftrightarrow H_2(\cdot)$ and since $H_2(f) = H(2f)G(f)$, we have $H_2(0) = H(0)G(0)$. Since $H(0) = \sum_l h_l = 0$, we have

$$H_2(0) = \sum_{l=0}^{L_2-1} h_{2,l} = 0,$$

as required. When $N \geq L_2$, the elements of $\{h_{2,l}\}$ form the nonzero elements of one of the rows of the DWT matrix $\mathcal{W}$ – since this matrix is orthonormal, each of its rows must have unit energy, and hence so must $\{h_{2,l}\}$. Alternatively, by Parseval's theorem (Equation (35c)),

$$\sum_{l=-\infty}^{\infty} h_{2,l}^2 = \int_{-1/2}^{1/2} \mathcal{H}_2(f)\,df = \int_{-1/2}^{1/2} \mathcal{H}(2f)\mathcal{G}(f)\,df = \frac{1}{2}\int_{-1}^{1} \mathcal{H}(f')\mathcal{G}(\tfrac{f'}{2})\,df'.$$

Following the same line of thought as we used to prove Exercise [23b], we split the last integral up into three parts and change the variables of integration to obtain

$$\sum_{l=-\infty}^{\infty} h_{2,l}^2 = \frac{1}{2}\int_{-1/2}^{1/2} \mathcal{H}(f)\mathcal{G}(\tfrac{f}{2}) + \mathcal{H}(f)\mathcal{G}(\tfrac{f}{2}+\tfrac{1}{2})\,df = \int_{-1/2}^{1/2} \mathcal{H}(f)\,df = \sum_{l=-\infty}^{\infty} h_l^2 = 1,$$

where we have used $\mathcal{G}(\frac{f}{2}) + \mathcal{G}(\frac{f}{2}+\frac{1}{2}) = 2$ and Parseval's theorem again.

To see that $\{h_{2,l}\}$ cannot be orthogonal to its even shifts, suppose that it is so. Since it has unit energy, its squared gain function must satisfy the orthonormality condition $\mathcal{H}_2(\frac{f}{2}) + \mathcal{H}_2(\frac{f}{2}+\frac{1}{2}) = 2$ for all f. However, for $f = 0$ and using $\mathcal{H}_2(f) = \mathcal{H}(2f)\mathcal{G}(f)$, we have

$$\mathcal{H}_2(0) + \mathcal{H}_2(\tfrac{1}{2}) = \mathcal{H}(0)\mathcal{G}(0) + \mathcal{H}(1)\mathcal{G}(\tfrac{1}{2}) = 0$$

since $\mathcal{H}(1) = \mathcal{H}(0) = 0$, which leads to a contradiction. Hence $\{h_{2,l}\}$ cannot be orthogonal to its even shifts.

The second level Haar wavelet filter is given by

$$\{h_{2,l}\} \equiv \{g * h_l^{\uparrow}\} = \{\tfrac{1}{\sqrt{2}}, \tfrac{1}{\sqrt{2}}\} * \{\tfrac{1}{\sqrt{2}}, 0, -\tfrac{1}{\sqrt{2}}\} = \{\tfrac{1}{2}, \tfrac{1}{2}, -\tfrac{1}{2}, -\tfrac{1}{2}\}.$$

This filter sums to zero and has unit energy, but

$$\sum_{l=0}^{1} h_{2,l}h_{2,l+2} = \tfrac{1}{2}\cdot -\tfrac{1}{2} + \tfrac{1}{2}\cdot -\tfrac{1}{2} = -\tfrac{1}{2} \neq 0,$$

so Equation (69c) is not satisfied.

Answer to Exercise [95]: We note that $\mathcal{D}_k = \mathcal{W}_k^T \mathbf{W}_k$ so that $\|\mathcal{D}_k\|^2 = \mathbf{W}_k^T \mathcal{W}_k \mathcal{W}_k^T \mathbf{W}_k$. Since $\mathcal{W}_k = \mathcal{B}_k \mathcal{A}_{k-1} \cdots \mathcal{A}_1$, we have

$$\begin{aligned}
\mathcal{W}_k\mathcal{W}_k^T &= \mathcal{B}_k\mathcal{A}_{k-1}\cdots\mathcal{A}_1\mathcal{A}_1^T\cdots\mathcal{A}_{k-1}^T\mathcal{B}_k^T \\
&= \mathcal{B}_k\mathcal{A}_{k-1}\cdots\mathcal{A}_2 I_{N_1}\mathcal{A}_2^T\cdots\mathcal{A}_{k-1}^T\mathcal{B}_k^T \\
&= \mathcal{B}_k\mathcal{A}_{k-1}\cdots\mathcal{A}_2\mathcal{A}_2^T\cdots\mathcal{A}_{k-1}^T\mathcal{B}_k^T \\
&\;\;\vdots \\
&= \mathcal{B}_k\mathcal{B}_k^T = I_{N_k}.
\end{aligned}$$

Hence

$$\|\mathcal{D}_k\|^2 = \mathbf{W}_k^T I_{N_k} \mathbf{W}_k = \mathbf{W}_k^T\mathbf{W}_k = \|\mathbf{W}_k\|^2.$$

Similarly $\|\mathcal{S}_j\|^2 = \|\mathbf{V}_j\|^2$, and the result follows from Equation (95d).

Answer to Exercise [96]: The result of Exercise [28a] says that the length of the equivalent filter is given by the sum of the lengths of the j individual filters minus $j-1$. Now the kth filter for $k = 1, \ldots, j$ has $2^{k-1} - 1$ zeros following the first $L-1$ nonzero elements, so it has $(L-1)(2^{k-1}-1)$ zero elements and L nonzero elements for a total of $(L-1)2^{k-1}+1$ elements. Hence, because

$$\sum_{k=1}^{j} 2^{k-1} = 2^j - 1,$$

the filter size is given by

$$\sum_{k=1}^{j} \left[(L-1)2^{k-1}+1\right] - (j-1) = (2^j-1)(L-1)+1 \equiv L_j.$$

The transfer function for a filter cascade is given by the product of the individual transfer functions. The first filter has transfer function $G(\cdot)$. The second filter is formed by upsampling the first, so Exercise [91] says that its transfer function is given by $G(2f)$; and the third filter is formed by upsampling the second, so a second application of Exercise [91] says that its transfer function is given by $G(4f)$. Continuing in this manner, the transfer function for the kth of the first $j-1$ filters is given by $G(2^{k-1}f)$. Similarly the transfer function for the jth filter is given by $H(2^{j-1}f)$, so the product of all these yields $H_j(f)$ as given by Equation (96b).

Answer to Exercise [97]: The final row $\mathcal{V}_J$ of the DWT matrix $\mathcal{W}$ is given by

$$[g^{\circ}_{J,N-1}, g^{\circ}_{J,N-2}, \ldots, g^{\circ}_{J,1}, g^{\circ}_{J,0}].$$

Consider

$$\sum_{l=0}^{N-1}\left(g^{\circ}_{J,l} - \tfrac{1}{\sqrt{N}}\right)^2 = \sum_{l=0}^{N-1}\left(g^{\circ}_{J,l}\right)^2 - \frac{2}{\sqrt{N}}\sum_{l=0}^{N-1} g^{\circ}_{J,l} + 1.$$

We claim that the above is equal to 0, from which it follows that $g^{\circ}_{J,l} = 1/\sqrt{N}$ for $l = 0, \ldots, N-1$. To see this, note that, because $\mathcal{V}_J$ is a row of an orthogonal matrix, we must have

$$\sum_{l=0}^{N-1}\left(g^{\circ}_{J,l}\right)^2 = 1, \text{ so } \sum_{l=0}^{N-1}\left(g^{\circ}_{J,l} - \tfrac{1}{\sqrt{N}}\right)^2 = 2 - \frac{2}{\sqrt{N}}\sum_{l=0}^{N-1} g^{\circ}_{J,l};$$

however,

$$\sum_{l=0}^{N-1} g^{\circ}_{J,l} = \sum_{l=0}^{L_J-1} g_{J,l} = G_J(0), \text{ yielding } \sum_{l=0}^{N-1}\left(g^{\circ}_{J,l} - \tfrac{1}{\sqrt{N}}\right)^2 = 2 - \frac{2G_J(0)}{\sqrt{N}}.$$

Recalling that

$$G_J(f) = \prod_{l=0}^{J-1} G(2^l f), \text{ we obtain } G_J(0) = \prod_{l=0}^{J-1} G(0) = \prod_{l=0}^{J-1}\left(\sum_{l=0}^{L-1} g_l\right) = 2^{J/2}.$$

Since $N = 2^J$, we obtain $G_J(0) = \sqrt{N}$, which establishes the claim.

Because $\mathcal{V}_J = [\frac{1}{\sqrt{N}}, \ldots, \frac{1}{\sqrt{N}}]$, it follows that

$$W_{N-1} = \mathcal{V}_J \mathbf{X} = \frac{1}{\sqrt{N}}\sum_{t=0}^{N-1} X_t = \overline{X}\sqrt{N}.$$

Since $\mathbf{V}_J$ is defined to be the 1×1 matrix whose single element is W_{N-1}, then $\mathcal{S}_J \equiv \mathcal{V}_J^T \mathbf{V}_J$ must equal the N dimensional column vector whose elements are all equal to $\overline{X}$.

Answer to Exercise [102a]: A comparison of Equations (95e) and (96e) shows that we can construct $\{h_{j,l}\}$ by convolving together $\{g_{j-1,l}\}$ and the filter $\{a_l\}$ whose impulse response sequence is

$$h_0, \underbrace{0,\ldots,0,}_{2^{j-1}-1 \text{ zeros}} h_1, \underbrace{0,\ldots,0,}_{2^{j-1}-1 \text{ zeros}} \ldots, h_{L-2}, \underbrace{0,\ldots,0,}_{2^{j-1}-1 \text{ zeros}} h_{L-1};$$

i.e., we have

$$a_l \equiv \begin{cases} h_{l/2^{j-1}}, & \text{if } l = 0, 2^{j-1}, \ldots, 2^{j-1}(L-1); \\ 0, & \text{otherwise.} \end{cases}$$

Hence

$$h_{j,l} = \sum_{m=-\infty}^{\infty} g_{j-1,m} a_{l-m}.$$

Making the substitution $l - m = 2^{j-1}k$ gives the first required result:

$$h_{j,l} = \sum_{k=-\infty}^{\infty} h_k g_{j-1,l-2^{j-1}k} = \sum_{k=0}^{L-1} h_k g_{j-1,l-2^{j-1}k}, \quad l = 0, \ldots, L_j - 1$$

(to construct $\{g_{j,l}\}$, we need only replace h_l with g_l when defining $\{a_l\}$).

For the second result, we know from Exercise [91] that $\{a_l\} \longleftrightarrow A(\cdot)$, where $A(f) \equiv H(2^{j-1}f)$. Hence, from Equations (36a) and (36b), the Fourier transform of the convolution $\{h_{j,l}\}$ of $\{g_{j-1,l}\}$ and $\{a_l\}$ is specified by

$$\{h_{j,l}\} \longleftrightarrow G_{j-1}(f)H(2^{j-1}f) = H_j(f).$$

(A similar argument yields $\{g_{j,l}\} \longleftrightarrow G_{j-1}(f)G(2^{j-1}f) = G_j(f)$.)

Answer to Exercise [102b]: Let $\{a_l\}$ be the filter formed by inserting a single zero between the elements of $\{h_{j-1,l}\}$:

$$a_l \equiv \begin{cases} h_{j-1,l/2}, & \text{if } l = 0, 2, \ldots, 2(L_{j-1} - 1); \\ 0, & \text{otherwise.} \end{cases}$$

We first show that the convolution of $\{a_l\}$ with $\{g_l\}$ yields $\{h_{j,l}\} \longleftrightarrow H_j(\cdot)$, where we know from Exercise [96] that

$$H_j(f) = H(2^{j-1}f) \prod_{l=0}^{j-2} G(2^l f).$$

To see this, note that

$$\{h_{j-1,l}\} \longleftrightarrow H_{j-1}(\cdot), \text{ where } H_{j-1}(f) = H(2^{j-2}f) \prod_{l=0}^{j-3} G(2^l f).$$

Using Exercise [91], the transfer function for $\{a_l\}$ is defined by

$$H_{j-1}(2f) = H(2^{j-2}[2f]) \prod_{l=0}^{j-3} G(2^l[2f]) = H(2^{j-1}f) \prod_{l=1}^{j-2} G(2^l f),$$

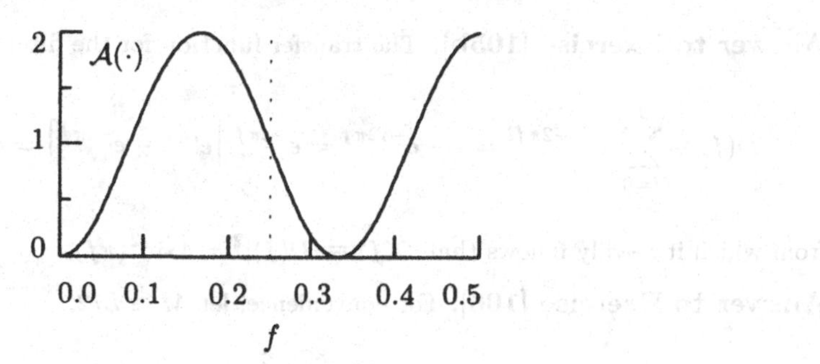

Figure 519. Squared gain function $\mathcal{A}(\cdot)$ for the filter $\{a_0 = -1/\sqrt{2}, a_1 = 0, a_2 = 0, a_3 = -1/\sqrt{2}\}$.

so the transfer function for the convolution of $\{a_l\}$ and $\{g_l\}$ is given by

$$H_{j-1}(2f)G(f) = \left[H(2^{j-1}f)\prod_{l=1}^{j-2} G(2^l f)\right] G(f) = H(2^{j-1}f)\prod_{l=0}^{j-2} G(2^l f) = H_j(f),$$

as claimed. Hence we can write

$$h_{j,l} = \sum_{m=-\infty}^{\infty} g_m a_{l-m}.$$

Making the substitution $l - m = 2k$ then gives

$$h_{j,l} = \sum_{k=-\infty}^{\infty} g_{l-2k} h_{j-1,k} = \sum_{k=0}^{L_{j-1}-1} g_{l-2k} h_{j-1,k}, \quad l = 0, \ldots, L_j - 1,$$

as required (we can use a similar argument to prove the formula for $\{g_{j,l}\}$).

Answer to Exercise [103]: Since

$$\sum_{l=-\infty}^{\infty} g_{j,l} = \sum_{l=0}^{2^j-1} \frac{1}{2^{j/2}} = 2^{j/2} \text{ while } \sum_{l=-\infty}^{\infty} g_{j,l}^2 = \sum_{l=0}^{2^j-1} \frac{1}{2^j} = 1,$$

Equation (103) yields

$$\text{width}_a \{g_{j,l}\} = \frac{\left(\sum_{l=-\infty}^{\infty} g_{j,l}\right)^2}{\sum_{l=-\infty}^{\infty} g_{j,l}^2} = 2^j,$$

as required.

Answer to Exercise [105a]: Since

$$\sum_{l=0}^{3} a_l = 0; \; \sum_{l=0}^{3} a_l^2 = 1; \text{ and } \sum_{l=0}^{3} a_l a_{l+2n} = 0 \text{ for all nonzero integers } n,$$

$\{a_l\}$ satisfies the definition of a wavelet filter (see Equations (69a), (69b) and (69c)). Its transfer function is given by

$$A(f) \equiv \sum_{l=0}^{3} a_l e^{-i2\pi fl} = -\tfrac{1}{\sqrt{2}} + \tfrac{1}{\sqrt{2}} e^{-i2\pi f3} = \tfrac{1}{\sqrt{2}} e^{-i3\pi f}\left[e^{-i3\pi f} - e^{i3\pi f}\right]$$

$$= \tfrac{1}{\sqrt{2}} e^{-i3\pi f}\left[-2i\sin(3\pi f)\right],$$

so the squared gain function is given by $\mathcal{A}(f) \equiv |A(f)|^2 = 2\sin^2(3\pi f)$. Figure 519 shows a plot of $\mathcal{A}(\cdot)$, from which it is evident that $\{a_l\}$ is not a high-pass filter.

Answer to Exercise [105b]: The transfer function for the filter $\{a_0 = 1, a_1 = -1\}$ is

$$D(f) \equiv \sum_{l=0}^{1} a_l e^{-i2\pi fl} = 1 - e^{-i2\pi f} = e^{-i\pi f}\left[e^{i\pi f} - e^{-i\pi f}\right] = e^{-i\pi f}\left[2i\sin(\pi f)\right],$$

from which it readily follows that $\mathcal{D}(f) \equiv |D(f)|^2 = 4\sin^2(\pi f)$.

Answer to Exercise [106]: For convenience, let $M \equiv L/2$,

$$C \equiv \cos^2(\pi f),\quad S \equiv \sin^2(\pi f), M_l \equiv \binom{M+l}{l} \text{ and } M_l^- \equiv \binom{M+l-1}{l},$$

so that

$$\mathcal{G}^{(\mathrm{D})}(f) = 2C^M \sum_{l=0}^{M-1} M_l^- S^l \text{ and } \mathcal{H}^{(\mathrm{D})}(f) = 2S^M \sum_{l=0}^{M-1} M_l^- C^l.$$

As noted in Equation (76b), the condition $\mathcal{H}^{(\mathrm{D})}(f)+\mathcal{H}^{(\mathrm{D})}(f+\frac{1}{2}) = 2$ is equivalent to $\mathcal{G}^{(\mathrm{D})}(f)+\mathcal{H}^{(\mathrm{D})}(f) = 2$, so we must show that

$$C^M \sum_{l=0}^{M-1} M_l^- S^l + S^M \sum_{l=0}^{M-1} M_l^- C^l = 1 \tag{520a}$$

for all positive integers M. We use proof by induction. The case $M = 1$ follows immediately because $M_0^- = 1$ for $M = 1$ and $C + S = \cos^2(\pi f) + \sin^2(\pi f) = 1$. For the inductive step, we need to show that, given Equation (520a), it follows that

$$C^{M+1} \sum_{l=0}^{M} M_l S^l + S^{M+1} \sum_{l=0}^{M} M_l C^l = 1. \tag{520b}$$

Now

$$\begin{aligned} C^{M+1}\sum_{l=0}^{M} M_l S^l &= C^M(1-S)\sum_{l=0}^{M} M_l S^l = C^M \sum_{l=0}^{M}\left[M_l S^l - M_l S^{l+1}\right] \\ &= C^M\left[\sum_{l=0}^{M} M_l S^l - \sum_{l=1}^{M+1} M_{l-1} S^l\right] \\ &= C^M\left[1 + \sum_{l=1}^{M}\left[M_l - M_{l-1}\right] S^l - M_M S^{M+1}\right]. \end{aligned}$$

Since

$$\begin{aligned} M_l - M_{l-1} &= \binom{M+l}{l} - \binom{M+l-1}{l-1} = \frac{(M+l)!}{M!l!} - \frac{(M+l-1)!}{M!(l-1)!} \\ &= \frac{(M+l)! - (M+l-1)!\cdot l}{M!l!} = \frac{(M+l-1)!\left[M+l-l\right]}{M!l!} \\ &= \frac{(M+l-1)!}{(M-1)!l!} = \binom{M+l-1}{l} = M_l^-, \end{aligned}$$

we have

$$C^{M+1}\sum_{l=0}^{M} M_l S^l = C^M\left[1+\sum_{l=1}^{M} M_l^- S^l - M_M S^{M+1}\right]$$
$$= C^M\left[\sum_{l=0}^{M-1} M_l^- S^l + M_M^- S^M - M_M S^{M+1}\right].$$

Likewise, the second term in Equation (520b) can be expressed as

$$S^{M+1}\sum_{l=0}^{M} M_l C^l = S^M\left[\sum_{l=0}^{M-1} M_l^- C^l + M_M^- C^M - M_M C^{M+1}\right].$$

Hence use of these two expressions and the induction hypothesis of Equation (520a) yields

$$C^{M+1}\sum_{l=0}^{M} M_l S^l + S^{M+1}\sum_{l=0}^{M} M_l C^l$$
$$= 1 + C^M\left(M_M^- S^M - M_M S^{M+1}\right) + S^M\left(M_M^- C^M - M_M C^{M+1}\right)$$
$$= 1 + 2C^M S^M M_M^- - C^M S^M M_M(C+S)$$
$$= 1 + 2C^M S^M M_M^- - C^M S^M M_M = 1 + C^M S^M\left(2M_M^- - M_M\right) = 1,$$

where we have used the fact that

$$2M_M^- - M_M = 2\binom{2M-1}{M} - \binom{2M}{M} = 2\frac{(2M-1)!}{M!(M-1)!} - \frac{(2M)!}{M!M!}$$
$$= \frac{2M\cdot(2M-1)! - (2M)!}{M!M!} = 0.$$

This establishes Equation (520b).

Answer to Exercise [111]: We have

$$U^{(\nu)}(f) = \sum_{l=-\infty}^{\infty} u_l^{(\nu)} e^{-i2\pi fl} = \sum_{l=-\infty}^{\infty} u_{l+\nu} e^{-i2\pi fl}$$
$$= \sum_{l=-\infty}^{\infty} u_l e^{-i2\pi f(l-\nu)}$$
$$= e^{i2\pi f\nu}\sum_{l=-\infty}^{\infty} u_l e^{-i2\pi fl} = e^{i2\pi f\nu} U(f),$$

as required.

Answer to Exercise [112]: Equation (76a) says that

$$G(f') = e^{-i2\pi f'(L-1)} H(\tfrac{1}{2} - f'),$$

so we have

$$H(\tfrac{1}{2} - f') = e^{i2\pi f'(L-1)} G(f').$$

Letting $f' = \frac{1}{2} - f$ and using the fact that $e^{i\pi(L-1)} = e^{i\pi}$ because $L-1$ is always an odd integer, we obtain

$$H(f) = e^{i2\pi(\frac{1}{2}-f)(L-1)} G(\tfrac{1}{2} - f) = e^{-i2\pi f(L-1)+i\pi} G(\tfrac{1}{2} - f),$$

as required.

Answer to Exercise [114]: Since $\theta^{(G)}(\cdot)$ is the phase function for $G(\cdot)$, we have $G(f) = |G(f)|e^{i\theta^{(G)}(f)}$. Hence

$$G_j(f) = \prod_{l=0}^{j-1} G(2^l f) = \prod_{l=0}^{j-1} \left[|G(2^l f)|e^{i\theta^{(G)}(2^l f)}\right] = \left[\prod_{l=0}^{j-1} |G(2^l f)|\right] e^{i\sum_{l=0}^{j-1}\theta^{(G)}(2^l f)},$$

so the phase function for $G_j(\cdot)$ is given by

$$\theta_j^{(G)}(f) \equiv \sum_{l=0}^{j-1} \theta^{(G)}(2^l f).$$

Since $\sum_{l=0}^{j-1} 2^l = 2^j - 1$, using $\theta^{(G)}(f) \approx 2\pi f\nu$ in the above yields

$$\theta_j^{(G)}(f) \approx \sum_{l=0}^{j-1} 2\pi(2^l f)\nu = 2\pi f\left(2^j - 1\right)\nu = 2\pi f \nu_j^{(G)} \tag{522}$$

with $\nu_j^{(G)} \equiv (2^j - 1)\nu$. Likewise, since $\theta^{(H)}(\cdot)$ is the phase function for $H(\cdot)$, we have $H(f) = |H(f)|e^{i\theta^{(H)}(f)}$. Hence

$$H_j(f) = H(2^{j-1}f)G_{j-1}(f) = |H(2^{j-1}f)||G_{j-1}(f)|e^{i\theta^{(H)}(2^{j-1}f)}e^{i\theta_{j-1}^{(G)}(f)},$$

so the phase function for $H_j(\cdot)$ is given by

$$\theta_j^{(H)}(f) \equiv \theta^{(H)}(2^{j-1}f) + \theta_{j-1}^{(G)}(f).$$

Using $\theta^{(H)}(f) \approx -2\pi f(L - 1 + \nu)$ and Equation (522) in the above yields

$$\theta_j^{(H)}(f) \approx -2\pi(2^{j-1}f)(L-1+\nu) + 2\pi f\left(2^{j-1} - 1\right)\nu = 2\pi f\left[-2^{j-1}(L-1) - \nu\right] = 2\pi f \nu_j^{(H)}$$

with $\nu_j^{(H)} \equiv -\left[2^{j-1}(L-1) + \nu\right]$.

Answer to Exercise [116]: The LA(8) scaling filter has width $L = 8$, so Equation (112e) says that $\nu = -\frac{L}{2} + 1 = -3$. Equation (114b) now says that

$$\nu_j^{(H)} = -\left(2^{j-1}[L-1] + \nu\right) = -7 \cdot 2^{j-1} + 3 = \begin{cases} -4, & j = 1; \\ -11, & j = 2; \\ -25, & j = 3; \end{cases}$$

on the other hand, Equation (114a) says that

$$\nu_3^{(G)} = (2^3 - 1)\nu = -21.$$

Since $\Delta t = 1$ here, the association between indices t and actual times for the wavelet coefficients $\{W_{1,t} : t = 0, 1, \ldots, 63\}$ making up $\mathbf{W}_1$ is given by the expression $t_0 + (2(t+1) - 1 + \nu_1^{(H)} \bmod N)$, so we have

$$\text{actual time} = 17 + (2t - 3 \bmod 128) = \begin{cases} 142, & t = 0; \\ 144, & t = 1; \\ 18, & t = 2; \\ 20, & t = 3; \\ 22, & t = 4; \\ \cdots & \\ 138, & t = 62; \\ 140, & t = 63; \end{cases}$$

the association between indices t and actual times for the wavelet coefficients $\{W_{2,t} : t = 0, 1, \ldots, 31\}$ making up $\mathbf{W}_2$ is given by the expression $t_0 + (4(t+1) - 1 + \nu_2^{(H)} \bmod N)$, so we have

$$\text{actual time} = 17 + (4t - 8 \bmod 128) = \begin{cases} 137, & t = 0; \\ 141, & t = 1; \\ 17, & t = 2; \\ 21, & t = 3; \\ 25, & t = 4; \\ \ldots & \\ 129, & t = 30; \\ 133, & t = 31; \end{cases}$$

the association between indices t and actual times for the wavelet coefficients $\{W_{3,t} : t = 0, 1, \ldots, 15\}$ making up $\mathbf{W}_3$ is given by the expression $t_0 + (8(t+1) - 1 + \nu_3^{(H)} \bmod N)$, so we have

$$\text{actual time} = 17 + (8t - 18 \bmod 128) = \begin{cases} 127, & t = 0; \\ 135, & t = 1; \\ 143, & t = 2; \\ 23, & t = 3; \\ 31, & t = 4; \\ \ldots & \\ 111, & t = 14; \\ 119, & t = 15; \end{cases}$$

and, finally, the association between indices t and actual times for the scaling coefficients $\{V_{3,t} : t = 0, 1, \ldots, 15\}$ making up $\mathbf{V}_3$ is given by the expression $t_0 + (8(t+1) - 1 + \nu_3^{(G)} \bmod N)$, so we have

$$\text{actual time} = 17 + (8t - 14 \bmod 128) = \begin{cases} 131, & t = 0; \\ 139, & t = 1; \\ 19, & t = 2; \\ 27, & t = 3; \\ 35, & t = 4; \\ \ldots & \\ 115, & t = 14; \\ 123, & t = 15. \end{cases}$$

Plots of the elements of $\mathbf{W}_1$, $\mathbf{W}_2$, $\mathbf{W}_3$ and $\mathbf{V}_3$ versus their actual times are shown in Figure 524.

Answer to Exercise [117]: Define the reversed filter by $a_l = g_{L-1-l}$, $l = 0, \ldots, L-1$. Then

$$A(f) \equiv \sum_{l=0}^{L-1} a_l e^{-i2\pi fl} = \sum_{l=0}^{L-1} g_{L-1-l} e^{-i2\pi fl} = \sum_{l=0}^{L-1} g_l e^{-i2\pi f(L-1-l)}$$

$$= e^{-i2\pi f(L-1)} \sum_{l=0}^{L-1} g_l e^{i2\pi fl} = e^{-i2\pi f(L-1)} G^*(f)$$

because $\{g_l\}$ is real-valued.

Answer to Exercise [118]: Since $\sum_l g_l^2 = \sum_l h_l^2 = 1$, we have

$$e\{g_l\} + e\{h_l\} = \sum_{l=0}^{L-1} l g_l^2 + \sum_{l=0}^{L-1} l h_l^2.$$

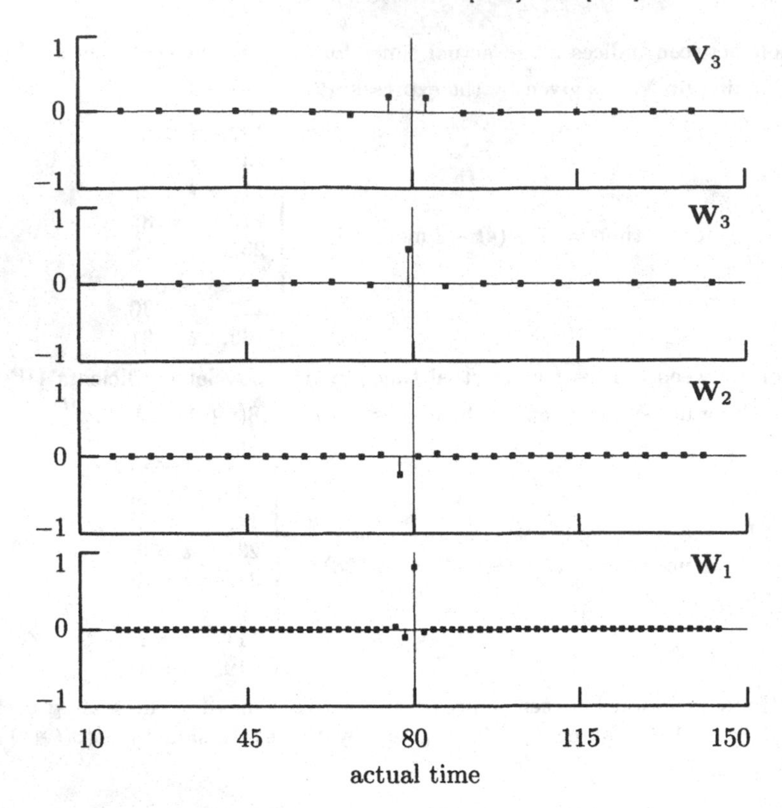

Figure 524. Wavelet and scaling coefficients for a level $J_0 = 3$ partial LA(8) DWT of a time series $\{X_t : t = 0, \ldots, 127\}$ that is zero everywhere for $X_{63} = 1$. The tth value of X_t is associated with actual time $t + 17$, so X_{63} is associated with actual time 80, which is indicated by the thin vertical lines in the plots above.

Using Equation (75a), we have

$$\sum_{l=0}^{L-1} l g_l^2 = \sum_{l=0}^{L-1} l h_{L-1-l}^2 = \sum_{l=0}^{L-1} (L-1-l) h_l^2,$$

which leads to

$$e\{g_l\} + e\{h_l\} = \sum_{l=0}^{L-1} (L-1-l) h_l^2 + \sum_{l=0}^{L-1} l h_l^2 = (L-1) \sum_{l=0}^{L-1} h_l^2 = L - 1,$$

as required.

Answer to Exercise [139]: The proportions of affected terms in $\mathcal{D}_j$ and in $\mathbf{W}_j$ are given by, respectively, $(2^j L_j' + L_j - 2^j)/N$ and $L_j'/N_j = 2^j L_j'/N$. For large j, we have $L_j' = L - 2$, so the ratio of proportions is

$$\frac{2^j L_j' + L_j - 2^j}{2^j L_j'} = \frac{2^j (L-2) + (2^j - 1)(L-1) + 1 - 2^j}{2^j (L-2)} = \frac{2^{j+1} - 1}{2^j} \approx 2,$$

as required.

Answer to Exercise [149]: Since $L'_j > 0$ if and only if $L \geq 4$, we have

$$L_j = (2^j - 1)(L - 1) + 1 \geq 3 \cdot 2^j - 2 > 2^j.$$

To show that $L_j > 2^j L'_j$, write $L'_j = r + \delta$ with $r = (L-2)(1 - \frac{1}{2^j})$ and $0 \leq \delta < 1$. Then

$$\begin{aligned} 2^j L'_j &= 2^j r + 2^j \delta = (L-2)(2^j - 1) + 2^j \delta \\ &= (2^j - 1)(L-1) + 1 - (2^j - 1) - 1 + 2^j \delta = L_j + 2^j(\delta - 1). \end{aligned}$$

Hence,

$$L_j = 2^j L'_j + 2^j(1 - \delta) > 2^j L'_j,$$

as required

Answer to Exercise [166]: Note that

$$\tfrac{1}{2}\mathcal{B}_1^T \mathbf{W}_1 = \tfrac{1}{2} \begin{bmatrix} h_1 & h_3 & \cdots \\ h_0 & h_2 & \cdots \\ 0 & h_1 & \cdots \\ 0 & h_0 & \cdots \\ \vdots & \vdots & \cdots \end{bmatrix} \begin{bmatrix} \sqrt{2}\widetilde{W}_{1,1} \\ \sqrt{2}\widetilde{W}_{1,3} \\ \vdots \end{bmatrix} = \begin{bmatrix} \tilde{h}_1 & \tilde{h}_3 & \cdots \\ \tilde{h}_0 & \tilde{h}_2 & \cdots \\ 0 & \tilde{h}_1 & \cdots \\ 0 & \tilde{h}_0 & \cdots \\ \vdots & \vdots & \cdots \end{bmatrix} \begin{bmatrix} \widetilde{W}_{1,1} \\ \widetilde{W}_{1,3} \\ \vdots \end{bmatrix}$$

and that

$$\tfrac{1}{2}\mathcal{B}_{\mathcal{T},1}^T \mathbf{W}_{\mathcal{T},1} = \tfrac{1}{2} \begin{bmatrix} h_0 & h_2 & \cdots \\ 0 & h_1 & \cdots \\ 0 & h_0 & \cdots \\ 0 & 0 & \cdots \\ \vdots & \vdots & \cdots \end{bmatrix} \begin{bmatrix} \sqrt{2}\widetilde{W}_{1,0} \\ \sqrt{2}\widetilde{W}_{1,2} \\ \vdots \end{bmatrix} = \begin{bmatrix} \tilde{h}_0 & \tilde{h}_2 & \cdots \\ 0 & \tilde{h}_1 & \cdots \\ 0 & \tilde{h}_0 & \cdots \\ 0 & 0 & \cdots \\ \vdots & \vdots & \cdots \end{bmatrix} \begin{bmatrix} \widetilde{W}_{1,0} \\ \widetilde{W}_{1,2} \\ \vdots \end{bmatrix}.$$

By interleaving the above, we can write

$$\tfrac{1}{2}\mathcal{B}_1^T \mathbf{W}_1 + \tfrac{1}{2}\mathcal{B}_{\mathcal{T},1}^T \mathbf{W}_{\mathcal{T},1} = \begin{bmatrix} \tilde{h}_0 & \tilde{h}_1 & \tilde{h}_2 & \tilde{h}_3 & \cdots \\ 0 & \tilde{h}_0 & \tilde{h}_1 & \tilde{h}_2 & \cdots \\ 0 & 0 & \tilde{h}_0 & \tilde{h}_1 & \cdots \\ 0 & 0 & 0 & \tilde{h}_0 & \cdots \\ \vdots & \vdots & \vdots & \vdots & \cdots \end{bmatrix} \begin{bmatrix} \widetilde{W}_{1,0} \\ \widetilde{W}_{1,1} \\ \widetilde{W}_{1,2} \\ \widetilde{W}_{1,3} \\ \vdots \end{bmatrix} = \widetilde{\mathcal{B}}_1 \widetilde{\mathbf{W}}_1.$$

A similar argument yields $\frac{1}{2}\mathcal{A}_1^T \mathbf{V}_1 + \frac{1}{2}\mathcal{A}_{\mathcal{T},1}^T \mathbf{V}_{\mathcal{T},1} = \widetilde{\mathcal{A}}_1 \widetilde{\mathbf{V}}_1$.

Answer to Exercise [167a]: Since $\widetilde{\mathcal{B}}_1^T \widetilde{\mathbf{W}}_1 = \frac{1}{2}(\mathcal{B}_1^T \mathbf{W}_1 + \mathcal{B}_{\mathcal{T},1}^T \mathbf{W}_{\mathcal{T},1})$ and since $\mathcal{B}_1 \mathcal{B}_1^T = I_{N/2}$, and $\mathcal{B}_{\mathcal{T},1}\mathcal{B}_{\mathcal{T},1}^T = I_{N/2}$ we have

$$\begin{aligned} \|\widetilde{\mathcal{D}}_1\|^2 &= \|\widetilde{\mathcal{B}}_1^T \widetilde{\mathbf{W}}_1\|^2 \\ &= \tfrac{1}{4}\|\mathcal{B}_1^T \mathbf{W}_1 + \mathcal{B}_{\mathcal{T},1}^T \mathbf{W}_{\mathcal{T},1}\|^2 \\ &= \tfrac{1}{4}\left(\mathcal{B}_1^T \mathbf{W}_1 + \mathcal{B}_{\mathcal{T},1}^T \mathbf{W}_{\mathcal{T},1}\right)^T \left(\mathcal{B}_1^T \mathbf{W}_1 + \mathcal{B}_{\mathcal{T},1}^T \mathbf{W}_{\mathcal{T},1}\right) \\ &= \tfrac{1}{4}(\mathbf{W}_1^T \mathcal{B}_1 \mathcal{B}_1^T \mathbf{W}_1 + \mathbf{W}_{\mathcal{T},1}^T \mathcal{B}_{\mathcal{T},1} \mathcal{B}_{\mathcal{T},1}^T \mathbf{W}_{\mathcal{T},1} + \mathbf{W}_1^T \mathcal{B}_1 \mathcal{B}_{\mathcal{T},1}^T \mathbf{W}_{\mathcal{T},1} + \mathbf{W}_{\mathcal{T},1}^T \mathcal{B}_{\mathcal{T},1} \mathcal{B}_1^T \mathbf{W}_1) \\ &= \tfrac{1}{4}(\mathbf{W}_1^T \mathbf{W}_1 + \mathbf{W}_{\mathcal{T},1}^T \mathbf{W}_{\mathcal{T},1} + \mathbf{W}_1^T \mathcal{B}_1 \mathcal{B}_{\mathcal{T},1}^T \mathbf{W}_{\mathcal{T},1} + \mathbf{W}_{\mathcal{T},1}^T \mathcal{B}_{\mathcal{T},1} \mathcal{B}_1^T \mathbf{W}_1) \\ &= \tfrac{1}{4}(\|\mathbf{W}_1\|^2 + \|\mathbf{W}_{\mathcal{T},1}\|^2 + \mathbf{W}_1^T \mathcal{B}_1 \mathcal{B}_{\mathcal{T},1}^T \mathbf{W}_{\mathcal{T},1} + \mathbf{W}_{\mathcal{T},1}^T \mathcal{B}_{\mathcal{T},1} \mathcal{B}_1^T \mathbf{W}_1). \end{aligned}$$

Since $\|\widetilde{\mathbf{W}}_1\|^2 = \frac{1}{2}\left(\|\mathbf{W}_1\|^2 + \|\mathbf{W}_{\mathcal{T},1}\|^2\right)$ and the scalar $\mathbf{W}_1^T \mathcal{B}_1 \mathcal{B}_{\mathcal{T},1}^T \mathbf{W}_{\mathcal{T},1}$ is equal to its transpose $\mathbf{W}_{\mathcal{T},1}^T \mathcal{B}_{\mathcal{T},1} \mathcal{B}_1^T \mathbf{W}_1$, we obtain

$$\|\widetilde{\mathcal{D}}_1\|^2 = \tfrac{1}{2}\left(\|\widetilde{\mathbf{W}}_1\|^2 + \mathbf{W}_1^T \mathcal{B}_1 \mathcal{B}_{\mathcal{T},1}^T \mathbf{W}_{\mathcal{T},1}\right),$$

as required.

Answer to Exercise [167b]: Let $\{\tilde{h}_l^\circ\}$ be $\{\tilde{h}_l\}$ periodized to length N:

$$\tilde{h}_l^\circ \equiv \begin{cases} \tilde{h}_l, & 0 \le l \le L-1; \\ 0, & L \le l \le N-1; \text{ and} \\ \tilde{h}_{l \bmod N}^\circ, & l < 0 \text{ or } l \ge N. \end{cases}$$

Define $\{\tilde{g}_l^\circ\}$ in an analogous manner. Then the (j,k)th element of $\widetilde{\mathcal{B}}_1^T \widetilde{\mathcal{B}}_1$ is given by

$$b_{j,k} = \sum_{l=0}^{N-1} \tilde{h}_{j+l}^\circ \tilde{h}_{k+l}^\circ = \sum_{l=0}^{N-1} \tilde{h}_l^\circ \tilde{h}_{k-j+l}^\circ = \sum_{l=0}^{L-1} \tilde{h}_l^\circ \tilde{h}_{k-j+l}^\circ, \quad 0 \le j,k \le N-1,$$

and there is an analogous expression (involving $\tilde{g}_l^\circ$'s) for the (j,k)th element $a_{j,k}$ of $\widetilde{\mathcal{A}}_1^T \widetilde{\mathcal{A}}_1$. The (j,k)th element of $\widetilde{\mathcal{B}}_1^T \widetilde{\mathcal{B}}_1 + \widetilde{\mathcal{A}}_1^T \widetilde{\mathcal{A}}_1$ thus is given by $b_{j,k} + a_{j,k}$, so we need to show that $b_{j,k} + a_{j,k} = 1$ when $j = k$ and is 0 otherwise. When $j = k$, we have

$$b_{k,k} = \sum_{l=0}^{L-1} [\tilde{h}_l^\circ]^2 = \sum_{l=0}^{L-1} \tilde{h}_l^2 = \tfrac{1}{2}$$

from the middle part of Equation (163a); since we similarly have $a_{k,k} = \frac{1}{2}$, it follows that $b_{k,k} + a_{k,k} = 1$. If $k - j = 2m$ for some nonzero integer m, then

$$b_{j,k} = \sum_{l=0}^{L-1} \tilde{h}_l^\circ \tilde{h}_{2m+l}^\circ = \sum_{l=0}^{L-1} \tilde{h}_l \tilde{h}_{2m+l \bmod N} = 0$$

because the second sum above can be rewritten as either one or two sums of the form

$$\sum_{l=0}^{L-1} \tilde{h}_l \tilde{h}_{2m'+l}$$

for some m', which is zero from the last part of Equation (163a). A similar argument holds for $a_{j,k}$, so $b_{j,k} + a_{j,k} = 0$ when $|k - j|$ is a positive even integer. Finally, if $k - j$ is an odd integer, note first that the first relationship of Equation (163d) and the definitions of $\{\tilde{g}_l^\circ\}$ and $\{\tilde{h}_l^\circ\}$ tell us that

$$g_l^\circ = (-1)^{l+1} \tilde{h}_{L-l-1}^\circ \text{ for all } l.$$

Hence we have

$$\begin{aligned} b_{j,k} + a_{j,k} &= \sum_{l=0}^{N-1} \tilde{h}_l \tilde{h}_{k-j+l} + \sum_{l=0}^{N-1} \tilde{g}_l \tilde{g}_{k-j+l} \\ &= \sum_{l=0}^{N-1} \tilde{h}_l \tilde{h}_{k-j+l} + \sum_{l=0}^{N-1} (-1)^{l+1} \tilde{h}_{L-l-1} (-1)^{k-j+l+1} \tilde{h}_{L-(k-j+l)-1} \\ &= \sum_{l=0}^{N-1} \tilde{h}_l \tilde{h}_{k-j+l} - \sum_{l=0}^{N-1} \tilde{h}_{L-l-1} \tilde{h}_{L-k+j-l-1} \\ &= \sum_{l=0}^{N-1} \tilde{h}_l \tilde{h}_{k-j+l} - \sum_{l=0}^{N-1} \tilde{h}_{-l} \tilde{h}_{-k+j-l} \\ &= \sum_{l=0}^{N-1} \tilde{h}_l \tilde{h}_{k-j+l} - \sum_{l=0}^{N-1} \tilde{h}_l \tilde{h}_{l-k+j} = \sum_{l=0}^{N-1} \tilde{h}_l \tilde{h}_{k-j+l} - \sum_{l=0}^{N-1} \tilde{h}_{l+k-j} \tilde{h}_l = 0, \end{aligned}$$

which proves the hint.

From Equation (165b), we have $\widetilde{\mathbf{W}}_1 = \widetilde{\mathcal{B}}_1\mathbf{X}$ and $\widetilde{\mathbf{V}}_1 = \widetilde{\mathcal{A}}_1\mathbf{X}$. Premultiplying both sides of the first equation by $\widetilde{\mathcal{B}}_1^T$ and both sides of the second by $\widetilde{\mathcal{A}}_1^T$ and then adding yields

$$\widetilde{\mathcal{B}}_1^T\widetilde{\mathbf{W}}_1 + \widetilde{\mathcal{A}}_1^T\widetilde{\mathbf{V}}_1 = \widetilde{\mathcal{B}}_1^T\widetilde{\mathcal{B}}_1\mathbf{X} + \widetilde{\mathcal{A}}_1^T\widetilde{\mathcal{A}}_1\mathbf{X} = (\widetilde{\mathcal{B}}_1^T\widetilde{\mathcal{B}}_1 + \widetilde{\mathcal{A}}_1^T\widetilde{\mathcal{A}}_1)\mathbf{X} = \mathbf{X},$$

where we have made use of the hint. Premultiplying both sides of the above by $\mathbf{X}^T$ yields

$$\mathbf{X}^T\widetilde{\mathcal{B}}_1^T\widetilde{\mathbf{W}}_1 + \mathbf{X}^T\widetilde{\mathcal{A}}_1^T\widetilde{\mathbf{V}}_1 = \mathbf{X}^T\mathbf{X};$$

however, since $\widetilde{\mathbf{W}}_1^T = \mathbf{X}^T\widetilde{\mathcal{B}}_1^T$ and $\widetilde{\mathbf{V}}_1^T = \mathbf{X}^T\widetilde{\mathcal{A}}_1^T$, the above becomes

$$\widetilde{\mathbf{W}}_1^T\widetilde{\mathbf{W}}_1 + \widetilde{\mathbf{V}}_1^T\widetilde{\mathbf{V}}_1 = \mathbf{X}^T\mathbf{X},$$

i.e., $\|\mathbf{W}_1\|^2 + \|\mathbf{V}_1\|^2 = \|\mathbf{X}\|^2$, as required.

Answer to Exercise [171a]: Let $\{\tilde{h}_j^\circ\}$ and $\{\tilde{g}_j^\circ\}$ be the filters obtained by periodizing $\{\tilde{h}_j\} = \{h_j/\sqrt{2}\}$ and $\{\tilde{g}_j\} = \{g_j/\sqrt{2}\}$ to length N. Note that $\{\tilde{h}_j^\circ\} \longleftrightarrow \{\widetilde{H}(\frac{k}{N})\}$ and $\{\tilde{g}_j^\circ\} \longleftrightarrow \{\widetilde{G}(\frac{k}{N})\}$. Now

$$\widetilde{W}_{1,t} = \sum_{l=0}^{N-1} \tilde{h}_l^\circ X_{t-l \bmod N} \text{ and } \widetilde{V}_{1,t} = \sum_{l=0}^{N-1} \tilde{g}_l^\circ X_{t-l \bmod N}.$$

Letting $\{\mathcal{X}_k\}$ be the DFT of $\{X_t\}$, we have

$$\{\widetilde{W}_{1,t}\} \longleftrightarrow \{\widetilde{H}(\tfrac{k}{N})\mathcal{X}_k\} \text{ and } \{\widetilde{V}_{1,t}\} \longleftrightarrow \{\widetilde{G}(\tfrac{k}{N})\mathcal{X}_k\}.$$

Parseval's theorem (Equation (36h)) yields

$$\|\widetilde{\mathbf{W}}_1\|^2 = \frac{1}{N}\sum_{k=0}^{N-1} |\widetilde{H}(\tfrac{k}{N})|^2|\mathcal{X}_k|^2 = \frac{1}{N}\sum_{k=0}^{N-1} \widetilde{\mathcal{H}}(\tfrac{k}{N})|\mathcal{X}_k|^2$$

and

$$\|\widetilde{\mathbf{V}}_1\|^2 = \frac{1}{N}\sum_{k=0}^{N-1} |\widetilde{G}(\tfrac{k}{N})|^2|\mathcal{X}_k|^2 = \frac{1}{N}\sum_{k=0}^{N-1} \widetilde{\mathcal{G}}(\tfrac{k}{N})|\mathcal{X}_k|^2.$$

Addition yields

$$\|\widetilde{\mathbf{W}}_1\|^2 + \|\widetilde{\mathbf{V}}_1\|^2 = \frac{1}{N}\sum_{k=0}^{N-1} |\mathcal{X}_k|^2 \left(\widetilde{\mathcal{H}}(\tfrac{k}{N}) + \widetilde{\mathcal{G}}(\tfrac{k}{N})\right) = \frac{1}{N}\sum_{k=0}^{N-1} |\mathcal{X}_k|^2 = \|\mathbf{X}\|^2,$$

where we have made use of (i) the fact that $\widetilde{\mathcal{H}}(f) + \widetilde{\mathcal{G}}(f) = 1$ for all f (see Equation (163c)) and (ii) Parseval's theorem again.

Answer to Exercise [171b]: The elements of $\widetilde{\mathbf{V}}_{J_0}$ are just $\{\widetilde{V}_{J_0,t}\}$, which is the result of circularly convolving $\{X_t\}$ with $\{\tilde{g}_{J_0,l}^\circ\}$ (see Equation (170a)). Since $\{\tilde{g}_{J_0,l}^\circ\}$ is the result of periodizing $\{\tilde{g}_{J_0,l}\}$ to length N and since $\{\tilde{g}_{J_0,l}\} \longleftrightarrow \widetilde{G}_{J_0}(\cdot)$, we have $\{\tilde{g}_{J_0,l}^\circ\} \longleftrightarrow \{\widetilde{G}_{J_0}(\frac{k}{N})\}$ (this follows from Exercise [33]). If we let $\{\mathcal{X}_k\}$ denote the DFT of $\{X_t\}$, then we have

$$\{\widetilde{V}_{J_0,t}\} \longleftrightarrow \{\widetilde{G}_{J_0}(\tfrac{k}{N})\mathcal{X}_k\}.$$

Next note that, if $\{a_t : t = 0, \ldots, N-1\} \longleftrightarrow \{A_k : k = 0, \ldots, N-1\}$, then $A_0 = N\bar{a}$, where $\bar{a}$ is the sample mean of $\{a_t\}$. It now follows that the sample mean of $\{\widetilde{V}_{J_0,t}\}$ is given by

$$\frac{1}{N}\widetilde{G}_{J_0}(0)\mathcal{X}_0 = \frac{1}{N}\left(\prod_{l=0}^{J_0-1} \widetilde{G}(0)\right) N\overline{X} = \overline{X}$$

since $\widetilde{G}(0) = G(0)/\sqrt{2} = 1$ because $G(0) = \sum_l g_l = \sqrt{2}$ from Exercise [76b]. Substituting $\{\widetilde{V}_{J_0,t}\}$ for $\{X_t\}$ in the relationship $\hat{\sigma}_X^2 = \frac{1}{N}\|\mathbf{X}\|^2 - \overline{X}^2$ now yields $\frac{1}{N}\|\widetilde{\mathbf{V}}_{J_0}\|^2 - \overline{X}^2$ as the sample variance for $\{\widetilde{V}_{J_0,t}\}$.

Answer to Exercise [172]: The elements of $\widetilde{\mathcal{D}}_1$ and $\widetilde{\mathcal{S}}_1$ are given by

$$\widetilde{\mathcal{D}}_{1,t} = \sum_{l=0}^{N-1} \tilde{h}_l^\circ \widetilde{W}_{1,t+l \bmod N} \text{ and } \widetilde{\mathcal{S}}_{1,t} = \sum_{l=0}^{N-1} \tilde{g}_l^\circ \widetilde{V}_{1,t+l \bmod N},$$

$t = 0, 1, \ldots, N-1$. As before, $\{\widetilde{\mathcal{D}}_{1,t}\}$ can be regarded as the result of circularly filtering $\{\widetilde{W}_{1,t}\}$ with a filter whose DFT is $\{\widetilde{H}^*(\frac{k}{N})\}$; on the other hand, $\{\widetilde{W}_{1,t}\}$ is the result of circularly filtering $\{X_t\}$ with the filter $\{\tilde{h}_l^\circ\}$ whose DFT is given by $\{\widetilde{H}(\frac{k}{N})\}$. Letting $\{\mathcal{X}_k\}$ represent the DFT of $\{X_t\}$ again, we have

$$\{\widetilde{\mathcal{D}}_{1,t}\} \longleftrightarrow \{\widetilde{H}^*(\tfrac{k}{N})\widetilde{H}(\tfrac{k}{N})\mathcal{X}_k\} = \{|\widetilde{H}(\tfrac{k}{N})|^2\mathcal{X}_k\} = \{\widetilde{\mathcal{H}}(\tfrac{k}{N})\mathcal{X}_k\},$$

and, likewise, $\{\widetilde{\mathcal{S}}_{1,t}\} \longleftrightarrow \{\widetilde{\mathcal{G}}(\frac{k}{N})\mathcal{X}_k\}$. Addition yields

$$\{\widetilde{\mathcal{D}}_{1,t} + \widetilde{\mathcal{S}}_{1,t}\} \longleftrightarrow \{(\widetilde{\mathcal{H}}(\tfrac{k}{N}) + \widetilde{\mathcal{G}}(\tfrac{k}{N}))\mathcal{X}_k\} = \{\mathcal{X}_k\}$$

because $\mathcal{H}(f) + \mathcal{G}(f) = 1$ for all f (see Equation (163b)). Since $\{X_t\} \longleftrightarrow \{\mathcal{X}_k\}$, we must have $X_t = \widetilde{\mathcal{D}}_{1,t} + \widetilde{\mathcal{S}}_{1,t}$ for all t, i.e., $\mathbf{X} = \widetilde{\mathcal{D}}_1 + \widetilde{\mathcal{S}}_1$.

Answer to Exercise [175]: Let $\{\widehat{W}_{j,k}\}$ be the DFT of $\{\widetilde{W}_{j,t}\}$, and let $\{\widehat{V}_{j,k}\}$ be the DFT of $\{\widetilde{V}_{j,t}\}$. Since Equation (175b) tells us that $\{\widetilde{W}_{j,t}\}$ is formed by circularly filtering $\{\widetilde{V}_{j-1,t}\}$ with the filter whose transfer function is given by $\widetilde{H}(2^{j-1}f)$, we have

$$\widehat{W}_{j,k} = \widetilde{H}(2^{j-1}\tfrac{k}{N})\widehat{V}_{j-1,k};$$

likewise, from Equation (175c),

$$\widehat{V}_{j,k} = \widetilde{G}(2^{j-1}\tfrac{k}{N})\widehat{V}_{j-1,k}.$$

The first summation on the right-hand side of Equation (175d) is a cross-correlation of $\{\widetilde{W}_{j,t}\}$ with the filter whose transfer function is defined by $\widetilde{H}(2^{j-1}f)$, so we have

$$\left\{\sum_{l=0}^{L-1} \tilde{h}_l \widetilde{W}_{j,t+2^{j-1}l \bmod N}\right\} \longleftrightarrow \{\widetilde{H}^*(2^{j-1}\tfrac{k}{N})\widehat{W}_{j,k}\};$$

likewise, for the second summation, we have

$$\left\{\sum_{l=0}^{L-1} \tilde{g}_l \widetilde{V}_{j,t+2^{j-1}l \bmod N}\right\} \longleftrightarrow \{\widetilde{G}^*(2^{j-1}\tfrac{k}{N})\widehat{V}_{j,k}\}.$$

The DFT of the right-hand side of Equation (175d) is thus given by

$$\begin{aligned}
&\widetilde{H}^*(2^{j-1}\tfrac{k}{N})\widehat{W}_{j,k} + \widetilde{G}^*(2^{j-1}\tfrac{k}{N})\widehat{V}_{j,k} \\
&\quad = \widetilde{H}^*(2^{j-1}\tfrac{k}{N})\widetilde{H}(2^{j-1}\tfrac{k}{N})\widehat{V}_{j-1,k} + \widetilde{G}^*(2^{j-1}\tfrac{k}{N})\widetilde{G}(2^{j-1}\tfrac{k}{N})\widehat{V}_{j-1,k} \\
&\quad = \left(|\widetilde{H}(2^{j-1}\tfrac{k}{N})|^2 + |\widetilde{G}(2^{j-1}\tfrac{k}{N})|^2\right)\widehat{V}_{j-1,k} = \widehat{V}_{j-1,k}
\end{aligned}$$

since $|\widetilde{H}(f)|^2 + |\widetilde{G}(f)|^2 = \widetilde{\mathcal{H}}(f) + \widetilde{\mathcal{G}}(f) = 1$ for all f (see Equation (163b)); i.e., the right-hand of Equation (175d) is equal to $\widetilde{V}_{j-1,t}$, as required.

Answer to Exercise [208]: We have

$$\begin{bmatrix} \mathcal{A}_2\mathcal{B}_1 \\ \mathcal{B}_2\mathcal{B}_1 \\ \mathcal{A}_1 \end{bmatrix} \begin{bmatrix} \mathcal{B}_1^T\mathcal{A}_2^T & \mathcal{B}_1^T\mathcal{B}_2^T & \mathcal{A}_1^T \end{bmatrix} = \begin{bmatrix} \mathcal{A}_2\mathcal{B}_1\mathcal{B}_1^T\mathcal{A}_2^T & \mathcal{A}_2\mathcal{B}_1\mathcal{B}_1^T\mathcal{B}_2^T & \mathcal{A}_2\mathcal{B}_1\mathcal{A}_1^T \\ \mathcal{B}_2\mathcal{B}_1\mathcal{B}_1^T\mathcal{A}_2^T & \mathcal{B}_2\mathcal{B}_1\mathcal{B}_1^T\mathcal{B}_2^T & \mathcal{B}_2\mathcal{B}_1\mathcal{A}_1^T \\ \mathcal{A}_1\mathcal{B}_1^T\mathcal{A}_2^T & \mathcal{A}_1\mathcal{B}_1^T\mathcal{B}_2^T & \mathcal{A}_1\mathcal{A}_1^T \end{bmatrix}$$

$$= \begin{bmatrix} \mathcal{A}_2 I_{\frac{N}{2}}\mathcal{A}_2^T & \mathcal{A}_2 I_{\frac{N}{2}}\mathcal{B}_2^T & \mathcal{A}_2 0_{\frac{N}{2}} \\ \mathcal{B}_2 I_{\frac{N}{2}}\mathcal{A}_2^T & \mathcal{B}_2 I_{\frac{N}{2}}\mathcal{B}_2^T & \mathcal{B}_2 0_{\frac{N}{2}} \\ 0_{\frac{N}{2}}\mathcal{A}_2^T & 0_{\frac{N}{2}}\mathcal{B}_2^T & I_{\frac{N}{2}} \end{bmatrix}$$

$$= \begin{bmatrix} \mathcal{A}_2\mathcal{A}_2^T & \mathcal{A}_2\mathcal{B}_2^T & 0_{\frac{N}{4},\frac{N}{2}} \\ \mathcal{B}_2\mathcal{A}_2^T & \mathcal{B}_2\mathcal{B}_2^T & 0_{\frac{N}{4},\frac{N}{2}} \\ 0_{\frac{N}{2},\frac{N}{4}} & 0_{\frac{N}{2},\frac{N}{4}} & I_{\frac{N}{2}} \end{bmatrix}$$

$$= \begin{bmatrix} I_{\frac{N}{4}} & 0_{\frac{N}{4}} & 0_{\frac{N}{4},\frac{N}{2}} \\ 0_{\frac{N}{4}} & I_{\frac{N}{4}} & 0_{\frac{N}{4},\frac{N}{2}} \\ 0_{\frac{N}{2},\frac{N}{4}} & 0_{\frac{N}{2},\frac{N}{4}} & I_{\frac{N}{2}} \end{bmatrix} = I_N,$$

as required.

Answer to Exercise [214]: Note first that, starting from Equation (214a),

$$W_{j,2n,t} = \sum_{l=0}^{L-1} u_{2n,l} W_{j-1,n,2t+1-l \bmod N_{j-1}}$$

while

$$W_{j,2n+1,t} = \sum_{l=0}^{L-1} u_{2n+1,l} W_{j-1,n,2t+1-l \bmod N_{j-1}}.$$

To establish the exercise, we just need to show that $u_{2n,l} = a_{n,l}$ and $u_{2n+1,l} = b_{n,l}$. In the first equation above, the first subscript in $u_{2n,l}$ is even, so

$$u_{2n,l} = \begin{cases} g_l, & \text{if } 2n \bmod 4 = 0, \text{ i.e., } n \text{ is even;} \\ h_l, & \text{if } 2n \bmod 4 = 2, \text{ i.e., } n \text{ is odd.} \end{cases}$$

Hence $u_{2n,l} = a_{n,l}$, as required. In the second equation, the first subscript in $u_{2n+1,l}$ is odd, so

$$u_{2n+1,l} = \begin{cases} h_l, & \text{if } 2n+1 \bmod 4 = 1, \text{ i.e., } n \text{ is even;} \\ g_l, & \text{if } 2n+1 \bmod 4 = 3, \text{ i.e., } n \text{ is odd.} \end{cases}$$

Hence $u_{2n+1,l} = b_{n,l}$, as required.

Answer to Exercise [223]: If we were to use $m(|W_{j,n,t}|) = |W_{j,n,t}|^2$, then, for any $\mathcal{C} \subset \mathcal{N}$, we would have

$$\sum_{(j,n)\in\mathcal{C}} M(\mathbf{W}_{j,n}) = \sum_{(j,n)\in\mathcal{C}} \|\mathbf{W}_{j,n}\|^2 = \|\mathbf{X}\|^2$$

because each transform is orthonormal. Thus, with this cost functional, every possible transform would satisfy the optimality criterion!

Answer to Exercise [230]: Since $\nu < 0$ for all LA and coiflet filters, we can write

$$\nu_{j,n} = -|\nu|(2^j - 1) - S_{j,n,1}(L - 1 - 2|\nu|),$$

so $\nu_{j,n} < 0$ if and only if

$$S_{j,n,1}(2|\nu| + 1 - L) < |\nu|(2^j - 1).$$

Since $|\nu|(2^j - 1) > 0$ and since $2^j - 1 \geq S_{j,n,1} \geq 0$, the above inequality holds if and only if $(2|\nu| + 1 - L) < |\nu|$, i.e., $|\nu| < L - 1$. From Equation (112e) we can deduce that $|\nu| \leq \frac{L}{2} - 2$ for all LA filters; since $\frac{L}{2} - 2 < L - 1$ is always true, it follows that $|\nu| < L - 1$. For the coiflet filters, we have $|\nu| = \frac{2L}{3} - 1 < L - 1$ always, which establishes the exercise.

Answer to Exercise [233]: If $j = 1$, we have

$$\begin{aligned}\sum_{n=0}^{1} |\widetilde{U}_{1,n}(f)|^2 &= |\widetilde{M}_{c_{1,0,0}}(f)|^2 + |\widetilde{M}_{c_{1,1,0}}(f)|^2 = |\widetilde{M}_0(f)|^2 + |\widetilde{M}_1(f)|^2 \\ &= |\widetilde{G}_0(f)|^2 + |\widetilde{H}_1(f)|^2 = 1,\end{aligned}$$

where we recall that $\mathbf{c}_{1,0} \equiv [0] = [c_{1,0,0}]$ and $\mathbf{c}_{1,1} \equiv [1] = [c_{1,1,0}]$, and the last equality above is Equation (163a). Thus the exercise holds when $j = 1$. To see that it holds for $j > 1$, we claim that

$$\sum_{n=0}^{2^j-1} |\widetilde{U}_{j,n}(f)|^2 = \sum_{n=0}^{2^{j-1}-1} |\widetilde{U}_{j-1,n}(f)|^2,$$

which – once proven – readily establishes the exercise via induction. To see that the claim holds, note that

$$\sum_{n=0}^{2^j-1} |\widetilde{U}_{j,n}(f)|^2 = \sum_{n=0}^{2^j-1} \prod_{m=0}^{j-1} |\widetilde{M}_{c_{j,n,m}}(2^m f)|^2, \tag{530}$$

and consider the first two terms in the summation:

$$\begin{aligned}&\prod_{m=0}^{j-1} |\widetilde{M}_{c_{j,0,m}}(2^m f)|^2 + \prod_{m=0}^{j-1} |\widetilde{M}_{c_{j,1,m}}(2^m f)|^2 \\ &= \left(\prod_{m=0}^{j-2} |\widetilde{M}_{c_{j,0,m}}(2^m f)|^2\right) |\widetilde{M}_{c_{j,0,j-1}}(2^m f)|^2 \\ &\quad + \left(\prod_{m=0}^{j-2} |\widetilde{M}_{c_{j,1,m}}(2^m f)|^2\right) |\widetilde{M}_{c_{j,1,j-1}}(2^m f)|^2 \\ &= \left(\prod_{m=0}^{j-2} |\widetilde{M}_{c_{j-1,0,m}}(2^m f)|^2\right) \left(|\widetilde{M}_0(2^m f)|^2 + |\widetilde{M}_1(2^m f)|^2\right) \\ &= \left(\prod_{m=0}^{j-2} |\widetilde{M}_{c_{j-1,0,m}}(2^m f)|^2\right) \left(|\widetilde{G}(2^m f)|^2 + |\widetilde{H}(2^m f)|^2\right) \\ &= \prod_{m=0}^{j-2} |\widetilde{M}_{c_{j-1,0,m}}(2^m f)|^2 = |\widetilde{U}_{j-1,0}(f)|^2;\end{aligned}$$

here we have used the fact that the first $j-1$ elements of $\mathbf{c}_{j,0}$ and $\mathbf{c}_{j,1}$ are the same as the elements of $\mathbf{c}_{j-1,0}$, while the last elements are, respectively, 0 and 1. In a similar manner, the next two terms in Equation (530) yield the following:

$$\prod_{m=0}^{j-1} |\widetilde{M}_{\mathbf{c}_{j,2,m}}(2^m f)|^2 + \prod_{m=0}^{j-1} |\widetilde{M}_{\mathbf{c}_{j,3,m}}(2^m f)|^2$$
$$= \left(\prod_{m=0}^{j-2} |\widetilde{M}_{\mathbf{c}_{j-1,1,m}}(2^m f)|^2\right)\left(|\widetilde{M}_1(2^m f)|^2 + |\widetilde{M}_0(2^m f)|^2\right)$$
$$= \left(\prod_{m=0}^{j-2} |\widetilde{M}_{\mathbf{c}_{j-1,1,m}}(2^m f)|^2\right)\left(|\widetilde{H}(2^m f)|^2 + |\widetilde{G}(2^m f)|^2\right)$$
$$= \prod_{m=0}^{j-2} |\widetilde{M}_{\mathbf{c}_{j-1,1,m}}(2^m f)|^2 = |\widetilde{U}_{j-1,1}(f)|^2,$$

where now we recall that the first $j-1$ elements of $\mathbf{c}_{j,2}$ and $\mathbf{c}_{j,3}$ are the same as the elements of $\mathbf{c}_{j-1,1}$, while the last elements are, respectively, 1 and 0. The number of terms in the summation in Equation (530) is a multiple of four, and each succeeding group of four can be treated in a manner similar to the first four terms, thus proving the claim.

Answer to Exercise [240]: We have

$$\langle \mathbf{X}^{(1)}, \mathbf{R}^{(1)}\rangle = (\mathbf{X}^{(1)})^T\mathbf{R}^{(1)} = \langle \mathbf{X}, \mathbf{d}_{\gamma_0}\rangle \mathbf{d}_{\gamma_0}^T(\mathbf{X} - \mathbf{X}^{(1)}) = \langle \mathbf{X}, \mathbf{d}_{\gamma_0}\rangle(\mathbf{d}_{\gamma_0}^T\mathbf{X} - \mathbf{d}_{\gamma_0}^T\mathbf{X}^{(1)}),$$

so the result holds if $\mathbf{d}_{\gamma_0}^T\mathbf{X} = \mathbf{d}_{\gamma_0}^T\mathbf{X}^{(1)}$. To see that this is true, note that

$$\mathbf{d}_{\gamma_0}^T\mathbf{X}^{(1)} = \mathbf{d}_{\gamma_0}^T(\langle \mathbf{X}, \mathbf{d}_{\gamma_0}\rangle \mathbf{d}_{\gamma_0}) = \langle \mathbf{X}, \mathbf{d}_{\gamma_0}\rangle \mathbf{d}_{\gamma_0}^T\mathbf{d}_{\gamma_0} = \langle \mathbf{X}, \mathbf{d}_{\gamma_0}\rangle = \mathbf{d}_{\gamma_0}^T\mathbf{X}$$

because $\mathbf{d}_{\gamma_0}^T\mathbf{d}_{\gamma_0} = \|\mathbf{d}_{\gamma_0}\|^2 = 1$. We thus have

$$\|\mathbf{X}\|^2 = \|\mathbf{X}^{(1)} + \mathbf{R}^{(1)}\|^2 = \|\mathbf{X}^{(1)}\|^2 + \|\mathbf{R}^{(1)}\|^2 + 2\langle \mathbf{X}^{(1)}, \mathbf{R}^{(1)}\rangle = \|\mathbf{X}^{(1)}\|^2 + \|\mathbf{R}^{(1)}\|^2;$$

additionally,

$$\|\mathbf{X}^{(1)}\|^2 = \|\langle \mathbf{X}, \mathbf{d}_{\gamma_0}\rangle \mathbf{d}_{\gamma_0}\|^2 = \left|\langle \mathbf{X}, \mathbf{d}_{\gamma_0}\rangle\right|^2 \|\mathbf{d}_{\gamma_0}\|^2 = \left|\langle \mathbf{X}, \mathbf{d}_{\gamma_0}\rangle\right|^2,$$

which completes the exercise.

Answer to Exercise [261]: We have

$$E\{(X-a)^2\} = E\{X^2\} - 2aE\{X\} + a^2 \equiv q(a),$$

which is a quadratic function of a. To determine the extremal value of this function, we differentiate the above with respect to a and set the resulting expression to 0, yielding

$$\frac{dq(a)}{da} = -2E\{X\} + 2a = 0$$

(since the second derivative is 2 the extremal value is in fact a minimum). The above equation is equivalent to $a = E\{X\}$.

Answer to Exercise [262a]: Let $u_0(x_0, x_1) \equiv x_0$ and $u_1(x_0, x_1) \equiv x_0 + x_1$. The Jacobian of this transform is given by

$$J = \begin{vmatrix} 1 & 0 \\ 1 & 1 \end{vmatrix} = 1.$$

With $y_0 = u_0(x_0, x_1)$ and $y_1 = u_1(x_0, x_1)$, we can take the inverse functions to be $v_0(y_0, y_1) = y_0$ and $v_1(y_0, y_1) = y_1 - y_0$. We thus have

$$f_{Y_0,Y_1}(y_0, y_1) = f_{X_0,X_1}(v_0(y_0, y_1), v_1(y_0, y_1)) = f_{X_0,X_1}(y_0, y_1 - y_0),$$

as claimed.

Answer to Exercise [262b]: Recall that the trace of a square matrix is the sum of its diagonal elements, so we have

$$\operatorname{tr}\{\Sigma_{\mathbf{X}}\} = \sum_{t=0}^{N-1} \operatorname{var}\{X_t\}.$$

We thus need to show that $\operatorname{tr}\{\Sigma_{\mathbf{Y}}\} = \operatorname{tr}\{\Sigma_{\mathbf{X}}\}$. If A and B are matrices such that their product AB is a square matrix, a standard result in matrix theory says that $\operatorname{tr}\{AB\} = \operatorname{tr}\{BA\}$ (see, for example, Graybill, 1983, Section 9.1). Hence, since $\mathcal{O}^T\mathcal{O} = I_N$ for an orthonormal transform,

$$\operatorname{tr}\{\Sigma_{\mathbf{Y}}\} = \operatorname{tr}\{\mathcal{O}\Sigma_{\mathbf{X}}\mathcal{O}^T\} = \operatorname{tr}\{\Sigma_{\mathbf{X}}\mathcal{O}^T\mathcal{O}\} = \operatorname{tr}\{\Sigma_{\mathbf{X}}I_N\} = \operatorname{tr}\{\Sigma_{\mathbf{X}}\},$$

as required.

Answer to Exercise [263]: If $\mu_{\mathbf{X}} = \mathbf{0}$, then it follows from Equation (262b) that $\mu_{\mathbf{Y}} = \mathcal{O}\mu_{\mathbf{X}} = \mathbf{0}$ also; likewise, if $\Sigma_{\mathbf{X}} = \sigma^2 I_N$, then it follows from Equation (262c) that, since $\mathcal{O}\mathcal{O}^T = I_N$ due to orthonormality, $\Sigma_{\mathbf{Y}} = \mathcal{O}\Sigma_{\mathbf{X}}\mathcal{O}^T = \sigma^2\mathcal{O}I_N\mathcal{O}^T = \sigma^2\mathcal{O}\mathcal{O}^T = \sigma^2 I_N$ also. Since the multivariate Gaussianity of $\mathbf{X}$ implies the multivariate Gaussianity of $\mathbf{Y}$, we can conclude that the statistical properties of $\mathbf{X}$ and $\mathbf{Y}$ are identical. Finally, if $\mu_{\mathbf{X}} \neq \mathbf{0}$, we still have $\Sigma_{\mathbf{Y}} = \Sigma_{\mathbf{X}}$, but now $\mu_{\mathbf{Y}}$ need not be equal to $\mu_{\mathbf{X}}$. For example, suppose $\mu_{\mathbf{X}} = [\mu, \mu]^T$, and consider the Haar DWT matrix

$$\mathcal{W} \equiv \begin{bmatrix} -\frac{1}{\sqrt{2}} & \frac{1}{\sqrt{2}} \\ \frac{1}{\sqrt{2}} & \frac{1}{\sqrt{2}} \end{bmatrix}, \text{ for which } \mu_{\mathbf{Y}} = \mathcal{W}\mu_{\mathbf{X}} = \begin{bmatrix} 0 \\ \mu\sqrt{2} \end{bmatrix}.$$

Answer to Exercise [265]: Using the definition of $r_d(\theta)$, we have

$$\begin{aligned} E_\theta\{r_d(\theta)\} = \int r_d(\theta_0) f_\theta(\theta_0)\, d\theta_0 &= \int \left(\int \ell(d(x), \theta_0) f_{X|\theta=\theta_0}(x)\, dx \right) f_\theta(\theta_0)\, d\theta_0 \\ &= \int\int \ell(d(x), \theta_0) f_{X|\theta=\theta_0}(x) f_\theta(\theta_0)\, dx\, d\theta_0. \end{aligned}$$

From Equation (260), we have $f_{X|\theta=\theta_0}(x) f_\theta(\theta_0) = f_{X,\theta}(x, \theta_0)$; a second appeal to this equation yields $f_{X,\theta}(x, \theta_0) = f_{\theta|X=x}(\theta_0) f_X(x)$, so we have

$$\begin{aligned} E_\theta\{r_d(\theta)\} &= \int\int \ell(d(x), \theta_0) f_{\theta|X=x}(\theta_0) f_X(x)\, dx\, d\theta_0 \\ &= \int \left(\int \ell(d(x), \theta_0) f_{\theta|X=x}(\theta_0)\, d\theta_0 \right) f_X(x)\, dx, \end{aligned}$$

as required.

Answer to Exercise [288a]: First note that, because $X_0 \equiv 0$ and because $E\{\varepsilon_u\} = 0$ for all u, we have $E\{X_0\} = 0$, whereas, for $t > 0$ and $t < 0$, we have, respectively,

$$E\{X_t\} = \sum_{u=1}^{t} E\{\varepsilon_u\} = 0 \text{ and } E\{X_t\} = -\sum_{u=0}^{|t|-1} E\{\varepsilon_{-u}\} = 0$$

for all $|t| > 0$. The first moment of a random walk process is thus independent of t (as is required for a stationary process). Let us now consider the variance of $\{X_t\}$. We have $\text{var}\{X_0\} = 0$, and, for $t > 0$,

$$\text{var}\{X_t\} = E\{X_t^2\} = E\left\{\left(\sum_{u=1}^{t} \varepsilon_u\right)\left(\sum_{u'=1}^{t} \varepsilon_{u'}\right)\right\} = \sum_{u=1}^{t}\sum_{u'=1}^{t} E\{\varepsilon_u \varepsilon_{u'}\} = t\sigma_\varepsilon^2,$$

since $E\{\varepsilon_u \varepsilon_{u'}\}$ is nonzero only if $u = u'$, which happens exactly $|t|$ times in the double summation. Since $\text{var}\{X_t\}$ depends on t, the random walk process $\{X_t\}$ is nonstationary. To establish part (b), let $Y_t \equiv X_t - X_{t-1}$ be the first order backward difference for $\{X_t\}$. For $t = 1$, we have $Y_1 = X_1 - X_0 = \varepsilon_1$; for $t > 1$,

$$Y_t = \sum_{u=1}^{t} \varepsilon_u - \sum_{u=1}^{t-1} \varepsilon_u = \varepsilon_t;$$

for $t = 0$, we have $Y_0 = X_0 - X_{-1} = \varepsilon_0$; and $t < 0$,

$$Y_t = -\sum_{u=0}^{|t|-1} \varepsilon_{-u} + \sum_{u=0}^{|t-1|-1} \varepsilon_{-u} = \varepsilon_t.$$

Thus $\{Y_t\}$ is the white noise process $\{\varepsilon_t\}$, so the first order backward difference of $\{X_t\}$ is a stationary process. To establish part (c), note that, since the SDF for the white noise process $\{Y_t\}$ is given by $S_Y(f) = \sigma_\varepsilon^2$, it follows that the SDF for $\{X_t\}$ is given by

$$S_X(f) = \frac{S_Y(f)}{4\sin^2(\pi f)} = \frac{\sigma_\varepsilon^2}{4\sin^2(\pi f)} \approx \frac{\sigma_\varepsilon^2}{4\pi^2 f^2},$$

where the approximation holds for small f (because $\sin(x) \approx x$ for small x).

Answer to Exercise [288b]: Since $E\{U_t\} = 0$ for all t, it follows that

$$E\{X_t\} = E\{a_0 + a_1 t + U_t\} = a_0 + a_1 t,$$

which is not independent of t, so $\{X_t\}$ is a nonstationary process. To establish (b), let $\{s_{U,\tau}\}$ denote the ACVS for $\{U_t\}$. Note that the first order backward difference

$$Y_t \equiv X_t - X_{t-1} = a_0 + a_1 t + U_t - (a_0 + a_1(t-1) + U_{t-1}) = a_1 + U_t - U_{t-1}$$

has mean value

$$E\{Y_t\} = E\{a_1 + U_t - U_{t-1}\} = a_1 + E\{U_t\} - E\{U_{t-1}\} = a_1 \neq 0$$

and covariance between RVs Y_t and $Y_{t+\tau}$

$$\begin{aligned}\text{cov}\{Y_t, Y_{t+\tau}\} &= E\{(Y_t - a_1)(Y_{t+\tau} - a_1)\} \\ &= E\{(U_t - U_{t-1})(U_{t+\tau} - U_{t+\tau-1})\} \\ &= E\{U_t U_{t+\tau}\} - E\{U_t U_{t+\tau-1}\} - E\{U_{t-1} U_{t+\tau}\} + E\{U_{t-1} U_{t+\tau-1}\} \\ &= s_{U,\tau} - s_{U,\tau-1} - s_{U,\tau+1} + s_{U,\tau} \\ &= 2s_{U,\tau} - s_{U,\tau-1} - s_{U,\tau+1} \equiv s_{Y,\tau}.\end{aligned}$$

Since $E\{Y_t\}$ and $\text{cov}\{Y_t, Y_{t+\tau}\}$ are both independent of t and finite, the process $E\{Y_t\}$ is stationary with a nonzero mean. To establish (c), let $\{Z_t\}$ be the first backward difference of $\{Y_t\}$, which is the same as the second backward difference of $\{X_t\}$. We have

$$E\{Z_t\} = E\{Y_t - Y_{t-1}\} = E\{Y_t\} - E\{Y_{t-1}\} = 0,$$

and we can use the same argument as above to establish that

$$\begin{aligned}\text{cov}\{Z_t, Z_{t+\tau}\} &= 2s_{Y,\tau} - s_{Y,\tau-1} - s_{Y,\tau+1} \\ &= 2(2s_{U,\tau} - s_{U,\tau-1} - s_{U,\tau+1}) \\ &\quad - (2s_{U,\tau-1} - s_{U,\tau-2} - s_{U,\tau}) \\ &\quad - (2s_{U,\tau+1} - s_{U,\tau} - s_{U,\tau+2}) \\ &= 6s_{U,\tau} - 4s_{U,\tau-1} - 4s_{U,\tau+1} + s_{U,\tau-2} + s_{U,\tau+2},\end{aligned}$$

which depends on τ, but not t. Thus $\{Z_t\}$ is a stationary process with mean zero.

Answer to Exercise [299]: Starting with Equation (299a), we have

$$\begin{aligned}E\{\hat{\sigma}_Y^2\} &= \frac{1}{N}\sum_{t=0}^{N-1} E\{[(Y_t - \mu_Y) - (\overline{Y} - \mu_Y)]^2\} \\ &= \frac{1}{N}\sum_{t=0}^{N-1} E\{(Y_t - \mu_Y)^2 - 2(Y_t - \mu_Y)(\overline{Y} - \mu_Y) + (\overline{Y} - \mu_Y)^2\} \\ &= \sigma_Y^2 + \text{var}\{\overline{Y}\} - \frac{2}{N} E\Big\{(\overline{Y} - \mu_Y)\sum_{t=0}^{N-1}(Y_t - \mu_Y)\Big\} \\ &= \sigma_Y^2 + \text{var}\{\overline{Y}\} - \frac{2}{N} E\{(\overline{Y} - \mu_Y)(N\overline{Y} - N\mu_Y)\} = \sigma_Y^2 - \text{var}\{\overline{Y}\},\end{aligned}$$

as required.

Answer to Exercise [300]: Let $\overline{Y}_N \equiv \frac{1}{N}\sum_{t=0}^{N-1} Y_t$. In this notation, we need to show that $\text{var}\{\overline{Y}_N\} = \sigma_Y^2 N^{2H-2}$. Since $\text{var}\{\overline{Y}_1\} = \text{var}\{Y_0\} = \sigma_Y^2$, the result holds for $N = 1$. Assume that it holds for $N - 1$; i.e., $\text{var}\{\overline{Y}_{N-1}\} = \sigma_Y^2 (N-1)^{2H-2}$. Using part (d) of Exercise [7.2] repeatedly, we have

$$\begin{aligned}\text{var}\{\overline{Y}_N\} &= \text{var}\Big\{\frac{N-1}{N}\overline{Y}_{N-1} + \frac{1}{N} Y_{N-1}\Big\} \\ &= \text{var}\Big\{\frac{N-1}{N}\overline{Y}_{N-1}\Big\} + \text{var}\Big\{\frac{1}{N} Y_{N-1}\Big\} + 2\,\text{cov}\Big\{\frac{N-1}{N}\overline{Y}_{N-1}, \frac{1}{N} Y_{N-1}\Big\} \\ &= \frac{(N-1)^2}{N^2}\sigma_Y^2 (N-1)^{2H-2} + \frac{\sigma_Y^2}{N^2} + \frac{2}{N^2}\,\text{cov}\Big\{\sum_{t=0}^{N-2} Y_t, Y_{N-1}\Big\}\end{aligned}$$

$$= \frac{1}{N^2}\left[\sigma_Y^2(N-1)^{2H} + \sigma_Y^2 + 2\sum_{t=0}^{N-2} \operatorname{cov}\{Y_t, Y_{N-1}\}\right]$$

$$= \frac{1}{N^2}\left[\sigma_Y^2(N-1)^{2H} + \sigma_Y^2 + 2\sum_{\tau=1}^{N-1} s_{Y,\tau}\right]$$

$$= \frac{\sigma_Y^2}{N^2}\left[(N-1)^{2H} + 1 + \sum_{\tau=1}^{N-1}\left(|\tau+1|^{2H} - 2|\tau|^{2H} + |\tau-1|^{2H}\right)\right]$$

$$= \frac{\sigma_Y^2}{N^2}\left[(N-1)^{2H} + 1 + \sum_{\tau=2}^{N}\tau^{2H} - 2\sum_{\tau=1}^{N-1}\tau^{2H} + \sum_{\tau=1}^{N-2}\tau^{2H}\right]$$

$$= \frac{\sigma_Y^2}{N^2}\left[(N-1)^{2H} + 1 + N^{2H} - (N-1)^{2H} - 1\right] = \sigma_Y^2 N^{2H-2},$$

thus completing the proof by induction.

Answer to Exercise [302]: Assume Equation (302) holds for some $J_0 \geq 2$:

$$\operatorname{var}\{Y_t\} = \operatorname{var}\{\overline{V}_{J_0-1,t}\} + \sum_{j=1}^{J_0-1} \nu_Y^2(\tau_j).$$

Now

$$\begin{aligned}
\operatorname{var}\{\overline{V}_{J_0-1,t}\} &= \int_{-1/2}^{1/2} \widetilde{\mathcal{G}}_{J_0-1}(f) S_Y(f)\,df \\
&= \int_{-1/2}^{1/2} \left(\prod_{l=0}^{J_0-2} \widetilde{\mathcal{G}}(2^l f)\right) S_Y(f)\,df \\
&= \int_{-1/2}^{1/2} \left(\left[\widetilde{\mathcal{G}}(2^{J_0-1} f) + \widetilde{\mathcal{H}}(2^{J_0-1} f)\right] \prod_{l=0}^{J_0-2} \widetilde{\mathcal{G}}(2^l f)\right) S_Y(f)\,df \\
&= \int_{-1/2}^{1/2} \left[\widetilde{\mathcal{G}}_{J_0}(f) + \widetilde{\mathcal{H}}_{J_0}(f)\right] S_Y(f)\,df \\
&= \operatorname{var}\{\overline{V}_{J_0,t}\} + \nu_Y^2(\tau_{J_0}),
\end{aligned}$$

so we have

$$\operatorname{var}\{Y_t\} = \operatorname{var}\{\overline{V}_{J_0,t}\} + \nu_Y^2(\tau_{J_0}) + \sum_{j=1}^{J_0-1} \nu_Y^2(\tau_j) = \operatorname{var}\{\overline{V}_{J_0,t}\} + \sum_{j=1}^{J_0} \nu_Y^2(\tau_j),$$

as required.

Answer to Exercise [304]: The proof for case $j = 1$ extends readily to general j if we can establish the claim that the filter $\{\tilde{h}_{j,l}\}$ is the equivalent filter for a filter cascade of two parts: the first part is a backward difference filter of order d (this turns $\{X_t\}$ into the stationary process $\{Y_t\}$), while the second is a filter of finite width. It readily follows from Equation (169b) and the proof for case $j = 1$ that the squared gain function $\widetilde{\mathcal{H}}_j^{(\mathrm{D})}(\cdot)$ for $\{\tilde{h}_{j,l}\}$, $j \geq 2$, can be expressed as

$$\widetilde{\mathcal{H}}_j^{(\mathrm{D})}(f) = \widetilde{\mathcal{H}}^{(\mathrm{D})}(2^{j-1} f)\widetilde{\mathcal{G}}_{j-1}^{(\mathrm{D})}(f) = \mathcal{D}^{\frac{L}{2}}(2^{j-1} f)\widetilde{\mathcal{A}}_L(2^{j-1} f)\widetilde{\mathcal{G}}_{j-1}^{(\mathrm{D})}(f).$$

Using the definition of $\mathcal{D}(\cdot)$ and the trigonometric identity in the hint, for $j \geq 2$

$$\begin{aligned}\mathcal{D}(2^{j-1}f) &= 4\sin^2(\pi 2^{j-1}f) \\ &= 4\sin^2(\pi 2^{j-2}f)\left[4\cos^2(\pi 2^{j-2}f)\right] \\ &= 4\sin^2(\pi 2^{j-3}f)\left[4\cos^2(\pi 2^{j-3}f)\right]\left[4\cos^2(\pi 2^{j-2}f)\right] \\ &\vdots \\ &= \mathcal{D}(f)\prod_{k=0}^{j-2} 4\cos^2(\pi 2^k f),\end{aligned}$$

where $4\cos^2(\pi f)$ defines the squared gain function for the backward sum filter $\{1, 1\}$. Hence the squared gain function defined by $\mathcal{D}^{\frac{L}{2}}(2^{j-1}f)$ can be decomposed into two filters, the first of which is an $L/2$ order backward difference filter. Since $L/2 \geq d$ we can now write

$$\widetilde{\mathcal{H}}_j^{(D)}(f) = \mathcal{D}^d(f)\mathcal{B}_j(f),$$

where

$$\mathcal{B}_j(f) \equiv \mathcal{D}^{\frac{L}{2}-d}(f)\left[\prod_{k=0}^{j-2} 4\cos^2(\pi 2^k f)\right]^{\frac{L}{2}} \widetilde{\mathcal{A}}_L(2^{j-1}f)\widetilde{\mathcal{G}}_{j-1}^{(D)}(f) \tag{536}$$

must be the squared gain function for a filter of finite width because the equivalent filter $\{\tilde{h}_{j,l}\}$ for the cascade has finite width.

Answer to Exercise [305]: The result established by Exercise [304] says that we can write

$$\overline{W}_{j,t} = \sum_{l=0}^{L_j-d-1} b_{j,l}Y_{t-l},$$

where $\{b_{j,l}\}$ is the filter whose squared gain function $\mathcal{B}_j(\cdot)$ is defined in Equation (536), and $\{Y_t\}$ is the stationary process with mean μ_Y formed by filtering $\{X_t\}$ with a dth order backward difference filter. Letting $B_j(\cdot)$ denote the transfer function for $\{b_{j,l}\}$ (so that $|B_j(f)|^2 = \mathcal{B}_j(f)$), we have

$$E\{\overline{W}_{j,t}\} = \sum_{l=0}^{L_j-d-1} b_{j,l}E\{Y_{t-l}\} = \mu_Y \sum_{l=0}^{L_j-d-1} b_{j,l} = \mu_Y B_j(0),$$

so $E\{\overline{W}_{j,t}\} = 0$ when $\mu_Y = 0$. Since $\mathcal{D}(0) = 4\sin^2(\pi \cdot 0) = 0$, Equation (536) tells us that $\mathcal{B}_j(0) = 0$ and hence $B_j(0) = 0$ when $L/2 > d$, in which case $E\{\overline{W}_{j,t}\} = 0$ no matter what μ_Y is. It remains to show that $E\{\overline{W}_{j,t}\} \neq 0$ when $\mu_Y \neq 0$ and $L/2 = d$, which is true if we can show that $B_j(0) \neq 0$ or, equivalently, $\mathcal{B}_j(0) > 0$. When $L/2 = d$,

$$\begin{aligned}\mathcal{B}_j(0) = \widetilde{\mathcal{A}}_L(0)\widetilde{\mathcal{G}}_{j-1}^{(D)}(0) = \widetilde{\mathcal{A}}_L(0) &= \frac{1}{2^L}\sum_{l=0}^{\frac{L}{2}-1}\binom{\frac{L}{2}-1+l}{l} \\ &= \begin{cases} 128/512, & L=2; \\ 96/512, & L=4; \\ 80/512, & L=6; \text{ etc.,}\end{cases}\end{aligned}$$

where we have used the fact that

$$\widetilde{\mathcal{G}}_{j-1}^{(D)}(0) = \prod_{l=0}^{j-2}\widetilde{\mathcal{G}}(0) = 1.$$

This establishes the exercise.

Answer to Exercise [312]: Suppose that the SDF estimator $\hat{S}_j(\cdot)$ is such that

$$\frac{\eta \hat{S}_j(f)}{S_j(f)} \stackrel{d}{=} \chi^2_\eta, \qquad 0 < |f| < 1/2.$$

Since $E\{\chi^2_\eta\} = \eta$ and $\operatorname{var}\{\chi^2_\eta\} = 2\eta$, we have

$$E\left\{\frac{\eta \hat{S}_j(f)}{S_j(f)}\right\} = \eta, \text{ i.e., } E\{\hat{S}_j(f)\} = S_j(f),$$

and

$$\operatorname{var}\left\{\frac{\eta \hat{S}_j(f)}{S_j(f)}\right\} = 2\eta, \text{ i.e., } \operatorname{var}\{\hat{S}_j(f)\} = \frac{2S_j^2(f)}{\eta}.$$

For any RV U with mean value $E\{U\}$, we have

$$\operatorname{var}\{U\} \equiv E\left\{(U - E\{U\})^2\right\} = E\{U^2\} - (E\{U\})^2,$$

so $E\{U^2\} = \operatorname{var}\{U\} + (E\{U\})^2$. Thus

$$E\{\hat{S}_j^2(f)\} = \operatorname{var}\{\hat{S}_j(f)\} + \left(E\{\hat{S}_j(f)\}\right)^2 = \frac{2S_j^2(f)}{\eta} + S_j^2(f) = S_j^2(f)\left(\frac{2}{\eta} + 1\right). \tag{537}$$

For the periodogram, we have $\eta = 2$, so Equation (537) tells us that

$$E\left\{\left[\hat{S}_j^{(p)}(f)\right]^2\right\} = 2S_j^2(f).$$

We thus have approximately

$$E\{\hat{A}_j\} = \frac{1}{2}\int_{-1/2}^{1/2} E\left\{\left[\hat{S}_j^{(p)}(f)\right]^2\right\} df = \frac{1}{2}\int_{-1/2}^{1/2} 2S_j^2(f)\, df = \int_{-1/2}^{1/2} S_j^2(f)\, df = A_j$$

(the contribution to the integral due to frequencies $f = 0, \pm 1/2$ is negligible); i.e., $\hat{A}_j$ is an approximately unbiased estimator of A_j.

Answer to Exercise [313a]: Since

$$E\{\hat{\nu}_X^2(\tau_j)\} = E\{a\chi^2_{\eta_1}\} = a\eta_1$$

and

$$\operatorname{var}\{\hat{\nu}_X^2(\tau_j)\} = \operatorname{var}\{a\chi^2_{\eta_1}\} = 2a^2\eta_1,$$

we have

$$\frac{2\left(E\{\hat{\nu}_X^2(\tau_j)\}\right)^2}{\operatorname{var}\{\hat{\nu}_X^2(\tau_j)\}} = \frac{2a^2\eta_1{}^2}{2a^2\eta_1} = \eta_1.$$

Using the large sample approximations for $E\{\hat{\nu}_X^2(\tau_j)\}$ and $\operatorname{var}\{\hat{\nu}_X^2(\tau_j)\}$ yields

$$\eta_1 = \frac{2\nu_X^4(\tau_j)}{2A_j/M_j} = \frac{M_j \nu_X^4(\tau_j)}{A_j}.$$

We also note that

$$a = \frac{E\{\hat{\nu}_X^2(\tau_j)\}}{\eta_1} = \frac{A_j}{M_j \nu_X^2(\tau_j)}$$

for completeness.

Answer to Exercise [313b]: From the definition of $Q_\eta(p)$, we have (for $0 \le p \le 1/2$)

$$\mathbf{P}\left[Q_\eta(p) \le \chi^2_\eta \le Q_\eta(1-p)\right] = 1 - 2p.$$

Since by assumption $\eta\hat{\nu}^2_X(\tau_j)/\nu^2_X(\tau_j)$ has the distribution of a χ^2_η RV, we have

$$\mathbf{P}\left[Q_\eta(p) \le \frac{\eta\hat{\nu}^2_X(\tau_j)}{\nu^2_X(\tau_j)} \le Q_\eta(1-p)\right] = 1 - 2p.$$

The event in the brackets occurs if and only if the events

$$Q_\eta(p) \le \frac{\eta\hat{\nu}^2_X(\tau_j)}{\nu^2_X(\tau_j)} \text{ and } \frac{\eta\hat{\nu}^2_X(\tau_j)}{\nu^2_X(\tau_j)} \le Q_\eta(1-p)$$

both occur; these two events are equivalent to

$$\nu^2_X(\tau_j) \le \frac{\eta\hat{\nu}^2_X(\tau_j)}{Q_\eta(p)} \text{ and } \frac{\eta\hat{\nu}^2_X(\tau_j)}{Q_\eta(1-p)} \le \nu^2_X(\tau_j).$$

Together these two events are equivalent to the single event

$$\frac{\eta\hat{\nu}^2_X(\tau_j)}{Q_\eta(1-p)} \le \nu^2_X(\tau_j) \le \frac{\eta\hat{\nu}^2_X(\tau_j)}{Q_\eta(p)},$$

which gives the confidence interval stated in the exercise.

Answer to Exercise [314a]: The large sample approximation to the distribution of $\hat{S}^{(\mathrm{p})}_j(f_k)$ tells us that $E\{\hat{S}^{(\mathrm{p})}_j(f_k)\} = S_j(f_k) = aC_j(f_k)$ and $\operatorname{var}\{\hat{S}^{(\mathrm{p})}_j(f_k)\} = S^2_j(f_k) = a^2C^2_j(f_k)$. Using Equation (314a), we have

$$E\{\hat{\nu}^2_X(\tau_j)\} \approx \frac{2}{M_j}\sum_{k=1}^{\lfloor (M_j-1)/2\rfloor} E\{\hat{S}^{(\mathrm{p})}_j(f_k)\} = \frac{2a}{M_j}\sum_{k=1}^{\lfloor (M_j-1)/2\rfloor} C_j(f_k);$$

using this equation again and the fact that the variance of a sum of independent RVs is the sum of the variances of the RVs, we have

$$\operatorname{var}\{\hat{\nu}^2_X(\tau_j)\} \approx \frac{4a^2}{M_j^2}\sum_{k=1}^{\lfloor (M_j-1)/2\rfloor} C^2_j(f_k).$$

Using the result of Exercise [313a], we obtain

$$\eta_2 = \frac{2\left(E\{\hat{\nu}^2_X(\tau_j)\}\right)^2}{\operatorname{var}\{\hat{\nu}^2_X(\tau_j)\}} = \frac{2\left(\frac{2a}{M_j}\sum_{k=1}^{\lfloor (M_j-1)/2\rfloor} C_j(f_k)\right)^2}{\frac{4a^2}{M_j^2}\sum_{k=1}^{\lfloor (M_j-1)/2\rfloor} C^2_j(f_k)},$$

from which the stated result follows.

Answer to Exercise [314b]: By assumption, we can write $S_j(f) \approx aC_j(f)$, where

$$C_j(f) \equiv \begin{cases} 1, & \frac{1}{2^{j+1}} < |f| \le \frac{1}{2^j}; \\ 0, & \text{otherwise.} \end{cases}$$

Now $\hat{\nu}_X^2(\tau_j)$ is asymptotically normal with mean $\nu_X^2(\tau_j)$ and large sample variance $2A_j/M_j$, where, from Equation (307a),

$$A_j = \int_{-1/2}^{1/2} S_j^2(f)\,df \approx 2\int_{-1/2}^{1/2} a^2 C_j^2(f)\,df = 2\int_{1/2^{j+1}}^{1/2^j} a^2\,df = \frac{a^2}{2^j}.$$

We thus have $\text{var}\{\hat{\nu}_X^2(\tau_j)\} \approx a^2/(2^{j-1}M_j)$. From Equation (305) we have

$$\nu_X^2(\tau_j) = \int_{-1/2}^{1/2} S_j(f)\,df \approx \int_{-1/2}^{1/2} aC_j(f)\,df = 2\int_{1/2^{j+1}}^{1/2^j} a\,df = \frac{a}{2^j},$$

and hence $E\{\hat{\nu}_X^2(\tau_j)\} \approx a/2^j$. From Exercise [313a],

$$\eta = \frac{2\left(E\{\hat{\nu}_X^2(\tau_j)\}\right)^2}{\text{var}\{\hat{\nu}_X^2(\tau_j)\}} \approx \frac{2(a/2^j)^2}{a^2/(2^{j-1}M_j)} = \frac{M_j}{2^j}.$$

If $M_j/2^j < 1$, we can argue that a better approximation is unity since the sum used to form $\hat{\nu}_X^2(\tau_j)$ always consists of at least one normally distributed RV with zero mean. This leads to $\eta_3 = \max\{M_j/2^j, 1\}$, as stated in the exercise.

Answer to Exercise [315]: Define $\eta \equiv 2N_B$ and

$$\widetilde{A}_j \equiv C\int_{-1/2}^{1/2} [\hat{S}_{(\text{sa})j}(f)]^2\,df,$$

where C is yet to be determined. From the assumption on the distribution of $\eta\hat{S}_{(\text{sa})j}(f)/S_j(f)$, we have $E\{\hat{S}_{(\text{sa})j}(f)\} = S_j(f)$ and $\text{var}\{\hat{S}_{(\text{sa})j}(f)\} = 2S_j^2(f)/\eta$. Since $E\{U^2\} = \text{var}\{U\} + (E\{U\})^2$ for any RV U, we obtain

$$E\{\widetilde{A}_j\} = C\int_{-1/2}^{1/2} E\left\{[\hat{S}_{(\text{sa})j}(f)]^2\right\} df = C\int_{-1/2}^{1/2} S_j^2(f)\left(\frac{2}{\eta}+1\right) df = C\left(\frac{2}{\eta}+1\right)A_j,$$

so we can obtain an unbiased estimator by letting

$$C \equiv \frac{1}{\frac{2}{\eta}+1} = \frac{\eta}{2+\eta}.$$

As N_B (and hence η) increases to infinity, we have $C \to 1$, so the estimator becomes

$$\widetilde{A}_j = \int_{-1/2}^{1/2} [\hat{S}_{(\text{sa})j}(f)]^2\,df,$$

as is reasonable from the definition of A_j.

Answer to Exercise [320a]: To obtain an unbiased MODWT estimator of the wavelet variance for scale τ_j, we must have $M_j \equiv N - L_j + 1 \geq 1$, where $L_j \equiv (2^j - 1)(L-1)+1$. With $N = 1026$, the largest j for which this inequality holds when $L = 2$, 4 and 6 is, respectively, $j = 10$, 8 and 7.

Answer to Exercise [320b]: The jth level MODWT Haar wavelet filter has an impulse response sequence $\{\tilde{h}_l\}$ consisting of the following $L_j = 2^j$ nonzero values:

$$\underbrace{\frac{1}{2^j}, \ldots, \frac{1}{2^j}}_{2^{j-1} \text{ of these}}, \underbrace{-\frac{1}{2^j}, \ldots, -\frac{1}{2^j}}_{2^{j-1} \text{ of these}}.$$

The corresponding MODWT coefficients for $\{X_t\}$ are thus given by

$$\begin{aligned}\overline{W}_{j,t} &= \sum_{l=0}^{L_j-1} \tilde{h}_l X_{t-l} = \frac{1}{2^j} \sum_{l=0}^{2^{j-1}-1} X_{t-l} - X_{t-2^{j-1}-l} \\ &= \frac{1}{2^j} \sum_{l=0}^{2^{j-1}-1} \Big[\big(X'_{t-l} + a_0 + a_1(t-l)\big) \\ &\qquad\qquad - \big(X'_{t-2^{j-1}-l} + a_0 + a_1(t - 2^{j-1} - l)\big)\Big] \\ &= \frac{1}{2^j} \sum_{l=0}^{2^{j-1}-1} \big(X'_{t-l} - X'_{t-2^{j-1}-l} + a_1 2^{j-1}\big) = \overline{W}'_{j,t} + \frac{a_1}{2}\tau_j,\end{aligned}$$

so

$$E\{\hat{\nu}^2_X(\tau_j)\} = E\{\overline{W}^2_{j,t}\} = E\{[\overline{W}'_{j,t}]^2\} + a_1\tau_j E\{\overline{W}'_{j,t}\} + \frac{a_1^2}{4}\tau_j^2 = \nu^2_{X'}(\tau_j) + \frac{a_1^2}{4}\tau_j^2$$

(the first result in Exercise [305] tells us that $E\{\overline{W}'_{j,t}\} = 0$). If the second term above dominates the first, we then have

$$\nu_X(\tau_j) \approx \frac{|a_1|}{2}\tau_j \text{ so } \log_{10}(\nu_X(\tau_j)) \approx -\log_{10}(2) + \log_{10}(|a_1|) + \log_{10}(\tau_j),$$

and thus a plot of $\log_{10}(\nu_X(\tau_j))$ versus $\log_{10}(\tau_j)$ should be approximately linear with unit slope. We should also have $\nu_X(\tau_1) \approx |a_1|/2$. A rough guess at a_1 from the top plot in Figure 318 is $-150\,000/1000 = -150$ nanoseconds, so we should have $\nu_X(\tau_1) \approx 75$ nanoseconds; in fact $\hat{\nu}_X(\tau_1) \doteq 71.38$ nanoseconds, so the appearance of the Haar wavelet variance of Figure 319 can be explained almost entirely in terms of the approximate linear drift in $\{X_t\}$.

Answer to Exercise [322]: As in the solution to Exercise [320b], we can express the level j MODWT Haar wavelet coefficient process for $\{\overline{Y}_t(\tau_1)\}$ as

$$\begin{aligned}\overline{W}_{j,t} &= \frac{1}{2^j} \sum_{l=0}^{2^{j-1}-1} \overline{Y}_{t-l}(\tau_1) - \overline{Y}_{t-2^{j-1}-l}(\tau_1) \\ &= \frac{1}{2\tau_j} \sum_{l=0}^{\tau_j-1} \overline{Y}_{t-l}(\tau_1) - \overline{Y}_{t-\tau_j-l}(\tau_1) = \frac{1}{2}\left[\overline{Y}_t(\tau_j) - \overline{Y}_{t-\tau_j}(\tau_j)\right].\end{aligned} \tag{540}$$

The assumption that the backward difference process $\{\overline{Y}_t(\tau_1) - \overline{Y}_{t-1}(\tau_1)\}$ has mean zero implies that $E\{\overline{W}_{j,t}\} = 0$ also (this follows because we can write

$$\begin{aligned}\overline{Y}_{t-l}(\tau_1) - \overline{Y}_{t-\tau_j-l}(\tau_1) = {} & [\overline{Y}_{t-l}(\tau_1) - \overline{Y}_{t-1-l}(\tau_1)] \\ & + [\overline{Y}_{t-1-l}(\tau_1) - \overline{Y}_{t-2-l}(\tau_1)] + \cdots \\ & + [\overline{Y}_{t-\tau_j+2-l}(\tau_1) - \overline{Y}_{t-\tau_j+1-l}(\tau_1)] \\ & + [\overline{Y}_{t-\tau_j+1-l}(\tau_1) - \overline{Y}_{t-\tau_j-l}(\tau_1)],\end{aligned}$$

with each term inside a set of square brackets having mean zero). Squaring both sides of Equation (540) and taking expectations yields the desired result since $\nu^2_{\overline{Y}}(\tau_j) = \text{var}\,\{\overline{W}_{j,t}\} = E\{\overline{W}^2_{j,t}\}$.

Answer to Exercise [345]: Because $W_{j,t}$ is not influenced by circularity, we can write

$$W_{j,t} = \sum_{l=0}^{L_j-1} h_{j,l} X_{2^j(t+1)-1-l},$$

with an analogous expression for $W_{j',t'}$ (the above is Equation (96d) with the 'mod N' dropped). Equation (345a) follows immediately from part (d) of Exercise [7.2] since

$$\text{cov}\,\{X_{2^j(t+1)-1-l}, X_{2^{j'}(t'+1)-1-l'}\} = s_{X,2^j(t+1)-l-2^{j'}(t'+1)+l'}.$$

When $j = j'$ and $t' = t + \tau$, we have

$$\text{cov}\,\{W_{j,t}, W_{j,t'}\} = \sum_{l=0}^{L_j-1}\sum_{l'=0}^{L_j-1} h_{j,l}h_{j,l'}s_{X,2^j\tau+l-l'}.$$

We can regard the right-hand side as summing all the elements in a symmetric $L_j \times L_j$ matrix, whose (l, l')th element is given by $h_{j,l}h_{j,l'}s_{X,2^j\tau+l-l'}$. We do this summation by going down each row, from $l = 0$ to $L_j - 1$; however, we can also do it by going down each diagonal instead. Consider the mth such diagonal, where $m = 0, 1, -1, \ldots$ are the indices for, respectively, the main diagonal, the first super- and sub-diagonals, and so forth. The elements of the mth diagonal satisfy $l - l' = m$ and hence are given by $h_{j,l}h_{j,m+l}s_{X,2^j\tau+m}$, where $l = 0, \ldots, L_j - m - 1$ when $m \geq 0$. By summing these $L_j - m$ elements and recalling that the matrix in question is symmetric, we obtain Equation (345b).

Answer to Exercise [348a]: Plugging the expression for $s_{X,\tau}$ given in Equation (267c) (with Δt taken to be unity so that $f_\mathcal{N} = 1/2$) into Equation (345a) yields

$$\begin{aligned}\text{cov}\,\{W_{j,t}, W_{j',t'}\} &= \sum_{l=0}^{L_j-1}\sum_{l'=0}^{L_{j'}-1} h_{j,l}h_{j',l'} \int_{-1/2}^{1/2} S_X(f)e^{i2\pi f(2^j(t+1)-l-2^{j'}(t'+1)+l')}\,df \\ &= \int_{-1/2}^{1/2}\left(\sum_{l=0}^{L_j-1} h_{j,l}e^{-i2\pi fl}\right)\left(\sum_{l'=0}^{L_{j'}-1} h_{j',l'}e^{-i2\pi fl'}\right)^* \\ &\qquad \times S_X(f)e^{i2\pi f(2^j(t+1)-2^{j'}(t'+1))}\,df \\ &= \int_{-1/2}^{1/2} H_j(f)H^*_{j'}(f)S_X(f)e^{i2\pi f(2^j(t+1)-2^{j'}(t'+1))}\,df.\end{aligned}$$

Since $\text{cov}\,\{W_{j,t}, W_{j',t'}\} = \text{cov}\,\{W_{j',t'}, W_{j,t}\}$, we can interchange j and t with j' and t' to obtain the stated result (this cosmetic swap allows us to eliminate some minus signs in some subsequent expressions).

Answer to Exercise [348b]: If we make the change of variable $f' = 2^j f$ on the right-hand side of Equation (348b), we obtain

$$\begin{aligned}\operatorname{cov}\{W_{j,t}, W_{j,t+\tau}\} &= \frac{1}{2^j}\int_{-2^{j-1}}^{2^{j-1}} e^{i2\pi f'\tau}\mathcal{H}_j(\tfrac{f'}{2^j})S_X(\tfrac{f'}{2^j})\,df' \\ &= \frac{1}{2^j}\int_0^{2^j} e^{i2\pi f'\tau}\mathcal{H}_j(\tfrac{f'}{2^j})S_X(\tfrac{f'}{2^j})\,df',\end{aligned}$$

where the last line follows because the integrand is periodic with a period of 2^j. If we partition the interval $[0, 2^j]$ into 2^j nonoverlapping intervals of unit length, we obtain

$$\begin{aligned}\operatorname{cov}\{W_{j,t}, W_{j,t+\tau}\} &= \frac{1}{2^j}\sum_{k=0}^{2^j-1}\int_k^{k+1} e^{i2\pi f'\tau}\mathcal{H}_j(\tfrac{f'}{2^j})S_X(\tfrac{f'}{2^j})\,df' \\ &= \frac{1}{2^j}\sum_{k=0}^{2^j-1}\int_0^1 e^{i2\pi(f+k)\tau}\mathcal{H}_j(\tfrac{f+k}{2^j})S_X(\tfrac{f+k}{2^j})\,df \\ &= \frac{1}{2^j}\int_0^1 e^{i2\pi f\tau}\sum_{k=0}^{2^j-1}\mathcal{H}_j(\tfrac{f+k}{2^j})S_X(\tfrac{f+k}{2^j})\,df,\end{aligned}$$

where we make the change of variable $f = f' - k$ and use the fact that $e^{i2\pi(f+k)\tau} = e^{i2\pi f\tau}$ because $k\tau$ is an integer. The desired result follows if we can argue that

$$A(f) \equiv \sum_{k=0}^{2^j-1}\mathcal{H}_j(\tfrac{f+k}{2^j})S_X(\tfrac{f+k}{2^j})$$

is periodic with unit period. To see that this is true, note that

$$\begin{aligned}A(f+1) &= \sum_{k=0}^{2^j-1}\mathcal{H}_j(\tfrac{f+1+k}{2^j})S_X(\tfrac{f+1+k}{2^j}) = \sum_{k=1}^{2^j}\mathcal{H}_j(\tfrac{f+k}{2^j})S_X(\tfrac{f+k}{2^j}) \\ &= \sum_{k=0}^{2^j-1}\mathcal{H}_j(\tfrac{f+k}{2^j})S_X(\tfrac{f+k}{2^j}) = A(f),\end{aligned}$$

where the last line follows from the fact that

$$\mathcal{H}_j(\tfrac{f+2^j}{2^j})S_X(\tfrac{f+2^j}{2^j}) = \mathcal{H}_j(\tfrac{f}{2^j})S_X(\tfrac{f}{2^j})$$

because both $\mathcal{H}_j(\cdot)$ and $S_X(\cdot)$ have unit periodicity.

Answer to Exercise [355]: In the current notation, Equation (302) reads as

$$\operatorname{var}\{X_t\} = \operatorname{var}\{\overline{V}_{J,0}\} + \sum_{j=1}^{J}\nu_X^2(\tau_j),$$

where $\overline{V}_{J,0}$ is a MODWT scaling coefficient that is related to the DWT scaling coefficient $V_{J,0}$ via $V_{J,0}/2^{J/2} = \overline{V}_{J,0}$, yielding $\operatorname{var}\{V_{J,0}\}/N = \operatorname{var}\{\overline{V}_{J,0}\}$; likewise, $\nu_X^2(\tau_j)$ is the variance of the MODWT wavelet coefficients $\{\widetilde{W}_{j,t}\}$, which are related to the DWT wavelet coefficients $\{W_{j,t}\}$ via $W_{j,t}/2^{j/2} = \widetilde{W}_{j,2^j(t+1)-1}$, yielding $\operatorname{var}\{W_{j,t}\}/2^j = \nu_X^2(\tau_j)$.

Answer to Exercise [362a]: We use the facts that, if A, B and C are square matrices, then $|ABC| = |BCA|$ and $|AB| = |A| \cdot |B|$ Thus

$$|\tilde{\Sigma}_{\mathbf{X}}| = |\mathcal{W}^T \Lambda_N \mathcal{W}| = |\Lambda_N \mathcal{W}\mathcal{W}^T| = |\Lambda_N I_N| = |\Lambda_N| \cdot |I_N| = |\Lambda_N|$$

because the determinant of the identity matrix is unity. Since the determinant of a diagonal matrix is the product of its diagonal elements and since the diagonal elements of Λ_N are as in Equation (355a), we have

$$|\Lambda_N| = C_{J+1} \prod_{j=1}^{J} C_j^{N_j} \text{ and hence } \log(|\Lambda_N|) = \log(C_{J+1}) + \sum_{j=1}^{J} N_j \log(C_j),$$

where $N_j \equiv N/2^j$ as usual.

Answer to Exercise [362b]: First note that

$$\tilde{\Sigma}_{\mathbf{X}}^{-1} = \left(\mathcal{W}^T \Lambda_N \mathcal{W}\right)^{-1} = \mathcal{W}^{-1} \Lambda_N^{-1} \left(\mathcal{W}^T\right)^{-1} = \mathcal{W}^T \Lambda_N^{-1} \mathcal{W},$$

where Λ_N^{-1} is a diagonal matrix whose elements are

$$\underbrace{\frac{1}{C_1}, \ldots, \frac{1}{C_1}}_{\frac{N}{2} \text{ of these}}, \underbrace{\frac{1}{C_2}, \ldots, \frac{1}{C_2}}_{\frac{N}{4} \text{ of these}}, \ldots, \underbrace{\frac{1}{C_j}, \ldots, \frac{1}{C_j}}_{\frac{N}{2^j} \text{ of these}}, \ldots, \underbrace{\frac{1}{C_{J-1}}, \frac{1}{C_{J-1}}}_{2 \text{ of these}}, \frac{1}{C_J}, \frac{1}{C_{J+1}}.$$

Thus

$$\begin{aligned}
\mathbf{X}^T \tilde{\Sigma}_{\mathbf{X}}^{-1} \mathbf{X} &= \mathbf{X}^T \mathcal{W}^T \Lambda_N^{-1} \mathcal{W} \mathbf{X} = \mathbf{W}^T \Lambda_N^{-1} \mathbf{W} \\
&= \left[\mathbf{W}_1^T, \mathbf{W}_2^T, \ldots, \mathbf{W}_J^T, \mathbf{V}_J^T\right] \begin{bmatrix} \mathbf{W}_1/C_1 \\ \mathbf{W}_2/C_2 \\ \vdots \\ \mathbf{W}_J/C_J \\ \mathbf{V}_J/C_{J+1} \end{bmatrix} \\
&= \frac{\mathbf{V}_J^T \mathbf{V}_J}{C_{J+1}} + \sum_{j=1}^{J} \frac{\mathbf{W}_j^T \mathbf{W}_j}{C_j} = \frac{V_{J,0}^2}{\sigma_\varepsilon^2 C'_{J+1}(\delta)} + \sum_{j=1}^{J} \frac{1}{\sigma_\varepsilon^2 C'_j(\delta)} \sum_{t=0}^{N_j - 1} W_{j,t}^2,
\end{aligned}$$

which – upon factoring out $1/\sigma_\varepsilon^2$ – yields the desired result.

Answer to Exercise [365a]: Because of the embedded differencing operation in the wavelet filter, the wavelet coefficients $W_{j,t}$ do not depend on the sample mean of the time series; on the other hand, the single scaling coefficient $V_{J,0}$ obviously does because, if we take the DWT of a time series $\mathbf{X}$ with sample mean $\overline{X}$, then $V_{J,0} = \overline{X}\sqrt{N}$. The desired result thus follows immediately from the result of Exercise [362b] by noting that the time series $\mathbf{X} - \mu\mathbf{1}$ has a sample mean of $\overline{X} - \mu$, and hence the corresponding scaling coefficient takes the form $V_{J,0} = (\overline{X} - \mu)\sqrt{N}$.

Answer to Exercise [365b]: The time series $\mathbf{X}' \equiv \mathbf{X} - \overline{X}\mathbf{1}$ has a sample mean of zero, so, while the wavelet coefficients for $\mathbf{X}$ and $\mathbf{X}'$ are identical, the scaling coefficients are, respectively, $\overline{X}\sqrt{N}$ and zero. When $V_{J,0}$ is set to zero, the MLE $\tilde{\sigma}_\varepsilon^2(\delta)$ of Equation (363b) reduces to that of Equation (365), which in turn leads to the correct reduced log likelihood for δ when the process mean μ is unknown.

Answer to Exercise [371]: The (rescaled and relocated) log likelihood function here is

$$\tilde{l}(\delta, \sigma_\varepsilon^2 \mid \mathbf{W}'_{nb}) \equiv -2\log(\tilde{L}(\delta, \sigma_\varepsilon^2 \mid \mathbf{W}'_{nb})) - N'\log(2\pi) = \sum_{j=1}^{J_0}\sum_{t=0}^{N'_j-1}\left(\frac{W^2_{j,t+L'_j}}{C_j} + \log(C_j)\right).$$

Recalling that $C_j = \sigma_\varepsilon^2 C'_j(\delta)$, we obtain

$$\tilde{l}(\delta, \sigma_\varepsilon^2 \mid \mathbf{W}'_{nb}) = \frac{1}{\sigma_\varepsilon^2}\sum_{j=1}^{J_0}\frac{1}{C'_j(\delta)}\sum_{t=0}^{N'_j-1} W^2_{j,t+L'_j} + N'\log(\sigma_\varepsilon^2) + \sum_{j=1}^{J_0} N'_j \log(C'_j(\delta)).$$

Differentiating the right-hand side with respect to σ_ε^2 and setting the resulting expression to zero yields $\tilde{\sigma}_\varepsilon^2(\delta)$ of Equation (371b). Substituting $\tilde{\sigma}_\varepsilon^2(\delta)$ into the above and subtracting off N' gives us the reduced likelihood function stated in Equation (371c).

Answer to Exercise [373]: From Equation (370a) we have $\mathbf{X} - \widehat{\mathbf{T}} = \widetilde{\mathbf{U}}$, where $\widetilde{\mathbf{U}} \equiv \mathcal{W}^T\mathbf{W}_{nb}$. Hence $\|\mathbf{X} - \widehat{\mathbf{T}}\|^2 = \|\mathcal{W}^T\mathbf{W}_{nb}\|^2 = \|\mathbf{W}_{nb}\|^2$ because $\mathcal{W}^T$ is an orthonormal transform.

Answer to Exercise [395]: The energy preserving property of orthonormal transforms says that $\|\mathbf{O}\|^2 = \|\mathbf{D}\|^2$. Because

$$\|\mathbf{O}\|^2 = \|\mathcal{I}_M\mathbf{O}\|^2 + \|(I_N - \mathcal{I}_M)\mathbf{O}\|^2$$

and

$$\|\mathbf{D} - \widehat{\mathbf{D}}_M\|^2 = \|\mathcal{O}^T\mathbf{O} - \mathcal{O}^T\mathcal{I}_M\mathbf{O}\|^2 = \|\mathcal{O}^T(I_N - \mathcal{I}_M)\mathbf{O}\|^2 = \|(I_N - \mathcal{I}_M)\mathbf{O}\|^2,$$

we have

$$C_{M-1} = \frac{\|\mathcal{I}_M\mathbf{O}\|^2}{\|\mathbf{O}\|^2} = \frac{\|\mathbf{O}\|^2 - \|(I_N - \mathcal{I}_M)\mathbf{O}\|^2}{\|\mathbf{O}\|^2} = 1 - \frac{\|\mathbf{D} - \widehat{\mathbf{D}}_M\|^2}{\|\mathbf{D}\|^2},$$

as required.

Answer to Exercise [404]: Note that

$$A^{(\delta)}(O_l) = -O_l 1_{[0,\delta^2)}(O_l^2) - \delta\,\mathrm{sign}\{O_l\}1_{[\delta^2,\infty)}(O_l^2).$$

Hence,

$$\begin{aligned}
\mathcal{R}(\sigma_{n_l}, O_l, \delta) &= \sigma_{n_l}^2 - 2\sigma_{n_l}^2 1_{[0,\delta^2)}(O_l^2) + O_l^2 1_{[0,\delta^2)}(O_l^2) + \delta^2 1_{[\delta^2,\infty)}(O_l^2) \\
&= -\sigma_{n_l}^2 1_{[0,\delta^2)}(O_l^2) + \sigma_{n_l}^2 1_{[\delta^2,\infty)}(O_l^2) + O_l^2 1_{[0,\delta^2)}(O_l^2) + \delta^2 1_{[\delta^2,\infty)}(O_l^2) \\
&= [O_l^2 - \sigma_{n_l}^2]1_{[0,\delta^2)}(O_l^2) + [\sigma_{n_l}^2 + \delta^2]1_{[\delta^2,\infty)}(O_l^2) \\
&= O_l^2 - \sigma_{n_l}^2 - [O_l^2 - \sigma_{n_l}^2]1_{[\delta^2,\infty)}(O_l^2) + [\sigma_{n_l}^2 + \delta^2]1_{[\delta^2,\infty)}(O_l^2) \\
&= O_l^2 - \sigma_{n_l}^2 + (2\sigma_{n_l}^2 - O_l^2 + \delta^2)1_{[\delta^2,\infty)}(O_l^2),
\end{aligned}$$

as required.

Answer to Exercise [407]: We have

$$E\{(R_l - a_l O_l)^2\} = E\{R_l^2\} - 2a_l E\{R_l O_l\} + a_l^2 E\{O_l^2\} \equiv f(a_l),$$

which is a quadratic function of a_l (assuming $E\{O_l^2\} \neq 0$). To determine the extremal value of this function, we differentiate the above with respect to a_l and set the resulting expression to 0, yielding

$$\frac{df(a_l)}{da_l} = -2E\{R_l O_l\} + 2a_l E\{O_l^2\} = 0$$

(since the second derivative is $2E\{O_l^2\}$ and hence positive if $E\{O_l^2\} \neq 0$, the extremal value is in fact a minimum). The above is equivalent to

$$E\{R_l O_l\} - a_l E\{O_l^2\} = E\{(R_l - a_l O_l) O_l\} = 0,$$

as required.

Answer to Exercise [441]: Let $\mathbf{b}^T$ be the row vector of $\mathcal{W}$ that yields the coefficient $n_{j,t}$ in $\mathbf{n} = \mathcal{W}\boldsymbol{\eta}$; i.e., $n_{j,t} = \mathbf{b}^T \boldsymbol{\eta}$. Letting $\mathcal{M} = \mathbf{b}^T$ in Equation (262c), using the decomposition for $\Sigma_{\boldsymbol{\eta}}$ stated prior to the exercise and recalling that $\mathbf{b}^T = \mathbf{b}^H$ because $\mathcal{W}$ is real-valued, we then have

$$\operatorname{var}\{n_{j,t}\} = \mathbf{b}^T \Sigma_{\boldsymbol{\eta}} \mathbf{b} = \mathbf{b}^H \mathcal{F}^H D \mathcal{F} \mathbf{b} = (\mathcal{F}\mathbf{b})^H D \mathcal{F} \mathbf{b}.$$

Now $\mathcal{F}\mathbf{b}$ is the ODFT of $\mathbf{b}$, which is related to the DFT, say $\{B_k\}$, of $\mathbf{b}$ via $\mathcal{F}\mathbf{b} = \frac{1}{\sqrt{M}}[B_0, B_1, \ldots, B_{M-1}]^T$ (compare Equations (46b) and (36g)). Hence

$$\begin{aligned}
\operatorname{var}\{n_{j,t}\} &= \frac{1}{M}[B_0^*, B_1^*, \ldots, B_{M-1}^*] \begin{bmatrix} S_0 & 0 & \cdots & 0 \\ 0 & S_1 & \cdots & 0 \\ \vdots & \vdots & \ddots & \vdots \\ 0 & 0 & \cdots & S_{M-1} \end{bmatrix} \begin{bmatrix} B_0 \\ B_1 \\ \vdots \\ B_{M-1} \end{bmatrix} \\
&= \frac{1}{M}[B_0^*, B_1^*, \ldots, B_{M-1}^*] \begin{bmatrix} S_0 B_0 \\ S_1 B_1 \\ \vdots \\ S_{M-1} B_{M-1} \end{bmatrix} = \frac{1}{M} \sum_{k=0}^{M-1} S_k |B_k|^2.
\end{aligned}$$

Equation (441b) follows by noting that $|B_k|^2 = |H_j(\frac{k}{N})|^2 = \mathcal{H}_j(\frac{k}{N})$.

Answer to Exercise [464]: Let $u \equiv 2^{-j}t - 2k$ so that

$$2^{-j}t - m = u + 2k - m, \quad 2^{-(j+1)}t - k = 2^{-1}u \text{ and } 2^{-j}dt = du. \tag{545}$$

Then

$$\begin{aligned}
\langle \phi_{j,m}(\cdot), \phi_{j+1,k}(\cdot) \rangle &\equiv \int_{-\infty}^{\infty} \phi_{j,m}(t)\phi_{j+1,k}(t)\, dt \\
&= \int_{-\infty}^{\infty} 2^{-j}\phi(2^{-j}t - m) 2^{-1/2}\phi(2^{-(j+1)}t - k)\, dt \\
&= \int_{-\infty}^{\infty} \phi(u + 2k - m) 2^{-1/2}\phi(2^{-1}u)\, du \\
&= \int_{-\infty}^{\infty} \phi_{0,m-2k}(u)\phi_{1,0}(u)\, du = \langle \phi_{0,m-2k}(\cdot), \phi_{1,0}(\cdot) \rangle = \bar{g}_{m-2k}.
\end{aligned}$$

To show that the corollary is true, note that, with $j' \equiv j - 1$, $m' = k$ and $k' = m$,

$$\langle \phi_{j,m}(\cdot), \phi_{j-1,k}(\cdot) \rangle = \langle \phi_{j-1,k}(\cdot), \phi_{j,m}(\cdot) \rangle = \langle \phi_{j',m'}(\cdot), \phi_{j'+1,k'}(\cdot) \rangle = \bar{g}_{m'-2k'} = \bar{g}_{k-2m},$$

where we have made use of the first part.

Answer to Exercise [466]: Fourier transforming both sides of Equation (466c) yields

$$\begin{aligned}\Phi_{j+1}(f) &= \sum_{l=-\infty}^{\infty} \bar{g}_l \int_{-\infty}^{\infty} \phi_{j,0}(t - 2^j l) e^{-i2\pi f t}\, dt \\ &= \sum_{l=-\infty}^{\infty} \bar{g}_l e^{-i2\pi 2^j f l} \int_{-\infty}^{\infty} \phi_{j,0}(t - 2^j l) e^{-i2\pi f(t-2^j l)}\, dt \\ &= \overline{G}(2^j f) \int_{-\infty}^{\infty} \phi_{j,0}(t') e^{-i2\pi f t'}\, dt' = \overline{G}(2^j f)\Phi_j(f),\end{aligned}$$

as required.

Answer to Exercise [468a]: First we claim that any nonzero integer m can be written as $m = 2^l[2j+1]$ for some $l \geq 0$ and $j \in \mathbb{Z}$. To see this, note that, by setting $l = 0$, we get $m = 2j + 1$, which yields all odd m; on the other hand, we can write any even m as the product of some odd integer $2j + 1$ and some 2^l with $l \geq 1$. Setting $n = l + 1$ in Equation (465c), we get

$$\Phi(m) = \Phi(\tfrac{2^l[2j+1]}{2^{l+1}}) \prod_{k=1}^{l+1} \frac{\overline{G}(\frac{2^l[2j+1]}{2^k})}{\sqrt{2}} = \Phi(\tfrac{2j+1}{2}) \prod_{k=1}^{l+1} \frac{\overline{G}(2^{l-k}[2j+1])}{\sqrt{2}}.$$

The $k = l + 1$ term in the product involves $\overline{G}(j + \frac{1}{2})$; however, $\overline{G}(\cdot)$ is periodic with period unity, so that $\overline{G}(j + \frac{1}{2}) = \overline{G}(\frac{1}{2}) = 0$ from Equation (468a). Hence $\Phi(m) = 0$. (The approach used here is based on Daubechies, 1992, pp. 174–5.)

Answer to Exercise [468b]: Note first that the two-scale relationship of Equation (465a) implies that

$$\phi(t+k) = \sqrt{2} \sum_{l'=-\infty}^{\infty} \bar{g}_{l'} \phi(2t + 2k - l')$$

and hence that

$$\phi(t)\phi(t+k) = 2 \sum_{l=-\infty}^{\infty} \sum_{l'=-\infty}^{\infty} \bar{g}_l \bar{g}_{l'} \phi(2t - l)\phi(2t + 2k - l').$$

Integrating both sides with respect to t yields

$$\begin{aligned}\delta_{k,0} &= 2 \sum_{l=-\infty}^{\infty} \sum_{l'=-\infty}^{\infty} \bar{g}_l \bar{g}_{l'} \int_{-\infty}^{\infty} \phi(2t - l)\phi(2t + 2k - l')\, dt \\ &= \sum_{l=-\infty}^{\infty} \sum_{l'=-\infty}^{\infty} \bar{g}_l \bar{g}_{l'} \int_{-\infty}^{\infty} \phi(t - l)\phi(t + 2k - l')\, dt = \sum_{l=-\infty}^{\infty} \bar{g}_l \bar{g}_{l+2k},\end{aligned}$$

as required.

Answer to Exercise [474a]:

$$\begin{aligned}
\int_{-\infty}^{\infty} \psi_{0,m}(t)\phi_{0,n}(t)\,dt
&= \int_{-\infty}^{\infty} \psi(t-m)\phi(t-n)\,dt \\
&= 2\int_{-\infty}^{\infty} \left[\sum_{l=-\infty}^{\infty} \bar{h}_l \phi(2t-2m-l)\right]\left[\sum_{l'=-\infty}^{\infty} \bar{g}_{l'}\phi(2t-2n-l')\right] dt \\
&= 2\sum_{l=-\infty}^{\infty}\sum_{l'=-\infty}^{\infty} \bar{h}_l \bar{g}_{l'} \int_{-\infty}^{\infty} \phi(2t-2m-l)\phi(2t-2n-l')\,dt \\
&= \sum_{l=-\infty}^{\infty}\sum_{l'=-\infty}^{\infty} \bar{h}_l \bar{g}_{l'} \int_{-\infty}^{\infty} \phi(t-2m-l)\phi(t-2n-l')\,dt \\
&= \sum_{l=-\infty}^{\infty} \bar{h}_l \bar{g}_{l+2m-2n} \\
&= \sum_{l=-\infty}^{\infty} \bar{h}_{l+n'} \bar{g}_{l-n'} \qquad (\text{letting } n' \equiv n-m) \\
&= \sum_{l=-\infty}^{\infty} (-1)^{l+n'} \bar{g}_{1-l-n'-L} \bar{g}_{l-n'} \\
&= \sum_{l=-\infty}^{-L/2} (-1)^{l+n'} \bar{g}_{1-l-n'-L} \bar{g}_{l-n'} + \sum_{l=-(L/2)+1}^{\infty} (-1)^{l+n'} \bar{g}_{1-l-n'-L} \bar{g}_{l-n'} \\
&= \sum_{l=L/2}^{\infty} (-1)^{-l+n'} \bar{g}_{1+l-n'-L} \bar{g}_{-l-n'} - \sum_{l=L/2}^{\infty} (-1)^{l+n'} \bar{g}_{-l-n'} \bar{g}_{1+l-n'-L} \\
&= 0,
\end{aligned}$$

where the final line is due to the fact that $(-1)^{-l+n'} = (-1)^{l+n'-L} = (-1)^{l+n'}$ for all l and n', since L is even. An alternative proof would be to note that

$$\sum_{l=-\infty}^{\infty} \bar{h}_l \bar{g}_{l+2m-2n} = \sum_{l=-\infty}^{\infty} h_{-l} g_{-l+2n-2m} = \sum_{l=-\infty}^{\infty} h_l g_{l+2(n-m)} = 0,$$

by invoking the result of Exercise [77].

Answer to Exercise [474b]: We have

$$\begin{aligned}
\int_{-\infty}^{\infty} \psi_{0,m}(t)\psi_{0,n}(t)\,dt = \int_{-\infty}^{\infty} \psi(t-m)\psi(t-n)\,dt
&= 2\int_{-\infty}^{\infty} \left[\sum_{l=-\infty}^{\infty} \bar{h}_l \phi(2t-2m-l)\right]\left[\sum_{l'=-\infty}^{\infty} \bar{h}_{l'}\phi(2t-2n-l')\right] dt \\
&= 2\sum_{l=-\infty}^{\infty}\sum_{l'=-\infty}^{\infty} \bar{h}_l \bar{h}_{l'} \int_{-\infty}^{\infty} \phi(2t-2m-l)\phi(2t-2n-l')\,dt
\end{aligned}$$

$$= \sum_{l=-\infty}^{\infty} \sum_{l'=-\infty}^{\infty} \bar{h}_l \bar{h}_{l'} \int_{-\infty}^{\infty} \phi(t - 2m - l)\phi(t - 2n - l')\, dt$$

$$= \sum_{l=-\infty}^{\infty} \bar{h}_l \bar{h}_{l+2m-2n} = \sum_{l=-\infty}^{\infty} (-1)^{2l+2m-2n} \bar{g}_{1-l} \bar{g}_{1-l-2m+2n}$$

$$= \sum_{l=-\infty}^{\infty} \bar{g}_{1-l} \bar{g}_{1-l-2m+2n}$$

$$= \sum_{l=-\infty}^{\infty} g_{l-1} g_{l-1-2n+2m} = \sum_{l=-\infty}^{\infty} g_l g_{l+2(m-n)} = \delta_{m,n},$$

because the unit energy filter $\{g_l\}$ is orthogonal to its nonzero even shifts (see Exercise [76b]).

Answer to Exercise [476]: Since $\bar{h}_l \equiv (-1)^l \bar{g}_{1-l-L}$ (Equation (472)), we have (letting $m \equiv 1 - l - L$)

$$\overline{H}(f) \equiv \sum_{l=-\infty}^{\infty} \bar{h}_l e^{-i2\pi f l} = \sum_{l=-(L-1)}^{0} (-1)^l \bar{g}_{1-l-L} e^{-i2\pi f l}$$

$$= \sum_{m=-(L-1)}^{0} (-1)^{1-m-L} \bar{g}_m e^{-i2\pi f(1-m-L)}$$

$$= -e^{-i2\pi f(1-L)} \sum_{m=-(L-1)}^{0} (-1)^{-m-L} \bar{g}_m e^{i2\pi f m}$$

$$= -e^{-i2\pi f(1-L)} \sum_{m=-(L-1)}^{0} (-1)^{-L} e^{-i\pi m} \bar{g}_m e^{i2\pi f m}$$

$$= -e^{-i2\pi f(1-L)} \sum_{m=-(L-1)}^{0} \bar{g}_m e^{i2\pi (f-\frac{1}{2})m}$$

$$= -e^{-i2\pi f(1-L)} \left(\sum_{m=-(L-1)}^{0} \bar{g}_m e^{-i2\pi (f-\frac{1}{2})m} \right)^*$$

$$= -e^{i2\pi f(L-1)} \overline{G}^*(f - \tfrac{1}{2}) = -e^{i2\pi f(L-1)} \overline{G}(\tfrac{1}{2} - f)$$

(we have used the fact that $(-1)^{-L} = 1$ since L is even).

Answer to Exercise [479]: We have

$$\langle \phi_{0,m-2k}(\cdot), \psi_{1,0}(\cdot) \rangle = \int_{-\infty}^{\infty} \phi(t - [m-2k]) \frac{\psi(\frac{t}{2})}{\sqrt{2}}\, dt$$

$$= \int_{-\infty}^{\infty} \phi(t - [m-2k]) \left[\sum_{l=-\infty}^{\infty} \bar{h}_l \phi(t - l) \right] dt$$

$$= \sum_{l=-\infty}^{\infty} \bar{h}_l \int_{-\infty}^{\infty} \phi(t - [m-2k]) \phi(t - l)\, dt$$

$$= \sum_{l=-\infty}^{\infty} \bar{h}_l \delta_{l,m-2k} = \bar{h}_{m-2k}.$$

As in Exercise [464], let $u \equiv 2^{-j}t - 2k$ so that Equation (545) holds. Then

$$\begin{aligned}
\langle \phi_{j,m}(\cdot), \psi_{j+1,k}(\cdot)\rangle &\equiv \int_{-\infty}^{\infty} \phi_{j,m}(t)\psi_{j+1,k}(t)\,dt \\
&= \int_{-\infty}^{\infty} 2^{-j}\phi(2^{-j}t - m)2^{-1/2}\psi(2^{-(j+1)}t - k)\,dt \\
&= \int_{-\infty}^{\infty} \phi(u + 2k - m)2^{-1/2}\psi(2^{-1}u)\,du \\
&= \int_{-\infty}^{\infty} \phi_{0,m-2k}(u)\psi_{1,0}(u)\,du \\
&= \langle \phi_{0,m-2k}(\cdot), \psi_{1,0}(\cdot)\rangle = \bar{h}_{m-2k},
\end{aligned}$$

as required.

Answer to Exercise [481]: We have

$$\begin{aligned}
v_{2,k} &= \langle s_1(\cdot), \phi_{2,k}(\cdot)\rangle = \langle x(\cdot) - d_1(\cdot), \phi_{2,k}(\cdot)\rangle \\
&= \langle x(\cdot), \phi_{2,k}(\cdot)\rangle - \langle d_1(\cdot), \phi_{2,k}(\cdot)\rangle \\
&= \langle x(\cdot), \phi_{2,k}(\cdot)\rangle - \sum_{m=-\infty}^{\infty} w_{1,m}\langle \psi_{1,m}(\cdot), \phi_{2,k}(\cdot)\rangle = \langle x(\cdot), \phi_{2,k}(\cdot)\rangle
\end{aligned}$$

since

$$\phi_{j,k}(\cdot) \perp \psi_{l,m}(\cdot) \quad \text{for} \quad j,k,l,m \in \mathbb{Z} \quad \text{such that} \quad l \le j.$$

Similarly,

$$\begin{aligned}
w_{2,k} &= \langle s_1(\cdot), \psi_{2,k}(\cdot)\rangle = \langle x(\cdot) - d_1(\cdot), \psi_{2,k}(\cdot)\rangle \\
&= \langle x(\cdot), \psi_{2,k}(\cdot)\rangle - \langle d_1(\cdot), \psi_{2,k}(\cdot)\rangle \\
&= \langle x(\cdot), \psi_{2,k}(\cdot)\rangle - \sum_{m=-\infty}^{\infty} w_{1,m}\langle \psi_{1,m}(\cdot), \psi_{2,k}(\cdot)\rangle = \langle x(\cdot), \psi_{2,k}(\cdot)\rangle,
\end{aligned}$$

since

$$\psi_{j,k}(\cdot) \perp \psi_{l,m}(\cdot) \text{ for all } j,k,l,m \in \mathbb{Z} \text{ such that } j \ne l.$$

For the general case we note that $x(t) = s_{j-1}(t) + d_{j-1}(t) + \cdots + d_1(t)$. Hence,

$$\begin{aligned}
v_{j,k} &= \langle s_{j-1}(\cdot), \phi_{j,k}(\cdot)\rangle = \langle x(\cdot) - \sum_{l=1}^{j-1} d_l(\cdot), \phi_{j,k}(\cdot)\rangle \\
&= \langle x(\cdot), \phi_{j,k}(\cdot)\rangle - \sum_{l=1}^{j-1}\langle d_l(\cdot), \phi_{j,k}(\cdot)\rangle \\
&= \langle x(\cdot), \phi_{j,k}(\cdot)\rangle - \sum_{l=1}^{j-1}\sum_{m=-\infty}^{\infty} w_{l,m}\langle \psi_{l,m}(\cdot), \phi_{j,k}(\cdot)\rangle = \langle x(\cdot), \phi_{j,k}(\cdot)\rangle,
\end{aligned}$$

where we have again used

$$\phi_{j,k}(\cdot) \perp \psi_{l,m}(\cdot) \quad \text{for} \quad j,k,l,m \in \mathbb{Z} \quad \text{such that} \quad l \le j,$$

as is the case here. The general case for $w_{j,k}$ follows analogously.

Answer to Exercise [482]: From the definition of a multiresolution analysis, we have $V_{j+1} \subset V_j$ for any j, which implies that

$$\bigcup_{l \in \mathbb{Z}} V_l = \bigcup_{-\infty < l \le j} V_l.$$

On the other hand, the fact that $V_{j-1} = V_j \oplus W_j$ implies that

$$\begin{aligned} V_j \cup V_{j-1} \cup V_{j-2} &= V_j \cup \left[V_j \oplus W_j\right] \cup \left[V_{j-1} \oplus W_{j-1}\right] \\ &= V_j \cup \left[V_j \oplus W_j\right] \cup \left[V_j \oplus W_j \oplus W_{j-1}\right]. \end{aligned}$$

From the definition of a direct sum, we have $A \subset A \oplus B$ for any sets A and B, which in turn implies that $A \cup [A \oplus B] = A \oplus B$. Hence

$$V_j \cup V_{j-1} \cup V_{j-2} = V_j \oplus W_j \oplus W_{j-1}.$$

We can thus argue that

$$L^2(\mathbb{R}) = \overline{\bigcup_{l \in \mathbb{Z}} V_l} = \overline{V_j \cup V_{j-1} \cup V_{j-2} \cup \cdots} = \overline{V_j \oplus W_j \oplus W_{j-1} \oplus \cdots},$$

as required.

Answer to Exercise [484]: We wish to show that $\Psi^{(m)}(0) = 0$ for $m = 0, \ldots, n$ implies $\overline{H}^{(m)}(0) = 0$ for $m = 0, \ldots, n$ also. We use the method of induction. The result is true for $n = 0$ and 1 since $\Psi(f) = \Phi(\frac{f}{2})\overline{H}(\frac{f}{2})/\sqrt{2}$, so that $\Psi(0) = 0$ implies $\overline{H}(0) = 0$, ($\Phi(0) = 1$), and we have just shown that $\Psi^{(1)}(0)$ implies $H^{(1)}(0) = 0$. Assume the result is true for $n-1$. Now suppose that $\Psi^{(m)}(0) = 0$ for $m = 0, \ldots, n$. From Leibniz's theorem for differentiation of a product,

$$\Psi^{(n)}(f) = \frac{1}{\sqrt{2}} \sum_{m=0}^{n} \binom{n}{m} \Phi^{(n-m)}(\tfrac{f}{2})\overline{H}^{(m)}(\tfrac{f}{2});$$

but, if $\Psi^{(m)}(0) = 0$ for $m = 0, \ldots, n$, then $\Psi^{(m)}(0) = 0$ for $m = 0, \ldots, n-1$. Since we are assuming the result is true for $n-1$, we have that $\overline{H}^{(m)}(0) = 0$ for $m = 0, \ldots, n-1$; however, from Leibniz's formula

$$\Psi^{(n)}(0) = \frac{1}{\sqrt{2}} \left[\sum_{m=0}^{n-1} \binom{n}{m} \Phi^{(n-m)}(0)\overline{H}^{(m)}(0) + \Phi(0)\overline{H}^{(n)}(0) \right].$$

Hence $\Psi^{(n)}(0) = 0 = \Phi(0)\overline{H}^{(n)}(0)$. Since $\Phi(0) = 1$, we have $\overline{H}^{(n)}(0) = 0$, which – together with $\overline{H}^{(m)}(0) = 0$ for $m = 0, \ldots, n-1$ – means that $\overline{H}^{(m)}(0) = 0$ for $m = 0, \ldots, n$, and the result is proved.

Answer to Exercise [485]: We have $\{\bar{g}_l\} \longleftrightarrow \overline{G}(\cdot)$, which implies that $\{\bar{g} \star \bar{g}_l\} \longleftrightarrow \overline{\mathcal{G}}(\cdot)$, where $\{\bar{g} \star \bar{g}_l\}$ is the autocorrelation of $\{\bar{g}_l\}$ (see Equations (25b) and (25c)). Because $\bar{g}_l$ is real-valued, $\bar{g} \star \bar{g}_l$ must be also, so we have

$$\bar{g} \star \bar{g}_l = (\bar{g} \star \bar{g}_l)^* = \left(\int_{-1/2}^{1/2} \overline{\mathcal{G}}(\cdot) e^{i2\pi fl}\, df \right)^* = \int_{-1/2}^{1/2} \overline{\mathcal{G}}(\cdot) e^{-i2\pi fl}\, df = \bar{g} \star \bar{g}_{-l},$$

so $\{\bar{g} \star \bar{g}_l\}$ is an even sequence. Hence

$$\overline{\mathcal{G}}(f) = \sum_{l=-(L-1)}^{L-1} \bar{g} \star \bar{g}_l e^{-i2\pi fl} = \bar{g} \star \bar{g}_0 + 2 \sum_{l=1}^{L-1} \bar{g} \star \bar{g}_l \cos(2\pi fl).$$

The relationship $\cos(2\pi fl) = 2\cos(2\pi f[l-1])\cos(2\pi f) - \cos(2\pi f[l-2])$ can be used repeatedly to reduce the summation above to a linear combination of powers of $\cos(2\pi f)$. Finally, because

$$\overline{\mathcal{G}}(f) = \cos^{2r}(\pi f)\mathcal{Q}(f) = \left(\frac{1+\cos(2\pi f)}{2}\right)^r \mathcal{Q}(f),$$

we can conclude that $\mathcal{Q}(f)$ must also be a polynomial in $\cos(2\pi f)$.

Answer to Exercise [488]: Note first that, since $z = e^{i2\pi f}$, it follows that $z^* = z^{-1}$ because $|z|^2 = zz^* = 1$. Thus

$$\left[\overline{\overline{Q}}_0(z)\right]^* = \prod_{l=1}^{\frac{L}{2}-1} (z^{-1} - z_l^*) = \prod_{l=1}^{\frac{L}{2}-1} (z^{-1} - z_l) = \overline{\overline{Q}}_0(z^{-1}),$$

where the middle equality holds because, in order for $\overline{\overline{Q}}_0(z)$ to be the z transform for a real-valued filter, the roots z_l must be either real-valued (in which case $z_l^* = z_l$) or occur in pairs that are the complex conjugates of each other.

Answer to Exercise [490]: Expanding $\overline{\overline{G}}(z)$, we obtain

$$\overline{\overline{G}}(z) = \frac{1-\sqrt{3}}{4\sqrt{2}}(b_1 z^3 + [1 + 2b_1]z^2 + [2 + b_1]z + 1);$$

however, comparison of the above with Equation (490b) shows that the coefficients of the polynomial are now reversed, from which it follows that the above leads to a filter that is the reverse of the D(4) scaling filter.

References

Abramovich, F., Sapatinas, T. and Silverman, B. W. (1998) Wavelet Thresholding via a Bayesian Approach. *Journal of the Royal Statistical Society, Series B*, 60, 725–49.

Abramowitz, M. and Stegun, I. A., editors (1964) *Handbook of Mathematical Functions.* Washington, DC: US Government Printing Office (reprinted in 1968 by Dover Publications, New York).

Abry, P. and Flandrin, P. (1994) On the Initialization of the Discrete Wavelet Transform Algorithm. *IEEE Signal Processing Letters*, 1, 32–4.

Abry, P., Gonçalvès, P. and Flandrin, P. (1993) Wavelet-Based Spectral Analysis of $1/f$ Processes. In *Proceedings of the IEEE International Conference on Acoustics, Speech and Signal Processing*, Minneapolis, 3, 237–40.

——— (1995) Wavelets, Spectrum Analysis and $1/f$ Processes. In *Wavelets and Statistics* (Lecture Notes in Statistics, Volume 103), edited by A. Antoniadis and G. Oppenheim. New York: Springer–Verlag, 15–29.

Abry, P. and Veitch, D. (1998) Wavelet Analysis of Long-Range-Dependent Traffic. *IEEE Transactions on Information Theory*, 44, 2–15.

Allan, D. W. (1966) Statistics of Atomic Frequency Standards. *Proceedings of the IEEE*, 54, 221–30.

Anderson, T. W. (1971) *The Statistical Analysis of Time Series.* New York: John Wiley & Sons.

Balek, J. (1977) *Hydrology and Water Resources in Tropical Africa*, Volume 8 of *Developments in Water Science.* New York: Elsevier Scientific Publishing Company.

Baliunas, S., Frick, P., Sokoloff, D. and Soon, W. (1997) Time Scales and Trends in the Central England Temperature Data (1659–1990): A Wavelet Analysis. *Geophysical Research Letters*, 24, 1351–4.

Balogh, A., Smith, E. J., Tsurutani, B. T., Southwood, D. J., Forsyth, R. J. and Horbury, T. S. (1995) The Heliospheric Magnetic Field over the South Polar Region of the Sun. *Science*, 268, 1007–10.

Barnes, J. A., Chi, A. R., Cutler, L. S., Healey, D. J., Leeson, D. B., McGunigal, T. E., Mullen, J. A., Jr., Smith, W. L., Sydnor, R. L., Vessot, R. F. C. and Winkler, G. M. R. (1971) Characterization of Frequency Stability. *IEEE Transactions on Instrumentation and Measurement*, 20, 105–20.

Bartlett, M. S. and Kendall, D. G. (1946) The Statistical Analysis of Variance-Heterogeneity and the Logarithmic Transformation. *Supplement to the Journal of the Royal Statistical Society*, 8, 128–38.

Beauchamp, K. G. (1984) *Applications of Walsh and Related Functions.* London: Academic Press.

Beran, J. (1994) *Statistics for Long-Memory Processes.* New York: Chapman & Hall.

Beran, J. and Terrin, N. (1996) Testing for a Change of the Long-Memory Parameter. *Biometrika*, 83, 627–38.

Best, D. J. and Roberts, D. E. (1975) The Percentage Points of the χ^2 Distribution: Algorithm AS 91. *Applied Statistics*, 24, 385–8.

Beylkin, G. (1992) On the Representation of Operators in Bases of Compactly Supported Wavelets. *SIAM Journal on Numerical Analysis*, 29, 1716–40.

Billingsley, P. (1968) *Convergence of Probability Measures.* New York: John Wiley & Sons.

Blackman, R. B. and Tukey, J. W. (1958) *The Measurement of Power Spectra.* New York: Dover Publications.

Bloomfield, P. (1976) *Fourier Analysis of Time Series: An Introduction.* New York: John Wiley & Sons.

Box, G. E. P., Jenkins, G. M. and Reinsel, G. C. (1994) *Time Series Analysis: Forecasting and Control* (3rd Edition). Englewood Cliffs, New Jersey: Prentice Hall.

Box, G. E. P. and Tiao, G. C. (1973) *Bayesian Inference in Statistical Analysis.* Reading, Massachusetts: Addison–Wesley.

Bracewell, R. N. (1978) *The Fourier Transform and Its Applications* (Second Edition). New York: McGraw–Hill.

Bradshaw, G. A. and Spies, T. A. (1992) Characterizing Canopy Gap Structure in Forests using Wavelet Analysis. *Journal of Ecology*, 80, 205–15.

Briggs, W. L. and Henson, V. E. (1995) *The DFT: An Owner's Manual for the Discrete Fourier Transform.* Philadelphia: SIAM.

Brillinger, D. R. (1981) *Time Series: Data Analysis and Theory* (Expanded Edition). San Francisco: Holden–Day.

——— (1994) Some River Wavelets. *Environmetrics*, 5, 211–20.

——— (1996) Some Uses of Cumulants in Wavelet Analysis. *Journal of Nonparametric Statistics*, 6, 93–114.

Brockwell, P. J. and Davis, R. A. (1991) *Time Series: Theory and Methods* (Second Edition). New York: Springer.

Brown, J. W. and Churchill, R. V. (1995) *Complex Variables and Applications* (Sixth Edition). New York: McGraw–Hill.

Brown, R. L., Durbin, J. and Evans, J. M. (1975) Techniques for Testing the Constancy of Regression Relationships over Time. *Journal of the Royal Statistical Society, Series B*, 37, 149–63.

Bruce, A. G., Donoho, D. L., Gao, H.–Y. and Martin, R. D. (1994) Denoising and Robust Non-Linear Wavelet Analysis. In *Wavelet Applications* (Proceedings of the SPIE 2242), edited by H. H. Szu. Bellingham, Washington: SPIE Press, 325–36.

Bruce, A. G. and Gao, H.–Y. (1996a) *Applied Wavelet Analysis with S-PLUS*. New York: Springer.

——— (1996b) Understanding WaveShrink: Variance and Bias Estimation. *Biometrika*, 83, 727–45.

Bruce, A. G., Gao, H.–Y. and Stuetzle, W. (1999) Subset-Selection and Ensemble Methods for Wavelet De-Noising. *Statistica Sinica*, 9, 167–82.

Calderón, A. P. (1964) Intermediate Spaces and Interpolation, the Complex Method. *Studia Mathematica*, 24, 113–90.

Carmona, R., Hwang, W.–L. and Torrésani, B. (1998) *Practical Time-Frequency Analysis*. San Diego: Academic Press.

Casella, G. and Berger, R. L. (1990) *Statistical Inference*. Pacific Grove, California: Wadswoth & Brooks/Cole.

Chambers, J. M., Cleveland, W. S., Kleiner, B. and Tukey, P. A. (1983) *Graphical Methods for Data Analysis*. Boston: Duxbury Press.

Chipman, H. A., Kolaczyk, E. D. and McCulloch, R. E. (1997) Adaptive Bayesian Wavelet Shrinkage. *Journal of the American Statistical Association*, 92, 1413–21.

Chui, C. K. (1997) *Wavelets: A Mathematical Tool for Signal Processing*. Philadelphia: SIAM.

Cleveland, W. S. and Parzen, E. (1975) The Estimation of Coherence, Frequency Response, and Envelope Delay. *Technometrics*, 17, 167–72.

Cohen, A., Daubechies, I. and Vial, P. (1993) Wavelets on the Interval and Fast Wavelet Transforms. *Applied and Computational Harmonic Analysis*, 1, 54–81.

Coifman, R. R. and Donoho, D. L. (1995) Translation-Invariant De-Noising. In *Wavelets and Statistics* (Lecture Notes in Statistics, Volume 103), edited by A. Antoniadis and G. Oppenheim. New York: Springer–Verlag, 125–50.

Coifman, R. R., Meyer, Y. and Wickerhauser, M. V. (1992) Size Properties of Wavelet Packets. In *Wavelets and Their Applications*, edited by M. B. Ruskai, G. Beylkin, R. R. Coifman, I. Daubechies, S. Mallat, Y. Meyer and L. Raphael. Boston: Jones and Bartlett, 453–70.

Coifman, R. R. and Wickerhauser, M. V. (1992) Entropy-Based Algorithms for Best Basis Selection. *IEEE Transactions on Information Theory*, 38, 713–8.

Craigmile, P. F., Percival, D. B. and Guttorp, P. (2000a) Wavelet-Based Parameter Estimation for Trend Contaminated Fractionally Differenced Processes. Submitted to *Journal of Time Series Analysis*.

——— (2000b) Trend Assessment Using the Discrete Wavelet Transform. Technical Report, National Research Center for Statistics and the Environment, University of Washington, Seattle.

Crouse, M. S., Nowak, R. D. and Baraniuk, R. G. (1998) Wavelet-Based Statistical Signal Processing Using Hidden Markov Models. *IEEE Transactions on Signal Processing*, 46, 886–902.

Dahlquist, G. and Björck, Å. (1974) *Numerical Methods*. Englewood Cliffs, New Jersey: Prentice Hall.

Daniels, H. E. (1987) Tail Probability Approximations. *International Statistical Review*, 55, 37–48.

Daubechies, I. (1988) Orthonormal Bases of Compactly Supported Wavelets. *Communications on Pure and Applied Mathematics*, 41, 909–96.

——— (1992) *Ten Lectures on Wavelets*. Philadelphia: SIAM.

——— (1993) Orthonormal Bases of Compactly Supported Wavelets II. Variations on a Theme. *SIAM Journal on Mathematical Analysis*, 24, 499–519.

David, H. A. (1985) Bias of S^2 Under Dependence. *The American Statistician*, 39, 201.

Davies, R. B. and Harte, D. S. (1987) Tests for Hurst Effect. *Biometrika*, 74, 95–101.

Davis, G., Mallat, S. G. and Zhang, Z. (1994) Adaptive Time-Frequency Approximations with Matching Pursuits. In *Wavelets: Theory, Algorithms, and Applications*, edited by C. K. Chui, L. Montefusco and L. Puccio. San Diego: Academic Press, 271–93.

Davison, A. C. and Hinkley, D. V. (1997) *Bootstrap Methods and their Application*. Cambridge, UK: Cambridge University Press.

Del Marco, S. and Weiss, J. (1997) Improved Transient Signal Detection Using a Wavepacket-Based Detector with an Extended Translation-Invariant Wavelet Transform. *IEEE Transactions on Signal Processing*, 45, 841–50.

Dijkerman, R. W. and Mazumdar, R. R. (1994) On the Correlation Structure of the Wavelet Coefficients of Fractional Brownian Motion. *IEEE Transactions on Information Theory*, 40, 1609–12.

Donoho, D. L. and Johnstone, I. M. (1994) Ideal Spatial Adaptation by Wavelet Shrinkage. *Biometrika*, 81, 425–55.

Doroslovački, M. I. (1998) On the Least Asymmetric Wavelets. *IEEE Transactions on Signal Processing*, 46, 1125–30.

Draper, N. R. and Smith, H. (1998) *Applied Regression Analysis* (Third Edition). New York: John Wiley & Sons.

Eltahir, E. A. B. and Wang, G. (1999) Nilometers, El Niño, and Climate Variability. *Geophysical Research Letters*, 26, 489–92.

Flandrin, P. (1989) On the Spectrum of Fractional Brownian Motions. *IEEE Transactions on Information Theory*, 35, 197–9.

——— (1992) Wavelet Analysis and Synthesis of Fractional Brownian Motion. *IEEE Transactions on Information Theory*, 38, 910–17.

Foufoula–Georgiou, E. and Kumar, P., editors (1994) *Wavelets in Geophysics*. San Diego: Academic Press.

Fuller, W. A. (1996) *Introduction to Statistical Time Series* (Second Edition). New York: John Wiley & Sons.

Gamage, N. K. K. (1990) Detection of Coherent Structures in Shear Induced Turbulence using Wavelet Transform Methods. In *Ninth Symposium on Turbulence and Diffusion*. Boston: American Meteorological Society, 389–92.

Gao, H.–Y. (1993) Wavelet Estimation of Spectral Densities in Time Series Analysis. Ph.D. dissertation, Department of Statistics, University of California, Berkeley.

——— (1997) Choice of Thresholds for Wavelet Shrinkage Estimate of the Spectrum. *Journal of Time Series Analysis*, 18, 231–51.

——— (1998) Wavelet Shrinkage Denoising using the Non-Negative Garrote. *Journal of Computational and Graphical Statistics*, 7, 469–88.

Gao, W. and Li, B. L. (1993) Wavelet Analysis of Coherent Structures at the Atmosphere-Forest Interface. *Journal of Applied Meteorology*, 32, 1717–25.

Gelman, A., Carlin, J. B., Stern, H. S. and Rubin, D. B. (1995) *Bayesian Data Analysis*. New York: Chapman & Hall.

Geweke, J. and Porter–Hudak, S. (1983) The Estimation and Application of Long Memory Time Series Models. *Journal of Time Series Analysis*, 4, 221–38.

Gneiting, T. (2000) Power-Law Correlations, Related Models for Long-Range Dependence, and Their Simulation. *Journal of Applied Probability*, 37, 1104–9.

Godfrey, R. and Rocca, F. (1981) Zero Memory Non-Linear Deconvolution. *Geophysical Prospecting*, 29, 189–228.

Gollmer, S. M., Harshvardhan, Cahalan, R. F. and Snider, J. B. (1995) Windowed and Wavelet Analysis of Marine Stratocumulus Cloud Inhomogeneity. *Journal of the Atmospheric Sciences*, 52, 3013–30.

González–Esparza, J. A., Balogh, A., Forsyth, R. J., Neugebauer, M., Smith, E. J. and Phillips, J. L. (1996) Interplanetary Shock Waves and Large-Scale Structures: Ulysses' Observations in and out of the Ecliptic Plane. *Journal of Geophysical Research*, 101, 17,057–71.

Goupillaud, P., Grossmann, A. and Morlet, J. (1984) Cycle-Octave and Related Transforms in Seismic Signal Analysis. *Geoexploration*, 23, 85–102.

Granger, C. W. J. and Hatanaka, M. (1964) *Spectral Analysis of Economic Time Series*. Princeton: Princeton University Press.

Granger, C. W. J. and Joyeux, R. (1980) An Introduction to Long-Memory Time Series Models and Fractional Differencing. *Journal of Time Series Analysis*, 1, 15–29.

Graybill, F. A. (1983) *Matrices with Applications in Statistics* (Second Edition). Belmont, California: Wadsworth.

Green, P. J. and Silverman, B. W. (1994) *Nonparametric Regression and Generalized Linear Models: A Roughness Penalty Approach*. London: Chapman & Hall.

Greenhall, C. A. (1991) Recipes for Degrees of Freedom of Frequency Stability Estimators. *IEEE Transactions on Instrumentation and Measurement*, 40, 994–9.

Greenhall, C. A., Howe, D. A. and Percival, D. B. (1999) Total Variance, an Estimator of Long-Term Frequency Stability. *IEEE Transactions on Ultrasonics, Ferroelectrics, and Frequency Control*, 46, 1183–91.

Grossmann, A. and Morlet, J. (1984) Decomposition of Hardy Functions into Square Integrable Wavelets of Constant Shape. *SIAM Journal on Mathematical Analysis*, 15, 723–36.

Haar, A. (1910) Zur Theorie der Orthogonalen Funktionensysteme. *Mathematische Annalen*, 69, 331–71.

Hamming, R. W. (1989) *Digital Filters* (Third Edition). Englewood Cliffs, New Jersey: Prentice Hall.

Hannan, E. J. (1970) *Multiple Time Series*. New York: John Wiley & Sons.

Härdle, W., Kerkyacharian, G., Picard, D. and Tsybakov, A. (1998) *Wavelets, Approximation, and Statistical Applications* (Lecture Notes in Statistics, Volume 129). New York: Springer.

Hess–Nielsen, N. and Wickerhauser, M. V. (1996) Wavelets and Time-Frequency Analysis. *Proceedings of the IEEE*, 84, 523–40.

Holden, A. V. (1976) *Models of the Stochastic Activity of Neurones* (Lecture Notes in Biomathematics, Volume 12). Berlin: Springer–Verlag.

Hosking, J. R. M. (1981) Fractional Differencing. *Biometrika*, 68, 165–76.

——— (1984) Modeling Persistence in Hydrological Time Series Using Fractional Differencing. *Water Resources Research*, 20, 1898–908.

Hsu, D. A. (1977) Tests for Variance Shift at an Unknown Time Point. *Applied Statistics*, 26, 279–84.

Hudgins, L. H. (1992) Wavelet Analysis of Atmospheric Turbulence. Ph.D. dissertation, University of California, Irvine.

Hudgins, L. H., Friehe, C. A. and Mayer, M. E. (1993) Wavelet Transforms and Atmospheric Turbulence. *Physical Review Letters*, 71, 3279–82.

Hurst, H. E. (1951) Long-Term Storage Capacity of Reservoirs. *Transactions of the American Society of Civil Engineers*, 116, 770–99.

Hurvich, C. M., Deo, R. and Brodsky, J. (1998) The Mean Squared Error of Geweke and Porter–Hudak's Estimator of the Memory Parameter of a Long-Memory Time Series. *Journal of Time Series Analysis*, 19, 19–46.

Hurvich, C. M. and Tsai, C.–L. (1998) A Crossvalidatory AIC for Hard Wavelet Thresholding in Spatially Adaptive Function Estimation. *Biometrika*, 85, 701–10.

Inclán, C. and Tiao, G. C. (1994) Use of Cumulative Sums of Squares for Retrospective Detection of Changes of Variance. *Journal of the American Statistical Association*, 427, 913–23.

Jansen, M., Malfait, M. and Bultheel, A. (1997) Generalized Cross Validation for Wavelet Thresholding. *Signal Processing*, 56, 33–44.

Jawerth, B. and Sweldens, W. (1994) An Overview of Wavelet Based Multiresolution Analyses. *SIAM Review*, 36, 377–412.

Jenkins, G. M. and Watts, D. G. (1968) *Spectral Analysis and Its Applications.* San Francisco: Holden–Day.

Jensen, M. J. (1999a) Using Wavelets to Obtain a Consistent Ordinary Least Squares Estimator of the Long-Memory Parameter. *Journal of Forecasting*, 18, 17–32.

——— (1999b) An Approximate Wavelet MLE of Short- and Long-Memory Parameters. *Studies in Nonlinear Dynamics and Econometrics*, 3, 239–53.

——— (2000) An Alternative Maximum Likelihood Estimator of Long-Memory Processes Using Compactly Supported Wavelets. *Journal of Economic Dynamics and Control*, 24, 361–87.

Jiang, J., Zhang, D. and Fraedrich, K. (1997) Historic Climate Variability of Wetness in East China (960–1992): A Wavelet Analysis. *International Journal of Climatology*, 17, 969–81.

Johnson, G. E. (1994) Constructions of Particular Random Processes. *Proceedings of the IEEE*, 82, 270–85.

Johnson, N. L., Kotz, S. and Balakrishnan, N. (1994) *Continuous Univariate Distributions, Volume 1* (Second Edition). New York: John Wiley & Sons.

Johnstone, I. M. and Silverman, B. W. (1997) Wavelet Threshold Estimators for Data with Correlated Noise. *Journal of the Royal Statistical Society, Series B*, 59, 319–51.

Kaplan, L. M. and Kuo, C.–C. J. (1993) Fractal Estimation from Noisy Data via Discrete Fractional Gaussian Noise (DFGN) and the Haar Basis. *IEEE Transactions on Signal Processing*, 41, 3554–62.

——— (1994) Extending Self-Similarity for Fractional Brownian Motion. *IEEE Transactions on Signal Processing*, 42, 3526–30.

Kashyap, R. L. and Eom, K.–B. (1988) Estimation in Long-Memory Time Series Model. *Journal of Time Series Analysis*, 9, 35–41.

Kay, S. M. (1981) Efficient Generation of Colored Noise. *Proceedings of the IEEE*, 69, 480–1.

——— (1988) *Modern Spectral Estimation: Theory & Application.* Englewood Cliffs, New Jersey: Prentice Hall.

Komm, R. W., Gu, Y., Hill, F., Stark, P. B. and Fodor, I. K. (1999) Multitaper Spectral Analysis and Wavelet Denoising Applied to Helioseismic Data. *Astrophysical Journal*, 519, 407–21.

Koopmans, L. H. (1974) *The Spectral Analysis of Time Series.* New York: Academic Press.

Kumar, P. and Foufoula–Georgiou, E. (1993) A Multicomponent Decomposition of Spatial Random Fields 1: Segregation of Large- and Small-Scale Features Using Wavelet Transforms. *Water Resources Research*, 29, 2515–32.

——— (1994) Wavelet Analysis in Geophysics: An Introduction. In *Wavelets in Geophysics*, edited by E. Foufoula–Georgiou and P. Kumar. San Diego: Academic Press, 1–43.

——— (1997) Wavelet Analysis for Geophysical Applications. *Reviews of Geophysics*, 35, 385–412.

Kuo, C., Lindberg, C. and Thomson, D. J. (1990) Coherence Established Between Atmospheric Carbon Dioxide and Global Temperature. *Nature*, 343, 709–14.

Lai, M.–J. (1995) On the Digital Filter Associated with Daubechies' Wavelets. *IEEE Transactions on Signal Processing*, 43, 2203–5.

Lang, M., Guo, H., Odegard, J. E., Burrus, C. S. and Wells, R. O. (1995) Nonlinear Processing of a Shift Invariant DWT for Noise Reduction. In *Wavelet Applications II* (Proceedings of the SPIE 2491), edited by H. H. Szu. Bellingham, Washington: SPIE Press, 640–51.

——— (1996) Noise Reduction using an Undecimated Discrete Wavelet Transform. *IEEE Signal Processing Letters*, 3, 10–12.

Lanning, E. N. and Johnson, D. M. (1983) Automated Identification of Rock Boundaries: An Application of the Walsh Transform to Geophysical Well-Log Analysis. *Geophysics*, 48, 197–205.

Lee, P. M. (1989) *Bayesian Statistics: An Introduction.* London: Edward Arnold.

Liang, J. and Parks, T. W. (1996) A Translation-Invariant Wavelet Representation Algorithm with Applications. *IEEE Transactions on Signal Processing*, 44, 225–32.

Lindsay, R. W., Percival, D. B. and Rothrock, D. A. (1996) The Discrete Wavelet Transform and the Scale Analysis of the Surface Properties of Sea Ice. *IEEE Transactions on Geoscience and Remote Sensing*, 34, 771–87.

Liu, P. C. (1994) Wavelet Spectrum Analysis and Ocean Wind Waves. In *Wavelets in Geophysics*, edited by E. Foufoula–Georgiou and P. Kumar. San Diego: Academic Press, 151–66.

Lugannani, R. and Rice, S. (1980) Saddle Point Approximation for the Distribution of the Sum of Independent Random Variables. *Advances in Applied Probability*, 12, 475–90.

Mak, M. (1995) Orthogonal Wavelet Analysis: Interannual Variability in the Sea Surface Temperature. *Bulletin of the American Meteorological Society*, 76, 2179–86.

Mallat, S. G. (1989a) Multiresolution Approximations and Wavelet Orthonormal Bases of $L^2(R)$. *Transactions of the American Mathematical Society*, 315, 69–87.

——— (1989b) A Theory for Multiresolution Signal Decomposition: The Wavelet Representation. *IEEE Transactions on Pattern Analysis and Machine Intelligence*, 11, 674–93.

——— (1989c) Multifrequency Channel Decompositions of Images and Wavelet Models. *IEEE Transactions on Acoustics, Speech, and Signal Processing*, 37, 2091–110.

——— (1998) *A Wavelet Tour of Signal Processing*. San Diego: Academic Press.

Mallat, S. G. and Zhang, Z. (1993) Matching Pursuits with Time-Frequency Dictionaries. *IEEE Transactions on Signal Processing*, 41, 3397–415.

Mandelbrot, B. B. and van Ness, J. W. (1968) Fractional Brownian Motions, Fractional Noises and Applications. *SIAM Review*, 10, 422–37.

Mandelbrot, B. B. and Wallis, J. R. (1968) Noah, Joseph, and Operational Hydrology. *Water Resources Research*, 4, 909–18.

Marple, S. L., Jr. (1987) *Digital Spectral Analysis with Applications*. Englewood Cliffs, New Jersey: Prentice–Hall.

Masry, E. (1993) The Wavelet Transform of Stochastic Processes with Stationary Increments and its Application to Fractional Brownian Motion. *IEEE Transactions on Information Theory*, 39, 260–4.

McCoy, E. J. and Walden, A. T. (1996) Wavelet Analysis and Synthesis of Stationary Long-Memory Processes. *Journal of Computational and Graphical Statistics*, 5, 26–56.

McCoy, E. J., Walden, A.T. and Percival, D. B. (1998) Multitaper Spectral Estimation of Power Law Processes. *IEEE Transactions on Signal Processing*, 46, 655–68.

McHardy, I. and Czerny, B. (1987) Fractal X-Ray Time Variability and Spectral Invariance of the Seyfert Galaxy NGC5506. *Nature*, 325, 696–8.

McLeod, A. I. and Hipel, K. W. (1978) Simulation Procedures for Box–Jenkins Models. *Water Resources Research*, 14, 969–75.

Medley, M., Saulnier, G. and Das, P. (1994) Applications of the Wavelet Transform in Spread Spectrum Communications Systems. In *Wavelet Applications* (Proceedings of the SPIE 2242), edited by H. H. Szu. Bellingham, Washington: SPIE Press, 54–68.

Meyers, S. D., Kelly, B. G. and O'Brien, J. J. (1993) An Introduction to Wavelet Analysis in Oceanography and Meteorology: With Applications to the Dispersion of Yanai Waves. *Monthly Weather Review*, 121, 2858–66.

Mood, A. M., Graybill, F. A. and Boes, D. C. (1974) *Introduction to the Theory of Statistics* (Third Edition). New York: McGraw–Hill.

Morettin, P. A. (1981) Walsh Spectral Analysis. *SIAM Review*, 23, 279–91.

Moulin, P. (1994) Wavelet Thresholding Techniques for Power Spectrum Estimation. *IEEE Transactions on Signal Processing*, 42, 3126–36.

——— (1995) Model Selection Criteria and the Orthogonal Series Method for Function Estimation. In *Proceedings of 1995 IEEE International Symposium on Information Theory*. New York: IEEE Press, 252.

Mulcahy, C. (1996) Plotting and Scheming with Wavelets. *Mathematics Magazine*, 69, 323–43.

Munk, W. H. and MacDonald, G. J. F. (1975) *The Rotation of the Earth*. Cambridge, UK: Cambridge University Press.

Musha, T. and Higuchi, H. (1976) The $1/f$ Fluctuation of a Traffic Current on an Expressway. *Japanese Journal of Applied Physics*, 15, 1271–5.

Nason, G. P. (1996) Wavelet Shrinkage using Cross-Validation. *Journal of the Royal Statistical Society, Series B*, 58, 463–79.

Nason, G. P. and Silverman, B. W. (1995) The Stationary Wavelet Transform and Some Statistical Applications. In *Wavelets and Statistics* (Lecture Notes in Statistics, Volume 103), edited by A. Antoniadis and G. Oppenheim. New York: Springer–Verlag, 281–99.

Ogden, R. T. (1997) *Essential Wavelets for Statistical Applications and Data Analysis.* Boston: Birkhäuser.

Oppenheim, A. V. and Schafer, R. W. (1989) *Discrete-Time Signal Processing.* Englewood Cliffs, New Jersey: Prentice Hall.

Papoulis, A. (1991) *Probability, Random Variables, and Stochastic Processes* (Third Edition). New York: McGraw–Hill.

Percival, D. B. (1983) The Statistics of Long Memory Processes. Ph.D. dissertation, Department of Statistics, University of Washington, Seattle.

——— (1991) Characterization of Frequency Stability: Frequency-Domain Estimation of Stability Measures. *Proceedings of the IEEE*, 79, 961–72.

——— (1992) Simulating Gaussian Random Processes with Specified Spectra. *Computing Science and Statistics*, 24, 534–8.

——— (1995) On Estimation of the Wavelet Variance. *Biometrika*, 82, 619–31.

Percival, D. B. and Constantine, W. (2000) Wavelet-Based Simulation of Stochastic Processes. Technical Report, University of Washington, Seattle.

Percival, D. B. and Guttorp, P. (1994) Long-Memory Processes, the Allan Variance and Wavelets. In *Wavelets in Geophysics*, edited by E. Foufoula–Georgiou and P. Kumar. San Diego: Academic Press, 325–44.

Percival, D. B. and Mofjeld, H. (1997) Analysis of Subtidal Coastal Sea Level Fluctuations Using Wavelets. *Journal of the American Statistical Association*, 92, 868–80.

Percival, D. B., Raymond, G. M. and Bassingthwaighte, J. B. (2000a) Approximation of Stationary Gaussian Processes via Spectral Synthesis Methods. Technical Report, University of Washington, Seattle.

Percival, D. B., Sardy, S. and Davison, A. C. (2000b) Wavestrapping Time Series: Adaptive Wavelet-Based Bootstrapping. In *Nonlinear and Nonstationary Signal Processing*, edited by W. J. Fitzgerald, R. L. Smith, A. T. Walden and P. C. Young. Cambridge, UK: Cambridge University Press, 442–71.

Percival, D. B. and Walden, A. T. (1993) *Spectral Analysis for Physical Applications: Multitaper and Conventional Univariate Techniques.* Cambridge, UK: Cambridge University Press.

Pesquet, J.–C., Krim, H. and Carfantan, H. (1996) Time-Invariant Orthonormal Wavelet Representations. *IEEE Transactions on Signal Processing*, 44, 1964–70.

Peterson, S. (1996) Filtering and Wavelet Regression Methods with Application to Exercise ECG. Ph.D. dissertation, Department of Mathematical Statistics, Lund University, Lund, Sweden.

Popper, W. (1951) *The Cairo Nilometer*, Volume 12 of *University of California Publications in Semitic Philology.* Berkeley, California: University of California Press.

Press, W. H., Teukolsky, S. A., Vetterling, W. T. and Flannery, B. P. (1992) *Numerical Recipes in Fortran* (Second Edition). Cambridge, UK: Cambridge University Press.

Priestley, M. B. (1981) *Spectral Analysis and Time Series.* London: Academic Press.

Rabiner, L. R. (1989) A Tutorial on Hidden Markov Models and Selected Applications in Speech Recognition. *Proceedings of the IEEE*, 77, 257–86.

Rao, C. R. and Mitra, S. K. (1971) *Generalized Inverse of Matrices and Its Applications.* New York: John Wiley & Sons.

Rao, K. R. and Yip, P. (1990) *Discrete Cosine Transform: Algorithms, Advantages, Applications.* Boston: Academic Press.

Reid, N. (1988) Saddlepoint Methods and Statistical Inference. *Statistical Science*, 3, 213–38.

Riedel, K. S. and Sidorenko, A. (1995) Minimum Bias Multiple Taper Spectral Estimation. *IEEE Transactions on Signal Processing*, 43, 188–95.

Rioul, O. (1992) Simple Regularity Criteria for Subdivision Schemes. *SIAM Journal on Mathematical Analysis*, 23, 1544–76.

Rioul, O. and Duhamel, P. (1992) Fast Algorithms for Discrete and Continuous Wavelet Transforms. *IEEE Transactions on Information Theory*, 38, 569–86.

Rioul, O. and Vetterli, M. (1991) Wavelets and Signal Processing. *IEEE Signal Processing Magazine*, 8, 14–38.

Robinson, E. A. (1980) *Physical Applications of Stationary Time-Series.* New York: MacMillan.

Robinson, E. A. and Treitel, S. (1980) *Geophysical Signal Analysis.* Englewood Cliffs, New Jersey: Prentice Hall.

Robinson, P. M. (1995) Log-Periodogram Regression of Time Series with Long Range Dependence. *Annals of Statistics*, 23, 1048–72.

Rutman, J. (1978) Characterization of Phase and Frequency Instabilities in Precision Frequency Sources: Fifteen Years of Progress. *Proceedings of the IEEE*, 66, 1048–75.

Scargle, J. D. (1997) Wavelet Methods in Astronomical Time Series Analysis. In *Applications of Time Series Analysis in Astronomy and Meteorology*, edited by T. Subba Rao, M. B. Priestley and O. Lessi. London: Chapman & Hall, 226–48.

Schick, K. L. and Verveen, A. A. (1974) $1/f$ Noise with a Low Frequency White Noise Limit. *Nature*, 251, 599–601.

Serroukh, A. and Walden, A. T. (2000a) Wavelet Scale Analysis of Bivariate Time Series I: Motivation and Estimation. *Journal of Nonparametric Statistics*, 13, 1–36.

——— (2000b) Wavelet Scale Analysis of Bivariate Time Series II: Statistical Properties for Linear Processes. *Journal of Nonparametric Statistics*, 13, 37–56.

Serroukh, A., Walden, A. T. and Percival, D. B. (2000) Statistical Properties and Uses of the Wavelet Variance Estimator for the Scale Analysis of Time Series. *Journal of the American Statistical Association*, 95, 184–96.

Shensa, M. J. (1992) The Discrete Wavelet Transform: Wedding the à Trous and Mallat Algorithms. *IEEE Transactions on Signal Processing*, 40, 2464–82.

Shimomai, T., Yamanaka, M. D. and Fukao, S. (1996) Application of Wavelet Analysis to Wind Disturbances Observed with MST Radar Techniques. *Journal of Atmospheric and Terrestrial Physics*, 58, 683–96.

Sinai, Y. G. (1976) Self-Similar Probability Distributions. *Theory of Probability and Its Applications*, 21, 64–80.

Stein, C. M. (1981) Estimation of the Mean of a Multivariate Normal Distribution. *Annals of Statistics*, 9, 1135–51.

Stoffer, D. S. (1991) Walsh–Fourier Analysis and Its Statistical Applications. *Journal of the American Statistical Association*, 86, 461–85.

Stoffer, D. S., Scher, M. S., Richardson, G. A., Day, N. L. and Coble, P. A. (1988) A Walsh–Fourier Analysis of the Effects of Moderate Maternal Alcohol Consumption on Neonatal Sleep-State Cycling. *Journal of the American Statistical Association*, 83, 954–63.

Stollnitz, E. J., DeRose, T. D. and Salesin, D. H. (1996) *Wavelets for Computer Graphics: Theory and Applications.* San Francisco: Morgan Kaufmann.

Strang, G. (1989) Wavelets and Dilation Equations: A Brief Introduction. *SIAM Review*, 31, 614–27.

——— (1993) Wavelet Transforms Versus Fourier Transforms. *Bulletin of the American Mathematical Society*, 28, 288–305.

Strang, G. and Nguyen, T. (1996) *Wavelets and Filter Banks.* Wellesley, Massachusetts: Wellesley–Cambridge Press.

Strichartz, R. S. (1993) How to Make Wavelets. *American Mathematical Monthly*, 100, 539–56.

——— (1994) Construction of Orthonormal Wavelets. In *Wavelets: Mathematics and Applications*, edited by J. J. Benedetto and M. W. Frazier. Boca Raton, Florida: CRC Press, 23–50.

Sweldens, W. (1996) The Lifting Scheme: A Custom-Design Construction of Biorthogonal Wavelets. *Applied and Computational Harmonic Analysis*, 3, 186–200.

Taswell, C. (1996) Satisficing Search Algorithms for Selecting Near-Best Bases in Adaptive Tree-Structured Wavelet Transforms. *IEEE Transactions on Signal Processing*, 44, 2423–38.

——— (2000) Constraint-Selected and Search-Optimized Families of Daubechies Wavelet Filters Computable by Spectral Factorization. *Journal of Computational and Applied Mathematics*, 121, 179–95.

Taswell, C. and McGill, K. (1994) Algorithm 735: Wavelet Transform Algorithms for Finite-Duration Discrete-Time Signals. *ACM Transactions on Mathematical Software*, 20, 398–412.

Teichroew, D. (1957) The Mixture of Normal Distributions with Different Variances. *Annals of Mathematical Statistics*, 28, 510–2.

Tewfik, A. H. and Kim, M. (1992) Correlation Structure of the Discrete Wavelet Coefficients of Fractional Brownian Motion. *IEEE Transactions on Information Theory*, 38, 904–9.

Tewfik, A. H., Kim, M. and Deriche, M. (1993) Multiscale Signal Processing Techniques: A Review. In *Signal Processing and its Applications,* Volume 10 of *Handbook of Statistics*, edited by N. K. Bose and C. R. Rao. Amsterdam: North–Holland, 819–81.

Thomson, D. J. (1982) Spectrum Estimation and Harmonic Analysis. *Proceedings of the IEEE*, 70, 1055–96.

Torrence, C. and Compo, G. P. (1998) A Practical Guide to Wavelet Analysis. *Bulletin of the American Meteorological Society*, 79, 61–78.

Toussoun, O. (1925) Mémoire sur l'Histoire du Nil. In *Mémoires a l'Institut d'Égypte*, 9, 366–404.

Tukey, J. W. (1977) *Exploratory Data Analysis.* Reading, Massachusetts: Addison–Wesley.

Unser, M. (1995) Texture Classification and Segmentation Using Wavelet Frames. *IEEE Transactions on Image Processing*, 4, 1549–60.

Van Trees, H. L. (1968) *Detection, Estimation, and Modulation Theory,* Part I. New York: John Wiley & Sons.

Vetterli, M. and Herley, C. (1992) Wavelets and Filter Banks: Theory and Design. *IEEE Transactions on Signal Processing*, 40, 2207–32.

Vidakovic, B. (1998) Nonlinear Wavelet Shrinkage with Bayes Rules and Bayes Factors. *Journal of the American Statistical Association*, 93, 173–9.

——— (1999) *Statistical Modeling by Wavelets.* New York: John Wiley & Sons.

Wahba, G. (1980) Automatic Smoothing of the Log Periodogram. *Journal of American Statistical Association*, 75, 122–32.

Walden, A. T. (1985) Non-Gaussian Reflectivity, Entropy, and Deconvolution. *Geophysics*, 50, 2862–88.

——— (1990) Variance and Degrees of Freedom of a Spectral Estimator Following Data Tapering and Spectral Smoothing. *Signal Processing*, 20, 67–79.

——— (1994) Interpretation of Geophysical Borehole Data via Interpolation of Fractionally Differenced White Noise. *Applied Statistics*, 43, 335–45.

Walden, A. T. and Contreras Cristan, A. (1998a) The Phase-Corrected Undecimated Discrete Wavelet Packet Transform and its Application to Interpreting the Timing of Events. *Proceedings of the Royal Society of London, Series A*, 454, 2243–66.

——— (1998b) Matching Pursuit by Undecimated Discrete Wavelet Transform for Non-Stationary Time Series of Arbitrary Length. *Statistics and Computing*, 8, 205–19.

Walden, A. T., McCoy, E. J. and Percival, D. B. (1995) The Effective Bandwidth of a Multitaper Spectral Estimator. *Biometrika*, 82, 201–14.

Walden, A. T., Percival, D. B. and McCoy, E. J. (1998) Spectrum Estimation by Wavelet Thresholding of Multitaper Estimators. *IEEE Transactions on Signal Processing*, 46, 3153–65.

Wang, Y. (1996) Function Estimation via Wavelet Shrinkage for Long-Memory Data. *Annals of Statistics*, 24, 466–84.

Welch, P. D. (1967) The Use of Fast Fourier Transform for the Estimation of Power Spectra: A Method Based on Time Averaging Over Short, Modified Periodograms. *IEEE Transactions on Audio and Electroacoustics*, 15, 70–3.

Whitcher, B. J (1998) Assessing Nonstationary Time Series Using Wavelets. Ph.D. dissertation, Department of Statistics, University of Washington, Seattle.

Whitcher, B. J, Byers, S. D., Guttorp, P. and Percival, D. B. (2000a) Testing for Homogeneity of Variance in Time Series: Long Memory, Wavelets and the Nile River. Technical Report, National Research Center for Statistics and the Environment, University of Washington, Seattle.

Whitcher, B. J, Guttorp, P. and Percival, D. B. (2000b) Wavelet Analysis of Covariance with Application to Atmospheric Time Series. *Journal of Geophysical Research*, 105, 14,941–62.

Wickerhauser, M. V. (1994) *Adapted Wavelet Analysis from Theory to Software*. Wellesley, Massachusetts: A K Peters.

Wood, A. T. A. and Chan, G. (1994) Simulation of Stationary Gaussian Processes in $[0,1]^d$. *Journal of Computational and Graphical Statistics*, 3, 409–32.

Wornell, G. W. (1990) A Karhunen–Loève-like Expansion for $1/f$ Processes via Wavelets. *IEEE Transactions on Information Theory*, 36, 859–61.

——— (1993) Wavelet-Based Representations for the $1/f$ Family of Fractal Processes. *Proceedings of the IEEE*, 81, 1428–50.

——— (1996) *Signal Processing with Fractals: A Wavelet-Based Approach*. Upper Saddle River, New Jersey: Prentice Hall.

Wornell, G. W. and Oppenheim, A. V. (1992) Estimation of Fractal Signals from Noisy Measurements using Wavelets. *IEEE Transactions on Signal Processing*, 40, 611–23.

Yaglom, A. M. (1958) Correlation Theory of Processes with Random Stationary nth Increments. *American Mathematical Society Translations* (Series 2), 8, 87–141.

Zurbuchen, Th., Bochsler, P. and von Steiger, R. (1996) Coronal Hole Differential Rotation Rate Observed With SWICS/Ulysses. In *Solar Wind Eight*, edited by D. Winterhalter, J. T. Gosling, S. R. Habbal, W. S. Kurth and M. Neugebauer. Woodbury, New York: American Institute of Physics, 273–6.

Author Index

Abramovich, F., 415, 426–7, 552
Abramowitz, M., 257, 552
Abry, P., 354, 374, 464, 552
Allan, D. W., 321, 354, 552
Anderson, T. W., 255, 552
Antoniadis, A., 552, 554, 559

Balakrishnan, N., 557
Balek, J., 193, 552
Baliunas, S., 6, 552
Balogh, A., 220–1, 236–7, 552, 556
Baraniuk, R. G., 554
Barnes, J. A., 321, 553
Bartlett, M. S., 270, 275–6, 376, 553
Bassingthwaighte, J. B., 560
Beauchamp, K. G., 41, 218, 553
Benedetto, J. J., 562
Beran, J., 192–3, 279–80, 284–5, 361, 367, 386, 388, 553
Berger, R. L., 256, 554
Best, D. J., 264, 553
Beylkin, G., 159, 553–4
Billingsley, P., 381, 553
Björck, Å, 280, 554
Blackman, R. B., 255, 269, 297, 317, 553
Bloomfield, P., 255, 269, 272, 553
Bochsler, P., 564
Boes, D. C., 559
Bose, N. K., 562
Box, G. E. P., 255, 266, 553
Bracewell, R. N., 4, 103, 216, 553
Bradshaw, G. A., 295, 553
Briggs, W. L., 20, 553
Brillinger, D. R., 255, 269, 368, 553
Brockwell, P. J., 363, 367, 553
Brodsky, J., 557
Brown, J. W., 20, 553
Brown, R. L., 380, 553
Bruce, A. G., 79, 106–7, 135, 141, 149–50, 159, 241, 253, 400, 429, 432, 553–4
Bultheel, A., 557
Burrus, C. S., 558
Byers, S. D., 563

Cahalan, R. F., 556
Calderón, A. P., 11, 554
Carfantan, H., 560
Carlin, J. B., 555
Carmona, R., 12, 255, 295, 554
Casella, G., 256, 554
Chambers, J. M., 264, 554
Chan, G., 290, 563
Chi, A. R., 553
Chipman, H. A., 410, 412–4, 426–7, 554
Chui, C. K., 492, 554–5
Churchill, R. V., 20, 553
Cleveland, W. S., 272, 554
Coble, P. A., 561
Cohen, A., 141, 554
Coifman, R. R., 144, 159, 204, 215, 223, 225, 374, 429, 554
Compo, G. P., 12, 295, 303, 562
Constantine, W., 360, 369, 560
Contreras Cristan, A., 217, 239, 247, 563
Craigmile, P. F., 368, 371, 554
Crouse, M. S., 450–2, 455, 554
Cutler, L. S., 553
Czerny, B., 340, 559

Dahlquist, G., 280, 554
Daniels, H. E., 434, 554
Das, P., 559
Daubechies, I., 4, 105–6, 108–9, 112–3, 117–8, 123, 153, 457, 461, 468, 472, 476, 483, 486–7, 492–3, 498, 546, 554

David, H. A., 299, 555
Davies, R. B., 290, 555
Davis, G., 239, 363, 367, 555
Davis, R. A., 553
Davison, A. C., 434, 555, 560
Day, N. L., 561
Del Marco, S., 159, 174, 228, 555
DeRose, T. D., 561
Deo, R., 557
Deriche, M., 562
Dijkerman, R. W., 354, 555
Donoho, D. L., 144, 159, 204, 374, 400, 418, 420, 429, 456, 553–5
Doroslovački, M. I., 119, 555
Draper, N. R., 376, 378, 555
Duhamel, P., 103, 561
Durbin, J., 553

Eltahir, E. A. B., 192, 555
Eom, K.-B., 364, 557
Evans, J. M., 553

Fitzgerald, W. J., 560
Flandrin, P., 280, 295, 321, 340, 354, 464, 552, 555
Flannery, B. P., 560
Fodor, I. K., 558
Forsyth, R. J., 552, 556
Foufoula-Georgiou, E., 4, 5, 6, 12, 295, 555, 558, 560
Fraedrich, K., 557
Frazier, M. W., 562
Frick, P., 552
Friehe, C. A., 557
Fukao, S., 561
Fuller, W. A., 16, 255, 299, 343, 441, 555

Gamage, N. K. K., 295, 555
Gao, H.-Y., 79, 106–7, 135, 141, 149–50, 159, 241, 253, 272, 400, 432, 435, 437, 439, 456, 553–5
Gao, W., 6, 295, 555
Gelman, A., 266, 555
Geweke, J., 379, 555
Gneiting, T., 290, 556
Godfrey, R., 411–2, 415, 556
Gollmer, S. M., 6, 556
Gonçalvès, P., 552
González-Esparza, J. A., 219–20, 556
Gosling, J. T., 564
Goupillaud, P., 4, 556
Granger, C. W. J., 277, 281, 285, 556
Graybill, F. A., 532, 556, 559
Green, P. J., 402, 556
Greenhall, C. A., 141, 159, 205, 310, 314, 321, 378, 556
Grossmann, A., 11, 556
Gu, Y., 558
Guo, H., 558
Guttorp, P., 159, 297, 310, 321, 554, 560, 563

Haar, A., 57, 556
Habbal, S. R., 564
Hamming, R. W., 20, 179, 556
Hannan, E. J., 255, 556
Härdle, W., 107, 556
Harshvardhan, 556
Harte, D. S., 290, 555
Hatanaka, M., 277, 556
Healey, D. J., 553
Henson, V. E., 20, 553
Herley, C., 457, 562
Hess-Nielsen, N., 118, 231, 556
Higuchi, H., 340, 559
Hill, F., 558
Hinkley, D. V., 434, 555
Hipel, K. W., 361, 559
Holden, A. V., 340, 556
Horbury, T. S., 219, 552
Hosking, J. R. M., 281, 285, 291, 367, 556
Howe, D. A., 556
Hsu, D. A., 380, 556
Hudgins, L. H., 295, 303, 557
Hurst, H. E., 191, 557
Hurvich, C. M., 379, 424, 557
Hwang, W.-L., 554

Inclán, C., 380–1, 557

Jansen, M., 424, 557
Jawerth, B., 461, 464, 557
Jenkins, G. M., 269, 317, 553, 557
Jensen, M. J., 366, 374–5, 378–9, 385, 557
Jiang, J., 6, 557
Johnson, D. M., 218, 558
Johnson, G. E., 290, 557
Johnson, N. L., 258, 557
Johnstone, I. M., 400–1, 406, 418, 420, 444, 450, 456, 555, 557
Joyeux, R., 281, 285, 556

Kaplan, L. M., 366–7, 557
Kashyap, R. L., 364, 557
Kay, S. M., 269, 292, 361, 557
Kelly, B. G., 559
Kendall, D. G., 270, 275–6, 376, 553
Kerkyacharian, G., 556
Kim, M., 354, 562
Kleiner, B., 554
Kolaczyk, E. D., 554
Komm, R. W., 450, 558
Koopmans, L. H., 119, 255, 269, 558
Kotz, S., 557
Krim, H., 560
Kumar, P., 4–6, 12, 295, 555, 560
Kuo, C., 273, 558
Kuo, C.-C. J., 366–7, 557
Kurth, W. S., 564

Lai, M.-J., 106, 558
Lang, M., 159, 429, 558
Lanning, E. N., 218, 558
Lee, P. M., 266, 558
Leeson, D. B., 553
Lessi, O., 561
Li, B. L., 6, 295, 555
Liang, J., 159, 174, 558
Lindberg, C., 558
Lindsay, R. W., 303, 558
Liu, P. C., 303, 558
Lugannani, R., 434, 558

MacDonald, G. J. F., 340, 559
Mak, M., 6, 558
Malfait, M., 557
Mallat, S. G., 11, 68, 100, 239–40, 457, 461, 554–5, 558–9
Mandelbrot, B. B., 191, 279–80, 559
Marple, S. L., Jr., 269, 559
Martin, R. D., 553
Masry, E., 280, 340, 354, 361, 559
Mayer, M. E., 557
Mazumdar, R. R., 354, 555
McCoy, E. J., 355, 379, 559, 563
McCulloch, R. E., 554
McGill, K., 141, 562
McGunigal, T. E., 553
McHardy, I., 340, 559
McLeod, A. I., 361, 559
Medley, M., 218, 559
Meyer, Y., 554
Meyers, S. D., 12, 559
Mitra, S. K., 204, 560
Mofjeld, H., 185, 244, 326, 560
Montefusco, L., 555
Mood, A. M., 256, 559
Morettin, P. A., 218, 559
Morlet, J., 11, 556
Moulin, P., 398, 432–7, 559
Mulcahy, C., 56, 559
Mullen, J. A., Jr., 553
Munk, W. H., 340, 559
Musha, T., 340, 559

Nason, G. P., 159, 402, 421–3, 559
Neugebauer, M., 556, 564
Nguyen, T., 465, 471, 562
Nowak, R. D., 554

O'Brien, J. J., 559
Odegard, J. E., 558
Ogden, R. T., 135, 255, 560
Oppenheim, A. V., 28, 106, 117, 366, 484, 560, 563
Oppenheim, G., 552, 554, 559

Papoulis, A., 256, 560
Parks, T. W., 159, 174, 558
Parzen, E., 272, 554
Percival, D. B., 159, 185, 244, 291–2, 297, 307, 309–10, 321, 326, 353–4, 360, 554, 556, 558–61, 563
Pesquet, J.-C., 159, 560
Peterson, S., 204, 560
Phillips, J. L., 556
Picard, D., 556
Popper, W., 193, 560
Porter-Hudak, S., 379, 555
Press, W. H., 56, 560
Priestley, M. B., 255–6, 268–9, 313, 560–1
Puccio, L., 555

Rabiner, L. R., 450–1, 560
Rao, C. R., 204, 560, 562
Rao, K. R., 54, 560
Raphael, L., 554
Raymond, G. M., 560
Reid, N., 434, 560
Reinsel, G. C., 553
Rice, S., 434, 558
Richardson, G. A., 561
Riedel, K. S., 273–4, 561
Rioul, O., 56, 102–3, 487, 494, 561
Roberts, D. E., 264, 553
Robinson, E. A., 488, 561
Robinson, P. M., 379, 561
Rocca, F., 411–2, 415, 556
Rothrock, D. A., 558
Rubin, D. B., 555
Ruskai, M. B., 554
Rutman, J., 7, 321, 561

Salesin, D. H., 561
Sapatinas, T., 552
Sardy, S., 560
Saulnier, G., 559
Scargle, J. D., 295, 561
Schafer, R. W., 28, 106, 117, 484, 560
Scher, M. S., 561
Schick, K. L., 340, 561
Serroukh, A., 303, 310, 334, 561
Shensa, M. J., 159, 561
Shimomai, T., 6, 561
Sidorenko, A., 273–4, 561
Silverman, B. W., 159, 401–2, 406, 444, 450, 552, 556–7, 559
Sinai, Y. G., 280, 561
Smith, E. J., 552, 556
Smith, H., 376, 378, 555
Smith, R. L., 560
Smith, W. L., 553
Snider, J. B., 556
Sokoloff, D., 552
Soon, W., 552
Southwood, D. J., 552
Spies, T. A., 295, 553
Stark, P. B., 558
Stegun, I. A., 257, 552
Stein, C. M., 403–4, 561
Stern, H. S., 555

Stoffer, D. S., 218, 561
Stollnitz, E. J., 1, 561
Strang, G., 56, 465, 471,
Strichartz, R. S., 80, 457, 562
Stuetzle, W., 554
Subba Rao, T., 561
Sweldens, W., 1, 461, 464, 557, 562
Sydnor, R. L., 553
Szu, H. H, 553, 558–9

Taswell, C., 119, 141, 226, 562
Teichroew, D., 265, 562
Terrin, N., 388, 553
Teukolsky, S. A., 560
Tewfik, A. H., 295, 354, 562
Thomson, D. J., 273, 275, 277, 558, 562
Tiao, G. C., 266, 380–1, 553, 557
Torrence, C., 12, 295, 303, 562
Torrésani, B., 554
Toussoun, O., 191, 562
Treitel, S., 488, 561
Tsai, C.-L., 424, 557
Tsurutani, B. T., 552
Tsybakov, A., 556
Tukey, J. W., 51, 255, 269, 297, 317, 553, 562
Tukey, P. A., 554

Unser, M., 159, 562

van Ness, J. W., 279–80, 559
Van Trees, H. L., 261, 562
Veitch, D., 374, 552
Verveen, A. A., 340, 561
Vessot, R. F. C., 553
Vetterli, M., 56, 457, 561–2
Vetterling, W. T., 560
Vial, P., 554
Vidakovic, B., 256, 413–4, 426–8, 452, 562
von Steiger, R., 564

Wahba, G., 272, 563
Walden, A. T., 217, 239, 247, 272–4, 303, 340, 355, 379, 411, 417, 442, 559–61, 563
Wallis, J. R., 191, 559
Wang, G., 192, 555
Wang, Y., 354, 563
Watts, D. G., 269, 317, 557
Weiss, J., 159, 174, 228, 555
Welch, P. D., 315, 563
Wells, R. O., 558
Whitcher, 303, 351, 353, 380, 382, 387, 391, 563
Wickerhauser, M. V., 118, 209, 223, 225, 231, 554, 556, 563
Winkler, G. M. R., 553
Winterhalter, D., 564
Wood, A. T. A., 290, 563
Wornell, G. W., 295, 340, 355, 361, 366, 563

Yaglom, A. M., 287, 563
Yamanaka, M. D., 561
Yip, P., 54, 560
Young, P. C., 560

Zhang, D., 557
Zhang, Z., 239–40, 555, 559
Zurbuchen, Th, 220, 564

Subject Index

1/f-type processes, 281
  nonstationary, 287

absolute error loss, 266, 415
absolute value of complex variable, 21
ACVS (*see* autocovariance sequence)
adaptive smoothing, 272
additive cost functional, 223, 251
  entropy, 223, 225
  entropy and cost table example, 225
  ℓ_p information, 223, 225, 253
  threshold, 223, 225, 253
admissibility condition, 4
aliasing, 35, 280, 287, 290
alignment of time series with LA(L) partial DWT coefficients, 115–6
Allan variance, 159, 297, 314, 321, 339, 354
  and Haar wavelet variance, 322
amplitude modulated complex exponential (Morlet wavelet function), 5
analysis of function
  scaling coefficients, 460, 495
analysis of signal
  scaling coefficients, 460, 495
analysis of time series, 43, 46, 53
  arbitrary disjoint dyadic decomposition, 213
  DWPT coefficients, 209–10, 212
  DWT coefficients, 57
  Fourier coefficients, 47
  MODWPT coefficients, 232
  MODWT coefficients, 169
  partial DWT coefficients, 80, 89
analysis of variance (ANOVA), 19, 267, 295–6, 307, 335
  by frequency band, 208, 214, 233, 250–1
  DWPT coefficients, 208, 250
  invalid using MODWT details, 182
  LA(8) MODWT for subtidal sea levels, 185
  matching pursuit, 240, 252
  MODWPT coefficients, 233, 251
  MODWT coefficients, 160, 166, 171
  padding/extending a sample for the DWT, 141
  scale-by-scale, 62, 151
  WP table decomposition, 214
ANOVA (*see* analysis of variance)
approximation space
  Haar scale one half, 461
  Haar scale two, 461
  Haar scale unity, 460–1
  scale λ_j, 460, 462, 495
  scale unity, 459
AR process (*see* autoregressive process)
AR(1) process
  ACS of nonboundary wavelet coefficients, 352–3
  and wavelet coefficients, 337
  decorrelation via DWT, 353
  SDFs, 351–2
AR(2) process, 436
  SDF, 436
  SDF estimation via thresholding
    of log multitaper, 446, 448
    of log periodogram, 437–8
  standard deviation of derived noise DWT coefficients, 442
AR(24) process, 272–3, 435
  SDF, 435
  SDF estimation, 273, 275
  SDF estimation via thresholding
    of log multitaper, 446–7
    of log periodogram, 437–8
ARE (asymptotic relative efficiency), 309

ARFIMA process (*see* autoregressive fractionally integrated moving average process)
argument of complex variable, 21
astronomical unit, 219
asymptotic properties of periodogram, 270
asymptotic relative efficiency of unbiased wavelet variance estimators, 309
atomic clock, 295, 318, 341
  wavelet variance estimates for, 317, 319, 320
atomic clock average fractional frequency deviates, 7–8, 11, 318, 321–2, 383, 392
  MLEs, 384
  MODWT, 17, 18
  partial DWT, 13, 14
  test for trend, 384
  wavelet variance estimates for, 322, 384
  WLSEs, 385
autocorrelation
  complex-valued dipole, 489
  DFT for finite sequence, 30, 37
  DFT for infinite sequence, 25, 36
  finite sequence (circular), 30, 37
  Fourier transform of, 38
  function, 38
  infinite sequence, 25, 36
  periodized wavelet filter, 72
  real-valued dipole, 488
  wavelet filter, 69
autocorrelation sequence
  FD process, 345–6
  nonboundary wavelet coefficients from AR(1) process, 352–3
  nonboundary wavelet coefficients from FD process, 345–6
  stationary process, 266
autocorrelation width, 12
  first difference of Gaussian, 12
  Haar (alternative formulation), 12
  Haar scaling filter of jth level, 103
  Mexican hat, 12
  scaling filter (DWT) of jth level, 103
  scaling filter (MODWT) of jth level, 174
autocovariance sequence, 266
  biased estimator, 269
  FD process, 284, 341
    Davies–Harte simulations, 359
    DWT-based simulations, 359
  FGN, 279
  long memory process, 279
  relationship to SDF, 267
  square summability condition, 267, 307
  symmetry, 266
  true versus simulated for FD process, 357
  white noise, 268
autoregressive (AR) process (*see* also AR(1), AR(2), AR(24)), 351
  of order p (AR(p)), 268
  simulation, 292
  spectral density function, 268
  variance, 294
autoregressive fractionally integrated moving average (ARFIMA) process, 285, 366
  DWT-based MLEs, 366
average
  air temperature, 6, 7
  fractional frequency deviates, 321–2
  rainfall, 6
  sea surface temperature, 6, 7
  signal over a scale, 6
  vertical wind velocity, 6
averages
  changes in, 7, 59

backward difference
  first, 60
  second, 60
backward difference filter, 304
  order $L/2$, 106–7
  squared gain function of, 304, 335
backward shift operator, 283, 287
band-pass filter, 5
  Daubechies wavelet filter of jth level, 96
  Daubechies wavelet filter of second level, 91
bandwidth, 445
  for smoothing, 269, 272
  of multitaper SDF estimator, 274, 277–8
  physical for MODWPT of solar magnetic field, 236
baseline drift, 125, 128–9, 134, 185, 197
basis
  approximation space for scale λ_j, 460, 495
  approximation space
    with scale one half, 461
    with scale two, 461
    with scale unity, 459, 461
  complex finite dimensional vector space, 46
  detail space
    with scale one half, 474–5, 496, 500
    with scale one quarter, 475
    with scale unity, 475
  real finite dimensional vector space, 43–4, 239
  subspace of real finite dimensional vector space, 45
basis vectors for LA(8) partial DWT, 162
Bayes estimator, 265
Bayes risk, 265–6
Bayesian decision rule, 265
Bayesian methods, 412, 426
Bayesian shrinkage estimator
  absolute error loss, 415
  squared error loss, 426
Bayesian shrinkage rule, 412–5, 454
  DWT-based, 426–8
  LA(8) partial DWT, 428
  squared error loss of estimator, 414

Bayesian statistics, 264, 394
best basis algorithm, 223, 225, 251
  boundary effects, 227
  circularly shifted series, 227
  contrasted with matching pursuit, 241
  shift-adapted, 228
  solar magnetic field, 226–8
best localized (BL) filter, 119
  scaling filter, 119
    plotted, 119
    reversal from Doroslovački, 119
  wavelet filter
    phase after advance, 120
best shift basis algorithm, 228
  solar magnetic field, 228
bias in the sample variance, 299–300
bias of periodogram, 298
bias-variance trade-off, 272
  MODWT and DWT, 430
biased ACVS estimator, 269
biased MODWT-based estimator of the wavelet variance, 307, 337, 351, 378
binary vector for sequency ordering at node (j, n), 215–7, 249, 253
binary-valued RV, 410, 454
BL filter (*see* best localized filter)
boundary coefficients
  DWPT
    delineated in plot after alignment, 221
    number of, 221
  DWT, 350, 370
    delineated in plot after alignment, 138
    number of, 136, 139, 146, 158
    time alignment, 137–8, 146–7
  MODWPT
    delineated in plot after alignment, 236
  MODWT
    delineated in plot after alignment, 182–3, 198
    number of, 198
boundary conditions (*see* circular or reflecting)
boundary details and smooths, 139
  DWT
    delineated in plot, 139–40, 196
    number of, 139, 149
  MODWPT
    delineated in plot, 237–8
  MODWT, 198–9
    delineated in plot, 183–4, 186, 192, 194–6, 199
    number of, 199
boundary wavelets, 141
box plots, 442–3
broad-band frequency characteristic, 396
bump signal, 395

C(L) scaling and wavelet filters, 123, 158
  advance required, 147, 156
  approximately linear phase, 124
  phase after advance, 124
  plotted, 123
  shift factor, 124, 198, 205, 229
C(L) wavelet packet filter for node (j, n)
  advance required, 230, 233
  phase function, 229
  shift factor, 230
C(6) scaling and wavelet filters
  filter coefficients, 109
  jth level filters, 125
  reversal from Daubechies, 123
Cartesian representation, 21
cascade of filters, 27
  D(4) wavelet filter, 68
  Daubechies wavelet filter, 105, 107
  equivalent filter, 28, 39
  flow diagram, 27–8
  for wavelet coefficients of second level, 90
  width of equivalent filter, 28
center of energy of a filter, 118, 231
Central Limit Theorem, 434
change point, 381–2, 391
  Nile River minima, 387–8
changes in averages, 7
chi-square distribution, 345
  scaled, 313, 336, 338
chi-square PDF, 263
  percentage points, 264
chi-square RV, 263, 271
  and IID Gaussian (normal) RVs, 293
  log transformation, 293
  mean for, 263, 270
  variance for, 263, 270
chirp, 157
Cholesky decomposition, 361
circular autocorrelation (*see* autocorrelation (circular))
circular boundary conditions for DWT, 60, 140
circular convolution (*see* convolution (circular))
circular covariance matrix, 441
circular cross-correlation (*see* cross-correlation (circular))
circular filtering
  boundary effects, 136, 139, 145, 197, 227
  DWPT coefficients, 210–2
  for node (j, n), 214–5, 249–50
  DWT coefficients, 70, 74, 77, 79, 80, 88, 92, 94–7, 152
  of a finite sequence, 31–2, 55, 268, 296, 336
  MODWPT coefficients, 232
  for node (j, n), 231, 251
  MODWT coefficients, 163, 168–70, 173, 175, 177, 200–1
  with periodogram, 433
circular shift
  random, 356, 358, 361, 389, 391
  rows of DWT matrix, 59
circular shift matrix, 52, 71
  inverse, 52
circular/reflecting boundary conditions
  electrocardiogram, 133, 140
circularity, 52

circularly shifted periodized filter, 111
scaling filter, 77, 79
wavelet filter, 71, 77, 92, 153
circularly shifted series, 52, 429
best basis algorithm, 227
discrete Fourier empirical power spectrum, 52
discrete wavelet empirical power spectrum, 63
Fourier coefficients, 52
Fourier details, 55
MODWT of, 179, 181
MRA via MODWT, 181
MRA via partial DWT, 160–1
partial DWT of, 160–1
wavelet details of jth level, 191
clustering of DWT coefficients, 450–1, 455
ECG, 450
coarse approximation
scale λ_j, 464, 495, 500
scale four, 463
scale two, 462, 495
coiflets (*see* also C(L)), 123
embedded backward differencing, 124, 156, 306
squared gain function, 124
complex conjugate
complex variable, 21
DFT, 22
complex demodulation, 22, 39
complex exponential, 21–2
complex orthonormal matrix (unitary), 45, 53
complex-valued transform, 41, 45, 53
complex-valued wavelet function, 4
Morlet, 5
complex variable, 20
absolute value, 21
argument, 21
Cartesian representation, 21
complex conjugate, 21
complex exponential, 21
de Moivre's theorem, 21
Euler relationship, 21, 48
imaginary component, 20–1
magnitude, 21
modulus, 21
polar representation, 21
real component, 20–1
conditional expectation (*see* conditional mean)
conditional mean, 260, 266, 394, 408–11, 414–5, 454, 456
estimator, 261
in geophysics, 411
shrinkage rule, 411–2, 424
LA(8) partial DWT, 425–6
conditional median, 266, 394, 408, 411, 415, 454
estimator, 261
shrinkage rule (approximate), 412, 415–6
conditional PDF, 260, 409
conditionally Gaussian, 451
confidence intervals
for SDF, 275, 278
for time-dependent wavelet variance, 325
for wavelet variance, 311, 313–4, 319–20, 327, 332–4, 336, 345, 385
atomic clock average fractional frequency deviates, 322–3
consistent estimator
and hard thresholding, 456
of SDF, 271–2
continuity of function and its derivatives, 486
continuous parameter, 266
continuous parameter nonstationary process, 340, 354
simulation, 361
continuous parameter pure power law process, 285
continuous RV, 256
continuous wavelet transform (CWT), 1, 457
energy decomposition, 11
first derivative of Gaussian, 10
grey-scaled, 8, 11
Haar, 10, 13
Mexican hat, 8, 11
reconstruction of a signal, 11
versus DWT, 1, 13
convolution and filtering, 25
convolution (circular) of finite sequences, 29–31, 37
DFT of, 30, 37, 52, 72
convolution of functions, 38
Fourier transform of, 38
convolution of infinite sequences, 24, 36
DFT of, 24, 36
coordinated universal time (UTC), 317
coronal mass ejection (CME), 220, 237
corotating interaction region (CIR), 219, 237
correlated Gaussian noise, 417
correlated Gaussian wavelet coefficients, 440
correlated noise, 393
correlated noise DWT coefficients, 440
level-dependent variance, 441, 456
correlation, 259
nonboundary wavelet coefficients at different levels, 346
Haar, 347
LA(8), 347
correlation matrix
DWT coefficients and FD process, 349
D(4), 351
Haar, 350
LA(8), 351
cost functional (*see* additive cost functional)
covariance, 259, 266, 293
log multitaper SDF estimator, 276, 277
nonboundary wavelet coefficients at different levels, 345, 348

at same level, 345, 348
periodogram, 270–1
covariance of stationary stochastic process, 266
covariance matrix, 259, 262
approximation via DWT coefficients, 355, 389
circular, 441
determinant of, 361
DWT coefficients, 349, 355
error in WLSE, 376
FD process, 355
IID noise, 393, 452
non-IID noise, 393, 453
signal vector, 393
simulated for FD process, 356
WLSE parameter estimates, 376
Cramér–Rao lower bound, 364, 386
cross-correlation (circular) of finite sequences, 30–1, 37, 234
DFT of, 30, 37, 39, 234
interleaving for wavelet details and smooths, 82
MODWPT details for node (j, n), 234, 251
MODWT details and smooths, 167, 172, 203
periodized scaling and wavelet filters, 77
upsampling for wavelet details and smooths, 82–3
cross-correlation of functions, 38
Fourier transform of, 38
cross-correlation of infinite sequences, 24, 36
DFT of, 25, 36, 39
cross-validation, 402, 453
leave-one-out, 402
leave-one-out using DWT, 422
two-fold, 402
two-fold using partial DWT, 422
cumulative distribution function
Gaussian (normal), 257
CWT (*see* continuous wavelet transform)
cycle spinning, 144, 204, 374, 429
and the MODWT, 429
cyclic (*see* circular)
cyclic filter (*see* circular filtering)

D(L) scaling and wavelet functions
D(4)
computation, 469, 477
connection to jth level filters, 471, 478
Hölder regularity, 487, 494
D(L) scaling filter, 106, 117, 155, 493
D(4) DWT matrix, 59–60
D(4) scaling filter, 75, 490
of jth level, 97–8
connection to scaling function, 471
squared gain function, 73, 76
of second level, 91
D(4) wavelet filter, 59, 155
as a filter cascade, 68
jth level, 97–8, 157
connection to wavelet function, 478
phase function, 158
phase function, 158
squared gain function, 73
of second level 91, 157
D(6) scaling filter, 106, 109, 491
D(8) scaling filter, 106, 109, 493
data tapers, 53, 273
orthonormal, 273
sine, 274, 277, 440
Daubechies scaling filter
BL method, 119
extremal phase, 108, 117
least asymmetric, 107, 112, 117
phase function, 106
squared gain function, 105, 116, 155
of time reversed, 486, 498
transfer function, 106
Daubechies wavelet filter
as a filter cascade, 105, 107
construction of, 457
embedded backward differencing, 306
from extremal phase scaling filter, 108
high-pass nature, 73–4, 151, 158
$L/2$ embedded differences, 106, 123–4, 155
low-pass filter component, 106, 155
of jth level
sidelobes of squared gain function, 123
squared gain function, 121–2
squared gain function, 105, 116, 153
Daubechies wavelets, 59, 442
daublets, 106
Davies–Harte simulation method, 281, 283, 290–1, 341
DCT (*see* discrete cosine transform)
de Moivre's theorem, 21
decibel (dB) scale, 121, 273
decorrelation via the DWT, 16, 19, 341
of AR(1) process, 353
of FD process, 340, 345, 350, 380, 388
of MA(1) process, 391
degrees of freedom, 270
denoising, 393
nuclear magnetic resonance
using LA(8) partial DWT, 431
using ODFT, 431
detail for finite energy function (signal), 479–80, 497
detail space
Haar
scale one half, 475
scale one quarter, 475
scale unity, 475
scale τ_j, 472, 496
details
boundary elements, 139, 147, 199
DWT, 64–5, 151, 158
D(4), 65–6
first level, 81
Haar, 64–5

jth level, 95, 191
second level, 89
Fourier, 50–1, 54–5
MODWPT
and zero phase, 234, 237, 252
for node (j, n) via cross-correlation (circular), 234, 251
MODWT
and DWT averaging, 204–5
and zero phase, 160, 168, 172, 189
jth level, 171, 173, 176–7, 191, 200, 203
via cross-correlation (circular), 172, 203
determinant
of approximate covariance matrix, 362
of covariance matrix, 361
deterministic signal, 393–4, 398, 417, 452–3
estimation via thresholding, 398
DFBM (*see* discrete fractional Brownian motion)
DFT (*see* discrete Fourier transform)
DHM (*see* Davies–Harte simulation method)
dictionary, 239, 252, 254
element of, 239
MODWT and constant vectors, 247
MODWT and ODFT vectors, 247
MODWT vectors, 242, 253
orthonormal, 240
partial DWT vectors, 240
redundant, 241–2
renormalized MODWT vectors, 242–3
WP table and ODFT vectors, 241
WP table vectors, 241, 254
differentiability (weak), 403–4, 456
digamma function, 275, 376, 390, 440
dilation, 457–8, 474
dyadic, 458, 494
inverse dyadic, 459
non-commuting with translation, 459
dipole filter, 488
direct spectrum estimator, 273
direct sum, 472
approximation and detail spaces, 482
illustrated for Haar scaling function, 472–3
discrete cosine transform (DCT), 41, 54
type II transform (DCT-II), 54
type IV transform (DCT-IV), 54
discrete Fourier empirical power spectrum, 48–9, 55
invariance to shifts, 52
discrete Fourier transform (DFT), 17, 22, 290–1
and ODFT, 52, 54
autocorrelation, 25, 36
circular autocorrelation, 30, 37
circular convolution, 30, 37, 52
circular cross-correlation, 30, 37, 39, 234
complex conjugate, 22
convolution, 24, 36
cross-correlation, 25, 36, 39
even indexed subsequence, 23
finite/infinite sequences, 28
finite sequence, 28, 36
finite sum of squared magnitudes, 21
infinite sequence, 22, 35, 39
inverse
for finite sequence, 29, 36, 78, 291
for infinite sequence, 22, 35, 70
odd indexed subsequence, 39
orthonormal, 41
periodicity
for finite sequence, 28, 36
for infinite sequence, 22, 35
periodized filter, 33–4, 37, 39
periodized scaling and wavelet filters
of jth level, 100
of second level, 92
discrete fractional Brownian motion (DFBM), 279
parameter and properties, 286
random walk, 288
SDF, 280
discrete parameter process, 266
pure power law process, 281
discrete RV, 256
discrete wavelet empirical power spectrum
DWT, 62–3, 66–7, 156, 160, 297
lack of invariance to shifts, 63
MODWT, 160, 181
discrete wavelet packet transform (DWPT), 15, 206, 278
advanced LA(8) coefficients for solar magnetic field, 221–2
ANOVA, 208, 250
defined, 209
energy decomposition, 208, 250
for solar magnetic field, 220
Haar filter and Walsh transform, 217–8
orthonormality, 206, 210, 221
parent node, 210
regular DWT, 218
second level, 208
table, 209, 212
time/frequency decomposition, 206
tree, 209, 212
vector for node (j, n), 250
vectors of jth level, 249
via modification of DWT, 207
discrete wavelet transform (DWT), 1, 41, 56
alternative definitions, 149–50
analysis, 94
and its time series, 15
and thresholding/shrinkage, 417
approximate covariance matrix of coefficients, 355, 389
atomic clock data, 14
autocorrelation implementation, 149
clustering of coefficients, 450–1, 455
coefficients, 13, 57
vector, 61, 67
via subsampling of MODWT, 174, 203

convolutional implementation, 149
correlation matrix from FD process, 349
D(4), 351
Haar, 350
LA(8), 351
covariance matrix of coefficients, 349, 355
D(4) matrix, 59–60
decomposition of sample variance, 62, 67
decorrelation of FD process, 340, 345, 350
energy decomposition, 19, 62
energy preservation, 57
Haar, 62–3, 156, 441
Haar matrix, 57–8
initialization, 464
inverse, 422
localization, 59
matrix, 57, 63, 99, 156
non-decimated, 159
orthonormality, 57, 67, 72
parameter estimation for FD process, 341
applied to trend, 368–9
partial, 55, 341, 350, 385, 418
persistence of coefficients, 450, 451, 455
power of two length, 57, 67, 141, 150
pyramid algorithm (*see* pyramid algorithm for DWT)
reconstruction of a time series, 16
scale, 58
sensitivity due to downsampling, 164
shift invariant, 159
stationary, 159
synthesis, 83, 94
time invariant, 159
time/scale decomposition, 206
translation invariant, 159, 174
undecimated, 159
versus CWT, 1, 13
disjoint dyadic decomposition
and WP table, 213, 250
best basis algorithm, 225
best basis from WP table, 223, 226, 251
example, 224
best shift basis from WP table, 228
downsampling, 70, 74, 80, 90, 92, 156, 164, 185, 211, 221, 463
and bandwidth
scaling and wavelet coefficients, 84–6
and filtering
magnitude squared DFTs, 209, 211
and frequency reversal
DWPT, 211
DWT wavelet coefficients, 85–6
energy preservation, 70
retaining even-indexed values, 75, 149
retaining odd-indexed values, 70, 149, 152
DWPT (*see* discrete wavelet packet transform)
DWT (*see* discrete wavelet transform)
dyadic dilation, 458
inverse, 459
dyadic scales, 13, 15
dynamic range, 273, 437

ECG (*see* electrocardiogram)
EDOF (*see* equivalent degrees of freedom)
efficiency (statistical), 17, 309
eigenspectra, 274–5, 444
El Niño, 192
electrocardiogram (ECG), 125, 56, 59, 456
advanced LA(8) coefficients
MODWT, 182–3
partial DWT, 127–8
clustering and persistence in DWT coefficients, 450
illustration of DWT tree structure, 451
instrumentation effects, 134
MRA for LA(8) MODWT, 182–4
MRA for partial DWT
C(6) wavelet, 129, 132
D(4) wavelet, 129, 131
Haar wavelet, 129, 130
LA(8) wavelet, 129, 133, 142
MRA using circular boundary conditions, 133, 140
MRA using reflecting boundary conditions, 140, 142
P waves, 125, 129, 135
partial DWTs, 126
R waves, 125, 129, 134
T waves, 125, 129, 135
wavelet smooths of 6th level, 196–7
embedded backward difference filters, 306
$L/2$ for Daubechies wavelet filter, 106, 123–4, 153, 188
$L/3$ for coiflet wavelet filter, 124, 156, 188
energy decomposition
across time, 43
frequency by frequency, 48, 54
of finite energy function (signal), 481, 497
via CWT, 11
via DWPT, 208, 250
via DWT, 19, 62
via matching pursuit, 240, 252
via MODWPT, 233, 251
via MODWT, 16–7, 166, 168–9, 171, 173, 200
via ODFT, 48, 54
via orthonormal transform, 43
via partial DWT, 15–6, 95, 104, 151
via smooths and details, 95, 104
via WP table disjoint dyadic decomposition, 214
energy of time series, 42
energy preservation

and downsampling, 70
via DWT, 57
via ODFT, 47
via orthonormal transform, 43, 53
via unitary transform, 46, 53
energy spectrum, 72–3, 268
entropy cost functional, 223, 225
best basis for solar magnetic field, 227–8
best shift basis for solar magnetic field, 228
cost table example, 225
equal in distribution, 257
equality in the mean square sense, 38
equivalent degrees of freedom (EDOF), 313–4, 336, 338–9, 376, 390
equivalent filter for a cascade of filters, 28, 39
equivalent width, 68
Euler relationship, 21–2, 48
Euler's constant, 270, 432
Euler–Maclaurin summation, 280
even indexed subsequence and DFT, 23
events
intersection, 435
union, 435
expected value (*see* mean)
exponential distribution (prior), 413, 426–7
exponential PDF, 264–5
extending/padding a sample for the DWT, 141, 422–3
extremal phase (minimum delay) filter (*see* also D(L)), 106, 117, 155, 493

false alarm probability, 435
fast Fourier transform (FFT), 17, 28, 68, 432, 440, 290
'father' wavelet filter, 80
FBM (*see* fractional Brownian motion)
FD process (*see* fractionally differenced process)
FFT (*see* fast Fourier transform)
FGN (*see* fractional Gaussian noise)
fidelity condition, 398
filter, 267
backward difference, 105, 304
band-pass, 5
center of energy, 118, 231
circular, 31–2, 55
dipole, 488
extremal phase (minimum delay), 106, 117, 155, 488, 493
first backward difference, 105
frequency response function, 25, 33–4
gain function, 25
generalized linear phase, 117
half-band, 76, 152
high-pass, 27
impulse response sequence, 25
infinite width, 79
least asymmetric (*see* LA and least asymmetric)
linear phase, 108, 111, 155, 158
linear time-invariant, 25
low-pass, 26, 185
maximum delay, 488
maximum phase, 488
minimum delay (extremal phase), 106, 117, 155, 488, 493
minimum phase, 488
MODWT scaling (*see* scaling filter (MODWT))
MODWT wavelet (*see* wavelet filter (MODWT))
periodized, 20, 32–3, 35, 37, 39–40
phase function, 25
scaling (*see* scaling filter)
squared gain function, 25
transfer function, 25, 33–4
transfer function for reversed form, 117
two-stage cascade, 304
wavelet (*see* wavelet filter)
wavelet packet (*see* wavelet packet filter)
width, 28, 39
zero phase, 108, 110, 158, 179, 234, 237, 252
filter cascade, 27, 89
D(4) wavelet filter, 68
Daubechies wavelet filter, 105, 107
flow diagram, 27–8
for wavelet coefficients of second level, 90
filter width, 90
filter with inserted zeros
transfer function, 91, 96–7
filtering, 20
circular, 136, 139, 145, 197, 268
finite sequences, 29
flow diagram, 25
high-pass example, 27
infinite sequences, 25
low-pass example, 26
filtering and convolution, 25
filtering and downsampling
magnitude squared DFTs, 209, 211
finite Fourier transform, 28
finite sequences
circular autocorrelation, 30, 37
circular convolution, 29, 37
circular cross-correlation, 30, 37
discrete Fourier transform of, 28, 36
finite support, 465, 472, 474, 496, 498
firm thresholding, 400, 411, 453
first backward difference, 60
filter, 105
first derivative of Gaussian CWT, 10
first derivative of Gaussian wavelet function, 3, 10
autocorrelation width, 12
first order Markov chain, 451
Fourier analysis, 1, 20, 22, 35–7, 47
Fourier coefficient, 46
Fourier details, 50–1, 54–5
circularly shifted series, 55
synthesis of time series, 50

Fourier frequencies, 271, 330
Fourier roughs, 50–1, 54
Fourier smooths, 50–1, 54
Fourier synthesis, 22, 29, 35–7, 49, 51
Fourier transform
  and derivatives of wavelet function, 483–4
  inverse for function, 37
  of autocorrelation, 38
  of convolution, 38
  of cross-correlation, 38
  of finite sequence (*see* DFT)
  of function, 37
  of infinite sequence (*see* DFT)
  of time reversed scaling filter, 465, 484, 496
  of time reversed wavelet filter, 476, 483, 498
  of scaling function, 465, 468–9, 496
  of wavelet function, 476, 498
  short time, 206
Fourier transform pair, 23–4, 29, 35–8, 307, 312
fractional Brownian motion (FBM), 279, 285, 354
  SDF, 280
  simulation, 361
fractional difference parameter, 286
fractional frequency deviates (*see* atomic clock average fractional frequency deviates)
fractional Gaussian noise (FGN), 191, 279, 299, 300, 354
  approximate SDF, 280
  autocovariance sequence, 279
  long memory property, 280
  parameter and properties, 285–6
  sample interval, 285
  SDF, 280, 282
  simulation, 283
fractionally differenced (FD) process, 281, 307, 349
  ACS, 345–6
  ACS of nonboundary wavelet coefficients, 345–6
  ACVS, 284, 341
  correlation between nonboundary wavelet coefficients, 347
  covariance matrix, 355
  decorrelation via DWT, 340, 345, 350, 380, 388
  DWT-based and exact MLEs, 363–4, 390
    with unknown mean, 366
  DWT-based log likelihood, 363
    with unknown mean, 365
  DWT-based MLEs, 364
    and scaling coefficient, 366
    with unknown mean, 365
  exact MLE, 367
  log multitaper WLSE, 379
  log periodogram LSE, 379
  long memory property, 284
  LSE, 374, 390
  MLE, 361–2
    versus WLSE, 378
  octave band average, 343, 353
  OLSE, 378
    versus WLSE, 378
  parameter and properties, 285–6
  parameter estimation via DWT coefficients, 341
  partial DWT, 342
  reduced log likelihood, 363, 389
  sample ACS, 341, 350
  SDF, 282, 284, 343–4, 348–9
  simulation, 283, 341
    covariance matrix, 356
    example, 358
    via inverse DWT, 341, 355, 389
  true versus simulated ACVS, 357
  WLSE, 374, 376, 391
    and exact MLEs, 377–9
frequency, 22
  negative, 22
  physical, 221
  positive, 22
frequency band ANOVA, 208, 214, 233, 250–1
frequency-based ANOVA, 267
frequency response function (transfer function), 25, 33–4, 267
frequency reversal property, 211, 348
frequentist view, 264
full-band series, 84–5, 211
function
  autocorrelation, 38
  dilation of, 457–8, 474, 494
  finite energy, 457, 459
    detail of, 479–80, 497
    MRA of, 478–9, 481–2, 496
    wavelet coefficients for, 479–80, 497
  Fourier transform of, 37
  inverse dilation of, 459
  regularity of, 457, 486
  scaling (*see* scaling function)
  square integrability condition (real-valued), 458, 494
  support of, 465
  translation of, 457–8, 474, 494
  wavelet (*see* wavelet function)
function estimation, 393
function of RVs
  mean of, 256, 258
functions
  convolution, 38
  cross-correlation, 38
  inner product of real-valued, 460

gain function, 25
gamma function, 257
garrote (nonnegative), 456
Gaussian noise, 440
Gaussian (normal) mixture model, 361, 410, 413, 415, 417, 451
Gaussian (normal) PDF, 3, 256–7, 456
  first derivative, 3
  percentage points for standard, 257

second derivative, 3
Gaussian (normal) RVs
cumulative distribution function for, 257
IID, 291–2
independent, 259
mean for, 257
PDF for jointly, 259
standard, 257
sum of two independent, 451
variance for, 257
uncorrelated, 259
Gaussian (normal) stationary processes
simulation
with known ACVS, 290–1
with known SDF, 291, 436
Gaussian spectral synthesis method (GSSM), 291, 357
generalized Cauchy (t) PDF, 258
mean for, 258
variance for, 258
generalized Gaussian distribution, 257, 417
geophysical signal analysis, 4
global thresholding, 399
grey-scaled CWT, 8, 11
GSSM (*see* Gaussian spectral synthesis method)

Haar approximation space
scale one half, 461
scale two, 461
scale unity, 460–1
Haar CWT, 10, 13
Haar detail space
scale one half, 475
scale one quarter, 475
scale unity, 475
Haar DWT, 62, 63, 156, 441
matrix, 57–8, 157
partial, 13–4, 55
Haar MODWT, 17–8
Haar scaling filter
and support of Haar scaling function, 465
autocorrelation width of jth level, 103
coefficients, 75, 463, 489
of second level, 40
squared gain function, 73, 76
of second level, 91
Haar scaling function, 460–3, 479
direct sum illustrated, 472–3
support via Haar scaling filter, 465
Haar wavelet filter, 60, 155–6, 192
coefficients, 75, 480
of second level, 40
phase function, 158
of jth level, 158
squared gain function, 73
of second level, 91
Haar wavelet function, 475, 479
alternative formulation, 2, 3, 9
autocorrelation width, 12
rescaled, 10
shifted, 9, 10
half-band filter, 76, 152
half-band series, 84–5, 211
hard thresholding, 399, 412, 430, 444, 453
and consistency, 456
D(4) and LA(8) partial DWTs, 418, 420
LA(8) MODWT, 430
LA(8) partial DWT, 419–20
Hermitian transpose, 45, 47, 53, 441
Hertz (Hz), 48
hidden Markov model (chain or tree), 450–1, 455
hidden state variable, 450
high-pass filter, 27
Daubechies wavelet filter, 73–4
example of impulse response, 26, 39
example of use, 27
transfer function, 27, 39
HMM (*see* hidden Markov model)
Hölder exponents, 487
Hölder regularity, 487
D(L) scaling and wavelet functions, 487
Hölder spaces, 487
homogeneous wavelet expansion, 482
Hosking simulation method, 291
Hurst coefficient (parameter), 279, 286, 299
hyperparameter, 264–5, 413–4, 426–7
Hz (Hertz), 48

ideal risk, 420
identity matrix, 42
IID (*see* independent and identically distributed)
imaginary component of complex variable, 20–1
imaginary number, 20
impulse response sequence, 25
high-pass example, 26, 39
low-pass example, 25, 39
inconsistency of periodogram, 270, 298
independence and uncorrelated Gaussian (normal) RVs, 259, 407
independent and identically distributed (IID) RVs, 262, 288, 393, 398, 452
Gaussian (normal), 291–2, 417, 453
and chi-square RV, 293
noise DWT coefficients, 418
wavelet coefficients, 417
independent RVs, 258, 260
independent signal and noise, 452
independent signal DWT coefficients, 451
indicator function, 404
infinite sequence
autocorrelation, 25, 36
convolution, 24, 36
cross-correlation, 24, 36
discrete Fourier transform of, 21, 35, 39
inhomogeneous wavelet expansion, 482

inner product
complex, 45
preservation via orthonormal transform, 54
real, 42
real functions, 460
instantaneous frequency, 158, 322
instantaneous long memory parameter estimation
LSE, 379, 391
variance, 392
MLE, 374, 391
integer part, 214
integers (set of $\mathbb{Z}$), 458
inter-quartile range, 442
interplanetary shock waves
corotating, 219
transient, 219
interpolation operator, 103
intersection of events, 435
inverse
DFT
finite sequence, 29, 36, 78, 291
infinite sequence, 22, 35, 70, 267
DWT, 422
use in simulation of FD process, 341
Fourier transform of function, 37
MODWT, 175, 429
transform matrix as Moore–Penrose generalized, 167, 204
ODFT, 49
partial DWT, 419, 422, 425, 433–4, 445, 454
isometric transform, 43, 46–7, 53

Jacobian, 258, 261
joint PDF, 258, 265, 409

Kronecker delta function, 43

ℓ_p information cost functional, 223, 225, 253
LA(L) scaling filter, 107, 112, 117, 155
advance required, 113, 156
approximately linear phase, 112, 117
of jth level
advance required, 156
shift factor, 114, 158, 198, 205
phase function, 112
plotted, 98, 112–3, 119
shift factor, 229
LA(L) scaling function
Hölder regularity, 494
LA(L) wavelet filter, 97–8
advance required, 113
approximately linear phase, 112, 188
of jth level
advance required, 146, 155
phase after advance, 114–5, 120
shift factor, 114, 198, 205
time shift function, 119, 121
phase function, 112
plotted, 98, 112–3
LA(L) wavelet function
Hölder regularity, 494
LA(L) wavelet packet filter for node (j, n)
advance required, 230, 233
phase function, 229
shift factor, 230
LA(8) partial DWT basis vectors, 162
LA(8) scaling filter
coefficients, 109, 112, 494
4th level
shift factor, 179
squared gain function, 97, 99
of jth level, 97–8
squared gain function, 157
LA(8) wavelet filter of jth level, 97–8
shift factor, 179
squared gain function, 97, 99, 157
LA(8) wavelet packet filter for node ($4, n$)
advance required, 253
shift factor, 221
LA(10) scaling filter
reversal from Daubechies, 117
LA(12) scaling filter coefficients, 109, 112
LA(14) scaling filter
reversal from Daubechies, 117
LA(16) scaling filter coefficients, 109, 112
Laplace (two-sided exponential) PDF, 257, 264, 413
mean for, 257
variance for, 257
leakage, 187–8, 193, 197
in periodogram, 272–3, 329, 437, 439
in wavelet variance, 303, 329–30
D(4), 332
Haar, 331
least asymmetric scaling filter (*see* LA(L) scaling filter)
least asymmetric wavelet filter (*see* LA(L) wavelet filter)
least squares estimation, 44, 53, 341
FD process, 374, 390
least squares estimation (ordinary) (*see* ordinary least squares)
least squares estimation (weighted) (*see* weighted least squares)
leave-one-out cross-validation, 402
DWT, 422
threshold, 423–4
length of series
analyzed via MODWPT, 231
analyzed via MODWT, 141, 159, 167, 199, 242
general length, 56, 141, 143–4
multiple of power of two, 104, 141, 150
power of two, 67, 141, 150
level, 65–7, 77
of partial DWT, 56
level-dependent thresholding
hard, 445–9
mid, 445–6

soft, 445–6
SURE-based, 406, 449–50, 453
universal, 406, 430, 444–5, 455
level-dependent variance
correlated noise DWT coefficients, 441, 456
level-independent thresholding
hard, 445–6
mid, 445–6
soft, 445–9
universal, 444–5, 455
likelihood function
for Gaussian stationary FD process, 361, 389
using Gaussian nonboundary DWT coefficients, 371
linear estimator of signal, 408
linear minimum mean square, 417
linear phase filter, 108, 111, 155, 158
generalized, 117
linear span, 459
linear time-invariant filter, 25
Lipschitz exponents, 487
local stationarity, 327, 333
log chi-square PDF versus Gaussian (normal) PDF, 276
log chi-square RV, 270, 272, 376, 390
distribution, 432–5, 437
log likelihood
DWT-based for Gaussian FD process, 363
with unknown mean, 365
log multitaper SDF estimator, 444
covariance of, 276–7
mean of, 275
thresholding, 440, 444
variance of, 275
WLSE for FD process, 379
log periodogram, 270, 432
LSE for FD process, 379
mean of, 270
smoothing of, 272
thresholding
and asymmetry, 456
for SDF estimation, 432, 435, 439
variance of, 270
log SDF estimator, 433
long memory process, 16, 191, 279, 299, 340
analysis, 17
autocovariance sequence, 279
defining property, 279
and FD process, 284
and FGN, 280
and PPL process, 281
and scaling coefficients from FD process, 351
instantaneous parameter estimation
LS, 379, 391
ML, 374, 391
nonstationary, 284
FD, 284, 288
PPL, 281
parameter estimation
LS, 374
ML, 361, 368
stationary, 281
synthesis, 17
loss, 265
absolute error, 266, 415
posterior expected, 265
squared error, 265–6, 414
low-pass filter, 26
Daubechies scaling filter, 76, 78
of second level, 91
example of impulse response, 25, 39
example of use, 26
tide removal, 185
transfer function, 26, 39
LSE (*see* least squares estimator)

MA process (*see* moving average process)
MAD (*see* median absolute deviation)
magnitude of complex variable, 21
magnitude squared DFTs
low/high pass filtering and downsampling, 209, 211
marginal PDF, 258, 265
Markov chain of first order, 451
Markovian dependencies, 451
matching pursuit, 239, 252, 254
algorithm, 239
ANOVA, 240, 252
approximation vector, 252–3
approximations of subtidal sea level fluctuations, 244–8
contrasted with best basis algorithm, 241
dictionary
of MODWT and constant vectors, 247
of MODWT and ODFT vectors, 247
of MODWT vectors, 242, 253
of partial DWT vectors, 240
of WP table and ODFT vectors, 241
of WP table vectors, 241, 254
energy decomposition, 240, 252
normalized residual sum of squares, 247–8
residual vector, 252–4
subtidal sea level fluctuations using MODWT dictionary, 243–5
matrix
circular shift, 52, 71
complex orthonormal (unitary), 45, 53
DWT, 57, 63, 99, 156
identity, 42
ODFT, 47
orthonormal, 42–3, 53
unitary, 46–7
maximal overlap discrete wavelet packet transform (MODWPT), 231, 251, 278
advanced LA(8) coefficients for solar magnetic field, 235–6
ANOVA, 233, 251
energy decomposition, 233, 251
general length series, 231
non-advanced LA(8) coefficients for solar magnetic field, 253

physical bandwidth for solar magnetic field, 236
physical time width for solar magnetic field, 236
zero phase filter for details, 234, 237, 252
maximal overlap discrete wavelet transform (MODWT), 17, 19, 159, 294, 374–5, 390
advanced LA(8) coefficients for electrocardiogram, 182–3
advanced LA(8) coefficients for subtidal sea levels, 189
analysis, 175
and cycle spinning, 429
ANOVA, 160, 166, 171
and circular shifts, 160, 180, 200
for subtidal sea levels, 185
atomic clock data, 17–8
choice of level, 188, 193, 199
circularly shifted series, 179, 181
discarded boundary coefficients, 310
DWT coefficients via subsampling, 174, 203
energy decomposition, 16–7, 166, 168–9, 171, 173, 200
general length series, 159, 167, 199, 242
Haar, 17–8
invalid ANOVA using details, 182
inverse, 429
LA(8), 182, 430
MRA, 166, 168–9, 172–3, 182, 200
nonorthogonality, 159, 200
pyramid algorithm, 164–5, 174–5, 177, 201
reflecting boundary conditions for, 310
renormalized vectors for waveform dictionary, 242–3
scaling filter (*see* scaling filter, MODWT)
synthesis, 175
upper bounds for level, 200
vectors for waveform dictionary, 242, 253
wavelet filter (*see* wavelet filter, MODWT)
zero phase filter for details and smooths, 160, 172, 168, 189
maximal overlap WP table, 232, 251
maximum delay dipole, 488, 490–1, 493–4, 500
maximum delay filter, 488
maximum likelihood estimator (MLE), 341
arbitrary sample size, 367, 373
DWT-based for Gaussian ARFIMA process, 366
DWT-based for Gaussian FD process, 363–4, 390
with unknown mean, 365–6
exact for Gaussian FD process, 363–4, 367, 372, 390
with unknown mean, 366
Gaussian nonboundary DWT coefficients, 371–2, 383–4, 390
Gaussian stationary FD process, 361–2
Nile River minima, 386, 392
nonstationary FD process, 368
scale restricted, 373, 385
maximum phase dipole, 488, 490–1, 493, 500
maximum phase filter, 488
mean
conditional, 260, 266, 394, 408–11, 414–5, 454, 456
of chi-square RV, 263, 270
of function of RVs, 256, 258
of Gaussian (normal) RV, 257
of generalized Cauchy (t) RV, 258
of Laplace (two-sided exponential) RV, 257
of log multitaper SDF estimator, 275
of log periodogram, 270
of periodogram, 270
of stationary stochastic process, 266
of RV, 256
of wavelet coefficients, 301, 305, 335
posterior, 266, 414, 454
sample (*see* sample mean)
mean absolute error, 261, 266, 411
minimization, 411
mean square
equality in, 38
error, 261, 265, 408, 411
minimization, 407, 409, 417
median
conditional, 266, 394, 408, 411, 415, 454
posterior, 266, 415–6, 454
median absolute deviation (MAD) standard deviation estimate, 420
DWT-based, 418–21, 426–7, 455
level-dependent, 450
MODWT-based, 429–30
Mexican hat CWT, 8, 11
Mexican hat wavelet function, 3, 10, 97
autocorrelation width, 12
microseconds, 318
mid thresholding, 399–400, 430, 444, 453
energy levels, 420
LA(8) MODWT, 430
LA(8) partial DWT, 419–20
minimization of Stein's unbiased risk, 405
minimum delay dipole, 488, 490, 494, 500
minimum delay (extremal phase) filter, 106, 117, 488
minimum mean square estimation, 394
minimum phase dipole, 488, 490, 500
minimum phase filter, 488
minimum unbiased risk, 402
mixing parameter, 413
mixture model (*see* Gaussian (normal) mixture model)
MLE (*see* maximum likelihood estimator)
mobile radio communications (MRC) process SDF, 436
estimation via thresholding
of log multitaper, 446–7, 449

of log periodogram, 437–8
modulo arithmetic, 30
modulus of complex variable, 21
MODWPT (*see* maximal overlap discrete wavelet packet transform)
MODWT (*see* maximal overlap discrete wavelet transform)
moment of inertia, 261
moments of wavelet function, 483
vanishing (*see* vanishing moments)
Moore–Penrose generalized inverse, 167, 204
Morlet (complex-valued) wavelet function, 5
'mother' wavelet filter, 80
moving average (MA) process, 391
decorrelation via DWT, 391
MRA (*see* multiresolution analysis)
MRC (*see* mobile radio communications)
multiresolution analysis (MRA), 2, 65, 151, 457
circular boundary conditions for electrocardiogram, 133, 140
finite energy function (signal), 478–9, 481–2, 496
function space, 461, 494
general length series, 143
Haar MODWT Nile river minima, 190, 192
LA(8) MODWT
for electrocardiogram, 183–4
for subtidal sea levels, 185–7
for vertical ocean shear, 193–5
MODWT and circular shifts, 160, 180–1, 200
partial DWT and circular shifts, 160–1
partial DWT for electrocardiogram
using C(6) wavelet, 129, 132
using D(4) wavelet, 129, 131
using Haar wavelet, 129–30
using LA(8) wavelet, 129, 133, 142
reflecting boundary conditions for electrocardiogram, 140, 142
via DWT, 65
via MODWT, 166, 168–9, 172–3, 182, 200
via partial DWT, 104, 129
multitaper SDF estimator, 269, 273, 329–30, 394, 439
applied to solar physics series, 278
bandwidth of, 274, 277–8
lack of leakage, 330, 446
logarithm of, 276, 444
multivariate RV, 393
Gaussian (normal), 400, 407, 441

nanoseconds, 318
nanoteslas (nT), 219
narrow-band frequency characteristic, 396
natural ordering for wavelet packet coefficients, 209–10
negative frequency, 22
nested subspaces, 460, 494
approximation and detail spaces, 472–3
convention for approximation spaces, 461
Nile River minima, 255, 295, 326, 341, 365, 386
change point, 387–8
MLE for, 386
using Haar, D(4) and LA(8) filters, 392
MRA for Haar MODWT, 190, 192
physical scales, 192
test for homogeneity of variance, 386–7
wavelet variance estimates for, 327
nilometer, 193
NMR (*see* nuclear magnetic resonance)
noise
correlated, 393
Gaussian (normal), 440
IID, 393, 398, 451–2
IID Gaussian (normal), 417
non-Gaussian, 432
non-IID, 393, 402
uncorrelated, 393
noise DWT coefficients
correlated, 440
IID Gaussian (normal), 418
non-IID, 432, 434
PDFs with level, 455
stochastic, 424, 426
uncorrelated, 432, 434
noise transform coefficients, 394, 406–8, 412, 452
IID Gaussian, 453
non-IID, 403
noise vector
IID, 452
non-IID, 452
nonboundary DWT coefficients, 370
MLE, 371–2, 383–4, 390
reduced log likelihood, 371, 383
non-decimated DWT, 159
non-Gaussian noise, 432
non-Gaussian signal DWT coefficients, 450
non-IID noise DWT coefficients, 432, 434
non-IID noise transform coefficients, 403
nonlinear conditional mean, 417
nonlinear estimator of signal, 408
nonlinear transform and shrinkage, 408, 411–2, 454
nonnegative garrote, 456
nonorthogonality of MODWT, 159
nonstationary process
$1/f$-type, 287
continuous parameter, 340
example simulation via DWT, 360
long memory, 284, 286
DWT-based MLEs for, 368
FD, 284, 288
PPL, 281
simulation, 289
via DWT, 360
with nonstationary first order backward difference, 294

with Pth order trend plus stationary process, 289, 294
with stationary dth order backward difference, 287, 368
SDF, 287
simulation, 292
with stationary first order backward difference
random walk, 288
trend plus stationary process, 288
with stationary second order backward difference
random run, 294
normal PDF (*see* Gaussian PDF)
normal RVs (*see* Gaussian RVs)
normalization and periodized filters, 34
normalization of scaling function, 468
normalized partial energy sequence (NPES), 129, 395
for electrocardiogram, 128–9
for signals, 395–6
for vertical ocean shear, 397
using DWT coefficients, 128–9, 395–6, 455
using ODFT coefficients, 128–9, 395–6
NPES (*see* normalized partial energy sequence)
nuclear magnetic resonance (NMR), 420, 424, 427
Bayesian shrinkage estimates of signal, 428
denoising
using LA(8) partial DWT, 431
using ODFT, 431
MODWT-based thresholding estimates of signal, 430
shrinkage estimates of signal, 425–6
thresholding estimates of signal, 418–9, 421, 423
Nyquist frequency, 22, 87, 267, 315

octave band average
FD process, 343, 353
octave band-pass filter, 102
octave band SDF estimation, 316, 329–30, 337, 343, 389
octave frequency band, 297
odd indexed subsequence and DFT, 39
ODFT (*see* orthonormal discrete Fourier transform)
OLSE (*see* ordinary least squares)
optimal discrete WP table transform, 223, 251
example of, 224
oracle, 420
ordinary least squares estimation for FD process, 378
orthogonal complement, 472, 496
orthogonality principle of linear mean square estimation, 407
orthonormal basis
approximation space
with scale λ_j, 460, 495
with scale one half, 461
with scale two, 461
with scale unity, 459
detail space
with scale one half, 474–5, 496, 500
with scale one quarter, 475
with scale unity, 475
orthonormal data tapers, 273
orthonormal discrete Fourier transform (ODFT), 41, 46, 49, 53, 395, 431
and DFT, 52, 54
energy preservation, 47
inverse, 49
matrix, 47
projection theorem, 52
orthonormal matrix, 42–3, 53
orthonormal transform, 41, 394, 452
DWT, 57, 67, 72
energy preservation, 43, 53
of IID Gaussian RVs, 263, 400
of noise coefficients, 394, 398, 406
of signal coefficients, 394, 398
projection theorem, 44, 53
orthonormality and DWT, 76–7, 79, 95
orthonormality property of wavelet filter, 69
outcome of RV, 256

PACS (partial autocorrelation sequence), 367
padding (extending a sample)
for the DWT, 141, 422–3
with sample mean, 338
with zeros, 308, 338, 382, 387
for the FFT, 440
paradigm for wavelet thresholding, 444, 449, 455
parameter and properties
of DFBM process, 286
of FD process, 285–6
of FGN process, 285–6
of PPL process, 285–6
parameter estimation for FD process via DWT coefficients, 341
parent node for DWPT, 210
Parseval's theorem, 23, 312, 314
one finite sequence, 29, 36, 72, 170, 233
one function, 38
one infinite sequence, 23, 35
two finite sequences, 29, 36, 233
two functions, 38, 467
two infinite sequences, 23, 35
partial autocorrelation sequence (PACS), 367
partial discrete wavelet transform (DWT), 19, 56, 80, 104, 211, 341, 350, 385, 418, 433, 445, 447, 451, 454
advanced LA(8) coefficients for electrocardiogram, 127–8
atomic clock data, 13
C(6), 126
choice of level, 145
circularly shifted series, 160–1

D(4), 126, 418, 420
decomposition of sample variance, 104
disjoint dyadic decomposition of WP table, 212
energy decomposition, 15, 16, 95, 104, 151
Haar, 13–4, 55, 126
inverse, 419, 422, 425, 433–4, 445, 454
LA(8), 126, 418–21, 423–5, 428, 431
basis vectors, 162
level of, 56
matrix, 79
multiple of power of two length, 104, 141, 150
of ECG data, 126
of FD process, 342
vectors for waveform dictionary, 240
partial DWT (*see* partial discrete wavelet transform)
partial energy sequence, 106
for maximum/minimum delay filter, 488, 493
PDF (*see* probability density function)
penalty term, 398
percentage points
for homogeneity of variance test, 381
for trend test, 373
of chi-square PDF, 264
of standard Gaussian (normal) PDF, 257
periodic extension, 30
periodicity of DFT
for finite sequence, 28, 36
for infinite sequence, 22, 35
periodized filter, 20, 32–3, 35, 37, 39–40
DFT of, 33–4, 37, 39
DWT scaling filter, 77, 79–80, 88
circularly shifted, 77, 79
DFT of, 92, 100
DWT wavelet filter, 71, 74, 80, 88, 92, 96–7, 153
autocorrelation, 72
circularly shifted, 71, 77, 92, 153
DFT of, 92, 100
example, 34
MODWT scaling and wavelet filter, 168–70, 201
normalization, 34
periodogram, 269, 298, 312, 314, 330, 394
asymptotic properties (independence and unbiasedness), 270
bias of, 298
covariances of, 270–1
inconsistency of, 270, 298
leakage in, 272–3, 329, 437, 439
logarithm of, 270, 432
mean of, 270
smoothing of, 271
variance of, 270
persistence of DWT coefficients, 450–1, 455
phase function, 25
D(4) wavelet filter, 158
Daubechies scaling filter, 106
Haar wavelet filter, 158
LA(L) and C(L) wavelet packet filter for node (j, n), 229
LA(L) scaling and wavelet filters, 112
wavelet packet filter for node $(3, n)$, 253
physical bandwidth
for MODWPT of solar magnetic field, 236
physical frequency, 221
physical scale, 15, 59, 128, 134, 185, 243, 297, 315, 328, 334
for MODWT details and smooth of jth level, 188, 192, 194
physical time width
for MODWPT of solar magnetic field, 236
piecewise power law process, 321, 323, 328
pilot spectrum estimator, 297
and unbiased Haar-based estimator, 317
point estimate, 265
polar representation of complex variable, 21
polynomial (trigonometric), 485
positive frequency, 22
posterior
expected loss, 265
mean, 266, 414, 454
median, 266, 415–6, 454
power law process, 288, 294, 297, 332, 337, 340
power of two
and DWT, 57, 67, 141, 150
multiple of, and partial DWT, 104, 141, 150
needed with periodogram, 437
power spectrum
discrete Fourier empirical, 48–9, 55
discrete wavelet empirical, 160, 181
DWT-based, 62–3, 66–7, 156, 297
MODWT-based, 160, 181
for stationary process, 267
PPL process (*see* pure power law process)
prior information, 264
prior PDF, 264–5
probability, 256
of false alarm, 435
probability density function (PDF), 256
chi-square, 263
conditional, 260, 409
exponential, 265
Gaussian (normal), 3, 256
generalized Cauchy (t), 258
generalized Gaussian, 257
joint, 258, 265, 409
jointly Gaussian (normal), 259
Laplace (two-sided exponential), 257, 264
marginal, 258, 265
prior, 264–5
sum of two independent RVs, 262, 409, 416

sum of two RVs, 262
t, 257
transformed RVs, 258, 261
projection onto subspace, 45
projection theorem, 41
ODFT, 52
orthonormal transform, 44, 53
unitary transform, 46
pseudo-code
for DWT
analysis, 101, 157
details and smooth, 102
pyramid algorithm, 100
synthesis, 101, 157
for MODWT
analysis, 177, 204
details and smooth, 179
pyramid algorithm, 177, 204
synthesis, 178, 204
pure power law process
continuous parameter, 285
discrete parameter, 281
long memory property, 281
parameter and properties, 285–6
simulation, 283
spectral density function (SDF), 281–2
purely random process, 268
pyramid algorithm for DWT, 17, 56, 68, 151, 157
first stage, 80
and DWPT, 207
summary, 86
general stage, 93
summary, 99
inverse, 95, 157
pseudo-code, 100, 157
for inverse, 101, 157
reformulation for general length series, 144
second stage, 88
and DWPT, 207
summary, 93
pyramid algorithm for MODWT, 175
first stage, 164–5
general stage, 174–5, 177, 201
inverse, 175, 177, 203
pseudo-code, 177, 204
pseudo-code for inverse, 178, 204

QMF (*see* quadrature mirror filter)
quadrature mirror filter (QMF), 75, 151, 163, 498
alternative definition, 79
for time reversed filters, 474, 497
quartiles, 442

rainfall average, 6
random circular shift, 356, 358, 361, 389, 391
random run, 288, 294
simulation, 289
random variable (RV), 256
binary-valued, 410, 454
continuous, 256
discrete, 256
mean of, 256
mean of function of, 256
multivariate, 393
Gaussian (normal), 400, 407, 441
outcome, 256
realization, 256
variance of, 256
random variables (RVs)
independent, 258
and identically distributed (IID), 262
random vector, 393, 258
random walk process, 288, 293, 385
and DFBM, 288
as first order backward difference of random run, 294
nonstationary process with stationary first order backward difference, 288
simulation of, 289
wavelet variance for, 337
real component of complex variable, 20–1
real numbers (set of $\mathbb{R}$), 457
realization of RV, 256
reciprocity relationships, 216
reduced log likelihood
for Gaussian FD process, 363, 389
for instantaneous estimation, 374
for Nile River minima, 386
using Gaussian nonboundary DWT coefficients, 371, 383
reflection boundary conditions, 19, 53, 140, 194
and best basis selection, 227–8
and best shift basis selection, 228
example with electrocardiogram, 140, 142
for DWPT, 227
for DWT, 140, 433, 439
for MODWT, 310
regular DWT (alternative name for DWPT), 218
regularity, 457
Hölder, 487
of function, 486
of signal, 486
regularization of the SDF, 297–8
relative approximation error, 395, 397
rescaled wavelet function (alternative formulation for Haar), 10
resolution of identity, 205
reversal of frequency order due to down-sampling
with DWPT, 211
with wavelet coefficients, 85–6
Riemann integral, 6
Riemann sum, 464
risk, 403, 408
minimum unbiased, 402
risk function, 265
RMSE (*see* root mean square error)
root mean square error (RMSE), 364, 372, 377, 379, 386, 436

thresholding of log multitaper estimator, 445–6
rotated cumulative variance, 189, 294, 382
for subtidal sea levels, 189–90
roughs
Fourier, 50–1, 54
wavelet, 66
D(4), 65–6
Haar, 64, 66
RV (*see* random variable)

saddlepoint method, 434, 436
sample ACS
FD process, 341, 350
time series, 14–5
wavelet coefficients, 14, 16, 19, 343
sample mean, 16, 48
and DCT-II, 55
and DCT-IV, 55
and scaling coefficient, 62, 67, 97, 146, 355
of DWPT coefficients for node (j, n), 253
of DWT coefficients, 158
of MODWT details and smooth, 205
of MODWT scaling coefficients, 171
padding with, 338
variance, 293
sample variance, 48
bias in, 299–300
decomposition of
via DWT, 62, 67
via MODWT, 171, 180, 205
via partial DWT, 104
via wavelet details, 66–7
of MODWT scaling coefficients of J_0th level, 171
of wavelet coefficients, 344
sampled sinusoid signal, 395
sampling interval, 48, 59, 266, 315
for FGN, 285
scale, 6–7, 58
dyadic, 13, 15
for scaling coefficients (λ_j)
DWT, 85, 103, 150
MODWT, 174
for wavelet coefficients (τ_j)
DWT, 59, 67–8, 85, 103, 150, 206
MODWT, 174
physical, 15, 59, 128, 134, 185
for Nile river minima, 192
for subtidal sea levels, 188, 243
for vertical ocean shear, 194
unitless/standardized, 59
scale-by-scale ANOVA, 62, 151
scale-restricted MLE, 373, 385
scaled chi-square distribution, 313, 336, 338
scaling coefficients, 13, 17, 59
and long memory property with FD process, 351
and sample mean, 62, 67, 97, 146, 355
boundary elements, 136, 146, 158
downsampling and bandwidth, 84, 86
MODWT
boundary elements, 198
compared to smooth of jth level, 180, 185
via circular filtering, 163, 169–70, 173, 175, 177, 200–1
scale λ_j, 464, 495
scale four, 463
scale two, 462, 495
scale unity, 460, 464, 495
variance of, 355
via circular filtering, 77, 79–80
for jth level, 94, 97, 152
for second level, 88
with DWT-based MLEs for Gaussian FD process, 366
scaling filter, 56, 75, 78, 151, 463, 495
advance required for LA(L), 113
and father wavelet filter, 80
approximately linear phase
for C(L), 124
for LA(L), 112, 117
BL(L) (*see* best localized filter)
C(L) (*see* C(L) scaling filter)
coefficients
for C(6), 109
for D(L), 493
for D(4), 75, 490
for D(6), 106, 109, 491
for D(8), 106, 109, 493
for Haar, 75, 463, 489
for LA(8), 109, 112, 494
for LA(12), 109, 112
for LA(16), 109, 112
coiflet (*see* coiflet scaling filter)
D(L) (*see* D(L) scaling filter)
Daubechies (*see* Daubechies scaling filter)
even width, 75, 78
Fourier transform of time reversed, 465, 484, 496
Haar squared gain function, 73, 76
key properties, 154, 499–500
LA(L) (*see* LA(L) scaling filter)
low-pass nature of Daubechies, 76, 78, 151
MODWT, 169, 200, 469, 476
autocorrelation width, 174
connection to scaling function, 469, 496
energy, 163
key properties, 202
non-zero sum, 163
orthogonality to even shifts, 163
periodized, 168, 170, 201
squared gain function, 163
transfer function, 163, 169, 173, 201, 469
width, 169, 201
nonzero sum, 76, 78, 151
of 4th level
shift factor for LA(8), 179

of Jth level
shift factor for LA(L), 158
of jth level, 96, 152
advance required for, 118, 156
alternative formulations, 102
autocorrelation width, 103, 174
connection to scaling function, 469, 496
for C(6), 125
for D(4), 97–8
for DWT/MODWT, 173
for LA(8), 97–8
shift factor for C(L), 124, 198, 205
shift factor for LA(L), 114, 198, 205
squared gain function, 157
transfer function, 97, 152, 157
width, 97
of second level, 91–2
Daubechies and low-pass nature, 91
Haar, 40
nonorthogonality to even shifts, 92
orthogonality to even shifted wavelet filter, 77, 151
orthogonality to even shifts, 76, 78, 151
periodized, 77, 79–80, 88
and circularly shifted, 77, 79
reversal
of BL(14) from Doroslovački, 119
of C(6) from Daubechies, 123
of LA(10), LA(14) and LA(16) from Daubechies, 117
shift factor
for C(L), 124, 229
for LA(L), 229
squared gain function, 76, 78, 151
of time reversed, 485, 498
support of scaling function, 465
time reversed, 463, 495
transfer function, 76, 78
unit energy, 76, 78
scaling function, 457, 459, 495
computation for D(4), 469
connection to scaling filter of jth level, 469, 496
finite support, 465, 472, 496
Fourier transform of, 465, 468–9, 496
Haar, 460–3, 479
integral of unity, 468
key properties, 499–500
normalization, 468
support via scaling filter, 465
two-scale relationship, 464, 466, 470, 495
SDF (*see* spectral density function)
sea surface temperature average, 6–7
seasonally dependent variation, 188–9
second backward difference, 60
second level details and smooths, 89
second order stationarity, 266
self-similarity parameter, 279
sequency (sequential frequency) ordering
binary vector for node (j, n), 215–7, 249, 253
for MODWPT coefficients, 232
for wavelet packet coefficients, 209–10, 212
shark's fin, 97, 124, 129, 196, 420, 471
shift invariant DWT, 159
shifted Haar wavelet function (alternative formulation), 9, 10
short-time Fourier transform, 206
shrinkage, 394, 408, 434, 453
and nonlinear transform, 408, 411–12
rules for, 411, 454
approximate conditional median, 412, 415–6
Bayesian, 412–3, 415, 426–8, 454
conditional mean, 411–2, 424–6
wavelet, 412, 417, 454
sign function, 400
signal
average value over a scale, 6
continuous time, 457
detail for finite energy, 479, 497
deterministic, 394, 398, 417, 452–3
dilation, 457–8, 474, 494
DWT coefficients
independent, 451
non-Gaussian, 450
estimation, 17, 393
finite energy, 457, 459
inverse dilation, 459
jth level detail for finite energy, 480, 497
jth level wavelet coefficients for finite energy, 480, 497
linear estimator, 408
MRA for finite energy, 478–9, 481–2, 496
nonlinear estimator, 408
plus noise model, 398, 402, 452
reconstruction via CWT, 11
regularity, 457, 486
representation via DWT, 394–5
sampled sinusoid, bump and sum, 395
sparse, 411, 415
square integrability condition (real-valued), 458, 494
stochastic, 393–4, 408, 412, 424, 452, 454
DWT coefficients, 424, 426, 450
Gaussian (normal) mixture model, 410
generalized Gaussian model, 417
transform coefficients, 407–8, 412, 452
support, 465
transform coefficients, 394
translation, 457–8, 474, 494
wavelet coefficients for finite energy, 479, 497
signal-to-noise ratio, 398
significance level for homogeneity of variance test, 381
simulation
Davies–Harte method, 281, 283, 290, 341
versus DWT method, 359

examples of
FD process via inverse DWT, 358
nonstationary process via DWT, 360
FD ACVS
by DWT method, 359
by Davies–Harte method, 359
Gaussian spectral synthesis method, 291
Hosking method, 291
localized nature of inverse DWT approach, 359
of autoregressive process, 292
of continuous parameter nonstationary process, 361
of FBM, 361
of FD process, 283
via Davies–Harte method, 341
via inverse DWT, 341, 355, 389
of FGN process, 283
of Gaussian (normal) stationary processes
with known ACVS, 290–1
with known SDF, 291, 436
of nonstationary process, 289
via DWT, 360
with stationary *d*th order backward differences, 292
of PPL process, 283
of random run, 289
of random walk, 289
of stationary processes, 290
of trend plus stationary process, 289
with stationary start-up values, 293
sine tapers, 274, 277, 440
small wave interpretation of wavelet function, 2, 476
smoothing
adaptive, 272
bandwidth, 269, 272
of periodogram, 271
spline, 272
smooths
DWT-based, 64, 66, 151, 158
artifact removal by averaging, 196–7
compared to MODWT-based smooths, 196
of *j*th level, 95
of first level, 81
of second level, 89
using Haar wavelet, 64, 66
Fourier, 50–1, 54
MODWT-based
and zero phase, 160, 168, 172
boundary elements, 199
compared to DWT-based smooths, 196
of *j*th level, 171, 173, 176–7, 200, 203
of *j*th level and DWT averaging, 204–5
of *j*th level via cross-correlation (circular), 172, 203
soft thresholding, 399–400, 430, 444, 453
cross-validation, 422
energy levels, 420
LA(8) MODWT, 430
LA(8) partial DWT, 419–21, 423–4
solar magnetic field, 207, 218, 234
advanced LA(8) DWPT coefficients, 221–2
advanced LA(8) MODWPT coefficients, 235–6
best basis
using entropy cost functional, 227
using reflection boundary conditions, 227–8
best shift basis using reflection boundary conditions, 228
details for LA(8) MODWPT, 237–8
DWPTs of 4th level, 220
modified time/frequency decomposition, 236–7
multitaper SDF estimate of, 278
non-advanced LA(8) MODWPT coefficients, 253
span
linear, 459
of subspace, 45
sparse signal model, 411, 415, 424, 454–5
spectral density function, 267
and ACVS, 267
confidence intervals for, 275, 278
consistent estimator, 271–2
via wavelet variance, 298
estimation via thresholding
of log multitaper estimator, 440, 444
of log periodogram, 432, 435, 439
statistical properties of, 434
DWT-based estimation of logarithm, 433
estimation via wavelet variance, 295, 297–8, 315–6, 329, 337
even function, 267
for AR process, 268
for AR(1) processes, 351–2
for AR(2) process, 436
estimated via log multitaper, 446, 448
estimated via log periodogram, 437–8
for AR(24) process, 435
estimated via log multitaper, 446–7
estimated via log periodogram, 437–8
estimated via multitaper, 275
estimated via periodogram, 273
for DFBM process, 280
for FBM process, 280
for FD process, 282, 284, 343–4, 348–9
for FGN process, 280, 282
for MRC process, 436
estimated via log multitaper, 446–7, 449
estimated via log periodogram, 437–8
for PPL process, 281–2
for white noise, 268
multitaper estimator, 269, 273–275, 394
nonboundary wavelet coefficients, 348–9, 352
nonstationary process with stationary *d*th order backward difference, 287

octave band estimation, 316, 329–30, 337, 343, 351, 389
periodogram (*see* periodogram)
piecewise power law, 321, 323
pilot estimate, 297
and unbiased Haar DWT-based estimator, 317
and unbiased Haar MODWT-based estimator, 317
power law, 297
square integrability condition, 307
spectral factorization, 106, 116, 486–7
D(L) scaling filter, 492, 500
D(4) scaling filter, 490–1
Haar scaling filter, 489
LA(L) scaling filter, 500
LA(8) scaling filter, 493
multiplicity of solutions, 489
spectral slope, 286
spectrum
energy, 268
power, (*see* power spectrum)
spline smoothing, 272
square integrability condition
for real-valued function, 458, 494
for SDF, 307
square summability condition for ACVS, 267, 307
squared error loss, 265–6, 414
squared gain function, 25, 268
associated with MODWT wavelet coefficients of jth level, 197
associated with MODWT wavelet details of jth level, 197
backward difference filter, 105, 280, 304, 335
coiflet wavelet filter, 124
D(4) scaling and wavelet filters, 73, 76
of second level, 91, 157
Daubechies scaling and wavelet filters, 105, 107, 116, 153, 155
of jth level, 121–2, 343, 391
DWT scaling and wavelet filters, 69, 74,, 76, 78, 151, 156
of jth level, 157
Haar scaling and wavelet filters, 73, 76
of second level, 91
LA(8) scaling filter
of 4th level, 97, 99
of jth level, 157
LA(8) wavelet filter of jth level, 97, 99
MODWT scaling and wavelet filters, 163
sidelobes for Daubechies wavelet filter of jth level, 123
time reversed scaling filter, 485–6, 498
wavelet packet filter for node (j, n), 215, 217, 233, 253
squared norm, 458
for complex-valued variables, 46
for real-valued variables, 42
standard deviation estimated
via level-dependent MAD, 450
via MAD and DWT, 418–21, 426–7, 455
via MAD and MODWT, 429–30
standard Gaussian (normal) RV, 257
state value, 451
state variable, 451
stationarity (second order), 266
local, 327, 333
stationary
backward differences, 287, 295, 301, 304–5, 335, 337–8, 424
DWT, 159
first difference series, 322, 328
long memory process, 280–1, 286
process, 268
and wavelet variance exponents, 323
covariance of, 266
mean of, 266
simulation of, 290
variance of, 266
second difference series, 319
start-up values for AR simulation, 293
statistical efficiency, 17, 309
Stein's unbiased risk estimator (SURE), 404, 420, 453, 456
level-dependent threshold, 449–50, 453
minimization, 405
threshold, 421, 454
step function, 5, 6
stochastic noise, 393, 398
DWT coefficients for, 424, 426
stochastic process, 255, 266
stochastic signal, 393–4, 408, 412, 424, 452, 454
DWT coefficients for, 424, 426, 450
estimation
via nonlinear shrinkage, 408
via scaling, 407
Gaussian (normal) mixture model, 410
generalized Gaussian model, 417
transform coefficients, 407–8, 412, 452
suboptimal basis selection
top down approach, 226
subsampling (*see* downsampling)
subspaces
convention for nesting approximation spaces, 461
nested, 460, 494
nesting for approximation and detail spaces, 472–3
projection onto, 45
span of, 45
subtidal sea level fluctuations, 295, 324
LA(8) MODWT of
advanced coefficients, 189
ANOVA, 185
MRA, 185–7
matching pursuit approximations, 244
using MODWT and ODFT dictionary, 247–8
using MODWT dictionary, 243–6
normalized residual sum of squares, 247–8
physical scales, 188

rotated cumulative variance, 189–90
sum
of linear trend and stationary process, 288
of Pth order polynomial trend and a stationary process, 289
of two independent RVs
PDF, 262, 409, 416
PDF for Gaussians, 451
of two RVs
PDF for, 262
support of a function, 465
SURE (*see* Stein's unbiased risk estimator)
symmetry of autocovariance sequence, 266
symmlets, 107
synthesis of function
scaling and wavelet coefficients, 481–2, 497
via approximation and details, 479, 481, 496
via scaling coefficients, 460
via wavelet coefficients, 482
synthesis of scaling coefficients of jth level
DWT, 95, 152
MODWT, 175, 177, 203
synthesis of signal
scaling and wavelet coefficients, 481–2, 497
via approximation and details, 479, 481, 496
via scaling coefficients, 460
via wavelet coefficients, 482
synthesis of time series, 43, 46, 53
from DWPT coefficients, 208, 250
from DWT coefficients, 57, 63
from Fourier coefficients, 49
from MODWPT coefficients, 234
from MODWT coefficients, 166, 168–9
from partial DWT coefficients, 80, 83, 90, 150
using matching pursuit, 240
via DWT smooths and roughs, 66–7
via DWT details and smooth, 65, 67, 129, 150
via Fourier details, 50
via Fourier roughs and smooth, 51, 54
via MODWPT details, 234, 251
via MODWT details and smooth, 166, 168–9, 173, 200

t distribution, 414, 427
tapers (*see* data tapers)
Taylor series expansion, 353
PDF (*see* also generalized Cauchy), 257
temperature average, 6–7
test for homogeneity of variance, 341
arbitrary sample size, 382, 386
DWT-based test statistic, 380
long memory processes, 379, 391
Nile River minima, 386–7
nonboundary MODWT-based, 381
test for trend, 373, 390
adjustment, 373
atomic clock average fractional frequency deviates, 384
mean removal, 373
threshold
cross-validation, 402
leave-one-out cross-validation, 423–4
two-fold cross-validation, 423–4
for SDF estimation via log periodogram, 436
level-dependent SURE-based, 406
universal, 400–1, 418–9, 443, 450, 453–4
via SURE, 404, 406, 421, 454
threshold cost functional, 223, 225, 253
threshold function, 399, 411, 454
thresholding, 394, 399, 419, 434, 453
firm, 400, 411, 453
global, 399
hard, 399, 412, 453
level-dependent, 445–9
level-independent, 445–6
using D(4) partial DWT, 418, 420
using LA(8) MODWT, 430
using LA(8) partial DWT, 418–20, 456
mid, 399–400, 453
and energy levels, 420
level-dependent, 445–6
level-independent, 445–6
using LA(8) MODWT, 430
using LA(8) partial DWT, 419–20, 456
paradigm, 444, 449, 455
soft, 399–40, 422, 453
and energy levels, 420
level-dependent, 445–6
level-independent, 445–9
using LA(8) MODWT, 430
using LA(8) partial DWT, 419–21, 423–4, 456
wavelet, 417, 454
thresholding and SDF estimation
log multitaper estimator, 440, 444
for AR(2) process, 446–8
for AR(24) process, 446–7
for MRC process, 447–9
RMSE, 445–6
log periodogram, 432, 435, 439
and asymmetry, 456
for AR(2) process, 437–8
for AR(24) process, 437–8
for MRC process, 437–8
RMSE, 446
time alignment, 17, 135, 333, 341, 374, 429
and zero phase, 182, 203, 234
for boundary scaling and wavelet coefficients, 137–8, 146–7
LA and coiflet wavelet filters, 136, 155, 188, 193

time-dependent wavelet variance, 296, 324, 333–4
  confidence intervals for, 325
time/frequency decomposition
  DWPT, 206
  matching pursuit, 239
  MODWPT of solar magnetic field, 236–7
time-independent wavelet variance, 296, 335
time invariant DWT, 159
time reversed filter
  scaling, 463, 495
    Fourier transform, 465, 484, 496
    key properties, 499
    squared gain function, 485–6, 498
  wavelet, 474, 479, 497
    Fourier transform, 476, 483, 498
    key properties, 499
time/scale decomposition
  DWT, 206
  matching pursuit, 239
time series, 1
  analysis (decomposition), 43, 46, 53, 393
  and its DWT, 15
  energy, 42
  reconstruction via DWT, 16
  sample ACS, 14–5
  synthesis (reconstruction), 43, 46, 53, 393
  vector form of, 42
time shift function
  LA(L) wavelet filter of jth level, 119, 121
time width
  of wavelet packet coefficients for node (j, n), 216
  physical, for MODWPT of solar magnetic field, 236
Toeplitz matrix, 356, 441
top down basis selection, 226
trace of a matrix, 442
trade-off between bias and variance, 272
transfer function (frequency response function), 25, 33–4, 267
  Daubechies scaling filter, 106
  DWPT wavelet packet filter for node (j, n), 215, 229, 250
  DWT scaling filter, 76, 78
  of jth level, 97, 152, 157
  DWT wavelet filter, 69
  jth level, 96, 152, 157
  of second level, 91
  filter with inserted zeros, 91, 96–7
  high- and low-pass examples, 26, 39
  MODWPT wavelet packet filter for node (j, n), 232
  MODWT scaling and wavelet filters, 163, 469, 476
  of jth level, 169, 173, 201, 469, 477
  reversed filter, 117
transform
  complex-valued (*see* complex-valued transform)
  discrete cosine (*see* discrete cosine transform)
  DWT (*see* discrete wavelet transform)
  isometric, 43, 46–7, 53
  ODFT (*see* orthonormal discrete Fourier transform)
  orthonormal (*see* orthonormal transform)
  Walsh (*see* Walsh transform)
  z, 484, 498
transform coefficient, 43
transformation, variance stabilizing, 270, 275
transformed RV
  PDF of, 258
transformed RVs
  joint PDF of, 261
transient, 396
translation, 457–8, 474, 494
  by minus one unit, 458
  by one unit, 457
  non-commuting with dilation, 459
translation invariant DWT, 159
  and MODWT, 174
transpose
  complex conjugate (Hermitian), 45
  ordinary, 42
tree structure of DWT, 451
trend, 317, 319–20, 323
  estimation with LA(8), D(4) and C(6) filters, 392
  partial DWT applied to, 368–9
  plus FD process, 368, 383
    estimation, 383
    filter length required, 370
  plus stationary process, 289
  test for, 373, 390
  and adjustment, 373
  and mean removal, 373
trigamma function, 275, 376, 391, 440, 442
trigonometric polynomial, 485
turbulence in the ocean, 332, 334
two-fold cross-validation, 402
  partial DWT, 422
  threshold, 423–4
two-scale difference equation, 465–6, 470, 495
two-scale relationship, 464, 466, 470, 495
two-sided exponential PDF, 257
two-stage cascade filter, 304

Ulysses spacecraft, 218, 237
  solar polar orbit, 219
unbiased MODWT estimator of the wavelet variance, 298, 306, 335, 337
uncorrelated Gaussian (normal) RVs, 259
  and independence, 259, 407
uncorrelated noise, 393
uncorrelated noise DWT coefficients, 432, 434
uncorrelated non-Gaussian noise, 417

uncorrelated non-Gaussian wavelet coefficients, 432
undecimated DWT, 159
union of events, 435
unitary matrix, 45–7, 53
unitary transform
  energy preservation, 46, 53
  projection theorem, 46
universal threshold, 400–1, 418–9, 443, 450, 453–4
  asymptotic considerations, 401, 439
  level-dependent, 430
universal thresholding
  level-dependent, 406, 444–5, 453, 455
  level-independent, 444–5, 455
Universal Time (UT), 219
upsampling, 82–3, 153, 156, 176–7, 201
UTC (coordinated universal time), 317

vanishing discrete moments of time reversed wavelet filter, 484, 498
vanishing moments, 123, 457
  of wavelet function, 483–7, 498
  relationship to Fourier transform and derivatives of wavelet function, 483–4
variance
  Allan (*see* Allan variance)
  inhomogeneity for Nile river minima, 193
  level-dependent for correlated noise DWT coefficients, 441, 456
  of autoregressive process, 294
  of chi-square RV, 263, 270
  of Gaussian (normal) RV, 257
  of generalized Cauchy (t) RV, 258
  of instantaneous LSE, 392
  of Laplace (two-sided exponential) RV, 257
  of log multitaper SDF estimator, 275
  of log periodogram, 270
  of periodogram, 270
  of RV, 256
  of sample mean, 293
  of scaling coefficient, 355
  of stationary stochastic process, 266
  rotated cumulative (*see* rotated cumulative variance)
  sample (*see* sample variance)
  stabilizing transformation, 270, 275
  subtraction of two estimated values of, 425, 427
  test for homogeneity of, 341, 379, 391
variance mixture of Gaussian (normal) distribution, 265, 413
vector space basis
  for complex finite dimensional, 46
  for real finite dimensional, 43–4, 239
vectors
  complex-valued
    inner product of, 45
    squared norm of, 46
  real-valued
    inner product of, 42
    squared norm of, 42
  transpose of, 42
Venn diagram, 473
vertical ocean shear, 193, 295, 327–8, 330, 396
  bursts in MRA, 194
  MRA for LA(8) MODWT, 193–5
  NPESs, 397
  physical scales, 194
  representation via LA(8) DWT coefficients, 397
  representation via ODFT coefficients, 397
  wavelet variance estimates for, 328–9, 332–3
vertical wind velocity average, 6
VisuShrink, 420

Walsh transform, 41
  and Haar DWPT, 217–8
wavelet, 2
  choice of, 56
wavelet-based analysis of covariance, 303
wavelet-based analysis of variance, 19, 296, 335
  proof of, 301
wavelet coefficients, 13, 17, 59
  boundary elements
    DWT, 136, 146, 158, 350
    MODWT, 198
  compared to MODWT details of jth level, 180, 185
  D(4), 60
  decorrelation property, 16, 19, 341
  downsampling
    and bandwidth, 85–6
    and frequency reversal, 85–6
  finite energy function (signal), 479, 497
  for AR(1) process, 337
  IID Gaussian (normal), 417
  jth level for finite energy function (signal), 480, 497
  mean of, 301, 335
  nonboundary elements
    correlation at different levels, 346–7
    covariance at different/same levels, 345, 348
    SDF for, 348–9, 352
  sample ACS, 14, 16, 19, 343
  statistical properties, 434
  synthesis of finite energy function (signal), 482
  time and scale, 19
  via circular filtering
    DWT, 70, 74, 80, 88, 92, 94, 152, 157
    MODWPT, 231, 251
    MODWT, 163, 168–70, 173, 175, 177, 200–1
wavelet correlation, 303
wavelet covariance, 303
wavelet details
  DWT, 64–5, 151, 158
    D(4), 65–6

decomposition of sample variance, 66–7
first level, 81
Haar, 64–5
jth level, 95, 191
pseudo-code, 102
MODWT (jth level), 171, 173, 176–7, 191, 200, 203
and DWT averaging, 204–5
pseudo-code, 179
via cross-correlation (circular), 172, 203
wavelet expansion
homogeneous, 482
inhomogeneous, 482
wavelet filter
advance required for LA(L), 113
approximately linear phase for LA(L), 112
autocorrelation, 69
choice of, 135, 195
coiflet (*see* coiflet wavelet filter)
connection to wavelet function, 477, 498
construction of Daubechies type, 457
D(4), 59, 73, 155
Daubechies (*see* Daubechies wavelet filter)
DWT, 56, 68, 151, 497
energy of, 163
even width, 68, 74
extremal phase, 61
'father,' 80
Fourier transform of time reversed, 476, 483, 498
Haar, 60, 73, 75, 155–6, 192
key properties, 154, 202, 499, 500
lacking high-pass nature, 105
least asymmetric (*see* LA(L))
MODWT, 163, 168–9, 200
applied circularly to finite sequence, 296, 336
applied to infinite sequence, 296, 335
'mother,' 80
of jth level, 95, 152, 343
advance required for C(L), 147, 156
advance required for general design, 118
advance required for LA(L), 146, 155
alternative formulations, 102
C(6), 125
connection to wavelet function, 477, 498
D(4), 97–8
Daubechies and band-pass nature, 96, 102
LA(8), 97–8
MODWT relationship to DWT, 173
periodized, 96–7
phase after advance for BL(L), 120
phase after advance for C(L), 124
phase after advance for LA(L), 114–5, 120
shift factor for C(L), 124, 198, 205
shift factor for LA(L), 114, 198, 205
shift factor for LA(8), 179
squared gain function, 157
time shift function for LA(L), 119, 121
transfer function, 96, 152, 157
width, 96
of second level, 90–2
Daubechies and band-pass nature, 91
Haar, 40
nonorthogonality to even shifts, 92
periodized, 92
transfer function, 91
orthogonality to even shifted scaling filter, 77, 163
orthogonality to even shifts, 69, 71, 74, 151, 163
orthonormality property, 69
periodized, 71, 74, 77, 80, 88, 153, 168–70, 201
squared gain function, 69, 74, 151, 156, 163
sum of zero, 69, 74, 151, 163
support of wavelet function, 474
time reversed, 474, 479, 497
transfer function, 69, 163, 169, 173, 201, 476–7
unit energy, 69, 71, 74
vanishing discrete moments of time reversed, 484, 498
width, 61, 135, 169, 201
and band-pass properties, 315
and mean of wavelet coefficients, 319
wavelet frames, 159
wavelet function, 457, 472, 496–7
as a small wave, 2, 476
complex-valued, 4
computation for D(4), 477
connection to MODWT wavelet filter of jth level, 477, 498
finite support, 474, 498
first derivative of Gaussian, 3, 10
Fourier transform of, 476, 498
Haar, 475, 479
alternative formulation, 2–3, 9
integral of zero, 476, 498
integration of, 2
integration of square of, 2
key properties, 499–500
Mexican hat, 3, 10
moments, 483
wavelet packet coefficients
for node (j, n)
time width, 216
via circular filtering, 214–5, 249–50
natural ordering, 209–10
sequential frequency (sequency) ordering, 209–10, 212, 232
via circular filtering, 210, 212, 232
wavelet packet cost table, 225
wavelet packet details (MODWPT)
LA(8) for solar magnetic field, 237–8
node (j, n) via cross-correlation (circular), 234, 251

wavelet packet filter for node (j, n), 214, 231, 250–1
  advance required
    for general design, 231
    for LA(L) and C(L), 230, 233, 253
  LA(L) and C(L) phase function, 229
  LA(8), 253
  nominal pass-band, 209, 216–8, 249
  phase function, 253
  shift factor for LA(L) and C(L), 221, 230
  squared gain function, 215, 217, 233, 253
  transfer function, 215, 229, 232, 250
  width, 214
wavelet packet table (WP table), 209, 212, 250
  arbitrary disjoint dyadic decomposition of, 213
  best basis, 223, 226, 251
    algorithm, 225
    example, 224
  best shift basis, 228
  disjoint dyadic decomposition
    ANOVA, 214
    energy decomposition, 214
  maximal overlap, 232, 251
  optimal transform, 223, 251
    example, 224
  partial DWT as disjoint dyadic decomposition of, 212
  vectors for waveform dictionary, 241, 254
wavelet packet tree (*see* wavelet packet table)
wavelet roughs, 66
  D(4), 65–6
  Haar, 64, 66
wavelet shrinkage (*see* shrinkage)
wavelet smooths
  DWT, 64, 66, 151, 158
    artifact removal by averaging, 197
    D(4), 65–6
    first level, 81
    Haar, 64, 66
    jth level, 95
    pseudo-code, 102
  MODWT
    and DWT averaging, 204–5
    jth level, 171, 173, 176–7, 200, 203
    pseudo-code, 179
    via cross-correlation (circular), 172, 203
wavelet spectrum, 295
wavelet thresholding (*see* thresholding)
wavelet variance, 15, 17, 295
  and Allan variance, 322
  and SDF, 296–7, 315–6, 329, 337, 343
  confidence intervals for, 311, 313–4, 319–20, 327, 332–4, 336, 345, 385
  estimates
    for Nile River minima, 327
    for atomic clock data, 317, 322, 319–20, 322–3
    for vertical ocean shear, 328–9, 332–3
  estimators
    asymptotic properties of, 295, 307, 311, 336
    based on Haar, D(4), D(6) and LA(8) filters, 320, 323, 328–9
    biased, 307, 308, 337, 351, 378
    leakage in, 303
    unbiased, 298, 306, 308, 335, 337, 375, 384, 390
  exponents and stationary/nonstationary processes, 323
  for FD process, 375
  for processes with stationary backward differences, 305
  for random walk process, 337
  for stationary processes, 305
  for white noise process, 337
  scale-by-scale ANOVA, 296, 335
    proof of, 301
  time-dependent, 296, 324, 333–4
  time-independent, 296, 335
waveshrink, 417
weak differentiability, 403–4, 456
weighted averages, 11
  changes in, 162
weighted least squares estimator (WLSE)
  for FD process, 374, 376, 391
    MODWT-based, 377–9, 385
  of atomic clock data, 385
weighted overlapped segment averaging (WOSA) spectral estimator, 315
white noise, 268, 286, 288, 293–4, 442
  ACVS of, 268
  Gaussian (normal), 358, 389
  spectral density function of, 268
  wavelet variance for, 337
width
  of equivalent filter for filter cascade, 28
  of filter, 28, 39, 90
  of scaling filter, 75, 78
    of jth level, 97, 169, 201
  of wavelet filter, 61, 68, 74, 135
    of jth level, 96, 169, 201
  of wavelet packet filter for node (j, n), 214
  time (*see* time width)
WLSE (*see* weighted least squares)
WOSA spectral estimator, 315
WP table (*see* wavelet packet table)

z transform of time reversed scaling filter, 484, 498
zero padding, 308, 338, 382, 387
zero phase filter, 108, 110, 158, 179, 184
  and time alignment, 182, 203, 234, 237, 252
  for MODWPT details for node (j, n), 234, 252
  for MODWT details, 189
    of jth level, 172, 168
  for MODWT smooths, 168
    of jth level, 172